Birkhäuser

Joan C. Artés • Jaume Llibre • Dana Schlomiuk
Nicolae Vulpe

Geometric Configurations of Singularities of Planar Polynomial Differential Systems

A Global Classification in the Quadratic Case

Birkhäuser

Joan C. Artés
Departament de Matemàtiques
Universitat Autònoma de Barcelona
Barcelona, Spain

Jaume Llibre
Departament de Matemàtiques
Universitat Autònoma de Barcelona
Barcelona, Spain

Dana Schlomiuk
Département de Mathématiques
et de Statistiques
Université de Montréal
Montréal, QC, Canada

Nicolae Vulpe
Vladimir Andrunachievici Institute of Mathematics
and Computer Science
Chisinau, Moldova

ISBN 978-3-030-50569-1 ISBN 978-3-030-50570-7 (eBook)
https://doi.org/10.1007/978-3-030-50570-7

Mathematics Subject Classification (2020): 58K45, 34C05, 34A34

This book is published under the imprint Birkhäuser, www.birkhauser-science.com by the registered company Springer Nature Switzerland AG
The registered company address is: Gewerbestrasse 11, 6330 Cham, Switzerland

We dedicate this book to the memory of the mathematician

Constantin Sibirschi (1928–1990)

on the occasion of the 90*th anniversary of his birth. Without the theory of algebraic invariants of polynomial differential equations, founded by Sibirschi, this book could not have been written.*

Contents

Preface **xi**

I Polynomial differential systems with emphasis on the quadratic ones **1**

1 Introduction **3**
1.1 Preliminaries . . . 3
1.2 Problems on planar polynomial differential systems . . . 6
1.2.1 The problem of the center . . . 6
1.2.2 Research arising from the work of Darboux and the problem on algebraic integrability . . . 6
1.2.3 The second part of Hilbert's 16th problem . . . 7
1.2.4 The general finiteness problem for limit cycles or the existential Hilbert's 16th problem . . . 9
1.2.5 The infinitesimal Hilbert's 16th problem and the Hilbert–Arnold problem . . . 10
1.3 The contents of this book . . . 10

2 Survey of results on quadratic differential systems **17**
2.1 Brief history of quadratic differential systems . . . 17
2.2 Some basic results obtained for quadratic differential systems . . . 20
2.3 Study of some subclasses of the family of quadratic differential systems . . . 23
2.4 Algebraic limit cycles in quadratic systems . . . 26
2.5 Finiteness problems for quadratic differential systems . . . 27
2.5.1 Basic concepts and results needed for studying the general finiteness problem . . . 27
2.5.2 The general finiteness problem for quadratic differential systems . . . 28
2.5.3 Application of Roussarie's ideas for the quadratic case . . . 29

2.5.4 The infinitesimal Hilbert's 16th problem and the Hilbert–Arnold problem 32
2.5.5 The infinitesimal Hilbert's 16th problem for quadratic differential systems 33
2.6 The initial steps in the global theory of quadratic differential systems 34

3 Singularities of polynomial differential systems 37
3.1 Compactification on the Poincaré sphere, Poincaré disc and projective plane 37
3.2 Classical definitions 41
3.3 New definitions 43
3.4 The blow-up technique 45
3.4.1 The polar blow-up 45
3.4.2 The blow-up using rational functions 46
3.4.3 The blow-up technique using only one direction 50
3.5 The borsec concept 64
3.6 Equivalence relations 75
3.7 Notations for singularities of polynomial differential systems 81
3.7.1 Elemental singularities 81
3.7.2 Non-elemental singularities 82
3.7.3 Lack of singularities and complex singularities 85
3.7.4 Infinite number of singularities 86

4 Invariants in mathematical classification problems 91
4.1 Basic concepts 91
4.2 Classification problems on planar polynomial vector fields 94
4.2.1 Equivalence relations for polynomial vector fields 94
4.2.2 Classifications of some families of polynomial vector fields . 96

5 Invariant theory of planar polynomial vector fields 99
5.1 Classical invariant theory 99
5.2 The work of Sibirschi's school 104
5.3 Basic concepts 109
5.3.1 Group actions on polynomial vector fields 110
5.3.2 Definition of invariant polynomials for polynomial differential systems 110
5.3.3 Assembling multiplicities of singularities in divisors of the line at infinity and in zero-cycles of the plane 113
5.3.4 Construction and geometric meaning of several basic invariant polynomials 114
5.4 Invariant polynomials associated to geometrical configurations . . . 116
5.4.1 Building blocks for the construction of the invariant polynomials needed for the classification theorems of **QS** . . 117

5.4.2 The set of all invariant polynomials which classify geometrically the global configurations of singularities in **QS** 120
5.4.3 The influence of complex singularities in the study of the geometrical global configurations of singularities in **QS** . . 131

6 Main results on classifications of singularities in QS 133
6.1 Finite singularities 133
6.2 Finite weak singularities 144
6.3 Singularities of **QS** with an integrable saddle 145
6.4 Infinite singularities 148

7 Classifications of quadratic systems with special singularities 163
7.1 A finite star node 164
7.1.1 Conditions for the existence of at least one finite star node 165
7.1.2 Configurations of singularities with a finite star node 168
7.2 An integrable saddle 179
7.3 A center 197
7.4 A star node at infinity and another special singularity 209
7.5 Three finite special singularities 221
7.6 A weak focus of order two or three 235
7.6.1 A weak focus of order two 236
7.6.2 A weak focus of order three 245
7.7 A weak saddle of order three 250

II Configurations of singularities of quadratic systems 261

8 QS with finite singularities of total multiplicity at most one 263
8.1 Systems without finite singularities 263
8.2 Systems with exactly one singularity 267

9 QS with finite singularities of total multiplicity two 281
9.1 Exactly one finite singularity 281
9.2 Two distinct real singularities 290
9.3 Two distinct complex singularities 323

10 QS with finite singularities of total multiplicity three 329
10.1 Exactly one singularity 329
10.2 Exactly two distinct singularities 334
10.3 Exactly three distinct singularities 345
10.3.1 Systems with zero-cycle $\mathcal{D}_{\mathbb{C}^2}(S) = p + q + r$ 346
10.3.2 Systems with zero-cycle $\mathcal{D}_{\mathbb{C}^2}(S) = p + q^c + r^c$ 371

11 QS with finite singularities of total multiplicity four **389**
11.1 Exactly one singularity 390
11.2 Exactly two distinct singularities 398
11.2.1 One triple and one simple real singularities 398
11.2.2 Two double real singularities 428
11.2.3 Two double complex singularities 433
11.3 Exactly three distinct singularities 437
11.3.1 One double and two elemental real singularities 437
11.3.2 One double real and two elemental complex singularities . . 484
11.4 Exactly four distinct finite singularities 489
11.4.1 Four real elemental singularities 489
11.4.2 Two real and two complex elemental singularities 546
11.4.3 Four complex elemental finite singularities 608

12 Degenerate quadratic systems ($m_f = \infty$) **613**

13 Conclusions **629**
13.1 New concepts 629
13.2 The classical versus the new way 630
13.3 Algorithm to study the singularities of quadratic differential systems 631
13.4 The topological configurations of singularities 632
13.5 The study of the quadratic differential systems modulo limit cycles 632

Appendix A Table of notation **635**

Appendix B Manual of Mathematica tools **645**
B.1 How to initiate the program 646
B.2 Notation 646
B.3 Examples 648

Bibliography **669**

Index **697**

Preface

In this book we consider planar polynomial differential systems, i.e. systems of the form

$$\frac{dx}{dt} = p(x,y), \qquad \frac{dy}{dt} = q(x,y)$$

where $p(x,y), q(x,y)$ are polynomials in x, y with real coefficients. To each such system there corresponds a point in $\mathbb{R}^N$ determined by its $N = (n+1)(n+2)$ coefficients, where n is the degree of the system, i.e. $n = \max(\deg(p), \deg(q))$. A system of degree 2 is called *quadratic*.

The study of these differential systems always begins with the study of their singularities, finite or infinite, followed by the study of separatrix connections and of limit cycles. Also in some particular cases, the study of first integrals, algebraic invariant curves and period function is of great interest.

Our main goal in this book is to classify in a geometrical way the global schemes of singularities, finite and infinite, of quadratic differential systems and to obtain their bifurcation diagram in the 12-dimensional space $\mathbb{R}^{12}$. This global classification and its bifurcation diagram is completely algebraic, and we provide the algorithm that computes, for every family of quadratic systems, the global bifurcation diagram of its corresponding schemes of singularities. The study of singularities is the first step in the topological classification of the phase portraits of these differential systems and their bifurcation diagram. The geometrical equivalence relation between singularities considered here, is deeper than the topological one, including features of an algebraic-geometric meaning that play a significant role in studying bifurcations of the systems.

This was a long-term project. Our work began seven or even eight years ago. Every year we met in the spring in Barcelona, then in the fall in Montreal, in order to work on the project. During the past three years, two of us met in late summer in Chişinău, Moldova. We were happy to have the opportunity to work together and in the acknowledgements we mention the institutions and grants that supported us.

Over the years, we published partial results such as the study of infinite singularities, then of quadratic systems with total multiplicity of finite singularities less than or equal to one, or with total multiplicity of finite singularities equal to two or three. From the class of quadratic differential systems with total multiplicity of finite singularities equal to four, those with total number of distinct finite singularities less than or equal to three, were also published. On one of these last published articles, we worked together with Alex C. Rezende, and we thank him for his contribution to our project.

The original results appearing in the book in Chapters 7, 12 and in Section 11.4 of Chapter 11 have never been published before and so they appear here for the first time. Section 11.4 contains the most generic and most difficult cases. This classification yielded 1765 distinct geometrical configurations of singularities, finite or infinite, plus at most 8 other such configurations (sharing the same finite part) that we conjecture are not realizable.

We give in the final chapter of the book some concluding comments with a view towards the future.

We are thankful to the editors and referees for the improvements they suggested and their advice was followed by us.

Joan Carles Artés
Jaume Llibre
Dana Schlomiuk
Nicolae Vulpe

Barcelona, Montréal, Chişinău, 2020

Acknowledgements

During the years we kept working on writing this book, in addition to the support of our universities, we received support from the CRM (Centre de Recherches Mathématiques) in Montreal and from the Academy of Sciences of Moldova. We mention the following grants that supported our work: MCYT/FEDER number MTM 2008-03437, MINECO number MTM2013-40998-P and MTM2016-77278-P (FEDER), ICREA Academia, the AGAUR grants 2009SGR 410 and 2014 SGR568, the European Community grants FP7-PEOPLE-2012-IRSES 316338 and 318999, several grants of NSERC of Canada, the last one being RGPIN-2015-04558 and the grants CRDF-MRDA CERIM-1006-06 and 12.839.08.05F-SCSTD-ASM from Moldova.

Part I

Polynomial differential systems with emphasis on the quadratic ones

Chapter 1

Introduction

1.1 Preliminaries

In this book we study the singularities of the simplest nonlinear 2polynomial differential equations, the planar quadratic differential systems. A *polynomial differential system* on the plane is a system of the form

$$\frac{dx}{dt} = p(x,y), \qquad \frac{dy}{dt} = q(x,y), \tag{1.1}$$

where p, $q \in \mathbb{R}[x,y]$, i.e. p, q are polynomials in the variables x and y over $\mathbb{R}$. To such a system one can associate the vector field

$$X = p(x,y)\frac{\partial}{\partial x} + q(x,y)\frac{\partial}{\partial y}.$$

We call *degree* of a system (1.1) the integer $n = \max(\deg(p), \deg(q))$. In particular we call *quadratic* a differential system (1.1) with degree $n = 2$. We denote by **QS** the whole class of real quadratic differential systems. A *singular point* of a differential system (1.1) is a common solution $(x_0, y_0) \in \mathbb{R}^2$ of $p(x,y) = 0$ and $q(x,y) = 0$.

These systems occur in many branches of applied mathematics, and they also have theoretical importance. Apart from their applications the major driving force in the development of their theory was a collection of very difficult problems, some of which are over one hundred years old. These problems also motivated the work we present in this book. For this reason in this introduction we briefly describe some of the main problems on these systems, the progress and also the difficulties encountered in solving them, and we indicate their present state.

The study of planar polynomial vector fields began to be pursued after the publication of the famous papers of Darboux [119] (1878) and of Poincaré [262] (1881), [263] (1885), [264] (1891), [265] (1897). Darboux's work was on integrability

J. C. Artés et al., *Geometric Configurations of Singularities of Planar Polynomial Differential Systems*, https://doi.org/10.1007/978-3-030-50570-7_1

in terms of the existence of *algebraic invariant curves* of complex polynomial differential equations over the complex projective plane. This very beautiful work, which could also be applied to compute first integrals of real polynomial differential equations, was very much admired by Poincaré and motivated his statement in 1891 (see [264]) of a very hard problem still open today, on planar polynomial differential systems (see Section 1.2). The main motivation of Poincaré for studying these systems came from his interest in the problem of the stability of the solar system. This problem asks if over a very long time the solar system will preserve its present state, or whether major changes, such as a planet escaping from the system, or a collision among bodies in the system, will occur. As this problem is very hard, Poincaré decided to begin by first studying the simplest nonlinear differential equations which are the planar polynomial ones, and he wrote the two seminal papers [262, 263] that founded the *qualitative study of differential equations*. By this we mean to determine the global behavior of the solutions of a differential system on the whole plane without the need of effectively integrating the system, which is in general not possible. Rather than emphasizing calculations of specific solutions, Poincaré took the global approach by considering the solutions in their totality. In these and other works he introduced many new notions, for example for special types of singularities such as *foci*, *nodes, saddles* and *centers*, which distinguished the behavior of solutions around singularities; the notion of *limit cycle*, namely an isolated periodic solution in the set of all periodic solutions of a system; the notion of the *Poincaré first return map*, ... In these papers Poincaré proved a number of theorems, among them the theorem saying that a necessary and sufficient condition for a polynomial vector field on the plane to have a center at a singular point with purely imaginary eigenvalues, is that the system admits a non-zero local analytic first integral in a neighborhood of this singular point. In his mémoire *General Problem of the Stability of Motion* [233], Lyapunov extended this theorem for analytic differential systems, (actually Lyapunov studied differential systems in n variables but when results are applied to two-dimensional systems, we obtain this more general theorem for analytic two-dimensional systems). In this mémoire Lyapunov developed the theory of stability of motion.

After its founding by Poincaré in the late nineteenth century, the qualitative study of differential equations was pursued in the Soviet Union around the middle of the twentieth century. A group of scientists centered around Andronov, who together with his collaborators Leontovich, Gordon and Maier wrote two books [2, 3] on the qualitative theory on differential equations. He also wrote together with Vitt and Khaikin a book on the theory of oscillations [4] (in Russian), [5] (in English) where apart from some results in qualitative theory of planar differential equations, the authors initiated the study of piecewise differential equations, important for their many applications. Another book on the qualitative theory of differential equations [241] was published by Nemytskii and Stepanov. In the second part of the twentieth century, the books of Arnold on differential equations and classical and celestial mechanics [7, 8, 9, 10, 11] highlighted the geometric, qualitative, aspects of the phenomena. Also in Moscow, Ilyashenko built a school on differential equations and published several books [174, 175, 178, 177, 179] on this subject. In Bielorussia, in Minsk, a group around Erugin and Cherkas also worked on the qualitative theory. Another center for the qualitative theory was Chişinău, Moldova, where Sibirschi and his student Vulpe and others worked in this direction. In Israel there is another school around Yakovenko, Binyamini and Novikov.

Lefschetz's book [200] on the geometric theory of differential equations includes results obtained by the schools in the Soviet Union and helped in making this literature known in the west.

In Poland a group around Żołądek and in Bulgaria a group around Horozov and Iliev are also active on planar differential systems. In Toulouse, Gavrilov, a former student of Horozov, is also very active in this area of research.

In the second part of the twentieth century, the qualitative theory of planar polynomial vector fields was an active research field also in China, mostly in two centers: Beijing and Nanjing. The Beijing group was lead by Zhang Zhifen, a scientist who received her Ph.D. from Moscow State University under the direction of Nemytskii. Zhang Zhifen and three other researchers published a book [362] on the qualitative theory. The group around Zhang Zhifen included Chengzhi Li and Weigu Li. The group in Nanjing centered around Ye Yan-Qian who together with collaborators published a book [355] on limit cycles. In Shanghai, at Shanghai Jiao Tong University, Xiang Zhang, Dongmei Xiao, Yiley Tang and Jiang Yu and, at Shanghai Normal University, Maoan Han also published articles on planar vector fields. At Quanzhou Normal University, Jibin Li was very active in differential equations and published several books on the subject [214, 215].

Another school on the qualitative theory of differential equations was the Brazilian school headed by M. Peixoto at IMPA (Instituto de Matemàtica Pura e Aplicada). One of the major works of Peixoto was on structural stability of differential equations in two dimensions. Another prominent member of this school is J. Sotomayor, who also wrote a book [318] on the qualitative theory of polynomial vector fields. Several people at IMPA around C. Camacho, in particular A. Lins Neto, V. Pereira, B. Scardua, work mainly on holomorphic foliations but occasionally publish articles on polynomial vector fields, see for example [76] and [210].

Other groups working in the qualitative theory of differential equations that need to be mentioned are in Europe: in Hasselt, Belgium around F. Dumortier; in Dijon, around R. Roussarie and P. Mardešić; in Paris, J. P. Françoise and his team; in Delft, Holland, J. Reyn; in Lleida, Spain, headed by Chavarriga and Giné; in Barcelona, headed by J. Llibre, A. Gasull and J. C. Artés; in Italy, R. Conti, M. Sabatini, G. Villari and M. Villarini; in United Kingdom, N. Lloyd and C. Christopher; in Germany, S. Walcher and in Slovenia, V. Romanovskii. In USA, several scientists such as J. Guckenheimer [153], D. Shafer, C. Chicone, L. Perko and T. Blows have also published in this area.

In Montreal C. Rousseau and D. Schlomiuk ran a seminar in the 1980s and 1990s and among the lectures quite a few were on the qualitative theory of differential equations. In 1992 the Séminaire de Mathématiques Supérieures (SMS) held at the Université de Montréal was on bifurcations and periodic orbits of vector fields. The lectures of this SMS appeared in the book [294] edited by D. Schlomiuk. Another SMS was held in the summer of 2002, and the book [177] edited by Y. Ilyashenko and C. Rousseau based on the lectures of this SMS was published in 2004.

Five other books on planar vector fields were written. The book of R. Roussarie [284] and the one of Dumortier, Llibre and Artés [128] are on qualitative theory of polynomial vector fields. The book [90] authored by C. Christopher and Chengzhi Li is divided into two parts: The first one is dedicated to the center-focus problem, and the second focuses on the Abelian integrals with special emphasis on the ones intervening in quadratic systems. J. Reyn published a book [279] on phase portraits of quadratic differential systems. The book [281] of Romanovski and Shafer is on planar polynomial vector fields and a computational algebra approach. Another book [297] has over 120 pages devoted to planar differential equations. Finally, chapters on the center-focus problem and isochronous centers are included in [214], while [361] contains chapters on Darboux and Liouvillian integrability.

The list of published works in this area is vast, and we cannot claim to have given it all. Our apologies are addressed to those authors not covered by our survey.

Planar polynomial differential systems are objects of a mixed nature involving analytic, geometric as well as algebraic features. Due to this, they form a fertile soil for intertwining diverse methods in a unified whole [293, 295].

1.2 Problems on planar polynomial differential systems

1.2.1 The problem of the center

This problem was formulated by Poincaré in 1885 [263]. One way in which we can state the problem of the center is the following: Given a positive integer n, find the necessary and sufficient conditions for a polynomial system of degree n to have a singularity that is a center. In fact Poincaré only considered singularities with a non-degenerate linear part of focus or center type. More explicitly he considered the case where the eigenvalues are purely imaginary, $\pm\beta i$, $\beta \in \mathbb{R}\setminus\{0\}$, and in this case the problem of the center is to give conditions for distinguishing between a center and a focus. The problem of the center was solved only for quadratic systems [123, 189, 190, 46]. The problem of the center is open for systems of degree $n \geq 3$. For cubic systems (i.e. $n = 3$), it is already a very hard problem.

1.2.2 Research arising from the work of Darboux and the problem on algebraic integrability

In 1878 in his remarkable paper [119] Darboux gave a geometric way of integrating a polynomial differential equation over the complex projective plane using invariant algebraic curves over $\mathbb{C}$. He proved a theorem which says that if we have a polynomial differential system of degree n and if we have a number m of invariant algebraic curves $f_i(x,y) = 0$, $i \in \{1,2,\dots m\}$, $f_i(x,y) \in \mathbb{C}[x,y]$ such that $m \geq n(n+1)/2+1$, then we can find a first integral of the form $f_1^{\lambda_1} \dots f_m^{\lambda_m}$ with $\lambda_i \in \mathbb{C}$ of the system. This work of Darboux can also be applied to polynomial differential systems over the reals as any such system can be complexified. Furthermore if the coefficients of such a system are real, then for any complex invariant curve of the system, its complex conjugate curve is also invariant. Using this fact and if we have enough such invariant curves, we can actually obtain a real first integral of the system. For an easily accessible introduction to the theory of Darboux, see [292] and also [128]. Poincaré was enthusiastic about this beautiful work of Darboux, which he called "oeuvre magistrale" in [264]. In this article he stated a problem now called the problem of Poincaré. He wrote in 1897 a second article on this problem [265]. There is a lot of ongoing work arising from the paper of Darboux. This work lies at the interface between dynamical systems and algebraic geometry, and it has many aspects. The literature on this topic is substantial and fast growing, and we cannot attempt to describe it in this introduction. We shall only consider here the problem stated by Poincaré on algebraic integrability.

This is the problem of recognizing when a planar polynomial system admits a rational first integral, i.e. $f(x,y)/g(x,y)$ where $f, g \in \mathbb{C}[x,y]$. We may suppose that f and g have no common factors over $\mathbb{C}$. This function is constant on solution curves, and so the solution curves lie on algebraic curves $f(x,y) - g(x,y)k = 0$ that are invariant under the flow, with $k \in \mathbb{R}$. These curves have a degree bounded by $\max(\deg(f), \deg(g))$. Poincaré remarks that if we have a bound N for the degrees of the invariant algebraic curves of the system, then it remains to perform some purely algebraic calculations in order to find the algebraic invariant curves and the rational first integral if it exists. For this reason the problem of Poincaré is sometimes considered to be the problem of bounding the degrees of invariant algebraic curves that a given polynomial differential system may have in the case that such curves exist. This problem is still unsolved even for quadratic systems. A theoretical answer to this problem has been given [91] using the ecstactic polynomial, but it is not a practical solution when the degree of the rational first integral is high. Carnicer [78] provided an upper bound of the degree of an invariant algebraic curve that has no dicritical singularities.

Both problems are of a global nature involving whole classes of polynomial differential systems, and this is one of the reasons that they are so hard.

1.2.3 The second part of Hilbert's 16th problem

A third famous problem on planar polynomial systems is *the second part of Hilbert's 16th problem* stated in the list of 23 problems posed by Hilbert in his address at the International Mathematical Congress in Paris in 1900. This problem asks to determine for any natural number n the maximum number of limit cycles that a planar polynomial differential system of degree n could have and their relative position. This statement needs some clarifications, which we indicate in what follows.

Does any polynomial differential system have a finite number of limit cycles? The *individual finiteness problem* asks to prove that we have a positive answer to the previous question. Poincaré proved the first finiteness result on planar polynomial differential systems. He showed that for some polynomial systems we have a finite number of limit cycles [263, Theorem XVII]. A note of warning: Poincaré defined the notion of "limit cycle of a semi-characteristic". This notion is not the same as the one defining a limit cycle as an isolated periodic orbit. This result of Poincaré inspired Chapter II of the book of Sotomayor [318], where he proves a version of this finiteness theorem of Poincaré.

In 1923 Dulac [124] claimed to have proved the individual finiteness theorem in the general case, i.e. for any planar polynomial differential system, but almost 60 years later in 1982, Ilyashenko found an error in Dulac's proof (see [170, 172]). Then Ilyashenko [171] proved that an analytic vector field on a closed analytic surface with a finite number of non-degenerate singular points has only a finite number of limit cycles. As a corollary we get the same theorem for a polynomial vector field on the plane via the Poincaré compactification in the sphere or the compactification of the line field on the projective plane. The general individual finiteness for polynomial vector fields thus became a conjecture, and work to prove it began with the treatment of the special case of quadratic differential systems. Chicone and Shafer [88] proved the finiteness conjecture in the special case of quadratic differential systems having only finite *graphics* (see Section 2.5.1), and Bamon [44] extended this theorem by including also infinite graphics, i.e. those with non-empty intersection with the line at infinity. In this work he used Ilyashenko's result [171].

In 1985 Moussu gave a survey lecture in the Séminaire Bourbaki on the problem of finiteness of the number of limit cycles in polynomial vector fields (see [238]). The Dulac conjecture for the general case of a polynomial differential system of degree n was first announced as proved in 1987 by Ilyashenko [173] and independently by Ecalle, Martinet, Moussu and Ramis [132]. In 1987 Yoccoz gave a lecture in the Séminaire Bourbaki summarizing the announced results just mentioned [356]. In 1989 Moussu and Roche submitted for publication a paper having the goal of showing how the Khovansky theory of Pfaffian varieties could be applied to solving the Dulac problem, and they applied this method to a particular case (see [239]). Two papers containing two independent proofs, one by Ilyashenko [173] and another one by Ecalle [130], were published. Later, two books including these proofs appeared, by Ilyashenko [174] in 1991, and by Ecalle [131] in 1992. Actually they proved a more general individual finiteness theorem, namely that any analytic vector field on a closed surface has a finite number of limit cycles. Regarding the proofs of these individual finiteness theorems, Smale said: "These two papers have yet to be thoroughly digested by the mathematical community" [317, Page 284]. In fact in the almost 30 years since these proofs were achieved, no other specialist on dynamical systems or differential equations, not even a specialist on Hilbert's 16th problem, has checked them. We observe that the situation is very different for the 1994 proof of Fermat's Last Theorem given by A. Wiles and R. Taylor [329, 349], or for the proof of Thurston's geometrization conjecture given by G. Perelman in 2002 [254, 255], based on the study of the Ricci flow introduced by R. Hamilton [157]. These proofs have been checked and fully understood by some people. We need however say that there is an effort on the part of some logicians to understand, if not the proofs of Ecalle and Ilyashenko, their ideas and notions such as the analyzable functions of Ecalle [36, 37, 38] and the regular functions of Ilyashenko [186]. These differential fields are of interest to some model theorists because of their algebraic and model-theoretic structure and their possible connection to o-minimality.[1]

The challenge posed by Hilbert's 16th problem prompted Smale to write the following phrase regarding this problem: "Except for the Riemann hypothesis, it seems to be the most elusive of Hilbert's problems" [317, Page 283]. A few survey papers were written on this problem (for example [176, 209]).

1.2.4 The general finiteness problem for limit cycles or the existential Hilbert's 16th problem

Once we have this individual finiteness theorem, the next question that comes to mind is the following: For fixed n, is there a finite bound for the number of limit cycles that a system of degree n could have? In contrast to the individual finiteness problem, we have the general finiteness problem for limit cycles, which asks for a proof that for any fixed n there exists a finite bound for the number of limit cycles which a polynomial differential system of degree n could have. In stating the second part of his 16th problem, Hilbert probably believed that we have such a finite bound, and in his problem he asked to find the least upper bound of the number of limit cycles occurring in systems of an arbitrary but fixed degree n. This presumably existent least upper bound is usually called in the literature the *Hilbert number* and it is denoted by $H(n)$. So far, however, we do not know that such a bound exists, not even in the case of quadratic differential systems. It is, however, generally believed that such a bound exists. So we have the *global finiteness conjecture*:

Conjecture 1.1. *For any fixed natural number n there exists a bound N for the number of limit cycles occurring in planar polynomial differential systems of degree n.*

Clearly this is a very hard open problem. One of the several reasons for which this problem is so hard is its global nature involving all phase portraits of polynomial differential systems of a fixed degree n.

[1] The authors thank van den Dries for this information.

1.2.5 The infinitesimal Hilbert's 16th problem and the Hilbert–Arnold problem

Hilbert's 16th problem gave birth to several other problems, among them the general finiteness problem indicated in the previous section and the so-called *infinitesimal Hilbert's 16th problem* (or *tangential Hilbert's 16th problem* or *weak Hilbert's 16th problem*) [176, 354]. This problem asks for an upper uniform bound with respect to F, A, B and the curves $\gamma(t)$, of the number of zeros of the integral $I(t) = \int_{\gamma(t)} \omega$ where $\omega = Adx + Bdy$ is a real 1-form with polynomial coefficients of degree at most n, $\gamma(t)$ is a closed connected component of a level curve $F = t$ and F is a real polynomial of degree $n + 1$. The problem of existence of such a bound was solved by Varchenko [330] using the fewnomial theory of Khovanskii [193]. A constructive proof of this theorem was given by Binyamini, Novikov and Yakovenko [51].

Another finiteness problem is the *Hilbert–Arnold problem*, which asks if it is true that for a generic finite parameter family of smooth vector fields on the 2-sphere, the number of limit cycles of the vector fields in the family is uniformly bounded with respect to the parameter, provided that the parameter set is compact (see [176, 188] and [297, Chapter 4] by Kaloshin). With the exception of the individual finiteness problem and of the infinitesimal Hilbert's 16th problem, none of the other finiteness problems have been solved. However these very challenging problems constituted a driving force for much of the work in the theory on polynomial, analytic and even C^∞ differential systems, and this is the reason that we mentioned them here.

Famous finiteness problems have been solved and their solutions fully understood in other areas of mathematics. Discussions of finiteness problems arising from Hilbert's 16th problem as well as some finiteness problems and results elsewhere in mathematics were published in the book [297]. We also mention here the book [177] on normal forms, bifurcations and finiteness problems in dynamical systems where, a collection of papers on these topics is given.

1.3 The contents of this book

In this section we describe the problem we solve in this book and its motivation. We also give a short statement about the results obtained, the new notion of geometrical equivalence of configurations of singularities and the background literature for our results. Finally we give a description of the contents of the chapters of the book.

In 1988 [283] Roussarie stated a *global finiteness conjecture* for any family of analytic oriented foliations defined on $S^2 \times S$ where S is any compact parameter manifold, and he proposed an approach for this problem. This approach was developed into a program for the solution of this problem for quadratic systems in 1994 [129]. A brief survey on this program is given in Section 2.5.2. Although in the

32 years since this approach was first formulated in [283] many papers appeared in which parts of this program were realized, the finiteness problem for quadratic systems is still open and for its solutions many more cases need to be dealt with.

Even harder than Hilbert's 16th problem for polynomial differential systems is the *problem of topologically classifying all phase portraits of polynomial systems of a given degree* n, that is hard even in the simplest case of quadratic differential systems. This last problem was solved for a number of subfamilies of the quadratic class. We refer to an article [296] that contains a three-page survey of *topological classifications* (see Section 2.6) of some families of polynomial vector fields, mainly subclasses of quadratic or cubic systems.

Our work in this book was partially motivated by the problem of topologically classifying the quadratic differential systems. For this problem it is necessary to first consider the singularities and determine all their possible global configurations.

By *global topological configuration of singularities* we understand a description of the local phase portraits around all the finite and infinite singularities. These singularities are considered in this book under the *geometrical equivalence relation* defined in Section 3.6. The geometrical equivalence relation is deeper and richer than the topological equivalence relation, involving also properties of singularities of an algebraic geometric nature, that play a major role in the bifurcations of the systems. It is the more important equivalence relation and for this reason it is frequently mentioned in this book. In order to lighten the language used in these frequent repetition we introduce a few shorter expressions as follows:

By *configuration* we will understand the global geometrical configuration. By *finite configuration* we will understand the global geometrical configuration of finite singularities. By *configuration at infinity* we will understand the global geometrical configuration of the singular points at infinity. In case we only mean to consider the phase portraits around finite and infinite singularities we will use the complete term *global topological configuration of singularities.*

Secondly we need to enumerate all possible *connections* among singularities. Then we need to find all phase portraits of quadratic systems *modulo limit cycles* (see this equivalence relation in Section 4.2.1). Finally the number of limit cycles and their relative positions occurring in quadratic systems need to be determined.

The first part of this program asks for the list of all possible global topological configurations of singularities occurring in quadratic differential systems. This part is entirely algebraic.

The second, third and fourth steps of this program are basically of an analytic nature and much harder to obtain. The second step involves a combinatorial discussion; we need to enumerate all the possibilities for obtaining a connection among the singularities. Then we need to see which ones of these are realizable and which ones can be proved not to exist. The proof of non-realizability of some hypothetical connections may be quite hard to obtain. Since there are many possibilities, this second step may take a long time. The third step is to find all phase portraits modulo limit cycles (see Section 4.2.1 in page 95 for definition of "modulo limit cycles") and the fourth one is to have the complete classification of phase portraits including limit cycles. The fourth step seems to be the hardest one of the four steps. For this fourth step the exact cyclicity of graphics would help. So far we have a rather small number of calculations of exact cyclicities of graphics in quadratic systems (see for example [167, 208]). The total number of distinct graphics in **QS** is not known. In fact, no notion of equivalence of graphics has ever been given, not even by Dumortier et al. [129] although the authors presented an infinity of graphics but affirmed that only 121 of them need to be studied. Clearly they identify certain graphics but without stating an equivalence relation. This reduction is important, and in fact the merit of the program in [129] is to have reduced the number of graphics in **QS** to be studied to this rather small number of 121 graphics. For a more ample discussion about this topic, see Section 2.5.

Here we need to point out, however, that if the general finiteness problem for quadratic systems were solved, the techniques used for this program cannot be of help for solving Hilbert's 16th problem or the topological classification problem for quadratic systems. Indeed, Hilbert asked for the *least upper bound* or the *maximum* of the number of limit cycles in systems of degree n and not for *a bound* for this number. Clearly Hilbert's problem for quadratic vector fields is a lot harder than the problem of finding an upper bound for the number of limit cycles that could occur in quadratic vector fields. And this last problem is a lot harder than showing the existence of such an upper bound, later described in more detail for the quadratic family [129], that still remains unsolved even in this simplest case.

In this book we deal with the first step of the problem of topologically classifying the quadratic class. The theorem that we prove gives the global classification of the global schemes of singularities according to the geometrical equivalence relation. This is a deeper equivalence relation for configurations of singularities than the topological one. For example it distinguishes among the weak foci, and also among weak saddles, of various orders. This distinction is important because weak foci of different orders produce different numbers of limit cycles close to these foci in perturbations. We also distinguish the various kinds of nodes such as one-direction or two-direction nodes or star nodes. These distinctions are also important since the star nodes and the one-direction nodes produce foci close to the nodes in perturbations. This geometrical equivalence relation, first introduced in 2015 [26], is defined in purely algebraic terms and is based on the fact that the blow-up of singularities is a purely algebraic process. The proofs of parts of our theorem were published in several papers [26, 21, 22, 23, 24, 18, 25]. These papers cover all the cases where the total number of finite singularities is at most equal to three. All cases with four distinct finite singularities are only covered in this book together with the cases with a multiplicity four singularity and the cases with an infinite number of finite singularities.

In this book we prove that there are 1765 distinct global geometrical configurations of singularities plus at most 8 other such configurations (sharing the same finite part) that we conjecture are not realizable (see Conjecture 11.1). The proof, which is entirely algebraic, ends the first part of the program.

We want to briefly sum up here the principal moments of our work on the geometrical classification of the configurations of singularities finite and infinite of systems in **QS**:

- Firstly Schlomiuk and Vulpe published an article [301] on the geometry of quadratic differential systems in the neighborhood of infinity where they used some geometrical concepts, earlier defined [299], for singularities at infinity. Using these concepts they gave the topological classification of quadratic systems in the neighborhood of infinity. This classification is done in terms of invariant polynomials, and it refines, using geometric meanings, the set of invariant polynomials previously used [244] for this problem.
- In 2008 [29] Artés, Llibre and Vulpe gave for the first time in terms of invariant polynomials, the topological classification of finite singular points, including the distinctions among singularities: nodes, foci and centers of systems in **QS** (see Table 6.2).
- In 2015 [26] the new notion of geometrical configuration of singularities and of geometrical equivalence of such configurations was given for the first time. It was then applied to obtain the geometrical classifications of geometrical configurations of singular points at infinity of systems in **QS**, in terms of invariant polynomials.
- In 2011 Vulpe published an important result [338] that yielded the algebraic classification of weak foci and saddles, centers and integrable saddles.
- In our papers [21, 22, 23, 24, 18, 25] (2013–2015) we gave the affine invariant classification of global geometrical configurations of singular points, finite and infinite, for all families of quadratic systems, with the exception of the case with four distinct finite singular points, real and/or complex, as well as the case of a unique finite singularity of multiplicity four. These last cases are treated here in Chapter 11. Finally, the cases of degenerate quadratic systems are presented in Chapter 12.

Part I begins with an introductory chapter on problems about polynomial differential systems, some of them open for more than a century. They provide part of the motivation for our work in this book. Part I contains the basic features of the theory of planar polynomial differential systems as well as the study of special singularities of quadratic differential systems. This theory is needed for the geometrical classification theorem of configurations of singularities of quadratic differential systems which forms the content of Part II of the book.

In Chapter 2 we give a brief survey on quadratic differential systems. There are more than one thousand papers written on these systems. For example there is a bibliography of some of these compiled by Reyn [278] that has 426 items plus references of 55 preprints and 10 reports published in the TUDelft series of reports in 1989. Our brief survey will only contain our choice of results, based on their significance in relation to our results presented here, or based on our view of their importance for the problems mentioned earlier in this introduction.

In Chapter 3 we start by defining the Poincaré compactification, next we recall the classical definitions of singularities and motivate the new terminology that we introduce. We continue with a description of the classical blow-up technique and the way we changed it so as to simplify the computations. We include some examples of using this simplification. We continue by defining a new concept called *borsec* (which stands for "border of a sector") that is needed to geometrize the study of singularities. We next define some equivalence relations of singularities, which we combine into the final notion of geometrical equivalence of singularities and the geometrical equivalence of configurations of singularities. Finally we assign a notation to every kind of singularity (finite and infinite). The reader is advised to consult Table A.1 in Appendix A, where a compendium of the notations is exposed with diagrams illustrating the neighborhood of each singularity.

Chapter 4 is devoted to the concept of invariant in mathematical classification problems, and several basic examples are given. Emphasis is placed on problems of classification on planar vector fields with respect to several equivalence relations and on the associated invariants such as the separatrix configuration, which is a topological invariant introduced by Markus [234] and developed by Peixoto [249] and Newmann [242].

Chapter 5 contains the essential tools and strategy for classifying families of planar polynomial differential systems in intrinsic ways, i.e. independent of the normal forms in which such families are presented. It is perhaps the most important chapter apart from the bifurcation diagrams of the global geometrical configurations of singularities (and their proofs) given in Part II as well as the new concepts on which these are based. The chapter starts with a section on the classical theory on invariants for binary and n-ary forms, a theory that inspired the formulation of the theory of invariants of polynomial differential systems of Sibirschi. The next section is devoted to the development of this theory by the Chişinău school founded by Sibirschi and the new developments of the joint work of Vulpe from the Chişinău school, Artés and Llibre from the Barcelona school and D. Schlomiuk. The remaining sections contain the basic concepts of this theory and the construction and geometrical meaning of several invariant polynomials as well as the basic tools needed for obtaining the result on quadratic systems presented in the book. A final section contains information about the role of complex singularities in classifying real quadratic differential systems according to the global geometrical configurations of singularities.

In Chapter 6 we include in a compact way the results of four published papers of ours. These papers form a foundation for all the new results in the book. As the results of each one of these four papers were obtained independently from the other papers, the relations of equivalence and notations included in these papers did not coincide, and in this book we assemble them in a consistent way.

These four works are presented as follows:

- In Section 6.1 we sum up the results in the paper [29] for the configurations of finite singularities. These configurations contain a bit more information than the local topological phase portraits, but they are still far from giving the full geometrical information.
- In Section 6.2 we present the main result of the paper [338] in which the affine invariant conditions for the number and order of weak foci and saddles are given. In this paper it is shown that geometrically the weak foci and saddles play a similar role. The invariant conditions for the order of weak singularities are the same for weak foci and weak saddles, and only the sign of an additional invariant distinguishes them.
- In Section 6.3 we present the main result of the paper [32] (with a minor correction) in which the authors classify by the affine group action, the configurations of singularities (finite and infinite) of quadratic systems with an integrable saddle. Already in this section some elements of the new geometric concepts and notations start to appear without being in their definitive form yet.
- In Section 6.4 the main results for configurations of infinite singularities [26] are given. In this paper already the notion of geometrical equivalence of singularities and the notations had been completely developed.

During the study of families of quadratic systems of total finite multiplicity of singularities 1 to 3, it became necessary to study in detail very specific configurations of singularities that contained a special kind of singularity, whether an integrable saddle or a star node or some combination of weak and/or one-direction node singularities.

In Chapter 7 we give a complete list of the realizable configurations of singularities containing some special kinds of singularities. This chapter contains entirely new, previously unpublished results.

In Part II we present the main results regarding the systematic geometrical classification of all the configurations of singularities, finite and infinite. This half of the book is divided into five chapters that cover all the possibilities of geometrical configurations of singularities. In Chapter 8 we consider the systems with total finite multiplicity of singularities less than or equal to 1. The complete results about this family were previously published [21]. In Chapter 9 we study the systems with total finite multiplicity of singularities equal to 2, results that were published in 2014 [23]. In Chapter 10 we study the systems with total finite multiplicity of singularities equal to 3. This section was published in two articles [24, 22]. In Chapter 11 we study the systems with total finite multiplicity of singularities equal to 4. The contents of some sections were published in two articles [18, 25], more precisely those configurations that have exactly two and three finite singular points, respectively. The remaining sections cover the most generic and difficult cases, for example the cases with four distinct finite singularities, real and/or complex. All these results appear for the first time in this book. In Chapter 12 we study the degenerate quadratic systems, which we conveniently denote by $m_f = \infty$. These results are also new.

We include a chapter with concluding comments in which we indicate the key points of this work and propose possible directions for advances towards completing the list of all phase portraits of quadratic systems modulo limit cycles.

We have included two appendices. In Appendix A we give a table that summarizes the notation we use for singularities and which is defined in Section 3.7. The table provides easy and fast access to the notation.

In Appendix B we offer to the mathematical community the computer tools that we have developed in Mathematica in order to calculate the invariant polynomials and determine the configurations of singularities of subfamilies of quadratic systems. We include a brief manual of these tools, which we hope may be useful for the interested reader.

Chapter 2

Survey of results on quadratic differential systems

We consider here polynomial quadratic differential systems of the form

$$\frac{dx}{dt} = p(x, y), \qquad \frac{dy}{dt} = q(x, y), \tag{2.1}$$

where p, $q \in \mathbb{R}[x, y]$ and $n = \max(\deg(p), \deg(q)) = 2$.

Quadratic differential systems occur often in many areas of applied mathematics, in population dynamics [145], nonlinear mechanics [236, 237, 69], chemistry, electrical circuits, neural networks, laser physics, hydrodynamics [347, 328, 183, 191], astrophysics [80] and others [280, 154, 102]. They are also of theoretical importance as a basic testing ground for the more general theory. Although these are the simplest nonlinear polynomial systems and their theory has been steadily advancing during the last part of the twentieth and the beginning of the twenty-first centuries, of the three classical problems formulated by Poincaré and Hilbert, only the problem of the center was solved for this class while the other two problems remain open, and they are very hard even in this case.

In this chapter we survey the history of quadratic differential systems, give an overview of the main results so far obtained for this family, point out stumbling blocks encountered in solving these very old open problems and look towards the future, indicating new directions of development.

2.1 Brief history of quadratic differential systems

Quadratic systems began to be studied at the beginning of the twentieth century. In 1996 Coppel [103] wrote that the first work which discusses quadratic systems seems to be the 1904 article [61] written by Büchel, but Coppel adds that this work mainly consists of a collection of examples. Coppel also gave a short survey

J. C. Artés et al., *Geometric Configurations of Singularities of Planar Polynomial Differential Systems*, https://doi.org/10.1007/978-3-030-50570-7_2

on quadratic systems. Another short survey on the early history of quadratic systems was written by Chicone and Tian in 1982 [89].

In 1908 Dulac published the first significant work [123] on quadratic differential systems. Using the method first described by Poincaré in his mémoire published in 1885 [263], Dulac succeeded in giving conditions for a complex quadratic system to have a center and he showed that in this case such a system is globally integrable. The notion of center introduced by Dulac is for complex differential systems. In general he calls *center* a singularity with non-zero eigenvalues λ_1, λ_2 whose quotient is negative and rational and around which there exists a local analytic first integral. In this paper on quadratic systems he takes $\lambda_1/\lambda_2 = -1$. After placing the singularity at the origin via a translation of axes, Dulac used the following normal form: $dx/dt = x + p_2(x, y)$, $dy/dt = -y + q_2(x, y)$ where p_2, q_2 are homogeneous polynomials over $\mathbb{C}$ of degree 2. In the case that the coefficients are real, the origin is a saddle and not a center. The conditions of Dulac for a center were thus not readily applicable to systems with real coefficients with a singular point having purely imaginary eigenvalues, and the search for a compact set of conditions for center for real systems in the normal form $dx/dt = -y + p_2(x, y)$, $dy/dt = x + q_2(x, y)$ began.

The first results in this direction were obtained in 1911 by Kapteyn [189, 190]. Kapteyn [189] gave a set of conditions on the coefficients of a real system in the form $dx/dt = -y+p_2(x, y)$, $dy/dt = x+q_2(x, y)$ to have a center at the origin, but these conditions are only sufficient. Later [190] he obtained necessary and sufficient conditions to have a center at the origin. Strangely enough, much later in 1934, Frommer [139] stated as necessary and sufficient for a center, conditions that were neither necessary nor sufficient (see [298] where counterexamples are given in both directions). Four other references [289, 47, 312, 313] appeared in Russian and were never translated. In 1986 Ye et al. [355] stated a "theorem" supposed to summarize Frommer's work. This theorem is different from Frommer's and states as necessary and sufficient, conditions that are only sufficient (see [298]). The development of these problems passed through stages marked by an abundance of errors with an occasional beautiful result such as the work of Bautin, first announced in a short note in 1939 followed by a longer article published in Russian in 1952, which later appeared in English translation in 1954 [46]. In this article Bautin gave necessary and sufficient conditions for a center of a real quadratic differential system, in a compact and neat form and also provided the first important result on limit cycles in quadratic perturbations of a quadratic system with a center. This theorem says that the cyclicity (see the definition in Section 2.5.1) of a center or a *weak focus* of a quadratic differential system is three.

Attempts to obtain all the phase portraits of all quadratic systems with a center were made by Frommer (1934), followed by others, but the complete set of phase portraits was obtained for the first time in 1983 by Vulpe [336]. The turbulent history of the problem of a center in quadratic differential systems was recounted in an article [298] where the global integrability of quadratic systems with a center was also proved.

Towards the end of the twentieth century, the bifurcation diagram of all the quadratic systems with a center was obtained [291, 364, 248]. In these bifurcation diagrams clearly two spaces intervene: the phase space where the phase portraits are drawn, and the global parameter space. In one article [364] the phase portraits are drawn on the affine plane and thus the behavior at infinity is missing. In two other articles [291, 248] the behavior at infinity is included, the phase portraits being drawn on the Poincaré disc. We denote the set of all quadratic systems with a center with **QSC**. Each system in **QS** is defined by twelve coefficients, and hence **QSC** is a subset of $\mathbb{R}^{12}$. The set **QSC** is actually an algebraic set in the parameter space that splits into four (irreducible) algebraic varieties. In all three papers [291, 364, 248] the bifurcation diagrams are drawn on separate pieces according to the normal forms of the four algebraic varieties of **QSC**. A normal form reduces the number of parameters, and in this case it slices the full parameter space on a linear subspace. The neighborhood of a point on such a slice corresponding to a specific normal form may be a limit point of points belonging to another normal form corresponding to another linear subspace, a fact that cannot be seen if we draw the bifurcation diagram on separate diagrams corresponding to the four irreducible components of the set **QSC**, not glued together. This fact was pointed out by Schlomiuk [296, Theorem 3.2]. To have the full picture, one would need to go beyond [291, 364, 248] and construct the *orbit space* (or the *moduli space*) (see Section 5.3.1) under the action of the affine group and time homotheties of the full space **QSC**. As we see, work on the problem of the center for the quadratic case, initiated by Dulac in 1908, spanned a full century, and the final touches on this topic are still lacking.

During the last quarter of the twentieth and the beginning of the twenty-first century, work on quadratic differential systems intensified and a large number of papers appeared on this subject. This work concentrated in several centers, among them Barcelona (the group around Llibre, Gasull and Artés); Beijing (the group around Zhang Zhifen and Chengzhi Li); Chişinău (the group around Sibirschi and Vulpe); Delft (the group around Reyn); Dijon (the group around Roussarie); Lleida (the group around Chavarriga and Giné); Minsk (the group around Erugin, Sadovskii, Rychov and Cherkas); Montreal (Schlomiuk, Rousseau and some of their students); Nanjing (the group around Ye Yan-Qian); Rio de Janeiro (the group at IMPA around Sotomayor); Toulouse (Gavrilov); UK (the group around Lloyd and Christopher); and Warsaw (Żołądek). In the U.S. some people (Chicone, Guckenheimer, Perko and Shafer) also had results on quadratic systems.

Work on quadratic systems was in part motivated by their many applications in various areas of applied mathematics. But the major motivation was the desire to solve the two classical problems remaining open for this family, the second part of Hilbert's 16th problem with its individual and general finiteness parts and Poincaré's problem on algebraic integrability, and also to obtain the topological classification of the global phase portraits of quadratic differential systems.

In 1955 Landis and Petrovski [257, 258] claimed to have proved that the maximum number of limit cycles occurring in quadratic systems is three. The methods used by Petrovki and Landis appeared to be interesting since they were over the complex field. In 1959 [260] they made some corrections on their paper, and later in 1967 [261] they acknowledged an error in the proof of Lemma 12 in the authors' paper [257], which was translated [259].

Chen and Wang [87], and independently Shi Songling [311], surprised the mathematical community by giving precise examples of quadratic systems having at least four limit cycles.

The class of quadratic differential systems forms a five-dimensional space modulo the action of the affine group and time rescaling. This is a large number of parameters, and so people started to work first on subclasses of this family depending on fewer parameters. During the last quarter of the twentieth century and the beginning of the twenty-first century, topological classification of numerous families of quadratic systems were given and bifurcation diagrams of some of these classes were obtained. While bifurcation diagrams were obtained for some three-dimensional families modulo the group action, so far no bifurcation diagrams were obtained for families that modulo the group action are four-dimensional. A survey of results on some classifications of subfamilies of **QS** is given in Section 2.2.

Another direction of work was also on classification problems but rather than consider subfamilies of **QS**, the classifications are of the whole quadratic class according to the specific geometrical global schemes of singularities of systems in **QS**. As indicated in Chapter 1 this is the beginning of the global theory of quadratic differential systems and it forms the topic of this book, where we obtain the global classification of all geometrical configurations of singularities encountered in **QS**.

Another line of work was on the individual finiteness problem for **QS**, solved jointly by Chicone and Shafer and by Bamon, for the graphics in the finite plane [88] and for infinite graphics [44].

The general finiteness problem, also called the existential Hilbert's 16th problem, for quadratic differential equations is still open. An approach to this problem given by Roussarie [283] was followed by a more elaborate program [129], reducing the finiteness part for the class **QS** to the proof of finite cyclicity of 121 limit periodic sets occurring in **QS**. This part will be discussed in Section 2.6.

Another work on the whole class **QS** is [285] where **QS** is viewed as a principal fiber bundle over the pencil of conics.

2.2 Some basic results obtained for quadratic differential systems

The following are some very basic theorems on these systems, some of which are given with proofs in the surveys of Coppel [103] and of Chicone and Tian [89].

(i) A straight line either has at most two (finite) contact points (which may include the singular points) with a quadratic system, or it is formed by trajectories of the system [355, Lemma 11.1]. We recall that by definition a *contact point* of a straight line L is a point of L where the vector field has the same direction as L, or it is zero.

(ii) If a straight line passing through two real finite singular points q_1 and q_2 of a quadratic system is not formed by trajectories, then it is divided by these two singular points in three segments $\overline{\infty q_1}$, $\overline{q_1 q_2}$ and $\overline{q_2 \infty}$ such that the trajectories cross $\overline{\infty q_1}$ and $\overline{q_2 \infty}$ in one direction, and they cross $\overline{q_1 q_2}$ in the opposite direction [355, Lemma 11.4].

(iii) If a finite singular point is situated on the straight line joining two opposite infinite singular points on the Poincaré disc (see Section 3.1) of a quadratic system, then this line is formed by trajectories, or it is a straight line without (finite) contact points except at that finite singular point [355, Lemma 11.5].

(iv) The interior of a closed curve is a convex region [103]. Here a *closed curve* means either a periodic orbit, or a finite graphic that separates the plane into two regions (the interior and exterior regions limited by the graphic).

(v) There exists a unique singular point in the interior region limited by a periodic orbit, and this point is either a focus or a center [103].

The following nice result due to Berlinskii [49] has a short proof given by Kukles and Hasanova [197] mentioned in both surveys [103, 89]. A more recent nice proof, using the Euler-Jacobi formula, was given by Cima, Gasull and Mañosas [93].

(vi) Suppose that a real quadratic system has exactly four real singular points. In this case if the quadrilateral formed by these points is convex, then two opposite singular points are anti-saddles (i.e. nodes, foci or centers) and the other two are saddles. If this quadrilateral is not convex, then either the three exterior vertices are saddles and the interior vertex is an anti-saddle, or the exterior vertices are anti-saddles and the interior vertex is a saddle. In particular, a quadratic system can have at most three saddles, respectively three anti-saddles.

(vii) Two periodic orbits are oppositely oriented if the bounded regions that they limit have no common points, and they have the same orientation if the bounded regions that they limit have a common point [103].

(viii) If a quadratic system has a limit cycle, then it surrounds a unique singular point, and this point is a focus [103].

(ix) If in a quadratic system the separatrix of an infinite saddle connects with the separatrix of the diametrically opposite infinite saddle, then this separatrix is an invariant straight line [319].

(x) If a quadratic system has a center, then it is integrable; i.e. there exists a non-constant analytic first integral defined in the whole real plane except perhaps on some invariant algebraic curves [230, 298, 293].

(xi) A quadratic system may have at most two foci. The proof is trivial using (ii).

The following are important global results on general quadratic systems.

(xii) A quadratic system with two nests of limit cycles has only one limit cycle in one of the two nests, in the case when the total multiplicity of the finite singularities of the system is equal to three [166].

(xiii) The following stronger result was obtained by Zhang Pingguang [358, 359]: If there are limit cycles surrounding two foci of a quadratic system, then around one of the foci there is at most one limit cycle. Like the results of Ecalle and Ilyashenko (mentioned in Chapter 1) on the finiteness of the number of limit cycles in an individual polynomial differential system, this result of Zhang Pingguang has so far not been checked by the mathematical community.

(xiv) A. Zegeling and R. Kooij proved that a system with four real finite singular points and two nests of limit cycles has only one limit cycle around each focus [357].

(xv) A corollary of this last result is that any configuration of limit cycles having two nests and more than two limit cycles can only happen in a situation with two foci and two complex singular points. Indeed, otherwise either we have a double finite singularity or a multiple infinite singularity coming from a coalescence of finite and infinite singularities, and by means of a perturbation, one could produce something contradicting statement (xiv).

The next results involve the notion of *weak focus* which is a focus with purely imaginary eigenvalues.

(xvi) There are no limit cycles in quadratic systems surrounding a weak focus of third order [202]. (A second proof of this result can be found in [358].)

(xvii) In quadratic systems there is at most one limit cycle surrounding a weak focus of second order, and when it exists it is hyperbolic [358].

(xviii) Quadratic systems with a weak focus of second order can have at most two limit cycles, and they must occur in a configuration $(1,1)$ [19].

(xix) If a quadratic differential system has two real invariant straight lines, then it has no limit cycles [45].

(xx) If a quadratic differential system has two complex invariant straight lines, meeting at a real point, then it has at most one limit cycle [326]. (The proof given in [326] is incomplete. A complete proof can be found in [308].)

(xxi) If a real quadratic system has one real invariant straight line, then it has at most one limit cycle [100, 104].

(xxii) If $H(x,y)$ is a real cubic polynomial with four distinct critical values, then all real quadratic vector fields sufficiently close to the Hamiltonian vector field $dx/dt = -\partial H/\partial y$, $dx/dt = \partial H/\partial x$ have at most two limit cycles [147].

Statement (xxii) was proved by L. Gavrilov [147] by complex analytic methods. This result settles the infinitesimal version of Hilbert's 16th problem for quadratic vector fields. In the review [353] of this work in MathSciNet, Yakovenko wrote "Yet the theorem proved by Gavrilov is one of the few results which are global with respect to the phase space. It covers all limit cycles appearing in the perturbation, and not just small neighborhoods of polycycles (in particular, singular points). The theorem proved in the paper concludes a long line of research and is a remarkable achievement". Another proof of this result using only real analysis was given by F. Chen, C. Li, J. Llibre and Z. Zhang [85].

2.3 Study of some subclasses of the family of quadratic differential systems

The study of the class **QS** has proved to be quite a challenge, as hard problems formulated more than a century ago are still open for this class. It is expected that we have a finite number of phase portraits in **QS**, but this number may even be greater than 2000. Although we have phase portraits for several subclasses of **QS**, listing all distinct phase portraits of the whole **QS** is still a very complex task, not within reach for the moment. This is partly due to the elusive nature of limit cycles and partly to the rather large number of parameters involved, five modulo the group action. For the moment only subclasses depending on at most three parameters have been studied globally, including their global bifurcation diagrams.

The first subclass of **QS**, which was studied during the last part of the twentieth century, is the family of all systems possessing a center. As indicated in Section 2.2, work on this problem spanned a century. A second family for which work spanned again almost a century was the class of Lotka–Volterra differential systems.

Work by Lotka [228] and Volterra [334] in population dynamics led to the study of these differential systems, which occur very often in applications. Due to their many applications, the literature is ample. On these systems we also have some theoretical studies. There were four attempts to topologically classify these systems [276, 350, 149, 77]. As explained by Schlomiuk and Vulpe [306, 307], each one of these papers has shortcomings and none gave the full topological classification of Lotka–Volterra systems. The lack of adequate global geometrical concepts at the time these papers were written explains in part this situation. The complete topological classification was obtained [307] using the geometrical global concept of configuration of invariant lines of a polynomial differential system (introduced in [300] and also used in [302, 304, 305, 306, 307]). This notion proved to be very efficient for topologically classifying the Lotka–Volterra systems leading to 112 topologically distinct phase portraits, exhibited [307] together with the bifurcation diagram of this family of systems. This notion also proved to be very helpful for topologically classifying the family of quadratic systems possessing invariant lines of total multiplicity at least four [303, 304].

We call *configuration of invariant straight lines* of a polynomial differential system (2.1), the set of all its invariant straight lines (real or complex), each endowed with its own multiplicity, together with all the real singular points of (2.1) located on these lines, each one endowed with its own multiplicity.

A survey on classification problems for polynomial differential systems [296, Section 3.2] was done and in particular, apart from the families mentioned above, several other families of quadratic differential systems are listed. The classifications are of two kinds: (A) without the use of topological invariants or invariant polynomials with respect to the group action of affine transformations and time homotheties, and (B) using both types of invariants. The earliest (correct) work of this kind was done for quadratic homogeneous systems, i.e. systems where p, q are quadratic homogeneous polynomials [344, 345]. All papers in the category (B) not only give the lists of phase portraits but also give, for each phase portrait, affine invariant necessary and sufficient conditions for its realization. The fact that these conditions are expressed in invariant form means that no matter how a system is presented, one can easily check if this system belongs to the family and, if it does, what its phase portrait is. Moreover, there are families of quadratic differential systems for which, modulo the group action of affine transformations and time homotheties, the bifurcation diagrams were done in the two-dimensional or three-dimensional projective spaces. These studies involve algebraic geometric methods, invariant polynomials and analytical and numerical studies. We will indicate those studies as (C). Here comes an extensive list of them:

- Quadratic systems that are structurally stable modulo limit cycles (i.e. all limit cycles surrounding the same singularity are coalesced to that singularity) have been topologically classified [12] (A).
- the global phase portraits of quadratic differential systems without finite singularities, also called quadratic foliations or chordal systems, have been topologically classified [142, 143] (A). Inside the quadratic foliations the ones that are structurally stable were also obtained [181] (A). The chordal quadratic systems have been classified [332] (B) using affine invariant polynomials.
- Semi-linear quadratic systems, i.e. one of the equations is defined by a polynomial of the form $ax + by$ [72, 226] (A).
- Systems with a unique finite singularity [98] (A).
- Systems with a focus and one anti-saddle modulo limit cycles [14](A,C).
- Quadratic and cubic systems with all points at infinity as singularities [94, 144] (A) and [305] (B).
- Quadratic systems with a higher-order singularity with two zero eigenvalues [180] (A,C).

- Quadratic systems of Darboux type [339] (B).
- Quadratic systems having a Darboux rational first integral (or a Darboux inverse integrating factor) [28] (B) and [74, 218, 219] (A).
- Quadratic systems having a Darboux polynomial first integral [30] (B) and [140] (A).
- Quadratic systems with a center [291, 248, 364] (A) and [336] (B).
- Quadratic Hamiltonian systems [13] (A,C), [15, 248] (A) and [187] (B).
- Bounded quadratic systems [97, 126, 207] (A).
- Reversible Darboux integrable quadratic systems [217] (A).
- Quadratic systems with a polynomial inverse integrating factor [95, 96] (A).
- Liouvillian integrable Lotka–Volterra quadratic systems [71] (A).
- homogeneous quadratic systems [232, 235] (A) and [120] (B).
- Quadratic systems with an integrable saddle [32] (B) and [52] (A).
- Quadratic systems with a symmetry center and simple infinite singular points [231, 340] (B).
- Quadratic systems with a unique finite singular point of multiplicity two, possessing two zero eigenvalues [243] (B).
- Quadratic systems with a single finite singularity that in addition is simple [333] (B).
- Quadratic systems with a finite singular point of multiplicity four [331, 342] (B)
- Quadratic systems with a singular point of multiplicity three [341] (B).
- Quadratic systems with invariant straight lines of total multiplicity greater than or equal to four [300, 302, 304, 303] (B).
- Quadratic Lotka–Volterra differential systems [73] (A) and [306, 307] (B).
- Homogeneous quadratic systems [344, 345] (B).
- Quadratic systems with a polynomial first integral [227] (A) and [30] (B).
- The bifurcation diagram of the class **QW3** of two-dimensional quadratic systems with a weak focus of order three in the two-dimensional real projective space [16, 222] (A,C) and [19] (B,C).
- The bifurcation diagram of the three-dimensional class **QW2**, modulo the group action, of quadratic systems with a weak focus of order two in the three-dimensional real projective space [19] (B,C) (this work revealed several interesting insights into the quadratic family, such as the fact that all limit cycles occurring in this class are topologically equivalent to perturbations within **QW2** of symmetric quadratic systems with a center).

- Quadratic systems with a weak focus and an invariant straight line [20] (B,C).
- Quadratic systems with a semi-elemental triple node [33] (B,C).
- Quadratic systems with a finite and an infinite saddle-node of multiplicity $\binom{0}{2}$ [34, 35] (B,C).

2.4 Algebraic limit cycles in quadratic systems

In 1958 Qin Yuan-Xun [275] proved that quadratic polynomial vector fields can have algebraic limit cycles of degree 2, and when such a limit cycle exists, then it is the unique limit cycle of the system.

Evdokimenko [133, 134, 135] proved that quadratic vector fields do not have algebraic limit cycles of degree 3. (For two different shorter proofs, see [82, 83, 216].)

In 1966 Yablonskii [352] found the first class of algebraic limit cycles of degree 4 inside the quadratic vector fields. The second class was found in 1973 by Filiptsov [136]. More recently two new classes have been found by Chavarriga et al. [83] where the authors proved that there are no other algebraic limit cycles of degree 4 for quadratic vector fields. The uniqueness of these limit cycles has been proved [81]. Some other results on the algebraic limit cycles of quadratic vector fields can be found in [223, 224].

By performing some convenient birational transformations on the plane to quadratic vector fields having algebraic limit cycles of degree 4, Christopher et al. [92] obtained algebraic limit cycles of degrees 5 and 6 for quadratic vector fields. Of course, in general a birational transformation does not preserve the degree of a polynomial vector field.

The following questions related to the algebraic limit cycles of quadratic polynomial vector fields remain open (see for instance [216]).

(i) What is the maximum degree of an algebraic limit cycle of a quadratic polynomial vector field?

(ii) Does there exist a chain of convenient rational transformations of the plane (as in [92]) that gives examples of quadratic systems with algebraic limit cycles of arbitrary degrees, or at least of degree greater than 6?

(iii) Is 1 the maximum number of algebraic limit cycles that a quadratic system can have?

For a quadratic differential system with a "generic" configuration of invariant algebraic solutions, the number of algebraic limit cycles is at most one. (For the term "generic" and for a proof, see [220, 221, 360].)

2.5 Finiteness problems for quadratic differential systems

As mentioned in Chapter 1, the *general finiteness problem for limit cycles* (or the *existential Hilbert's 16th problem*) asks for a proof that for any fixed n there exists a finite bound for the number of limit cycles that a polynomial differential system of degree n could have. So far the general finiteness problem is not even solved for the "simplest" case of quadratic differential systems. However, a lot of work has been done for this special case. For a brief description of this work, we need some concepts and preliminary results.

2.5.1 Basic concepts and results needed for studying the general finiteness problem

Françoise and Pugh [137] defined the concept of *limit periodic set* and of *graphic*. To introduce these concepts, we first compactify the set of all planar polynomial vector fields of degree n as follows: If

$$\begin{aligned} X =& p(x,y)\frac{\partial}{\partial x} + q(x,y)\frac{\partial}{\partial y}, \\ \text{with } p(x,y) =& \sum_{i+j\le n} p_{ij}x^i y^j, \\ \text{and } q(x,y) =& \sum_{i+j\le n} q_{ij}x^i y^j, \end{aligned}$$

then via a time rescaling of positive constant $\sum_{i,j}(p_{ij}^2 + q_{ij}^2) = 1$. So the set of all such vector fields is the unit sphere $\mathbb{S}^N$ with $N = (n+1)(n+2) - 1$.

A compact subset γ_0 of $\mathbb{S}^2$ is a *limit periodic set* for a vector field X_0 if there exists a sequence of vector fields X_k with periodic solutions $\{\gamma_k\}$ such that $X_k \to X_0$ in $\mathbb{S}^N$ and $\{\gamma_k\}$ converges to γ_0, in the topology given by the Hausdorff distance of compact sets K_1, K_2:

$$\operatorname{dist}_H(K_1,K_2) = \sup_{x\in K_1, y\in K_2} (\operatorname{dist}(x,K_2), \operatorname{dist}(y,K_1)).$$

A limit periodic set is *degenerate* if it contains non-isolated singularities.

A *graphic* is a loop formed by singular points p_i, $i = 1,2,\ldots,m+1$, where $m \ge 0$, $p_{m+1} = p_1$, and separatrices s_i, $i = 1,2,\ldots,m$ (if $m > 0$) connecting them, such that p_j is the α-limit set of the orbit s_j and p_{j+1} is the ω-limit set of s_j.

Remark 2.1. We point out that this definition includes the case of a graphic with only one singular point and no other orbit ($m = 0$).

A *degenerate graphic* is a loop formed by singular points p_i, $i = 1,2,\ldots,m+1$, where $p_{m+1} = p_1$ and orbits s_j, $j \in S$, where S is a strict subset of $\{1,\ldots,m\}$

such that for each orbit s_j the α-limit set of s_j is p_j, the ω-limit set of s_j is p_{j+1} and arcs of curves s_k connecting p_k with p_{k+1} are filled with singular points for k belonging to the set $\{1, \dots, m\}\backslash S$.

An *irreducible graphic* is a graphic with only hyperbolic or semi-hyperbolic singular points.

The *hyperbolicity ratio* of a hyperbolic saddle with eigenvalues $\lambda_1 < 0 < \lambda_2$ is the number $r = -\lambda_1/\lambda_2$. A graphic is said to be *generic* when all its singular points are hyperbolic and the product of all its hyperbolicity ratios is different from 1.

A graphic that admits a first return map on one of its sides is called a *polycycle*.

It is well known that the Poincaré–Bendixon Theorem implies the next result.

Proposition 2.1. *Limit periodic sets of planar polynomial vector fields are graphics, periodic solutions, singular points or degenerate graphics.*

Another notion needed here is the concept of cyclicity of a graphic. The word *cyclicity* was used by Bautin [46] for the case of a graphic reduced to a single singular point. The concept of *finite cyclicity* was introduced by Roussarie [283], whose definition we give next:

Let X_λ be a family of analytic vector fields on $\mathbb{S}^2$ with $\lambda \in S$ where S is a compact parameter set in $\mathbb{R}^k$ for some $k \geq 1$. We say that a *compact subset* Γ *of* $\mathbb{S}^2$*, which is invariant by* X_{λ_0}*, has finite cyclicity in the family* X_λ if and only if there exist $N < \infty$, $\varepsilon > 0$ and a neighborhood V of $\lambda_0 \in S$ such that, for any $\lambda \in V$, the number of limit cycles γ of X_λ with $\text{dist}_H(\gamma, \Gamma) \leq \varepsilon$ is less than N. The *cyclicity* of Γ is the infimum of such N when ε and the diameter of V go to zero.

2.5.2 The general finiteness problem for quadratic differential systems

The work on the *general finiteness conjecture* was initiated by Roussarie [283]. His paper can be summed up as follows.

Roussarie first generalized the global finiteness conjecture stated for polynomial vector fields in Chapter 1, to a global conjecture stated for analytic families of vector fields on the sphere, arising from analytic families of polynomial vector fields on the plane.

Conjecture 2.1 (Global finiteness conjecture). *Let* X_λ *be a family of analytic vector fields on* $\mathbb{S}^2 \times S$ *arising from a family of planar polynomial vector fields by compactification, with* $\lambda \in S$ *where* S *is in an analytic compact manifold of parameters. Then there exists a bound* C *such that for any* $\lambda \in S$ *the number of limit cycles of* X_λ *is less than* C*.*

Roussarie then states the following local finiteness conjecture:

Conjecture 2.2 (Local finiteness conjecture). *Let X_λ be a family of analytic vector fields on $\mathbb{S}^2 \times S$ arising from a family of planar polynomial vector fields by compactification, with $\lambda \in S$ where S is an analytic compact manifold of parameters. Then each limit periodic set of a vector field in the family has finite cyclicity in the family.*

Roussarie then proves the following:

Proposition 2.2. *Let X_λ be any family of analytic vector fields on $\mathbb{S}^2$ with $\lambda \in S$ where S an analytic compact manifold of parameters. Suppose that each limit periodic set has finite cyclycity in X_λ. Then there exists a uniform bound C such that for any $\lambda \in S$ the number of limit cycles of X_λ is less than C.*

In the last section of his paper [283], Roussarie discusses the reduction of the local conjecture to studying the specific cases of limit periodic sets, such as the weak foci or centers, cusp points, periodic orbits, hyperbolic loops, hyperbolic graphics and irreducible graphics, then the general graphics and isolated singular points and finally the degenerate limit periodic sets.

Roussarie did not mention specifically quadratic differential systems, but it is clear that the breaking up of the local finiteness conjecture into a finite number of possible specific cases of limit periodic sets, and proving finite cyclicity for these cases, must start with the consideration of quadratic systems. The general case is far more complicated and beyond our reach at this state. However, even for this case, proving finite cyclicity for all such possible limit periodic sets is not a simple endeavour.

2.5.3 Application of Roussarie's ideas for the quadratic case

The ideas of Roussarie were further applied [129] for quadratic differential systems[1]. In the initial paper of Roussarie [283] the global finiteness problem is reduced to the problem of proving finite cyclicity of limit periodic sets. The list of limit periodic sets that need to be investigated must be as short and as simple as possible. A specific list of 121 limit periodic sets was given [129] in order to prove the local finiteness conjecture for **QS**. We point out that all 121 cases have limit periodic sets that surround a focus or a center or are reduced to a singular point. For this reduction to just 121 graphics, extensive use of the hypothesis that we work on the specific family of quadratic systems was made. This is a rather small number of limit periodic sets, degenerate or not, occurring in the family **QS**, and this is the merit of the paper [129]. The program of reducing the proof of the general finiteness to the proof of finite cyclicity of only 121 graphics in quadratic systems was presented at the SMS of Université de Montréal in the summer of 1992 by Rousseau (see [286]).

[1]The authors are grateful to Roussarie for discussions and clarifications regarding the paper [129].

The core of the work in the paper [129] was however not entirely understood, even by some of the most prominent mathematicians in the field. Thus referring to [129], Ilyashenko wrote: "*a complete list of 121 polycycles that may occur for quadratic vector fields is presented*" [176, Page 324].

Firstly we point out that the list of 121 graphics is very far from being a complete list of graphics in quadratic systems. Actually Dumortier et al. [129] exhibited an infinite number of graphics occurring in quadratic systems but claimed that it suffices to prove finite cyclicity only for 121 of them. Secondly not all the 121 graphics are polycycles as it is stated by Ilyashenko. Indeed, for example the graphics H_6^3, H_7^3 as well as many others in [129] are not polycycles since they do not admit a Poincaré return map. Thirdly, we point out that the authors considered only the limit periodic sets surrounding a focus or a center. In **QS** we have many other graphics, but the authors claim that these other graphics are irrelevant for the proof of the local conjecture. Examples of graphics not listed among the 121 graphics [129] can easily be found, see for example [302, Pictures 5.25 to 5.26 and 5.29]. The merit of the program [129] is exactly the fact that we do not have to check finite cyclicity for *all the graphics* occurring in quadratic differential systems. We only have to check finite cyclicity for the 121 graphics listed.

The basic idea of Roussarie [283] is to compactify both the phase space and the parameter space. The quadratic systems on the plane are compactified by using the Poincaré compactification of the sphere. Indeed, for every vector field X on the plane we can construct an analytic vector field $\hat{X}$ on the sphere such that its restriction to the upper hemisphere and the vector field X have topologically equivalent phase portraits. We can also compactify the parameter space that for a system of degree n is $\mathbb{R}^N$ with $N = (n+1)(n+2)$. Indeed, suppose that the coefficients of a system of degree n are p_{ij}, q_{ij}; then we may suppose that these coefficients are such that $\sum_{i+j\leq n}(p_{ij}^2 + q_{ij}^2) = 1$ and hence the parameter space is a sphere $\mathbb{S}^{N-1}$ in $\mathbb{R}^N$. In the quadratic case $N = 12$ and $\mathbb{S}^{N-1}$ is the 11-dimensional sphere. So both the parameter space and the phase space can be assumed compact.

The fact that this space $\mathbb{S}^{N-1}$ is compact plays an essential role in solving the global finiteness problem for quadratic vector fields. This is done by using the following result (which is Theorem 2.6 of [129] indicated as having been proven previously in [283]). We point out however that an essential ingredient, namely that the family X_λ is analytic, is missing in the statement that appeared earlier [129], where the result is stated as follows:

Theorem 2.1. *Let $\{X_\lambda\}_{\lambda\in S}$ be a family of vector fields defined on the Poincaré sphere $\mathbb{S}^2$ and indexed by λ in a compact set S. Suppose that each limit periodic set has finite cyclicity. Then there exists a uniform bound K, independent of λ, for the number of limit cycles of any vector fields X_λ.*

The next thing to use is the fact that a quadratic system with a limit cycle has a unique singular point inside the limit cycle, which is a focus (see [103]). Due to this such a system can clearly be brought to the following normal form:

$$\frac{dx}{dt} = \lambda x - \mu y + \epsilon_1 x^2 + \epsilon_2 xy + \epsilon_3 y^2, \qquad \frac{dy}{dt} = \mu x + \lambda y + \delta_1 x^2 + \delta_2 xy + \delta_3 y^2, \quad (2.2)$$

with $(\epsilon_1, \epsilon_2, \epsilon_3, \delta_1, \delta_2, \delta_3) \neq 0$. Using a time rescaling, (λ, μ) can be assumed to be in $\mathbb{S}^1$. Then a homothety on the axes $(x, y) \mapsto (x/u, y/u)$ allows one to pass from $(\epsilon_1, \epsilon_2, \epsilon_3, \delta_1, \delta_2, \delta_3) \in \mathbb{R}^6$ to the real projective space $\mathbb{PR}^5$. Hence it is sufficient to study phase portraits for systems (2.2) for parameter values in $\mathbb{S}^1 \times \mathbb{PR}^5$.

Thus we only need to prove the local conjecture for this family of systems. Furthermore, by the above mentioned argument, it suffices to consider only limit cycles surrounding the origin for the normal form (2.2).

Dumortier et al. [129] claim that they can reduce the problem even further by considering that it is only necessary to prove the finite cyclicity for limit periodic sets that surround the origin in (2.2), and they produce the list of such limit periodic sets.

A survey of the program [283, 129] up to the year 2004 was given by Rousseau [287]. In a section titled "The status of all graphics" [287, Section 4.7], the author's intention was to summarize the progress on the finiteness program for quadratic systems since the beginning of the 1990s and up to 2004 when her paper was published.

The graphics surrounding a center are of particular importance, especially those surrounding a symmetric center. How many are they and for how many of them was finite cyclicity proved? This summary [287] does not answer these questions, and no other summary has appeared since 2004.

Moreover this summary contains some typos. For example the graphic I^2_{16a} [129] occurs in this summary in two clearly disjoint families, the family of graphics already studied and the family of graphics to be studied, while I^2_{16b} is completely missing. One of these two occurrences of I^2_{16a} must be the missing I^2_{16b}.

Furthermore we point out that in [287], we see listed as degenerated graphics (DF_{3a}), (DF_{3b}), (DF_{4a}) and (DF_{4b}). However, these last notations do not appear in the original list of 121 graphics [129], where all the DF graphics mentioned are $DF_{1a}, DF_{1b}, DF_{2a}, DF_{2b}$.

Finally, Rousseau [287] expresses her belief that the list of 121 graphics for proving finite cyclicity is incomplete, and she presents four graphics that should be studied twice because they present two distinct kinds of blow-ups of their singularities. This observation is significant because from the graphics in [129] which contain an elliptic-saddle, Rousseau realized [287] that they had considered only the possibility of having a generic nilpotent elliptic-saddle (the one that we later denote by $\widehat{es}_{(3)}$, see Section 3.7). She also realized that four of the cases had to be considered also under the possibility of the elliptic-saddle being non-generic (the one that we later denote $\widehat{es}^{-}_{(3)}$). They are geometrically different elliptic-saddles since their respective blow-ups are not equivalent, so perturbing the system containing the graphic may have an impact on its cyclicity. Moreover, perhaps other cases such as H^3_8 could also occur with a non-generic elliptic-saddle. This observation of Rousseau has greatly helped us not to commit the same mistake in our classification.

In 2016 the authors [288] say that a total of 88 graphics of the 121 graphics listed in [129] have been proved to have finite cyclicity.

2.5.4 The infinitesimal Hilbert's 16th problem and the Hilbert–Arnold problem

An *Abelian integral* is a real analytic function $I(h)$ obtained integrating a polynomial 1-form $\omega = p(x,y)dx + q(x,y)dy$ over closed ovals of a real algebraic curve $H(x,y) = h$, where h is a constant.

The infinitesimal Hilbert's 16th problem, is the problem of obtaining upper bounds for the number of isolated zeros of Abelian integrals. More precisely, for fixed integers m and n find the maximum number $Z(m,n)$ of (isolated) zeros of the Abelian integrals $I(h) = \oint_{H=h} \omega$, where $m = \deg(H)$ and n is the maximum degree of the polynomials p and q.

Khovanskii [192] and Varchenko [330] proved that for given m and n the number $Z(m,n)$ is uniformly bounded with respect to the choice of the polynomial H, the family of ovals γ_h and the 1-form ω. In other words, they proved the existence of a bound without giving an explicit value for this bound.

In 1997 Petrov [256] provided an asymptotic upper bound for the maximal number of zeros $Z(m,n)$. An explicit estimate for $Z(m,n)$ was given in 2010 by Binyamini, Novikov and Yakovenko in the form of a double exponential $\exp(\exp(P(m,n)))$, where $P(m,n)$ is a certain polynomial of degree less than or equal to 60 (see [51]).

The infinitesimal Hilbert 16th problem is sometimes also called the weak Hilbert 16th problem, or the tangential Hilbert 16th problem as Binyamini et al. say:

> The central result of this paper is an explicit upper bound for the number of limit cycles born from nonsingular (smooth compact) energy level ovals in a non-conservative polynomial perturbation of a polynomial Hamiltonian vector field on the plane. This problem was repeatedly posed in various sources under different names as the weakened, infinitesimal or tangential Hilbert 16th problem. In this introductory section we briefly outline some connections between different problems concerning limit cycles of polynomial vector fields on the plane. [51]

Much more complete expositions can be found in the surveys [176, 354] and the books [179, 365].

Finally in 2012 Binyamini and Dor established a fully explicit upper bound for $Z(m,n)$ that is linear in m (see [50]).

At first sight the above problem does not appear to be related to Hilbert's 16th problem. However these two problems are related as follows: The function $I(h)$ given by the Abelian integral is the first approximation with respect to a small parameter ε of the displacement function on a transversal segment to the ovals $H(x,y) = h$ for the system

$$\dot{x} = -\frac{\partial H(x,y)}{\partial y} + \varepsilon f(x,y), \quad \dot{y} = x\frac{\partial H(x,y)}{\partial x} + \varepsilon g(x,y),$$

where H, f and g are polynomials in x, y with real coefficients. The number of isolated zeros of $I(h)$ taking into account their multiplicities gives an upper bound of the number of limit cycles of this polynomial differential system with ε small.

The Hilbert–Arnold problem is stated as follows:

> Is it true that for a generic finite parameter family of smooth vector fields on the 2-sphere the number of limit cycles of the equations in the family is uniformly bounded with respect to the parameter, provided that the parameter set is compact? [176]

According to Ilyashenko [176], only a restricted version of the Hilbert–Arnold problem was solved before the publication of his survey in 2002.

We call an *elementary polycycle* a polycycle whose singular points are elementary. Denoting by $E(k)$ the maximum number of limit cycles that may bifurcate from an elementary polycycle in a typical k-parameter family of smooth planar vector fields, a theorem of Ilyashenko and Yakovenko [178] says: For any k, the number $E(k)$ exists. The following result was proved by Kaloshin [188]: For any k, $E(k) < 2^{25k^2}$.

2.5.5 The infinitesimal Hilbert's 16th problem for quadratic differential systems

Let us consider a polynomial differential system $dx/dt = p(x,y)$, $dy/dt = q(x,y)$, with $\max(\deg(p), \deg(q)) = 2$ and its associated differential form $\omega = q(x,y)dx - p(x,y)dy$. Let H be a first integral of a quadratic polynomial differential system having a center. Then the weak Hilbert's 16th problem for quadratic differential systems is the study of the zeros of the Abelian integral $I(h)$ (see [90]).

Gavrilov and Horozov [148] initiated the study of quadratic perturbations of a Hamiltonian quadratic system with four distinct real finite singularities, three of them saddles and the fourth one a center. They gave an upper bound of three limit cycles arising in such perturbations.

In 1998 Iliev published a paper [168] where he studied the bifurcations of limit cycles from the periodic orbits surrounding a center of a quadratic system under general quadratic perturbations. For any of the cases, he determined the essential perturbations, which turned out to have at most three parameters, and computed the corresponding bifurcation function $I(h)$. In view of the fact that the space of quadratic vector fields is 12-dimensional and that modulo the group of affine transformations and time rescaling it is five-dimensional, this reduction to at most three parameters is of great value. After the publication of this work, almost all results obtained for perturbations of quadratic centers used the normal forms of Iliev.

In 2001 Gavrilov [147] proved that if $H(x, y)$ is a real cubic polynomial with four distinct real critical values, then all quadratic vector fields sufficiently close to the Hamiltonian vector field $dx/dt = -\partial H/\partial y$, $dy/dt = \partial H/\partial x$ have at most two limit cycles. The result was proved by using complex analysis methods.

A more general result saying that within the class of quadratic vector fields, the perturbations of any Hamiltonian quadratic system possessing a center, and different from the systems possessing three invariant straight lines intersecting transversally, have at most two limit cycles, was proved by Chen, Li, Llibre and Zhang [85] using only real analysis methods. The proof summarizes several previous results, including the one of Gavrilov just mentioned, but also settles the infinitesimal version of Hilbert's 16th problem for Hamiltonian quadratic vector fields.

In the second part of the book of Christopher and Li [90], written by Li, there is a good survey of results about the weak Hilbert's 16th problem for quadratic differential systems published before 2005. This book contains 190 references on Hilbert's 16th infinitesimal problem, in particular many references on quadratic systems.

In 2012 Li published a survey paper [203] that is a follow-up to the book [90]. In the first part of this survey, methods about the application of Abelian integrals to the study of limit cycles and to Hilbert's 16th problem are given. Due to these two surveys of Li, here we only give the references we know on results for quadratic systems from 2011 on.

All the published papers after 2011 on perturbations of quadratic systems with a center using Abelian integrals that we found, are on quadratic reversible systems with a center [86, 99, 127, 162, 165, 163, 164, 169, 204, 205, 212, 213, 250, 251, 252, 253, 310, 351, 363]. The reversible systems are much harder to study than the Hamiltonian ones because the first integrals are in this case not given by polynomials or by rational functions but by Darboux functions. These studies produced results for isolated points, or for segments, or for arcs in the two-dimensional parameter space of a normal form of quadratic reversible differential systems. In [206] systems in a class of non-Hamiltonian centers are perturbed inside the quadratic class, and the authors prove that the cyclicity of the period annulus is one.

2.6 The initial steps in the global theory of quadratic differential systems

Another direction of work is on classifying the whole quadratic class according to the global configurations of singularities of the systems. We obtain in this book the global geometrical classification theorem of all geometrical configurations of singularities, finite and infinite, encountered in **QS**. This completes the first part (algebraic) of the global theory of quadratic differential systems.

At this stage we are very far from having a substantial global theory of the class of quadratic differential systems. Some results listed in the previous sections have, however, a certain global aspect: for example the topological classification of structurally stable quadratic systems, modulo limit cycles [12], and the topological classification of structurally unstable quadratic systems of codimension one, modulo limit cycles [17], or the topological classification of all quadratic systems with invariant straight lines of total multiplicity greater than or equal to four [302, 303]. The work on Lotka–Volterra systems [306, 307] is also of a global character. These theories are all for strict subfamilies of the quadratic class.

Work on the global theory of singularities of quadratic differential systems began in the last part of the twentieth century and at the beginning of the twenty-first century. This work was initiated in 1987 by Coll [94] who characterized all possible phase portraits in a neighborhood of infinity in terms of the coefficients of the normal forms for all quadratic systems. In 1997 Nikolaev and Vulpe [244] classified topologically the configurations of singularities at infinity in terms of invariant polynomials. Using geometrical concepts [299] and additional concepts [301], Schlomiuk and Vulpe simplified the invariant polynomials in [244] to render more transparent the classification. To reduce the number of phase portraits by half, in both cases the *topological equivalence relation* was taken to mean the existence of a homeomorphism of the phase plane carrying orbits to orbits and *preserving or reversing* the orientation. By *topological classification* of a family of polynomial vector fields we mean the classification with respect to topological equivalence. Artés et al. [29] classified topologically (adding also the distinction between nodes and foci) the whole quadratic class, according to configurations of their finite singularities.

In the topological classification no distinction was made among the various types of foci or saddles, strong or weak of various orders. However these distinctions of an algebraic nature are very important in the study of perturbations of systems possessing such singularities. Indeed, the maximum number of limit cycles that can be produced close to the weak foci in perturbations depends on the orders of the foci.

The distinction among weak saddles is also important because, for example, when a loop is formed using two separatrices of one weak saddle, the maximum number of limit cycles that can be obtained close to the loop in perturbations depends on the order of the weak saddle. (For example see [201, 1, 282].)

The geometrical equivalence relation for finite or infinite singularities (introduced in [26] and used in [21]) takes into account such distinctions. This equivalence relation is finer and deeper than the *qualitative equivalence relation* introduced by Jiang and Llibre [182] because it distinguishes among the foci (or saddles) of different orders and among the various types of nodes. This equivalence relation also induces a finer distinction among the more complicated degenerate singularities.

We recall that a focus (or saddle) with trace zero is called a weak focus (weak saddle). In quadratic systems these could be of orders one, two or three [46]. For details on Poincaré–Lyapunov constants and weak foci of various orders, we refer to [293, 222]. Necessary and sufficient conditions for a quadratic system to have weak foci (saddles) of orders i, $i = 1, 2, 3$ have been given in invariant form [338].

Algebraic information may not be significant for the local (topological) phase portrait around a singularity. For example, topologically there is no distinction between a focus and a node or between a weak and a strong focus. However, as indicated before, algebraic information plays a fundamental role in the study of perturbations of systems possessing such singularities.

Although we now know that in trying to understand these systems, there is a limit to the power of algebraic methods, these methods have not been used far enough.

In this book we obtain this global geometrical classification by purely algebraic tools. This classification has yielded 1765 distinct geometrical configurations of singularities and 8 additional configurations (sharing the same finite configuration) that we conjecture are not realizable. The first step in this direction is [26] where the global classification of singularities at infinity of *the whole class* **QS**, according to the *geometrical equivalence relation* of configurations of infinite singularities, was done by using only algebraic methods. This work was extended to also incorporate all the finite singularities. We initiated the work in this direction [21] where this classification was done for all global configurations of singularities that have a total finite multiplicity $m_f \leq 1$, and we continued the classification for $m_f = 2$ [23]. In this book we give the full geometrical classification of global configurations of singularities, which, as we indicated in Chapter 1, is the first step in obtaining the topological classification of **QS**.

Chapter 3

Singularities of polynomial differential systems

We consider here polynomial differential systems of the form

$$\frac{dx}{dt} = p(x,y), \qquad \frac{dy}{dt} = q(x,y), \tag{3.1}$$

where p, $q \in \mathbb{R}[x,y]$, i.e. p, q are polynomials in x, y over $\mathbb{R}$. We recall that the degree of a system (3.1) is the integer $n = \max(\deg(p), \deg(q))$. In particular we call *quadratic* a differential system with $n = 2$.

Since we are going to talk about finite and infinite singularities, we must first describe the compactified space in which we are going to work.

3.1 Compactification on the Poincaré sphere, Poincaré disc and projective plane

The main idea for the compactification of a plane polynomial vector field was given by Poincaré who stated it in the opening section of his article [262]. This is the idea of replacing the study of the vector field on the plane with a vector field on the sphere. The theorem giving the compactification based on this idea of Poincaré was proved in [150]. We describe here the Poincaré compactification, following mainly [150] with a bit more of the details.

Let $\mathcal{X} = P\partial/\partial x_1 + Q\partial/\partial x_2$ be a polynomial vector field of degree d associated to the polynomial differential system

$$\frac{dx_1}{dt} = P(x_1, x_2), \qquad \frac{dx_2}{dt} = Q(x_1, x_2).$$

We consider the *Poincaré sphere* $\mathbb{S}^2 = \{y = (y_1, y_2, y_3) \in \mathbb{R}^3 : y_1^2 + y_2^2 + y_3^2 = 1\}$, where its tangent plane to the point $(0,0,1)$ is identified with $\mathbb{R}^2$.

J. C. Artés et al., *Geometric Configurations of Singularities of Planar Polynomial Differential Systems*, https://doi.org/10.1007/978-3-030-50570-7_3

Poincaré's idea was to use *central projection* sending this plane to the sphere. To describe the central projection we consider the line passing through the point $(x_1, x_2, 1)$ and the center of the sphere, i.e. the origin of coordinates. This line cuts the sphere at two symmetric points with respect to the origin, $y = (y_1, y_2, y_3)$ on the upper hemisphere $U_3 = \{y \in \mathbb{S}^2 | y_3 > 0\}$ and the other $-(y_1, y_2, y_3)$ on the lower hemisphere $V_3 = \{y \in \mathbb{S}^2 | y_3 < 0\}$ of $\mathbb{S}^2$. The point $(x_1, x_2, 1)$, the origin and (y_1, y_2, y_3) being collinear, there exists $\lambda \in \mathbb{R}\backslash\{0\}$ such that $(y_1, y_2, y_3) = \lambda(x_1, x_2, 1)$ hence $y_3 = \lambda$. Since $y_1^2 + y_2^2 + y_3^2 = 1$ we have that $\lambda^2(x_1^2 + x_2^2 + 1) = 1$. Denoting $\Delta(x) = \sqrt{x_1^2 + x_2^2 + 1}$ we have $\lambda^2 = 1/\Delta(x)^2$, so for $y_3 > 0$, $y_3 = \lambda = 1/\Delta(x)$. Therefore $(y_1, y_2, y_3) = (\frac{x_1}{\Delta(x)}, \frac{x_2}{\Delta(x)}, \frac{1}{\Delta(x)})$. We get the differentiable function $f_+ : \mathbb{R}^2 \to U_3$

$$y = f_+(x_1, x_2) = (\frac{x_1}{\Delta(x)}, \frac{x_2}{\Delta(x)}, \frac{1}{\Delta(x)}) = (y_1, y_2, y_3). \tag{3.2}$$

Analogously if y is in V_3, i.e. $y_3 < 0$, then $y_3 = \lambda = -1/\Delta(x)$ and we have a function $f_- : \mathbb{R}^2 \to V_3$

$$y = f_-(x_1, x_2) = -(\frac{x_1}{\Delta(x)}, \frac{x_2}{\Delta(x)}, \frac{1}{\Delta(x)}) = (y_1, y_2, y_3). \tag{3.3}$$

The points $f_+(x_1, x_2)$ and $f_-(x_1, x_2)$ are called the central projections of (x_1, x_2) on the sphere. These maps are differentiable, transferring the vector field $\mathcal{X}$ on $\mathbb{R}^2$ to a vector field $\mathcal{X}^*$ defined on $\mathbb{S}^2\backslash\mathbb{S}^1$ where $\mathbb{S}^1$ is the equator, i.e. $\{y \in \mathbb{S}^2 | y_3 = 0\}$. Clearly these functions have inverses $\phi_3 : U_3 \to \mathbb{R}^2$, $\psi_3 : V_3 \to \mathbb{R}^2$ where $x = \phi_3(y) = (y_1/y_3, y_2/y_3)$ for $y_3 > 0$ and $x = \psi_3(y) = (y_1/y_3, y_2/y_3)$ for $y_3 < 0$. These maps define two charts (U_3, ϕ_3) and (V_3, ψ_3) of the differentiable manifold $\mathbb{S}^2$. Four more charts are needed to cover all the sphere which are (U_i, ϕ_i) and (V_i, ψ_i), $i \in \{1, 2\}$, $U_i = \{y \in \mathbb{S}^2 | y_i > 0\}$, $V_i = \{y \in \mathbb{S}^2 | y_i < 0\}$, $\phi_1(y) = (y_2/y_1, y_3/y_1)$ for $y_1 > 0$, $\psi_1(y) = (y_2/y_1, y_3/y_1)$ for $y_1 < 0$, $\phi_2(y) = (y_1/y_2, y_3/y_2)$ for $y_2 > 0$, $\psi_2(y) = (y_1/y_2, y_3/y_2)$ for $y_2 < 0$. We make the convention to denote in all charts the coordinates by $z = (z_1, z_2)$.

We need to construct a vector field on the whole sphere which has the same phase portraits on U_3 as $\mathcal{X}^*$. We need to cover the equator and for this we use the other four charts (U_i, ϕ_i) with and (V_i, ψ_i), $i \in \{1, 2\}$ of the sphere.

We first calculate the vector field $\mathcal{X}^*$ on the chart (U_1, ϕ_1). We observe that U_1 is split in $U_1 \bigcap U_3$, on which we can use $\phi_1 \circ f_+$, $U_1 \bigcap V_3$ on which we can use $\phi_1 \circ f_-$, and the equator of the sphere $\{y_3 = 0\}$.

We first calculate f_+ in the chart (U_1, ϕ_1), i.e. we consider the map $\phi_1 \circ f_+$. In order that this composition be well defined we need to have $y = f_+(x) \in U_1$, i.e. $y_1 > 0$ and hence, in view of (3.2), $x_1 > 0$. On the other hand f_+ sends (x_1, x_2), $x_1 > 0$, to a point y in U_3 so that $y_3 > 0$. Then $z = (z_1, z_2) = \phi_1 \circ f_+(x_1, x_2) = (y_2/y_1, y_3/y_1) = (x_2/x_1, 1/x_1)$. This map, defined for $x_1 > 0$, is analytic and we have $z_2 = 1/x_1 > 0$. Therefore $\phi_1 \circ f_+ : \{x \in \mathbb{R}^2 | x_1 > 0\} \to \{z \in \mathbb{R}^2 | z_2 > 0\}$. Another part of U_1 is $U_1 \bigcap V_3$ for which we need to compute $\phi_1 \circ f_-$. We have $z = (z_1, z_2) = \phi_1 \circ f_-(x_1, x_2) = \phi_1(-(\frac{x_1}{\Delta(x)}, \frac{x_2}{\Delta(x)}, \frac{1}{\Delta(x)})) = (y_2/y_1, y_3/y_1) = (x_2/x_1, 1/x_1)$ which has the same expression as for $x_1 > 0$ but which is here defined for $x_1 = y_1/y_3 < 0$ and hence $z_2 = 1/x_1 < 0$. Therefore the differentials of f_+ and f_- have in the chart (U_1, ϕ_1) the matrix

$$\begin{pmatrix} -x_2/x_1^2 & 1/x_1 \\ -1/x_1^2 & 0 \end{pmatrix}$$

sending our vector field in the x plane for $x_1 \neq 0$ to the vector field in the z plane for $z_2 \neq 0$,

$$((1/x_1)^2)(x_1 Q(x_1, x_2) - x_2 P(x_1, x_2), -P(x_1, x_2)). \tag{3.4}$$

Written in the z coordinates, for $z_2 \neq 0$, this vector field is

$$z_2(-z_1 P(1/z_2, z_1/z_2) + Q(1/z_2, z_1/z_2), -z_2 P(1/z_2, z_1/z_2)), \tag{3.5}$$

expressed by rational functions, analytic for $z_2 \neq 0$ and it cannot be extended on $z_2 = 0$.

In U_1 only half of the equator is covered, namely $\{y \in \mathbb{S}^1 | y_1 > 0\}$. The part of the equator $\{y \in \mathbb{S}^1 | y_1 < 0\}$ is covered using the chart (V_1, ψ_1). Using the maps $\psi_1 \circ f_+$ and $\psi_1 \circ f_-$ on $V_1 \bigcap U_3$ and $V_1 \bigcap V_3$ respectively, we obtain the same expressions for both maps $\psi_1 \circ f_\pm$ on V_1 as for U_1, and we have (3.5) for the vector field in V_1 for $z_2 \neq 0$.

We want to obtain a vector field defined on the whole sphere, including the equator, and which is closely related to our vector field $\mathcal{X}^*$ on $\mathbb{S}^2 \backslash \mathbb{S}^1$. For this reason we write the vector field (3.4) in terms of the y coordinates on the sphere and we obtain:

$$\left(\frac{y_3}{y_1}\right)^2 \left[\frac{y_1}{y_3} Q\left(\frac{y_1}{y_3}, \frac{y_2}{y_3}\right) - \frac{y_2}{y_3} P\left(\frac{y_1}{y_3}, \frac{y_2}{y_3}\right), -P\left(\frac{y_1}{y_3}, \frac{y_2}{y_3}\right)\right] \tag{3.6}$$

The two components of (3.6) are of degrees $1 - d$ and $2 - d$ respectively, in the variable y_3. By multiplying both components by y_3^{d-1}, the two components become polynomial.

In the chart (U_1, ϕ_1), for $\phi_1(U_1 \bigcap U_3)$, we have $z_2 = y_3/y_1 > 0$. Since $y_1 = 1/\Delta(z)$ we have $y_3 = z_2/\Delta(z)$ and hence $y_3^{d-1} = (\frac{z_2}{\Delta(z)})^{d-1}$. So that this multiplication leads to the vector field:

$$\frac{z_2^d}{\Delta(z)^{d-1}}(Q(1/z_2, z_1/z_2) - z_1 P(1/z_2, z_1/z_2), -z_2 Q(1/z_2, z_1/z_2)) \tag{3.7}$$

for $z_2 > 0$ and this vector field has the same phase portrait as (3.5).

In $U_1 \bigcap V_3$ we have $y_3 < 0$ but $y_1 > 0$ and hence $y_1 = -1/\Delta(z)$. This leads to $z_2 = y_3/y_1 = -y_3\Delta(z) < 0$. Thus $y_3 = -z_2/\Delta(z) < 0$ and hence multiplying the corresponding vector field expressed in y coordinates by $y_3^{d-1} = (-1)^{d-1}z_2^{d-1}/\Delta(z)^{d-1}$ we obtain in the z coordinates the vector field (3.7) multiplied by $(-1)^{d-1}$ for $z_2 < 0$. If d is odd and $d > 1$, then this factor is 1 and we have again the expression (3.7). But in case d is even then this factor is -1 and hence (3.7) needs to be multiplied by -1 to obtain the vector field for $z_2 < 0$. This vector field has exactly the same phase portrait as (3.5) on $z_2 \neq 0$ whenever d is odd. If d is even then the two phase portraits (3.5), (3.7) coincide for $z_2 > 0$ and they have the same phase curves for $z_2 < 0$ but with opposite orientation.

The remaining points on the equator $(0, 1, 0)$ and $(0, -1, 0)$ are covered respectively by charts (U_2, ϕ_2) and (V_2, ψ_2). Analogous calculations for these charts yield a vector field in the plane $\mathbb{R}^2$ identified with the tangent plane to the sphere at the point $(0, 1, 0)$. Using again the same coordinates $z = (z_1, z_2)$ on this new plane, we obtain the vector field for $z_2 \neq 0$:

$$z_2(P(z_1/z_2, 1/z_2) - z_1Q(z_1/z_2, 1/z_2), -z_2Q(z_1/z_2, 1/z_2)), \tag{3.8}$$

expressed by rational functions for $z_2 \neq 0$ and it cannot be extended to the equator $z_2 = 0$.

We consider the chart (U_2, ϕ_2) and the map $z = (z_1, z_2) = \phi_2 \circ f_+(x_1, x_2) = (y_1/y_2, y_3/y_2) = (x_1/x_2, 1/x_2)$. Here $f_+(x_1, x_2) \in U_3$ we have $y_3 > 0$ and $y_2 > 0$ so $y_2 = 1/\Delta(z)$. Then $z_2 = y_3/y_2 > 0$, $z_2 = 1/x_1 > 0$ so $x_1 > 0$ and $y_3 = z_2y_2 = z_2/\Delta(z)$ and hence $y_3^{d-1} = z_2^{d-1}/\Delta(z)^{d-1}$. Then multiplying the vector field in y by y_3^{d-1} we obtain, for $z_2 > 0$, the vector field

$$\frac{z_2^d}{\Delta(z)^{d-1}}[P(z_1/z_2, 1/z_2) - z_1Q(z_1/z_2, 1/z_2), -z_2Q(z_1/z_2, 1/z_2)]. \tag{3.9}$$

In the same chart (U_2, ϕ_2) using now the map f_-, $f_-(x_1, x_2) \in V_3$, then $(z_1, z_2) = \phi_2 \circ f_-(x_1, x_2) = (y_1/y_2, y_3/y_2) = (x_1/x_2, 1/x_2)$ and we obtain $z_2 = y_3/y_2 < 0$ since $y_3 < 0$ and $y_2 > 0$, so $y_2 = 1/\Delta(z)$. Hence $x_2 = 1/z_2 < 0$ and the differential of the map $\phi_2 \circ f_-$ sends our vector field $\mathcal{X}$, for $x_2 < 0$ to the vector field (3.8) for $z_2 < 0$. Hence everywhere in the chart (U_2, ϕ_2) the vector field is (3.8) for $z \neq 0$. We have $y_3 = z_2/\Delta(z)$ and hence $y_3^{d-1} = z_2^{d-1}/\Delta(z)^{d-1}$ and by multiplication with $y_3^{d-1} = z_2^{d-1}/\Delta(z)^{d-1}$ and we obtain again the vector field (3.9). This vector field has the same phase portrait as (3.8) for $z > 0$ for any d and in case d is odd it has the same phase portrait as (3.8) on $z \neq 0$. In case d is even, it has the same phase curves as (3.8) for $z_2 < 0$ but with opposite orientation. In addition this vector field is analytic and it is also defined on $z_2 = 0$.

We also obtain a vector field on the chart (U_3, ϕ_3) and using again the coordinates (z_1, z_2) we have $y_3 = 1/\Delta(z)$ and hence the usual multiplication by $y_3^{d-1} = 1/\Delta(z)^{d-1}$ leads to the vector field

$$\frac{1}{\Delta(z)^{d-1}}[P(z_1, z_2), Q(z_1, z_2)], \tag{3.10}$$

which is analytic in the whole plane. In an analogous way we define a corresponding analytic vector field for the chart (V_3, ψ_3). In this way we obtain an analytic vector field $p(\mathcal{X})$ defined on the whole sphere, including the equator. This vector field has exactly the same phase portrait as the vector field $\mathcal{X}^*$ in the chart (U_1, ϕ_1).

We call compactification on the sphere of the vector field $\mathcal{X}$, the restriction of the vector field $p(\mathcal{X})$ on $U_3 \bigcup \mathbb{S}^1$. We observe that this vector field is invariant on the equator. Indeed, we note that the second component of this vector field has the factor z_2 and hence for $z_2 = 0$ this component is zero. The *infinite singular points* of $\mathcal{X}$ are the singular points of $p(\mathcal{X})$ that lie on the equator $\mathcal{S}^1$. Note that if $y \in \mathbb{S}^1$ is an infinite singular point, then $-y$ is also an infinite singular point.

We call *Poincaré disk* the image of the upper hemisphere completed with the equator, i.e. $\pi(U_3 \bigcup \mathcal{S}^1)$ under the projection $\pi(y_1, y_2, y_3) = (y_1, y_2, 0) \equiv (y_1, y_2)$. We denote the Poincaré disk by $\mathbb{D}$.

To a polynomial differential system (3.1) of degree n with real coefficients, we associate the differential equation $\omega_1 = q(x, y)dx - p(x, y)dy = 0$. This equation defines two foliations with singularities, one on the real and one on the complex affine planes. We can compactify these foliations with singularities on the real (respectively complex) projective plane with homogeneous coordinates X, Y, Z. This is done by introducing the homogeneous coordinates via the equations $x = X/Z, y = Y/Z$ and taking the pull-back of the form ω_1 into the projective plane. We obtain a foliation with singularities on $\mathbb{PK}^2$ ($\mathbb{K}$ equal to $\mathbb{R}$ or $\mathbb{C}$) defined by the equation $\omega = A(X, Y, Z)dX + B(X, Y, Z)dY + C(X, Y, Z)dZ = 0$ on the projective plane over $\mathbb{K}$, which is called the *compactification on the projective plane* of the foliation with singularities defined by $\omega_1 = 0$ on the affine plane $\mathbb{K}^2$. Here A, B, C are homogeneous polynomials over $\mathbb{K}$, defined by $A(X, Y, Z) = ZQ(X, Y, Z)$, $Q(X, Y, Z) = Z^n q(X/Z, Y/Z)$, $B(X, Y, Z) = ZP(X, Y, Z)$, $P(X, Y, Z) = Z^n p(X/Z, Y/Z)$ and $C(X, Y, Z) = YP(X, Y, Z) - XQ(X, Y, Z)$. The *singular points at infinity* of the foliation defined by $\omega_1 = 0$ on the affine plane are the singular points of the type $[X : Y : 0] \in \mathbb{PK}^2$, and the line $Z = 0$ is called the *line at infinity* of this foliation. The singular points of the foliation F are the solutions of the three equations $A = 0$, $B = 0$, $C = 0$. In view of the definitions of A, B, C, it is clear that the singular points at infinity are the points of intersection of $Z = 0$ with $C = 0$, (for more details see [222], or [26] or [21]).

3.2 Classical definitions

We recall here the following terminology used in the classical literature:

We call *regular* a point (x_0, y_0) such that $p(x_0, y_0)^2 + q(x_0, y_0)^2 \neq 0$.

We call *singular* a point (x_0, y_0) such that $p(x_0, y_0) = q(x_0, y_0) = 0$.

A *degenerate* singular point is a point whose Jacobian matrix has determinant equal to zero.

A *hyperbolic* singular point is a point such that the real part of its eigenvalues is not zero.

A *semi-hyperbolic* singular point is a point such that exactly one of its eigenvalues is zero.

Hyperbolic and semi-hyperbolic singularities are also called *elementary.*

A *nilpotent* singular point is a point with its two eigenvalues zero but with its Jacobian matrix not zero.

A *linearly zero* singular point is a point which has its Jacobian matrix zero.

The non-degenerate singular points can also be classified according to the following criteria. Let $\lambda_1, \lambda_2 \in \mathbb{C}$ be the eigenvalues of the Jacobian matrix at a singular point (x_0, y_0).

A singular point such that $\lambda_1\lambda_2 < 0$ is called a *saddle* (*strong saddle* if $\lambda_1 + \lambda_2 \neq 0$ or *weak saddle* if $\lambda_1 + \lambda_2 = 0$).

A singular point such that $\lambda_1, \lambda_2 \in \mathbb{R}$ and $\lambda_1\lambda_2 > 0$ is called a *node* (stable if $\lambda_1 < 0$ and unstable if $\lambda_1 > 0$).

A singular point such that $\lambda_1, \lambda_2 \notin \mathbb{R}$ and $\Re(\lambda_1) \neq 0$ is called a *strong focus* (stable if $\Re(\lambda_1) < 0$ and unstable if $\Re(\lambda_1) > 0$).

These three singularities are all hyperbolic.

A singular point such that $\lambda_1, \lambda_2 \notin \mathbb{R}$ and $\Re(\lambda_i) = 0$ (for $i = 1, 2$) that possesses a neighborhood filled with periodic orbits is called a *center.* Among the centers we distinguish the *isochronous centers,* for which the period of every periodic orbit of the center is the same, and *uniform isochronous centers,* which are isochronous centers with a constant angular velocity for every orbit.

A singular point such that $\lambda_1, \lambda_2 \notin \mathbb{R}$ and $\Re(\lambda_i) = 0$ (for $i = 1, 2$) that is not a center is called a *weak focus.*

Weak foci or weak saddles can be classified according to different orders using the algebraic concept of Poincaré–Lyapunov constants. For details on Poincaré–Lyapunov constants and weak foci/saddles of various orders, we refer to [293, 222].

We note that in a perturbation of a system with a hyperbolic singular point, the local phase portraits around this singularity would stay the same by the Hartman–Grobman Theorem [128]. This was the main reason for introducing the notion of hyperbolic singular point, so as to separate the weak foci and centers from strong foci, since in perturbations the former could produce limit cycles close to them and the latter could not.

A linearly zero singularity is either topologically equivalent to a focus or a center or its neighborhood is a union of *hyperbolic, elliptic* and/or *parabolic sectors* (in a finite number if the system is polynomial). We refer the reader to [128, page 18] for a complete definition of sectors; see Figure 3.1.

Among the semi-hyperbolic singularities we may have the ones that are topologically equivalent to elementary nodes or saddles, and for these we keep the same name. We can also have a new type that is called a *saddle-node* and its local topological phase portrait is formed by two hyperbolic sectors and one parabolic sector.

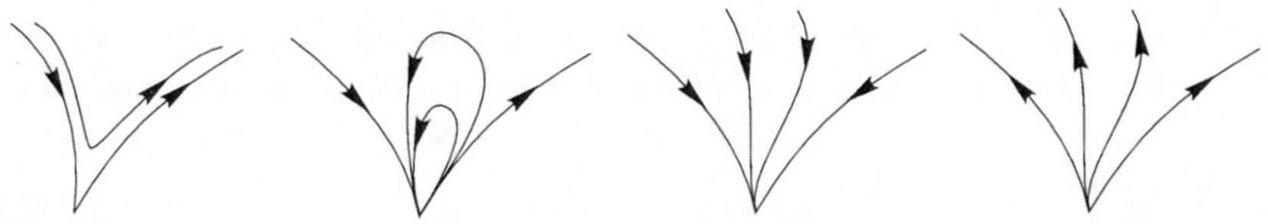

Figure 3.1: Sectors near a singular point.

Among the nilpotent singularities we may have the ones that are topologically equivalent to elementary nodes or saddles, to semi-hyperbolic saddle-nodes or to foci or centers and for these we keep the same name. And we can also have two new types: (1) a *cusp* whose local topological phase portrait is formed by two hyperbolic sectors with the two *local separatrices* (which form the border of the hyperbolic sectors) being tangent to the same half line at the singularity and (2) an *elliptic-saddle* whose local topological phase portrait is formed by one hyperbolic, one elliptic plus two parabolic sectors. See the drawings in the table in Appendix A.

However some time after the definitions of hyperbolic singularities and of other notions were introduced and adopted by the mathematical community, it was discovered that the weak saddles, whose traces of their Jacobian matrices are zero, are also relevant in perturbations and they are closely related to weak foci. But according to the classical definitions the weak foci are non-hyperbolic singularities and the weak saddles are hyperbolic singularities. We also point out that the invariant polynomials (see Chapter 5) that rule weak foci are the same ones that rule weak saddles. Finally we note that in his (1908) work on centers [123], Dulac places weak saddles and weak foci on the same footing.

We thus observe that the classical definitions are not adequate for the deeper classification that we present in this book. In order to produce a coherent systematic classification of all the singularities, we have had to introduce the new notions defined in the next section that are related exclusively to the Jacobian matrix and its eigenvalues (without considering their reals parts).

3.3 New definitions

Here are the new definitions:

We call *elemental* a singular point with its two eigenvalues not zero.

We call *semi-elemental* a singular point with exactly one of its eigenvalues equal to zero.

We maintain the name of *nilpotent* for a singular point with its two eigenvalues zero but with its Jacobian matrix at this singularity not identically zero.

We call *intricate* a singular point with its Jacobian matrix identically zero.

The definition of a nilpotent singular point remains exactly the same, and the linearly zero singularities are renamed intricate so as to use a single word. The semi-hyperbolic are now called semi-elemental. And finally the non-degenerate singularities become the elemental ones.

In this section we use the same concepts that we previously considered in [26, 21] such as *orbit* γ *tangent to a half line* L *at* p, *well defined angle at* p, *characteristic orbit at a singular point* p, *characteristic angle at a singular point* and *characteristic direction at* p. Since these are basic concepts for the notion of *geometrical equivalence relation*, we recall here these notions as well as a few others.

We assume that we have an isolated singularity p, i.e. suppose that there exists a neighborhood U of p where there is no other singularity. Consider an orbit γ in U defined by a solution $\Gamma(t) = (x(t), y(t))$ such that $\lim_{t\to+\infty}\Gamma(t) = p$ (or $\lim_{t\to-\infty}\Gamma(t) = p$). For a fixed t consider the unit vector defined by $C(t) = (\overrightarrow{\Gamma(t) - p})/\|\overrightarrow{\Gamma(t) - p}\|$. Let L be a half line ending at p. We shall say that *the orbit* γ *is tangent to the half line* L *at* p if $\lim_{t\to+\infty} C(t)$ (or $\lim_{t\to-\infty} C(t)$) exists and L contains this limit point on the unit circle centered at p. In this case we call *well defined angle of* Γ *at* p the angle between the positive x-axis and the half line L measured in the counterclockwise sense. We may also say that *the solution curve* $\Gamma(t)$ *tends to* p *with a well defined angle.* A *characteristic orbit at a singular point* p is an orbit of a solution curve $\Gamma(t)$ that tends to p with a well defined angle. We call *characteristic angle at the singular point* p a well defined angle of a solution curve $\Gamma(t)$. The line through p extending the half line L is called a *characteristic direction*.

Assume that the singularity is placed at the point $(0,0)$. Then the polynomial $PCD(x,y) = yp_m(x,y) - xq_m(x,y)$, where m is the starting degree of a polynomial differential system of the form (3.1), is called the *Polynomial of Characteristic Directions* of (3.1) (see page 51). In fact, in the case that $PCD(x,y) \not\equiv 0$, the factorization of $PCD(x,y)$ gives the characteristic directions at the origin.

If a singular point has an infinite number of characteristic directions, we will call it a *star-like* singularity. A star-like singularity can either be a node or something much more complicated with elliptic and hyperbolic sectors included. In the case when there are hyperbolic sectors, they must be *cusp-like* (for the definition see page 74). Elliptic sectors can either be cusp-like or normal ones. We call *special characteristic angle* any well defined angle at a star-like singularity, in which either none or more than one solution curve tends to p within this well defined angle. We will call *special characteristic directions at singular point* p any line such that at least one of the two angles defining it, is a special characteristic angle.

Since the study of the most degenerate singular points (nilpotent and intricate) needs the technique of blow-up, we are going to introduce it now.

3.4 The blow-up technique

To draw the phase portrait around an elemental hyperbolic singularity of a smooth planar vector field, we just need to use the Hartman–Grobman Theorem (see [128]). For an elemental non-hyperbolic singularity the system can be brought by an affine change of coordinates and time rescaling to the form $dx/dt = -y + \ldots$, $dy/dt = x + \ldots$, and it is well known that in this case the singularity is either a center or a focus. One way to see this is by the Poincaré–Lyapunov theory (see [263, 233]). In the quadratic case we can actually determine using the Poincaré–Lyapunov constants if it is a focus or a center, so the local phase portrait is known. For systems of higher degree, we have the *center-focus problem*; in general we can only say that the phase portrait around the singularity is of a center or of a focus but we cannot determine with certainty which one of the two it is.

In order to study semi-elemental singularities, one simply needs to apply the corresponding theorems of the monograph by Andronov et al. [2]. Nilpotent singularities can also be studied applying the corresponding theorem of Andronov et al. [2], but this gives only the topological behavior in the vicinity of the singularity, which it is not enough for the geometrical classification that is the main topic of this book.

In the case of a more complicated singularity, such as a degenerate one, we need to use the *blow-up technique*. This is a well known technique but since it plays such a crucial role in this work and also in order to make this chapter as self-contained as possible, we shall briefly describe it here. Another reason why we need to insist on describing this technique here is because we are going to use it in a slightly modified (actually simplified) way so as to lighten the calculations. For this modified way to be perfectly clear, we show that it is in complete agreement with the usual blow-up procedure.

The idea behind the blow-up technique is to replace the plane by a surface and the singular point p on the plane by a line or by a circle on the surface such that the singular point p splits (ideally) into a finite number of simpler singularities p_i on the line or the circle. For this idea to work, we need to construct a new surface on which we have a diffeomorphic copy of the vector field on $\mathbb{R}^2 \backslash \{p\}$ or at least on the complement of a line passing through p, and whose associated singularities lie on the circle (or on a line) that replaces the singularity p on the new surface.

3.4.1 The polar blow-up

The *polar blow-up* uses polar coordinates. Clearly we may assume that the singularity is placed at the origin. Consider the map $\phi : \mathbb{S}^1 \times \mathbb{R} \longrightarrow \mathbb{R}^2$ defined by $\phi(\theta, r) \mapsto (r\cos\theta, r\sin\theta)$. This map is a diffeomorphism for $r \in (0, \infty)$ and for

$r \in (-\infty, 0)$ onto $\mathbb{R}^2 \backslash \{(0,0)\}$ but $\phi^{-1}(0,0)$ is the circle $\mathbb{S}^1 \times \{0\}$. This application defines a diffeomorphic vector field on the upper part of the cylinder $\mathbb{S}^1 \times \mathbb{R}$. In fact this is the conversion to polar coordinates. The resulting smooth vector field extends to the whole cylinder just by allowing r to be negative or zero. This full vector field on the cylinder has either a finite number of singularities on the circle (this occurs only when the initial singular point is nilpotent) or singularities that fill up the circle (when we start with a linearly zero singularity). In this latter case we need to make a time rescaling $T = r^k t$ of the vector field with an adequate k to obtain a finite number of singularities. We thus "decompose" the singularity in $\mathbb{R}^2$ into several singularities located on the circle $\mathbb{S}^1 \times \{0\}$ of the cylinder. The map ϕ collapses the circle on the cylinder (and hence the singularities located on this circle) to the origin of coordinates in the plane. If the phase portraits around the singularities on the circle can be drawn, then the inverse process of *blowing down* the upper side of the cylinder completed with the circle allows us to draw the portrait around the origin of $\mathbb{R}^2$. In the case that the singularities on the circle are still degenerate, we need to repeat the process a finite number of times. This is guaranteed by the theorem of de-singularization of singularities (see [48, 125]).

3.4.2 The blow-up using rational functions

The blow-up by polar coordinates is simple, leading to a simple surface (the cylinder), on which a diffeomorphic copy of the vector field on $\mathbb{R}^2 \backslash \{(0,0)\}$ extends to a vector field on the full cylinder. The visualization of this blow-up on the cylinder surface is easy. But this process has the disadvantage of using the transcendental functions: cosine and sine. When several such blow-ups are needed, this is computationally very inconvenient.

It would be more advantageous to use a construction involving rational functions. A little more difficult to visualize, this algebraic blow-up is computationally simpler, using only rational transformations. To blow-up a singularity of the plane means to construct a surface playing the role of the cylinder in the polar case and to replace the singularity on the plane with a line (directional blow-up) or with a circle (viewed as the space of directions of $\mathbb{R}^2$ at this singularity) on this surface. If the singularity is replaced by a line, then the surface will not be a cylinder but an algebraic surface, a quadric in $\mathbb{R}^3$. If the singularity is replaced by a circle, then the surface will be a Möbius band.

To visualize the construction, we take the surface $A = \{(x, y, z) \in \mathbb{R}^3 |\, y = zx\}$ in $\mathbb{R}^3$, which is a smooth algebraic surface (a hyperbolic paraboloid). This manifold is analytically diffeomorphic to the plane via the canonical projection $\pi_{1,3}$: $A \longrightarrow \mathbb{R}^2$, $\pi_{1,3}(x, y, z) = (x, z)$ with inverse $(\pi_{1,3})^{-1} : \mathbb{R}^2 \longrightarrow A$, $(\pi_{1,3})^{-1}(x, z) = (x, xz, z)$. The map $\pi_{1,3}$ gives a global chart of this surface A in $\mathbb{R}^3$ that contains both the x- and z-axes but only contains the origin from the y-axis. It is easy to visualize it as it is a ruled surface generated by the line $L = \{(ux_0, uy_0, z_0)|\, y_0 = z_0x_0,\ u \in \mathbb{R}\}$ where (x_0, y_0) describes the half circle $\{(x, y) \in \mathbb{R}^2 |\, x > 0, x^2 + y^2 = 1\}$. The line L is parallel to the line $L_{(x_0,y_0)}$ joining in the (x, y)-plane the origin with (x_0, y_0) and it cuts the vertical axis at the point $(0, 0, z_0)$, where z_0 is the slope of $L_{(x_0,y_0)}$, $z_0 = y_0/x_0$. When the point (x_0, y_0) describes the half circle, L turns around, describing the "spiraling" surface A.

The surface A is the blow-up surface since it plays here the role of the cylinder. On A the singular point at the origin of the plane is replaced by the z-axis $L_z = \{(0, 0, z)|\, z \in \mathbb{R}\}$. From the line $x = 0 = z$ in $\mathbb{R}^3$, A only contains the origin. We firstly show that we could place on A a diffeomorphic copy of the vector field on $\mathbb{R}^2 \backslash \{x = 0\}$. This is done as follows:

We consider the set $A_x = \{(x, y, z) \in A|\, x \neq 0\}$. This is an open set in A and its closure in $\mathbb{R}^3$ is A. We have $A = A_x \bigcup L_z$. The diffeomorphism $\pi_{1,2} : A_x \longrightarrow (\mathbb{R}^2 \backslash \{x = 0\})$, $\pi_{1,2}(x, y, z) = (x, y)$ (with inverse $(\pi_{1,2})^{-1}(x, y) = (x, y, z)$ with $z = y/x$, defines a diffeomorphic vector field on $A_x = A \backslash L_z$ of the vector field on $\mathbb{R}^2 \backslash \{x = 0\}$. At this moment we cannot extend this vector field to the whole of A. But we can consider the foliation with singularities generated by the planar vector field with components $p(x, y), q(x, y)$, associated to the form $\omega = p(x, y)dy - q(x, y)dx$, and we can pull back this foliation along $\pi_{1,2}$ to obtain a foliation with singularities on the whole surface A that is associated to the form $\omega' = p(x, zx)(zdx + xdz) - q(x, zx)dx = (p(x, zx)z - q(x, zx))dx + p(x, zx)xdz$ on A, written here in the coordinates of the global chart $A \to \mathbb{R}^2$ $(x, xz, z) \to (x, z)$. This form yields a foliation with singularities that has an infinite number of singular points on L_z. Indeed, write $p(x, y) = p_0(x, y) + p_1(x, y) + \ldots + p_n(x, y)$ and $q(x, y) = q_0(x, y) + q_1(x, y) + \ldots + q_n(x, y)$ where p_i, q_i collect all the terms of p, respectively q, containing terms $x^{l_1}y^{l_2}$ with $l_1 + l_2 = i$. Since the origin is a singular point, $p_0 + q_0 = 0$. Let m be an integer such that $p_i = 0 = q_i$ for all $i < m$ and $(p_m, q_m) \neq 0$ in $\mathbb{R}[x, y]$. We may call this m the *starting degree of system* (3.1), and we have $m \geq 1$. Then $p(x, zx) = x^m(p_m(1, z) + xp_{m+1}(1, z) + \ldots + x^{n-m}p_n(1, z))$ and $q(x, zx) = x^m(q_m(1, z) + xq_{m+1}(1, z) + \ldots + x^{n-m}q_n(1, z))$. Therefore the form ω' becomes $\omega' = x^m\omega''$ where

$$\begin{aligned}\omega'' = {} & [(p_m(1, z) + xp_{m+1}(1, z) + \ldots + x^{n-m}p_n(1, z))z - (q_m(1, z) + xq_{m+1}(1, z) \\ & + \ldots + x^{n-m}q_n(1, z))]dx \\ & + [(p_m(1, z) + xp_{m+1}(1, z) + \ldots + x^{n-m}p_n(1, z)]dz.\end{aligned}$$

We call ω'' the *reduced form* of ω'. Clearly the foliation with singularities given by $\omega' = 0$ has the line $x = 0$ filled up with singularities and elsewhere it is determined by the equation $\omega'' = 0$. This last foliation has a finite number of singularities on $x = 0$ given by the polynomial equation $p_m(1, z)z - q_m(1, z) = 0$. These are the singularities into which the singular point at the origin of the plane has been decomposed on A, after reduction by a factor x^m.

The diffeomorphism $\mathbb{R}^2 \setminus \{x = 0\} \longrightarrow \mathbb{R}^2 \setminus \{x = 0\}$ given by $(x, y) \hookrightarrow (x, z)$ where $z = y/x$ sends the system on the (x, y)-plane for $x \neq 0$ to a system on the

(x,z)-plane for $x \neq 0$. This resulting system is

$$\begin{aligned} dx/dt =& p(x,xz) = x^m(p_m(1,z) + xp_{m+1}(1,z) + \ldots + x^{n-m}p_n(1,z)),\\ dz/dt =& (q(x,xz)x - p(x,zx)z)/x^2 = x^{m-1}[(q_m(1,z) + xq_{m+1}(1,z) + \ldots \\ &+ x^{n-m}q_n(1,z)) - (p_m(1,z) + xp_{m+1}(1,z) + \ldots + x^{n-m}p_n(1,z))z]. \end{aligned}$$

With the time rescaling $T = x^{m-1}t$ we obtain a differential system whose associated differential form is defined over the whole $\mathbb{R}^2$ and it is exactly the form ω''.

We define the *directional blow-up or affine blow-up of the affine plane* $\mathbb{R}^2$ *at the point* $(0,0)$ *in the direction of the* y*-axis*, the surface A on which we have a foliation with singularities extending the foliation determined on the plane by $\omega = 0$. The *directional blow-up of the point* $(0,0)$ *of* $\mathbb{R}^2$ *in the direction of the* y*-axis* is in this case the affine line $\{(0,0,z)|\, z \in \mathbb{R}\}$ belonging to A.

The *blow-down* map is the map $\xi : A \to \mathbb{R}^2$ defined as follows: For $x \neq 0$ we define $\xi(x,y,z) = (x,y)$ where $z = y/x$ and $\xi(0,0,z) = (0,0)$. It is worth drawing attention to the fact that depending on the parity of $m-1$, the orbits on the half-plane $x < 0$ will remain with the same direction, or they may change to the opposite direction.

We remark that on $A\backslash L_z$ we have a vector field diffeomorphic to the vector field on $\mathbb{R}^2\backslash\{x=0\}$. Via $\pi_{1,3}$ this vector field is transferred diffeomorphically on the plane of coordinates (x,z) on $\mathbb{R}^2\backslash L_z$. The foliation with singularities on A is transferred diffeomorphically to a foliation with singularities on the (x,z)-plane. We therefore could also call the *directional blow-up or affine blow-up of the affine plane* $\mathbb{R}^2$ *at the point* $(0,0)$ *in the direction of the* y*-axis*, the (x,z)-plane and the *directional blow-up or affine blow-up of the singular point* $(0,0)$ *in the direction of the* y*-axis*, the z-axis on the (x,z)-plane on which $(0,0)$ splits into a number of singularities of the foliation.

Analogously we define the surfaces $B = \{(x,y,u) \in \mathbb{R}^3|\, x = uy\}$, $B_y = \{(x,y,u) \in B|\, y \neq 0\}$ in $\mathbb{R}^3$. Clearly B_y is an open subset in B and we have $B = B_y \bigcup \{\{(0,0)\} \times \mathbb{R}\}$. Surface B is a smooth algebraic surface analytically diffeomorphic with the (y,u)-plane via the canonical map $\pi_{2,3} : B \longrightarrow \mathbb{R}^2$, $\pi_{2,3}(x,y,u) = (y,u)$ with inverse $(\pi_{2,3})^{-1}(y,u) = (uy,y,u)$ and it is a surface in $\mathbb{R}^3$ that contains both the y- and u-axes but only contains the origin from the x-axis. On B the singular point at the origin of the (x,y)-plane is replaced by the u-axis $L_u = \{(0,0,u)|\, u \in \mathbb{R}\}$. We can place on B a diffeomorphic copy of the vector field on $\mathbb{R}^2\backslash\{y=0\}$ in a similar way to the previous construction on the surface A. This is done as follows: We have $B = B_y \cup L_u$. The diffeomorphism $\pi_{2,3} : B_y \to \mathbb{R}^2\backslash\{y=0\}$, $\pi_{2,3}(x,y,u) = (y,u)$ with inverse $\pi_{2,3}^{-1}(y,u) = (yu,y,u)$ defines a diffeomorphic vector field on $B_y = B\backslash L_u$ of the vector field on $\mathbb{R}^2\backslash\{y=0\}$. Analogously to how we obtained a foliation with singularities on A from the one on A_x, we obtain here a foliation with singularities on B.

We define the *directional blow-up or affine blow-up of the affine plane* $\mathbb{R}^2$ *at the point* $(0,0)$ *in the direction of the* x*-axis*, the surface B on which we replace the point $(0,0)$ of $\mathbb{R}^2$ with the line $\{(0,0,u)|\, u \in \mathbb{R}\}$ and on which we have a foliation with singularities extending the foliation determined on the plane, by $\omega = 0$ with $y \neq 0$. The *directional blow-up of the point* $(0,0)$ *of* $\mathbb{R}^2$ *in the direction of the* x*-axis* is in this case the affine line $\{(0,0,u)|\, u \in \mathbb{R}\}$ in B.

The two foliations with singularities can be glued on a single surface, a Möbius band on which we replace the singular point $(0,0)$ by the circle $\{(0,0) \times \mathbb{PR}^1\} \cong \mathbb{PR}^1$ viewed as the space of directions on $\mathbb{R}^2$. We describe next how this is done following Arnold's description of σ-process (another name for blow-up) [9].

As before we assume the singular point of the planar vector field to be placed at the origin. Let ζ be the canonical map $\zeta : \mathbb{R}^2\backslash\{(0,0)\} \longrightarrow \mathbb{PR}^1$, $\zeta(x,y) = [x : y]$ and consider its graph $\Gamma = \{(x,y,[x:y])|\,(x,y) \in \mathbb{R}^2\backslash\{(0,0)\}\} \subset (\mathbb{R}^2\backslash\{(0,0)\}) \times \mathbb{PR}^1$ and also the closure $\bar{\Gamma}$ of Γ in the space $\mathbb{R}^2 \times \mathbb{PR}^1$. We have $\bar{\Gamma} = \Gamma \bigcup (\{(0,0)\} \times \mathbb{PR}^1)$. The surface $\bar{\Gamma}$ clearly contains the copy $\{(0,0,u)|\, u \in \mathbb{PR}^1\}$ of the circle $\mathbb{PR}^1$. We shall show that the surface $\bar{\Gamma}$, which is topologically a Möbius band, is endowed with an analytic structure, and that we can define on it a foliation with singularities corresponding to the foliation with singularities defined on the plane by the vector field. The *blow-up of the plane at the point* $(0,0)$ is the surface $\bar{\Gamma}$ and the *blow-up of the point* $(0,0)$ is the circle $\{(0,0)\} \times \mathbb{PR}^1\}$.

$\mathbb{PR}^1$ is covered by the two open sets:

$$
\begin{aligned}
U_1 =& \{[x:y]|\, x \neq 0\} = \{[1 : y/x]|\,(x,y) \in (\mathbb{R}^2\backslash\{x=0\})\} = \{[1:z]|\, z \in \mathbb{R}\} \cong \mathbb{R} \\
U_2 =& \{[x:y]|\, y \neq 0\} = \{[x/y : 1]|\,(x,y) \in (\mathbb{R}^2\backslash\{y=0\})\} = \{[u:1]|\, u \in \mathbb{R}\} \cong \mathbb{R}.
\end{aligned}
$$

The open set U_1 is $\mathbb{PR}^1$ with the point $[0 : 1]$ removed, U_2 is $\mathbb{PR}^1$ with $[1 : 0]$ removed and clearly $U_1 \bigcap U_2$ is $\mathbb{PR}^1$ with both points $[1:0], [0:1]$ removed.

Let $\Gamma_x = \Gamma \bigcap (\mathbb{R}^2 \times U_1)$ and $\Gamma_y = \Gamma \bigcap (\mathbb{R}^2 \times U_2)$. Then $\Gamma = \Gamma_x \bigcup \Gamma_y$. The closure $\bar{\Gamma}_x$ of Γ_x in $\mathbb{R}^2 \times U_1$ is $\Gamma_x \bigcup \{(0,0,[x:y])|\, x \neq 0\} = \Gamma_x \bigcup \{(0,0,[1:z])|\, z \in \mathbb{R}\}$. The closure $\bar{\Gamma}_y$ of Γ_y in $\mathbb{R}^2 \times U_2$ is $\Gamma_y \bigcup \{(0,0,[x:y])|\, y \neq 0\} = \Gamma_y \bigcup \{(0,0,[u:1])|\, u \in \mathbb{R}\}$. We have: $\bar{\Gamma} = \bar{\Gamma}_x \bigcup \{(0,y,[0:1])|\, y \neq 0\} = \bar{\Gamma}_y \bigcup \{(x,0,[1:0])|\, x \neq 0\}$. We note that both $\bar{\Gamma}_x$ and $\bar{\Gamma}_y$ are open sets of $\bar{\Gamma}$ and that $\{\bar{\Gamma}_x, \bar{\Gamma}_y\}$ is an open covering of $\bar{\Gamma}$.

The map $\tau : \Gamma_x \longrightarrow A_x$, $\tau(x,y,[x:y]) = (x,y,z)$ with $z = y/x$ is a bijection with inverse $\tau^{-1}(x,zx,z) = (x,zx,[x:zx])$. The map $\bar{\tau} : \bar{\Gamma}_x \longrightarrow A$ defined by $\bar{\tau}(x,y,[x:y]) = (x,y,y/x)$ for $x \neq 0$ and $\bar{\tau}(0,0,[x:y]) = (0,0,y/x)$, $x \neq 0$, is a bijection of $\bar{\Gamma}_x$ into A. The map $\phi = \pi_{1,3} \circ \bar{\tau} : \bar{\Gamma}_x \longrightarrow \mathbb{R}^2$, $\phi(x,y,[x:y]) \mapsto (x,z)$ whenever $x \neq 0$ and $y = zx$ and $\phi(0,0,[x:y]) = (0,z)$ for $x \neq 0$ and $y = zx$, is a homeomorphism giving a global chart of $\bar{\Gamma}_x$.

The map $\nu : \Gamma_y \longrightarrow B_y$ defined by $\nu(x,y,[x:y]) = (x,y,u)$ with $y \neq 0$ and $u = x/y$ is a bijection with inverse $\nu^{-1}(x,y,u) = (uy,y,[uy:y])$. The map $\bar{\nu} : \bar{\Gamma}_y \longrightarrow B$ such that $\bar{\nu}(x,y,[x:y]) = (y,u)$ when $y \neq 0$ and $x = uy$ and $\bar{\nu}(0,0,[x:y]) = (0,0,x/y)$, where $y \neq 0$, is a bijection of $\bar{\Gamma}_y$ onto B. The map $\psi = \pi_{2,3} \circ \bar{\nu} : \bar{\Gamma}_y \longrightarrow \mathbb{R}^2$ is a homeomorphism. The maps ϕ, ψ define an analytic structure on $\bar{\Gamma}$. Furthermore $\bar{\Gamma}$ is a Möbius band. We first note that $\bar{\Gamma} = \bar{\Gamma}_x \bigcup \{(0, y, [0:1]) | \, y \neq 0\}$, so the only thing missing from $\bar{\Gamma}$ when we take $\bar{\Gamma}_x \simeq A$ is the set $\{(0, y, [0:1]) | \, y \neq 0\}$, which belongs to the domain of the second chart of $\bar{\Gamma}$. The surface $\bar{\Gamma}_x$ is diffeomorphic to A, which looks like a spiraling surface, (for more information see [9, page 21 and Figures 9 and 10]). Furthermore $\bar{\Gamma}$ is a non-orientable surface as can easily be checked by computing the Jacobian determinant function of the maps $\phi \circ \psi^{-1}$ and $\psi \circ \phi^{-1}$.

The *(algebraic) blow-up of* $\mathbb{R}^2$ *at the point* $(0, 0)$ is the surface $\bar{\Gamma}$ on which the blow-up of the point $(0, 0)$ is the circle $\{(0, 0)\} \times \mathbb{PR}^1$. The singularities on the circle of this foliation with singularities, are blown-down to the origin of coordinates by the map $(0, 0, [x : y]) \mapsto (0, 0)$.

If the singularities on the circle $\{(0, 0)\} \times \mathbb{PR}^1$ in this new surface are elemental or semi-elemental, then their phase portraits are easily drawn. If they are not, then after a finite number of applications of this process (σ-process [9, 10]), the singularity splits into elemental and semi-elemental ones. This is guaranteed by the *Desingularization Theorem* (see [48, 125, 128]), which leads to the *finite sectorial decomposition* (see [128]). The finite sectorial decomposition says that the neighborhood of any singular point (different from a focus or center) of a polynomial differential system may be decomposed into the union of parabolic (P), hyperbolic (H) and elliptic (E) sectors (see Appendix A for its portraits). Using the phase portraits around the simpler singularities and the so-called blow-down procedure, we can construct the phase portrait of the planar vector field around the singular point at the origin. The study of the simpler singularities on the circle $\{(0, 0)\} \times \mathbb{PR}^1$ is done in the two local charts. In fact we study the singularities in the chart $\phi : \bar{\Gamma}_x \longrightarrow \mathbb{R}^2$ and then we study the only possibly still missing singularity in $\bar{\Gamma}$, which is in the direction of the y-axis in the chart $\psi : \bar{\Gamma}_y \longrightarrow \mathbb{R}^2$.

3.4.3 The blow-up technique using only one direction

An alternative procedure is to use just one chart but in which we are sure that all singularities on the circle are included. This could be done by applying first a very simple linear transformation so that in the resulting chart there is no missing singularity and then studying the singularities in just this chart, which may be taken to be $\bar{\Gamma}_x \longrightarrow \mathbb{R}^2$. Clearly here A can be replaced by the (x, z)-plane. We will use this second approach here, which we call the *one-directional blow-up*.

Assume that we have a polynomial differential system (3.1) with a nilpotent or intricate singular point placed at the origin $(0, 0)$ and we want to blow it up in the y-axis direction so that the blow-up along the x-axis direction is unnecessary. We first note that the characteristic directions (see Section 3.3) at the origin of the (x, y)-plane correspond bijectively to the singularities on the blow-up circle. Indeed, let $(x(t), y(t))$ $(x(t) \neq 0)$ be a solution curve that tends to $(0, 0)$ as t tends to ∞ (or to $-\infty$) and whose characteristic direction has the finite slope z_0. Then $z_0 = \lim z(t) = \lim \big(y(t)/x(t)\big)$. In the (x, z)-plane clearly $(0, z_0)$ is a singular point since the curve $(x(t), z(t))$ has z_0 as its α-limit (or ω-limit). If the slope of a solution curve $(x(t), y(t))$ tends to $\pm\infty$ as t tends to ∞, it means that the characteristic direction is the y-axis. This means that at infinity of the z-axis we have a singularity, and to study it we need to use the other chart $B_y \longrightarrow \mathbb{R}^2$. Since we have generically a finite number of characteristic directions, we can make a very simple linear transformation so as to avoid having the y-axis as a characteristic direction. In this case all the singularities will be on the blow-up (x, z)-plane and hence we can study all of them. We need to compute these singularities, which are the singularities of the reduced form $\omega'' = q''(x, zx) - p''(x, zx)z$ where $p'' = p/x^m$ and q''/x^m.

The singular point at $(0, 0)$ in the (x, y)-plane has split into singularities $(0, z)$ on the z-axis. We will call this z-axis the *blow-up line*. These are the points $(0, z)$ with z a solution of the equation $p_m(1, z)z - q_m(1, z) = 0$, and going back to the (x, y)-plane by replacing $z = y/x$ for $x \neq 0$, we get the equation $(p_m(x, y)y - q_m(x, y)x)/x^m = 0$ and hence $p_m(x, y)y - q_m(x, y)x = 0$. We recall that this was previously called the Polynomial of Characteristic Directions $PCD(x, y) = p_m(x, y)y - q_m(x, y)x$ and the equation of the characteristic directions on the (x, y)-plane is $PCD(x, y) = 0$. The factors of the homogeneous polynomial $PCD(x, y)$ provide the characteristic directions at the origin of the (x, y)-plane. In order to be sure that the y-axis is not a characteristic direction we only need to check that x is not a factor of $PCD(x, y)$. If it is, we need to make a linear change of variables that moves this direction out of the y-axis and which does not move any other characteristic direction into the y-axis.

If all the directions are characteristic, then the degenerate singularity will be a star-like singularity and at least two blow-ups must be done to obtain the desingularization (see Example 3.4). Otherwise, the number of characteristic directions is finite and there exists the possibility to make such a linear change. We will use changes of the type $(x, y) \to (x + ky, y)$ where k is some number (usually 1). It seems natural to call this linear change a *k-twist* as the y-axis is twisted by some angle depending on k.

Once we are sure that we have no characteristic direction on the y-axis, we perform the directional blow-up $(x, y) = (x, xz)$.

For this transformation the x-axis remains invariant and the origin of coordinates goes to the z-axis on the (x, z)-plane.

The old orbits that arrived at $(0, 0)$ with a well defined angle of slope k now arrive at the singular point $(0, k)$ of the new system. Studying these new singular points, one can determine their local behavior and separatrices, which after the blow-down describe the behavior of the orbits around the original singular point. Often one needs to do several blow-ups in chain (combined with some translation and/or k–twists) if the singularities that appear after the first blow-up are either nilpotent or intricate.

In the process of studying a singularity that is either nilpotent or intricate, we will then use transformations of the following types: blow-ups, k-twist linear changes and translations (when we need to move a newly created singular point) to the origin. In order to keep track of these changes, we will subindex the variables starting with (x_0, y_0) (or whichever other couple of letters) in the original system, and (x_i, y_i) the variables corresponding to the i-th change.

The whole procedure is described in a sequence of systems and figures that explains the transformations done up to the final unfolding of the singular point initially nilpotent or intricate. In the final unfolding we determine the complete phase portrait of the desingularized systems. Finally, a new chain of figures in which we undo the changes, carrying the newly found separatrices, sectors and orbits up to the original system will give the desired phase portrait.

After doing a *vertical blow-up* (i.e. a blow-up in the direction of the y-axis), it is possible that the newly created vertical axis is filled with singular points, that is, the system is degenerate with x_i^j being a common factor of both polynomials. If this happens, then we must study the reduced system obtained by removing this common factor. This is done with a change on the time variable $t \to x_i^j t$. If j is odd then the orientation of the orbits on the half plane $x_i < 0$ changes. If j is even (including the case $j = 0$), then there is no change of orientation. Nilpotent singularities never need the reduction ($j = 0$) and generically, intricate singularities have $j = 1$. We find cases with $j > 1$ in highly degenerate singular points starting with minimal degree 3, which will not appear here.

When we study the characteristic directions of a singular point, we may find that $PCD(x, y) = x(ax - by)G(x, y)$, so we are forced to do a twist to move the vertical characteristic direction so as to become a direction in the chart. However, if $a^2 + b^2 \neq 0$ we do not want to move this direction into the vertical axis. Since we can chose the change $(x, y) \to (x + ky, y)$ with k freely, we can always do it. Moreover, if one does the correct changes and blow-ups (that is, we never blow-up along a characteristic direction), the final result must be independent of the k (or k's) used. Thus, unless $a = -b$, one can always assume that $x + y = 0$ is not a characteristic direction and use $k = 1$. If it is a characteristic direction, then we take $k = 2$ (assuming that $x + 2y = 0$ is not a characteristic direction) and so on.

Now we will present in full detail one simple blow-up and one complicated blow-up that can be found when studying intricate singularities, which may appear as infinite singular points of quadratic differential systems. Afterwards we will study a highly degenerate singular point with a simple topological phase portrait but which has geometrical properties that make it quite special. Some of the studies are done using parameters that prove the power of the method when studying bifurcations. And finally we will study a singular point of a fifth-degree system whose $PCD(x, y) \equiv 0$.

Example 3.1. We want to study the systems:

$$\begin{aligned} \dot{v}_0 &= z_0 + v_0^2 + cv_0z_0 + az_0^2 - bv_0z_0^2, \\ \dot{z}_0 &= (1 - g)v_0z_0 - bz_0^3. \end{aligned} \tag{3.11}$$

For these systems the origin $(0, 0)$ is a nilpotent singular point, and $PCD(v_0, z_0) = z_0^2$. So there is no need of twist and we can blow up directly with $j = 0$ (no change on time). After the blow-up we obtain the systems:

$$\begin{aligned}\dot v_1 &= v_1^2 + (cv_1^2 + v_1)z_1 + (av_1^2 - bv_1^3)z_1^2,\\ \dot z_1 &= z_1(-gv_1 - (cv_1 + 1)z_1 - av_1z_1^2),\end{aligned}$$

for which $(0,0)$ is an intricate singular point. Now $PCD(v_1, z_1) = v_1z_1((g+1)v_1 + 2z_1)$ and we need to make a k-twist. In principle one should need to consider five possibilities: either (a) $g < -1$, (b) $g = -1$, (c) $-1 < g < 1$, (d) $g = 1$ and (e) $g > 1$. We will see in detail in this example that the last three cases can be considered all together with a single twist.

Even though we have the possibility of $g = 1$, we make a 1-twist and we get

$$\begin{aligned}\dot v_2 &= bz_2^5 + 2az_2^4 - 3bv_2z_2^4 + 3bv_2^2z_2^3 + 2cz_2^3 - 3av_2z_2^3 - bv_2^3z_2^2\\ &\quad + av_2^2z_2^2 + gz_2^2 - 3cv_2z_2^2 - z_2^2 + cv_2^2z_2 - gv_2z_2 - v_2z_2 + v_2^2,\\ \dot z_2 &= z_2\left(az_2^3 + cz_2^2 - av_2z_2^2 + gz_2 - cv_2z_2 - z_2 - gv_2\right).\end{aligned}$$

Now we do a blow-up with $j = 1$ (time reduction of odd order) to obtain:

$$\begin{aligned}\dot v_3 &= v_3(bv_3^3z_3^5 - 3bv_3^3z_3^4 + 2av_3^2z_3^4 + 3bv_3^3z_3^3 - 3av_3^2z_3^3\\ &\quad + 2cv_3z_3^3 - bv_3^3z_3^2 + av_3^2z_3^2 + gz_3^2 - 3cv_3z_3^2 - z_3^2 - gz_3 + cv_3z_3 - z_3 + 1),\\ \dot z_3 &= -bv_3^3z_3^6 + 3bv_3^3z_3^5 - 2av_3^2z_3^5 - 3bv_3^3z_3^4 + 4av_3^2z_3^4 - 2cv_3z_3^4\\ &\quad + bv_3^3z_3^3 - 2av_3^2z_3^3 - gz_3^3 + 4cv_3z_3^3 + z_3^3 + 2gz_3^2 - 2cv_3z_3^2 - gz_3 - z_3.\end{aligned}$$

The singular points on the straight line $v_3 = 0$ are the solutions of the equation $(z_3 - 1)z_3(z_3(g - 1) - g - 1) = 0$. So we have three singular points with Jacobian matrices:

$$(0,0) \Rightarrow \begin{pmatrix} 1 & 0 \\ 0 & -1-g \end{pmatrix};\ (0,1) \Rightarrow \begin{pmatrix} -1 & 0 \\ 0 & 2 \end{pmatrix};\ \left(0, \frac{g+1}{g-1}\right) \Rightarrow \begin{pmatrix} 1 & 0 \\ \star & \frac{2+2g}{1-g} \end{pmatrix}.$$

Notice that if $g = 1$, one of the singular points of the blown-up system has escaped to infinity. Also notice that if $g > 1$, the singularity has positive ordinate greater than 1, while if $-1 < g < 1$ it has negative ordinate.

Then if $-1 < g$, the singularities $(0,0)$ and $(0,1)$ are saddles. Moreover if $-1 < g < 1$ then $(0, (g + 1)/(g - 1))$ is an unstable node. Thus the blow-down procedure is described in Figure 3.2 in full detail. In this figure (and from now on) the final phase portrait is described with a notation that will be defined in Section 3.7. We describe the changes done one by one up to the complete desingularization of the point, and next going back step by step in the opposite direction, we retrieve the local phase portrait. Notice that when we blow down every pair of adjacent sectors, it becomes a single one according to the simple rules: $P \cup H \equiv P$ (see Figure 3.2), $H \cup H \equiv H$ (see Figure 3.3) and $P \cup P \equiv E$ (see Figure 3.9, second row). The top sector in the first quadrant is adjacent to the bottom sector in the third, and the top sector in the second quadrant is adjacent to the bottom sector

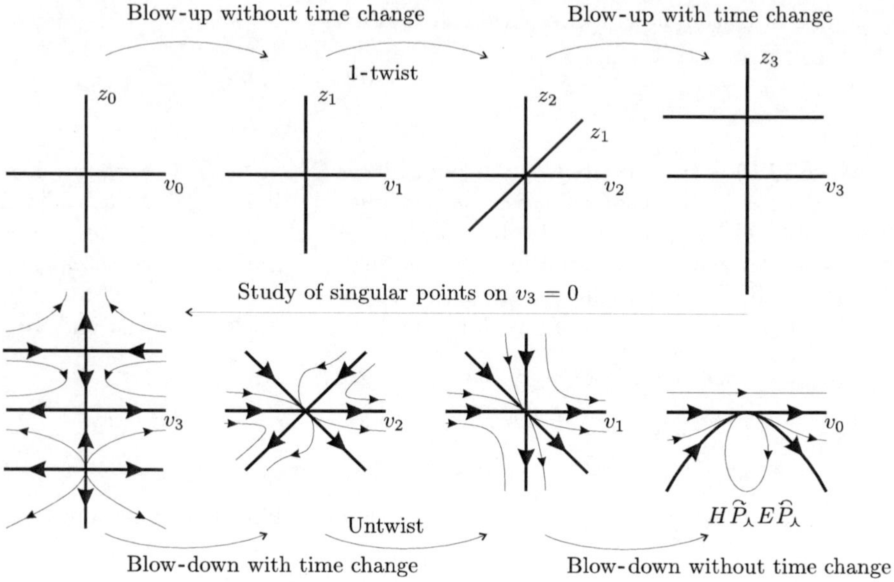

Figure 3.2: Blow-up and phase portrait of (3.11) with $-1 < g < 1$.

in the fourth through the Möbius band. We always add the label for the horizontal axis v_i on the straight line $z_i = 0$.

If $g > 1$ then $(0, (g+1)/(g-1))$ has positive ordinate greater than 1, and the singularity is now a saddle. The unfolding can be seen in Figure 3.3.

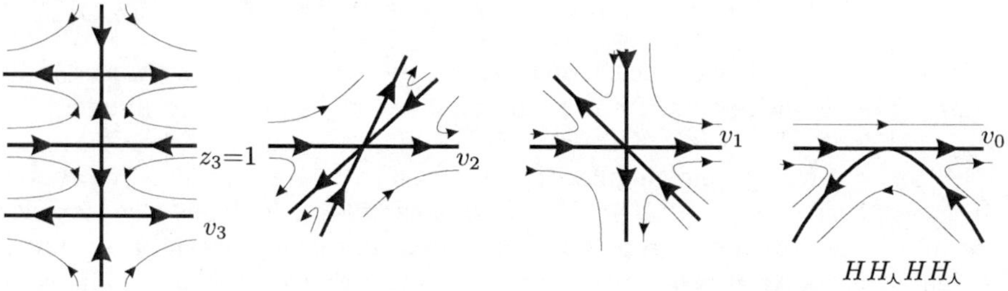

Figure 3.3: Blow-up and phase portrait of (3.11) with $g > 1$.

Moreover if $g = 1$, then the singularity $(0, (g+1)/(g-1))$ is undefined; it has gone to infinity. In fact this is the bifurcation point between the two cases $-1 < g < 1$ and $g > 1$. The unstable node of Figure 3.2 tends to $-\infty$ and reappears as a saddle in Figure 3.3. For the case $g = 1$ we cannot make a 1-twist because we lose the information of the direction $z_1 = -v_1$, but we can do either a 1/2-twist or a 2-twist to get the unfolding shown in Figure 3.4.

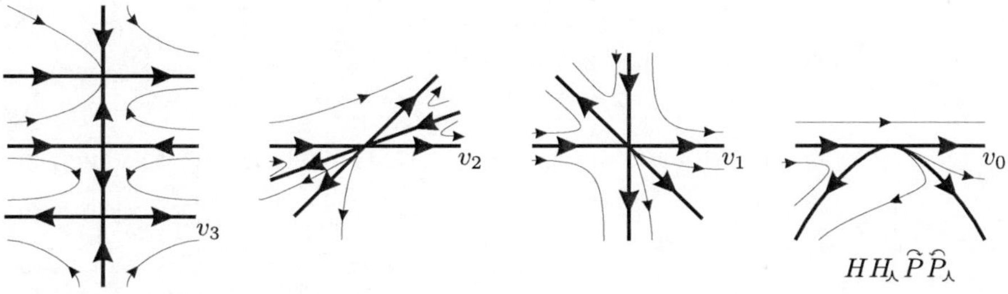

Figure 3.4: Blow-up and phase portrait of (3.11) with $g = 1$.

To understand the continuity among the final phase portraits obtained in Figures 3.2, 3.4 and 3.3, one has to notice that the first and the third represent singularities of odd multiplicity (in this case three) while the middle one portrays a singularity of even multiplicity (in this case four). In order to move from the phase portrait in Figure 3.3 to the phase portrait in Figure 3.2, it is necessary that a singular point with index +1, which is the α-limit of one of the non-straight separatrices of a hyperbolic sector, collides with the nilpotent singular point that we are studying (producing the saddle-node we see in Figure 3.4). Afterwards, a saddle separates from the nilpotent singularity along the other separatrix (see Figure 3.5 where for a better understanding we name the singularities as if they were finite).

If $g < -1$, the singular point $(0, 0)$ is an unstable node, $(0, 1)$ is a saddle and $(0, (g + 1)/(g - 1))$ has positive ordinate (lower than 1) and is also a saddle. Thus the blow-down procedure is described in Figure 3.6.

If $g = -1$, the singular points $(0, 1)$ and $(0, 0)$ are respectively a saddle and a semi-elemental singularity of multiplicity two, which is a saddle-node unstable with the null eigenvalue associated to the vertical direction. Thus the blow-down procedure is described in Figure 3.7.

To understand the continuity among the final phase portraits obtained on Figures 3.2, 3.7 and 3.6 one has to notice that when $g > -1$ we have a saddle at $(0, 0)$ and an unstable node at $(0, (g + 1)/(g - 1))$ and just the opposite happens if $g < -1$. Moreover the singular point $(0, (g + 1)/(g - 1))$ has moved from a negative ordinate to a positive. So in the middle we have a saddle-node. The parabolic behavior of $(0, (g + 1)/(g - 1))$ has been transferred to the origin and the hyperbolic behavior belongs to the traveling singularity. The result is that the three phase portraits are topologically equivalent but with qualitative different properties. In fact the elliptic-saddle in the middle will fall in a different geometrical equivalence class, as we will see in the definition given in Section 3.6 (see Figure 3.8 where we

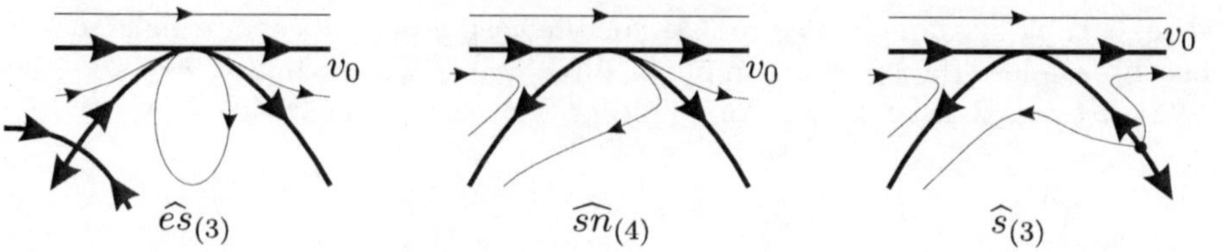

Figure 3.5: Transition from nilpotent elliptic-saddle to nilpotent saddle.

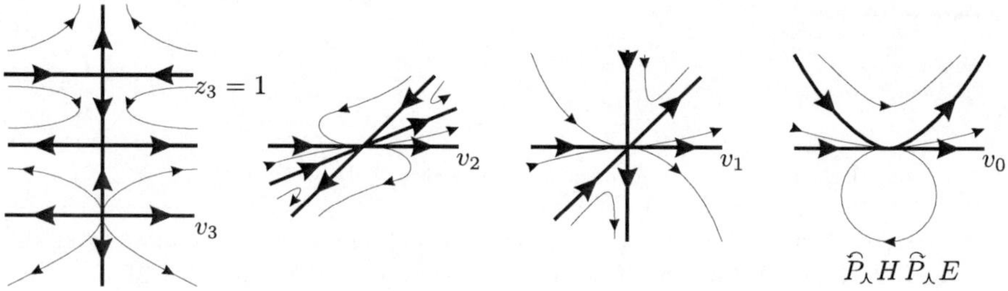

Figure 3.6: Blow-up and phase portrait of (3.11) with $g < -1$.

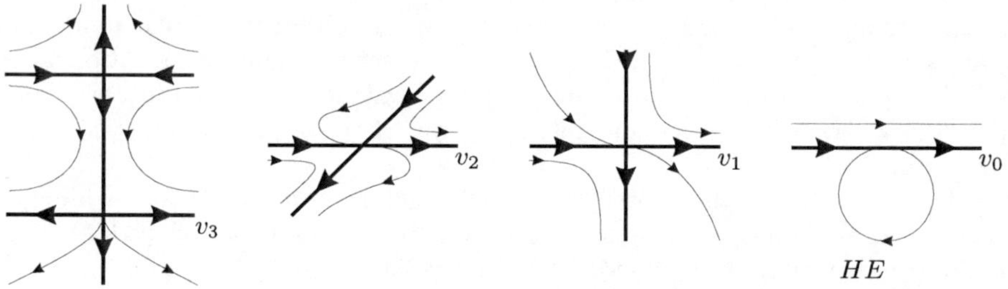

Figure 3.7: Blow-up and phase portrait of (3.11) with $g = -1$.

call the singularities as if they were finite for a better understanding).

Example 3.2. We want to study the systems

$$\begin{aligned} \dot{v}_0 &= -v_0^2 + v_0 z_0^2, \\ \dot{z}_0 &= (g-1)v_0 z_0 + z_0^3. \end{aligned} \tag{3.12}$$

For these systems if $g = 0$ there is an infinite number of finite singular points along the curve $z_0^2 - v_0 = 0$. In this case we need to find the reduced system and then the origin becomes elemental. If $g \neq 0$ then the origin $(0, 0)$ is an intricate

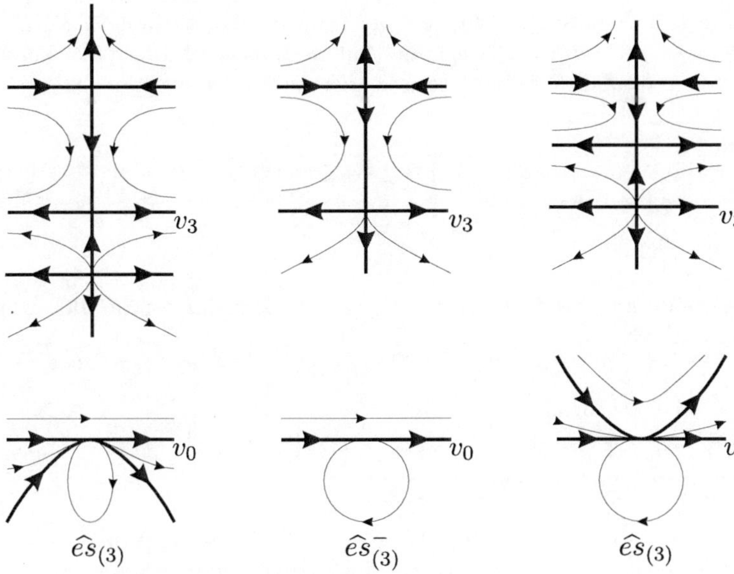

Figure 3.8: Transition of nilpotent elliptic-saddles.

singular point, and $PCD(v_0, z_0) = -gv_0^2 z_0$. Then we need a 1-twist and we obtain the systems:

$$\begin{aligned}\dot{v}_1 &= -v_1^2 + v_1 z_1 + g v_1 z_1 - g z_1^2 + v_1 z_1^2,\\ \dot{z}_1 &= z_1(-v_1 + g v_1 + z_1 - g z_1 + z_1^2).\end{aligned}$$

Now we do the blow-up with $j = 1$ (time reduction of odd order) to get

$$\begin{aligned}\dot{v}_2 &= v_2(-1 + z_2 + g z_2 - g z_2^2 + v_2 z_2^2),\\ \dot{z}_2 &= g z_2 - 2g z_2^2 + g z_2^3.\end{aligned}$$

The singular points on the straight line $v_2 = 0$ are the solutions of the equation $g(z_2 - 1)^2 z_2 = 0$. So we have two singular points with Jacobian matrices:

$$(0,0) \Rightarrow \begin{pmatrix} -1 & 0 \\ 0 & g \end{pmatrix}; \qquad (0,1) \Rightarrow \begin{pmatrix} 0 & 0 \\ 0 & 0 \end{pmatrix}.$$

The singular point $(0,0)$ can be a saddle or a node (we have assumed $g \neq 0$) while $(0,1)$ is again intricate.

We need to translate this singular point to the origin:

$$\begin{aligned}\dot{v}_3 &= v_3(v_3 + z_3 - g z_3 + 2v_3 z_3 - g z_3^2 + v_3 z_3^2),\\ \dot{z}_3 &= g z_3^2(1 + z_3).\end{aligned}$$

We check now the characteristic directions that are the solutions of $PCD(v_3, z_3) = v_3 z_3(v_3 + z_3 - 2gz_3)$, and we determine that we must make a 1-twist knowing that we must not move another characteristic direction to the vertical axis since $g \neq 0$.

We get

$$\begin{aligned}\dot{v}_4 &= v_4^2 - v_4 z_4 - g v_4 z_4 + 2v_4^2 z_4 + 2g z_4^2 - 4v_4 z_4^2 - g v_4 z_4^2 + v_4^2 z_4^2 + 2z_4^3 \\ &\quad + 2g z_4^3 - 2v_4 z_4^3 + z_4^4, \\ \dot{z}_4 &= g z_4^2(1 + z_4).\end{aligned}$$

Now we do another blow-up with $j = 1$ (time reduction of odd order) to get

$$\begin{aligned}\dot{v}_5 &= v_5(1 - z_5 - g z_5 + 2v_5 z_5 + 2g z_5^2 - 4v_5 z_5^2 - g v_5 z_5^2 + v_5^2 z_5^2 + 2v_5 z_5^3 + 2g v_5 z_5^3 \\ &\quad - 2v_5^2 z_5^3 + v_5^2 z_5^4), \\ \dot{z}_5 &= -z_5 + z_5^2 + 2g z_5^2 - 2v_5 z_5^2 - 2g z_5^3 + 4v_5 z_5^3 + 2g v_5 z_5^3 - v_5^2 z_5^3 - 2v_5 z_5^4 \\ &\quad - 2g v_5 z_5^4 + 2v_5^2 z_5^4 - v_5^2 z_5^5.\end{aligned}$$

The singular points on the straight line $v_5 = 0$ are the solutions of the equation $-(-1 + z_5)z_5(-1 + 2cgz_5) = 0$, so we have three singular points with Jacobian matrices:

$$(0,0) \Rightarrow \begin{pmatrix} 1 & 0 \\ 0 & -1 \end{pmatrix}; \ (0,1) \Rightarrow \begin{pmatrix} g & 0 \\ 0 & 1-2g \end{pmatrix}; \ \left(0, \frac{1}{2g}\right) \Rightarrow \begin{pmatrix} \frac{1}{2} & 0 \\ \star & \frac{2g-1}{2g} \end{pmatrix}.$$

Then the singular point $(0,0)$ is always a saddle. If $g < 0$ the singular point $(0,1)$ is also a saddle, and $\left(0, \frac{1}{2g}\right)$ has negative ordinate and is an unstable node. Moreover the singular point $(0,0)$ of the second system is a stable node. Thus the blow-down procedure is described in Figure 3.9.

If $0 < g < 1/2$ the singular point $(0,1)$ is an unstable node and $(0, 1/(2g))$ has positive ordinate (greater than 1) and it is a saddle. Moreover the singular point $(0,0)$ of the second system is a saddle. Thus the blow-down procedure is described in Figure 3.10.

If $g = 1/2$ the singular points $(0,1)$ is a semi-elemental singularity of multiplicity two, which is an unstable saddle-node with the null eigenvalue associated to the vertical direction. Thus the blow-down procedure is described in Figure 3.11.

As a curiosity, notice that this phase portrait is topologically equivalent to a semi-elemental saddle-node but not qualitatively equivalent (regardless of multiplicity), since the latter must be described (in sector notation) as $H\widehat{P}\widehat{P}H$. For a definition of the qualitative equivalence, we refer the reader to [182]. We will later define even finer equivalence relations in Section 3.6.

If $1/2 < g < 1$ or $g > 1$, the singular point $(0,1)$ is a saddle and $(0, 1/(2g))$ has positive ordinate (lower than 1) and it is a stable node. Moreover the singular point $(0,0)$ of the second system is a saddle. Thus the blow-down procedure is described in Figure 3.12.

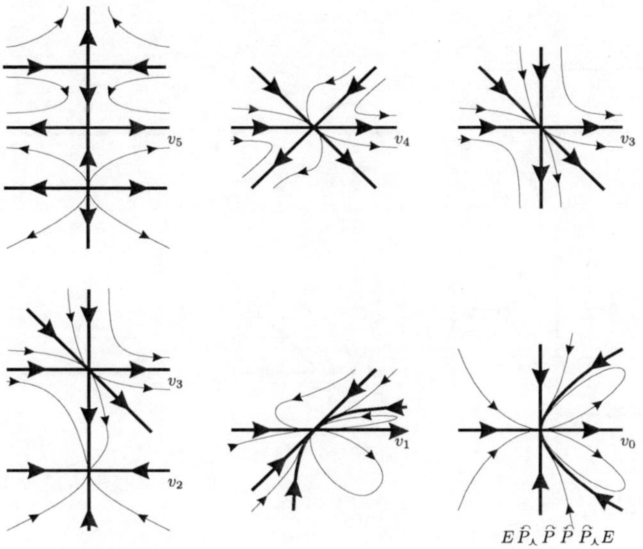

Figure 3.9: Blow-up and phase portrait of (3.12) with $g < 0$.

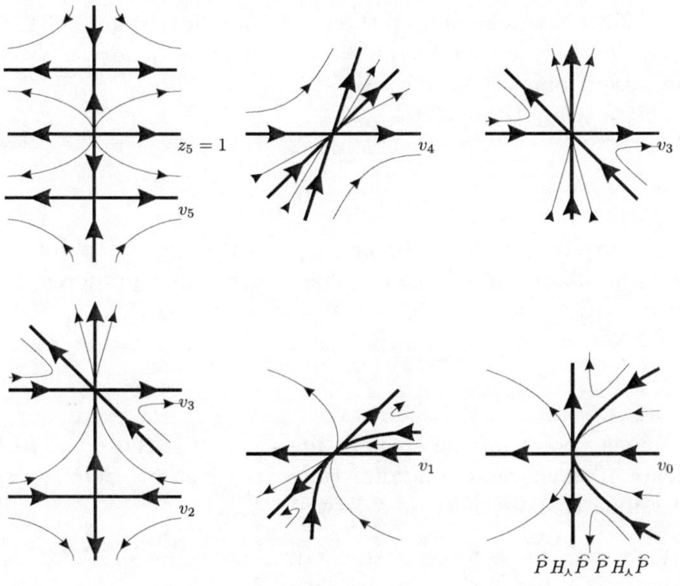

Figure 3.10: Blow-up and phase portrait of (3.12) with $0 < g < 1/2$.

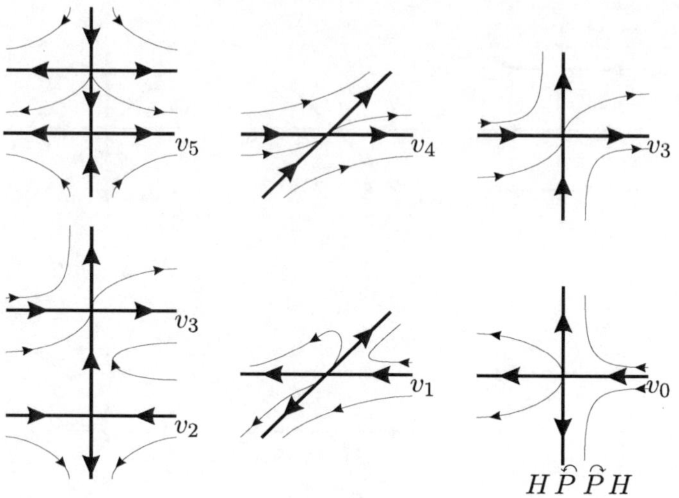

Figure 3.11: Blow-up and phase portrait of (3.12) with $g = 1/2$.

Example 3.3. This is an example of the most complicated kind of the desingularization process of infinite singular points appearing in quadratic systems. It is of degree 3 because it appears at infinity.

We want to study the systems:

$$\begin{aligned}\dot{v}_0 &= -v_0^3 - (1 - bv_0)z_0^2,\\ \dot{z}_0 &= -v_0^2 z_0 + bz_0^3.\end{aligned} \tag{3.13}$$

For these systems the origin $(0,0)$ is an intricate singular point, and we have $PCD(v_0, z_0) = -z_0^3$. So there is no need of a twist and we proceed directly to the blow-up with $j = 1$ (time reduction of odd order) to get

$$\begin{aligned}\dot{v}_1 &= v_1(-v_1 - z_1^2 + bv_1 z_1^2),\\ \dot{z}_1 &= z_1^3.\end{aligned}$$

The only singular point on the straight line $v_1 = 0$ is the point $(0,0)$, which is again intricate. The characteristic directions are given by $PCD(v_1, z_1) = -v_1^2 z_1$, and we determine that we must do a 1-twist. We get

$$\begin{aligned}\dot{v}_2 &= -v_2^2 + 2v_2 z_2 - z_2^2 - v_2 z_2^2 + bv_2^2 z_2^2 + 2z_2^3 - 2bv_2 z_2^3 + bz_2^4,\\ \dot{z}_2 &= z_2^3.\end{aligned}$$

Now we do another blow-up with $j = 1$ (time reduction of odd order) to get

$$\begin{aligned}\dot{v}_3 &= v_3(-1 + 2z_3 - z_3^2 - v_3 z_3^2 + bv_3^2 z_3^2 + 2v_3 z_3^3 - 2bv_3^2 z_3^3 + bv_3^2 z_3^4),\\ \dot{z}_3 &= z_3 - 2z_3^2 + z_3^3 + 2v_3 z_3^3 - bv_3^2 z_3^3 - 2v_3 z_3^4 + 2bv_3^2 z_3^4 - bv_3^2 z_3^5.\end{aligned}$$

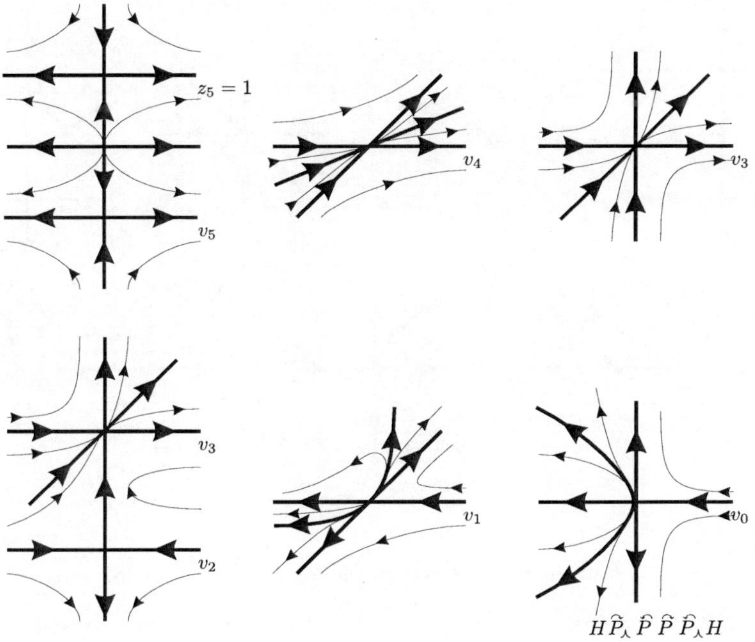

Figure 3.12: Blow-up and phase portrait of (3.12) with $1/2 < g < 1$ or $g > 1$.

The singular points on the straight line $v_3 = 0$ are the solutions of the equation $(-1+z_3)^2 z_3 = 0$. So we have two singular points with Jacobian matrices:

$$(0,0) \Rightarrow \begin{pmatrix} -1 & 0 \\ 0 & 1 \end{pmatrix}; \qquad (0,1) \Rightarrow \begin{pmatrix} 0 & 0 \\ 0 & 0 \end{pmatrix}.$$

The singular point $(0,1)$ is also intricate. We need to translate this singularity to the origin:

$$\begin{aligned} \dot{v}_4 &= v_4(v_4 + 4v_4 z_4 - z_4^2 + 5v_4 z_4^2 + bv_4^2 z_4^2 + 2v_4 z_4^3 + 2bv_4^2 z_4^3 + bv_4^2 z_4^4), \\ \dot{z}_4 &= -z_4(1+z_4)(2v_4 - z_4 + 4v_4 z_4 + bv_4^2 z_4 + 2v_4 z_4^2 + 2bv_4^2 z_4^2 + bv_4^2 z_4^3). \end{aligned}$$

We check now the characteristic directions, which are the solutions of $PCD(v_4, z_4) = v_4(3v_4 - z_4)z_4$, and determine that we must do a 1-twist:

$$\begin{aligned} \dot{v}_5 = {} & v_5^2 - 4v_5 z_5 + 4v_5^2 z_5 + 4z_5^2 - 15v_5 z_5^2 + 5v_5^2 z_5^2 - bv_5^2 z_5^2 + bv_5^3 z_5^2 + 12z_5^3 - 16v_5 z_5^3 \\ & + 2bv_5 z_5^3 + 2v_5^2 z_5^3 - 6bv_5^2 z_5^3 + 2bv_5^3 z_5^3 + 11z_5^4 - bz_5^4 - 6v_5 z_5^4 + 9bv_5 z_5^4 - 9bv_5^2 z_5^4 \\ & + bv_5^3 z_5^4 + 4z_5^5 - 4bz_5^5 + 12bv_5 z_5^5 - 4bv_5^2 z_5^5 - 5bz_5^6 + 5bv_5 z_5^6 - 2bz_5^7, \\ \dot{z}_5 = {} & -z_5(1+z_5)(2v_5 - 3z_5 + 4v_5 z_5 + bv_5^2 z_5 - 4z_5^2 + 2v_5 z_5^2 - 2bv_5 z_5^2 \\ & + 2bv_5^2 z_5^2 - 2z_5^3 + bz_5^3 - 4bv_5 z_5^3 + bv_5^2 z_5^3 + 2bz_5^4 - 2bv_5 z_5^4 + bz_5^5). \end{aligned}$$

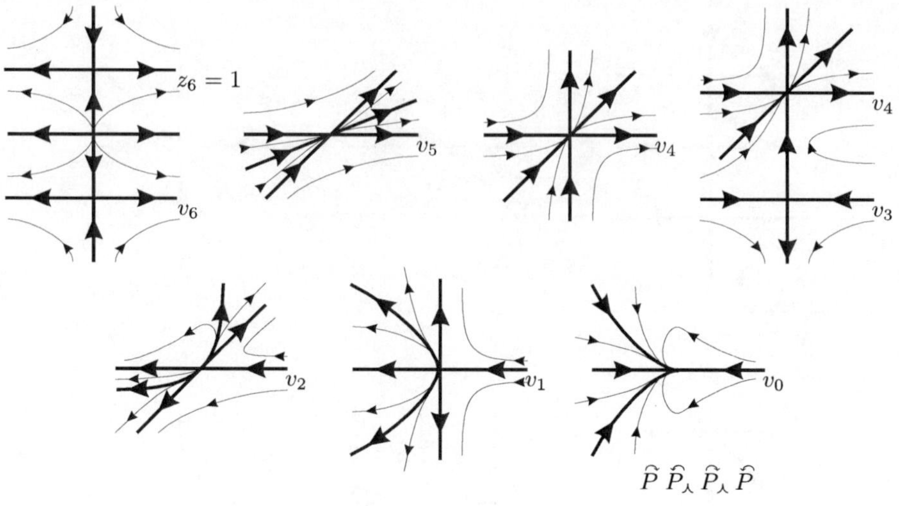

Figure 3.13: Blow-up and phase portrait of (3.13).

Now we do another blow-up with $j = 1$ (time reduction of odd order) to get

$$
\begin{aligned}
\dot{v}_6 = {} & -v_6(-1 + 4z_6 - 4v_6z_6 - 4z_6^2 + 15v_6z_6^2 - 5v_6^2z_6^2 + bv_6^2z_6^2 - bv_6^3z_6^2 - 12v_6z_6^3 \\
& + 16v_6^2z_6^3 - 2bv_6^2z_6^3 - 2v_6^3z_6^3 + 6bv_6^3z_6^3 - 2bv_6^4z_6^3 - 11v_6^2z_6^4 + bv_6^2z_6^4 + 6v_6^3z_6^4 \\
& - 9bv_6^3z_6^4 + 9bv_6^4z_6^4 - bv_6^5z_6^4 - 4v_6^3z_6^5 + 4bv_6^3z_6^5 - 12bv_6^4z_6^5 + 4bv_6^5z_6^5 + 5bv_6^4z_6^6 \\
& - 5bv_6^5z_6^6 + 2bv_6^5z_6^7), \\
\dot{z}_6 = {} & -3z_6 + 7z_6^2 - 10v_6z_6^2 - bv_6^2z_6^2 - 4z_6^3 + 22v_6z_6^3 - 11v_6^2z_6^3 + 3bv_6^2z_6^3 - 4bv_6^3z_6^3 \\
& - 12v_6z_6^4 + 22v_6^2z_6^4 - 3bv_6^2z_6^4 - 4v_6^3z_6^4 + 12bv_6^3z_6^4 - 5bv_6^4z_6^4 - 11v_6^2z_6^5 + bv_6^2z_6^5 \\
& + 8v_6^3z_6^5 - 12bv_6^3z_6^5 + 15bv_6^4z_6^5 - 2bv_6^5z_6^5 - 4v_6^3z_6^6 + 4bv_6^3z_6^6 - 15bv_6^4z_6^6 + 6bv_6^5z_6^6 \\
& + 5bv_6^4z_6^7 - 6bv_6^5z_6^7 + 2bv_6^5z_6^8.
\end{aligned}
$$

The singular points on the straight line $v_6 = 0$ are the solutions of the equation $-(-1 + z_6)z_6(-3 + 4z_6) = 0$. So we have three singular points with Jacobian matrices:

$$
(0,0) \Rightarrow \begin{pmatrix} 1 & 0 \\ 0 & -3 \end{pmatrix}; \quad (0,1) \Rightarrow \begin{pmatrix} 1 & 0 \\ 0 & -1 \end{pmatrix}; \quad \left(0, \frac{3}{4}\right) \Rightarrow \begin{pmatrix} \frac{1}{4} & 0 \\ \star & \frac{3}{4} \end{pmatrix}.
$$

Then the singular points $(0,0)$ and $(0,1)$ are always saddles and $(0,3/4)$ is an unstable node. Thus the blow-down procedure is described in Figure 3.13.

Example 3.4. Now we will study an intricate star-like singularity, that is a singular point such that for every slope there is a phase curve arriving at that point with that slope. We note that such a singularity could not appear in quadratic systems.

We want to study the system

$$\begin{aligned}\dot{x}_0 &= -x_0^5 - 3x_0^4 y_0 - 2x_0^3 y_0^2 + 2x_0^2 y_0^3 + x_0^2 y_0 + 3x_0 y_0^4 + y_0^5,\\ \dot{y}_0 &= x_0^5 + 3x_0^4 y_0 + 2x_0^3 y_0^2 - 2x_0^2 y_0^3 - 3x_0 y_0^4 + x_0 y_0^2 - y_0^5.\end{aligned} \tag{3.14}$$

For this system the origin $(0,0)$ is an intricate singular point, and $PCD(x_0, y_0) \equiv 0$. In this case there is an infinite number of characteristic directions, and even though it is theoretically possible to be lucky and choose a non-special characteristic direction to do the blow-up, one can never be sure to have done it. We need to detect the special characteristic directions to be sure that the desingularization is complete, and thus we need a minimum of two blow-ups.

We proceed directly to the blow-up with $j = 1$ (the standard time reduction for linearly zero singularities) since we cannot know a priori which directions may be of special interest, and what will be the needed reduction on time. We get

$$\begin{aligned}\dot{x}_1 &= x_1^2(x_1^2 y_1^5 + 3x_1^2 y_1^4 + 2x_1^2 y_1^3 - 2x_1^2 y_1^2 - 3x_1^2 y_1 + y_1 - x_1^2),\\ \dot{y}_1 &= -x_1^3(y_1 - 1)(y_1 + 1)^5.\end{aligned}$$

So the line $x_1 = 0$ is filled up with singularities as we already expected, and we must make a reduction to a new system without such degeneracy by dividing by x_1^2. This new system has only a singular point at $(0,0)$. With the previous blow-up we may have thrown to infinity some important information related to the direction $x_0 = 0$, but we know for sure that there is no important information related to the direction $y_0 = -x_0$ (as well as with many others directions).

We start again from system (3.14) and do first a 1-twist to get

$$\begin{aligned}\dot{v}_1 &= -v_1^2 + 2v_1 z_1 - z_1^2 - v_1 z_1^2 + b v_1^2 z_1^2 + 2z_1^3 - 2b v_1 z_1^3 + b z_1^4,\\ \dot{z}_1 &= z_1^3.\end{aligned}$$

Now we do another blow-up with $j = 1$ (time reduction of odd order) to get

$$\begin{aligned}\dot{v}_2 &= v_2^2 z_2 - v_2^2 z_2^2,\\ \dot{z}_2 &= v_2^3 - 2v_2^3 z_2.\end{aligned}$$

Finally we remove the degeneracy $v_2^2 = 0$ and get the system:

$$\begin{aligned}\dot{v}_2 &= z_2 - z_2^2,\\ \dot{z}_2 &= v_2(1 - 2z_2).\end{aligned}$$

The singular points on the straight line $v_2 = 0$ are the solutions of the equation $(-1 + z_2)z_2 = 0$. So we have two singular points with Jacobian matrices:

$$(0,0) \Rightarrow \begin{pmatrix} 0 & 1 \\ 1 & 0 \end{pmatrix}; \qquad (0,1) \Rightarrow \begin{pmatrix} 0 & -1 \\ -1 & 0 \end{pmatrix}.$$

Therefore both singularities are saddles and the axis $v_2 = 0$ is not invariant. It appears that in the first attempt we have already thrown away some important information to infinity and so this was not a good blow-up. This second blow-up is the good one. Thus the blow-down procedure is described in Figure 3.14.

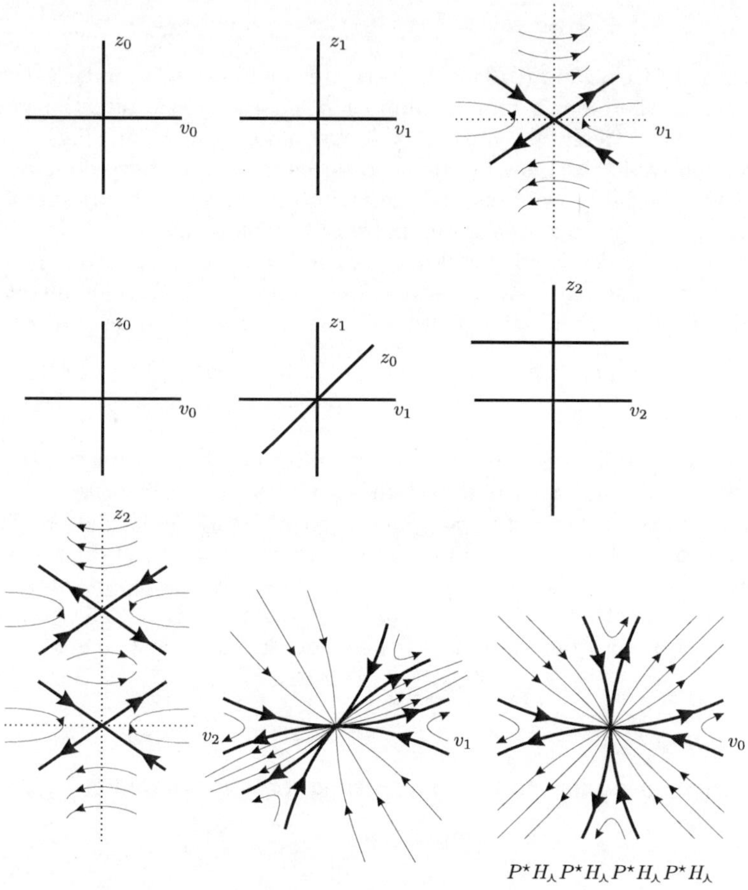

Figure 3.14: Blow-up and phase portrait of (3.14).

3.5 The borsec concept

It is known that the neighborhood of any isolated singular point of a polynomial vector field, which is not a focus or a center, is formed by a finite number of sectors that can only be of three types: parabolic, hyperbolic and elliptic (see [128]). These notions are of topological nature. It is also known that any nilpotent or intricate singular point can be decomposed by means of a finite number of changes of variables, called blow-ups, into elemental and semi-elemental singular points (for more details see [26, Section 3] and [128, Section 3.1]).

Consider the three singular points given in Figure 3.15. All three are topologically equivalent, and their neighborhoods can be described as having two elliptic sectors and two parabolic ones. But we can easily detect some geometric features which distinguish them. For example (a) and (b) have three characteristic directions and (c) has only two. Moreover in (a) the solution curves of the parabolic sectors are tangent to only one characteristic direction and in (b) they are tangent to two characteristic directions. All these properties can be determined algebraically.

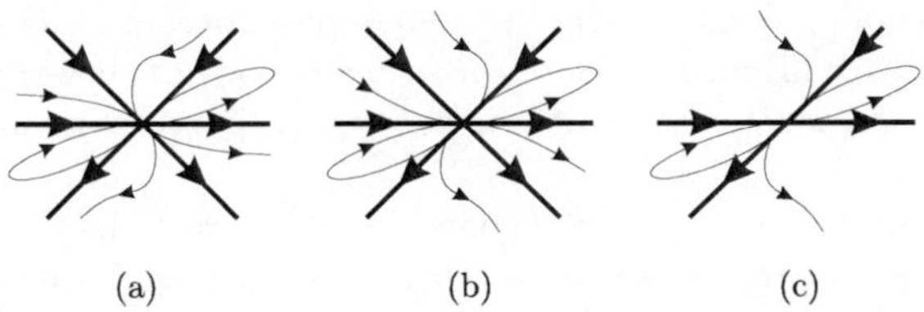

Figure 3.15: Some topologically equivalent singular points.

The usual definition of a sector is of topological nature and is local (*topological local sector*), i.e. it is given with respect to a neighborhood around the singular point. We work with a new notion, namely of *geometrical local sector* (introduced in [26]), that distinguishes the phase portraits of Figure 3.15. As we shall see later, this notion is characterized in algebraic terms.

We begin with the elemental singular points having characteristic directions. These are either two-direction nodes, one-direction nodes, star nodes or saddles. The first three cases are distinguished algebraically using their eigenvalues (see Figure 3.16). In the case of saddles the notion of geometrical local sector coincides with the usual notion of topological sector.

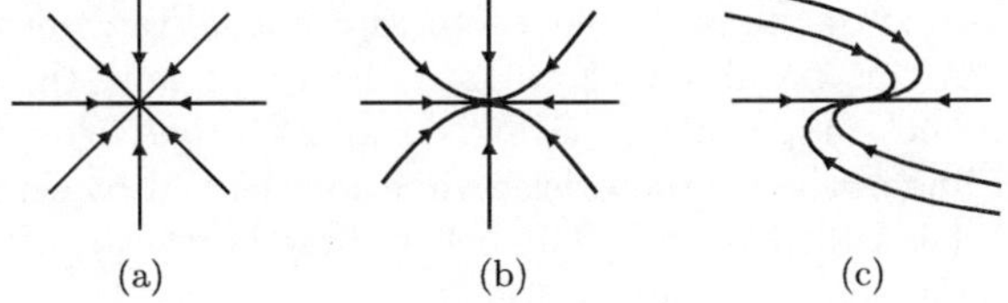

Figure 3.16: Different types of nodes.

We consider now the semi-elemental singular points. These could be saddles,

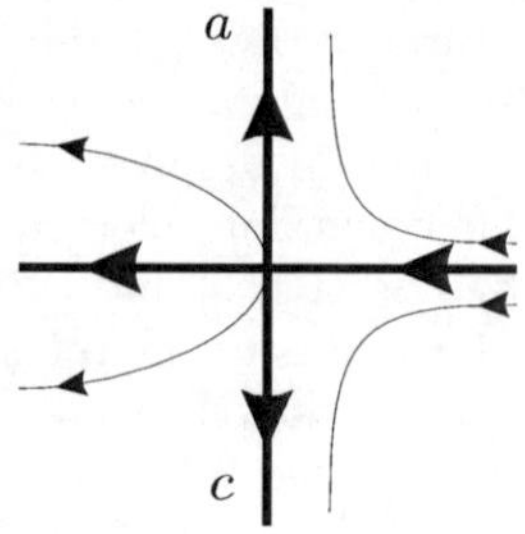

Figure 3.17: Local phase portrait of a non-semi-elemental saddle-node.

nodes or saddle-nodes. Each saddle has four separatrices and four hyperbolic sectors. Here again we call *geometrical local sector* anyone of these hyperbolic sectors and we call *borsec* (contraction of "border" with "sector") anyone of the four separatrices.

A semi-elemental node has two characteristic directions generating four half lines. For each one of these half lines, there exists at least one orbit tangent to that half line, and we pick an orbit tangent to that half line. Removing these four orbits (which we call *borsecs*) together with the singular point, we are left with four sectors that we call *geometrical local sectors*.

Consider now a semi-elemental saddle-node. Such a singular point has three separatrices and three topological sectors, two hyperbolic ones and one parabolic sector. Such a singular point has four characteristic half lines, and one of them separates the parabolic sector in two. By removing an orbit tangent to a half line for each half line as well as the singular point, we obtain four sectors that we call geometrical local sectors. We call borsecs these four orbits.

We now proceed to extend the notion of geometrical local sector and of borsec for nilpotent and intricate singular points.

The introduction of the concept of borsec in the general case will play a role in distinguishing a semi-elemental saddle-node from an intricate saddle-node such as the one indicate in Figure 3.17. In the semi-elemental saddle-node all orbits inside the parabolic sector are tangent to the same half line, but in the saddle-node of Figure 3.17 the orbits in the parabolic sector are not all tangent to the same half line. The orbits in this parabolic sector are of three kinds: the ones tangent to separatrix a, the ones tangent to separatrix c and a single orbit which is tangent to the other half line of the characteristic direction defined by separatrix b. In this case this last orbit is called borsec. The other three borsecs are separatrices as in the case of the semi-elemental node.

We start introducing some terminology.

Let δ be the border of a sufficiently small open disc D centered at an isolated singular point p so that δ intersects all the elliptic, parabolic and hyperbolic sectors of a nilpotent or intricate singular point p.

Consider a solution $\Gamma : (a,b) \to \mathbb{R}^2$ where (a,b) is its maximal interval of definition, and let γ be the orbit of Γ, i.e. $\gamma = \{\Gamma(t) \,|\, t \in (a,b)\}$. We call a *half orbit* of γ at a singular point p a subset $\gamma' \subseteq \gamma$ such that there exists $t_1 \in (a,b)$ for which we have either $\gamma' = \{\Gamma(t) \,|\, t \in (a,t_1)\}$, in which case we have $a = -\infty$, $\lim_{t\to-\infty} \Gamma(t) = p$, $\Gamma(t_1) \in \delta$ and $\Gamma(t) \in D$ for $t \in (-\infty, t_1)$, or have $\gamma' = \{\Gamma(t) \,|\, t \in (t_1,b)\}$, $b = +\infty$, $\lim_{t\to+\infty} \Gamma(t) = p$, $\Gamma(t_1) \in \delta$ and $\Gamma(t) \in D$ for $t \in (t_1, \infty)$.

We note that in the case of elliptic sectors there may exist orbits that are divided exactly in two half orbits.

Let $\Omega_p = \{\gamma' \,|\, \gamma' \text{ is a half orbit at } p\}$.

We shall define a relation of equivalence on Ω_p by using the complete desingularization of the singular point p in the case that this singularity is nilpotent or intricate (see [128, Section 1.5 and Theorem 3.3] and [48]).

In order to desingularize a singular point, we will use the technique of the one-directional blow-up described in Section 3.4 and apply k-twist changes if needed.

Since the complete desingularization of a nilpotent or an intricate singular point in general needs more than one blow-up, we have as many blow-up lines as we have blow-ups. As indicated above a blow-up line may be transformed by means of linear changes and through other blow-ups in other straight lines. We will call such straight lines *blow-up lines of higher order.*

We now introduce an equivalence relation on Ω_p. We say that two half orbits $\gamma_1', \gamma_2' \in \Omega_p$ are *equivalent* if and only if (1) for both γ_1' and γ_2' we have $\lim_{t\to-\infty} \Gamma_1(t) = p = \lim_{t\to-\infty} \Gamma_2(t)$ or $\lim_{t\to+\infty} \Gamma_1(t) = p = \lim_{t\to+\infty} \Gamma_2(t)$, (2) after the complete desingularization, these orbits lifted to the final stage are tangent to the same half line at the same singular point, or end as orbits of a star node on the same half plane defined by the blow-up line, and (3) both orbits remain in the same half plane defined by the blow-up line in all the successive blow-ups.

We recall that after a complete desingularization all singular points are elemental or semi-elemental. We now single out two types of equivalence classes:

(a) Suppose that an equivalence class $\mathcal{C} \in \Omega_p/\sim$ is such that its half orbits lifted to the last stage in the desingularization process lead to orbits that possess the following properties: (A) They belong to an elemental two-direction node or to a semi-elemental saddle-node, and (B) they are all tangent to the same half line which lies on the blow-up line.

(b) Suppose that an equivalence class $\mathcal{C} \in \Omega_p/\sim$ is such that (A) its half orbits lifted to the final stage of the desingularization process, are tangent to a blow-up line of higher order, and (B) its lifted orbits blown down to the previous stage of the desingularization, form a part of an elliptic sector.

Let $\Omega_p'/\sim$ be the set of all equivalence classes that are of type (a) or (b). Then consider the complement $\mathcal{B}_p = (\Omega_p/\sim) \setminus (\Omega_p'/\sim)$ and consider a set of representatives of $\mathcal{B}_p$. We call *borsec* anyone of these representatives.

Note that the definition of borsec is independent of the choice of the disc D with boundary δ if D is sufficiently small.

We call a *geometrical local sector* of a singular point p with respect to a neighborhood V, a region in V delimited by two consecutive borsecs.

In order to illustrate the definitions of borsec and geometrical local sector we discuss the following example given in Figures 3.18, 3.19(a) and 3.19(b).

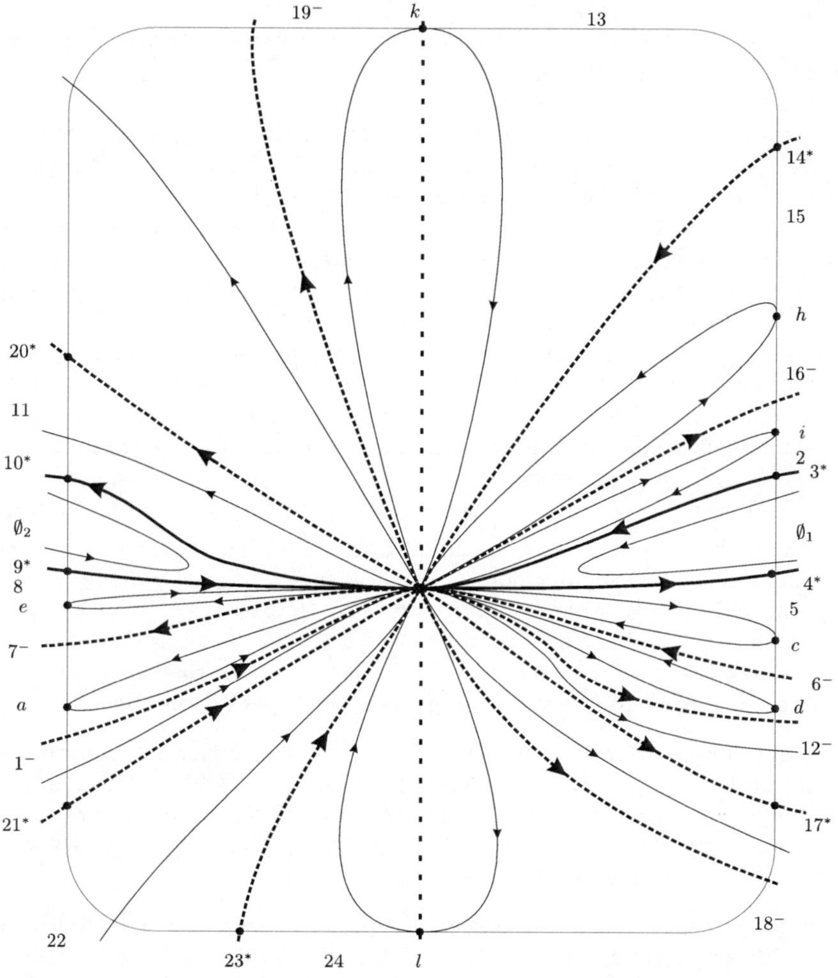

Figure 3.18: Local phase portrait of an intricate singular point.

In Figure 3.18 we portray an intricate singular point p whose desingularization needs a chain of two blow-ups, where all different kinds of elemental singular

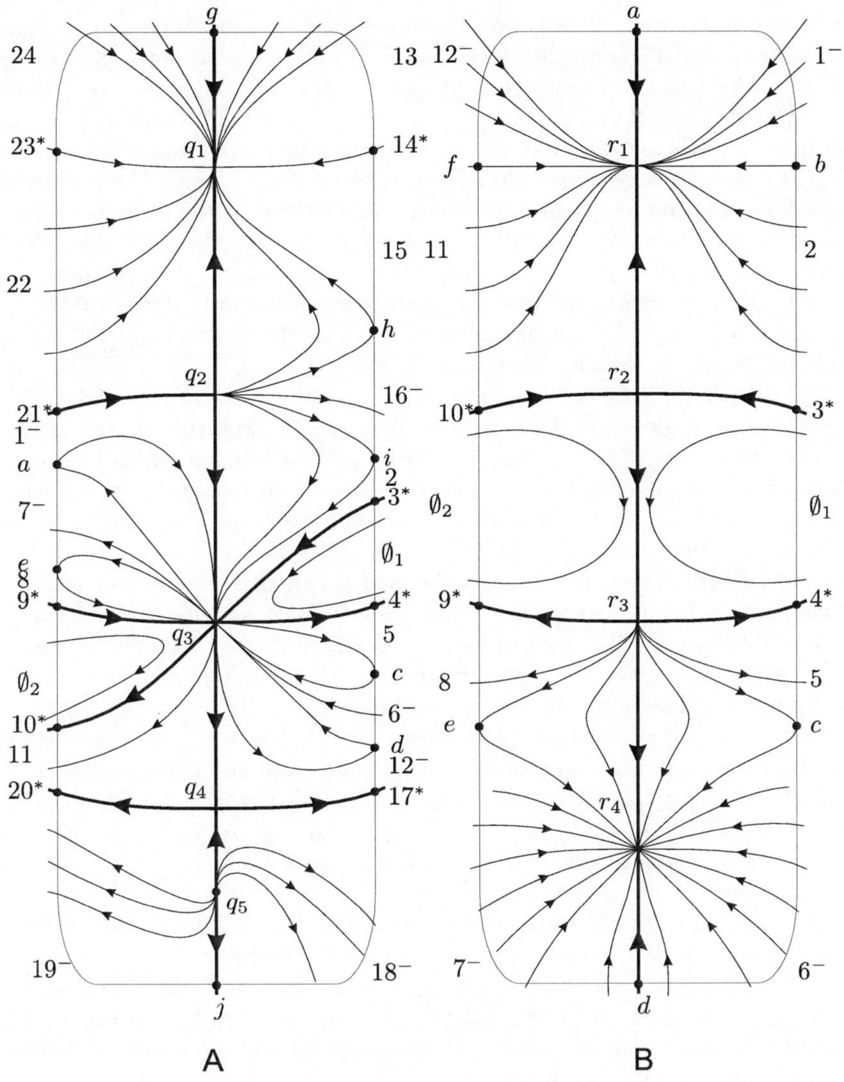

Figure 3.19: The two needed blow-ups for the singularity of Figure 3.18.

points (which can appear in a blow-up) are present, and where semi-elemental saddle-nodes appear in both possible positions with respect of the blow-up line.

We have taken a small enough neighborhood of the singular point p of boundary δ. We split the boundary δ into different arcs and points that correspond to the different equivalence classes of orbits. We have enumerated them from 1 to 24.

The arcs of δ denoted with $\emptyset_1$ and $\emptyset_2$ correspond to hyperbolic sectors that are not considered in the equivalence classes since the orbits do not tend to p. Some of these equivalence classes have a unique orbit that is then a borsec (like 14^* or 4^*); we add an asterisk superscript to denote these equivalence classes. Other equivalence classes are arcs, like 16^- or 12^-, and one representative of each one of them is taken as a borsec; we add a dash superscript to denote these equivalence classes. The remaining equivalence classes, just denoted by their number, are those that do not produce a borsec by the exceptions given in the definition. We have drawn the separatrices (which are always borsecs) with a bold continuous line. We have drawn the borsecs that are not separatrices with bold dashed lines. Other orbits are drawn as thin continuous lines. Finally, the vertical dashed line is the y-direction in which the first blow-up was done.

We describe a little the blow-ups of the phase portrait of the intricate singularity p given in Figure 3.18. Its first blow-up is given in Figure 3.19(a). In it we see, from the upper part of the figure to its lower part, q_1, an elemental two-direction node with all but two orbits tangent to the blow-up line; q_2, a semi-elemental saddle-node with direction associated to the non-zero eigenvalue being the blow-up line; q_3, another intricate singular point that needs another blow-up portrayed in Figure 3.19(b); q_4, an elemental saddle; and q_5, an elemental one-direction node that necessarily has its characteristic direction coinciding with the blow-up line.

In order to make the vertical blow-up of the intricate singularity q_3, we must first do an ε-twist since the vertical direction that corresponds to the previous blow-up line is a characteristic direction of q_3.

In this second blow-up given in Figure 3.19(b), we see, going down from its upper part, the following elemental or semi-elemental singular points: r_1, a two-direction node with only two orbits tangent to the blow-up line (this singular point corresponds to the characteristic direction given by the previous blow-up line); r_2, a saddle; r_3, a saddle-node with the direction associated to the zero eigenvalue being the blow-up line; and r_4, a star node.

Now we describe all the classes of equivalence that we obtain in order to clarify the definitions of borsec and geometrical local sector.

We must move from the second blow-up to the first and after that to the original phase portrait. We enumerate the arcs in the boundary of Figure 3.19(b) (following the clockwise sense) that will correspond to the classes of equivalence of orbits in Figure 3.18 as follows:

(1^-) The arc 1^- goes from the point a on the vertical axis to the point b without including any of them.

(2) The arc 2 goes from the point b to the point 3^* without including any of them.

The orbit that ends at point b corresponds to the blow-up line in Figure 3.19(a), and so does not survive in the original phase portrait. Thus the orbits associated to arc 1^- cannot belong to the same equivalence class as the orbits associated to arc 2 since in Figure 3.19(a) they are in different half planes defined by the blow-up line.

(3^*) The point 3^* belongs to the orbit that is a separatrix of the saddle r_2.

($\emptyset_1$) The open arc $\emptyset_1$ goes between the points 3^* and 4^*, is associated to a hyperbolic sector and plays no role.

(4^*) The point 4^* belongs to the orbit that is a separatrix of the saddle-node r_3.

(5) The arc 5 goes from the point 4^* to the point c including only the second.

(6^-) The arc 6^- goes from the point c to the point d on the vertical axis, including the point c.

The point c belongs to both arcs 5 and 6^-. In fact it is just a point of partition of the boundary, splitting the orbits that come from r_3 from the orbits that go to r_4. Since the equivalence classes are defined regarding the half orbits, there is no contradiction.

(7^-) The arc 7^- goes from the point d on the vertical axis to the point e including the point e (i.e. $7^- = (d, e]$).

(8) The arc 8 goes from the point e to the point 9^* including the point e.

The same comment made for the point c applies to point e.

(9^*) The point 9^* belongs to the orbit that is a separatrix of the saddle-node r_3.

($\emptyset_2$) The open arc $\emptyset_2$ between the points 9^* and 10^* is associated to a hyperbolic sector and plays no role.

(10^*) The point 10^* belongs to the orbit that is a separatrix of the saddle r_2.

(11) The arc 11 goes from the point 10^* to the point f without including any of them.

(12^-) The arc 12^- goes from the point f to the point a in the vertical axis without including any of them (i.e. $12^- = (d, e)$).

The same comment about the point b applies to point f.

Now we translate these notations to Figure 3.19(a) and complete the notation of the arcs on the boundary of this figure (again following the clockwise sense).

(13) The arc 13 goes from the point g on the vertical axis to the point 14^* without including any of them.

(14^*) The point 14^* belongs to the orbit that is tangent to the eigenvector associated to the greatest eigenvalue of the node q_1.

(15) The arc 15 goes from the point 14^* to the point h including only the second.

(16^-) The arc 16^- goes from the point h to the point i including both (i.e. $16^- = [d, e]$).

The following arcs and points from the point i to the point 17^* have already received their names when we did the blow-down from Figure 3.19(b) to Figure 3.19(a).

The arcs 6^- and 12^- of Figure 3.19(b) become adjacent in Figure 3.19(a) and the points a and d are glued together and correspond to the point that after the $-\varepsilon$-twist goes to the vertical axis. The region defined by these arcs forms now an elliptic sector.

(17^*) The point 17^* belongs to the orbit that is a separatrix of the saddle q_4.

(18^-) The arc 18^- goes from the point 17^* to the point j without including any of them (i.e. $18^- = (d, e)$).

(19^-) The arc 19^- goes from the point j to the point 20^* without including any of them (i.e. $19^- = (j, 12^*)$).

(20^*) The point 20^* belongs to the orbit that is a separatrix of the saddle q_4.

The following arcs and points from the point 20^* to the point 21^* have already received their names when we did the blow-down from Figure 3.19(b) to Figure 3.19(a).

(21^*) The point 21^* belongs to the orbit that is a separatrix of the saddle-node q_2.

(22) The arc 22 goes from the point 21^* to the point 23^* without including any of them.

(23^*) The point 23^* belongs to the orbit that is tangent to the eigenvector associated to the greatest eigenvalue of the node q_1.

(24) The arc 24 goes from the point 23^* to the point g in the vertical axis without including any of them.

Now we move to the original phase portrait in Figure 3.18. For clarity it is convenient to start the description with a hyperbolic sector.

The orbit associated to the point 4^* defines an equivalent class with a single element, and this element is a borsec. Moreover it is a global separatrix.

The orbits associated to the points of the arc 5 form a class of equivalence but define no borsec since in the final desingularization (Figure 3.19(b)) these orbits end at a saddle-node tangent to the blow-up line; thus these orbits are in a class of equivalence of type (a) that does not produce a borsec.

The orbits associated to the points of the arc 6^- form a class of equivalence defining a borsec that splits the two local geometrical elliptic sectors that we see in Figure 3.18. This borsec is not a separatrix.

The orbits associated to the points of the arc 12^- form a class of equivalence defining a borsec that splits a local elliptic sector from a parabolic local sector that we can see in Figure 3.18. Even though the class 12^- has been split from class 11 by the blow-up line of higher order (the straight line passing through point r_1 and going from point b to point f in Figure 3.19(b)), we see that class

12^- corresponds to the part of an elliptic sector with its characteristic direction tangent to the blow-up line. So this class of equivalence is not of type (b), and we must define a borsec there. The type (b) however will occur later on in our discussion, more precisely when we consider the arc 11.

The orbit associated to the point 17^* defines an equivalent class with a single element, and this element is a borsec. This borsec is not a separatrix. It is just part of a global parabolic sector but locally distinguishes the three different characteristic directions of the orbits in the arc of δ going from d to l.

The orbits associated to the points of the arc 18^- form a class of equivalence defining a borsec that splits a local elliptic sector from a parabolic one that we can see in Figure 3.18.

The orbits associated to the points of the arc 24 form a class of equivalence, but this does not define a borsec because in the final desingularization, the corresponding orbits end at a two-direction node tangent to the blow-up line (this class of equivalence is of type (a)).

The orbit associated to the point 23^* defines an equivalent class with a single element, and this element is a borsec that splits a local elliptic sector from a parabolic one that we can see in Figure 3.18.

The orbits associated to the points of the arc 22^- form a class of equivalence, but this does not define a borsec because in the final desingularization, the corresponding orbits end at a two-direction node tangent to the blow-up line (this class of equivalence is of type (a)).

The orbit associated to the point 21^* defines an equivalent class with a single element, and this element is a borsec.

The orbits associated to the points of the arc 1^- form a class of equivalence defining a borsec that splits a local elliptic sector from a parabolic one that we can see in Figure 3.18. Even though the class 1^- has been split from class 2 by the blow-up line of higher order, in Figure 3.19(b), we see that class 1^- corresponds to a part of an elliptic sector with its characteristic direction tangent to the blow-up line. So this is not a class of equivalence of type (b), and we must define a borsec here.

The orbits associated to the points of the arc 7^- form a class of equivalence defining a borsec that splits the two local elliptic sectors that we see in Figure 3.18. As in the case of arc 6^-, this borsec is not a separatrix.

The orbits associated to the points of the arc 8 form a class of equivalence but define no borsec since in the final desingularization (Figure 3.19(b)) these orbits end at a saddle-node tangent to the blow-up line (this equivalence class is of type (a)).

The orbit associated to the point 9^* defines an equivalent class with a single element, and this element is a borsec. Moreover it is a global separatrix.

The orbits associated to the open arc $\emptyset_2$ form a hyperbolic sector and are not associated to any equivalence class since they do not end at the singular point.

The orbit associated to the point 10^* defines an equivalent class with a single element, and this element is a borsec. Moreover it is a global separatrix.

The orbits associated to the points of the arc 11 form a class of equivalence but define no borsec since class 11 is of type (b). In this case we are in a similar situation as with the arc 12^-, but now, since the point r_2 is a saddle, the arc 11 in Figure 3.19(a) defines a parabolic sector, so there is no need of a borsec, which would otherwise be needed if the sector were elliptic.

The orbit associated to the point 20^* defines an equivalent class with a single element, and this element is a borsec. This is similar to the case 17^*.

The orbits associated to the points of the arc 19^- form a class of equivalence defining a borsec that splits a local elliptic sector from a parabolic one that we can see in Figure 3.18. This is similar to the case 18^-.

The orbits associated to the points of the arc 13 form a class of equivalence, but this does not define a borsec analogously with the case 24.

The orbit associated to the point 14^* defines an equivalent class with a single element, and this element is a borsec.

The orbits associated to the points of the arc 15 form a class of equivalence that does not define a borsec analogously to the case 13.

The orbits associated to the points of the arc 16^- form a class of equivalence defining a borsec that splits two local elliptic sectors. This is similar to the case 7^-.

The orbits associated to the points of arc 2 form a class of equivalence but define no borsec by the same arguments used for the arc 11.

The orbit associated to the point 3^* defines an equivalent class with a single element, and this element is a borsec. Moreover it is a separatrix.

Generically a *geometrical local sector* is defined by two consecutive borsecs arriving at the singular point with two different well defined angles. If this sector is parabolic, then the solutions can arrive at the singular point with one of the two characteristic angles, and this is geometrical information that can be revealed with the blow-up.

There is also the possibility that two borsecs defining a geometrical local sector tend to the singular point with the same well defined angle. Such a sector will be called a *cusp-like sector* and can be either hyperbolic, elliptic or parabolic, denoted by $H_\curlywedge$, $E_\curlywedge$ and $P_\curlywedge$ respectively. The reader can find a more detailed explanation of the notations used to describe a singular point in Section 3.7.

In the case of parabolic sectors, we want to include the information of the orbits that arrive tangent to one or to the other borsec. We distinguish the two cases by writing $\widehat{P}$ if they arrive tangent to the borsec limiting the previous sector in clockwise sense or $\widetilde{P}$ if they arrive tangent to the borsec limiting the next sector. In the case of a cusp-like parabolic sector, all orbits must arrive with only one well defined angle, but the distinction between $\widehat{P}$ and $\widetilde{P}$ is still valid because it occurs at some stage of the desingularization, and we must keep this information since they are geometrically different according to our definition. Thus complicated intricate singular points like the two we see in Figure 3.20 may be described as $\widetilde{P}E\widehat{P}HHH$ (case (a)) and $E\widehat{P}_\curlywedge HH\widetilde{P}_\curlywedge E$ (case (b)), respectively.

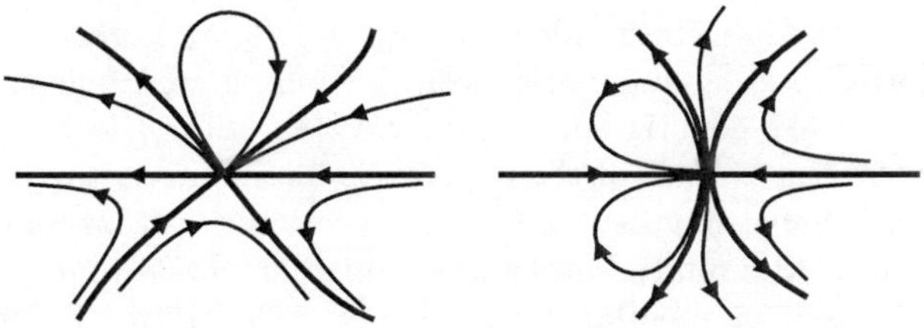

(a) (b)

Figure 3.20: Two phase portraits of intricate singular points.

The phase portrait of the intricate singularity of Figure 3.18 could be described as

$$H_{\curlywedge} E_{\curlywedge} E_{\curlywedge} \widehat{P}\widehat{P} E \widehat{P}\widehat{P} E_{\curlywedge} E_{\curlywedge} H_{\curlywedge} \widehat{P}\widehat{P} EEE$$

starting with the hyperbolic sector $\emptyset_1$ and going in the clockwise direction.

3.6 Equivalence relations for singularities and for geometrical configurations of singularities of planar polynomial vector fields

We first recall the *topological equivalence relation* as it is used in most of the literature. Two singularities p_1 and p_2 are *topologically equivalent* if there exist open neighborhoods N_1 and N_2 of these singularities and a homeomorphism $\Psi : N_1 \to N_2$ carrying orbits to orbits and preserving their orientations. To reduce the number of cases, by *topological equivalence* we shall mean here that the homeomorphism Ψ *preserves or reverses* the orientation. In this book we use this second notion, sometimes also used elsewhere in the literature (see [182, 19]).

In 2005 [182] Jiang and Llibre introduced another equivalence relation for singularities that is finer than the topological equivalence: We say that p_1 and p_2 are *qualitatively equivalent* if (1) they are topologically equivalent through a homeomorphism Ψ and (2) two orbits are tangent to the same straight line at p_1 if and only if the corresponding two orbits are tangent to the same straight line at p_2.

We say that two simple finite nodes, with the respective eigenvalues λ_1, λ_2 and σ_1, σ_2, of planar polynomial vector fields are *tangent equivalent* if and only if they satisfy one of the following three conditions: (1) $(\lambda_1 - \lambda_2)(\sigma_1 - \sigma_2) \neq 0$; (2) $\lambda_1 - \lambda_2 = 0 = \sigma_1 - \sigma_2$ and both linearization matrices at the two singularities are diagonal; (3) $\lambda_1 - \lambda_2 = 0 = \sigma_1 - \sigma_2$ and the corresponding linearization matrices are not diagonal.

We say that two infinite simple nodes P_1 and P_2 are *tangent equivalent* if and only if their corresponding singularities on the sphere are tangent equivalent and in addition, if they are generic nodes, we have $(|\lambda_1| - |\lambda_2|)(|\sigma_1| - |\sigma_2|) > 0$ where λ_1 and σ_1 are the eigenvalues of the eigenvectors tangent to the line at infinity.

Finite and infinite singular points may be either real or complex. We point out that the sum of the multiplicities of all singular points of a quadratic system (with a finite number of singular points) is always 7. (Here of course we refer to the compactification on the complex projective space $\mathbb{P}\mathbb{C}^2$ of the foliation with singularities associated to the complexification of the vector field.) The sum of the multiplicities of the infinite singular points is always at least 3; more precisely it is always 3 plus the sum of the multiplicities of the finite singularities that have gone to infinity.

For the sake of completeness of this section, we repeat the basic definitions concerning singularities already given in Section 3.3:

We call *elemental* a singular point with both its eigenvalues not zero.

We call *semi-elemental* a singular point with exactly one of its eigenvalues equal to zero.

We call *nilpotent* a singular point with both its eigenvalues zero but with its Jacobian matrix at that singularity not identically zero.

We call *intricate* a singular point with its Jacobian matrix identically zero.

The *multiplicity* of a finite singularity of a given polynomial vector field of degree n is the maximum number of singularities that can bifurcate from it when we perturb the system inside the whole class of polynomial vector fields of degree n. In the case of an infinite singularity, we define the multiplicity as the couple $\binom{a}{b}$ where a is the maximum number of finite singularities and b is the maximum number of infinite singularities that can bifurcate from it when we perturb the system inside the whole class of polynomial vector fields of degree n.

We will say that two foci (or saddles) are *order equivalent* if they are both strong or both weak of the same order. We will say that two centers are of *same level of isochronicity* if they are both non-isochronous, both isochronous but not uniform isochronous or both uniform isochronous.

Semi-elemental saddle-nodes are always topologically equivalent.

To define the notion of *geometrical equivalence relation* of singularities, we first define the notion of *blow-up equivalence*, necessary for nilpotent and intricate singular points. We start by considering such a singular point p_1 at the origin of the plane (x_0, y_0) with a finite number of characteristic directions. We define an ε-twist as a k-twist with k small enough so that no characteristic direction (or special characteristic direction in the case of a star singularity) with negative slope is moved to positive slope. Then if $x_0 = 0$ is a characteristic direction, we do an ε-twist. After the blow-up $(x_0, y_0) = (x_1, y_1 x_0)$, the singular point is replaced by the straight line $x_1 = 0$ in the plane (x_1, y_1). The neighborhood of the straight line $x_1 = 0$ in the projective plane obtained identifying the opposite points of the border of the Poincaré disc is a Möbius band M_1.

The straight line $x_1 = 0$ will be invariant and may be formed by a continuous set of singular points. In that case, with a time change, this degeneracy may be removed and the y_1-axis will remain invariant.

Now we have a number k_1 of singularities located on the axis $x_1 = 0$. We do not include the infinite singular point at the origin of local chart U_2 because we already know that it does not play any role to understand the local phase portrait of the singularity p_1. We can then list the k_1 singularities as $p_{1,1}, p_{1,2}, \ldots, p_{1,k_1}$ with decreasing order of the y_1 coordinate. The $p_{1,i}$ is adjacent to $p_{1,i+1}$ in the usual sense, and p_{1,k_1} is also adjacent to $p_{1,1}$ by the Möbius band.

Assume now that we have an intricate singular point p_1 at the origin of the plane (x_0, y_0) with an infinite number of characteristic directions. Then if $x_0 = 0$ is a special characteristic direction, we do an ε-twist. After the blow-up $(x_0, y_0) = (x_1, y_1 x_0)$, the singular point is replaced by the straight line $x_1 = 0$ in the plane (x_1, y_1). The neighborhood of the straight line $x_1 = 0$ in the projective plane obtained identifying the opposite points of the border of the Poincaré disc is a Möbius band M_1.

The straight line $x_1 = 0$ will be invariant and formed by a continuous set of singular points. In that case, with a time change, this degeneracy may be removed and the y_1-axis is not anymore invariant.

Now we have a set of cardinality k_1 formed by singularities located on the axis $x_1 = 0$ plus contact points of the flow with the axis $x_1 = 0$. Again we do not include the infinite singular point at the origin of local chart U_2 because we already know that it does not play any role to understand the local phase portrait of the singularity p_1. We list again the k_1 points as $p_{1,1}, p_{1,2}, \ldots, p_{1,k_1}$ with decreasing order of the y_1 coordinate. The $p_{1,i}$ is adjacent to $p_{1,i+1}$ in the usual sense, and p_{1,k_1} is also adjacent to $p_{1,1}$ by the Möbius band.

Let p_2 be another intricate singularity located at the origin of another plane $(\bar{x}_0, \bar{y}_0)$ and let M_2 be the analogue of M_1 for p_2.

The next definition works whether the singular points are star-like or not. We say that p_1 and p_2 are *one-step blow-up equivalent* if modulo a rotation with center p_2 (before the blow-up) and a symmetry (if needed), we have the following:

(i) The cardinality k_1 from p_1 equals the cardinality k_2 from p_2.

(ii) There exists a homeomorphism $\phi^1_{p_1} : M_1 \to M_2$ such that $\phi^1_{p_1}(\{x_1 = 0\}) = \{\bar{x}_1 = 0\}$, $\phi^1_{p_1}$ sends the singularities $p_{1,i}$ to $p_{2,i}$ and the flow on a neighborhood U of the axis $x_1 = 0$ is topologically equivalent to the flow on $\phi^1_{p_1}(U)$.

(iii) $\phi^1_{p_1}$ sends an elemental (respectively semi-elemental, nilpotent or intricate) singular point to an elemental (respectively semi-elemental, nilpotent or intricate) singular point.

(iv) $\phi^1_{p_1}$ sends a contact point to a contact point.

Assuming that $p_{1,j}$ and $\phi^1_{p_1}(p_{1,j}) = p_{2,j}$ are both intricate or both nilpotent, the process of desingularization (blow-up) must be continued.

We do exactly the same study we did before for p_1 and p_2 now for $p_{1,j}$ and $p_{2,j}$. We move them to the respective origins of the planes (x_1, y_1) and $(\bar{x}_1, \bar{y}_1)$, and we can determine whether they are one-step blow-up equivalent or not.

If successive nilpotent or intricate singular points appear from desingularization of p_1 we do the same kind of changes that we did for $p_{1,j}$ and apply the corresponding definition of one-step blow-up equivalence. This is repeated until after a finite number of blow-ups all the singular points that appear are elemental or semi-elemental.

We say that two singularities, both nilpotent or both intricate p_1 and p_2 of two polynomial vector fields χ_1 and χ_2 are *blow-up equivalent* if and only if all of the following hold:

(i) They are one-step blow-up equivalent.

(ii) The tree of blow-up steps that we must do to desingularize each one of the two singularities is the same.

(iii) At each level j in the process of desingularization of p_1 and of p_2, two singularities that are related via the corresponding homeomorphism are one-step blow-up equivalent.

Definition 3.1. Two finite singularities p_1 and p_2 of two polynomial vector fields are *geometrically equivalent* if and only if they are topologically equivalent, they have the same multiplicity and one of the following conditions is satisfied:

- p_1 and p_2 are order equivalent foci (or saddles),
- p_1 and p_2 are tangent equivalent simple nodes,
- p_1 and p_2 are both elemental centers with the same level of isochronicity,
- p_1 and p_2 are both semi-elemental singularities,
- p_1 and p_2 are blow-up equivalent nilpotent or intricate singularities.

Definition 3.2. We say that two infinite singularities P_1 and P_2 of two polynomial vector fields are *geometrically equivalent* if they are blow-up equivalent finite singularities in the corresponding local charts at infinity and the number, type and ordering of sectors on each side of the line at infinity of P_1 coincide with those of P_2.

Definition 3.3. By the *geometry of a singularity* we understand its local topological phase portrait, together with its multiplicity, its order if it is a weak focus or a weak saddle, its type of tangent equivalence if it is a node, its level of isochronicity if it is a center, and its geometrical sectorial decomposition.

Definition 3.4. We call a *geometrical configuration of singularities of a polynomial system* an ordered set of symbols encoding the complete geometry of all its singularities. This set encodes the geometry of the isolated singularities (finite and infinite) and occasionally may include a subset encoding the geometry of the whole set of non-isolated singularities if an infinite number of singularities occurs.

The symbols we will use to describe the singularities have been progressively introduced in this chapter, but a complete description is given in Section 3.7.

Definition 3.5. (i) Let χ_1 and χ_2 be two non-degenerate polynomial vector fields with a finite number of singularities, finite or infinite. We say that χ_1 and χ_2 have *geometrical equivalent configurations of singularities* if and only if we have a bijection σ of the set of singularities of χ_1 onto the set of singularities of χ_2 (sending finite singularities to finite singularities and infinite ones to infinite ones) such that for every singularity s of χ_1, $\sigma(s)$ is geometrically equivalent to s and furthermore the restriction σ_∞ to the subset of infinite singularities listed in the counterclockwise direction on the Poincaré disc respects, for each one of the infinite singularities, the number and types of its sectors as well as the order of these sectors when we move from a point on the line at infinity on a curve around the singularity sufficiently close to the line at infinity.

(ii) Let χ_1 and χ_2 be two non-degenerate polynomial vector fields with a finite number of finite singularities and an infinite number of infinite singularities. We say that the configurations of the two systems are *geometrically equivalent* if the configurations of the finite singularities are geometrically equivalent and the configurations of singularities that appear from the reductions obtained after the removal of the line at infinity are also geometrically equivalent.

(iii) Let χ_1 and χ_2 be two degenerate polynomial vector fields. We say that the configurations of these two systems are *geometrically equivalent* if (a) there is a one-to-one correspondence σ of the set of isolated singularities of χ_1 onto the set of isolated singularities of χ_2 that yields geometrical equivalent configurations of their isolated singularities, and (b) the two curves $f_1 = 0$ and $f_2 = 0$, where f_1 is the greatest common divisor over $\mathbb{R}$ of the polynomials $p_1, q_1 \in \mathbb{R}[x, y]$, defining χ_1, and f_2 is the greatest common divisor over $\mathbb{R}$ of the polynomials $p_2, q_2 \in \mathbb{R}[x, y]$, defining χ_2, are affinely equivalent. Moreover, the singularities that appear from the reduction after the removal of the curve of degeneracy must also be geometrically equivalent.

Remark 3.1. Important: We are perfectly aware that our definition of geometrically equivalent singularities has been done without taking into account the time, that is, we consider geometrically equivalent a strong stable focus to a strong unstable focus. Furthermore, our definition of geometrically equivalent configurations of singularities also does not take into account the possibility of different stabilities among the singularities. We consider that two configurations of singularities are equivalent if the singularities are geometrically equivalent one to one, plus a detail about the relative position of infinite saddle-nodes. This definition does not take into account that a couple of two finite stable anti-saddles can never be transformed by means of a homeomorphism (after composing it with the solution curves) to a couple of finite anti-saddles with different stabilities. The simplest example is the case with two finite foci and one infinite saddle not having a separatrix connection with its opposite point on the Poincaré disc. We may denote this case with $a, a; S$ (where a stands simply for anti-saddle). The most common situation is that there are no limit cycles, and both foci have different stability, and then the stable infinite separatrix comes from the unstable focus and the unstable separatrix goes to the stable focus. But it may also happen that one of the foci changes its stability after passing through a Hopf bifurcation, and then both foci are again strong and with the same stability. According to our definition of geometrical equivalence of configurations of singularities, these two configurations are equivalent. To distinguish them we would need to make a deeper definition that takes into account the stabilities and considers them as distinct configurations.

As we go deeper into the subject, definitions need to be expanded. This already happened before; for example, our definition of equivalence of singularities in [29] did not distinguish among weak foci or weak saddles whereas the present definition does distinguish them.

The feature we signaled in the above paragraph can be detected by means of algebraic tools. This is not the main reason why we do not include this in our definition. In fact, there can be found invariant polynomials to distinguish the different combinations of stabilities. The problem is that the cases we need to consider are not only $a, a; S$ but many others much more complicated. If there are only two anti-saddles, the different possibilities are just two: Either they have the same stability or they have different stability. The fact of both being stable or both unstable is not a real problem since we can always consider the classification modulo a reversal of time. Also in the case with three anti-saddles like $a, a, a, s; S$, there are only two possibilities, either all three have the same stability or one has the opposite stability to the other two. But in a bit more complicated case like $a, a, sn; S$ the number of possibilities would start to increase. Just in this case we have the possibility of (1) the two anti-saddles having the same stability of the nodal part of the saddle-node, (2) the two anti-saddles having the same stability opposite to the one of the nodal part of the saddle-node and (3) the two anti-saddles having different stabilities (and obviously the nodal part of the saddle-node coincides with one of them). And in an even more complicated case with a finite anti-saddle, a finite saddle-node and an infinite saddle-node coming from the coalescence of a finite node and an infinite saddle, we would have up to four possibilities to consider. There are no richer cases since all the situations can be derived from the case a, a, a, s when some anti-saddles collide with other singularities (finite or infinite).

In short, for every situation in which there is more that one singular point containing parabolic sectors, we would need to describe in detail all the possible combinations of stabilities of these parabolic sectors. In some cases like $n, n; S$ it is clear that both must have different stabilities, but in others like $f, f; S$ we have already seen that two cases are possible. This fact may imply also the existence of limit cycles in some cases, or may yield other implications for the global phase portrait.

Even though a deeper geometrical classification can be done, this could double the number of elements in the classification. Presently our goal is to succeed in giving a complete topological classification, modulo limit cycles, of the whole class **QS**. The first step in this problem beyond this work is to determine the α- or ω-limit points of every separatrix in each one of the configurations. It is at this point that the different stabilities need to be taken into account.

3.7 Notations for singularities of polynomial differential systems

In this work we encounter all the possibilities we have for the geometric features of the infinite singularities in the whole quadratic class as well as the way they assemble in systems of this class. Since we want to describe precisely these geometric features and in order to facilitate understanding, it is important to have a clear, compact and congenial notation that conveys easily the information. Of course this notation must be compatible with the one used to describe finite singularities, and thus we must start with the finite ones. The notation we use, even though it is used here to describe finite and infinite singular points of quadratic systems, can easily be extended to general polynomial systems.

We start by first describing the finite and infinite singularities, denoting the first ones with lowercase letters and the second with capital letters. When describing in a sequence both finite and infinite singular points, we will always place first the finite ones and only later the infinite ones, separating them by a semicolon ";". Finite and infinite singular points may either be real of complex. When we have a complex singular point, we will specify this with the symbols © and Ⓒ for finite and infinite singular points respectively.

3.7.1 Elemental singularities

We use the letters s, S for saddles; n, N for nodes; f for foci and c for centers. In order to augment the level of precision we will distinguish the finite nodes as follows:

n for a node with two distinct eigenvalues,

n^d (a one-direction node) for a node with two identical eigenvalues whose Jacobian matrix cannot be diagonal,

n^* (a star node) for a node with two identical eigenvalues whose Jacobian matrix is diagonal.

Moreover, in the case of an elemental infinite node, we want to distinguish whether the absolute value of the eigenvalue associated to the eigenvector directed towards the affine plane is greater or less than the absolute value of the eigenvalue associated to the eigenvector tangent to the line at infinity. This is relevant because this determines if all the orbits except one arrive at infinity tangent to the line at infinity or transversal to this line. We will denote them as N^∞ and N^f respectively.

Finite elemental foci and saddles are classified as strong or weak. When the trace of the Jacobian matrix evaluated at those singular points is not zero, we call them strong saddles and strong foci and we maintain the standard notations s and f. But when the trace is zero, except for centers and saddles of infinite order (i.e. saddles with all their Poincaré–Lyapunov constants equal to zero), it is known that the foci and saddles, in the quadratic case, may have up to three orders. We denote them by $s^{(i)}$ and $f^{(i)}$ where $i = 1, 2, 3$ is the order. In addition we have the centers, which we denote by c and saddles of infinite order (integrable saddles), which we denote by $\$$. Moreover, we will denote by $c^{\odot}$ the isochronous centers and by $c^{\oplus}$ the uniform isochronous centers.

Foci and centers cannot appear as singular points at infinity, and hence there is no need to introduce their order in this case. In the case of saddles, we can have weak saddles at infinity, but it is premature at this stage to describe them since the maximum order of weak singularities in cubic systems is not yet known.

3.7.2 Non-elemental singularities

All non-elemental singular points are multiple singularities, in the sense that there are perturbations that have at least two elemental singular points as close as we wish to the multiple singularity. For finite singular points we denote with a subindex their multiplicity, as in $\overline{sn}_{(2)}$ or in $\widehat{es}_{(3)}$ (the meanings of $^-$ and $\widehat{}$ will be explained in the next two subsections). In order to describe the various kinds of multiplicity for infinite singular points, we use the concepts and notations introduced by Schlomiuk and Vulpe [301]. Thus we denote by $\binom{a}{b}$ the maximum number a (respectively b) of finite (respectively infinite) singularities that can be obtained by perturbation of the multiple singularity. For example $\overline{\binom{1}{1}}SN$ means a saddle-node at infinity produced by the collision of one finite singularity with an infinite one; $\overline{\binom{0}{3}}S$ means a saddle produced by the collision of three infinite singularities.

3.7.2.1 Semi-elemental singularities

They can be either nodes, saddles or saddle-nodes, finite or infinite. We will denote the semi-elemental ones always with an overline, for example $\overline{n}$, $\overline{s}$ and $\overline{sn}$ with the corresponding multiplicity. In the case of infinite singularities, we will put the overline on top of the parenthesis of multiplicities.

Moreover, in cases explained in this section (see page 84), an infinite saddle-node may be denoted by $\overline{\binom{1}{1}}NS$. Semi-elemental nodes could never be n^d or n^* since the eigenvalues are always different. In the case of an infinite semi-elemental node, the type of collision determines whether the singular point is tangent equivalent to N^f or to N^∞ where $\overline{\binom{2}{1}}N$ is an N^f and $\overline{\binom{0}{3}}N$ is an N^∞.

3.7.2.2 Nilpotent singularities

They can either be saddles, nodes, saddle-nodes, elliptic-saddles, cusps, foci or centers.

The first four of these could be at infinity. We denote a nilpotent singular points with a hat, as in $\widehat{es}_{(3)}$ for a finite nilpotent elliptic-saddle of multiplicity three and $\widehat{cp}_{(2)}$ for a finite nilpotent cusp singularity of multiplicity two. In the case of nilpotent infinite singularities, we will put the hat on top of the parenthesis of multiplicities, for example $\widehat{\binom{1}{2}}PEP-H$ (the meaning of $PEP-H$ will be explained in Section 3.7.2.3). The relative position of the sectors of an infinite nilpotent singularity, with respect to the line at infinity, can produce topologically different phase portraits. This forces us to enlarge the notation for these singularities similar to the one that we will use for the intricate singular points.

Moreover, in the case of a finite nilpotent elliptic-saddle, it may happen that it can belong to two different geometrical equivalence classes according to our definition of blow-up equivalence. A nilpotent elliptic-saddle of a quadratic system needs a first directional blow-up followed by a 1-twist and a second blow-up. After this second blow-up it generically produces two saddles and one node. The four separatrices of the saddles (outside the y-axis) become borsecs of the original systems, but only two of them are in fact separatrices. The other two borsecs are situated on the same half plane delimited by the separatrices, and there are local parabolic sectors for each one of the couples formed by a separatrix and a borsec. But in some special cases, it may happen that after the second blow-up we obtain a saddle plus a saddle-node. In this case there are just two borsecs in the original system and they are both separatrices. The reader could look at Figure 3.8 of Example 3.1 where the two different elliptic-saddles are portrayed with their corresponding blow-ups.

Since the non-generic case is an elliptic-saddle with less borsecs than the generic one, for the generic one we will maintain the notation $\widehat{es}_{(3)}$ and denote by $\widehat{es}^{-}_{(3)}$ the non-generic one.

Possibly for higher-degree systems, the notation of all finite nilpotent singularities will need to be done in sectorial notation (or may even need additional concepts) since the features of geometric possibilities may become more complex.

3.7.2.3 Intricate singularities

As we have already said, the neighborhood of any singular point of a polynomial vector field (except for foci and centers) is formed by a finite number of sectors that can only be of three types: parabolic, hyperbolic and elliptic. Then a reasonable way to describe intricate and nilpotent singularities at infinity is to use a sequence formed by the types of their sectors. Here we use the definition of geometrical local sector (see Section 3.5). The description we give is the one that appears in the clockwise direction once the blow-down of the desingularization is done. Thus in quadratic systems we have just seven possibilities for finite intricate singular points of multiplicity four (see [29]) which are the following ones:

- $phpphp_{(4)}$,
- $phph_{(4)}$,
- $hh_{(4)}$,
- $hhhhhh_{(4)}$,
- $peppep_{(4)}$,
- $pepe_{(4)}$,
- $ee_{(4)}$.

In agreement with our notation, we use lowercase letters because these singularities are finite, and we add the subindex (4) since they are all of multiplicity four.

For infinite intricate and nilpotent singular points, we insert a minus sign (en dash) between two of the sectors to split those that appear on one side or the other of the equator of the sphere. In this way we will distinguish between $\binom{2}{2}PHP - PHP$ and $\binom{2}{2}PPH - PPH$.

When describing a single finite nilpotent or intricate singular point, one can always apply an affine change of coordinates to the system, so it does not really matter which sector starts the sequence, or the direction (clockwise or counter-clockwise) we choose. If it is an infinite nilpotent or intricate singular point, then we will always start with a sector bordering the infinity (to avoid using two dashes). However, when one needs to describe several singular points at infinity together, then the relative positions of one with respect to those other singularities, may be relevant. This will happen when trying to describe at the same time at least two non-elemental singular points together. In this book this situation only occurs once for systems having two $\binom{1}{1}$ semi-elemental saddle-nodes at infinity and a third singular point that is an elemental node. In this case we have just two possibilities (which appear in [301, Figures 3 and 4 in Table 5]) that correspond to 3 and 4 in Figure 6.1. In Config. 3 the node is represented by two opposite points on the circle of the Poincaré disc. Each one of these two points lies between a hyperbolic sector of one of the saddle-nodes and the parabolic sector of the other saddle-node. We make the convention to denote this situation by $\overline{\binom{1}{1}}SN$, $\overline{\binom{1}{1}}SN$, N. In Config. 4 we have a different situation. The node yields two opposite points on the circle, one of them lying between two hyperbolic sectors belonging to distinct saddle-nodes, whereas the other lies between the parabolic sectors of both saddle-nodes. We make the convention to denote this situation by $\overline{\binom{1}{1}}SN$, $\overline{\binom{1}{1}}NS$, N. These two configurations contain the same kind of singular points, but their relative position produces topologically different portraits around the line at infinity.

For the description of the topological phase portraits around the isolated singular points, the information described above is sufficient. However we are interested in additional geometrical features such as the number of characteristic directions, which figure in the final global picture of the desingularization. In order to add this information we need to introduce more notation. If the two limiting orbits of a sector arrive at the singular point with the same slope and tangent to the same half line, then the sector will be denoted by $H_\curlywedge$, $E_\curlywedge$ or $P_\curlywedge$. The index in this notation refers to the cusp-like form of limiting trajectories of the sectors.

Moreover in the case of parabolic sectors, we want to make precise whether the orbits arrive tangent to one or the other limiting trajectories. We distinguish the two cases by $\overset{\curvearrowright}{P}$ if they arrive tangent to the trajectory limiting the previous sector in the clockwise sense or $\overset{\curvearrowleft}{P}$ if they arrive tangent to the trajectory limiting the next sector. Clearly a parabolic sector denoted by P^* would correspond to a sector in which orbits arrive with all possible slopes between the slopes of the limiting trajectories. In the case of a cusp-like parabolic sector, all orbits must arrive with only one slope, but the distinction between $\overset{\curvearrowright}{P}$ and $\overset{\curvearrowleft}{P}$ is still valid if we consider the different desingularizations that we obtain from the singularities having such sectors. Thus, complicated intricate singular points like the two we see in Figure 3.20 may be described as $\binom{4}{2}\ \overset{\curvearrowleft}{P}E\overset{\curvearrowright}{P}-HHH$ (case (a)) and $\binom{4}{3}\ E\overset{\curvearrowright}{P}_\curlywedge H-H\overset{\curvearrowleft}{P}_\curlywedge E$ (case (b)), respectively.

In the definition of finite intricate singular points, we may need to use arrows over parabolic sectors. However in the case of quadratic systems, for each arrangement of sectors only one possibility is realizable, and so we avoid the use of arrows.

Examples of all these possibilities can be seen in Appendix A.

If the intricate singularities are formed by star-like sectors we will use the notations P^*, E^* and $H_\curlywedge^*$. A hyperbolic sector of a star-like singularity is always a cusp-like sector.

3.7.3 Lack of singularities and complex singularities

The lack of finite singular points will be encapsulated in the notation $\emptyset$. If we need to point out the lack of an infinite singular point, we will use the symbol $\emptyset$.

As we said in the beginning of this section, we will denote the presence of complex singularities with the symbols © and © for finite and infinite singular complex points respectively.

Sometimes we need to say more than just that we have complex singularities. We may need for example to say that a trace of the Jacobian matrix of the singularity is zero. We denote this by writing $©^\rho$. Other times we may need to add the information that a complex singularity has the discriminant of the characteristic equation of its Jacobian matrix equal to zero. We denote this by $©^\tau$. For a detailed description and motivation of the importance of these singularities, see Section 5.4.3. In particular, this distinction is never needed for infinite complex singularities.

3.7.4 Infinite number of singularities

Finally there is also the possibility that we have an infinite number of finite or of infinite singular points. In the first case, this means that the polynomials defining the differential system are not coprime. Their common factor may produce a line or conic with real coefficients filled up with singular points.

3.7.4.1 Line at infinity filled up with singularities

It is known that any such system has in a sufficiently small neighborhood of infinity one of six topologically distinct phase portraits (see [305]). The way to determine these portraits is by studying the reduced systems on the infinite local charts after removing the degeneracy of the systems within these charts. In the case that a singular point still remains on the line at infinity, we study such a singularity. In previous work [305] the tangential behavior of the solution curves was not considered in the case of a node. If a node remains after the removal of the degeneracy in the local charts at infinity, this could either be of the type N^d, N and $N^\star$ (this last case does not occur in quadratic systems as we will see in Diagram 6.8). Since no eigenvector of such a resulting node N (for quadratic systems) will have the direction of the line at infinity, we do not need to distinguish N^f and N^∞. Other singular points at infinity of quadratic systems, after removal of the degeneracy, can be a saddle, a center, a semi-elemental saddle-node or a nilpotent elliptic-saddle. We also have the possibility of no singularities after the removal of the degeneracy. To convey the way these singularities were obtained as well as their nature, we use the notation $[\infty; \emptyset]$, $[\infty; N]$, $[\infty; N^d]$, $[\infty; S]$, $[\infty; C]$, $[\infty; \overline{\binom{2}{0}} SN]$ or $[\infty; \widehat{\binom{3}{0}} ES]$. These are all the possibilities in **QS**.

3.7.4.2 Degenerate systems

We will denote with the symbol $\ominus$ the case when the polynomials defining the system have a common factor. The symbol stands for the most generic of these cases, which corresponds to a real line filled up of singular points. The degeneracy can also be produced by a common quadratic factor that defines a conic. It is well known that by an affine transformation any conic over $\mathbb{R}$ can be brought to one of the following forms: $x^2 + y^2 - 1 = 0$ (real ellipse), $x^2 + y^2 + 1 = 0$ (complex ellipse, i.e. an irreducible conic over $\mathbb{R}$ that has only complex points), $x^2 - y^2 = 1$ (hyperbola), $y - x^2 = 0$ (parabola), $x^2 - y^2 = 0$ (pair of intersecting real lines), $x^2 + y^2 = 0$ (pair of intersecting complex lines), $x^2 - 1 = 0$ (pair of parallel real lines), $x^2 + 1 = 0$ (pair of parallel complex lines) or $x^2 = 0$ (double line).

We will indicate each case by the following symbols:

- $\ominus[|]$ for a real straight line;
- $\ominus[\circ]$ for an ellipse over $\mathbb{R}$;

- $\ominus[©]$ for a complex ellipse, i.e. an irreducible conic over $\mathbb{R}$ that has only complex points;
- $\ominus[)(]$ for a hyperbola;
- $\ominus[\cup]$ for a parabola;
- $\ominus[\times]$ for two real straight lines intersecting at a finite point;
- $\ominus[\cdot]$ for two complex straight lines that intersect at a real finite point;
- $\ominus[||]$ for two real parallel lines;
- $\ominus[||^c]$ for two complex parallel lines that intersect at a real infinite point;
- $\ominus[|_2]$ for a double real straight line.

It is worth noticing that the degeneracy $\ominus[©]$ implies the non-existence of real singularities, so we have a chordal system whose phase portrait is equivalent to the case when there are four complex finite singularities and one real infinite singularity. Moreover, the degeneracy $\ominus[\cdot]$ implies the existence of one real finite singularity, and its local phase portrait is the same as that of a singularity $hh_{(4)}$. And finally, we point out that the degeneracy $\ominus[||^c]$ produces one real singularity at infinity with configuration $H - H$.

We also want to determine whether after removing the common factor of the polynomials, the singular points remain on the curve defined by this common factor. If the reduced system has no finite singularity on this curve, we will use the symbol $\emptyset$ to describe this situation. If some singular points remain, we will use the corresponding notation of their types. As an example we complete the notation above as follows:

- $(\ominus\,[|];\emptyset)$ denotes the presence of a real straight line filled up with singular points such that the reduced system has no singularity on this line.
- $(\ominus\,[|];f)$ denotes the presence of a real straight line such that the reduced system has a strong focus on this line.
- $(\ominus\,[\cup];\emptyset)$ denotes the presence of a parabola filled up with singularities such that no singular point of the reduced system is situated on this parabola.

3.7.4.3 Degenerate systems with non-isolated singular points at infinity, isolated on the line at infinity

The existence of a common factor of the polynomials defining the differential system also affects the infinite singular points. We point out that the projective completion of a real affine line filled up with singular points has a point on the line at infinity that will also be a non-isolated singularity.

In order to describe correctly the singularities at infinity, we must mention also this kind of phenomena and describe what happens to such singular points at infinity after the removal of the common factor. To show the existence of a common factor, we will use the same symbol as before, $\ominus$, and for the type of degeneracy we use the symbols introduced above. We will use the symbol $\emptyset$ to denote the non-existence of real infinite singular points after the removal of the degeneracy. We will use the corresponding capital letters to describe the singularities that remain there. We take note that a simple straight line, two parallel lines (real or complex), one double line or one parabola defined by the common factor (all taken over the reals) imply the existence of one real non-isolated singular point at infinity in the original degenerate system. However, a hyperbola and two real straight lines intersecting at a finite point imply the presence of two real non-isolated singular points at infinity in the original degenerate system. Finally, a complex ellipse and two complex straight lines that intersect at a real finite point imply the presence of two complex non-isolated singular points at infinity in the original degenerate system. Thus, in the reduced system these singular points may disappear as singularities, and if they remain, they must be described. In the first five cases we will give the description of the corresponding infinite singular point. In the next five cases we give the description of the corresponding two singular points. As agreed, we use capital letters to denote them since they are on the line at infinity. We give below some examples:

- $N^f, S, \big(\ominus\, [|]; \emptyset\big)$ means that the system has a node at infinity such that an infinite number of orbits arrive tangent to the eigenvector in the affine part, a saddle, that one non-isolated singular point is part of a real finite straight line filled up with singularities and that the reduced linear system has no infinite singular point in that position.
- $S, \big(\ominus\, [|]; N^*\big)$ means that the system has a saddle at infinity, that one non-isolated singular point is part of a real finite straight line filled up with singularities and that the reduced linear system has a star node in that position.
- $S, \big(\ominus\, [)(]; \emptyset, \emptyset\big)$ means that the system has a saddle at infinity, that two non-isolated singular points are part of a hyperbola filled up with singularities, and that the reduced constant system has no singularity in those positions.
- $\big(\ominus\, [\times]; N^*, \emptyset\big)$ means that the system has two non-isolated singular points at infinity, which are part of two real intersecting straight lines filled up with singularities, and that the reduced constant system has a star node in one of those positions and no singularity in the other.
- $S, \big(\ominus\, [\circ]; \emptyset, \emptyset\big)$ means that the system has a saddle at infinity, that two non-isolated (complex) singular points which are located on the complexification of an ellipse over $\mathbb{R}$, which has no real point at infinity and that the reduced constant system has no singularity in those positions.

When there is a non-isolated infinite singular point such that the reduced system has a singularity at that position, it may happen that one or several characteristic directions at this singular point, directed towards the affine plane, could coincide with a tangent line to the curve of singularities at this singular point. This situation could produce many different geometrical (or even topological) combinations, but in the quadratic case we only have a few of them for which we introduce a coherent notation. This notation can be adapted and further developed for higher-degree systems. In quadratic systems we only need to distinguish among some situations in which a characteristic direction of the infinite singular point, after the removal of the degeneracy, may coincide or not with a tangent line to the curve of singularities at this singular point. We show in Figure 3.21 two cases that need to be distinguished (cases (a) and (b)). Here we will use a numerical subscript that denotes the cardinal number $\mathcal{K}$ of the union of the characteristic directions, together with the tangent lines to the curve of singularities at this singular point, all of them considered in a neighborhood of the point at infinity on the Poincaré sphere. The singularities at infinity of the examples of Figure 3.21 would then be denoted by S, $\left(\ominus\,[|];N_3^{\infty}\right)$ (case (a)) and S, $\left(\ominus\,[|];N_2^{\infty}\right)$ (case (b)).

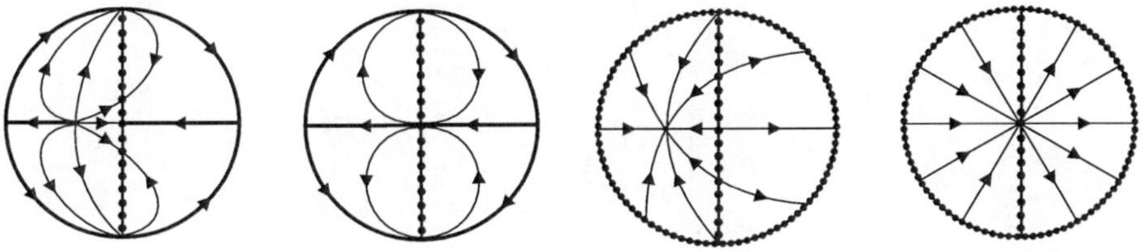

(a) (b) (c) (d)

Figure 3.21: Most degenerate quadratic systems.

3.7.4.4 Degenerate systems with the line at infinity filled up with singularities

For a quadratic system this implies that the polynomials must have a common linear factor, and there are only two possible phase portraits, which can be seen in Figure 3.21 (the portraits (c) and (d)). In order to be consistent with our notation and considering generalization to higher-degree systems, we describe the two cases in a coherent way with what we have done up to now.

The case (c) is denoted by $[\infty;\,(\,\ominus\,[|];\,\emptyset_3)]$, which means that

- the line at infinity is filled up with singular points;
- the reduced quadratic system on the infinite local charts has a non-isolated singular point on the line at infinity due to the affine line of degeneracy;
- once the original system is reduced to a linear one by removing the common factor, the infinity continues to be filled up with singular points;
- once the system on a local chart at infinity around the singularity that is common to both lines filled up with singular points, is reduced by completely removing the degeneracy, there is no singular point on that intersection;

- the cardinal number $\mathcal{K}$ is 3, indicating that apart from the line of singularities and the line at infinity, we have another characteristic direction pointing towards the affine plane at that point.

The second case (d) is denoted by $[\infty; (\ominus\, [|]; \emptyset_2)]$, which means exactly the same items as above with the exception that the cardinal number $\mathcal{K}$ is 2. That is, beyond the line of singularities and the line at infinity, we have no other characteristic direction.

We have included for easy access a table with the notation in Appendix A.

Chapter 4

Invariants in mathematical classification problems

4.1 Basic concepts

In general in a *classification problem* one starts with a set of objects X and an equivalence relation $\sim$ on X. One way of stating the classification problem of the elements of X with respect to $\sim$ is to ask for a *complete list of representatives* of all equivalence classes of X, which is *minimal* in the sense that any two representatives in the list are non-equivalent. Since the objects of X could be quite complex, in constructing this list, there is the possibility of ending up with repetitions, i.e. with distinct representatives for the same equivalence class. It is therefore necessary to produce a proof that two elements in the list belong to distinct equivalence classes. Such a proof usually involves the notion of *invariant*.

Given a set X endowed with an equivalence relation $\sim$, an *invariant* on X with respect to $\sim$ is a map $I : X \to Y$ such that if $x_1 \sim x_2$ then $I(x_1) = I(x_2)$.

For an invariant $I : X \to Y$, Y could for example be a set of numbers. Thus the map associating to every ellipse its largest axis is an invariant with respect to the action of the group of rigid motions and $Y = \mathbb{R}$ in this case. In general the construction of invariants could involve rather complicated mathematical objects. For example this is the case for Donaldson's invariants used to distinguish between different smooth differential structures on four-dimensional topological manifolds (see [196]). For this work Donaldson was awarded the Fields medal in 1986.

In proving non-equivalence, a major role is played by invariants. Indeed, to show that two elements a, b of X are not equivalent, it suffices to find and invariant I on X with respect to $\sim$ such that $I(a) \neq I(b)$.

By a *complete set of invariants* we mean a set of invariants $I_j : X \to Y_j$, $j \in J$ with respect to $(X, \sim)$ such that the set of invariants separates the equivalence

J. C. Artés et al., *Geometric Configurations of Singularities of Planar Polynomial Differential Systems*, https://doi.org/10.1007/978-3-030-50570-7_4

classes, i.e. for all a, b such that $a \not\simeq b$ there exists an invariant I_j such that $I_j(a) \neq I_j(b)$.

The elements of X could for example be algebraic curves, topological spaces, or smooth manifolds. In topology one of the equivalence relations that is considered is the homeomorphism of two topological spaces, and a main problem is to establish if two topological spaces X, Y are homeomorphic or not. To see that they are, it suffices to construct a homeomorphism $f : X \to Y$. To prove that they are not homeomorphic appears to be much harder because this would mean proving that any bijection $g : X \to Y$ is not a homeomorphism. A proof of non-equivalence is produced once we have a topological invariant I such that $I(X) \neq I(Y)$.

Given a set X endowed with an equivalence relation $\sim$, we can form the quotient set $X/\sim$ of equivalence classes $[a] = \{b|\, b \sim a\}$ of X. The canonical map $X \to X/\sim$ is clearly an invariant, the trivial invariant, and any other invariant on X factors through this map.

If X is a set endowed with structure, for example a topology, the quotient set may also be endowed with structure, in this case the quotient topology. This quotient object plays an important role in modern invariant theory, where following Mumford, for the classification problem with respect to an equivalence relation $\sim$, one also asks for the structure of the quotient object $X/\sim$ (see [240]). In fact in the modern theory, to solve a classification problem one needs to:

(i) find a complete set of invariants so as to be able to distinguish the classes,

(ii) find canonical representatives of the equivalence classes of $(X, \sim)$,

(iii) describe the moduli space $X/\sim$.

An important family of classification problems are those generated by *group actions*. By an action of a group G on a set X, we mean a map $A : G \times X \to X$ that respects the group law of G, i.e. if we denote by gx the element $A(g, x)$, then we ask that (i) $g_2(g_1x) = (g_2g_1)x$ and (ii) $ex = x$ for every x in X, e being the neutral element of G. Such an action is called a *left group action*. Analogously we can define also a *right group action*.

We say that two elements a, b of X *are equivalent with respect to the group action* of G on X if and only if there exists an element g of G such that $b = ga$. In this case we say that *we classify* X *with respect to (or modulo) the group action*. The equivalence class $[a]$ of a is then called the *orbit* of a under the action of G. To a group action of G on X we can associate the corresponding quotient set $X/\sim = \{[x]|\, x \in X\}$, where $[x] = Orb(x) = \{gx|\, g \in G\}$. In this case the invariants are called *invariants of the action of the group* G on X, or simply *invariants of the group*.

As an example we consider the one-dimensional real projective space that is obtained as the space of orbits of the action on the space $\mathbb{R}^2\backslash\{(0, 0)\}$ by the group $\mathbb{R}^*$ of invertible real numbers.

Assuming that we have a group action $A : G \times X \to X$ on a set X, associating to each $g \in G$ and each $x \in X$ an element $gx = A(g, x)$, then for $g \in G$, we have a map $A(g,-) : X \to X$ sending an element $x \in X$ to $gx = A(g,x) \in X$. This map is a bijection. Indeed, supposing that $gx_1 = gx_2$, then $x_1 = g^{-1}(gx_1) = g^{-1}(gx_2) = x_2$ so this map is injective. The map $A(g,-)$ is also surjective because for every $y \in X$ we have $y = g(g^{-1}y) = gx$ where $x = g^{-1}y$. We have thus obtained a map $G \to \text{Bij}(X)$, where $\text{Bij}(X)$ is the set of all bijections $X \to X$. Clearly $\text{Bij}(X)$ forms a group with respect to the composition of maps. The map $G \to \text{Bij}(X)$ is a group homomorphism. Alternatively each group homomorphism $G \to \text{Bij}(X)$ yields a group action of G on X. So group actions $G \times X \to X$ are in one-to-one correspondence with group homomorphisms $G \to \text{Bij}(X)$. In particular if $X = V$ where V is a vector space over a field $\mathbb{K}$ and if we have a homomorphism $G \to GL(V) \subset \text{Bij}(X)$ of a group G into the group $GL(V)$ of all linear transformations $V \to V$ of V, then the associated action is said to be linear. In this case the homomorphism $G \to GL(V)$ is called a *representation of the group* G into the group $GL(V)$.

The group actions that induce group representations play an important role in classification problems of planar polynomial vector fields.

We give here some examples.

Example 4.1. Let X be the set of affine conics over $\mathbb{R}$. On this set we have the following two group actions:

(1) The action of the group of rigid transformations of the plane, i.e. for every rigid motion M and conic C, we get the conic MC obtained after applying to C the rigid motion M. It is well known that each such transformation is a combination of rotations, reflections and translations. A set of representatives of the classification of real non-empty conics with respect to this group action is formed by all ellipses $x^2/a^2 + y^2/b^2 = 1$, all hyperbolas $x^2/a^2 - y^2/b^2 = 1$ with $a, b > 0$, all parabolas $y - kx^2 = 0$ and the reducible conics $x(y-kx) = 0$, $x(x-a) = 0$, $x^2 = 0$, $x^2 + y^2 = 0$.

(2) The action of the group on the set X of affine conics over $\mathbb{R}$ is the action of the group $\mathit{Aff}(2,\mathbb{R})$ of affine transformations of the plane, i.e. for every affine transformation A we get the conic AC obtained after applying to C the affine transformation A. A set of representatives of this group action on X is formed by the following seven conics: $x^2 + y^2 = 1$, $x^2 - y^2 = 1$, $y - x^2 = 0$, $xy = 0$, $x(x-1) = 0$, $x^2 = 0$ and $x^2 + y^2 = 0$. The moduli space $X/\sim$ is the discrete space of these seven conics.

Example 4.2. Let X be the set of projective conics over $\mathbb{R}$. On this set acts the group $PGL(3,\mathbb{R})$ of all projective transformations on the projective plane $\mathbb{PR}^2$. A set of representatives is formed by the conic $Z^2 - XY = 0$, the reducible conics $XY = 0$ and $X^2 = 0$, and $X/\sim$ is the set of the equivalence classes of these three conics.

Example 4.3. Consider the set S of two-dimensional connected, compact, orientable, topological manifolds. The topological classification of these manifolds with respect to the homeomorphism equivalence relation is given by the following theorem:

> *Every compact, connected, orientable two-dimensional topological manifold is a sphere with g handles.*

Here the number g of handles is a topological invariant, the number of holes of the compact surface. The singleton $\{g\}$ is a complete set of invariants. Apparently this classification was known since the nineteenth century. A proof can be found in the book of Seifert and Trelfall [309]. The set $S/\sim$ is denumerable, the enumeration being done by $g = 1, 2, \ldots$.

Example 4.4. A greatly more complicated example is the classification of smooth, compact, connected, three-dimensional manifolds, which was only recently completed by G. Perelman based on the geometrization conjecture of W. Thurston and on work done by R. Hamilton on the Ricci flow (see [327]). The proof of this conjecture also solves the problem of Poincaré that any three-dimensional compact topological manifold for which any loop can be continuously deformed to a point is homeomorphic to the three-dimensional sphere.

The problem of listing all smooth structures on four-dimensional topological manifolds is open. It is known that some four-dimensional topological manifolds do not admit smooth structures (see [327]).

Other examples of classification problems that have been solved are: the classification of compact Riemann surfaces with respect to biholomorphical equivalence relation and the classification of finite simple groups with respect to isomorphism.

4.2 Classification problems on planar polynomial vector fields

4.2.1 Equivalence relations for polynomial vector fields

Let X be a specific family of planar polynomial vector fields that we consider equipped with the topological equivalence relation, defined as follows:

> *Two polynomial differential systems S_1 and S_2 are topologically equivalent if and only if there exists a homeomorphism of the plane carrying the oriented phase curves of S_1 to the oriented phase curves of S_2 and preserving the orientation.*

To every polynomial differential system we associate its *phase portrait* that is the partition of the plane into oriented phase curves. Clearly if we reverse the time, we get the same phase curves but with reversed orientation. When listing all phase portraits of a given family of systems, it is then convenient to list only one of the two portraits that both induce the same foliation with singularities.

By reversing the orientation, another phase portrait of the family is obtained. For this reason sometimes authors prefer to use the following equivalence relation:

> *Two polynomial differential systems S_1 and S_2 are topologically equivalent if and only if there exists a homeomorphism of the plane carrying the phase curves of S_1 to the phase curves of S_2 and preserving or reversing (globally) the orientation.*

The first definition is the one commonly used, and it is the better notion as it preserves stability, or instability, of singularities. But people working on classifying systems belonging to specific families prefer the second notion because it cuts in half the number of phase portraits to be exhibited.

We also have other equivalence relations on polynomial vector fields. In particular we have the following:

- The *qualitative equivalence relation* of Jiang and Llibre [182] is an equivalence relation for singularities. We say that two singularities p_1 and p_2 are qualitatively equivalent if (i) they are topologically equivalent through a local homeomorphism Ψ and (ii) two orbits are tangent to the same straight semi-line at p_1 if and only if the corresponding two orbits are tangent to the same straight semi-line at p_2.

- The *equivalence relation modulo the group action of the affine transformations and time rescaling*, which will be considered in Section 5.3.1, is finer than the topological equivalence relation, i.e. if two systems S_1 and S_2 are affinely equivalent, then they are also topologically equivalent. This is a very important equivalence relation because it opens up the possibility of using invariant polynomials, which form an invaluable tool for constructing algebraic invariants and for effective computer calculations (see Section 5.3.2).

- The *equivalence relation modulo limit cycles* is defined for **QS** follows: Two systems S_1, S_2 *are equivalent modulo limit cycles* $S_1 \sim_{LC} S_2$, if and only if by identifying the unique focus inside each eye of limit cycles with each one of the points inside the closed region bounded by the largest one of the limit cycles of an eye of limit cycles, the two phase portraits become topologically equivalent with the resulting quotient topology on the plane. Then the *classification problem modulo limit cycles* is the problem of describing the complete list of phase portraits of the quadratic class obtained in this way, i.e. one phase portrait for each equivalence class.

When the systems are of higher degree, the equivalence modulo limit cycles should be redefined so as to cover other possibilities such as having several singularities inside the limit cycles, or even several nests of limit cycles inside a larger one.

4.2.2 Classifications of some families of polynomial vector fields

The first successful articles on classifying families of planar polynomial systems were done by Sibirschi and Vulpe, one published in 1975 [344] and another in 1977 [345]. A few unsuccessful attempts to classify quadratic systems with a center were done before 1975 (see [298] for a history of this problem and for references). Vulpe's article [336] published in 1983 contained the first correct classification for this important class of all quadratic systems with a center. Vulpe not only listed all 31 phase portraits of this family but also gave necessary and sufficient conditions in terms of invariant polynomials for the realization of each one of the phase portraits, when the center is placed at the origin. Thirty seven years have passed since the publication of this article, and in this period of time over 50 articles appeared solving classification problems for quadratic or cubic differential systems. As already mentioned in Chapter 2, these articles are of two kinds: (A) those listing phase portraits but without proving with the help of topological invariants that the lists contain topologically distinct phase portraits and without characterizing the phase portraits (whenever possible) in terms of invariant polynomials (for this concept see the next section) and (B) those proving that the phase portraits are indeed distinct and characterizing each phase portrait in terms of invariant polynomials (see [296]) whenever this was possible. Recall that a survey of papers on topologically classifying quadratic or cubic polynomial differential systems is included in Section 2.3.

In the case of algebraic families of systems such as the class of quadratic systems with center, the classification is completely done in terms of purely geometric properties of the systems, and the bifurcation diagram in $\mathbb{R}^{12}$ can be described by equalities and inequalities expressed in terms of invariant polynomials to be defined in Section 5.4.2. This is a family of integrable systems. A complete topological classification was also obtained for the class of Lotka–Volterra differential systems. Not all the systems in this class are integrable; however, necessary and sufficient conditions in terms of invariant polynomials have been given for all phase portraits of systems in this class [307].

In general when we need to classify families of non-integrable differential systems, equalities and inequalities expressed in terms of algebraic expressions (invariant polynomial) do not suffice but they necessarily appear in the initial stages of such classifications. The study of a specific family of polynomial systems, for example the quadratic family, implies the use of a number of charts that eventually need to be glued together to produce the whole picture. We therefore need to use distinct presentations of the systems, and we need to be able to easily pass from one to another. The invariant polynomials are of help here. Although they could not do the whole work, they are a necessary first part of the work. Classifying non-integrable differential systems implies the use of not only algebraic equalities and inequalities of invariant polynomials, but also analytic or smooth ones and a lot of analytic work along with some numerical analysis. Furthermore, in a number of problems of an algebraic or algebro-geometric nature, the theory of invariant polynomials with respect to some subgroup of the affine group acting on differential equations, intervenes substantially. Polynomial differential systems with specific algebraic properties, for example those possessing algebraic invariant curves, or those with a focus of specific order, are interesting in their own right and have potential for applications. A specific example is provided by the study of quadratic differential systems with a third order focus: The bifurcation diagram (see [16, 222]) of this family is basically obtained by using the algebraic varieties of bifurcation of singularities and by perturbation of systems with a center. In addition we also have an analytic bifurcation curve.

The general theory of planar polynomial vector fields is in the process of growing, and some of its connections with algebraic geometry are emerging. Much work in this area was done from the viewpoint of bifurcation theory. Bifurcation theory is one of the basic tools in this area, but confining the study to bifurcation theory only, is a limitation.

Planar polynomial vector fields form an area of mathematics presenting enormous difficulties because their limit cycles are such elusive objects and also because of other phenomena (such as detecting connections of separatrices and multiple limit cycles) of non-algebraic nature that intervene in this study. Up to now understanding these vector fields, including their limit cycles, was possible only in a small number of cases. We observe that the proof of the *individual finiteness theorem* for limit cycles is still not understood by the mathematical community (see [317]). For the moment, dealing with limit cycles within global families of polynomial vector fields is not an easy task, even in the case of quadratic systems. It is however conceivable to obtain the full classification of some families of polynomial vector fields modulo limit cycles, for example for the quadratic class. Work on this classification problem is in full steam. Thus we have the initial work [12] where the authors classify modulo limit cycles the class of all quadratic vector fields that are structurally stable. Work by Artés, Llibre and Rezende on the classification modulo limit cycles of all quadratic systems that have a *first degree of degeneracy* is already completed [17].

Strictly speaking the articles do not actually list phase portraits since these would be impossible to draw, containing *all* oriented phase curves. Instead the articles list the *separatrix configurations* as defined by Markus [234]. But giving the separatrix configuration suffices since from it the corresponding phase portrait is immediately grasped, so we can actually say that the phase portraits are listed.

Definition 4.1. By a separatrix configuration of a C^1 planar differential system we mean a collection of curves consisting of separatrices and one oriented phase curve from each canonical region.

The separatrix configuration is a *topological invariant*, i.e. it is an invariant with respect to the topological equivalence relation. It could contain many curves, which makes it sometimes hard to see if two separatrix configurations are or are not homeomorphic. Finer topological invariants are necessary for this (see [296]). As we mentioned in the preceding section, one way to solve a classification problem is to present a list of representatives for which a proof is given that any two objects in the list are not equivalent. Since some of the articles do not explicitly prove that the phase portraits are in fact non-equivalent, in the lists occasionally a separatrix configuration appears more than once. Also, erratum or corrigendum papers appeared adjusting the results.

The classifications are based on specific normal forms and are done in terms of equalities and inequalities expressed in terms of the coefficients of these normal forms, which are a basic tool for obtaining these classifications. They are so chosen as to be well adapted for the specific families that we want to classify. In general, for studying globally a family of systems, more than just one normal form is usually needed. Thus there arises the problem of gluing any two such normal forms. But this gluing is usually not done. Furthermore this results in listing several times the same phase portrait, once for each one of the normal forms, and the domains of definition of these normal forms overlap (see for example [276, 77]) so that we do not end up with a single global list of distinct phase portraits. When studies are done in invariant form, i.e. are *independent* of the normal forms in which the systems are presented, the results obtained from distinct normal forms can be glued together, i.e. they can be integrated. While some topological invariants are difficult if not even impossible to compute, others such as the invariant polynomials (see Section 5.4) can easily be computed, allowing us to glue results. During the past twelve years collaborations of the Chişinău school founded by Sibirschi, the Barcelona school and Dana Schlomiuk from Montreal were made possible, and classifications where both geometric methods and algebraic ones involving invariant polynomials were obtained. Work on this topic is in full progress. The global classification theorem of the global schemes of singularities according to the geometrical equivalence relation (see Part II) proved in this book is an example of this kind of work.

Chapter 5

Invariant theory of planar polynomial vector fields

The roots of the invariant theory of polynomial vector fields lie in the classical invariant theory. The idea to adapt to polynomial vector fields the concepts of classical invariant theory is due to C. S. Sibirschi, the founder of the Chişinău school of qualitative theory of differential equations. In this chapter we first describe briefly the classical invariant theory and then explain how the ideas of this theory were used by Sibirschi and his school to built an analogous theory for polynomial vector fields.

5.1 Classical invariant theory

In the introduction of his book [247], Olver defines classical invariant theory as *the study of the intrinsic or geometrical properties of polynomials.* By *geometric property* we mean a property that is not affected by changes of coordinates. These changes are usually assembled in groups such as for example the group $GL(n, \mathbb{C})$ of linear transformations of $\mathbb{C}^n$ or its subgroup $SL(n, \mathbb{C})$ of special linear transformations of determinant 1. In his famous Erlangen program [195], Felix Klein called a property *geometric* if it is invariant under a group action. Depending on the groups we consider, we could have the Euclidean geometry for the group of isometries, the affine geometry for the group of affine transformations or the projective geometry for the group of projective transformations.

We can trace the beginning of classical invariant theory to Gauss, who in 1801 published in Latin his book *Disquisitiones arithmeticae* (its English translation is [146]) where Gauss observed invariant behavior in the theory of quadratic forms over $\mathbb{C}$. Gauss considered binary quadratic forms $Q(x, y) = ax^2 + 2bxy + cy^2$ over $\mathbb{C}$. On these forms acts the group $GL(2, \mathbb{C})$ of linear transformations, i.e. for every 2×2 non-singular matrix M over $\mathbb{C}$ and every quadratic form $Q(x, y)$, we can

J. C. Artés et al., *Geometric Configurations of Singularities of Planar Polynomial Differential Systems*, https://doi.org/10.1007/978-3-030-50570-7_5

associate a quadratic form $\tilde{Q}(\tilde{x},\tilde{y}) = A\tilde{x}^2 + 2B\tilde{x}\tilde{y} + C\tilde{y}^2$ where $(x,y)^t = M(\tilde{x},\tilde{y})^t$. The discriminant of a quadratic form Q over $\mathbb{C}$ is the number $\Delta = b^2 - ac$. For the transformed form we have $\tilde{\Delta} = B^2 - AC$, and calculations yield $\tilde{\Delta} = (\det(M))^2\Delta = K\Delta$ with $K = (\det(M))^2 \in \mathbb{C}$, K depending only on the element M of the group $GL(2,\mathbb{R})$ and not on the quadratic form. If $\det(M)) = 1$ then Δ is an invariant. Gauss [146] was interested in number theory and used quadratic forms over $\mathbb{Z}$ with the action of the group $SL(2,\mathbb{Z})$ on these forms.

The classical theory of invariants deals with transformations of forms, i.e. homogeneous polynomials in n variables $x = (x_1,\dots,x_n)$ of degree m with coefficients in $\mathbb{C}$ (in particular, for the special case of binary forms we have $f(a,x) = \sum_{i=0}^m \binom{m}{i} a_i x_1^{m-i} x_2^i$ with $a = (a_0,\dots,a_m)$). If $x = Mx'$ is a non-singular linear transformation, then this induces a transformation of f, i.e. $f(a,x) = f(a,Mx') = f'(a',x')$ where f' a new n-form with new coefficients denoted by $a' = (a'_0,\dots a'_m)$. Then the coefficients a' of f' are related to those of f by a linear transformation $a = \nu(M)a'$. Furthermore if $M = M_1M_2$ then $\nu(M) = \nu(M_1M_2) = \nu(M_1)\nu(M_2)$, so the map $M \to \nu M$ is a homomorphism of the group $GL(n,\mathbb{C})$ into $GL(N,\mathbb{C})$ with N the number of coefficients of the form f. If we only consider M in a subgroup G of $GL(n,\mathbb{C})$, then this map is a group homomorphism $G \to GL(N,\mathbb{C})$. In general, a homomorphism of a group G into $GL(N,\mathbb{C})$ is called a *linear representation of G in $GL(N,\mathbb{C})$*, and N is said to be the *degree of the representation*.

A polynomial $I(a)$ in the a_i is a *relative invariant of a form f (or of a family of such forms f)* if $I(a') = (\det(M))^\mu I(a)$ for all $M \in GL(n,\mathbb{C})$ and μ is called the weight of the relative invariant I. If $I(a') = I(a)$ then I is called an *invariant* (or an *absolute invariant*).

A polynomial $u(a,x)$ in the x_j such that $u'(a',x') = (\det(M))^\mu u(a,x)$ for all $M \in GL(n,\mathbb{C})$ is called a *covariant* of f. The classical theory of invariants is the study of these invariants and covariants of n-forms over $\mathbb{C}$ and finding all of them.

Thomas Hawking [158] observed that in the first phase of the development of classical invariant theory, the fact that the map $G \to GL(n,\mathbb{C})$, given by $M \to \nu(M)$, is actually a homomorphism was ignored. The role of groups was later brought into focus by S. Lie and by F. Klein in his Erlangen program of 1872 [195]. Regarding the Erlangen program Hermann Weyl said:

> *The adjustment of invariant theory to this standpoint has been slow; it could not be made without recognizing that the study of the groups themselves and their representations necessarily has to precede the study of their invariants* [346].

The slow adjustment mentioned in Weyl's remark could be explained by comparing the time span of the developments of classical invariant theory (1840–1893) with that of group theory and group representations, which began to be developed only at the end of nineteenth century.

Groups began to be studied in connection with algebraic equations of degree greater than four. The beginning of group theory was marked by the contributions of Cauchy, Abel and especially Galois in the first half of the nineteenth century. It was Galois who first used the word *group* and the groups encountered in his work were groups of permutations. In the second part of the nineteenth century several works, among them the treaty of Camille Jordan "Traité des substitutions et des équations algébriques" (1870) [184], appeared. As already mentioned, Klein used groups in his 1872 Erlangen program [195], for classifying geometries. These were groups of linear transformations. W. Burnside wrote a book [68] (1897) on groups of finite order that was the first English book on the theory of groups. In general in the nineteenth century, the groups that were considered were concrete: permutation groups, groups of linear transformations, matrix groups, etc. The notion of an abstract group only appeared later, in the twentieth century. Representation theory began with the pioneer work of G. Frobenius, W. Burnside, I. Schur and R. Brauer (see [115]). Frobenius published the first paper on representation theory [138] in 1897, where for the first time the notion of representation was introduced. As we shall see, invariant theory began to be developed in the 1840s, and it became a major area of mathematics research in the second part of the nineteenth century, while both group theory and representation theory began to be developed as independent subjects only at the end of the nineteenth century century.

The British mathematician George Boole launched the classic invariant theory by publishing the treatise "Exposition of a General Theory of Linear Transformations" [53] in 1841. The basic example considered by Boole was the set of monic polynomials $Q(x) = x^2 + 2b_1x + b_2$ on which acts the group of translations, $Q(x+c) = x^2 + 2(b_1+c)x + c^2 + 2b_1c + b_2$. The discriminant of $Q(x)$ is $\Delta = b_1^2 - b_2$ and calculations yield that the discriminant Δ' of $Q(x+c)$ coincides with Δ. Thus the discriminant is an invariant of monic quadratic polynomials under the action of translations. Boole used elimination theory to construct an invariant of homogeneous polynomials of degree m in n variables, namely the discriminant K. In 1845 [79] Cayley gave a technique for generating invariants, distinct from that of Boole. Cayley's goal was to elaborate a theory that would allow for the production of some sort of minimal set of invariants for a given form and also of a minimal set of relations or syzygies among invariants. For example Cayley constructed two simpler invariants I, J than the ones of Boole and it turned out that K, I, J are related by the syzygy $K = I^3 - 27J^2$. The term *syzygy* was introduced in 1853 by James Joseph Sylvester, who was also a leader of the British school of invariants. Another member of the British school in invariant theory was Salmon.

The German school led by Otto Hesse included his student S. Aronhold, also A. Clebsch and P. Gordan. Paul Gordan proved the first fundamental theorem of invariant theory in 1868 [151]. Gordan considered only binary forms over $\mathbb{C}$ under the action of the group $GL(2, \mathbb{C})$. His theorem says that *all covariants are explicitely constructible as polynomials in a finite number of them over the underlying field.* The covariants are said in this case to have a finite basis. The syzygy problem, i.e. find a finite ideal basis for syzygies, remained unsolved.

For over twenty years, extending the theorem of Gordan to n-ary forms proved to be a difficult task, but in 1890 Hilbert, who obtained his doctorate in

1885 with a thesis on invariant theory and with an Habilitationsschrift (1886) also on invariant theory, succeeded in generalizing the result of Gordan to n–ary forms over $\mathbb{C}$, i.e. homogeneous polynomials in n variables $F(x_1, \ldots, x_n)$ with coefficients in $\mathbb{C}$, of any degree [159]. But while Gordan presented a finite complete system of invariants, Hilbert's first proof of the more general result for n-forms of degree m was existential, i.e. he proved the existence of a finite basis of invariants by reductio ad absurdum. It was only in 1893 that he gave a more constructive proof for this theorem (see [160]). A collection of invariants $I_1, \ldots, I_s$ of n-forms $f_1, \ldots, f_l$ of degree m such that any invariant I is a polynomial function $I = P(I_1, \ldots, I_s)$ is called a *Hilbert basis*. Similarly a finite collection of covariants $J_1, \ldots, J_k$ forms a Hilbert basis if every covariant J is a polynomial function $J = P(J_1, \ldots, J_k)$.

In 1893 Hilbert solved also the problem of finding a finite basis for syzygies of invariants of n-ary forms. Hilbert's work went beyond invariant theory since in these papers he proved results of modern algebra that are now at the basis of commutative algebra: Hilbert's basis theorem and the Nullstellensatz. Thus invariant theory played an important role in the development of abstract algebra at the beginning of the twentieth century. Referring to Hilbert, Hermann Weyl said:

> *His papers mark a turning point in the history of invariant theory. He solves the main problem and thus almost kills the subject. But its life lingers on, however flickering, during the next decades* [346, page 27].

Hermann Weyl said:

> *After the formal investigations from Cayley and Sylvester to Gordan, Hilbert inaugurated a new epoch in the theory of invariants. Indeed, by discovering new ideas and introducing new powerful methods he not only brought the subject up to the new level set for algebra by Kronecker and Dedekind, but made such a thorough job of it that he all but finished it, at least as far as the full linear group is concerned* [348].

In 1900, at the International Congress of Mathematicians in Paris, Hilbert included in his list of problems for the twentieth century [161] also a problem on invariant theory. This is Hilbert's 14th problem. However, after his work [160], Hilbert abandoned invariant theory.

Work on invariant theory continued to appear sporadically after the turn of the century. Emmy Noether's thesis, under the direction of Paul Gordan, published in 1908 [245] was on invariant theory, on biquadratic forms in three variables. She had other contributions to invariant theory (for example [246]), but she was to be mainly remembered for her absolutely major role in the development of abstract algebra. Her name appears in the modern statement of Hilbert's basis theorem which says that if A is a Noetherian ring then the polynomial ring $A[x]$ is also Noetherian. Recall that a *Noetherian ring* is a commutative ring in which every ascending chain of ideals is stationary. This statement is equivalent to saying that every ideal I of the ring A is of finite type, i.e. there exist a finite number

$i_1, \ldots, i_r$ of elements of I such that $I = Ai_1 + \ldots + Ai_r$. If k is a field then $k[x]$ (and hence $k[x_1, \ldots, x_n]$) is Noetherian, and thus any subset of the polynomial ring $k[x_1, \ldots, x_n]$ has a finite ideal basis.

With few exceptions, after the turn of the twentieth century, invariant theory remained rather stagnant until around 1935 when there occurred the first revival. In his book [122], Dieudonné said that this revival was due "*to the developments (I. Schur, H. Weyl, E. Cartan) of the global theory of semi-simple groups and their representations around 1935 when it was realized that classical invariant theory was really a special case of that new theory*". But Dieudonné also said that, after the publication of Weyl's famous book *The Classical Groups* [346], "*a lack of outstanding problems was probably the reason why important new developments failed to materialize*". It was only in the second half of the twentieth century that, infused with new ideas, invariant theory gave birth to modern invariant theory or geometric invariant theory as it was called by Mumford [240], nowadays a thriving subject.

Summing up the ingredients occurring in classical invariant theory, we have:

- a group G that is $GL(n, \mathbb{C})$ or a subgroup G of $GL(n, \mathbb{C})$;
- a finite-dimensional vector space V that is $\mathbb{C}^n$ over $\mathbb{C}$;
- a finite-dimensional vector space W over $\mathbb{C}$, for example the vector space of binary or more generally n-forms of degree m;
- a group action of G on V that is linear, equivalently a representation of G in V, i.e. a group homomorphism $\pi : G \to GL(V)$ where $GL(V)$ is the group of linear transformations on V;
- the action on W induced by the group action of G on V, which is again linear, a representation of G in W, i.e. a homomorphism $G \to GL(W)$;
- the invariants and covariants of this action.

The covariants of the above-mentioned action are assembled in a structured object, an algebra over $\mathbb{C}$ and more generally an algebra over a field k as follows: The coordinate-free analog $k[V]$ of a polynomial ring over an infinite field k, with V a finite-dimensional vector space over k, is defined as being the subring generated by the dual space V^* of the ring of all functions $V \to k$. Explicitly if $k[x_1, \ldots, x_n]$ is the polynomial ring in $x_1, \ldots, x_n$ over k, then we can view x_i as the function $x_i : V \to k$ associating to every $(a_1, \ldots, a_n) \in k^n$ the value $x_i(a) = a_i$. Then $k[V]$ is the subalgebra of the k-algebra of all the k-valued function on V, formed by all polynomials in the functions x_i. Let $GL(V)$ be the group of all invertible linear transformations of V. We define an action $A : GL(V) \times k[V] \to k(V)$ of $GL(V)$ in $k[V]$ as follows: If $g \in GL(V)$ and $f \in k[V]$, let $A(g, f)(w) = f(g^{-1}w)$ for all $w \in k[V]$. Clearly $A(g_2, (g_1(f)) = A(g_1 g_2)(f)$. If G is a subgroup of $GL(V)$, we say that $f \in k[V]$ is invariant by G if $A(g, f) = f$ for every $g \in G$. Let $k[V]^G$ be the subalgebra of $k[V]$ over k of all invariant elements of $k[V]$.

Invariant theory is the study of such subalgebras and the first question of invariant theory is whether $k[V]^G$ is a finitely generated algebra over k. One way to formulate Hilbert's 14th problem is to ask for a description of all subgroups G of $GL(V)$ for which $k[V]^G$ is finitely generated. Actually in his address at the ICM in Paris in 1900, Hilbert formulated this problem in a different way (see [161]). In 1958 Nagata gave an example of such a subgroup G for which $k[V]^G$ is not finitely generated. (For a more general form of Hilbert's 14th problem, see [273, 274].)

In the modern invariant theory or *geometric invariant theory*, X is an algebraic variety over a field k, and $\sim$ is an equivalence relation modulo the action of an algebraic group on X. In this case the algebra of invariants $k[V]^G$ is a subalgebra of the *coordinate ring* $k[X]$ of polynomial functions $f : X \to k$ on X. For the classification problem to be complete, apart from constructing a complete systems of invariants and choosing canonical representatives, one also needs to endow the quotient space (*moduli space*) with the structure of algebraic variety. For this to become possible, some orbits would need to be removed as indicated by Mumford [240]. Mumford used from the start the language of *schemes.* (For easy access to quotients, see [274], a source book on the modern developments of invariant theory. For a presentation of the classical concepts and results, in modern and very elegant form, see [320].)

Analogously, when classifying families of polynomial differential systems with respect to some group action such as a subgroup of the group $Aff(2,\mathbb{R})$, we introduce invariant polynomials to distinguish orbits (see the next two sections of this chapter). We are also interested in the construction of the moduli space, which may or may not be Hausdorff. As an example we mention here the construction of the moduli space of the space **QW3** of quadratic differential systems possessing a third-order weak focus, modulo the action of the group $G = Aff(2,\mathbb{R}) \times \mathbb{R}^*$ of the affine transformations and time rescaling. This topological space turns out to be very simple. It is formed by two open discs having quadratic differential systems with a center on their borders, and all systems with limit cycles are contained in only one of the two discs [121].

5.2 The work of Sibirschi's school and collaborators in invariant theory of planar polynomial vector fields

Classical invariant theory involved vector spaces of n-forms over $\mathbb{C}$ of degree m, where n and m are natural numbers. It was a natural problem to try to develop an analogous theory in which the vector space of n-forms is replaced by a vector space of differential equations such as the vector space of polynomial differential systems of degree m. Although algebraic invariants in the theory of differential equations were used in work of some nineteenth-century mathematicians (i.e. Laguerre [198, 199]), the development of a full invariant theory for polynomial differential systems only began in the second part of the twentieth century. This theory sprang directly from the classical invariant theory, and it was for this reason that the previous section of this chapter was devoted to this classical theory. The initial steps in this direction were made by Constantin Sibirschi, who wrote several books and published numerous works on this theme. He established the connection between the algebraic invariants of the classical groups (rotation groups, $O(2,\mathbb{R})$, $GL(2,\mathbb{R})$, $\mathit{Aff}(2,\mathbb{R})$) and the qualitative theory of planar polynomial differential systems. Sibirschi was the founder of the Chişinău school in the theory of invariants of planar polynomial vector fields. He had many students, among them V. Baltag, D. Boularas, I. Calin, Dang Din' Bik, V. I. Daniljuc, F. Gasinskaja-Kirnickaja, V. Lunchevich, A. V. Marincuk, I. I. Pleshkan, M. Popa, A. M. Stakhi, A. Şubă, V. D. Taku and N. Vulpe. Most of them had students and they continue to be active.

In his book [315], Sibirschi wrote that the first works on the theory of invariants of differential equations were done by Laguerre [198, 199]. These papers, published in 1879, are on linear differential equations. From 1879 on, other articles on invariants of linear differential equations were published, among them the work of Halphen [156]. Beginning with the work of Liouville [211] and of Appell [6], the theory of invariants was extended to nonlinear differential equations, and new contacts between this theory and algebraic geometry appeared (see [314, 315] for more information).

Sibirschi began to work on developing the invariant theory of polynomial vector fields during the 1950s. Here we give a very brief survey of results obtained by Sibirschi and by members of the Chişinău school and their collaborators. The source book used by the members of this school was the book of Gurevich [155].

With his students, Sibirschi constructed algebraic invariants and comitants (analogs in this theory of the covariants in the classical theory used by Gurevich [155]) for polynomial vector fields under the action of the groups $O(2,\mathbb{R})$, $GL(2,\mathbb{R})$ and $\mathit{Aff}(2,\mathbb{R})$.

The basic problem of giving polynomial bases for invariants and comitants for the whole class of quadratic and some families of cubic differential systems for the action of the group $GL(2,\mathbb{R})$ was solved by Sibirschi and his students Boularas, Calin, Daniljuc, Dang Din' Bik, Gasinskaja–Kirnickaja and Vulpe (see [141, 116, 117, 335, 59]). In addition to the construction of polynomial bases, Calin constructed algorithms for creating bases of the $GL(2,\mathbb{R})$-comitants in rational form for polynomial differential systems (see [75]).

The problem of giving polynomial bases for invariants and comitants for the quadratic differential systems with respect to the group $\mathit{Aff}(2,\mathbb{R})$ was solved by Boularas in [54, 55]. He obtained 36 affine comitants such that every invariant or comitant with respect to $\mathit{Aff}(2,\mathbb{R})$ is obtained as a polynomial in these 36 elements. Using these elements as well as the elements of polynomial bases of $GL(2,\mathbb{R})$-comitants, a minimal polynomial basis for T-comitants (see definition in Section 5.3.2) was constructed up to and including the degree 12 [58].

For cubic differentials systems the problem of giving polynomial bases for invariants and comitants, for the action of the group $\mathit{Aff}(2,\mathbb{R})$, is still unsolved.

These bases play a fundamental role in the study of families of quadratic or cubic differential systems. In particular they were used for the purpose of classifying some families of polynomial differential systems. This work was an important contribution because, unlike other classification results on these systems that were non-intrinsic, the results obtained by Sibirschi and his students were intrinsic, i.e. they were invariant under the corresponding coordinate changes and hence independent of the specific presentation of the systems (normal forms) used in the classifications.

The first invariant classifications were done by Vulpe and Sibirschi for homogeneous quadratic differential systems. More precisely [343] the authors classified this family with respect to the $GL(2, \mathbb{R})$ group action. This work was refined by Boularas, who reduced the number of canonical forms for the systems [56].

The analogous classification for quadratic differential systems without constant terms was done by Popa and Sibirschi in [271]. Sibirschi together with Dang Din' Bik gave a classification of a family of cubic differential systems with respect to the $GL(2, \mathbb{R})$ group action in [118].

In 1975 [344] the authors gave necessary and sufficient $GL(2, \mathbb{R})$-invariant conditions for distinguishing topologically the phase portraits in the family of homogenous quadratic systems. This classification was later refined in 1977 [345], where the authors used a finer equivalence relation that distinguished the singular points at infinity geometrically (see Section 3.6).

Sibirschi determined necessary and sufficient conditions for the existence of a center at the origin in terms of algebraic invariants with respect to the $GL(2, \mathbb{R})$ action, for a system with only quadratic or cubic nonlinearities (see [312, 313]). Later [194] analogous conditions were constructed for quadratic differential systems for the existence of two centers, one of which being placed at the origin.

For a general quadratic system in the plane (the point $(0, 0)$ is not necessarily a singularity), Boularas, Vulpe and Sibirschi [60] obtained necessary and sufficient conditions for the existence of one center, and also of two centers. These conditions are expressed in the form of equalities and inequalities between affine-invariant polynomials in the coefficients of the right-hand sides of the system.

The Poincaré–Lyapunov constants intervene in the problem stated by Poincaré of distinguishing a focus from a center (see [263, 293]). Calin worked on the expression of the general Poincaré–Lyapunov constants for a weak focus or center placed at the origin, in terms of invariants of the $GL(2, \mathbb{R})$ group action. (In this case the singular point is placed at the origin.)

Another theme on which Sibirschi and his former student Popa worked was the problem of determining $GL(2, \mathbb{R})$ invariant necessary and sufficient conditions for the existence of an invariant straight line passing through the origin [272]. This theme was further pursued by Schlomiuk and Vulpe in a series of papers where they gave (i) necessary and sufficient conditions, in terms of invariant polynomials, for a quadratic differential system to have invariant straight lines with total multiplicity four (respectively five or six); (ii) the complete list of phase portraits of all such systems, which they proved to be integrable; and (iii) the bifurcation diagram in the space $\mathbb{R}^{12}$ of the coefficients of the systems [300, 304, 305]. Schlomiuk and

Vulpe also gave necessary and sufficient invariant conditions for a quadratic system to be a Lotka–Volterra system [306]. For these systems they also gave the full list of phase portraits as well as the bifurcation diagram of Lotka–Volterra systems in the space $\mathbb{R}^{12}$ of their coefficients [307].

Şubă and his former students Puţuntică and Repeşco gave necessary and sufficient conditions for cubic systems to have invariant lines of total parallel multiplicity 7 (see [324, 325]). But their conditions are given in terms of the coefficients of special normal forms. They obtained a total of 17 distinct configurations of invariant lines. With only one exception, each one of these configurations leads to just one phase portrait.

Şubă and Cozma have published a series of papers [110, 111, 112, 113, 114, 321, 322, 323] on the family of cubic systems possessing invariant straight lines. In these papers they have investigated cubic systems having a singularity at the origin $O(0,0)$ of coordinates, which has pure imaginary eigenvalues. In addition they forced the systems to possess three or four invariant straight lines and constructed the necessary and sufficient conditions for the singularity $O(0,0)$ to be a center.

Cozma considered cubic systems with the singularity $O(0,0)$ of the same type as above [105, 106, 107, 108, 109]. For such systems he solved the problem of a center in the case when the systems possess two invariant straight lines and one invariant conic.

Llibre and Vulpe classified all cubic systems possessing the maximum number nine of invariant straight lines, including the line at infinity, and taking into account their multiplicities [225]. They detected 23 topologically distinct configurations of invariant lines for such systems and proved that modulo the group $\mathit{Aff}(2,\mathbb{R})$ action and time rescaling, each configuration leads to just one point in the coefficient space $\mathbb{R}^{20}$. Moreover, every configuration is characterized by a set of affine invariant conditions. This classification was completed by Bujac [62]. The first integrals and the phase portraits of the systems in this class were given in [63].

Vulpe and his former student Cristina Bujac have characterized in invariant terms the cubic differential systems having invariant straight lines, including the line at infinity, of total multiplicity 8 (see [64, 65, 66, 67]). They obtained a total of 51 distinct possible configurations of invariant lines.

Baltag together with Calin proved that planar systems $dx/dt = cx + dy + xC(x,y)$, $dy/dt = ex+fy+yC(x,y)$, where $C(x,y) \neq 0$ is a non-zero homogeneous polynomial of degree $r \geq 0$, under the condition $R(x,y) = -ex^2+(c-f)xy+dy^2 \neq 0$, has a first integral of generalized Darboux type. The main idea is to express an invariant algebraic curve of the system using comitants and invariants of the system (see [39]).

Baltag worked together with Vulpe to construct invariant polynomials helpful in the qualitative theory of polynomial differential systems. Some invariant polynomials with respect to the group of affine transformations which they constructed in [42, 43] give affine invariant conditions for specifying the total multiplicity of finite singularities for generic polynomial differential systems (see the polynomials

μ_i, $i = 0, \ldots, n^2$, in Section 5.3.4). In particular, for the quadratic differential systems with real coefficients they found necessary and sufficient conditions in invariant form, for any number $k \leq 4$ of finite singularities to be with prescribed multiplicities (see [40, 41]). These invariant polynomials were used in many works in classification problems of quadratic differential systems (for example [19, 20]).

In another paper [41] Baltag and Vulpe provide a classification of planar quadratic differential systems, according to the number and multiplicities of the singular points on the line at infinity. The results of these papers are essentially based on the classification of the family of quadratic systems according to the total multiplicity of all finite singular points of systems. This classification (in terms of the coefficients of normal forms of the systems) was given for the first time by J. W. Reyn [277]. The authors emphasized that the methods applied were based on the knowledge of the minimal polynomial basis of $GL(2,\mathbb{R})$-comitants for quadratic systems, and thus did not admit a generalization for the case of polynomial differential systems of higher degree.

Mihail Popa and his students (A. Braicov, O. Diaconescu, N. Gherstega, Anca-Veronica, P. Makar', E. Naidenova, V. Orlov, S. Port and V. Pricop) studied the finite-dimensional graded algebras of comitants and invariants, which Popa called the *Sibirschi algebras* [269], and they obtained characteristics of these algebras (the Krull dimension and the type and number of generators for autonomous polynomial differential systems, Hilbert series for the graded algebras of comitants and invariants of two-dimensional autonomous polynomial differential systems) [57, 267, 268]. They also obtained invariant classifications of $GL(2,\mathbb{R})$ and $Aff(2,\mathbb{R})$-orbits according to their dimensions for some classes of polynomial differential systems.

Popa and Pricop obtained a bound for the number of algebraically independent Poincaré–Lyapunov constants for polynomial vector fields possessing a singular point of center or focus type. More exactly [269] (see also [270]) the authors proved the following result.

Let $\{m_0, m_1, \ldots m_l\}$ with $m_0 = 1$ be an ordered sequence of natural numbers in strictly increasing order. We consider differential systems of the form

$$\begin{aligned}\frac{dx}{dt} &= P_{m_0}(x,y) + P_{m_1}(x,y) + \ldots P_{m_l}(x,y),\\ \frac{dy}{dt} &= Q_{m_0}(x,y) + Q_{m_1}(x,y) + \ldots Q_{m_l}(x,y),\end{aligned} \tag{5.1}$$

where by

$$P_{m_i}(x,y), \quad Q_{m_i}(x,y), \quad 1 < m_i < m_l \quad \text{for} \quad 1 \leq i < l$$

we denote the homogeneous parts of degree m_i of the two equations (5.1) such that at least one of them is not zero. Then the number of algebraically independent focal values of systems (5.1) at the origin is bounded by $\rho = 2(\sum_{i=0}^{l} m_i + l) + 3$.

The work of C. S. Sibirschi and the Chişinău school has implications that extend to the most recent geometric research in this area. For example, in the quadratic case we need to classify topologically the set **QS** of quadratic differential systems. This set modulo the group action of the group $G = Aff(2,\mathbb{R}) \times \mathbb{R}^*$ of affine transformations and time rescaling is five-dimensional. The first step in this direction is to classify the global schemes of singularities, and for this, the invariant theory approach is essential. This is what is done in this book. Of course the invariants and comitants are helpful in topologically classifying families of systems in **QS** defined by specific algebraic-geometric properties of the systems. This has allowed us to classify topologically some classes of systems which have algebraic-geometric properties such as the Lotka–Volterra differential systems (see [306, 307]) as well as other families of quadratic or cubic systems.

5.3 Basic concepts

We consider planar polynomial differential equations of degree n. Such systems can be written in the form

$$(S)\qquad \begin{aligned} \dot{x} &= p_0 + p_1(x,y) + p_2(x,y) + \ldots + p_n(x,y) \equiv p(x,y),\\ \dot{y} &= q_0 + q_1(x,y) + q_2(x,y) + \ldots + q_n(x,y) \equiv q(x,y), \end{aligned} \tag{5.2}$$

where $p_i(x,y)$ are homogeneous polynomials, which we write in the classical form as

$$p_i(x,y) = \sum_{k+l=i} \binom{i}{l} a_{kl} x^k y^l, \quad q_i(x,y) = \sum_{k+l=i} \binom{i}{l} b_{kl} x^k y^l$$

where $p_i(x,y) = a_{i0}x^i + \ldots + a_{0i}y^i$ and $q_i(x,y) = b_{i0}x^i + \ldots + b_{0i}y^i$ of degree i $(i = 0,\ldots,n)$ whenever either p_i or q_i or both are not identically zero and $\max(\deg(p_n), \deg(q_n)) = n$.

We note that these systems can be presented in tensorial form (see for instance [316]):

$$\frac{dx^j}{dt} = a^j + \sum_{\omega=1}^{n} a^j_{j_1 j_2 \cdots j_\omega} x^{j_1} x^{j_2} \ldots x^{j_\omega} \quad (j, j_1, \ldots, j_\omega = 1,2), \tag{5.3}$$

where the coefficients $a^j_{j_1 j_2 \cdots j_\omega}$ are symmetrical with respect to the lower indices (i.e. their values are not dependent on the order of succession of these indices) and

$$x = x^1,\ y = x^2, \quad a_{kl} = a^1_{\underbrace{11\ldots1}_{k}\underbrace{22\ldots2}_{l}}, \quad b_{kl} = a^2_{\underbrace{11\ldots1}_{k}\underbrace{22\ldots2}_{l}}.$$

Let $\mathbf{PS}_n$ be the set of all polynomial differential systems (5.2) of degree n. In the invariant theory of polynomial differential systems, the set $\mathbf{PS}_n$ plays the role of the set of n-forms in classical invariant theory. We also need a group action on $\mathbf{PS}_n$.

5.3.1 Group actions on polynomial vector fields

On the set $\mathbf{PS}_n$ of all degree-n polynomial differential systems acts the group $Aff(2,\mathbb{R})$ of affine transformations of $\mathbb{R}^2$. For each $g \in Aff(2,\mathbb{R})$, $g: \mathbb{R}^2 \longrightarrow \mathbb{R}^2$ we have:

$$g: \begin{pmatrix}\tilde{x}\\ \tilde{y}\end{pmatrix} = M\begin{pmatrix}x\\ y\end{pmatrix} + B; \qquad g^{-1}: \begin{pmatrix}x\\ y\end{pmatrix} = M^{-1}\begin{pmatrix}\tilde{x}\\ \tilde{y}\end{pmatrix} - M^{-1}B.$$

where $M = ||M_{ij}||$ is a 2×2 nonsingular matrix and B is a 2×1 matrix over $\mathbb{R}$. For every $S \in \mathbf{PS}_n$ we can form its transformed system $\tilde{S} = gS$:

$$(\tilde{S}) \qquad \frac{d\tilde{x}}{dt} = \tilde{p}(\tilde{x},\tilde{y}), \qquad \frac{d\tilde{y}}{dt} = \tilde{q}(\tilde{x},\tilde{y}),$$

where

$$\begin{pmatrix}\tilde{p}(\tilde{x},\tilde{y})\\ \tilde{q}(\tilde{x},\tilde{y})\end{pmatrix} = M\begin{pmatrix}(p\circ g^{-1})(\tilde{x},\tilde{y})\\ (q\circ g^{-1})(\tilde{x},\tilde{y})\end{pmatrix}.$$

The map

$$\begin{array}{rcl} Aff(2,\mathbb{R})\times \mathbf{PS}_n & \longrightarrow & \mathbf{PS}_n\\ (g,S) & \longrightarrow & \tilde{S} = gS \end{array}$$

verifies the axioms for a left group action. For every subgroup $G \subseteq Aff(2,\mathbb{R})$ we have an induced action of G on $\mathbf{PS}_n$. We can identify the set $\mathbf{PS}_n$ of systems (5.2) with a subset of $\mathbb{R}^N$, where $N=(n+2)(n+1)$, via the embedding $\mathbf{PS}_n \hookrightarrow \mathbb{R}^N$ that associates to each system (5.2) the N-tuple $\tilde{a} = (a_{00}, a_{10}, \dots, a_{0n}, b_{00}, b_{10}, \dots, b_{0n})$ of its coefficients. Whenever we consider a particular point in $\mathbb{R}^N$, we shall denote it in boldface: $\boldsymbol{a} = (\boldsymbol{a}_{00}, \boldsymbol{a}_{10}, \dots, \boldsymbol{a}_{0n}, \boldsymbol{b}_{00}, \boldsymbol{b}_{10}, \dots, \boldsymbol{b}_{0n}) \in \mathbb{R}^N$.

On systems (S) such that $\max(\deg(p),\deg(q)) \le n$, we consider the action of the group $Aff(2,\mathbb{R})$, which yields an action of this group on $\mathbb{R}^N$. For every $g \in Aff(2,\mathbb{R})$ let $r_g: \mathbb{R}^N \longrightarrow \mathbb{R}^N$ be the map that corresponds to g via this action. We know (see [315]) that r_g is linear and that the map $r: Aff(2,\mathbb{R}) \longrightarrow GL(N,\mathbb{R})$ thus obtained is a group homomorphism. For every subgroup G of $Aff(2,\mathbb{R})$, r induces a representation of G onto a subgroup $\mathcal{G}$ of $GL(12,\mathbb{R})$.

5.3.2 Definition of invariant polynomials for polynomial differential systems

We now define the concept of *comitant*, which in this theory plays the role of the *covariant* in classical invariant theory. The term *comitant* is short for *concomitant*, i.e. a polynomial in both the coefficients of (5.2) and in the variables x, y.

Definition 5.1. A polynomial $U(\tilde{a}, x, y) \in \mathbb{R}[\tilde{a}, x, y]$ is called a comitant of systems (5.2) with respect to a subgroup G of $Aff(2,\mathbb{R})$ if there exists $\chi \in \mathbb{Z}$ such that for every $(g, \boldsymbol{a}) \in G\times\mathbb{R}^N$ with $N=(n+1)(n+2)$ the identity

$$U(r_g(\boldsymbol{a}),\ g(x,y)) \equiv (\det g)^{-\chi}\, U(\boldsymbol{a}, x, y),$$

holds in $\mathbb{R}[x,y]$, where $\det g$ is the determinant of the linear matrix associated to the transformation g. If the polynomial U does not explicitly depend on x and y, then it is called invariant. The number $\chi \in \mathbb{Z}$ is called the weight of the comitant $U(a,x,y)$. If $G = GL(2,\mathbb{R})$ (or $G = Aff(2,\mathbb{R})$) then the comitant $U(\tilde{a},x,y)$ of systems (5.2) is called a GL-comitant (respectively, affine comitant).

Definition 5.2. A subset $X \subset \mathbb{R}^N$ will be called G-invariant if for every $g \in G$ we have $r_g(X) \subseteq X$.

Let $T(2,\mathbb{R})$ be the subgroup of $Aff(2,\mathbb{R})$ formed by translations. Consider the linear representation of $T(2,\mathbb{R})$ into its corresponding subgroup $\mathcal{T} \subset GL(N,\mathbb{R})$, i.e. for every $\tau \in T(2,\mathbb{R})$, $\tau: \ x = \tilde{x} + \alpha, y = \tilde{y} + \beta$, we consider as above $r_\tau: \mathbb{R}^N \longrightarrow \mathbb{R}^N$.

Definition 5.3. A GL-comitant $U(\tilde{a},x,y)$ of systems (5.2) is called a T-comitant if for every $(\tau,\, \boldsymbol{a}) \in T(2,\mathbb{R}) \times \mathbb{R}^N$ the relation $U(r_\tau \cdot \boldsymbol{a},\ \tilde{x},\ \tilde{y}) = U(\boldsymbol{a},\ \tilde{x},\ \tilde{y})$ holds in $\mathbb{R}[\tilde{x},\tilde{y}]$.

Let

$$U_i(\tilde{a},x,y) = \sum_{j=0}^{d_i} U_{ij}(\tilde{a}) x^{d_i-j} y^j, \quad i = 1,\ldots,s,$$

be a set of GL-comitants of systems (5.2) where d_i denotes the degree of the binary form $U_i(\tilde{a},x,y)$ in x and y with coefficients in $\mathbb{R}[\tilde{a}]$, where $\mathbb{R}[\tilde{a}] = \mathbb{R}[a_{00},\ldots,b_{0n}]$. We denote by

$$\mathcal{U} = \{\, U_{ij}(\tilde{a}) \in \mathbb{R}[\tilde{a}] \mid i = 1,\ldots,s,\ j = 0,1,\ldots,d_i \,\}$$

the set of the coefficients in $\mathbb{R}[a]$ of the GL-comitants $U_i(\tilde{a},x,y)$, $i = 1,\ldots,s$, and by $V(\mathcal{U})$ its associated algebraic set:

$$V(\mathcal{U}) = \left\{ \boldsymbol{a} \in \mathbb{R}^N \mid U_{ij}(\boldsymbol{a}) = 0 \ \forall\ U_{ij}(\tilde{a}) \in \mathcal{U} \right\}.$$

Definition 5.4. A GL-comitant $U(\tilde{a},x,y)$ of systems (5.2) is called a conditional T-comitant (or CT-comitant) modulo a set of polynomials $\langle U_1, U_2, \ldots, U_s \rangle$ if the following two conditions are satisfied:

(i) The algebraic subset $V(\mathcal{U}) \subset \mathbb{R}^N$ is affinely invariant (see Definition 5.2).

(ii) For every $(\tau,\ \boldsymbol{a}) \in T(2,\mathbb{R}) \times V(\mathcal{U})$ we have $U(r_\tau \cdot \boldsymbol{a},\ \tilde{x},\ \tilde{y}) = U(\boldsymbol{a},\ \tilde{x},\ \tilde{y})$ in $\mathbb{R}[\tilde{x},\ \tilde{y}]$.

In other words, a CT-comitant $U(\tilde{a},x,y)$ modulo $\langle U_1, U_2, \ldots, U_s \rangle$ is a T-comitant on the algebraic subset $V(\mathcal{U}) \subset \mathbb{R}^N$.

Definition 5.5. The polynomial $U(\tilde{a},x,y) \in \mathbb{R}[\tilde{a},x,y]$ has well determined sign on $V \subset \mathbb{R}^N$(where $N = (n+1)(n+2)$) with respect to x, y if for every fixed $\boldsymbol{a} \in V$, the sign of the polynomial function $U(\boldsymbol{a},x,y)$ on $\mathbb{R}^2$ is constant where this function is not zero.

Observation 5.1. *We draw attention to the fact that if a CT-comitant $U(a, x, y)$ of even weight is a binary form in x, y, it is of even degree in the coefficients of* (5.2) *and has well determined sign on some affine invariant algebraic subset $V(\mathcal{U})$, then this property is conserved by any affine transformation and the sign is conserved.*

From the classical theory of invariants it is known (see for instance [152, 155, 247]) that there exist different methods for constructing the invariant polynomials with respect to linear transformations: tensorial, symbolic and applying differential operators (Ω-operator, transvectant, polarization, annihilators and so on).

In order to construct the invariant polynomials that we need for the classification of the global configurations of singularities, we chose two tools: the transvectant $(f, g)^{(k)} : \mathbb{R}[\tilde{a}, x, y] \to \mathbb{R}[\tilde{a}, x, y]$ and the differential operator $\mathcal{L} : \mathbb{R}[\tilde{a}, x, y] \to \mathbb{R}[\tilde{a}, x, y]$, which are defined below.

Definition 5.6. Let f, $g \in \mathbb{R}[\tilde{a}, x, y]$ and

$$(f, g)^{(k)} = \sum_{h=0}^{k} (-1)^h \binom{k}{h} \frac{\partial^k f}{\partial x^{k-h} \partial y^h} \frac{\partial^k g}{\partial x^h \partial y^{k-h}}.$$

Then the operator $(*, *)^{(k)}$ which associates to (f, g) the polynomial $(f, g)^{(k)} \in \mathbb{R}[\tilde{a}, x, y]$ is called the transvectant of index k of (f, g).

According to Olver [247], if polynomials f, $g \in \mathbb{R}[\tilde{a}, x, y]$ are invariant polynomials of systems (5.2) with respect to the group $GL(2, \mathbb{R})$, then $(f, g)^{(k)}$ $(k \leq \max(\deg(f), \deg(g)))$ is also a GL-invariant polynomial of these systems.

We consider the two differential operators $\mathbf{L}_i : \mathbb{R}[\tilde{a}] \to \mathbb{R}[\tilde{a}]$ of the form (see [42])

$$\begin{aligned} \mathbf{L}_1 &= \sum_{k=1}^{n} k \sum_{l=0}^{n-k} \frac{n+1-k-l}{k+l} \Big(a_{k-1,\,l} \frac{\partial}{\partial a_{k\,l}} + b_{k-1,\,l} \frac{\partial}{\partial b_{k\,l}} \Big), \\ \mathbf{L}_2 &= \sum_{l=1}^{n} l \sum_{k=0}^{n-l} \frac{n+1-k-l}{k+l} \Big(a_{k,\,l-1} \frac{\partial}{\partial a_{k\,l}} + b_{k,\,l-1} \frac{\partial}{\partial b_{k\,l}} \Big). \end{aligned} \tag{5.4}$$

We then define the differential operator $\mathcal{L} : \mathbb{R}[\tilde{a}, x, y] \to \mathbb{R}[\tilde{a}, x, y]$, where $\mathcal{L} = x \cdot \mathbf{L}_2 - y \cdot \mathbf{L}_1$.

It has been proved [42] that if a polynomial $U \in \mathbb{R}[\tilde{a}, x, y]$ is an invariant polynomial of systems (5.2) with respect to the group $GL(2, \mathbb{R})$, then $\mathcal{L}(U)$ is also a GL-invariant polynomial of these systems.

In this subsection we introduced several terms: comitant with respect to a subgroup G of $\mathit{Aff}(2, \mathbb{R})$ (in the special case when $G = GL(2, \mathbb{R})$, we use the term GL-comitant), T-comitant and CT-comitant. These are all polynomials in the ring $\mathbb{R}[\tilde{a}, x, y]$ and each one of them is in a certain sense invariant. For this reason we call them invariant polynomials. We construct in Sections 5.3.4 and 5.4 a number of such invariant polynomials that are needed for our classification theorem of global geometrical configurations of singularities, finite and infinite.

Although the majority of polynomials used in the classification that we obtain in this book, are T-comitants (or CT-comitants), and they are not invariant with respect to the group $Aff(2,\mathbb{R})$ (except T-comitants that do not depend on x and y), our classification of the global configurations of singularities is affine invariant.

5.3.3 Assembling multiplicities of singularities in divisors of the line at infinity and in zero-cycles of the plane

An isolated singular point p at infinity of a polynomial vector field of degree n has two types of multiplicities: the maximum number m of finite singularities that can split from p, in small perturbations of the system within polynomial systems of degree n, and the maximum number m' of infinite singularities that can split from p, in small such perturbations of the system. We encode the two in the column $(m, m')^t$. We then encode the global information about all isolated singularities at infinity using formal sums called *cycles* and *divisors* (as defined in [248] or in [222] and used in [222, 301, 21, 26]).

We briefly recall here the definition of cycle. Let V be an irreducible algebraic variety over a field $\mathbb{K}$. A *cycle* of dimension r, or r-cycle on V, is a formal sum $\sum_W n_W W$, where W is a subvariety of V of dimension r that is not contained in the singular locus of V, $n_W \in \mathbb{Z}$, and only a finite number of the coefficients n_W are non-zero. The *degree* $\deg(J)$ of a cycle J is defined by $\sum_W n_W$. An $(n-1)$-cycle is called a *divisor* on V. These notions have been used for classification purposes of planar quadratic differential systems [248, 20, 222, 301].

We have two formal sums (*divisors on the line at infinity* $Z = 0$ of the complex affine plane) $D_S(P,Q;Z) = \sum_w I_w(P,Q)w$ and $D_S(C,Z) = \sum_w I_w(C,Z)w$, where $w \in \{Z = 0\}$, P, Q are the homogenizations of the polynomials p, q i.e. $P(X,Y,Z) = Z^n p(X/Z, Y/Z)$, $Q(X,Y,Z) = Z^n q(X/Z, Y/Z)$ and n is the degree of the system. Furthermore here by $I_w(F,G)$ we mean the intersection multiplicity of two algebraic curves $F(X,Y,Z) = 0$, $G(X,Y,Z) = 0$ on the complex projective plane.

$$D_S = \sum_{\omega\in\{Z=0\}} \begin{pmatrix} I_w(P,Q) \\ I_w(C,Z) \end{pmatrix} w.$$

This divisor encodes the total number of singularities at infinity of a system (S) as well as the two kinds of multiplicities that each singularity has. The meaning of these two kinds of multiplicities are described in the definition of the two divisors $D_S(P,Q;Z)$ and $D_S(C,Z)$ on the line at infinity.

For a system (S) with isolated finite singularities, we consider the formal sum (*zero-cycle on the plane*) $D_{\mathbb{C}^2}(S) = \sum_{\omega\in\mathbb{R}^2} I_w(p,q)w$ encoding the multiplicities of all finite singularities. (For more details see [222, 19].)

5.3.4 Construction and geometric meaning of several basic invariant polynomials

We construct the following polynomials

$$\begin{aligned} C_i(x,y) &= yp_i(x,y) - xq_i(x,y),\ i=0,1,\dots,n, \\ D_i(x,y) &= \frac{\partial}{\partial x}p_i(x,y) + \frac{\partial}{\partial y}q_i(x,y),\ i=1,2,\dots,n. \end{aligned} \tag{5.5}$$

These polynomials are in fact invariant polynomials with respect to the group $GL(2,\mathbb{R})$ of linear transformations and are of degree 1 with respect to the coefficients of systems (5.2). Indeed for these systems written in tensorial form (5.3), we can accordingly present the above polynomials as

$$\begin{aligned} C_0 = a^{\beta}x^{\gamma}\varepsilon_{\beta\gamma},\ \ C_i(x^1,x^2) &= a^{\beta}_{\alpha_1\alpha_2\dots\alpha_i}x^{\alpha_1}x^{\alpha_2}\dots x^{\alpha_i}x^{\gamma}\varepsilon_{\beta\gamma}, \\ D_i(x^1,x^2) &= a^{\beta}_{\alpha_1\alpha_2\dots\alpha_{i-1}\beta}x^{\alpha_1}x^{\alpha_2}\dots x^{\alpha_{i-1}}, \end{aligned}$$

where the bi-vector $\varepsilon_{\beta\gamma}$ has the coordinates $\varepsilon_{11}=\varepsilon_{22}=0$ and $\varepsilon_{12}=-\varepsilon_{21}=1$. Therefore according to [155], the polynomials (5.5) are GL-comitants of systems (5.2).

We draw attention to the fact that the polynomials (5.5) serve as building bricks for constructing any invariant polynomial of any degree for polynomial systems (5.2) by using $+$, $-$, $\times$ and operator $(*,*)^{(k)}$.

We have the next result (see [337]):

Theorem 5.1. [337] *Any GL-invariant polynomial of systems (5.2) can be constructed from the elements of the set (5.5) by using the operations: $+$, $-$, $\times$ and by applying the differential operation $(*,*)^{(k)}$.*

Remark 5.1. We point out that the elements of the set (5.5) generate the whole set of GL-invariant polynomial of systems (5.2) and hence also the set of affine comitants as well as the set of the T-comitants of these systems.

Some geometric properties of the basic invariant polynomials (5.5) are revealed in the next theorem.

Theorem 5.2. [30] *If a polynomial system (5.2) verifies the conditions*

$$D_i(x,y)=0\ (i=1,2,\dots,n)\ \left(\Leftrightarrow \frac{\partial}{\partial x}p(x,y)+\frac{\partial}{\partial y}q(x,y)\equiv 0\right), \tag{5.6}$$

i.e. it is Hamiltonian, then this system possesses the following polynomial first integral

$$H(x,y)=\sum_{j=0}^{n}\frac{1}{j+1}C_j(x,y). \tag{5.7}$$

Proof. Assuming that the conditions (5.6) hold, we show that the polynomial $H(x,y)$ of the form (5.7) is a first integral of systems (5.2). Indeed considering (5.7) and applying the Euler's formula for homogeneous polynomials we have

$$H_x = \frac{\partial H}{\partial x} = \sum_{j=0}^{n} \frac{1}{j+1}\left(y\frac{\partial p_j}{\partial x} - q_j - x\frac{\partial q_j}{\partial x}\right) \overset{(5.6)}{=} \sum_{j=0}^{n} \frac{1}{j+1}(-jq_j - q_j) =$$
$$= -\sum_{j=0}^{n} q_j = -q(x,y);$$
$$H_y = \frac{\partial H}{\partial y} = \sum_{j=0}^{n} \frac{1}{j+1}\left(p_j + y\frac{\partial p_j}{\partial y} - x\frac{\partial q_j}{\partial y}\right) \overset{(5.6)}{=} \sum_{j=0}^{n} \frac{1}{j+1}(p_j + jp_j) =$$
$$= \sum_{j=0}^{n} p_j = p(x,y).$$

Therefore

$$H_x\frac{dx}{dt} + H_y\frac{dy}{dt} = -q(x,y)p(x,y) + p(x,y)q(x,y) = 0$$

and this completes the proof of the theorem. □

Using the homogeneous parts of degree n of polynomial systems (5.2) we construct the following polynomial:

$$\mu_0 = \text{Resultant}[p_n(x,y), q_n(x,y), \gamma], \quad \text{where} \quad \gamma = \frac{x}{y} \text{ or } \gamma = \frac{y}{x}.$$

This polynomial is in fact an affine invariant and serves as a basis for constructing a set of $GL(2,\mathbb{R})$-invariant polynomials, which are responsible for the total multiplicity of finite singularities of polynomials systems (5.2). These invariant polynomials are defined by using a classical differential operator $\mathcal{L}$ (defined on page 112).

So, by using the operator $\mathcal{L}$ we construct the following polynomials:

$$\mu_i(\tilde{a},x,y) = \frac{1}{i!}\mathcal{L}^{(i)}(\mu_0(\tilde{a})), \ i = 1,2,\ldots,n^2, \tag{5.8}$$

where $\mathcal{L}^{(i)}(\mu_0) = \mathcal{L}(\mathcal{L}^{(i-1)}(\mu_0))$, $\mathcal{L}^{(0)}(\mu_0) = \mu_0$ and $\mu_i(\tilde{a},x,y) \in \mathbb{R}[\tilde{a},x,y]$.

As mentioned above these polynomials are in fact invariant polynomials of systems (5.2) with respect to the group $GL(2,\mathbb{R})$.

To reveal the geometrical meaning of the comitants $\mu_i(\tilde{a},x,y)$ $(i = 0,1,\ldots,n^2)$ we use the homogenization of the polynomials $p(x,y)$ and $q(x,y)$, i.e.

$$P(X,Y,Z) = Z^n p(X/Z, Y/Z), \quad Q(X,Y,Z) = Z^n q(X/Z, Y/Z),$$

and we calculate the following resultants:

$$\mathrm{Res}_X(P,Q) = \mu_0 Y^{n^2} + \mu_{10} Y^{n^2-1} Z + \ldots + \mu_{n^2-1,0} Y Z^{n^2-1} + \mu_{n^2,0} Z^{n^2}, \quad (5.9)$$
$$\mathrm{Res}_Y(P,Q) = \mu_0 X^{n^2} + \mu_{01} X^{n^2-1} Z + \ldots + \mu_{0,n^2-1} X Z^{n^2-1} + \mu_{0,n^2} Z^{n^2}, \quad (5.10)$$

where $\mu_{ij} = \mu_{ij}(\tilde{a}) \in \mathbb{R}[a_{00}, \ldots, b_{0n}]$.

Additionally, for μ_i, $i = 0, 1, \ldots, n^2$ from (5.8), we have

$$\begin{aligned}
\mu_0(\tilde{a}) &= \mu_0,\\
\mu_1(\tilde{a}, x, y) &= \mu_{10}x + \mu_{01}y,\\
\mu_2(\tilde{a}, x, y) &= \mu_{20}x^2 + \mu_{11}xy + \mu_{02}y^2,\\
\mu_3(\tilde{a}, x, y) &= \mu_{30}x^3 + \mu_{21}x^2y + \mu_{12}xy^2 + \mu_{03}y^3,\\
&\vdots\\
\mu_{n^2-1}(\tilde{a}, x, y) &= \mu_{n^2-1,0}x^{n^2-1} + \mu_{n^2-2,1}x^{n^2-2}y + \ldots + \mu_{0,n^2-1}y^{n^2-1},\\
\mu_{n^2}(\tilde{a}, x, y) &= \mu_{n^2,0}x^{n^2} + \mu_{n^2-1,1}x^{n^2-1}y + \ldots + \mu_{1,n^2-1}xy^{n^2-1} + \mu_{0,n^2}y^{n^2}.
\end{aligned}$$

We observe that the leading coefficients of the comitants μ_i $(i = 0, 1, \ldots, n^2)$ with respect to x are the corresponding coefficients in (5.9) and with respect to y are the corresponding coefficients in (5.10).

We draw attention to the fact that if the comitant $\mu_i(\tilde{a}, x, y)$ $(i = 0, 1, \ldots, n^2)$ is not equal to zero, then we may assume that its leading coefficients are both non-zero, as this can be obtained by applying a rotation of the phase plane of the systems (5.2). From this, (5.9), (5.10) and the above values of μ_i, $i = 0, 1, \ldots, n^2$ we have the following:

Lemma 5.1. *The system $P(X,Y,Z) = Q(X,Y,Z) = 0$ possesses m $(1 \le m \le n^2)$ solutions $[X_i : Y_i : Z_i]$ with $Z_i = 0$ $(i = 1, \ldots, m)$ (considered with multiplicities) if and only if for every $i \in \{0, 1, \ldots, m-1\}$ we have $\mu_i(\tilde{a}, x, y) = 0$ in $\mathbb{R}[\tilde{a}, x, y]$ and $\mu_m(\tilde{a}, x, y) \neq 0$.*

From the above lemma it is clear that the annihilation of $\mu_i(\tilde{a}, x, y)$, $i \in \{0, 1, \ldots, m-1\}$, leads to the total multiplicity of finite singularities being equal to $n^2 - m$.

5.4 Invariant polynomials associated to geometrical configurations of singularities for quadratic differential systems

Our goal in this book is to give an intrinsic geometric classification of the global geometrical configurations of singularities for the whole class of quadratic differential systems, i.e. for systems of the form

$$\begin{aligned}\frac{dx}{dt} &= p_0 + p_1(x,y) + p_2(x,y) \equiv p(x,y),\\ \frac{dy}{dt} &= q_0 + q_1(x,y) + q_2(x,y) \equiv q(x,y)\end{aligned} \tag{5.11}$$

where $\max(\deg(p), \deg(q)) = 2$ and p_i, q_i $(i = 0, 1, 2)$ are homogeneous polynomials of degree i in x, y in the case that they are not identically zero:

$$\begin{aligned} p_0 &= a_{00}, \quad p_1(x,y) = a_{10}x + a_{01}y, \quad p_2(x,y) = a_{20}x^2 + 2a_{11}xy + a_{02}y^2,\\ q_0 &= a_{00}, \quad q_1(x,y) = a_{10}x + a_{01}y, \quad q_2(x,y) = a_{20}x^2 + 2a_{11}xy + a_{02}y^2.\end{aligned}$$

To obtain this classification we use invariant polynomials.

Assume that $M_i(x_i, y_i)$ $(1 \le i \le 4)$ is a singular point of systems (5.11) with the corresponding Jacobian matrix

$$\begin{pmatrix} \dfrac{\partial p}{\partial x}(x_i,y_i) & \dfrac{\partial p}{\partial y}(x_i,y_i)\\ \dfrac{\partial q}{\partial x}(x_i,y_i) & \dfrac{\partial q}{\partial y}(x_i,y_i)\end{pmatrix} \equiv \begin{pmatrix} \hat{p}_x^i & \hat{p}_y^i \\ \hat{q}_x^i & \hat{q}_y^i\end{pmatrix}.$$

We set the notations

$$\rho_i = \hat{p}_x^i + \hat{q}_y^i, \quad \Delta_i = \hat{p}_x^i\hat{q}_y^i - \hat{p}_y^i\hat{q}_x^i, \quad \tau_i = \rho_i^2 - 4\Delta_i,$$

for the traces, determinants and discriminants, respectively, corresponding to the finite singularity M_i and its characteristic equation.

In this section we construct all the invariant polynomials which intervene in our classification theorems of global geometrical configurations of singularities, finite and infinite.

5.4.1 Building blocks for the construction of the invariant polynomials needed for the classification theorems of QS

In this subsection we shall define several building blocks to be later used in the construction of our invariant polynomials that will allow us to find for any family of quadratic systems, given in whatever normal form, their global geometrical configuration of singularities, finite or infinite. The following polynomials in $\mathbb{R}[\tilde{a}, x, y]$ are the simplest invariant polynomials of degree 1 with respect to the coefficients of the differential systems (5.11) that are GL-comitants:

$$\begin{aligned} C_i(x,y) &= yp_i(x,y) - xq_i(x,y), \ i = 0,1,2,\\ D_i(x,y) &= \frac{\partial}{\partial x}p_i(x,y) + \frac{\partial}{\partial y}q_i(x,y), \ i = 1,2.\end{aligned} \tag{5.12}$$

According to Theorem 5.1 we deduce that any GL-comitant (as well as an affine comitant or a T-comitant of any degree) could be constructed from the elements

(5.12) by using the operations $+$, $-$, $\times$ and by applying the differential operation $(*,*)^{(k)}$. Moreover the geometrical meaning of the invariant polynomials (5.12) is revealed by Theorem 5.2, setting $n=2$.

Apart from these simple invariant polynomials, we shall also make use of nine other GL-invariant polynomials, all of which help in the construction of the invariant polynomials we need to define. The GL-invariants are the following:

$$\begin{array}{lll} T_1=(C_0,C_1)^{(1)}, & T_2=(C_0,C_2)^{(1)}, & T_3=(C_0,D_2)^{(1)}, \\ T_4=(C_1,C_1)^{(2)}, & T_5=(C_1,C_2)^{(1)}, & T_6=(C_1,C_2)^{(2)}, \\ T_7=(C_1,D_2)^{(1)}, & T_8=(C_2,C_2)^{(2)}, & T_9=(C_2,D_2)^{(1)}. \end{array}$$

These are of degree two with respect to the coefficients of systems (5.11).

We next define a list of T-comitants:

$$\begin{aligned}
\hat{A}(\tilde{a}) &= (C_1, T_8 - 2T_9 + D_2^2)^{(2)}/144,\\
\widehat{B}(\tilde{a},x,y) &= \Big\{16D_1(D_2,T_8)^{(1)}(3C_1D_1 - 2C_0D_2 + 4T_2) + 32C_0(D_2,T_9)^{(1)}(9T_7\\
&\quad - 5T_6 + 3D_1D_2) + 2(D_2,T_9)^{(1)}(27C_1T_4 - 18C_1D_1^2 - 32D_1T_2\\
&\quad + 32(C_0,T_5)^{(1)}) + 6(D_2,T_7)^{(1)}\big[8C_0(T_8 - 12T_9) - 12C_1(D_1D_2 + T_7)\\
&\quad + D_1(26C_2D_1 + 32T_5) + C_2(9T_4 + 96T_3)\big] + 6(D_2,T_6)^{(1)}\big[32C_0T_9\\
&\quad - C_1(12T_7 + 52D_1D_2) - 32C_2D_1^2\big] + 48D_2(D_2,T_1)^{(1)}(2D_2^2 - T_8)\\
&\quad + 6D_1D_2T_4(T_8 - 7D_2^2 - 42T_9) - 32D_1T_8(D_2,T_2)^{(1)} + 9D_2^2T_4(T_6 - 2T_7)\\
&\quad - 16D_1(C_2,T_8)^{(1)}(D_1^2 + 4T_3) + 12D_1(C_1,T_8)^{(2)}(C_1D_2 - 2C_2D_1)\\
&\quad + 12D_1(C_1,T_8)^{(1)}(T_7 + 2D_1D_2) + 96D_2^2\big[D_1(C_1,T_6)^{(1)} + D_2(C_0,T_6)^{(1)}\big]\\
&\quad - 4D_1^3D_2(D_2^2 + 3T_8 + 6T_9) - 16D_1D_2T_3(2D_2^2 + 3T_8)\\
&\quad + 6D_1^2D_2^2(7T_6 + 2T_7) - 252D_1D_2T_4T_9\Big\}/(2^8 3^3),\\
\widehat{D}(\tilde{a},x,y) &= \big[2C_0(T_8 - 8T_9 - 2D_2^2) + C_1(6T_7 - T_6) - (C_1,T_5)^{(1)} - 9D_1^2C_2\\
&\quad + 6D_1(C_1D_2 - T_5)\big]/36,\\
\widehat{E}(\tilde{a},x,y) &= \big[D_1(2T_9 - T_8) - 3(C_1,T_9)^{(1)} - D_2(3T_7 + D_1D_2)\big]/72,\\
\widehat{F}(\tilde{a},x,y) &= \big[6D_1^2(D_2^2 - 4T_9) + 4D_1D_2(T_6 + 6T_7) + 48C_0(D_2,T_9)^{(1)} - 9D_2^2T_4\\
&\quad + 288D_1\widehat{E} - 24(C_2,\widehat{D})^{(2)} + 120(D_2,\widehat{D})^{(1)} - 36C_1(D_2,T_7)^{(1)}\\
&\quad + 8D_1(D_2,T_5)^{(1)}\big]/144,\\
\widehat{K}(\tilde{a},x,y) &= (T_8 + 4T_9 + 4D_2^2)/72,\\
\widehat{H}(\tilde{a},x,y) &= (-T_8 + 8T_9 + 2D_2^2)/72,
\end{aligned}$$

which together with the T-comitants $C_2(\tilde{a},x,y)$ and $D_2(\tilde{a},x,y)$ helped us to define the following affine invariants (constructed in [58]):

$$
\begin{aligned}
A_1(\tilde{a}) &= \hat{A},\\
A_2(\tilde{a}) &= (C_2, \widehat{D})^{(3)}/12,\\
A_3(\tilde{a}) &= \big[\!\big[C_2, D_2)^{(1)}, D_2\big)^{(1)}, D_2\big)^{(1)}/48,\\
A_4(\tilde{a}) &= (\widehat{H}, \widehat{H})^{(2)},\\
A_5(\tilde{a}) &= (\widehat{H}, \widehat{K})^{(2)}/2,\\
A_6(\tilde{a}) &= (\widehat{E}, \widehat{H})^{(2)}/2,\\
A_7(\tilde{a}) &= \big[\!\big[C_2, \widehat{E})^{(2)}, D_2\big)^{(1)}/8,\\
A_8(\tilde{a}) &= \big[\!\big[\widehat{D}, \widehat{H})^{(2)}, D_2\big)^{(1)}/8,\\
A_9(\tilde{a}) &= \big[\!\big[\widehat{D}, D_2)^{(1)}, D_2\big)^{(1)}, D_2\big)^{(1)}/48,\\
A_{10}(\tilde{a}) &= \big[\!\big[\widehat{D}, \widehat{K})^{(2)}, D_2\big)^{(1)}/8,\\
A_{11}(\tilde{a}) &= (\widehat{F}, \widehat{K})^{(2)}/4,\\
A_{12}(\tilde{a}) &= (\widehat{F}, \widehat{H})^{(2)}/4,\\
A_{13}(\tilde{a}) &= \big[\!\big[C_2, \widehat{H})^{(1)}, \widehat{H}\big)^{(2)}, D_2\big)^{(1)}/24,\\
A_{14}(\tilde{a}) &= (\widehat{B}, C_2)^{(3)}/36,\\
A_{15}(\tilde{a}) &= (\widehat{E}, \widehat{F})^{(2)}/4,\\
A_{16}(\tilde{a}) &= \big[\!\big[\widehat{E}, D_2)^{(1)}, C_2\big)^{(1)}, \widehat{K}\big)^{(2)}/16,\\
A_{17}(\tilde{a}) &= \big[\!\big[\widehat{D}, \widehat{D})^{(2)}, D_2\big)^{(1)}, D_2\big)^{(1)}/64,\\
A_{18}(\tilde{a}) &= \big[\!\big[\widehat{D}, \widehat{F})^{(2)}, D_2\big)^{(1)}/16,\\
A_{19}(\tilde{a}) &= \big[\!\big[\widehat{D}, \widehat{D})^{(2)}, \widehat{H}\big)^{(2)}/16,\\
A_{20}(\tilde{a}) &= \big[\!\big[C_2, \widehat{D})^{(2)}, \widehat{F}\big)^{(2)}/16,\\
A_{21}(\tilde{a}) &= \big[\!\big[\widehat{D}, \widehat{D})^{(2)}, \widehat{K}\big)^{(2)}/16,\\
A_{22}(\tilde{a}) &= \big[\!\big[C_2, \widehat{D})^{(1)}, D_2\big)^{(1)}, D_2\big)^{(1)}, D_2\big)^{(1)} D_2\big)^{(1)}/1152,\\
A_{23}(\tilde{a}) &= \big[\!\big[\widehat{F}, \widehat{H})^{(1)}, \widehat{K}\big)^{(2)}/8,\\
A_{24}(\tilde{a}) &= \big[\!\big[C_2, \widehat{D})^{(2)}, \widehat{K}\big)^{(1)}, \widehat{H}\big)^{(2)}/32,\\
A_{25}(\tilde{a}) &= \big[\!\big[\widehat{D}, \widehat{D})^{(2)}, \widehat{E}\big)^{(2)}/16,\\
A_{26}(\tilde{a}) &= (\widehat{B}, \widehat{D})^{(3)}/36,\\
A_{27}(\tilde{a}) &= \big[\!\big[\widehat{B}, D_2)^{(1)}, \widehat{H}\big)^{(2)}/24,\\
A_{28}(\tilde{a}) &= \big[\!\big[C_2, \widehat{K})^{(2)}, \widehat{D}\big)^{(1)}, \widehat{E}\big)^{(2)}/16,\\
A_{29}(\tilde{a}) &= \big[\!\big[\widehat{D}, \widehat{F})^{(1)}, \widehat{D}\big)^{(3)}/96,
\end{aligned}
$$

$$A_{30}(\tilde{a}) = [\![C_2, \widehat{D})^{(2)}, \widehat{D})^{(1)}, \widehat{D})^{(3)}/288,$$
$$A_{31}(\tilde{a}) = [\![\widehat{D}, \widehat{D})^{(2)}, \widehat{K})^{(1)}, \widehat{H})^{(2)}/64,$$
$$A_{32}(\tilde{a}) = [\![\widehat{D}, \widehat{D})^{(2)}, D_2)^{(1)}, \widehat{H})^{(1)}, D_2)^{(1)}/64,$$
$$A_{33}(\tilde{a}) = [\![\widehat{D}, D_2)^{(1)}, \widehat{F})^{(1)}, D_2)^{(1)}, D_2)^{(1)}/128,$$
$$A_{34}(\tilde{a}) = [\![\widehat{D}, \widehat{D})^{(2)}, D_2)^{(1)}, \widehat{K})^{(1)}, D_2)^{(1)}/64,$$
$$A_{35}(\tilde{a}) = [\![\widehat{D}, \widehat{D})^{(2)}, \widehat{E})^{(1)}, D_2)^{(1)}, D_2)^{(1)}/128,$$
$$A_{36}(\tilde{a}) = [\![\widehat{D}, \widehat{E})^{(2)}, \widehat{D})^{(1)}, \widehat{H})^{(2)}/16,$$
$$A_{37}(\tilde{a}) = [\![\widehat{D}, \widehat{D})^{(2)}, \widehat{D})^{(1)}, \widehat{D})^{(3)}/576,$$
$$A_{38}(\tilde{a}) = [\![C_2, \widehat{D})^{(2)}, \widehat{D})^{(2)}, \widehat{D})^{(1)}, \widehat{H})^{(2)}/64,$$
$$A_{39}(\tilde{a}) = [\![\widehat{D}, \widehat{D})^{(2)}, \widehat{F})^{(1)}, \widehat{H})^{(2)}/64,$$
$$A_{40}(\tilde{a}) = [\![\widehat{D}, \widehat{D})^{(2)}, \widehat{F})^{(1)}, \widehat{K})^{(2)}/64,$$
$$A_{41}(\tilde{a}) = [\![C_2, \widehat{D})^{(2)}, \widehat{D})^{(2)}, \widehat{F})^{(1)}, D_2)^{(1)}/64,$$
$$A_{42}(\tilde{a}) = [\![\widehat{D}, \widehat{F})^{(2)}, \widehat{F})^{(1)}, D_2)^{(1)}/16.$$

In the above list the bracket "$[\![$" means a succession of two or up to five parentheses "(" depending on the row in which it appears.

5.4.2 The set of all invariant polynomials which classify geometrically the global configurations of singularities in QS

Diagrams 8.1, 8.2, 9.2–9.4, 10.2–10.5 and 11.2–11.10 classify the global geometrical configurations of singularities, finite and infinite, of quadratic differential systems, yielding 1765 such configurations and 8 additional configurations (related to a single finite configuration) that we conjecture are not realizable.

For any family of quadratic systems, in whatever normal form, we can then determine from these diagrams the geometrical global configurations of singularities of the systems in the family. Each one of these configurations is characterized using some of the following 83 invariant polynomials occurring in these diagrams:

$$\begin{gathered}\Big\{\mu_0,\ldots,\mu_4,\ \mathbf{D},\ \mathbf{P},\ \mathbf{R},\ \mathbf{S},\ \mathbf{T},\ \mathbf{U},\ W_1,\ldots,W_{11},\ H_1,\ldots,H_5,\\ V_1,\ldots,V_4,\ E_1,\ E_3,\ U_1,\ldots,U_6,\ Y_1,\ Y_2,\ \mathcal{T}_1,\ldots,\mathcal{T}_4,\ \mathcal{H},\\ \mathcal{F},\ \mathcal{F}_1,\ldots,\mathcal{F}_4,\ \mathcal{B},\mathcal{B}_1,\ldots,\mathcal{B}_5,\ \mathcal{I}_1,\ \mathcal{I}_2,\ \sigma,\ \eta,\ \widetilde{M},\ C_2,\ \kappa,\ \kappa_1,\\ \theta,\ \theta_1,\ldots,\theta_6,\ \widetilde{D},\ \widetilde{K},\ \widetilde{N},\ \widetilde{R},\ \tilde{L},\ L_1,\ L_2,\ L_3,K_1,\ K_2,\ K_3\Big\}.\end{gathered} \tag{5.13}$$

In what follows we define these elements. Recalling the formulas (5.4) for the quadratic systems (5.11), we obtain the differential operator $\mathcal{L} = x\cdot\mathbf{L}_2 - y\cdot\mathbf{L}_1$

acting on $\mathbb{R}[\tilde{a}, x, y]$ with

$$\mathbf{L}_1 = 2a_{00}\frac{\partial}{\partial a_{10}} + a_{10}\frac{\partial}{\partial a_{20}} + \frac{1}{2}a_{01}\frac{\partial}{\partial a_{11}} + 2b_{00}\frac{\partial}{\partial b_{10}} + b_{10}\frac{\partial}{\partial b_{20}} + \frac{1}{2}b_{01}\frac{\partial}{\partial b_{11}},$$
$$\mathbf{L}_2 = 2a_{00}\frac{\partial}{\partial a_{01}} + a_{01}\frac{\partial}{\partial a_{02}} + \frac{1}{2}a_{10}\frac{\partial}{\partial a_{11}} + 2b_{00}\frac{\partial}{\partial b_{01}} + b_{01}\frac{\partial}{\partial b_{02}} + \frac{1}{2}b_{10}\frac{\partial}{\partial b_{11}}.$$

Using this operator and the affine invariant $\mu_0 = \operatorname{Res}_x\big(p_2(\tilde{a}, x, y), q_2(\tilde{a}, x, y)\big)/y^4$, according to the formulas (5.8) we construct the following polynomials:

$$\mu_i(\tilde{a}, x, y) = \frac{1}{i!}\mathcal{L}^{(i)}(\mu_0),\ i = 1, \ldots, 4, \quad \text{where } \mathcal{L}^{(i)}(\mu_0) = \mathcal{L}(\mathcal{L}^{(i-1)}(\mu_0)).$$

As mentioned earlier (see page 112), if a polynomial $U \in \mathbb{R}[\tilde{a}, x, y]$ is an invariant polynomial of systems (5.11) with respect to the group $GL(2,\mathbb{R})$, then $\mathcal{L}(U)$ is also a GL-invariant polynomial of these systems. So since $\mu_0 = -A_4/2$ is a GL-invariant (even more, it is an affine invariant), it follows that each one of the polynomials $\mu_i(\tilde{a}, x, y)$ $(i = 1, \ldots, 4)$ is a GL-comitant.

The very important geometrical meaning of these invariant polynomials is revealed in the next lemma (see [42, 301] and also Lemma 5.1).

Lemma 5.2. *Assume that a quadratic system (S) with coefficients $\boldsymbol{a} \in \mathbb{R}^{12}$ belongs to the family* (5.11)*. Then:*

(i) *The total multiplicity of the finite singularities of this system is $4 - k$ if and only if for every i such that $0 \le i \le k-1$ we have $\mu_i(\boldsymbol{a}, x, y) = 0$ in $\mathbb{R}[x, y]$ and $\mu_k(\boldsymbol{a}, x, y) \neq 0$. In this case the factorization $\mu_k(\boldsymbol{a}, x, y) = \prod_{i=1}^{k}(u_i x - v_i y) \neq 0$ over $\mathbb{C}$ yields the coordinates $[v_i : u_i : 0]$ of points at infinity that have multiplicity greater than one, this being the result of coalescence of finite and infinite singularities. Moreover the number of distinct expressions $u_i x - v_i y$ in this factorization is less than or equal to three (the maximum number of infinite singularities of a quadratic system), and the multiplicity of each one of the expressions $u_i x - v_i y$ gives us the number of the finite singularities of the system (S) that have coalesced with the infinite singular point $[v_i : u_i : 0]$.*

(ii) *Let the point $M_0(0, 0)$ be a singular point for the quadratic system (S). Then the point $M_0(0, 0)$ is a singular point of multiplicity k $(1 \le k \le 4)$ if and only if for every i such that $0 \le i \le k - 1$ we have $\mu_{4-i}(\boldsymbol{a}, x, y) = 0$ in $\mathbb{R}[x, y]$ and $\mu_{4-k}(\boldsymbol{a}, x, y) \neq 0$.*

(iii) *The system (S) is degenerate (i.e. $\gcd(p, q) \neq$ constant) if and only if $\mu_i(\boldsymbol{a}, x, y) = 0$ in $\mathbb{R}[x, y]$ for every $i = 0, 1, 2, 3, 4$.*

Using the GL-comitants $\mu_0, \ldots, \mu_4$, we define the following invariant polynomials:

$$\begin{aligned}
\mathbf{D} &= \Big[3\big((\mu_3, \mu_3)^{(2)}, \mu_2\big)^{(2)} - \big(6\mu_0\mu_4 - 3\mu_1\mu_3 + \mu_2^2, \mu_4\big)^{(4)}\Big]/48,\\
\mathbf{P} &= 12\mu_0\mu_4 - 3\mu_1\mu_3 + \mu_2^2,\\
\mathbf{R} &= 3\mu_1^2 - 8\mu_0\mu_2,\\
\mathbf{S} &= \mathbf{R}^2 - 16\mu_0^2\mathbf{P},\\
\mathbf{T} &= 18\mu_0^2(3\mu_3^2 - 8\mu_2\mu_4) + 2\mu_0(2\mu_2^3 - 9\mu_1\mu_2\mu_3 + 27\mu_1^2\mu_4) - \mathbf{PR},\\
\mathbf{U} &= \mu_3^2 - 4\mu_2\mu_4.
\end{aligned} \tag{5.14}$$

With the exception of $\mathbf{U}$, these invariant polynomials are T-comitants because they are GL-comitants, and their coefficients in variables x and y are absolute invariants under translations, as it can be checked directly. The polynomial $\mathbf{U}$ is a CT-comitant modulo $\langle \mu_0, \mu_1 \rangle$ because when $\mu_0 = \mu_1 = 0$ it can be checked that the coefficients of $\mathbf{U}$ remains invariant under translations.

We remark that the invariant polynomials (5.14) have a very important geometrical meaning. More exactly, together with $\mu_0, \ldots, \mu_4$ they are responsible for the number, kinds (real or/and complex) and the multiplicities of finite singularities of systems (5.11) (see Table 5.1).

According to [41, 29], we have the next proposition.

Proposition 5.1. *The value of the zero-cycle $\mathcal{D}_{\mathbb{C}^2}(S)$ for non-degenerate quadratic systems* (5.11) *is determined by the corresponding conditions in Table* 5.1*, where we write $p + q + r^c + s^c$ if two of the finite points, i.e. r^c, s^c are complex but not real.*

Remark 5.2. We point out that $\mathbf{D} = 0$, $\mathbf{P} = 0$, ... $\mathbf{U} = 0$ are the hypersurfaces in the parameter space whose union is the bifurcation set for the number and multiplicity of finite singularities.

Next we define the invariant polynomials that govern the number of the traces that vanish. We denote

$$\sigma(\tilde{a}, x, y) = \frac{\partial p}{\partial x} + \frac{\partial q}{\partial y} = \sigma_0(\tilde{a}) + \sigma_1(\tilde{a}, x, y)\ (\equiv D_1(\tilde{a}) + D_2(\tilde{a}, x, y)),$$

and observe that the polynomial $\sigma(\tilde{a}, x, y)$ is an affine comitant of systems (5.11). If (x_i, y_i) is a singular point of a system (5.11), then for the trace ρ_i of its respective Jacobian matrix, we have $\rho_i = \sigma(\tilde{a}, x_i, y_i)$.

Applying the differential operators $\mathcal{L}$ and $(*, *)^{(k)}$ (i.e. transvectant of index k), we define the following polynomial function that governs the values of the traces for finite singularities of systems (5.11).

Table 5.1

No.	Zero-cycle $\mathcal{D}_{\mathbb{C}^2}(S)$	Invariant criteria	No.	Zero-cycle $\mathcal{D}_{\mathbb{C}^2}(S)$	Invariant criteria
1	$p+q+r+s$	$\mu_0\neq 0, \mathbf{D}<0, \mathbf{R}>0, \mathbf{S}>0$	10	$p+q+r$	$\mu_0=0, \mathbf{D}<0, \mathbf{R}\neq 0$
2	$p+q+r^c+s^c$	$\mu_0\neq 0, \mathbf{D}>0$	11	$p+q^c+r^c$	$\mu_0=0, \mathbf{D}>0, \mathbf{R}\neq 0$
3	$p^c+q^c+r^c+s^c$	$\mu_0\neq 0, \mathbf{D}<0, \mathbf{R}\leq 0$; $\mu_0\neq 0, \mathbf{D}<0, \mathbf{S}\leq 0$	12	$2p+q$	$\mu_0=\mathbf{D}=0, \mathbf{PR}\neq 0$
4	$2p+q+r$	$\mu_0\neq 0, \mathbf{D}=0, \mathbf{T}<0$	13	$3p$	$\mu_0=\mathbf{D}=\mathbf{P}=0, \mathbf{R}\neq 0$
5	$2p+q^c+r^c$	$\mu_0\neq 0, \mathbf{D}=0, \mathbf{T}>0$	14	$p+q$	$\mu_0=\mathbf{R}=0, \mathbf{P}\neq 0, \mathbf{U}>0$
6	$2p+2q$	$\mu_0\neq 0, \mathbf{D}=\mathbf{T}=0, \mathbf{PR}>0$	15	p^c+q^c	$\mu_0=\mathbf{R}=0, \mathbf{P}\neq 0, \mathbf{U}<0$
7	$2p^c+2q^c$	$\mu_0\neq 0, \mathbf{D}=\mathbf{T}=0, \mathbf{PR}<0$	16	$2p$	$\mu_0=\mathbf{R}=0, \mathbf{P}\neq 0, \mathbf{U}=0$
8	$3p+q$	$\mu_0\neq 0, \mathbf{D}=\mathbf{T}=0, \mathbf{P}=0, \mathbf{R}\neq 0$	17	p	$\mu_0=\mathbf{R}=\mathbf{P}=0, \mathbf{U}\neq 0$
9	$4p$	$\mu_0\neq 0, \mathbf{D}=\mathbf{T}=0, \mathbf{P}=\mathbf{R}=0$	18	0	$\mu_0=\mathbf{R}=\mathbf{P}=0, \mathbf{U}=0, \ \mu_4\neq 0$

Definition 5.7 ([338]). We call *trace polynomial* $\mathfrak{T}(w)$ over the ring $\mathbb{R}[\tilde{a}]$ the polynomial defined as follows

$$\mathfrak{T}(w)=\sum_{i=0}^{4}\frac{1}{(i!)^2}\left(\sigma_1^i,\ \frac{1}{i!}\mathcal{L}^{(i)}(\mu_0)\right)^{(i)}w^{4-i}=\sum_{i=0}^{4}\mathcal{G}_i(\tilde{a})\,w^{4-i},$$

where the coefficients

$$\mathcal{G}_i(\tilde{a})=\frac{1}{(i!)^2}(\sigma_1^i,\ \mu_i)^{(i)}\in\mathbb{R}[\tilde{a}],\quad i=0,1,2,3,4\quad \big(\mathcal{G}_0(\tilde{a})\equiv\mu_0(\tilde{a})\big) \tag{5.15}$$

are GL-invariants of systems (5.11).

Using the polynomial $\mathfrak{T}(w)$, we can construct the following four polynomials $\mathcal{T}_4, \mathcal{T}_3, \mathcal{T}_2, \mathcal{T}_1$, which are responsible for the weak singularities:

$$\mathcal{T}_{4-i}(\tilde{a})=\frac{1}{i!}\frac{d^i\mathfrak{T}}{dw^i}\bigg|_{w=\sigma_0}\in\mathbb{R}[\tilde{a}],\quad i=0,1,2,3\quad \big(\mathcal{T}_4\equiv\mathfrak{T}(\sigma_0)\big).$$

These polynomials are in fact affine invariants since they can be expressed through the elements $A_1, \dots, A_{42}$ constructed earlier, which are affine invariants:

$$\begin{aligned}
\mathcal{T}_1(\tilde{a}) &= -4A_6,\\
\mathcal{T}_2(\tilde{a}) &= 5A_8 - 18A_1^2 - A_{10} - 3A_{11} + 9A_{12},\\
\mathcal{T}_3(\tilde{a}) &= 8A_{15} - 4A_1A_2,\\
\mathcal{T}_4(\tilde{a}) &= (5A_2^2 - 22A_{17} - 16A_{18} + 21A_{19} - 2A_{20} - 3A_{21})/9.
\end{aligned}$$

The geometric meaning of these invariants is revealed in the next lemma. The equations in the lemma could be checked directly by evaluating the affine invariant polynomials $\mathcal{T}_1, \dots, \mathcal{T}_4$ for each one of the canonical systems that will be given in Table 6.1 of Section 6.1.

Lemma 5.3. *Consider a non-degenerate system (5.11) and let $\boldsymbol{a} \in \mathbb{R}^{12}$ be its 12-tuple of coefficients. Let ρ_i be the trace corresponding to the finite singular point M_i, $1 \le i \le 4$ (real or complex, simple or multiple). Then the following equalities hold, respectively:*

(i) *For $\mu_0(\boldsymbol{a}) \neq 0$ (total multiplicity of finite singularities equal to four),*

$$\begin{aligned}
\mathcal{T}_4(\boldsymbol{a}) &= \mathcal{G}_0(\boldsymbol{a})\rho_1\rho_2\rho_3\rho_4,\\
\mathcal{T}_3(\boldsymbol{a}) &= \mathcal{G}_0(\boldsymbol{a})(\rho_1\rho_2\rho_3 + \rho_1\rho_2\rho_4 + \rho_1\rho_3\rho_4 + \rho_2\rho_3\rho_4),\\
\mathcal{T}_2(\boldsymbol{a}) &= \mathcal{G}_0(\boldsymbol{a})(\rho_1\rho_2 + \rho_1\rho_3 + \rho_1\rho_4 + \rho_2\rho_3 + \rho_2\rho_4 + \rho_3\rho_4),\\
\mathcal{T}_1(\boldsymbol{a}) &= \mathcal{G}_0(\boldsymbol{a})(\rho_1 + \rho_2 + \rho_3 + \rho_4).
\end{aligned}$$

(ii) *For $\mu_0(\boldsymbol{a}) = 0$, $\mu_1(\boldsymbol{a}, x, y) \neq 0$ (total multiplicity of finite singularities equal to three),*

$$\begin{aligned}
\mathcal{T}_4(\boldsymbol{a}) &= \mathcal{G}_1(\boldsymbol{a})\rho_1\rho_2\rho_3, & \mathcal{T}_3(\boldsymbol{a}) &= \mathcal{G}_1(\boldsymbol{a})(\rho_1\rho_2 + \rho_1\rho_3 + \rho_2\rho_3),\\
\mathcal{T}_2(\boldsymbol{a}) &= \mathcal{G}_1(\boldsymbol{a})(\rho_1 + \rho_2 + \rho_3), & \mathcal{T}_1(\boldsymbol{a}) &= \mathcal{G}_1(\boldsymbol{a}),
\end{aligned}$$

(iii) *For $\mu_0(\boldsymbol{a}) = \mu_1(\boldsymbol{a}, x, y) = 0$, $\mu_2(\boldsymbol{a}, x, y) \neq 0$ (total multiplicity of finite singularities equal to two),*

$$\begin{aligned}
\mathcal{T}_4(\boldsymbol{a}) &= \mathcal{G}_2(\boldsymbol{a})\rho_1\rho_2, & \mathcal{T}_3(\boldsymbol{a}) &= \mathcal{G}_2(\boldsymbol{a})(\rho_1 + \rho_2),\\
\mathcal{T}_2(\boldsymbol{a}) &= \mathcal{G}_2(\boldsymbol{a}), & \mathcal{T}_1(\boldsymbol{a}) &= 0.
\end{aligned}$$

(iv) *For $\mu_0(\boldsymbol{a}) = \mu_1(\boldsymbol{a}, x, y) = \mu_2(\boldsymbol{a}, x, y) = 0$, $\mu_3(\boldsymbol{a}, x, y) \neq 0$ (one simple singularity),*

$$\mathcal{T}_4(\boldsymbol{a}) = \mathcal{G}_3(\boldsymbol{a})\rho_1, \quad \mathcal{T}_3(\boldsymbol{a}) = \mathcal{G}_3(\boldsymbol{a}), \quad \mathcal{T}_2(\boldsymbol{a}) = \mathcal{T}_1(\boldsymbol{a}) = 0.$$

Next we list the invariant polynomials that together with $\mathcal{T}_1, \dots, \mathcal{T}_4$ are responsible for the number of finite weak singularities as well as for their kinds (foci, centers or saddles) and their orders (see Theorem 6.2), and we specify their definitions.

$$\begin{aligned}
\mathcal{F}(\tilde{a}) &= A_7,\\
\mathcal{F}_1(\tilde{a}) &= A_2,\\
\mathcal{F}_2(\tilde{a}) &= -2A_1^2A_3 + 2A_5(5A_8 + 3A_9) + A_3(A_8 - 3A_{10} + 3A_{11} + A_{12})\\
&\quad - A_4(10A_8 - 3A_9 + 5A_{10} + 5A_{11} + 5A_{12}),\\
\mathcal{F}_3(\tilde{a}) &= -10A_1^2A_3 + 2A_5(A_8 - A_9) - A_4(2A_8 + A_9 + A_{10} + A_{11} + A_{12})\\
&\quad + A_3(5A_8 + A_{10} - A_{11} + 5A_{12}),\\
\mathcal{F}_4(\tilde{a}) &= 20A_1^2A_2 - A_2(7A_8 - 4A_9 + A_{10} + A_{11} + 7A_{12}) + A_1(6A_{14} - 22A_{15})\\
&\quad - 4A_{33} + 4A_{34},\\
\mathcal{H}(\tilde{a}) &= -(A_4 + 2A_5),\\
\widetilde{K}(\tilde{a}, x, y) &= 4\widehat{K} \equiv \text{Jacob}\big(p_2(\tilde{a}, x, y), q_2(\tilde{a}, x, y)\big),\\
\mathcal{B}(\tilde{a}) &= -(3A_8 + 2A_9 + A_{10} + A_{11} + A_{12}),\\
\mathcal{B}_1(\tilde{a}, x, y) &= \Big\{(T_7, D_2)^{(1)}\big[12D_1T_3 + 2D_1^3 + 9D_1T_4 + 36(T_1, D_2)^{(1)}\big] - 2D_1(T_6, D_2)^{(1)}\\
&\quad \times (D_1^2 + 12T_3) + D_1^2\big[D_1(T_8, C_1)^{(2)} + 6\big((T_6, C_1)^{(1)}, D_2\big)^{(1)}\big]\Big\}/144,\\
\mathcal{B}_2(\tilde{a}, x, y) &= \Big\{(T_7, D_2)^{(1)}\big[8T_3(T_6, D_2)^{(1)} - D_1^2(T_8, C_1)^{(2)} - 4D_1\big((T_6, C_1)^{(1)}, D_2\big)^{(1)}\big]\\
&\quad + \big[(T_7, D_2)^{(1)}\big]^2(8T_3 - 3T_4 + 2D_1^2)\Big\}/384,\\
\mathcal{B}_3(\tilde{a}, x, y) &= -D_1^2(4D_2^2 + T_8 + 4T_9) + 3D_1D_2(T_6 + 4T_7) - 24T_3(D_2^2 - T_9),\\
\mathcal{B}_4(\tilde{a}, x, y) &= D_1(T_5 + 2D_2C_1) - 3C_2(D_1^2 + 2T_3),\\
\mathcal{B}_5(\tilde{a}) &= \mathcal{T}_1^2 - 4\mu_0\mathcal{T}_2,\\
\mathcal{I}_1(\tilde{a}) &= 4A_8 - 5A_{10} - 11A_{11} + 10A_{12},\\
\mathcal{I}_2(\tilde{a}) &= A_8 - 2A_{10}.
\end{aligned}$$

In what follows we define a set of invariant polynomials which are responsible for the local topological structures of neighborhoods of the real finite singularities of systems (5.11) (see Theorem 6.1, Table 6.2):

$$\begin{aligned}
W_1(\tilde{a}) &= \big[A_1^2(A_4 - 2A_5) - 4A_1A_{16} - 4A_5A_8 - 2A_4(3A_8 + 2A_{11})\\
&\quad + 4A_6(A_7 + 7A_6)\big]/2,\\
W_2(\tilde{a}) &= \big[144A_1A_2A_6 - 23A_2^2A_4 - 72A_3(A_{19} + A_{21}) + 72A_5(A_{17} + 4A_{19})\\
&\quad + A_4(14A_{20} - 50A_{17} - 32A_{18} - 75A_{19} + 93A_{21}) - 288A_6A_{15}\big]/9,
\end{aligned}$$

$$\begin{aligned}
W_3(\tilde{a}) = \big[&9A_1^2(36A_{18} - 19A_2^2 + 134A_{17} + 165A_{19}) + 102A_3(3A_{30} - 14A_{29})\\
&+ 3A_{11}(42A_{18} - 102A_{17} + 195A_{19}) - 63A_6(17A_{25} + 30A_{26})\\
&+ 2A_2^2(A_{10} + 3A_{11}) + 3A_{10}(14A_{18} - 118A_{17} + 153A_{19} + 120A_{21})\\
&+ 6A_7(329A_{25} - 108A_{26}) + 3A_8(164A_{18} + 153A_{19} - 442A_{17})\\
&+ 9A_{12}(2A_{20} - 160A_{17} - 2A_{18} - 59A_{19}) + 18A_{21}(21A_9 - 5A_{11})\\
&+ 302A_2A_{34} + 3A_1(77A_2A_{14} + 235A_2A_{15} - 54A_{36}) - 366A_{14}^2\\
&- 12A_{15}(71A_{14} + 80A_{15})\big]/9,\\
W_4(\tilde{a}) = \big[&1512A_1^2(A_{30} - 2A_{29}) - 648A_{15}A_{26} + 72A_1A_2(49A_{25} + 39A_{26})\\
&+ 6A_2^2(23A_{21} - 1093A_{19}) - 87A_2^4 + 4A_2^2(61A_{17} + 52A_{18} + 11A_{20})\\
&- 6A_{37}(352A_3 + 939A_4 - 1578A_5) + 72A_{29}(17A_{12} - 38A_9 - 109A_{11})\\
&- 36A_8(396A_{29} + 265A_{30}) - 4A_{17}(158A_{17} + 30A_{20} + 87A_{21})\\
&+ 12A_{30}(76A_9 - 189A_{10} - 273A_{11} - 651A_{12}) - 24A_{18}(3A_{20} + 31A_{17})\\
&- 648A_{14}(23A_{25} + 5A_{26}) + 36A_{19}(63A_{20} + 478A_{21})\\
&+ 18A_{21}(2A_{20} + 137A_{21}) - 18A_{19}(238A_{17} + 669A_{19})\big]/81,\\
W_5(\tilde{a}) = {}&3A_1A_2(A_1A_2 + 6A_{14}) + 2A_2^2A_8 - 24A_{19}(A_8 + A_{12}),\\
W_6(\tilde{a}) = {}&2A_1^2(2A_{29} - A_{30}) - A_{26}(A_1A_2 + 2A_{14}),\\
W_7(\tilde{a}) = {}&12A_{26}(A_{26} - 2A_{25}) + (2A_{29} - A_{30})(A_2^2 - 20A_{17} - 12A_{18} + 6A_{19}\\
&+ 6A_{21}) + 48A_{37}(A_1^2 - A_8 - A_{12}),\\
W_8(\tilde{a}) = {}&64D_1\big[\big((T_6, C_1)^{(1)}, D_2\big)^{(1)}\big]^2\big[16(C_0, T_6)^{(1)} - 37(D_2, T_1)^{(1)} + 12D_1T_3\big]\\
&+ 4(108D_1^4 - 3T_4^2 - 128T_3T_4 + 42D_1^2T_4)\big[\big((T_6, C_1)^{(1)}, D_2\big)^{(1)}\big]^2\\
&+ 36D_1\big((T_6, C_1)^{(1)}, D_2\big)^{(1)}(C_1, T_8)^{(2)}\big[4D_1(C_0, T_6)^{(1)} - D_1^2(4T_3 + T_4)\\
&+ 24T_3^2\big] + 64\big[\big((T_6, C_1)^{(1)}, D_2\big)^{(1)}\big]^2\big[27T_3^2 + +16\big((T_6, C_1)^{(1)}, C_0\big)^{(1)}\big]\\
&- 54\big((T_6, C_1)^{(1)}, T_6\big)^{(1)}(C_1, T_8)^{(2)}\big[8D_1^4 + D_1^2T_4 - 8D_1(C_0, T_6)^{(1)}\\
&+ 8D_1^2T_3 + 8T_3^2\big] + 108D_1T_3\big[(C_1, T_8)^{(2)}\big]^2\big[D_1T_3 - 2(C_0, T_6)^{(1)}\big]\\
&+ 576\big((T_6, C_1)^{(1)}, D_2\big)^{(1)}\big((T_6, C_1)^{(1)}, T_6\big)^{(1)}\big[2(D_2, T_1)^{(1)} - 5D_1T_3\big]\\
&- 27\big[(C_1, T_8)^{(2)}\big]^2\big[T_4^4/8 + (C_0, T_1)^{(1)}\big],\\
W_9(\tilde{a}) = {}&-3T_4(T_4 + 4D_1^2) - 32D_1(C_0, T_6)^{(1)} - 32\Big((T_6, C_1)^{(1)}, C_0\Big)^{(1)},\\
W_{10}(\tilde{a}) = {}&H_1(\tilde{a})H_2(\tilde{a})H_3(\tilde{a}) \text{ (see } H_1\text{–}H_3 \text{ below)},\\
W_{11}(\tilde{a}) = {}&2\widehat{K}(C_2, \widehat{D})^{(2)} + 12\widehat{F}(\widehat{K} - 2\widehat{H}) - D_2(\widehat{D}, \widehat{K})^{(1)},
\end{aligned}$$

$$
\begin{aligned}
H_1(\tilde a) &= A_1^2(16A_3 - A_4 + 22A_5) - 16A_7^2 - 2A_4(8A_8 - A_{12}) + 4A_6(2A_6 + 17A_7)\\
&\quad - 4A_1A_{16} + 4A_5(A_8 - A_{10} - A_{11} - 4A_{12}),\\
H_2(\tilde a) &= 2A_5 - 3A_4,\\
H_3(\tilde a) &= -3A_1^2(A_4 + 8A_5) + A_4(9A_{10} + A_{11} + 2A_{12}) - 8A_3(A_{11} + 2A_{10})\\
&\quad + 2A_5(7A_{10} - A_{11} + 6A_{12}) + 2A_6(8A_7 - 3A_6) - 8A_2A_{13},\\
H_4(\tilde a) &= 2A_3(A_{10} + A_{11}) + 2A_5A_{10} - 12A_6A_7 + A_4(13A_8 + 3A_{11} + 3A_{12}),\\
H_5(\tilde a) &= 4A_2A_{13} + 6A_3(5A_{10} + A_{11}) - 3A_4(3A_8 + 4A_{10} + A_{11} + A_{12})\\
&\quad - 18A_5A_{10} - 12A_6A_7,\\
V_1(\tilde a) &= W_1\left(W_2W_3 - W_1W_4\right) - \mu_0^2W_3^2,\\
V_2(\tilde a) &= W_1^3 - 4\mu_0^2\left(W_1W_2 - 2\mu_0^2W_3\right),\\
V_3(\tilde a) &= W_2^2 + W_1W_3 - 4\mu_0^2W_4,\\
V_4(\tilde a) &= W_1^2 - 4\mu_0^2W_2,\\
E_1(\tilde a) &= A_{25},\\
E_3(\tilde a) &= 2A_1^2 - A_8 - A_{12},\\
\widetilde{D}(\tilde a, x, y) &= \widehat{D}.
\end{aligned}
$$

Except for W_8, W_9 and W_{11} which are CT-comitants and $\widetilde{D}$, which is T-comitant, all other invariant polynomials above are affine invariants. Moreover the invariant polynomials $W_1, \dots, W_{11}$ govern the discriminants of the characteristic equations corresponding to finite singularities of systems (5.11), similarly as the invariant polynomials $\mathcal{T}_1, \dots, \mathcal{T}_4$ are responsible for the traces corresponding to these singularities. More exactly in the generic case we have the following lemma. The equalities in the lemma could be verified by straightforward calculation of the values of the affine invariant polynomials $W_1, \dots W_4$ for each one of the canonical systems given in Table 6.1 of Section 6.1.

Lemma 5.4. *Consider a non-degenerate system* (5.11) *and let* $\boldsymbol{a} \in \mathbb{R}^{12}$ *be its 12-tuple of coefficients. Let* τ_i *be the discriminant corresponding to the finite singular point* M_i, $1 \le i \le 4$ *(real or complex, simple or multiple). Then the following equalities hold, respectively (the values of the invariant polynomials* $\mathcal{G}_0, \dots, \mathcal{G}_3$ *are given in formulas* (5.15)*):*

(i) *For* $\mu_0(\boldsymbol{a}) \neq 0$ *(total multiplicity four of finite singularities),*

$$
\begin{aligned}
W_4(\boldsymbol{a}) &= \mathcal{G}_0^2(\boldsymbol{a})\tau_1\tau_2\tau_3\tau_4,\\
W_3(\boldsymbol{a}) &= \mathcal{G}_0^2(\boldsymbol{a})(\tau_1\tau_2\tau_3 + \tau_1\tau_2\tau_4 + \tau_1\tau_3\tau_4 + \tau_2\tau_3\tau_4),\\
W_2(\boldsymbol{a}) &= \mathcal{G}_0^2(\boldsymbol{a})(\tau_1\tau_2 + \tau_1\tau_3 + \tau_1\tau_4 + \tau_2\tau_3 + \tau_2\tau_4 + \tau_3\tau_4),\\
W_1(\boldsymbol{a}) &= \mathcal{G}_0^2(\boldsymbol{a})(\tau_1 + \tau_2 + \tau_3 + \tau_4).
\end{aligned}
$$

(ii) *For* $\mu_0(\boldsymbol{a})=0,\ \mu_1(\boldsymbol{a},x,y)\neq 0$ *(total multiplicity three),*

$$\begin{array}{ll} W_4(\boldsymbol{a})=\mathcal{G}_1^2(\boldsymbol{a})\tau_1\tau_2\tau_3, & W_3(\boldsymbol{a})=\mathcal{G}_1^2(\boldsymbol{a})(\tau_1\tau_2+\tau_1\tau_3+\tau_2\tau_3),\\ W_2(\boldsymbol{a})=\mathcal{G}_1^2(\boldsymbol{a})(\tau_1+\tau_2+\tau_3). & W_1(\boldsymbol{a})=\mathcal{G}_1^2(\boldsymbol{a}). \end{array}$$

(iii) *For* $\mu_0(\boldsymbol{a})=\mu_1(\boldsymbol{a},x,y)=0,\ \mu_2(\boldsymbol{a},x,y)\neq 0$ *(total multiplicity two),*

$$\begin{array}{ll} W_4(\boldsymbol{a})=\mathcal{G}_2^2(\boldsymbol{a})\tau_1\tau_2, & W_3(\boldsymbol{a})=\mathcal{G}_2^2(\boldsymbol{a})(\tau_1+\tau_2),\\ W_2(\boldsymbol{a})=\mathcal{G}_2^2(\boldsymbol{a}), & W_1(\boldsymbol{a})=0. \end{array}$$

(iv) *For* $\mu_0(\boldsymbol{a})=\mu_1(\boldsymbol{a},x,y)=\mu_2(\boldsymbol{a},x,y)=0,\ \mu_3(\boldsymbol{a},x,y)\neq 0$ *(one simple singularity),*

$$W_4(\boldsymbol{a})=\mathcal{G}_3^2(\boldsymbol{a})\tau_1,\quad W_3(\boldsymbol{a})=\mathcal{G}_3^2(\boldsymbol{a}),\quad W_2(\boldsymbol{a})=W_1(\boldsymbol{a})=0.$$

Thus if for an elemental singular point its corresponding discriminant vanishes, then this point is a node with coinciding eigenvalues. However as we mentioned earlier, we would like to distinguish a one-direction node from a star node. The existence and the number of star nodes for a quadratic system (5.11) are detected by the following set of invariant polynomials (see Theorem 7.1):

$$\begin{aligned} U_1(\tilde a,x,y)&=\widetilde K-4\widehat H,\\ U_2(\tilde a,x,y)&=(C_1,-4\widehat H-\widetilde K)^{(1)}-2D_1U_1,\\ U_3(\tilde a,x,y)&=3\widehat D(D_2^2-16\widetilde K)+C_2\big[(C_2,\widehat D)^{(2)}-5(D_2,\widehat D)^{(1)}+6\,\widehat F\big],\\ U_4(\tilde a,x,y)&=2T_5+C_1D_2,\\ U_5(\tilde a,x,y)&=3C_1D_1+4T_2-2C_0D_1,\\ U_6(\tilde a,x,y)&=-4\widehat H,\\ Y_1(\tilde a)&=A_1,\\ Y_2(\tilde a,x,y)&=2D_1^2+8T_3-T_4. \end{aligned}$$

The following invariant polynomials govern the number, kind (real or complex) and multiplicities of infinite singularities of systems (5.11) and furthermore their geometrical configurations:

$$\begin{aligned} C_2(\tilde a,x,y)&=yp_2(\tilde a,x,y)-xq_2(\tilde a,x,y),\\ \widetilde M(\tilde a,x,y)&=(C_2,C_2)^{(2)}\equiv 2\text{Hess}\,\big(C_2(\tilde a,x,y)\big),\\ \eta(\tilde a)&=(\widetilde M,\widetilde M)^{(2)}/384\equiv \text{Discrim}\,\big(C_2(\tilde a,x,y)\big),\\ \kappa(\tilde a)&=(\widetilde M,\widetilde K)^{(2)}/4,\\ \kappa_1(\tilde a)&=(\widetilde M,C_1)^{(2)},\\ \widetilde N(\tilde a,x,y)&=\widetilde K-4\widehat H, \end{aligned}$$

$$
\begin{aligned}
\theta(\tilde{a}) &= -(\widetilde{N},\widetilde{N})^{(2)}/2 \equiv \text{Discrim}\big(\widetilde{N}(\tilde{a},x,y)\big),\\
\theta_1(\tilde{a}) &= 16\eta(\tilde{a}) + \kappa(\tilde{a}),\\
\theta_2(\tilde{a}) &= A_1,\\
\theta_3(\tilde{a}) &= A_8 + A_{11},\\
\theta_4(\tilde{a}) &= A_7,\\
\theta_5(\tilde{a},x,y) &= 2C_2(T_6,T_7)^{(1)} - (T_5 + 2D_2C_1)(C_1,D_2^2)^{(2)},\\
\theta_6(\tilde{a},x,y) &= C_1T_8 - 2C_2T_6,\\
K_1(\tilde{a},x,y) &= (T_5 - 3C_2D_1 + 2C_1D_2)/6 \;\equiv\; p_1q_2 - p_2q_1,\\
K_2(\tilde{a},x,y) &= 4(T_2,\widetilde{M} - 2\widetilde{K})^{(1)} + 3D_1(C_1,\widetilde{M} - 2\widetilde{K})^{(1)}\\
&\quad - (\widetilde{M} - 2\widetilde{K})\big(16T_3 - 3T_4/2 + 3D_1^2\big),\\
K_3(\tilde{a},x,y) &= C_2^2(4T_3 + 3T_4) + C_2(3C_0\widetilde{K} - 2C_1T_7) + 2K_1(3K_1 - C_1D_2),\\
\tilde{L}(\tilde{a},x,y) &= 4\widetilde{K} - 32\widehat{H} - \widetilde{M},\\
L_1(\tilde{a},x,y) &= (C_2,\widetilde{D})^{(2)},\\
L_2(\tilde{a},x,y) &= (C_2,\widetilde{D})^{(1)},\\
L_3(\tilde{a},x,y) &= C_1^2 - 4C_0C_2,\\
\widetilde{R}(\tilde{a},x,y) &= \tilde{L} + 8\widetilde{K}.
\end{aligned}
$$

The geometrical meaning of the above invariant polynomials is revealed in Theorem 6.4. We point out here several properties of some of these polynomials as well as some polynomials from the set (5.13).

Lemma 5.5. *Consider the family of non-degenerate quadratic systems* (5.11) *and assume that a quadratic system* (S) *with coefficients* $\boldsymbol{a} \in \mathbb{R}^{12}$ *belongs to this family.*

(i) *The invariant polynomials* $C_2(\tilde{a},x,y)$, $\widetilde{M}(\tilde{a},x,y)$ *and* $\eta(\tilde{a})$ *govern the number and kind (real and/or complex) of infinite singularities. More precisely the type of the divisor* $D_S(C,Z)$ *for the system* (S) *is determined by the following conditions:*

$$
\begin{array}{rcl}
D_S(C,Z) = & w_1 + w_2 + w_3 & \Leftrightarrow \quad \eta(\boldsymbol{a}) > 0;\\
D_S(C,Z) = & w_1 + w_2^c + w_3^c & \Leftrightarrow \quad \eta(\boldsymbol{a}) < 0;\\
D_S(C,Z) = & 2w_1 + w_2 & \Leftrightarrow \quad \eta(\boldsymbol{a}) = 0,\ \widetilde{M}(\boldsymbol{a},x,y) \neq 0;\\
D_S(C,Z) = & 3w & \Leftrightarrow \quad \widetilde{M}(\boldsymbol{a},x,y) = 0,\ C_2(\boldsymbol{a},x,y) \neq 0;\\
D_S(C,Z) = & \mathit{undefined} & \Leftrightarrow \quad C_2(\boldsymbol{a},x,y) = 0,
\end{array}
$$

where we write $w_1 + w_2^c + w_3^c$ *if two of the points, i.e.* w_2^c, w_3^c, *are complex but not real.*

(ii) *The invariant polynomials* $\mu_0(\tilde{a})$ *and* $\widetilde{K}(\tilde{a},x,y)$ *are responsible for the number of directions that indicate where finite singularities have coalesced with* *infinite ones (these directions are defined by the common factor of homogeneities $p_2(\tilde{a},x,y)$ and $q_2(\tilde{a},x,y)$ of systems (5.11)). More exactly we have:*

$$deg\,\mathrm{gcd}(p_2(\boldsymbol{a},x,y),q_2(\boldsymbol{a},x,y)) = \begin{cases} 0 & \Leftrightarrow \quad \mu_0(\boldsymbol{a})\neq 0; \\ 1 & \Leftrightarrow \quad \mu_0(\boldsymbol{a})=0,\ \widetilde{K}(\boldsymbol{a},x,y)\neq 0; \\ 2 & \Leftrightarrow \quad \widetilde{K}(\boldsymbol{a},x,y)=0. \end{cases}$$

Moreover in the case of $\widetilde{K}(\boldsymbol{a},x,y)=0$, we have the following factorization of $\mathrm{gcd}(p_2(\boldsymbol{a},x,y),q_2(\boldsymbol{a},x,y))$ over $\mathbb{C}$: $(\alpha x+\beta y)(\gamma x+\delta y)$, $\alpha\delta-\beta\gamma\neq 0$ if $\widehat{H}(\boldsymbol{a},x,y)\neq 0$ and $(\alpha x+\beta y)^2$ if $\widehat{H}(\boldsymbol{a},x,y)=0$.

(iii) *The invariant polynomials $\theta(\tilde{a})$ and $\widetilde{N}(\tilde{a},x,y)$ are related to the couples of parallel invariant lines of systems (5.11). More exactly, a necessary condition for the existence of one couple (respectively, two couples) of parallel invariant straight lines of the system (S) is the condition $\theta(\boldsymbol{a})=0$ (respectively, $\widetilde{N}(\boldsymbol{a},x,y)=0$).*

For the proof of this lemma see [300, 301].

Remark 5.3. We point out that apart from the set (5.13) of invariant polynomials that give a complete classification of the global geometrical configurations of singularities, finite and infinite, of systems (5.11), we also use in this book some additional invariant polynomials that are needed for the proofs of these classification theorems and not in the statements of these theorems. We give as follows the definition of these additional 16 invariant polynomials:

$$\begin{aligned}
B_1(\tilde{a},x,y) &= T_6+T_7+3D_1D_2,\\
B_2(\tilde{a},x,y) &= T_5-3C_2D_1+2C_1D_2,\\
B_3(\tilde{a},x,y) &= T_2+C_0D_2,\\
F_1(\tilde{a},x,y) &= \widehat{D},\\
F_2(\tilde{a},x,y) &= \big[288\big(\widetilde{K},B_3\big)^{(1)}-8\big(B_1,\,B_2\big)^{(1)}-\big(B_2,\,B_2\big)^{(2)}\big]/864,\\
F_3(\tilde{a},x,y) &= 3D_1^2D_2+2D_1(T_6+T_7)+3\left(C_1,T_7\right)^{(1)},\\
F_4(\tilde{a},x,y) &= \big[3\,B_2(T_4+2D_1^2)-8\,B_1B_3+288(T_1+C_0D_1)\widetilde{K}\big]/144,\\
F_5(\tilde{a},x,y) &= T_5+2C_1D_2-3C_2D_1,\\
\mathcal{G}(\tilde{a},x,y) &= \widetilde{M}-32\widehat{H},\\
E_2(\tilde{a}) &= A_6,\\
G_1(\tilde{a}) &= (5A_2^2-22A_{17}-16A_{18}+21A_{19}-2A_{20}-3A_{21})/9,\\
G_3(\tilde{a}) &= A_2,\\
G_8(\tilde{a}) &= (5A_8-18A_1^2-A_{10}-3A_{11}+9A_{12})/2,
\end{aligned}$$

$$G_9(\tilde{a}) = (A_4 + 2A_5)/4,$$
$$G_{10}(\tilde{a}) = A_{26},$$
$$G_{13}(\tilde{a}) = \left[4(D_2, T_1)^{(1)} - D_1(T_4 + 2D_1^2 + 4T_3)\right]/8.$$

Observation 5.2. *The results leading to this complete classification were published separately in a sequence of papers* [21, 22, 23, 24, 25, 26], *some on infinite, others on finite singularities. As we were advancing in our study, we were introducing new invariant polynomials to deal with for example the order of weak foci or saddles or star nodes. It was only when the work was complete that we realized that we had some repetitions, for example an invariant used for infinite singularities appeared also when dealing with some finite singularities. Examples are:* $\widetilde{N} = U_1$, $\theta_2 = Y_1$ *and* $\theta_4 = \mathcal{F}$. *We keep these notations as they appeared in the previous papers, for coherence in the full work and compatibility with previous papers and because they may intervene in some algebraic relations among polynomials.*

5.4.3 The influence of complex singularities in the study of the geometrical global configurations of singularities in QS

The ultimate goal in the study of **QS** is to obtain the topological classification of this family. The first step in this direction is the global topological classification of its singularities. For this we use the geometrical classification, which is deeper and more helpful than the topological one. Clearly in the phase portraits the only singularities that appear are the real ones. Although hidden from view, the complex singularities appear to play an important role, and this is what we intend to explain in this section.

Apart from the trivial fact that a quadratic system has always singular points (finite and infinite) of total multiplicity exactly seven in the projective plane, some real and maybe some complex, the complex singularities affect the study of geometrical configurations of singularities in a deeper way. The main impact of the complex singularities occurs in the detection of zero traces or zero discriminants of the characteristic equations of singularities.

Let us present an example.

Assume that we have a quadratic system with two complex finite singularities and two real ones. From Lemma 5.3 we have the invariants $\mathcal{T}_4(\boldsymbol{a}) = \mathcal{G}_0(\boldsymbol{a})\rho_1\rho_2\rho_3\rho_4$ and $\mathcal{T}_3(\boldsymbol{a}) = \mathcal{G}_0(\boldsymbol{a})(\rho_1\rho_2\rho_3 + \rho_1\rho_2\rho_4 + \rho_1\rho_3\rho_4 + \rho_2\rho_3\rho_4)$, where ρ_i are the traces of the finite singularities. It is clear that if $\mathcal{T}_4 = 0 \neq \mathcal{T}_3$ then only one trace may be zero, and this must be a real singularity, which will be weak. But if both are zero, then one cannot know at this level if the zero traces correspond to the complex singularities or to the real ones. If at least one of the real singularities is a node, then both complex must have zero trace. But if both real singularities are saddle or focus, then the doubt will prevail. We need to define a new invariant to deal with this problem, i.e. to separate these cases. This new invariant is $\mathcal{B}_5$ (see page 125). A similar phenomenon occurs with two vanishing discriminants when the invariant polynomials W_4 and W_3 are zero, which leads to a similar situation. In this case, the existence of a finite saddle or focus (which cannot have discriminant zero) clears the question, but if both real singularities are nodes, then a set of new conditions that we call $(\mathcal{C}^{\tau}_{nn})$ and $(\mathcal{C}^{\tau}_{n^d n^d})$ will help us to decide whether the complex singularities or the real ones have discriminant zero.

Thus in some configurations of singularities, we indicate that complex singularities have trace or discriminant zero. However these special complex singularities marked with zero trace or zero discriminant were not distinguished in the geometrical classification from generic complex singularities (see Definition 3.1 in Section 3.6). It may happen that the same real parts of two configurations are identical but the complex part is generic in one case and special in the other. These configurations could appear in our bifurcation diagram in two (or three) distinct branches of the bifurcation tree, one branch leading to the generic configuration and the others to special cases (with either traces or discriminants zero). All of them are needed for the completeness of the diagram. We place within brackets the configurations of singularities including special complex singularities that are geometrically equivalent to the corresponding generic ones, so as not to count them several times. For example $[f,\ ©^{\tau}, ©^{\tau};\ \widehat{\binom{1}{2}}\ \widehat{P}_{\curlywedge} E \widehat{P}_{\curlywedge} - H,\ S]$ is geometrically equivalent to $f,\ ©, ©;\ \widehat{\binom{1}{2}}\ \widehat{P}_{\curlywedge} E \widehat{P}_{\curlywedge} - H,\ S$.

However it may happen that some properties of the real singularities force the complex singularities to have trace or discriminant zero. This produces a configuration that cannot exist with generic complex singularities. For example, configuration $c^{\odot}, c^{\odot}, ©, ©;\ S, Ⓒ, Ⓒ$ with two isochronous centers does not exist because the condition of presence of the isochronous centers forces the complex singularities to have discriminant zero. Thus, the right configuration is $c^{\odot}, c^{\odot}, ©^{\tau}, ©^{\tau};\ S, Ⓒ, Ⓒ$. So we do not put it in brackets since this configuration must be counted. There are another three examples of the same kind.

We might have decided to add the concept of complex singularities with trace or discriminant zero to be geometrically different from the generic ones to all **QS**. But then this would have meant to apply this distinction to all systems with four finite complex singularities which would have added superfluous information in the theory of **QS**.

Chapter 6

Main results on classifications of finite, infinite and weak singularities in quadratic systems

In this chapter we sum up results on finite singularities of quadratic differential systems previously obtained by us and that were published prior to the publication of this book (see [41, 29, 338, 301, 26, 32]). Roughly speaking these results give us global information about the possibilities for the number and multiplicity of finite singularities (see [41, 29]), the canonical forms for these possibilities, the weak singularities that may occur as well as the cases of integrable saddles and information on the infinite singularities.

The description of the singularities that we give in this chapter is not the definitive one, which will be given in Part II after improvement.

6.1 Finite singularities

Consider real quadratic systems of the form:

$$\begin{aligned} \frac{dx}{dt} &= p_0 + p_1(x,y) + p_2(x,y) \equiv p(x,y), \\ \frac{dy}{dt} &= q_0 + q_1(x,y) + q_2(x,y) \equiv q(x,y), \end{aligned} \tag{6.1}$$

with homogeneous polynomials p_i and q_i $(i = 0, 1, 2)$ of degree i in x, y:

$$\begin{aligned} p_0 &= a_{00}, \quad p_1(x,y) = a_{10}x + a_{01}y, \quad p_2(x,y) = a_{20}x^2 + 2a_{11}xy + a_{02}y^2, \\ q_0 &= a_{00}, \quad q_1(x,y) = a_{10}x + a_{01}y, \quad q_2(x,y) = a_{20}x^2 + 2a_{11}xy + a_{02}y^2. \end{aligned}$$

J. C. Artés et al., *Geometric Configurations of Singularities of Planar Polynomial Differential Systems*, https://doi.org/10.1007/978-3-030-50570-7_6

In Chapter 5 it is mentioned that the number, multiplicities and kinds (real or/and complex) of finite singularities of these systems are governed by the invariant polynomials $\mu_0, \ldots, \mu_4$ as well as by the polynomials (5.14) (see Proposition 5.1 and Table 5.1 on page 123).

In view of the expressions (5.14), we point out the following:

Remark 6.1. If $\mu_0 = 0$ then the condition $\mathbf{R} = 0$ (respectively $\mathbf{R} = \mathbf{P} = 0$; $\mathbf{R} = \mathbf{P} = \mathbf{U} = 0$; $\mathbf{R} = \mathbf{P} = \mathbf{U} = \mathbf{V} = 0$) is equivalent to $\mu_1 = 0$ (respectively $\mu_1 = \mu_2 = 0$; $\mu_1 = \mu_2 = \mu_3 = 0$; $\mu_1 = \mu_2 = \mu_3 = \mu_4 = 0$).

Using the above remark and Proposition 5.1, we present in the next proposition the bifurcation diagram in $\mathbb{R}^{12}$ of the type of the zero-cycle $\mathcal{D}_{\mathbb{C}^2}(S)$), for non-degenerate quadratic systems (6.1).

Proposition 6.1. *The number and multiplicities of the finite singular points (described by the zero-cycle $\mathcal{D}_{\mathbb{C}^2}(S)$) for non-degenerate quadratic systems* (6.1) *is given by Diagram* 6.1.

Based on the results in [29, 338], we obtain here the following proposition.

Proposition 6.2. *Any non-degenerate quadratic system can be brought via an affine transformation and time rescaling to one of the canonical forms in Table* 6.1*, according to the type of the zero-cycle $\mathcal{D}_{\mathbb{C}^2}(S)$.*

Next we focus our attention on the local topological structure of the vicinities of the finite singularities of a quadratic system. According to [29] we have the next result.

Theorem 6.1. *The number and the topological types of the finite singular points of a quadratic system are contained in Table* 6.2*. The conditions $(\mathfrak{D}_j)$ $(j = 1, 2)$ are given by the following formulas and all the invariant polynomials used in Table* 6.2 *are presented in Section* 5.4*:*

$$\begin{aligned} (\mathfrak{D}_1):\quad & W_4 > 0,\ V_1 \le 0,\ V_2 = 0, \\ (\mathfrak{D}_2):\quad & W_4 = W_3 = 0,\ V_4 \equiv W_1^2 - 4\mu_0^2 W_2 = 0. \end{aligned} \tag{6.2}$$

Remark 6.2. We note that in Table 6.2 we made the following modifications with respect to [29, Table 1]:

- Since we have obtained stronger tools (see Theorem 6.2) for classifying weak singularities, at this moment it is convenient to eliminate the distinction between foci and centers by joining these cases and using the notation $(f \vee c)$ for "focus or center" when both cases are possible. As we advance in this chapter, we will recover the description with centers and weak singularities using these new tools.
- We have replaced the old notations for the finite singularities (which were in capital letters) with the notations explicitly introduced in Section 3.7.

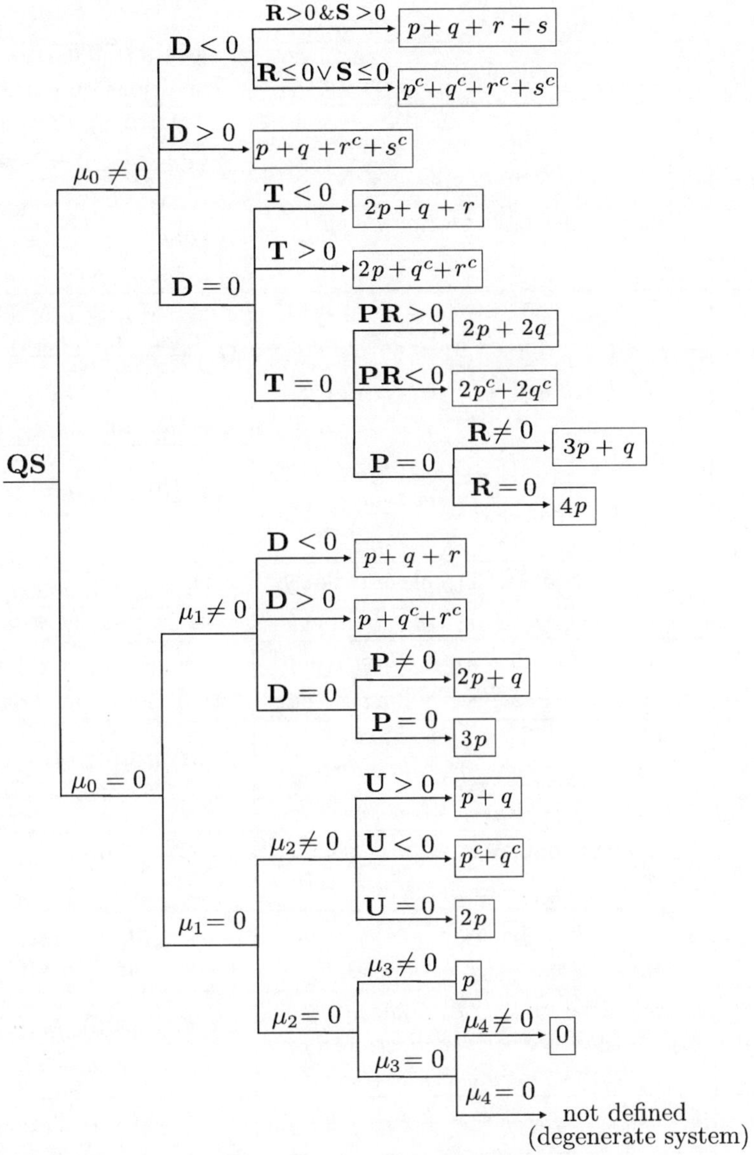

Diagram 6.1: The bifurcation diagram of quadratic systems according to the number and multiplicity of finite singularities.

Table 6.1

Type of $\mathcal{D}_{\mathbb{C}^2}(S)$		Canonical form	Finite singularities and conditions on parameters
$p+q+$ $r+s$	1	$\dot{x}=cx+dy-cx^2+2hxy-dy^2,$ $\dot{y}=ex+fy-ex^2+2mxy-fy^2$	$M_1(0,0), M_2(1,0),$ $M_4\left(\frac{\beta(\beta-2\gamma)}{\beta^2-4\alpha\gamma}, \frac{\beta(\beta-2\alpha)}{\beta^2-4\alpha\gamma}\right),$ $M_3(0,1),$ where $\alpha=eh-cm,\ \beta=cf-de,$ $\gamma=dm-fh;\ \beta^2-4\alpha\gamma\neq 0$
$p+q+$ r^c+s^c	2	$\dot{x}=a-(a+g)x+gx^2+2hxy+ay^2,$ $\dot{y}=b-(b+l)x+lx^2+2mxy+by^2$	$M_{1,2}(0,\pm i),\ M_3(1,0),$ $M_4\left(\frac{\beta^2+4\gamma^2}{\beta^2-4\alpha\gamma}, \frac{2\beta(\alpha+\gamma)}{4\alpha\gamma-\beta^2}\right),$ where $\alpha=gm-hl,\ \beta=bg-al,$ $\gamma=bh-am;\ \beta^2-4\alpha\gamma\neq 0$
p^c+q^c+ r^c+s^c	3a	$\dot{x}=\frac{g(u^2+1)}{v^2+1}(1+y^2)-2(gu+hv)x$ $+gx^2+2hxy,\ \dot{y}=\frac{l(u^2+1)}{v^2+1}(1+y^2)$ $-2(lu+mv)x+lx^2+2mxy$	$M_{1,2}(0,\pm i),\ M_{3,4}(u\pm i,v);$ $gm-lh\neq 0$
	3b	$\dot{x}=a+[a(u^2-1)-g]x+gx^2+ay^2,$ $\dot{y}=b+[b(u^2-1)-l]x+lx^2+by^2$	$M_{1,2}(0,\pm i),\ M_{3,4}(1,\pm iu);$ $al-bg\neq 0$
$2p+q+r$	4	$\dot{x}=cx+cuy-cx^2+2hxy-cuy^2,$ $\dot{y}=ex+euy-ex^2+2mxy-euy^2$	$M_{1,2}(0,0), M_3(1,0),$ $M_4(0,1);\ u(cm-eh)\neq 0$
$2p+$ q^c+r^c	5	$\dot{x}=cm\,x+2cn\,y+g\,x^2-2cn\,xy$ $+(g+cm)\,y^2,\ \dot{y}=em\,x+2en\,y$ $+l\,x^2-2en\,xy+(l+em)\,y^2$	$M_{1,2}(0,0),\ M_{3,4}(1,\pm i);$ $(cl-eg)(m^2+n^2)\neq 0$
$2p+2q$	6	$\dot{x}=cx+cuy-cx^2+2cvxy+ky^2,$ $\dot{y}=ex+euy-ex^2+2evxy+ny^2$	$M_{1,2}(0,0),\ M_{3,4}(1,0);$ $cn-ek\neq 0$
$2p^c+2q^c$	7	$\dot{x}=a+aux+gx^2+2avxy+ay^2,$ $\dot{y}=b+bux+lx^2+2bvxy+by^2$	$M_{1,2}(0,i),\ M_{3,4}(0,-i);$ $al-bg\neq 0$
$3p+q$	8	$\dot{x}=cx+cuy-cx^2+2hxy$ $+(2hu+cv)y^2,\ \dot{y}=ex+euy$ $-ex^2+2mxy+(2mu+ev)y^2$	$M_{1,2,3}(0,0),\ M_4(1,0);$ $(u^2-v)(cm-eh)\neq 0$
$4p$	9a	$\dot{x}=y^2,\ \dot{y}=y+x^2+2mxy+ny^2$	$M_{1,2,3,4}(0,0)$-semi-elemen.
	9b	$\dot{x}=y+x^2+ky^2,\ \dot{y}=y^2$	$M_{1,2,3,4}(0,0)$-nilpotent
	9c	$\dot{x}=gx^2+2hxy,$ $\dot{y}=lx^2+2mxy+ny^2$	$M_{1,2,3,4}(0,0)$-intricate; $g^2n^2-4hn(gm-hl)\neq 0$

Table 6.1 *(continued)*

Type of $\mathcal{D}_{\mathbb{C}^2}(S)$	Canonical form		Finite singularities and conditions on parameters
$p+q+r$	10	$\dot{x}=cx+dy-cx^2+2hxy$, $\dot{y}=ex+fy-ex^2+2mxy$	$M_1(0,0)$, $M_2(1,0)$, $M_3\left(\frac{\beta}{2\alpha},\frac{\beta(2\alpha-\beta)}{4\alpha\gamma}\right)$, where $\alpha=eh-cm$, $\beta=cf-de$, $\gamma=dm-fh$; $\alpha\beta\gamma(2\alpha-\beta)\neq0$
$p+q^c+r^c$	11	$\dot{x}=2(h-gu)x+g(u^2+1)y$ $+gx^2-2hxy$, $\dot{y}=2(m-lu)x$ $+l(u^2+1)y+lx^2-2mxy$	$M_1(0,0)$, $M_{2,3}(u\pm i,1)$; $gm-hl\neq0$
$2p+q$	12	$\dot{x}=cx+cy-cx^2+2hxy$, $\dot{y}=ex+ey-ex^2+2mxy$	$M_{1,2}(0,0)$, $M_3(1,0)$; $cm-eh\neq0$
$3p$	13	$\dot{x}=gy+gx^2+2hxy$, $\dot{y}=ly+lx^2+2mxy$	$M_{1,2,3}(0,0)$; $gm-lh\neq0$
$p+q$	14a	$\dot{x}=cx+dy-cx^2+2duxy$, $\dot{y}=ex+fy-ex^2+2fuxy$	$M_1(0,0)$, $M_2(1,0)$; $(cf-de)(2u+1)u\neq0$
	14b	$\dot{x}=-(g+ku^2)x-2hy+gx^2$ $+2hxy+ky^2$, $\dot{y}=ux+y$	$M_1(0,0)$, $M_2(1,-u)$; $g-2hu+ku^2\neq0$
p^c+q^c	15a	$\dot{x}=a+hux+2hxy+ay^2$, $\dot{y}=b+mux+2mxy+by^2$	$M_{1,2}(0,\pm i)$; $am-bh\neq0$
	15b	$\dot{x}=a+cx+gx^2+2hxy+ay^2$, $\dot{y}=x$	$M_{1,2}(0,\pm i)$; $a\neq0$
$2p$	16a	$\dot{x}=dy+gx^2+2dxy$, $\dot{y}=fy+lx^2+2fxy$	$M_{1,2}(0,0)$; $fg-dl\neq0$
	16b	$\dot{x}=cx+dy$, $\dot{y}=lx^2+2mxy+ny^2$	$M_{1,2}(0,0)$; $c^2n-2cdm+d^2l\neq0$
p	17a	$\dot{x}=cx+dy+(2c+d)x^2+2dxy$, $\dot{y}=ex+fy+(2e+f)x^2+2fxy$	$M_1(0,0)$; $cf-de\neq0$
	17b	$\dot{x}=x+dy$, $\dot{y}=ex+fy$ $+lx^2+2mxy+(2dm-ld^2)y^2$	$M_1(0,0)$; $(de-f)(l^2+m^2)\neq0$
0	18a	$\dot{x}=h+gx^2+2hxy$, $\dot{y}=m+lx^2+2mxy$	$hl-gm\neq0$
	18b	$\dot{x}=y$, $\dot{y}=1+fy+2mxy+ny^2$	$m^2+n^2\neq0$
	18c	$\dot{x}=x$, $\dot{y}=1+ex+lx^2+2mxy$	$l^2+m^2\neq0$
	18d	$\dot{x}=1$, $\dot{y}=ex+fy$ $+lx^2+2mxy+ny^2$	$(e,f\in\{0,1\})$ $l^2+m^2+n^2\neq0$

Table 6.2

$\mathcal{D}_{\mathbb{C}^2}(S)$ $\Updownarrow$ Affine invariant condit.	Additional conditions for determining the characters of singular points		N	Type
$p+q+r+s$ $\Updownarrow$ $\mu_0 \neq 0$, $\mathbf{D}<0$, $\mathbf{R}>0$, $\mathbf{S}>0$	$\mu_0<0$, $\widetilde{K}<0$	$W_4 \geq 0$	1	s, s, s, n
		$W_4<0$	2	s, s, s, $(f \vee c)$
	$\mu_0>0$	$W_4>0, W_2>0, W_1W_3>0$	4	s, s, n, n
		$W_4=0,\ W_3 \geq 0$	5	
		$W_4<0$	6	s, s, n, f
		$W_4=0,\ W_3<0$	7	
		$W_4>0,\ W_2<0$	8	s, s, $(f \vee c)$, $(f \vee c)$
		$W_4>0, W_2 \geq 0, W_1W_3<0$	9	
	$\mu_0<0$, $\widetilde{K}>0$	$W_4>0, W_2>0, W_1W_3>0$	12	s, n, n, n
		$W_4=0,\ W_3>0$	13	
		$W_4=W_3=0,\ W_2>0$	14	
		$W_4<0$	15	s, n, n, $(f \vee c)$
		$W_4=0, W_3<0$	16	
		$W_4=W_3=0,\ W_2<0$	17	
		$W_4>0,\ W_2<0$	20	s, n, f, f
		$W_4>0, W_2 \geq 0, W_1W_3<0$	21	
$p+q+r^c+s^c$ $\Updownarrow$ $\mu_0 \neq 0$, $\mathbf{D}>0$	$\mu<0,\ \widetilde{K}<0$		22	s, s
	$\mu_0>0$	$W_4>0$	23	s, n
		$W_4=0,\ W_3 \neq 0$	24	
		$W_4=W_3=0,\ W_2 \geq 0$	25	
		$W_4<0$	26	s, $(f \vee c)$
		$W_4=W_3=0,\ W_2<0$	27	
	$\mu_0<0$, $\widetilde{K}>0$	$W_4>0, V_1>0, W_1>0$	30	n, n
		$W_4>0, V_1<0, V_2>0$	31	
		$W_4>0, V_1=0, V_2V_3 \neq 0, W_1>0$	32	
		$W_4>0, V_1=V_3=0, V_2>0$	33	
		$W_4=0,\ W_3>0$	34	
		$W_4=W_3=0, W_2>0,\ V_4<0$	35	
		$W_4=W_3=0, W_2 \geq 0,$ $V_4>0, W_1>0$	36	

Table 6.2 (continued)

$\mathcal{D}_{\mathbb{C}^2}(S)$ $\Updownarrow$ Affine invariant condit.	Additional conditions for determining the characters of singular points			N	Type
$p+q+r^c+s^c$ $\Updownarrow$ $\mu_0 \neq 0,$ $\mathbf{D} > 0$	$\mu_0 < 0,$ $\widetilde{K} > 0$		$H_1 \neq 0,\ W_{10} > 0,\ (\mathfrak{D}_1)$	37	n, n
			$H_1 = 0,\ H_4 H_5 > 0,\ (\mathfrak{D}_1)$	38	
			$H_1 = H_4 = 0,\ H_2 > 0,\ (\mathfrak{D}_1)$	39	
			$H_1 \neq 0,\ W_1 > 0,\ W_{10} \geq 0,\ (\mathfrak{D}_2)$	40	
			$H_1 \neq 0,\ W_1 W_{10} < 0,\ (\mathfrak{D}_2)$	41	
			$H_1 = 0, H_4 H_5 > 0, W_1 > 0,\ (\mathfrak{D}_2)$	42	
			$H_1 = 0, H_4 H_5 W_1 < 0,\ (\mathfrak{D}_2)$	43	
			$W_1 = 0,\ (\mathfrak{D}_2)$	44	
			$W_4 < 0$	45	$n, (f \vee c)$
			$W_4 = 0,\ W_3 < 0$	46	
			$W_4 = W_3 = 0,\ W_2 < 0$	47	
			$W_4 = W_3 = W_2 = 0,\ W_1 < 0$	48	
			$W_4 > 0, V_1 > 0, W_1 < 0$	56	$(f \vee c), (f \vee c)$
			$W_4 > 0, V_1 < 0, V_2 < 0$	57	
			$W_4 > 0,\ V_1 = 0,\ V_3 \neq 0,$ $W_1 < 0,\ \neg(\mathfrak{D}_1)$	58	
			$W_4 > 0, V_1 = V_3 = 0, V_2 < 0$	59	
			$W_4 = W_3 = 0,\ W_2 > 0,$ $V_4 > 0,\ W_1 < 0$	60	
			$H_1 \neq 0,\ W_{10} \leq 0,\ (\mathfrak{D}_1)$	61	
			$H_1 = 0,\ H_4 H_5 < 0,\ (\mathfrak{D}_1)$	62	
			$H_1 = H_4 = 0,\ H_2 < 0,\ (\mathfrak{D}_1)$	63	
			$H_1 \neq 0, W_1 < 0, W_{10} < 0, (\mathfrak{D}_2)$	64	
			$H_1 = 0, H_4 H_5 < 0, W_1 < 0,\ (\mathfrak{D}_2)$	65	
$p^c + q^c + r^c + s^c \ \Leftrightarrow\ \mu_0 \neq 0,\ \mathbf{D} < 0\ \&\ (\mathbf{R} \leq 0$ or $\mathbf{S} \leq 0)$					—
$2p+q+r$ $\Updownarrow$ $\mu_0 \neq 0,$ $\mathbf{D} = 0,$ $\mathbf{T} < 0$	$\mu_0 < 0,$ $\widetilde{K} > 0$	$E_1 \neq 0$	$W_4 > 0, W_2 > 0, W_1 W_3 > 0$	66	$n, n, \overline{sn}_{(2)}$
			$W_4 = 0,\ W_3 \geq 0$	67	
			$W_4 < 0$	68	$n, (f \vee c),$ $\overline{sn}_{(2)}$
			$W_4 = 0,\ W_3 < 0$	69	
			$W_4 > 0, W_2 \geq 0, W_1 W_3 < 0$	71	$f, f, \overline{sn}_{(2)}$
			$W_4 > 0,\ W_2 < 0$	72	

Table 6.2 (continued)

$\mathcal{D}_{\mathbb{C}^2}(S)$ $\Updownarrow$ Affine invariant condit.	Additional conditions for determining the characters of singular points			N	Type
$2p+q+r$ $\Updownarrow$ $\mu_0 \neq 0$, $\mathbf{D}=0$, $\mathbf{T}<0$	$\mu_0<0$, $\widetilde{K}>0$	$E_1=0$	$W_2 \geq 0$	*73*	$\text{n, n}, \widehat{cp}_{(2)}$
			$W_2<0$	*74*	$\text{n, f}, \widehat{cp}_{(2)}$
	$\mu_0>0$	$E_1 \neq 0$	$W_4 \geq 0$	*75*	$\text{s, n}, \overline{sn}_{(2)}$
			$W_4<0$	*76*	$\text{s, f}, \overline{sn}_{(2)}$
		$E_1=0$	$W_2 \geq 0$	*77*	$\text{s, n}, \widehat{cp}_{(2)}$
			$W_2<0$	*78*	$\text{s}, (f \vee c), \widehat{cp}_{(2)}$
	$\mu_0<0$, $\widetilde{K}<0$	$E_1 \neq 0$		*80*	$\text{s, s}, \overline{sn}_{(2)}$
		$E_1=0$		*81*	$\text{s, s}, \widehat{cp}_{(2)}$
$2p+q^c+r^c$ $\Leftrightarrow$ $\mu_0 \neq 0$, $\mathbf{D}=0, \mathbf{T}>0$	$E_1 \neq 0$			*82*	$\overline{sn}_{(2)}$
	$E_1=0$			*83*	$\widehat{cp}_{(2)}$
$2p+2q$ $\Leftrightarrow$ $\mu_0 \neq 0$, $\mathbf{D}=\mathbf{T}=0$, $\mathbf{PR}>0$	$G_1 \neq 0$			*84*	$\overline{sn}_{(2)}, \overline{sn}_{(2)}$
	$G_1=0,\ E_2 \neq 0$			*85*	$\overline{sn}_{(2)}, \widehat{cp}_{(2)}$
	$G_1=0,\ E_2=0$			*86*	$\widehat{cp}_{(2)}, \widehat{cp}_{(2)}$
$2p^c+2q^c \Leftrightarrow \mu_0 \neq 0, \mathbf{D}=\mathbf{T}=0, \mathbf{PR}<0$					—
$3p+q$ $\Updownarrow$ $\mu_0 \neq 0$, $\mathbf{D}=0$, $\mathbf{T}=\mathbf{P}=0$, $\mathbf{R} \neq 0$	$\mu_0>0$	$E_3<0$	$G_{10} \neq 0$	*87*	$\overline{n}_{(3)},\ s$
			$G_{10}=0$	*88*	$\widehat{es}_{(3)},\ s$
		$E_3>0$	$G_{10} \neq 0, W_4 \geq 0$	*89*	$\overline{s}_{(3)},\ n$
			$G_{10}=0, W_1 \geq 0$	*90*	$\widehat{s}_{(3)},\ n$
			$G_{10} \neq 0, W_4<0$	*91*	$\overline{s}_{(3)}, (f \vee c)$
			$G_{10}=0, W_1<0$	*92*	$\widehat{s}_{(3)}, (f \vee c)$
	$\mu_0<0$	$E_3>0$		*94*	$(\overline{s}_{(3)} \vee \widehat{s}_{(3)}),\ s$
		$E_3<0$	$G_{10} \neq 0, W_4 \geq 0$	*95*	$\overline{n}_{(3)},\ n$
			$G_{10} \neq 0, W_4<0$	*96*	$\overline{n}_{(3)}, f$
			$G_{10}=0,\ W_1 \geq 0$	*97*	$\widehat{es}_{(3)},\ n$
			$G_{10}=0,\ W_1<0$	*98*	$\widehat{es}_{(3)}, (f \vee c)$
$4p$ $\Updownarrow$ $\mu_0 \neq 0$, $\mathbf{D}=\mathbf{T}=0$, $\mathbf{P}=\mathbf{R}=0$	$F_1 \neq 0$			*100*	$\overline{sn}_{(4)} \vee \widehat{sn}_{(4)}$
	$F_1=0$	$\mu_0>0$	$\eta>0$	*101*	$phpphp_{(4)}$
			$\eta=0, \widetilde{M} \neq 0$	*102*	$phph_{(4)}$
			$\eta=\widetilde{M}=0$	*103*	$hh_{(4)}$
			$\eta<0$	*104*	

Table 6.2 (continued)

$\mathcal{D}_{\mathbb{C}^2}(S)$ $\Updownarrow$ Affine invariant condit.	Additional conditions for determining the characters of singular points			N	Type
$4p$ $\Updownarrow$ $\mu_0 \neq 0$, $\mathbf{D}=\mathbf{T}=0$, $\mathbf{P}=\mathbf{R}=0$	$F_1 = 0$	$\mu_0 < 0$	$\eta > 0, \kappa > 0$	*105*	$hhhhhh_{(4)}$
			$\eta > 0, \kappa < 0$	*106*	$peppep_{(4)}$
			$\eta = 0, \widetilde{M} \neq 0$	*107*	$pepe_{(4)}$
			$\eta = \widetilde{M} = 0$	*108*	$ee_{(4)}$
			$\eta < 0$	*109*	
$p+q+r$ $\Updownarrow$ $\mu_0 = 0$, $\mathbf{D} < 0$, $\mathbf{R} \neq 0$	$\widetilde{K} < 0$	$G_9 \neq 0,\ W_4 \geq 0$		*110*	s, s, n
		$G_9 = 0,\ W_7 \geq 0$		*111*	
		$G_9 \neq 0,\ W_4 < 0$		*112*	$s, s, (f \vee c)$
		$G_9 = 0,\ W_7 < 0$		*113*	
	$\widetilde{K} > 0$	$G_9 \neq 0$	$W_4 > 0, W_3 > 0, W_2 > 0$	*115*	s, n, n
			$W_4 = 0, W_3 \geq 0$	*116*	
		$G_9 = 0$	$W_7 \geq 0$	*117*	
		$G_9 \neq 0$	$W_4 < 0$	*118*	s, n, f
			$W_4 = 0, W_3 < 0$	*119*	
		$G_9 = 0$	$W_7 < 0$	*120*	
		$G_9 \neq 0$	$W_4 > 0, W_3 < 0$	*121*	s, f, f
			$W_4 > 0, W_3 \geq 0, W_2 < 0$	*122*	
$p+q^c+r^c$ $\Updownarrow$ $\mu_0 = 0$, $\mathbf{D} > 0$, $\mathbf{R} \neq 0$	$\widetilde{K} < 0$			*123*	s
	$\widetilde{K} > 0$	$G_9 \neq 0$	$W_4 > 0$	*124*	n
			$W_4 = 0,\ W_3 \neq 0$	*125*	
			$W_4 = 0,\ W_3 = 0,\ W_2 \geq 0$	*126*	
		$G_9 = 0$	$W_7 > 0$	*127*	
			$W_7 = 0,\ W_6 \neq 0$	*128*	
			$W_7 = 0, W_6 = 0, W_5 \geq 0$	*129*	
		$G_9 \neq 0$	$W_4 < 0$	*130*	$(f \vee c)$
			$W_4 = W_3 = 0,\ W_2 < 0$	*131*	
		$G_9 = 0$	$W_7 < 0$	*132*	
			$W_7 = W_6 = 0,\ W_5 < 0$	*133*	

Table 6.2 (continued)

$\mathcal{D}_{\mathbb{C}^2}(S)$ $\Updownarrow$ Affine invariant condit.	Additional conditions for determining the characters of singular points			N	Type
$2p+q$ $\Updownarrow$ $\mu_0=\mathbf{D}=0$, $\mathbf{PR}\neq 0$	$\widetilde{K}<0$	$E_1\neq 0$		*136*	$\overline{sn}_{(2)}, s$
		$E_1=0$		*137*	$\widehat{cp}_{(2)}, s$
	$\widetilde{K}>0$	$E_1\neq 0$	$G_9\neq 0,\ W_4\geq 0$	*138*	$\overline{sn}_{(2)}, n$
			$G_9=0,\ W_7\geq 0$	*139*	
			$G_9\neq 0,\ W_4<0$	*140*	$\overline{sn}_{(2)}, f$
			$G_9=0,\ W_7<0$	*141*	
		$E_1=0$	$G_9\neq 0,\ W_2\geq 0$	*142*	$\widehat{cp}_{(2)}, n$
			$G_9=0$	*143*	
			$G_9\neq 0,\ W_2<0$	*144*	$\widehat{cp}_{(2)}, f$
$3p \Leftrightarrow$ $\mu_0=\mathbf{D}=0$, $\mathbf{P}=0, \mathbf{R}\neq 0$	$\widetilde{K}<0$			*145*	$\overline{s}_{(3)} \vee \widehat{s}_{(3)}$
	$\widetilde{K}>0$	$G_{10}\neq 0$		*146*	$\overline{n}_{(3)}$
		$G_{10}=0$		*147*	$\widehat{es}_{(3)}$
$p+q$ $\Updownarrow$ $\mu_0=\mathbf{R}=0$, $\mathbf{P}\neq 0$, $\mathbf{U}>0$	$\widetilde{K}<0,\ F_2<0$			*148*	s, s
	$\widetilde{K}\neq 0$	$G_8>0, W_4\geq 0$		*149*	s, n
		$G_8=0,\ F_2>0$		*150*	
	$\widetilde{K}=0$	$\mathbf{D}\neq 0$	$G_8\neq 0,\ W_4\geq 0$	*151*	
			$G_8=0,\ W_7\geq 0$	*152*	
		$\mathbf{D}=0$	$G_3\neq 0,\ W_8\geq 0$	*153*	
			$G_3=0,\ D_2\neq 0$	*154*	
			$G_3=D_2=0,\ W_9\geq 0$	*155*	
	$\widetilde{K}\neq 0$	$G_8>0,\ W_4<0$		*156*	s, $(f\vee c)$
	$\widetilde{K}=0$	$\mathbf{D}\neq 0$	$G_8\neq 0, W_4<0$	*157*	
			$G_8=0,\ W_7<0$	*158*	
		$\mathbf{D}=0$	$G_3\neq 0,\ W_8<0$	*159*	
			$G_3=D_2=0,\ W_9<0$	*160*	
	$\widetilde{K}>0$	$G_8=0,\ F_2<0$		*164*	n, n

Table 6.2 *(continued)*

$\mathcal{D}_{\mathbb{C}^2}(S)$ $\Updownarrow$ Affine invariant condit.	Additional conditions for determining the characters of singular points			N	*Type*
$p+q$ $\Updownarrow$ $\mu_0=\mathbf{R}=0,$ $\mathbf{P}\neq 0,$ $\mathbf{U}>0$	$\widetilde{K}>0$	$G_8<0$	$W_4>0,\ W_3>0$	*165*	n, n
			$W_4=0,\ W_3\geq 0$	*166*	
			$W_4<0$	*167*	n, f
			$W_4=0,\ W_3<0$	*168*	
			$W_4>0,\ W_3<0$	*169*	$(f\vee c)$,$(f\vee c)$
$p^c+q^c \ \Leftrightarrow\ \mu_0=\mathbf{R}=0,\ \mathbf{P}\neq 0,\ \mathbf{U}<0$					—
$2p$ $\Updownarrow$ $\mu_0=\mathbf{R}=0,$ $\mathbf{P}\neq 0,$ $\mathbf{U}=0$	$G_8\neq 0,\quad G_1\neq 0$			*171*	$\overline{sn}_{(2)}$
	$G_8=0$	$\widetilde{K}\neq 0$		*172*	
		$\widetilde{K}=0, \widetilde{N}\neq 0,\ G_{10}\neq 0$		*173*	
		$\widetilde{K}=\widetilde{N}=0,\ F_3\neq 0$		*174*	
	$G_8\neq 0,\quad G_1=0$			*175*	$\widehat{cp}_{(2)}$
	$G_8=0$	$\widetilde{K}=G_{10}=0, \widetilde{N}\neq 0$		*176*	
		$\widetilde{K}=\widetilde{N}=F_3=0$		*177*	
p $\Updownarrow$ $\mu_0=\mathbf{R}=0,$ $\mathbf{P}=0,$ $\mathbf{U}\neq 0$	$\widetilde{K}<0$			*178*	s
	$\widetilde{K}=0,\ F_4F_5<0$			*179*	
	$\widetilde{K}>0,\ W_4\geq 0$			*180*	n
	$\widetilde{K}=0,$ $F_4F_5>0$	$G_3\neq 0,\ W_4\geq 0$		*181*	
		$\widetilde{N}\neq 0,\ G_3=0,\ W_7\geq 0$		*182*	
		$\widetilde{N}=0,\ W_8\geq 0$		*183*	
	$\widetilde{K}>0,\ W_4<0$			*184*	$(f\vee c)$
	$\widetilde{K}=0,$ $F_4F_5>0$	$G_3\neq 0,\ W_4<0$		*185*	
		$\widetilde{N}\neq 0,\ G_3=0,\ W_7<0,\ G_{10}\neq 0$		*186*	
		$\widetilde{N}=0,\ W_8<0,\ G_{13}\neq 0$		*187*	
$0 \ \Leftrightarrow\ \begin{cases}\mu_0=\mathbf{R}=\mathbf{P}=0,\\ \mathbf{U}=0, \mathbf{V}\neq 0\end{cases}$	There are no finite singular points				
$\mathcal{D}_{\mathbb{C}^2}(S)$ undefined $\Leftrightarrow \begin{cases}\mu_0=\mathbf{R}=\mathbf{P}=0,\\ \mathbf{U}=\mathbf{V}=0\end{cases}$	Degenerate systems				

- In several of our articles we did not use everywhere the same notation for invariant polynomials. In this book we have unified the notation in Section 5.4. In particular the polynomials $K(x,y)$, $H(x,y)$ and $N(x,y)$ in Table 1 of [29] are renamed in other papers $\widetilde{K}(x,y)$, $\widetilde{H}(x,y)$ and $\widetilde{N}(x,y)$, respectively. Here we only use these invariant polynomials with tilde; see Table 6.2.

6.2 Finite weak singularities

Necessary and sufficient conditions for a real quadratic system to possess a given number of weak singularities of specific orders have been determined [338]. The geometrical meaning of the invariant polynomials $\mathcal{T}_i$ for $i \in \{1,2,3,4\}$ appears in Lemma 5.3. We present here the main result proved in [338].

Theorem 6.2. *Consider a non-degenerate quadratic system. The symbols* $(\mathcal{T}_1, \ldots, \mathbf{P}, \mathbf{R})$ *denote invariant polynomials defined in Section* 5.4.

(a) *If* $\mathcal{T}_4 \neq 0$, *then this system has no weak singularities.*

(b) *If* $\mathcal{T}_4 = 0$ *and* $\mathcal{T}_3 \neq 0$, *then the system has exactly one weak singularity. Moreover this singularity is either a weak focus (respectively a weak saddle) of the indicated order below, or a center (respectively an integrable saddle) if and only if* $\mathcal{T}_3\mathcal{F} < 0$ *(respectively* $\mathcal{T}_3\mathcal{F} > 0$*) and the following corresponding conditions hold:*

$$
\begin{array}{llll}
(\mathrm{b}_1) & f^{(1)}\ \big(\textit{respectively } s^{(1)}\big) & \Leftrightarrow & \mathcal{F}_1 \neq 0,\\
(\mathrm{b}_2) & f^{(2)}\ \big(\textit{respectively } s^{(2)}\big) & \Leftrightarrow & \mathcal{F}_1 = 0,\ \mathcal{F}_2 \neq 0,\\
(\mathrm{b}_3) & f^{(3)}\ \big(\textit{respectively } s^{(3)}\big) & \Leftrightarrow & \mathcal{F}_1 = \mathcal{F}_2 = 0,\ \mathcal{F}_3\mathcal{F}_4 \neq 0,\\
(\mathrm{b}_4) & c\quad \big(\textit{respectively } \$\big) & \Leftrightarrow & \mathcal{F}_1 = \mathcal{F}_2 = \mathcal{F}_3\mathcal{F}_4 = 0.
\end{array}
$$

(c) *If* $\mathcal{T}_4 = \mathcal{T}_3 = 0$ *and* $\mathcal{T}_2 \neq 0$, *then the system could possess two and only two weak singularities and none of them is of order two or three. Moreover this system possesses two weak singularities, which are of the types indicated below, if and only if* $\mathcal{F} = 0$ *and one of the following conditions holds:*

$$
\begin{array}{llcl}
(\mathrm{c}_1) & s^{(1)},\ s^{(1)} & \Leftrightarrow & \mathcal{F}_1 \neq 0,\ \mathcal{T}_2 < 0,\ \mathcal{B} \leq 0,\ \mathcal{H} > 0,\\
(\mathrm{c}_2) & s^{(1)},\ f^{(1)} & \Leftrightarrow & \mathcal{F}_1 \neq 0,\ \mathcal{T}_2 > 0,\ \mathcal{B} < 0,\\
(\mathrm{c}_3) & f^{(1)},\ f^{(1)} & \Leftrightarrow & \mathcal{F}_1 \neq 0,\ \mathcal{T}_2 < 0,\ \mathcal{B} < 0,\ \mathcal{H} < 0,\\
(\mathrm{c}_4) & \$,\quad \$ & \Leftrightarrow & \mathcal{F}_1 = 0,\ \mathcal{T}_2 < 0,\ \mathcal{B} < 0,\ \mathcal{H} > 0,\\
(\mathrm{c}_5) & \$,\quad c & \Leftrightarrow & \mathcal{F}_1 = 0,\ \mathcal{T}_2 > 0,\ \mathcal{B} < 0,\\
(\mathrm{c}_6) & c,\quad c & \Leftrightarrow & \mathcal{F}_1 = 0,\ \mathcal{T}_2 < 0,\ \mathcal{B} < 0,\ \mathcal{H} < 0.
\end{array}
$$

(d) *If* $\mathcal{T}_4 = \mathcal{T}_3 = \mathcal{T}_2 = 0$ *and* $\mathcal{T}_1 \neq 0$, *then the system could possess one and only one weak singularity (which is of order one). Moreover this system has one weak singularity of the type indicated below if and only if* $\mathcal{F} = 0$ *and one of the following conditions holds:*

(d$_1$) $s^{(1)}$ $\Leftrightarrow$ $\mathcal{F}_1 \neq 0,\ \mathcal{B} < 0,\ \mathcal{H} > 0,$
(d$_2$) $f^{(1)}$ $\Leftrightarrow$ $\mathcal{F}_1 \neq 0,\ \mathcal{B} < 0,\ \mathcal{H} < 0.$

(e) *If* $\mathcal{T}_4 = \mathcal{T}_3 = \mathcal{T}_2 = \mathcal{T}_1 = 0$ *and* $\sigma(a, x, y) \neq 0$, *then the system could possess one and only one weak singularity. Moreover this system has one weak singularity, which is of the type indicated below, if and only if one of the following conditions holds:*

(e$_1$) $s^{(1)}$ $\Leftrightarrow$ $\mathcal{F}_1 \neq 0,\ \mathcal{H} = \mathcal{B}_1 = 0,\ \mathcal{B}_2 > 0,$
(e$_2$) $f^{(1)}$ $\Leftrightarrow$ $\mathcal{F}_1 \neq 0,\ \mathcal{H} = \mathcal{B}_1 = 0,\ \mathcal{B}_2 < 0,$

$$(\mathrm{e}_3)\quad \$ \ \Leftrightarrow \begin{cases} [\alpha]\ \mathcal{F}_1 = 0,\ \mathcal{F} = 0,\ \mathcal{B} < 0,\ \mathcal{H} > 0,\ \textit{or} \\ [\beta]\ \mathcal{F}_1 = 0,\ \mathcal{H} = \mathcal{B}_1 = 0,\ \mathcal{B}_2 > 0,\ \textit{or} \\ [\gamma]\ \mathcal{F}_1 = 0,\ \mathcal{H} = \mathcal{B} = \mathcal{B}_1 = \mathcal{B}_2 = \mathcal{B}_3 = \mu_0 = 0,\ \widetilde{K}(\mu_2^2 + \mu_3^2) \neq 0,\ \textit{or} \\ [\delta]\ \mathcal{F}_1 = 0,\ \mathcal{H} = \mathcal{B} = \mathcal{B}_1 = \mathcal{B}_2 = \mathcal{B}_3 = \widetilde{K} = 0,\ \mu_2 \mathcal{G} \neq 0,\ \textit{or} \\ [\varepsilon]\ \mathcal{F}_1 = 0,\ \mathcal{H} = \mathcal{B} = \mathcal{B}_1 = \mathcal{B}_2 = \mathcal{B}_3 = \mathcal{B}_4 = \widetilde{K} = \mu_2 = 0,\ \mu_3 \neq 0, \end{cases}$$

$$(\mathrm{e}_4)\quad c \ \Leftrightarrow \begin{cases} [\alpha]\ \mathcal{F}_1 = 0,\ \mathcal{F} = 0,\ \mathcal{B} < 0,\ \mathcal{H} < 0,\ \textit{or} \\ [\beta]\ \mathcal{F}_1 = 0,\ \mathcal{H} = \mathcal{B}_1 = 0,\ \mathcal{B}_2 < 0. \end{cases}$$

(f) *If* $\sigma(a, x, y) = 0$ *(this implies* $\mathcal{T}_4 = \mathcal{T}_3 = \mathcal{T}_2 = \mathcal{T}_1 = 0$*), then the system is Hamiltonian and it possesses* i $(1 \leq i \leq 4)$ *weak singular points of the types indicated below if and only if one of the following conditions holds:*

(f$_1$) $\$, \$, \$, c$ $\Leftrightarrow$ $\mu_0 < 0,\ \mathbf{D} < 0,\ \mathbf{R} > 0,\ \mathbf{S} > 0,$
(f$_2$) $\$, \$, c, c$ $\Leftrightarrow$ $\mu_0 > 0,\ \mathbf{D} < 0,\ \mathbf{R} > 0,\ \mathbf{S} > 0,$
(f$_3$) $\$, \$, c$ $\Leftrightarrow$ $\mu_0 = 0,\ \mathbf{D} < 0,\ \mathbf{R} \neq 0,$

$$(\mathrm{f}_4)\quad \$, \$ \ \Leftrightarrow \begin{cases} [\alpha]\ \mu_0 < 0,\ \mathbf{D} > 0,\ \textit{or} \\ [\beta]\ \mu_0 < 0,\ \mathbf{D} = 0,\ \mathbf{T} < 0,\ \textit{or} \\ [\gamma]\ \mu_0 = \mathbf{R} = 0,\ \mathbf{P} \neq 0,\ \mathbf{U} > 0,\ \widetilde{K} \neq 0, \end{cases}$$

$$(\mathrm{f}_5)\quad \$, c \ \Leftrightarrow \begin{cases} [\alpha]\ \mu_0 > 0,\ \mathbf{D} > 0,\ \textit{or} \\ [\beta]\ \mu_0 > 0,\ \mathbf{D} = 0,\ \mathbf{T} < 0,\ \textit{or} \\ [\gamma]\ \mu_0 = \mathbf{R} = 0,\ \mathbf{P} \neq 0,\ \mathbf{U} > 0,\ \widetilde{K} = 0, \end{cases}$$

$$(\mathrm{f}_6)\quad \$ \ \Leftrightarrow \begin{cases} [\alpha]\ \mu_0 < 0,\ \mathbf{D} = \mathbf{T} = \mathbf{P} = 0,\ \mathbf{R} \neq 0,\ \textit{or} \\ [\beta]\ \mu_0 = 0,\ \mathbf{D} > 0,\ \mathbf{R} \neq 0,\ \textit{or} \\ [\gamma]\ \mu_0 = 0,\ \mathbf{D} = 0,\ \mathbf{PR} \neq 0,\ \textit{or} \\ [\delta]\ \mu_0 = \mathbf{R} = \mathbf{P} = 0,\ \mathbf{U} \neq 0, \end{cases}$$

(f$_7$) c $\Leftrightarrow$ $\mu_0 > 0,\ \mathbf{D} = \mathbf{T} = \mathbf{P} = 0,\ \mathbf{R} \neq 0.$

6.3 Singularities of QS with an integrable saddle

In this section we present the affine invariant necessary and sufficient conditions for anyone of the topological configurations of singularities that contain at least one finite integrable saddle. These conditions were obtained in [32, Table 1]. We note that there is an error in line 8 of that table. Instead of configuration $c,\ s;\ N,\ \overline{\binom{1}{1}}SN,\ \overline{\binom{1}{1}}SN$, we must have $\$,\ n;\ N,\ \overline{\binom{1}{1}}SN,\ \overline{\binom{1}{1}}SN$.

We underline that many of the configurations provided by the next theorem will be improved so as to contain more geometrical information.

Theorem 6.3 ([32]). *Assume that a non-degenerate quadratic differential system possesses an integrable saddle, i.e. one of the corresponding sets of conditions provided by Theorem 6.2 is satisfied. Then all the types of its possible combinations with other singularities (finite and infinite) are given by one of the corresponding set of conditions indicated below on the right:*

($\mathcal{A}$) *In the case of* $\mathcal{T}_3 \neq 0$, $\mathcal{T}_4 = \mathcal{F}_1 = \mathcal{F}_2 = \mathcal{F}_3\mathcal{F}_4 = 0$ *and* $\mathcal{T}_3\mathcal{F} > 0$,

$$
\begin{array}{lll}
(\mathcal{A}_1) & \$, n, n, n;\ S, S, N & \Leftrightarrow\ \mu_0 < 0; \\
(\mathcal{A}_2) & \$, s, n, n;\ S, N, N & \Leftrightarrow\ \mu_0 > 0,\ \mathbf{D} < 0; \\
(\mathcal{A}_3) & \$, f;\ N & \Leftrightarrow\ \mu_0 > 0,\ \mathbf{D} > 0,\ W_4 < 0; \\
(\mathcal{A}_4) & \$, n;\ N & \Leftrightarrow\ \mu_0 > 0,\ \mathbf{D} > 0,\ W_4 \geq 0; \\
(\mathcal{A}_5) & \$,\ n,\ \overline{sn}_{(2)};\ N,\ \overline{\binom{0}{2}}SN & \Leftrightarrow\ \mu_0 > 0,\ \mathbf{D} = 0; \\
(\mathcal{A}_6) & \$,\ f;\ N & \Leftrightarrow\ \mu_0 = 0,\ \eta < 0,\ W_4 \neq 0; \\
(\mathcal{A}_7) & \$, n;\ N & \Leftrightarrow\ \mu_0 = 0,\ \eta < 0,\ W_4 = 0; \\
(\mathcal{A}_8) & \$,\ n;\ N,\ \overline{\binom{1}{1}}SN,\ \overline{\binom{1}{1}}SN & \Leftrightarrow\ \mu_0 = 0,\ \eta > 0.
\end{array}
$$

($\mathcal{B}$) *In the case of* $\mathcal{T}_4 = \mathcal{T}_3 = \mathcal{F} = \mathcal{F}_1 = 0$, $\mathcal{B} < 0$ *and* $(\mathcal{T}_2 > 0) \cup (\mathcal{T}_2 < 0, \mathcal{H} > 0)$,

$$
\begin{array}{lll}
(\mathcal{B}_1) & \$, c, n, n;\ S & \Leftrightarrow\ \mu_0 < 0,\ \mathbf{D} < 0,\ \eta < 0; \\
(\mathcal{B}_2) & \$, c, s, s;\ N, N, N & \Leftrightarrow\ \mu_0 < 0,\ \mathbf{D} < 0,\ \eta > 0,\ \widetilde{K} < 0; \\
(\mathcal{B}_3) & \$, c, n, n;\ S, S, N & \Leftrightarrow\ \mu_0 < 0,\ \mathbf{D} < 0,\ \eta > 0,\ \widetilde{K} > 0; \\
(\mathcal{B}_4) & \$, c, n, n;\ \overline{\binom{0}{3}}S & \Leftrightarrow\ \mu_0 < 0,\ \mathbf{D} < 0,\ \eta = 0; \\
(\mathcal{B}_5) & \$, \$;\ N, N, N & \Leftrightarrow\ \mu_0 < 0,\ \mathbf{D} > 0; \\
(\mathcal{B}_6) & \$, \$, n, n;\ N & \Leftrightarrow\ \mu_0 > 0,\ \mathbf{D} < 0,\ \eta < 0; \\
(\mathcal{B}_7) & \$, \$, n, n;\ S, N, N & \Leftrightarrow\ \mu_0 > 0,\ \mathbf{D} < 0,\ \eta > 0; \\
(\mathcal{B}_8) & \$, \$, n, n; \overline{\binom{0}{3}}N & \Leftrightarrow\ \mu_0 > 0,\ \mathbf{D} < 0,\ \eta = 0; \\
(\mathcal{B}_9) & \$, c;\ N & \Leftrightarrow\ \mu_0 > 0,\ \mathbf{D} > 0,\ \eta < 0; \\
(\mathcal{B}_{10}) & \$, c;\ S, N, N & \Leftrightarrow\ \mu_0 > 0,\ \mathbf{D} > 0,\ \eta > 0; \\
(\mathcal{B}_{11}) & \$, c;\ \overline{\binom{0}{3}}N & \Leftrightarrow\ \mu_0 > 0,\ \mathbf{D} > 0,\ \eta = 0; \\
(\mathcal{B}_{12}) & \$, c;\ \overline{\binom{2}{1}}N & \Leftrightarrow\ \mu_0 = 0,\ \eta < 0,\ \widetilde{K} \neq 0; \\
(\mathcal{B}_{13}) & \$, c;\ \overline{N} & \Leftrightarrow\ \mu_0 = 0,\ \eta < 0,\ \widetilde{K} = 0; \\
(\mathcal{B}_{14}) & \$, c;\ \overline{\binom{2}{1}}S, N, N & \Leftrightarrow\ \mu_0 = 0,\ \eta > 0,\ \widetilde{K} \neq 0,\ W_4 < 0; \\
(\mathcal{B}_{15}) & \$, \$;\ \overline{\binom{2}{1}N, N, N} & \Leftrightarrow\ \mu_0 = 0,\ \eta > 0,\ \widetilde{K} \neq 0,\ W_4 > 0; \\
(\mathcal{B}_{16}) & \$, c;\ N, \overline{\binom{1}{1}}SN,\ \overline{\binom{1}{1}}SN & \Leftrightarrow\ \mu_0 = 0,\ \eta > 0,\ \widetilde{K} = 0.
\end{array}
$$

$(\mathcal{C})$ *In the case of* $\mathcal{T}_4 = \mathcal{T}_3 = \mathcal{T}_2 = 0$, $\sigma \neq 0$ *and the conditions* (e_3) *of Theorem* 6.2,

$(\mathcal{C}_1)$ $\$, \bar{s}_{(3)}; N, N, N \Leftrightarrow \mu_0 < 0;$
$(\mathcal{C}_2)$ $\$, \widehat{es}_{(3)}; N \Leftrightarrow \mu_0 > 0,\ \eta < 0;$
$(\mathcal{C}_3)$ $\$, \widehat{es}_{(3)};\ \underline{S}, N, N \Leftrightarrow \mu_0 > 0,\ \eta > 0;$
$(\mathcal{C}_4)$ $\$, \widehat{es}_{(3)};\ \overline{\binom{0}{3}} N \Leftrightarrow \mu_0 > 0,\ \eta = 0;$
$(\mathcal{C}_5)$ $\$, n, n;\ S,\ \widehat{\binom{1}{2}}\ PEP{-}H \Leftrightarrow \begin{cases}\mu_0 = 0,\ \mu_1 \neq 0,\ \mathbf{D} < 0,\\ C_2 \neq 0,\ \tilde{L} < 0,\ \widetilde{N} \leq 0;\end{cases}$
$(\mathcal{C}_6)$ $\$, n, n;\ S,\ \widehat{\binom{1}{2}} PEP{-}PHP \Leftrightarrow \begin{cases}\mu_0 = 0,\ \mu_1 \neq 0,\ \mathbf{D} < 0,\\ C_2 \neq 0,\ \tilde{L} < 0,\ \widetilde{N} > 0;\end{cases}$
$(\mathcal{C}_7)$ $\$, n, n;\ N,\ \widehat{\binom{1}{2}}\ HHH - H \Leftrightarrow \begin{cases}\mu_0 = 0,\ \mu_1 \neq 0,\ \mathbf{D} < 0,\\ C_2 \neq 0,\ \tilde{L} > 0;\end{cases}$
$(\mathcal{C}_8)$ $\$, n, n;\ \underline{[\infty,\ \emptyset]} \Leftrightarrow \mu_0 = 0,\ \mu_1 \neq 0,\ \mathbf{D} < 0,\ C_2 = 0;$
$(\mathcal{C}_9)$ $\$;\ N,\ \widehat{\binom{1}{2}}\ PEP - H \Leftrightarrow \mu_0 = 0,\ \mu_1 \neq 0,\ \mathbf{D} > 0;$
$(\mathcal{C}_{10})$ $\$, s;\ N, \binom{2}{2} PEP - PEP \Leftrightarrow \mu_0 = \mu_1 = 0,\ \mu_2 < 0;$
$(\mathcal{C}_{11})$ $\$, n;\ S, \binom{2}{2}\ PPE - PPE \Leftrightarrow \mu_0 = \mu_1 = 0,\ \mu_2 > 0,\ \tilde{L} < 0;$
$(\mathcal{C}_{12})$ $\$, n;\ N, \binom{2}{2}\ PPH - PPH \Leftrightarrow \mu_0 = \mu_1 = 0,\ \mu_2 > 0,\ \tilde{L} > 0;$
$(\mathcal{C}_{13})$ $\$, n;\ \binom{2}{3}\ PEPH - P \Leftrightarrow \mu_0 = \mu_1 = 0,\ \mu_2 > 0,\ \tilde{L} = 0,\ C_2 \neq 0;$
$(\mathcal{C}_{14})$ $\$, n;\ [\infty,\ N] \Leftrightarrow \mu_0 = \mu_1 = 0,\ \mu_2 > 0,\ \tilde{L} = 0,\ C_2 = 0;$
$(\mathcal{C}_{15})$ $\$;\ N, \binom{3}{2}\ PEP - PH \Leftrightarrow \mu_0 = \mu_1 = \mu_2 = 0,\ \widetilde{K} \neq 0;$
$(\mathcal{C}_{16})$ $\$;\ N, \binom{3}{2} PEP - H \Leftrightarrow \mu_0 = \mu_1 = \mu_2 = 0,\ \widetilde{K} = 0,\ \tilde{L} \neq 0;$
$(\mathcal{C}_{17})$ $\$;\ \widehat{\binom{1}{2}} PEP - H, \overline{\binom{2}{1}}\ N \Leftrightarrow \mu_0 = \mu_1 = \mu_2 = 0,\ \widetilde{K} = 0,\ \tilde{L} = 0,\ \kappa_1 \neq 0;$
$(\mathcal{C}_{18})$ $\$;\ \binom{1}{1}\ SN, \binom{2}{2} PEP - PEP \Leftrightarrow \mu_0 = \mu_1 = \mu_2 = 0,\ \widetilde{K} = 0,\ \tilde{L} = 0,\ \kappa_1 = 0.$

$(\mathcal{D})$ *In the case of Hamiltonian systems, i.e.* $\sigma = 0$, *and the conditions* $\cup_{j=1}^{6}(f_j)$ *of Theorem* 6.2,

$(\mathcal{D}_1)$ $\$, \$, \$, c;\ N, N, N \Leftrightarrow \mu_0 < 0,\ \mathbf{D} < 0;$
$(\mathcal{D}_2)$ $\$, \$;\ N, N, N \Leftrightarrow \mu_0 < 0,\ \mathbf{D} > 0;$
$(\mathcal{D}_3)$ $\$, \$, \widehat{cp}_{(2)};\ N, N, N \Leftrightarrow \mu_0 < 0,\ \mathbf{D} = 0,\ \mathbf{T} \neq 0;$
$(\mathcal{D}_4)$ $\$, \widehat{s}_{(3)};\ N, N, N \Leftrightarrow \mu_0 < 0,\ \mathbf{D} = 0,\ \mathbf{T} = 0;$
$(\mathcal{D}_5)$ $\$, \$, c, c;\ N \Leftrightarrow \mu_0 > 0,\ \mathbf{D} < 0;$
$(\mathcal{D}_6)$ $\$, c;\ N \Leftrightarrow \mu_0 > 0,\ \mathbf{D} > 0;$
$(\mathcal{D}_7)$ $\$, c, \widehat{cp}_{(2)};\ N \Leftrightarrow \mu_0 > 0,\ \mathbf{D} = 0;$
$(\mathcal{D}_8)$ $\$, \$, c;\ N,\ \widehat{\binom{1}{2}}\ PEP - H \Leftrightarrow \mu_0 = 0,\ \mu_1 \neq 0,\ \mathbf{D} < 0;$
$(\mathcal{D}_9)$ $\$;\ N,\ \widehat{\binom{1}{2}}\ PEP - H \Leftrightarrow \mu_0 = 0,\ \mu_1 \neq 0,\ \mathbf{D} > 0;$
$(\mathcal{D}_{10})$ $\$, \widehat{cp}_{(2)};\ N,\ \widehat{\binom{1}{2}}\ PEP - H \Leftrightarrow \mu_0 = 0,\ \mu_1 \neq 0,\ \mathbf{D} = 0;$
$(\mathcal{D}_{11})$ $\$, \$;\ N, \binom{2}{2} PEP - PEP \Leftrightarrow \mu_0 = \mu_1 = 0,\ \mu_2 \neq 0,\ \widetilde{K} \neq 0;$
$(\mathcal{D}_{12})$ $\$, c;\ \widehat{\binom{2}{3}} N \Leftrightarrow \mu_0 = \mu_1 = 0,\ \mu_2 \neq 0,\ \widetilde{K} = 0;$
$(\mathcal{D}_{13})$ $\$;\ \binom{3}{3}\ PEPEP - P \Leftrightarrow \mu_0 = \mu_1 = \mu_2 = 0.$

6.4 Infinite singularities

When considering a quadratic system, we shall use the abbreviation **ISP** for its *infinite singular points*. We recall that the number and kinds of infinite singular points of a quadratic system (6.1) are governed by the invariant polynomials $\eta(a)$, $\widetilde{M}(a,x,y)$ and $C_2(a,x,y)$ (see Lemma 5.5). More precisely according to [301], we have the next lemma.

Lemma 6.1. *The type of the divisor $D_S(C,Z)$ for systems (6.1) is determined by the corresponding conditions indicated in Table 6.3, where we write $w_1^c+w_2^c+w_3$ if two of the points, i.e. w_1^c, w_2^c, are complex but not real. Moreover, for each type of the divisor $D_S(C,Z)$ given in Table 6.3, the quadratic systems (6.1) can be brought via a linear transformation to one of the following canonical systems $(\mathbf{S}_I)$–$(\mathbf{S}_V)$ corresponding to their behavior at infinity:*

Table 6.3

Case	Type of $D_S(C,Z)$	Necessary and sufficient conditions on the comitants	Canonical systems
1	$w_1+w_2+w_3$	$\eta>0$	$(\mathbf{S}_I)$
2	$w_1^c+w_2^c+w_3$	$\eta<0$	$(\mathbf{S}_{II})$
3	$2w_1+w_2$	$\eta=0,\quad \widetilde{M}\neq 0$	$(\mathbf{S}_{III})$
4	$3w$	$\widetilde{M}=0,\quad C_2\neq 0$	$(\mathbf{S}_{IV})$
5	$D_S(C,Z)$ undefined	$C_2=0$	$(\mathbf{S}_V)$

$$\begin{cases} \dot{x} = a+cx+dy+gx^2+(h-1)xy, \\ \dot{y} = b+ex+fy+(g-1)xy+hy^2; \end{cases} \qquad (\mathbf{S}_I)$$

$$\begin{cases} \dot{x} = a+cx+dy+gx^2+(h+1)xy, \\ \dot{y} = b+ex+fy-x^2+gxy+hy^2; \end{cases} \qquad (\mathbf{S}_{II})$$

$$\begin{cases} \dot{x} = a+cx+dy+gx^2+hxy, \\ \dot{y} = b+ex+fy+(g-1)xy+hy^2; \end{cases} \qquad (\mathbf{S}_{III})$$

$$\begin{cases} \dot{x} = a+cx+dy+gx^2+hxy, \\ \dot{y} = b+ex+fy-x^2+gxy+hy^2, \end{cases} \qquad (\mathbf{S}_{IV})$$

$$\begin{cases} \dot{x} = a+cx+dy+x^2, \\ \dot{y} = b+ex+fy+xy. \end{cases} \qquad (\mathbf{S}_V)$$

In article [301] the topological classification of quadratic differential systems according to their phase portraits in the neighborhood of infinity was done. According to this result and [305], we have the following proposition:

Proposition 6.3. *There exist exactly 46 topologically distinct phase portraits around infinity for the family of non-degenerate quadratic systems, given in Figure* 6.1. *The necessary and sufficient conditions for the realization of each one of these portraits could be assembled from Diagrams* 6.2–6.4. *Moreover in these diagrams the necessary and sufficient conditions for the number and multiplicities of infinite singular points are given.*

The geometrical configurations of singularities at infinity for quadratic differential systems have been studied [26]. For the particular case of degenerate systems, in [26] it was stated that there were 30 different possible geometrical configurations of singularities. Actually the real number is 31, since the case of systems having two complex conjugate straight lines intersecting at infinity was missed. This case forces the existence of a real singularity at infinity, which was not noticed in that paper. So we state the result (which includes the correction) for the case of degenerate systems (i.e. when the polynomials defining the differential system are not coprime).

Proposition 6.4. *For the family of degenerate quadratic systems, there exist exactly 31 topologically distinct phase portraits around infinity, given in Figure* 6.2. *Moreover necessary and sufficient conditions for the realization of each one of these portraits are given in Diagrams* 6.6–6.8 *(see Theorem* 6.4*). These are the cases occurring for* $\mu_i = 0$ *for all* $i \in \{0, 1, 2, 3, 4\}$.

The next theorem was proved in [26]:

Theorem 6.4. (a) *The configurations of singularities at infinity of all quadratic vector fields are classified in Diagrams* 6.5–6.8 *according to the geometrical equivalence relation. Necessary and sufficient conditions for each one of the* 167 *different equivalence classes can be assembled from these diagrams in terms of* 27 *invariant polynomials with respect to the action of the affine group and time rescaling, given in Section* 5.4.

(b) *Diagrams* 6.5–6.8 *actually contain the bifurcation diagram in the* 12*-dimensional space of parameters of the global geometrical configurations of singularities at infinity of quadratic differential systems.*

We remark that the geometrical meaning of some of the conditions given in terms of invariant polynomials in Diagrams 6.5–6.8 appear in Diagrams 6.2–6.4.

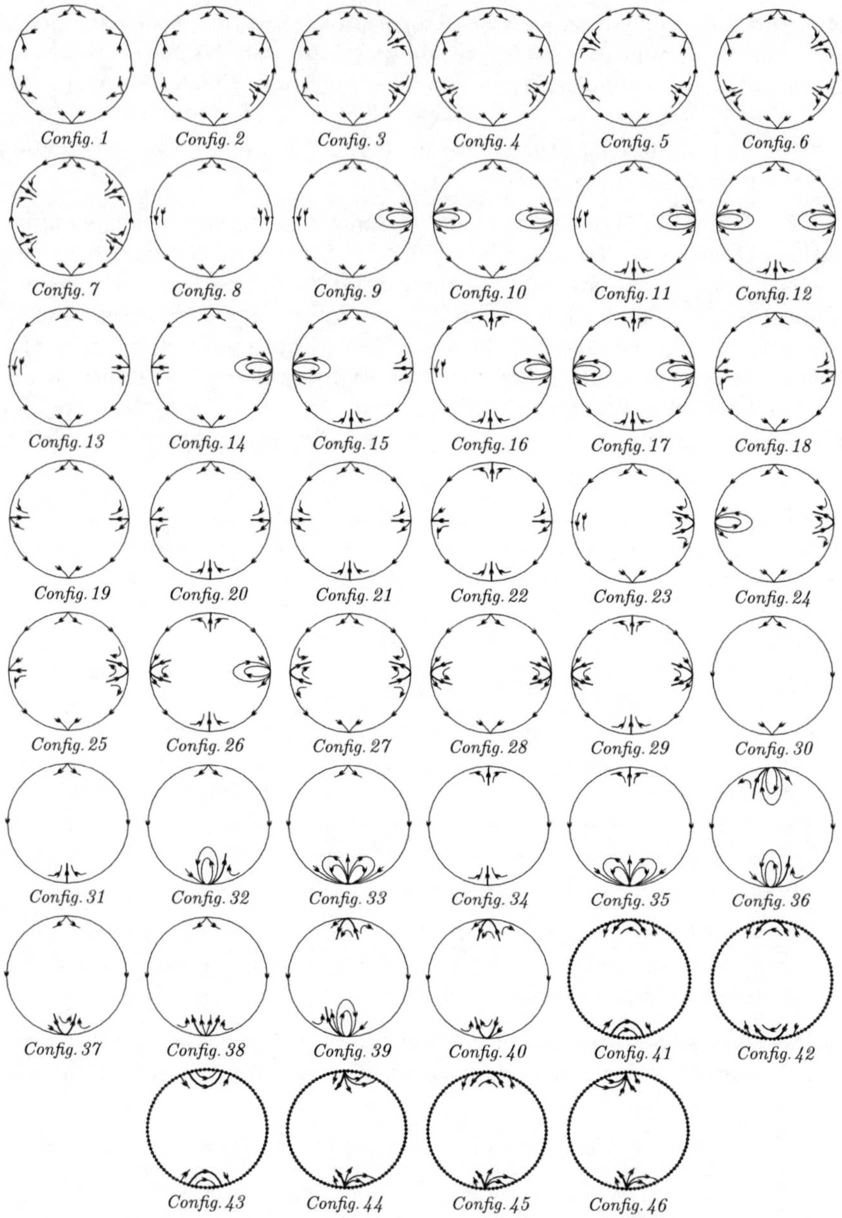

Figure 6.1: Topologically distinct local configurations of **ISP**s for non-degenerate quadratic systems.

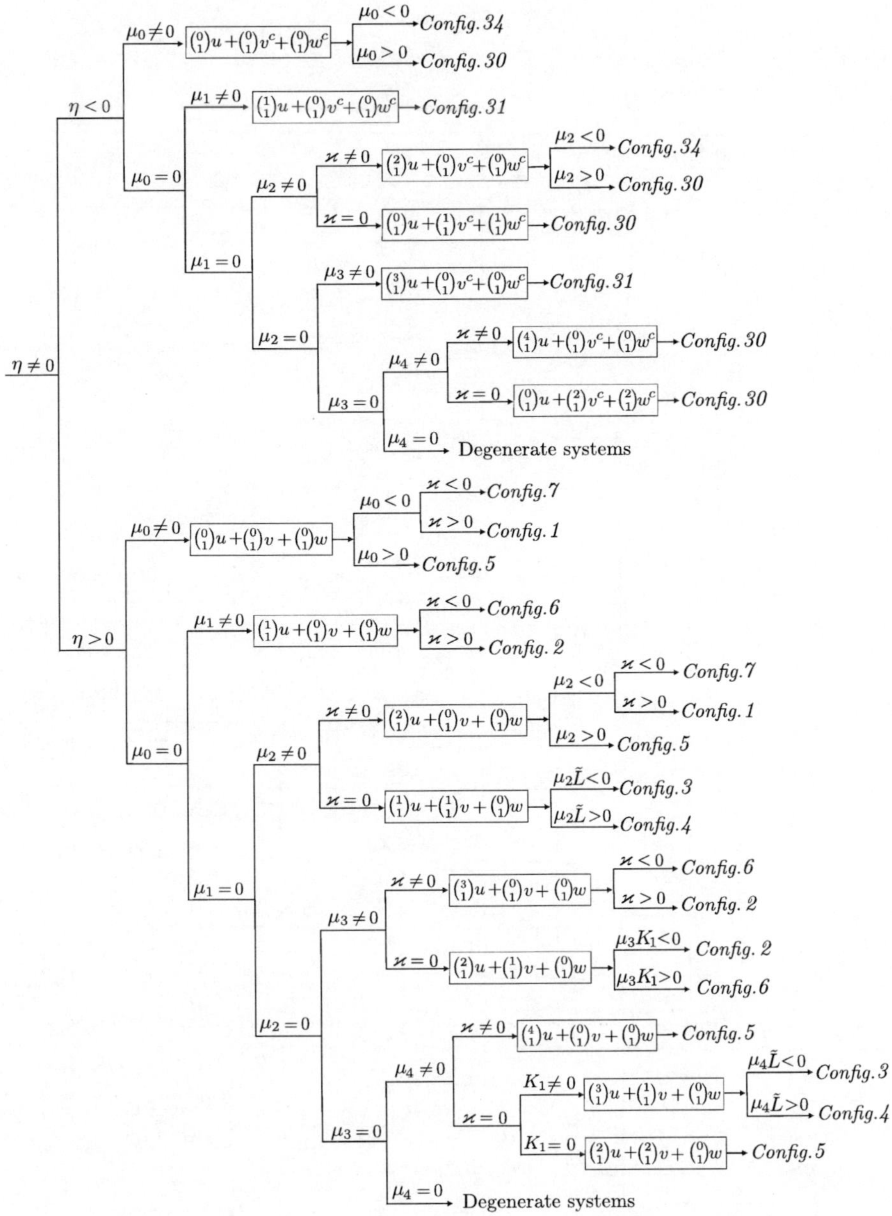

Diagram 6.2: Multiplicities of **ISP**s in the case $\eta \neq 0$.

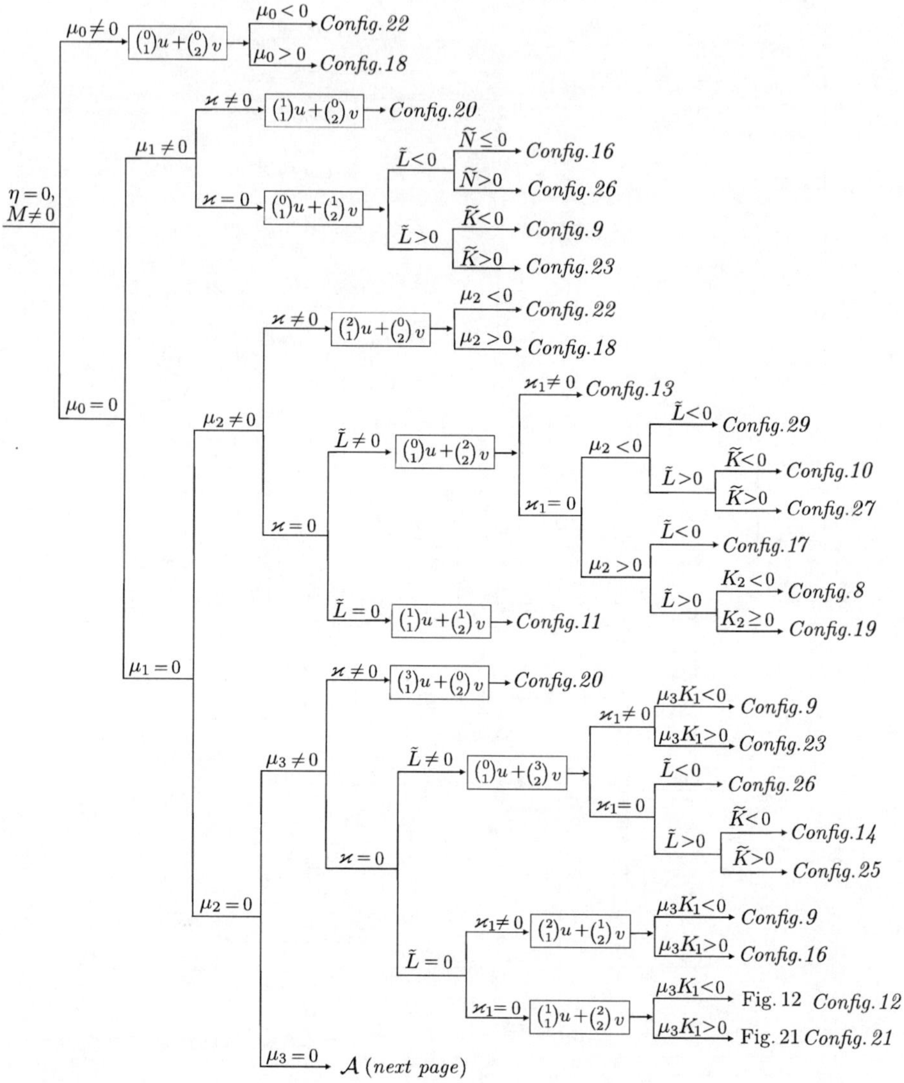

Diagram 6.3: Multiplicities of **ISP**s in the case $\eta = 0$, $\widetilde{M} \neq 0$.

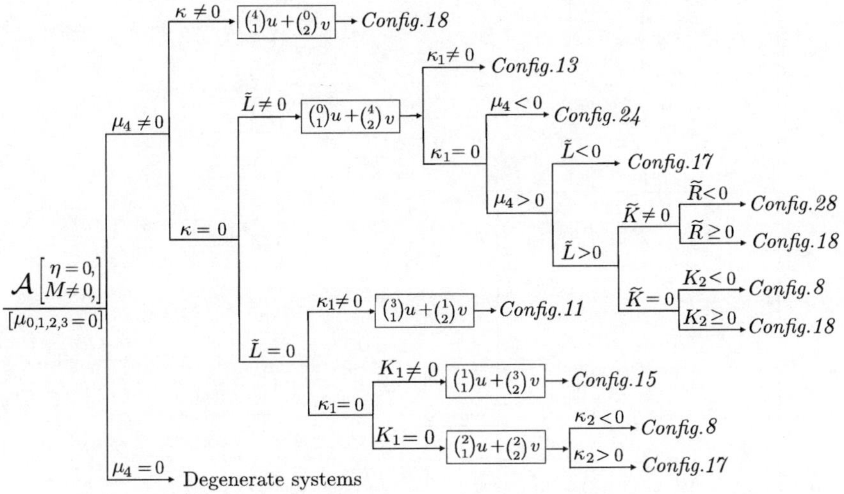

Diagram 6.3 (*cont.*): Multiplicities of **ISP**s in the case $\eta = 0$, $\widetilde{M} \neq 0$.

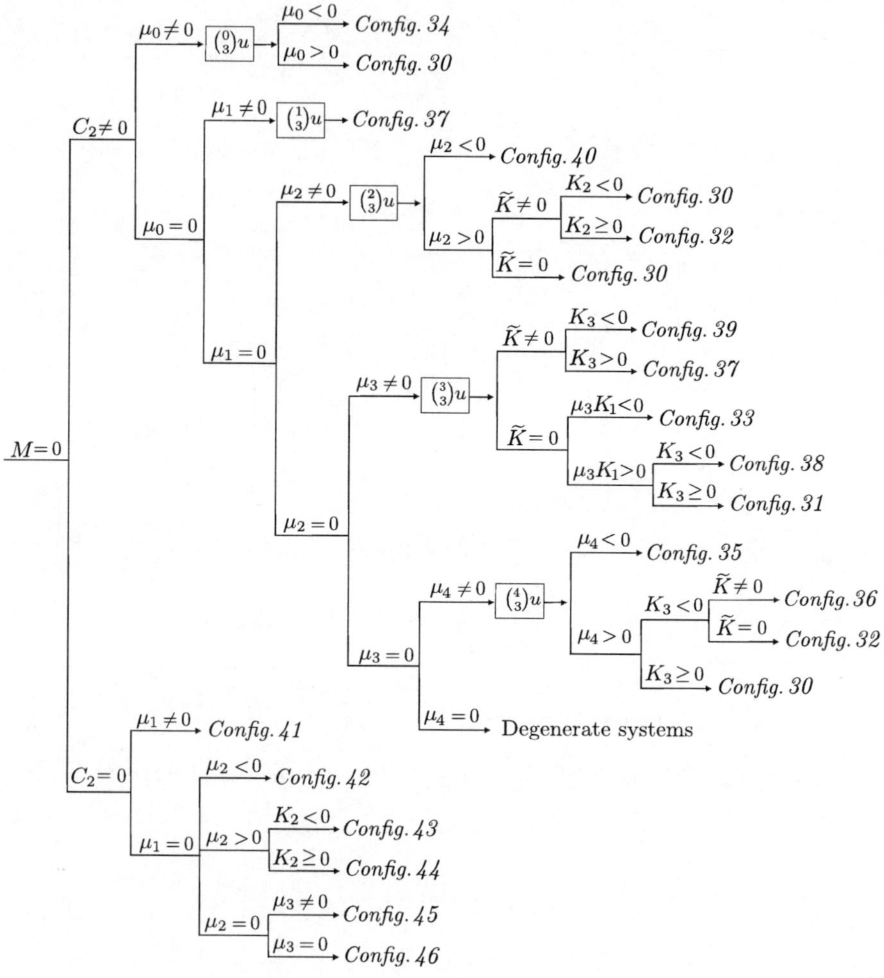

Diagram 6.4: Multiplicities of **ISP**s in the case $\widetilde{M} = 0$.

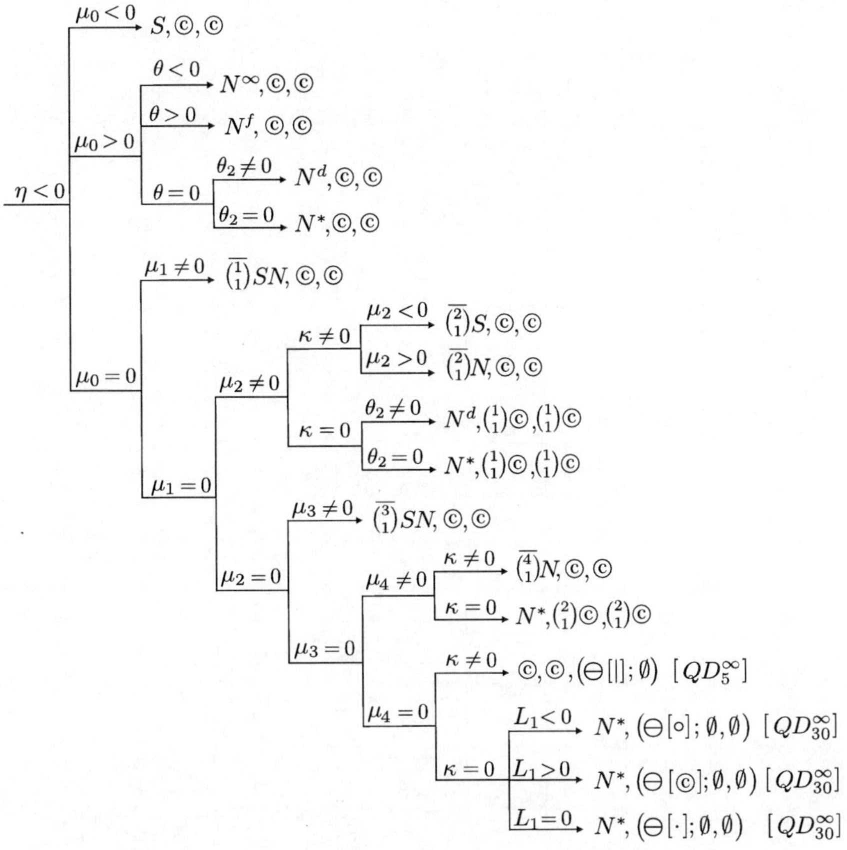

Diagram 6.5: Configurations of **ISPs** in the case $\eta<0$.

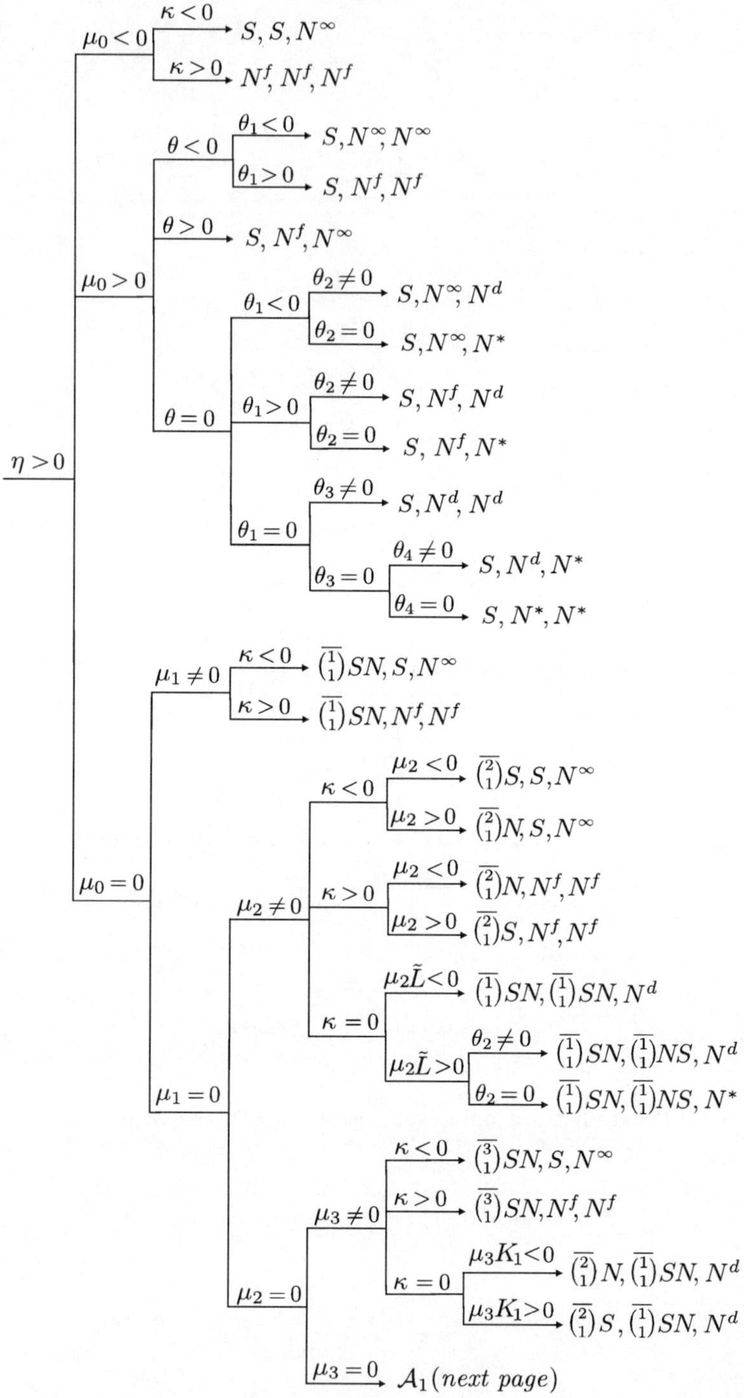

Diagram 6.6: Configurations of **ISPs** in the case $\eta>0$.

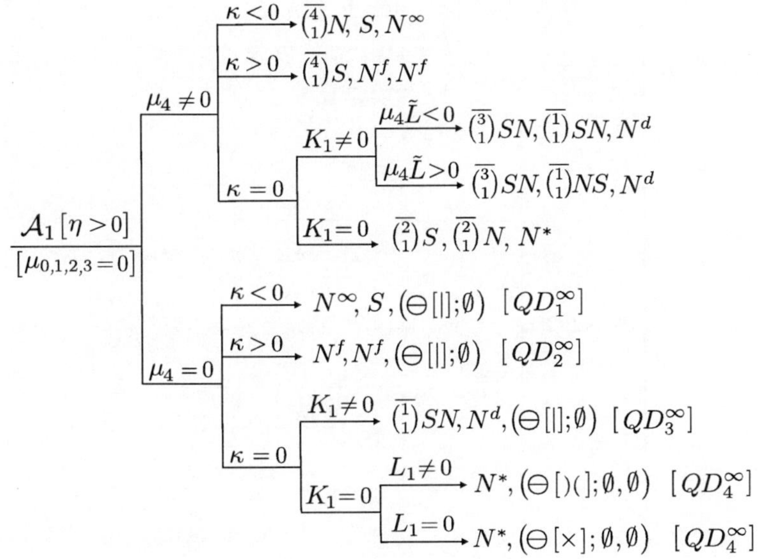

Diagram 6.6 *(cont.)*: Configurations of **ISPs** in the case $\eta > 0$.

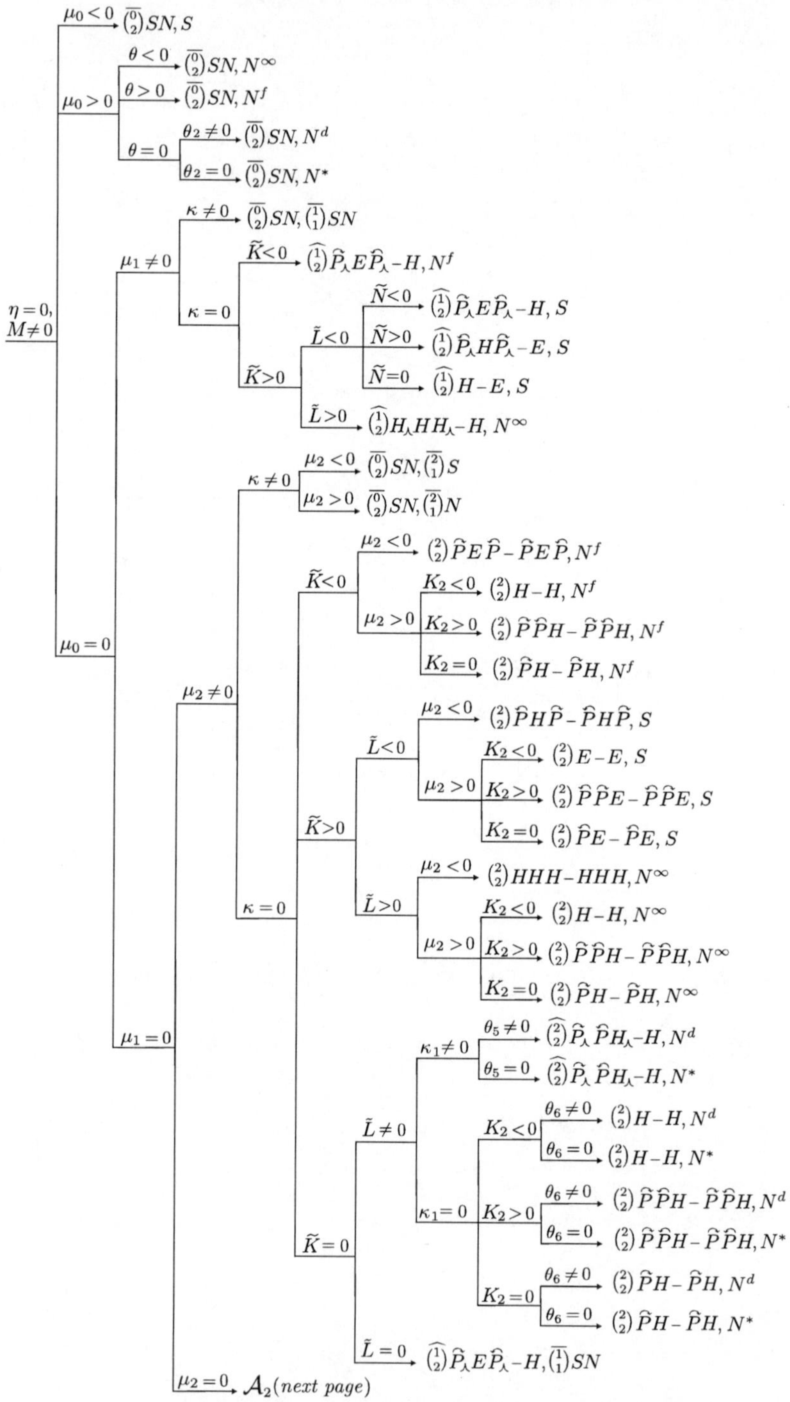

Diagram 6.7: Configurations of **ISP**s in the case $\eta=0$, $\widetilde{M}\neq 0$.

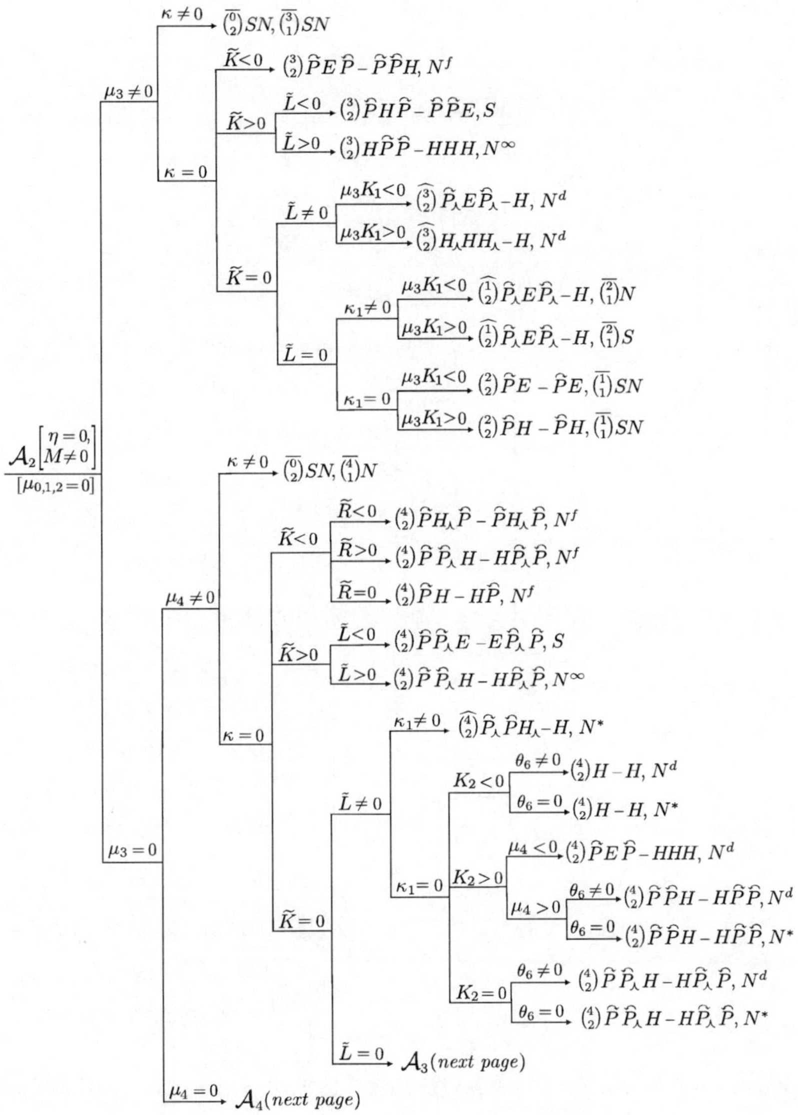

Diagram 6.7 *(cont.)*: Configurations of **ISP**s in the case $\eta = 0$, $\widetilde{M} \neq 0$.

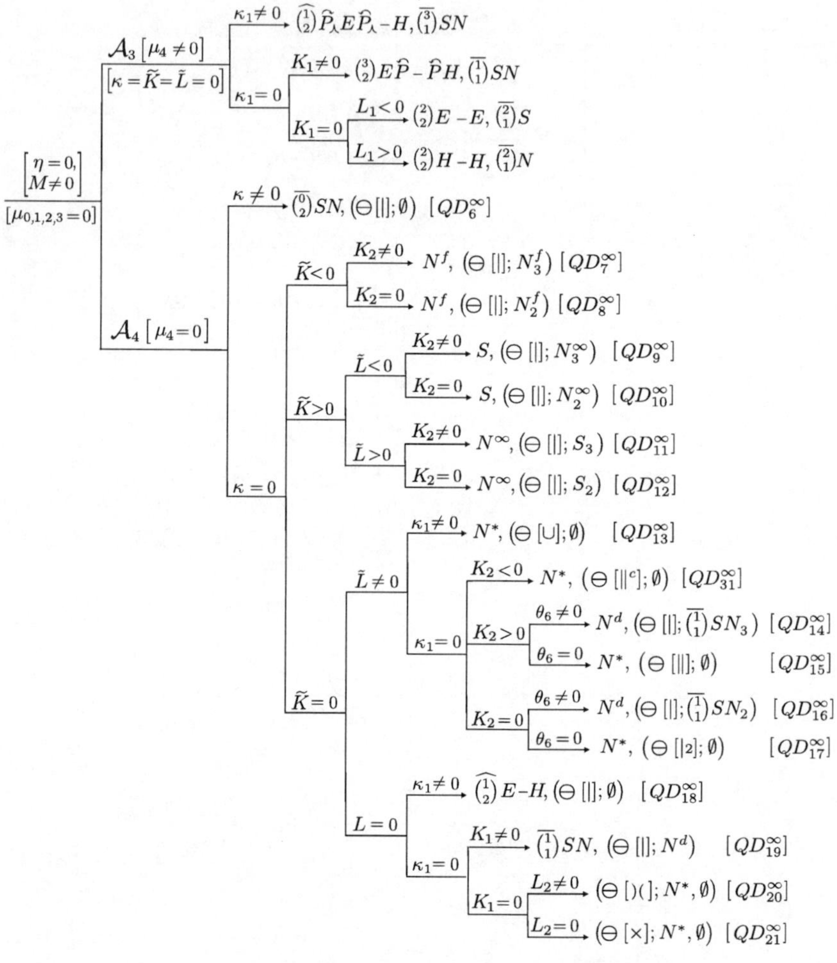

Diagram 6.7 *(cont.)*: Configurations of **ISP**s in the case $\eta = 0$, $\widetilde{M} \neq 0$.

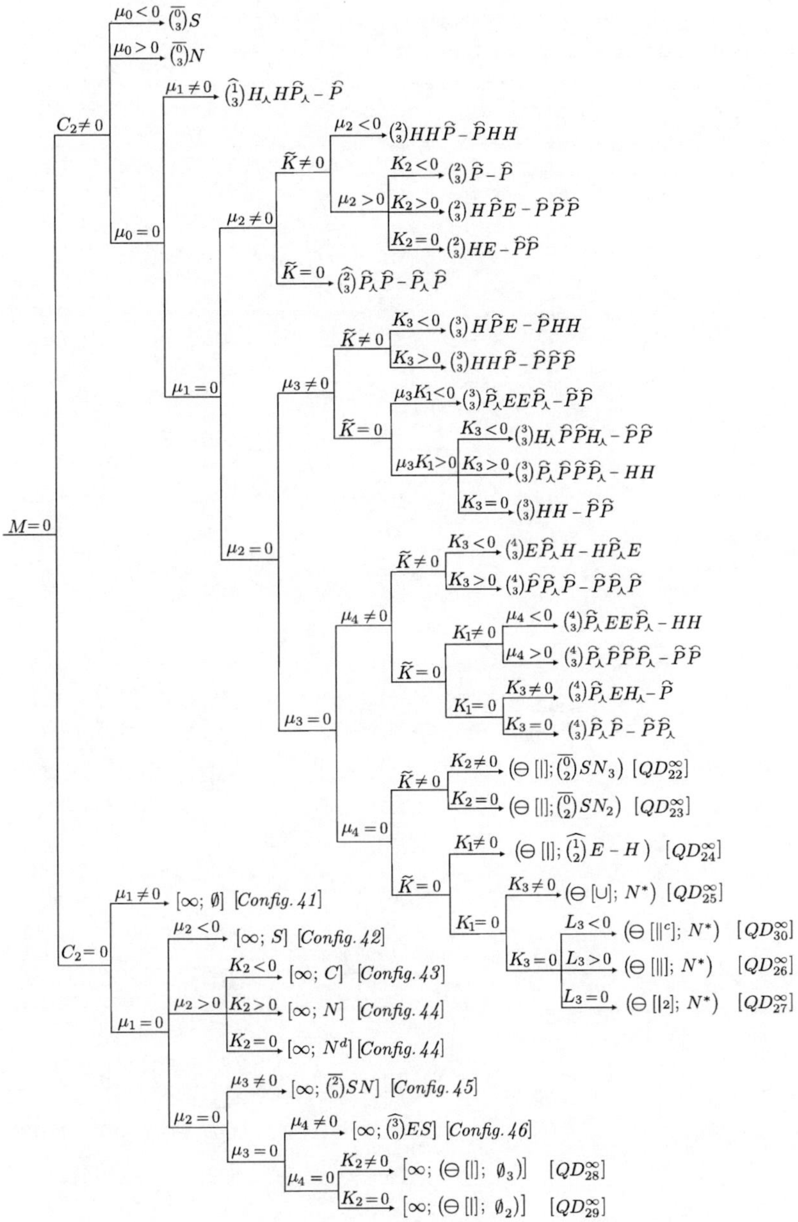

Diagram 6.8: Configurations of **ISP**s in the case $\eta = 0$, $\widetilde{M} = 0$.

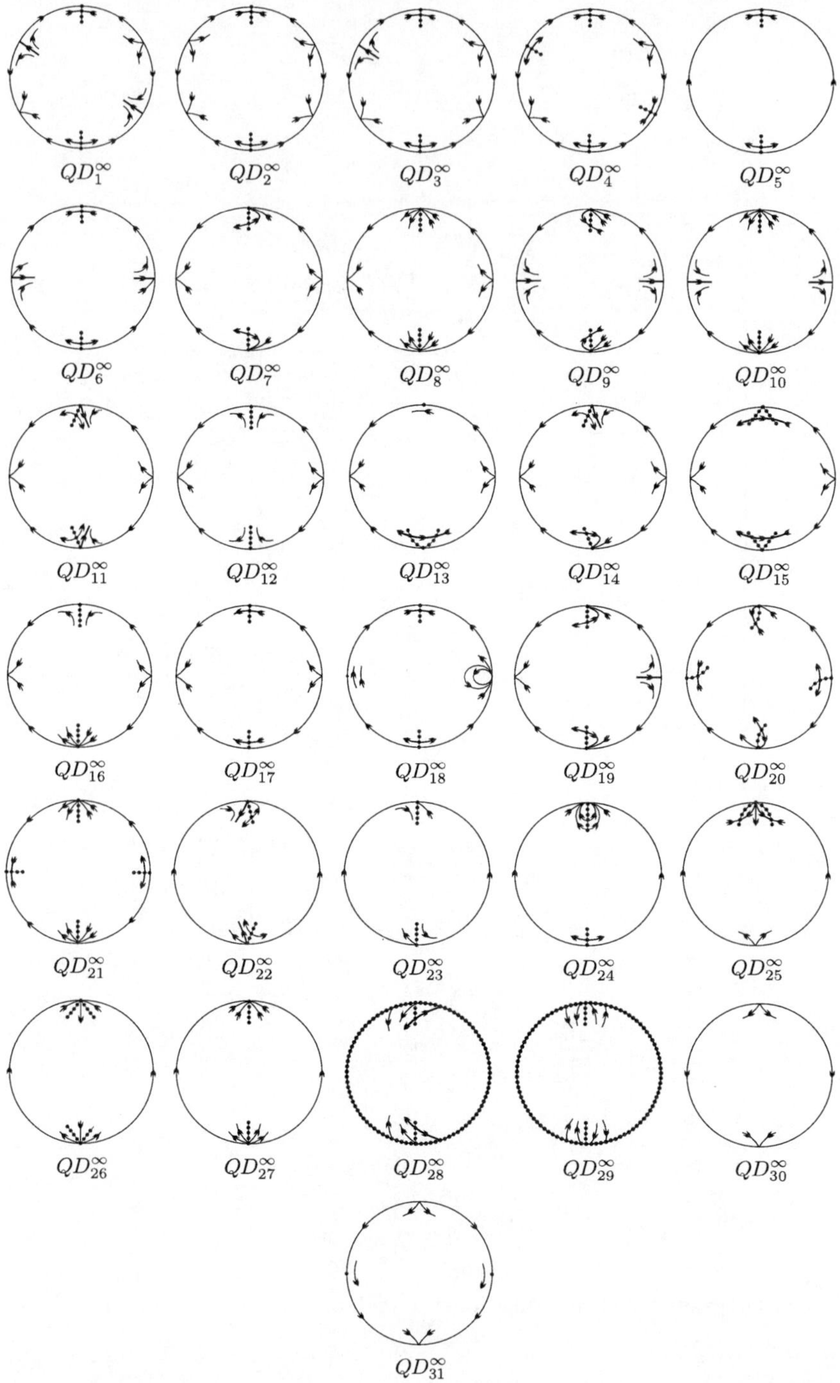

Figure 6.2: Topologically distinct local configurations of **ISP**s for degenerate quadratic systems.

Chapter 7

Classifications of quadratic systems with special singularities

One of the goals of this book is to enumerate all possible geometrical configurations of singularities, finite and infinite, of real quadratic differential systems. The foliations with singularities of the complexified systems can be compactified over the complex projective space. We then have a total of seven singularities counted with multiplicities. The total multiplicity of finite singularities is at most four, and there is the possibility of having a unique finite singular point of such multiplicity. A singular point at infinity could have higher multiplicity, which could even be seven. This occurs when all finite singularities coalesce with a triple point at infinity. Hence the local phase portraits around a singular point at infinity could be much more complicated than around anyone of the finite singularities.

In principle we could write all possible combinations of finite and infinite singularities starting firstly with the elemental singularities, for example the combinations $s, s^{(1)}, n, n; N^*, N, S$, or $f^{(2)}, s, n, n; N^d, S, S$. We need then to list all cases that could include degenerate singularities, for example we could have $s, n, f^{(1)}$; S, SN, S. Listing of all such combination is a tedious but not difficult task. It is just a matter of keeping track of all such possibilities. The much more difficult task is to determine which ones among these combinations can actually be realized and which must be discarded. Some combinations could easily be discarded by using results already proved about quadratic systems. For example we know that we could have at most two foci and also at most two centers in quadratic systems. So clearly the combinations $f, f, f^{(2)}, s; S, N, S$ and $c, c, c; S, SN, S$ cannot be realized. Moreover the total index of infinite singularities is directly related with that of finite singularities. But the number of combinations we get by enumerating all possibilities is very large, and by comparison the number of combinations that

J. C. Artés et al., *Geometric Configurations of Singularities of Planar Polynomial Differential Systems*, https://doi.org/10.1007/978-3-030-50570-7_7

could be discarded by using the classical literature on quadratic systems is rather small. For this reason we need to adopt a strategy that will simplify our task.

Among the elemental singular points we have some special singularities such as the finite star nodes, star nodes at infinity, weak foci of various orders, weak saddles of various orders, centers and integrable saddles. We propose first to consider the combinations where such special singularities occur. This will simplify our work when we incorporate these blocks into the full geometry of the configurations of singularities.

For weak foci of order two and three, we are lucky because we can use previous studies of the class of all quadratic systems possessing a weak focus of order two [19] and of order three [222]. The singularities were not classified up to geometrical equivalence as we do here, but in any case these results have eased our task. We do not have such luck for the case of systems with a saddle of order two or three, and this missing part causes some difficulties as we see later in the book when we need to confront these cases (see Part II). We can also use the article [32] where all systems with an integrable saddle were studied. Another class for which the topological study was done in detail is the family of quadratic systems with a center (see [336, 291, 364, 248, 31]). Here we add the geometrical information in our list of configurations (see Table 7.1 in Section 7.3).

The goal of this chapter is to enumerate all possible geometrical configurations of singularities where at least one singularity is special and of the following kinds: finite star node, weak focus of orders two or three, weak saddle of order three, center or integrable saddle. Secondly we consider the case of an infinite star node together with another special singularity of the following special kinds: either an elemental node (infinite or finite) with equal eigenvalues or a weak finite singularity. Thirdly we consider the particular case of quadratic differential systems possessing at least three finite elemental singular points of a special type, and namely each one of them is either a node n^d or a weak singularity. More precisely we will investigate the existence of the triplets (n^d, n^d, n^d), (n^d, n^d, w), or (n^d, w, w), where by w we denote a weak singularity, which could be a focus or a saddle.

Since one configuration may appear in several of these families, we start with what appears to be the basic, simplest case among these families, namely the configurations that include a finite star node. Whenever a star node will appear in a family, we will apply the results already obtained for this first family, and we proceed in this way for the remaining families.

7.1 A finite star node

We take real planar quadratic differential systems that can be written in the form

$$\begin{aligned}\frac{dx}{dt} &= a + cx + dy + gx^2 + 2hxy + ky^2 \equiv p(x, y),\\ \frac{dy}{dt} &= b + ex + fy + lx^2 + 2mxy + ny^2 \equiv q(x, y).\end{aligned} \tag{7.1}$$

7.1.1 Conditions for the existence of at least one finite star node

Theorem 7.1. *A quadratic system* (7.1) *possesses a single star node if and only if one of the following two sets of conditions holds:*

$$\begin{aligned}
&(\mathcal{S}^1_{n^*}) \quad U_1 \neq 0,\ U_2 \neq 0,\ U_3 = Y_1 = 0;\\
&(\mathcal{S}^2_{n^*}) \quad U_1 = U_4 = U_5 = 0,\ \ Y_2 > 0,\ \ U_6 = 0.
\end{aligned}$$

It possesses two star nodes if and only if

$$(\mathcal{S}^3_{n^*}) \quad U_1 = U_4 = U_5 = 0,\ \ Y_2 > 0,\ \ U_6 \neq 0,$$

where the invariant polynomials used are given in Section 5.4.

Proof. If a system (7.1) possesses a star node $M_0(x_0, y_0)$, then via a translation of axes we can bring it to the origin of coordinates, and hence the system can be brought to the form:

$$\begin{aligned}
\dot{x} &= \lambda x + p_2(x, y),\\
\dot{y} &= \lambda y + q_2(x, y),
\end{aligned} \tag{7.2}$$

with $\lambda \neq 0$. So we apply to systems (7.1) the translation: $x = \bar{x} + x_0$, $y = \bar{y} + y_0$. Then the coefficients $\bar{c}$, $\bar{e}$, $\bar{a}$ of the resulting new system are expressed in terms of the old parameters of (7.1) and x_0, y_0 as follows:

$$\begin{aligned}
\bar{c} &= c + 2gx_0 + 2hy_0, & \bar{d} &= d + 2hx_0 + 2ky_0,\\
\bar{e} &= e + 2lx_0 + 2my_0, & \bar{f} &= f + 2mx_0 + 2ny_0,\\
\bar{a} &= p(x_0, y_0), & \bar{b} &= q(x_0, y_0).
\end{aligned} \tag{7.3}$$

To have the linear matrix of the new systems of the form $\begin{pmatrix} \lambda & 0 \\ 0 & \lambda \end{pmatrix}$, the conditions $\bar{d} = \bar{e} = \bar{c} - \bar{f} = 0$ have to be fulfilled. Introducing these values in equations (7.3), we obtain the linear system of equations in the unknowns x_0 and y_0,

$$\begin{aligned}
2hx_0 + 2ky_0 + d &= 0,\\
2lx_0 + 2my_0 + e &= 0,\\
2(g - m)x_0 + 2(h - n)y_0 + (c - f) &= 0.
\end{aligned} \tag{7.4}$$

We denote by Δ_{ij} $(1 \leq i < j \leq 3)$ the 2×2 minors of the 3×2 matrix of this system in x_0, y_0, constructed with the lines i and j,

$$\Delta_{12} = 4(hm - kl), \quad \Delta_{13} = 4(h^2 - hn - gk + km), \quad \Delta_{23} = 4(hl - ln - gm + m^2),$$

and by Δ the determinant of the 3×3 matrix, extension of the previous 3×2 matrix by the column of the free terms of the system in x_0, y_0. A straightforward calculation yields

$$U_1 = -\Delta_{23}x^2 + 2\Delta_{12}xy - \Delta_{13}y^2, \quad 4Y_1 = -\Delta.$$

We consider two possibilities: $U_1 \neq 0$ and $U_1 = 0$.

1: $U_1 \neq 0$. Then clearly the rank of the 3×2 matrix of system (7.4) equals 2. Hence this system has a unique solution if the rank of the extended matrix equals 2 (i.e. $Y_1 = 0$), and it does not have any solution if this rank equals 3 (i.e. $Y_1 \neq 0$).

Thus in the case that $U_1 \neq 0$ and $Y_1 = 0$, there exists a unique point (x_0, y_0) such that the system of linear equations (7.4) has a unique solution x_0, y_0. In this case if the point (x_0, y_0) is a star node, then $\bar{a} = p(x_0, y_0) = 0$, $\bar{b} = q(x_0, y_0) = 0$ and since the eigenvalues are equal and not zero, the form of the new system is the one given in (7.2).

For the systems:

$$\dot{x} = a + \lambda x + p_2(x, y), \quad \dot{y} = b + \lambda y + q_2(x, y),$$

we calculate

$$U_2 = -4\lambda U_1, \quad U_3 = 3(bx - ay)U_1^2.$$

Therefore since $U_1 \neq 0$, the condition $\lambda \neq 0$ is equivalent to $U_2 \neq 0$, and $a = b = 0$ if and only if $U_3 = 0$. So the conditions $(\mathcal{S}_{n^*}^1)$ of the theorem are all verified.

2: $U_1 = 0$. We claim that in this case the condition $C_2 \neq 0$ must hold, otherwise we get a linear differential system. Indeed, supposing $C_2 = 0$ we have

$$C_2 = -lx^3 + (g - 2m)x^2y + (2h - n)xy^2 + ky^3 = 0.$$

For systems (7.1) we obtain

$$l = k = 0, \quad g = 2m, \quad n = 2h \quad \Rightarrow \quad U_1 = 4(mx + hy)^2.$$

The condition $U_1 = 0$ yields $m = h = 0$, and we obtain a linear differential system, i.e. our claim is proved.

So we have $C_2 \neq 0$, then without loss of generality we may assume $lk \neq 0$ due to a linear transformation, and via a translation we impose $d = e = 0$. Moreover we can consider $k = 1$ due to a time rescaling, and therefore we compute

$$U_1 = -4(lh - gm + m^2 - ln)x^2 - 8(l - hm)xy + 4(g - h^2 - m + hn)y^2.$$

So it could easily be detected that the condition $U_1 = 0$ is equivalent to the conditions $l = hm \neq 0$ and $g = h^2 + m - hn$. This leads to the systems:

$$\begin{aligned} \dot{x} &= a + cx + (h^2 + m - hn)x^2 + 2hxy + y^2, \\ \dot{y} &= b + fy + hmx^2 + 2mxy + ny^2, \quad hm \neq 0, \end{aligned} \tag{7.5}$$

for which the system of linear equations (7.4) becomes

$$\begin{aligned} 2(hx_0 + y_0) &= 0, \\ 2m(hx_0 + y_0) &= 0, \\ 2(h - n)(hx_0 + y_0) + (c - f) &= 0. \end{aligned}$$

Thus we have $hx_0 + y_0 = 0$ and hence $c - f = 0$. We obtain the line of solutions $y_0 = -hx_0$ for the solutions (x_0, y_0). As for systems (7.5) we have

$$U_4 = 6(c-f)(hx+y)(mx^2+y^2),$$

and we conclude that the condition $c - f = 0$ is given by $U_4 = 0$.

So in what follows we assume $U_4 = 0$ (i.e. $f = c$). However in order to have a star node, according to (7.3) the additional conditions $\bar{a} = p(x_0, y_0) = 0$ and $\bar{b} = q(x_0, y_0) = 0$ have to be verified. So setting $y_0 = -hx_0$ for systems (7.5), we obtain the equations

$$\bar{a} = (m-hn)x_0^2 + cx_0 + a = 0, \quad \bar{b} = -h\big[(m-hn)x_0^2 + cx_0\big] + b = 0. \tag{7.6}$$

It is clear that for the existence of at least one solution the condition $b = -ah$ is necessary. For systems (7.5) with $f = c$, calculation yields

$$U_5 = 12(b+ah)(mx^2 - hxy + nxy - y^2), \tag{7.7}$$

and the condition $b = -ah$ is clearly governed by the invariant polynomial U_5. Assuming $U_5 = 0$ we get the following family of systems:

$$\begin{aligned} \dot{x} &= a + cx + (h^2+m-hn)x^2 + 2hxy + y^2, \\ \dot{y} &= -ah + cy + hmx^2 + 2mxy + ny^2, \quad hm \neq 0, \end{aligned} \tag{7.8}$$

for which the relations (7.6) become

$$\bar{b} = -h\bar{a} = -h\big[(m-hn)x_0^2 + cx_0 + a\big] = 0. \tag{7.9}$$

For systems (7.8) we compute

$$U_6 = -4(m-hn)(mx^2 - hxy + nxy - y^2),$$

and we consider two cases: $U_6 \neq 0$ and $U_6 = 0$.

2.1: $U_6 \neq 0$. Then $m - hn \neq 0$ and the equation (7.9) could have real solutions if and only if its discriminant $c^2 - 4a(m-hn) \geq 0$. Then since the star nodes must be located on the line $y = -hx$, the coordinates of the star nodes are

$$x_0 = \frac{-c \pm \sqrt{c^2 - 4a(m-hn)}}{2(m-hn)}, \quad y_0 = -hx_0.$$

Therefore replacing this singular point at the origin of coordinates (via the translation $x = X + x_0$, $y = Y + y_0$ in systems (7.8)), we arrive at the following family of systems:

$$\begin{aligned} \dot{X} &= \pm\sqrt{c^2 - 4a(m-hn)}\; X + (h^2+m-hn)X^2 + 2hXY + Y^2, \\ \dot{Y} &= \pm\sqrt{c^2 - 4a(m-hn)}\; Y + hmX^2 + 2mXY + nY^2, \quad hm(m-hn) \neq 0. \end{aligned}$$

Clearly (0,0) will be a real star node for these systems if and only if $c^2 - 4a(m - hn) > 0$. Additionally for these systems we calculate

$$Y_2 = 8(c^2 - 4am + 4ahn),$$

and hence in the case $U_6 \neq 0$ we could have exactly two star nodes, which happens if and only if $Y_2 > 0$. So the conditions $(\mathcal{S}^3_{n^*})$ of the theorem are verified.

2.2: $U_6 = 0$. Considering (7.7) this implies $m = hn \neq 0$, and systems (7.8) become

$$\begin{aligned} \dot{x} &= a + cx + (hx + y)^2, \\ \dot{y} &= -ah + cy + n(hx + y)^2, \quad hn \neq 0. \end{aligned} \tag{7.10}$$

At the same time the relations (7.9) take the form

$$\bar{b} = -h\bar{a} = -h(cx_0 + a) = 0,$$

i.e. we could have a solution (which is unique) if and only if $c \neq 0$. Since for systems (7.10) we have $Y_2 = 8c^2$, we conclude that for the existence of a star node, it is necessary to have $Y_2 \neq 0$, which is equivalent to $Y_2 > 0$. Thus we obtain the following coordinates of the star node: $x_0 = -a/c$, $y_0 = ah/c$. Replacing this singular point at the origin of coordinates (via the translation $x = X + x_0$, $y = Y + y_0$ in systems (7.10)), we arrive at the following family of systems:

$$\begin{aligned} \dot{X} &= cX + (hX + Y)^2, \\ \dot{Y} &= cY + n(hX + Y)^2. \end{aligned}$$

Since $c \neq 0$ (due to $Y_2 \neq 0$), the singular point (0,0) is a star node for these systems. So the conditions $(\mathcal{S}^2_{n^*})$ of the theorem are verified. □

7.1.2 Configurations of singularities with a finite star node

Theorem 7.2. *There are exactly* 40 *geometrically distinct configurations of singularities, finite and infinite, of quadratic vector fields possessing a finite star node, and all of them are contained in the following list, enumerated and grouped according to the values of the type of the zero-cycle* $\mathcal{D}_{\mathbb{C}^2}(S)$*:*

	(α_1) $s, s, s, n^*; N^f, N^f, N^f;$	(α_2) $s, n, n, n^*; S, S, N^\infty;$
	(α_3) $s, n, n^d, n^*; S, S, N^\infty;$	(α_4) $s, n^d, n^d, n^*; S, S, N^\infty;$
$\mathcal{D}_{\mathbb{C}^2}(S) =$	(α_5) $s, s, n, n^*; S, N^\infty, N^\infty;$	(α_6) $s, s, n, n^*; S, N^f, N^f;$
$p + q + r + s$	(α_7) $s, s, n, n^*; S, N^f, N^\infty;$	(α_8) $s, s, n^d, n^*; S, N^f, N^\infty;$
	(α_9) $s, \$, n, n^*; S, N^\infty, N^*;$	(α_{10}) $s, \$, n, n^*; S, N^f, N^*;$
	(α_{11}) $\$, \$, n^*, n^*; S, N^*, N^*;$	

	(α_{12}) n, n^*, ⓒ, ⓒ; S, Ⓒ, Ⓒ;	(α_{13}) n^d, n^*, ⓒ, ⓒ; S, Ⓒ, Ⓒ;
$\mathcal{D}_{\mathbb{C}^2}(S) =$	(α_{14}) n^*, n^*, ⓒ$^\rho$, ⓒ$^\rho$; S, Ⓒ, Ⓒ;	(α_{15}) s, n^*, ⓒ, ⓒ; N^∞, Ⓒ, Ⓒ;
$p + q + r^c + s^c$	(α_{16}) s, n^*, ⓒ, ⓒ; N^f, Ⓒ, Ⓒ;	(α_{17}) s, n^*, ⓒ$^\tau$, ⓒ$^\tau$; N^∞, Ⓒ, Ⓒ;
	(α_{18}) $\$, n^*$, ⓒ, ⓒ; N^*, Ⓒ, Ⓒ;	

$\mathcal{D}_{\mathbb{C}^2}(S) = 2p+q+r$	$(\alpha_{19})\ \overline{sn}_{(2)}, n, n^*; \overline{\binom{0}{2}} SN,\ S;$ $(\alpha_{20})\ \overline{sn}_{(2)}, n^d, n^*; \overline{\binom{0}{2}} SN,\ S;$ $(\alpha_{21})\ \overline{sn}_{(2)}, s, n^*; \overline{\binom{0}{2}} SN,\ N^\infty;$ $(\alpha_{22})\ \overline{sn}_{(2)}, s, n^*; \overline{\binom{0}{2}} SN,\ N^f;$ $(\alpha_{23})\ \overline{sn}_{(2)}, \$, n^*; \overline{\binom{0}{2}} SN,\ N^*;$
$\mathcal{D}_{\mathbb{C}^2}(S) = 3p+q$	$(\alpha_{24})\ \overline{n}_{(3)}, n^*; \overline{\binom{0}{3}} S;$ $(\alpha_{25})\ \overline{s}_{(3)}, n^*; \overline{\binom{0}{3}} N;$
$\mathcal{D}_{\mathbb{C}^2}(S) = p+q+r$	$(\alpha_{26})\ s, s, n^*; \overline{\binom{1}{1}} SN,\ N^f,\ N^f;$ $(\alpha_{27})\ s, n, n^*; \overline{\binom{1}{1}} SN,\ S,\ N^\infty;$ $(\alpha_{28})\ s, n^d, n^*; \overline{\binom{1}{1}} SN,\ S,\ N^\infty;$
$\mathcal{D}_{\mathbb{C}^2}(S) = p+q^c+r^c$	$(\alpha_{29})\ n^*, \copyright, \copyright; \overline{\binom{1}{1}} SN,\ \copyright, \copyright;$
$\mathcal{D}_{\mathbb{C}^2}(S) = 2p+q$	$(\alpha_{30})\ \overline{sn}_{(2)}, n^*; \overline{\binom{0}{2}} SN, \overline{\binom{1}{1}} SN;$
$\mathcal{D}_{\mathbb{C}^2}(S) = p+q$	$(\alpha_{31})\ n, n^*; \binom{2}{2}\ \widehat{P}H\widehat{P} - \widehat{P}H\widehat{P},\ S;$ $(\alpha_{32})\ n^*, n^*; \binom{2}{2}\ \widehat{P}H\widehat{P} - \widehat{P}H\widehat{P},\ S,;$ $(\alpha_{33})\ s, n^*; \binom{2}{2}\ \widehat{P}\widehat{P}H - \widehat{P}\widehat{P}H,\ N^f;$ $(\alpha_{34})\ s, n^*; \binom{2}{2}\ \widehat{P}\widehat{P}H - \widehat{P}\widehat{P}H,\ N^\infty;$ $(\alpha_{35})\ \$, n^*; N^*, \binom{1}{1} \copyright, \binom{1}{1} \copyright;$ $(\alpha_{36})\ \$, n^*; \overline{\binom{1}{1}} SN, \overline{\binom{1}{1}} NS,\ N^*;$ $(\alpha_{37})\ \$, n^*; \binom{2}{2}\ \widehat{P}\widehat{P}H - \widehat{P}\widehat{P}H,\ N^*;$
$\mathcal{D}_{\mathbb{C}^2}(S) = p$	$(\alpha_{38})\ n^*; \binom{3}{3}\ HH\widehat{P} - \widehat{P}\widehat{P}\widehat{P};$ $(\alpha_{39})\ n^*; \binom{2}{2}\ \widehat{P}H - \widehat{P}H, \overline{\binom{1}{1}} SN;$ $(\alpha_{40})\ n^*; \binom{3}{3}\ \widehat{P}_{\lambda}\widehat{P}\widehat{P}\widehat{P}_{\lambda} - HH.$

The realizability of each one of these configurations will be proved by means of an example, which will be provided in Part II.

In order to prove the theorem, we will first prove the following lemma.

Lemma 7.1. *No quadratic systems could simultaneously possess a star node and a focus (or a center).*

Proof. Assume that a system (7.1) possesses a node n^* and a focus. Then due to an affine transformation, we may assume that these singularities are located at $M_1(0,0)$ and $M_2(1,0)$, respectively. Then for systems (7.1) the conditions $a = b = 0$, $g = -c$ and $l = -e$ must be satisfied. Moreover, since $M_1(0,0)$ is a star node, we have $d = e = 0$, $f = c \neq 0$, and we arrive at the family of systems:

$$\dot{x} = cx - cx^2 + 2hxy + ky^2, \quad \dot{y} = y(c + 2mx + ny).$$

Since these systems possess the invariant line $y = 0$, we deduce that the singular point $M_2(1,0)$ could be neither a focus nor a center, and this completes the proof of the lemma. □

Proof of Theorem 7.2. We shall prove this theorem considering step by step all the possible values of the zero-cycle $\mathcal{D}_{\mathbb{C}^2}(S)$, taking into account Proposition 6.2 (see Table 6.1) and examining only the cases when there exists at least one real elemental singularity. Along the whole proof of this theorem, any finite anti-saddle is a node according to Lemma 7.1.

1: $\mathcal{D}_{\mathbb{C}^2}(S) = p + q + r + s$. In this case we have four real finite singular points, and according to Table 6.1 we consider the canonical form 1. Moreover without loss of generality we may assume that $M_1(0,0)$ is a star node, i.e. the conditions $d = e = 0$ and $c = f = 1$, after a rescaling, are satisfied. We thus get the family of systems:

$$\dot{x} = x - x^2 + 2hxy, \quad \dot{y} = y + 2mxy - y^2 \tag{7.11}$$

for which calculations yield

$$\begin{aligned} \mu_0 &= 1 - 4hm, \ \widetilde{K} = -4(mx^2 - xy + hy^2), \ Y_1 = U_3 = W_4 = 0, \\ U_2 &= 16m(1+m)x^2 - 32hmxy + 16h(1+h)y^2 = -4U_1, \\ W_3 &= 64(1+h)^2(1+m)^2(h+m+4hm)^2, \ \mathcal{T}_4 = -16hm(1+h+m), \\ \mathbf{D} &= -48(1+2h)^2(1+2m)^2, \ \eta = (1+2h)^2(1+2m)^2. \end{aligned} \tag{7.12}$$

Remark 7.1. We observe that the change $(x, y, h, m) \mapsto (y, x, m, h)$ preserves the form of systems (7.11).

By Table 5.1 on page 123 the condition $\mathbf{D} < 0$, holds and this implies $\eta > 0$, i.e. at infinity systems (7.11) have three real distinct singularities.

1.1: $\mu_0 < 0$. According to Theorem 6.1 (see Table 6.2), we have three saddles and one anti-saddle if $\widetilde{K} < 0$ and one saddle and three anti-saddles if $\widetilde{K} > 0$.

1.1.1: $\widetilde{K} < 0$. Then $m > 0$, $h > 0$, $1 - 4hm < 0$ and by (7.12) this implies $\mathcal{T}_4 \neq 0$. According to Theorem 6.2 systems (7.11) could not possess a weak singularity (which in this case must be a saddle). In addition, by Theorem 6.4, at infinity we have the configuration of singularities N^f, N^f, N^f. So we get the global configuration of singularities (α_1).

1.1.2: $\widetilde{K} > 0$. Since for systems (7.11) the conditions $W_4 = 0$ and $W_3 \geq 0$ hold, according to Table 6.2 of Theorem 6.1, we must have a saddle and three nodes.

The conditions $\widetilde{K} > 0$ and $\mu_0 < 0$ imply $m < 0$, $h < 0$ and $1 - 4hm < 0$. We claim that also in this case we could not have a weak saddle. Indeed the condition $\mathcal{T}_4 = 0$ gives $m = -(h+1)$, and then $\mu_0 = (1+2h)^2 \geq 0$, which contradicts $\mu_0 < 0$. So our claim is proved.

We remark that we could have additional nodes with coinciding eigenvalues. For this it is necessary to have $W_3 = 0$, i.e. considering (7.12) we have $(1+h)(1+m)(h+m+4hm) = 0$. We claim that in the case under consideration, we could not have two star nodes. Indeed according to Theorem 7.1 for this it is necessary that $U_1 = 0$. However by (7.12) this condition implies $hm = 0$, which contradicts the condition $1 - 4hm < 0$.

So the conditions $W_3 = 0$ and $W_2 \neq 0$ lead to one node with one direction, whereas if $W_3 = W_2 = 0$ we get two such nodes.

Taking into consideration that by Theorem 6.4 in the case $\mu_0 < 0$, $\widetilde{K} > 0$ and $\eta > 0$, at infinity we have the configuration of singularities S, S, N^∞, we obtain the three configurations of singularities (α_2), (α_2) and (α_4).

1.2: $\mu_0 > 0$. In this case $1 - 4hm > 0$, and according to Table 6.2 (see Theorem 6.1), systems (7.11) possess two saddles and two nodes.

To determine the types of infinite singularities, according to Theorem 6.4 we calculate

$$\theta = -64hm(1 + h + m) = 4\mathcal{T}_4,$$

$$\theta_1 = 64(h + h^2 + m + 7hm + 6h^2m + m^2 + 6hm^2 + 4h^2m^2).$$

1.2.1: $\mathcal{T}_4 \neq 0$. Then $\theta \neq 0$ and by Theorem 6.4 in the case $\theta < 0$ at infinity we have S, N^∞, N^∞ if $\theta_1 < 0$ and S, N^f, N^f if $\theta_1 > 0$. In the case $\theta > 0$ we have S, N^f, N^∞. Regarding finite singularities, we shall consider two possibilities: $W_3 \neq 0$ and $W_3 = 0$.

1. $W_3 \neq 0$. As it was mentioned above, apart from the star node, we have a generic node and hence we get the following three global configurations of singularities:

   (α_5) if $\theta < 0$ and $\theta_1 < 0$, (α_6) if $\theta < 0$ and $\theta_1 > 0$ and (α_7) if $\theta > 0$.

2. $W_3 = 0$. Then by (7.12) we have $(1+h)(1+m)(h+m+4hm) = 0$, and since $\theta = 4\mathcal{T}_4 \neq 0$, we have $hm(1 + h + m) \neq 0$.

   We observe that in this case $U_1 \neq 0$ (as $hm \neq 0$) and by Theorem 7.1 systems (7.11) could not have two star nodes.

   Regarding Remark 7.1, in order to examine the condition $W_3 = 0$, it is sufficient to consider two relations: $h = -1$ and $h = -m/(4m + 1)$ (we observe that in the case $4m + 1 = 0$ the condition $W_3 = 0$ is equivalent to $h = -1$).

   So assuming $h = -1$ (respectively $h = -m/(4m + 1)$), we obtain $\theta = 64m^2$ (respectively $\theta = 64m^2(1 + 2m)^2/(1 + 4m)^2$). Therefore due to $\mathcal{T}_4 \neq 0$ in both cases we get $\theta > 0$. Thus considering Theorem 6.4 we obtain the configuration of singularities (α_8).

1.2.2: $\mathcal{T}_4 = 0$. We shall consider again the two cases $W_3 \neq 0$ and $W_3 = 0$.

1. $W_3 \neq 0$. For systems (7.11) we have

$$\mathcal{F}_1 = 0,\ \mathcal{F}_i = \mathcal{T}_4\Phi_i(m, h)\ (i = 2, 3),$$

where $\Phi_i(m,h)$ are some polynomials in m and h. Therefore the condition $\mathcal{T}_4 = 0$ implies $\mathcal{F}_2 = \mathcal{F}_3 = 0$. We claim that due to $W_3 \neq 0$ the condition $\mathcal{T}_4 = 0$ implies $\mathcal{T}_3 \neq 0$. Indeed, by Remark 7.1 we shall consider two relations: $h = 0$ and $h = -(m+1)$. In the first case we get

$$\mathcal{T}_4 = \mathcal{F}_1 = \mathcal{F}_2 = \mathcal{F}_3 = 0, \quad \mathcal{T}_3 = -8m(1+m),$$
$$W_3 = 64m^2(1+m)^2, \ \theta = \theta_2 = 0, \ \theta_1 = 64m(1+m).$$

Assuming $h = -(m+1)$, we obtain

$$\mathcal{T}_4 = \mathcal{F}_1 = \mathcal{F}_2 = \mathcal{F}_3 = 0, \ \mathcal{T}_3 = -8m(1+m)(1+2m)^2,$$
$$W_3 = 64m^2(1+m)^2(1+2m)^4, \ \theta = \theta_2 = 0, \ \theta_1 = 64m(1+m)(1+2m)^2.$$

Thus the condition $W_3 \neq 0$ yields $\mathcal{T}_3 \neq 0$, and this proves our claim. According to Theorem 6.2 one of the saddles of systems (7.11) is integrable, i.e. we get the singularities $s, \$, n, n^*$.

Additionally, in both cases considered above, we have $\theta = \theta_2 = 0$ and $\theta_1 \neq 0$. Then according to Theorem 6.4, at infinity we have the singularities S, N^∞, N^* if $\theta_1 < 0$ and S, N^f, N^* if $\theta_1 > 0$. Therefore we arrive at the configurations (α_9) and (α_{10}), respectively.

2. $W_3 = 0$. The condition $\mathcal{T}_4 = 0 = W_3$ implies

$$(1+h)(1+m)(h+m+4hm) = hm(1+h+m) = 0.$$

Herein we get four solutions (h,m): $(-1,0)$, $(-1/2,-1/2)$, $(0,-1)$ and $(0,0)$. Taking into consideration Remark 7.1 and the condition $\mu_0 > 0$, we shall consider two of them: $(-1,0)$ and $(0,0)$. In both cases for systems (7.11) calculations yield

$$\begin{gathered}\mathcal{T}_4 = \mathcal{T}_3 = \mathcal{F}_1 = \mathcal{F}_2 = \mathcal{F}_3 = \mathcal{F} = 0, \ \mathcal{T}_2 = -4, \ \mathcal{B} = -8, \ \mathcal{H} = 4, \\ W_3 = 0, \ W_2 = 16, \ \eta = 1, \ \mu_0 = 1, \ \theta = \theta_1 = \theta_3 = \theta_4 = 0, \\ U_1 = U_4 = U_5 = 0, \ U_6 = -4xy - 4\varepsilon y^2, \ Y_2 = 8,\end{gathered} \tag{7.13}$$

where $\varepsilon = 1$ if $(h,m) = (-1,0)$ and $\varepsilon = 0$ if $(h,m) = (0,0)$. So according to Theorems 7.1, 6.2 and 6.4, we obtain the global configuration of singularities (α_{11}).

2: $\mathcal{D}_{\mathbb{C}^2}(S) = p+q+r^c+s^c$. Consider systems 2 from Table 6.1 possessing two real and two complex finite singularities. Assuming that the singular point $M_3(1,0)$ for these systems is a star node, we obtain $h = 0$, $l = b$ and $g = a + 2m$. In this case we have $\Delta_3 = 4m^2 \neq 0$, and we may consider $m = 1/2$ (due to a time rescaling). Therefore we get the 2-parameter family of systems:

$$\dot{x} = a - (1+2a)x + (1+a)x^2 + ay^2, \quad \dot{y} = b - 2bx + bx^2 + xy + by^2 \tag{7.14}$$

which, apart from the complex singularities $M_{1,2}(0, \pm i)$, possess the star node $M_3(1,0)$. For these systems calculations yield

$$\begin{gathered}\mu_0 = a + a^2 + b^2,\ \widetilde{K} = 2(1+a)x^2 + 4bxy - 2ay^2,\ W_4 = 0,\\ Y_1 = U_3 = 0,\ \ U_1 = (1+2a+4b^2)x^2 - 8abxy + 2a(1+2a)y^2 = -U_2/4,\\ W_3 = (a+2a^2+2b^2)^2[(1-2a)^2+4b^2]^2,\ \mathcal{T}_4 = -2a[(1+2a)^2+4b^2],\\ \mathbf{D} = 192(a^2+b^2)^2 > 0,\ \mathcal{F}_1 = 0,\ \ \eta = -4(a^2+b^2)^2.\end{gathered} \tag{7.15}$$

We observe that $\eta < 0$ (due to $\mathbf{D} > 0$), and hence at infinity there are one real and two complex singularities.

2.1: $\mu_0 < 0$. Then $a + a^2 + b^2 < 0$ and this implies $\widetilde{K} > 0$. According to Theorem 6.1 (see Table 6.2), apart from the star node, systems (7.14) possess an anti-saddle, which due to Lemma 7.1 must be a node. Regarding the infinite singularities since $\mu_0 < 0$, according to Theorem 6.4 we have the configuration S, ©, ©. We shall consider two possibilities: $W_3 \neq 0$ and $W_3 = 0$.

1. $W_3 \neq 0$. Then the second node is generic, i.e. a node with two directions. Since by (7.15) and $\mu_0 < 0$ the condition $W_3 \neq 0$ implies $\mathcal{T}_4 \neq 0$, we arrive in this case at the configuration of singularities (α_{12}).
2. $W_3 = 0$. In this case due to $\mu_0 < 0$ the condition $W_3 = 0$ yields $b^2 = -(a + 2a^2)/2$, and then $W_2 = a^2(6a-1)^2/4 \neq 0$ due to $a < 0$. So the eigenvalues of the second node of systems (7.14) coincide. According to Theorem 7.1, in order to have two star nodes, the condition $U_1 = 0$ is necessary. By (7.15) this condition is equivalent to $a = -1/2$ and $b = 0$, and in this case it is not too hard to convince ourselves that the condition $(\mathcal{S}^3_{n^*})$ of Theorem 7.1 is satisfied, i.e. the second node is a star node, too. In addition, from (7.15) we observe that the condition $U_1 = 0$ is equivalent to $\mathcal{T}_4 = 0$ (and clearly in this case $\mathcal{T}_3 = 0$ because both complex points must have traces zero).

   In such a way we get the following two global configurations of singularities: (α_{13}) if $U_1 \neq 0$ and (α_{14}) if $U_1 = 0$.

2.2: $\mu_0 > 0$. Then by Theorem 6.1 (see Table 6.2), apart from the star node, systems (7.14) possess one saddle.

1. $\mathcal{T}_4 \neq 0$. According to Theorem 6.2 the saddle cannot be a weak one, i.e. on the phase plane we have the real singularities s, n^*. For systems (7.14) we have
$$\theta = -8a[(1+2a)^2 + 4b^2] = 4\mathcal{T}_4,$$
   and hence the condition $\mathcal{T}_4 \neq 0$ implies $\theta \neq 0$.

   If $W_3 \neq 0$ then the corresponding discriminants of complex singularities are not zero. So considering Theorem 6.4, we obtain the configuration of singularities (α_{15}) if $\theta < 0$ and (α_{16}) if $\theta > 0$.

Assume now $W_3 = 0$, i.e. by (7.15) we have $a = 1/2$ and $b = 0$, which implies $W_2 = 0$. Then, apart from the star node $M_3(0,0)$, we have the strong saddle $M_4(1,0)$. Since for the system (7.14) with the conditions $a = 1/2$ and $b = 0$ we have $\mu_0 = 3/4 > 0$, $\eta = -1/4 < 0$ and $\theta = -16$, by Theorem 6.4 we arrive at the configuration (α_{17}).

2. $\mathcal{T}_4 = 0$. As $\mu_0 > 0$ the condition $\mathcal{T}_4 = 0$ is equivalent to $a = 0$ (and then $b \neq 0$ due to $\mathbf{D} \neq 0$). So we obtain

$$\mathcal{T}_4 = \mathcal{F}_1 = \mathcal{F}_2 = \mathcal{F}_3 = 0,\ \mathcal{T}_3\mathcal{F} = b^4(1+4b^2)^2/2 > 0,\ W_3 = 4b^4(1+4b^2)^2,$$

and by Theorem 6.2, apart from the star node, systems (7.14) possess an integrable saddle. For these systems in the case $a = 0$, we obtain $\theta = \theta_2 = 0$, and according to Theorem 6.4, apart from two complex singularities, at infinity we have a star node. So we get the global configuration of singularities given by (α_{18}).

3: $\mathcal{D}_{\mathbb{C}^2}(S) = 2p+q+r$. In this case we consider the normal form 4 (see Table 6.1) and assume that the simple singular point $M_4(0,1)$ is a star node. Then taking into account the conditions from Table 6.1 corresponding to this normal form, we obtain $c = 0$, $h = -eu/2$ and $m = -e/2$. Then $\mu_0 = e^4u^3 \neq 0$, and due to a time rescaling we may assume $e = 1$. So we get the 1-parameter family of systems:

$$\dot{x} = -uxy, \quad \dot{y} = x + uy - x^2 - xy - uy^2, \tag{7.16}$$

for which calculations yield

$$\begin{gathered} \mu_0 = u^3,\ \widetilde{K} = 2u(-x^2+uy^2),\ W_4 = 0,\ W_3 = u^{10}(1+u)^2, \\ E_1 = -u^7/2,\ Y_1 = U_3 = 0,\ \ U_1 = (2u-1)x^2 + 2uxy + u^2y^2, \\ \mathcal{T}_4 = 2(1-u)u^6,\ \eta = 0,\ \widetilde{M} = -8x^2. \end{gathered} \tag{7.17}$$

3.1: $\mu_0 < 0$. Then $u < 0$ and this implies $\widetilde{K} > 0$ and $E_1 \neq 0$. As $W_4 = 0$ and $W_3 \geq 0$, according to Theorem 6.1 (see Table 6.2), apart from the star node, systems (7.16) possess a node and a semi-elemental saddle-node. As by (7.17) we have $U_1 \neq 0$, considering Theorem 7.1 there could not be two star nodes. Therefore we could have only two possibilities: $\overline{sn}_{(2)}, n, n^*$ if $W_3 \neq 0$ and $\overline{sn}_{(2)}, n^d, n^*$ if $W_3 = 0$.

Additionally, as $\eta = 0$ and $\widetilde{M} \neq 0$, according to Theorem 6.4, in the case $\mu_0 < 0$, systems (7.16) have at infinity a saddle and a saddle-node (semi-elemental). Therefore we get the following two global configurations of singularities: (α_{19}) if $W_3 \neq 0$ and (α_{20}) if $W_3 = 0$.

3.2: $\mu_0 > 0$. Then $u > 0$ and as $E_1 \neq 0$ and $W_4 = 0$, according to Theorem 6.1 (see Table 6.2), apart from the star node and the semi-elemental saddle-node, systems (7.16) possess a saddle.

If $\mathcal{T}_4 \neq 0$ then by Theorem 6.2 this saddle is strong. For systems (7.16) we have $\theta = 8(1-u)u^2$, and due to $\mathcal{T}_4 \neq 0$ we get $\theta \neq 0$. Therefore since $\mu_0 > 0$, by Theorem 6.4 we arrive at the configuration (α_{21}) if $\theta < 0$, and (α_{22}) if $\theta > 0$.

Assume now $\mathcal{T}_4 = 0$. Then $u = 1$ and this implies

$$\mathcal{F}_1 = \mathcal{F}_2 = \mathcal{F}_3 = 0,\ \mathcal{T}_3\mathcal{F} = 1/2 > 0,$$

and hence by Theorem 6.2 we obtain an integrable saddle. Additionally we have $\theta = \theta_2 = 0$, the infinite node is a star node and we get the configuration (α_{23}).

4: $\mathcal{D}_{\mathbb{C}^2}(S) = 3p + q$. In this case we consider the normal form 8 (see Table 6.1) and force the elemental singular point $M_4(1,0)$ to be a star node. Then we obtain $e = 0$, $h = -cu/2$ and $m = -c/2 \neq 0$. Due to a time rescaling we may assume $c = 1$, and this leads to the 2-parameter family of systems:

$$\dot{x} = x + uy - x^2 - uxy + (v - u^2)y^2, \quad \dot{y} = -y(x + uy), \tag{7.18}$$

for which calculations yield

$$\mu_0 = u^2 - v,\ E_3 = (u^2 - v)/4,\ W_4 = 0,\ G_{10} = (v - u^2)/8.$$

As $\mu_0 \neq 0$ (see Table 5.1 on page 123) we obtain $E_3 G_{10} \neq 0$ and $\text{sign}\,(E_3) = \text{sign}\,(\mu_0)$. Therefore according to Theorem 6.1 (see Table 6.2), apart from the star node, systems (7.18) possess a semi-elemental triple node if $\mu_0 < 0$ and a semi-elemental triple saddle if $\mu_0 > 0$. Regarding infinite singularities, we calculate

$$\eta = \widetilde{M} = 0,\ C_2 = (v - u^2)y^3 \neq 0.$$

Therefore considering Theorem 6.4 we obtain the configuration (α_{24}) if $\mu_0 < 0$ and (α_{25}) if $\mu_0 > 0$.

5: $\mathcal{D}_{\mathbb{C}^2}(S) = p + q + r$. Assuming that the singular point $M_1(0,0)$ for canonical form 10 (see Table 6.1) is a star node, we obtain $d = e = 0$ and $f = c$. Then considering the conditions from Table 6.1 under the parameters corresponding to systems 10, we get $chm \neq 0$, and due to the rescaling $(x, y, t) \mapsto (x, cy/h, t/c)$, we may assume $c = 1 = h$. Therefore we get the 1-parameter family of systems:

$$\dot{x} = x(1 - x + 2y), \quad \dot{y} = y(1 + 2mx), \quad m \neq 0, \tag{7.19}$$

for which calculations yield

$$\begin{aligned} &\mu_0 = W_4 = 0,\ \mu_1 = 4mx,\ \eta = 4(1+2m)^2,\ W_3 = 64(1+m)^2(1+4m)^2,\\ &\widetilde{K} = -4mx^2,\ \mathcal{T}_4 = -16m,\ \mathbf{D} = -192(1+2m)^2 < 0,\ Y_1 = U_3 = 0,\\ &U_2 = 16\big[m(1+m)x^2 - 2mxy + y^2\big] = -4U_1,\ G_9 = -m,\ \kappa = 128m. \end{aligned} \tag{7.20}$$

As $m \neq 0$ we have $G_9 \neq 0$, and since $W_4 = 0$ and $W_3 \geq 0$, according to Theorem 6.1 (see Table 6.2), apart from the star node, systems (7.19) possess two saddles if $\widetilde{K} < 0$, and one node and one saddle if $\widetilde{K} > 0$. We observe that by Theorem 6.2 these systems could not possess weak singularities as $\mathcal{T}_4 \neq 0$.

5.1: $\widetilde{K} < 0$. Then $m > 0$, and this implies $\eta > 0$ and $\kappa > 0$. Therefore considering Theorem 6.4 we get the configuration (α_{26}).

5.2: $\widetilde{K} > 0$. We have $m < 0$, and apart from the star node, systems (7.19) possess a node and a saddle.

We observe that due to $\mathbf{D} \neq 0$ we have $\eta \neq 0$, and then $\eta > 0$. Since $\kappa < 0$, by Theorem 6.4 we have at infinity the configuration $\overline{\binom{1}{1}}SN,\ S,\ N^{\infty}$.

It remains to examine the condition $W_3 = 0$, which could provide one more finite node with coinciding eigenvalues.

1. $W_3 \neq 0$. Then the second finite node is generic (with two directions), and we arrive at the configuration (α_{27}).
2. $W_3 = 0$. By (7.20) we obtain $(m+1)(4m+1) = 0$, and we have either $m = -1$ or $m = -1/4$. In both cases we have $U_1 \neq 0$, and by Theorem 7.1 we could not have two star nodes. This leads to a single configuration, namely (α_{28}).

6: $\mathcal{D}_{\mathbb{C}^2}(S) = p + q^c + r^c$. In this case we consider the normal form 11 (see Table 6.1) and force the elemental real singular point $M_1(0,0)$ to be a star node. Then we obtain $g = 0$, $m = lu$ and $h = l(1+u^2)/2 \neq 0$. Due to a time rescaling we may assume $l = 1$, and we get the family of systems:

$$\dot{x} = (1+u^2)x(1-y), \quad \dot{y} = (1+u^2)y + x^2 - 2uxy,$$

for which calculations yield

$$\mu_0 = 0,\ \mu_1 = (1+u^2)^3 x \neq 0,\ \eta = -4(1+u^2)^2 < 0.$$

Therefore considering Theorem 6.4 we get a single configuration given by (α_{29}). It remains to observe that for the complex singularities $M_{2,3}(u \pm i, 1)$ we have $\rho_{2,3} = -(u \pm i)^2$ and $\tau_{2,3} = (u \pm i)^2(u \mp 3i)^2$. Since $u \in \mathbb{R}$ the above systems could not have, apart from a star node, complex singularities with zero traces or zero discriminants.

7: $\mathcal{D}_{\mathbb{C}^2}(S) = 2p + q$. Assuming that the singular point $M_3(1,0)$ for the canonical form 12 (see Table 6.1) is a star node, we obtain $e = 0$, $h = -c/2$ and $m = -c/2 \neq 0$. Then due to a time rescaling we may assume $c = 1$, and we get the system

$$\dot{x} = (1-x)(x+y), \quad \dot{y} = -xy,$$

for which calculations yield

$$\mu_0 = 0,\ \mu_1 = -x \neq 0,\ W_4 = 0,\ W_3 = 1,\ \widetilde{K} = 2x^2 > 0,$$
$$E_1 = -1/2,\ G_9 = 1/8,\ \eta = 0,\ \widetilde{M} = -8y^2,\ \kappa = -16.$$

So we have $G_9E_1 \neq 0$, and since $W_4 = 0$ and $\widetilde{K} > 0$, according to Theorem 6.1 (see Table 6.2), apart from the star node, the above systems possess one semi-elemental saddle-node. Additionally at infinity we have two real singularities, and considering Theorem 6.4, we arrive at the configuration (α_{30}).

8: $\mathcal{D}_{\mathbb{C}^2}(S) = p + q$. In these cases we have to consider two canonical systems from Table 6.1: 14a (when quadratic parts $p_2(x, y)$ and $q_2(x, y)$ are not proportional, i.e. $\widetilde{K} = \text{Jacob}\,(p_2, q_2) \neq 0$) and 14b (when p_2 and q_2 are proportional, i.e. $\widetilde{K} = 0$).

8.1: Systems 14a. Forcing the elemental singular point $M_1(0, 0)$ to be a star node for these systems (see Table 6.1), we obtain $d = e = 0$ and $f = c \neq 0$. Then due to a time rescaling we may assume $c = 1$, and we get the 1-parameter family of systems:

$$\dot{x} = x(1 - x), \quad \dot{y} = y(1 + 2ux), \quad u(2u + 1) \neq 0, \tag{7.21}$$

for which calculations yield

$$\begin{gathered}\mu_0 = \mu_1 = 0, \ \mu_2 = (1 + 2u)x^2, \ \widetilde{K} = -4ux^2, \ W_4 = Y_1 = U_3 = G_8 = 0,\\ U_2 = 16u(1 + u)x^2 = -4U_1, \ F_2 = (1 + 2u)x^2, \ \eta = \kappa = \kappa_1 = 0,\\ \widetilde{M} = -8(1 + 2u)^2x^2, \ \tilde{L} = 8(1 + 2u)x^2, \ K_2 = 96(1 + 3u + 4u^2)x^2.\end{gathered} \tag{7.22}$$

According to Theorem 6.4 (see Diagram 6.7), in the case $\mu_0 = \mu_1 = 0$ and $\mu_2\widetilde{K} \neq 0$, the configurations of infinite singularities depend on the sign of the invariant polynomial μ_2. So we consider two cases: $\mu_2 < 0$ and $\mu_2 > 0$.

1. $\mu_2 < 0$. Then $u < -1/2$ and this implies $\widetilde{K} > 0$ and $F_2 < 0$. As $G_8 = 0$, according to Theorem 6.1 (see Table 6.2, line 164), apart from the star node, systems (7.21) possess the node $M_2(1, 0)$ for which we have $\tau_2 = 4(1 + u)^2$ (τ_2 is the respective discriminant). So we have a simple node if $u \neq -1$ and a node with two coinciding eigenvalues if $u = -1$. However in the second case we calculate

   $$U_1 = U_4 = U_5 = 0, \ U_6 = x^2 \neq 0, \ Y_2 = 2 > 0,$$

   and by Theorem 7.1 the second point is also a star node.

   Additionally, for $u < -1/2$ considering (7.22) we obtain $\tilde{L} < 0$, and as $\eta = 0$, $\widetilde{M} \neq 0$, $\mu_2 < 0$ and $\kappa = \kappa_1 = 0$, according to Theorem 6.4 we get the configuration (α_{31}) if $U_1 \neq 0$ and (α_{32}) if $U_1 = 0$.

2. $\mu_2 > 0$. In this case $u > -1/2$ and this yields $F_2 > 0$. As $G_8 = 0$ and $\widetilde{K} \neq 0$, according to Theorem 6.1 (see Table 6.2, line 150), apart from the star node, systems (7.21) possess the saddle $M_2(1, 0)$ for which we have $\rho_2 = u$ (ρ_2 is the corresponding trace). So as $u \neq 0$ (due to $\widetilde{K} \neq 0$) this saddle could not be weak.

Additionally, we have $\mu_2 > 0$, $\tilde{L} > 0$, $K_2 > 0$ and either $\widetilde{K} < 0$ (if $u > 0$) or $\widetilde{K} > 0$ (if $-1/2 < u < 0$). So according to Theorem 6.4, we arrive at the configuration (α_{33}) if $\widetilde{K} < 0$ and (α_{34}) if $\widetilde{K} > 0$.

8.2: Systems 14b. Assuming that the singular point $M_1(0,0)$ for this canonical form (see Table 6.1) is a star node, we obtain $u = h = 0$ and $g = -1$. Then we get the family of systems:

$$\dot{x} = x - x^2 + ky^2, \quad \dot{y} = y, \tag{7.23}$$

for which calculations yield

$$\mu_0 = \mu_1 = 0, \ \mu_2 = x^2 - ky^2 \neq 0, \ \widetilde{K} = 0, \ W_4 = 0, \ \mathbf{D} = -192k, \ G_3 = 0,$$
$$G_8 = -2k, \ W_7 = 108k^2, \ \eta = 4k, \ \widetilde{M} = -8(x^2 + 3ky^2),$$
$$\kappa = \theta_2 = 0, \ \mu_2\tilde{L} = 8(x^2 - ky^2)^2.$$

1. $\eta \neq 0$. Then $k \neq 0$ and this implies $\mathbf{D}G_8 \neq 0$. As $W_4 = \widetilde{K} = 0$, according to Theorem 6.1 (see Table 6.2, line 151), apart from the star node, systems (7.23) possess the saddle $M_2(1,0)$. Since for these systems we have

$$\mathcal{T}_4 = \mathcal{F}_1 = \mathcal{F}_2 = \mathcal{F}_3 = 0, \ \mathcal{T}_3\mathcal{F} = 8k^2 > 0,$$

by Theorem 6.2 it follows that this saddle is integrable. Additionally as $\kappa = \theta_2 = 0$ and $\mu_2\tilde{L} > 0$, considering Theorem 6.4 we get the configuration (α_{35}) if $\eta < 0$, and (α_{36}) if $\eta > 0$.

2. $\eta = 0$. Then $k = 0$ and we have $\mathbf{D} = W_7 = 0$. As $G_3 = \widetilde{K} = 0$ and $D_2 = -2x \neq 0$, according to Theorem 6.1 (see Table 6.2, line 154), we again have a star node and a saddle. Moreover, the saddle $M_2(1,0)$ is again integrable. Indeed, for systems (7.23) with $k = 0$, we calculate

$$\mathcal{T}_i = 0\,(i = 1,2,3,4), \ \sigma = 2 - 2x, \ \mathcal{G} = -8x^2,$$
$$\mathcal{F}_1 = \mathcal{H} = \mathcal{B} = 0, \ \mathcal{B}_j = 0\,(j = 1,2,3),$$

and therefore by Theorem 6.2 (see statement $(e_3)[\delta]$) this system has an integrable saddle.

Additionally, in order to determine the configuration of singularities at infinity for the system under examination, we have

$$\eta = 0, \ \widetilde{M} \neq 0, \ \kappa = \kappa_1 = \widetilde{K} = \theta_6 = 0, \ \mu_2 \neq 0, \ \tilde{L} = 8x^2, \ K_2 = 96x^2 > 0,$$

and considering Theorem 6.4 we arrive at the global configuration of singularities given by (α_{37}).

9: $\mathcal{D}_{\mathbb{C}^2}(S) = p$. Similarly to the previous case we have to consider two canonical systems from Table 6.1, namely systems 17a and systems 17b.

9.1: Systems 17a. Assuming that the singular point $M_1(0,0)$ for this canonical form (see Table 6.1) is a star node, we obtain $d = e = 0$ and $f = c \neq 0$. Then due to a time rescaling we may assume $c = 1$, and we get the system

$$\dot{x} = x(1+2x), \quad \dot{y} = y + x^2 + 2xy,$$

for which we have

$$\mu_0 = \mu_1 = \mu_2 = 0,\ \mu_3 = x^3,\ \eta = \widetilde{M} = 0,\ C_2 = -x^3,\ \widetilde{K} = 8x^2,\ K_3 = 6x^6.$$

So according to Theorem 6.4 this leads to the configuration (α_{38}).

9.2: Systems 17b. Forcing the singular point $M_1(0,0)$ for this canonical form (see Table 6.1) to be a star node, we obtain $d = e = 0$ and $f = 1$. Then we get the family of systems:

$$\dot{x} = x, \quad \dot{y} = y + lx^2 + 2mxy, \tag{7.24}$$

for which calculations yield

$$\mu_0 = \mu_1 = \mu_2 = \eta = 0,\ \mu_3 = x^2(lx+2my) \neq 0,\ \widetilde{M} = -32m^2x^2,\ C_{2|_{m=0}} = -lx^3,$$
$$\widetilde{K} = \kappa = \tilde{L} = \kappa_1 = 0,\ \mu_3 K_1 = x^4(lx+2my)^2 > 0,\ K_{3|_{m=0}} = 6l^2x^6.$$

So systems (7.24) possess two real distinct infinite singularities if $\widetilde{M} \neq 0$ and only one if $\widetilde{M} = 0$. Considering Theorem 6.4, we arrive at the configuration (α_{39}) if $\widetilde{M} \neq 0$ and (α_{40}) if $\widetilde{M} = 0$.

This completes the proof of Theorem 7.2. □

7.2 An integrable saddle

Assume that a quadratic system (7.1) possesses a weak saddle. Then according to [70] (see also [185]), via an affine transformation this system can be written as

$$\dot{x} = x + ax^2 + bxy + cy^2, \quad \dot{y} = -y - kx^2 - lxy - my^2, \tag{7.25}$$

with the weak saddle at the origin. For these systems Cai Sui Lin [70] has calculated the first three saddle quantities:

$$\begin{aligned} L_1^* &= lm - ab, \\ L_2^* &= kb(2m-b)(m+2b) - cl(2a-l)(a+2l), \\ L_3^* &= (ck-lb)\big[acl(2a-l) - bkm(2m-b)\big]. \end{aligned}$$

Clearly in order to have an integrable saddle, the condition $L_1^* = L_2^* = L_3^* = 0$ needs to be satisfied.

We observe that the condition $b = l = 0$ implies $L_1^* = L_2^* = L_3^* = 0$, and we get the family of systems:

$$\dot{x} = x + ax^2 + cy^2, \quad \dot{y} = -y - kx^2 - my^2, \tag{7.26}$$

which possess an integrable saddle at the origin of coordinates.

In the case $b^2 + l^2 \neq 0$ we have the next result.

Lemma 7.2. *If a quadratic system (7.25) with $b^2 + l^2 \neq 0$ possesses an integrable saddle, then via a linear transformation and a time rescaling this system could be brought to one of the following canonical forms if and only if the corresponding conditions are satisfied:*

$$\mathcal{T}_3 \neq 0 \quad \Leftrightarrow \quad \begin{cases} \dot{x} = x - 2ckx^2 + xy + cy^2, \\ \dot{y} = -y - kx^2 - ckxy + 2y^2; \end{cases} \tag{7.27}$$

$$\mathcal{T}_3 = 0,\ \mathcal{T}_2 \neq 0 \quad \Leftrightarrow \quad \begin{cases} \dot{x} = x + mx^2 + xy + cy^2, \\ \dot{y} = -y - cx^2 - xy - my^2; \end{cases} \tag{7.28}$$

$$\mathcal{T}_3 = \mathcal{T}_2 = 0,\ \sigma \neq 0 \quad \Leftrightarrow \quad \begin{cases} \dot{x} = x + lmx^2 + xy + cy^2, \\ \dot{y} = -y - cl^3x^2 - lxy - my^2, \\ l(1 + 3cl - m)(1 + cl + m) = 0; \end{cases} \tag{7.29}$$

$$\mathcal{T}_3 = \mathcal{T}_2 = \sigma = 0 \quad \Leftrightarrow \quad \begin{cases} \dot{x} = x + lx^2/2 + xy + cy^2, \\ \dot{y} = -y - kx^2 - lxy - y^2/2. \end{cases} \tag{7.30}$$

Proof. Assume that for systems (7.25) the condition $b^2 + l^2 \neq 0$ holds. Then without loss of generality we may consider $b \neq 0$ due to the change $(x, y, t, a, b, c, k, l, m) \mapsto (y, x, -t, m, l, k, c, b, a)$. Moreover we may consider $b = 1$ due to the rescaling $y \to y/b$, and then the condition $L_1^* = lm - a = 0$ gives $a = lm$. Herein we calculate

$$\begin{aligned} &L_2^* = (k - cl^3)(2 + m)(2m - 1), \\ &L_3^* = (l - ck)(k - cl^3)m(2m - 1), \\ &\mathcal{T}_4 = 0, \quad \mathcal{T}_3 = (k - cl^3)(2m - 1))^3, \\ &\mathcal{T}_2 = (2m - 1)^2(2k + 3ckl + l^2 + 2cl^3 + km + cl^3m - l^2m^2), \\ &\mathcal{T}_1 = (k - cl^3)(2m - 1)(1 + 2m), \quad \sigma = (2m - 1)(lx - y), \end{aligned} \tag{7.31}$$

and we shall force the condition $L_2^* = 0$.

1. If $\mathcal{T}_3 \neq 0$ then we have $(k - cl^3)(2m - 1) \neq 0$. The condition $L_2^* = 0$ gives $m = -2$, and then $L_3^* = 10(l - ck)(k - cl^3) = 0$. As $\mathcal{T}_3 \neq 0$ (i.e. $k \neq cl^3$) we obtain $l = ck$, and this leads to the family of systems (7.27).

2. Assume now $\mathcal{T}_3 = 0$ and $\mathcal{T}_2 \neq 0$. Taking into account (7.31) the condition $L_2^* = 0$ yields $k = cl^3$, and this implies $L_3^* = 0$. In this case we have

$$\mathcal{T}_2 = l^2(1 + 3cl - m)(1 + cl + m)(2m - 1)^2, \tag{7.32}$$

and since $\mathcal{T}_2 \neq 0$ this implies $l \neq 0$. Then we may assume $l = 1$ due to the changes $x \to x/l$ and $c \to c/l$, and this leads to the family of systems (7.28).

3. In the case $\mathcal{T}_3 = \mathcal{T}_2 = 0$ and $\sigma \neq 0$ considering (7.31) the condition $\sigma \neq 0$ gives $2m - 1 \neq 0$, and then $\mathcal{T}_3 = 0$ implies $k = cl^3$. Since $\mathcal{T}_2 = 0$ considering (7.32) we arrive at the family of systems (7.29) for which the indicated condition must be satisfied.

4. Suppose finally $\mathcal{T}_3 = \mathcal{T}_2 = \sigma = 0$. Then by (7.31) we obtain $m = 1/2$ and this leads to the family of Hamiltonian systems (7.30).

This completes the proof of Lemma 7.2. □

Theorem 7.3. *We consider here all the configurations of singularities, finite and infinite, of quadratic vector fields possessing an integrable saddle. There are exactly 77 geometrically distinct configurations of singularities, finite and infinite, and all of them are contained in the following list, enumerated and grouped according to the conditions* $(\mathcal{A})$, $(\mathcal{B})$, $(\mathcal{C})$ *and* $(\mathcal{D})$ *of Theorem 6.3:*

$(\mathcal{A})$

(β_1) $\$, f, \textcircled{c}, \textcircled{c};\ N^\infty, \textcircled{C}, \textcircled{C};$ (β_2) $\$, f, \textcircled{c}, \textcircled{c};\ N^f, \textcircled{C}, \textcircled{C};$
(β_3) $s, \$, n, n;\ S, N^\infty, N^\infty;$ (β_4) $s, \$, n, n;\ S, N^f, N^f;$
(β_5) $s, \$, n, n;\ S, N^\infty, N^f;$ (β_6) $s, \$, n, n^*;\ S,\ N^\infty,\ N^*;$
(β_7) $s, \$, n, n^*;\ S,\ N^f,\ N^*;$ (β_8) $\$, n^*, \textcircled{c}, \textcircled{c};\ N^*, \textcircled{C}, \textcircled{C};$
(β_9) $\$, n^d, \overline{sn}_{(2)};\ \overline{\binom{0}{2}}SN,\ N^\infty;$ (β_{10}) $\$, n^d, \overline{sn}_{(2)};\ \overline{\binom{0}{2}}SN,\ N^f;$
(β_{11}) $\$, n^*, \overline{sn}_{(2)};\ \overline{\binom{0}{2}}SN,\ N^*;$ (β_{12}) $\$, f;\ N^d,\ \binom{1}{1}\textcircled{C}, \binom{1}{1}\textcircled{C};$
(β_{13}) $\$, n^*;\ N^*,\ \binom{1}{1}\textcircled{C}, \binom{1}{1}\textcircled{C};$ (β_{14}) $\$, n;\ \overline{\binom{1}{1}}SN,\ \overline{\binom{1}{1}}NS,\ N^d;$
(β_{15}) $\$, n^*;\ \overline{\binom{1}{1}}SN,\ \overline{\binom{1}{1}}NS,\ N^*;$ (β_{16}) $\$, n, n, n;\ S,\ S,\ N^\infty;$
(β_{17}) $\$, n, \textcircled{c}, \textcircled{c};\ N^\infty,\ \textcircled{C}, \textcircled{C};$

$(\mathcal{B})$

(β_{18}) $\$, \$, n, n;\ S, N^\infty, N^\infty;$ (β_{19}) $\$, \$, n, n;\ S, N^f, N^f;$
(β_{20}) $\$, \$, n^*, n^*;\ S, N^*, N^*;$ (β_{21}) $\$, c, \textcircled{c}, \textcircled{c};\ N^\infty,\ \textcircled{C}, \textcircled{C};$
(β_{22}) $\$, c, \textcircled{c}, \textcircled{c};\ N^f,\ \textcircled{C}, \textcircled{C};$ (β_{23}) $\$, c;\ \overline{\binom{1}{1}}SN,\ \overline{\binom{1}{1}}SN,\ N^d;$
(β_{24}) $\$, n, n, c;\ S, \textcircled{C}, \textcircled{C};$ (β_{25}) $\$, n^d, n^d, c;\ S, \textcircled{C}, \textcircled{C};$
(β_{26}) $s, s, \$, c;\ N^f, N^f, N^f;$ (β_{27}) $\$, n, n, c;\ S,\ S,\ N^\infty;$
(β_{28}) $\$, n, n, c;\ \binom{0}{3}\overline{S};$ (β_{29}) $\$, \$, \textcircled{c}, \textcircled{c};\ N^f,\ N^f,\ N^f;$
(β_{30}) $\$, \$, n, n;\ N^\infty,\ \textcircled{C}, \textcircled{C};$ (β_{31}) $\$, \$, n^d, n^d;\ S, N^d, N^d;$
(β_{32}) $\$, \$, n, n; \overline{\binom{0}{3}}N;$ (β_{33}) $\$, c, \textcircled{c}, \textcircled{c};\ S, N^\infty, N^\infty;$
(β_{34}) $\$, c, \textcircled{c}, \textcircled{c};\ S, N^f, N^f;$ (β_{35}) $\$, c, \textcircled{c}^\tau, \textcircled{c}^\tau;\ S, N^d, N^d;$
(β_{36}) $\$, c, \textcircled{c}, \textcircled{c}; \overline{\binom{0}{3}}N;$ (β_{37}) $\$, c;\ \overline{\binom{2}{1}}N,\ \textcircled{C}, \textcircled{C};$
(β_{38}) $\$, c;\ \overline{\binom{2}{1}}S, N^f, N^f;$ (β_{39}) $\$, \$;\ \overline{\binom{2}{1}}N, N^f, N^f;$
(β_{40}) $\$, c;\ N^d,\ \binom{1}{1}\textcircled{C}, \binom{1}{1}\textcircled{C};$

(β_{41}) $\$,\ \widehat{es}_{(3)};\ \overline{\binom{0}{3}}N;$ (β_{42}) $\$, n^d; \widehat{\binom{2}{2}}\, \widetilde{P}\widehat{P}H - \widetilde{P}\widehat{P}H, N^d;$
(β_{43}) $\$, n^*; \binom{2}{2}\widetilde{P}\widehat{P}H - \widetilde{P}\widehat{P}H, N^*;$ (β_{44}) $\$;\ \widehat{\binom{1}{2}}\, \widetilde{P}_{\curlywedge}E\widehat{P}_{\curlywedge} - H, \overline{\binom{2}{1}}N;$
(β_{45}) $\$,\ \overline{s}_{(3)}; N^f, N^f, N^f;$ (β_{46}) $\$,\ \widehat{es}_{(3)};\ N^\infty, ©, ©;$
(β_{47}) $\$,\ \widehat{es}_{(3)};\ S,\ N^\infty,\ N^\infty;$ (β_{48}) $\$,\ \widehat{es}_{(3)};\ S,\ N^f,\ N^f;$

(C)

(β_{49}) $\$,\ \widehat{es}_{(3)};\ S,\ N^d,\ N^d;$ (β_{50}) $\$, n, n; \widehat{\binom{1}{2}}H_{\curlywedge}HH_{\curlywedge} - H, N^\infty;$
(β_{51}) $\$, n, n; \widehat{\binom{1}{2}}\widetilde{P}_{\curlywedge}H\widehat{P}_{\curlywedge} - E, S;$ (β_{52}) $\$, n, n;\ \widehat{\binom{1}{2}}\ \widetilde{P}_{\curlywedge}E\widehat{P}_{\curlywedge} - H,\ S;$
(β_{53}) $\$, n^d, n^d;\ \widehat{\binom{1}{2}}\ H - E,\ S;$ (β_{54}) $\$, n, n;\ [\infty;\ \emptyset];$
(β_{55}) $\$, ©, ©; \widehat{\binom{1}{2}}\widetilde{P}_{\curlywedge}E\widehat{P}_{\curlywedge} - H, N^f;$ (β_{56}) $s, \$; \binom{2}{2}\widetilde{P}E\widehat{P} - \widetilde{P}E\widehat{P}, N^f;$
(β_{57}) $\$, n; \binom{2}{2}\ \widehat{P}\widetilde{P}E - \widehat{P}\widetilde{P}E, S;$ (β_{58}) $\$, n; \binom{2}{2}\widetilde{P}\widehat{P}H - \widetilde{P}\widehat{P}H, N^\infty;$
(β_{59}) $\$, n; \binom{2}{2}\widetilde{P}\widehat{P}H - \widetilde{P}\widehat{P}H, N^f;$ (β_{60}) $\$, n;\ \binom{2}{3}\ H\widehat{P}E - \widehat{P}\widehat{P}\widehat{P};$
(β_{61}) $\$, n;\ [\infty,\ N];$ (β_{62}) $\$;\ \binom{3}{2}\ \widetilde{P}E\widehat{P} - \widetilde{P}\widehat{P}H, N^f;$
(β_{63}) $\$;\ \widehat{\binom{3}{2}}\ \widetilde{P}_{\curlywedge}E\widehat{P}_{\curlywedge} - H,\ N^d;$ (β_{64}) $\$;\ \binom{2}{2}\ \widetilde{P}E - \widetilde{P}E, \overline{\binom{1}{1}}SN;$

(D)

(β_{65}) $\$, c, ©^\rho, ©^\rho;\ N^f,\ ©, ©;$ (β_{66}) $\$;\ \binom{3}{3}\ \widetilde{P}_{\curlywedge}EE\widehat{P}_{\curlywedge} - \widetilde{P}\widehat{P};$
(β_{67}) $\$, \$, \$, c;\ N^f, N^f, N^f;$ (β_{68}) $\$, \$, ©^\rho, ©^\rho;\ N^f, N^f, N^f;$
(β_{69}) $\$, \$, \widehat{cp}_{(2)};\ N^f, N^f, N^f;$ (β_{70}) $\$, \widehat{s}_{(3)};\ N^f, N^f, N^f;$
(β_{71}) $\$, \$, c, c;\ N^f, ©, ©;$ (β_{72}) $\$, c, \widehat{cp}_{(2)};\ N^f, ©, ©;$
(β_{73}) $\$, \$, c;\ \widehat{\binom{1}{2}}\ \widetilde{P}_{\curlywedge}E\widehat{P}_{\curlywedge} - H,\ N^f;$ (β_{74}) $\$, ©^\rho, ©^\rho; \widehat{\binom{1}{2}}\widetilde{P}_{\curlywedge}E\widehat{P}_{\curlywedge} - H, N^f;$
(β_{75}) $\$, \widehat{cp}_{(2)}; \widehat{\binom{1}{2}}\widetilde{P}_{\curlywedge}E\widehat{P}_{\curlywedge} - H, N^f;$ (β_{76}) $\$, \$;\ \binom{2}{2}\ \widetilde{P}E\widehat{P} - \widetilde{P}E\widehat{P},\ N^f;$
(β_{77}) $\$, c;\ \widehat{\binom{2}{3}}\ \widetilde{P}_{\curlywedge}\widehat{P} - \widetilde{P}_{\curlywedge}\widehat{P}.$

The realizability of each one of these configurations will be proved by means of an example provided in Part II.

Proof. Assume that a quadratic system (7.1) possesses an integrable saddle. Then for this system one of the sets of the conditions $(\mathcal{A})$, $(\mathcal{B})$, $(\mathcal{C})$ or $(\mathcal{D})$ of Theorem 6.3 must be satisfied. In what follows we consider each one of these sets of conditions.

1: Case $(\mathcal{A})$. Since the system has an integrable saddle and $\mathcal{T}_3 \neq 0$ considering Lemma 7.2, we have to examine each one of the canonical forms (7.26) and (7.27).

1.1: Systems (7.26). For these systems we calculate

$$\begin{aligned}&\mathcal{T}_3 = -8(a^3c - km^3), \quad \eta = a^2m^2 - 4a^3c - 27c^2k^2 + 18ackm - 4km^3,\\ &\mathbf{D} = -48\eta, \quad W_4 = 256c^2k^2\eta, \quad \mu_0 = (ck - am)^2, \quad \theta = 64ck(ck - am).\end{aligned} \tag{7.33}$$

1.1.1: $\mu_0 \neq 0$. Then $\mu_0 > 0$ and we consider three subcases: $W_4 < 0$, $W_4 > 0$ and $W_4 = 0$.

1.1.1.1: $W_4 < 0$. Then from (7.33) it follows that $\mathbf{D} > 0$, and we arrive at the case $(\mathcal{A}_3)$ of Theorem 6.3. Since in this case we have $\theta \neq 0$, according to Theorem 6.4 we obtain the configuration (β_1) if $\theta < 0$ and (β_2) if $\theta > 0$.

1.1.1.2: $W_4 > 0$. In this case we get $\mathbf{D} < 0$, and this leads to the case $(\mathcal{A}_2)$ of Theorem 6.3. Since $W_4 \neq 0$ we conclude that both finite nodes are generic. Additionally we have again $\theta \neq 0$, and according to Theorem 6.4 we arrive at the following three configurations: (β_3) and (β_4) if $\theta < 0$ and (β_5) if $\theta > 0$.

1.1.1.3: $W_4 = 0$. If $\mathbf{D} \neq 0$ (then $\eta \neq 0$) we obtain $ck = 0$, and without loss of generality we may assume $k = 0$ due to the change $(x, y, t, a, c, k, m) \mapsto (y, x, -t, m, k, c, a)$. Then we calculate

$$\begin{gathered} \mathcal{T}_3 = -8a^3c, \quad \eta = -a^2(4ac - m^2), \quad \mathbf{D} = -48\eta, \\ W_3 = 64a^6c^2, \quad \mu_0 = a^2m^2, \quad \theta = \theta_2 = 0, \quad \theta_1 = -64a^3c, \\ U_1 = 4acy^2, \quad U_2 = 16acy^2, \quad U_3 = Y_1 = 0. \end{gathered} \tag{7.34}$$

In the case $\mathbf{D} < 0$ we arrive at the case $(\mathcal{A}_2)$ of Theorem 6.3, where due to $W_4 = 0$ we have exactly one node with coinciding eigenvalues because the condition $\mathcal{T}_3 \neq 0$ implies $W_3 \neq 0$. Considering (7.34) we deduce that this node is a star node since the conditions $(\mathcal{S}^1_{n^*})$ of Theorem 7.1 are satisfied.

So considering the existence of a finite star node and of exactly one weak singularity (due to $\mathcal{T}_3 \neq 0$), according to Theorem 7.2 we get two configurations: (β_6) and (β_7), which coincide with (α_9) and (α_{10}), respectively.

If $\mathbf{D} > 0$ we get the case $(\mathcal{A}_4)$ of Theorem 6.3, and due to the existence of a star node, we could apply again Theorem 7.2, which leads to the configuration (β_8) $(\equiv (\alpha_{18}))$. We point out that according to Theorem 7.2 we could not have, apart from a star node, one integrable saddle and two complex singularities with zero discriminants.

Assume now that the condition $\mathbf{D} = 0$ holds. Then considering (7.33) we get $\eta = 0$, and we have the case $(\mathcal{A}_5)$ of Theorem 6.3 where, apart from the integrable saddle, we have a node with coinciding eigenvalues and a double semi-elemental saddle-node.

We claim that if the node is n^d then the node at infinity could only be generic. Indeed, to have at infinity a node with coinciding eigenvalues, by Theorem 6.4 the condition $\theta = 0$ is necessary. Taking into account conditions (7.33), since $\mu_0 \neq 0$, the condition $\theta = 0$ implies $ck = 0$ (then we may assume $k = 0$ due to a linear change of variables and renaming the parameters), and we arrive at the relations (7.34). As it was mentioned above, we must have a star node, and moreover, the condition $\theta = 0$ implies $\theta_2 = 0$, i.e. by Theorem 6.4 at infinity we could not have a node N^d.

Thus we arrive at the following three configurations: (β_9) if $\theta < 0$, (β_{10}) if $\theta > 0$ and (β_{11}) $(\equiv (\alpha_{21}))$ if $\theta = 0$.

1.1.2: $\mu_0 = 0$. Since $c^2 + k^2 \neq 0$ (as $\mathcal{T}_3 \neq 0$) we may assume $k = 1$ (due to a linear change of variables and renaming the parameters), and then the condition $\mu_0 = (c - am)^2 = 0$ gives $c = am$. So we obtain the family of systems:

$$\dot{x} = x + ax^2 + amy^2, \quad \dot{y} = -y - x^2 - my^2, \tag{7.35}$$

for which calculations yield

$$\mathcal{T}_3 = -8m(a^2 - m)(a^2 + m) \neq 0, \quad \eta = -4m(a^2 + m)^2,$$
$$\mu_0 = \mu_1 = 0, \quad \mu_2 = (a^2 + m)(x^2 + my^2), \quad W_4 = -1024a^2m^3(a^2 + m)^2,$$
$$\theta = \kappa = 0, \quad \theta_1 = -64m(a^2 + m)^2, \quad \theta_2 = -2am.$$

The condition $\mathcal{T}_3 \neq 0$ implies $\eta\theta_1\mu_2 \neq 0$, and since $\mu_0 = \mu_1 = 0$, by Lemma 5.2 we conclude that two finite singular points have coalesced with infinite ones. We consider two subcases: $\eta < 0$ and $\eta > 0$.

1.1.2.1: $\eta < 0$. Considering Theorem 6.3, we arrive at the case $(\mathcal{A}_6)$ if $W_4 \neq 0$ and $(\mathcal{A}_7)$ if $W_4 = 0$.

If $W_4 \neq 0$ we obtain $\theta_2 \neq 0$, and since $\theta = \kappa = 0$, by Theorem 6.4 we get the configuration (β_{12}).

Assume now $W_4 = 0$. Due to $\mathcal{T}_3 \neq 0$ we get $a = 0$, and this implies $\theta_2 = 0$. So by Theorem 6.4 at infinity we have a star node. Additionally for systems (7.35) with $a = 0$, calculations yield

$$U_1 = 4mx^2, \quad U_2 = -16mx^2, \quad U_3 = Y_1 = 0,$$

and by Theorem 7.1 we have a finite star node, too. So we get the configuration (β_{13}) $(\equiv (\alpha_{35}))$.

1.1.2.2: $\eta > 0$. This leads to the case $(\mathcal{A}_8)$ of Theorem 6.3, and it remains to determine what kind of nodes (finite and infinite) we have.

If $W_4 \neq 0$ (this implies $\theta_2 \neq 0$) then the finite node is generic, whereas by Theorem 6.4 at infinity we have a node N^d. So we arrive at the configuration (β_{14}).

In the case $W_4 = 0$, as it was shown above, we have a finite star node as well as an infinite one. As a result we get the configuration (β_{15}) $(\equiv (\alpha_{36}))$.

1.2: Systems (7.27). So we consider the family of systems:

$$\dot{x} = x - 2ckx^2 + xy + cy^2, \quad \dot{y} = -y - kx^2 - ckxy + 2y^2,$$

for which calculations yield

$$\mu_0 = -2k(1 - c^2k)^2 = -\eta/2 = -\theta/36, \quad \mathbf{D} = -5184k(1 + c^2k)^2,$$
$$\mathcal{T}_3 = 125k(c^2k - 1)(1 + c^2k) \neq 0, \quad W_4 = 4k^2(1 - c^2k)^2(1 + c^2k)^2.$$

We observe that the condition $\mathcal{T}_3 \neq 0$ implies μ_0 and $W_4 > 0$, and we consider two cases: $\mu_0 < 0$ and $\mu_0 > 0$.

1.2.1: $\mu_0 < 0$. By Theorem 6.3 we have the case $(\mathcal{A}_1)$, i.e. apart from the integrable saddle, we have three finite elemental nodes and at infinity we have two saddles and one node. According to Theorem 6.4, at infinity we have one node N^∞.

Additionally by $W_4 \neq 0$ we deduce that all three finite nodes are generic, and this gives the configuration (β_{16}).

1.2.2: $\mu_0 > 0$. This condition implies $\mathbf{D} > 0$, and since $W_4 > 0$ we arrive at the case $(\mathcal{A}_4)$ of Theorem 6.3. So the finite node is generic, and as $\eta < 0$ at infinity, apart from the node, we have two complex singular points. Moreover the condition $\mu_0 > 0$ implies $\theta < 0$, and considering Theorem 6.4 we get the configuration (β_{17}).

2: Case $(\mathcal{B})$. Since we have an integrable saddle and $\mathcal{T}_3 = 0$ and $\mathcal{T}_2 \neq 0$ considering Lemma 7.2, we have to examine each one of the canonical forms (7.26) and (7.28).

2.1: Systems (7.26). In this case for systems (7.26) the conditions

$$\mathcal{T}_3 = -8(a^3c - km^3) = 0, \quad \mathcal{T}_2 = 4(a^3c + 3ackm - a^2m^2 + km^3) \neq 0$$

have to be satisfied. We claim that these conditions imply $am \neq 0$. Indeed suppose the contrary, that $am = 0$. We observe that the condition $\mathcal{T}_2 \neq 0$ implies $a^2 + m^2 \neq 0$, and we may assume $m \neq 0$ due to the change $(x, y, t, a, c, k, m) \mapsto (y, x, -t, m, k, c, a)$. Then $a = 0$ and therefore the condition $\mathcal{T}_3 = 0$ gives $k = 0$. However this implies $\mathcal{T}_2 = 0$, and this contradiction proves our claim.

So $am \neq 0$ and we may assume $a = m = 1$ due to the rescaling $(x, y) \mapsto (x/a, y/m)$. Hence the condition $\mathcal{T}_3 = 0$ gives $k = c$, and we arrive at the 1-parameter family of systems:

$$\dot{x} = x + x^2 + cy^2, \quad \dot{y} = -y - cx^2 - y^2, \tag{7.36}$$

for which we calculate

$$\begin{gathered}\mathcal{T}_2 = 4(1+c)(3c-1) \neq 0, \quad \eta = -(1+c)(3c-1)^3, \\ \mu_0 = (c-1)^2(c+1)^2, \quad \mathbf{D} = 48(1+c)(3c-1)^3, \\ W_4 = -256c^4(1+c)(3c-1)^3, \quad \theta = 64c^2(c-1)(1+c).\end{gathered}$$

We observe that $\mu_0 \geq 0$, and the condition $\mathcal{T}_2 \neq 0$ implies $\eta\mathbf{D} < 0$.

2.1.1: $\mu_0 \neq 0$. Then $\mu_0 > 0$ and we consider two subcases: $\mathbf{D} < 0$ and $\mathbf{D} > 0$.

2.1.1.1: $\mathbf{D} < 0$. In this case $\eta > 0$ and this leads to the case $(\mathcal{B}_7)$ of Theorem 6.3, i.e. systems (7.36) have the singularities $\$, \$, n, n;\ S, N, N$. So we need to detect the types of the nodes (finite and infinite).

If $W_4 \neq 0$ then both finite nodes are generic. We observe that the conditions $W_4 \neq 0$ and $\mu_0 \neq 0$ imply $\theta \neq 0$. Moreover the condition $\mathbf{D} < 0$ gives $-1 < c < 1/3$, and this implies $\theta < 0$. So by Theorem 6.4 we get the two global configurations of singularities: (β_{18}) and (β_{19}).

Assume now $W_4 = 0$. Since $\mathcal{T}_2 \neq 0$ we have $c = 0$, and then we calculate

$$\theta = \theta_1 = \theta_3 = \theta_4 = 0, \quad U_1 = U_4 = U_5 = 0, \quad U_6 = 4xy \neq 0, \quad Y_2 = 8 > 0.$$

Therefore by Theorems 6.4 and 7.1 we obtain that both finite as well as both infinite nodes are star nodes. So we get the configuration (β_{20}).

2.1.1.2: $\mathbf{D} > 0$. Then $\eta < 0$ and in this case systems (7.36) possess the singularities described in the case $(\mathcal{B}_9)$ of Theorem 6.3, i.e. the singularities $\$, c;\ N$ where we must add the complex singularities: two finite and two infinite. Since $c \neq 0$ (otherwise we get $\mathbf{D} = -48 < 0$), the condition $\theta \neq 0$ holds. Therefore by Theorem 6.4 we get the configuration of singularities: (β_{21}) if $\theta < 0$, and (β_{22}) if $\theta > 0$.

2.1.2: $\mu_0 = 0$. Considering $\mathcal{T}_2 \neq 0$ we get $c = 1$, and then we obtain $\eta = -16 < 0$ and $\widetilde{K} = 0$. This leads to the case $(\mathcal{B}_{13})$ of Theorem 6.3, i.e. we have the singularities $\$, c;\ N$, coinciding with the singularities in the above mentioned case $(\mathcal{B}_9)$. However in this case calculations yield

$$\mu_0 = \mu_1 = 0,\ \ \mu_2 = 2(x^2 + y^2),\ \ \kappa = 0,\ \ \theta_2 = -2,$$

and the complex finite singular points coalesce with the complex infinite ones. According to Theorem 6.4 we have the configuration (β_{40}).

It remains to remark that the systems (7.36) could not have the complex singularities with zero discriminants. Indeed due to $\mathcal{T}_2 \neq 0$ the condition $W_4 = 0$ implies $c = 0$, and then $\mathbf{D} = -48 < 0$, i.e. by Proposition 5.1 we could not have complex singularities.

2.2: Systems (7.28). In this case we consider the family of systems:

$$\dot{x} = x + mx^2 + xy + cy^2, \quad \dot{y} = -y - cx^2 - xy - my^2, \tag{7.37}$$

for which calculations yield

$$\begin{aligned}
&\mathcal{T}_2 = (1 + 3c - m)(1 + c + m)(2m - 1)^2 \neq 0,\\
&\mu_0 = (c - m)^2(c - 1 + m)(1 + c + m),\\
&\widetilde{K} = 2(c - m)\left[x^2 + 2(c + m)xy + y^2\right],\\
&\mathbf{D} = 48(1 + 3c - m)^3(c - 1 + m),\\
&\eta = (1 + c + m)(1 - 3c + m)^3,\\
&W_4 = -16(1 - 2c)^4(1 + 3c - m)^3(1 + c + m),\\
&\theta = 16(1 - 2c)^2(c - m)(1 + c + m),\ \ \theta_2 = (1 - 4c^2)/2.
\end{aligned} \tag{7.38}$$

First we prove the next lemma.

Lemma 7.3. *Assume that for a system* (7.37) *the condition* $\mathcal{T}_2 \neq 0$ *holds. Then for this system the following hold.*

(i) *The condition* $\mu_0 \neq 0$ *implies* $\mathbf{D} \neq 0$, *and* $\mu_0 W_4 \neq 0$ *implies* $\theta \neq 0$.

(ii) *If* $W_4 = 0$ *then* $\theta = 0$ *and* $\mu_0 = \eta$. *Moreover the system could not possess a star node.*

(iii) *The conditions* $\mu_0 > 0$ *and* $\theta > 0$ *imply* $\mathbf{D} > 0$ *and* $\eta < 0$.

Proof. Statement (i) follows directly from (7.38) considering the values of the corresponding invariant polynomials.

For (ii) assume $W_4 = 0$. Then due to the condition $\mathcal{T}_2 \neq 0$ this yields $c = 1/2$, and we calculate

$$U_1 = 0, \;\; U_4 = 6(y-x)(x+y)^2 \neq 0, \;\; \mu_0 = (2m-1)^3(3+2m)/16 = \eta, \;\; \theta = 0.$$

So by Theorem 7.1 systems (7.37) could not have a star node.

Considering (7.38) the conditions of (iii) $\mu_0 > 0, \theta > 0$, give $(c-m)(c+m-1) > 0$.

If $c-m > 0$ then $c+m-1 > 0$, and setting $c-m = u^2$ and $c+m-1 = v^2$, we get $c = (1+u^2+v^2)/2$ and $m = (1+u^2+v^2)/2$. Calculations yield

$$\mu_0 = u^4v^2(2+v^2) > 0, \;\; \mathbf{D} = 48v^2(2+2u^2+v^2)^3 > 0, \;\; \eta = -(2+v^2)(2u^2+v^2)^3 < 0.$$

Suppose now that $c-m < 0$ and $c+m-1 < 0$. Then setting $c-m = -u^2$ and $c+m-1 = -v^2$, we get $c = (1-u^2-v^2)/2$ and $m = (1+u^2-v^2)/2$, and we calculate

$$\mu_0 = u^4v^2(v^2-2), \;\; \mathbf{D} = 48v^2(v^2-2+2u^2)^3, \;\; \eta = -(v^2-2)(2u^2+v^2)^3.$$

Clearly the condition $\mu_0 > 0$ implies $\mathbf{D} > 0$ and $\eta < 0$.

This completes the proof of Lemma 7.3. □

In order to determine the global configurations of singularities for systems (7.37), we consider three cases: $\mu_0 < 0$, $\mu_0 > 0$ and $\mu_0 = 0$.

2.2.1: $\mu_0 < 0$. As μ_0 is the discriminant of the binary form $\widetilde{K}$, we have an invariant polynomial with well defined sign, namely $\operatorname{sign}(\widetilde{K}) = \operatorname{sign}(c-m)$. Since by Lemma 7.3 the condition $\mathbf{D} \neq 0$ holds, we consider two subcases: $\mathbf{D} < 0$ and $\mathbf{D} > 0$.

2.2.1.1: $\mathbf{D} < 0$. We examine three possibilities: $\eta < 0$, $\eta > 0$ and $\eta = 0$.

1. $\eta < 0$. In this case we arrive at the case $(\mathcal{B}_1)$ of Theorem 6.3, i.e. systems (7.37) have the singularities $\$, c, n, n;\ S$ (and additionally two complex points at infinity due to $\eta < 0$). So we need to detect the types of the finite nodes.

   Assume first $W_4 \neq 0$. Then both finite nodes are generic, and clearly we get the configuration (β_{24}).

   If $W_4 = 0$ then due to the condition $\mathcal{T}_2 \neq 0$ this yields $c = 1/2$, and this implies $W_3 = 0$, i.e. we have two finite nodes with coinciding eigenvalues. However by Lemma 7.3 (see statement (ii)) systems (7.37) could not have a star node. So we have two n^d, and this leads to the configuration (β_{25}).

2. $\eta > 0$. According to Theorem 6.3, we arrive at the case $(\mathcal{B}_2)$ if $\widetilde{K} < 0$, and at the case $(\mathcal{B}_3)$ if $\widetilde{K} > 0$.

If $\widetilde{K} < 0$ systems (7.37) have the singularities $\$, c, s, s;\ N, N, N$, and due to the conditions $\eta > 0$ and $\mu_0 < 0$, according to Theorem 6.4 we get the configuration (β_{26}).

Assume now $\widetilde{K} > 0$. Then systems (7.37) have the singularities $\$, c, n, n;$ S, S, N, and as $\eta > 0$ and $\mu_0 < 0$, by Theorem 6.4, at infinity, apart from two saddles, we have a node N^∞. So it remains to detect what kind of finite nodes the systems have.

Since $\eta > 0$ and $\mu_0 < 0$, by Lemma 7.3 we get $W_4 \neq 0$, i.e. both finite nodes are generic, and this leads to the configuration (β_{27}).

3. $\eta = 0$. Since $\mathbf{D} < 0$, according to Theorem 6.3 (see case $(\mathcal{B}_4)$) systems (7.37) possess the singularities $\$, c, n, n;\ \overline{\binom{0}{3}}\, S$. So in order to detect the type of the finite nodes, we observe that the condition $\eta = 0$ (due to $\mathcal{T}_2 \neq 0$) gives $1 - 3c + m = 0$ (i.e. $m = 3c - 1$), and we calculate $\mu_0 = 8c(2c-1)^3$ and $W_4 = -512c(2c-1)^4$. Therefore $W_4 \neq 0$ (as $\mu_0 \neq 0$), i.e. both finite nodes are generic, and we arrive at the configuration (β_{28}).

2.2.1.2: $\mathbf{D} > 0$. This immediately leads to the case $(\mathcal{B}_5)$ of Theorem 6.3, i.e. we have the singularities $\$, \$;\ N, N, N$ (and additionally two finite complex singular points). Therefore by Theorem 6.4 we get the configuration (β_{29}).

2.2.2: $\mu_0 > 0$. According to Lemma 7.3 we have $\mathbf{D} \neq 0$.

2.2.2.1: $\mathbf{D} < 0$. As above we consider three possibilities: $\eta < 0$, $\eta > 0$ and $\eta = 0$.

1. $\eta < 0$. We arrive at the case $(\mathcal{B}_6)$ of Theorem 6.3, i.e. systems (7.37) have the singularities $\$, \$, n, n;\ N$ (and additionally two complex points at infinity). So we need to detect the types of the nodes, finite and infinite.

   By Lemma 7.3 (see statement (ii)) the conditions $\mu_0 > 0$ and $\eta < 0$ imply that $W_4 \neq 0$ and that the finite nodes are generic. By statement (i) of the same lemma the condition $\theta \neq 0$ must hold, and moreover by statement (iii) of Lemma 7.3 the conditions $\mu_0 > 0$ and $\mathbf{D} < 0$ imply $\theta < 0$. Therefore by Theorem 6.4 we obtain the configuration (β_{30}).

2. $\eta > 0$. In this case systems (7.37) possess the singularities described by the case $(\mathcal{B}_7)$ of Theorem 6.3, i.e. $\$, \$, n, n;\ S, N, N$.

   If $W_4 \neq 0$ the finite nodes are generic. By Lemma 7.3 (see statement (i)) the condition $\theta \neq 0$ must hold. Moreover from statement (iii) of the same lemma the condition $\theta < 0$ must hold. So by Theorem 6.4 we get the configurations (β_{18}) and (β_{19}).

   Assume now $W_4 = 0$. As it was shown above we get $c = 1/2$, and we calculate

$$\begin{gathered} W_4 = W_3 = \theta = \theta_1 = 0,\ \theta_3 = (1-2m)^2/2, \\ \mu_0 = (2m-1)^3(3+2m)/16 = \eta. \end{gathered} \tag{7.39}$$

So considering Lemma 7.3 we conclude that both finite nodes are n^d. Since $\theta_3 \neq 0$ (due to $\mu_0 \neq 0$), according to Theorem 6.4, at infinity we also have two N^d, and we arrive at the configuration (β_{31}).

3. $\eta = 0$. We have the case $(\mathcal{B}_8)$ of Theorem 6.3, i.e. the singularities of systems (7.37) are $\$, \$, n, n; \overline{\binom{0}{3}}N$. Considering the condition $\mu_0 \neq 0$ we get $m = 3c-1$, and then $W_4 = -512c(2c-1)^4 \neq 0$ due to $\mu_0 = 8c(2c-1)^3 \neq 0$. Therefore both finite nodes are generic, and we arrive at the configuration (β_{32}).

2.2.2.2: D > 0. In this case, apart from the two real finite singularities, we have two complex ones.

1. $\eta < 0$. We arrive at the case $(\mathcal{B}_9)$ of Theorem 6.3, i.e. systems (7.37) have the singularities $\$, c;\ N$ (and additionally two complex points at infinity).

   Since $\mu_0 \eta < 0$ by (7.38), the condition $\theta \neq 0$ must be satisfied (otherwise $c = 1/2$, which implies $\mu_0 = \eta$). Therefore by Theorem 6.4 we get the configuration (β_{21}) if $\theta < 0$ and (β_{22}) if $\theta > 0$.

2. $\eta > 0$. In this case systems (7.37) possess the singularities described by the case $(\mathcal{B}_{10})$ of Theorem 6.3, i.e. $\$, c;\ S, N, N$. Since $\mu_0 > 0$ and $\eta > 0$, by Lemma 7.3 (see statement (iii)) the condition $\theta \leq 0$ must hold. As a result in the case $\theta \neq 0$, considering Theorem 6.4 we obtain two configurations: (β_{33}) and (β_{34}).

   Assume now $\theta = 0$. Since $\mu_0 \neq 0$ we get $c = 1/2$, and this leads to the relation (7.39). So, by Theorem 6.4, at infinity we have two nodes N^d. Additionally the conditions $W_4 = W_3 = 0$ imply the existence of two finite singularities with the vanishing corresponding discriminants, and clearly these singularities could only be complex. Therefore we get the configuration (β_{35}).

3. $\eta = 0$. In this case we have the case $(\mathcal{B}_{11})$ of Theorem 6.3, i.e. the singularities of systems (7.37) are $\$, c;\ \overline{\binom{0}{3}}N$. Due to $\mu_0 \neq 0$ by Lemma 7.3, the condition $W_4 \neq 0$ must hold. As a result we arrive at the configuration (β_{36}).

2.2.3: $\mu_0 = 0$. Then considering (7.38) and the condition $\mathcal{T}_2 \neq 0$ we obtain $(c-m)(c-1+m) = 0$, and we consider two subcases: $\widetilde{K} \neq 0$ and $\widetilde{K} = 0$.

2.2.3.1: $\widetilde{K} \neq 0$. Then $c - m \neq 0$ and therefore $m = 1-c$, and this leads to the 1-parameter family of systems:

$$\dot{x} = x + (1-c)x^2 + xy + cy^2, \quad \dot{y} = -y - cx^2 - xy + (c-1)y^2. \tag{7.40}$$

Calculations yield

$$\mathcal{T}_2 = 8c(1-2c)^2 \neq 0, \quad \mu_0 = \mu_1 = 0, \quad \mu_2 = 2c(x+y)^2,$$
$$\eta = 16(1-2c)^3, \quad W_4 = -2048c^3(1-2c)^4, \quad \kappa = -64(2c-1)^3.$$

We observe that the condition $\mathcal{T}_2 \neq 0$ implies $\mu_2\eta W_4\kappa \neq 0$.

Since $\widetilde{K} \neq 0$ in the case $\eta < 0$ we arrive at the case $(\mathcal{B}_{12})$ of Theorem 6.3, i.e. systems (7.40) have the real singularities $\$, c;\ \overline{\binom{2}{1}}N$. Therefore we get the global configuration (β_{37}).

Assume now $\eta > 0$. By Theorem 6.3 the systems (7.40) possess the real singularities $\$, c;\ \overline{\binom{2}{1}}S, N, N$ if $W_4 < 0$ (see statement $(\mathcal{B}_{14})$) and the real singularities $\$, \$;\ \overline{\binom{2}{1}}N, N, N$ if $W_4 > 0$ (see statement $(\mathcal{B}_{15})$).

Additionally for systems (7.40) we have $\text{sign}\,(\eta) = \text{sign}\,(\kappa)$ and $\text{sign}\,(\mu_2) = -\text{sign}\,(W_4)$, and since $\eta > 0$ we obtain the configuration (β_{38}) if $W_4 < 0$ and (β_{39}) if $W_4 > 0$.

2.2.3.2: $\widetilde{K} = 0$. In this case we get $m = c$, and we obtain the 1-parameter family of systems:

$$\dot{x} = x + cx^2 + xy + cy^2, \quad \dot{y} = -y - cx^2 - xy - cy^2.$$

For these systems we calculate

$$\begin{gathered}\mathcal{T}_2 = (1-2c)^2(1+2c)^2 \neq 0, \quad \eta = (1-2c)^3(1+2c),\\ \mu_2\tilde{L} = 8(2c-1)(1+2c)(cx^2+xy+cy^2)^2,\\ \mu_0 = \mu_1 = \kappa = 0, \quad \theta_2 = (1-2c)(1+2c)/2,\end{gathered}$$

and clearly the condition $\mathcal{T}_2 \neq 0$ implies $\mu_2\tilde{L}\eta\theta_2 \neq 0$. Considering Theorem 6.3, we arrive at the case $(\mathcal{B}_{13})$ (the real singularities $\$, c;\ N$) if $\eta < 0$ and at the case $(\mathcal{B}_{16})$ (the real singularities $\$, c;\ N, \overline{\binom{1}{1}}SN, \overline{\binom{1}{1}}SN$) if $\eta > 0$.

We observe that $\text{sign}\,(\mu_2\tilde{L}) = -\text{sign}\,(\eta)$. By Lemma 5.2 the conditions $\mu_0 = \mu_1 = 0$ mean that two finite singularities coalesced with two infinite ones, and if $\eta < 0$ these singularities are complex. Therefore considering Theorem 6.4 we obtain the global configurations (β_{40}) if $\eta < 0$, and (β_{23}) if $\eta > 0$.

3: Case $(\mathcal{C})$. Since we have an integrable saddle and $\mathcal{T}_3 = \mathcal{T}_2 = 0$ and $\sigma \neq 0$ considering Lemma 7.2, we have to examine each one of the canonical forms (7.26) and (7.29).

3.1: Systems (7.26). In this case the conditions $\mathcal{T}_3 = \mathcal{T}_2 = 0$ and $\sigma \neq 0$ hold, where

$$\mathcal{T}_3 = -8(a^3c - km^3), \quad \mathcal{T}_2 = 4(a^3c + 3ackm - a^2m^2 + km^3), \quad \sigma = 2(ax - my).$$

So the condition $\sigma \neq 0$ implies $a^2 + m^2 \neq 0$, and we may assume $m \neq 0$ due to the change $(x, y, t, a, c, k, m) \mapsto (y, x, -t, m, k, c, a)$. Moreover we may consider $m = 1$ due to the rescaling $y \to y/m$. Therefore the condition $\mathcal{T}_3 = 0$ gives $k = a^3c$, and then we get

$$\begin{gathered}\mathcal{T}_2 = 4a^2(1+ac)(3ac-1), \quad \mu_0 = a^2(ac-1)^2(1+ac)^2,\\ \mu_1 = 2a(1-ac)(1+ac)(ax-y),\\ \mu_2 = (1+ac)\big[a^2x^2 + 3a(ac-1)xy + y^2\big].\end{gathered}$$

3.1.1: $\mu_0 \neq 0$. Then $a \neq 0$ and we may assume $a = 1$ due to the change $x \to x/a$ and $c \to c/a$. Therefore as $\mu_0 \neq 0$ the condition $\mathcal{T}_2 = 0$ implies $c = 1/3$. This leads to the system

$$\dot{x} = x + x^2 + y^2/3, \quad \dot{y} = -y - x^2/3 - y^2,$$

for which we calculate

$$\mathcal{T}_2 = 0, \quad \sigma = 2(x - y) \neq 0, \; \mu_0 = 64/81, \quad \eta = 0.$$

Since $\mu_0 > 0$ and $\eta = 0$ we have the case $(\mathcal{C}_4)$ of Theorem 6.3, i.e. the above system has the singularities $\$, \widehat{es}_{(3)}$; $\overline{\binom{0}{3}}N$. Therefore we get the global configuration (β_{41}).

3.1.2: $\mu_0 = 0$. We observe that this condition implies $\mu_1 = 0$, and we shall examine two subcases: $\mu_2 \neq 0$ and $\mu_2 = 0$.

3.1.2.1: $\mu_2 \neq 0$. Then $(1 + ac) \neq 0$ and hence the condition $\mathcal{T}_2 = \mu_0 = 0$ yields $a = 0$. So we get the 1-parameter family of systems

$$\dot{x} = x + cy^2, \quad \dot{y} = -y - y^2, \tag{7.41}$$

for which calculations yield

$$\begin{gathered}\eta = 0, \quad \widetilde{M} = -8y^2, \quad \mu_0 = \mu_1 = 0, \quad \mu_2 = y^2, \quad \widetilde{K} = 0, \\ \tilde{L} = 8y^2, \quad \kappa = \kappa_1 = 0, \quad K_2 = 96y^2, \quad \theta_6 = 16cy^4.\end{gathered} \tag{7.42}$$

Due to $\mu_0 = \mu_1 = 0$, $\mu_2 > 0$ and $\tilde{L} > 0$, according to Theorem 6.3 we arrive at the case $(\mathcal{C}_{12})$, i.e. systems (7.41) possess the singularities $\$, n$; N, $\binom{2}{2}$ $PPH - PPH$. So we need to distinguish the types of the elemental nodes (finite and infinite) and to complete the geometric description of the intricate infinite singularity.

We observe that, apart from the integrable saddle $M_1(0,0)$, systems (7.41) possess the node $M_1(-c, 1)$ with the following corresponding Jacobian matrix: $\begin{pmatrix} 1 & -2c \\ 0 & 1 \end{pmatrix}$. Therefore the finite node has the coinciding corresponding eigenvalues, and it is a star node if and only if $c = 0$.

Additionally considering (7.42), by Theorem 6.4 (see Diagram 6.7) we deduce that the infinite node is a node N^d if $\theta_6 \neq 0$ (i.e. $c \neq 0$) and is a star node if $\theta_6 = 0$ (i.e. $c = 0$). As a result we get the configurations (β_{42}) and (β_{43}), respectively.

3.1.2.2: $\mu_2 = 0$. In this case $1 + ac = 0$ and clearly $a \neq 0$. As it was mentioned in the case $\mu_0 \neq 0$, we may consider $a = 1$, and therefore we get $c = -1$. This leads to the system

$$\dot{x} = x + x^2 - y^2, \quad \dot{y} = -y + x^2 - y^2,$$

for which we have

$$\begin{gathered}\mu_0 = \mu_1 = \mu_2 = 0, \quad \mu_3 = (y - x)(x + y)^2 = -K_1, \\ \eta = \kappa = \widetilde{K} = \tilde{L} = 0, \quad \kappa_1 = -256, \quad \widetilde{M} = -32(x - y)^2 \neq 0.\end{gathered}$$

Therefore we have the case $(\mathcal{C}_{17})$ of Theorem 6.3, i.e. the above system has the singularities $\$;\ \widehat{\binom{1}{2}}PEP-H,\overline{\binom{2}{1}}\ N$, which improved by Theorem 6.4 gives the global configuration (β_{44}).

3.2: Systems (7.29). Since for these systems the conditions $\mathcal{T}_2=0$ and $\sigma\neq 0$ must be satisfied, we calculate

$$\begin{aligned}\mathcal{T}_2 &= l^2(1+3cl-m)(1+cl+m)(1-2m)^2, \quad \sigma=(2m-1)(lx-y),\\ \mu_0 &= l^2(cl-m)^2(cl+m-1)(1+cl+m),\\ \mu_1 &= l(m-cl)(cl+m-1)(1+2m)(lx-y).\end{aligned} \tag{7.43}$$

We consider two cases: $\mu_0\neq 0$ and $\mu_0=0$.

3.2.1: $\mu_0\neq 0$. Then $l\neq 0$ and we may assume $l=1$ due to the change $x\to x/l$ and $c\to c/l$. Therefore as $\sigma\neq 0$ (i.e. $2m-1\neq 0$) the condition $\mathcal{T}_2=0$ implies $m=1+3c$. This leads to the 1-parameter family of systems:

$$\dot{x}=x+(1+3c)x^2+xy+cy^2,\quad \dot{y}=-y-cx^2-xy-(1+3c)y^2, \tag{7.44}$$

for which we calculate

$$\begin{aligned}\sigma=(1+6c)(x-y)\neq 0,\quad \mu_0=8c(1+2c)^3\neq 0,\quad \eta=16(1+2c),\\ \theta=-32(1-2c)^2(1+2c)^2,\quad \theta_1=64(1-2c)(1+2c)(5+6c),\\ \theta_3=c(1+2c)(5+12c-12c^2).\end{aligned}$$

If $\mu_0<0$ then we have the case $(\mathcal{C}_1)$ of Theorem 6.3, i.e. the above systems possess the singularities $\$,\overline{s}_{(3)};N,N,N$. This implies $\eta>0$, and by Theorem 6.4 we get the configuration (β_{45}).

Assume now $\mu_0>0$. Then according to Theorem 6.3, we arrive at the case $(\mathcal{C}_2)$ if $\eta<0$, at the case $(\mathcal{C}_3)$ if $\eta>0$ and at the case $(\mathcal{C}_4)$ if $\eta=0$.

Thus for $\eta<0$ systems (7.44) possess the real singularities $\$,\widehat{es}_{(3)};N$ and additionally two complex singularities at infinity. We observe that the condition $\eta<0$ gives $c<-1/2$, and this implies $\theta<0$. So considering Theorem 6.4 we get the configuration (β_{46}).

If $\eta>0$ then we have the singularities $\$,\widehat{es}_{(3)};\ S,N,N$. Since in this case the condition $c>-1/2$ holds, then $\theta\leq 0$, and we could have $\theta=0$ for $c=1/2$ and in this case $\theta_3\neq 0$. Moreover in the case $\theta<0$ we obtain $\theta_1\neq 0$ and $\operatorname{sign}(\theta_1)=\operatorname{sign}(2c-1)$.

Thus considering Theorem 6.4, we arrive at the following three configurations of singularities: (β_{47}) and (β_{48}) if $\theta\neq 0$, and (β_{49}) if $\theta=0$.

Suppose finally $\eta=0$. As it was mentioned above, we are in the case $(\mathcal{C}_4)$, i.e. systems (7.44) $\$,\widehat{es}_{(3)};\ \overline{\binom{0}{3}}N$, and this immediately leads to the configuration (β_{41}).

3.2.2: $\mu_0=0$. We consider two subcases: $\mu_1\neq 0$ and $\mu_1=0$.

3.2.2.1: $\mu_1 \neq 0$. Then considering (7.43) we have $l \neq 0$, and similarly as above we may assume $l = 1$. Since $\sigma \neq 0$ the conditions $\mathcal{T}_2 = \mu_0 = 0$ evidently yield $m = -1 - c$, and this leads to the family of systems

$$\dot{x} = x - (1+c)x^2 + xy + cy^2, \quad \dot{y} = -y - cx^2 - xy + (1+c)y^2, \tag{7.45}$$

for which we calculate

$$\sigma = (3+2c)(y-x) \neq 0, \quad \mu_0 = 0, \quad \mu_1 = 2(1+2c)^2(y-x),$$
$$\mathbf{D} = -768(1+2c)^3, \quad \tilde{L} = -16c(x-y)^2, \quad C_2 = c(x-y)^2(x+y),$$
$$\widetilde{N} = (1-2c)(1+2c)(x-y)^2, \quad \eta = \kappa = 0, \quad \widetilde{M} = -32c^2(x-y)^2,$$
$$\widetilde{K} = 2(1+2c)(x-y)^2.$$

We observe that the condition $\mu_1 \neq 0$ implies $\mathbf{D} \neq 0$ and the condition $C_2 = 0$ is equivalent to $\tilde{L} = 0$ as well as to $\widetilde{M} = 0$.

1. $\mathbf{D} < 0$. According to Theorem 6.3 the types of real singularities of the above systems are described in the cases $(\mathcal{C}_5)$–$(\mathcal{C}_8)$. Moreover in all these cases we have the finite singularities $\$, n, n$, whereas the types of the singularities at infinity depend on the values of the invariant polynomials $\tilde{L}$, $\widetilde{N}$ and C_2.

   In order to detect the conditions for the existence of finite nodes with coinciding eigenvalues, since $\mathbf{D} < 0$ (i.e. $1 + 2c > 0$) we set a new parameter u: $1 + 2c = u^2 > 0$. Then $c = (u^2-1)/2$ and systems (7.45), apart from the integrable saddle $M_1(0,0)$, possess the nodes $M_{2,3}$ with the coordinates

$$(x_2, y_2) = \left(\frac{1+u}{2u^2}, \frac{1-u}{2u^2}\right), \quad (x_3, y_3) = \left(\frac{1-u}{2u^2}, \frac{1+u}{2u^2}\right).$$

   For these singularities we calculate $\tau_2 = (u^2-2)^2/u^2 = \tau_3$. Since $u^2 - 2 = 2c - 1$ we conclude that, in the case under consideration, both nodes of systems (7.45) have coinciding eigenvalues if and only if $c = 1/2$. Since in this case the conditions $U_1 = 0$ and $U_4 = 6(y-x)(x+y)^2 \neq 0$ hold, by Theorem 7.1 these systems could not have a star node.

   Additionally as the condition $\mathbf{D} < 0$ $(c > -1/2)$ implies $\widetilde{K} > 0$, considering Theorem 6.4, at infinity we have the next configurations of singularities: $\widehat{\binom{1}{2}}\, H_\curlywedge H H_\curlywedge - H,\ N^\infty$ if $-1/2 < c < 0$; $\widehat{\binom{1}{2}}\, \widehat{P}_\curlywedge H \widehat{P}_\curlywedge - E,\ S$ if $0 < c < 1/2$; $\widehat{\binom{1}{2}}\, \widehat{P}_\curlywedge E \widehat{P}_\curlywedge - H,\ S$ if $c > 1/2$; $\widehat{\binom{1}{2}}\, H - E,\ S$ if $c = 1/2$ and $[\infty; \emptyset]$ if $c = 0$ (then $C_2 = 0$). As a result we get the configurations of singularities (β_{50}), (β_{51}), (β_{52}), (β_{53}) and (β_{54}), respectively.

2. $\mathbf{D} > 0$. This leads to the case $(\mathcal{C}_9)$ of Theorem 6.3, i.e. systems (7.45) possess the real singularities $\$;\ N,\ \widehat{\binom{1}{2}}\, PEP - H$, and in addition they possess two complex finite singularities.

   Additionally since the condition $\mathbf{D} > 0$ $(c < -1/2)$ implies $\widetilde{K} < 0$, by Theorem 6.4, we arrive at the configuration (β_{55}).

3.2.2.2: $\mu_1 = 0$. Then at least two finite singularities have gone to infinity.

3.2.2.2.1: $\mu_2 \neq 0$. We claim that in this case for a non-degenerate system (7.29) the condition $l = 0$ must hold. Indeed supposing the contrary, we may consider $l = 1$ (due to a rescaling), and taking into account (7.43) and $\sigma \neq 0$, the conditions $\mathcal{T}_2 = \mu_0 = \mu_1 = 0$ imply $c = -1/2 = m$ (because the case $c = 0, m = 1$ gives a degenerate system). However in this case we get $\mu_2 = 0$, and this contradiction proves our claim.

So $l = 0$ and we obtain the family of systems

$$\dot{x} = x + xy + cy^2, \quad \dot{y} = -y - my^2, \tag{7.46}$$

for which we calculate

$$\begin{aligned} \sigma = (1-2m)y \neq 0, \quad \mu_0 = \mu_1 = 0, \quad \mu_2 = m(m-1)y^2, \quad \widetilde{K} = -2my^2, \\ \tilde{L} = 8m(1+m)y^2, \quad \eta = \kappa = \kappa_1 = 0, \quad \widetilde{M} = -8(1+m)^2y^2, \\ C_2 = (1+m)xy^2 + cy^3, \quad K_2 = 48(2+3m+2m^2)y^2 > 0. \end{aligned} \tag{7.47}$$

3.2.2.2.1.1: $\mu_2 < 0$. Then by Theorem 6.3 we have the case $(\mathcal{C}_{10})$, i.e. the systems (7.46) possess the singularities $\$, s;\ N, \binom{2}{2}PEP - PEP$. Additionally the condition $\mu_2 < 0$ (i.e. $0 < m < 1$) implies $\widetilde{K} < 0$ and $\widetilde{M} \neq 0$, and according to Theorem 6.4 we obtain the configuration (β_{56}).

3.2.2.2.1.2: $\mu_2 > 0$. According to Theorem 6.3 the types of real singularities of the systems (7.46) are described in the cases $(\mathcal{C}_{11})$–$(\mathcal{C}_{14})$. In all these cases we have the finite singularities $\$, n$, whereas the types of the singularities at infinity depend on the values of the invariant polynomials $\tilde{L}$ and C_2.

We observe that the finite node could not be a node with coinciding eigenvalues because for the node $M_2(c/(m(1-m)), -1/m)$ we have $\tau_2 = 1/m^2 \neq 0$.

In order to detect the geometrically distinct configurations of the infinite singularities, we apply Theorem 6.4 considering three subcases: $\tilde{L} < 0$, $\tilde{L} > 0$ and $\tilde{L} = 0$.

1. $\tilde{L} < 0$. Then $-1 < m < 0$, and this implies $\widetilde{K} > 0$ and $\widetilde{M} \neq 0$. Since $\eta = \kappa = 0$ and $K_2 > 0$, by Theorem 6.4 we have the configuration (β_{57}).

2. $\tilde{L} > 0$. In this case $m(m+1) > 0$, and as $\mu_2 > 0$ (i.e. $m(m-1) > 0$) we have either $m < 0$ (i.e. $\widetilde{K} > 0$) or $m > 0$ (i.e. $\widetilde{K} < 0$). As a result of Theorem 6.4, we arrive at the configuration (β_{58}) if $\widetilde{K} > 0$ and at the configuration (β_{59}) if $\widetilde{K} < 0$.

3. $\tilde{L} = 0$. Since $m \neq 0$ (due to $\mu_2 > 0$) we get $m = -1$, and this implies $\widetilde{M} = 0 \neq \widetilde{K}$ and $C_2 = cy^3$. So according to Theorem 6.4, we obtain the configuration (β_{60}) if $C_2 \neq 0$ and the configuration (β_{61}) if $C_2 = 0$.

3.2.2.2.2: $\mu_2 = 0$. Then $m(m-1) = 0$, and considering the cases $(\mathcal{C}_{15})$–$(\mathcal{C}_{18})$ of Theorem 6.3, we consider two subcases: $\widetilde{K} \neq 0$ and $\widetilde{K} = 0$.

1. $\widetilde{K} \neq 0$. We claim that in this case for a non-degenerate system (7.29), the condition $l = 0$ must hold. Indeed if $l \neq 0$ then due to a rescaling we may consider $l = 1$, and taking into account (7.43) and $\sigma \neq 0$, the conditions $\mathcal{T}_2 = \mu_0 = \mu_1 = 0$ imply $c = -1/2 = m$ (the case $c = 0, m = 1$ gives a degenerate system). However in this case we get $\widetilde{K} = 0$, and this contradiction proves our claim.

   So $l = 0$ and this leads to systems (7.46), for which the condition $\mu_2 = 0$ due to $\widetilde{K} \neq 0$ gives $m = 1$. Then $\widetilde{M} \neq 0$ and $\widetilde{K} < 0$, and by Theorem 6.4 we have the configuration (β_{62}).

2. $\widetilde{K} = 0$. If $\tilde{L} \neq 0$ we have $l \neq 0$, otherwise setting $l = 0$ we get again systems (7.46), for which, according to (7.47), the condition $\widetilde{K} = 0$ implies $\tilde{L} = 0$.

   So $l \neq 0$ (then $l = 1$ due to a rescaling), and it was shown above that in this case we get $c = -1/2 = m$. Then we obtain

$$\mu_0 = \mu_1 = \mu_2 = 0, \quad \mu_3 = -1/2(x-y)^3, \quad \tilde{L} = 8(x-y)^2, \; \eta = \kappa = 0,$$
$$\widetilde{M} = -8(x-y)^2, \quad K_1 = (x-y)^3/2,$$

   and as $\tilde{L} \neq 0$ and $\mu_3 K_1 = -(x-y)^6/4 < 0$, according to Theorem 6.4 this leads to the configuration (β_{63}).

   Assume now $\tilde{L} = 0$. In this case clearly we must have $l = 0$, and for systems (7.46) the condition $\widetilde{K} = \tilde{L} = 0$ yields $m = 0$. Therefore we calculate

$$\mu_3 = -y^2(x+cy), \quad \widetilde{M} = -8y^2, \quad \tilde{L} = \widetilde{K} = 0, \quad K_1 = y^2(x+cy), \quad \kappa_1 = 0,$$

   and since $\mu_3 K_1 = -y^4(x+cy)^2 < 0$ and $\kappa_1 = 0$, by Theorem 6.4 we arrive at the configuration (β_{64}).

4: Case $(\mathcal{D})$. Since $\sigma = 0$ the systems become Hamiltonian, and considering Lemma 7.2 we have to examine each one of the canonical forms (7.26) and (7.30). We point out that for the Hamiltonian systems in the case of the existence of two finite complex singularities, they must have zero traces, and therefore we could not have complex singularities with zero discriminants.

4.1: Systems (7.26) In this case for systems (7.26) the condition $\sigma = 2(ax - my) = 0$ gives $a = m = 0$, and we arrive at the family of Hamiltonian systems:

$$\dot{x} = x + cy^2, \quad \dot{y} = -y - kx^2,$$

with $c^2 + k^2 \neq 0$. Then due to the change $(x, y, t, c, k) \mapsto (y, x, -t, k, c)$, we may assume $c \neq 0$, and by applying the rescaling $x \to cx$, we get $c = 1$. For these systems we calculate

$$\mu_0 = k^2, \quad \mathbf{D} = 1296k^2, \quad \eta = -27k^2, \quad \theta = 64k^2.$$

If $k \neq 0$ we arrive at the case $(\mathcal{D}_6)$ of Theorem 6.3, i.e. the above systems possess the singularities $\$, c$; N. Moreover, in addition we have four complex singularities: two finite with zero traces (as systems are Hamiltonian) and two infinite ones.

Due to the condition $\mu_0 > 0$ $\eta < 0$ and $\theta > 0$, according to Theorem 6.4 we obtain the global configuration (β_{65}).

Assuming $k = 0$ for the above system, we obtain:

$$\mu_0 = \mu_1 = \mu_2 = 0, \quad \mu_3 = -y^3, \quad \eta = 0 = \widetilde{M}, \quad C_2 = y^3 = K_1.$$

So the unique finite singularity of the systems under consideration is the integrable saddle, and since $\mu_3 K_1 = -y^6 < 0$, by Theorem 6.4 we get the configuration (β_{66}).

4.2: Systems (7.30). For these systems calculations yield

$$\begin{aligned} \mu_0 &= \big[16c^2k^2 + 8cl(l^2 - 3k) + 8k - 3l^2\big]/16, \quad \eta = -27\mu_0, \\ \mathbf{D} &= 3\big[432c^2k^2 - 8cl(l^2 + 9k) - 8k - l^2\big], \quad \theta = 64\mu_0, \quad \kappa = -144\mu_0. \end{aligned} \tag{7.48}$$

Following Theorem 6.3 we consider three cases: $\mu_0 < 0$, $\mu_0 > 0$ and $\mu_0 = 0$.

4.2.1: $\mu_0 < 0$. According to Theorem 6.3, the types of real singularities of systems (7.30) are described in the cases $(\mathcal{D}_1)$–$(\mathcal{D}_4)$. We observe that in all these cases at infinity there are three nodes, and due to $\mu_0 < 0$, by Theorem 6.4 all the nodes are of type N^f.

Thus considering Theorem 6.3, we arrive at the following configurations: (β_{67}) in the case $(\mathcal{D}_1)$, (β_{68}) in the case $(\mathcal{D}_2)$, (β_{69}) in the case $(\mathcal{D}_3)$ and (β_{70}) in the case $(\mathcal{D}_4)$.

4.2.2: $\mu_0 > 0$. Then we arrive at the three cases $(\mathcal{D}_5)$, $(\mathcal{D}_6)$ and $(\mathcal{D}_7)$ of Theorem 6.3, which are defined by the conditions $\mathbf{D} < 0$, $\mathbf{D} > 0$ and $\mathbf{D} = 0$, respectively. We observe that in each one of these cases at infinity we have one simple node (apart from the two complex singularities). Moreover the last condition implies also $\theta > 0$, and according to Theorem 6.4 in all these cases we have a node N^f. As a result we get the configuration (β_{71}) if $\mathbf{D} < 0$, (β_{65}) if $\mathbf{D} > 0$ and (β_{72}) if $\mathbf{D} = 0$.

4.2.3: $\mu_0 = 0$. We consider two subcases: $\mu_1 \neq 0$ and $\mu_1 = 0$.

4.2.3.1: $\mu_1 \neq 0$. In this case by Theorem 6.3 we have the cases $(\mathcal{D}_8)$–$(\mathcal{D}_{10})$, which are distinguished by the invariant polynomial $\mathbf{D}$. We observe that in all three cases at infinity we have the singularities N, $\widehat{\binom{1}{2}}PEP-H$. So there are two distinct real singularities, i.e. $\eta = 0$ and $\widetilde{M} \neq 0$. Taking into consideration Theorem 6.4, we deduce that at infinity we could have only the configuration $\widehat{\binom{1}{2}}\ \widehat{P}_\lambda E\widehat{P}_\lambda - H$, N^f.

Thus we arrive at the following geometrically distinct configurations of singularities: (β_{73}) if $\mathbf{D} < 0$, (β_{74}) if $\mathbf{D} > 0$ and (β_{75}) if $\mathbf{D} = 0$.

4.2.3.2: $\mu_1 = 0$. If $\mu_2 \neq 0$ then by Theorem 6.3 we have the singularities $\$, \$$; N, $\binom{2}{2}PEP-PEP$ if $\widetilde{K} \neq 0$ (see the case $(\mathcal{D}_{11})$) and $\$, c$; $\widehat{\binom{2}{3}}N$ if $\widetilde{K} = 0$ (see the case $(\mathcal{D}_{12})$).

Assume first $\widetilde{K} \neq 0$. Then we have two infinite singularities (i.e. $\widetilde{M} \neq 0$). According to (7.48) the condition $\mu_0 = 0$ implies $\kappa = 0$, and considering Theorem 6.4 we obtain the configuration (β_{76}).

Consider now the case $\widetilde{K} = 0$. As we observe all three infinite singular points coalesced (i.e. $\widetilde{M} = 0$). Moreover two finite singularities also coalesced with this point. As a result we have obtained a nilpotent infinite node of total multiplicity five, and we need to describe its sectors.

For systems (7.30) the condition

$$\widetilde{K} = (2k - l^2)x^2 + (4ck - l)xy + (2cl - 1)y^2 = 0$$

must be satisfied. Since $c \neq 0$ (otherwise $2cl - 1 \neq 0$), we may assume $c = 1$ due to the rescaling $x \to cx$. Then $\widetilde{K} = 0$ yields $k = 1/8$ and $l = 1/2$. Herein we calculate

$$\eta = \mu_0 = \mu_1 = 0, \quad \mu_2 = (x + 2y)^2/4 > 0, \quad \widetilde{M} = 0, \quad C_2 = (x + 2y)^3/8 \neq 0.$$

So according to Theorem 6.4, we arrive at the configuration (β_{77}).

Suppose now $\mu_2 = 0$. Then by Theorem 6.3 (see the case $(\mathcal{D}_{13})$) systems (7.30) possess the singularities $\$$; $\binom{3}{3}$ $PEPEP - P$. So the conditions $\eta = 0 = \widetilde{M}$ and $C_2 \neq 0$ have to be fulfilled. We observe that in this case $\widetilde{K} \neq 0$, otherwise as it was shown above, we get $\mu_2 \neq 0$, which contradicts our assumption. So considering these conditions as well as the conditions $\mu_0 = \mu_1 = \mu_2 = 0$, by Theorem 6.4 we get the configuration (β_{66}).

This completes the proof of Theorem 7.3. □

7.3 A center

Consider a quadratic system (7.1) having a center. We have already given [31] a classification of these systems with a center, using what we called the *congruent equivalence relation.* We were then at a stage in our work preceding our introduction of the geometrical equivalence relation. Also the phase portraits were determined using the affine invariant polynomials that we also use in this book. Even though at that time we already used some of the concepts of the geometrical equivalence relation, this was not done with the same depth as it is in this book. Some concepts such as the nodes with discriminant zero, the complex finite singularities with trace or discriminant zero, the different types of nodes at infinity or a complete sectorial decomposition of nilpotent singularities were not done at that time. Moreover there are some minor mistakes in Table 6.1 of [31], which will be corrected here together with the improvements mentioned.

So, we will state again the points (i) and (ii) of Theorem 1.1 of [31], and we will redo its proof but will skip the cases already done and pay attention only to those steps in which the geometrical equivalence relation brings some new data to the table.

We need here the following results concerning isochronous and uniformly isochronous centers (see [84, 229, 101]).

Lemma 7.4 ([229]). *A quadratic system* (7.1) *has an isochronous center on its phase plane if and only if this system can be brought via an affine change of coordinates and a rescaling of the time to one of the following four systems:*

$$(S_1) \qquad \dot{x} = -y + x^2 - y^2, \quad \dot{y} = x + 2xy;$$
$$(S_2) \qquad \dot{x} = -y + x^2, \quad \dot{y} = x + xy;$$
$$(S_3) \qquad \dot{x} = -y - 4x^2/3, \quad \dot{y} = x - 16xy/3;$$
$$(S_4) \qquad \dot{x} = -y + 16x^2/3 - 4y^2/3, \quad \dot{y} = x + 8xy/3.$$

Lemma 7.5 ([101]). *Assume that a polynomial differential system* (5.2) *has a center at the origin of coordinates. Then, this center is uniformly isochronous if and only if due to a linear change of variables and a rescaling of time this system can be written under the form*

$$x' = -y + xR(x,y), \quad y' = x + yR(x,y), \tag{7.49}$$

where $R(x,y)$ is a polynomial in x and y of degree $m-1$ and $R(0,0) = 0$.

Considering Lemmas 7.4 and 7.5, we arrive at the next remark.

Remark 7.2. (i) Only the system (S_2) is of the form (7.49) (with $R(x,y) = x$), and hence the center of this system is uniformly isochronous.

(ii) Each one of the systems (S_1)–(S_4) possesses on the phase plane two complex singularities and either two centers (systems (S_1) and (S_4) or one center (systems (S_2) and (S_3)). Moreover both centers of system (S_1) (respectively (S_4)) are isochronous because making the change $(x,y,t) \mapsto (x,-y-1,t)$ (respectively $(x,y,t) \mapsto (x,-y-3/4,t)$), which interchanges the centers, we obtain the same system.
Furthermore for systems (S_1)–(S_4) we have, respectively,

$$\begin{aligned}
&(S_1): \ \mathcal{T}_2 = -16, \ \mathbf{D} = 192, \ \mu_0 = \eta = -4, \ W_4 = \mathcal{I}_1 = 0;\\
&(S_2): \ \mathcal{T}_2 = \mu_0 = 0, \ \mathbf{D} = 192, \ \mu_1 = -x, \ \tilde{L} = 0;\\
&(S_3): \ \mathcal{T}_2 = \mu_0 = \mathcal{I}_2 = 0, \ \mathbf{D} = 2^{20}/27, \ \mu_1 = 2^{10}x/27,\\
&\qquad \tilde{L} = -128x^2/3 = 3\widetilde{N};\\
&(S_4): \ \mathcal{T}_2 = -25600/81, \ \mathbf{D} = 2^{16}/27, \ \mu_0 = -4096/81, \ \eta = 8192/81,\\
&\qquad E_3 = \mathcal{I}_1 = 0.
\end{aligned} \tag{7.50}$$

Next we prove the following statement:

Theorem 7.4. *Consider a quadratic system of differential equations* (7.1).

(i) *This system possesses a center and the configuration of all its singularities (finite and infinite) up to geometrical equivalence is the one given in Table 7.1 if and only if the corresponding affine invariant conditions described in this table hold.*

(ii) *We have a total of 50 geometrically distinct configurations of singularities for quadratic systems possessing a center.*

We have maintained the same enumeration of cases from Table 6.1 of [31] to avoid introducing a new enumeration but use this enumeration as subindices of the letter γ. Thus for the cases in which one configuration of singularities has split in two due to new geometric features, we have named them, for example, γ_{7a} and γ_{7b} (there are several cases that split into three geometrically distinct configurations, and then the letter c is used). We have also exchanged the order of the invariants $\widetilde{N}$ and $\tilde{L}$ in the cases from 28 to 31, so as to coincide with what we have in Diagram 6.7 in this book. This has helped us to discover that case 28 can have three geometrically distinct possibilities. Moreover we have split the cases in which the center could be isochronous and even uniformly isochronous.

The minor mistakes corrected from Table 6.1 of [31] (apart from the geometrical improvements) are as follows:

(i) In case γ_{26} the finite saddle was written as a semi-elemental triple saddle, and it is in fact a nilpotent triple saddle.

(ii) In case γ_{29} the infinite nilpotent point was written as $\widehat{\binom{1}{2}}PEP-PHP$, and it must be as it is indicated in the table here.

(iii) In case γ_{32} the infinite multiple point was written as intricate and must be nilpotent.

(iv) In case γ_{41} the infinite multiple point was written without sectorial decomposition, which is included here.

The next proposition follows directly from [31] and gives a complete set of normal forms to study the quadratic systems with a center.

Proposition 7.1. *Assume that a quadratic system has a center at* $(0,0)$. *Then this system could be brought to one of the following canonical forms via a linear transformation:*

$$(S_1^{(c)}) \quad \begin{cases} \dot{x} = -y + gx^2 - xy, \ (g \neq 0), \\ \dot{y} = x + x^2 + 3gxy - 2y^2; \end{cases}$$

$$(S_2^{(c)}) \quad \begin{cases} \dot{x} = y + 2nxy, \ (wm \neq 0), \\ \dot{y} = -x + lx^2 + 2mxy - ly^2; \end{cases}$$

$$(S_3^{(c)}) \quad \begin{cases} \dot{x} = y + 2(1-e)xy, \\ \dot{y} = -x + dx^2 + ey^2; \end{cases}$$

$$(S_4^{(c)}) \quad \begin{cases} \dot{x} = y + 2cxy + by^2, \\ \dot{y} = -x - ax^2 - cy^2, \ c \in \{0, 1/2\}, \end{cases}$$

where $w = m^2(2n-l) - (n-l)^2(2n+l)$.

Before proving Theorem 7.4, we prove first the next lemma, concerning the isochronous centers.

Lemma 7.6. *A quadratic system* (7.1) *possesses an isochronous center if and only if one of the sets of conditions given in Table* 7.1 *for the configuration* γ_{11c}, γ_{12b} *or* γ_{28b} *is satisfied. And this system possesses an uniformly isochronous center if and only if the conditions defining the configuration* γ_{31} *from Table* 7.1 *are fulfilled.*

Proof. From the proof of Theorem 1.1 of [31] we conclude that all four configurations mentioned in the statement of the lemma could be produced only by systems $(S_3^{(c)})$. More precisely these systems produce the configurations γ_{11c} and γ_{12b} in the case $\mathcal{T}_2 \neq 0$, whereas the configurations γ_{28b} and γ_{31} correspond to the case $\mathcal{T}_2 = 0$.

The *necessity* follows directly from Remark 7.2 and Table 7.1, comparing the conditions defining the configuration γ_{11c} (respectively γ_{12b}, γ_{28b}, γ_{28c}) with the conditions (7.50) for the system (S_1) (respectively (S_4), (S_3), (S_2)).

To prove *sufficiency*, for systems $(S_3^{(c)})$ we calculate

$$\begin{aligned} \mathcal{T}_2 &= 4d(2+d-2e), \ E_3 = 2e(e-1)(1+d-e)^2, \\ W_4 &= -256d(2+d-2e)^3(1-2e)^4, \ \mu_0 = 4d(e-1)^2 e, \end{aligned} \tag{7.51}$$

and we consider two cases: $\mathcal{T}_2 \neq 0$ and $\mathcal{T}_2 = 0$.

1: $\mathcal{T}_2 \neq 0$. Then according to Table 7.1 we need to prove that the conditions for the configurations γ_{11c} and γ_{12b} imply the existence of isochronous centers.

It remains to observe that this system via the change $(x, y, t) \mapsto (2y, 2x, t)$ can be brought to the system (S_1), which by Lemma 7.4 and Remark 7.2 possesses two isochronous centers.

Table 7.1

				Configuration of singularities	No.
Conditions for the existence of a center [statement (*b*) of Theorem 6.2]: $\mathcal{T}_3 \neq 0,\ \mathcal{T}_4 = \mathcal{F}_1 = \mathcal{F}_2 = \mathcal{F}_3\mathcal{F}_4 = 0,\ \mathcal{T}_3\mathcal{F} < 0,\ (b_4)$					
Additional conditions for configurations				*Configuration of singularities*	No.
$\mu_0 \neq 0$	$\mathbf{D} < 0$	$\widetilde{K} < 0$		$s, s, s, c;\ N^f, N^f, N^f$	γ_1
		$\widetilde{K} > 0$		$s, n, n, c;\ S, S, N^{\infty}$	γ_2
	$\mathbf{D} > 0$	$\eta < 0$		$f, c,$ ©,©$;\ S,$ ©, ©	γ_3
		$\eta > 0$		$n, c,$ ©,©$;\ S, S, N^{\infty}$	γ_4
	$\mathbf{D} = 0$			$n^d, c, \overline{sn}_{(2)};\ \overline{\binom{0}{2}}\ SN, S$	γ_5
$\mu_0 = 0$				$s, c;\ \overline{\binom{1}{1}}\ SN, \overline{\binom{1}{1}}\ SN, N^d$	γ_6
Conditions for the existence of a center [statement (*c*) of Theorem 6.2]: $\mathcal{T}_4 = \mathcal{T}_3 = 0,\ \mathcal{T}_2 \neq 0,\ (c_5) \cup (c_6)$					
$\mu_0 < 0$	$\mathbf{D} < 0$	$\eta < 0$	$W_4 \neq 0$	$\$, n, n, c;\ S,$ ©, ©	γ_{7a}
			$W_4 = 0$	$\$, n^d, n^d, c;\ S,$ ©, ©	γ_{7b}
		$\eta > 0$	$\widetilde{K} < 0$	$s, s, \$, c;\ N^f, N^f, N^f$	γ_8
			$\widetilde{K} > 0$	$\$, n, n, c;\ S, S, N^{\infty}$	γ_9
		$\eta = 0$		$\$, n, n, c;\ \overline{\binom{0}{3}}\ S$	γ_{10}
	$\mathbf{D} > 0$	$\eta < 0$	$W_4 \neq 0$	$c, c,$ ©,©$;\ S,$ ©, ©	γ_{11a}
			$W_4 = 0 \neq \mathcal{I}_1$	$c, c,$ ©$^{\tau}$, ©$^{\tau};\ S,$ ©, ©	γ_{11b}
			$W_4 = 0 = \mathcal{I}_1$	$c^{\odot}, c^{\odot},$ ©$^{\tau}$, ©$^{\tau};\ S,$ ©, ©	γ_{11c}
		$\eta > 0$	$E_3^2 + \mathcal{I}_1^2 \neq 0$	$c, c,$ ©,©$;\ S, S, N^{\infty}$	γ_{12a}
			$E_3 = 0 = \mathcal{I}_1$	$c^{\odot}, c^{\odot},$ ©,©$;\ S, S, N^{\infty}$	γ_{12b}
		$\eta = 0$		$c, c,$ ©,©$;\ \overline{\binom{0}{3}}\ S$	γ_{13}
$\mu_0 > 0$	$\mathbf{D} < 0$			$s, s, c, c;\ N^f,$ ©, ©	γ_{14}
	$\mathbf{D} > 0$	$\eta < 0$	$\theta < 0$	$\$, c,$ ©,©$;\ N^{\infty},$ ©, ©	γ_{15a}
			$\theta > 0$	$\$, c,$ ©,©$;\ N^f,$ ©, ©	γ_{15b}
		$\eta > 0$	$\theta_1 < 0$	$\$, c,$ ©,©$;\ S, N^{\infty}, N^{\infty}$	γ_{16a}
			$\theta_1 > 0$	$\$, c,$ ©,©$;\ S, N^f, N^f$	γ_{16b}
			$\theta_1 = 0$	$\$, c,$ ©$^{\tau}$, ©$^{\tau};\ S, N^d, N^d$	γ_{16c}
		$\eta = 0$		$\$, c,$ ©,©$;\ \overline{\binom{0}{3}}\ N$	γ_{17}
$\mu_0 = 0$	$\eta < 0$	$\widetilde{K} \neq 0$	$\mu_2 < 0$	$c, c;\ \overline{\binom{2}{1}}\ S,$ ©, ©	γ_{18}
			$\mu_2 > 0$	$\$, c;\ \overline{\binom{2}{1}}\ N,$ ©, ©	γ_{19}
		$\widetilde{K} = 0$		$\$, c;\ N^d, \overline{\binom{1}{1}}$©$, \overline{\binom{1}{1}}$©	γ_{20}
	$\eta > 0$	$\widetilde{K} \neq 0$		$\$, c;\ \overline{\binom{2}{1}}\ S, N^f, N^f$	γ_{21}
		$\widetilde{K} = 0$		$\$, c;\ \overline{\binom{1}{1}}\ SN, \overline{\binom{1}{1}}\ SN, N^d$	γ_{22}

Table 7.1 *(continued)*

Additional conditions for configurations					Configuration of singularities	No.
Conditions for the existence of a center [statement (e) of Theorem 6.2]: $\mathcal{T}_4=\mathcal{T}_3=\mathcal{T}_2=\mathcal{T}_1=0,\ \sigma\neq 0,\ (e_4)$						
$\mu_0<0$	$\eta<0$				$\widehat{es}_{(3)}, c;\ S, \copyright, \copyright$	γ_{23}
	$\eta>0$				$\widehat{es}_{(3)}, c;\ S, S, N^{\infty}$	γ_{24}
	$\eta=0$				$\widehat{es}_{(3)}, c;\ \overline{\binom{0}{3}}\ S$	γ_{25}
$\mu_0>0$					$\widehat{s}_{(3)}, c;\ N^f, \copyright, \copyright$	γ_{26}
$\mu_0=0$	$\mu_1\neq 0$	$\mathbf{D}<0$			$s, s, c;\ \widehat{\binom{1}{2}}\ \widehat{P}_{\curlywedge} E \widehat{P}_{\curlywedge}-H,\ N^f$	γ_{27}
		$\mathbf{D}>0$	$\tilde{L}<0$	$\widetilde{N}<0\neq\mathcal{I}_2$	$c, \copyright, \copyright;\ \widehat{\binom{1}{2}}\ \widehat{P}_{\curlywedge} E \widehat{P}_{\curlywedge}-H,\ S$	γ_{28a}
				$\widetilde{N}<0=\mathcal{I}_2$	$c^{\odot}, \copyright, \copyright;\ \widehat{\binom{1}{2}}\ \widehat{P}_{\curlywedge} E \widehat{P}_{\curlywedge}-H, S$	γ_{28b}
				$\widetilde{N}>0$	$c, \copyright, \copyright;\ \widehat{\binom{1}{2}}\ \widehat{P}_{\curlywedge} H \widehat{P}_{\curlywedge}-E, S$	γ_{29}
				$\widetilde{N}=0$	$c, \copyright, \copyright;\ \widehat{\binom{1}{2}}\ E-H, S$	γ_{28c}
			$\tilde{L}>0$		$c, \copyright, \copyright;\ \widehat{\binom{1}{2}} H_{\curlywedge} H H_{\curlywedge}-H, N^{\infty}$	γ_{30}
			$\tilde{L}=0$		$c^{\oplus}, \copyright, \copyright;\ [\infty; S]$	γ_{31}
	$\mu_1=0$	$\mu_3\neq 0$	$\tilde{L}\neq 0$		$c;\ \widehat{\binom{3}{2}} H_{\curlywedge} H H_{\curlywedge}-H,\ N^d$	γ_{32}
			$\tilde{L}=0$		$c;\ \widehat{\binom{1}{2}}\ \widehat{P}_{\curlywedge} E \widehat{P}_{\curlywedge}-H, \overline{\binom{2}{1}} S$	γ_{33}
		$\mu_3=0$			$c, (\ominus\,[\vert]; \emptyset);\ \copyright, \copyright, (\ominus\,[\vert]; \emptyset)$	γ_{34}
Conditions for the existence of a center [statement (f) of Theorem 6.2]: Hamiltonian systems $\Rightarrow \sigma=0,\ (f_1)$–$(f_3),\ (f_5),\ (f_7)$						
$\mu_0<0$					$\$, \$, \$, c;\ N^f, N^f, N^f$	γ_{35}
$\mu_0>0$	$\mathbf{D}<0$				$\$, \$, c, c;\ N^f, \copyright, \copyright$	γ_{36}
	$\mathbf{D}>0$				$\$, c, \copyright^{\rho}, \copyright^{\rho};\ N^f, \copyright, \copyright$	γ_{37}
	$\mathbf{D}=0$	$\mathbf{T}\neq 0$			$\widehat{cp}_{(2)}, \$, c;\ N^f, \copyright, \copyright$	γ_{38}
		$\mathbf{T}=0$			$\widehat{s}_{(3)}, c;\ N^f, \copyright, \copyright$	γ_{39}
$\mu_0=0$	$\mu_1\neq 0$				$\$, \$, c;\ \widehat{\binom{1}{2}}\ \widehat{P}_{\curlywedge} E \widehat{P}_{\curlywedge}-H,\ N^f$	γ_{40}
	$\mu_1=0$				$\$, c;\ \widehat{\binom{2}{3}}\ \widehat{P}_{\curlywedge} \widehat{P}-\widehat{P}_{\curlywedge} \widehat{P}$	γ_{41}

Consider first the configuration γ_{11c}. In this case by Table 7.1 the condition $W_4=0$ is necessary. Since $\mathcal{T}_2\neq 0$ we get $e=1/2$, and then for the systems $(S_3^{(c)})|_{\{e=1/2\}}$ we obtain $\mathcal{I}_1=-(1+2d)^2/8$. Therefore the condition $\mathcal{I}_1=0$ gives $d=-1/2$, and then the system $(S_3^{(c)})|_{\{e=1/2=-d\}}$ becomes

$$\dot{x}=(1+x)y,\quad \dot{y}=-x-x^2/2+y^2/2.$$

We examine now the configuration γ_{12b}. Due to $\mu_0\neq 0$ the condition $E_3=0$ gives $e=d+1$, and then for systems $(S_3^{(c)})$ we calculate

$$\mathcal{I}_1=4d(1+2d)(1+5d),\quad \eta=-4d(1+3d)^3.$$

According to Table 7.1 for the configuration γ_{12b} the condition $\eta > 0$ is necessary, and this implies $-1/3 < d < 0$. Hence the condition $\mathcal{I}_1 = 0$ yields $d = -1/5$, and we arrive at the system

$$\dot{x} = (5+2x)y/5, \quad \dot{y} = -x - x^2/5 + (4y^2)/5.$$

Applying the change $(x, y, t) \mapsto (20y/3, 20x/3, t)$, this system will be brought to the system (S_4), which by Lemma 7.4 and Remark 7.2 possesses two isochronous centers.

2: $\mathcal{T}_2 = 0$. By Table 7.1 we need to examine the conditions for the configurations γ_{28b} and γ_{28c}. According to this table for both configurations the condition $\mu_0 = 0$ holds. Then from (7.51) the conditions $\mathcal{T}_2 = \mu_0 = 0$ imply $d = 0$, and then for systems $(S_3^{(c)})|_{\{d=0\}}$ we calculate

$$\mu_1 = 4e(e-1)^2 y, \quad \mathcal{I}_2 = 3(e-1)e(3e-1), \quad \tilde{L} = 8e(3e-2)y^2.$$

According to Table 7.1 for the configuration γ_{28b} the conditions $\mathcal{I}_2 = 0$ and $\mu_1 \neq 0$ are necessary. Therefore we get $e = 1/3$, and this leads to the system

$$\dot{x} = (3+4x)y/3, \quad \dot{y} = -x + y^2/3,$$

which via the change $(x, y, t) \mapsto (-4y, -4x, t)$ becomes the system (S_3). So by Lemma 7.4 this system possesses one isochronous center.

Consider now the configuration γ_{28c}, for which by Table 7.1 the condition $\tilde{L} = 0$ has to be satisfied. Since $\mu_1 \neq 0$ we get $e = 2/3$, and we arrive at the system

$$\dot{x} = (3+2x)y/3, \quad \dot{y} = -x + (2y^2)/3.$$

Applying the change $(x, y, t) \mapsto (-3y/2, -3x/2, t)$, this system will be brought to the system (S_2), which by Lemma 7.4 and Remark 7.2 possesses one uniformly isochronous center.

This completes the proof of the lemma. □

Proof of Theorem 7.4. From the proof of Theorem 1.1 of [31], we conclude that systems $(S_1^{(c)})$ can only produce the configuration γ_4, systems $(S_2^{(c)})$ can only produce the configurations from γ_1 to γ_6 except γ_4 and systems $(S_3^{(c)})$ can only produce the configurations from γ_7 (now split into γ_{7a} and γ_{7b}) to γ_{22} if $\mathcal{T}_2 \neq 0$, and from γ_{23} to γ_{34} if $\mathcal{T}_2 = 0$. Finally, systems $(S_4^{(c)})$ can only produce the configurations from γ_{35} to γ_{41}.

In order to complete the proof, we must check whether the new geometrical point of view can provide some new features in each one of the configurations. In case there are several possibilities, we must check which are realizable and which are impossible. We recall that by Lemma 7.1 a center is incompatible with a finite star node. Moreover as it will be proved in Lemma 7.7 (see statement (ii) on page 210), a center is also incompatible with an infinite star node.

1: The normal form $(S_1^{(c)})$. This normal form gives only the configuration 4 $c, n; \underset{\smile}{S}, S, N$ of [31].

The configuration of the singularities (and the normal form itself) says that $\mu_0 < 0$ and $\eta > 0$, which implies that the configuration of infinite singularities S, S, N must be improved here to S, S, N^∞ by Theorem 6.4. Regarding the finite singularities, apart from the two real ones, we have two complex ones. The traces of the complex ones must be different from zero since $\mathcal{T}_3 \neq 0$. But some discriminants may be zero, and this could allow the node to be n^d (it cannot be n^* by Lemma 7.1), or the complex points to have discriminant zero, depending on how many discriminants are zero. But normal form $(S_1^{(c)})$ gives $W_4 = -4g^2(1+g^2)^2$, and since $g \neq 0$ then $W_4 < 0$, the node is generic and the discriminants of the complex singularities also are not zero. So we end up with configuration γ_4.

2: The normal form $(S_2^{(c)})$. This normal form gives configurations from γ_1 to γ_6 except γ_4.

Remark 7.3. (i) The normal form $(S_2^{(c)})$ has the condition $wm = \mathcal{T}_3 \neq 0$, so only one finite singular point has trace zero and that must be the center.

(ii) We have that $\mathbf{D} = -192(l+2n)^2(l^2+m^2+2ln)$ and $W_4 = -256(l+2n)^2(l^2+m^2+2ln)(l^2+m^2+2ln+n^2)^2$. So if $\mathbf{D} \neq 0$ the only possibility to have $W_4 = 0$ would be $m = 0$ and $l = -n$, but this implies $wm = \mathcal{T}_3 = 0$. So if $\mathbf{D} \neq 0$ then sign$(\mathbf{D}) =$ sign(W_4).

(iii) We have $\mu_0 = -4l^2n^2 \leq 0$ and $\kappa = 64ln(2l^2+2m^2+2n^2+5ln)$. So $\mu_0 = 0$ implies $\kappa = 0$.

(iv) We have $\eta = 4(l+2n)^2(l^2+m^2+2ln)$ so sign$(\mathbf{D}) = -$sign(η).

2.1: γ_1. Since the infinite part is formed by three nodes, Theorem 6.4 says that all of them must be N^f. This configuration is completely determined according to geometrical equivalence.

2.2: γ_2. Since the infinite part is formed by two saddles and one node, Theorem 6.4 says that the node must be N^∞. By Remark 7.3(ii) since $\mathbf{D} < 0$ then $W_4 < 0$, and this configuration is completely determined according to geometrical equivalence.

2.3: γ_3. With $\mu_0 < 0$, $\mathbf{D} > 0$ and $\eta < 0$, the infinite part must be formed by one saddle and two complex points, which is geometrically complete. Moreover by Remark 7.3(ii) $W_4 > 0$, and this configuration is completely determined according to geometrical equivalence.

2.4: γ_5. With $\mu_0 < 0$ and $\mathbf{D} = 0$, it implies $\eta = W_4 = 0$ by Remarks 7.3(ii) and (iv). At infinity we have one saddle and one saddle-node formed by the coalescence of two infinite singularities. We have that $W_4 = 0$ forces one finite singularity to have discriminant zero, and this can only happen if the node is n^d.

2.5: γ_6. In this case two finite points have gone to infinity, so we have $\mu_0 = \mu_1 = \kappa = 0 \neq \mu_2$. Since in the case $\mu_0 = -4l^2n^2 = 0$ (i.e. $n = 0$, or $l = 0$) we get $\mu_2\tilde{L} < 0$, and by Theorem 6.4 we arrive univocally at the configuration of infinite singularities $\overline{\binom{1}{1}}\ SN, \overline{\binom{1}{1}}\ SN, N^d$.

3: The normal form $(S_3^{(c)})$ with $\mathcal{T}_2 \neq 0$. This normal form gives configurations from γ_7 (now split into γ_{7a} and γ_{7b}) to γ_{22}.

For normal form $(S_3^{(c)})$ we have the following invariants and comitants:

$$\begin{aligned}
&\mathcal{T}_2 = 4d(2 + d - 2e) \neq 0, \quad W_4 = -256d(2 + d - 2e)^3(-1 + 2e)^4,\\
&\mu_0 = 4d(-1 + e)^2e, \quad \mu_1 = 4(-1 + e)e(-1 + d + e)y,\\
&\mathbf{D} = 192(2 + d - 2e)^3e, \quad \eta = -4d(-2 + 3e)^3,\\
&\kappa = -64d(-1 + e)(-2 + 3e)(-1 + 3e),\\
&\mathcal{T}_3 = 0, \quad \mathcal{B} = -2 < 0,\\
&\theta = 64d(-1 + e)(-1 + 2e)^2,\\
&\theta_1 = -64d(-1 + 2e)(-2 + 3e)(-5 + 6e).
\end{aligned}$$

3.1: γ_{7a} ($\equiv (\beta_{24})$) and γ_{7b} ($\equiv (\beta_{25})$). This is the first case in which one configuration from Table 6.1 of [31] gives two geometrically distinct configurations. The reason is quite simple. The configuration of infinite singularities must be completed with two simple complex ones. The finite singularities are the center, a saddle and two nodes. Since $\mathcal{T}_3 = 0$ the saddle must be integrable by Theorem 6.2. And then the nodes can either be generic or have discriminant zero. Since $\mathcal{T}_2 \neq 0$ the only way to have $W_4 = 0$ is $e = 1/2$, and then also $W_3 = 0$. So, either both nodes are generic, or both are n^d (they cannot be n^* by Lemma 7.1). And both configurations are possible, as it will be proved in the corresponding examples in Part II.

3.2: γ_8 ($\equiv (\beta_{26})$). The infinite singularities must be all N^f, and the rest of the configuration has no other geometrical option to be improved. One saddle must be integrable by $\mathcal{T}_3 = 0$ and Theorem 6.2, and the other saddles must be strong by the same theorem.

3.3: γ_9 ($\equiv (\beta_{27})$). The infinite node must be N^∞ since this is the only possibility with two infinite saddles. If we try to force $W_4 = 0$ as in γ_{7b} with $e = 1/2$, then we have $\mu_0 = \eta = -d/2$, and we reach a contradiction. Thus the finite nodes must be generic.

3.4: γ_{10} ($\equiv (\beta_{28})$). For the same reason as above, the finite nodes must be generic, and the rest of the configuration is already geometrically fixed.

3.5: γ_{11a}, γ_{11b} and γ_{11c}. The configuration of infinite singularities must be completed with two simple complex ones. The finite singularities must be completed also with two simple complex ones, which cannot have traces zero if $\mathcal{T}_2 \neq 0$ but can have discriminants zero, as it will be shown with the corresponding examples in Part II. Moreover according to Lemma 7.6 in the case of discriminants of the complex singularities being zero, we have the possibility for the centers to be isochronous.

3.6: γ_{12a} and γ_{12b}. The infinite node must be N^∞ since this is the only possibility with two infinite saddles. As we have seen for configuration γ_9, trying to force $W_4 = 0$ gives an incompatibility in the signs of μ_0 and η, so in this case the finite complex points must be generic. Moreover by Lemma 7.6 we have the possibility for the centers to be isochronous.

3.7: γ_{13}. For the same reason as above the finite complex points must be generic, and the rest of the configuration is already geometrically fixed.

3.8: γ_{14}. The finite singularities are geometrically fixed since $\mathcal{T}_2 \neq 0$. And the invariants force that if $\mu_0 > 0$ and $\mathbf{D} < 0$, then $\theta > 0$ and the infinite node must be N^f, which is accompanied by two complex singularities.

3.9: γ_{15a} ($\equiv (\beta_{21})$) and γ_{15b} ($\equiv (\beta_{22})$). In this case it is not true that θ is forced to be positive; it can also be negative, but it cannot be zero (with $\mu_0 > 0$) since this implies also $\eta > 0$ and we are not in this case. Thus the infinite node may be N^f or N^∞ (as the examples will prove) but neither N^d nor N^*, and it is accompanied by two complex singularities. For the same reason W_4 cannot be zero, and the complex finite singularities are generic.

3.10: γ_{16a} ($\equiv (\beta_{33})$), γ_{16b} ($\equiv (\beta_{34})$) and γ_{16c} ($\equiv (\beta_{35})$). In this case if $W_4 \neq 0$, then $\theta < 0$, and we may have two possibilities according to the sign of θ_1. But if $W_4 = 0$ then $\theta = \theta_1 = 0$. So we can use the invariant θ_1 to split the three cases, which will be proved realizable by the corresponding examples in Part II. Moreover if $\theta_1 \neq 0$ the finite complex points are generic, and if $\theta_1 = 0$ the finite complex points have discriminant zero.

3.11: γ_{17} ($\equiv (\beta_{36})$). As in γ_{10} or γ_{13} the invariant $W_4 \neq 0$, and so the configuration must be completed just adding two simple finite complex singularities.

3.12: γ_{18}. From this case up to γ_{22} the fact that $\mu_0 = 0$ implies also $\mu_1 = 0$. That is, two finite singular points have gone to infinity, and thus the finite singularities are completely determined. The infinite singularity is also geometrically determined in this case.

3.13: γ_{19} ($\equiv (\beta_{37})$). There is nothing more that can be geometrically said about the real infinite singularity.

3.14: γ_{20} ($\equiv (\beta_{40})$). We must complete the description of the infinite singularities by adding two complex double singularities, which are the coalescence of two couples of simple complex singularities (one finite and one infinite). Moreover since the node at infinity cannot be N^*, by Theorem 6.4 it must be N^d.

3.15: γ_{21} ($\equiv (\beta_{38})$). We must complete the description of the infinite singularities by saying that by Theorem 6.4 both infinite nodes must be N^f.

3.16: γ_{22} ($\equiv (\beta_{23})$). Same as above, from Theorem 6.4 the only possibility for the infinite node is N^d. This case implies $e = 1$ and $\eta = -4d$. So the condition $\eta > 0$ yields $d < 0$ and hence $\mu_2 \tilde{L} = 8d(dx^2 + y^2)^2 < 0$, and by Theorem 6.4 we get the configuration γ_{22}.

4: The normal form $(S_3^{(c)})$ with $\mathcal{T}_2 = 0$. This normal form gives configurations from γ_{23} up to γ_{34}. For normal form $(S_3^{(c)})$ we have $\mathcal{T}_2 = 4d(2+d-2e) = 0$, which implies $d = 2e - 2$ if $\mu_0 \neq 0$.

4.1: γ_{23}. The finite configuration was already geometrically complete, and the infinite one must be completed with two complex singularities.

4.2: γ_{24}. Again the finite configuration was already geometrically complete, and by Theorem 6.4 the infinite node must be N^∞.

4.3: γ_{25}. The configuration given in [31] was already geometrically correct.

4.4: γ_{26}. In this case there is a minor mistake in the configuration given in [31] since the triple saddle was written to be semi-elemental while it must be nilpotent. Moreover this case implies $\theta > 0$, and then the infinite node must be N^f and accompanied by two complex simple singularities.

In the following configurations (up to γ_{33}) there appear some nilpotent infinite points whose correct geometrical description has been improved with respect to [31].

If $\mu_1 \neq 0 = \mu_0$ (corresponding to the configurations from γ_{27} to γ_{31}), then $d = 0$ and we have $\tilde{L} = 8e(-2+3e)y^2$ and $\widetilde{N} = -4(-1+e)(-1+2e)y^2$. Computing the traces of the Jacobian matrices on the finite singular points, it appears that (apart from the origin, which is the center) the others are $\pm\sqrt{2}/\sqrt{(e-1)e}$, and whether the singularities are real or complex, they cannot have trace zero.

4.5: γ_{27}. By Theorem 6.2 there must be just one finite singular point with trace zero, one finite singular point has gone to infinity, and the nilpotent infinite point has now its improved description from Theorem 6.4.

In [31] we gave priority to comitant $\widetilde{N}$ over $\tilde{L}$, but in [26] the priorities were reversed. Since all later papers have used the second option, we will rearrange the table of centers accordingly, and this splits the cases $\widetilde{N} < 0$ and $\widetilde{N} = 0$ into two geometrically distinct possibilities.

4.6: γ_{28a}, γ_{28b}, γ_{28c}. Since $\mu_1 \neq \mu_0 = 0 < \mathbf{D}$ one finite singular point has gone to infinity, and there are two finite complex singularities whose discriminant will only be zero if $W_7 = -192(-1+e)^4(-1+2e)^4$ is equal to zero, which is equivalent to $\widetilde{N} = 0$. It also implies $\tilde{L} < 0$. So if $\tilde{L} < 0$ and $\widetilde{N} < 0$ then the finite complex singularities are generic, and the configuration at infinity comes from Theorem 6.4. If $\tilde{L} < 0$ and $\widetilde{N} = 0$ then the finite complex singularities have discriminant zero, and the infinite nilpotent point is topologically equivalent to the previous case but is geometrically different. Moreover from Lemma 7.6 we know that if $\mathcal{I}_2 = 0$ the center is isochronous, which forces $\widetilde{N} < 0$.

4.7: γ_{29}. Since $\widetilde{N} > 0$ then $W_7 \neq 0$, and the complex finite singularities are generic. We simply improve the description of the infinite nilpotent point.

4.8: γ_{30}. The case $\tilde{L} > 0$ makes $\widetilde{N}$ irrelevant since W_7 cannot be zero. So the complex finite singularities are generic. We simply improve the description of the infinite nilpotent point and add that in this case the infinite node must be N^{∞}.

4.9: γ_{31}. The configuration given in [31] was already geometrically correct (we have simply changed a "," to a ";" in the description of the degenerate infinity). Moreover by Lemma 7.6 the center is uniformly isochronous.

4.10: γ_{32}. Now $\mu_1 = \mu_2 = 0$, hence two more singular points have gone to infinity, thus leaving the center as the only finite singularity. In [31] we forgot to add the "hat" that describes the infinite multiple point as nilpotent, and it seemed to be intricate. We now repair this, improving the sectorial description of this point, and since $\kappa = \widetilde{K} = 0$, from Theorem 6.4 the infinite node must be N^d.

4.11: γ_{33}. Again the center is the only finite singularity, and the configuration at infinity is uniquely determined and only needed the addition of geometrical improvements.

4.12: γ_{34}. The configuration given in [31] was already geometrically correct. It is worth to note that this configuration corresponds to a degenerate quadratic system since there is an infinite continuous set of singularities.

5: The normal form $(S_4^{(c)})$. This normal form gives configurations from γ_{35} up to γ_{41}. These are all Hamiltonian systems, and thus the traces of all finite singularities (real or complex) are zero and the elemental saddles are integrable. So the main improvements must be done to the description of infinite singularities.

For this normal form we have the affine invariants:

$$\mu_0 = a(ab^2 + 4c^3), \quad \eta = -27a(ab^2 + 4c^3), \quad \theta = 64a(ab^2 + 4c^3).$$

5.1: γ_{35}. Three infinite elemental nodes must always be N^f.

5.2: γ_{36} to γ_{39}. Since $\mu_0 > 0$ implies $\eta < 0$ and $\theta > 0$, the only needed improvement in these cases is to confirm the infinite node as N^f and to add the two complex infinite singularities. Moreover in γ_{37} we must add two complex finite singularities with zero traces.

5.3: γ_{40}. Since $\mu_0 = 0$ implies $\eta = 0$, the only needed improvement in this case is to confirm the infinite node as N^f and to improve the description of the infinite nilpotent point.

5.4: γ_{41}. In [31] we described the infinite singular point as a nilpotent node of multiplicity five (coalescence of two finite and three infinite singularities). According to Theorem 6.4 we have given in Table 7.1 a completely correct geometrical description. □

7.4 A star node at infinity and another special singularity

Theorem 7.5. *We consider here all the configurations of singularities, finite and infinite, of quadratic vector fields possessing an infinite star node and either an elemental node (infinite or finite) with coinciding eigenvalues or a weak finite singularity. We prove that there are exactly* 42 *geometrically distinct configurations of singularities, finite and infinite, and all of them are contained in the following list, enumerated and grouped according to the values of the divisor on the line at infinity* $D_S(C,Z)$*:*

$D_S(C,Z) = u+v+w$

$(\delta_1)\ s, \$, n, n^*;\ S,\ N^{\infty},\ N^*;$ $(\delta_2)\ s, \$, n, n^*;\ S,\ N^f,\ N^*;$
$(\delta_3)\ \$, \$, n^*, n^*;\ S,\ N^*,\ N^*;$ $(\delta_4)\ \$, n^*;\ \overline{\binom{1}{1}}SN,\ \overline{\binom{1}{1}}NS,\ N^*$
$(\delta_5)\ s, s, n, n;\ S, N^*, N^*;$ $(\delta_6)\ \copyright, \copyright, \copyright, \copyright;\ S, N^*, N^*;$
$(\delta_7)\ \copyright_{(2)}, \copyright_{(2)};\ S, N^*, N^*;$ $(\delta_8)\ \overline{sn}_{(2)}, \overline{sn}_{(2)};\ S, N^*, N^*;$
$(\delta_9)\ phpphp_{(4)};\ S, N^*, N^*;$ $(\delta_{10})\ s, s, n, n;\ S, N^d, N^*;$
$(\delta_{11})\ s, s^{(1)}, n, n^d;\ S, N^d, N^*;$ $(\delta_{12})\ \copyright, \copyright, \copyright, \copyright;\ S, N^d, N^*;$
$(\delta_{13})\ s, n, \copyright, \copyright;\ S, N^d, N^*;$ $(\delta_{14})\ s^{(1)}, n^d, \copyright, \copyright;\ S, N^d, N^*;$
$(\delta_{15})\ \overline{sn}_{(2)}, s, n;\ S, N^d, N^*;$ $(\delta_{16})\ \overline{sn}_{(2)}, s^{(1)}, n^d;\ S, N^d, N^*;$
$(\delta_{17})\ \overline{sn}_{(2)}, \copyright, \copyright;\ S, N^d, N^*;$ $(\delta_{18})\ \copyright_{(2)}, \copyright_{(2)};\ S, N^d, N^*;$
$(\delta_{19})\ \overline{sn}_{(2)}, \overline{sn}_{(2)};\ S, N^d, N^*;$ $(\delta_{20})\ \widehat{sn}_{(4)};\ S, N^d, N^*;$
$(\delta_{21})\ s, s^{(1)}, n, n^d;\ S, N^{\infty}, N^*;$ $(\delta_{22})\ s, s^{(1)}, n, n^d;\ S, N^f, N^*;$
$(\delta_{23})\ s^{(1)}, s^{(1)}, n^d, n^d;\ S, N^{\infty}, N^*;$ $(\delta_{24})\ s^{(1)}, s^{(1)}, n^d, n^d;\ S, N^f, N^*;$
$(\delta_{25})\ s^{(1)}, n^d, \copyright, \copyright;\ S, N^{\infty}, N^*;$ $(\delta_{26})\ s^{(1)}, n^d, \copyright, \copyright;\ S, N^f, N^*;$
$(\delta_{27})\ \overline{sn}_{(2)}, s^{(1)}, n^d;\ S, N^{\infty}, N^*;$ $(\delta_{28})\ \overline{sn}_{(2)}, s^{(1)}, n^d;\ S, N^f, N^*;$
$(\delta_{29})\ s^{(1)}, n^d;\ \overline{\binom{1}{1}}SN, \overline{\binom{1}{1}}NS, N^*.$

$D_S(C,Z) = u+v^c+w^c$

$(\delta_{30})\ \$, n^*, \copyright, \copyright;\ N^*, \copyright, \copyright;$ $(\delta_{31})\ \$, n^*;\ N^*,\ \binom{1}{1}\copyright, \binom{1}{1}\copyright;$
$(\delta_{32})\ s, s^{(1)}, n, n^d;\ N^*, \copyright, \copyright;$ $(\delta_{33})\ s^{(1)}, s^{(1)}, n^d, n^d;\ N^*, \copyright, \copyright;$
$(\delta_{34})\ s^{(1)}, n^d, \copyright, \copyright;\ N^*, \copyright, \copyright;$ $(\delta_{35})\ \overline{sn}_{(2)}, s^{(1)}, n^d;\ N^*, \copyright, \copyright;$
$(\delta_{36})\ s^{(1)}, n^d;\ N^*, \binom{1}{1}\copyright, \binom{1}{1}\copyright.$

$D_S(C,Z) = 2u+v$

$(\delta_{37})\ \overline{sn}_{(2)}, \$, n^*;\ \overline{\binom{0}{2}}SN,\ N^*;$ $(\delta_{38})\ \$, n^*;\ \binom{2}{2}\ \widehat{P}\widehat{P}H - \widehat{P}\widehat{P}H, N^*;$
$(\delta_{39})\ s, s^{(1)}, n, n^d;\ \overline{\binom{0}{2}}SN,\ N^*;$ $(\delta_{40})\ s^{(1)}, s^{(1)}, n^d, n^d;\ \overline{\binom{0}{2}}SN, N^*;$
$(\delta_{41})\ s^{(1)}, n^d, \copyright, \copyright;\ \overline{\binom{0}{2}}SN,\ N^*;$ $(\delta_{42})\ s^{(1)}, n^d;\ \widehat{\binom{2}{2}}\ \widehat{P}_{\curlywedge}\widehat{P}\,H_{\curlywedge} - H, N^*.$

The realizability of each one of these configurations will be proved by means of an example, which will be provided in Part II.

Considering real planar quadratic differential systems (7.1), we first prove the next lemma concerning this family of systems.

Lemma 7.7. *If a quadratic system (7.1) possesses an infinite star node N^*, then*

(i) *for this system the condition $W_4 = 0$ is equivalent to $\mathcal{T}_4 = 0$;*

(ii) *this system could possess neither a focus (or a center) nor a weak saddle of order two or three;*

(iii) *if in addition this system possesses a finite one-direction node n^d, then it also possesses a weak saddle of order one.*

Proof. Consider a quadratic system (7.1) and assume that it has an infinite star node. Without loss of generality we may suppose that this point is located at $N[0:1:0]$, i.e. the parameter $k = 0$. So we detect the following Jacobian matrix corresponding to this singularity: $\begin{pmatrix} 2h-n & d \\ 0 & -n \end{pmatrix}$. Evidently we have a star node if and only if $h = d = 0$, $n \neq 0$, and then we obtain $W_4 = \mathcal{T}_4^2$, i.e. the statement (i) is valid.

Suppose now that in addition we have a finite singularity. Then doing a translation we may consider that it is located at $(0,0)$, i.e. the condition $a = b = 0$ holds. Since $n \neq 0$ via a time rescaling we may force $n = 1$, and as a result we get the systems:

$$\dot{x} = cx + gx^2, \quad \dot{y} = ex + fy + lx^2 + 2mxy + y^2.$$

We observe that these systems possess the invariant line $x = 0$, and hence they could not have a focus (or a center) at the origin of coordinates.

Assume that this finite singularity is a weak saddle, i.e. the condition $f = -c \neq 0$ is satisfied. Then doing a rescaling we may consider $c = 1$, and in this case for the above systems we calculate

$$\mathcal{T}_4 = 0, \ \mathcal{T}_3 = 8\varphi, \ \mathcal{F}_1 = 2(e+2m),$$
$$\mathcal{F}_2 = 12g(e+2m)\varphi, \ \mathcal{F}_3 = -4g(e+2m)\varphi,$$

where $\varphi(e,g,l,m) = l - eg - gm - m^2$. According to Theorem 6.2 for the existence of a weak saddle of order two (respectively three), the conditions $\mathcal{F}_1 = 0$ and $\mathcal{F}_2 \neq 0$ (respectively $\mathcal{F}_1 = \mathcal{F}_2 = 0$ and $\mathcal{F}_3 \neq 0$) are necessary. However taking into account the above values of the invariant polynomials, we deduce that the condition $\mathcal{F}_1 = 0$ implies $\mathcal{F}_2 = \mathcal{F}_3 = 0$, and this completes the proof of statement (ii) of the lemma.

Assume finally that a system (7.1), apart from an infinite star node, possesses a node n^d, which could be considered to be at the origin. So for the above systems the conditions $f = c \neq 0$ and $e \neq 0$ must hold, and doing a rescaling we may assume $c = 1$. Therefore we arrive at the family of systems:

$$\dot{x} = x(1+gx), \quad \dot{y} = ex + y + lx^2 + 2mxy + y^2, \quad e \neq 0, \tag{7.52}$$

for which calculations yield

$$\mathcal{T}_4 = 0, \ \mathcal{T}_3 = 8(gm - m^2 + l - eg) = \mathcal{F}/8, \ \mathcal{F}_1 = 2e.$$

So $\mathcal{F}_1 \neq 0$ and by Theorem 6.2 we conclude that the above systems could not have a weak saddle of order ≥ 2, nor an integrable saddle.

We show now that this system possesses necessarily a weak saddle of order one. Indeed, if $\mathcal{T}_3 \neq 0$ then $\mathcal{T}_3\mathcal{F} > 0$, and by Theorem 6.2 systems (7.52) possess a weak saddle of order one.

Assume now $\mathcal{T}_3 = 0$, i.e. $l = m^2 - gm + eg$. We calculate

$$\mathcal{T}_4 = \mathcal{T}_3 = \mathcal{F} = 0, \ \mathcal{T}_2 = -4g^2, \ \mathcal{B} = -2(e+2g)^2, \ \mathcal{H} = 4g^2.$$

If $\mathcal{T}_2 \neq 0$ then we have $\mathcal{T}_2 < 0$, $\mathcal{B} \leq 0$, $\mathcal{H} > 0$, and since $\mathcal{F}_1 \neq 0$ according to Theorem 6.2 (see statement (c_1)), systems (7.52) have two weak saddles both of order one.

Supposing $\mathcal{T}_2 = 0$ (i.e. $g = 0$), we obtain

$$\mathcal{T}_4 = \mathcal{T}_3 = \mathcal{T}_2 = \mathcal{T}_1 = \mathcal{F} = 0, \ \sigma = 2(1 + mx + y) \neq 0,$$
$$\mathcal{F}_1 = 2e, \ \mathcal{H} = \mathcal{B}_1 = 0, \ \mathcal{B}_2 = 4e^2.$$

So we have $\mathcal{F}_1 \neq 0$ and this implies $\mathcal{B}_2 > 0$. Therefore by Theorem 6.2 (see statement (e_1)) systems (7.52) have one saddle of order one, and this completes the proof of the lemma. □

Proof of Theorem 7.5. We shall prove this theorem considering canonical systems associated to infinite singular points (see Lemma 6.1, $(\mathbf{S}_I)$–$(\mathbf{S}_V)$). More precisely we examine the following three canonical forms that possess at least one real elemental infinite singularity and are defined by the values of the divisor $D_S(C,Z)$:

$$D_S(C,Z) = u + v + w \ \Rightarrow \begin{cases} \dot{x} = a + cx + dy + gx^2 + (h-1)xy, \\ \dot{y} = b + ex + fy + (g-1)xy + hy^2; \end{cases} \qquad (\mathbf{S}_I)$$

$$D_S(C,Z) = u + v^c + w^c \ \Rightarrow \begin{cases} \dot{x} = a + cx + dy + gx^2 + (h+1)xy, \\ \dot{y} = b + ex + fy - x^2 + gxy + hy^2; \end{cases} \qquad (\mathbf{S}_{II})$$

$$D_S(C,Z) = 2u + v \ \Rightarrow \begin{cases} \dot{x} = a + cx + dy + gx^2 + hxy, \\ \dot{y} = b + ex + fy + (g-1)xy + hy^2. \end{cases} \qquad (\mathbf{S}_{III})$$

1: $D_S(C,Z) = u+v+w$. In this case we consider the family of systems $(\mathbf{S}_I)$, which possess three real infinite singularities: $R_1[1:0:0]$, $R_2[0:1:0]$ and $R_3[1:1:0]$. Without loss of generality we may assume that the star node is located at the point R_1, for which the respective Jacobian matrix has the form: $\begin{pmatrix} 1 & -e \\ 0 & g \end{pmatrix}$. Therefore this singular point is a star node if and only if $e = 0$ and $g = 1$. Then via a translation we may assume $c = 0$, and we get the following family of systems:

$$\dot{x} = a + dy + x^2 + (h-1)xy, \quad \dot{y} = b + fy + hy^2, \qquad (7.53)$$

possessing at infinity the star node $R_1[1:0:0]$.

Next we ask for the existence of a second special singularity (infinite of finite), which considering Lemma 7.7 could be N^*, N^d, n^d, n^*, $s^{(1)}$ or $\$$.

Considering Theorems 7.2 and 7.3 we deduce that if, apart from an infinite star node N^*, a quadratic system possesses a finite star node, then necessarily it possesses an integrable saddle, too. Conversely the existence of a node N^* and of an integrable saddle implies also the existence of a finite star node n^*. So by Theorem 7.2, assuming that, apart from a node N^*, a system possess a node n^* (or a saddle $\$$), in the case $D_S(C, Z) = u + v + w$ we arrive at the following four configurations of singularities: $(\delta_1) \equiv (\alpha_9)$, $(\delta_2) \equiv (\alpha_{10})$, $(\delta_3) \equiv (\alpha_{11})$ and $(\delta_4) \equiv (\alpha_{35})$.

Additionally by Lemma 7.7 (see statement (iii)) for the family of systems under consideration, the existence of a one-direction finite node implies the existence of one weak saddle of order one. So it remains to examine three cases of an additional special singular point: N^*, N^d and $s^{(1)}$.

1.1: N^* and N^*. Without loss of generality we may assume that the second infinite star node of systems (7.53) is $R_2[0 : 1 : 0]$, for which the corresponding Jacobian matrix has the form $\begin{pmatrix} 1 & -d \\ 0 & h \end{pmatrix}$. So R_2 is a star node if and only if $d = 0$ and $h = 1$. Then doing a translation we may consider $f = 0$, and we get the following 2-parameter family of systems:

$$\dot{x} = a + x^2, \quad \dot{y} = b + y^2. \tag{7.54}$$

For these systems we calculate

$$\begin{gathered} \eta = 1 = \mu_0,\ \mathbf{D} = -12288a^2b^2,\ W_4 = 256(a-b)^4, \\ W_3 = -256(a-b)^2(a+b),\ W_2 = 32(3a^2 + 2ab + 3b^2), \\ W_1 = -16(a+b),\ \mathcal{T}_4 = 16(a-b)^2. \end{gathered} \tag{7.55}$$

As $\eta > 0$ and $\mu_0 > 0$, by Theorem 6.4 apart from the two star nodes, these systems have at infinity one saddle.

1.1.1: $\mathbf{D} \neq 0$. By Proposition 5.1 (see Table 5.1 on page 123) systems (7.54) could possess either four real finite singular points (and this implies $a < 0$ and $b < 0$) or four complex finite singular points (in which case we have either $a > 0$ or $b > 0$).

1. Assume first $a < 0$ and $b < 0$. If $W_4 \neq 0$ then considering (7.55) we get $a - b \neq 0$, $a + b \neq 0$, and we obtain

$$W_4 > 0,\ W_2 > 0,\ W_1 W_3 > 0,\ \mathcal{T}_4 \neq 0.$$

Therefore by Theorem 6.1 (see Table 6.2, line 4) and Theorem 6.2 (there are no weak singularities), we arrive at the configuration (δ_5).

If $W_4 = 0$ we get $a = b < 0$, and clearly due to a rescaling we obtain a system (7.54) with $a = b = -1$. This system possesses five invariant affine lines ($x = \pm 1$, $y = \pm 1$ and $y = x$), and this implies the existence of two star nodes $M_1(-1,-1)$ and $M_2(1,1)$ as well as of two saddles $M_1(-1,1)$ and $M_2(1,-1)$. Moreover by Theorem 6.2 the saddles are integrable because for this system calculations yield

$$\mathcal{T}_4 = \mathcal{T}_3 = \mathcal{F}_1 = 0, \quad \mathcal{T}_2 = -16 < 0, \quad \mathcal{B} = -32 < 0, \quad \mathcal{H} = 4 > 0.$$

This leads to the configuration (δ_3).

2. Suppose now that that either the condition $a > 0$ or $b > 0$ holds, i.e. we have four distinct complex finite singularities. Clearly we get the configuration given by (δ_6).

1.1.2: $\mathbf{D} = 0$. In this case $ab = 0$, and making a linear change of variables and a rescaling, we may assume in systems (7.54) $b = 0$ and $a \in \{-1, 0, 1\}$. Then we get the three systems:

$$\dot{x} = a + x^2, \quad \dot{y} = y^2, \quad a \in \{-1, 0, 1\},$$

for which we calculate

$$\mathbf{T} = 0, \ \mathbf{P} = 16a^2y^4, \ \mathbf{R} = -16ay^2,$$

and $\text{sign}\,(a) = -\text{sign}\,(\mathbf{PR})$.

If $a = 1$ we have $\mathbf{PR} < 0$, and by Proposition 5.1 (see Table 5.1) we have two double complex finite singular points. This leads to the configuration (δ_7).

Suppose $a = -1$. Then $\mathbf{PR} > 0$ and $G_1 = 16 \neq 0$, and considering Theorem 6.1 (see Table 6.2, line 84) we have two double real finite singular points that are saddle-nodes. So we get the configuration (δ_8).

Assume finally $a = 0$. In this case we get a homogeneous quadratic system for which $\mu_0 > 0$ and $\eta > 0$. Therefore by Theorem 6.1 (see Table 6.2, line 101) we have the intricate finite singularity $phpphp_{(4)}$ and this leads to the configuration (δ_9).

1.2: N^* and N^d. Assume now that the second infinite singular point $R_2[0:1:0]$ of systems (7.53) is N^d. Considering the corresponding Jacobian matrix given in the previous case of two star nodes, we must have $h = 1$ and $d \neq 0$. Then doing a translation and a rescaling we get the 2-parameter family of systems:

$$\dot{x} = a + y + x^2, \quad \dot{y} = b + y^2, \tag{7.56}$$

for which we calculate

$$\begin{gathered} \eta = 1 = \mu_0, \ \mathbf{D} = -12288b^2(a^2 + b), \ \mathbf{R} = -16(bx^2 + ay^2), \\ \mathbf{S} = 192by^2(4ax^2 - y^2), \ \sigma = 2(x + y), \ W_4 = 256\big[(a - b)^2 + b\big]^2, \\ W_3 = -256\big[(a - b)^2(a + b) + (a - 3b)b\big], \ W_1 = -16(a + b), \\ W_2 = 32(3a^2 + b + 2ab + 3b^2), \ \mathcal{T}_4 = 16\big[(a - b)^2 + b\big], \ \mathcal{T}_3 = 32b. \end{gathered} \tag{7.57}$$

As $\eta > 0$ and $\mu_0 > 0$, by Theorem 6.4 the above systems, apart from the star node and one-direction node, have at infinity one saddle.

1.2.1: $\mathbf{D} < 0$. By Proposition 5.1 (see Table 5.1) systems (7.56) could possess either four real finite singular points if $\mathbf{R} > 0$ and $\mathbf{S} > 0$ or four complex finite singular points if $\mathbf{R} \leq 0$ or $\mathbf{S} \leq 0$.

1. Assume first that $\mathbf{R} > 0$ and $\mathbf{S} > 0$, and considering (7.57) this gives $a < 0$ and $b < 0$. If $W_4 \neq 0$ (this implies $W_4 > 0$ and $\mathcal{T}_4 \neq 0$), then considering the graphics of the polynomial functions $W_3 = 0$, $W_2 = 0$ and $W_1 = 0$ and (7.57), we obtain $W_2 > 0$ and $W_1 W_3 > 0$. Therefore by Theorem 6.1 (see Table 6.2, line 4) and Theorem 6.2 (there are no weak singularities), we arrive at the configuration (δ_{10}).

   Assume now $W_4 = 0$. Since $b < 0$, setting $b = -u^2 \neq 0$, the condition $W_4 = 0$ yields $a = -u(u \pm 1)$, and for systems (7.56) in this case we calculate

$$\begin{aligned} &W_4 = 0, \ W_3 = 1024u^4 > 0, \ \mathcal{T}_4 = 0, \\ &\mathcal{T}_3 = -32u^2, \ \mathcal{F} = -1, \ \mathcal{F}_1 = \mp 4u. \end{aligned} \tag{7.58}$$

   By Theorem 6.1 (see Table 6.2, line 5) we have two saddles and two nodes. Moreover, as $W_4 = 0$ and $W_3 \neq 0$, one node has coinciding eigenvalues.

   Additionally, as $\mathcal{T}_3\mathcal{F} > 0$ and $\mathcal{F}_1 \neq 0$, by Theorem 6.2 one of the finite saddle is of order one. Since by Theorem 7.2 we could not have a star node, this leads to the configuration (δ_{11}).

2. Suppose now that $\mathbf{R} \leq 0$ or $\mathbf{S} \leq 0$, i.e. systems (7.56) possess four complex finite singular points. We claim that among these four singularities we could not have singularities with zero discriminants nor singularities with zero traces. Indeed, supposing $W_4 = 0$ (which is equivalent to $\mathcal{T}_4 = 0$) we arrive at (7.58), and since $W_3\mathcal{T}_3 \neq 0$ we deduce that our claim is proved. So in this case we have the configuration given by (δ_{12}).

1.2.2: $\mathbf{D} > 0$. By Proposition 5.1 systems (7.56) possess two real and two complex finite singularities. Considering (7.57) we obtain $b < -a^2 < 0$.

If $W_4 \neq 0$ then $W_4 > 0$, and by Theorem 6.1 we have one node and one saddle. Moreover the saddle is strong since the condition $W_4 \neq 0$ implies $\mathcal{T}_4 \neq 0$, and we get the configuration (δ_{13}).

In the case $W_4 = 0$ we get $\mathcal{T}_4 = 0$, i.e. we have a node with coinciding eigenvalues, which could not be a star node due to Theorem 7.2. In addition, considering (7.58) we have a weak saddle of order one, and moreover we conclude that we could not have complex singularities with zero traces or discriminants. This leads to the configuration (δ_{14}).

1.2.3: $\mathbf{D} = 0$. For systems (7.56) we calculate

$$\mathbf{T} = 96by^2\left[4abx^4 + 4(a^2 + 2b)x^2y^2 - ay^4\right],$$

and we shall consider two subcases: $\mathbf{T} \neq 0$ and $\mathbf{T} = 0$.

1.2.3.1: Assume first $\mathbf{T} \neq 0$. Then $b \neq 0$ and hence the condition $\mathbf{D} = 0$ yields $b = -a^2 < 0$. In this case we obtain $\mathbf{T} = 96a^3y^2(2ax^2 + y^2)^2$ and therefore $\text{sign}(\mathbf{T}) = \text{sign}(a)$.

1. If $\mathbf{T} < 0$ then $a < 0$, and by Proposition 5.1 systems (7.56) with $b = -a^2$ possess two elemental and one double real singularities.

   In the case $W_4 \neq 0$ we have $W_4 > 0$, and as $E_1 = -32a^4 \neq 0$, by Theorem 6.1 apart from the semi-elemental saddle-node, we have one node and one saddle. Moreover by (7.57) the condition $W_4 \neq 0$ implies $\mathcal{T}_4 \neq 0$, and considering Theorem 6.2 the saddle is strong, i.e. we get the configuration (δ_{15}).

   Suppose $W_4 = 0$. As for systems (7.56) with $b = -a^2$ we have $W_4 = 256a^6(2 + a)^2 = 0$, then due to $\mathbf{T} \neq 0$ (i.e. $a \neq 0$) we get $a = -2$. Then $\mathcal{T}_4 = 0$ and $\mathcal{T}_3\mathcal{F}_1 \neq 0$ (i.e. the weak saddle has order one). So in this case we arrive at the configuration (δ_{16}).

2. In the case $\mathbf{T} > 0$ (then $a > 0$), by Theorem 6.1 systems (7.56) with $b = -a^2$ possess one double real singularity (which is a semi-elemental saddle-node) and two complex ones. So we obtain the configuration (δ_{17}).

1.2.3.2: Admit now $\mathbf{T} = 0$. Then $b = 0$ and we calculate $\mathbf{P} = 16a^2y^4$, $\mathbf{R} = -16ay^2$ and hence $\text{sign}(\mathbf{PR}) = -\text{sign}(a)$.

If $a > 0$ then $\mathbf{PR} < 0$, and by Proposition 5.1 systems (7.56) with $b = 0$ possess two double complex singularities. This leads to the configuration (δ_{18}).

If $a < 0$ we have $\mathbf{PR} > 0$, and as for systems (7.56) with $b = 0$ we have $G_1 = 16a^2 \neq 0$. Then by Theorem 6.1 (see Table 6.2, line 84) we have two double real singularities, which are semi-elemental saddle-nodes. So we obtain the configuration (δ_{19}).

Suppose now $a = 0$. Then $\mathbf{P} = \mathbf{R} = 0$, and as for systems (7.56) with $a = b = 0$ we have $F_1 = -y^3 \neq 0$. By Theorem 6.1 (see Table 6.2, line 100) we have one saddle-node of multiplicity four. This saddle-node is nilpotent as it is shown by the linear part of the system (7.56), and hence we get the configuration (δ_{20}).

1.3: N^* and $s^{(1)}$. We consider systems (7.53) and assume that we have an infinite star node N^* and no other special point at infinity: N^* or N^d. We assume also that these systems possess a finite saddle $s^{(1)}$. Doing a translation we may consider that this saddle is located at the origin of coordinates, i.e. $a = b = 0$, and since we have a saddle, the condition $f = -c$ must hold. By Theorem 6.2 in order to have a saddle $s^{(1)}$ the condition $\mathcal{F}_1 = 2c(c + d - ch) \neq 0$ is necessary for these systems. So $c \neq 0$ and we may assume $c = 1$ doing a rescaling. This leads to the 2-parameter family of systems:

$$\dot{x} = x + dy + x^2 + (h - 1)xy, \quad \dot{y} = -y + hy^2, \tag{7.59}$$

for which we calculate

$$\begin{aligned} &\eta = 1,\ \mu_0 = h^2,\ \mathbf{D} = -48\big[(1-2h)^2 - 4dh\big],\ W_4 = \mathcal{T}_4 = 0,\\ &W_3 = 4(1-4h-4dh+3h^2)^2,\ \theta = \theta_2 = 0,\ \theta_1 = 16(1-h^2),\\ &\mathcal{T}_3 = 2(1-4h-4dh+3h^2) = 8\mathcal{F},\ \mathcal{F}_1 = 2(1+d-h) \neq 0. \end{aligned} \tag{7.60}$$

1.3.1: $\mu_0 \neq 0$. Then $\mu_0 > 0$ and by Theorem 6.4 systems (7.59) possess at infinity one saddle and two nodes. Moreover as one node is a star node and $\theta = 0$, we arrive at the next remark.

Remark 7.4. By Theorem 6.4 in the case $\mu_0 > 0$ and $\theta = 0$, we have the following configurations of infinite singularities:

$$\begin{array}{ll} S, N^\infty, N^* \ \Leftrightarrow\ \theta_1 < 0; & S, N^f, N^* \ \Leftrightarrow\ \theta_1 > 0;\\ S, N^d, N^* \ \Leftrightarrow\ \theta_1 = 0,\ \theta_4 \neq 0; & S, N^*, N^* \ \Leftrightarrow\ \theta_1 = \theta_4 = 0. \end{array}$$

However, as the last two cases have already been considered, in what follows we assume that the condition $\theta_1 \neq 0$ is fulfilled.

We shall consider three subcases: $\mathbf{D} < 0$, $\mathbf{D} > 0$ and $\mathbf{D} = 0$.

1.3.1.1: $\mathbf{D} < 0$. As there exists a real finite singularity (a weak saddle) and $\mu_0 > 0$, according to Theorem 6.1 systems (7.59) possess four real finite singularities: two saddles and two anti-saddles. Moreover by Lemma 7.7 the anti-saddles are both nodes.

1. $W_3 \neq 0$. Then only one node has coinciding eigenvalues, and another is a generic node (with two directions). By Theorem 7.2 we could not have a star node (otherwise necessarily we must have an integrable saddle). So we have the finite singularities $s, s^{(1)}, n, n^d$. Considering Remark 7.4 (i.e. the condition $\theta_1 \neq 0$), we arrive at the configuration (δ_{21}) if $\theta_1 < 0$ and (δ_{22}) if $\theta_1 > 0$.

2. $W_3 = 0$. Then $1 - 4h - 4dh + 3h^2 = 0$, and as $h \neq 0$ (due to $\mu_0 \neq 0$) we obtain $d = (h-1)(3h-1)/(4h)$. In this case for systems (7.59) we calculate

$$\begin{aligned} &\mathbf{D} = -48h^2,\ W_4 = W_3 = 0,\ \theta = \theta_2 = 0,\ \theta_1 = 16(1-h^2),\ \mathcal{T}_2 = -4h^2,\\ &\mathcal{T}_4 = \mathcal{T}_3 = \mathcal{F} = 0,\ \mathcal{F}_1 = (1-h^2)/(2h),\ \mathcal{B} = -\frac{(9h^2-1)^2}{8h^2} \leq 0,\ \mathcal{H} = 4h^2 > 0. \end{aligned}$$

Since $W_4 = W_3 = 0$ both finite nodes have coinciding eigenvalues, and it was mentioned above that due to Theorem 7.2 we could not have a finite star node. Additionally considering the above conditions, by Theorem 6.2 systems (7.59) possess two weak saddles both being of order one. Thus we get the following two configurations for systems (7.59): (δ_{23}) if $\theta_1 < 0$ and (δ_{24}) if $\theta_1 > 0$.

1.3.1.2: $\mathbf{D} > 0$. Then by Proposition 5.1 systems (7.59) possess two real and two complex finite singularities. Since $\mu_0 > 0$, considering Theorem 6.1, apart from the weak saddle of order one, we must have one anti-saddle, which by Lemma 7.7 is a node. Since the condition $\mathbf{D} > 0$ implies $1 - 4h - 4dh + 4h^2 < 0$, we conclude that $1 - 4h - 4dh + 3h^2 \neq 0$, i.e. $W_3 \neq 0$, and therefore we could not have more than one singularity with coinciding eigenvalues. Moreover by Theorem 7.2 this singularity cannot be a star node (incompatible with a weak saddle of order one), i.e. we have a node n^d. Additionally by (7.60) the condition $W_3 \neq 0$ implies $\mathcal{T}_3 \neq 0$, and therefore we also cannot have complex singularities with zero traces.

Thus considering Remark 7.4, we arrive at the configuration (δ_{25}) if $\theta_1 < 0$ and (δ_{26}) if $\theta_1 > 0$.

1.3.1.3: $\mathbf{D} = 0$. This yields $1 - 4h - 4dh + 4h^2 = 0$, and as $h \neq 0$ we obtain $d = (2h-1)^2/(4h)$. Thus we arrive at the 1-parameter family of systems:

$$\dot{x} = x + (2h-1)^2 y/(4h) + x^2 + (h-1)xy, \quad \dot{y} = -y + hy^2.$$

Besides the weak saddle of order one $M_1(0,0)$, these systems possess one elemental singular point $M_2(-1,0)$ and one double semi-elemental singularity. Since the corresponding Jacobian matrix of the singularity M_2 is $\begin{pmatrix} -1 & \frac{1}{4h} \\ 0 & -1 \end{pmatrix}$, we conclude that it is a one-direction node. Additionally the double point is a saddle-node because of $E_1 = -h^2/2 \neq 0$ and Theorem 6.1 (see Table 6.2, line 75).

Considering the infinite singularities described in Remark 7.4, we get the configuration (δ_{27}) if $\theta_1 < 0$ and (δ_{28}) if $\theta_1 > 0$.

1.3.2: $\mu_0 = 0$. Then $h = 0$ and we get the 1-parameter family of systems:

$$\dot{x} = x + dy + x^2 - xy, \quad \dot{y} = -y,$$

which possess two real finite singularities: $M_1(0,0)$ (a weak saddle of order one, i.e. $\mathcal{F}_1 = 2(d+1) \neq 0$) and $M_2(-1,0)$, which is a node with Jacobian matrix $\begin{pmatrix} -1 & d+1 \\ 0 & -1 \end{pmatrix}$. So we have a node n^d because $d + 1 \neq 0$.

Additionally for these systems we calculate

$$\mu_0 = \mu_1 = 0, \ \mu_2 = x(x-y), \ \kappa = \theta_2 = 0, \ \tilde{L} = 8x(x-y).$$

Therefore since $\mu_2 \tilde{L} > 0$, according to Theorem 6.4, at infinity we have the configuration of singularities $\binom{1}{1} SN$, $\binom{1}{1} NS$, N^*. Thus in this case we arrive at the configuration (δ_{29}).

2: $D_S(C,Z) = u + v^c + w^c$. In this case we consider the family of systems $(\mathbf{S}_{II})$ that possess two complex $(R_{1,2}[1 : \pm i : 0])$ and one real $(R_3[1 : 1 : 0])$ singularities. For the real singularity we have the corresponding Jacobian matrix $\begin{pmatrix} -1 & -d \\ 0 & h \end{pmatrix}$. Therefore this singular point is a star node if and only if $d = 0$ and $h = -1$.

We remark that by Theorem 7.2, assuming that apart from a node N^*, a quadratic system possesses a node n^* (or a saddle $), in the case $D_S(C, Z) = u + v^c + w^c$ we get two configurations of singularities: $(\delta_{30}) \equiv (\alpha_{18})$ and $(\delta_{31}) \equiv (\alpha_{34})$.

Since we could not have a second special singular point at infinity in this case, considering Lemma 7.7 it remains to examine the case when, apart from N^*, there exists a finite singularity $s^{(1)}$. Then via a translation we may assume that this saddle is located at the origin of coordinates (i.e. $a = 0 = b$), and we get the following family of systems:

$$\dot{x} = cx + gx^2, \quad \dot{y} = ex + fy - x^2 + gxy - y^2.$$

Since the singular point $M_1(0, 0)$ is a weak saddle, we must have $f = -c \neq 0$, and via a rescaling we set $c = 1$. This leads to the 2-parameter family of systems:

$$\dot{x} = x + gx^2, \quad \dot{y} = ex + y - x^2 + gxy - y^2, \quad g - e \neq 0, \tag{7.61}$$

because the condition $\mathcal{F}_1 = 2(e - g) \neq 0$ must be satisfied in order to have a weak saddle of order one. For these systems we calculate

$$\begin{gathered} \eta = -4, \ \mu_0 = g^2, \ \theta = \theta_2 = 0, \ W_3 = 4(4 + 4eg - 3g^2)^2, \\ W_4 = 0, \ \mathbf{D} = 192(1 + eg - g^2), \ \mathcal{T}_3 = 2(4 + 4eg - 3g^2), \\ \mathcal{T}_4 = 0, \ \mathcal{F} = (4 + 4eg - 3g^2)/4. \end{gathered}$$

2.1: $\mu_0 \neq 0$. This implies $\mu_0 > 0$, and by Proposition 5.1 systems (7.61) possess finite singular points of total multiplicity four. We examine three subcases: $\mathbf{D} < 0$, $\mathbf{D} > 0$ and $\mathbf{D} = 0$.

1. $\mathbf{D} < 0$. As there exists a real finite singularity (a weak saddle) and $\mu_0 > 0$, according to Theorem 6.1 systems (7.61) possess four real finite singularities: two saddles and two anti-saddles. Moreover by Lemma 7.7 the anti-saddles are both nodes.

   If $W_3 \neq 0$ then only one node has coinciding eigenvalues, and the other is a generic node (with two directions). Moreover by Theorem 7.2 we could not have a star node apart from a weak saddle of order one.

   Additionally the condition $W_3 \neq 0$ implies $\mathcal{T}_3 \neq 0$, and by Theorem 6.2 we could not have two weak saddles. Since at infinity we have one star node and two complex singularities, we arrive at the configuration (δ_{32}).

   Assume now $W_3 = 0$. Then $4 + 4eg - 3g^2 = 0$, and as $g \neq 0$ (due to $\mu_0 \neq 0$) we obtain $e = (3g^2 - 4)/(4g)$. In this case for systems (7.61) we calculate

$$\begin{gathered} \mathbf{D} = -48g^2, \ W_4 = W_3 = 0, \quad \mathcal{T}_4 = \mathcal{T}_3 = \mathcal{F} = 0, \ \mathcal{T}_2 = -4g^2, \\ \mathcal{F}_1 = -(g^2 + 4)/(2g), \ \mathcal{B} = -\frac{(9g^2 + 4)^2}{8g^2} \leq 0, \ \mathcal{H} = 4g^2 > 0. \end{gathered}$$

As $\mathcal{F}_1 \neq 0$, according to Theorem 6.2 systems (7.61) possess two weak saddles both of order one. Moreover as $W_3 = 0$ both nodes are one-direction nodes, and we get the configuration (δ_{33}).

2. $\mathbf{D} > 0$. Then by Proposition 5.1 we have two real and two complex finite singularities. Moreover since $\mu_0 > 0$, by Theorem 6.1, apart from the weak saddle, systems (7.61) possess an anti-saddle, which considering Lemma 7.7 and the condition $W_4 = 0$ is a node n^d.

   We observe that the condition $\mathbf{D} > 0$ implies $W_3\mathcal{T}_3 \neq 0$; otherwise for $W_3 = 0$ (this implies $\mathcal{T}_3 = 0$), as it was shown above we get $\mathbf{D} = -48g^2 < 0$. So we get the unique global configuration (δ_{34}).

3. $\mathbf{D} = 0$. This yields $1 + eg - g^2 = 0$, and as $g \neq 0$ we obtain $e = (g^2 - 1)/g$. We arrive at the 1-parameter family of systems:

$$\dot{x} = x + gx^2, \quad \dot{y} = (g^2 - 1)x/g - y - x^2 + gxy - y^2.$$

   Besides the weak saddle $M_1(0,0)$, these systems possess one elemental singularity $M_2(0,-1)$ and one double semi-elemental singular point. The Jacobian matrix corresponding to M_2 has the form $\begin{pmatrix} 1 & 0 \\ -1/g & 1 \end{pmatrix}$. So M_2 is a node with one direction, whereas the double point is a saddle-node as $E_1 = g^2/2 \neq 0$ (see Table 6.2, line 75). Taking into account that $M_1(0,0)$ is a weak saddle of order one, we arrive at the configuration (δ_{35}).

2.2: $\mu_0 = 0$. In this case we get $g = 0$, and systems (7.61) possess two real finite singularities: $M_1(0,0)$ (a weak saddle) and $M_2(0,-1)$, which is a node with Jacobian matrix $\begin{pmatrix} 1 & 0 \\ e & 1 \end{pmatrix}$. So this node is n^d as $e \neq 0$ due to $g - e \neq 0$.

Additionally for these systems we calculate

$$\mu_0 = \mu_1 = 0, \ \mu_2 = x^2 + y^2, \ \kappa = \theta_2 = 0,$$

and according to Theorem 6.4 we have the following configuration of infinite singularities: N^*, $\binom{1}{1}$ ©, $\binom{1}{1}$ ©. This leads to the configuration (δ_{36}).

3: $D_S(C,Z) = 2u + v$. In this case we consider the family of systems $(\mathbf{S}_{III})$ that possess two real singularities: $R_1[1:0:0]$ (simple) and $R_2[0:1:0]$ (double). For $R_1[1:0:0]$ we have the corresponding Jacobian matrix $\begin{pmatrix} 1 & -e \\ 0 & g \end{pmatrix}$, and clearly this singular point is a star node if and only if $e = 0$ and $g = 1$.

Additionally by Theorem 7.2, assuming that, apart from a node N^*, a system possess a node n^* (or a saddle $), in the case $D_S(C,Z) = 2u+v$ we get the following two configurations of singularities: $(\delta_{37}) \equiv (\alpha_{23})$ and $(\delta_{38}) \equiv (\alpha_{36})$.

Evidently we could not have a second special singular point at infinity in this case, and considering Lemma 7.7 it remains to examine the case when, apart from

N^*, there exists a finite singularity $s^{(1)}$. Then via a translation we may assume that this saddle is located at the origin of coordinates (i.e. $a = 0 = b$), and we get the following family of systems:

$$\dot{x} = cx + dy + x^2 + hxy, \quad \dot{y} = fy + hy^2.$$

As $M_1(0,0)$ has to be a weak saddle, it is necessary that $f = -c \neq 0$, and via a rescaling (to force $c = 1$) we obtain the 2-parameter family of systems:

$$\dot{x} = x + dy + x^2 + hxy, \quad \dot{y} = -y + hy^2, \quad d - h \neq 0, \tag{7.62}$$

since for these systems we have $\mathcal{F}_1 = 2(d-h) \neq 0$. Calculations yield

$$\begin{aligned} &\eta = 0, \ \widetilde{M} = -8x^2, \ \mu_0 = h^2, \ \mathbf{D} = 192h(d-h), \ W_4 = 0, \\ &W_3 = 4h^2(4d-3h)^2, \ \theta = \theta_2 = 0, \ \mathcal{T}_4 = 0, \ \mathcal{T}_3 = -2h(4d-3h). \end{aligned} \tag{7.63}$$

3.1: $\mu_0 \neq 0$. Then $\mu_0 > 0$ and by Theorem 6.4 systems (7.62) possess at infinity one star node and one double semi-elemental saddle-node: $\overline{\binom{0}{2}}SN, N^*$. We observe that in this case the condition $\mu_0\mathcal{F}_1 \neq 0$ implies $\mathbf{D} \neq 0$, and we examine two subcases: $\mathbf{D} < 0$ and $\mathbf{D} > 0$.

1. $\mathbf{D} < 0$. As there exists a real finite singularity (a weak saddle) and $\mu_0 > 0$, according to Theorem 6.1 systems (7.61) possess four real finite singularities: two saddles and two anti-saddles. Moreover by Lemma 7.7 these systems could not have a focus, i.e. the anti-saddles are both nodes.

   If $W_3 \neq 0$ then only one node has coinciding eigenvalues and another one is a generic node. Additionally by Theorem 7.2 we could not have a star node, apart from a weak saddle of order one, and this implies the existence of a node n^d. Thus we get the configuration given by (δ_{39}).

   Assuming $W_3 = 0$ we obtain $d = 3h/4$, and in this case for systems (7.61) we calculate

   $$\begin{gathered} \mathbf{D} = -48h^2 < 0, \ W_4 = W_3 = 0, \quad \mathcal{T}_4 = \mathcal{T}_3 = \mathcal{F} = 0, \\ \mathcal{T}_2 = -4h^2 < 0, \ \mathcal{B} = -81h^2/8 < 0, \ \mathcal{H} = 4h^2 > 0. \end{gathered}$$

   As $\mathcal{F}_1 \neq 0$, according to Theorem 6.2 systems (7.62) possess two weak saddles both of order one. Moreover as $W_3 = 0$ both nodes are one-direction nodes, and this leads to the configuration (δ_{40}).

2. $\mathbf{D} > 0$. Then by Proposition 5.1 systems (7.62) possess two complex and two real elemental singularities. Since $\mu_0 > 0$, by Theorem 6.1 and Lemma 7.7, apart from the weak saddle, we have a node, which is n^d due to $W_4 = 0$ and Theorem 7.2.

   Considering (7.63) we observe that the condition $\mathbf{D} > 0$ implies $W_3\mathcal{T}_3 \neq 0$; otherwise for $W_3 = 0$ (this implies $\mathcal{T}_3 = 0$) as it was shown above, we get $\mathbf{D} = -48h^2 < 0$. Therefore we arrive at the unique global configuration (δ_{41}).

3.2: $\mu_0 = 0$. This condition implies $h = 0$ (then $d \neq 0$), and then systems (7.62) possess two real finite singularities: $M_1(0,0)$ (a weak saddle of order one) and $M_2(-1,0)$, which is a node with Jacobian matrix $\begin{pmatrix} -1 & d \\ 0 & -1 \end{pmatrix}$. Since $d \neq 0$ this node is n^d.

Additionally for systems (7.62) with $h = 0$ we calculate

$$\mu_0 = \mu_1 = 0, \ \mu_2 = x^2, \ \kappa = \theta_5 = 0 = \widetilde{K}, \ \tilde{L} = 8x^2, \ \kappa_1 = -32d.$$

Therefore considering Theorem 6.4 we obtain the configuration (δ_{42}).

As all the possibilities have been examined, Theorem 7.5 is proved. □

7.5 Three finite special singularities

Theorem 7.6. *We consider here all the configurations of singularities, finite and infinite, of quadratic vector fields possessing three finite elemental singular points of a special type, namely each one of them is either a node n^d or a weak singularity (i.e. the triplets (n^d, n^d, n^d), (n^d, n^d, w) or (n^d, w, w), where by "w" we denote a weak singularity, which could be a focus or a saddle). We prove that there are exactly 42 geometrically distinct configurations of singularities, finite and infinite, and all of them are contained in the following list, enumerated and grouped according to the above mentioned triplets:*

(n^d, n^d, n^d)

(λ_1) $s, n^d, n^d, n^d;\ \underline{S,}\ \copyright, \copyright;$ (λ_2) $s, n^d, n^d, n^d;\ S, S, N^\infty;$
(λ_3) $s, n^d, n^d, n^d;\ \overline{\binom{0}{2}} SN, S.$

(n^d, n^d, w)

(λ_4) $s^{(1)}, s^{(1)}, n^d, n^d; S, N^\infty, N^*;$ (λ_5) $s^{(1)}, s^{(1)}, n^d, n^d;\ \underline{S,}\ N^f, N^*;$
(λ_6) $s^{(1)}, s^{(1)}, n^d, n^d;\ N^*, \copyright, \copyright;$ (λ_7) $s^{(1)}, s^{(1)}, n^d, n^d;\ \overline{\binom{0}{2}} SN, N^*;$
(λ_8) $s^{(1)}, n, n^d, n^d;\ \underline{S,}\ \copyright, \copyright;$ (λ_9) $s^{(1)}, n, n^d, n^d;\ S, S, N^\infty;$
(λ_{10}) $s^{(1)}, n, n^d, n^d;\ \overline{\binom{0}{2}} SN, S;$ (λ_{11}) $s^{(2)}, n, n^d, n^d;\ \underline{S,}\ \copyright, \copyright;$
(λ_{12}) $s^{(2)}, n, n^d, n^d;\ S, S, N^\infty;$ (λ_{13}) $s^{(2)}, n, n^d, n^d;\ \overline{\binom{0}{2}} SN, S;$
(λ_{14}) $s, s^{(1)}, n^d, n^d;\ \underline{N^f,}\ \copyright, \copyright;$ (λ_{15}) $s, s^{(1)}, n^d, n^d;\ \underline{S,}\ N^f, N^\infty;$
(λ_{16}) $s, s^{(1)}, n^d, n^d;\ \overline{\binom{0}{2}} SN, N^f;$ (λ_{17}) $s, s^{(2)}, n^d, n^d;\ N^f, \copyright, \copyright;$
(λ_{18}) $s^{(1)}, n^d, n^d;\ \overline{\binom{1}{1}} SN,\ \underline{\copyright,}\ \copyright;$ (λ_{19}) $s^{(1)}, n^d, n^d;\ \overline{\binom{1}{1}} SN,\ S,\ N^\infty;$
(λ_{20}) $s^{(1)}, n^d, n^d;\ \overline{\binom{1}{1}} SN, \overline{\binom{0}{2}} SN;$ (λ_{21}) $s^{(2)}, n^d, n^d;\ \overline{\binom{1}{1}} SN, \copyright, \copyright;$
(λ_{22}) $\$, n^d, n^d, c;\ \underline{S,}\ \copyright, \copyright;$ (λ_{23}) $\$, \$, n^d, n^d;\ S, N^d, N^d;$
(λ_{24}) $\$, n^d, n^d;\ \widehat{\binom{1}{2}}\ H\!-\!E, S.$

$$
\begin{array}{lll}
 & (\lambda_{25})\ s, s^{(1)}, s^{(1)}, n^d;\ N^f, N^f, N^f; & (\lambda_{26})\ s^{(1)}, n, n^d, f^{(1)};\ S, ©, ©; \\
 & (\lambda_{27})\ s^{(1)}, n, n^d, f^{(1)};\ S, S, N^\infty; & (\lambda_{28})\ s^{(1)}, n, n^d, f^{(1)};\ \overline{\binom{0}{2}}SN, S; \\
 & (\lambda_{29})\ s^{(1)}, s^{(1)}, n^d, f;\ N^f, ©, ©; & (\lambda_{30})\ s^{(1)}, s^{(1)}, n^d, f;\ S, N^f, N^\infty; \\
 & (\lambda_{31})\ s^{(1)}, s^{(1)}, n^d, f;\ \overline{\binom{0}{2}}SN, N^f; & (\lambda_{32})\ s, s^{(1)}, n^d, f^{(1)};\ N^f, ©, ©; \\
(n^d, w, w) & (\lambda_{33})\ s, s^{(1)}, n^d, f^{(1)};\ S, N^f, N^\infty; & (\lambda_{34})\ s, s^{(1)}, n^d, f^{(1)};\ \overline{\binom{0}{2}}SN, N^f; \\
 & (\lambda_{35})\ s^{(1)}, s^{(1)}, n, n^d;\ N^\infty, ©, ©; & (\lambda_{36})\ s^{(1)}, s^{(1)}, n, n^d;\ S, N^\infty, N^\infty; \\
 & (\lambda_{37})\ s^{(1)}, s^{(1)}, n, n^d;\ S, N^f, N^f; & (\lambda_{38})\ s^{(1)}, s^{(1)}, n, n^d;\ \overline{\binom{0}{2}}SN, N^\infty; \\
 & (\lambda_{39})\ s^{(1)}, s^{(1)}, n^d;\ \overline{\binom{1}{1}}SN, N^f, N^f; & (\lambda_{40})\ s^{(1)}, n^d, f^{(1)};\ \overline{\binom{1}{1}}SN, ©, ©; \\
 & (\lambda_{41})\ s^{(1)}, n^d, f^{(1)};\ \overline{\binom{1}{1}}SN, S, N^\infty; & (\lambda_{42})\ s^{(1)}, n^d, f^{(1)};\ \overline{\binom{1}{1}}SN, \overline{\binom{0}{2}}SN.
\end{array}
$$

The realizability of each one of these configurations will be proved by means of an example, which will be provided in Part II.

Proof. Consider a real quadratic system (7.1) possessing three finite elemental singular points. Then without loss of generality we may assume that these points are located at $M_1(0,0)$, $M_2(1,0)$ and $M_3(0,1)$, and we get the family of systems:

$$\dot{x} = cx + dy - cx^2 + 2hxy - dy^2, \quad \dot{y} = ex + fy - ex^2 + 2mxy - fy^2. \tag{7.64}$$

The Jacobian matrices for these three finite singularities are respectively

$$M_1 \Rightarrow \begin{pmatrix} c & d \\ e & f \end{pmatrix}, \; M_2 \Rightarrow \begin{pmatrix} -c & d+2h \\ -e & f+2m \end{pmatrix}, \; M_3 \Rightarrow \begin{pmatrix} c+2h & -d \\ e+2m & -f \end{pmatrix}. \tag{7.65}$$

We need to force each one of the singularities $M_1(0,0)$, $M_2(1,0)$ and $M_3(0,1)$ to be either a node n^d or a weak singularity. Since we could not have three weak singularities (see for example Theorem 6.2), we examine three possibilities, depending on the number of the nodes n^d: three, two or one.

1: *Three nodes n^d.* Since systems (7.64) must have three nodes n^d, the following conditions have to be satisfied:

$$
\begin{aligned}
\tau_1 &= (c-f)^2 + 4de = 0, \\
\tau_2 &= (c+f+2m)^2 - 4e(d+2h) = 0, \\
\tau_3 &= (c+f+2h)^2 - 4d(e+2m) = 0.
\end{aligned} \tag{7.66}
$$

We distinguish two cases: $e \neq 0$ and $e = 0$.

1.1: $e \neq 0$. Then we may assume $e = 1/4$ (due to a time rescaling), and therefore the condition $\tau_1 = 0$ implies $d = -(c-f)^2$. Thus we get

$$\tau_2 = 2(c^2 + f^2 - h + 2cm + 2fm + 2m^2) = 0 \;\Rightarrow\; h = (c+m)^2 + (f+m)^2.$$

To simplify the computations as well as the normal form of the systems, we introduce two new parameters u and v as follows: $u = f - c$ and $v = c + f + 2m$. Thus we obtain the 3-parameter family of systems:

$$
\begin{aligned}
\dot{x} &= cx - u^2y - cx^2 + (u^2+v^2)xy + u^2y^2, \\
\dot{y} &= x/4 + (c+u)y - x^2/4 + (v-u-2c)xy - (c+u)y^2,
\end{aligned} \tag{7.67}
$$

for which the following condition holds:

$$\tau_3 = 4c^2 + 4c(v^2 - u^2 + u) + u^4 - 2u^3 + 2u^2(1+v)^2 + v^2(2u+v^2) = 0. \quad (7.68)$$

Computing the discriminant of the quadratic polynomial τ_3 with respect to the parameter c, we obtain

$$\text{Discrim}\,[\tau_3, c] = -16u^2(1+2v)^2 \le 0.$$

Therefore the condition $\tau_3 = 0$ could be satisfied only if either $u = 0$ or $v = -1/2$.

1.1.1: Assume first $u = 0$. Then we obtain $\tau_3 = (2c + v^2)^2 = 0$, which gives $c = -v^2/2$. This leads to the 1-parameter family of systems:

$$\dot{x} = v^2 x(-1 + x + 2y)/2,$$
$$\dot{y} = \left[x - 2v^2 y - x^2 + 4v(1+v)xy + 2v^2y^2\right]/4,$$

for which we calculate

$$\mu_0 = -v^6(2 + 4v + 3v^2)/16, \quad \eta = v^6(2 + 4v + v^2)/16,$$
$$\mathbf{D} = -3v^{20}(1+v)^4/256, \ \mathcal{T}_4 = v^{14}(1+v)^2/16,$$
$$\widetilde{M} = -v^2(5 + 8v + 2v^2)x^2 + 2v^3(2+v)xy - 2v^4y^2.$$

We observe that $v \neq 0$ because these systems are non-degenerate. Then $\mu_0 < 0$ and we have $\mathbf{D} \neq 0$ (otherwise we obtain a double nilpotent singularity). So we have four distinct finite singularities, and clearly the fourth one is a saddle. Additionally as $v(v+1) \neq 0$ we get $\mathcal{T}_4 \neq 0$, and by Theorem 6.2 the saddle is strong.

We remark that the total index of the finite singularities is $+2$, and hence the total index of infinite singularities must be -1 (here and in what follows, we calculate the total index of singularities only on half of the Poincaré sphere, skipping the symmetric ones). Considering the condition $\mu_0 < 0$ and Theorem 6.4, we obtain the following three global configurations of singularities: (λ_1) if $\eta < 0$, (λ_2) if $\eta > 0$ and (λ_3) if $\eta = 0$.

1.1.2: Suppose now that $u \neq 0$, i.e. in this case we must have $v = -1/2$, and we obtain $\tau_3 = (1 + 8c + 4u - 4u^2)^2/16 = 0$, which implies $c = (-1 - 4u + 4u^2)/8$. This leads to the 1-parameter family of systems:

$$\dot{x} = (4u^2 - 4u - 1)x/8 - u^2y + (1 + 4u - 4u^2)x^2/8 + (1 + 4u^2)xy/4 + u^2y^2,$$
$$\dot{y} = x/4 + (-1 + 4u + 4u^2)y/8 - x^2/4 - (1 + 4u^2)xy/4 + (1 - 4u - 4u^2)y^2/8,$$

for which we calculate

$$\mu_0 = -(1-2u)^6(3 + 4u + 12u^2)/4096, \quad \eta = (1-2u)^6(1 + 12u + 4u^2)/4096,$$
$$\mathbf{D} = -2^{-32}3(1-2u)^{20}(1+2u)^4, \quad \mathcal{T}_4 = 2^{-20}(1-2u)^{14}(1+2u)^2,$$
$$\widetilde{M} = -\frac{1}{8}(1-2u)^2\left[(3 + 12u + 4u^2)x^2 + (3 + 28u + 12u^2)xy + (1 + 12u + 12u^2)y^2\right].$$

We observe that $(2u-1) \neq 0$ because these systems are non-degenerate. Then $\mu_0 < 0$ and we have $\mathbf{D} \neq 0$ (otherwise we obtain a double singularity). So we have four distinct finite singularities, and clearly the fourth one is a saddle. Additionally as $(2u-1)(2u+1) \neq 0$ we get $\mathcal{T}_4 \neq 0$, and by Theorem 6.2 the saddle is a strong one. As a result (due to $\widetilde{M} \neq 0$) we get the same configurations as in the previous case.

1.2: $e = 0$. In this case the condition $\tau_1 = 0$ gives $f = c$, and then we have $\tau_2 = 4(c+m)^2 = 0$, i.e. $m = -c$. As M_1 is an elemental singularity we have $c \neq 0$, and doing a time rescaling we may assume $c = 1$. Then the condition $\tau_3 = 4(1+2d+2h+h^2) = 0$ yields $d = -(h+1)^2/2$, and we get the 1-parameter family of systems:

$$\dot{x} = x - (h+1)^2 y/2 - x^2 + 2hxy + (h+1)^2 y^2/2, \quad \dot{y} = y - 2xy - y^2.$$

For these systems we calculate

$$\mu_0 = -(2h^2+1) < 0, \ \eta = 2h^2 - 1, \ \mathcal{T}_4 = 16h^2, \ \mathbf{D} = -48h^2,$$
$$\widetilde{M} = -8x^2 - 8(1+2h)xy - 4(-1+2h+5h^2)y^2.$$

Clearly the fourth singular point is a saddle and $h \neq 0$, by $\mathbf{D} \neq 0$. Then $\mathcal{T}_4 \neq 0$ and by Theorem 6.2 the saddle M_4 could not be a weak saddle.

Since $\widetilde{M} \neq 0$, considering Theorem 6.4 we arrive at the same three global configurations of singularities, namely (λ_1) if $\eta < 0$, (λ_2) if $\eta > 0$ and (λ_3) if $\eta = 0$.

2: *Two nodes n^d and one weak singularity.* As we have three elemental singularities we shall consider systems (7.64) possessing the singularities $M_1(0,0)$, $M_2(1,0)$ and $M_3(0,1)$ with the respective Jacobian matrices (7.65). We shall follow the same way as in the previous part of the proof forcing M_1 and M_2 to be nodes n^d and M_3 to be a weak singularity. We again consider two cases: $e \neq 0$ and $e = 0$.

2.1: $e \neq 0$. In this case we consider the family of systems (7.67), which posses two nodes n^d ($M_1(0,0)$ and $M_2(1,0)$) and a third singularity $M_3(0,1)$. We impose the singular point $M_3(0,1)$ to be a weak singularity, i.e. instead of the condition $\tau_3 = 0$ we force the condition $\rho_3 = v^2 + u(u-1) = 0$.

We point out the following restrictions under the parameters of these systems. We prove a lemma for them.

Lemma 7.8. *Assume that for systems (7.67) the condition $\rho_3 = u^2 + v^2 - u = 0$ is satisfied. Then for these systems*

(i) *the condition $\widetilde{M}(\mathcal{F}_1^2 + \mathcal{F}_2^2) \neq 0$ holds,*

(ii) *the condition $\mathcal{T}_3 = 0$ implies the existence at infinity of a node N^*,*

(iii) *the weak singular point could not be a focus or a center.*

Proof. (i) We first assume that for these systems the condition $\widetilde{M} = 0$ holds. Calculations yield $\widetilde{M} = -2(\widetilde{M}_1x^2 + \widetilde{M}_2xy + 4\widetilde{M}_3y^2)$, where

$$\begin{aligned}\widetilde{M}_1 &= 4c^2 + c(8u - 8v - 3) + u^2 + v^2 - 3u - 8uv,\\ \widetilde{M}_2 &= 4c^2 + 4c(2u + u^2 - v + v^2) + 4(u - v)(u^2 + v^2) - u(5u + 4v),\\ \widetilde{M}_3 &= c^2 + c(2u - u^2 + 2v^2) + (u^2 + v^2)^2 + u^2 - u^3 + 3u^2v + 2uv^2,\end{aligned}$$

and we have to force the condition $\widetilde{M}_1 = \widetilde{M}_2 = \widetilde{M}_3 = 0$. We observe that $\operatorname{Res}_c(\widetilde{M}_1, \widetilde{M}_2) = \mathcal{R}$, where $\mathcal{R}$ can be presented in the form

$$\begin{aligned}\mathcal{R} &= -12\big[\rho_3\mathcal{R}_1 + (1 + 2v)\mathcal{R}_2\big],\\ \mathcal{R}_1 &= 16u(1 + u)(1 + u^2) + 43 + 156v + 48u(1 + u)v + 4(39 + 4u + 8u^2)v^2\\ &\quad + 16v^3(3 + v),\\ \mathcal{R}_2 &= u(43 + 70v) - v^2(43 + 54v).\end{aligned}$$

It is clear that due to $\rho_3 = u^2 + v^2 - u = 0$, the condition $\mathcal{R} = 0$ is equivalent to $(1 + 2v)\mathcal{R}_2 = 0$.

Assume first $1 + 2v = 0$, i.e. $v = -1/2$. Then we get $\rho_3 = (1 - 2u)^2/4 = 0$, which gives $u = 1/2$, and calculations yield

$$\widetilde{M} = -(1 + 4c)\big[4(1 + c)x^2 + (7 + 4c)xy + 4(1 + c)y^2\big]/2 = 0.$$

However this contradicts $\Delta_1 = (1+4c)^2/16 \neq 0$ (as M_1 is an elemental singularity).

Suppose now $1+2v \neq 0$, i.e. in this case the condition $\mathcal{R}_2 = 0$ must hold. We observe that in this case $v \neq -43/70$, and we have $u = v^2(43 + 54v)/(43 + 70v)$, which implies

$$\rho_3 = v^3(1 + 2v)(688 + 1593v + 1458v^2)/(43 + 70v)^{20} = 0.$$

Clearly the last condition gives $v = 0$, and this implies $u = 0$. However we again arrive at the contradiction:

$$\widetilde{M} = -2c\big[(4c - 3)x^2 + 4cxy + 4cy^2\big] = 0, \quad \Delta_1 = c^2 \neq 0.$$

Thus we proved that for systems (7.67) that possess two nodes n^d and a weak singularity (which is a saddle), we could not have $\widetilde{M} = 0$.

Suppose now that the conditions $\mathcal{F}_1 = \mathcal{F}_2 = 0$ hold for systems (7.67), for which the condition $\rho_3 = 0$ is satisfied. Then the polynomials $\mathcal{F}_1$ and $\mathcal{F}_2$ must have a common solution depending on c, and we calculate

$$\operatorname{Res}_c(\mathcal{F}_1, \mathcal{F}_2) = u^2(u + v)^{12}(1 + 2v)^2(u^2 - 2u - v - 2uv - v^2)(u^2 + u + 2uv - v^2)\widetilde{\mathcal{R}},$$

where $\widetilde{\mathcal{R}}$ is a polynomial of degree 36 in u and v. Using the function of the computer algebra system Mathematica "PolynomialReduce[$\widetilde{\mathcal{R}},\ u^2 + v^2 - u,\ \{u, v\}$]″ the last polynomial can be presented in the form

$$\widetilde{\mathcal{R}} = \big[\rho_3\widetilde{\mathcal{R}}_1 + 2^{-29}(1 + 2v)^{12}\widetilde{\mathcal{R}}_2\big],$$

where $\widetilde{\mathcal{R}}_1$ is a polynomial in u and v and

$$\begin{aligned}\widetilde{\mathcal{R}}_2 = &(17+12v)(u-v^2)+12v^{12}(u+v)+420(1+u)v^9-2uv^2(91+60v)\\&+8uv^4(90+54v-161v^2-84v^3-9v^7)+5v^8(203u-99v^2)\\&+3v^4(55+36v)-4v^6(143+84v)-6v^{10}(47u+30v)+77v^8(11+v^4).\end{aligned}$$

Clearly due to $\rho_3 = u^2+v^2-u = 0$, the condition $\widetilde{\mathcal{R}} = 0$ is equivalent to $\widetilde{\mathcal{R}}_2 = 0$. Additionally we calculate

$$\text{Res}_{\,u}(\widetilde{\mathcal{R}}_2, \rho_3) = v^{24}(85+264v+288v^2),$$

and Discrim $[85+264v+288v^2, v] = -28224 < 0$. Therefore considering the above expression for the polynomial $\text{Res}_{\,c}(\mathcal{F}_1, \mathcal{F}_2)$, which must vanish, we deduce that, apart from the equality $\rho_3 = 0$, the following condition is necessary, too:

$$u(u+v)(1+2v)(u^2-2u-v-2uv-v^2)(u^2+u+2uv-v^2) = 0.$$

However since the condition $u = 0$ implies $v = 0$ (due to $\rho_3 = 0$) and

$$\text{Res}_{\,u}(u^2-2u-v-2uv-v^2, \rho_3) = v(1+2v)^3, \quad \rho_3\big|_{\{v=-1/2\}} = (1-2u)^2/4,$$
$$\text{Res}_{\,u}(u^2-2u-v-2uv-v^2, \rho_3) = 4v^3(1+2v),$$

it is sufficient to consider the following two relations: $v = 0 = u$ and $v = -1/2 = -u$. Calculations yield

$$\mathcal{F}_1 = c^3/2, \quad \Delta_1 = c^2 \ \text{ if } \ v = 0 = u;$$
$$\mathcal{F}_1 = (1+4c)^3/128, \quad \Delta_1 = (1+4c)^2/16 \ \text{ if } \ v = -1/2 = -u.$$

Clearly since $\Delta_1 \neq 0$ (as M_1 is an elemental singularity), this completes the proof of statement (i) of Lemma 7.8.

(ii) Suppose now that for systems (7.67) with $\rho_3 = u^2+v^2-u = 0$, the condition $\mathcal{T}_3 = 0$ holds. We calculate

$$\begin{aligned}\mathcal{T}_3 &= (2c+u)(2c-v)(u+v)\widehat{T}_3/16,\\ \Delta_1\Delta_2 &= (2c+u)^2(2c-v)^2/16 \neq 0,\end{aligned} \tag{7.69}$$

where the polynomial $\widehat{T}_3$ could be presented in the form

$$\begin{aligned}\widehat{T}_3 &= \rho_3\widehat{U}_1 + (2c+u)(2c-v)\widehat{U}_2,\\ \widehat{U}_1 &= 32c^3(u-v)+4c^2(u+3u^2-2v-16uv+3v^2)+2c(3u^3-u^2-7uv\\&\quad -7u^2v+7uv^2-3v^3)-u(2u^2+5uv+7u^2v+8uv^2+7v^3),\\ \widehat{U}_2 &= 4c^2+8cu+5u^2-8cu^2-4u^3+4u^2v.\end{aligned}$$

We claim that the condition $\mathcal{T}_3 = 0$ is equivalent to $u+v = 0$. Indeed suppose that $u+v \neq 0$. Then considering (7.69) the condition $\mathcal{T}_3 = 0$ implies $\widehat{T}_3 = 0$, which due to $\rho_3 = 0$ gives $\widehat{U}_2 = 0$. However due to $u^2 - u = -v^2$, we have

$$\text{Discrim}\,[\widehat{U}_2, c] = 16u^2(-1 - 4u + 4u^2 - 4v) = -16u^2(1+2v)^2 < 0$$

since the equality $u(1+2v) = 0$ contradicts $u+v \neq 0$ due to the condition $\rho_3 = u^2 + v^2 - u = 0$, because $\rho_3|_{u=0} = v^2$ and $\rho_3|_{v=-1/2} = (2u-1)^2/4$.

So our claim is proved and we examine the condition $u+v = 0$. In this case we calculate

$$\begin{gathered}\mathcal{T}_3 = 0, \quad \rho_3 = v(1+2v) = 0, \quad \Delta_1 = (2c-v)^2/2,\\ \theta = 4(2c-v)^2 v^2 (1+2v)^2 = 0 = \theta_2, \quad \mu_0 = (2c-v)^4/16 > 0,\end{gathered}$$

and we have

$$\mu_0 = c^4, \quad \theta_1 = -16c^3, \quad \eta = (c-1)c^3, \quad \widetilde{M} = -2c(-3x^2 + 4cx^2 + 4cxy + 4cy^2),$$

if $v = 0$ and

$$\begin{gathered}\mu_0 = (1+4c)^4/256 > 0, \quad \theta_1 = (1+4c)^3/4, \quad \eta = (1+4c)^3(5+4c)/256,\\ \widetilde{M} = -(1+4c)(4x^2 + 4cx^2 + 7xy + 4cxy + 4y^2 + 4cy^2)/2,\end{gathered}$$

if $v = -1/2$. So we obtain $\theta = \theta_2 = 0$, $\theta_1 \neq 0$, $\mu_0 > 0$ and either $\eta \neq 0$ or $\eta = 0$ and $\widetilde{M} \neq 0$. According to Theorem 6.4 in all these cases systems (7.67) possess at infinity one N^*.

(iii) Suppose that systems (7.67) possess a weak focus or a center. Then without loss of generality we may assume that this singularity is located at $M_3(0,1)$. Then we must have $\tau_3 < 0$. However considering the expression for the discriminant τ_3 given in (7.68), we have $\text{Discrim}\,[\tau_3, c] = -16u^2(1+2v)^2 \leq 0$, and we conclude that $\tau_3 \geq 0$, i.e. we get a contradiction.

This completes the proof of the lemma. □

So according to Lemma 7.8 we could have two nodes n^d and two weak saddles together with an infinite star node N^* in the case $\mathcal{T}_3 = 0$. So considering Theorem 7.5, we arrive at the next four global configurations of singularities: $(\lambda_4) \equiv (\delta_{23})$, $(\lambda_5) \equiv (\delta_{24})$, $(\lambda_6) \equiv (\delta_{33})$ and $(\lambda_7) \equiv (\delta_{40})$.

Next we examine the configurations of singularities of systems (7.67) assuming $\mathcal{T}_3 \neq 0$ and considering three subcases: $\mu_0 < 0$, $\mu_0 > 0$ and $\mu_0 = 0$.

2.1.1: $\mu_0 < 0$. As we have two nodes, by Theorem 6.1 systems (7.67) possess a saddle and three anti-saddles. However by Lemma 7.8 we could not have a focus, and therefore, apart from the weak saddle, we have three nodes (two of them being n^d). Additionally it was proved earlier that systems with three nodes n^d could not possess a weak saddle, and therefore we conclude that the third node is generic.

For the infinite singularities according to Theorem 6.4 and taking into account that the total index of finite singularities in this case equals +2 as well as Lemma 7.8, we get three configurations: $S, ©, ©$ (if $\eta < 0$), S, S, N^{∞} (if $\eta > 0$) and $\overline{\binom{0}{2}} SN, S$ (if $\eta = 0$).

Additionally since systems (7.67) have three nodes, we must have $\mathcal{T}_3 \neq 0$, and by Theorem 6.2 (see statement (b)) the weak saddle has order one if $\mathcal{F}_1 \neq 0$ and order two if $\mathcal{F}_1 = 0$, because by Lemma 7.8 in this case the condition $\mathcal{F}_2 \neq 0$ has to be fulfilled. This leads to six geometrically distinct global configurations of singularities. More exactly we have three configurations in the case $\mathcal{F}_1 \neq 0$, (λ_8), (λ_9) and (λ_{10}), and three in the case $\mathcal{F}_1 = 0$, (λ_{11}), (λ_{12}) and (λ_{13}).

2.1.2: $\mu_0 > 0$. Since we have two nodes, by Theorem 6.1 systems (7.67) possess two saddles, one of them being weak. Moreover as the condition $\mathcal{T}_3 \neq 0$ is satisfied, by Theorem 6.2 in this case only one saddle is weak, and its order is governed by $\mathcal{F}_i$, $i = 1, \ldots, 4$ (see Theorem 6.2, statement (b)).

Additionally by Lemma 7.8 the condition $\mathcal{F}_1^2 + \mathcal{F}_2^2 \neq 0$ is satisfied for systems (7.67), and hence we could have a weak saddle only of order one (if $\mathcal{F}_1 \neq 0$) or two (if $\mathcal{F}_1 = 0$).

1. $\mathcal{F}_1 \neq 0$. In this case the weak saddle is of order one, and we examine the types of infinite singularities. For systems (7.67) we calculate

$$\begin{aligned}\theta = {} & 8c^2(u^2+v^2)(1-2u+2u^2+2v+2v^2) + 4u^4(1-u+v) \\ & + 8c\big[u^3(1-u)^2 - v(u^2+v^2)^2 + 2uv^2(1-u+u^2) + uv(u+4v^2+v^3)\big] \\ & + 2u^6 + 8u^2v(u+v) + 6u^3v^2(2+u) + 20u^2v^3 + 6u^2v^4 + 2v^6,\end{aligned}$$

and we observe that the discriminant of the quadratic polynomial θ with respect to the parameter c due to the condition $u^2 + v^2 - u = 0$ equals

$$\widehat{\mathcal{D}} = \text{Discrim}\,[\theta, c] = -64u^2(u+v)^4(1+2v)^2 \leq 0.$$

If $\widehat{\mathcal{D}} = 0$ then due to the same condition $u^2 + v^2 - u = 0$, this implies $u = -v$ because in the case $u = 0$ (respectively $v = -1/2$) we get $v = 0$ (respectively $u = 1/2$), i.e. we get particular cases of the relation $u = -v$, but this implies $\mathcal{T}_3 = 0$. Hence $\widehat{\mathcal{D}}$ cannot be zero (see (7.69)), i.e. $\widehat{\mathcal{D}} < 0$.

We observe that the sign of the coefficient in front of c^2 in the expression for θ depends on the sign of the polynomial

$$1 - 2u + 2u^2 + 2v + 2v^2 = 1 + 2v + 2(u^2 + v^2 - u) = 1 + 2v.$$

Additionally the condition $u^2 + v^2 - u = 0$ implies $0 \leq u \leq 1$ and $-1/2 \leq v \leq 1/2$, and hence $1 + 2v \geq 0$. However due to $\widehat{\mathcal{D}} < 0$ we have $1 + 2v \neq 0$, and this leads again to the condition $\theta > 0$. Then considering the conditions $\mu_0 > 0$, $\mathcal{F}_1 \neq 0$ and Lemma 7.8, by Theorem 6.4 we get the following three global configurations of singularities: (λ_{14}) (if $\eta < 0$), (λ_{15}) (if $\eta > 0$) and (λ_{16}) (if $\eta = 0$).

2. $\mathcal{F}_1 = 0$. Since by Lemma 7.8 in this case the condition $\mathcal{F}_2 \neq 0$ must hold for systems (7.67), according to Theorem 6.2 we have a weak saddle of order two.

   Applying the Mathematica function FindInstance (or Reduce) we detect that the conditions $\rho_3 = u^2 + v^2 - u = 0$, $\mathcal{F}_1 = 0$, $\mu_0 > 0$ and either $\eta \geq 0$ or $\theta \leq 0$ are incompatible. So for systems (7.67) only the condition $\eta < 0$ is satisfied, and we get the configuration (λ_{17}).

2.1.3: $\mu_0 = 0$. In this case a finite singularity has gone to infinity, and systems (7.67) possess three finite singularities: two nodes n^d and a weak saddle. Here the condition $\mu_1 \neq 0$ holds.

1. $\mathcal{F}_1 \neq 0$. In this case by Theorem 6.2 the saddle is of order one. We observe that the total index of finite singularities is $+1$, and hence the total index of infinite singularities must be zero. Therefore in the case $\eta \neq 0$, considering Theorem 6.4 we arrive at (λ_{18}) if $\eta < 0$, and at (λ_{19}) if $\eta > 0$.

   Assume now that the condition $\eta = 0$ is satisfied. In this case by Lemma 7.8 the condition $\widetilde{M} \neq 0$ holds. So at infinity we have two distinct real singularities.

   We claim that in this case the condition $\kappa \neq 0$ holds. Indeed, considering the condition $\rho_3 = u^2 + v^2 - u = 0$, for systems (7.67) we calculate

$$\mathrm{Res}_c(\mu_0, \kappa) = 256(u+v)^8 \left[\rho_3 \Psi(u,v) + 2u^3(1+2v)^3\right]^2,$$

   where $\Psi(u,v)$ is a polynomial in variables u and v. Therefore due to the condition $\rho_3 = 0$, we obtain that the condition $\mathrm{Res}_c(\mu_0, \kappa) = 0$ is equivalent to $u(u+v)(1+2v) = 0$. As it was mentioned earlier, we may assume $u+v=0$ because the other two possibilities lead to the particular cases of this relation. However for $u = -v$ we get the contradiction $\mu_0 = (2c-v)^4/16 = 0$ and $\Delta_1 = (2c-v)^4/4 \neq 0$, and this completes the proof of our claim.

   So $\kappa \neq 0$, and as $\eta = 0$ and $\mu_0 = 0$, by Theorem 6.4 at infinity we get two saddle-nodes, which lead to the configuration (λ_{20}).

2. $\mathcal{F}_1 = 0$. In this case since by Lemma 7.8 the condition $\mathcal{F}_2 \neq 0$ holds, we conclude that the saddle is of order two.

   Applying the Mathematica function FindInstance (or Reduce) we detect that the conditions $\rho_3 = u^2 + v^2 - u = 0$, $\mathcal{F}_1 = 0$, $\mu_0 = 0$ and $\eta > 0$ are incompatible. Additionally the condition $\eta = 0$, together with the indicated ones, leads to degenerate systems. So in this case for systems (7.67) only the condition $\eta < 0$ is satisfied, and we get the configuration (λ_{21}).

2.2: $e = 0$. Then considering (7.66) for systems (7.64) the condition $\tau_1 = \tau_2 = 0$ gives $f = c$ and $m = -c$, and therefore the condition $\rho_3 = 2h = 0$ yields $h = 0$.

As $M_1(0,0)$ is an elemental singularity, we have $\Delta_1 = c^2 \neq 0$, and doing a time rescaling we may assume $c = 1$. So we get the 1-parameter family of systems:

$$\dot{x} = x + dy - x^2 - dy^2, \quad \dot{y} = y - 2xy - y^2, \quad d(2d+1) \neq 0, \tag{7.70}$$

since $\Delta_3 = -(1+2d) \neq 0$ and $M_1(0,0)$ has the Jacobian matrix $\begin{pmatrix} 1 & d \\ 0 & 1 \end{pmatrix}$, i.e. $d \neq 0$ as M_1 must be a node n^d. For these systems we calculate

$$\begin{aligned} &\mu_0 = 1 + 4d = \eta, \ \mathcal{T}_4 = \mathcal{T}_3 = \mathcal{F}_1 = \mathcal{F} = \theta = \theta_1 = 0, \\ &\mathcal{T}_2 = -4(1+4d), \ \theta_3 = 2d^2, \ \mathcal{B} = -8(1+2d)^2 < 0, \\ &\mathcal{H} = 4(1+4d) = 0, \ \widetilde{M} = -8\big[x^2 + xy + (1+3d)y^2\big], \end{aligned} \tag{7.71}$$

and we consider three possibilities: $\mu_0 < 0$, $\mu_0 > 0$ and $\mu_0 = 0$.

1. $\mu_0 < 0$. This implies $\eta < 0$, and considering (7.71) we obtain $\mathcal{T}_2 > 0$ and $\mathcal{B} < 0$. So according to Theorem 6.2 in this case systems (7.70), apart from two nodes n^d, possess an integrable saddle and a center.

   Since $\eta < 0$ and $\mu_0 < 0$, considering Theorem 6.4 we arrive at the configuration $(\lambda_{22}) \equiv (\beta_{25})$.

2. $\mu_0 > 0$. In this case we have two saddles and two anti-saddles. Moreover from (7.71) we have $\mathcal{T}_2 < 0$, $\mathcal{F}_1 = \mathcal{F} = 0$, $\mathcal{B} < 0$ and $\mathcal{H} > 0$, and according to Theorem 6.2 both saddles are integrable.

   Additionally at infinity we have three real singularities, and as $\mu_0 > 0$ they are a saddle and two nodes. Since for systems (7.70) we have $\theta = \theta_1 = 0$ and $\theta_3 \neq 0$, by Theorem 6.4 both infinite nodes are N^d. So we obtain the configuration $(\lambda_{23}) \equiv (\beta_{31})$.

3. $\mu_0 = 0$. In this case we get $d = -1/4$, and then

   $$\mathcal{T}_4 = \mathcal{T}_3 = \mathcal{T}_2 = \mathcal{T}_1 = \mathcal{F}_1 = \mathcal{H} = \mathcal{B}_1 = 0, \ \mathcal{B}_2 = 2 > 0, \ \sigma = 2 - 4x - 2y \neq 0.$$

   According to Theorem 6.2 (see statement $(e_3), [\beta]$) the system has one integrable saddle. To determine the configuration of singularities at infinity, we calculate

   $$\begin{aligned} &\mu_0 = \eta = \kappa = \widetilde{N} = 0, \ \mu_1 = (2x+y)/2 \neq 0, \\ &\widetilde{K} = (2x+y)^2 > 0, \ \tilde{L} = -2(2x+y)^2 < 0. \end{aligned}$$

   Therefore by Theorem 6.4 we have at infinity one saddle and a triple nilpotent singular point $\widehat{\binom{1}{2}}\ H-E$, and this leads to the configuration $(\lambda_{24}) \equiv (\beta_{53})$.

3: *One node n^d and two weak singularities.* As we have three elemental singularities, we shall consider systems (7.64) possessing the singularities $M_1(0,0)$, $M_2(1,0)$ and $M_3(0,1)$ with the respective Jacobian matrices (7.65). We proceed in the same way as in the second part of the proof, forcing M_1 to be a node n^d and M_2 and M_3 to be weak singularities. So we get the relations

$$\tau_1 = (c-f)^2 + 4de = 0, \ \rho_2 = f - c + 2m = 0, \ \rho_3 = c - f + 2h = 0. \qquad (7.72)$$

In order to set the condition $\tau_1 = 0$, we consider again two cases: $e \neq 0$ and $e = 0$.

3.1: $e \neq 0$. Then due to a time rescaling we may assume $e = 1$, and considering (7.72) we obtain $d = -h^2, \ f = c + 2h, \ m = -h$. This leads to the 2-parameter family of systems:

$$\begin{aligned} \dot{x} &= cx - h^2y - cx^2 + 2hxy + h^2y^2, \\ \dot{y} &= x + (c+2h)y - x^2 - 2hxy - (c+2h)y^2. \end{aligned} \qquad (7.73)$$

For these systems we calculate the following invariant polynomials that are responsible for their finite singularities:

$$\begin{aligned} \mu_0 &= c^4 + 4c^3h + 2c(2+5c)h^2 + 4(2+3c)h^3 - (3+4c)h^4, \\ \mathbf{D} &= -48(c+h)^4(c^2 - 2h + h^2)^2(c^2 + 4ch + 5h^2 - 2h^3)^2, \\ \widetilde{K} &= 4\big[h(1+c)x^2 + (c+h)^2xy + h(h^2 - 2h - c)y^2\big], \\ W_2 &= 16(c^2 - 2h + h^2)^3(c^2 + 4ch + 5h^2 - 2h^3)^3, \\ W_3 &= -128h(c+h)^2(c + 3h + ch - h^2)^2(c^2 - 2h + h^2)^2 \\ &\quad \times (c^2 + 4ch + 5h^2 - 2h^3)^2, \quad W_4 = 0, \\ \mathcal{T}_2 &= -4(c+h)^2(c^2 - 2h + h^2)(c^2 + 4ch + 5h^2 - 2h^3), \\ \mathcal{T}_4 &= \mathcal{T}_3 = \mathcal{F} = 0, \ \mathcal{F}_1 = 2(c+h)(c + 3h + ch - h^2)^2, \\ \mathcal{B} &= -2(c+h)^4(1 + 2c + 2h - h^2)^2, \\ \mathcal{H} &= 4(c+h)^2(c^2 - h + 2ch + 3h^2 - h^3). \end{aligned} \qquad (7.74)$$

We also calculate the invariant polynomials that are responsible for their infinite singularities:

$$\begin{aligned} \eta &= c^4 + 4c^3(h^2 + h - 1) - 6c^2h(8 + 5h + 4h^2) + 4ch^2(12h^2 - 17h - 48) \\ &\quad - h^3(256 - 181h + 32h^2), \\ \widetilde{M} &= 8(3c - c^2 + 12h + 4ch - 4h^2)x^2 + 8(c+h)^2xy \\ &\quad - 8(c^2 + 8ch + 16h^2 + 3ch^2 - 6h^3)y^2, \\ \kappa &= -16\big[c^4 + 4c^3h + 2c^2h(4 + 13h + 4h^2) + 4ch^2(13 + 11h - 5h^2) \\ &\quad + h^3(80 - 51h + 8h^2)\big], \ \ \theta = 64h(c + 3h + ch - h^2)^2, \\ G_9 &= -(c+h)^2(c^2 - h + 2ch + 3h^2 - h^3), \\ \theta_2 &= h(1-h)(c + 3h + ch - h^2). \end{aligned} \qquad (7.75)$$

Since systems (7.73) possess at least three real elemental singularities, by Proposition 5.1 we must have $\mathbf{D} < 0$. Considering the above expressions we have the next lemma.

Lemma 7.9. *Assume that for a system (7.73) the condition $\mathbf{D} < 0$ holds. Then*

(i) *$\mathcal{T}_2\widetilde{M} \neq 0$ and the condition $W_3 \neq 0$ gives $W_3\theta < 0$ and $\mathcal{F}_1 \neq 0$,*

(ii) *the conditions $\mu_0 \leq 0$ and $\widetilde{K} > 0$ imply $h > 0$,*

(iii) *the conditions $\mu_0 = 0$ implies $\kappa\mathcal{F}_1 \neq 0$.*

Proof. (i) The first statement follows directly from (7.74) and (7.75).

(ii) Since $\mu_0 = \text{Discrim}\,[\widetilde{K}, x]/(16y^2)$ considering (7.74) we conclude that the binary form $\widetilde{K}$ has either two complex factors or one real double. Hence in both cases the condition $h(1+c)(h^2-2h-c) \neq 0$ is necessary, and moreover we have

$$\text{sign}\,(\widetilde{K}) = \text{sign}\,\big(h(1+c)\big) = \text{sign}\,\big(h(h^2-2h-c)\big).$$

So $h \neq 0$ and suppose the contrary, that $h < 0$. Then as $\widetilde{K} > 0$ we get $c < -1$, and this implies $h(h^2-2h-c) < 0$. The contradiction obtained completes the proof of statement (ii) of the lemma.

(iii) Assume $\mu_0 = 0$ and suppose the contrary, that $\kappa\mathcal{F}_1 = 0$. If $\mathcal{F}_1 = 0$ then considering (7.74) and the condition $\mathbf{D} \neq 0$, we obtain $c(1+h) - h(h-3) = 0$. Clearly $h+1 \neq 0$ and we get $c = h(h-3)/(1+h)$. Then calculations yield

$$\mathcal{F}_1 = 0, \quad \mu_0 = -4h^3(1-h)^6/(1+h)^4, \quad \mathbf{D} = -12288h^{10}(1-h)^{16}/(1+h)^{12},$$

and evidently the condition $\mathbf{D} \neq 0$ implies $\mu_0 \neq 0$.

Suppose now $\kappa = 0$. Then considering (7.74) the following condition is necessary:

$$\text{Res}\,_c(\mu_0, \kappa) = 2^{32}h^{10}(1-h)^{12} = 0.$$

However for $h = 0$ (respectively $h = 1$) we calculate $\mu_0 = c^4 = 0$ (respectively $\mu_0 = (1+c)^2(5+2c+c^2) = 0$), and the condition either $c = 0$ or $c = -1$ gives $\mathbf{D} = 0$. This contradiction completes the proof of the lemma. □

Since $\mathcal{T}_2 \neq 0$, according to Theorem 6.2 both weak singularities are of first order if $\mathcal{F}_1 \neq 0$, and we have either an integrable saddle or a center if $\mathcal{F}_1 = 0$.

3.1.1: $\mu_0 < 0$. This implies $\widetilde{K} \neq 0$, and according to Theorem 6.1 we have either three saddles and one anti-saddle (if $\widetilde{K} < 0$) or three anti-saddles and one saddle (if $\widetilde{K} > 0$).

If $\widetilde{K} < 0$, apart from the node n^d, we have three saddles, and this implies $W_3 \neq 0$. Hence by Lemma 7.9 the condition $\mathcal{F}_1 \neq 0$ holds, and we have two weak singularities of order one.

Additionally, as the total index of finite singularities equals -2, at infinity we must have three nodes, and considering Theorem 6.4 we arrive at the unique global configuration (λ_{25}).

Assume now $\widetilde{K} > 0$. In this case we have a saddle and three anti-saddles. By Lemma 7.9 the conditions $\mu_0 < 0$ and $\widetilde{K} > 0$ imply $h > 0$ and we examine two cases: $W_3 \neq 0$ and $W_3 = 0$.

1. $W_3 \neq 0$. Then due to $h > 0$ we get $W_3 < 0$ and since $W_4 = 0$, by Theorem 6.1 (see Table 6.2, line 16), apart from the weak saddle, we have a focus and two nodes. Clearly the saddle and the focus are weak, and by Lemma 7.9 we have $\mathcal{F}_1 \neq 0$. So we deduce that both weak singularities are of order one and the second node is generic. It remains to examine the infinite singularities.

   Since $\mu_0 < 0$ and $\widetilde{M} \neq 0$, considering Theorem 6.4 we get three configurations of infinite singular points: S, ©, © (if $\eta < 0$), S, S, N^{∞} (if $\eta > 0$), and $\binom{0}{2}SN, S$ (if $\eta = 0$). So considering Theorem 7.3, we arrive at the three global configurations (λ_{26}), (λ_{27}) and (λ_{28}), respectively.

2. $W_3 = 0$. Then due to $\mathbf{D} \neq 0$ and $h \neq 0$ we have $c(1+h) - h(h-3) = 0$, and since $h + 1 \neq 0$ we get $c = h(h-3)/(1+h)$. Herein we obtain

   $$W_2 = -1024h^9(1-h)^{18}/(1+h)^{12}, \quad \mu_0 = -4h^3(1-h)^6)/(1+h)^4, \quad \mathcal{F}_1 = 0,$$

   and as the condition $\mu_0 < 0$ yields $h > 0$, this implies $W_2 < 0$. So by Theorem 6.1 (see Table 6.2, line 17), apart from the weak saddle, we again have a focus and two nodes. But in this case due to $\mathcal{F}_1 = 0$ we have an integrable saddle and a center, and we again arrive at the configuration $(\lambda_{22}) \equiv (\beta_{25})$.

3.1.2: $\mu_0 > 0$. We consider three possibilities: $W_3 < 0$, $W_3 > 0$ and $W_3 = 0$.

3.1.2.1: $W_3 < 0$. Then by Lemma 7.9 we have $\mathcal{F}_1 \neq 0$ and $\theta > 0$. Since $W_4 = 0$, according to Theorem 6.1 (see Table 6.2, line 7) in this case systems (7.73) possess two saddles, a node (which is n^d) and a focus. Clearly one saddle is weak and the type of the second weak singularity according to Theorem 6.2 is determined by the invariant polynomial $\mathcal{T}_2 \neq 0$.

1. $\mathcal{T}_2 < 0$. Considering (7.74) and Theorem 6.2 we have two weak saddles (both of the first order due to $\mathcal{F}_1 \neq 0$).

   Since $\mu_0 > 0$, $\widetilde{M} \neq 0$ and $\theta > 0$, by Theorem 6.4 we get the following three configurations of singularities: (λ_{29}) if $\eta < 0$, (λ_{30}) if $\eta > 0$ and (λ_{31}) if $\eta = 0$.

2. $\mathcal{T}_2 > 0$. Then by Theorem 6.2 systems (7.73) possess a weak saddle and a weak focus (both of the first order).

   Thus since $\theta > 0$, $\mu_0 > 0$ and $\widetilde{M} \neq 0$, by Theorem 6.4 we arrive at the following three configurations: (λ_{32}) (if $\eta < 0$), (λ_{33}) (if $\eta > 0$) and (λ_{34}) (if $\eta = 0$).

3.1.2.2: $W_3 > 0$. Then due to $W_4 = 0$, according to Theorem 6.1 (see Table 6.2, line 5) in this case systems (7.73) possess two saddles and two nodes. This means that the two singular points with zero traces are weak saddles of order one due to $\mathcal{F}_1 \neq 0$ (see Lemma 7.9). So we have the configuration of finite singularities $s^{(1)}, s^{(1)}, n, n^d$.

Since by Lemma 7.9 the condition $\theta < 0$ holds, considering the conditions $\mu_0 > 0$ and $\widetilde{M} \neq 0$ and Theorem 6.4, we get the following four configurations: (λ_{35}) if $\eta < 0$, (λ_{36}) and (λ_{37}) if $\eta > 0$ and (λ_{38}) if $\eta = 0$.

3.1.2.3: $W_3 = 0$. Since $W_4 = 0$, by Theorem 6.1 (see Table 6.2, line 5) systems (7.73) possess two weak saddles and two nodes, which are both n^d.

If $\mathcal{F}_1 \neq 0$ then both weak saddles have order one. In this case by (7.74) due to $\mathbf{D} \neq 0$ we obtain $h = 0$ and then $\theta = 0$. As it was proved earlier for the family of systems with two such nodes and a weak singularity, in this case we could have only configurations belonging to the set $\{(\lambda_4), (\lambda_5), (\lambda_6), (\lambda_7)\}$.

Assume now $\mathcal{F}_1 = 0$. Then the weak saddles are integrable, and we arrive at the unique global configuration (λ_{23}).

3.1.3: $\mu_0 = 0$. Since the three finite singularities are fixed (i.e. $m_f = 3$), by Lemma 5.2 we conclude that $\mu_1 \neq 0$, and this implies $\widetilde{K} \neq 0$. Additionally by Lemma 7.9 the condition $\mathcal{F}_1 \neq 0$ must hold in this case.

If $\widetilde{K} < 0$ then according to Theorem 6.1 systems (7.73) possess one anti-saddle (which is a node n^d) and two weak saddles both of order one. Since the total index of finite singularities is -1, we deduce that the total index of infinite singularities must be $+2$, i.e. we have at least two nodes. This implies $\eta > 0$, and following Theorem 6.4 we get the configuration given by (λ_{39}).

Assume now $\widetilde{K} > 0$. Then, apart from a one-direction node, we have a saddle and an anti-saddle, which must be a focus in order to have two weak singularities (i.e. we have n^d, $s^{(1)}$ and $f^{(1)}$).

In the case $\eta < 0$ considering the conditions $\mu_0 = 0$, $\mu_1 \neq 0$, by Theorem 6.4 we obtain the configuration (λ_{40}).

Suppose now $\eta > 0$. We observe that the total index of finite singularities is $+1$, and then the total index of infinite singularities must be zero. So by Theorem 6.4 in this case we get the configuration (λ_{41}).

Assume finally $\eta = 0$. Since $\widetilde{M} \neq 0$ and by Lemma 7.9 (iii) we have $\kappa \neq 0$, and considering Theorem 6.4, we arrive at the configuration (λ_{42}).

3.2: $e = 0$. Considering the conditions (7.72) we get $f = c$ and $m = h = 0$. As $M_1(0,0)$ is an elemental singularity, we have $c \neq 0$, and doing a time rescaling we may assume $c = 1$. Then we get the 1-parameter family of systems:

$$\dot{x} = x + dy - x^2 - dy^2, \quad \dot{y} = y - y^2,$$

possessing the fourth finite singularity $M_4(1,1)$, which is a node with the Jacobian matrix $\begin{pmatrix} -1 & -d \\ 0 & -1 \end{pmatrix}$. Since the Jacobian matrix corresponding to the one-direction node $M_1(0,0)$ is $\begin{pmatrix} 1 & d \\ 0 & 1 \end{pmatrix}$, the condition $d \neq 0$ must hold. Then we obtain $\mathcal{F}_1 = -2d \neq 0$. Therefore the above systems possess two nodes n^d, apart from two weak saddles of order one. Hence as it was shown in the case of two nodes, the global configurations of singularities that possess the above systems belong to the set $\{(\lambda_4), (\lambda_5), (\lambda_6), (\lambda_7)\}$.

As all the cases have been considered, Theorem 7.6 is proved. □

7.6 A weak focus of order two or three

We join in one section the cases with weak foci of order two and three because there exists a nice normal form for the weak focus of order two, which includes the weak focus of order three as a limiting case.

Next we introduce a couple of lemmas that will be useful in this section and in later ones.

Lemma 7.10. *Assume that a quadratic system with $\mu_0 > 0$ possesses two foci or a focus plus a node n^d. Then $\theta > 0$.*

Proof. We prove first the impossibility of $\theta = 0$. Assume that we have a system with two finite singularities. We can move them to $M_1(0,0)$ and $M_2(1,0)$. The condition $\mu_0 > 0$ implies $C_2 \neq 0$, and therefore the number of infinite singular points is finite and at least one infinite singularity is real. If at least one of the finite singular points is a focus, then the real infinite singularity cannot be located at the end of the x-axis, otherwise the line $y = 0$ must be invariant on which a focus could not be located. So by means of a linear change (which keeps the finite singularities M_1 and M_2), we can locate the real infinite singularity at $N_1[0:1:0]$, and we get the following family of systems:

$$\dot{x} = cx + dy - cx^2 + 2hxy,$$
$$\dot{y} = ex + fy - ex^2 + 2mxy + ny^2.$$

We obtain that the discriminants corresponding to the finite singularities M_1 and M_2 are

$$\tau_1 = (c-f)^2 + 4de, \quad \tau_2 = (c - f - 2m)^2 + 4(de - cf + 2eh - 2cm).$$

Additionally for the above systems we have

$$\theta = 64h(eh^2 - chm - 2ehn + cmn + m^2n + en^2),$$

and the Jacobian at the point $N_1[0:1:0]$ is $\begin{pmatrix} n-2h & -d \\ 0 & n \end{pmatrix}$. According to Theorem 6.4 (see Diagrams 6.5 - 6.7) in the case $\mu_0 > 0$, the condition $\theta = 0$ implies the existence of an infinite node with coinciding eigenvalues. So we may consider that the point $N_1[0:1:0]$ is a such node, i.e. the condition $h=0$ is satisfied. Since we have at least one focus, we may assume $\tau_1<0$, which implies $d\neq 0$, and doing a time rescaling we fix $d=1$. We set a new parameter u as follows: $\tau_1=-u^2<0$. This gives $e=-((c-f)^2+u^2)/4$. Then we obtain

$$\tau_2 = 2(c+m)^2 + 2(f+m)^2 + u^2 > 0,$$

i.e. M_2 could be neither a focus nor n^d.

The impossibility of $\theta<0$ has not been obtained by direct algebraic computation. Indirectly, the Mathematica tool "Reduce" gave us the reply that this is impossible. □

Notice that Lemma 7.10 does not require the existence of a weak focus of order two, not even a weak focus. So this lemma will be very useful when studying more generic families.

Lemma 7.11. *Assume that a quadratic system* (7.1) *possesses a triple finite singularity and a weak singularity of finite order greater than one (i.e. not a center or an integrable saddle). Then this weak singularity could only be a saddle of order two.*

Proof. If a quadratic system (7.1) possesses one simple and one triple singularities, then by Proposition 6.2 via an affine transformation we can bring this system to the canonical form 8 from Table 6.1. For these systems we calculate

$$\mathcal{T}_4 = 4(cm-eh)^2(c-2m-eu)(c+eu)^3,$$
$$\mathcal{T}_3 = 8(cm-eh)^2(c-3m-2eu)(c+eu)^2(v-u^2).$$

In order to have a weak singularity of order greater than one, by Theorem 6.2 the conditions $\mathcal{T}_4=\mathcal{F}_1=0$ and $\mathcal{T}_3\neq 0$ are necessary. So $\mathcal{T}_4=0$ and $\mathcal{T}_3\neq 0$ give $c=2m+eu$, and then we calculate

$$\mathcal{F}_1 = 12e(m+eu)(eh-2m^2-emu)(u^2-v),$$
$$\mathcal{T}_3 = 32(m+eu)^3(eh-2m^2-emu)^2(u^2-v).$$

Clearly the conditions $\mathcal{F}_1=0$ and $\mathcal{T}_3\neq 0$ imply $e=0$, and we obtain

$$\mathcal{F}_2 = -1536m^10(u^2-v)^2, \quad \mathcal{T}_3 = 128m^7(u^2-v), \quad \mathcal{F} = 4m^5(u^2-v).$$

Therefore the condition $\mathcal{T}_3\neq 0$ implies $\mathcal{F}_2\neq 0$, and moreover since $\mathcal{T}_3\mathcal{F}>0$, according to Theorem 6.2 we have a saddle of order two.

This completes the proof of the lemma. □

7.6.1 A weak focus of order two

According to [19] a quadratic system with a second order weak focus can be brought via an affine transformation and time rescaling to the normal form

$$\begin{aligned} \dot{x} &= -y + (b-v)x^2/3 + 2axy + ny^2, \\ \dot{y} &= x + ax^2 + (b+2v)xy/3 - ay^2, \end{aligned} \tag{7.76}$$

for which the following condition holds:

$$ab(b+3n-v)(2b+15n-5v) \neq 0. \tag{7.77}$$

We remark that the condition $ab(b+3n-v)=0$ implies the existence of a center instead of a weak focus, whereas the conditions $(2b+15n-5v)=0 \neq ab(b+3n-v)$ imply that the weak focus is of order three.

Lemma 7.12. *For a system* (7.76) *under the condition* (7.77), *the condition* $\widetilde{M} \neq 0$ *is satisfied.*

Proof. For systems (7.76) we calculate

$$\widetilde{M} = -8\big[(9a^2+v^2)x^2 + 3a(3n-v)xy + 3(3a^2+nv)y^2\big],$$

and therefore the condition $\widetilde{M}=0$ implies $a=0$, which contradicts the condition (7.77). This proves the lemma. □

Lemma 7.13. *A quadratic system having a weak point (saddle or focus) of order two or three cannot have a nilpotent finite singular point.*

Proof. This is an immediate consequence of Theorem 6.2, which shows that a weak singularity of order two (or three) yields $\mathcal{T}_3 \neq 0$. This implies that we cannot have another singular point with trace zero. □

Lemma 7.14. *Assume that a quadratic system possesses a weak focus of order two or three. Then the conditions* $\mathbf{D}=\theta=\theta_1=0$ *are incompatible for this system.*

Proof. We shall use the canonical systems (7.76) for which the singular point $M_1(0,0)$ is a weak focus of order two or three. As it was mentioned earlier, in this case the condition $ab(b+3n-v)\neq 0$ is necessary.

Assume that for systems (7.76) the conditions $\mathbf{D}=\theta=\theta_1=0$ hold. For these systems we calculate

$$\theta = 8\big[216a^2n^2 - 18a^2(48a^2-b^2+4v^2) + n(-b^3-432a^2v+6b^2v-32v^3)\big]/27,$$

and since $a\neq 0$ the solution of the equation $\theta=0$ must depend on the parameter n. Then for these systems we calculate

$$\operatorname{Res}_n(\theta,\theta_1) = 2^{16}3^{-4}a^4b^4(72a^2-b^2+2bv+8v^2)^2 \equiv 2^{16}3^{-4}a^4b^4\big[\phi_1(a,b,v)\big]^2,$$

and due to $ab\neq 0$ we deduce that the equations $\theta=\theta_1=0$ have a common solution (depending on n) if only if $\phi_1(a,b,v)=0$. Additionally we observe that the last equation has solutions depending on b. So for systems (7.76) we calculate

$$\operatorname{Res}_b\Big(\phi_1, \operatorname{Res}_n(\mathbf{D},\theta)\Big) = 2^{56}3^{16}a^{16}(9a^2+v^2)^6 = 0,$$

and since $a\neq 0$ we conclude that the conditions $\mathbf{D}=\theta=\theta_1=0$ are incompatible. □

Lemma 7.15. *Assume that a quadratic system possesses a weak focus of order two or three. Then the conditions from each one of the following sets are incompatible:*

(i) $\mathbf{D} < 0$, $W_4 > 0$ *and* $\eta \geq 0$;

(ii) $\mathbf{D} < 0$, $W_4 < 0$, $\eta < 0$ *and* $\theta \leq 0$;

(iii) $\mathbf{D} < 0$, $\theta \leq 0$ *and* $\theta_1 < 0$;

(iv) $\mathbf{D} > 0$, $\mu_0 > 0$, *and* $W_4 = 0$;

(v) $\mathbf{D} > 0$, $W_4 = 0$, $W_3 \neq 0$ *and* $\eta = 0$;

(vi) $\mathbf{D} \geq 0$, $\mu_0 < 0$, $W_4 < 0$ *and* $\eta = 0$;

(vii) $\mathbf{D} = 0$, $\mu_0 < 0$, $W_4 \geq 0$ *and* $\eta \geq 0$;

(viii) $\mathbf{D} = 0$, $\eta < 0$ *and* $\theta \leq 0$;

(ix) $\mathbf{D} = 0$, $\eta > 0$, $\theta \leq 0$ *and* $\theta_1 < 0$;

(x) $\mathbf{D} = \eta = 0$, $\mu_0 > 0$ *and* $\theta \leq 0$;

(xi) $\mu_0 = 0$, $W_4 \geq 0$ *and* $\eta \geq 0$;

(xii) $W_4 = W_3 = W_2 = 0$.

Proof. We have not been able to prove all these statements directly. However the Mathematica tool "Reduce" gives incompatibility of each one of the above sets of conditions for systems (7.76). Since the condition $(2b + 15n - 5v) \neq 0$ from (7.77) is not used in the proof of the statements, we conclude that the lemma holds for both weak foci of order two and three. We observe that some of these sets have also been proved to be impossible in the bifurcation diagram done in [19]. □

Theorem 7.7. *There are exactly* 56 *geometrically distinct configurations of singularities, finite and infinite, for quadratic vector fields possessing one weak focus of order exactly two. All of them are contained in the following list, enumerated and grouped according to the value of the invariant polynomial* μ_0:

$\mu_0 < 0$

(ν_1) $s, s, s, f^{(2)};\ N^f, N^f, N^f$;
(ν_2) $s, n, n, f^{(2)};\ S, ©, ©$;
(ν_3) $s, n, n, f^{(2)};\ S, S, N^\infty$;
(ν_4) $s, n, n, f^{(2)};\ \overline{\binom{0}{2}} SN, S$;
(ν_5) $s, n, f, f^{(2)};\ S, ©, ©$;
(ν_6) $s, n, n^d, f^{(2)};\ S, ©, ©$;
(ν_7) $n, f^{(2)}, ©, ©;\ S, ©, ©$;
(ν_8) $n, f^{(2)}, ©, ©;\ S, S, N^\infty$;
(ν_9) $f, f^{(2)}, ©, ©;\ S, ©, ©$;
(ν_{10}) $f, f^{(2)}, ©, ©;\ S, S, N^\infty$;
(ν_{11}) $f, f^{(2)}, ©, ©;\ \overline{\binom{0}{2}} SN, S$;
(ν_{12}) $n^d, f^{(2)}, ©, ©;\ S, ©, ©$;
(ν_{13}) $n^d, f^{(2)}, ©, ©;\ S, S, N^\infty$;
(ν_{14}) $n, f^{(2)}, ©^\tau, ©^\tau;\ S, ©, ©$;
(ν_{15}) $n, f^{(2)}, ©^\tau, ©^\tau;\ S, S, N^\infty$;
(ν_{16}) $n, f^{(2)}, ©^\tau, ©^\tau;\ \overline{\binom{0}{2}} SN, S$;
(ν_{17}) $f, f^{(2)}, ©^\tau, ©^\tau;\ S, ©, ©$;
(ν_{18}) $f, f^{(2)}, ©^\tau, ©^\tau;\ S, S, N^\infty$;
(ν_{19}) $f, f^{(2)}, ©, ©;\ \overline{\binom{0}{2}} SN, S$;
(ν_{20}) $\overline{sn}_{(2)}, n, f^{(2)};\ S, ©, ©$;
(ν_{21}) $\overline{sn}_{(2)}, n, f^{(2)};\ S, S, N^\infty$;
(ν_{22}) $\overline{sn}_{(2)}, f, f^{(2)};\ S, ©, ©$;
(ν_{23}) $\overline{sn}_{(2)}, n^d, f^{(2)};\ S, ©, ©$.

$\mu_0 > 0$

(ν_{24}) $s, s, n, f^{(2)};\ N^f, \copyright, \copyright;$ (ν_{25}) $s, s, n, f^{(2)};\ S, N^f, N^f;$
(ν_{26}) $s, s, n, f^{(2)};\ S, N^\infty, N^f;$ (ν_{27}) $s, s, n, f^{(2)};\ S, N^f, N^d;$
(ν_{28}) $s, s, n, f^{(2)};\ \overline{\binom{0}{2}}SN, N^f;$ (ν_{29}) $s, s, f, f^{(2)};\ N^f, \copyright, \copyright;$
(ν_{30}) $s, s, n^d, f^{(2)};\ N^f, \copyright, \copyright;$ (ν_{31}) $s, f^{(2)}, \copyright, \copyright;\ N^\infty, \copyright, \copyright;$
(ν_{32}) $s, f^{(2)}, \copyright, \copyright;\ N^f, \copyright, \copyright;$ (ν_{33}) $s, f^{(2)}, \copyright, \copyright;\ N^d, \copyright, \copyright;$
(ν_{34}) $s, f^{(2)}, \copyright, \copyright;\ S, N^\infty, N^\infty;$ (ν_{35}) $s, f^{(2)}, \copyright, \copyright;\ S, N^f, N^f;$
(ν_{36}) $s, f^{(2)}, \copyright, \copyright;\ S, N^\infty, N^f;$ (ν_{37}) $s, f^{(2)}, \copyright, \copyright;\ S, N^\infty, N^d;$
(ν_{38}) $s, f^{(2)}, \copyright, \copyright;\ S, N^f, N^d;$ (ν_{39}) $s, f^{(2)}, \copyright, \copyright;\ S, N^d, N^d;$
(ν_{40}) $s, f^{(2)}, \copyright, \copyright;\ \overline{\binom{0}{2}}SN, N^\infty;$ (ν_{41}) $s, f^{(2)}, \copyright, \copyright;\ \overline{\binom{0}{2}}SN, N^f;$
(ν_{42}) $s, f^{(2)}, \copyright, \copyright;\ \overline{\binom{0}{2}}SN, N^d;$ (ν_{43}) $\overline{sn}_{(2)}, s, f^{(2)};\ N^f, \copyright, \copyright;$
(ν_{44}) $\overline{sn}_{(2)}, s, f^{(2)};\ S, N^f, N^f;$ (ν_{45}) $\overline{sn}_{(2)}, s, f^{(2)};\ S, N^\infty, N^f;$
(ν_{46}) $\overline{sn}_{(2)}, s, f^{(2)};\ S, N^f, N^d;$ (ν_{47}) $\overline{sn}_{(2)}, s, f^{(2)};\ \overline{\binom{0}{2}}SN, N^f.$

$\mu_0 = 0$

(ν_{48}) $s, n, f^{(2)};\ \overline{\binom{1}{1}}SN, \copyright, \copyright;$ (ν_{49}) $s, n, f^{(2)};\ \overline{\binom{1}{1}}SN, S, N^\infty;$
(ν_{50}) $s, n, f^{(2)};\ \overline{\binom{1}{1}}SN, \overline{\binom{0}{2}}SN;$ (ν_{51}) $s, f, f^{(2)};\ \overline{\binom{1}{1}}SN, \copyright, \copyright;$
(ν_{52}) $s, n^d, f^{(2)};\ \overline{\binom{1}{1}}SN, \copyright, \copyright;$ (ν_{53}) $s, s, f^{(2)};\ \overline{\binom{1}{1}}SN, N^f, N^f;$
(ν_{54}) $f^{(2)}, \copyright, \copyright;\ \overline{\binom{1}{1}}SN, S, N^\infty;$ (ν_{55}) $\overline{sn}_{(2)}, f^{(2)};\ \overline{\binom{1}{1}}SN, S, N^\infty;$
(ν_{56}) $s, f^{(2)};\ \overline{\binom{2}{1}}S, N^f, N^f.$

The realizability of each one of these configurations will be proved by means of an example, which will be provided in Part II.

Proof. The bifurcation diagram of the phase portraits of all quadratic systems with a second order weak focus was studied in [19], where in addition the distinguishing between a focus and a node was included. In [19] we did not however look for geometrical configurations of singularities. This will be covered in this book.

As explained in [19], systems (7.76), which depend on four parameters, can be split into two cases: $n \neq 0$ (and then take $n = 1$ by a time change) or $n = 0$. In this last case one may take $a = 1$ also by a time change since $a = 0$ implies that the weak focus becomes a center.

Remark 7.5. The following results always apply when we have a focus of order two:

- By Theorem 6.2 we could not have a second weak singularity.
- By Lemma 7.1 we cannot have an n^*.
- By Lemma 7.7(*ii*) we cannot have an N^*.
- By Lemma 7.12 we have always $\widetilde{M} \neq 0$.
- By Lemma 7.13 any double point must be a saddle-node.

In what follows we will avoid repeating the application of these results in every needed step.

1: $\mu_0 < 0$. Following Theorem 6.1 we consider three subcases: $\mathbf{D} < 0$, $\mathbf{D} > 0$ and $\mathbf{D} = 0$.

1.1: $\mathbf{D} < 0$. As systems (7.76) have the real singular point $(0, 0)$, by Proposition 5.1 in this case we obtain four real distinct finite singularities.

1.1.1: $\widetilde{K} < 0$. By Theorem 6.1 (see Table 6.2), apart from the weak focus, systems (7.76) possess three finite saddles, and as a consequence at infinity there must be three nodes. As $\mu_0 < 0$, by Theorem 6.4 (see Diagram 6.6), each one is of the type N^f. We arrive at the configuration (ν_1).

1.1.2: $\widetilde{K} > 0$. According to Theorem 6.1 (see Table 6.2) systems (7.76) possess one saddle and three anti-saddles.

1. $W_4 < 0$. Then, apart from the weak focus, on the phase plane we have one saddle and two nodes. Since the nodes are generic (due to $W_4 \neq 0$), considering Diagrams 6.5, 6.6 and 6.7 (see Theorem 6.4), we arrive at the following three configurations of singularities: (ν_2) if $\eta < 0$, (ν_3) if $\eta > 0$ and (ν_4) if $\eta = 0$.

2. $W_4 > 0$. According to Table 6.2 (see Theorem 6.1), apart from the weak focus, on the phase plane we have one saddle, one node and one focus.

   Additionally by Lemma 7.15(i) in this case for a system (7.76) the condition $\eta < 0$ holds. So at infinity we have a saddle and two complex singularities, and we get the configuration (ν_5).

3. $W_4 = 0$. By Theorem 7.6 we see that we cannot have a weak focus of order two and two nodes with coinciding eigenvalues, so $W_3 \neq 0$. Moreover $\eta < 0$ must hold for systems (7.76). The direct algebraic proof is not easy to do, but this fact is already proved by the bifurcation diagram in [19], and moreover, it is confirmed by the Mathematica tool "Reduce".

   So considering Theorems 6.1 and 6.4 (see Diagram 6.5), we get the unique global configuration (ν_6).

1.2: $\mathbf{D} > 0$. Then according to Proposition 5.1 systems (7.76) have two real and two complex finite singularities, and moreover due to $\mu_0 < 0$ the real singular points have the same index. So since we have a focus, by Theorem 6.1 (see Table 6.2) the condition $\widetilde{K} > 0$ must hold, and we consider three possibilities: $W_4 < 0$, $W_4 > 0$ and $W_4 = 0$.

1.2.1: $W_4 < 0$. According to Theorem 6.1, apart from the weak focus of order two, we have a generic node. Considering Lemma 7.15(vi), we arrive at the configuration (ν_7) if $\eta < 0$ and (ν_8) if $\eta > 0$.

1.2.2: $W_4 > 0$. In this case by Theorem 6.1, apart from the weak, focus of order two, we have a focus, which must be strong due to Theorem 6.2. So considering

Diagrams 6.5, 6.6 and 6.7 (see Theorem 6.4), we arrive at the following three configurations of singularities: (ν_9) if $\eta < 0$, (ν_{10}) if $\eta > 0$ and (ν_{11}) if $\eta = 0$.

1.2.3: $W_4 = 0$. Since we have a focus, according to Theorem 6.1 the condition $W_3 \le 0$ must be fulfilled.

1. $W_3 \neq 0$. Then $W_3 < 0$ and hence, apart from a weak focus of order two, there exists a node with coinciding eigenvalues. So taking into account the types of infinite singularities (see Theorem 6.4) as well as Lemma 7.15(v), we get the configuration (ν_{12}) if $\eta < 0$ and (ν_{13}) if $\eta > 0$.
2. $W_3 = 0$. Here Lemma 7.15(xii) says that $W_2 \neq 0$. In this case we must have exactly two finite singularities with discriminant zero. But we only have two real finite singularities, and one of them is a focus whose corresponding discriminant is negative. So, these two points must be the complex ones. Then, using Theorem 6.4, if $W_2 < 0$ we get the configuration (ν_{14}) if $\eta < 0$, (ν_{15}) if $\eta > 0$ and (ν_{16}) if $\eta = 0$. And if $W_2 > 0$ we get the configuration (ν_{17}) if $\eta < 0$, (ν_{18}) if $\eta > 0$ and (ν_{19}) if $\eta = 0$. These six geometrical configurations correspond to one singularity of the surface $W_4 = 0$ in the parameter space $\mathbb{PR}^3$. These possibilities were not taken into account in [19] since we were not interested in complex singularities. Anyway it does not modify the main statement in [19], which dealt with phase portraits.

1.3: $\mathbf{D} = 0$. Then according to Proposition 5.1 systems (7.76) have either two simple real and one double, or one simple and one triple finite singularities. However the second case could not be realizable by Lemma 7.11. We observe that due to $\mu_0 < 0$ the simple real singular points have the same index. Moreover, the existence of the weak focus of order two forces the double point to be a semi-elemental saddle-node (see Lemma 7.13).

So we consider three possibilities: $W_4 < 0$, $W_4 > 0$ and $W_4 = 0$.

1. $W_4 < 0$. According to Theorem 6.1, apart from the weak focus of order two (and the saddle-node), we have a generic node. Considering Lemma 7.15(vi) and Theorem 6.4, we arrive at the configuration (ν_{20}) if $\eta < 0$ and (ν_{21}) if $\eta > 0$.
2. $W_4 > 0$. In this case by Theorem 6.1, apart from the weak focus of order two (and the saddle-node), we have a focus, which must be strong due to Theorem 6.2. Considering Lemma 7.15(vii), the only possible configuration is (ν_{22}), which is possible for $\eta < 0$.
3. $W_4 = 0$. Since we have a focus, according to Theorem 6.1 the condition $W_3 \neq 0$ (then $W_3 < 0$) must be fulfilled, and we have a node with discriminant zero apart from the weak focus and the saddle-node. Considering Lemma 7.15(vii), the only possible configuration is (ν_{23}), which has $\eta < 0$.

2: $\mu_0 > 0$. Following Theorem 6.1 we consider three subcases: $\mathbf{D} < 0$, $\mathbf{D} > 0$ and $\mathbf{D} = 0$.

2.1: $\mathbf{D} < 0$. Since systems (7.76) have the real singular point $(0,0)$, by Proposition 5.1 in this case we obtain four real distinct finite singularities. According to Theorem 6.1 (see Table 6.2) systems (7.76) possess two saddles and two anti-saddles, one of them being the weak focus of order two. We must split the study according to the type of the second anti-saddle. Thus the value of the affine invariant W_4 must be studied.

2.1.1: $W_4 < 0$. Then, apart from the weak focus, on the phase plane we have two saddles and one node. The node is generic (due to $W_4 \neq 0$). We must split the study according to the number of real infinite singularities, which is governed by the invariant η.

1. $\eta < 0$. Considering Diagram 6.5 (see Theorem 6.4) we could have several configurations, but Lemma 7.15(ii) says that $\theta > 0$. Then only the configuration (ν_{24}) is possible.

2. $\eta > 0$. Considering Diagram 6.6 (see Theorem 6.4) we must study now the invariant θ. If $\theta < 0$ then Lemma 7.15(iii) says that $\theta_1 > 0$, and only the configuration (ν_{25}) is possible. If $\theta > 0$ then we get directly the configuration (ν_{26}). And if $\theta = 0$ then again Lemma 7.15(iii) says that $\theta_1 > 0$, and only the configuration (ν_{27}) is possible.

3. $\eta = 0$. Considering Diagram 6.7 (see Theorem 6.4) we can have several possibilities according to θ being negative, positive or zero. We claim that in these conditions θ can only be positive.

   Start from (7.76) and consider two possibilities: (1) $a = 1, n = 0$, or (2) $n = 1$. In the first case $\eta = 81 + 54v + 9v^2 + 4v^3$, and this gives a single real solution for v. Substituting this value in $\{\theta, \mu_0, \mathbf{D}\}$, we find that they depend on a single variable b, and it is immediate to check that the combination $\{\theta \leq 0, \mu_0 > 0, \mathbf{D} < 0\}$ is not possible.

   In the second case $\eta = 108a^4 - 27a^2 + 54a^2v + 9a^2v^2 + 4v^3$, and θ, μ_0 and $\mathbf{D}$ all depend on a^2. So $\eta = 0$ can be solved in terms of a^2, giving the discriminant $3 - 10v + 3v^2$. Then $\{\theta, \mu_0, \mathbf{D}\}|_{\eta=0}$ depend on the parameters b and v, and plotting the graphics of these functions, we see that the combination $\{\theta \leq 0, \mu_0 > 0, \mathbf{D} < 0, 3 - 10v + 3v^2 > 0\}$ is not possible.

   So $\theta > 0$ and only the configuration (ν_{28}) is possible.

2.1.2: $W_4 > 0$. Then by Theorem 6.1 (see Table 6.2, line 9), apart from the weak focus, on the phase plane we have two saddles and a strong focus.

We must split the study according to the number of real infinite singularities, that is, the invariant η must be studied. From [19] this configuration of finite singularities can only happen if $\eta < 0$. Then by Lemma 7.10 we have that $\theta > 0$, and only the configuration (ν_{29}) is possible.

2.1.3: $W_4 = 0$. Then, apart from the weak focus, on the phase plane we have two saddles and an n^d.

By exactly the same arguments as in the previous case, this configuration of finite singularities can only happen if $\eta < 0$, and by Lemma 7.10 only the configuration (ν_{30}) is possible.

2.2: $\mathbf{D} > 0$. Then according to Proposition 5.1 systems (7.76) have two real and two complex finite singularities, and moreover due to $\mu_0 > 0$ the real singular points have different index. So since we have a focus, the other point is a saddle. Since we have already a weak point of order two, the saddle must be strong, and we consider two possibilities: $W_4 < 0$ and $W_4 = 0$. But this last possibility is eliminated by Lemma 7.15(iv). So we must split the study according to the number of real infinite singularities, hence the invariant η must be studied.

1. $\eta < 0$. All cases from Diagram 6.5 of Theorem 6.4 (except the infinite star node, see Lemma 7.7 (ii)) are possible, and we arrive at the configuration (ν_{31}) if $\theta < 0$, (ν_{32}) if $\theta > 0$ and (ν_{33}) if $\theta = 0$.
2. $\eta > 0$. Again all cases from Diagram 6.6 of Theorem 6.4 (except the ones including an infinite star node) are possible, and we arrive at the configuration (ν_{34}) if $\theta < 0$ and $\theta_1 < 0$, (ν_{35}) if $\theta < 0$ and $\theta_1 > 0$, (ν_{36}) if $\theta > 0$, (ν_{37}) if $\theta = 0$ and $\theta_1 < 0$, (ν_{38}) if $\theta = 0$ and $\theta_1 > 0$ and (ν_{39}) if $\theta = 0$ and $\theta_1 = 0$.
3. $\eta = 0$. Again all cases from Diagram 6.7 of Theorem 6.4 (except the ones including an infinite star node) are possible, and we arrive at the configuration (ν_{40}) if $\theta < 0$, (ν_{41}) if $\theta > 0$ and (ν_{42}) if $\theta = 0$.

2.3: $\mathbf{D} = 0$. Then according to Proposition 5.1 systems (7.76) have two simple real and one double finite singularities, or one simple and one triple finite singularities. However the second case could not be realizable by Lemma 7.11. Then by Theorem 6.1 due to $\mu_0 > 0$ the simple real singular points have different index. Moreover the existence of the weak focus of order two forces the double point to be a semi-elemental saddle-node (see Lemma 7.13). This already implies $W_4 < 0$, and the saddle must be strong since we already have a weak point of order two. We must split the study according to the invariant η.

1. $\eta < 0$. From Diagram 6.5 of Theorem 6.4 and according to Lemma 7.15(viii) $\theta > 0$, hence we arrive at the configuration (ν_{43}).
2. $\eta > 0$. From Diagram 6.6 of Theorem 6.4 and according to Lemma 7.14 and Lemma 7.15 (statements (ix)) if $\theta \leq 0$ then $\theta_1 > 0$, and we arrive at the configuration (ν_{44}) if $\theta < 0$, (ν_{45}) if $\theta > 0$ and (ν_{46}) if $\theta = 0$.
3. $\eta = 0$. From Diagram 6.7 of Theorem 6.4 and according to Lemma 7.15(x) $\theta > 0$, hence we arrive at the configuration (ν_{47}).

3: $\mu_0 = 0$. We prepare a simple normal form to compute some invariant polynomials for this case in an easier way. We take a generic quadratic system with a weak focus

$$\dot{x} = -y + gx^2 + 2hxy + ky^2, \quad \dot{y} = x + lx^2 + 2mxy + ny^2.$$

By means of a rotation we can move any infinite singular point (particularly, the point that has coalesced with a finite singular point) to the origin of local chart U_2, that is, $k = 0$. The Jacobian matrix at this point is

$$\begin{pmatrix} 2h - n & -1 \\ 0 & n \end{pmatrix}.$$

Since we want the second eigenvalue to be zero (in order to have the coalescence of a finite singular point with this one), the condition $n = 0$ must hold. Then we get the normal form

$$\dot{x} = -y + gx^2 + 2hxy, \quad \dot{y} = x + lx^2 + 2mxy,$$

for which calculations yield

$$\begin{aligned} \mu_1 &= 4m(hl - gm)x, \quad \mathbf{D} = 192m^2(8gm - 4h^2 - 4hl - l^2), \\ \kappa &= 128h^2(hl - gm), \quad \widetilde{K} = -4(hl - gm)x^2, \\ \mathcal{T}_3 &= -8h(gh^2 + ghl - 2g^2m + 2h^2m + hlm - 4gm^2 - 2m^3), \\ \mathcal{F}_1 &= 2(lg - gh + lm). \end{aligned} \tag{7.78}$$

Since $\mathcal{T}_3 \neq 0$, we have $h \neq 0$, and we can assume $h = 1$ doing a time rescaling. Then the condition $\mathcal{F}_1 = 0$ implies either $l = g = 0$ (forcing $\mu_1 = 0$) or $l \neq 0$ and $m = g(l-1)/l$.

Following Theorem 6.1 we consider two subcases: $\mu_1 \neq 0$ and $\mu_1 = 0$.

3.1: $\mu_1 \neq 0$. From Proposition 6.1 (see Diagram 6.1) we have finite multiplicity of singularities equal to three (exactly one singularity has moved to infinity). From (7.78) we have that the condition $h\mu_1 \neq 0$ implies $\kappa \neq 0$ and then $\operatorname{sign}(\kappa) = -\operatorname{sign}(\widetilde{K})$. We consider three possibilities: $\mathbf{D} < 0$, $\mathbf{D} > 0$ and $\mathbf{D} = 0$.

3.1.1: $\mathbf{D} < 0$. Then according to Proposition 6.1 systems (7.76) have three simple real finite singularities. Due to $\mu_0 = 0$ the invariant polynomial $\widetilde{K}$ is a well defined binary form, and according to Theorem 6.1 we must continue the study according to its sign. Additionally by (7.78) we have $\operatorname{sign}(\kappa) = -\operatorname{sign}(\widetilde{K})$, and we examine two cases: $\kappa < 0$ and $\kappa > 0$.

1. $\kappa < 0$. Then $\widetilde{K} > 0$ and by Theorem 6.1 systems (7.76) have a saddle and an anti-saddle apart from the weak focus. The behavior of the anti-saddle is ruled by invariant W_4. If $W_4 < 0$, according to Theorem 6.1 (see Table 6.2, line 118) we have a saddle and a node, which is generic. So by Theorem 6.4 we arrive at the configurations (ν_{48}), (ν_{49}) and (ν_{50}) if respectively $\eta < 0$, $\eta > 0$ and $\eta = 0$.

   If $W_4 \geq 0$ then $\eta < 0$ (according to [19] and also Lemma 7.15(xi)), and the only configurations are (ν_{51}) if $W_4 > 0$ and (ν_{52}) if $W_4 = 0$.

2. $\kappa > 0$. Then according to Theorem 6.1 systems (7.76) have two saddles apart from the weak focus. By [19] we have that $\eta > 0$, and Diagram 6.5 completes the study, giving the configuration (ν_{53}).

3.1.2: $\mathbf{D} > 0$. In this case by Theorem 6.1 we must have two complex finite singularities and a real one, which will be a saddle or an anti-saddle according to the sign of $\widetilde{K}$ $(= -\text{sign}\,(\kappa))$. As we are in the family (7.76), we have the weak focus and $\widetilde{K}$ becomes irrelevant. By [19] we have that $\eta > 0$, and considering Diagram 6.6 we get the configuration (ν_{54}).

3.1.3: $\mathbf{D} = 0$. According to Theorem 6.1 systems (7.76) have either one double and one simple singularities, or one triple. Since we have a simple point (the weak focus), the second possibility is eliminated. By [19] we have $\eta > 0$, and Diagram 6.6 completes the study, giving the configuration (ν_{55}).

3.2: $\mu_1 = 0$. Following Theorem 6.1 we could have several cases, but considering the unique phase portrait, which could exist by [19], taking into account Theorem 6.4 we have only the configuration (ν_{56}).

This completes the proof of Theorem 7.7. □

7.6.2 A weak focus of order three

The normal form of this family is obtained from (7.76) by taking $(2b+15n-5v) = 0$ and $ab(b+3n-v) \neq 0$. This family was first studied [16, 222] using Li's normal form, but was also studied [19] under Bautin's normal form as it stays in the closure of systems with a weak focus of order two. So, the possibilities for configurations of singularities with a weak focus of order three will come from those studied in the previous subsection.

So we only need to determine which ones of the 56 possibilities from Theorem 7.7 are not realizable if we replace the weak focus of order two by a weak focus of order three.

Theorem 7.8. *We consider here all the configurations of singularities, finite and infinite, of quadratic vector fields possessing one weak focus of order exactly three. We prove that there are exactly 21 geometrically distinct configurations of singularities, finite and infinite, and all of them are contained in the following list, enumerated and grouped according to the value of the invariant polynomial* μ_0*:*

$$\mu_0 < 0 \quad \begin{array}{ll} (\chi_1)\ s, s, s, f^{(3)};\ N^f, N^f, N^f; & (\chi_2)\ s, n, n, f^{(3)};\ S, S, N^\infty; \\ (\chi_3)\ n, f^{(3)}, \copyright, \copyright;\ S, S, N^\infty; & (\chi_4)\ f, f^{(3)}, \copyright, \copyright;\ \underline{S, \copyright, \copyright}; \\ (\chi_5)\ f, f^{(3)}, \copyright, \copyright;\ S, S, N^\infty; & (\chi_6)\ f, f^{(3)}, \copyright, \copyright;\ \overline{\tbinom{0}{2}}SN, S; \\ (\chi_7)\ n^d, f^{(3)}, \copyright, \copyright;\ S, S, N^\infty; & (\chi_8)\ \overline{sn}_{(2)}, n, f^{(3)};\ S, S, N^\infty. \end{array}$$

$\mu_0 > 0$
(χ_9) $s, s, n, f^{(3)}$; S, N^f, N^f; (χ_{10}) $s, s, n, f^{(3)}$; S, N^∞, N^f;
(χ_{11}) $s, s, n, f^{(3)}$; S, N^f, N^d; (χ_{12}) $s, f^{(3)}, ©, ©$; S, N^f, N^f;
(χ_{13}) $s, f^{(3)}, ©, ©$; S, N^f, N^∞; (χ_{14}) $s, f^{(3)}, ©, ©$; S, N^f, N^d;
(χ_{15}) $\overline{sn}_{(2)}, s, f^{(3)}$; S, N^f, N^f; (χ_{16}) $\overline{sn}_{(2)}, s, f^{(3)}$; S, N^∞, N^f;
(χ_{17}) $\overline{sn}_{(2)}, s, f^{(3)}$; S, N^f, N^d.

$\mu_0 = 0$
(χ_{18}) $s, n, f^{(3)}$; $\overline{\binom{1}{1}}SN, S, N^\infty$; (χ_{19}) $s, s, f^{(3)}$; $\overline{\binom{1}{1}}SN, N^f, N^f$;
(χ_{20}) $f^{(3)}, ©, ©$; $\overline{\binom{1}{1}}SN, S, N^\infty$; (χ_{21}) $\overline{sn}_{(2)}, f^{(3)}$; $\overline{\binom{1}{1}}SN, S, N^\infty$.

The realizability of each one of these configurations will be proved by means of an example, which will be provided in Part II.

Proof. In order to prove this theorem, we work mainly with the normal form (7.76) studied in [19] setting $(2b + 15n - 5v) = 0 \neq ab(b + 3n - v)$. The parameter space is included in $\mathbb{PR}^2$, and we may split it in the plane $n = 1$ plus the line $n = 0$ and $a = 1$.

Thus the normal form is

$$\begin{aligned} \dot{x} &= -y + (v - 5n)x^2/2 + 2axy + ny^2, \\ \dot{y} &= x + ax^2 + (3v - 5n)xy/2 - ay^2, \end{aligned} \tag{7.79}$$

for which the following condition holds:

$$a(v - 3n)(2a^2 - n^2 + nv) \neq 0.$$

In the plane $n = 1$ the invariants we need are:

$$\begin{aligned} W_4 =& (33750 - 40365a^2 - 636336a^4 + 2045952a^6 - 1769472a^8 - 74250v \\ &- 238302a^2v + 1995840a^4v - 2543616a^6v + 28350v^2 + 535005a^2v^2 \\ &- 1184544a^4v^2 - 165888a^6v^2 + 24030v^3 - 196452a^2v^3 - 174528a^4v^3 \\ &- 5886v^4 - 54243a^2v^4 - 7344a^4v^4 - 5022v^5 - 5598a^2v^5 - 918v^6 \\ &- 109a^2v^6 - 54v^7)/16, \\ W_3 =& 25(-3 + v)^2(-2250 + 10773a^2 - 21600a^4 + 18432a^6 + 4950v - 14292a^2v \\ &+ 14400a^4v - 2700v^2 + 774a^2v^2 + 3744a^4v^2 - 396v^3 + 2652a^2v^3 + 342v^4 \\ &+ 109a^2v^4 + 54v^5)/64, \\ \eta =& - 27a^2 + 108a^4 + 54a^2v + 9a^2v^2 + 4v^3, \\ \mathbf{D} =& - 324(-5 - 52a^2 + 16a^4 + 16v + 40a^2v - 18v^2 - 4a^2v^2 + 8v^3 - v^4), \\ \mu_0 =& (-125 + 158a^2 - 32a^4 + 175v - 116a^2v - 75v^2 + 14a^2v^2 + 9v^3)/8, \\ \mathcal{F}_3\mathcal{F}_4 =& - 78125a^3(-3 + v)^9(-1 + 2a^2 + v)/128, \\ \theta =& 125 + 364a^2 - 256a^4 - 25v - 328a^2v - 25v^2 + 12a^2v^2 - 3v^3, \\ \theta_1 =& 4(-369a^2 + 576a^4 - 75v + 438a^2v + 50v^2 + 23a^2v^2 + 13v^3). \end{aligned}$$

And on the line $n = 0$, $a = 1$ the invariants are

$$\begin{aligned} W_4 =& -(1769472 + 165888v^2 + 7344v^4 + 109v^6)/16, \\ W_3 =& \, 25v^2(18432 + 3744v^2 + 109v^4)/64, \ \eta = 9(12 + v^2), \\ \mathbf{D} =& \, 324(-16 + 4v^2 + v^4), \ \mu_0 = (7v^2 - 16)/4, \\ \mathcal{F}_3\mathcal{F}_4 =& -78125v^9/64, \ \theta = 4(3v^2 - 64), \ \theta_1 = 4(576 + 23v^2). \end{aligned}$$

From these two sets of invariants in two and one parameters respectively, it becomes easy to determine the incompatibilities, studying their intersections and their joint graphics. The bifurcation diagram in $\mathbb{PR}^2$ can be seen in Figure 7.1. This bifurcation was done in [19], and here we have added the graphics of the invariants $\theta = 0$ and $\theta_1 = 0$, the names of each component and a positive or negative sign to describe the regions bounded by them.

1: $\mu_0 < 0$. Following Theorem 6.1 we consider three subcases: $\mathbf{D} < 0$, $\mathbf{D} > 0$ and $\mathbf{D} = 0$.

1.1: $\mathbf{D} < 0$. As systems (7.76) have the real singular point $(0, 0)$, by Proposition 5.1 in this case we obtain four real distinct finite singularities.

Considering Figure 7.1 we observe that we have two domains defined by the conditions $\mu_0 < 0$ and $\mathbf{D} < 0$: the domain between the upper and the middle branches of the graphic $\mu_0 = 0$, and the triangular domain formed by the lower branch of $\mu_0 = 0$, lower branch of $\mathbf{D} = 0$ and the axis $a = 0$. We detect that in any point of the first (respectively second) domain the condition $\widetilde{K} < 0$ (respectively $\widetilde{K} > 0$) holds. So we examine these two possibilities.

1. $\widetilde{K} < 0$. By Theorem 6.1 (see Table 6.2), apart from the weak focus, systems (7.79) possess three finite saddles and as a consequence at infinity must have three nodes. As $\mu_0 < 0$, by Theorem 6.4 (see Diagram 6.6) each node is of the type N^f. We arrive at the configuration (χ_1).

2. $\widetilde{K} > 0$. According to Theorem 6.1 (see Table 6.2) systems (7.79) possess one saddle and three anti-saddles. But from Figure 7.1 it results that $W_4 < 0$ and $\eta > 0$, so the only possible configuration is (χ_2).

1.2: $\mathbf{D} > 0$. Then according to Proposition 5.1 systems (7.79) have two real and two complex finite singularities, and moreover due to $\mu_0 < 0$ the real singular points have the same index. So since we have a focus the condition $\widetilde{K} > 0$ must hold, and we consider three possibilities: $W_4 < 0$, $W_4 > 0$ and $W_4 = 0$.

1. $W_4 < 0$. According to Theorem 6.1, apart from the weak focus of order two, we have a generic node. From Figure 7.1 it comes that $\eta > 0$, so the only possible configuration is (χ_3).

2. $W_4 > 0$. In this case by Theorem 6.1, apart from the weak focus of order two, we have a focus that must be strong due to Theorem 6.2. So considering

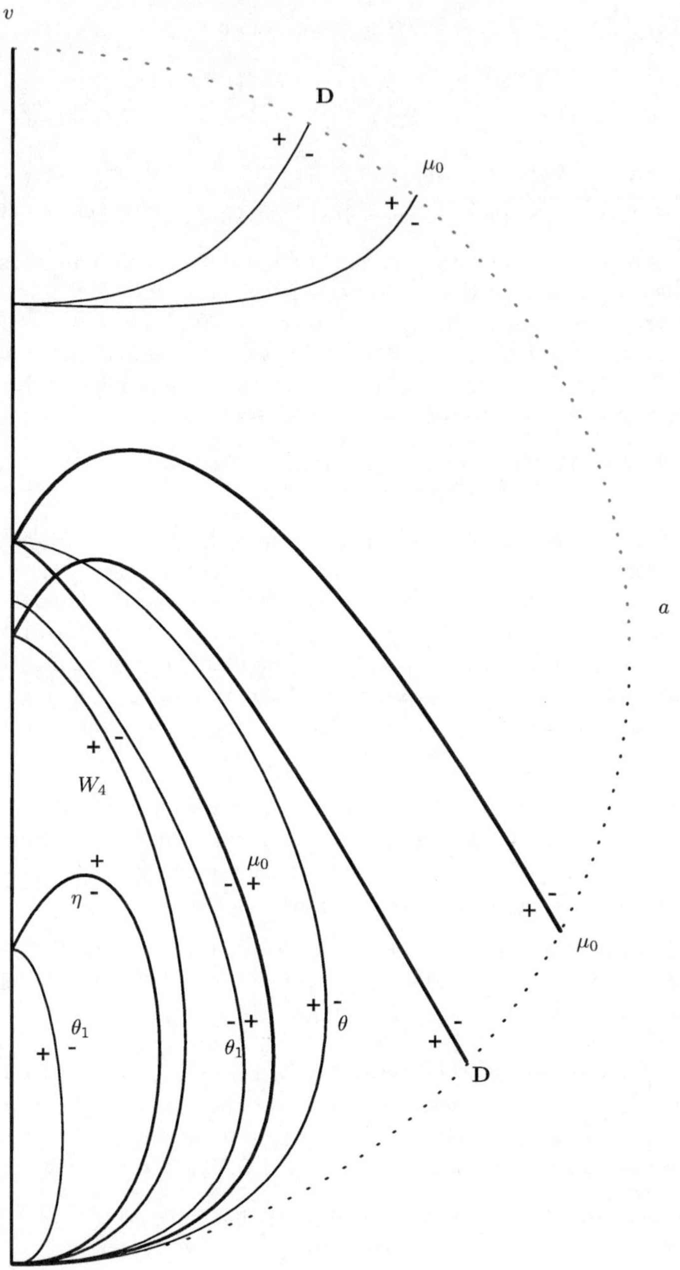

Figure 7.1: The algebraic bifurcation set of systems with a weak focus of order three.

Diagrams 6.5, 6.6 and 6.7 (see Theorem 6.4), we arrive at the following three configurations: (χ_4) if $\eta < 0$, (χ_5) if $\eta > 0$ and (χ_6) if $\eta = 0$.

3. $W_4 = 0$. Since we have a focus, according to Theorem 6.1 the condition $W_3 \leq 0$ must be fulfilled. However the Mathematica tool "Reduce" gives that $W_4 = W_3 = 0$ imply $a = 0$, and then the weak point is a center. Then $W_3 < 0$ and from Figure 7.1 it results that $\eta > 0$, so the only possible configuration is (χ_7).

1.3: $\mathbf{D} = 0$. Then according to Proposition 5.1 systems (7.79) have two simple real and a double finite singularities, or one simple and one triple finite singularities. However the second case could not be realizable by Lemma 7.11. So due to $\mu_0 < 0$ the simple real singular points have the same index. Moreover, the existence of the weak focus of order three forces that the double point is a semi-elemental saddle-node (see Lemma 7.13). From Figure 7.1 it results that $W_4 < 0$ and $\eta > 0$, so the only possible configuration is (χ_8).

2: $\mu_0 > 0$. Following Theorem 6.1 we consider three subcases: $\mathbf{D} < 0$, $\mathbf{D} > 0$ and $\mathbf{D} = 0$.

2.1: $\mathbf{D} < 0$. As systems (7.79) have the real singular point $(0,0)$, by Proposition 5.1 we obtain four real distinct finite singularities. According to Theorem 6.1 (see Table 6.2) systems (7.79) possess two saddles and two anti-saddles, and one of them is the weak focus of order three. We must split the study according to the type of the second anti-saddle. From Figure 7.1 it comes that $W_4 < 0$. Then, apart from the weak focus, on the phase plane we have two saddles and one node. The node is generic (due to $W_4 \neq 0$). From Figure 7.1 it comes that $\eta > 0$, and we have the equivalent possibilities that we had for a focus of order two, that is, (χ_9) if $\theta < 0$, (χ_{10}) if $\theta > 0$ and (χ_{11}) if $\theta = 0$,

2.2: $\mathbf{D} > 0$. Then according to Proposition 5.1 systems (7.79) have two real and two complex finite singularities, and moreover due to $\mu_0 > 0$ the real singular points have different index. So since we have a focus, the other point is a saddle. Since we have already a weak singular point of order three, the saddle must be strong. If $W_4 = 0$ then also $W_3 = 0$, and this implies $a = 0$ (i.e. we get a center). So the only possibility is $W_4 < 0$. And from Figure 7.1 it comes that $\eta > 0$. At this point, for the weak focus of order two we had several possibilities, but when the focus is of order three, from Figure 7.1 there are three possible configurations: if $\theta < 0$ (then $\theta_1 > 0$) it must be (χ_{12}), if $\theta > 0$ it must be (χ_{13}), and if $\theta = 0$ (then again $\theta_1 > 0$) it must be (χ_{14}).

2.3: $\mathbf{D} = 0$. Then according to Proposition 5.1 systems (7.79) have two simple real and a double finite singularities, or one simple and one triple finite singularities. However the second case could not be realizable by Lemma 7.11. So due to $\mu_0 > 0$ the simple real singular points have different index. The existence of the weak focus of order three forces that the double point is a semi-elemental saddle-node (see Lemma 7.13). This already implies $W_4 < 0$, and the saddle must be strong since we already have a weak singular point of order three. And from Figure 7.1 it comes that $\eta > 0$. So we have the same options as with a focus of order two, that is, configuration (χ_{15}) if $\theta < 0$, (χ_{16}) if $\theta > 0$ and (χ_{17}) if $\theta = 0$.

3: $\mu_0 = 0$. It is easy to see that $\mu_1 = 3(3-v)(25-8a^2-20v+3v^2)x/8+3a(3-v)^2y/4$. Thus $\mu_1 = 0$ implies $v = 3$, and then $\mu_0 = -4(1+a^2)^2 \neq 0$. Following Theorem 6.1 we only need to consider the case $\mu_1 \neq 0$, and we have finite multiplicity of singularities equal to three (one has moved to infinity). From Figure 7.1 it comes that $\eta > 0$. We consider three subcases: $\mathbf{D} < 0$, $\mathbf{D} > 0$ and $\mathbf{D} = 0$.

3.1: $\mathbf{D} < 0$. We have two possibilities according to the sign of the invariant polynomial κ. It is easy to see from Figure 7.1 that if $\mu_0 = 0$ then $W_4 < 0$. Therefore if $\kappa < 0$ we get the configuration (χ_{18}), whereas for $\kappa > 0$ the configuration is (χ_{19}).

3.2: $\mathbf{D} > 0$. Since we have a single real finite point and we know that this must be the weak focus, the only configuration is (χ_{20}).

3.3: $\mathbf{D} = 0$. Since we have a single simple real finite point (the weak focus) and a double one that must be a saddle-node, the only configuration is (χ_{21}).

This completes the proof of Theorem 7.8. □

7.7 A weak saddle of order three

We first create a convenient normal form to work with this family. We start from a generic 12 parameters system in which we have moved a singular point (which will become the weak saddle of order three) to the origin, and by means of a linear change of variables, we can set the eigenvectors on the axes. As the point is weak, we can set the trace equal to zero. Finally by a time change we can set the eigenvalues to $1, -1$. Thus we have the system

$$\dot{x} = x + ax^2 + bxy + cy^2, \quad \dot{y} = -y - kx^2 - lxy - ny^2. \tag{7.80}$$

For systems (7.80) we obtain the first Lyapunov constant, which is $L_1 = -(ab - ln)$, and we must consider two possibilities: either $b = l = 0$ or $b^2 + l^2 \neq 0$. In the second case, if $b = 0$ we can make a change $(x, y, t) \to (y, x, -t)$, and thus we have that we may always consider $b \neq 0$. And after, by a rescaling we can set $b = 1$. If we take $b = l = 0$, it comes immediately that the rest of the Lyapunov constants are also zero, and then the weak saddle is integrable. So we must concentrate on the case $b = 1$.

In order to obtain a weak saddle of order equal to or greater than two, forcing the first Lyapunov constant to be zero we get $a = ln$. Doing this, the second and third Lyapunov constants are $L_2 = -(k - cl^3)^2(2+n)(1-2n)^4$ and $L_3 = (ck - l)(cl^3 - k)^3 n(2n-1)^7$. So, the conditions $L_2 = 0 \neq L_3$ imply $n = -2$. Thus we arrive at the 3-parameter family

$$\dot{x} = x - 2lx^2 + xy + cy^2, \quad \dot{y} = -y - kx^2 - lxy + 2y^2 \tag{7.81}$$

with $l \in \{1, 0\}$ due to the rescaling $x \to lx$ for $l \neq 0$. We claim that in both cases ($l = 1$ or $l = 0$) we may assume $k \neq 0$. Indeed for these systems we have $\mathcal{T}_3 = 125(cl^3 - k) \neq 0$, and in the case $l = 0$ we get $k \neq 0$. If $l = 1$ we have $c^2 + k^2 \neq 0$, and due to the change $(x, y, c, k, t) \mapsto (y, x, k, c, -t)$ we may consider $k \neq 0$ and our claim is proved. We observe that in the case $l = 0$ we may assume $k = \pm 1$ due to the rescaling $x \to |k|^{-1/2}x$.

We begin by proving the next lemma.

Lemma 7.16. *Assume that a quadratic system possesses a weak saddle of order three. Then the following statements hold:*

(i) $\mu_0 < 0 \Rightarrow \widetilde{K} > 0$ *and* $\mathbf{D} < 0$.

(ii) $\widetilde{M}(\theta^2 + \theta_1^2) \neq 0$.

(iii) $\mu_0 = 0 \Rightarrow \mu_1 \neq 0$, $\kappa < 0$ *and* $\mathbf{D} < 0$.

(iv) $W_4 = W_3 = 0 \Rightarrow W_2 > 0$.

Proof. We shall use the canonical systems (7.81) for which we will consider first the sign of μ_0.

(i) Assume that $\mu_0 < 0$. Since $\mu_0 = \text{Discrim}[\widetilde{K}]/16$, we deduce that the quadratic form $\widetilde{K}$ has a well defined sign, i.e. its coefficients in front of x^2 and y^2 must have the same sign and it coincides with the sign of $\widetilde{K}$.

For systems $(7.81)|_{\{l=0\}}$ we calculate

$$\widetilde{K} = 2kx^2 + 4ckxy + 4y^2,$$

and evidently the condition $\mu_0 < 0$ implies $\widetilde{K} > 0$ as $\text{Coefficient}[\widetilde{K}, y^2] = 4 > 0$.

Consider now systems $(7.81)|_{\{l=1\}}$ for which we calculate

$$\widetilde{K} = 2(2 + k)x^2 + 4(ck - 4)xy + 2(2 + c)y^2.$$

Suppose the contrary, that $\widetilde{K} < 0$. Then we must have $2 + k < 0$ and $2 + c < 0$. Therefore setting $2 + c = -u^2$ and $2 + k = -v^2$, we get $c = -2 - u^2$, $k = -2 - v^2$. We calculate

$$\mu_0 = (2u^2 - uv + 2v^2 + u^2v^2)(2u^2 + uv + 2v^2 + u^2v^2) > 0$$

because both quadratic forms $2u^2 \pm uv + 2v^2$ are positive, i.e. the condition $\mu_0 < 0$ implies $\widetilde{K} > 0$.

Consider now the condition $\mathbf{D} < 0$ and suppose the contrary, that $\mathbf{D} \geq 0$.

If $\mathbf{D} > 0$, according to Theorem 6.1 (see Table 6.2) we must have two real and two complex singularities, i.e. $\mathcal{D}_{\mathbb{C}^2}(S) = p + q + r^c + s^c$. However due to the condition $\widetilde{K} > 0$, by Table 6.2 both real finite singularities must be anti-saddles, and this contradicts the existence of a weak saddle of order three.

Assume now $\mathbf{D} = 0$. Since there exists at least one real finite elemental singularity and $\mu_0 \neq 0$, by Table 6.2 we could only have either $\mathcal{D}_{\mathbb{C}^2}(S) = 2p+q+r$ or $\mathcal{D}_{\mathbb{C}^2}(S) = 3p+q$. In the first case due to $\mu_0 < 0$ and $\widetilde{K} > 0$, according to Table 6.2 we obtain again that both elemental finite singularities must be anti-saddles.

Let's examine the second case, when $\mathcal{D}_{\mathbb{C}^2}(S) = 3p+q$. According to Proposition 6.2 in this case a quadratic system could be brought via an affine transformation to the canonical form 8 (see Table 6.1) for which we calculate

$$\mu_0 = 4(eh-cm)^2(u^2-v), \quad E_3 = 2(eh-cm)^3(u^2-v),$$
$$\widetilde{K} = 4(eh-cm)(x^2+2uxy+vy^2).$$

We observe that the conditions $\mu_0 < 0$ and $\widetilde{K} > 0$ imply $E_3 < 0$, and then by Table 6.2 the elemental finite singularity is an anti-saddle. This completes the proof of the statement (i) of the lemma.

(ii) Suppose that for a system (7.81) we have $\mu_0 < 0$ but $\widetilde{M}(\theta^2+\theta_1^2) = 0$.

1. Assume first $\widetilde{M} = 0$. In this case $l \neq 0$, otherwise for systems $(7.81)|_{\{l=0\}}$ we get Coefficient$[\widetilde{M}, y^2] = -8 \neq 0$.

   Therefore we consider systems $(7.81)|_{\{l=1\}}$ for which calculations yield

$$\widetilde{M} = -8\big[(1+3k)x^2 + (-1+9ck)xy - (1+3c)y^2\big].$$

   Thus the condition $\widetilde{M} = 0$ is equivalent to $c = -1/3 = k$, however this leads to the contradiction: $\mu_0 = 1000/81 > 0$.

2. Suppose now $\theta = \theta_1 = 0$. Then for systems (7.81) it is necessary that $l \neq 0$, otherwise we get

$$\theta = 8k(9+8c^2k), \quad \theta_1 = -48k(1+12c^2k), \quad \mathcal{T}_3 = -125k,$$

   and due to $\mathcal{T}_3 \neq 0$ it is clear that $\theta^2+\theta_1^2 \neq 0$.

   For systems $(7.81)|_{\{l=1\}}$ we have $\mathcal{T}_3 = 125(c-k) \neq 0$ and

$$\theta = 8(-4+9c+9k-22ck+8c^2k^2), \quad \theta_1 = -16(7+3c+3k-49ck+36c^2k^2).$$

   Since $k \neq 0$ we calculate

$$\text{Res}_c[\theta, \theta_1] = 2^{16}3^25^4(k-1)^2k^2(2k-1)^2$$

   and clearly in order to have a common solution in c of the equations $\theta = 0$ and $\theta_1 = 0$, the condition $(k-1)(2k-1) = 0$ has to be fulfilled. However calculations yield

$$\theta|_{\{k=1\}} = 8(c-1)(8c-5), \quad \theta_1|_{\{k=1\}} = -32(c-1)(18c-5),$$
$$\theta|_{\{k=1/2\}} = 4(2c-1)^2, \quad \theta_1|_{\{k=1/2\}} = -8(2c-1)(9c-17).$$

We conclude that the conditions $\theta = \theta_1 = 0$ could be satisfied only if $c = k$, but this implies $\mathcal{T}_3 = 0$. This contradiction completes the proof of the statement (ii) of Lemma 7.16.

(iii) Assume first that the condition $\mu_0 = 0$ holds for systems $(7.81)|_{\{l=0\}}$. Calculations yield

$$\mu_0 = k(c^2k - 2),\ \mu_1 = -3k(x + cy),\ \kappa = -16k(7 + 9c^2k),\ \mathbf{D} = 1296k(c^2k - 4),$$

and since $k \neq 0$ we have $\mu_1 \neq 0$ and the condition $\mu_0 = 0$ gives $c \neq 0$ and $k = 2/c^2$. Then we get $\kappa = -800/c^2 < 0$ and $\mathbf{D} = -5184/c^2 < 0$.

Next we consider systems $(7.81)|_{\{l=1\}}$ and each one of the following conditions: $\mu_1 \neq 0$, $\kappa < 0$ and $\mathbf{D} < 0$. For these systems we calculate

$$\mu_0 = (ck - 4)^2 - (2 + k)(2 + c),\quad \mu_1 = 3(ck - 6 - k)x - 3(ck - 6 - c)y.$$

Suppose the contrary, that $\mu_1 = 0$. In this case we obtain either $c = k = -2$ or $c = k = 3$, which implies $\mu_0 = 0$. However in both cases we get $\mathcal{T}_3 = 0$, and this leads to an integrable saddle. Therefore the condition $\mu_1 \neq 0$ must be satisfied for systems (7.81) in the case $\mu_0 = 0$.

Consider now the conditions $\kappa < 0$ and $\mathbf{D} < 0$. For systems $(7.81)|_{\{l=1\}}$ we have

$$\kappa = -16(8 + 7c + 7k - 31ck + 9c^2k^2),\quad \mathbf{D} = -1296(3 + 4c + 4k + 6ck - c^2k^2).$$

Since $k \neq 0$ the condition $\mu_0 = 0$ yields

$$c^{\pm} = \frac{2 + 9k \pm (2 + k)\sqrt{1 + 8k}}{2k^2},$$

and we deduce that the condition $0 \neq k > -1/8$ has to be satisfied. Then we calculate

$$\kappa|_{\{c=c^{\pm}\}} = -\frac{200(2 + k)}{k^2}\left[(1 + 6k + 2k^2) \pm (1 + 2k)\sqrt{1 + 8k}\,\right],$$
$$\mathbf{D}|_{\{c=c^{\pm}\}} = -\frac{648}{k^2}\left[(4 + 12k + 3k^2 + 4k^3) \pm (3k^2 + 4k - 4)\sqrt{1 + 8k}\,\right].$$

We observe that for $k > -1/8$ we get $2 + k > 0$, $1 + 6k + 2k^2 > 0$ and $4 + 12k + 3k^2 + 4k^3 > 0$.

Additionally we calculate

$$(1 + 6k + 2k^2)^2 - \left[(1 + 2k)\sqrt{1 + 8k}\,\right]^2 = 4k^2(k - 1)^2 \geq 0,$$

and

$$(4 + 12k + 3k^2 + 4k^3)^2 - \left[(3k^2 + 4k - 4)\sqrt{1 + 8k}\,\right]^2$$
$$= 16k^2(k - 3)^2(3 + 3k + k^2) \geq 0.$$

Due to $k \neq 0$ we could have equality in the first (respectively second) relation only if $k = 1$ (respectively $k = 3$).

If $k = 1$ then $c^+ = 10$ and $c^- = 1$. Since $c - k \neq 0$ (otherwise we get $\mathcal{T}_3 = 0$), we obtain the couple $(c, k) = (10, 1)$ for which we get $\kappa = -14512 < 0$.

Analogously for $k = 3$ we obtain $c^+ = 3$ and $c^- = 2/9$. Since $c \neq k$, we get the couple $(c, k) = (2/, 3)$ for which we obtain $\mathbf{D} = -25200 < 0$. The statement (iii) of the lemma is proved.

(iv) Assume that for a system (7.81) the conditions $W_4 = W_3 = 0$ hold. If $l = 0$ then we calculate

$$W_4 = -4k^2(-1 + 432c^2k + 1728c^4k^2), \quad W_3 = -25k^2(-1 + 288c^2k),$$

and since $k \neq 0$ the condition $W_3 = 0$ (then $c \neq 0$) gives $k = 1/(288c^2)$. However in this case we get $W_4 = -25/(995328c^4) \neq 0$.

So we have to examine systems $(7.81)|_{\{l=1\}}$ for which we obtain

$$\operatorname{Res}_c[W_4, W_3] = -2^{16}3^7 5^{12}(k-1)^4 k^4 (1+k)^4 (2k-1)^4 (11 + 36k + 12k^2)^2 = 0.$$

Since $k \neq 0$ in order to have a common solution (which depends on c), the condition $(k-1)(1+k)(2k-1)(11+36k+12k^2) = 0$ must hold. A straightforward computation shows that in the case $k = 1$ (respectively $k = -1$; $k = 1/2$) the common solution of the equations $W_4 = 0$ and $W_3 = 0$ is $c = 1$ (respectively $c = -1$; $c = 1/2$). So we get $c = k$ and this leads to the contradiction: $\mathcal{T}_3 = 0$.

It remains to examine the identity

$$11 + 36k + 12k^2 = 0 \quad \Rightarrow \quad k_0^{\pm} = (-9 \pm 4\sqrt{3})/6.$$

Calculations yield

$$\begin{aligned} W_4^{\pm}(c) &= \big[(7171 + 13104c - 118248c^2 - 504576c^3 - 515088c^4) \\ &\quad \pm 8\sqrt{3}(-327 - 1406c + 7956c^2 + 36216c^3 + 37152c^4)\big]/3, \\ W_3^{\pm}(c) &= 25\big[(5011 + 24480c + 45660c^2 + 20016c^3) \\ &\quad \mp 8\sqrt{3}(183 + 1766c + 3372c^2 + 1464c^3)\big]/12, \end{aligned}$$

and we obtain that the unique common solution of $W_4^+(c) = 0 = W_3^+(c)$ (respectively $W_4^-(c) = 0 = W_3^-(c)$) is $c_0^+ = (-9 - 4\sqrt{3})/6$ (respectively $c_0^- = (-9 + 4\sqrt{3})/6$). Then we obtain $W_2(k_0^+, c_0^+) = W_2(k_0^-, c_0^-) = 15625/144 > 0$, and this completes the proof of the statement (iv) of Lemma 7.16. □

We need also other statements provided by the next lemma, which is proved with the Mathematica tool "Reduce".

Lemma 7.17. *Assume that a quadratic system possesses a weak saddle of order three. Then the following statements hold:*

(i) $\mathbf{D} < 0 \Rightarrow$ *the system cannot have two foci; moreover if* $\mu_0 < 0$ *then it is impossible even to have* n^d, f.

(ii) $\mu_0 < 0$ *and* $W_4 \geq 0 \Rightarrow \eta > 0$.

(iii) $\mu_0 > 0$ *and* $\mathbf{D} \geq 0 \Rightarrow \eta < 0$.

(iv) $W_4 = 0 \Rightarrow W_3 \geq 0$; *moreover*

(a) $\mu_0 < 0 \Rightarrow W_3 > 0$,

(b) $\mu_0 > 0$ *and* $\mathbf{D} < 0 \Rightarrow W_3 > 0$.

(v) $W_4 \geq 0$ *and* $\eta \leq 0 \Rightarrow \theta < 0$.

(vi) $W_4 \geq 0$ *and* $\theta = 0 \Rightarrow \theta_1 > 0$.

(vii) $\mu_0 = 0$ *and* $W_4 \geq 0 \Rightarrow \eta > 0$.

Since we have already studied the families with a finite star node, an infinite star node, a center, an integrable saddle, a weak focus of order two, a weak focus of order three and three special points, we already know that a saddle of order three cannot appear in combination with any of them. So, the remaining possible cases are given in the next theorem.

Theorem 7.9. *We consider here all the configurations of singularities, finite and infinite, of quadratic vector fields possessing one weak saddle of order three. We prove that there are exactly* 44 *geometrically distinct configurations of singularities, finite and infinite, and all of them are contained in the following list, enumerated and grouped according to the value of the invariant polynomial* μ_0*:*

$\mu_0 < 0$:
- (ψ_1) $s^{(3)}, n, n, f;\ S, \textcircled{c}, \textcircled{c}$;
- (ψ_2) $s^{(3)}, n, n, f;\ S, S, N^\infty$;
- (ψ_3) $s^{(3)}, n, n, f;\ \binom{0}{2}SN, S$;
- (ψ_4) $s^{(3)}, n, n, n;\ S, S, N^\infty$;
- (ψ_5) $s^{(3)}, n, n, n^d;\ S, S, N^\infty$.

$\mu_0 = 0$:
- (ψ_{40}) $s^{(3)}, n, f;\ \overline{\binom{1}{1}}SN, \textcircled{c}, \textcircled{c}$;
- (ψ_{41}) $s^{(3)}, n, f;\ \overline{\binom{1}{1}}SN, S, N^\infty$;
- (ψ_{42}) $s^{(3)}, n, f;\ \overline{\binom{1}{1}}SN, \overline{\binom{0}{2}}SN$;
- (ψ_{43}) $s^{(3)}, n, n;\ \overline{\binom{1}{1}}SN, S, N^\infty$;
- (ψ_{44}) $s^{(3)}, n, n^d;\ \overline{\binom{1}{1}}SN, S, N^\infty$.

$\mu_0 > 0$

$(\psi_6)\ s, s^{(3)}, n, f;\ N^\infty, ©, ©;$	$(\psi_7)\ s, s^{(3)}, n, f;\ N^f, ©, ©;$
$(\psi_8)\ s, s^{(3)}, n, f;\ N^d, ©, ©;$	$(\psi_9)\ s, s^{(3)}, n, f;\ S, N^\infty, N^\infty;$
$(\psi_{10})\ s, s^{(3)}, n, f;\ S, N^f, N^f;$	$(\psi_{11})\ s, s^{(3)}, n, f;\ S, N^\infty, N^f;$
$(\psi_{12})\ s, s^{(3)}, n, f;\ S, N^\infty, N^d;$	$(\psi_{13})\ s, s^{(3)}, n, f;\ S, N^f, N^d;$
$(\psi_{14})\ s, s^{(3)}, n, f;\ \overline{\binom{0}{2}}SN, N^\infty;$	$(\psi_{15})\ s, s^{(3)}, n, f;\ \overline{\binom{0}{2}}SN, N^f;$
$(\psi_{16})\ s, s^{(3)}, n, f;\ \overline{\binom{0}{2}}SN, N^d;$	$(\psi_{17})\ s, s^{(3)}, n, n;\ N^\infty, ©, ©;$
$(\psi_{18})\ s, s^{(3)}, n, n;\ S, N^\infty, N^\infty;$	$(\psi_{19})\ s, s^{(3)}, n, n;\ S, N^f, N^f;$
$(\psi_{20})\ s, s^{(3)}, n, n;\ S, N^\infty, N^f;$	$(\psi_{21})\ s, s^{(3)}, n, n;\ S, N^f, N^d;$
$(\psi_{22})\ s, s^{(3)}, n, n;\ \overline{\binom{0}{2}}SN, N^\infty;$	$(\psi_{23})\ s, s^{(3)}, n, n^d;\ N^\infty, ©, ©;$
$(\psi_{24})\ s, s^{(3)}, n, n^d;\ S, N^\infty, N^\infty;$	$(\psi_{25})\ s, s^{(3)}, n, n^d;\ S, N^f, N^f;$
$(\psi_{26})\ s, s^{(3)}, n, n^d;\ S, N^\infty, N^f;$	$(\psi_{27})\ s, s^{(3)}, n, n^d;\ S, N^f, N^d;$
$(\psi_{28})\ s, s^{(3)}, n, n^d;\ \overline{\binom{0}{2}}SN, N^\infty;$	$(\psi_{29})\ s^{(3)}, f, ©, ©;\ N^\infty, ©, ©;$
$(\psi_{30})\ s^{(3)}, f, ©, ©;\ N^f, ©, ©;$	$(\psi_{31})\ s^{(3)}, f, ©, ©;\ N^d, ©, ©;$
$(\psi_{32})\ s^{(3)}, n, ©, ©;\ N^\infty, ©, ©;$	$(\psi_{33})\ s^{(3)}, n^d, ©, ©;\ N^\infty, ©, ©;$
$(\psi_{34})\ s^{(3)}, n, ©, ©;\ N^\infty, ©, ©;$	$(\psi_{35})\ \overline{sn}_{(2)}, s^{(3)}, f;\ N^\infty, ©, ©;$
$(\psi_{36})\ \overline{sn}_{(2)}, s^{(3)}, f;\ N^f, ©, ©;$	$(\psi_{37})\ \overline{sn}_{(2)}, s^{(3)}, f;\ N^d, ©, ©;$
$(\psi_{38})\ \overline{sn}_{(2)}, s^{(3)}, n;\ N^\infty, ©, ©;$	$(\psi_{39})\ \overline{sn}_{(2)}, s^{(3)}, n^d;\ N^\infty, ©, ©.$

We have ordered these blocks so as to obtain a better paging, even if the numerical order is not respected. The realizability of each one of these configurations will be proved by means of an example, which will be provided in Part II.

Proof. In order to prove the above theorem, we work mainly with the normal form (7.81) with $(cl^3 - k)(ck - l) \neq 0$ (otherwise we get an integrable saddle). As we mentioned earlier, for these systems we may consider $l \in \{1, 0\}$.

In the case $l = 0$ (then we can assume $k = \pm 1$), the invariants are

$$\begin{aligned}
W_4 &= -4k^2(-1 + 432c^2k + 1728c^4k^2),\\
W_2 &= 6k^2(7 - 162c^2k + 48c^4k^2),\\
W_1W_3 &= 25k^4(-1 + 288c^2k)(-25 - 164c^2k + 32c^4k^2),\\
\eta &= k(4 - 27c^2k),\ \mathbf{D} = 1296k(c^2k - 4),\\
\mu_0 &= k(c^2k - 2),\ \mathcal{F}_3\mathcal{F}_4 = -312500ck^4,\\
\theta &= 8k(9 + 8c^2k),\ \theta_1 = -48k(1 + 12c^2k),\ \kappa = -16k(7 + 9c^2k).
\end{aligned}$$

And in the case $l = 1$ the invariants we need are

$$\begin{aligned}
W_4 = &-4(324c - 108 - c^2 + 324k - 862ck - 756c^2k - k^2 - 756ck^2\\
&+ 3564c^2k^2 + 432c^3k^2 + 432c^2k^3 - 4320c^3k^3 + 1728c^4k^4),\\
W_2 = &\ 3(2691 + 980c + 14c^2 + 980k - 2170ck - 656c^2k + 14k^2\\
&- 656ck^2 - 253c^2k^2 - 324c^3k^2 - 324c^2k^3 - 392c^3k^3 + 96c^4k^4),
\end{aligned}$$

$$\begin{aligned}
W_1W_3 =& \, 25(-144 + 180c - c^2 + 180k + 362ck - 468c^2k - k^2 - 468ck^2 \\
& - 504c^2k^2 + 288c^3k^2 + 288c^2k^3 + 288c^3k^3)(-2376 - 804c \\
& - 25c^2 - 804k + 1840ck + 1032c^2k - 25k^2 + 1032ck^2 \\
& + 780c^2k^2 - 164c^3k^2 - 164c^2k^3 - 354c^3k^3 + 32c^4k^4), \\
\eta =& \, 1 + 4c + 4k + 18ck - 27c^2k^2, \\
\mathbf{D} =& \, 1296(-3 - 4c - 4k - 6ck + c^2k^2), \\
\mu_0 =& \, 12 - 2c - 2k - 9ck + c^2k^2, \\
\mathcal{F}_3\mathcal{F}_4 =& \, 312500(c - k)^3(ck - 1), \\
\theta =& \, 8(-4 + 9c + 9k - 22ck + 8c^2k^2), \\
\theta_1 =& - 16(7 + 3c + 3k - 49ck + 36c^2k^2), \\
\kappa =& - 16(8 + 7c + 7k - 31ck + 9c^2k^2).
\end{aligned}$$

From these two sets of invariants in two and one parameters, it becomes relatively easy to determine the incompatibilities of the corresponding sets of conditions by simply studying the domains defined by the curves obtained by the annihilation of the invariants.

The corresponding algebraic bifurcation set of systems with a weak saddle of order three can be simply drawn in a half disc, as we have done with Figure 7.1. However, in the case of a saddle we also need the curves defined by $W_i = 0$ for $i = 1, 2, 3$, and the bifurcation set is a bit more complicated and difficult to see without colors. We prefer to omit it because it is not critical for the proof of Theorem 7.9.

1: $\mu_0 < 0$. By Lemma 7.16(i) we have $\mathbf{D} < 0$ and $\widetilde{K} > 0$. As systems (7.81) have the real singular point $(0, 0)$, by Proposition 5.1 we obtain four real distinct finite singularities. Since $\widetilde{K} > 0$, according to Theorem 6.1 (see Table 6.2, lines 12–21) systems (7.81) possess one saddle and three anti-saddles. In order to detect the types of the anti-saddles, we must consider three subcases: $W_4 < 0$, $W_4 > 0$ and $W_4 = 0$.

1.1: $W_4 < 0$. According to Theorem 6.1 (see Table 6.2, line 15) systems (7.81) possess one saddle, one focus and two nodes. Since the total index of the finite singular points is +2, from Theorem 6.4 there are four possibilities at infinity, but Lemma 7.16(ii) eliminates the condition $\widetilde{M} = 0$. So we get the configurations (ψ_1), (ψ_2) and (ψ_3) if $\eta < 0$, $\eta > 0$ and $\eta = 0$, respectively.

1.2: $W_4 > 0$. According to Theorem 6.1 (see Table 6.2) systems (7.81) possess one saddle and either three nodes or one node and two foci. However Lemma 7.17(i) eliminates the second case. So all are generic nodes. And Lemma 7.17(ii) says that now $\eta > 0$, and hence (ψ_4) is the only possible configuration.

1.3: $W_4 = 0$. Then one of the three anti-saddles is a node with discriminant zero. Additionally Lemma 7.17(iv(a)) says that $W_3 > 0$, and hence by Table 6.2 (line

13) the other two anti-saddles must be generic nodes. Again by Lemma 7.17(ii) we have that $\eta > 0$, and so we get the unique global configuration (ψ_5).

2: $\mu_0 > 0$. Following Theorem 6.1 we consider three subcases: $\mathbf{D} < 0$, $\mathbf{D} > 0$ and $\mathbf{D} = 0$.

2.1: $\mathbf{D} < 0$. As systems (7.81) have the real singular point $(0, 0)$, by Proposition 5.1 in this case we obtain four real distinct finite singularities. According to Theorem 6.1 (see Table 6.2) systems (7.81) possess two saddles and two anti-saddles. We must split the study into the cases $W_4 < 0$, $W_4 > 0$ and $W_4 = 0$.

2.1.1: $W_4 < 0$. Then we have a focus and a node. The node is generic (due to $W_4 \neq 0$). We must divide the study into three cases: $\eta < 0$, $\eta > 0$ and $\eta = 0$.

1. $\eta < 0$. From Theorem 6.4 there are four possibilities at infinity, but from Theorem 7.5 we know that an N^* is incompatible with a weak saddle of order three. So we get configurations (ψ_6), (ψ_7) and (ψ_8) if $\theta < 0$, $\theta > 0$ and $\theta = 0$, respectively.

2. $\eta > 0$. By Theorem 6.4 there are 10 possibilities at infinity, but from Theorem 7.5 we know that an N^* is incompatible with a weak saddle of order three. And Lemma 7.17(ii) eliminates the possibility $\theta = \theta_1 = 0$. So we get the following configurations: (ψ_9) if $\theta < 0$ and $\theta_1 < 0$, (ψ_{10}) if $\theta < 0$ and $\theta_1 > 0$, (ψ_{11}) if $\theta > 0$, (ψ_{12}) if $\theta = 0$ and $\theta_1 < 0$ and (ψ_{13}) if $\theta = 0$ and $\theta_1 > 0$.

3. $\eta = 0$. According to Theorem 6.4 there are four possibilities at infinity, but from Theorem 7.5 we know that an N^* is incompatible with a weak saddle of order three. So we get the configurations (ψ_{14}), (ψ_{15}) and (ψ_{16}) if $\theta < 0$, $\theta > 0$ and $\theta = 0$, respectively.

2.1.2: $W_4 > 0$. According to Theorem 6.1 (see Table 6.2), apart from two saddles, we have either two foci or two nodes. By Lemma 7.17(i) we cannot have two foci. Then we must have two nodes, which are generic due to $W_4 \neq 0$. We must divide the study into the cases $\eta < 0$, $\eta > 0$ and $\eta = 0$.

1. $\eta < 0$. From Lemma 7.17(viii) we have $\theta < 0$, and the only possible configuration is (ψ_{17}).

2. $\eta > 0$. From Theorem 6.4 there are 10 possibilities at infinity, but from Theorem 7.5 we know that an N^* is incompatible with a weak saddle of order three. And Lemma 7.17(ix) eliminates the possibility $\theta = \theta_1 = 0$. Moreover Lemma 7.17(xi) eliminates the possibility $\theta = 0$ and $\theta_1 \geq 0$. So we get the following configurations: (ψ_{18}) if $\theta < 0$ and $\theta_1 < 0$, (ψ_{19}) if $\theta < 0$ and $\theta_1 > 0$, (ψ_{20}) if $\theta > 0$ and (ψ_{21}) if $\theta = 0$.

3. $\eta = 0$. From Lemma 7.17(v) we have $\theta < 0$, and the only possible configuration is (ψ_{22}).

2.1.3: $W_4 = 0$. By Lemma 7.17(vi(b)) we have $W_3 > 0$. We must divide the study into the cases $\eta < 0$, $\eta > 0$ and $\eta = 0$.

1. $\eta < 0$. From Lemma 7.17(iv) we have $\theta < 0$, and the only possible configuration is (ψ_{23}).
2. $\eta > 0$. From Theorem 6.4 there are 10 possibilities at infinity, but from Theorem 7.5 we know that an N^* is incompatible with a weak saddle of order three. And Lemma 7.16(ii) eliminates the possibility $\theta = \theta_1 = 0$. Moreover Lemma 7.17(vi) eliminates the possibility $\theta = 0$ and $\theta_1 \leq 0$. So we get the following configurations: (ψ_{24}) if $\theta < 0$ and $\theta_1 < 0$, (ψ_{25}) if $\theta < 0$ and $\theta_1 > 0$, (ψ_{26}) if $\theta > 0$ and (ψ_{27}) if $\theta = 0$.
3. $\eta = 0$. From Lemma 7.17(v) we have $\theta < 0$, and the only possible configuration is (ψ_{28}).

2.2: $\mathbf{D} > 0$. Then according to Proposition 5.1 systems (7.81) have two real and two complex finite singularities, and moreover due to $\mu_0 > 0$ the real singular points have different index. So since we have a saddle, the other point is an anti-saddle. Since we have already a weak point of order three, the anti-saddle must be a node or a strong focus. Lemma 7.17(iii) implies $\eta < 0$. We must split the study into the cases $W_4 < 0$, $W_4 > 0$ and $W_4 = 0$.

2.2.1: $W_4 < 0$. Then we have a focus and a saddle. From Theorem 6.4 there are four possibilities at infinity, but from Theorem 7.5 we know that an N^* is incompatible with a weak saddle of order three. So we get the configurations (ψ_{29}), (ψ_{30}) and (ψ_{31}) if $\theta < 0$, $\theta > 0$, and $\theta = 0$, respectively.

2.2.2: $W_4 > 0$. Now we have a saddle and a node. The node is generic (due to $W_4 \neq 0$). From Lemma 7.17(v) we have $\theta < 0$, and the only possible configuration is (ψ_{32}).

2.2.3: $W_4 = 0$. By Lemma 7.17(iv) we have $W_3 \geq 0$, but this time the possibility $W_3 = 0$ exists.

1. $W_3 > 0$. Now we have a saddle and a node with discriminant zero, which cannot be a star node by Theorem 7.2. From Lemma 7.17(v) we have $\theta < 0$, and the only possible configuration is (ψ_{33}).
2. $W_3 = 0$. Now Lemma 7.16(iv) says that $W_2 > 0$, we have a saddle and a generic node, and the ones with discriminant zero are the complex finite singularities. Considering the condition $\eta < 0$ from Lemma 7.17(v) we have $\theta < 0$, and the only possible configuration is (ψ_{34}).

2.3: $\mathbf{D} = 0$. Since we already have a simple finite singularity, according to Proposition 5.1 systems (7.81) have either two simple real and a double, or one simple and one triple finite singularities. However by Lemma 7.11 we could not have a triple point and a weak saddle of order three.

Thus we have two simple real and one double singularities. Moreover due to $\mu_0 > 0$ the simple real singular points have different index. Furthermore the existence of the weak saddle of order three forces that the double point is a semi-elemental saddle-node (see Lemma 7.13). Additionally by Lemma 7.17(iii) the condition $\eta < 0$ has to be fulfilled. We must split the study into the cases $W_4 < 0$, $W_4 > 0$ and $W_4 = 0$.

1. $W_4 < 0$. Then we have a focus and a saddle. From Theorem 6.4 there are four possibilities at infinity, but from Theorem 7.5 we know that an N^* is incompatible with a weak saddle of order three. So we get configurations (ψ_{35}), (ψ_{36}) and (ψ_{37}) if $\theta < 0$, $\theta > 0$, and $\theta = 0$, respectively.

2. $W_4 > 0$. Now we have a saddle and a node. The node is generic (due to $W_4 \neq 0$). From Lemma 7.17(v) we have $\theta < 0$, and as $\eta < 0$ the only possible configuration is (ψ_{38}).

3. $W_4 = 0$. Now $W_3 \neq 0$ since we have finite multiplicity four and no other finite singularity can have discriminant zero (even $W_3 > 0$ since the other elemental point is a saddle). From Lemma 7.17(v) we have $\theta < 0$, and the only possible configuration is (ψ_{39}).

3: $\mu_0 = 0$. From Lemma 7.16(iii) we have $\mu_1 \neq 0$. Then following Theorem 6.1 we only need to consider the case $\mu_1 \neq 0$ (i.e. $\mathbf{R} \neq 0$, see Remark 6.1), and we have finite multiplicity of singularities equal to three (one has moved to infinity). From Lemma 7.16(iii) we also have $\mathbf{D} < 0$ and $\kappa < 0$. According to Theorem 6.4 the total index of infinite singularities is zero (independently of the value of η). So the total index of the finite singularities must be $+1$, and this implies the existence of a saddle and two anti-saddles.

We divide the study into the cases $W_4 < 0$, $W_4 > 0$ and $W_4 = 0$.

1. $W_4 < 0$. Then we have a focus and a node apart from the saddle of order three. From Theorem 6.4 there are three possibilities at infinity, giving configurations (ψ_{40}), (ψ_{41}) and (ψ_{42}) if $\eta < 0$, $\eta > 0$ and $\eta = 0$, respectively.

2. $W_4 > 0$. Then we have two nodes apart from the saddle of order three. The nodes are generic (due to $W_4 \neq 0$). From Lemma 7.17(vii) we have $\eta > 0$, and the only possible configuration is (ψ_{43}).

3. $W_4 = 0$. We cannot have $W_3 = 0$ because in this case we get three special singularities and by Theorems 7.2 and 7.6 we cannot have a weak saddle of order three and two nodes with coinciding eigenvalues. Moreover we cannot have $W_3 < 0$ by Lemma 7.17(iv). So the second anti-saddle is a generic node. Finally by Lemma 7.17(vii) we have $\eta > 0$, and the only possible configuration is (ψ_{44}).

This completes the proof of Theorem 7.9. □

Part II

Configurations of singularities of quadratic systems

Chapter 8

Quadratic systems with finite singularities of total multiplicity at most one

In this and the next four chapters, we consider real planar quadratic differential systems that can be written in the form

$$\begin{aligned}\frac{dx}{dt} &= a + cx + dy + gx^2 + 2hxy + ky^2 \equiv p(x, y),\\ \frac{dy}{dt} &= b + ex + fy + lx^2 + 2mxy + ny^2 \equiv q(x, y).\end{aligned} \tag{8.1}$$

We will give lots of examples of quadratic systems. Some sections will contain only examples from one specific normal form (having some parameters), and we will give the values of those parameters. But in other sections where we need to use several normal forms, we have preferred to indicate the 12 coefficients of the system (8.1) in the form $(a, c, d, g, h, k), (b, e, f, l, m, n)$.

8.1 Systems without finite singularities

Theorem 8.1. (a) *The global configurations of singularities of all quadratic vector fields with finite singularities of total multiplicity $m_f = 0$ are classified in Diagram* 8.1 *according to the geometrical equivalence relation. We have a total of* 32 *geometrically distinct global configurations of singularities.*

(b) *Necessary and sufficient conditions for each one of the* 32 *different equivalence classes can be assembled from Diagram* 8.1 *in terms of* 18 *invariant*

J. C. Artés et al., *Geometric Configurations of Singularities of Planar Polynomial Differential Systems*, https://doi.org/10.1007/978-3-030-50570-7_8

polynomials with respect to the action of the affine group and time rescaling. These invariant polynomials, which appear in this diagram, are given in Section 5.4.

(c) *Diagram* 8.1 *actually contains the global bifurcation diagram in the* 12*-dimensional space of parameters of the global configurations of singularities of this family of quadratic differential systems and provides an algorithm for finding for any given system in the family considered, its respective geometrical global configuration of singularities.*

Proof. The total multiplicity m_f of finite singularities of every system in this family is zero. Theorem 6.4 gives the full global geometrical classification of the whole class of quadratic systems according to their singularities at infinity. Since only infinite singularities occur in this family ($m_f = 0$), we can extract from Diagrams 6.6–6.8 (see Theorem 6.4) the classification of the configurations of singularities of this family. In fact from this theorem we obtain more. Indeed, we extract from Diagrams 6.6–6.8 the part of the *global bifurcation diagram* of configurations of singularities at infinity of **QS**, the fragment covering the case we need here, i.e. $m_f = 0$. We obtain the bifurcation diagram (see Diagram 8.1) of global configurations of singularities of this class, done in the 12-parameter space of coefficients and obtained with the help of invariant polynomials. The proof for this diagram is completely covered in the article [26], and thus there is no need for a proof here. We shall only give here examples, one for each kind of distinct geometrical global configurations occurring in this family, which we indicate with "⇒".

1. Systems with $\eta < 0$:
   (i) $\overline{\binom{4}{1}}N$, ©, © $\Rightarrow (\dot{x} = 1 + xy;\ \dot{y} = -x^2)$;
   (ii) N^*, $\binom{2}{1}$ ©, $\binom{2}{1}$ © $\Rightarrow (\dot{x} = 1;\ \dot{y} = -x^2 - y^2)$.
2. Systems with $\eta > 0$:
   (i) $\overline{\binom{4}{1}}N$, S, N^∞ $\Rightarrow (\dot{x} = -1 + xy;\ \dot{y} = 1 - xy + 2y^2)$;
   (ii) $\overline{\binom{4}{1}}S$, N^f, N^f $\Rightarrow (\dot{x} = 1 - xy;\ \dot{y} = 2 - 2xy + y^2)$;
   (iii) $\overline{\binom{3}{1}}SN$, $\overline{\binom{1}{1}}SN$, N^d $\Rightarrow (\dot{x} = 1 + x - xy;\ \dot{y} = 1 - xy)$;
   (iv) $\overline{\binom{3}{1}}SN$, $\overline{\binom{1}{1}}NS$, N^d $\Rightarrow (\dot{x} = 1 - x + xy;\ \dot{y} = 1 + xy)$;
   (v) $\overline{\binom{2}{1}}S$, $\overline{\binom{2}{1}}N$, N^* $\Rightarrow (\dot{x} = 1 - xy;\ \dot{y} = -xy)$.
3. Systems with $\eta = 0$, $\widetilde{M} \neq 0$:
   (i) $\overline{\binom{0}{2}}SN$, $\overline{\binom{4}{1}}N$ $\Rightarrow (\dot{x} = 1 + xy;\ \dot{y} = -1 - xy + y^2)$;

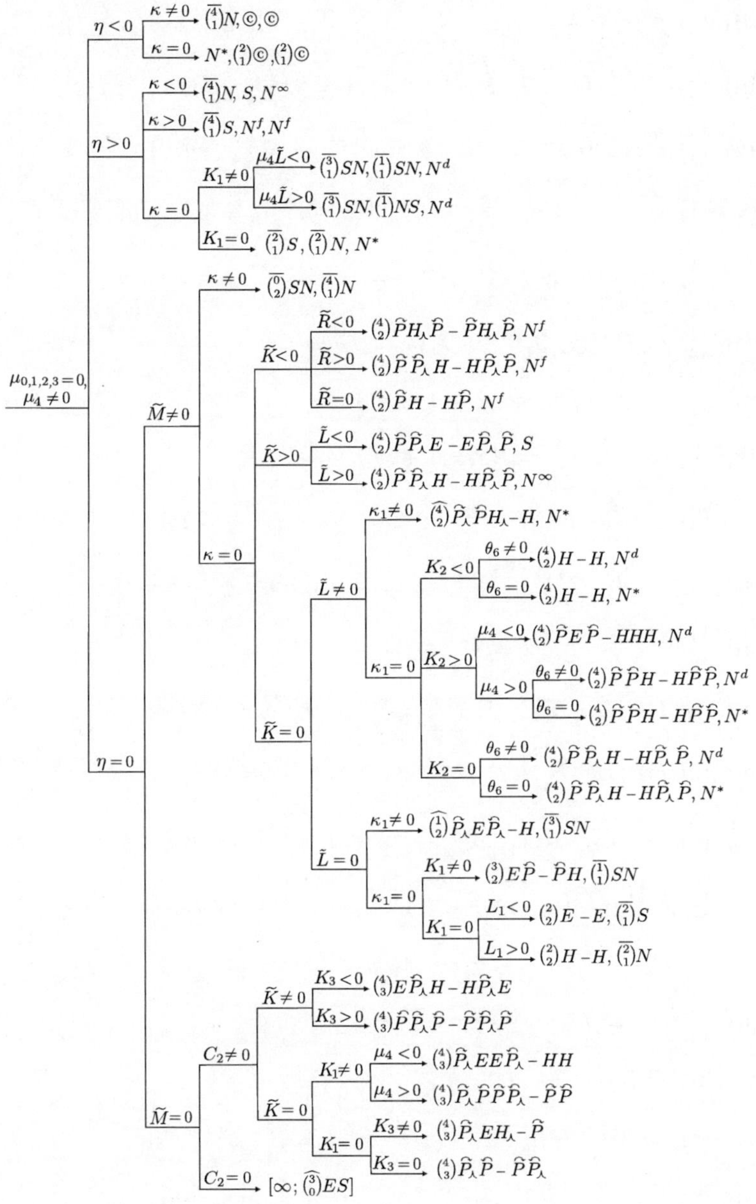

Diagram 8.1: Global configurations: case $\mu_0 = \mu_1 = \mu_2 = \mu_3 = 0$, $\mu_4 \neq 0$.

(ii) $\binom{4}{2}\ \widehat{P}H_{\curlywedge}\widehat{P}-\widehat{P}H_{\curlywedge}\widehat{P},\ N^f \Rightarrow (\dot{x}=x^2/4;\ \dot{y}=1-3xy/4)$;

(iii) $\binom{4}{2}\ \widehat{P}\,\widehat{P}_{\curlywedge}H-H\,\widehat{P}_{\curlywedge}\widehat{P},\ N^f \Rightarrow (\dot{x}=2x^2/3;\ \dot{y}=1-xy/3)$;

(iv) $\binom{4}{2}\ \widehat{P}H-H\widehat{P},\ N^f \Rightarrow (\dot{x}=x^2/2;\ \dot{y}=1-xy/2)$;

(v) $\binom{4}{2}\ \widehat{P}\,\widehat{P}_{\curlywedge}E-E\widehat{P}_{\curlywedge}\widehat{P},\ S \Rightarrow (\dot{x}=-x^2;\ \dot{y}=1-2xy)$;

(vi) $\binom{4}{2}\ \widehat{P}\,\widehat{P}_{\curlywedge}H-H\,\widehat{P}_{\curlywedge}\widehat{P},\ N^{\infty} \Rightarrow (\dot{x}=2x^2;\ \dot{y}=1+xy)$;

(vii) $\widehat{\binom{4}{2}}\ \widehat{P}_{\curlywedge}\widehat{P}H_{\curlywedge}-H,\ N^{*} \Rightarrow (\dot{x}=y+x^2;\ \dot{y}=1)$;

(viii) $\binom{4}{2}\ H-H,\ N^d \Rightarrow (\dot{x}=1+x^2;\ \dot{y}=x)$;

(ix) $\binom{4}{2}\ H-H,\ N^{*} \Rightarrow (\dot{x}=1+x^2;\ \dot{y}=1)$;

(x) $\binom{4}{2}\ \widehat{P}E\widehat{P}-HHH,\ N^d \Rightarrow (\dot{x}=-2+x^2;\ \dot{y}=1+x)$;

(xi) $\binom{4}{2}\ \widehat{P}\widehat{P}H-H\widehat{P}\widehat{P},\ N^d \Rightarrow (\dot{x}=-1+x^2;\ \dot{y}=2+x)$;

(xii) $\binom{4}{2}\ \widehat{P}\widehat{P}H-H\widehat{P}\widehat{P},\ N^{*} \Rightarrow (\dot{x}=-1+x^2;\ \dot{y}=1)$;

(xiii) $\binom{4}{2}\ \widehat{P}\widehat{P}_{\curlywedge}H-H\,\widehat{P}_{\curlywedge}\widehat{P},\ N^d \Rightarrow (\dot{x}=x^2;\ \dot{y}=1+x)$;

(xiv) $\binom{4}{2}\ \widehat{P}\widehat{P}_{\curlywedge}H-H\,\widehat{P}_{\curlywedge}\widehat{P},\ N^{*} \Rightarrow (\dot{x}=x^2;\ \dot{y}=1)$;

(xv) $\widehat{\binom{1}{2}}\ \widehat{P}_{\curlywedge}E\widehat{P}_{\curlywedge}-H,\ \overline{\binom{3}{1}}SN \Rightarrow (\dot{x}=y;\ \dot{y}=1-xy)$;

(xvi) $\binom{3}{2}\ E\widehat{P}-\widehat{P}H,\ \overline{\binom{1}{1}}SN \Rightarrow (\dot{x}=x;\ \dot{y}=1-xy)$;

(xvii) $\binom{2}{2}\ E-E,\ \overline{\binom{2}{1}}S \Rightarrow (\dot{x}=-1;\ \dot{y}=1-xy)$;

(xviii) $\binom{2}{2}\ H-H,\ \overline{\binom{2}{1}}N \Rightarrow (\dot{x}=1.\ \dot{y}=1-xy)$;

4. Systems with $\eta=\widetilde{M}=0$:

(i) $\binom{4}{3}\ E\widehat{P}_{\curlywedge}H-H\widehat{P}_{\curlywedge}E \Rightarrow (\dot{x}=x^2;\ \dot{y}=1-x^2+xy)$;

(ii) $\binom{4}{3}\ \widehat{P}\,\widehat{P}_{\curlywedge}\widehat{P}-\widehat{P}\,\widehat{P}_{\curlywedge}\widehat{P} \Rightarrow (\dot{x}=x^2;\ \dot{y}=-1-x^2+xy)$;

(iii) $\binom{4}{3}\ \widehat{P}_{\curlywedge}EE\widehat{P}_{\curlywedge}-HH \Rightarrow (\dot{x}=x;\ \dot{y}=1-x^2)$;

(iv) $\binom{4}{3}\ \widehat{P}_{\curlywedge}\widehat{P}\,\widehat{P}\,\widehat{P}_{\curlywedge}-\widehat{P}\,\widehat{P} \Rightarrow (\dot{x}=1+x;\ \dot{y}=-x^2)$;

(v) $\binom{4}{3}\widehat{P}_{\curlywedge}EH_{\curlywedge}-\widehat{P} \Rightarrow (\dot{x}=1;\ \dot{y}=y-x^2);$

(vi) $\binom{4}{3}\widehat{P}_{\curlywedge}\widehat{P}-\widehat{P}\widehat{P}_{\curlywedge} \Rightarrow (\dot{x}=1;\ \dot{y}=-x^2);$

(vii) $\left[\infty;\ \widehat{\binom{3}{0}}\,ES\right] \Rightarrow (\dot{x}=x^2;\ \dot{y}=1+xy).$ □

8.2 Systems with exactly one singularity

Theorem 8.2. (a) *The global configurations of singularities of all quadratic vector fields with finite singularities of total multiplicity* $m_f=1$ *are classified in Diagram* 8.2 *according to the geometrical equivalence relation of configurations. We have a total of* 52 *geometrically distinct global configurations of singularities.*

(b) *Necessary and sufficient conditions for each one of the* 52 *different equivalence classes can be assembled from Diagram* 8.2 *in terms of* 26 *invariant polynomials with respect to the action of the affine group and time rescaling. These invariant polynomials, which appear in this diagram, are given in Section* 5.4.

(c) *Diagram* 8.2 *actually contains the global bifurcation diagram in the* 12*-dimensional space of parameters of the global configurations of singularities of this family of quadratic differential systems and provides an algorithm for finding for any given system in the family considered, its respective geometrical global configuration of singularities.*

Proof. We consider all quadratic vector fields with total multiplicity m_f of finite singularities equal to 1. Since we have only one finite singular point, this point is of course real. To obtain the full global classification of global configurations of singularities with respect to the *geometrical equivalence relation* for this family, we need to (i) deepen the *topological* classification of all configurations of finite singularities done by Theorem 6.1 (see Table 6.2, lines 178–187) by using the finer *geometrical equivalence relation*, (ii) integrate this with the *geometric* classification of infinite singularities given by Theorem 6.4 and (iii) search for a minimal set of invariants that allows us to obtain for this family the bifurcation diagram with respect to the *geometrical equivalence relation* of global configurations of singularities in the 12-dimensional space of parameters.

According to Proposition 5.1 (see Table 6.2) in this case the conditions $\mu_0=\mathbf{R}=\mathbf{P}=0$ and $\mathbf{U}\neq 0$ must be satisfied, and due to Remark 6.1 these conditions are equivalent to $\mu_0=\mu_1=\mu_2=0$ and $\mu_3\neq 0$. So according to Theorem 6.4 the following lemma is valid.

Lemma 8.1. *The configurations of singularities at infinity of the family of quadratic systems possessing one elemental (real) finite singularity (i.e.* $\mu_0=\mu_1=\mu_2=0$

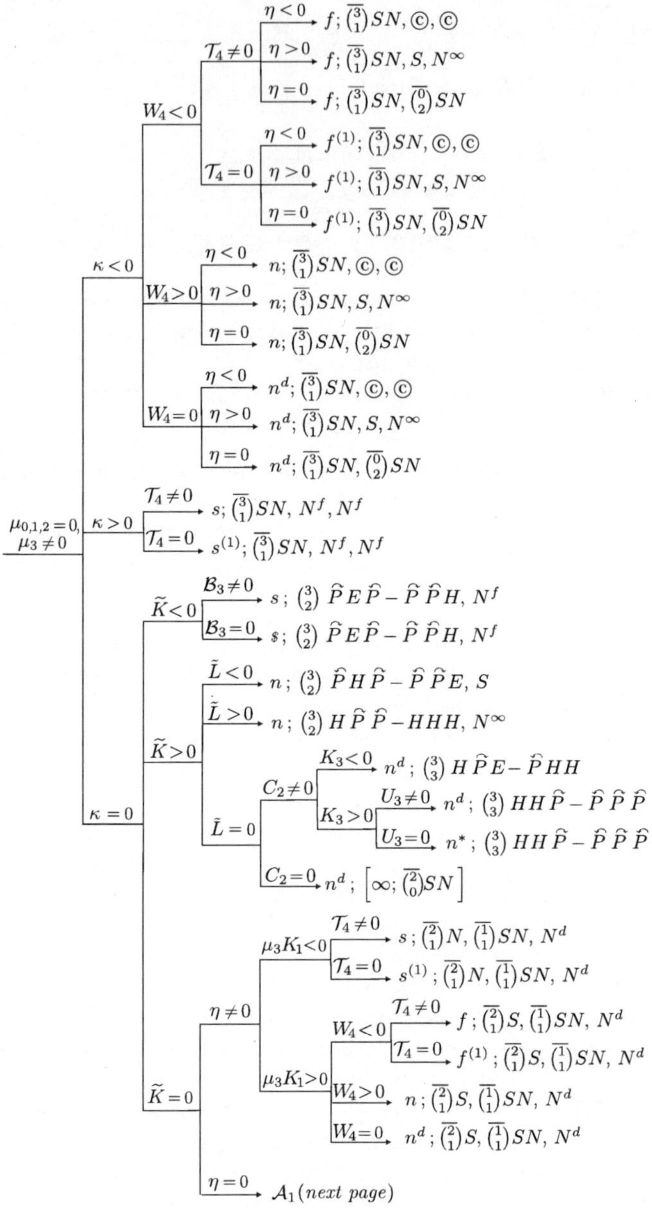

Diagram 8.2: Global configurations: case $\mu_0=\mu_1=\mu_2=0\neq\mu_3$.

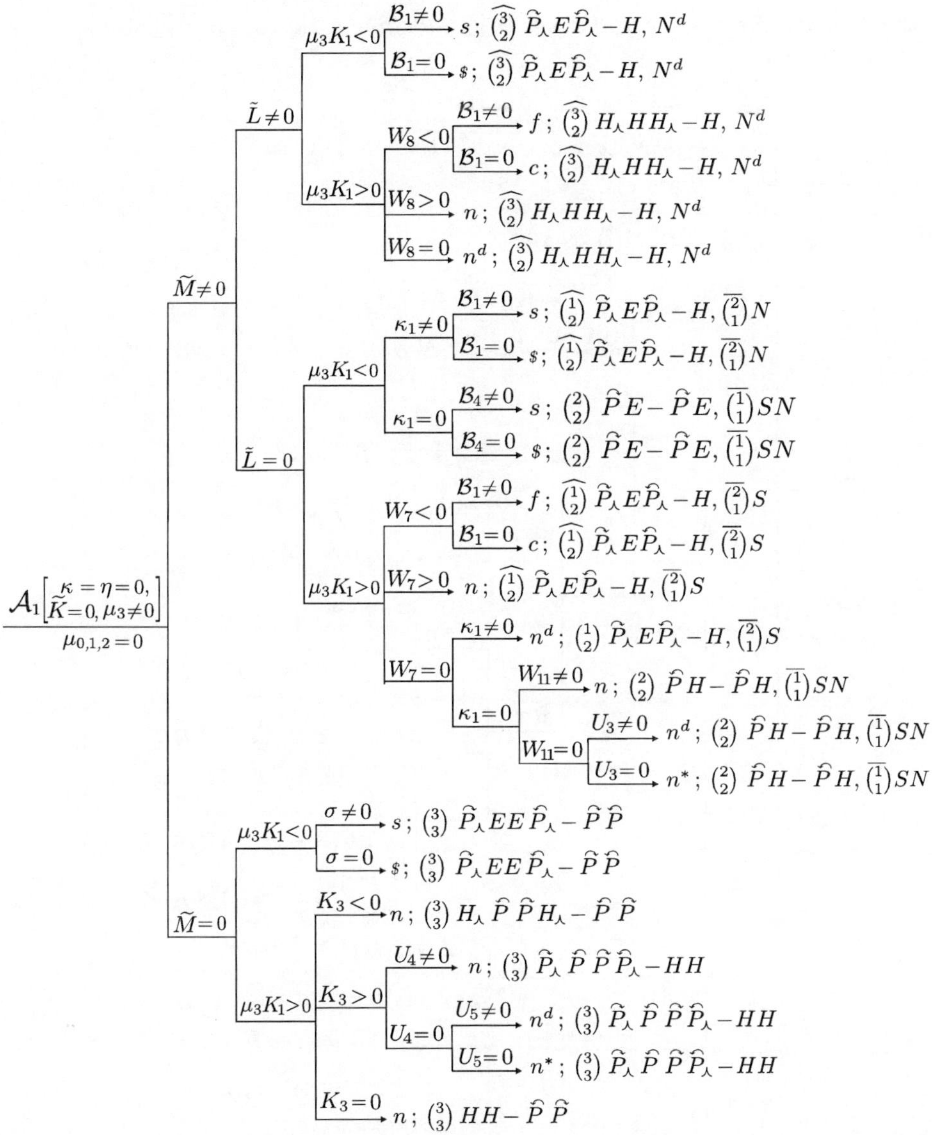

Diagram 8.2 (*cont.*): Global configurations: case $\mu_0 = \mu_1 = \mu_2 = 0 \neq \mu_3$.

and $\mu_3 \neq 0$) are classified in Diagram 8.3 according to the geometrical equivalence relation. Necessary and sufficient conditions for each one of the 22 different equivalence classes can be assembled from this diagram in terms of 14 invariant *polynomials with respect to the action of the affine group and time rescaling, given in Section* 5.4.

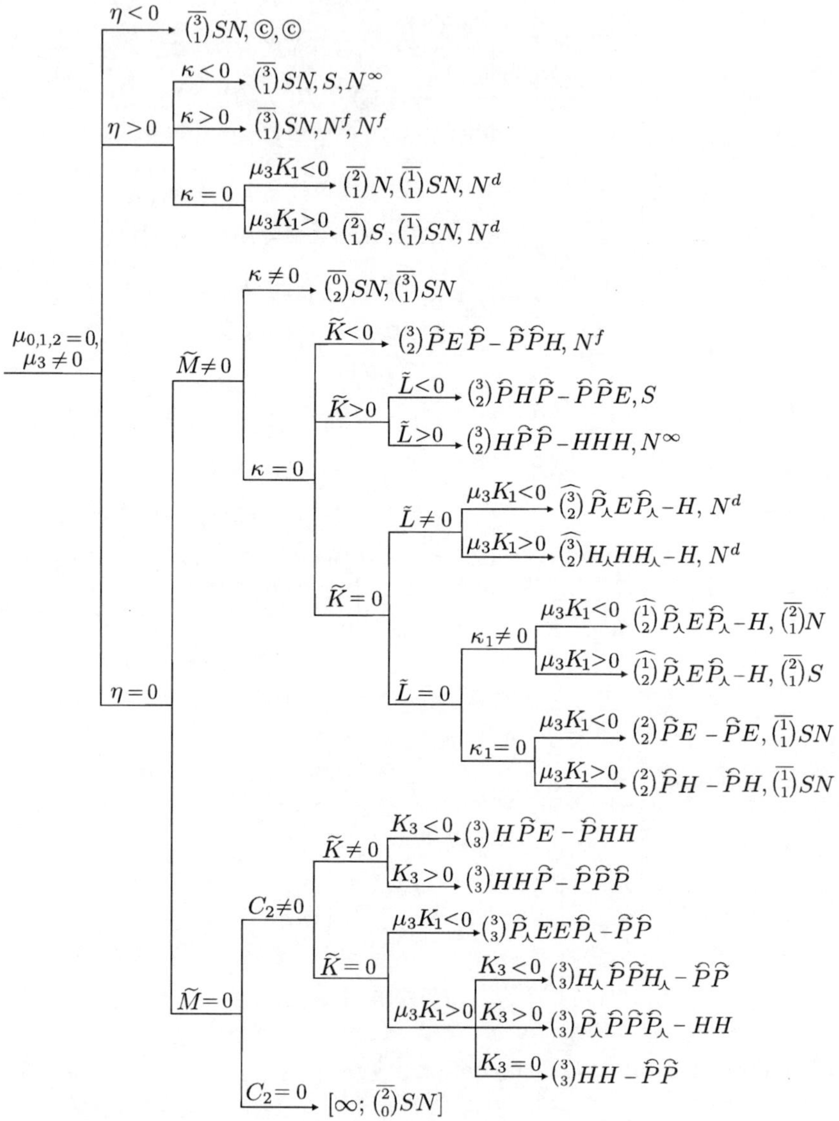

Diagram 8.3: Configurations of infinite singularities: the case $\mu_0 = \mu_1 = \mu_2 = 0 \neq \mu_3$.

According to Proposition 6.2 the family of quadratic systems with one elemental finite singularity could be brought via an affine transformation to one of the two canonical forms in Table 6.1 (see systems 17a and 17b), governed by the invariant polynomial $\widetilde{K} \neq 0$. We consider two cases: $\widetilde{K} \neq 0$ and $\widetilde{K} = 0$.

1: $\widetilde{K} \neq 0$. In this case by Proposition 6.2 via an affine transformation, quadratic systems in this family could be brought to the systems:

$$\begin{aligned}\dot{x} &= cx + dy + (2c+d)x^2 + 2dxy,\\ \dot{y} &= ex + fy + (2e+f)x^2 + 2fxy,\end{aligned}$$

possessing the singular point $M_1(0,0)$. For these systems calculations yield

$$\mu_0 = \mu_1 = \mu_2 = 0, \quad \mu_3 = (cf-de)^2x^3, \quad \kappa = 256d^2(de-cf).$$

We remark that for these systems we have $\mu_3 \neq 0$, and therefore in what follows we assume that the condition $cf - de \neq 0$ holds (i.e. the singular point $M_1(0,0)$ is elemental).

1.1: $\kappa \neq 0$. Then $d \neq 0$ and due to a time rescaling we may assume $d = 1$. So we consider the 3-parameter family of systems:

$$\begin{aligned}\dot{x} &= cx + y + (2c+1)x^2 + 2xy,\\ \dot{y} &= ex + fy + (2e+f)x^2 + 2fxy, \quad cf - e \neq 0,\end{aligned} \tag{8.2}$$

for which calculations yield

$$\begin{aligned}\mu_0 = \mu_1 = \mu_2 = 0, \quad \mu_3 = (cf-e)^2x^3, \quad \widetilde{K} = 8(cf-e)x^2,\\ \eta = 4\big[(2c+1+2f)^2 + 16(e-cf)\big], \quad \kappa = 256(e-cf),\\ \mathcal{T}_4 = -8(c+f)(cf-e)^2, \quad \mathcal{T}_3 = -8(cf-e)^2, \quad \mathcal{F}_1 = 6(e-cf),\\ W_4 = 64(cf-e)^4\big[(c-f)^2+4e\big] = 64(cf-e)^4\big[(c+f)^2+4(e-cf)\big],\\ \widetilde{M} = -8\big[(1+2c-2f)^2 + 6(2e+f)\big]x^2 - 16(1+2c-2f)xy - 32y^2.\end{aligned} \tag{8.3}$$

Considering (8.3) we make the following remark:

Remark 8.1. Assume that the condition $\kappa \neq 0$ holds. Then

(i) $\mathcal{T}_3\mathcal{F}_1 \neq 0$ and $\mathrm{sign}\,(\widetilde{K}) = -\mathrm{sign}\,(\kappa)$;

(ii) the condition $\kappa > 0$ implies $\eta > 0$ and $W_4 > 0$;

(iii) in the case $\mathcal{T}_4 = 0$ we have $W_4 \neq 0$ and $\mathrm{sign}\,(W_4) = \mathrm{sign}\,(\kappa)$.

The first two statements follow obviously from (8.3). In the case $\mathcal{T}_4 = 0$ we get $f = -c$ and then $\kappa = 256(c^2+e)$ and $W_4 = 256(c^2+e)^5$. This proves the last assertion.

1.1.1: $\kappa < 0$. Then by Remark 8.1 we obtain $\widetilde{K} > 0$.

1.1.1.1: $W_4 < 0$. In this case considering the condition $\widetilde{K} > 0$, according to Theorem 6.1 (see Table 6.2, line 184) the finite singularity is a focus or a center.

1.1.1.1.1: $\mathcal{T}_4 \neq 0$. Then by Theorem 6.2 we have a strong focus. Since $\kappa < 0$ according to Lemma 8.1 we get the following three global configurations:

- $f;\ \overline{\binom{3}{1}}SN, ©, © \Rightarrow (c=0,\ e=-1,\ f=1)$ (if $\eta < 0$);
- $f;\ \overline{\binom{3}{1}}SN, S, N^{\infty} \Rightarrow (c=0,\ e=-1,\ f=7/4)$ (if $\eta > 0$);
- $f;\ \overline{\binom{0}{2}}SN, \overline{\binom{3}{1}}SN \Rightarrow (c=0,\ e=-1,\ f=3/2)$ (if $\eta = 0$).

1.1.1.1.2: $\mathcal{T}_4 = 0$. Then $f = -c$ and since by Remark 8.1 we have $\mathcal{T}_3\mathcal{F}_1 \neq 0$, by Theorem 6.2 the finite singularity is a first order weak focus. Considering the types of the infinite singularities given in Diagram 8.3, we obtain the following three global configurations:

- $f^{(1)};\ \overline{\binom{3}{1}}SN, ©, © \Rightarrow (c=0,\ e=-1,\ f=0)$ (if $\eta < 0$);
- $f^{(1)};\ \overline{\binom{3}{1}}SN, S, N^{\infty} \Rightarrow (c=0,\ e=-1/18,\ f=0)$ (if $\eta > 0$);
- $f^{(1)};\ \overline{\binom{0}{2}}SN, \overline{\binom{3}{1}}SN \Rightarrow (c=0,\ e=-1/16,\ f=0)$ (if $\eta = 0$).

1.1.1.2: $W_4 > 0$. Since $\widetilde{K} > 0$, according to Theorem 6.1 (see Table 6.2, line 180) systems (8.2) possess a node, which is generic (due to $W_4 \neq 0$). So considering Lemma 8.1 we have the following three global configurations:

- $n;\ \overline{\binom{3}{1}}SN, ©, © \Rightarrow (c=0,\ e=-1,\ f=-9/4)$ (if $\eta < 0$);
- $n;\ \overline{\binom{3}{1}}SN, S, N^{\infty} \Rightarrow (c=0,\ e=-1,\ f=-3)$ (if $\eta > 0$);
- $n;\ \overline{\binom{0}{2}}SN, \overline{\binom{3}{1}}SN \Rightarrow (c=0,\ e=-1,\ f=-5/2)$ (if $\eta = 0$).

1.1.1.3: $W_4 = 0$. Then the singular point $M_1(0,0)$ of systems (8.2) is a node with coinciding eigenvalues, which could not be a star node (due to the respective linear matrix). Considering the types of the infinite singularities given by Lemma 8.1, we get the next three global configurations:

- $n^d;\ \overline{\binom{3}{1}}SN, ©, © \Rightarrow (c=0,\ e=-1/4,\ f=-1)$ (if $\eta < 0$);
- $n^d;\ \overline{\binom{3}{1}}SN, S, N^{\infty} \Rightarrow$ y$(c=0,\ e=-1/4,\ f=1)$ (if $\eta > 0$);
- $n^d;\ \overline{\binom{0}{2}}SN, \overline{\binom{3}{1}}SN \Rightarrow (c=0,\ e=-1/64,\ f=-1/4)$ (if y$\eta = 0$).

1.1.2: $\kappa > 0$. According to Remark 8.1 we obtain $\widetilde{K} < 0$, and according to Theorem 6.1 (see Table 6.2, line 178) the finite singularity is a saddle. By Remark 8.1, in this case we have $\eta > 0$, and considering Lemma 8.1 we have the unique configuration of infinite singularities $\overline{\binom{3}{1}}SN,\ N^f, N^f$.

1.1.2.1: $\mathcal{T}_4 \neq 0$. In this case by Theorem 6.2 the saddle is strong, and we arrive at the global configuration

• s; $\overline{\binom{3}{1}}SN,\ N^f, N^f \Rightarrow (c=0,\ e=1,\ f=1)$.

1.1.2.2: $\mathcal{T}_4=0$. Then $f=-c$ and by Remark 8.1 we have $\mathcal{T}_3\mathcal{F}_1\neq 0$. Considering Theorem 6.2 we deduce that the finite singularity is a weak saddle of the first order. So we obtain the global configuration

• $s^{(1)}$; $\overline{\binom{3}{1}}SN,\ N^f, N^f \Rightarrow (c=0,\ e=1,\ f=0)$.

1.2: $\kappa=0$. Then by (8.2) we have $d=0$, and considering the condition $\mu_3=c^2f^2x^3\neq 0$, we obtain $cf\neq 0$. So doing a time rescaling we may assume $f=1$, and we consider the 2-parameter family of systems:

$$\dot{x}=cx+2cx^2,\quad \dot{y}=ex+y+(2e+1)x^2+2xy,\quad c\neq 0,$$

for which calculations yield

$$\begin{aligned}\mu_0=\mu_1=\mu_2=0,\ \mu_3=c^2x^3,\ \widetilde{K}=8cx^2,\ \eta=\kappa=0,\\ \widetilde{M}=-32(c-1)^2x^2,\ \sigma=1+c+2(1+2c)x,\\ C_2=-(1+2e)x^3+2(c-1)x^2y,\ W_4=0,\\ \mathcal{T}_i=0,\ i=1,2,3,4,\ \mathcal{F}_1=\mathcal{H}=\mathcal{B}=\mathcal{B}_1=\mathcal{B}_2=0,\\ \mathcal{B}_3=-288c^3(1+c)x^2,\ \tilde{L}=32c(c-1)x^2.\end{aligned}\tag{8.4}$$

Remark 8.2. We observe that the corresponding matrix for the singular point $M_1(0,0)$ is $\begin{pmatrix} c & 0\\ e & 1\end{pmatrix}$, and hence this singular point is a saddle if $c<0$, a node with two directions if $c>0$ and $c\neq 1$, a node with one direction if $c=1$ and $e\neq 0$, and a star node if $c=1$ and $e=0$.

1.2.1: $\widetilde{K}<0$. Then $c<0$ and by the above remark the finite singularity is a saddle. Considering (8.4) according to Theorem 6.2 (see the statement $e_3[\gamma]$) the saddle is weak if and only if $\mathcal{B}_3=0$. Moreover in this case we have an integrable saddle.

Since $c<0$ we have $\widetilde{M}\neq 0$. Then according to Lemma 8.1, we arrive at the next two global global configurations:

• s; $\binom{3}{2}\ \widehat{P}E\widehat{P}-\widehat{P}\,\widehat{P}H,\ N^f \Rightarrow (c=-2,\ e=0)$ (if $\mathcal{B}_3\neq 0$);

• $\$$; $\binom{3}{2}\ \widehat{P}E\widehat{P}-\widehat{P}\,\widehat{P}H,\ N^f \Rightarrow (c=-1,\ e=0)$ (if $\mathcal{B}_3=0$).

1.2.2: $\widetilde{K}>0$. Then $c>0$ and by Remark 8.2 the finite singularity is a node. We observe that due to $\mu_3\neq 0$ the condition $c=1$ is equivalent to $\tilde{L}=0$.

1.2.2.1: $\tilde{L}\neq 0$. Then $\widetilde{M}\neq 0$ and by Remark 8.2 we have a generic node. Additionally, as $\widetilde{K}>0$ we obtain $\operatorname{sign}(\tilde{L})=\operatorname{sign}(c-1)$, and considering Lemma 8.1 we get the following two global configurations:

• n; $\binom{3}{2}\ \widehat{P}H\widehat{P}-\widehat{P}\,\widehat{P}E,\ S \Rightarrow (c=1/2,\ e=0)$ (if $\tilde{L}<0$);

• n ; $\binom{3}{2}$ $H\widehat{P}\,\widehat{P}-HHH$, $N^\infty \Rightarrow (c=2,\ e=0)$ (if $\tilde{L}>0$).

1.2.2.2: $\tilde{L}=0$. Then $c=1$ and this implies $\widetilde{M}=0$. By Remark 8.2 we have a node with coinciding eigenvalues. Additionally for $c=1$ we obtain $C_2=-(1+2e)x^3$, $U_3=-24ex^5$.

1.2.2.2.1: $C_2\neq 0$. Then we have a single real infinite singularity of multiplicity six, and according to Lemma 8.1 the type of this singularity depends on the sign of the invariant polynomial $K_3=6(1+2e)x^6$, which is non-zero due to $C_2\neq 0$.

Thus taking into consideration Remark 8.2 and Lemma 8.1, we arrive at the next three global configurations:

• n^d ; $\binom{3}{3}$ $H\widehat{P}E-\widehat{P}HH \Rightarrow (c=1,\ e=-1)$ (if $K_3<0$);

• n^d ; $\binom{3}{3}$ $HH\widehat{P}-\widehat{P}\,\widehat{P}\,\widehat{P} \Rightarrow (c=1,\ e=1)$ (if $K_3>0$, $U_3\neq 0$);

• n^* ; $\binom{3}{3}$ $HH\widehat{P}-\widehat{P}\,\widehat{P}\,\widehat{P} \Rightarrow (c=1,\ e=0)$ (if $K_3>0$, $U_3=0$).

1.2.2.2.2: $C_2=0$. Then $e=-1/2$ and we get the system

$$\dot{x}=x(1+2x),\quad \dot{y}=-x/2+y+2xy,$$

possessing a node n^d and the infinite line filled up with singularities. Considering Lemma 8.1 we obtain the global configuration

• n^d ; $\left[\infty;\ \overline{\binom{2}{0}}SN\right] \Rightarrow (c=1,\ e=-1/2)$.

2: $\widetilde{K}=0$. In this case, according to Proposition 6.2 we consider the family of systems:

$$\begin{aligned}\dot{x}&=x+dy,\\ \dot{y}&=ex+fy+lx^2+2mxy-d(dl-2m)y^2,\end{aligned}\tag{8.5}$$

possessing the singular point $M_1(0,0)$. For these systems calculations yield

$$\eta=4d^2(dl-m)^2(dl-2m)^2,\quad \tilde{L}=8d(2m-dl)(x+dy)\big[lx-(dl-2m)y\big].\tag{8.6}$$

We consider two cases: $\eta\neq 0$ and $\eta=0$.

2.1: $\eta\neq 0$. Then $d(dl-m)(dl-2m)\neq 0$, and we may assume $d=l=1$ and $m=0$ due to the transformation

$$x_1=(dl-2m)x,\quad y_1=-\frac{(dl-2m)m}{dl-m}x+\frac{d(dl-2m)^2}{dl-m}y,\quad t_1=\frac{dl-m}{dl-2m}t.$$

So we consider the 2-parameter family of systems:

$$\dot{x}=x+y,\quad \dot{y}=ex+fy+x^2-y^2,$$

for which calculations yield

$$\mu_0 = \mu_1 = \mu_2 = 0, \quad \mu_3 = (f-e)(x-y)(x+y)^2, \quad \widetilde{K} = \kappa = 0,$$
$$\eta = 4, \quad K_1 = (x-y)(x+y)^2, \quad F_4F_5 = 6(f-e)(x-y)^2(x+y)^4,$$
$$G_3 = 2(e-f), \quad W_4 = 64(e-f)^2\big[(f-1)^2+4e\big],$$
$$\mathcal{T}_4 = 8(f-e)(1+f), \quad \mathcal{T}_3 = 8(f-e), \quad \mathcal{F}_1 = 2(e-f).$$

Remark 8.3. In the case $\eta \neq 0$ the condition $\mu_3 \neq 0$ implies $\mathcal{T}_3\mathcal{F}_1F_4F_5G_3 \neq 0$ and sign $(\mu_3K_1) =$ sign (F_4F_5).

2.1.1: $\mu_3K_1 < 0$. By Remark 8.3 we have $F_4F_5 < 0$, and according to Theorem 6.1 (see Table 6.2, line 179) the finite singularity is a saddle. Clearly this saddle is weak if and only if $f = -1$, and this is equivalent to $\mathcal{T}_4 = 0$. Additionally by Remark 8.3 we have $\mathcal{T}_3\mathcal{F}_1 \neq 0$, and according to Theorem 6.2 the weak saddle could be only of the first order. So considering Lemma 8.1 we get the following two global configurations:

- $s\,;\,\overline{\binom{2}{1}}N,\ \overline{\binom{1}{1}}SN,\ N^d \Rightarrow (e=2,\ f=1)\quad (\text{if } \mathcal{T}_4 \neq 0);$
- $s^{(1)}\,;\,\overline{\binom{2}{1}}N,\ \overline{\binom{1}{1}}SN,\ N^d \Rightarrow (e=2,\ f=-1)\quad (\text{if } \mathcal{T}_4 = 0).$

2.1.2: $\mu_3K_1 > 0$. In this case we have $F_4F_5 > 0$, and as $G_3 \neq 0$ by Theorem 6.1 we have a focus or a center if $W_4 < 0$ and a node if $W_4 \geq 0$.

2.1.2.1: $W_4 < 0$. Then we have a focus, which is strong if $\mathcal{T}_4 \neq 0$ and weak of the first order if $\mathcal{T}_4 = 0$ (due to Theorem 6.2 and $\mathcal{T}_3\mathcal{F}_1 \neq 0$, see Remark 8.3). Considering Lemma 8.1, we arrive at the next two global configurations:

- $f\,;\,\overline{\binom{2}{1}}S,\ \overline{\binom{1}{1}}SN,\ N^d \Rightarrow (e=-2,\ f=1)\quad (\text{if } \mathcal{T}_4 \neq 0);$
- $f^{(1)}\,;\,\overline{\binom{2}{1}}S,\ \overline{\binom{1}{1}}SN,\ N^d \Rightarrow (e=-2,\ f=-1)\quad (\text{if } \mathcal{T}_4 = 0).$

2.1.2.2: $W_4 > 0$. In this case we have a generic node (as $W_4 \neq 0$), and hence we get

- $n\,;\,\overline{\binom{2}{1}}S,\ \overline{\binom{1}{1}}SN,\ N^d \Rightarrow (e=0,\ f=2).$

2.1.2.3: $W_4 = 0$. Then we have a node with coinciding eigenvalues, and due to the linearization matrix at the singularity $M_1(0,0)$, this is a one-direction node. So we have the global configuration

- $n^d\,;\,\overline{\binom{2}{1}}S,\ \overline{\binom{1}{1}}SN,\ N^d \Rightarrow (e=-1/4,\ f=0).$

2.2: $\eta = 0$. Then $d(dl-m)(dl-2m) = 0$, and we consider two subcases: $\tilde{L} \neq 0$ and $\tilde{L} = 0$.

2.2.1: $\tilde{L} \neq 0$. Considering (8.6) we obtain $d(dl-2m) \neq 0$, and then the condition $\eta = 0$ gives $m = dl$. In this case we have $\tilde{L} = 8d^2l^2(x+dy)^2 \neq 0$, and then via the rescaling $(x, y) \mapsto \big(x/(dl), y/(d^2l)\big)$ we obtain the following 2-parameter family of systems:

$$\dot{x} = x + y, \quad \dot{y} = ex + fy + (x + y)^2.$$

For these systems calculations yield

$$\begin{gathered}
\mu_0 = \mu_1 = \mu_2 = \eta = 0, \quad \mu_3 = (f - e)(x + y)^3, \quad \widetilde{K} = \widetilde{N} = \kappa = 0,\\
\tilde{L} = 8(x + y)^2 = -\widetilde{M}, \quad K_1 = (x + y)^3, \quad F_4F_5 = 6(f - e)(x + y)^6,\\
G_3 = 0, \quad W_8 = 2^{14}3^3(e - f)^4\big[(f - 1)^2 + 4e\big], \quad \mathcal{T}_i = 0, \ i = 1, 2, 3, 4, \\
\sigma = 1 + f + 2x + 2y, \quad \mathcal{F}_1 = \mathcal{H} = 0,\\
\mathcal{B}_1 = 4(e - f)^2(1 + f), \quad \mathcal{B}_2 = 4(e - f)^3
\end{gathered} \tag{8.7}$$

and we again have sign $(\mu_3K_1) =$ sign (F_4F_5).

2.2.1.1: $\mu_3K_1 < 0$. Then we have $F_4F_5 < 0$, and according to Theorem 6.1 (see Table 6.2, line 179) the finite singularity is a saddle. Clearly this saddle is weak if and only if $f = -1$, and this is equivalent to $\mathcal{B}_1 = 0$. Additionally considering (8.7) we obtain $\mathcal{B}_2 > 0$, and according to Theorem 6.2 the weak saddle is an integrable one. So considering Lemma 8.1 we get the following two global configurations:

- $s\,;\ \widehat{\binom{3}{2}}\ \widehat{P}_{\curlywedge} E \widehat{P}_{\curlywedge} - H,\ N^d \Rightarrow (e = 2,\ f = 1)$ (if $\mathcal{B}_1 \neq 0$);
- $\$\,;\ \widehat{\binom{3}{2}}\ \widehat{P}_{\curlywedge} E \widehat{P}_{\curlywedge} - H,\ N^d \Rightarrow (e = 2,\ f = -1)$ (if $\mathcal{B}_1 = 0$).

2.2.1.2: $\mu_3K_1 > 0$. In this case we have $F_4F_5 > 0$, and as $G_3 = \widetilde{N} = 0$ by Theorem 6.1 we have a focus or a center if $W_8 < 0$ and a node if $W_8 \geq 0$.

2.2.1.2.1: $W_8 < 0$. Then we have a focus, which is strong if $\mathcal{B}_1 \neq 0$. Considering (8.7) we have $\mathcal{B}_2 < 0$, and according to Theorem 6.2 in the case $\mathcal{B}_1 = 0$ we have a center. So considering Lemma 8.1, we arrive at the next two global configurations:

- $f\,;\ \widehat{\binom{3}{2}}\ H_{\curlywedge} H H_{\curlywedge} - H,\ N^d \Rightarrow (e = -2,\ f = 1)$ (if $\mathcal{B}_1 \neq 0$);
- $c\,;\ \widehat{\binom{3}{2}}\ H_{\curlywedge} H H_{\curlywedge} - H,\ N^d \Rightarrow (e = -2,\ f = -1)$ (if $\mathcal{B}_1 = 0$).

2.2.1.2.2: $W_8 > 0$. In this case we have a generic node (as the condition $W_8 \neq 0$ implies $\tau_1 = (f - 1)^2 + 4e \neq 0$), and hence we get the global configuration

- $n\,;\ \widehat{\binom{3}{2}}\ H_{\curlywedge} H H_{\curlywedge} - H,\ N^d \Rightarrow (e = 0,\ f = 2)$.

2.2.1.2.3: $W_8 = 0$. Then we have a node with coinciding eigenvalues, and due to the matrix of the linearization of the system at the singularity $M_1(0, 0)$, this is a one-direction node, providing the global configuration

- $n^d\,;\ \widehat{\binom{3}{2}}\ H_{\curlywedge} H H_{\curlywedge} - H,\ N^d \Rightarrow (e = -1/4,\ f = 0)$.

2.2.2: $\tilde{L} = 0$. Considering (8.6) we obtain $d(dl - 2m) = 0$, and as for systems (8.5) we have

$$\widetilde{M} = -32m^2x^2 - 8d(dl - 2m)(3lx^2 - 2mxy + d^2ly^2 - 2dmy^2).$$

The above condition gives $\widetilde{M} = -32m^2x^2$. We consider two possibilities: $\widetilde{M} \neq 0$ and $\widetilde{M} = 0$.

2.2.2.1: $\widetilde{M} \neq 0$. Then $m \neq 0$ and as the condition $d(dl - 2m) = 0$ holds, applying the transformation

$$x_1 = dlx, \quad y_1 = dl(x + dy),$$

when $d \neq 0$ (then $m = dl/2 \neq 0$ due to $\widetilde{M} \neq 0$), or the transformation

$$x_1 = 2mx, \quad y_1 = lx/(2m) + y,$$

when $d = 0$, we arrive at the following family of systems:

$$\begin{aligned} \dot{x} &= \varepsilon_1 x + \varepsilon_2 y, & \varepsilon_1\varepsilon_2 &= 0, \\ \dot{y} &= ex + fy + xy, & \varepsilon_1 + \varepsilon_2 &= 1. \end{aligned} \tag{8.8}$$

For these systems calculations yield

$$\begin{gathered} \mu_0 = \mu_1 = \mu_2 = 0, \;\; \mu_3 = (\varepsilon_1 f - \varepsilon_2 e)xy(\varepsilon_1 x + \varepsilon_2 y), \;\; \widetilde{K} = \kappa = \tilde{L} = 0, \\ \widetilde{N} = -x^2, \;\; \eta = 0, \;\; \widetilde{M} = -8x^2, \;\; \kappa_1 = -32\varepsilon_2, \;\; K_1 = xy(\varepsilon_1 x + \varepsilon_2 y), \\ F_4F_5 = 6(\varepsilon_1 f - \varepsilon_2 e)x^2y^2(\varepsilon_1 x + \varepsilon_2 y)^2, \quad W_7 = 3\varepsilon_2 e^2(4\varepsilon_2 e + f^2)/16, \\ G_3 = 0, \;\; \mathcal{T}_i = 0, \; i = 1,2,3,4, \;\; \sigma = \varepsilon_1 + f + x, \;\; \mathcal{F}_1 = \mathcal{H} = 0, \\ \mathcal{B}_1 = -\varepsilon_2 ef, \;\; \mathcal{B}_2 = \varepsilon_2 e/4. \end{gathered} \tag{8.9}$$

So we obtain again sign $(\mu_3 K_1)$ = sign (F_4F_5), and we consider two cases: $\mu_3 K_1 < 0$ and $\mu_3 K_1 > 0$.

2.2.2.1.1: $\mu_3 K_1 < 0$. Then we have $F_4F_5 < 0$, and according to Theorem 6.1 (see Table 6.2, line 179) the finite singularity is a saddle. Clearly this saddle is weak if and only if $\varepsilon_1 + f = 0$.

2.2.2.1.1.1: $\kappa_1 \neq 0$. Then by (8.9) we have $\varepsilon_2 = 1$, $\varepsilon_1 = 0$, and the condition $\mu_3 K_1 < 0$ yields $e > 0$. So $\mathcal{B}_2 > 0$ and we have $\mathcal{B}_1 = 0$ if and only if $f = 0$. In this case according to Theorem 6.2 we have an integrable saddle. Therefore considering the condition $\kappa_1 \neq 0$ and Lemma 8.1, we get the following two global configurations:

- $s;\ \widehat{\binom{1}{2}}\ \widehat{P}_\curlywedge E\widehat{P}_\curlywedge - H,\ \overline{\binom{2}{1}}N \Rightarrow (\varepsilon_2 = 1,\ e = 1,\ f = 1)$ (if $\mathcal{B}_1 \neq 0$);
- $\$;\ \widehat{\binom{1}{2}}\ \widehat{P}_\curlywedge E\widehat{P}_\curlywedge - H,\ \overline{\binom{2}{1}}N \Rightarrow (\varepsilon_2 = 1,\ e = 1,\ f = 0)$ (if $\mathcal{B}_1 = 0$).

2.2.2.1.1.2: $\kappa_1 = 0$. Then we have $\varepsilon_2 = 0$, $\varepsilon_1 = 1$, and the condition $\mu_3 K_1 < 0$ yields $f < 0$. We observe that in this case the saddle is a weak one if and only if $f + 1 = 0$. Additionally calculations yield

$$\mathcal{F}_1 = \mathcal{H} = \mathcal{B}_1 = \mathcal{B}_2 = \mathcal{B}_3 = 0, \quad \mathcal{B}_4 = 6(1 + f)x^2y.$$

So according to Theorem 6.2 in the case of weak saddle (i.e. $f=-1$), we have an integrable saddle. Therefore considering the condition $\kappa_1=0$ and Lemma 8.1, we obtain the following two global configurations:

- $s;\ \binom{2}{2}\ \widehat{P}E-\widehat{P}E,\ \overline{\binom{1}{1}}SN \Rightarrow (\varepsilon_2=0,\ e=0,\ f=-2)$ (if $\mathcal{B}_4\neq 0$);
- $\$;\ \binom{2}{2}\ \widehat{P}E-\widehat{P}E,\ \overline{\binom{1}{1}}SN \Rightarrow (\varepsilon_2=0,\ e=0,\ f=-1)$ (if $\mathcal{B}_4=0$).

2.2.2.1.2: $\mu_3K_1>0$. In this case we have $F_4F_5>0$, and as $G_3=0$ and $\widetilde{N}\neq 0$, according to Theorem 6.1 (see Table 6.2, lines 182 and 186) we have a focus or a center if $W_7<0$ and a node if $W_7\geq 0$.

2.2.2.1.2.1: $W_7<0$. Then we have $\varepsilon_2=1$, $\varepsilon_1=0$ (i.e. $\kappa_1\neq 0$), and $e<0$. So $\mathcal{B}_2=e/4\leq 0$, $\mathcal{B}_1=-ef$ and according to Theorem 6.2 we have a strong focus if $\mathcal{B}_1\neq 0$ and a center if $\mathcal{B}_1=0$.

Thus considering Lemma 8.1, we arrive at the following two global configurations:

- $f;\ \widehat{\binom{1}{2}}\ \widehat{P}_{\lambda}E\widehat{P}_{\lambda}-H,\ \overline{\binom{2}{1}}S \Rightarrow (\varepsilon_2=1,\ e=-1,\ f=1)$ (if $\kappa_1\neq 0,\ \mathcal{B}_1\neq 0$);
- $c;\ \widehat{\binom{1}{2}}\ \widehat{P}_{\lambda}E\widehat{P}_{\lambda}-H,\ \overline{\binom{2}{1}}S \Rightarrow (\varepsilon_2=1,\ e=-1,\ f=0)$ (if $\kappa_1\neq 0,\ \mathcal{B}_1=0$).

2.2.2.1.2.2: $W_7>0$. Then we again have $\varepsilon_2=1, \varepsilon_1=0$, and hence $\kappa_1\neq 0$. So the singular point is a generic node, and by Lemma 8.1 we get the global configuration

- $n;\ \widehat{\binom{1}{2}}\ \widehat{P}_{\lambda}E\widehat{P}_{\lambda}-H,\ \overline{\binom{2}{1}}S \Rightarrow (\varepsilon_2=1,\ e=-1,\ f=3)$.

2.2.2.1.2.3: $W_7=0$. Then by (8.9) we have $\varepsilon_2e(4\varepsilon_2e+f^2)=0$, and we consider two subcases: $\kappa_1\neq 0$ and $\kappa_1=0$.

2.2.2.1.2.3.1: $\kappa_1\neq 0$. Then we have $\varepsilon_2=1, \varepsilon_1=0$, and the condition $W_7=0$ gives $e=-f^2/4$. Considering the linearization matrix of the singularity $M_1(0,0)$, we conclude that systems (8.8) possess a node n^d. So by Lemma 8.1 we have the global configuration

- $n^d;\ \widehat{\binom{1}{2}}\ \widehat{P}_{\lambda}E\widehat{P}_{\lambda}-H,\ \overline{\binom{2}{1}}S \Rightarrow (\varepsilon_2=1,\ e=-1,\ f=2)$.

2.2.2.1.2.3.2: $\kappa_1=0$. In this case we have $\varepsilon_2=0, \varepsilon_1=1$, and the linearization matrix of the singularity $M_1(0,0)$ is $\begin{pmatrix}1 & 0\\ e & f\end{pmatrix}$ with $f>0$ due to $\mu_3K_1>0$. So systems (8.8) possess a generic node if $f\neq 1$, a one-direction node if $f=1$ and $e\neq 0$ and a star node if $f=1$ and $e=0$. Additionally for these systems in the considered case we have

$$W_{11}=3(f-1)x^4,\quad U_3=-3x^4\big[efx+(1-f)y\big],$$

and clearly these invariant polynomials govern the possibilities mentioned above. So considering the condition $\kappa_1=0$ and Lemma 8.1, we get the following three global configurations:

- n ; $\binom{2}{2}\ \widehat{P}H-\widehat{P}H, \overline{\binom{1}{1}}SN \Rightarrow (\varepsilon_2=0,\ e=1,\ f=2)$ (if $W_{11}\neq 0$);
- n^d ; $\binom{2}{2}\ \widehat{P}H-\widehat{P}H, \overline{\binom{1}{1}}SN \Rightarrow (\varepsilon_2=0,\ e=1,\ f=1)$ (if $W_{11}=0,\ U_3\neq 0$);
- n^* ; $\binom{2}{2}\ \widehat{P}H-\widehat{P}H, \overline{\binom{1}{1}}SN \Rightarrow (\varepsilon_2=0,\ e=0,\ f=1)$ (if $W_{11}=0,\ U_3=0$).

2.2.2.2: $\widetilde{M}=0$. In this case $m=0$ and then the condition $\tilde{L}=0$ yields $dl=0$. As $l\neq 0$ (due to $\mu_3\neq 0$) we get $d=0$, and then via the rescaling $(x,y)\mapsto(x/l,\ y/l)$ we may assume $l=1$. Therefore we obtain the family of systems:

$$\dot{x}=x,\quad \dot{y}=ex+fy+x^2,$$

for which calculations yield

$$\begin{aligned}\mu_0=\mu_1=\mu_2=0,\quad \mu_3=fx^3,\quad \eta=\widetilde{M}=0,\quad C_2=-x^3,\quad K_1=x^3,\\ \widetilde{K}=\kappa=\tilde{L}=\widetilde{N}=0,\quad G_3=W_8=0,\quad K_3=6(2-f)fx^6,\\ F_4F_5=6fx^6,\quad \mathcal{T}_i=0,\ i=1,2,3,4,\quad \sigma=1+f.\end{aligned}\tag{8.10}$$

2.2.2.2.1: $\mu_3K_1<0$. Then we have $F_4F_5<0$, and according to Theorem 6.1 (see Table 6.2, line 179) the finite singularity is a saddle. Clearly this saddle is weak if and only if $f=-1$, and this is equivalent to $\sigma=0$. However in the last case we get Hamiltonian systems, and hence the weak saddle is an integrable one. So considering Lemma 8.1, we arrive at the next two global configurations:

- s ; $\binom{3}{3}\ \widehat{P}_{\curlywedge}EE\widehat{P}_{\curlywedge}-\widehat{P}\widehat{P} \Rightarrow (e=0,\ f=-2)$ (if $\sigma\neq 0$);
- $\$$; $\binom{3}{3}\ \widehat{P}_{\curlywedge}EE\widehat{P}_{\curlywedge}-\widehat{P}\widehat{P} \Rightarrow (e=0,\ f=-1)$ (if $\sigma=0$).

2.2.2.2.2: $\mu_3K_1>0$. In this case we have $F_4F_5>0$ (i.e. $f>0$), and considering the matrix of the linearization at the singular point, we conclude that the singular point $M_1(0,0)$ is a node. Moreover, this node is generic if $f\neq 1$, a one-direction node if $f=1$ and $e\neq 0$ and a star node if $f=1$ and $e=0$. Additionally for these systems in the considered case we have

$$U_4=-6(f-1)x^3,\quad U_5\big|_{f=1}=-6ex^2.$$

The behavior of the trajectories in the vicinity of the infinite singularity (which is of multiplicity six) according to Lemma 8.1 is governed by the invariant polynomial K_3. By (8.10) as $f>0$ we have $\operatorname{sign}(K_3)=\operatorname{sign}(2-f)$. Thus we arrive at the following five global configurations:

- n ; $\binom{3}{3}\ H_{\curlywedge}\widehat{P}\widehat{P}H_{\curlywedge}-\widehat{P}\widehat{P} \Rightarrow (e=0,\ f=3)$ (if $K_3<0$);
- n ; $\binom{3}{3}\ HH-\widehat{P}\widehat{P} \Rightarrow (e=0,\ f=2)$ (if $K_3=0$);
- n ; $\binom{3}{3}\ \widehat{P}_{\curlywedge}\widehat{P}\widehat{P}\widehat{P}_{\curlywedge}-HH \Rightarrow (e=0,\ f=1/2)$ (if $K_3>0,\ U_4\neq 0$);
- n^d ; $\binom{3}{3}\ \widehat{P}_{\curlywedge}\widehat{P}\widehat{P}\widehat{P}_{\curlywedge}-HH \Rightarrow (e=1,\ f=1)$ (if $K_3>0,\ U_4=0,\ U_5\neq 0$);

• n^* ; $\binom{3}{3}\, \widetilde{P}_{\curlywedge}\, \widehat{P}\, \widehat{P}\widehat{P}_{\curlywedge} - HH \Rightarrow (e = 0,\; f = 1)$ (if $K_3 > 0$, $U_4 = 0$, $U_5 = 0$).

As all the cases have been considered we have obtained 52 possible geometrically distinct global configurations of singularities of the family of quadratic systems with only one finite singularity, that in addition is elemental. □

Chapter 9

Quadratic systems with finite singularities of total multiplicity two

According to Proposition 6.1 for the quadratic systems having the finite singularities of total multiplicity two, the conditions $\mu_0 = \mu_1 = 0$ and $\mu_2 \neq 0$ must be satisfied. So by Theorem 6.4 the following lemma is valid.

Lemma 9.1. *The configurations of singularities at infinity of the family of quadratic systems possessing finite singularities (real or complex) of total multiplicity two (i.e.* $\mu_0 = \mu_1 = 0$ *and* $\mu_2 \neq 0$*) are classified in Diagram* 9.1 *according to the geometrical equivalence relation. Necessary and sufficient conditions for each one of the* 43 *different equivalence classes can be assembled from this diagram in terms of* 14 *invariant polynomials with respect to the action of the affine group and time rescaling, given in Section* 5.4.

9.1 Exactly one finite singularity

Theorem 9.1. (a) *We consider here all the global configurations of singularities of quadratic vector fields with finite singularities of total multiplicity* $m_f = 2$ *and exactly one finite singularity. These global configurations are classified in Diagram* 9.2 *according to the geometrical equivalence relation of configurations. We have a total of* 30 *geometrically distinct global configurations of singularities with one real finite singularity of multiplicity two.*

(b) *Necessary and sufficient conditions for each one of the* 30 *different equivalence classes can be assembled from this diagram in terms of* 17 *invariant*

J. C. Artés et al., *Geometric Configurations of Singularities of Planar Polynomial Differential Systems*, https://doi.org/10.1007/978-3-030-50570-7_9

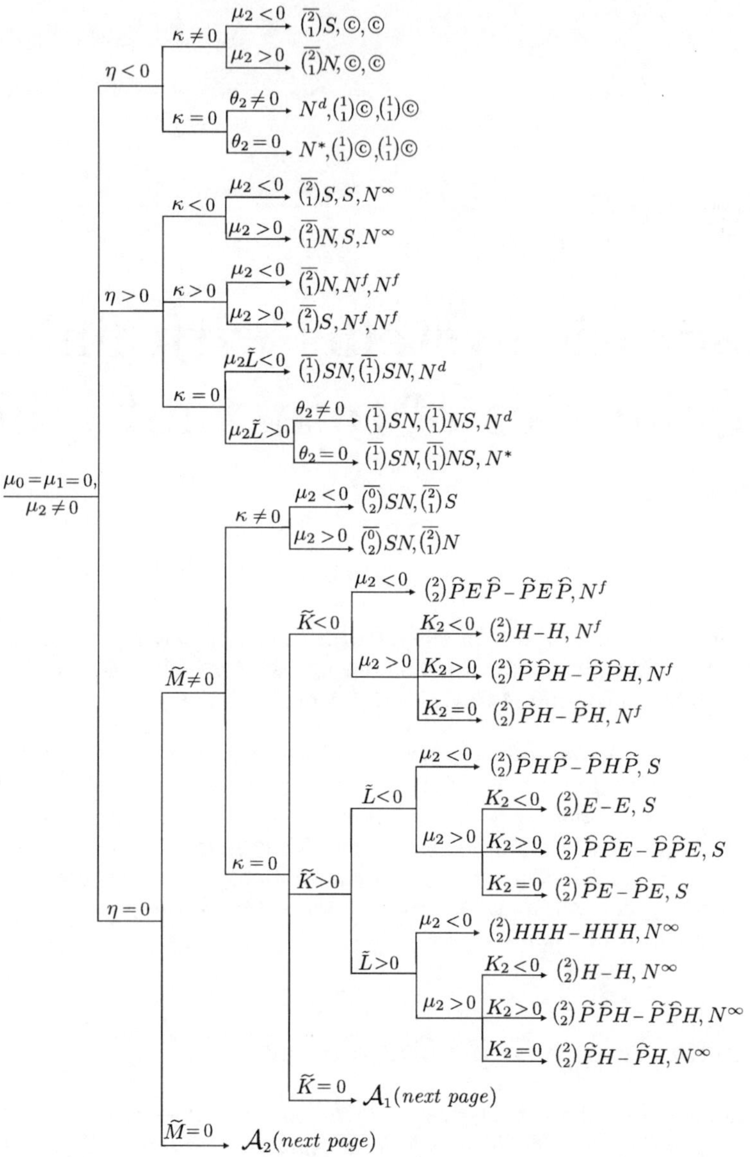

Diagram 9.1: Configurations of infinite singularities: the case $\mu_0=\mu_1=0$, $\mu_2\neq 0$.

polynomials with respect to the action of the affine group and time rescaling. These invariant polynomials, which appear in Diagram 9.2, *are given in Section* 5.4.

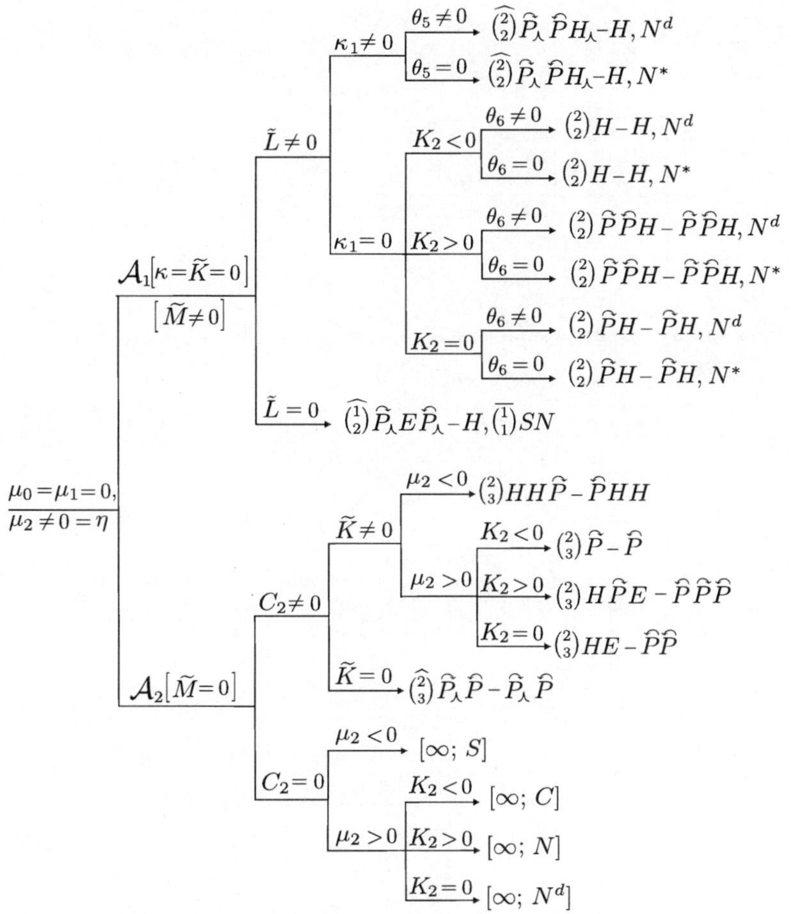

Diagram 9.1 (*cont.*): Configurations of infinite singularities: the case $\mu_0 = 0$, $\mu_1 = 0, \mu_2 \neq 0$.

(c) *Diagram 9.2 actually contains the global bifurcation diagram in the* 12*-dimensional space of parameters of the global configurations of singularities of this family of quadratic differential systems and provides an algorithm for finding for any given system in the family considered, its respective geometrical global configuration of singularities.*

Proof. Assuming that the quadratic systems (8.1) possess a double singular point, taking into account [338] and Proposition 6.2 (see Table 6.1) we have to consider two families of systems: 16a if $\widetilde{K} \neq 0$ and 16b if $\widetilde{K} = 0$.

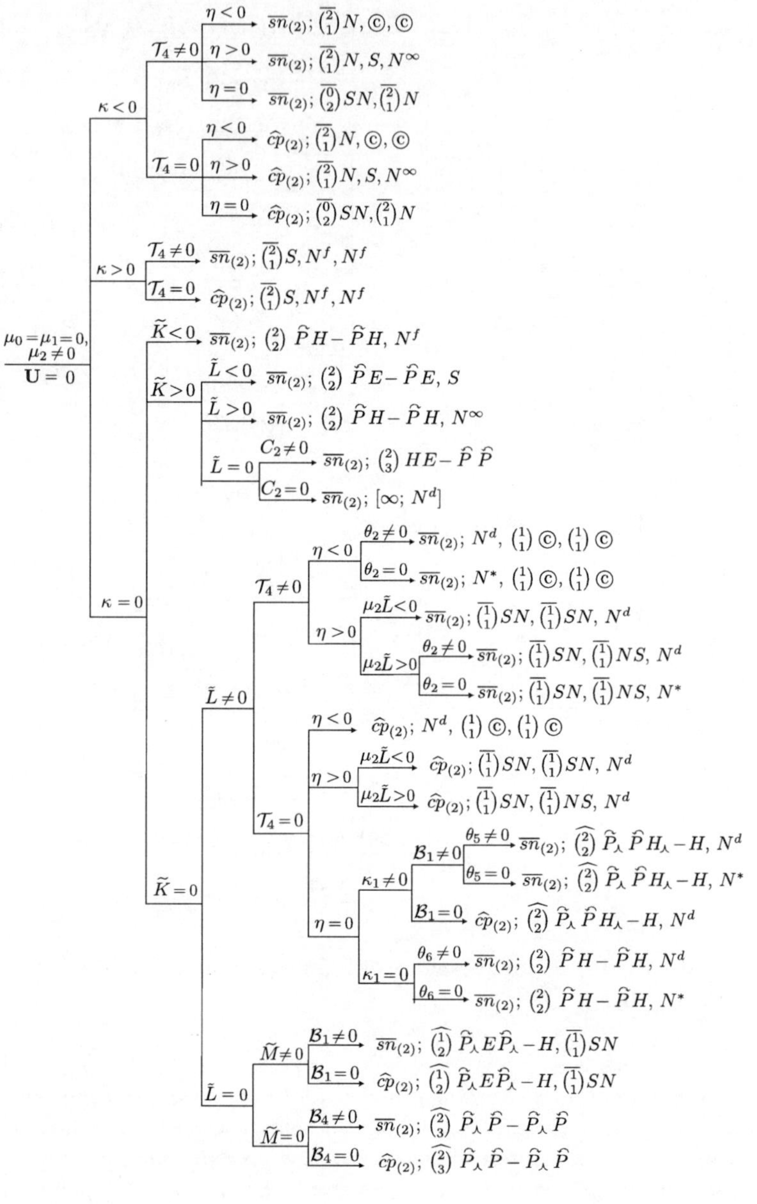

Diagram 9.2: Global configurations: the case $\mu_0 = \mu_1 = 0$, $\mu_2 \neq 0$, $\mathbf{U} = 0$.

1: $\widetilde{K} \neq 0$. In this case, following Proposition 6.2 (see Table 6.1), we consider the family of systems:

$$\dot{x} = dy + gx^2 + 2dxy, \quad \dot{y} = fy + lx^2 + 2fxy, \tag{9.1}$$

possessing the double singular point $M_{1,2}(0,0)$. For these systems calculations yield

$$\begin{gathered} \mu_0 = \mu_1 = 0, \quad \mu_2 = (dl - fg)^2 x^2, \\ \kappa = 128d^2(dl - fg), \quad \mathcal{T}_4 = 4d^2 f^2 (dl - fg)^2. \end{gathered} \tag{9.2}$$

Remark 9.1. We observe that the family of systems (9.1) depends on four parameters. However due to a rescaling we can reduce the number of the parameters to two. More precisely since by the condition $\mu_2 \neq 0$ we have $d^2 + f^2 \neq 0$, we may assume $(d,f) \in \{(1,1),(1,0),(0,1)\}$ due to the rescaling (i) $(x,y,t) \mapsto (x, fy/d, t/f)$ if $df \neq 0$, (ii) $(x,y,t) \mapsto (x, y/d, t)$ if $f = 0$ and (iii) $(x,y,t) \mapsto (x,y,t/f)$ if $d = 0$.

Considering (9.2) and $\mu_2 \neq 0$, the condition $d = 0$ is equivalent to $\kappa = 0$, and in the case $\kappa \neq 0$ the condition $f = 0$ is equivalent to $\mathcal{T}_4 = 0$.

1.1: $\kappa \neq 0$.

1.1.1: $\mathcal{T}_4 \neq 0$. Then we have $df \neq 0$, and considering Remark 9.1 we may assume $d = f = 1$. So we obtain the 2-parameter family of systems:

$$\dot{x} = y + gx^2 + 2xy, \quad \dot{y} = y + lx^2 + 2xy,$$

for which calculations yield

$$\begin{gathered} \mu_0 = \mu_1 = 0, \quad \mu_2 = (g-l)^2 x^2, \quad \widetilde{K} = 4(g-l)x^2, \quad G_8 = 2(g-l)^2, \\ \mathcal{T}_4 = 4(g-l)^2 = G_1, \quad \kappa = 128(l-g), \quad \eta = 4\big[(g+2)^2 + 8(l-g)\big], \\ \widetilde{M} = -8\big[(g-2)^2 + 6l\big]x^2 + 16(2-g)xy - 32y^2. \end{gathered}$$

Remark 9.2. We observe that $\widetilde{M} \neq 0$ and $\mu_2 > 0$. Moreover the condition $\kappa > 0$ implies $\eta > 0$.

As $G_8 G_1 \neq 0$ according to Theorem 6.1 (see Table 6.2, line 171) the double finite singular point is a saddle-node.

1.1.1.1: $\kappa < 0$. As $\widetilde{M} \neq 0$ and $\mu_2 > 0$, by Lemma 9.1 we get the following three global configurations:

- $\overline{sn}_{(2)};\ \overline{\binom{2}{1}}N, ©, © \Rightarrow (g = 0,\ l = -1)$ (if $\eta < 0$);
- $\overline{sn}_{(2)};\ \overline{\binom{2}{1}}N, S, N^{\infty} \Rightarrow (g = 0,\ l = -1/4)$ (if $\eta > 0$);
- $\overline{sn}_{(2)};\ \overline{\binom{0}{2}}SN, \overline{\binom{2}{1}}N \Rightarrow (g = 0,\ l = -1/2)$ (if $\eta = 0$).

1.1.1.2: $\kappa > 0$. By Remark 9.2 we have $\eta > 0$, and according to Lemma 9.1 we arrive at the global configuration

• $\overline{sn}_{(2)}$; $\overline{\binom{2}{1}}S, N^f, N^f \Rightarrow (g = 0,\ l = 1)$.

1.1.2: $\mathcal{T}_4 = 0$. Then we have $d \neq 0$, $f = 0$, and considering Remark 9.1 we may assume $d = 1$. So we obtain the 2-parameter family of systems:

$$\dot{x} = y + gx^2 + 2xy, \quad \dot{y} = lx^2,$$

for which calculations yield

$$\mu_0 = \mu_1 = 0, \quad \mu_2 = l^2x^2, \quad \widetilde{K} = -4lx^2, \quad G_8 = 2l^2, \quad G_1 = 0,$$
$$\kappa = 128l, \quad \eta = 4(8l + g^2), \quad \widetilde{M} = -8(6cl + g^2)x^2 - 16gxy - 32y^2.$$

As $G_8 \neq 0$ and $G_1 = 0$, according to Theorem 6.1 (see Table 6.2, line 175) the double finite singular point is a cusp.

1.1.2.1: $\kappa < 0$. As $\widetilde{M} \neq 0$ and $\mu_2 > 0$, by Lemma 9.1 we get the following three global configurations:

• $\widehat{cp}_{(2)}$; $\overline{\binom{2}{1}}N$, ©, © $\Rightarrow (g = 2,\ l = -1)$ (if $\eta < 0$);
• $\widehat{cp}_{(2)}$; $\overline{\binom{2}{1}}N, S, N^{\infty} \Rightarrow (g = 2,\ l = -1/4)$ (if $\eta > 0$);
• $\widehat{cp}_{(2)}$; $\overline{\binom{0}{2}}SN, \overline{\binom{2}{1}}N \Rightarrow (g = 2,\ l = -1/2)$ (if $\eta = 0$).

1.1.2.2: $\kappa > 0$. We observe that the condition $\kappa > 0$ implies $l > 0$ and then $\eta = 4(8l+g^2) > 0$. So considering Lemma 9.1 we arrive at the global configuration

• $\widehat{cp}_{(2)}$; $\overline{\binom{2}{1}}S, N^f, N^f \Rightarrow (g = 1,\ l = 1)$.

1.2: $\kappa = 0$. Then for systems (9.1) we have $d = 0$, and by Remark 9.1 we may assume $f = 1$. So we get the family of systems:

$$\dot{x} = gx^2, \quad \dot{y} = y + lx^2 + 2xy,$$

for which calculations yield

$$\mu_0 = \mu_1 = 0, \quad \mu_2 = g^2x^2, \quad \widetilde{K} = 4gx^2, \quad \eta = \kappa = G_8 = 0, \quad K_2 = 0,$$
$$\tilde{L} = 8g(g-2)x^2, \quad \widetilde{M} = -8(g-2)^2x^2, \quad C_2 = -lx^3 + (g-2)x^2y.$$

As $G_8 = 0$ and $\widetilde{K} \neq 0$ according to Theorem 6.1 (see Table 6.2, line 172), the double finite singular point is a saddle-node.

1.2.1: $\widetilde{K} < 0$. Then $g < 0$ and this implies $\widetilde{M} \neq 0$. As $\mu_2 > 0$ and $K_2 = 0$, considering Lemma 9.1 we obtain the global configuration

• $\overline{sn}_{(2)}$; $\binom{2}{2}\ \widehat{P}H - \widehat{P}H,\ N^f \Rightarrow (g = -1,\ l = 0)$.

1.2.2: $\widetilde{K} > 0$. In view of Lemma 9.1 we consider two possibilities: $\tilde{L} \neq 0$ and $\tilde{L} = 0$.

1.2.2.1: $\tilde{L} \neq 0$. Then $g - 2 \neq 0$ and we have $\widetilde{M} \neq 0$. So taking into account the conditions $\mu_2 > 0$ and $K_2 = 0$, by Lemma 9.1 we get the following two global configurations:

• $\overline{sn}_{(2)}$; $\binom{2}{2}$ $\widehat{P}E-\widehat{P}E$, $S \Rightarrow (g=1,\ l=0)$ (if $\tilde{L}<0$);
• $\overline{sn}_{(2)}$; $\binom{2}{2}$ $\widehat{P}H-\widehat{P}H$, $N^{\infty} \Rightarrow (g=3,\ l=0)$ (if $\tilde{L}>0$).

1.2.2.2: $\tilde{L}=0$. In this case $g=2$ and this implies $\widetilde{M}=0$. As $\mu_2>0$ and $K_2=0$, by Lemma 9.1 we arrive at the following two global configurations:
• $\overline{sn}_{(2)}$; $\binom{2}{3}$ $HE-\widehat{P}\,\widehat{P} \Rightarrow (g=2,\ l=1)$ (if $C_2\neq 0$);
• $\overline{sn}_{(2)}$; $[\infty;\ N^d] \Rightarrow (g=2,\ l=0)$ (if $C_2=0$).

2: $\widetilde{K}=0$. As it was mentioned earlier, in this case we consider the family of systems 16b (see Table 6.1), i.e. the systems:

$$\dot{x}=cx+dy,\quad \dot{y}=lx^2+2mxy+ny^2,\quad 0\neq c^2n-2cdm+d^2l\equiv \mathcal{Z}, \tag{9.3}$$

for which we calculate

$$\begin{gathered}\mu_0=\mu_1=\kappa=0,\quad \mu_2=\mathcal{Z}(lx^2+2mxy+ny^2),\quad \widetilde{K}=0,\\ \tilde{L}=8n(lx^2+2mxy+ny^2),\quad \eta=4n^2(m^2-ln),\\ \theta_2=-d(m^2-ln),\quad \widetilde{M}=-8(4m^2-3ln)x^2-16mnxy-8n^2y^2,\\ G_8=2n(ln-m^2)\mathcal{Z},\quad G_1=4c^2n(ln-m^2)\mathcal{Z}=\mathcal{T}_4.\end{gathered} \tag{9.4}$$

2.1: $\tilde{L}\neq 0$. We consider two subcases: $\eta\neq 0$ and $\eta=0$.

2.1.1: $\eta\neq 0$. In this case $G_8\neq 0$ and we examine two possibilities: $\mathcal{T}_4\neq 0$ and $\mathcal{T}_4=0$.

2.1.1.1: $\mathcal{T}_4\neq 0$. Then we have $cn\neq 0$, and due to the rescaling $(x,y,t)\mapsto(x,cy/n,t/c)$, we may assume $c=n=1$. So we get the family of systems:

$$\dot{x}=x+dy,\quad \dot{y}=lx^2+2mxy+y^2,\quad 0\neq 1-2dm+d^2l\equiv \mathcal{Z}',$$

where we may consider $d\in\{0,1\}$ due to the rescaling $x\to dx$ if $d\neq 0$. For these systems we have

$$\begin{gathered}\mathcal{T}_4=4(l-m^2)\mathcal{Z}'=G_1=2G_8,\quad \eta=4(m^2-l),\quad \theta_2=d(l-m^2),\\ \mu_2=\mathcal{Z}'(lx^2+2mxy+y^2),\quad \tilde{L}=8(lx^2+2mxy+y^2).\end{gathered}$$

As $G_8G_1\neq 0$ according to Theorem 6.1 (see Table 6.2, line 171) the finite singular point is a saddle-node.

2.1.1.1.1: $\eta<0$. Considering Lemma 9.1 we get the following two global configurations:
• $\overline{sn}_{(2)}$; N^d, $\binom{1}{1}$ ©, $\binom{1}{1}$ © $\Rightarrow (d=1,\ l=1,\ m=0)$ (if $\theta_2\neq 0$);
• $\overline{sn}_{(2)}$; N^*, $\binom{1}{1}$ ©, $\binom{1}{1}$ © $\Rightarrow (d=0,\ l=1,\ m=0)$ (if $\theta_2=0$).

2.1.1.1.2: $\eta > 0$. We observe that $\operatorname{sign}(\mu_2\tilde{L}) = \operatorname{sign}(\mathcal{Z}')$. Moreover if $d = 0$ we obtain $\mathcal{Z}' = 1 > 0$. So considering Lemma 9.1 we arrive at the following three global configurations:

- $\overline{sn}_{(2)};\ \overline{\binom{1}{1}}SN,\ \overline{\binom{1}{1}}SN,\ N^d \Rightarrow (d=1,\ l=0,\ m=1)\quad (\text{if } \mu_2\tilde{L} < 0)$;
- $\overline{sn}_{(2)};\ \overline{\binom{1}{1}}SN,\ \overline{\binom{1}{1}}NS,\ N^d \Rightarrow (d=1,\ l=0,\ m=-1)\quad (\text{if } \mu_2\tilde{L} > 0,\ \theta_2 \neq 0)$;
- $\overline{sn}_{(2)};\ \overline{\binom{1}{1}}SN,\ \overline{\binom{1}{1}}NS,\ N^* \Rightarrow (d=0,\ l=0,\ m=1)\quad (\text{if } \mu_2\tilde{L} > 0,\ \theta_2 = 0)$.

2.1.1.2: $\mathcal{T}_4 = 0$. By (9.4) due to the condition $\eta \neq 0$ we obtain $c = 0$ and then $dln \neq 0$. So via the rescaling $(x, y, t) \mapsto (dx/n, y/n, t)$ we may assume $n = d = 1$, and we arrive at the family of systems:

$$\dot{x} = y, \quad \dot{y} = lx^2 + 2mxy + y^2.$$

We observe that we may assume $m \in \{0, 1\}$ due to the rescaling $(x, y, t) \mapsto (x, my, t/m)$ if $m \neq 0$. For these systems we calculate

$$G_1 = 0, \quad G_8 = 2l(l - m^2), \quad \eta = 4(m^2 - l), \quad \theta_2 = l - m^2,$$
$$\mu_2 = l(lx^2 + 2mxy + y^2), \quad \tilde{L} = 8(lx^2 + 2mxy + y^2).$$

So the condition $\eta\mu_2 \neq 0$ implies $G_8 \neq 0$, and due to $G_1 = 0$ by Theorem 6.1 (see Table 6.2, line 175) the double finite singular point is a cusp.

Additionally we have $\theta_2 \neq 0$ and $\operatorname{sign}(\mu_2\tilde{L}) = \operatorname{sign}(l)$. So considering Lemma 9.1 we get the following three global configurations:

- $\widehat{cp}_{(2)};\ N^d,\ \binom{1}{1}$ ©, $\binom{1}{1}$ © $\Rightarrow (l=1,\ m=0)\quad (\text{if } \eta < 0)$;
- $\widehat{cp}_{(2)};\ \overline{\binom{1}{1}}SN,\ \overline{\binom{1}{1}}SN,\ N^d \Rightarrow (l=-1,\ m=1)\quad (\text{if } \eta > 0,\ \mu_2\tilde{L} < 0)$;
- $\widehat{cp}_{(2)};\ \overline{\binom{1}{1}}SN,\ \overline{\binom{1}{1}}NS,\ N^d \Rightarrow (l=1,\ m=2)\quad (\text{if } \eta > 0,\ \mu_2\tilde{L} > 0)$.

2.1.2: $\eta = 0$. As $n \neq 0$ (due to $\tilde{L} \neq 0$) we may assume $n = 1$ due to a rescaling. So considering (9.4) the condition $\eta = 0$ gives $m^2 - l = 0$, and we obtain $l = m^2$. Then for systems (9.3) we have

$$\mu_2 = (c - dm)^2(mx + y)^2, \quad \kappa_1 = 32m(c - dm),$$

and as $\mu_2 \neq 0$ the condition $\kappa_1 = 0$ is equivalent to $m = 0$.

2.1.2.1: $\kappa_1 \neq 0$. Then $m \neq 0$ and we may assume $m = 1$ due to the rescaling $x \to x/m$. Therefore we arrive at the family of systems:

$$\dot{x} = cx + dy, \quad \dot{y} = (x + y)^2,$$

for which we calculate

$$\begin{gathered}\eta = \kappa = G_8 = 0, \quad \mu_2 = (c - d)^2(x + y)^2, \quad \widetilde{K} = \widetilde{N} = 0,\\ \tilde{L} = 8(x + y)^2, \quad \widetilde{M} = -8(x + y)^2, \quad \theta_5 = 96(c - d)d(x + y)^3,\\ \kappa_1 = 32(c - d), \quad F_3 = 24c(c - d)(x + y), \quad \mathcal{B}_1 = 2c^2(c - d)^2.\end{gathered} \tag{9.5}$$

2.1.2.1.1: $\mathcal{B}_1 \neq 0$. Then $c \neq 0$ and we observe that in this case $F_3 \neq 0$, and as $G_8 = \widetilde{K} = \widetilde{N} = 0$, according to Theorem 6.1 (see Table 6.2, line 174), the finite singular point is a saddle-node.

Additionally we have $\eta = \kappa = \widetilde{K} = 0$ and $\widetilde{M}\tilde{L}\kappa_1 \neq 0$. Therefore considering Lemma 9.1 we obtain the following two global configurations:

- $\overline{sn}_{(2)};\ \widehat{\binom{2}{2}}\ \widehat{P}_{\curlywedge}\ \widehat{P}H_{\curlywedge}-H,\ N^d \Rightarrow (c=1,\ d=2)$ (if $\theta_5 \neq 0$);
- $\overline{sn}_{(2)};\ \widehat{\binom{2}{2}}\ \widehat{P}_{\curlywedge}\ \widehat{P}H_{\curlywedge}-H,\ N^* \Rightarrow (c=1,\ d=0)$ (if $\theta_5 = 0$).

2.1.2.1.2: $\mathcal{B}_1 = 0$. Then $c = 0$ and this implies $F_3 = 0$. So by Theorem 6.1 (see Table 6.2, line 177) the double finite singular point is a cusp. As in this case $\theta_5 = -96d^2(x+y)^3 \neq 0$ (due to $\mu_2 = d^2(x+y)^2 \neq 0$), considering (9.5) and Lemma 9.1 we get the global configuration

- $\widehat{cp}_{(2)};\ \widehat{\binom{2}{2}}\ \widehat{P}_{\curlywedge}\ \widehat{P}H_{\curlywedge}-H,\ N^d \Rightarrow (c=0,\ d=1)$.

2.1.2.2: $\kappa_1 = 0$. In this case we have $m = 0$, and this leads to the family of systems:

$$\dot{x} = cx + dy, \quad \dot{y} = y^2,$$

for which we calculate $\mu_2 = c^2y^2,\ F_3 = 24c^2y,\ K_2 = 0,\ \theta_6 = -8dy^4$. Therefore the condition $\mu_2 \neq 0$ implies $F_3 \neq 0$. So the double finite singularity is a saddle-node, and considering Lemma 9.1 we obtain the following two global configurations:

- $\overline{sn}_{(2)};\ \binom{2}{2}\ \widehat{P}H - \widehat{P}H,\ N^d \Rightarrow (c=1, d=1)$ (if $\theta_6 \neq 0$);
- $\overline{sn}_{(2)};\ \binom{2}{2}\ \widehat{P}H - \widehat{P}H,\ N^* \Rightarrow (c=1, d=0)$ (if $\theta_6 = 0$).

2.2: $\tilde{L} = 0$. Then for systems (9.3) we have $n = 0$ and then $d \neq 0$ (otherwise we get degenerate systems). So we may assume $d = 1$ (due to a rescaling), and we obtain the family of systems:

$$\dot{x} = cx + y, \qquad \dot{y} = lx^2 + 2mxy, \tag{9.6}$$

for which we calculate

$$\begin{aligned}\eta = \kappa = G_8 = \widetilde{K} = \tilde{L} = 0,\ \ \mu_2 = (l-2cm)x(lx+2my),\ \ \widetilde{N} = -4m^2x^2,\\ \widetilde{M} = -32m^2x^2,\ \ G_{10} = c^2m^3(l-2cm),\ \ \mathcal{B}_1 = 2c^2m(l-2cm).\end{aligned} \tag{9.7}$$

2.2.1: $\widetilde{M} \neq 0$. Then $m \neq 0$ and we may assume $m = 1$ due to the rescaling $(x, y, t) \mapsto (x/m, y/m, t)$. Moreover for systems (9.6) with $m = 1$, we may consider $c \in \{0, 1\}$ due to the rescaling $(x, y, t) \mapsto (cx, c^2y, t/c)$ if $c \neq 0$.

Therefore we have $\widetilde{N} \neq 0$, and by Theorem 6.1 (see Table 6.2, lines 173 and 176) the finite singular point is a saddle-node if $G_{10} \neq 0$ and a cusp if $G_{10} = 0$. We observe that the condition $G_{10} = 0$ is equivalent to $\mathcal{B}_1 = 0$. So as $\widetilde{M} \neq 0$ and $\tilde{L} = 0$, considering Lemma 9.1 we obtain the following two global configurations:

- $\overline{sn}_{(2)};\ \widehat{\binom{1}{2}}\ \widehat{P}_{\curlywedge}E\widehat{P}_{\curlywedge}-H,\ \overline{\binom{1}{1}}SN \Rightarrow (c=1,\ m=1,\ l=1)$ (if $\mathcal{B}_1 \neq 0$);

- $\widehat{cp}_{(2)}$; $\widehat{\binom{1}{2}}\ \widehat{\widetilde{P}}_{\curlywedge} E \widehat{P}_{\curlywedge} - H,\ \overline{\binom{1}{1}}SN \Rightarrow (c=0,\ m=1,\ l=1)$ (if $\mathcal{B}_1 = 0$).

2.2.2: $\widetilde{M} = 0$. Then $m = 0$ and $l \neq 0$ (otherwise we get degenerate systems). Hence we may assume $l = 1$ due to the rescaling $(x, y, t) \mapsto (x/l, y/l, t)$. Then for systems (9.6) with $m = 0$ and $l = 1$, considering (9.7) we obtain

$$\widetilde{M} = \widetilde{K} = \widetilde{N} = G_8 = 0, \quad C_2 = -x^3, \quad F_3 = -24cx, \quad \mathcal{B}_4 = 6cx^2(cx + y).$$

So by Theorem 6.1 (see Table 6.2, lines 174 and 177) the finite singular point is a saddle-node if $F_3 \neq 0$ and a cusp if $F_3 = 0$. We observe that the condition $F_3 = 0$ is equivalent to $\mathcal{B}_4 = 0$. Considering the above conditions, according to Lemma 9.1 we obtain the following two global configurations:

- $\overline{sn}_{(2)}$; $\widehat{\binom{2}{3}}\ \widehat{\widetilde{P}}_{\curlywedge} \widehat{P} - \widehat{\widetilde{P}}_{\curlywedge} \widehat{P} \Rightarrow (c=1,\ m=0,\ l=1)$ (if $\mathcal{B}_4 \neq 0$);
- $\widehat{cp}_{(2)}$; $\widehat{\binom{2}{3}}\ \widehat{\widetilde{P}}_{\curlywedge} \widehat{P} - \widehat{\widetilde{P}}_{\curlywedge} \widehat{P} \Rightarrow (c=0,\ m=0,\ l=1)$ (if $\mathcal{B}_4 = 0$). □

9.2 Two distinct real singularities

Theorem 9.2. (a) *We consider here all the global configurations of singularities of quadratic vector fields with exactly two distinct real finite singularities that in addition are elemental. These global configurations are classified in Diagram* 9.3 *according to the geometrical equivalence relation of configurations. We have* 151 *geometrically distinct global configurations of singularities.*

(b) *Necessary and sufficient conditions for each one of the* 151 *different equivalence classes can be assembled from this diagram in terms of* 31 *invariant polynomials with respect to the action of the affine group and time rescaling. These invariant polynomials, which appear in Diagram* 9.3, *are given in Section* 5.4.

(c) *Diagram* 9.3 *actually contains the global bifurcation diagram in the* 12*-dimensional space of parameters of the global configurations of singularities of this family of quadratic differential systems and provides an algorithm for finding for any given system in the family considered, its respective geometrical global configuration of singularities.*

Proof. Assume that the quadratic systems (8.1) possess two real finite singular points and both are elemental, i.e. by Proposition 5.1 and Remark 6.1 the conditions $\mu_0 = \mu_1 = 0$, $\mu_2 \neq 0$ and $\mathbf{U} > 0$ hold. Taking into account [338] and Proposition 6.2 (see Table 6.1) we have to consider two families of systems: 14a if $\widetilde{K} \neq 0$ and 14b if $\widetilde{K} = 0$.

1: $\widetilde{K} \neq 0$. In this case we consider the family of systems 14a, i.e. the systems:

$$\dot{x} = cx + dy - cx^2 + 2duxy, \qquad \dot{y} = ex + fy - ex^2 + 2fuxy, \tag{9.8}$$

which possess the singular points $M_1(0,0)$ and $M_2(1,0)$.

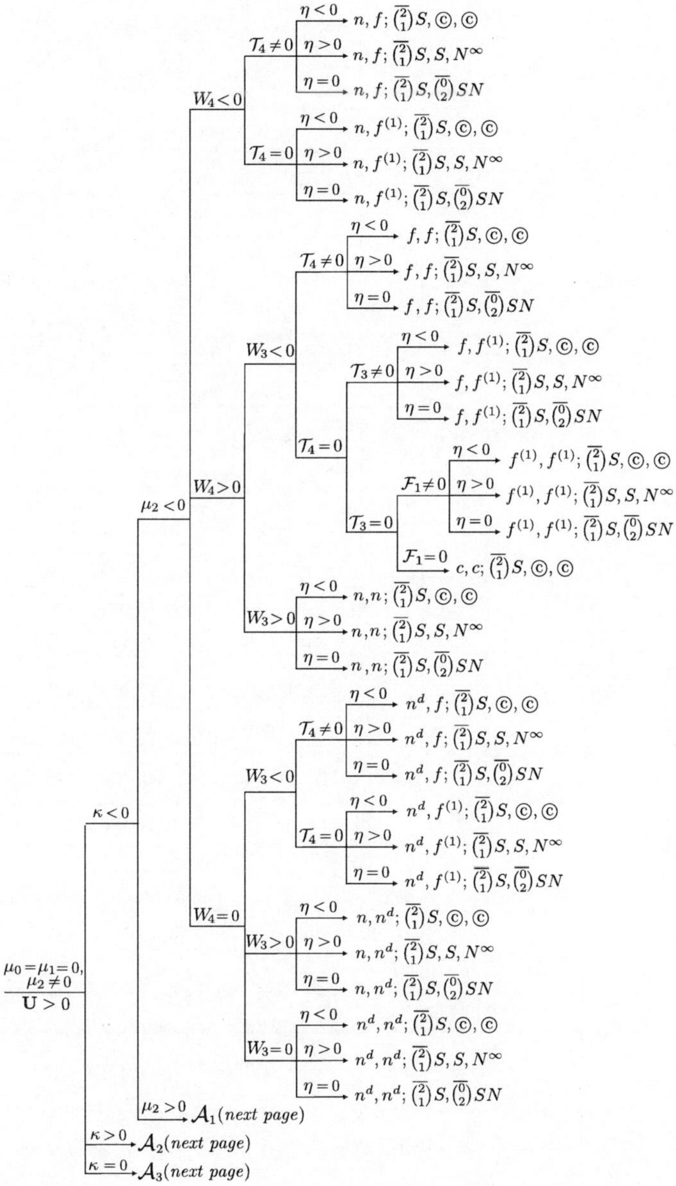

Diagram 9.3: Global configurations: the case $\mu_0 = \mu_1 = 0$, $\mu_2 \neq 0$, $\mathbf{U} > 0$.

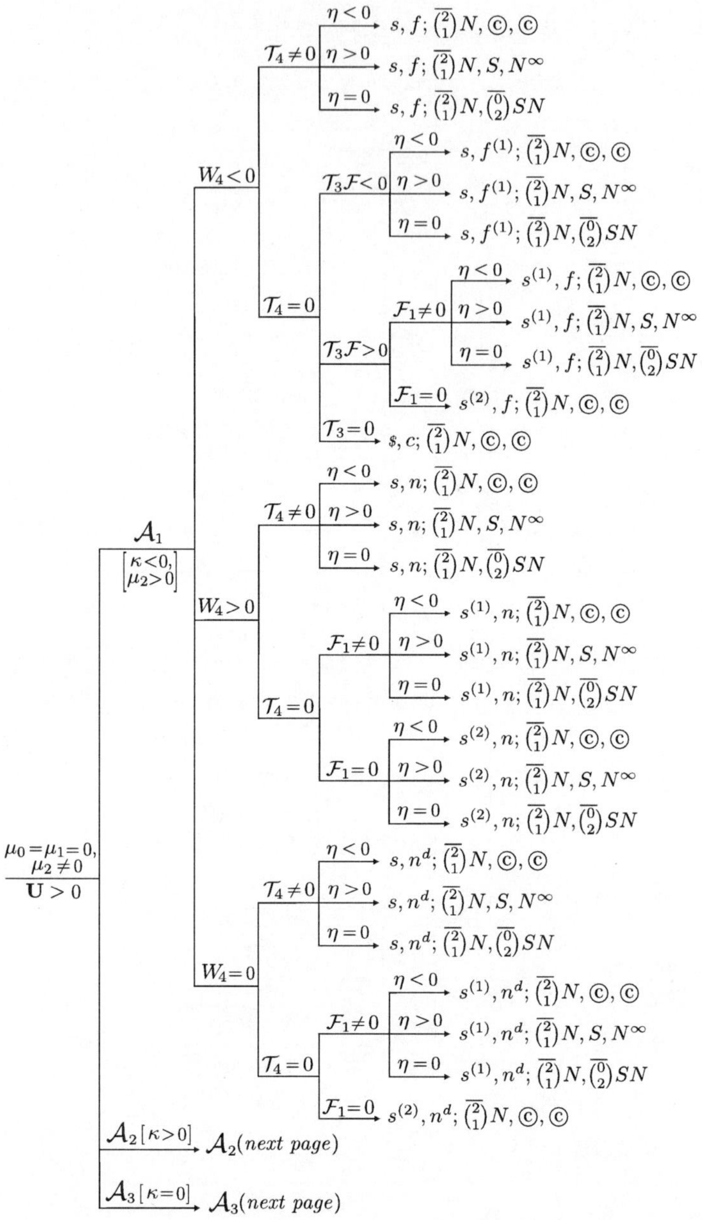

Diagram 9.3 (*cont.*): Global configurations: the case $\mu_0=\mu_1=0$, $\mu_2\neq0$, $\mathbf{U}>0$.

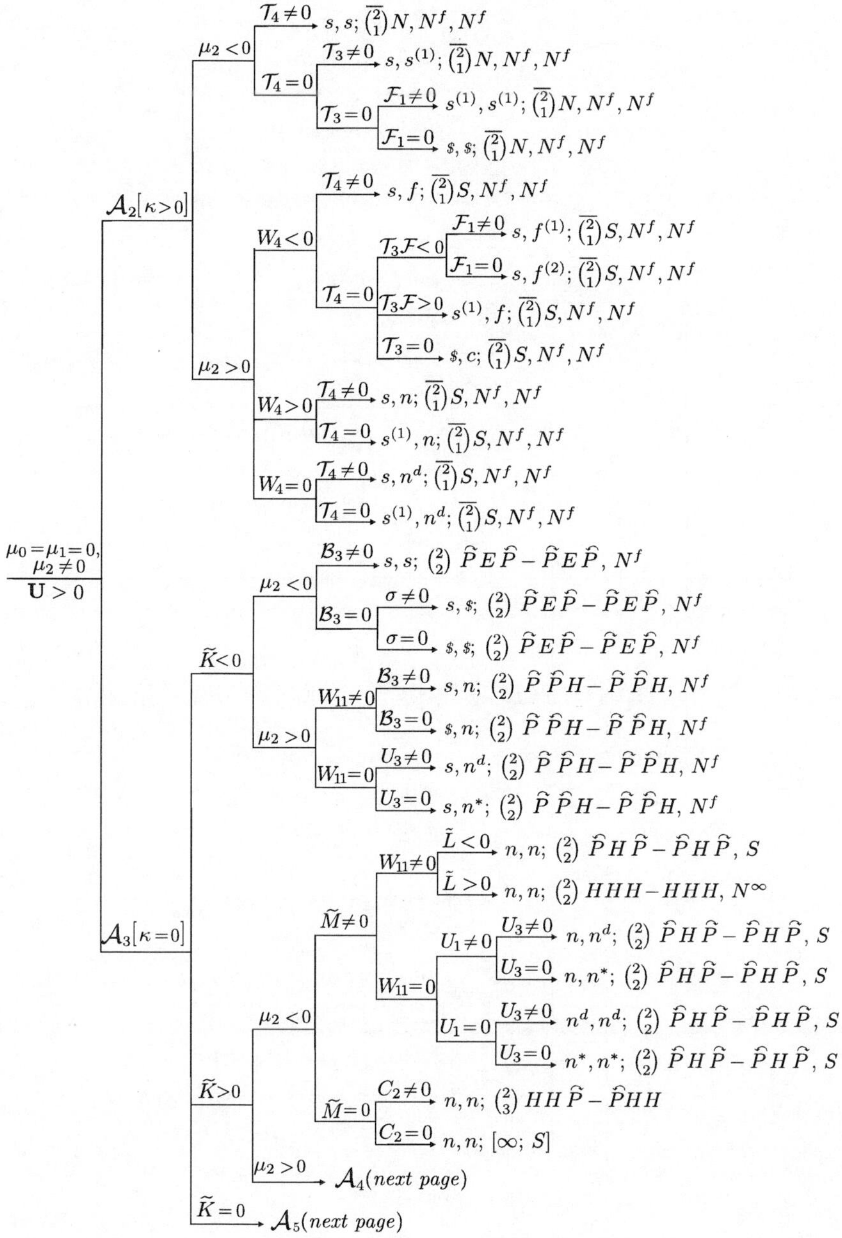

Diagram 9.3 (*cont.*): Global configurations: the case $\mu_0=\mu_1=0,\ \mu_2\neq 0,\ \mathbf{U}>0$.

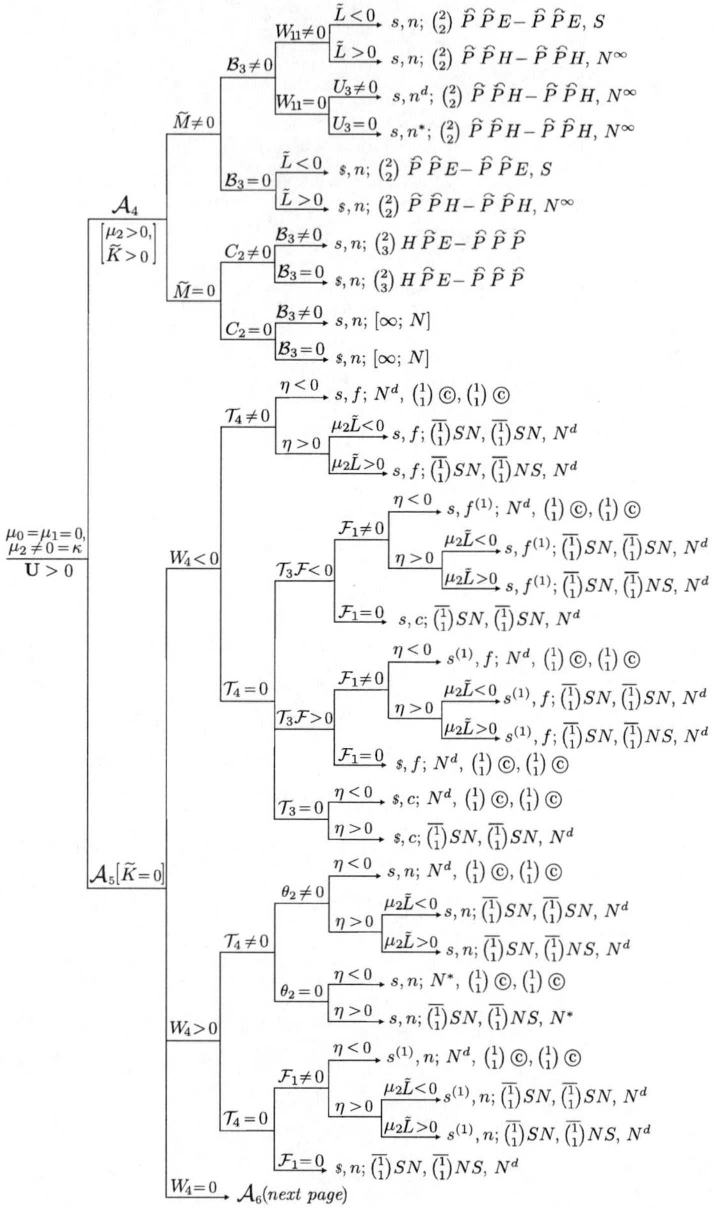

Diagram 9.3 (*cont.*): Global configurations: the case $\mu_0=\mu_1=0,\ \mu_2\neq 0,\ \mathbf{U}>0$.

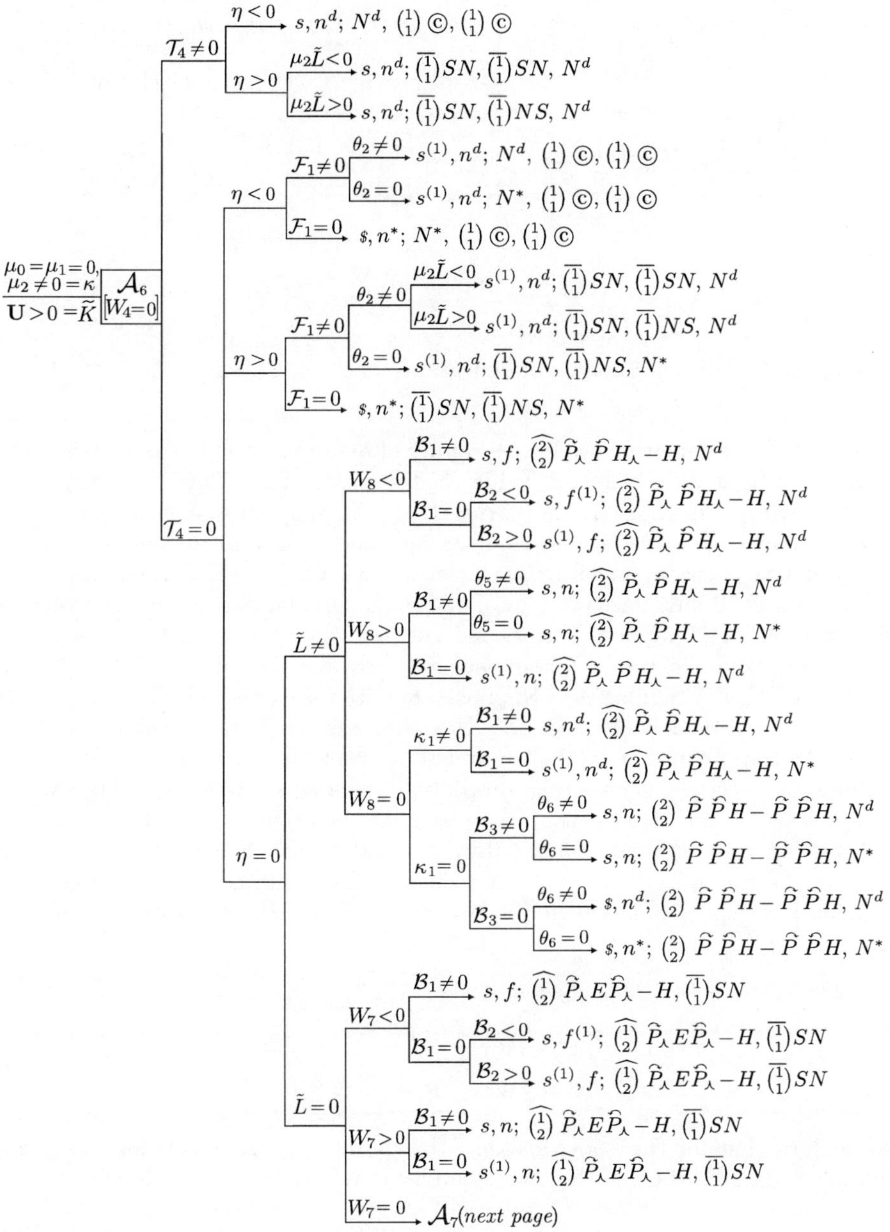

Diagram 9.3 (*cont.*): Global configurations: the case $\mu_0=\mu_1=0,\ \mu_2\neq 0,\ \mathbf{U}>0$.

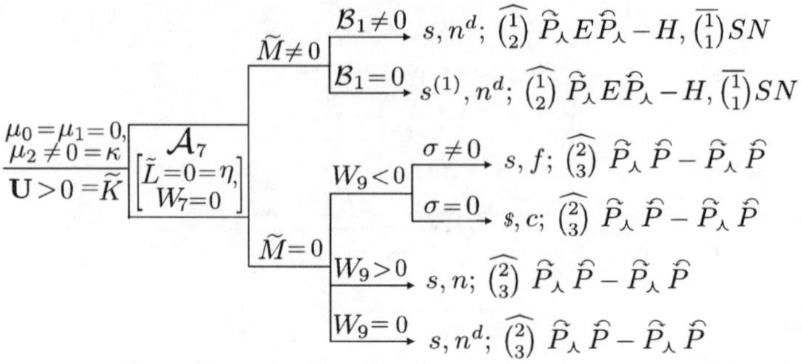

Diagram 9.3 (*cont.*): Global configurations: the case $\mu_0=\mu_1=0$, $\mu_2\neq 0$, $\mathbf{U}>0$.

Remark 9.3. Assume that we have a family of quadratic systems possessing a real elemental singular point for all values of the parameters. Then by a translation of axes (which depends on the parameters), we may suppose this point to be placed at the origin. In the case that we have one other real elemental singularity (or even two such singularities), we can always use a linear transformation to place the second singularity (or even two such singularities) in specific positions (for example at $(1,0)$) and if a second such singular point exists, to place it at $(0,1)$. We arrive thus at a certain normal form for the family, dictated by the position of these singularities. Suppose that in the course of the study of this family, under certain conditions on parameters expressed in invariant form, i.e. in terms of invariant polynomials, we find that an elemental real singularity of the systems has a certain geometrical property, for example it is a node. Then we may always suppose this singularity to be placed at the origin. This is clear if we have just one real elemental singularity. If we have other real elemental singular points, then by the argument above we can exchange its position with one of the other elemental singular points via an affine transformation without changing the aspect of the normal form.

For systems (9.8) calculations yield

$$\begin{aligned} &\mu_0 = \mu_1 = 0, \quad \mu_2 = (cf-de)^2(1+2u)x^2, \\ &\widetilde{K} = 4(de-cf)ux^2, \quad \kappa = 128d^2(cf-de)u^3. \end{aligned} \tag{9.9}$$

We remark that for the above systems the condition $\mu_2\widetilde{K}\neq 0$ holds. So in what follows we assume that the following condition is satisfied:

$$(cf-de)(1+2u)u \neq 0. \tag{9.10}$$

Remark 9.4. We observe that the family of systems (9.8) depends on five parameters. However due to a rescaling we can reduce the number of the parameters to three. More precisely since according to condition (9.10) we have $d^2+f^2\neq 0$, we may assume $(d, f) \in \{(1,1),(1,0),(0,1)\}$ due to the rescaling (i) $(x, y, t) \mapsto (x, fy/d, t/f)$ if $df \neq 0$, (ii) $(x, y, t) \mapsto (x, y/d, t)$ if $f = 0$ and (iii) $(x, y, t) \mapsto (x, y, t/f)$ if $d = 0$.

Considering (9.9) and (9.10), the condition $d \neq 0$ is equivalent to $\kappa \neq 0$.

1.1: $\kappa \neq 0$. Then considering Remark 9.4 we examine the subfamilies of systems (9.8) with $(d, f) = (1, 1)$ and $(d, f) = (1, 0)$.

1.1.1: $(d, f) = (1, 1)$. We consider the 3-parameter family of systems:

$$\begin{aligned} \dot{x} &= cx + y - cx^2 + 2uxy, \quad (c-e)(1+2u)u \neq 0, \\ \dot{y} &= ex + y - ex^2 + 2uxy, \end{aligned} \tag{9.11}$$

for which calculations yield

$$\begin{aligned} &\mu_0 = \mu_1 = 0, \ \mu_2 = (c-e)^2(1+2u)x^2, \ \widetilde{K} = 4(e-c)ux^2, \\ &\kappa = 128(c-e)u^3, \ \eta = 4u^2\big[(c-2u)^2 + 8(c-e)u\big], \\ &F_2 = (c-e)^2(1+2u)x^2, \ G_8 = 2(c-e)^2u^2(1+2u), \\ &W_4 = 16(c-e)^4u^4(1+2u)^2\big[(1-c+2u)^2 + 4(c-e)(1+2u)\big] \\ &\times\big[(c-1)^2 + 4e\big], \ \mathcal{T}_4 = 4(1+c)(c-e)^2(1-c+2u)u^2(1+2u), \\ &W_3 = 32(c-e)^4u^4(1+2u)^2(1+c^2+2u+2cu-4eu+2u^2), \\ &\widetilde{M} = -8\big[(c+2u)^2 - 6eu\big])x^2 + 16u(c+2u)xy - 32u^2y^2. \end{aligned} \tag{9.12}$$

Remark 9.5. We observe that the condition $\mu_2\kappa \neq 0$ gives $\widetilde{M}F_2G_8 \neq 0$ and the condition $\kappa > 0$ implies $\eta > 0$. Moreover we have

$$\operatorname{sign}(\widetilde{K}) = -\operatorname{sign}(\kappa), \quad \operatorname{sign}(G_8) = \operatorname{sign}(\mu_2) = \operatorname{sign}(F_2).$$

1.1.1.1: $\kappa < 0$. Then by Remark 9.5 we obtain $\widetilde{K} > 0$.

1.1.1.1.1: $\mu_2 < 0$. Then $1 + 2u < 0$, i.e. $u < -1/2$, and considering (9.12) we obtain $G_8 < 0$. So since $\widetilde{K} > 0$, according to Theorem 6.1 (see Table 6.2, lines 165–169) both finite singularities are anti-saddles.

1.1.1.1.1.1: $W_4 < 0$. Then we have a node and a focus, and whether the focus is a weak one or not depends on the invariant polynomial $\mathcal{T}_4$. Additionally due to $W_4 \neq 0$ we have a generic node.

1.1.1.1.1.1.1: $\mathcal{T}_4 \neq 0$. Then by Theorem 6.2 the focus is strong.

1. $\eta < 0$. Then at infinity we have one real and two complex singularities, and as $\mu_2 < 0$ and $\kappa \neq 0$, considering Lemma 9.1 we get the global configuration
   - $n, f; \overline{\binom{2}{1}}S, ©, © \Rightarrow (c = 5, \ e = -1, \ u = -2)$.

2. $\eta > 0$. In this case at infinity we have three real singularities. As $\kappa < 0$ and $\mu_2 < 0$, by Lemma 9.1 we get the global configuration

   - $n, f; \overline{\binom{2}{1}}S, S, N^{\infty} \Rightarrow (c = 1/2,\ e = -1/5,\ u = -2)$.

3. $\eta = 0$. In this case considering Remark 9.5 we have $\widetilde{M} \neq 0$. As $\kappa \neq 0$ and $\mu_2 < 0$, considering Lemma 9.1 we get the global configuration

   - $n, f; \overline{\binom{0}{2}}SN, \overline{\binom{2}{1}}S \Rightarrow (c = 3,\ e = -1/16,\ u = -2)$.

1.1.1.1.1.1.2: $\mathcal{T}_4 = 0$. Then by Theorem 6.2 the focus is weak. Considering (9.12) and the condition (9.10), the condition $\mathcal{T}_4 = 0$ gives $(c+1)(2u+1-c) = \rho_1\rho_2 = 0$. By Remark 9.3 we may assume without loss of generality that $\rho_1 = 0$, i.e. $c = -1$. Then for systems (9.11) we calculate

$$\begin{aligned} \mathcal{T}_3 &= 8(1+e)^2u^2(1+u)(1+2u), \quad \mathcal{F}_1 = 2(1+e)(u-1)(1+2u),\\ \mu_2 &= (1+e)^2(1+2u)x^2, \quad \kappa = -128(1+e)u^3, \\ W_4 &= -256(1+e)^5u^4(1+2u)^2(e+2eu-u^2). \end{aligned} \tag{9.13}$$

We observe that the condition $\mu_2 < 0$ implies $\mathcal{F}_1 \neq 0$. Thus by Theorem 6.2 the weak focus has order one. Therefore according to Lemma 9.1 we get the following three global configurations:

- $n, f^{(1)}; \overline{\binom{2}{1}}S, ©, © \Rightarrow (c = -1, e = -3/2, u = -6/10)$ (if $\eta < 0$);
- $n, f^{(1)}; \overline{\binom{2}{1}}S, S, N^{\infty} \Rightarrow (c = -1, e = -51/50, u = -7/10)$ (if $\eta > 0$);
- $n, f^{(1)}; \overline{\binom{0}{2}}SN, \overline{\binom{2}{1}}S \Rightarrow (c = -1, e = -36/35, u = -7/10)$ (if $\eta = 0$).

1.1.1.1.1.2: Suppose now $W_4 > 0$. In this case as $\widetilde{K} > 0$ and $G_8 < 0$, according to Theorem 6.1 the systems (9.11) possess two nodes if $W_3 > 0$ and two foci or/and centers if $W_3 < 0$.

1.1.1.1.1.2.1: $W_3 < 0$.

1. $\mathcal{T}_4 \neq 0$. Then by Theorem 6.2 both foci are strong. Thus considering the conditions $\mu_0 = \mu_1 = 0$, $\mu_2 < 0$, $\kappa \neq 0$ and Lemma 9.1, we arrive at the following three global configurations:

   - $f, f; \overline{\binom{2}{1}}S, ©, © \Rightarrow (c = 1,\ e = -1,\ u = -2)$ (if $\eta < 0$);
   - $f, f; \overline{\binom{2}{1}}S, S, N^{\infty} \Rightarrow (c = 1,\ e = -1/2,\ u = -2)$ (if $\eta > 0$);
   - $f, f; \overline{\binom{0}{2}}SN, \overline{\binom{2}{1}}S \Rightarrow (c = 1,\ e = -9/16,\ u = -2)$ (if $\eta = 0$).

2. $\mathcal{T}_4 = 0$. As it was mentioned earlier, we may assume $c = -1$. Then at least one focus is weak.

(a) $\mathcal{T}_3 \neq 0$. In this case only one focus is weak. Moreover considering (9.13) we observe that the condition $\mu_2 < 0$ implies $\mathcal{F}_1 \neq 0$, and the weak focus could only be of the first order. So in view of the arguments above and Lemma 9.1, we get the following three global configurations:
- $f, f^{(1)};\ \overline{\binom{2}{1}}S, \copyright, \copyright \Rightarrow (c = -1,\ e = -2,\ u = -3/2)$ (if $\eta < 0$);
- $f, f^{(1)};\ \overline{\binom{2}{1}}S, S, N^{\infty} \Rightarrow (c = -1, e = -5/4, u = -3/2)$ (if $\eta > 0$);
- $f, f^{(1)};\ \overline{\binom{0}{2}}SN, \overline{\binom{2}{1}}S \Rightarrow (c = -1, e = -4/3, u = -3/2)$ (if $\eta = 0$).

(b) $\mathcal{T}_3 = 0$. Considering (9.13) we get $u = -1$, and then we calculate

$$\mathcal{T}_4 = \mathcal{T}_3 = \mathcal{F} = 0, \quad \mathcal{T}_2 = -4(1+e)^2, \quad \mathcal{F}_1 = 4(1+e),$$
$$\mu_2 = -(1+e)^2x^2, \quad \kappa = 128(1+e), \quad \eta = 4(9+8e).$$

So the condition $\mu_2 \neq 0$ gives $\mathcal{F}_1 \neq 0$, and according to Theorem 6.2 we have two first order weak singularities, which in this case are foci. Thus considering Lemma 9.1 we obtain the following three global configurations:
- $f^{(1)}, f^{(1)};\ \overline{\binom{2}{1}}S, \copyright, \copyright \Rightarrow (c = -1,\ e = -2,\ u = -1)$ (if $\eta < 0$);
- $f^{(1)}, f^{(1)};\ \overline{\binom{2}{1}}S, S, N^{\infty} \Rightarrow (c = -1,\ e = -17/16,\ u = -1)$ (if $\eta > 0$);
- $f^{(1)}, f^{(1)};\ \overline{\binom{0}{2}}SN, \overline{\binom{2}{1}}S \Rightarrow (c = -1,\ e = -9/8,\ u = -1)$ (if $\eta = 0$).

1.1.1.1.1.2.2: $W_3 > 0$. According to Theorem 6.1 (see Table 6.2, line 165) we have two nodes and both are generic (due to $W_4 \neq 0$). According to Lemma 9.1 we arrive at the following three global configurations:
- $n, n;\ \overline{\binom{2}{1}}S, \copyright, \copyright \Rightarrow (c = 6,\ e = -1/3,\ u = -2)$ (if $\eta < 0$);
- $n, n;\ \overline{\binom{2}{1}}S, S, N^{\infty} \Rightarrow (c = 6,\ e = -1/5,\ u = -2)$ (if $\eta > 0$);
- $n, n;\ \overline{\binom{0}{2}}SN, \overline{\binom{2}{1}}S \Rightarrow (c = 6,\ e = -1/4,\ u = -2)$ (if $\eta = 0$).

1.1.1.1.1.3: $W_4 = 0$. Then we have a node with coinciding eigenvalues, and by Remark 9.3 without loss of generality we may assume that $M_1(0,0)$ is such a point. So $\tau_1 = (c-1)^2 + 4e = 0$, we obtain $e = -(c-1)^2/4$ and in this case calculations yield

$$\begin{aligned} &\mu_2 = (1+c)^4(1+2u)x^2/16, \ \eta = 4u^2(1+2u)(c^2+2u), \\ &W_4 = 0, \ W_3 = (1+c)^8u^4(1+u)(1+2u)^2(1+c^2+2u)/8, \\ &\kappa = 32(1+c)^2u^3, \ \mathcal{T}_4 = -(1+c)^5(c-1-2u)u^2(1+2u)/4. \end{aligned} \tag{9.14}$$

1.1.1.1.1.3.1: $W_3 < 0$. According to Theorem 6.1 (see Table 6.2, line 168) we have one focus and one node, which is n^d (as the Jacobian matrix is not diagonal).

1. $\mathcal{T}_4 \neq 0$. Then the focus is strong, and we get the following three global configurations:

- $n^d, f; \overline{\binom{2}{1}}S, ©, ©$ ⇒ $(c = 2,\ e = -1/4,\ u = -3/2)$ (if $\eta < 0$);
- $n^d, f; \overline{\binom{2}{1}}S, S, N^\infty$ ⇒ $(c = 2,\ e = -1/4,\ u = -9/4)$ (if $\eta > 0$);
- $n^d, f; \overline{\binom{0}{2}}SN, \overline{\binom{2}{1}}S$ ⇒ $(c = 2,\ e = -1/4,\ u = -2)$ (if $\eta = 0$).

2. $\mathcal{T}_4 = 0$. Considering (9.14) and the condition $\mu_2\kappa \neq 0$, we obtain $(c-1-2u) = 0$, i.e. $c = 2u + 1$. Then we calculate

$$\begin{aligned}\mathcal{T}_3 &= 8u^2(1+u)^5(1+2u),\quad \mathcal{F}_1 = -2(1+u)^2(1+2u)(1+3u),\\ \eta &= 4u^2(1+2u)(1+6u+4u^2),\quad \mu_2 = (1+u)^4(1+2u)x^2,\\ \kappa &= 128u^3(1+u)^2,\quad W_4 = 0,\quad W_3 = 64u^4(1+u)^{10}(1+2u)^3.\end{aligned}\tag{9.15}$$

So the conditions $W_3 \neq 0$ and $\mu_2 < 0$ imply $\mathcal{T}_3\mathcal{F}_1 \neq 0$, and according to Theorem 6.2 the focus $M_2(1,0)$ is a weak focus of the first order. Therefore considering Lemma 9.1 we arrive at the following three global configurations:

- $n^d, f^{(1)}; \overline{\binom{2}{1}}S, ©, ©$ ⇒ $(c = -2,\ e = -9/4,\ u = -3/2)$ (if $\eta < 0$);
- $n^d, f^{(1)}; \overline{\binom{2}{1}}S, S, N^\infty$ ⇒ $(c = -7/5,\ e = -36/25,\ u = -6/5)$ (if $\eta > 0$);
- $n^d, f^{(1)}; \overline{\binom{0}{2}}SN, \overline{\binom{2}{1}}S$ ⇒ $(c = 2u+1,\ e = -(c-1)^2/4,\ u = -(3+\sqrt{5})/4)$ (if $\eta = 0$).

1.1.1.1.1.3.2: $W_3 > 0$. According to Theorem 6.1 (see Table 6.2, line 166) we have two nodes (one of them being n^d).

Therefore considering Lemma 9.1 we get the following three global configurations:

- $n, n^d; \overline{\binom{2}{1}}S, ©, ©$ ⇒ $(c = 4/3,\ e = -1/36,\ u = -8/10)$ (if $\eta < 0$);
- $n, n^d; \overline{\binom{2}{1}}S, S, N^\infty$ ⇒ $(c = 4/3,\ e = -1/36,\ u = -17/18)$ (if $\eta > 0$);
- $n, n^d; \overline{\binom{0}{2}}SN, \overline{\binom{2}{1}}S$ ⇒ $(c = 4/3,\ e = -1/36,\ u = -8/9)$ (if $\eta = 0$).

1.1.1.1.1.3.3: $W_3 = 0$. Considering (9.14) and the condition $\mu_2 < 0$, we obtain $(1+u)(1+c^2+2u) = 0$, and in this case we have two nodes n^d (as none of the Jacobian matrices is diagonal).

1. $u = -1$. We have

$$\eta = 4(2-c^2),\quad \mu_2 = -(1+c)^4x^2/16 < 0,\quad \kappa = -32(1+c)^2 < 0.$$

So considering Lemma 9.1 we obtain the following three global configurations:

- $n^d, n^d; \overline{\binom{2}{1}}S, ©, ©$ ⇒ $(c = 2,\ e = -1/4,\ u = -1)$ (if $\eta < 0$);
- $n^d, n^d; \overline{\binom{2}{1}}S, S, N^\infty$ ⇒ $(c = 1/2,\ e = -1/16,\ u = -1)$ (if $\eta > 0$);
- $n^d, n^d; \overline{\binom{0}{2}}SN, \overline{\binom{2}{1}}S$ ⇒ $(c = \sqrt{2},\ e = -(c-1)^2/4,\ u = -1)$ (if $\eta = 0$).

2. $u = -(c^2+1)/2$. We calculate

$$\eta = c^2(1+c^2)^2, \quad \mu_2 = -c^2(1+c)^4x^2/16 < 0, \quad \kappa = -4(1+c)^2(1+c^2)^3 < 0.$$

In this case we could only have $\eta > 0$, and such a global configuration was detected above.

1.1.1.1.2: $\mu_2 > 0$. Then by (9.12) we have $1+2u > 0$, and this implies $G_8 > 0$. Since $\widetilde{K} > 0$, according to Theorem 6.1 the systems (9.11) possess a saddle and a focus (or a center) if $W_4 < 0$ and a saddle and a node if $W_4 \geq 0$.

1.1.1.1.2.1: $W_4 < 0$. Then we have a saddle and a focus (or a center).

1.1.1.1.2.1.1: $\mathcal{T}_4 \neq 0$. Then by Theorem 6.2 we could not have weak singularities, i.e. we have a strong focus and a strong saddle. So considering Remark 9.5 and Lemma 9.1, we get the following three global configurations:

- $s, f; \overline{\binom{2}{1}}N, \copyright, \copyright \Rightarrow (c=-8,\ e=2,\ u=2)$ (if $\eta < 0$);
- $s, f; \overline{\binom{2}{1}}N, S, N^\infty \Rightarrow (c=-8,\ e=1/2,\ u=2)$ (if $\eta > 0$);
- $s, f; \overline{\binom{0}{2}}SN, \overline{\binom{2}{1}}N \Rightarrow (c=-8,\ e=1,\ u=2)$ (if $\eta = 0$).

1.1.1.1.2.1.2: $\mathcal{T}_4 = 0$. As it was shown earlier, we can assume $c=-1$, and we calculate

$$\mathcal{T}_3\mathcal{F} = 8(1+e)^3u^4(1+u)^2(1+2u)^2.$$

Therefore considering (9.13) we conclude that the conditions $\kappa < 0$ and $\mu_2 > 0$ imply $\mathcal{T}_3\mathcal{F} \neq 0$. Moreover, as $(1+e)u > 0$ (due to $\kappa < 0$), from (9.13) we have

$$\operatorname{sign}(\mathcal{T}_3\mathcal{F}) = \operatorname{sign}(u), \quad \operatorname{sign}(W_4) = \operatorname{sign}(u(e+2eu-u^2)).$$

1. $\mathcal{T}_3\mathcal{F} < 0$. Then $u < 0$ and according to Theorem 6.2 the systems (9.11) with $c=-1$ possess a weak focus. As $-1/2 < u < 0$, then $\mathcal{F}_1 \neq 0$ and we have a weak first order focus. So considering Lemma 9.1 we get the following three global configurations:

   - $s, f^{(1)}; \overline{\binom{2}{1}}N, \copyright, \copyright \Rightarrow (c=-1,\ e=-2,\ u=-1/4)$ (if $\eta < 0$);
   - $s, f^{(1)}; \overline{\binom{2}{1}}N, S, N^\infty \Rightarrow (c=-1,\ e=-17/16,\ u=-1/4)$ (if $\eta > 0$);
   - $s, f^{(1)}; \overline{\binom{0}{2}}SN, \overline{\binom{2}{1}}N \Rightarrow (c=-1,\ e=-9/8,\ u=-1/4)$ (if $\eta = 0$).

2. $\mathcal{T}_3\mathcal{F} > 0$. In this case $u > 0$ and according to Theorem 6.2 the systems (9.11) with $c=-1$ possess a weak saddle.

   (a) $\mathcal{F}_1 \neq 0$. The weak saddle has order one, and considering Lemma 9.1 we obtain the following three global configurations:
   - $s^{(1)}, f; \overline{\binom{2}{1}}N, \copyright, \copyright \Rightarrow (c=-1,\ e=1,\ u=1/5)$ (if $\eta < 0$);
   - $s^{(1)}, f; \overline{\binom{2}{1}}N, S, N^\infty \Rightarrow (c=-1,\ e=1/5,\ u=1/5)$ (if $\eta > 0$);
   - $s^{(1)}, f; \overline{\binom{0}{2}}SN, \overline{\binom{2}{1}}N \Rightarrow (c=-1,\ e=9/40,\ u=1/5)$ (if $\eta = 0$).

(b) $\mathcal{F}_1 = 0$. Then $u = 1$ and we calculate

$$\begin{gathered} \mathcal{T}_3\mathcal{F} = 288(1+e)^3, \quad \mathcal{F}_1 = 0, \quad \mathcal{F}_2 = -432(1+e)^2, \\ \kappa = -128(1+e), \quad \eta = 4(1-8e), \quad W_4 = 2304(1+e)^5(1-3e). \end{gathered} \tag{9.16}$$

As $\kappa < 0$ we get $e+1 > 0$ and hence $\mathcal{F}_2 \neq 0$. Therefore as $\mathcal{T}_3\mathcal{F} > 0$ and $\mathcal{F}_1 = 0$, by Theorem 6.2 we obtain a weak saddle of order two. Additionally the condition $W_4 < 0$ due to $e+1 > 0$ gives $e > 1/3$, and this implies $\eta < 0$.
Thus considering Lemma 9.1 we obtain the global configuration
• $s^{(2)}, f; \overline{\binom{2}{1}}N, ©, © \Rightarrow (c = -1,\ e = 1,\ u = 1)$.

1.1.1.1.2.2: $W_4 > 0$. By Theorem 6.1 we have a saddle and a node.

1.1.1.1.2.2.1: $\mathcal{T}_4 \neq 0$. Then by Theorem 6.2 we could not have weak singularities, i.e. the saddle is strong.

So considering Lemma 9.1 we obtain the following three global configurations:
• $s, n; \overline{\binom{2}{1}}N, ©, © \Rightarrow (c = -2,\ e = 2/5,\ u = 2)$ (if $\eta < 0$);
• $s, n; \overline{\binom{2}{1}}N, S, N^{\infty} \Rightarrow (c = -2,\ e = 1/5,\ u = 2)$ (if $\eta > 0$);
• $s, n; \overline{\binom{0}{2}}SN, \overline{\binom{2}{1}}N \Rightarrow (c = -2,\ e = 1/4,\ u = 2)$ (if $\eta = 0$).

1.1.1.1.2.2.2: $\mathcal{T}_4 = 0$. As it was shown earlier, we can assume $c = -1$, and considering (9.13) we conclude that the conditions $\kappa < 0$ and $\mu_2 > 0$ imply $\mathcal{T}_3 \neq 0$.

1. $\mathcal{F}_1 \neq 0$. Then the weak saddle has order one, and considering Lemma 9.1 we obtain the following three global configurations:

   • $s^{(1)}, n; \overline{\binom{2}{1}}N, ©, © \Rightarrow (c = -1,\ e = 5/8,\ u = 2)$ (if $\eta < 0$);

   • $s^{(1)}, n; \overline{\binom{2}{1}}N, S, N^{\infty} \Rightarrow (c = -1,\ e = 1/2,\ u = 2)$ (if $\eta > 0$);

   • $s^{(1)}, n; \overline{\binom{0}{2}}SN, \overline{\binom{2}{1}}N \Rightarrow (c = -1,\ e = 9/16,\ u = 2)$ (if $\eta = 0$).

2. $\mathcal{F}_1 = 0$. In this case we get $u = 1$, and considering (9.16) we deduce that the condition $\kappa < 0$ implies $\mathcal{F}_2 \neq 0$. Therefore according to Theorem 6.2 we have a second order weak saddle, and we arrive at the following three global configurations:

   • $s^{(2)}, n; \overline{\binom{2}{1}}N, ©, © \Rightarrow (c = -1,\ e = 1/4,\ u = 1)$ (if $\eta < 0$);

   • $s^{(2)}, n; \overline{\binom{2}{1}}N, S, N^{\infty} \Rightarrow (c = -1,\ e = 1/9,\ u = 1)$ (if $\eta > 0$);

   • $s^{(2)}, n; \overline{\binom{0}{2}}SN, \overline{\binom{2}{1}}N \Rightarrow (c = -1,\ e = 1/8,\ u = 1)$ (if $\eta = 0$).

1.1.1.1.2.3: $W_4 = 0$. As it was mentioned above, in this case we may assume $e = -(c-1)^2/4$. Additionally as $G_8 > 0$ and $\widetilde{K} > 0$, according to Theorem 6.1 the systems (9.11) possess a saddle and a node, which is n^d (due to $W_4 = 0$ and the fact that the Jacobian matrix is not diagonal). According to (9.14) the conditions $\kappa < 0$ and $\mu_2 > 0$ yield $-1/2 < u < 0$.

1.1.1.1.2.3.1: $\mathcal{T}_4 \neq 0$. Then by Theorem 6.2 we could not have weak singularities, i.e. the saddle is strong.

So considering Lemma 9.1 we obtain the following three global configurations:

- $s, n^d; \overline{\binom{2}{1}}N, ©, © \Rightarrow (c = 1/2,\ e = -1/16,\ u = -1/5)$ (if $\eta < 0$);
- $s, n^d; \overline{\binom{2}{1}}N, S, N^\infty \Rightarrow (c = 1/2,\ e = -1/16,\ u = -1/9)$ (if $\eta > 0$);
- $s, n^d; \overline{\binom{0}{2}}SN, \overline{\binom{2}{1}}N \Rightarrow (c = 1/2,\ e = -1/16,\ u = -1/8)$ (if $\eta = 0$).

1.1.1.1.2.3.2: $\mathcal{T}_4 = 0$. Considering (9.14) and the condition $\mu_2\kappa \neq 0$, we obtain $c - 1 - 2u = 0$, i.e. $c = 2u + 1$. Then we obtain (9.15) and we can observe that $\mathcal{T}_3 \neq 0$ (due to $\mu_2\kappa \neq 0$).

1. $\mathcal{F}_1 \neq 0$. Then by Theorem 6.2 the weak saddle has order one, and considering Lemma 9.1 we obtain the following three global configurations:
   - $s^{(1)}, n^d; \overline{\binom{2}{1}}N, ©, © \Rightarrow (c = 1/2,\ e = -1/16,\ u = -1/4)$ (if $\eta < 0$);
   - $s^{(1)}, n^d; \overline{\binom{2}{1}}N, S, N^\infty \Rightarrow (c = 2/3,\ e = -1/36,\ u = -1/6)$ (if $\eta > 0$);
   - $s^{(1)}, n^d; \overline{\binom{0}{2}}SN, \overline{\binom{2}{1}}N \Rightarrow (c = 2u + 1,\ e = -(c-1)^2/4,\ u = (\sqrt{5} - 3)/4)$ (if $\eta = 0$).
2. $\mathcal{F}_1 = 0$. Considering (9.15) we obtain $u = -1/3$, and then we have

$$\mathcal{T}_4 = \mathcal{F}_1 = 0,\quad \mathcal{T}_3 = 256/6561,\quad \mathcal{F}_2 = -256/19683,\quad \eta = -20/243.$$

   Hence by Theorem 6.2 the saddle is of the second order. As $\eta < 0$ considering Lemma 9.1 we obtain the global configuration
   - $s^{(2)}, n^d; \overline{\binom{2}{1}}N, ©, © \Rightarrow (c = 1/3,\ e = -1/9,\ u = -1/3)$.

1.1.1.2: $\kappa > 0$. According to (9.12) and Remark 9.5, the condition $\kappa > 0$ implies $\eta > 0$ and $\widetilde{K} < 0$, and we consider two possibilities: $\mu_2 < 0$ and $\mu_2 > 0$.

1.1.1.2.1: $\mu_2 < 0$. Then $u < -1/2$, and considering (9.12) we obtain $F_2 < 0$ and $G_8 < 0$. As $\widetilde{K} < 0$ according to Theorem 6.1 (see Table 6.2, line 148) both finite singularities are saddles.

Additionally according to Lemma 9.1, due to $\eta > 0$, $\kappa > 0$ and $\mu_2 < 0$, at infinity we have the configuration $\overline{\binom{2}{1}}N, N^f, N^f$.

1.1.1.2.1.1: $\mathcal{T}_4 \neq 0$. Then both saddles are strong, and this leads to the global configuration

- $s, s; \overline{\binom{2}{1}}N, N^f, N^f \Rightarrow (c = -2,\ e = 0,\ u = -2)$.

1.1.1.2.1.2: Suppose now $\mathcal{T}_4 = 0$. As it was shown earlier, we can assume $c = -1$ and we consider (9.13).

1.1.1.2.1.2.1: $\mathcal{T}_3 \neq 0$. In this situation only one saddle is weak, and as $\mathcal{F}_1 \neq 0$ (due to $\mu_2 < 0$), according to Theorem 6.2 the order of the weak saddle is one. Thus we get the global configuration
• $s, s^{(1)}; \overline{\binom{2}{1}}N, N^f, N^f \Rightarrow (c = -1,\ e = 1,\ u = -2)$.

1.1.1.2.1.2.2: $\mathcal{T}_3 = 0$. Then $u = -1$ and we have

$$\mathcal{T}_4 = \mathcal{T}_3 = \mathcal{F} = 0,\quad \mathcal{T}_2 = -4(1+e)^2,\quad \mathcal{F}_1 = 4(1+e),\quad \mu_2 = -(1+e)^2x^2 \neq 0.$$

Hence $\mathcal{T}_2 < 0$ and as $\mathcal{F}_1 \neq 0$ according to Theorem 6.2 we have two weak saddles each one of the first order. This gives us the global configuration
• $s^{(1)}, s^{(1)}; \overline{\binom{2}{1}}N, N^f, N^f \Rightarrow (c = -1,\ e = 1,\ u = -1)$.

1.1.1.2.2: $\mu_2 > 0$. In this case we have $u > -1/2$, and considering (9.12) we obtain $F_2 > 0$ and $G_8 > 0$. As $\widetilde{K} < 0$ according to Theorem 6.1 (see Table 6.2, lines 149 and 156) we have a saddle and an anti-saddle. The type of the anti-saddle is governed by the invariant polynomial W_4.

Additionally, due to the conditions $\eta > 0$, $\kappa > 0$, $\mu_2 > 0$ and considering Remark 9.5, according to Lemma 9.1 at infinity we have the configuration $\overline{\binom{2}{1}}S, N^f, N^f$.

1.1.1.2.2.1: $W_4 < 0$. Then we have a saddle and a focus (or a center).

1.1.1.2.2.1.1: $\mathcal{T}_4 \neq 0$. Then by Theorem 6.2 we could not have weak singularities i.e. the saddle and the focus are both strong ones. Thus we get the global configuration
• $s, f; \overline{\binom{2}{1}}S, N^f, N^f \Rightarrow (c = 1,\ e = -1,\ u = 2)$.

1.1.1.2.2.1.2: $\mathcal{T}_4 = 0$. As it was shown earlier, we can assume $c = -1$, and considering (9.13) we conclude that the conditions $\kappa > 0$ and $\mu_2 > 0$ imply $\mathcal{T}_3 \neq 0$. Moreover as $(1+e)u < 0$ (due to $\kappa > 0$) we have $\text{sign}\,(\mathcal{T}_3\mathcal{F}) = -\text{sign}\,(u)$.

1. $\mathcal{T}_3\mathcal{F} < 0$. Then $u > 0$ and in this case according to Theorem 6.2 the systems (9.11) with $c = -1$ possess a weak focus.
   (a) $\mathcal{F}_1 \neq 0$. In this case by Theorem 6.2 the weak focus is of order one, and this leads to the global configuration
   • $s, f^{(1)}; \overline{\binom{2}{1}}S, N^f, N^f \Rightarrow (c = -1,\ e = -2,\ u = 2)$.
   (b) $\mathcal{F}_1 = 0$. Then $u = 1$, $e < -1$ and this implies $\mathcal{F}_2 = -432(1+e)^2 \neq 0$. So by Theorem 6.2 we could have a weak focus of order at most two, and this leads to the global configuration
   • $s, f^{(2)}; \overline{\binom{2}{1}}S, N^f, N^f \Rightarrow (c = -1,\ e = -2,\ u = 1)$.
2. $\mathcal{T}_3\mathcal{F} > 0$. Then $u < 0$ and in this case according to Theorem 6.2 the systems (9.11) with $c = -1$ possess a weak saddle. As $-1/2 < u < 0$, then $\mathcal{F}_1 \neq 0$ and we have a weak saddle of order one. Thus we get the global configuration
   • $s^{(1)}, f; \overline{\binom{2}{1}}S, N^f, N^f \Rightarrow (c = -1,\ e = 1/2,\ u = -1/4)$.

1.1.1.2.2.2: $W_4 > 0$. By Theorem 6.1 apart from the saddle, we have a node (which is generic due to $W_4 \neq 0$).

1.1.1.2.2.2.1: $\mathcal{T}_4 \neq 0$. Then by Theorem 6.2 we could not have weak singularities, and this leads to the global configuration

• $s, n; \overline{\binom{2}{1}}S, N^f, N^f \Rightarrow (c = 2,\ e = 1,\ u = 2)$.

1.1.1.2.2.2.2: $\mathcal{T}_4 = 0$. As it was shown earlier, we can assume $c = -1$, and considering (9.13) we conclude that the conditions $\kappa > 0$ and $\mu_2 > 0$ imply $\mathcal{T}_3 \neq 0$. Moreover we claim that in this case the condition $\mathcal{F}_1 \neq 0$ holds. Indeed assuming $\mathcal{F}_1 = 0$, by (9.13) we obtain $u = 1$, and then the conditions

$$\kappa = -128(1+e) > 0, \quad W_4 = 2304(1+e)^5(1-3e) > 0$$

are incompatible. Thus $\mathcal{F}_1 \neq 0$ and we have a weak saddle of order one, i.e. we get the global configuration

• $s^{(1)}, n; \overline{\binom{2}{1}}S, N^f, N^f \Rightarrow (c = -1,\ e = -1/2,\ u = -1/3)$.

1.1.1.2.2.3: $W_4 = 0$. As it was mentioned above, in this case we may assume $e = -(c-1)^2/4$ (i.e. $\tau_1 = 0$). Additionally as $G_8 > 0$, $F_2 > 0$ and $\widetilde{K} < 0$, according to Theorem 6.1 (see Table 6.2, line 149) systems (9.11) possess a saddle and a node, which is n^d (due to $W_4 = 0$ and the non-diagonal corresponding matrix of the linearization). According to (9.14) the condition $\kappa > 0$ yields $u > 0$.

1.1.1.2.2.3.1: $\mathcal{T}_4 \neq 0$. Then by Theorem 6.2 we could not have weak singularities, i.e. the saddle is strong. So we get the global configuration

• $s, n^d; \overline{\binom{2}{1}}S, N^f, N^f \Rightarrow (c = 1,\ e = 0,\ u = 1)$.

1.1.1.2.2.3.2: $\mathcal{T}_4 = 0$. Considering (9.14) and the condition $\mu_2\kappa \neq 0$, we get $c - 1 - 2u = 0$, i.e. $c = 2u+1$. Then we obtain (9.15), and we observe that $\mathcal{T}_3 \neq 0$ (due to $\mu_2\kappa \neq 0$). Moreover in this case we obtain $\mathcal{F}_1 = -2(1+u)^2(1+2u)(1+3u) \neq 0$ due to $u > 0$. Therefore we have a weak saddle of order one, and this leads to the global configuration

• $s^{(1)}, n^d; \overline{\binom{2}{1}}S, N^f, N^f \Rightarrow (c = 3,\ e = -1,\ u = 1)$.

As all the cases have been examined the investigation of systems (9.11) is completed.

1.1.2: $(d, f) = (1, 0)$. In this case for systems (9.8) we calculate

$$\kappa = -128eu^3, \quad \mu_2 = e^2(1+2u)x^2, \quad \mathcal{T}_4 = -4c^2e^2u^2(1+2u).$$

We observe that due to $\mu_2\kappa \neq 0$ the condition $c \neq 0$ is equivalent to $\mathcal{T}_4 \neq 0$.

1.1.2.1: $\mathcal{T}_4 \neq 0$. Then $c \neq 0$ and we may assume $c = 1$ due to the rescaling $(x, y, t) \mapsto (x, cy, t/c)$. So we get the 2-parameter family of systems:

$$\dot{x} = x + y - x^2 + 2uxy, \quad \dot{y} = ex - ex^2, \quad eu(1+2u) \neq 0, \tag{9.17}$$

for which we calculate

$$\begin{gathered}\mu_0=\mu_1=0,\quad \mu_2=e^2(1+2u)x^2,\quad \widetilde{K}=4eux^2,\quad \kappa=-128eu^3,\\ \eta=4u^2(1-8eu),\quad F_2=e^2(1+2u)x^2,\quad G_8=2e^2u^2(1+2u),\\ W_4=16e^4(1+4e)u^4(1+2u)^2(1-4e-8eu),\\ W_3=32e^4u^4(1+2u)^2(1-4eu),\quad \widetilde{M}=8(6eu-1)x^2+16uxy-32u^2y^2.\end{gathered}\tag{9.18}$$

Remark 9.6. We observe that the condition $\mu_2\kappa\neq 0$ gives $\widetilde{M}F_2G_8\neq 0$. Moreover we have

$$\operatorname{sign}(\widetilde{K})=-\operatorname{sign}(\kappa),\quad \operatorname{sign}(G_8)=\operatorname{sign}(\mu_2)=\operatorname{sign}(F_2).$$

So, for systems (9.17) the same relations between the signs of invariant polynomials as in the case of systems (9.11) hold. Moreover as we shall use the same Table 6.2 of Theorem 6.1 and the same conditions for infinite singularities (i.e. Lemma 9.1), we only need to detect if for systems (9.17) we could obtain some global configurations that we have not obtained for systems (9.11).

For this goal we only need to detect if some logically possible global configurations of singularities in the case $\mathcal{T}_4\neq 0$ could not be realized for systems (9.11), to examine the respective case for systems (9.17) and to find out if such detected global configuration could be realized for systems (9.17).

1.1.2.1.1: $\kappa<0$. We observe that all the logically possible global configurations for systems (9.8) in the case $\kappa<0$ and $\mathcal{T}_4\neq 0$ are realized. More precisely we have the following number of global configurations in this case:

$$\begin{array}{llll}\mu_2<0,\ W_4<0 & \Rightarrow\ 3; & \mu_2<0,\ W_4>0 & \Rightarrow\ 6;\\ \mu_2<0,\ W_4=0 & \Rightarrow\ 9; & \mu_2>0,\ W_4<0 & \Rightarrow\ 3;\\ \mu_2>0,\ W_4>0 & \Rightarrow\ 3; & \mu_2>0,\ W_4=0 & \Rightarrow\ 3.\end{array}$$

1.1.2.1.2: $\kappa>0$. In this case by (9.12) the condition $\kappa>0$ implies $\eta>0$ for systems (9.11), i.e. the global configurations with $\eta\leq 0$ do not exist for these systems.

The question is: *Could we obtain such global configurations for systems* (9.17)*?*

The answer is *no*, as from (9.18) we observe that the condition $\kappa>0$ gives $eu<0$ and then clearly $\eta>0$.

Considering Remark 9.6 and Theorem 6.1, we conclude that all logical possibilities for global configurations in the case $\kappa>0$ and $\mathcal{T}_4\neq 0$ are realized for systems (9.11). More precisely we have the following number of global configurations in this case:

$$\begin{array}{llll}\mu_2<0 & \Rightarrow\ 1; & \mu_2>0,\ W_4<0 & \Rightarrow\ 1;\\ \mu_2>0,\ W_4>0 & \Rightarrow\ 1; & \mu_2>0,\ W_4=0 & \Rightarrow\ 1.\end{array}$$

Thus we consider now the next possibility.

1.1.2.2: $\mathcal{T}_4 = 0$. Then $c = 0$ and we get the systems:

$$\dot{x} = (1 + 2ux)y, \quad \dot{y} = ex - ex^2, \quad eu(1 + 2u) \neq 0. \tag{9.19}$$

Moreover due to the rescaling $(x, y, t) \mapsto (x, |e|^{1/2}y, |e|^{-1/2}t)$, we can assume $e \in \{-1, 1\}$.

For these systems we calculate

$$\begin{gathered}
\mu_0 = \mu_1 = 0, \ \mu_2 = e^2(1 + 2u)x^2, \ \widetilde{K} = 4eux^2, \quad \kappa = -128eu^3, \\
\eta = -32eu^3, \ \widetilde{M} = 16u(3ex^2 - 2uy^2), \ F_2 = e^2(1 + 2u)x^2, \\
G_8 = 2e^2u^2(1 + 2u), \ W_4 = -256e^6u^4(1 + 2u)^3, \\
W_3 = -128e^5u^5(1 + 2u)^2, \ \mathcal{T}_4 = \mathcal{T}_3 = \mathcal{F}_1 = \mathcal{F} = 0, \\
\mathcal{T}_2 = 4e^2u^2(1 + 2u), \ \ \mathcal{B} = -2e^2u^4, \ \ \mathcal{H} = -4eu^3.
\end{gathered}$$

As $\mathcal{T}_4 = \mathcal{T}_3 = \mathcal{F} = 0$, according to Theorem 6.2 the systems (9.19) possess two weak singularities. Moreover, since $\mathcal{F}_1 = 0$ these singularities could be only centers and/or integrable saddles. We observe that due to $\mu_2\widetilde{K} \neq 0$ we have

$$\begin{gathered}
\text{sign}\,(\eta) = \text{sign}\,(\mathcal{H}) = \text{sign}\,(\kappa) = -\text{sign}\,(\widetilde{K}), \\
\text{sign}\,(G_8) = \text{sign}\,(\mu_2) = \text{sign}\,(\mathcal{T}_2) = \text{sign}\,(F_2).
\end{gathered} \tag{9.20}$$

1.1.2.2.1: $\kappa < 0$. In this case we get $\eta < 0$ and $\mathcal{H} < 0$.

1.1.2.2.1.1: $\mu_2 < 0$. This implies $\mathcal{T}_2 < 0$, and as $\mathcal{H} < 0$ and $\mathcal{B} < 0$, according to Theorem 6.2 we have two centers. Therefore considering Lemma 9.1 we get the global configuration

- $c, c; \overline{\binom{2}{1}} S, ©, © \Rightarrow (e = -1, \ u = -1)$.

1.1.2.2.1.2: $\mu_2 > 0$. Then $\mathcal{T}_2 > 0$, and as $\mathcal{B} < 0$, according to Theorem 6.2 we have one saddle and one center. At infinity we have the same configuration, and we get

- $\$, c; \overline{\binom{2}{1}} N, ©, © \Rightarrow (e = 1, \ u = 1)$.

1.1.2.2.2: $\kappa > 0$. In this case by (9.20) we have $\eta > 0$ and $\mathcal{H} > 0$.

1.1.2.2.2.1: $\mu_2 < 0$. Then $\mathcal{T}_2 < 0$, and since $\mathcal{H} > 0$ and $\mathcal{B} < 0$, according to Theorem 6.2 we have two integrable saddles. Additionally considering the signs of the invariant polynomials μ_2, κ and η, according to Lemma 9.1 we get the global configuration

- $\$, \$; \overline{\binom{2}{1}} N, N^f, N^f \Rightarrow (e = 1, \ u = -1)$.

1.1.2.2.2.2: $\mu_2 > 0$. Then we obtain $\mathcal{T}_2 > 0$, and as $\mathcal{B} < 0$, according to Theorem 6.2 we have one saddle and one center. So considering Lemma 9.1 we obtain the global configuration

- $\$, c;\ \overline{\binom{2}{1}}S, N^f, N^f \Rightarrow (e=-1,\ u=1)$.

Thus in the case $\kappa \neq 0$ all the possibilities are examined for systems (9.8).

We observe that the four global configurations detected for systems (9.19) are not realizable for systems (9.11). In order to insert these global configurations in the global diagram, we use the next remark.

Remark 9.7. For the last four global configurations the following conditions are satisfied, respectively:

$$\begin{array}{lcl}
c, c;\ \overline{\binom{2}{1}}S, ©, © & \Rightarrow & \kappa<0, \mu_2<0, W_4>0, W_3<0, \mathcal{T}_4=\mathcal{T}_3=\mathcal{F}_1=0;\\
\$, c;\ \binom{2}{\underline{1}}N, ©, © & \Rightarrow & \kappa<0, \mu_2>0, W_4<0, \mathcal{T}_4=\mathcal{T}_3=\mathcal{F}_1=0;\\
\$, \$;\ \binom{2}{\underline{1}}N, N^f, N^f & \Rightarrow & \kappa>0, \mu_2<0, W_4>0, W_3>0, \mathcal{T}_4=\mathcal{T}_3=\mathcal{F}_1=0;\\
\$, c;\ \binom{2}{1}S, N^f, N^f & \Rightarrow & \kappa>0, \mu_2>0, W_4<0, \mathcal{T}_4=\mathcal{T}_3=\mathcal{F}_1=0.
\end{array}$$

1.2: $\kappa=0$. Considering (9.9), due to $\widetilde{K}\neq 0$ we obtain $d=0$, and then by Remark 9.4 we may assume $f=1$. Thus we arrive at the systems:

$$\begin{aligned}
\dot{x} &= cx(1-x), \qquad cu(1+2u)\neq 0,\\
\dot{y} &= ex+y-ex^2+2uxy,
\end{aligned} \tag{9.21}$$

for which, due to the rescaling $(x,y,t)\mapsto(x,\ ey,\ t)$ (if $e\neq 0$), we can assume $e\in\{0,1\}$.

For these systems we calculate

$$\begin{gathered}
\mu_0=\mu_1=\kappa=\kappa_1=0,\ \mu_2=c^2(1+2u)x^2,\ \widetilde{K}=-4cux^2,\\
\tilde{L}=8c(c+2u)x^2,\ K_2=96c^2(c^2+3cu+4u^2)x^2,\\
\eta=0,\ \widetilde{M}=-8(c+2u)^2x^2,\ \ C_2=ex^3-(c+2u)x^2y,\ G_8=0,\\
F_2=c^2(1+2u)x^2,\ \sigma=1+c-2(c-u)x,\\
\mathcal{T}_i=0\ (i=1,2,3,4),\ \mathcal{F}_1=\mathcal{H}=\mathcal{B}=\mathcal{B}_1=\mathcal{B}_2=0,\\
\mathcal{B}_3=72c^2(1+c)(1-c+2u)x^2\equiv 72c^2\rho_1\rho_2x^2.
\end{gathered} \tag{9.22}$$

Considering the values of the above invariant polynomials, according to Theorem 6.2 we arrive at the following remark.

Remark 9.8. Systems (9.21) possess at least one weak singularity if and only if $\mathcal{B}_3=0$. More exactly as $\widetilde{K}\neq 0$, by Theorem 6.2 we have one integrable saddle in the case $\sigma\neq 0$ (statement $(e_3)[\gamma]$), and we have two integrable saddles in the case $\sigma=0$ (statement $(f_4)[\gamma]$).

1.2.1: $\widetilde{K}<0$. As $G_8=0$, according to Theorem 6.1 (see Table 6.2, lines 148 and 150) we have two saddles if $F_2<0$ and a saddle and a node if $F_2>0$. Additionally from (9.22) it follows that sign (F_2) = sign (μ_2).

1.2.1.1: $\mu_2 < 0$. Then $F_2 < 0$ and systems (9.21) possess two saddles.

However by (9.22) the condition $\widetilde{K} < 0$ gives $cu > 0$ and then $\widetilde{M} \neq 0$. Therefore as $\widetilde{K} < 0$ and $\mu_2 < 0$, according to Lemma 9.1 and Remark 9.8, we arrive at the following three global configurations:

- s, s; $\binom{2}{2}$ $\widehat{P}E\widehat{P} - \widehat{P}E\widehat{P}$, $N^f \Rightarrow (c = -2,\ e = 0,\ u = -1)$ (if $\sigma\mathcal{B}_3 \neq 0$);
- $\$, s$; $\binom{2}{2}$ $\widehat{P}E\widehat{P} - \widehat{P}E\widehat{P}$, $N^f \Rightarrow (c = -1,\ e = 0,\ u = -2)$ (if $\sigma \neq 0,\ \mathcal{B}_3 = 0$);
- $\$, \$$; $\binom{2}{2}$ $\widehat{P}E\widehat{P} - \widehat{P}E\widehat{P}$, $N^f \Rightarrow (c = -1,\ e = 0,\ u = -1)$ (if $\sigma = 0$).

1.2.1.2: $\mu_2 > 0$. Then $F_2 > 0$ and systems (9.22) possess one saddle and one node. We observe that the Jacobian matrices for the singularities $M_1(0,0)$ and $M_2(1,0)$ are:

$$M_1 \Rightarrow \begin{pmatrix} c & 0 \\ e & 1 \end{pmatrix}, \qquad M_2 \Rightarrow \begin{pmatrix} -c & 0 \\ -e & 1+2u \end{pmatrix}.$$

Therefore systems (9.21) possess a node with coinciding eigenvalues if and only if $(c-1)(c+2u+1) = 0$, and this node is a star node if and only if $(c-1)(c+2u+1) = e = 0$.

Additionally for these systems we have $W_{11} = 96cu^3(1-c)(1+c+2u)x^4$ and

$$U_3\big|_{c=1} = -24eu(1+u)^2x^5, \quad U_3\big|_{c=-1-2u} = 24eu(1+u)^2(1+2u)x^5.$$

Since the conditions $\widetilde{K} \neq 0$ and $\mu_2 > 0$ imply $cu(1+u)(1+2u) \neq 0$, we have the next remark.

Remark 9.9. Systems (9.21) with $\mu_2 > 0$ possess a node with coinciding eigenvalues if and only if $W_{11} = 0$. Moreover this node is n^d if $U_3 \neq 0$, and it is a star node if $U_3 = 0$.

1.2.1.2.1: $W_{11} \neq 0$. Then the node is generic. We observe that in this case $\sigma \neq 0$, otherwise we get $c = u = -1$, which contradicts $\mu_2 > 0$.

1.2.1.2.1.1: $\mathcal{B}_3 \neq 0$. Then by Remark 9.8 the saddle is strong, and we shall examine the infinite singularities. We have again $\widetilde{M} \neq 0$ (due to $cu > 0$), and considering (9.22) we obtain $K_2 > 0$. So according to Lemma 9.1 we get the global configuration

- s, n; $\binom{2}{2}$ $\widehat{P}\widehat{P}H - \widehat{P}\widehat{P}H$, $N^f \Rightarrow (c = 2,\ e = 0,\ u = 1)$.

1.2.1.2.1.2: $\mathcal{B}_3 = 0$. Since $\sigma \neq 0$, by Remark 9.8 we have an integrable saddle, and we arrive at the global configuration

- $\$, n$; $\binom{2}{2}$ $\widehat{P}\widehat{P}H - \widehat{P}\widehat{P}H$, $N^f \Rightarrow (c = -1,\ e = 0,\ u = -1/3)$.

1.2.1.2.2: $W_{11} = 0$. Then one of the finite singularities is a node with coinciding eigenvalues. Due to Remark 9.3 without loss of generality we may assume that such a node is $M_1(0,0)$, i.e. the condition $c = 1$ holds.

1.2.1.2.2.1: $U_3 \neq 0$. Then $e \neq 0$ and by Remark 9.9, apart from the saddle, we have a node n^d. Additionally if $c = 1$ then $u > 0$ (due to $\widetilde{K} < 0$), and we obtain $\mathcal{B}_3 \neq 0$, i.e. the saddle is strong. Thus we obtain the global configuration

- $s, n^d;\ \binom{2}{2}\ \widehat{P}\,\widehat{P}H - \widehat{P}\,\widehat{P}H,\ N^f \Rightarrow (c = 1,\ e = 1,\ u = 1)$.

1.2.1.2.2.2: $U_3 = 0$. In this case we have a star node and a strong saddle, and this leads to the global configuration

- $s, n^*;\ \binom{2}{2}\ \widehat{P}\,\widehat{P}H - \widehat{P}\,\widehat{P}H,\ N^f \Rightarrow (c = 1,\ e = 0,\ u = 1)$.

1.2.2: $\widetilde{K} > 0$. Since $G_8 = 0$, according to Theorem 6.1, the types of the finite singularities of systems (9.21) are governed by the polynomial F_2.

1.2.2.1: $\mu_2 < 0$. Then $F_2 < 0$ and as $G_8 = 0$ according to Theorem 6.1 (see Table 6.2, line 164) systems (9.21) possess two nodes.

1.2.2.1.1: $\widetilde{M} \neq 0$. Then at infinity we have two real distinct singularities.

1.2.2.1.1.1: $W_{11} \neq 0$. Then by Remark 9.9 both nodes are generic. Additionally according to Lemma 9.1 the configuration of the infinite singularities depends on the sign of the invariant polynomial $\tilde{L}$ (we note that $\tilde{L} \neq 0$ due to $\widetilde{M} \neq 0$). So we get the following two global configurations:

- $n, n;\ \binom{2}{2}\ \widehat{P}H\widehat{P} - \widehat{P}H\widehat{P},\ S \Rightarrow (c = 2,\ e = 0,\ u = -2)$ (if $\tilde{L} < 0$);
- $n, n;\ \binom{2}{2}\ HHH - HHH,\ N^{\infty} \Rightarrow (c = 2,\ e = 0,\ u = -2/3)$ (if $\tilde{L} > 0$).

1.2.2.1.1.2: $W_{11} = 0$. Then we may assume $c = 1$, and hence the singular point M_1 of systems (9.21) is a node with coinciding eigenvalues, whereas for the second singularity M_2 we have $\lambda_1 = -1$ and $\lambda_2 = 2u + 1$. Therefore the second node will be a node with coinciding eigenvalues if and only if $u = -1$. For $c = 1$ for systems (9.21) we calculate

$$U_1 = -4u(1+u)x^2,\quad U_3 = -24eu(1+u)^2x^5,\quad \mu_2 = (1+2u)x^2,$$
$$\widetilde{K} = -4ux^2,\quad U_5\big|_{u=-1} = -6ex^2,\quad \tilde{L} = 8(2u+1), \tag{9.23}$$

and due to $\widetilde{K} \neq 0$ the condition $u = -1$ is equivalent to $U_1 = 0$. Moreover if $U_1 \neq 0$ the condition $e = 0$ (to have a star node) is equivalent to $U_3 = 0$. In the case $U_1 = 0$ (i.e. $u = -1$) the condition $e = 0$ is equivalent to $U_5 = 0$, and in this case we have two star nodes.

Additionally due to $\mu_2\widetilde{K} \neq 0$ we get $\tilde{L} < 0$. Thus considering Lemma 9.1, we arrive at the following four global configurations:

- $n, n^d;\ \binom{2}{2}\ \widehat{P}H\widehat{P} - \widehat{P}H\widehat{P},\ S \Rightarrow (c = 1,\ e = 1,\ u = -2)$ (if $U_1 \neq 0,\ U_3 \neq 0$);
- $n, n^*;\ \binom{2}{2}\ \widehat{P}H\widehat{P} - \widehat{P}H\widehat{P},\ S \Rightarrow (c = 1,\ e = 0,\ u = -2)$ (if $U_1 \neq 0,\ U_3 = 0$);
- $n^d, n^d;\ \binom{2}{2}\ \widehat{P}H\widehat{P} - \widehat{P}H\widehat{P},\ S \Rightarrow (c = 1,\ e = 1,\ u = -1)$ (if $U_1 = 0,\ U_5 \neq 0$);
- $n^*, n^*;\ \binom{2}{2}\ \widehat{P}H\widehat{P} - \widehat{P}H\widehat{P},\ S \Rightarrow (c = 1,\ e = 0,\ u = -1)$ (if $U_1 = 0,\ U_5 = 0$).

1.2.2.1.2: $\widetilde{M} = 0$. Then we have $c = -2u$, and this gives $C_2 = ex^3$ and $W_{11} = -192u^4(1+2u)x^4 \neq 0$ (due to $\mu_2 \neq 0$). Therefore at infinity we have one real singularity of multiplicity five if $C_2 \neq 0$, and the infinite line is filled up with singularities if $C_2 = 0$. At the same time due to Remark 9.9 the systems (9.21) possess two generic nodes.

1.2.2.1.2.1: $C_2 \neq 0$. As $\widetilde{K} \neq 0$ and $\mu_2 < 0$, according to Lemma 9.1 we arrive at the global configuration
• $n, n;\ \binom{2}{3}\ HH\widehat{P} - \widehat{P}HH \Rightarrow (c = 2,\ e = 1,\ u = -1)$.

1.2.2.1.2.2: $C_2 = 0$. Then we have $c + 2u = e = 0$, and as the nodes are generic, considering Lemma 9.1 we obtain the global configuration
• $n, n;\ [\infty;\ S] \Rightarrow (c = 2,\ e = 0,\ u = -1)$.

1.2.2.2: $\mu_2 > 0$. Then $F_2 > 0$ and as $G_8 = 0$ according to Theorem 6.1 (see Table 6.2, line 150) systems (9.22) possess a saddle and a node. We observe that due to $\widetilde{K} > 0$ and $\mu_2 > 0$, in this case the conditions $2u + 1 > 0$ and $cu < 0$ hold.

1.2.2.2.1: $\widetilde{M} \neq 0$. Then at infinity we have two real distinct singularities.

1.2.2.2.1.1: $\mathcal{B}_3 \neq 0$. Then by Remark 9.8 the saddle is strong.

1.2.2.2.1.1.1: $W_{11} \neq 0$. In this case the node is generic, and as $\widetilde{K} \neq 0$, $K_2 > 0$ and $\mu_2 > 0$, considering Lemma 9.1 we arrive at the following two global configurations:
• $\mathrm{s}, n;\ \binom{2}{2}\ \widehat{P}\,\widehat{P}E - \widehat{P}\,\widehat{P}E,\ S \Rightarrow (c = -2,\ e = 0,\ u = 2)$ (if $\tilde{L} < 0$);
• $\mathrm{s}, n;\ \binom{2}{2}\ \widehat{P}\,\widehat{P}H - \widehat{P}\,\widehat{P}H,\ N^{\infty} \Rightarrow (c = -2,\ e = 0,\ u = 1/4)$ (if $\tilde{L} > 0$).

1.2.2.2.1.1.2: $W_{11} = 0$. Then we may assume $c = 1$, and considering (9.23) the condition $\mu_2 > 0$ implies $\tilde{L} = 8(1 + 2u)x^2 > 0$. Moreover in this case the condition $e = 0$ is equivalent to $U_3 = 0$. As a result we get the following two global configurations:
• $\mathrm{s}, n^d;\ \binom{2}{2}\ \widehat{P}\,\widehat{P}H - \widehat{P}\,\widehat{P}H,\ N^{\infty} \Rightarrow (c = 1,\ e = 1,\ u = -1/4)$ (if $U_3 \neq 0$);
• $\mathrm{s}, n^*;\ \binom{2}{2}\ \widehat{P}\,\widehat{P}H - \widehat{P}\,\widehat{P}H,\ N^{\infty} \Rightarrow (c = 1,\ e = 0,\ u = -1/4)$ (if $U_3 = 0$).

1.2.2.2.1.2: $\mathcal{B}_3 = 0$. By Remark 9.8 we have an integrable saddle, and we may assume that it is located at $M_1(0,0)$, i.e. $c = -1$. Then we have $W_{11} = -384u^4x^4 \neq 0$, and by Remark 9.9 the node is generic. Considering Lemma 9.1 we arrive at the following two global configurations:
• $\$, n;\ \binom{2}{2}\ \widehat{P}\,\widehat{P}E - \widehat{P}\,\widehat{P}E,\ S \Rightarrow (c = -1,\ e = 0,\ u = 2)$ (if $\tilde{L} < 0$);
• $\$, n;\ \binom{2}{2}\ \widehat{P}\,\widehat{P}H - \widehat{P}\,\widehat{P}H,\ N^{\infty} \Rightarrow (c = -1,\ e = 0,\ u = 1/4)$ (if $\tilde{L} > 0$).

1.2.2.2.2: $\widetilde{M} = 0$. Then we have $c = -2u$, and we obtain

$$C_2 = ex^3,\quad \mathcal{B}_3 = -288u^2(-1 + 2u)(1 + 4u)x^2,\quad W_{11} = -192u^4(1 + 2u)x^4.$$

1.2.2.2.2.1: $C_2 \neq 0$. Due to $\mu_2 > 0$ we have $u > -1/2$, and this implies $W_{11} \neq 0$, i.e. the node is generic. So since $\widetilde{K} \neq 0$, $\mu_2 > 0$ and $K_2 > 0$, according to Lemma 9.1 we arrive at the following two global configurations:

• s, n; $\binom{2}{3}$ $H\,\widehat{P}\,E-\widehat{P}\,\widehat{P}\,\widehat{P}$ $\Rightarrow (c=-2,\ e=1,\ u=1)$ (if $\mathcal{B}_3 \neq 0$);
• $\$, n$; $\binom{2}{3}$ $H\,\widehat{P}\,E-\widehat{P}\,\widehat{P}\,\widehat{P}$ $\Rightarrow (c=-1,\ e=1,\ u=1/2)$ (if $\mathcal{B}_3 = 0$).

1.2.2.2.2.2: $C_2 = 0$. Then we have $c + 2u = e = 0$, and as the node is generic, considering Lemma 9.1 we obtain the following two global configurations:
• s, n; $[\infty; N] \Rightarrow (c=-2,\ e=0,\ u=1)$ (if $\mathcal{B}_3 \neq 0$);
• $\$, n$; $[\infty; N] \Rightarrow (c=-1,\ e=0,\ u=1/2)$ (if $\mathcal{B}_3 = 0$).

2: $\widetilde{K} = 0$. Since $\widetilde{K}(\tilde{a},x,y) = \text{Jacob}\,(p_2(\tilde{a},x,y), q_2(\tilde{a},x,y))$ (see Section 5.4), the condition $\widetilde{K} = 0$ means that the homogeneous quadratic parts of generic quadratic systems are proportional, say $q_2 = \lambda p_2$. Therefore clearly the transformation $x_1 = x,\ y_1 = -\lambda x + y$ leads to quadratic systems, of which the second equation is linear. Applying a translation (as these systems must have two real finite distinct singularities), we get the systems:

$$\dot{x} = cx + dy + gx^2 + 2hxy + ky^2, \quad \dot{y} = ex + fy,$$

for which we calculate

$$\mathbf{U} = \alpha^2(ex+fy)^2(gx^2+2hxy+ky^2)^2, \quad \mu_2 = \beta(gx^2+2hxy+ky^2),$$

where $\alpha = cf - de$ and $\beta = f^2g - 2efh + e^2k$. These systems possess the singularities $M_1(0,0)$ and $M_2(-f\alpha/\beta,\ e\alpha/\beta)$, which are distinct due to $\mathbf{U} > 0$ and $\mu_2 \neq 0$. We observe that the condition $e = 0$ implies the existence of the invariant line $y = 0$ for these systems. So we consider two cases: $e \neq 0$ and $e = 0$.

In the first case ($e \neq 0$) we apply the transformation $x_1 = ex + fy$, $y_1 = \beta y/(e\alpha)$ and $t_1 = \beta t/(e\alpha)$, which places the point M_2 at the point $(0,1)$. This leads to the family of systems (we keep the old notation of variables)

$$\dot{x} = cx + dy + gx^2 + 2hxy - dy^2, \quad \dot{y} = x. \tag{9.24}$$

If $e = 0$ then $cfg \neq 0$ (as $\alpha\beta \neq 0$), and after the rescaling $(x,y,t) \mapsto (-cx/g, y, t/f)$ (which places $M_2(-c/g, 0)$ to the point $(1,0)$), we arrive at the family of systems:

$$\dot{x} = cx + dy - cx^2 + 2hxy + ky^2, \quad \dot{y} = y. \tag{9.25}$$

In what follows we consider each one of the families of systems we just obtained.

2.1: Systems (9.24). We observe that the Jacobian matrices for the singularities $M_1(0,0)$ and $M_2(0,1)$ of these systems are $\begin{pmatrix} c & d \\ 1 & 0 \end{pmatrix}$ and $\begin{pmatrix} c+2h & -d \\ 1 & 0 \end{pmatrix}$ respectively. So the next remark becomes obvious.

Remark 9.10. The family of systems (9.24) could not have a finite star node.

For systems (9.24) we calculate

$$\begin{aligned}
\mu_0 &= \mu_1 = \kappa = \widetilde{K} = 0, \ \eta = 4g^2(dg + h^2), \ \theta_2 = dg + h^2, \\
\mu_2 &= -d(gx^2 + 2hxy - dy^2), \ \mathbf{U} = d^2x^2(gx^2 + 2hxy - dy^2)^2, \\
G_8 &= 2dg(dg + h^2), \ \tilde{L} = 8g(gx^2 + 2hxy - dy^2), \\
W_4 &= 16d^2g^2(dg + h^2)^2(c^2 + 4d)(c^2 - 4d + 4ch + 4h^2) \\
&= 16d^2g^2(dg + h^2)^2\tau_1\tau_2, \ \mathbf{D} = -192d^4(dg + h^2), \\
\mathcal{F}_1 &= 2d(cg + h + gh), \ \mathcal{T}_4 = 4cdg(c + 2h)(dg + h^2).
\end{aligned} \tag{9.26}$$

Remark 9.11. The condition $W_4 \neq 0$ implies $G_8\mathbf{D}\theta_2\eta \neq 0$ for systems (9.24).

2.1.1: $W_4 < 0$. Then by Remark 9.11 we have $G_8\mathbf{D} \neq 0$, and due to $\widetilde{K} = 0$, according to Theorem 6.1 (see Table 6.2, line 157) we have a saddle and either a focus or a center.

2.1.1.1: $\mathcal{T}_4 \neq 0$. By Theorem 6.2 we have a strong saddle and a strong focus. Additionally as by Remark 9.11 the condition $\eta\theta_2 \neq 0$ holds, considering Lemma 9.1 we obtain the following three global configurations:

- $s, f;\ N^d,\ \binom{1}{1}\copyright,\ \binom{1}{1}\copyright \Rightarrow (c = 1,\ d = -2,\ g = 1,\ h = 1)$ (if $\eta < 0$);
- $s, f;\ \overline{\binom{1}{1}}SN,\ \overline{\binom{1}{1}}SN,\ N^d \Rightarrow (c = 1,\ d = -2,\ g = -1,\ h = 1)$ (if $\eta > 0,\ \mu_2\tilde{L} < 0$);
- $s, f;\ \overline{\binom{1}{1}}SN,\ \overline{\binom{1}{1}}NS,\ N^d \Rightarrow (c = 1,\ d = 3,\ g = -1/4,\ h = 1)$ (if $\eta > 0\ \mu_2\tilde{L} > 0$).

2.1.1.2: $\mathcal{T}_4 = 0$. Then one of the finite singularities is weak, and by Remark 9.3 without loss of generality we may assume $\rho_1 = c = 0$, i.e. this weak singularity is $M_1(0,0)$. In this case for systems (9.24) we calculate

$$\begin{aligned}
\mathcal{T}_3 &= 8dgh(dg + h^2), \ \ \mathcal{F} = gh(dg + h^2), \ \ \mathcal{F}_1 = 2dh(1 + g), \\
\mathcal{F}_2 &= \mathcal{F}_3 = 0, \ \ W_4 = -256d^3g^2(d - h^2)(dg + h^2)^2.
\end{aligned} \tag{9.27}$$

2.1.1.2.1: $\mathcal{T}_3\mathcal{F} < 0$. According to Theorem 6.2 the systems possess a weak focus, the order of which is determined by the invariant polynomial $\mathcal{F}_1$.

2.1.1.2.1.1: $\mathcal{F}_1 \neq 0$. Then we have a first order weak focus, and considering Lemma 9.1 and the conditions $\eta\theta_2 \neq 0$ and $\kappa = 0$, we obtain the following three global configurations:

- $s, f^{(1)};\ N^d,\ \binom{1}{1}\copyright,\ \binom{1}{1}\copyright \Rightarrow (c = 0,\ d = -1,\ g = 2,\ h = 1)$ (if $\eta < 0$);
- $s, f^{(1)};\ \overline{\binom{1}{1}}SN,\ \overline{\binom{1}{1}}SN,\ N^d \Rightarrow (c = 0,\ d = -1,\ g = -2,\ h = 1)$ (if $\eta > 0$, $\mu_2\tilde{L} < 0$);
- $s, f^{(1)};\ \overline{\binom{1}{1}}SN,\ \overline{\binom{1}{1}}NS,\ N^d \Rightarrow (c = 0,\ d = -1,\ g = 1/2,\ h = 1)$ (if $\eta > 0$, $\mu_2\tilde{L} > 0$).

2.1.1.2.1.2: $\mathcal{F}_1 = 0$. Then by (9.27) due to $\mathcal{T}_3 \neq 0$ we get $g = -1$, and then we have

$$\mu_2\tilde{L} = 8d(x^2 - 2hxy + dy^2)^2, \quad \mathcal{T}_3\mathcal{F} = 8dh^2(d - h^2)^2,$$
$$\eta = 4(h^2 - d), \quad W_4 = 256d^3(h^2 - d)^3.$$

So we arrive at the next remark:

Remark 9.12. If $\mathcal{T}_4 = 0$ and $\mathcal{T}_3 \neq 0$, then the condition $\mathcal{F}_1 = 0$ implies sign $(\mathcal{T}_3\mathcal{F})$ = sign $(\mu_2\tilde{L})$ = sign (ηW_4).

Since by (9.27) we have $\mathcal{F}_2 = \mathcal{F}_3 = 0$, according to Theorem 6.2 the systems (9.24) possess a center. By Remark 9.12 the conditions $\mathcal{T}_3\mathcal{F} < 0$ and $W_4 < 0$ imply $\mu_2\tilde{L} < 0$ and $\eta > 0$. So considering Lemma 9.1 we obtain the global configuration

- $s, c;\ \overline{\binom{1}{1}}SN,\ \overline{\binom{1}{1}}SN,\ N^d \Rightarrow (c = 0,\ d = -2,\ g = -1,\ h = 1)$.

2.1.1.2.2: $\mathcal{T}_3\mathcal{F} > 0$. In this case by Theorem 6.2 the systems possess a weak saddle.

2.1.1.2.2.1: $\mathcal{F}_1 \neq 0$. Then we have a first order weak saddle, and considering Lemma 9.1 and the conditions $\eta\theta_2 \neq 0$ and $\kappa = 0$, we obtain the following three global configurations:

- $s^{(1)}, f;\ N^d,\ \binom{1}{1}$ ©, $\binom{1}{1}$ © $\Rightarrow (c = 0,\ d = 2,\ g = -2,\ h = 1)$ (if $\eta < 0$);
- $s^{(1)}, f;\ \overline{\binom{1}{1}}SN,\ \overline{\binom{1}{1}}SN,\ N^d \Rightarrow (c = 0,\ d = 2,\ g = 1,\ h = 1)$ (if $\eta > 0,\ \mu_2\tilde{L} < 0$);
- $s^{(1)}, f;\ \overline{\binom{1}{1}}SN,\ \binom{1}{1}NS,\ N^d \Rightarrow (c = 0,\ d = 2,\ g = -1/3,\ h = 1)$ (if $\eta > 0\ \mu_2\tilde{L} > 0$).

2.1.1.2.2.2: $\mathcal{F}_1 = 0$. Then the saddle is integrable. Additionally by Remark 9.12 the condition $W_4 < 0$ implies $\eta < 0$, and this leads to the global configuration

- \$$, f;\ N^d,\ \binom{1}{1}$ ©, $\binom{1}{1}$ © $\Rightarrow (c = 0,\ d = 2,\ g = -1,\ h = 1)$.

2.1.1.2.3: $\mathcal{T}_3 = 0$. By (9.27) due to $W_4 \neq 0$ the condition $\mathcal{T}_3 = 0$ implies $h = 0$, and then we obtain

$$\mathcal{T}_4 = \mathcal{T}_3 = \mathcal{F}_1 = 0, \quad \mathcal{T}_2 = 4d^2g^2, \quad \mathcal{B} = -2d^2g^4,$$
$$W_4 = -256d^6g^4, \quad \eta = 4dg^3, \quad \mu_2\tilde{L} = -8dg(gx^2 - dy^2)^2.$$

Remark 9.13. If $W_4 \neq 0$ then the condition $\mathcal{T}_4 = \mathcal{T}_3 = 0$ implies $W_4 < 0$.

The condition $W_4 \neq 0$ implies $\mathcal{B} < 0$ and $\mathcal{T}_2 > 0$. Therefore by Theorem 6.2 we have a center and an integrable saddle. As the condition $\eta > 0$ implies $\mu_2\tilde{L} < 0$, taking into consideration Lemma 9.1 we get the following two global configurations:

- \$$, c;\ N^d,\ \binom{1}{1}$ ©, $\binom{1}{1}$ © $\Rightarrow (c = 0,\ d = 1,\ g = -1,\ h = 0)$ (if $\eta < 0$);
- \$$, c;\ \overline{\binom{1}{1}}SN,\ \overline{\binom{1}{1}}SN,\ N^d \Rightarrow (c = 0,\ d = 1,\ g = 1,\ h = 0)$ (if $\eta > 0$).

2.1.2: $W_4 > 0$. Since $\widetilde{K} = 0$ and $G_8\mathbf{D} \neq 0$ (see Remark 9.11), according to Theorem 6.1 (see Table 6.2, line 151) systems (9.24) possess a saddle and a node.

2.1.2.1: $\mathcal{T}_4 \neq 0$. Then by Theorem 6.2 the saddle is strong, and as $W_4 \neq 0$ the node is generic and $\theta_2 \neq 0$. Therefore considering Lemma 9.1 we obtain the following three global configurations:
- $s, n;\ N^d,\ \binom{1}{1}$ ©, $\binom{1}{1}$ © $\Rightarrow (c=1,\ d=1/2,\ g=-3,\ h=1)$ (if $\eta<0$);
- $s, n;\ \binom{\overline{1}}{1}SN,\ \binom{\overline{1}}{1}SN,\ N^d \Rightarrow (c=1,\ d=1/2,\ g=1,\ h=1)$ (if $\eta>0,\ \mu_2\tilde{L}<0$);
- $s, n;\ \binom{\overline{1}}{1}SN,\ \binom{\overline{1}}{1}NS,\ N^d \Rightarrow (c=1,\ d=1/2,\ g=-1,\ h=1)$ (if $\eta>0,\ \mu_2\tilde{L}>0$).

2.1.2.2: $\mathcal{T}_4 = 0$. Then the saddle is weak, and by Remark 9.3 without loss of generality we may assume that this saddle is located at $M_1(0,0)$, i.e. $c=0$. We consider the relations (9.27).

We observe that by Remark 9.13 the condition $W_4>0$ implies $\mathcal{T}_3 \neq 0$.

2.1.2.2.1: $\mathcal{F}_1 \neq 0$. In this case by Theorem 6.2 the weak saddle is of order one, and considering Lemma 9.1 we arrive at the following three global configurations:
- $s^{(1)}, n;\ N^d,\ \binom{1}{1}$ ©, $\binom{1}{1}$ © $\Rightarrow (c=0,\ d=1/2,\ g=-3,\ h=1)$ (if $\eta<0$);
- $s^{(1)}, n;\ \binom{\overline{1}}{1}SN,\ \binom{\overline{1}}{1}SN,\ N^d \Rightarrow (c=0,\ d=1/2,\ g=1,\ h=1)$ (if $\eta>0,\ \mu_2\tilde{L}<0$);
- $s^{(1)}, n;\ \binom{\overline{1}}{1}SN,\ \binom{\overline{1}}{1}NS,\ N^d \Rightarrow (c=0,\ d=1/2,\ g=-1/2,\ h=1)$ (if $\eta>0$, $\mu_2\tilde{L}>0$).

2.1.2.2.2: $\mathcal{F}_1 = 0$. As $\mathcal{F}_2=\mathcal{F}_3=0$, according to Theorem 6.2 the systems (9.24) possess an integrable saddle. Additionally by Remark 9.12 the condition $W_4>0$ implies $\eta>0$, and this leads to the global configuration
- $\$, n;\ \binom{\overline{1}}{1}SN,\ \binom{\overline{1}}{1}NS,\ N^d \Rightarrow (c=0,\ d=1/2,\ g=-1,\ h=1)$.

2.1.3: $W_4 = 0$. Taking into account (9.26) we consider two possibilities: $\eta \neq 0$ and $\eta = 0$.

2.1.3.1: $\eta \neq 0$. Then due to $\mu_2 \neq 0$ the condition $W_4=0$ implies $\tau_1\tau_2 = (c^2+4d)(c^2-4d+4ch+4h^2)=0$, i.e. one of the singularities is a node with coinciding eigenvalues. Moreover by Remark 9.10 we have a node with one direction. According to Remark 9.3 we may assume that the singularity $M_1(0,0)$ is such a node, and this implies $\tau_1 = c^2+4d=0$, i.e. $d=-c^2/4\neq 0$. So we may assume $c=1$ due to the rescaling $(x,y,t)\mapsto(cx,y,t/c)$, and this leads to the family of systems:

$$\dot{x} = x - y/4 + gx^2 + 2hxy + y^2/4, \quad \dot{y} = x,$$

for which we calculate

$$\mu_2\tilde{L} = 2g(4gx^2+8hxy+y^2)^2/8,\ \eta = g^2(4h^2-g),$$
$$\mathcal{T}_4 = g(1+2h)(g-4h^2)/4,\ \theta_2 = (4h^2-g)/4.$$

2.1.3.1.1: $\mathcal{T}_4 \neq 0$. Then by Theorem 6.2 we have a strong saddle. We observe that the condition $\eta \neq 0$ implies $\theta_2 \neq 0$, and considering Lemma 9.1 we obtain the following three global configurations:

- $s, n^d;\ N^d,\ \binom{1}{1}$ ©, $\binom{1}{1}$ © $\Rightarrow (c=1,\ d=-1/4,\ g=1,\ h=0)$ (if $\eta<0$);
- $s, n^d;\ \overline{\binom{1}{1}}SN,\ \overline{\binom{1}{1}}SN,\ N^d \Rightarrow (c=1,\ d=-1/4,\ g=-1,\ h=0)$ (if $\eta>0$, $\mu_2\tilde{L}<0$);
- $s, n^d;\ \overline{\binom{1}{1}}SN,\ \overline{\binom{1}{1}}NS,\ N^d \Rightarrow (c=1,\ d=-1/4,\ g=1,\ h=1)$ (if $\eta>0$, $\mu_2\tilde{L}>0$).

2.1.3.1.2: $\mathcal{T}_4=0$. Then the saddle M_2 is weak. As $\eta\neq 0$ we get $h=-1/2$, and then

$$\eta=g^2(1-g),\quad \mathcal{F}_1=(1-g)/4,\quad \theta_2=(1-g)/4.$$

So due to $\eta\neq 0$ we have $\theta_2\mathcal{F}_1\neq 0$, i.e. the weak saddle is of order one. Therefore we get the following three global configurations:

- $s^{(1)}, n^d;\ N^d,\ \binom{1}{1}$ ©, $\binom{1}{1}$ © $\Rightarrow (c=1,\ d=-1/4,\ g=2,\ h=-1/2)$ (if $\eta<0$);
- $s^{(1)}, n^d;\ \overline{\binom{1}{1}}SN,\ \overline{\binom{1}{1}}SN,\ N^d \Rightarrow (c=1,\ d=-1/4,\ g=-1,\ h=-1/2)$ (if $\eta>0$, $\mu_2\tilde{L}<0$);
- $s^{(1)}, n^d;\ \overline{\binom{1}{1}}SN,\ \overline{\binom{1}{1}}NS,\ N^d \Rightarrow (c=1,\ d=-1/4,\ g=1/2,\ h=-1/2)$ (if $\eta>0$, $\mu_2\tilde{L}>0$).

2.1.3.2: $\eta=0$. Considering (9.26) we have $g(dg+h^2)=0$, and we consider two possibilities: $\tilde{L}\neq 0$ (then $g\neq 0$) and $\tilde{L}=0$ (then $g=0$).

2.1.3.2.1: $\tilde{L}\neq 0$. In this case $g\neq 0$ and as $d\neq 0$ we obtain $g=-h^2/d\neq 0$. Hence we may assume $h=1$ due to the rescaling $(x,y,t)\mapsto(hx,y,t/h)$ (and by introducing new parameters $\tilde{c}=c/h$, $\tilde{d}=d/h^2$). So we arrive at the family of systems (we keep the old parameters)

$$\dot{x}=cx+dy-x^2/d+2xy-dy^2,\quad \dot{y}=x, \tag{9.28}$$

for which we calculate

$$\begin{gathered}
\kappa=\eta=\widetilde{K}=\mathbf{D}=0,\ \kappa_1=-32(1+c-d)/d,\ \mu_2=(x-dy)^2,\\
\theta_5=96(1+c-d)(x-dy)^3)/d^2,\ \widetilde{M}=-8(x-dy)^2/d^2=-\tilde{L},\\
G_3=-2(1+c-d),\ \mathcal{T}_i=0,\ i=1,2,3,4,\ \sigma=c-2x/d+2y,\\
W_8=2^{12}3^3(1+c-d)^2(c^2+4d)\big[(c+2)^2-4d\big]\\
=2^{12}3^3(1+c-d)^2\tau_1\tau_2,\ \mathcal{F}_1=-2(1+c-d),\\
\mathcal{H}=0,\ \mathcal{B}_2=4(1+c)(1+c-d)^2/d,\\
\mathcal{B}_1=-2c(2+c)(1+c-d)=-2(1+c-d)\rho_1\rho_2.
\end{gathered} \tag{9.29}$$

Remark 9.14. The condition $W_8\neq 0$ implies $G_3\kappa_1\theta_5\mathcal{F}_1\neq 0$ for systems (9.28). Moreover in this case the condition $\mathcal{B}_1=0$ is equivalent to $\rho_1\rho_2=0$.

Since $\widetilde{K}=\mathbf{D}=0$, according to Theorem 6.1 the types of the finite singularities are governed by the invariant polynomials W_8, W_9, G_3 and D_2.

2.1.3.2.1.1: $W_8 < 0$. Then by Remark 9.14 we have $G_3 \neq 0$, and by Theorem 6.1 (see Table 6.2, line 159) systems (9.28) possess a saddle and a focus (or a center).

2.1.3.2.1.1.1: $\mathcal{B}_1 \neq 0$. Then $\rho_1\rho_2 \neq 0$ and we do not have weak singularities. Additionally as $\widetilde{M}\tilde{L}\kappa_1\theta_5 \neq 0$, considering Lemma 9.1, we obtain the global configuration

- $s, f;\ \widehat{\binom{2}{2}}\ \widehat{P}_{\curlywedge}\ \widehat{P}\,H_{\curlywedge}-H,\ N^d \Rightarrow (c=1,\ d=-1,\ g=1,\ h=1)$.

2.1.3.2.1.1.2: $\mathcal{B}_1 = 0$. Then one of the finite singularities is weak, and by Remark 9.3 without loss of generality we may assume such a point to be $M_1(0,0)$, i.e. for systems (9.28) the condition $\rho_1 = c = 0$ holds. So we have

$$\mathcal{B}_1 = 0,\quad \mathcal{B}_2 = 4(1-d)^2/d,\quad W_8 = 2^{16}3^3 d(1-d)^3.$$

Therefore the condition $W_8 < 0$ implies $\mathcal{B}_2 \neq 0$, and by Theorem 6.2 the type of weak singularity (which is of order one due to $\mathcal{F}_1 \neq 0$) depends of the sign of $\mathcal{B}_2$. Thus considering Lemma 9.1 we get the following two global configurations:

- $s, f^{(1)};\ \widehat{\binom{2}{2}}\ \widehat{P}_{\curlywedge}\ \widehat{P}\,H_{\curlywedge}-H,\ N^d \Rightarrow (c=0,\ d=-1,\ g=1,\ h=1)$ (if $\mathcal{B}_2 < 0$);
- $s^{(1)}, f;\ \widehat{\binom{2}{2}}\ \widehat{P}_{\curlywedge}\ \widehat{P}\,H_{\curlywedge}-H,\ N^d \Rightarrow (c=0,\ d=2,\ g=-1/2,\ h=1)$ (if $\mathcal{B}_2 > 0$).

2.1.3.2.1.2: $W_8 > 0$. Then by Theorem 6.1 (see Table 6.2, line 153) systems (9.28) possess a saddle and a node, which is generic (due to $W_8 \neq 0$).

2.1.3.2.1.2.1: $\mathcal{B}_1 \neq 0$. Then the saddle is strong, and considering Remark 9.14 and Lemma 9.1, we arrive at the global configuration

- $s, n;\ \widehat{\binom{2}{2}}\ \widehat{P}_{\curlywedge}\ \widehat{P}\,H_{\curlywedge}-H,\ N^d \Rightarrow (c=1,\ d=1,\ g=-1,\ h=1)$.

2.1.3.2.1.2.2: $\mathcal{B}_1 = 0$. It was mentioned earlier that in this case we may assume $c = 0$, i.e. the weak saddle of the first order (as $\mathcal{F}_1 \neq 0$) is located at the origin of coordinates. Considering Remark 9.14 and Lemma 9.1, we obtain the global configuration

- $s^{(1)}, n;\ \widehat{\binom{2}{2}}\ \widehat{P}_{\curlywedge}\ \widehat{P}\,H_{\curlywedge}-H,\ N^d \Rightarrow (c=0,\ d=1/2,\ g=-2,\ h=1)$.

2.1.3.2.1.3: $W_8 = 0$. Then $(1+c-d)\tau_1\tau_2 = 0$, and considering (9.29) we have to distinguish two subcases given by the invariant polynomial κ_1.

2.1.3.2.1.3.1: $\kappa_1 \neq 0$. Then $(1+c-d) \neq 0$, and the condition $W_8 = 0$ gives $\tau_1\tau_2 = 0$. Therefore we have a node with coinciding eigenvalues, and by Remark 9.3 we may assume that this node is located at the origin of coordinates, i.e. the condition $\tau_1 = c^2 + 4d = 0$ holds. We note that by Remark 9.10 this node could not be a star node. So we have $d = -c^2/4 \neq 0$, and we calculate

$$\mathcal{B}_1 = -c(2+c)^3/2,\ \kappa_1 = 32(2+c)^2/c^2,\ \ \theta_5 = -6(2+c)^2(4x+c^2y)^3/c^4.$$

The condition $\kappa_1 \neq 0$ implies $\mathcal{B}_1 \neq 0$, and by Theorem 6.2, the saddle is strong. Additionally the condition $\kappa_1 \neq 0$ implies $\theta_5 \neq 0$, and considering Lemma 9.1, we arrive at the global configuration

- s, n^d; $\widehat{\binom{2}{2}}\, \widehat{P}_{\curlywedge}\, \widehat{P}\, H_{\curlywedge} - H,\ N^d \Rightarrow (c = 1,\ d = -1/4,\ g = 4,\ h = 1)$.

2.1.3.2.1.3.2: $\kappa_1 = 0$. Then by (9.29) we have $c = d - 1$, and for systems (9.28) we calculate

$$\begin{gathered}\mu_0 = \mu_1 = \widetilde{K} = \eta = \mathbf{D} = \kappa = \kappa_1 = 0,\ \ \widetilde{M} = -8(x - dy)^2/d^2 = \mathcal{G},\\ \mu_2 = (x - dy)^2,\ \ K_2 = 96(x - dy)^2/d^2,\ \ \theta_6 = 8(x - dy)^4/d^2,\\ G_3 = 0,\ \ D_2 = 2(x - dy)/d,\ \ \mathbf{U} = x^2(x - dy)^4,\\ \mathcal{T}_i = 0,\ i = 1, 2, 3, 4,\ \ \sigma = d - 1 - 2x/d + 2y,\ \ \mathcal{F}_1 = \mathcal{H} = 0,\\ \mathcal{B} = \mathcal{B}_1 = \mathcal{B}_2 = 0,\ \ \mathcal{B}_3 = 72(d - 1)(1 + d)(x - dy)/d^2.\end{gathered} \tag{9.30}$$

Since $\mathbf{D} = \widetilde{K} = G_3 = 0$ and $D_2 \neq 0$, according to Theorem 6.1 (see Table 6.2, line 154) we have a saddle and a node. Considering (9.30), by Theorem 6.2 (see statement $(e_3)[\delta]$) the saddle will be weak (more precisely it will be an integrable one) if and only if $\mathcal{B}_3 = 0$. Moreover for the singular points $M_1(0,0)$ and $M_2(0,1)$, we have respectively

$$\rho_1 = d - 1,\ \ \tau_1 = (d + 1)^2,\ \ \ \rho_2 = d + 1,\ \ \tau_2 = (d - 1)^2.$$

Therefore we observe that if one of the singular points is a weak saddle, the second one becomes a node with coinciding eigenvalues, which by Remark 9.10 is a node n^d.

Since $K_2 > 0$ and $\theta_6 \neq 0$, considering Lemma 9.1, we obtain the following two global configurations:

- s, n; $\binom{2}{2}\, \widehat{\widetilde{P}}\, \widehat{P}\, H - \widehat{\widetilde{P}}\, \widehat{P}\, H,\ N^d \Rightarrow (c = 1,\ d = 2,\ g = -1/2,\ h = 1)$ (if $\mathcal{B}_3 \neq 0$);
- $\$, n^d$; $\binom{2}{2}\, \widehat{\widetilde{P}}\, \widehat{P}\, H - \widehat{\widetilde{P}}\, \widehat{P}\, H,\ N^d \Rightarrow (c = 0,\ d = 1,\ g = -1,\ h = 1)$ (if $\mathcal{B}_3 = 0$).

2.1.3.2.2: $\tilde{L} = 0$. Then $g = 0$ and we obtain the family of systems:

$$\dot{x} = cx + dy + 2hxy - dy^2,\quad \dot{y} = x, \tag{9.31}$$

for which we calculate

$$\begin{gathered}\mu_0 = \mu_1 = \widetilde{K} = \kappa = \eta = G_8 = 0,\ \mu_2 = dy(-2hx + dy),\\ \mathbf{D} = -192d^4h^2,\ \mathbf{U} = d^2x^2y^2(2hx - dy)^2,\ \widetilde{M} = -32h^2y^2,\\ W_7 = 12d^2h^6(c^2 + 4d)\big[(c + 2h)^2 - 4d\big] = 12d^2h^6\tau_1\tau_2,\\ \mathcal{T}_i = 0,\ i = 1, 2, 3, 4,\ \ \sigma = c + 2hy,\ \ \mathcal{F}_1 = 2dh,\ \ \mathcal{H} = 0,\\ \mathcal{B}_1 = 2cdh(c + 2h) = 2dh\rho_1\rho_2,\ \ \mathcal{B}_2 = 4dh^3(c + h).\end{gathered} \tag{9.32}$$

For the singular points $M_1(0,0)$ and $M_2(0,1)$ of the above systems, we have $\Delta_2 = d = -\Delta_1$, and hence these systems possess a saddle and an anti-saddle.

2.1.3.2.2.1: $W_7 < 0$. Then we have a saddle and a focus (or a center), and considering (9.32) we observe that the condition $\mathcal{B}_1 = 0$ is equivalent to $\rho_1\rho_2 = 0$.

2.1.3.2.2.1.1: $\mathcal{B}_1 \neq 0$. Then both singularities are strong, and considering Lemma 9.1, we arrive at the global configuration

- $s, f;\ \widehat{\binom{1}{2}}\ \widehat{P}_\curlywedge E\widehat{P}_\curlywedge\!-\!H, \overline{\binom{1}{1}}SN \Rightarrow (c=1,\ d=-1,\ g=0,\ h=1)$.

2.1.3.2.2.1.2: $\mathcal{B}_1 = 0$. Then one of the finite singularities is weak, and by Remark 9.3 without loss of generality we may assume such a point is $M_1(0,0)$, i.e. for systems (9.31) the condition $\rho_1 = c = 0$ holds. In this case we calculate

$$\mathcal{B}_1 = 0,\quad \mathcal{B}_2 = 4dh^4,\quad \mathcal{F}_1 = 2dh,\quad W_7 = -192d^3h^6(d-h^2).$$

Therefore the condition $W_7 \neq 0$ implies $\mathcal{F}_1\mathcal{B}_2 \neq 0$, and by Theorem 6.2 the type of the weak singularity (which is of order one due to $\mathcal{F}_1 \neq 0$) depends on the sign of $\mathcal{B}_2$. Thus considering Lemma 9.1 we get the following two global configurations:

- $s, f^{(1)};\ \widehat{\binom{1}{2}}\ \widehat{P}_\curlywedge E\widehat{P}_\curlywedge\!-\!H, \overline{\binom{1}{1}}SN \Rightarrow (c=0,\ d=-1,\ g=0,\ h=1)\quad (\text{if } \mathcal{B}_2<0)$;
- $s^{(1)}, f;\ \widehat{\binom{1}{2}}\ \widehat{P}_\curlywedge E\widehat{P}_\curlywedge\!-\!H, \overline{\binom{1}{1}}SN \Rightarrow (c=0,\ d=2,\ g=0,\ h=1)\quad (\text{if } \mathcal{B}_2>0)$.

2.1.3.2.2.2: $W_7 > 0$. Then by Theorem 6.1 (see Table 6.2, line 152) systems (9.31) possess a saddle and a node, which is generic (due to $W_7 \neq 0$).

Taking into account the fact that the saddle is weak if and only if $\mathcal{B}_1 = 0$, we get the following two global configurations:

- $s, n;\ \widehat{\binom{1}{2}}\ \widehat{P}_\curlywedge E\widehat{P}_\curlywedge\!-\!H, \overline{\binom{1}{1}}SN \Rightarrow (c=1,\ d=1/2,\ g=0,\ h=1)\quad (\text{if } \mathcal{B}_1 \neq 0)$;
- $s^{(1)}, n;\ \widehat{\binom{1}{2}}\ \widehat{P}_\curlywedge E\widehat{P}_\curlywedge\!-\!H, \overline{\binom{1}{1}}SN \Rightarrow (c=0,\ d=1/2,\ g=0,\ h=1)\quad (\text{if } \mathcal{B}_1 = 0)$.

2.1.3.2.2.3: $W_7 = 0$. Since $\mu_2 \neq 0$, by (9.32) we have $h\tau_1\tau_2 = 0$, and we consider two subcases: $\widetilde{M} \neq 0$ and $\widetilde{M} = 0$.

2.1.3.2.2.3.1: $\widetilde{M} \neq 0$. Then $h \neq 0$ and we have a node with coinciding eigenvalues. By Remark 9.3 we may assume that this node is located at the origin of coordinates, i.e. we have $\tau_1 = c^2+4d = 0$. So we have a node n^d (see Remark 9.10), and setting $d = -c^2/4 \neq 0$ we may assume $c = 1$ due to the rescaling$(x,y,t) \mapsto (cx, yt/c)$. This leads to the family of systems:

$$\dot{x} = x - y/4 + 2hxy + y^2/4,\quad \dot{y} = x,$$

and we calculate

$$\sigma = 1+2hy,\ \mathcal{B}_1 = -h(1+2h)/2,\ \mathcal{B}_2 = -h^3(1+h),\ \mathcal{F}_1 = -h/2,\ \widetilde{M} = -32h^2y^2.$$

Since $\widetilde{M} \neq 0$ we could have a weak saddle if and only if $h = -1/2$, and the weak saddle is of order one due to $\mathcal{F}_1 \neq 0$ (see Theorem 6.2, statement (e_1)). So considering Lemma 9.1 we get the following two global configurations:

- $s, n^d;\ \widehat{\binom{1}{2}}\ \widehat{P}_\curlywedge E\widehat{P}_\curlywedge\!-\!H, \overline{\binom{1}{1}}SN \Rightarrow (c=1,\ d=-1/4,\ g=0,\ h=1)\quad (\text{if } \mathcal{B}_1 \neq 0)$;
- $s^{(1)}, n^d;\ \widehat{\binom{1}{2}}\ \widehat{P}_\curlywedge E\widehat{P}_\curlywedge\!-\!H, \overline{\binom{1}{1}}SN \Rightarrow (c=1,\ d=-1/4,\ g=0,\ h=-1/2)$
  (if $\mathcal{B}_1 = 0$).

2.1.3.2.2.3.2: $\widetilde{M} = 0$. Then $h = 0$ and we get the family of systems:

$$\dot{x} = cx + dy - dy^2, \quad \dot{y} = x.$$

We calculate

$$\begin{aligned} \widetilde{M} = 0, \ \ C_2 = -dy^3, \ \ \mu_2 = d^2y^2, \ \ W_9 = 12(c^2 + 4d)(c^2 - 4d) = 12\tau_1\tau_2, \\ \mathcal{T}_i = 0, \ i = 1, 2, 3, 4, \ \ \sigma = c = \rho_1 = \rho_2, \ \ \Delta_2 = d = -\Delta_1. \end{aligned} \tag{9.33}$$

So the above systems possess a saddle and an anti-saddle, and clearly the type of the anti-saddle is governed by the invariant polynomial W_9.

1. $W_9 < 0$. In this case we have a saddle and a focus (or a center), and considering (9.33) we observe that we could have a weak singularity if and only if we have $\sigma = 0$. However in this case we get Hamiltonian systems possessing a center and an integrable saddle. As $C_2 \neq 0$ (due to $\mu_2 \neq 0$), considering Lemma 9.1 we obtain the following two global configurations:

   - $s, f;\ \widehat{\binom{2}{3}}\ \widehat{P}_{\curlywedge}\, \widehat{P} - \widehat{P}_{\curlywedge}\, \widehat{P} \Rightarrow (c = 1,\ d = 1,\ g = 0,\ h = 0) \quad (\text{if } \sigma \neq 0);$

   - $\$, c;\ \widehat{\binom{2}{3}}\ \widehat{P}_{\curlywedge}\, \widehat{P} - \widehat{P}_{\curlywedge}\, \widehat{P} \Rightarrow (c = 0,\ d = 1,\ g = 0,\ h = 0) \quad (\text{if } \sigma = 0).$

2. $W_9 \geq 0$. Then we have a saddle and a node, and in this case we have $\sigma \neq 0$; otherwise if $c = 0$ we get $W_9 = -192d^2 < 0$. So the saddle is strong and the node is generic if $W_9 > 0$, and it is a node with one direction (see Remark 9.10) if $W_9 = 0$. Therefore considering Lemma 9.1 we obtain the following two global configurations:

   - $s, n;\ \widehat{\binom{2}{3}}\ \widehat{P}_{\curlywedge}\, \widehat{P} - \widehat{P}_{\curlywedge}\, \widehat{P} \Rightarrow (c = 3,\ d = 1,\ g = 0,\ h = 0) \quad (\text{if } W_9 > 0);$

   - $s, n^d;\ \widehat{\binom{2}{3}}\ \widehat{P}_{\curlywedge}\, \widehat{P} - \widehat{P}_{\curlywedge}\, \widehat{P} \Rightarrow (c = 2,\ d = 1,\ g = 0,\ h = 0) \quad (\text{if } W_9 = 0).$

As all the possibilities have been investigated, we have completed the examination of systems (9.24).

2.2: Systems (9.25). We consider systems (9.25), i.e. the systems of the form

$$\dot{x} = cx + dy - cx^2 + 2hxy + ky^2, \quad \dot{y} = y, \tag{9.34}$$

which possess the finite singularities $M_1(0,0)$ and $M_2(1,0)$. We observe that the Jacobian matrices corresponding to these singular points are respectively $\begin{pmatrix} c & d \\ 0 & 1 \end{pmatrix}$ and $\begin{pmatrix} -c & d + 2h \\ 0 & 1 \end{pmatrix}$, and therefore we have

$$\rho_1 = c + 1, \ \ \Delta_1 = c, \ \ \tau_1 = (c - 1)^2; \ \ \rho_2 = 1 - c, \ \ \Delta_2 = -c, \ \ \tau_2 = (c + 1)^2.$$

So obviously we have the next remark.

Remark 9.15. The family of systems (9.34) have a finite node and a finite saddle. The node has coinciding eigenvalues if and only if $(c-1)(c+1)=0$, and in this case the systems simultaneously have a weak saddle. Moreover these systems have a star node if and only if either $c-1=d=0$, or $c+1=d+2h=0$.

For systems (9.34) we calculate

$$\begin{gathered}\mu_0=\mu_1=\kappa=\widetilde{K}=0,\quad \eta=4c^2(h^2+ck),\quad \mu_2=c(cx^2-2hxy-ky^2),\\ \theta_2=0,\quad \mathbf{U}=c^2y^2(cx^2-2hxy-ky^2)^2,\quad G_8=-2c^2(h^2+ck),\\ W_4=16c^4(c-1)^2(c+1)^2(h^2+ck)^2,\\ \mathbf{D}=-192c^4(h^2+ck),\quad \tilde{L}=8c(cx^2-2hxy-ky^2),\\ \mathcal{F}_1=-2c(cd-h+ch),\quad \mathcal{T}_4=4c^2(c+1)(c-1)(h^2+ck).\end{gathered}\tag{9.35}$$

We observe that for systems (9.34) the following conditions hold:

$$W_4\geq 0,\quad \mu_2\tilde{L}>0,\quad \theta_2=0.\tag{9.36}$$

2.2.1: $W_4\neq 0$. In this case by (9.36) we have $W_4>0$, and this implies $\eta\mathcal{T}_4\neq 0$. Hence obviously the node is generic and the saddle is strong.

Considering (9.36) and Lemma 9.1, we obtain the following two global configurations:

- $s, n;\ N^*,\ \binom{1}{1}$ ©, $\binom{1}{1}$ © $\Rightarrow (c=2,\ d=0,\ h=1,\ k=-1)$ (if $\eta<0$);
- $s, n;\ \overline{\binom{1}{1}}SN,\ \overline{\binom{1}{1}}NS,\ N^* \Rightarrow (c=2,\ d=0,\ h=1,\ k=1)$ (if $\eta>0$).

2.2.2: $W_4=0$. Taking into account (9.35), we consider two possibilities: $\eta\neq 0$ and $\eta=0$.

2.2.2.1: $\eta\neq 0$. Then the condition $W_4=0$ implies $\tau_1\tau_2=(c-1)^2(c+1)^2=0$, i.e. one of singularities is a node with coinciding eigenvalues. According to Remark 9.3 we may assume that the singularity $M_1(0,0)$ is such a node, and this implies $\tau_1=(c-1)^2=0$, i.e. $c=1$. Then we calculate

$$\mathcal{T}_4=\mathcal{F}_2=\mathcal{F}_3=0,\quad \mathcal{T}_3\mathcal{F}=8(h^2+k)^2,\quad \mathcal{F}_1=-2d,\quad \eta=4(h^2+k).$$

We observe that the condition $\eta\neq 0$ implies $\mathcal{T}_3\mathcal{F}>0$, and by Theorem 6.2 we have a weak saddle of order one if $\mathcal{F}_1\neq 0$ and an integrable saddle if $\mathcal{F}_1=0$. Additionallyt by Remark 9.15 we have a node n^d if $\mathcal{F}_1\neq 0$ and a star node if $\mathcal{F}_1=0$.

Thus considering (9.36) and Lemma 9.1, we obtain the following four global configurations:

- $s^{(1)}, n^d;\ N^*,\ \binom{1}{1}$ ©, $\binom{1}{1}$ © $\Rightarrow (c=1,\ d=1,\ h=1,\ k=-2)$ (if $\eta<0,\ \mathcal{F}_1\neq 0$);
- $\$, n^*;\ N^*,\ \binom{1}{1}$ ©, $\binom{1}{1}$ © $\Rightarrow (c=1,\ d=0,\ h=1,\ k=-2)$ (if $\eta<0,\ \mathcal{F}_1=0$);
- $s^{(1)}, n^d;\ \overline{\binom{1}{1}}SN,\ \overline{\binom{1}{1}}NS,\ N^* \Rightarrow (c=1,\ d=1,\ h=1,\ k=0)$ (if $\eta>0,\ \mathcal{F}_1\neq 0$);

• $\$, n^*; \binom{\overline{1}}{1}SN, \binom{\overline{1}}{1}NS, N^* \Rightarrow (c=1,\ d=0,\ h=1,\ k=0)$ (if $\eta>0$, $\mathcal{F}_1=0$).

2.2.2.2: $\eta=0$. Considering (9.35) due to $\mu_2\neq 0$ (i.e. $c\neq 0$), we obtain $k=-h^2/c\neq 0$, and this implies $W_4=0$. Then we may assume $h=1$ due to the rescaling $(x,y,t)\mapsto(x,y/h,t)$. So we get the family of systems:

$$\dot{x}=cx+dy-cx^2+2xy-y^2/c,\quad \dot{y}=y, \tag{9.37}$$

for which we calculate

$$\begin{gathered}\kappa=\eta=\widetilde{K}=\mathbf{D}=\theta_5=0,\quad \kappa_1=-32c(c-1+cd)=16G_3,\\ \widetilde{M}=-8(cx-y)^2=-\tilde{L},\quad \mu_2=(cx-y)^2,\quad \sigma=1+c-2cx+2y,\\ W_8=2^{12}3^3c^2(c-1)^2(1+c)^2(c-1+cd)^2,\quad \mathcal{H}=0,\\ \mathcal{T}_i=0,\ i=1,2,3,4,\quad \mathcal{F}_1=-2(c-1+cd),\quad \mathcal{B}_2=4c^2(c-1+cd)^2,\\ \mathcal{B}_1=2c(c-1+cd)(c+1)(c-1)=2c(c-1+cd)\rho_1\rho_2.\end{gathered} \tag{9.38}$$

We note that by Remark 9.15 the above systems possess a node and a saddle.

2.2.2.2.1: $W_8\neq 0$. Then by (9.38) we have $W_8>0$, and this implies $\kappa_1\mathcal{B}_1\neq 0$. Hence obviously the node is generic and the saddle is strong. Since $\kappa_1\widetilde{ML}\neq 0$ and $\theta_5=0$, considering Lemma 9.1 we obtain the global configuration

• $s, n;\ \widehat{\binom{2}{2}}\ \widehat{P}_{\curlywedge}\ \widehat{P}H_{\curlywedge}-H,\ N^* \Rightarrow (c=2,\ d=0,\ h=1,\ k=-1/2)$.

2.2.2.2.2: $W_8=0$. Then $(c-1)(1+c)(c-1+cd)=0$ (as $c\neq 0$), and considering (9.38) we have to distinguish two subcases: $\kappa_1\neq 0$ and $\kappa_1=0$.

2.2.2.2.2.1: $\kappa_1\neq 0$. Then $(c-1+cd)\neq 0$, and hence we get $(c-1)(1+c)=0$. So one of the singularities is a node with coinciding eigenvalues, and by Remark 9.3 we may assume that the singularity $M_1(0,0)$ is such a node, i.e. $c=1$. Then we calculate

$$\mathcal{H}=\mathcal{B}_1=0,\quad \mathcal{F}_1=-2d,\quad \mathcal{B}_2=4d^2,\quad \kappa_1=-32d.$$

Therefore the condition $\kappa_1\neq 0$ implies $\mathcal{F}_1\neq 0$ and $\mathcal{B}_2>0$. By Theorem 6.2 (see statement (e_1)) and by Remark 9.15, systems (9.37) possess a weak saddle of order one and a node n^d.

Thus considering the condition $\theta_5=0$, by Lemma 9.1 we get the global configuration

• $s^{(1)}, n^d;\ \widehat{\binom{2}{2}}\ \widehat{P}_{\curlywedge}\ \widehat{P}H_{\curlywedge}-H,\ N^* \Rightarrow (c=1,\ d=1,\ h=1,\ k=-1)$.

2.2.2.2.2.2: $\kappa_1=0$. Then due to $\mu_2\neq 0$ (i.e. $c\neq 0$), by (9.38) we have $d=(1-c)/c$. So for systems (9.37) we calculate

$$\begin{gathered}\widetilde{K}=\kappa_1=\theta_6=0,\quad \widetilde{M}=-8(cx-y)^2=\mathcal{G},\quad K_2=96c^2(cx-y)^2,\\ \mu_2=(cx-y)^2,\quad \sigma=1+c-2cx+2y,\quad \mathcal{F}_1=\mathcal{H}=\mathcal{B}=\mathcal{B}_1=\mathcal{B}_2=0,\\ \mathcal{B}_3=-72(c-1)(1+c)(cx-y)^2=-72\rho_1\rho_2(cx-y)^2.\end{gathered} \tag{9.39}$$

Considering (9.39) by Theorem 6.2 (see statement $(e_3)[\delta]$) the saddle will be weak (more precisely will be an integrable one) if and only if $\mathcal{B}_3 = 0$. Moreover by Remark 9.15, apart from the integrable saddle, we have a star node. Since $K_2 > 0$ and $\theta_6 = 0$, considering Lemma 9.1 we obtain the following two global configurations:

- s, n; $\binom{2}{2}\ \widehat{P}\,\widehat{P}H - \widehat{P}\,\widehat{P}H,\ N^* \Rightarrow (c = 2,\ d = -1/2,\ h = 1,\ k = -1/2)$ (if $\mathcal{B}_3 \neq 0$);
- $\$, n^*$; $\binom{2}{2}\ \widehat{P}\,\widehat{P}H - \widehat{P}\,\widehat{P}H,\ N^* \Rightarrow (c = 1,\ d = 0,\ h = 1,\ k = -1)$ (if $\mathcal{B}_3 = 0$).

As all possible cases are examined, we have proved that the family of systems with two distinct real finite singularities possesses exactly 151 geometrically distinct global configurations of singularities. □

9.3 Two distinct complex singularities

Theorem 9.3. (a) *We consider here all the global configurations of singularities of quadratic vector fields with two distinct finite singularities, that in addition are complex and elemental. These global configurations are classified in Diagram* 9.4 *according to the geometrical equivalence relation of configurations. We have exactly* 16 *global configurations with two distinct complex finite singularities.*

(b) *Necessary and sufficient conditions for each one of the* 16 *different equivalence classes can be assembled from this diagram in terms of* 11 *invariant polynomials with respect to the action of the affine group and time rescaling. These invariant polynomials, which appear in Diagram* 9.4*, are given in Section* 5.4*.*

(c) *Diagram* 9.4 *actually contains the global bifurcation diagram in the* 12*-dimensional space of parameters of the global configurations of singularities of this family of quadratic differential systems and provides an algorithm for finding for any given system in the family considered, its respective geometrical global configuration of singularities.*

Proof. Assume that the quadratic systems (8.1) possess two complex finite singular points and both are elemental, i.e. by Proposition 5.1 and Remark 6.1 the conditions $\mu_0 = \mu_1 = 0$, $\mu_2 \neq 0$ and $\mathbf{U} < 0$ hold. Taking into account [338] and Proposition 6.2 (see Table 6.1), we have to consider two families of systems: 15a if $\widetilde{K} \neq 0$ and 15b if $\widetilde{K} = 0$.

1: $\widetilde{K} \neq 0$. In this case according to Proposition 6.2 (see Table 6.1) we shall consider the family of systems 15a, i.e. the systems:

$$\dot{x} = a + hux + 2hxy + ay^2, \quad \dot{y} = b + mux + 2mxy + by^2, \tag{9.40}$$

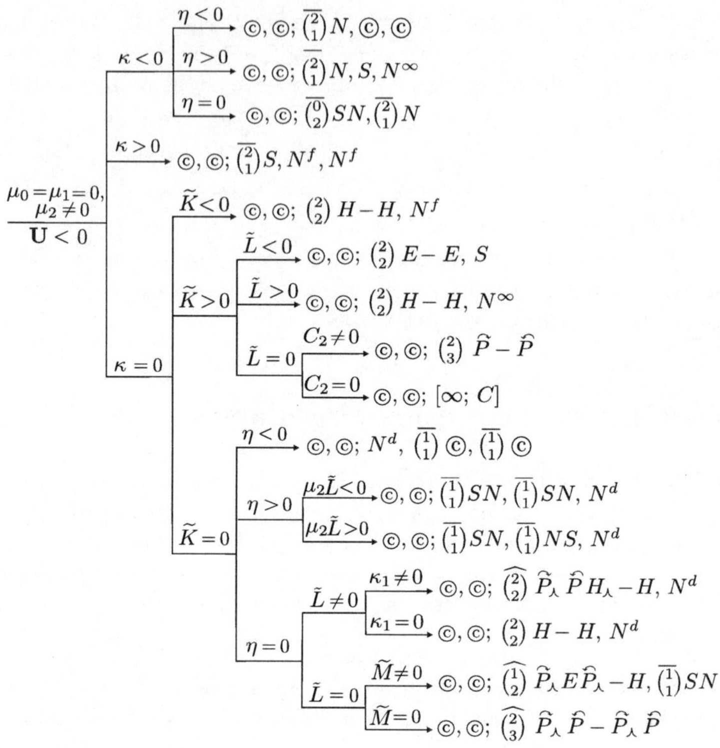

Diagram 9.4: Global configurations: the case $\mu_0 = \mu_1 = 0$, $\mu_2 \neq 0$, $\mathbf{U} < 0$.

possessing the singular points $M_{1,2}(0, \pm i)$. For these systems calculations yield

$$\begin{aligned}
&\mu_0 = \mu_1 = 0,\ \mu_2 = (bh - am)^2(4 + u^2)y^2,\ \widetilde{K} = 4(bh - am)y^2,\\
&\kappa = -128m^2(bh - am),\ \eta = 4m^2\big[(b + 2h)^2 - 8(bh - am)\big],\\
&\widetilde{M} = -32m^2x^2 - 16(b - 2h)mxy - 8\big[(b - 2h)^2 + 6am\big]y^2.
\end{aligned} \tag{9.41}$$

Remark 9.16. We observe that $\mu_2 > 0$ and if $\kappa \neq 0$ then $\widetilde{M} \neq 0$. Moreover the condition $\kappa > 0$ implies $\eta > 0$.

Remark 9.17. The family of systems (9.40) depends on five parameters. However due to a rescaling we can reduce the number of the parameters to three. More precisely since by the condition $\widetilde{K} \neq 0$ (i.e. $bh - am \neq 0$) we have $m^2 + h^2 \neq 0$, then we may assume $(m, h) \in \{(1, 1), (1, 0), (0, 1)\}$ due to the rescaling (i) $(x, y, t) \mapsto (hx/m, y, t/h)$ if $mh \neq 0$, (ii) $(x, y, t) \mapsto (x/m, y, t)$ if $h = 0$ and (iii) $(x, y, t) \mapsto (x, y, t/h)$ if $m = 0$.

Considering (9.41) and $\widetilde{K} \neq 0$, we deduce that the condition $m \neq 0$ is equivalent to $\kappa \neq 0$.

1.1: $\kappa \neq 0$. Then considering Remark 9.17 we shall examine the subfamilies of systems (9.40) with $(m,h) = (1,1)$ and $(m,h) = (1,0)$.

1.1.1: $(m,h) = (1,1)$. We consider the 3-parameter family of systems:

$$\dot{x} = a + ux + 2xy + ay^2, \quad \dot{y} = b + ux + 2xy + by^2, \tag{9.42}$$

for which calculations yield

$$\mu_0 = \mu_1 = 0,\ \mu_2 = (a-b)^2(4+u^2)y^2,\ \widetilde{K} = 4(b-a)y^2,$$
$$\kappa = 128(a-b),\ \eta = 4\big[(b-2)^2 + 8a\big],\ \theta = 64(b-a).$$

1.1.1.1: $\kappa < 0$. Since $\mu_2 > 0$ and $\widetilde{M} \neq 0$ (see Remark 9.16), according to Lemma 9.1 we get the following three global configurations:

- ©, ©; $\overline{\binom{2}{1}}N$, ©, © $\Rightarrow (a=-1,\ b=0,\ u=0)$ (if $\eta < 0$);
- ©, ©; $\overline{\binom{2}{1}}N, S, N^{\infty}$ $\Rightarrow (a=0,\ b=1,\ u=0)$ (if $\eta > 0$);
- ©, ©; $\overline{\binom{0}{2}}SN, \overline{\binom{2}{1}}N$ $\Rightarrow (a=0,\ b=2,\ u=0)$ (if $\eta = 0$).

1.1.1.2: $\kappa > 0$. By Remark 9.16 we have $\eta > 0$, and considering Lemma 9.1, we arrive at the global configuration

- ©, ©; $\overline{\binom{2}{1}}S, N^f, N^f$ $\Rightarrow (a=1,\ b=0,\ u=0)$.

1.1.2: $(m,h) = (1,0)$. We consider the 3-parameter family of systems:

$$\dot{x} = a + ay^2, \quad \dot{y} = b + ux + 2xy + by^2,$$

where we may assume $b \in \{0,1\}$ due to the rescaling $(x,y,t) \mapsto (bx,\ y,\ t/b)$ (if $b \neq 0$). For these systems calculations yield

$$\mu_0 = \mu_1 = 0,\ \mu_2 = a^2(4+u^2)y^2,\ \widetilde{K} = -4ay^2,\ \kappa = 128a,\ \eta = 4(8a+b^2).$$

Considering Remark 9.16 we conclude that the above systems could not possess new global configurations different from the global configurations of systems (9.42).

1.2: $\kappa = 0$. Considering (9.41), due to $\widetilde{K} \neq 0$ we obtain $m = 0$, and then by Remark 9.17 we may assume $h = 1$. Thus we arrive at the systems:

$$\dot{x} = a + ux + 2xy + ay^2, \quad \dot{y} = b + by^2,$$

where due to the rescaling $(x,y,t) \mapsto (ax,y,t)$ (if $a \neq 0$) we can assume $a \in \{0,1\}$.

For these systems we calculate

$$\mu_0 = \mu_1 = \kappa = \kappa_1 = 0,\ \mu_2 = b^2(4+u^2)y^2,\ \widetilde{K} = 4by^2,$$
$$\widetilde{L} = 8b(b-2)y^2,\ K_2 = -384b^2(4-3b+b^2)y^2,$$
$$\eta = 0,\ \widetilde{M} = -8(b-2)^2y^2,\ C_2 = (2-b)xy^2 + ay^3.$$

1.2.1: $\widetilde{K} < 0$. Then $b < 0$ and this implies $\widetilde{M} \neq 0$. We observe that $K_2 < 0$, $\mu_2 > 0$, and considering Lemma 9.1 we obtain the global configuration
• ©, ©; $\binom{2}{2}$ $H-H$, $N^f \Rightarrow (a = 0,\ b = -2,\ u = 0)$.

1.2.2: $\widetilde{K} > 0$. We consider two possibilities: $\tilde{L} \neq 0$ and $\tilde{L} = 0$.

1.2.2.1: $\tilde{L} \neq 0$. In this case $\widetilde{M} \neq 0$, and taking into account the conditions $K_2 < 0$ and $\mu_2 > 0$, by Lemma 9.1 we arrive at the following two global configurations:
• ©, ©; $\binom{2}{2}$ $E-E$, $S \Rightarrow (a = 0,\ b = 1,\ u = 0)$ (if $\tilde{L} < 0$);
• ©, ©; $\binom{2}{2}$ $H-H$, $N^\infty \Rightarrow (a = 0,\ b = 3,\ u = 0)$ (if $\tilde{L} > 0$).

1.2.2.2: $\tilde{L} = 0$. Since $b \neq 0$ (due to $\mu_2 \neq 0$) we have $b = 2$ and then $\widetilde{M} = 0$ and $C_2 = ay^3$. So considering Lemma 9.1 we obtain the following two global configurations:
• ©, ©; $\binom{2}{3}$ $\widehat{P}-\widehat{P} \Rightarrow (a = 1,\ b = 2,\ u = 0)$ (if $C_2 \neq 0$);
• ©, ©; $[\infty; C] \Rightarrow (a = 0,\ b = 2,\ u = 0)$ (if $C_2 = 0$).

2: $\widetilde{K} = 0$. In this case according to Proposition 6.2 (see Table 6.1) we shall consider the family of systems 15b, i.e. the systems:

$$\dot{x} = a + cx + gx^2 + 2hxy + ay^2, \quad \dot{y} = x, \quad (a \neq 0)$$

for which we calculate

$$\begin{aligned} \mu_0 = \mu_1 = \kappa = 0, \quad \mu_2 = a(gx^2 + 2hxy + ay^2)x^2, \quad \widetilde{K} = 0, \\ \tilde{L} = 8g(gx^2 + 2hxy + ay^2), \quad \eta = 4g^2(h^2 - ag), \\ \theta_2 = h^2 - ag, \quad \widetilde{M} = -8g^2x^2 - 16ghxy + 8(3ag - 4h^2)y^2. \end{aligned} \tag{9.43}$$

As $\mu_2 \neq 0$ we have $\operatorname{sign}(\mu_2\tilde{L}) = \operatorname{sign}(ag)$.

2.1: $\eta < 0$. Then $\theta_2 \neq 0$ and considering Lemma 9.1, we arrive at the global configuration
• ©, ©; N^d, $\binom{1}{1}$ ©, $\binom{1}{1}$ © $\Rightarrow (a = 2,\ c = 0,\ g = 1,\ h = 1)$.

2.2: $\eta > 0$. As $\theta_2 \neq 0$, considering Lemma 9.1 we obtain the following two global configurations:
• ©, ©; $\overline{\binom{1}{1}}SN$, $\overline{\binom{1}{1}}SN$, $N^d \Rightarrow (a = 1,\ c = 0,\ g = -1,\ h = 1)$ (if $\mu_2\tilde{L} < 0$);
• ©, ©; $\overline{\binom{1}{1}}SN$, $\overline{\binom{1}{1}}NS$, $N^d \Rightarrow (a = 1,\ c = 0,\ g = 1/2,\ h = 1)$ (if $\mu_2\tilde{L} > 0$).

2.3: $\eta = 0$. We must consider to cases whether $\tilde{L} \neq 0$ or $\tilde{L} = 0$.

2.3.1: $\tilde{L} \neq 0$. Then $g \neq 0$ and we obtain $h^2 - ag = 0$. As $a \neq 0$ we get $g = h^2/a$. Calculations yield

$$\begin{aligned} \tilde{L} = 8h^2(hx + ay)^2/a^2, \quad \widetilde{M} = -8h^2(hx + ay)^2/a^2, \\ \kappa_1 = 32h^2(a + ch)/a, \quad \theta_5 = -96h(a + ch)(hx + ay)^3/a^2. \end{aligned}$$

We observe that the condition $\tilde{L} \neq 0$ implies $\widetilde{M} \neq 0$, i.e. at infinity we have two distinct singularities.

2.3.1.1: $\kappa_1 \neq 0$. Then $\theta_5 \neq 0$ and considering Lemma 9.1 we obtain the global configuration
- $©, ©;\ \widehat{\binom{2}{2}}\ \widehat{P}_\lambda\,\widehat{P}H_\lambda - H,\ N^d \Rightarrow (a=1,\ c=0,\ g=1,\ h=1)$.

2.3.1.2: $\kappa_1 = 0$. As $\tilde{L} \neq 0$ we get $a = -ch \neq 0$, and then we have

$$K_2 = -384h^4(x-cy)^2/c^2,\quad \theta_6 = 8h^2(x-cy)^4/c^2,\quad \mu_2 = h^2x^2(x-cy)^2 \neq 0.$$

So we obtain $K_2 < 0$, $\theta_6 \neq 0$, and considering Lemma 9.1 we obtain the global configuration
- $©, ©;\ \widehat{\binom{2}{2}}\ H - H,\ N^d \Rightarrow (a=-1,\ c=1,\ g=-1,\ h=1)$.

2.3.2: $\tilde{L} = 0$. In this case considering (9.43) we get $g = 0$, and then we calculate

$$\eta = \tilde{L} = 0,\ \widetilde{M} = -32h^2y^2,\ C_2 = y^2(2hx+ay),$$
$$\kappa_1 = 128h^2,\ \mu_2 = ay(2hx+ay) \neq 0.$$

We observe that the condition $\mu_2 \neq 0$ implies $C_2 \neq 0$. Therefore since $\tilde{L} = 0$, according to Lemma 9.1 we obtain the following two global configurations:
- $©, ©;\ \widehat{\binom{1}{2}}\ \widehat{P}_\lambda E\widehat{P}_\lambda - H,\ \overline{\binom{1}{1}}SN \Rightarrow (a=1,\ c=0,\ g=0,\ h=1)$ (if $\widetilde{M} \neq 0$);
- $©, ©;\ \widehat{\binom{2}{3}}\ \widehat{P}_\lambda\,\widehat{P} - \widehat{P}_\lambda\,\widehat{P} \Rightarrow (a=1,\ c=0,\ g=0,\ h=0)$ (if $\widetilde{M} = 0$).

As all possible cases are examined, we have proved that the family of systems with two complex distinct finite singularities possesses exactly 16 geometrically distinct global configurations of singularities. □

Chapter 10

Quadratic systems with finite singularities of total multiplicity three

According to Proposition 5.1, for a quadratic system to have finite singularities of total multiplicity three (i.e. $m_f = 3$), the conditions $\mu_0 = 0$ and $\mu_1 \neq 0$ must be satisfied. Then by Theorem 6.4 the following lemma is valid.

Lemma 10.1. *The geometrical configurations of singularities at infinity of the family of quadratic systems possessing finite singularities of total multiplicity three (i.e. $\mu_0 = 0$, $\mu_1 \neq 0$) are classified in Diagram* 10.1 *according to the geometrical equivalence relation.*

10.1 Exactly one singularity

Theorem 10.1. (a) *We consider here all the global configurations of singularities of quadratic vector fields possessing exactly one real singularity that in addition is triple. These configurations are classified in Diagram* 10.2 *according to the geometrical equivalence relation of configurations. We have a total of* 20 *geometrically distinct global configurations of singularities.*

(b) *Necessary and sufficient conditions for each one of the* 20 *different geometrical equivalence classes can be assembled from this diagram in terms of* 11 *invariant polynomials with respect to the action of the affine group and time rescaling. These invariant polynomials, which appear in Diagram* 10.2, *are given in Section* 5.4.

(c) *Diagram* 10.2 *actually contains the global bifurcation diagram in the* 12-*dimensional space of parameters of the global geometrical configurations, of*

J. C. Artés et al., *Geometric Configurations of Singularities of Planar Polynomial Differential Systems*, https://doi.org/10.1007/978-3-030-50570-7_10

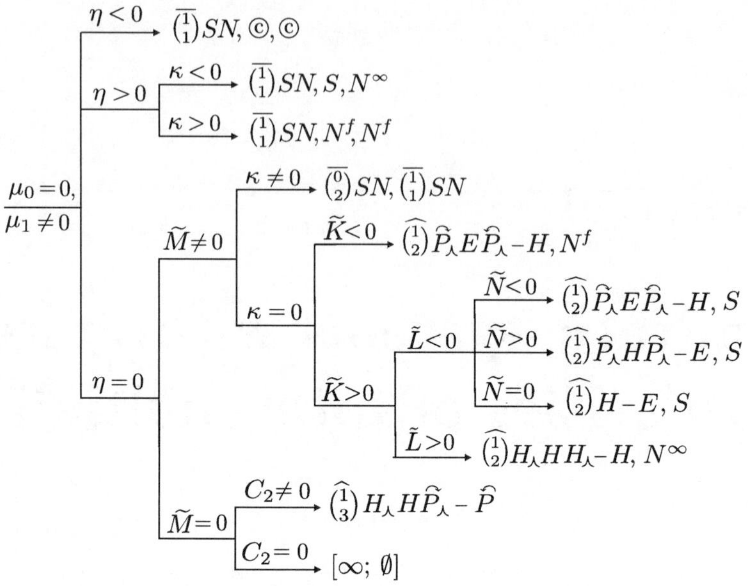

Diagram 10.1: Configurations of infinite singularities: the case $\mu_0 = 0$, $\mu_1 \neq 0$.

this subclass of quadratic differential systems and provides an algorithm for finding for any given system in the family considered, its respective geometrical global configuration.

Proof. Assume that the quadratic systems (8.1) possess a triple finite singularity. In this case according to Proposition 6.2 (see Table 6.1) we shall consider the family of systems 13, i.e. the systems:

$$\dot{x} = gy + gx^2 + 2hxy, \quad \dot{y} = ly + lx^2 + 2mxy \tag{10.1}$$

possessing only one finite singularity: the triple point $M_{1,2,3}(0,0)$. For these systems calculations yield

$$\begin{aligned} \mu_0 = 0, \ \ \mu_1 = 4(gm - hl)^2 x, \ \ \kappa = 128h^2(hl - gm), \\ E_3 = -2g^3m^3, \ \theta_2 = g(g - m)m. \end{aligned} \tag{10.2}$$

Since for the above systems we have $\mu_1 \neq 0$ (i.e. $gm - hl \neq 0$), we observe that the condition $\kappa = 0$ is equivalent to $h = 0$.

We need first a technical lemma to control the existence of non-generic finite elliptic-saddles.

Lemma 10.2. *If a system of the family* (10.1) *has an elliptic-saddle, then it is generic if* $\theta_2 \neq 0$. *Otherwise it is non-generic.*

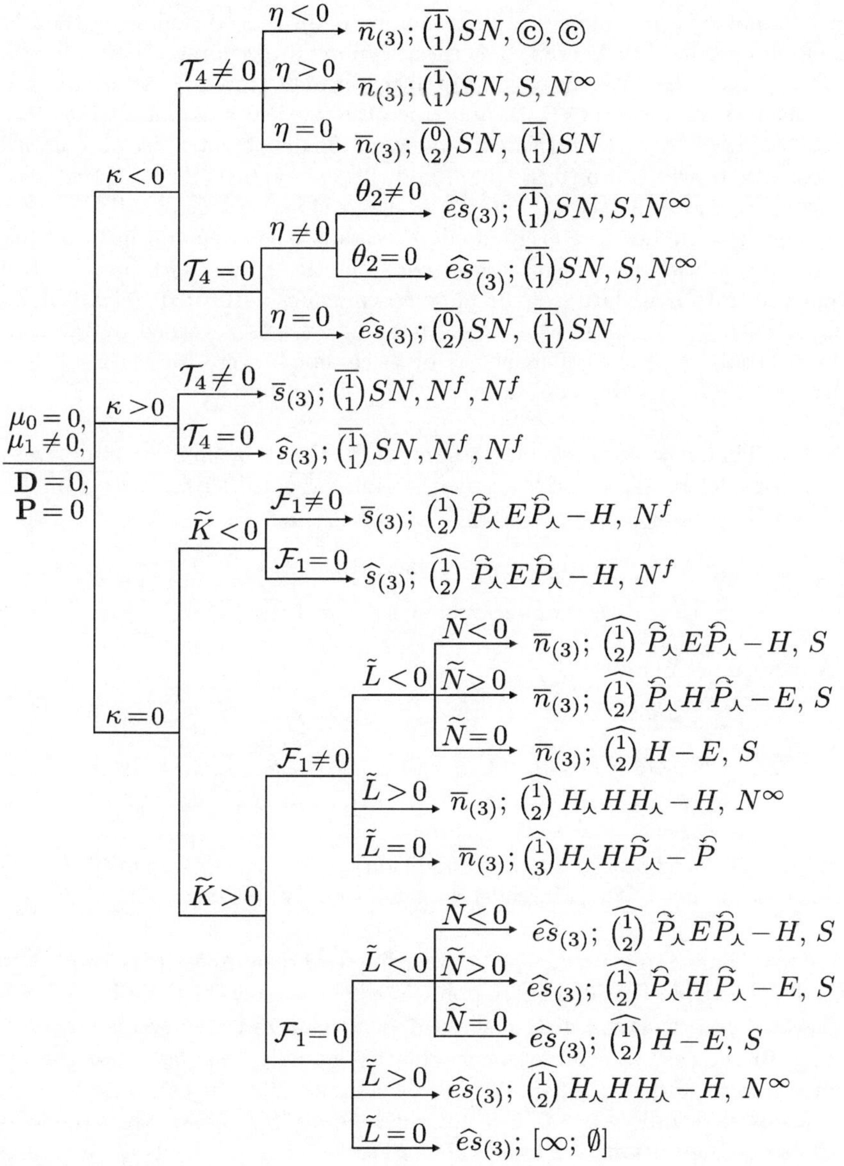

Diagram 10.2: Global configurations: the case $\mu_0 = 0$, $\mu_1 \neq 0$, $\mathbf{D} = \mathbf{P} = 0$.

Proof. We first force that a system of the family (10.1) has a nilpotent singularity at the origin by making its trace zero, that is, $l = 0$ and $g \neq 0$. The line

$x = 0$ cannot be a characteristic direction at $(0, 0)$. Indeed, following the blow-up technique described in Section 3.4, the polynomial of characteristic directions is $PCD(x, y) = -gy^2$. We make a vertical directional blow-up. After this we have an intricate singularity at $(0, 0)$, and since the vertical axis is invariant, we do a 1-twist followed by a second vertical blow-up. Then we get three singularities on the axis $x = 0$, which are $(0, 0)$, $(0, 1)$ and $(0, (m - g)/m)$ (if $m = 0$ we just make a 2-twist instead of a 1-twist, and we arrive at the same result). If the three singularities are different, the original elliptic-saddle is generic and has four borsecs. Then, the condition to obtain a non-generic elliptic-saddle (with just two borsecs) is that the third singularity of the blow-up coincides with $(0, 0)$ (clearly it cannot coincide with $(0, 1)$). Moreover the expression $m - g$ is governed by the invariant θ_2 from (10.2) since the other factors of θ_2 cannot be zero due to $E_3 < 0$, which is needed to have an elliptic-saddle. □

1: $\kappa \neq 0$. Then $h \neq 0$ and we may assume $h = 1$ due to a time rescaling. Moreover we can consider $m \in \{0, 1\}$ due to the rescaling $(x, y, t) \mapsto (mx, m^2y, t/m^2)$ in the case $m \neq 0$. So we get the systems:

$$\dot{x} = gy + gx^2 + 2xy, \quad gm - l \neq 0,$$
$$\dot{y} = ly + lx^2 + 2mxy, \quad m \in \{0, 1\},$$

for which we calculate

$$\kappa = 128(l - gm), \ \widetilde{K} = 4(gm - l)x^2, \ G_{10} = l^3(gm - l)^3,$$
$$\mathcal{T}_4 = -8l^3(gm - l)^2, \ \eta = 4\big[(g - 2m)^2 + 8l\big], \ \theta_2 = (g - m)(gm - l),$$
$$\widetilde{M} = -8\big[(g - 2m)^2 + 6l\big]x^2 - 16(g - 2m)xy - 32y^2.$$

Remark 10.1. We observe that the condition $G_{10} = 0$ is equivalent to $\mathcal{T}_4 = 0$, and in this case we get $\eta \geq 0$. Moreover we have $\operatorname{sign}(\widetilde{K}) = -\operatorname{sign}(\kappa)$.

1.1: $\kappa < 0$. This implies $\widetilde{K} > 0$, and according to Theorem 6.1 (see Table 6.2, lines 146,147) the triple finite singular point is a semi-elemental node if $G_{10} \neq 0$ and a nilpotent elliptic-saddle if $G_{10} = 0$ (in this case $l = 0$ due to $\kappa \neq 0$). We have $\widetilde{M} \neq 0$. In the case of a nilpotent singularity we have that $\theta_2 = gm(g - m) = 0$ implies $g = m$ and $\eta = 4m^2 > 0$. So, by Lemma 10.2 we can only have a non-generic elliptic-saddle if $\eta > 0$. Considering Lemma 10.1, we arrive at the following six global configurations:

- $\overline{n}_{(3)}; \overline{\binom{1}{1}}SN, ©, ©\ \Rightarrow (g = 1,\ l = -1,\ m = 1)$ (if $\mathcal{T}_4 \neq 0$, $\eta < 0$);
- $\overline{n}_{(3)}; \overline{\binom{1}{1}}SN, S, N^{\infty}\ \Rightarrow (g = 1,\ l = 1/2,\ m = 1)$ (if $\mathcal{T}_4 \neq 0$, $\eta > 0$);
- $\overline{n}_{(3)}; \overline{\binom{0}{2}}SN, \overline{\binom{1}{1}}SN\ \Rightarrow (g = 1,\ l = -1/8,\ m = 1)$ (if $\mathcal{T}_4 \neq 0$, $\eta = 0$);
- $\widehat{es}_{(3)}; \overline{\binom{1}{1}}SN, S, N^{\infty}\ \Rightarrow (g = 1,\ l = 0,\ m = 2)$ (if $\mathcal{T}_4 = 0$, $\eta > 0$, $\theta_2 \neq 0$);
- $\widehat{es}^{-}_{(3)}; \overline{\binom{1}{1}}SN, S, N^{\infty}\ \Rightarrow (g = 1,\ l = 0,\ m = 1)$ (if $\mathcal{T}_4 = 0$, $\eta > 0$, $\theta_2 = 0$);

• $\widehat{es}_{(3)}$; $\overline{\binom{0}{2}}SN$, $\overline{\binom{1}{1}}SN$ $\Rightarrow (g=2,\ l=0,\ m=1)$ (if $\mathcal{T}_4=0,\ \eta=0$).

1.2: $\kappa>0$. In this case we have $\widetilde{K}<0$, and according to Theorem 6.1 (see Table 6.2, line 145) the triple finite singular point is a saddle. As its eigenvalues are $\lambda_1=0$ and $\lambda_2=l$, we conclude that this triple saddle is semi-elemental if $l\neq 0$ and nilpotent if $l=0$.

We claim that the condition $\kappa>0$ implies $\eta>0$. Indeed, as $\kappa=128(l-gm)>0$, we can set $l=gm+\varepsilon$ ($\varepsilon>0$) and obtain

$$\eta=4\big[(g-2m)^2+8l\big]=4\big[(g+2m)^2+8\varepsilon\big]>0.$$

Thus considering Lemma 10.1 we obtain the following two global configurations:
• $\overline{s}_{(3)}$; $\overline{\binom{1}{1}}SN, N^f, N^f$ $\Rightarrow (g=0,\ l=1,\ m=1)$ (if $\mathcal{T}_4\neq 0$);
• $\widehat{s}_{(3)}$; $\overline{\binom{1}{1}}SN, N^f, N^f$ $\Rightarrow (g=-1,\ l=0,\ m=1)$ (if $\mathcal{T}_4=0$).

2: $\kappa=0$. Then $h=0$ and $gm\neq 0$ due to $\mu_1=4g^2m^2x\neq 0$. So we may assume $g=1$ due to the rescaling $(x,y)\mapsto(x/g,\ y/g^2)$, and we get the family of systems:

$$\dot{x}=y+x^2,\quad \dot{y}=ly+lx^2+2mxy,\quad m\neq 0,$$

for which we calculate

$$\begin{gathered}\mu_0=\kappa=\eta=0,\quad \mu_1=4m^2x,\quad \widetilde{K}=4mx^2,\quad G_{10}=l^3m^3,\\ \mathcal{F}_1=6lm,\quad \tilde{L}=8(1-2m)x^2,\quad \widetilde{N}=4m(1-m)x^2,\ \theta_2=m(1-m),\\ \widetilde{M}=-8(2m-1)^2x^2,\quad C_2=-lx^3+(1-2m)x^2y.\end{gathered}$$

2.1: $\widetilde{K}<0$. According to Theorem 6.1 (see Table 6.2, line 145) the triple finite singular point is a saddle. As it was mentioned earlier, this triple saddle is semi-elemental if $G_{10}\neq 0$ and nilpotent if $G_{10}=0$, which is equivalent to $\mathcal{F}_1=0$.

Additionally the condition $\widetilde{K}<0$ implies $m<0$ and then $\widetilde{M}\neq 0$. Therefore considering Lemma 10.1 we obtain the following two global configurations:
• $\overline{s}_{(3)}$; $\widehat{\binom{1}{2}}\ \widehat{P}_{\curlywedge}E\widehat{P}_{\curlywedge}-H,\ N^f$ $\Rightarrow (l=1,\ m=-1)$ (if $\mathcal{F}_1\neq 0$);
• $\widehat{s}_{(3)}$; $\widehat{\binom{1}{2}}\ \widehat{P}_{\curlywedge}E\widehat{P}_{\curlywedge}-H,\ N^f$ $\Rightarrow (l=0,\ m=-1)$ (if $\mathcal{F}_1=0$).

2.2: $\widetilde{K}>0$. As we have mentioned earlier, the triple finite singular point is a semi-elemental node if $G_{10}\neq 0$ and a nilpotent elliptic-saddle if $G_{10}=0$.

1. $\widetilde{M}\neq 0$. Then $2m-1\neq 0$, and as $m>0$ (due to $\widetilde{K}>0$), we have

$$\operatorname{sign}(\tilde{L})=\operatorname{sign}(1-2m)\quad \operatorname{sign}(\widetilde{N})=\operatorname{sign}(1-m).$$

We have that $\theta_2=0\Leftrightarrow \widetilde{N}=0$. So if there is an elliptic-saddle, by Lemma 10.2 it will be non-generic if and only if $\widetilde{N}=0$. Therefore according to Lemma 10.1, we arrive at the following eight global configurations:

- $\overline{n}_{(3)}$; $\widehat{\binom{1}{2}}\ \widehat{P}_\lambda E\widehat{P}_\lambda - H,\ S\ \Rightarrow (l=1,\ m=2)$ (if $\mathcal{F}_1 \neq 0,\ \tilde{L}<0,\ \widetilde{N}<0$);
- $\overline{n}_{(3)}$; $\widehat{\binom{1}{2}}\ \widehat{P}_\lambda H\widehat{P}_\lambda - E,\ S\ \Rightarrow (l=1,\ m=2/3)$ (if $\mathcal{F}_1 \neq 0,\ \tilde{L}<0,\ \widetilde{N}>0$);
- $\overline{n}_{(3)}$; $\widehat{\binom{1}{2}}\ H-E,\ S\ \Rightarrow (l=1,\ m=1)$ (if $\mathcal{F}_1 \neq 0,\ \tilde{L}<0,\ \widetilde{N}=0$);
- $\overline{n}_{(3)}$; $\widehat{\binom{1}{2}}\ H_\lambda H H_\lambda - H,\ N^\infty\ \Rightarrow (l=1,\ m=1/3)$ (if $\mathcal{F}_1 \neq 0,\ \tilde{L}>0$);
- $\widehat{es}_{(3)}$; $\widehat{\binom{1}{2}}\ \widehat{P}_\lambda E\widehat{P}_\lambda - H,\ S\ \Rightarrow (l=0,\ m=2)$ (if $\mathcal{F}_1 = 0,\ \tilde{L}<0,\ \widetilde{N}<0$);
- $\widehat{es}_{(3)}$; $\widehat{\binom{1}{2}}\ \widehat{P}_\lambda H\widehat{P}_\lambda - E,\ S\ \Rightarrow (l=0,\ m=2/3)$ (if $\mathcal{F}_1 = 0,\ \tilde{L}<0,\ \widetilde{N}>0$);
- $\widehat{es}_{(3)}$; $\widehat{\binom{1}{2}}\ H-E,\ S\ \Rightarrow (l=0,\ m=1)$ (if $\mathcal{F}_1 = 0,\ \tilde{L}<0,\ \widetilde{N}=0$);
- $\widehat{es}_{(3)}$; $\widehat{\binom{1}{2}}\ H_\lambda H H_\lambda - H,\ N^\infty\ \Rightarrow (l=0,\ m=1/3)$ (if $\mathcal{F}_1 = 0,\ \tilde{L}>0$).

2. $\widetilde{M}=0$. Then $m=1/2$ ($\theta_2 \neq 0$) and we obtain $C_2=-lx^3$. So the condition $\mathcal{F}_1=0$ is equivalent to $C_2=0$, and considering Lemma 10.1 we get the following two global configurations:

   - $\overline{n}_{(3)}$; $\widehat{\binom{1}{3}} H_\lambda H\widehat{P}_\lambda - \widehat{P}\ \Rightarrow (l=1,\ m=1/2)$ (if $\mathcal{F}_1 \neq 0$);
   - $\widehat{es}_{(3)}$; $[\infty;\ \emptyset]\ \Rightarrow (l=0,\ m=1/2)$ (if $\mathcal{F}_1 = 0$).

Thus we obtained that a quadratic system with a triple finite singularity possesses one of the 20 global configurations given in Diagram 10.2. □

10.2 Exactly two distinct singularities

Theorem 10.2. (a) *We consider here all the global configurations of singularities of quadratic vector fields with finite singularities of total multiplicity $m_f=3$ possessing two real finite singularities: one elemental and one double. These configurations are classified in Diagram 10.3 according to the geometrical equivalence relation of configurations. We have a total of 62 geometrically distinct global configurations.*

(b) *Necessary and sufficient conditions for each one of the 62 different geometrical equivalence classes can be assembled from these diagrams in terms of 22 invariant polynomials with respect to the action of the affine group and time rescaling. These invariant polynomials, which appear in Diagram 10.3, are given in Section 5.4.*

(c) *Diagram 10.3 actually contains the global bifurcation diagram in the 12-dimensional space of parameters of the global geometrical configurations, of this subclass of quadratic differential systems and provides an algorithm for finding for any given system in the family considered, its respective geometrical global configuration.*

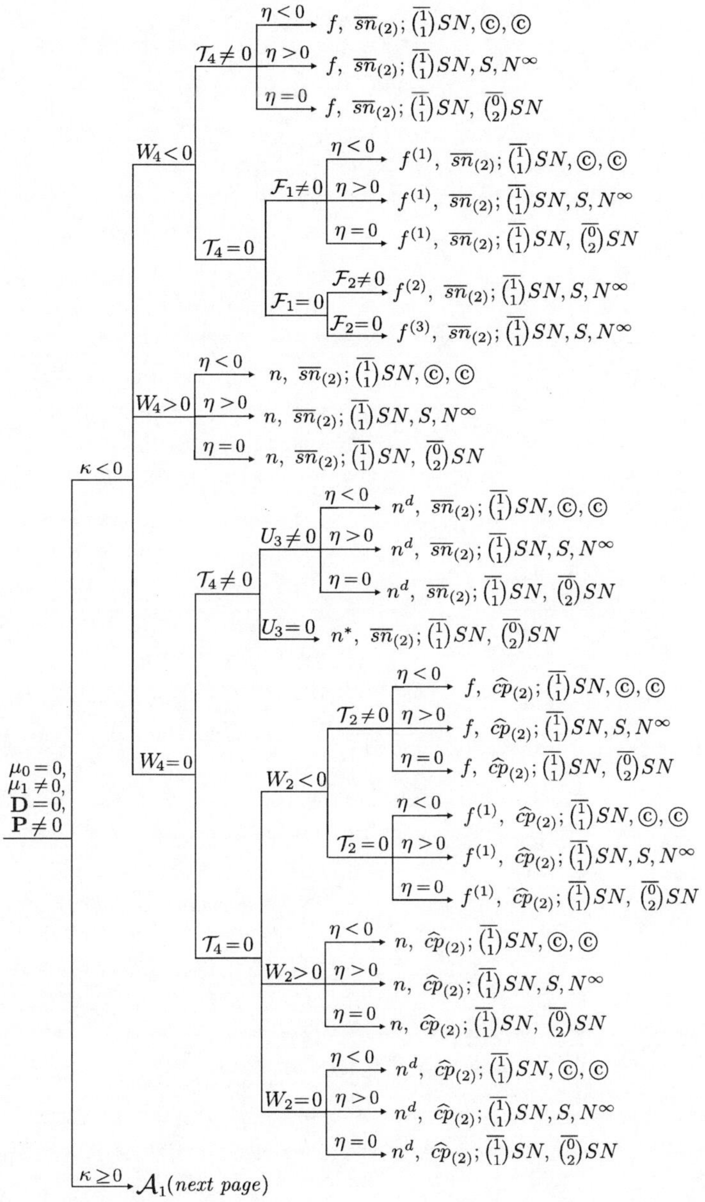

Diagram 10.3: Global configurations: the case $\mu_0 = 0$, $\mu_1 \neq 0$, $\mathbf{D} = 0$, $\mathbf{P} \neq 0$.

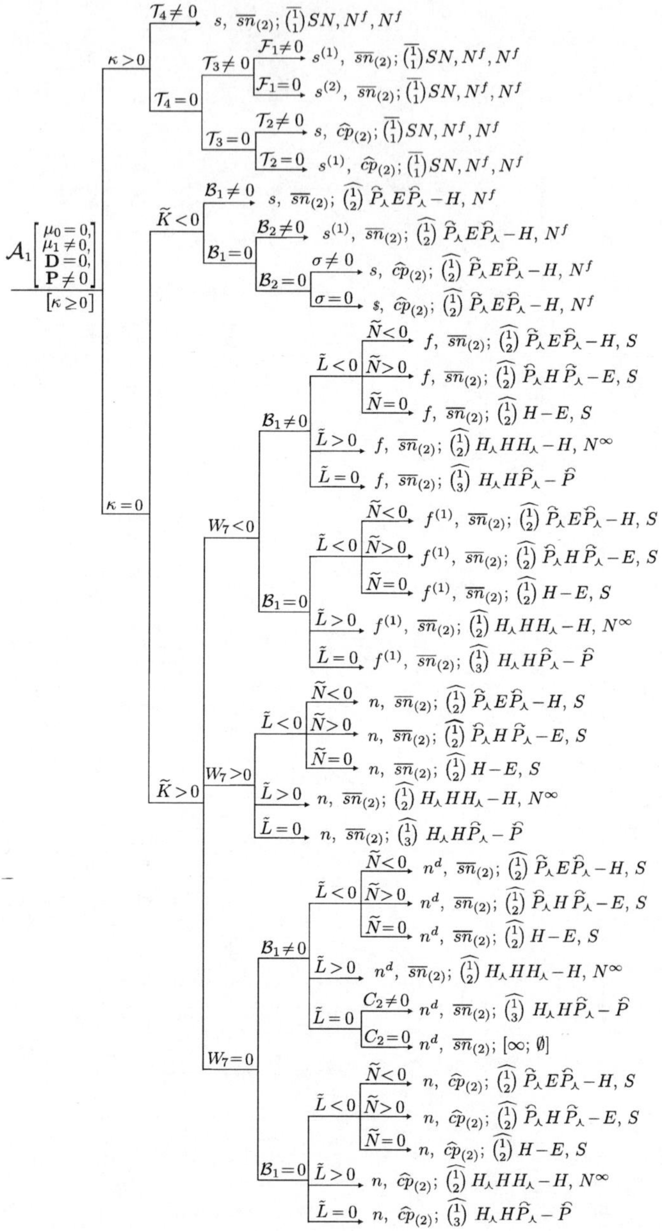

Diagram 10.3 (*cont.*): Global configurations: the case $\mu_0=0$, $\mu_1\neq 0$, $\mathbf{D}=0$, $\mathbf{P}\neq 0$.

Proof. Assume that the quadratic systems (8.1) possess one double and one simple finite singularities. Then according to Proposition 6.2 (see Table 6.1) via an affine transformation these systems could be brought to the systems 12, i.e. to systems of the form:

$$\dot{x} = cx + cy - cx^2 + 2hxy, \quad \dot{y} = ex + ey - ex^2 + 2mxy \tag{10.3}$$

possessing the double singular point $M_{1,2}(0,0)$ and the elemental singularity $M_3(1,0)$. For these systems calculations yield

$$\mu_0 = 0, \quad \mu_1 = -4(cm - eh)^2 x, \quad \kappa = 128h^2(cm - eh).$$

Since for the above systems we have $\mu_1 \neq 0$ (i.e. $gm - hl \neq 0$), we deduce that the condition $\kappa = 0$ is equivalent to $h = 0$.

1.1: $\kappa \neq 0$. Then $h \neq 0$ and we may assume $h = 1$ due to a time rescaling. So we get the systems:

$$\dot{x} = cx + cy - cx^2 + 2xy, \quad \dot{y} = ex + ey - ex^2 + 2mxy, \tag{10.4}$$

possessing the singular points $M_{1,2}(0,0)$ and $M_3(1,0)$. For these singularities we have the following values of the traces ρ_i, of the determinants Δ_i and of the discriminants τ_i:

$$\rho_1 = c + e, \quad \Delta_1 = 0, \quad \tau_1 = (c+e)^2;$$
$$\rho_3 = e - c + 2m, \quad \Delta_3 = 2(e - cm), \quad \tau_3 = (e - c + 2m)^2 + 8(cm - e).$$

Then for the above systems we calculate

$$\begin{aligned}
\kappa = -64\Delta_3, \quad \widetilde{K} = 2\Delta_3 x^2, \quad E_1 = -\rho_1\Delta_3^4/2, \quad G_9 = \Delta_3/2,\\
\mathcal{T}_4 = -2\rho_1^2\rho_3\Delta_3^2, \quad \mathcal{T}_3 = 2\rho_1(\rho_1 + 2\rho_3)\Delta_3^2,\\
\eta = 4\big[(c - 2m)^2 - 4\Delta_3\big], \quad W_4 = 4\rho_1^4\Delta_3^4\tau_3,\\
\widetilde{M} = -8\big[(c+2m)^2 - 6e\big]x^2 + 16(c+2m)xy - 32y^2.
\end{aligned} \tag{10.5}$$

Remark 10.2. We observe that $\widetilde{M} \neq 0$ and the condition $\kappa \neq 0$ implies $\widetilde{K}G_9 \neq 0$. Moreover we have sign $(\widetilde{K}) = -$sign (κ).

1.1: $\kappa < 0$. This implies $\widetilde{K} > 0$, and according to Theorem 6.1 (see Table 6.2, lines 138–144) the elemental singular point is an anti-saddle and its type is governed by the invariant polynomials W_4 and W_2 (as in the considered case we have $G_9 \neq 0$).

1.1.1: $W_4 < 0$. Considering (10.5) we have $\tau_3 < 0$, i.e. the elemental singular point is a focus. Additionally as the condition $W_4 \neq 0$ implies $E_1 \neq 0$ and $G_9 \neq 0$, according to Theorem 6.1 (see Table 6.2, line 140) the double point is a semi-elemental saddle-node.

1.1.1.1: $\mathcal{T}_4 \neq 0$. Then the focus is strong, and considering Lemma 10.1 and $\widetilde{M} \neq 0$ from (10.5), we arrive at the following three global configurations:

- $f,\ \overline{sn}_{(2)};\ \overline{\binom{1}{1}}SN, \copyright, \copyright\ \Rightarrow (c=2,\ e=1,\ m=0)\quad (\text{if } \eta<0);$
- $f,\ \overline{sn}_{(2)};\ \overline{\binom{1}{1}}SN, S, N^{\infty}\ \Rightarrow (c=2,\ e=5/12,\ m=0)\quad (\text{if } \eta>0);$
- $f,\ \overline{sn}_{(2)};\ \overline{\binom{0}{2}}SN,\ \overline{\binom{1}{1}}SN\ \Rightarrow (c=2,\ e=1/2,\ m=0)\quad (\text{if } \eta=0).$

1.1.1.2: Suppose now $\mathcal{T}_4 = 0$. Then considering (10.5) we get $\rho_3 = 0$, i.e. the focus is weak. In this case we have $e = c - 2m$, and we calculate

$$\begin{aligned} \mathcal{T}_3 &= 32(c-m)^2(c-2m-cm)^2,\ \mathcal{F} = -2(c-m)^2(c-2m-cm),\\ \mathcal{T}_4 &= 0,\ \mathcal{F}_1 = 4(c+2c^2-2m-2cm)(c-2m-cm),\\ W_4 &= -8192(c-m)^4(c-2m-cm)^5. \end{aligned} \tag{10.6}$$

So the condition $W_4 < 0$ gives $\mathcal{T}_3\mathcal{F} < 0$, and by Theorem 6.2 the order of the weak focus is governed by the invariant polynomial $\mathcal{F}_1$.

1. $\mathcal{F}_1 \neq 0$. Then the simple singular point of systems (10.4) is a first order weak focus, and considering Lemma 10.1 we arrive at the following three global configurations:

   - $f^{(1)}, \overline{sn}_{(2)};\ \overline{\binom{1}{1}}SN, \copyright, \copyright\ \Rightarrow (c=0,\ e=2,\ m=-1)\quad (\text{if } \eta<0);$
   - $f^{(1)}, \overline{sn}_{(2)};\ \overline{\binom{1}{1}}SN, S, N^{\infty}\ \Rightarrow (c=0,\ e=10,\ m=-5)\quad (\text{if } \eta>0);$
   - $f^{(1)}, \overline{sn}_{(2)};\ \overline{\binom{0}{2}}SN,\ \overline{\binom{1}{1}}SN\ \Rightarrow (c=0,\ e=8,\ m=-4)\quad (\text{if } \eta=0).$

2. $\mathcal{F}_1 = 0$. By (10.6) the condition $\mathcal{F}_1 = 0$ gives $c+2c^2-2m-2cm = 0$ and $c+1 \neq 0$, otherwise we get $\mathcal{F}_1 = -4(1+m) \neq 0$ due to the condition $\kappa = 128(1+m) \neq 0$. So $c+1 \neq 0$ and we obtain $m = c(1+2c)/(2(1+c))$. Then we calculate

$$\begin{aligned} \mathcal{F}_2 &= -\frac{c^8(3+2c)^2(6+5c)}{(1+c)^5}, \quad \eta = 4\frac{c^2(12+20c+9c^2)}{(1+c)^2},\\ W_4 &= 16\frac{c^{14}(3+2c)^5}{(1+c)^9}, \quad \kappa = 64\frac{c^2(3+2c)}{1+c}, \end{aligned}$$

   and we observe that due to $\kappa \neq 0$ the condition $\eta > 0$ holds. Moreover, we get $\mathcal{F}_2 = 0$ for $c = -6/5$, and in this case the weak focus is of order three as $\mathcal{F}_3\mathcal{F}_4 = 2^{23}3^{21}5^{-13} \neq 0$ (see Theorem 6.2). Thus considering Lemma 10.1 we get the following two global configurations:

   - $f^{(2)}, \overline{sn}_{(2)};\ \overline{\binom{1}{1}}SN, S, N^{\infty}\ \Rightarrow (c=-4/3,\ e=16/3,\ m=-10/3)$ $(\text{if } \mathcal{F}_2 \neq 0);$
   - $f^{(3)}, \overline{sn}_{(2)};\ \overline{\binom{1}{1}}SN, S, N^{\infty}\ \Rightarrow (c=-6/5,\ e=36/5,\ m=-21/5)$ $(\text{if } \mathcal{F}_2 = 0).$

1.1.2: $W_4 > 0$. Then the anti-saddle is a generic node, and we get the next three global configurations:

- $n,\ \overline{sn}_{(2)};\ \overline{\binom{1}{1}}SN, ©, ©\ \Rightarrow (c=0,\ e=1,\ m=1)$ (if $\eta<0$);
- $n,\ \overline{sn}_{(2)};\ \overline{\binom{1}{1}}SN, S, N^{\infty}\ \Rightarrow (c=0,\ e=1/3,\ m=1)$ (if $\eta>0$);
- $n,\ \overline{sn}_{(2)};\ \overline{\binom{0}{2}}SN,\ \overline{\binom{1}{1}}SN\ \Rightarrow (c=0,\ e=1/2,\ m=1)$ (if $\eta=0$).

1.1.3: $W_4 = 0$. Then as $\Delta_3 \neq 0$ by (10.5) we have $\rho_1\tau_3 = 0$, and clearly $\rho_1 = 0$ if and only if $W_4 = \mathcal{T}_4 = 0$.

1.1.3.1: $\mathcal{T}_4 \neq 0$. In this case $\rho_1 \neq 0$ and then the condition $W_4 = 0$ gives $\tau_3 = 0$. So we have a node with coinciding eigenvalues. We claim that this node will be a star node if and only if $U_3 = 0$. Indeed the linearization matrix corresponding to $M_3(1,0)$ is $\mathcal{M}_3 = \begin{pmatrix} -c & 2+c \\ -e & e+2m \end{pmatrix}$, and clearly we have a star node if and only if $e=0$, $c=-2$ and $m=1$. In this case a straightforward computation gives us $U_3 = 0$.

Conversely, assume that $U_3 = 0$. As Coefficient$[U_3, y^5] = 12c^2(2+c)$, we have $c(2+c) = 0$. We claim that we could not have $c = 0$, otherwise we get Coefficient$[U_3, xy^4] = 24e^2 = 0$, which contradicts $\kappa \neq 0$. So we have $c = -2$, and then we calculate Coefficient$[U_3, xy^4] = 24(e-2+2m)(e+2m) = 0$. As $\kappa = -128(e+2m) \neq 0$, we get $e = 2(1-m)$. Finally we calculate Coefficient$[U_3, x^2y^3] = 288(m-1) = 0$, and we have $m = 1$, which implies $U_3 = 0$. So we arrive at the conditions above (i.e. the elemental singularity is a star node), and this proves our claim.

Additionally when we have a star node, we obtain $\eta = 0$. So considering Lemma 10.1 we arrive at the following four global configurations:

- $n^d, \overline{sn}_{(2)};\ \overline{\binom{1}{1}}SN, ©, ©\ \Rightarrow (c=1,\ e=3,\ m=1)$ (if $U_3 \neq 0,\ \eta<0$);
- $n^d, \overline{sn}_{(2)};\ \overline{\binom{1}{1}}SN, S, N^{\infty}\ \Rightarrow (c=-3,\ e=-1,\ m=1)$ (if $U_3 \neq 0,\ \eta>0$);
- $n^d, \overline{sn}_{(2)};\ \overline{\binom{0}{2}}SN,\ \overline{\binom{1}{1}}SN\ \Rightarrow (c=6,\ e=8,\ m=1)$ (if $U_3 \neq 0,\ \eta=0$);
- $n^*, \overline{sn}_{(2)};\ \overline{\binom{0}{2}}SN,\ \overline{\binom{1}{1}}SN\ \Rightarrow (c=-2,\ e=0,\ m=1)$ (if $U_3 = 0$).

1.1.3.2: $\mathcal{T}_4 = 0$. As it was mentioned above, in this case we have $\rho_1 = 0$ and then $E_1 = 0$. As $\widetilde{K} > 0$ and $G_9 \neq 0$, according to Theorem 6.1 (see Table 6.2, lines 142 and 144) the double point is a nilpotent cusp, whereas the type of the anti-saddle is governed by the invariant polynomial W_2. Setting $e = -c$ for the elemental singular point, we have

$$\begin{aligned} &\mathcal{M}_3 = \begin{pmatrix} -c & 2+c \\ c & 2m-c \end{pmatrix}, \quad \rho_3 = 2(m-c), \\ &\Delta_3 = -2c(1+m), \quad \tau_3 = 4(c^2+2c+m^2), \end{aligned} \tag{10.7}$$

and calculations yield

$$
\begin{gathered}
\kappa = -64\Delta_3, \quad E_1 = \mathcal{T}_4 = \mathcal{T}_3 = 0, \quad G_9 = \Delta_3/2, \quad \mathcal{T}_2 = 2\rho_3\Delta_3^2, \\
\mathcal{T}_1 = 2\Delta_3^2, \quad \mathcal{F} = (1+c-m)\rho_3\Delta_3/2, \quad \mathcal{F}_1 = -2c(1+c-m)\Delta_3, \\
\eta = 4\big[(c+2m)^2 + 8c\big], \quad W_2 = 4\Delta_3^4\tau_3.
\end{gathered} \tag{10.8}
$$

1.1.3.2.1: $W_2 < 0$. This gives $\tau_3 < 0$, and the elemental singularity is a focus.

1. $\mathcal{T}_2 \neq 0$. Then $\rho_3 \neq 0$ and the focus is strong. So considering Lemma 10.1, we arrive at the next three global configurations:
   - $f,\ \widehat{cp}_{(2)};\ \overline{\binom{1}{1}}SN, ©, ©\ \Rightarrow (c = -1/12,\ e = 1/12,\ m = -9/25)$ (if $\eta < 0$);
   - $f,\ \widehat{cp}_{(2)};\ \overline{\binom{1}{1}}SN, S, N^{\infty}\ \Rightarrow (c = -1/13,\ e = 1/13,\ m = -9/25)$ (if $\eta > 0$);
   - $f,\ \widehat{cp}_{(2)};\ \overline{\binom{0}{2}}SN,\ \overline{\binom{1}{1}}SN\ \Rightarrow (c = -2/25,\ e = 2/25,\ m = -9/25)$ (if $\eta = 0$).
2. $\mathcal{T}_2 = 0$. We have a weak focus and the condition $\rho_3 = 0$ gives $m = c$. Then by (10.8) we obtain

$$
\begin{gathered}
\mathcal{T}_2 = \mathcal{F} = 0, \quad \mathcal{T}_1 = 8c^2(1+c)^2, \quad \mathcal{F}_1 = 4c^2(1+c), \\
\mathcal{B} = -2c^2, \quad \mathcal{H} = 4c(1+c), \quad \kappa = 128c(1+c).
\end{gathered} \tag{10.9}
$$

   So the condition $\kappa \neq 0$ yields $\mathcal{T}_1\mathcal{F}_1 \neq 0$ (moreover $\kappa < 0$ implies $\mathcal{H} < 0$), and according to Theorem 6.2 (see statement (d_2)) the weak focus could only be of order one. So we get the next three global configurations:
   - $f^{(1)},\ \widehat{cp}_{(2)};\ \overline{\binom{1}{1}}SN, ©, ©\ \Rightarrow (c = -1/2,\ e = 1/2,\ m = -1/2)$ (if $\eta < 0$);
   - $f^{(1)},\ \widehat{cp}_{(2)};\ \overline{\binom{1}{1}}SN, S, N^{\infty} \Rightarrow (c = -17/18, e = 17/18, m = -17/18)$ (if $\eta > 0$);
   - $f^{(1)},\ \widehat{cp}_{(2)};\ \overline{\binom{0}{2}}SN,\ \overline{\binom{1}{1}}SN\ \Rightarrow (c = -8/9,\ e = 8/9,\ m = -8/9)$ (if $\eta = 0$).

1.1.3.2.2: $W_2 > 0$. Then we have $\tau_3 > 0$ and the elemental singularity is a generic node:

- $n,\ \widehat{cp}_{(2)};\ \overline{\binom{1}{1}}SN, ©, ©\ \Rightarrow (c = -2,\ e = 2,\ m = 2)$ (if $\eta < 0$);
- $n,\ \widehat{cp}_{(2)};\ \overline{\binom{1}{1}}SN, S, N^{\infty}\ \Rightarrow (c = -1,\ e = 1,\ m = 2)$ (if $\eta > 0$);
- $n,\ \widehat{cp}_{(2)};\ \overline{\binom{0}{2}}SN,\ \overline{\binom{1}{1}}SN\ \Rightarrow (c = -8/9,\ e = 8/9,\ m = 16/9)$ (if $\eta = 0$).

1.1.3.2.3: $W_2 = 0$. Considering (10.8) we have $\tau_3 = 0$, and systems (10.4) with $e = -c$ possess a node with coinciding eigenvalues. As the linearization matrix has the form given in (10.7) and $c \neq 0$ (due to $\kappa \neq 0$), this node could not be a star node. So considering Lemma 10.1 we arrive at the following three global configurations:

- $n^d,\ \widehat{cp}_{(2)};\ \overline{\binom{1}{1}}SN, ©, ©\ \Rightarrow (c = -9/5,\ e = 9/5,\ m = -3/5)$ (if $\eta < 0$);

• n^d, $\widehat{cp}_{(2)}$; $\overline{\binom{1}{1}}SN, S, N^\infty$ $\Rightarrow (c = -1/5,\ e = 1/5,\ m = -3/5)$ (if $\eta > 0$);
• n^d, $\widehat{cp}_{(2)}$; $\overline{\binom{0}{2}}SN$, $\overline{\binom{1}{1}}SN$ $\Rightarrow (c = -32/25,\ e = 32/25,\ m = -24/25)$ (if $\eta = 0$).

1.2: $\kappa > 0$. This implies $\widetilde{K} < 0$, and according to Theorem 6.1 (see Table 6.2, lines 136 and 137) the elemental singular point is a saddle. At the same time the type of the double point depends on the value of the invariant polynomial E_1.

Additionally considering (10.5) we observe that the condition $\kappa > 0$ implies $\eta > 0$, and according to Lemma 10.1, at infinity there could be the unique global configuration $\overline{\binom{1}{1}}SN, N^f, N^f$.

1.2.1: $\mathcal{T}_4 \neq 0$. By (10.5) this implies $\rho_1\rho_3 \neq 0$ and then $E_1 \neq 0$, i.e. the double point is a semi-elemental saddle-node. Additionally as $\rho_3 \neq 0$, the saddle is strong, and we obtain the global configuration
• s, $\overline{sn}_{(2)}$; $\overline{\binom{1}{1}}SN, N^f, N^f$ $\Rightarrow (c = 0,\ e = -1,\ m = 0)$.

1.2.2: $\mathcal{T}_4 = 0$. Considering (10.5) we observe that $\rho_1 = 0$ if and only if $\mathcal{T}_4 = \mathcal{T}_3 = 0$.

1.2.2.1: $\mathcal{T}_3 \neq 0$. In this case we get $\rho_1 \neq 0$ (then $E_1 \neq 0$) and $\rho_3 = 0$. We get $e = c - 2m$, and by (10.6) and Theorem 6.2, we obtain a weak saddle, the order of which is governed by the invariant polynomial $\mathcal{F}_1$.

1. $\mathcal{F}_1 \neq 0$. Clearly we arrive at the global configuration

   • $s^{(1)}$, $\overline{sn}_{(2)}$; $\overline{\binom{1}{1}}SN, N^f, N^f$ $\Rightarrow (c = 0,\ e = -2,\ m = 1)$.

2. $\mathcal{F}_1 = 0$. It was shown above (for the possibility $W_4 < 0$ on page 339), in this case the condition $c + 1 \neq 0$ holds, and we obtain $m = c(1 + 2c)/(2(1 + c))$. Then calculations yield

$$\mathcal{F}_2 = -\frac{c^8(3 + 2c)^2(6 + 5c)}{(1 + c)^5}, \qquad \kappa = 64\frac{c^2(3 + 2c)}{1 + c},$$

   and due to $\kappa > 0$ the condition $\mathcal{F}_2 \neq 0$ holds. So by Theorem 6.2 we could only have a weak saddle of order two, and this leads to the global configuration

   • $s^{(2)}$, $\overline{sn}_{(2)}$; $\overline{\binom{1}{1}}SN, N^f, N^f$ $\Rightarrow (c = 1,\ e = -1/2,\ m = 3/4)$.

1.2.2.2: $\mathcal{T}_3 = 0$. In this case we have $\rho_1 = 0$ and then $E_1 = 0$. As $\widetilde{K} < 0$, according to Theorem 6.1 (see Table 6.2, line 137), apart from the saddle, we have a nilpotent cusp.

1. $\mathcal{T}_2 \neq 0$. Considering (10.8) we get $\rho_3 \neq 0$, and therefore we have a strong saddle:

   • s, $\widehat{cp}_{(2)}$; $\overline{\binom{1}{1}}SN, N^f, N^f$ $\Rightarrow (c = 1,\ e = -1,\ m = 0)$.

2. $\mathcal{T}_2 = 0$. Then $m = c$, and taking into account (10.9), since $\mathcal{F}_1 \neq 0$ (due to $\kappa \neq 0$), by Theorem 6.2 we conclude that the weak saddle could be only of the first order:
   - $s^{(1)}$, $\widehat{cp}_{(2)}$; $\overline{\binom{1}{1}}SN, N^f, N^f \Rightarrow (c = 1,\ e = -1,\ m = 1)$.

1.3: $\kappa = 0$. In this case $h = 0$ and then $c \neq 0$, otherwise systems (10.3) become degenerate. So we may assume $c = 1$ due to a time rescaling, and we obtain the 2-parameter family of systems:

$$\dot{x} = x + y - x^2, \quad \dot{y} = ex + ey - ex^2 + 2mxy, \tag{10.10}$$

for the singular points $M_{1,2}(0,0)$ and $M_3(1,0)$ of which we have the following linearization matrices, traces, determinants and discriminants:

$$\begin{aligned} \mathcal{M}_1 &= \begin{pmatrix} 1 & 1 \\ e & e \end{pmatrix}, \ \rho_1 = \rho_2 = 1 + e,\ \Delta_1 = \Delta_2 = 0,\ \tau_1 = \tau_2 = (1+e)^2; \\ \mathcal{M}_3 &= \begin{pmatrix} -1 & 1 \\ -e & e + 2m \end{pmatrix}, \ \rho_3 = e - 1 + 2m,\ \Delta_3 = -2m,\ \tau_3 = \rho_3^2 - 4\Delta_3. \end{aligned} \tag{10.11}$$

For the above systems calculations yield

$$\begin{aligned} &\mu_0 = 0,\ \ \mu_1 = -\Delta_3^2 x,\ \ \widetilde{K} = 2\Delta_3 x^2,\ \ E_1 = -\rho_1\Delta_3^4/2,\ \ G_9 = 0, \\ &\mathcal{T}_i = 0,\ i = 1,2,3,4,\ \ \sigma = \rho_1 + 2(m-1)x,\ \ W_7 = 3\rho_1^4\Delta_3^6\tau_3/16, \\ &\mathcal{B}_1 = \rho_1^2\rho_3\Delta_3,\ \ \mathcal{B}_2 = -\rho_1(m-1)^2(3e - 1 + 4m)\Delta_3. \end{aligned} \tag{10.12}$$

We observe that the condition $\mu_1 \neq 0$ implies $\widetilde{K} \neq 0$. According to Theorems 6.1 and 6.2, the polynomials above are responsible for the types of the finite singularities of systems (10.10). In order to describe the infinite singularities, considering Lemma 10.1 we calculate the additional invariant polynomials

$$\begin{aligned} &\kappa = \eta = 0,\ \ \widetilde{M} = -8(1+2m)^2x^2,\ \ C_2 = ex^3 - (1+2m)x^2y, \\ &\tilde{L} = 8(1+2m)x^2,\ \ \widetilde{N} = -4m(1+m)x^2. \end{aligned} \tag{10.13}$$

So considering Lemma 10.1 we have the next remark.

Remark 10.3. Systems (10.10) could possess at infinity only one of the following seven global configurations:

$$\begin{aligned} \widetilde{M} \neq 0,\ \widetilde{K} < 0 \quad &\Rightarrow \quad \widehat{\binom{1}{2}}\ \widehat{P}_{\curlywedge}E\widehat{P}_{\curlywedge} - H,\ N^f; \\ \widetilde{M} \neq 0,\ \widetilde{K} > 0,\ \tilde{L} < 0,\ \widetilde{N} < 0 \quad &\Rightarrow \quad \widehat{\binom{1}{2}}\ \widehat{P}_{\curlywedge}E\widehat{P}_{\curlywedge} - H,\ S; \\ \widetilde{M} \neq 0,\ \widetilde{K} > 0,\ \tilde{L} < 0,\ \widetilde{N} > 0 \quad &\Rightarrow \quad \widehat{\binom{1}{2}}\ \widehat{P}_{\curlywedge}H\widehat{P}_{\curlywedge} - E,\ S; \\ \widetilde{M} \neq 0,\ \widetilde{K} > 0,\ \tilde{L} < 0,\ \widetilde{N} = 0 \quad &\Rightarrow \quad \widehat{\binom{1}{2}}\ H - E,\ S; \\ \widetilde{M} \neq 0,\ \widetilde{K} > 0,\ \tilde{L} > 0 \quad &\Rightarrow \quad \widehat{\binom{1}{2}}\ H_{\curlywedge}HH_{\curlywedge} - H,\ N^{\infty}; \\ \widetilde{M} = 0,\ C_2 \neq 0 \quad &\Rightarrow \quad \widehat{\binom{1}{3}}\ H_{\curlywedge}H\widehat{P}_{\curlywedge} - \widehat{P}; \\ \widetilde{M} = C_2 = 0 \quad &\Rightarrow \quad [\infty;\ \emptyset]. \end{aligned}$$

1.3.1: $\widetilde{K} < 0$. Then $\Delta_3 < 0$ and the elemental singularity is a saddle.

1.3.1.1: $\mathcal{B}_1 \neq 0$. In this case we have $\rho_1\rho_3 \neq 0$, and therefore the double point is a semi-elemental saddle-node and the saddle is strong. We observe that the condition $\widetilde{K} < 0$ implies $m > 0$ and then $\widetilde{M} \neq 0$. So considering the above remark we arrive at the global configuration

- $s,\ \overline{sn}_{(2)};\ \widehat{\binom{1}{2}}\ \widehat{P}_{\curlywedge} E \widehat{P}_{\curlywedge} - H,\ N^f \ \Rightarrow (e = 0,\ m = 1).$

1.3.1.2: $\mathcal{B}_1 = 0$. Then $\rho_1\rho_3 = 0$ and we shall consider two cases: $\mathcal{B}_2 \neq 0$ and $\mathcal{B}_2 = 0$

1. $\mathcal{B}_2 \neq 0$. Considering (10.12) we have $\rho_1 \neq 0$ and then $\rho_3 = 0$ (i.e. $e = 1-2m$) and the saddle is weak. We calculate

$$\begin{gathered}\mathcal{B}_1 = \mathcal{H} = 0,\ \mathcal{B}_2 = 8m(m-1)^4,\ \mathcal{F}_1 = 8m(m-1),\\ \sigma = 2(m-1)(x-1),\end{gathered} \tag{10.14}$$

   and clearly the conditions $\mathcal{B}_2 \neq 0$ and $\widetilde{K} < 0$ imply $\sigma\mathcal{F}_1 \neq 0$ and $\mathcal{B}_2 > 0$. So according to Theorem 6.2 (see statement (e_1)) the weak saddle is of order one. Considering Remark 10.3 we get the global configuration

   - $s^{(1)},\ \overline{sn}_{(2)};\ \widehat{\binom{1}{2}}\ \widehat{P}_{\curlywedge} E \widehat{P}_{\curlywedge} - H,\ N^f \ \Rightarrow (e = 0,\ m = 1/2).$

2. $\mathcal{B}_2 = 0$. We examine two possibilities: $\sigma \neq 0$ and $\sigma = 0$.

   (a) $\sigma \neq 0$. We claim that in this case the condition $\mathcal{B}_1 = \mathcal{B}_2 = 0$ is equivalent to $\rho_1 = 0$. Indeed suppose the contrary. Then $\rho_3 = 0$, and since $\widetilde{K} \neq 0$, considering (10.14) the condition $\mathcal{B}_2 = 0$ implies $\sigma = 0$. This contradiction proves our claim. So $\rho_1 = 0$ (i.e. $e = -1$ and then $E_1 = 0$), and by Theorem 6.1 (see Table 6.2, line 137) the double singular point is a nilpotent cusp. In this case $\rho_3 = 2(m-1) \neq 0$ due to $\sigma = 2(m-1)x \neq 0$. By Remark 10.3 we get the global configuration

   - $s,\ \widehat{cp}_{(2)};\ \widehat{\binom{1}{2}}\ \widehat{P}_{\curlywedge} E \widehat{P}_{\curlywedge} - H,\ N^f \ \Rightarrow (e = -1,\ m = 2).$

   (b) $\sigma = 0$. Then $e = -1$, $m = 1$ and the system (10.10) is Hamiltonian, i.e. the saddle is integrable:

   - $\$,\ \widehat{cp}_{(2)};\ \widehat{\binom{1}{2}}\ \widehat{P}_{\curlywedge} E \widehat{P}_{\curlywedge} - H,\ N^f \ \Rightarrow (e = -1,\ m = 1).$

1.3.2: $\widetilde{K} > 0$. In this case according to Theorem 6.1 (see Table 6.2, lines 139, 141 and 143) the elemental singular point is an anti-saddle, and its type is governed by the invariant polynomial W_7 (as in the considered case we have $G_9 = 0$).

Remark 10.4. Considering (10.13) we observe that the condition $\tilde{L} = 0$ implies $\widetilde{M} = 0$. Moreover as in this case $W_7 = 3e(e-4)(1+e)^4/16$, we obtain that the condition $W_7 \neq 0$ yields $C_2 \neq 0$.

1.3.2.1: $W_7 < 0$. By (10.12) we have $\rho_1 \neq 0$ and $\tau_3 < 0$, and we conclude that the double point is a semi-elemental saddle-node, whereas the simple point is a focus.

1. $\mathcal{B}_1 \neq 0$. Then $\rho_3 \neq 0$ and the focus is strong. Considering Remarks 10.3 and 10.4, we arrive at the next five global configurations:

   - $f,\ \overline{sn}_{(2)};\ \widehat{\binom{1}{2}}\ \widehat{P}_{\curlywedge} E \widehat{P}_{\curlywedge} - H,\ S\ \Rightarrow (e=1,\ m=-3/2)$ (if $\tilde{L}<0,\ \widetilde{N}<0$);
   - $f,\ \overline{sn}_{(2)};\ \widehat{\binom{1}{2}}\ \widehat{P}_{\curlywedge} H \widehat{P}_{\curlywedge} - E,\ S\ \Rightarrow (e=1,\ m=-2/3)$ (if $\tilde{L}<0,\ \widetilde{N}>0$);
   - $f,\ \overline{sn}_{(2)};\ \widehat{\binom{1}{2}}\ H - E,\ S\ \Rightarrow (e=1,\ m=-1)$ (if $\tilde{L}<0,\ \widetilde{N}=0$);
   - $f,\ \overline{sn}_{(2)};\ \widehat{\binom{1}{2}}\ H_{\curlywedge} H H_{\curlywedge} - H,\ N^{\infty}\ \Rightarrow (e=1,\ m=-1/3)$ (if $\tilde{L}>0$);
   - $f,\ \overline{sn}_{(2)};\ \widehat{\binom{1}{3}}\ H_{\curlywedge} H \widehat{P}_{\curlywedge} - \widehat{P}\ \Rightarrow (e=1,\ m=-1/2)$ (if $\tilde{L}=0$).

2. $\mathcal{B}_1 = 0$. As $W_7 \neq 0$ (i.e. $\rho_1 \neq 0$) we have $\rho_3 = 0$, and this gives $e = 1 - 2m$. Then we obtain $W_7 = 1536(m-1)^4 m^7$, and considering (10.14) we observe that the condition $W_7 < 0$ implies $\sigma\mathcal{F}_1 \neq 0$ and $\mathcal{B}_2 < 0$. According to Theorem 6.2 (see statement (e_2)) we have a first order weak focus. Considering Remarks 10.3 and 10.4, we get the following five global configurations:

   - $f^{(1)},\ \overline{sn}_{(2)};\ \widehat{\binom{1}{2}}\ \widehat{P}_{\curlywedge} E \widehat{P}_{\curlywedge} - H,\ S\ \Rightarrow (e=4,\ m=-3/2)$ (if $\tilde{L}<0,\ \widetilde{N}<0$);
   - $f^{(1)},\ \overline{sn}_{(2)};\ \widehat{\binom{1}{2}}\ \widehat{P}_{\curlywedge} H \widehat{P}_{\curlywedge} - E,\ S\ \Rightarrow (e=7/3,\ m=-2/3)$ (if $\tilde{L}<0,\ \widetilde{N}>0$);
   - $f^{(1)},\ \overline{sn}_{(2)};\ \widehat{\binom{1}{2}}\ H - E,\ S\ \Rightarrow (e=3,\ m=-1)$ (if $\tilde{L}<0,\ \widetilde{N}=0$);
   - $f^{(1)},\ \overline{sn}_{(2)};\ \widehat{\binom{1}{2}}\ H_{\curlywedge} H H_{\curlywedge} - H,\ N^{\infty}\ \Rightarrow (e=5/3,\ m=-1/3)$ (if $\tilde{L}>0$);
   - $f^{(1)},\ \overline{sn}_{(2)};\ \widehat{\binom{1}{3}}\ H_{\curlywedge} H \widehat{P}_{\curlywedge} - \widehat{P}\ \Rightarrow (e=2,\ m=-1/2)$ (if $\tilde{L}=0$).

1.3.2.2: $W_7 > 0$. By (10.12) we have $\rho_1 \neq 0$ and $\tau_3 > 0$, and we conclude that the double point is a semi-elemental saddle-node, whereas the simple point is a generic node. So by Remark 10.3, we arrive at the next five global configurations:

- $n,\ \overline{sn}_{(2)};\ \widehat{\binom{1}{2}}\ \widehat{P}_{\curlywedge} E \widehat{P}_{\curlywedge} - H,\ S\ \Rightarrow (e=-2,\ m=-3/2)$ (if $\tilde{L}<0,\ \widetilde{N}<0$);
- $n,\ \overline{sn}_{(2)};\ \widehat{\binom{1}{2}}\ \widehat{P}_{\curlywedge} H \widehat{P}_{\curlywedge} - E,\ S\ \Rightarrow (e=-2,\ m=-2/3)$ (if $\tilde{L}<0,\ \widetilde{N}>0$);
- $n,\ \overline{sn}_{(2)};\ \widehat{\binom{1}{2}}\ H - E,\ S\ \Rightarrow (e=-2,\ m=-1)$ (if $\tilde{L}<0,\ \widetilde{N}=0$);
- $n,\ \overline{sn}_{(2)};\ \widehat{\binom{1}{2}}\ H_{\curlywedge} H H_{\curlywedge} - H,\ N^{\infty}\ \Rightarrow (e=-2,\ m=-1/3)$ (if $\tilde{L}>0$);
- $n,\ \overline{sn}_{(2)};\ \widehat{\binom{1}{3}}\ H_{\curlywedge} H \widehat{P}_{\curlywedge} - \widehat{P}\ \Rightarrow (e=-2,\ m=-1/2)$ (if $\tilde{L}=0$).

1.3.2.3: $W_7 = 0$. By (10.12) we have $\rho_1\tau_3 = 0$.

1. $\mathcal{B}_1 \neq 0$. Then we obtain $\rho_1 \neq 0$ and $\tau_3 = 0$. So the simple singularity is a node with coinciding eigenvalues. Additionally considering the linearization matrix $\mathcal{M}_3$ from (10.11), we observe that we could not have a star node. Therefore considering Remark 10.3 we get the next six global configurations:

   - $n^d,\ \overline{sn}_{(2)};\ \widehat{\binom{1}{2}}\ \widehat{P}_{\curlywedge} E \widehat{P}_{\curlywedge} - H,\ S\ \Rightarrow (e = 1,\ m = -2)\quad (\text{if } \tilde{L} < 0,\ \widetilde{N} < 0)$;
   - $n^d,\ \overline{sn}_{(2)};\ \widehat{\binom{1}{2}}\ \widehat{P}_{\curlywedge} H \widehat{P}_{\curlywedge} - E,\ S\ \Rightarrow (e = 1/25,\ m = -18/25)\quad (\text{if } \tilde{L} < 0,\ \widetilde{N} > 0)$;
   - $n^d,\ \overline{sn}_{(2)};\ \widehat{\binom{1}{2}}\ H - E,\ S\ \Rightarrow (e = 3 + 2\sqrt{2},\ m = -1)\quad (\text{if } \tilde{L} < 0,\ \widetilde{N} = 0)$;
   - $n^d,\ \overline{sn}_{(2)};\ \widehat{\binom{1}{2}}\ H_{\curlywedge} H H_{\curlywedge} - H,\ N^{\infty}\ \Rightarrow (e = 1/4,\ m = -1/8)\quad (\text{if } \tilde{L} > 0)$;
   - $n^d,\ \overline{sn}_{(2)};\ \widehat{\binom{1}{3}}\ H_{\curlywedge} H \widehat{P}_{\curlywedge} - \widehat{P}\ \Rightarrow (e = 4,\ m = -1/2)\quad (\text{if } \tilde{L} = 0,\ C_2 \neq 0)$;
   - $n^d,\ \overline{sn}_{(2)};\ [\infty; \emptyset]\ \Rightarrow (e = 0,\ m = -1/2)\quad (\text{if } \tilde{L} = 0,\ C_2 = 0)$.

2. $\mathcal{B}_1 = 0$. Then $\rho_1 \rho_3 = 0$ and this gives $\rho_1 = 0$, otherwise we get $\rho_3 = \tau_3 = 0$ and M_3 becomes a non-elemental singularity. So $\rho_1 = 0$ (i.e. $e = -1$ and then $E_1 = 0$) and the double point is a cusp. As in this case we have $\tau_3 = 4(1 + m^2) > 0$, the simple singular point is a generic node. So as $e \neq 0$ (i.e. $C_2 \neq 0$), we get the following five global configurations:

   - $n,\ \widehat{cp}_{(2)};\ \widehat{\binom{1}{2}}\ \widehat{P}_{\curlywedge} E \widehat{P}_{\curlywedge} - H,\ S\ \Rightarrow (e = -1,\ m = -3/2)\quad (\text{if } \tilde{L} < 0,\ \widetilde{N} < 0)$;
   - $n,\ \widehat{cp}_{(2)};\ \widehat{\binom{1}{2}}\ \widehat{P}_{\curlywedge} H \widehat{P}_{\curlywedge} - E,\ S\ \Rightarrow (e = -1,\ m = -2/3)\quad (\text{if } \tilde{L} < 0,\ \widetilde{N} > 0)$;
   - $n,\ \widehat{cp}_{(2)};\ \widehat{\binom{1}{2}}\ H - E,\ S\ \Rightarrow (e = -1,\ m = -1)\quad (\text{if } \tilde{L} < 0,\ \widetilde{N} = 0)$;
   - $n,\ \widehat{cp}_{(2)};\ \widehat{\binom{1}{2}}\ H_{\curlywedge} H H_{\curlywedge} - H,\ N^{\infty}\ \Rightarrow (e = -1,\ m = -1/3)\quad (\text{if } \tilde{L} > 0)$;
   - $n,\ \widehat{cp}_{(2)};\ \widehat{\binom{1}{3}}\ H_{\curlywedge} H \widehat{P}_{\curlywedge} - \widehat{P}\ \Rightarrow (e = -1,\ m = -1/2)\quad (\text{if } \tilde{L} = 0)$.

Thus we obtain a total of 62 geometrically distinct configurations of singularities for the family of quadratic systems with one double and one simple finite singularities. □

10.3 Exactly three distinct singularities

If a quadratic system possesses three distinct finite singularities, it is clear that either all three singularities are real (i.e. $\mathcal{D}_{\mathbb{C}^2}(S) = p + q + r$) or one is real and two are complex (i.e. $\mathcal{D}_{\mathbb{C}^2}(S) = p + q^c + r^c$). By Proposition 5.1 in the first case the conditions $\mu_0 = 0$, $\mu_1 \neq 0$ and $\mathbf{D} < 0$ hold, whereas in the second case we have $\mu_0 = 0$, $\mu_1 \neq 0$ and $\mathbf{D} > 0$. We consider each one of these possibilities.

10.3.1 Systems with zero-cycle $\mathcal{D}_{\mathbb{C}^2}(S) = p + q + r$

Theorem 10.3. (a) *We consider here all the geometrical global configurations of singularities of quadratic vector fields with three distinct real finite singularities. These configurations are classified in Diagram* 10.4 *according to the geometrical equivalence relation of configurations. The total number of geometrically distinct global configurations of singularities is* 147.

(b) *Necessary and sufficient conditions for each one of the* 147 *different geometrical equivalence classes can be assembled from this diagram in terms of* 21 *invariant polynomials with respect to the action of the affine group and time rescaling. These invariant polynomials, which appear in Diagram* 10.4, *are given in Section* 5.4.

(c) *Diagram* 10.4 *actually contains the global bifurcation diagram in the* 12*-dimensional space of parameters of the global geometrical configurations, of this family of quadratic differential systems and provides an algorithm for deciding for any given system which is its respective global configuration.*

Proof. Assume that the quadratic systems (8.1) possess three real elemental finite singularities. Then according to Proposition 6.2 (see Table 6.1) via an affine transformation these systems could be brought to the systems 10. However applying (for some geometrical reasons) the change $(x, y, c, d, h, e, f, m) \rightarrow (y, x, e, f, m, c, d, h)$, we consider here the family of systems:

$$\dot{x} = cx + dy + 2hxy - dy^2, \quad \dot{y} = ex + fy + 2mxy - fy^2. \tag{10.15}$$

We remark that these systems possess three distinct real finite singularities: $M_1(0,0)$, $M_2(0,1)$ and $M_3\left(\dfrac{\alpha(\alpha-2\gamma)}{4\beta\gamma}, \dfrac{\alpha}{2\gamma}\right)$, where $\alpha = cf - de$, $\beta = cm - eh$ and $\gamma = dm - fh$. For the above systems calculations yield

$$\mu_0 = 0, \quad \mu_1 = 4\beta\gamma y, \quad \kappa = -128m^2\gamma.$$

Since for the above systems we have $\mu_1 \neq 0$ (i.e. $\gamma \neq 0$), we observe that the condition $\kappa = 0$ is equivalent to $m = 0$.

By Lemma 10.1 the sign of the invariant polynomial κ is relevant, so in what follows we consider three cases: $\kappa < 0$, $\kappa > 0$ and $\kappa = 0$.

If $\kappa \neq 0$ we have $m \neq 0$, and we may assume $m = 1$ due to a time rescaling. So we get the systems:

$$\dot{x} = cx + dy + 2hxy - dy^2, \quad \dot{y} = ex + fy + 2xy - fy^2, \tag{10.16}$$

and for the singular points M_i we have the following values for the traces ρ_i, for the determinants Δ_i, for the discriminants τ_i and for the linearization matrices

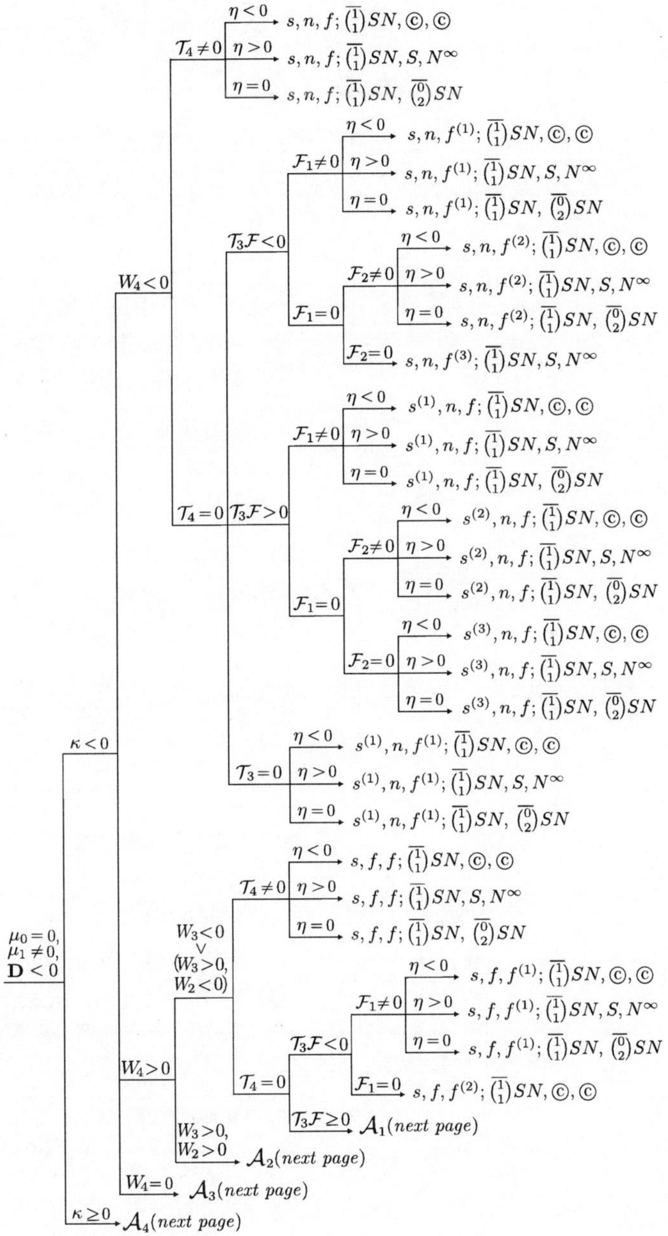

Diagram 10.4: Global configurations: the case $\mu_0 = 0$, $\mu_1 \neq 0$, $\mathbf{D} < 0$.

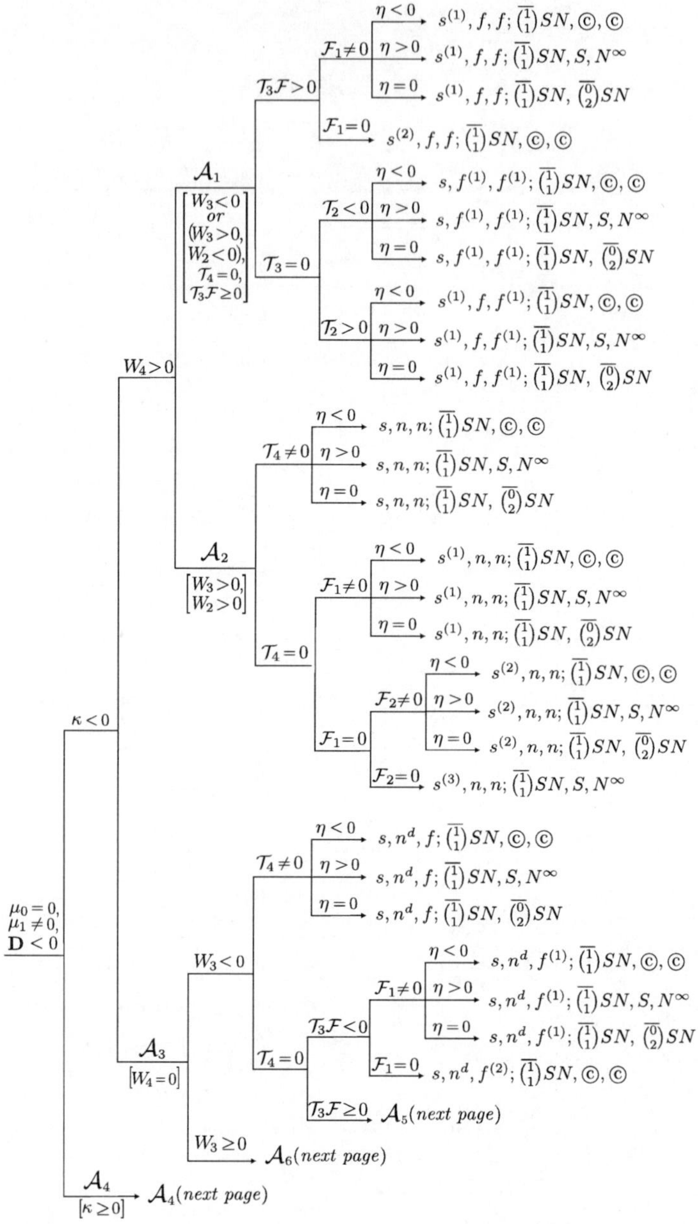

Diagram 10.4 (*cont.*): Global configurations: the case $\mu_0 = 0$, $\mu_1 \neq 0$, $\mathbf{D} < 0$.

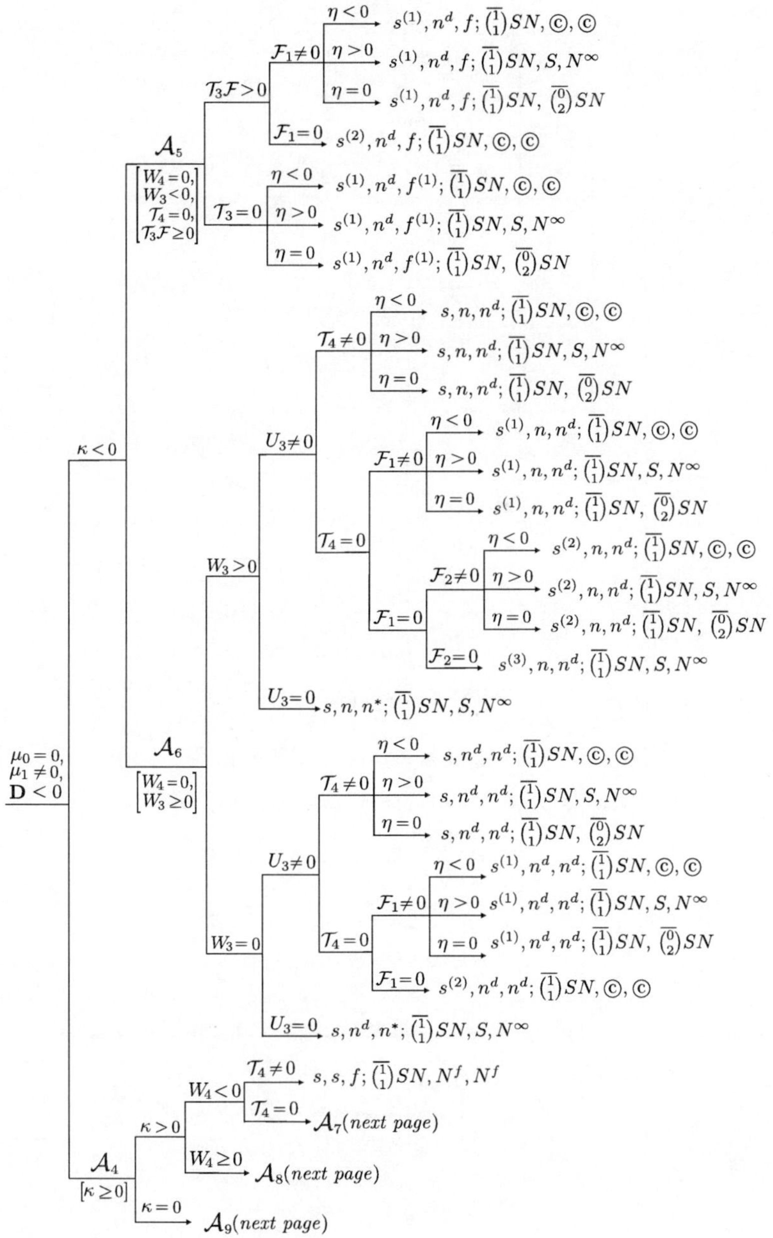

Diagram 10.4 (*cont.*): Global configurations: the case $\mu_0 = 0$, $\mu_1 \neq 0$, $\mathbf{D} < 0$.

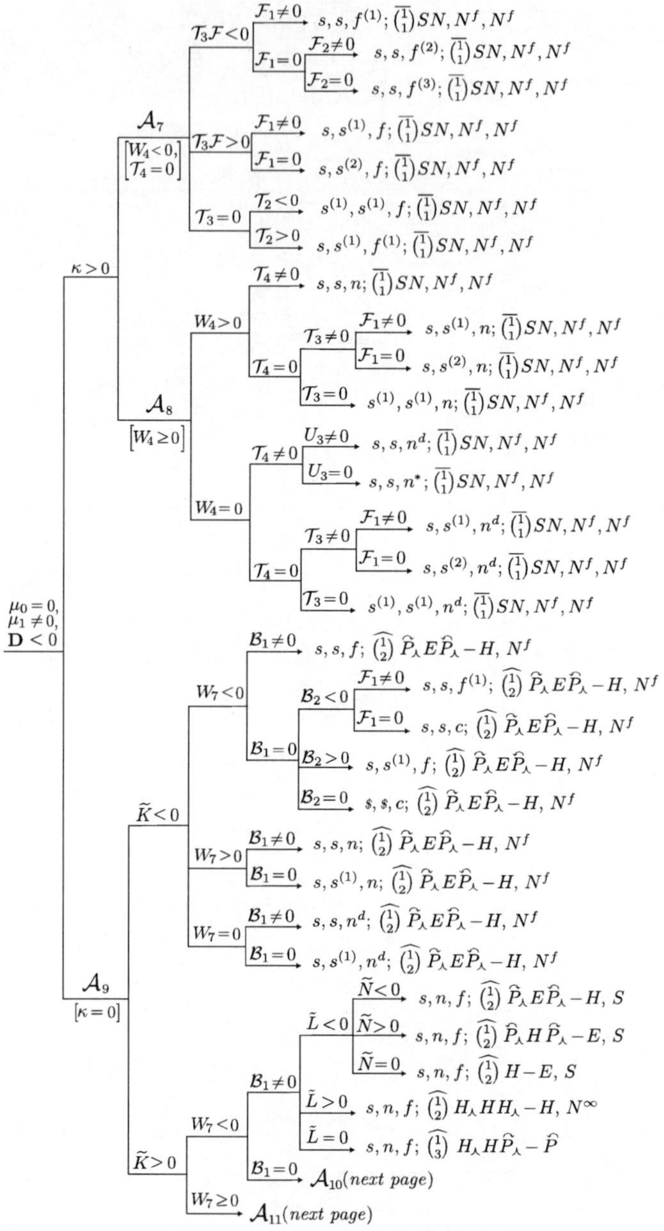

Diagram 10.4 (*cont.*): Global configurations: the case $\mu_0=0$, $\mu_1\neq 0$, $\mathbf{D}<0$.

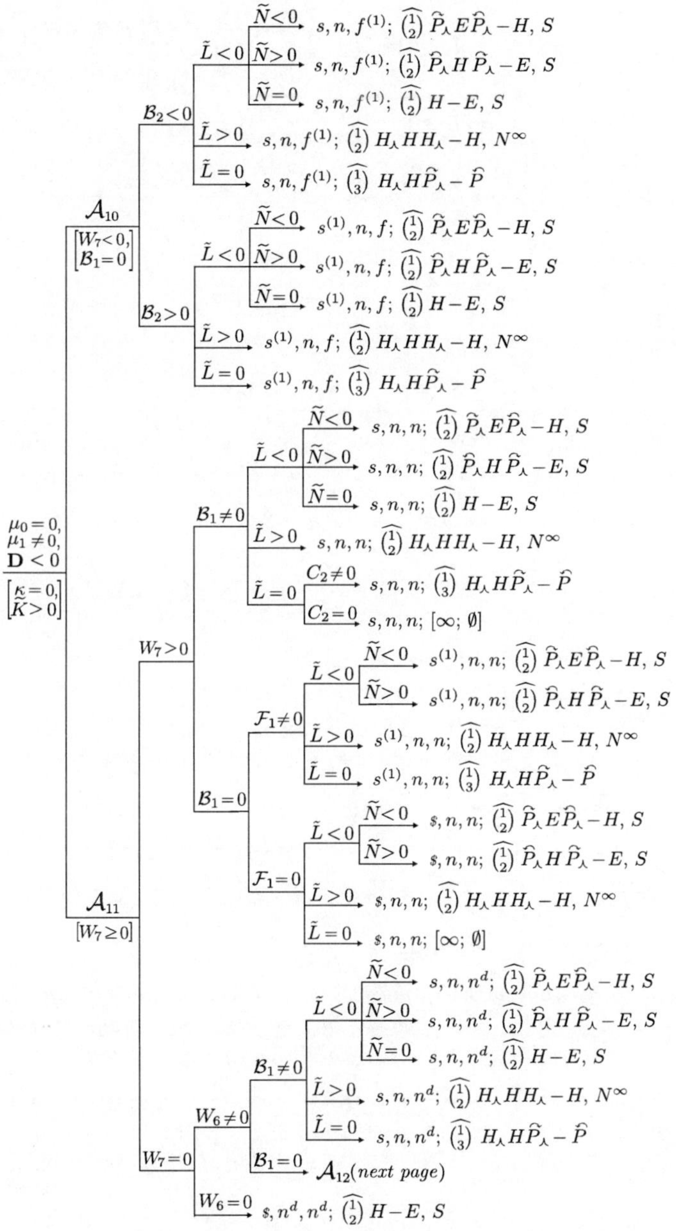

Diagram 10.4 (*cont.*): Global configurations: the case $\mu_0 = 0$, $\mu_1 \neq 0$, $\mathbf{D} < 0$.

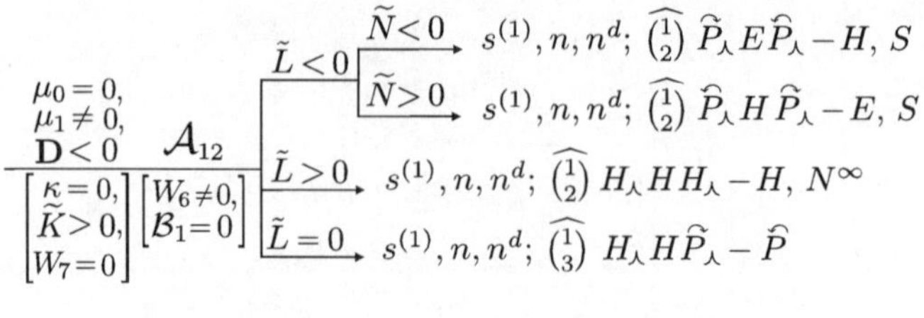

Diagram 10.4 (*cont.*): Global configurations: the case $\mu_0=0$, $\mu_1\neq 0$, $\mathbf{D}<0$.

$\mathcal{M}_1$ and $\mathcal{M}_2$:

$$\mathcal{M}_1=\begin{pmatrix} c & d\\ e & f\end{pmatrix},\quad \mathcal{M}_2=\begin{pmatrix} c+2h & -d\\ e+2 & -f\end{pmatrix},$$
$$\begin{aligned}\rho_1=c+f,\ \ \Delta_1=\alpha,\quad \rho_2=c-f+2h,\ \ \Delta_2=2\gamma-\alpha,\\ \rho_3=\frac{1}{2\beta\gamma}\big[d^2e(2+e)+2d(c^2-2ceh-2efh-e^2fh+e^2h^2)-\\ -f^2(c^2-2ceh-2eh^2)\big],\ \ \Delta_3=\frac{\alpha(\alpha-2\gamma)}{2\gamma},\quad \tau_i=\rho_i^2-4\Delta_i,\ \ i=1,2,3.\end{aligned}\tag{10.17}$$

Then for systems (10.16) we calculate

$$\begin{aligned}\kappa=-128\gamma,\ \ \widetilde{K}=4\gamma y^2,\ \ \mu_1=4\beta\gamma y,\ \ \eta=4\big[(f-2h)^2-8\gamma\big],\\ G_9=\gamma,\ \ \widetilde{M}=-32x^2+16(f+2h)xy-8\big[(f+2h)^2-6d\big]y^2,\\ \mathcal{T}_4=8\beta\gamma\rho_1\rho_2\rho_3,\ \ \mathcal{T}_3=8\beta\gamma\big[\rho_1(\rho_2+\rho_3)+\rho_2\rho_3\big],\\ \mathcal{T}_2=8\beta\gamma(\rho_1+\rho_2+\rho_3),\ \ W_4=64\beta^2\gamma^2\tau_1\tau_2\tau_3,\\ W_2=64\beta^2\gamma^2(\tau_1+\tau_2+\tau_3),\ \ W_3=64\beta^2\gamma^2\big[\tau_1\tau_2+\tau_1\tau_3+\tau_2\tau_3\big].\end{aligned}\tag{10.18}$$

First we prove the following lemma.

Lemma 10.3. *A system* (10.16) *possesses a star node if and only if the conditions* $W_4=U_3=0$ *hold. Moreover in this case the star node is unique, and the conditions* $W_3\geq 0$ *and* $\mathcal{T}_4\neq 0$ *hold. In addition if* $W_3=0$ *then* $\eta>0$.

Proof. Assuming that a system (10.16) possesses a star node, without loss of generality we can consider that such a singular point is $M_1(0,0)$. In this case considering (10.17) we must have $d=e=0$ and $f=c$, and this implies $W_4=U_3=0$.

Conversely, assume that the conditions $W_4=U_3=0$ hold. Since $\mu_1\neq 0$ (i.e. $\beta\gamma\neq 0$), the relation $W_4=0$ is equivalent to $\tau_1\tau_2\tau_3=0$, i.e. at least one of the discriminants corresponding to finite singularities vanishes. Then without loss of generality we can consider that the singular point with vanishing discriminant is $M_1(0,0)$, and hence the relation $(c-f)^2+4de=0$ holds.

Assume first $e = 0$. Then $f = c$ and we have

$$\text{Coefficient}[U_3, x^3y^2] = -72cd, \quad \mu_1 = 4c(d - ch)y.$$

Since $\mu_1 \neq 0$ the condition $U_3 = 0$ implies $d = 0$ (then U_3 vanishes), and we get a star node. We note that in this case we obtain

$$W_3 = 64c^6(c+h)^2(c+4h)^2 \geq 0, \quad \eta = 4(c+2h)^2, \quad \mathcal{T}_4 = -16c^5h, \quad \mu_1 = -4c^2hy.$$

It is clear that the condition $\mu_1 \neq 0$ gives $\mathcal{T}_4 \neq 0$ and that the condition $W_3 = 0$ implies $\eta > 0$.

Suppose now $e \neq 0$. In this case we obtain $d = -(c-f)^2/(4e)$, and we calculate Coefficient$[U_3, x^5] = -12ef(e+2)$. Since $e \neq 0$ we consider two cases: $e + 2 = 0$ and $e + 2 \neq 0$.

1. If $e = -2$, then calculations yield

$$\text{Coefficient}[U_3, x^4y] = -24(c+2h)(c+f+2h),$$
$$\mu_1 = (c+2h)(c^2 - 2cf + f^2 - 8fh)y/2.$$

   Due to $\mu_1 \neq 0$ the condition $U_3 = 0$ implies $f = -(c+2h)$, and we obtain Coefficient$[[U_3, x^3y^2] = 36(c+h)^2(c+2h) = 0$. Since $c + 2h \neq 0$ we get $h = -c$. In this case we have $U_3 = 0$, and at the same time considering the matrix $\mathcal{M}_2$ from (10.17), we observe that the singular point $M_2(0,1)$ is a star node. We remark that in this case $M_1(0,0)$ is a node n^d, and the condition $W_3 = 0$ is fulfilled. Additionally we have $\eta = 4c^2 > 0$ and $\mathcal{T}_4 = 16c^6 \neq 0$ due to $\mu_1 = -4c^3y \neq 0$.

2. Suppose now $e + 2 \neq 0$. Then $f = 0$ and we calculate $\mu_1 = -c^2(c - eh)y/e$ and

$$\begin{aligned}\text{Coefficient}[U_3, x^4y] &= -\,3(10c^2 + c^2e - 16ceh + 8e^2h^2) \equiv -3\Phi_1,\\ \text{Coefficient}[U_3, x^3y^2] &= 6(3c^3 + 3c^3e + 5c^2eh - c^2e^2h - 8ce^2h^2 + 4e^3h^3)/e\\ &\equiv 6\Phi_2/e.\end{aligned}$$

   Since $U_3 = 0$ we necessarily have $\text{Res}_h(\Phi_1, \Phi_2) = 144c^6e^6(2+e)(4+e)^2 = 0$, and as $e(2+e) \neq 0$ and $\mu_1 \neq 0$ (then $c \neq 0$), we get $e = -4$. Thus we obtain

$$\begin{aligned}U_3 = 3(3c+8h)y(8x+cy)\big[&-256(c+8h)x^3 + 32(7c^2+16ch+64h^2)x^2y\\ &-8c(c+4h)(c+8h)xy^2 - c^2(11c^2 - 8ch - 64h^2)y^3\big]/1024,\end{aligned}$$

   and it is easy to determine that the condition $U_3 = 0$ is equivalent to $3c + 8h = 0$. Hence $h = -3c/8$ and we obtain that in this case the singular point $M_3(-c/4, 2)$ is a star node (apart from the node n^d at $M_1(0,0)$) and the condition $W_3 = 0$ holds. We note that in this case $\eta = c^2/4 > 0$ and $\mathcal{T}_4 = c^6/16 \neq 0$ due to $\mu_1 = -c^3y/8 \neq 0$.

To prove the uniqueness of the star node, it is enough to apply Theorem 7.1. For systems (10.16) we calculate

$$U_1 = -4x^2 + 8hxy + 4(d - fh - h^2)y^2 \neq 0,$$

and by Theorem 7.1 (see statement $(\mathcal{S}^3_{n^*})$) these systems could not have two star nodes.

Since all the cases are examined, Lemma 10.3 is proved. □

Remark 10.5. We observe that for systems (10.16) the invariant polynomial $\widetilde{M} \neq 0$, and in addition the condition $\kappa \neq 0$ implies $\widetilde{K}G_9 \neq 0$. Moreover we have $\text{sign}\,(\widetilde{K}) = -\text{sign}\,(\kappa)$.

Now we continue to study the cases $\kappa < 0$, $\kappa > 0$ and $\kappa = 0$.

1: $\kappa < 0$. This implies $\widetilde{K} > 0$, and according to Theorem 6.1 (see Table 6.2, lines 115–122) systems (10.16) possess one saddle and two anti-saddles. Moreover the types of the anti-saddles are governed by the invariant polynomials W_i, $i = 4, 3, 2$ (because in the case we consider here we have $G_9 \neq 0$).

1.1: $W_4 < 0$. By Theorem 6.1, apart from the saddle, we have a node (which is generic) and a focus.

1.1.1: $\mathcal{T}_4 \neq 0$. Then by Theorem 6.2 the focus is strong, and considering the condition $\widetilde{M} \neq 0$ (see Remark 10.5), by Lemma 10.1 we arrive at the following three global configurations:

- $s, n, f; \overline{\binom{1}{1}}SN, ©, ©$ $\Rightarrow (c = 1, h = 3, e = -1, d = 10, f = 1)$ (if $\eta < 0$);
- $s, n, f; \overline{\binom{1}{1}}SN, S, N^{\infty}$ $\Rightarrow (c = 1,\ h = 3,\ e = -1,\ d = 4,\ f = 1/5)$ (if $\eta > 0$);
- $s, n, f; \overline{\binom{1}{1}}SN, \overline{\binom{0}{2}}SN$ $\Rightarrow (c = 1, h = 3, e = -1, d = 961/200, f = 1/5)$ (if $\eta = 0$).

1.1.2: $\mathcal{T}_4 = 0$, $\mathcal{T}_3 \neq 0$. According to Theorem 6.2 we have either a weak focus or a weak saddle depending on the sign of the product $\mathcal{T}_3\mathcal{F}$ (in the case $\mathcal{T}_3 \neq 0$).

1.1.2.1: $\mathcal{T}_3\mathcal{F} < 0$. Then we have a weak focus, the order of which is governed by the invariant polynomials $\mathcal{F}_i \neq 0$ $(i = 1, 2, 3, 4)$ (see Theorem 6.2).

1.1.2.1.1: $\mathcal{F}_1 \neq 0$. In this case considering Lemma 10.1 and the condition $\kappa < 0$, we get the next three global configurations:

- $s, n, f^{(1)}; \overline{\binom{1}{1}}SN, ©, ©$ $\Rightarrow (c = 1,\ h = 3,\ e = -1,\ d = 5,\ f = -1)$ (if $\eta < 0$);
- $s, n, f^{(1)}; \overline{\binom{1}{1}}SN, S, N^{\infty}$ $\Rightarrow (c = 1,\ h = 3,\ e = -1,\ d = 2,\ f = -1)$ (if $\eta > 0$);
- $s, n, f^{(1)}; \overline{\binom{1}{1}}SN, \overline{\binom{0}{2}}SN \Rightarrow (c = 1, h = 3, e = -1, d = 25/8, f = -1)$ (if $\eta = 0$).

1.1.2.1.2: $\mathcal{F}_1 = 0$. Then the weak focus is of order at least two, and this depends on $\mathcal{F}_2$.

1. $\mathcal{F}_2 \neq 0$. In this case the weak focus has order two, and we arrive at the following three global configurations (to construct the examples below we use the normal form

$$\dot{x} = -y+(b-v)x^2/3+2axy+y^2, \quad \dot{y} = x+ax^2+(b+2v)xy/3-ay^2) : \quad (10.19)$$

   - $s, n, f^{(2)}; \overline{\binom{1}{1}}SN, ©, ©$ $\Rightarrow (b = -23/7, v = -6/7, a = \sqrt{1338 - 4\sqrt{2559}}/42)$ (if $\eta < 0$);
   - $s, n, f^{(2)}; \overline{\binom{1}{1}}SN, S, N^{\infty}$ $\Rightarrow (b = -163/49,\ v = -6/7,$ $a = \sqrt{66534 - 26\sqrt{127374}}/294)$ (if $\eta > 0$);
   - $s, n, f^{(2)}; \overline{\binom{1}{1}}SN,\ \overline{\binom{0}{2}}SN$ $\Rightarrow (b = -162/49,\ v = -6/7,\ a = 4\sqrt{2}/7)$ (if $\eta = 0$).

2. $\mathcal{F}_2 = 0$. According to Theorem 7.4 we could not have a center apart from a node and a saddle. So we have a focus of order three. Additionally in the case we consider, according to Theorem 7.8 we could have only three real distinct infinite singularities (i.e. $\eta > 0$). Hence we have one global configuration (see systems (10.19)):

   - $s, n, f^{(3)}; \overline{\binom{1}{1}}SN, S, N^{\infty}$: $\Rightarrow (b = -15/4, v = 3/2, a = \sqrt{(31-3\sqrt{57})/2}/8)$.

1.1.2.2: $\mathcal{T}_3\mathcal{F} > 0$. Then we have a weak saddle, and we consider two subcases: $\mathcal{F}_1 \neq 0$ and $\mathcal{F}_1 = 0$.

1.1.2.2.1: $\mathcal{F}_1 \neq 0$. Then by Theorem 6.2 the weak saddle is of order one, and considering Lemma 10.1 we get the following three global configurations:

- $s^{(1)}, n, f; \overline{\binom{1}{1}}SN, ©, ©$ $\Rightarrow (c = 1,\ h = 3,\ e = -1,\ d = 5,\ f = -9/5)$ (if $\eta < 0$);
- $s^{(1)}, n, f; \overline{\binom{1}{1}}SN, S, N^{\infty}$ $\Rightarrow (c = 1,\ h = 3,\ e = -1,\ d = 2,\ f = 2(3 - 8\sqrt{6})/25)$ (if $\eta > 0$);
- $s^{(1)}, n, f; \overline{\binom{1}{1}}SN,\ \overline{\binom{0}{2}}SN$ $\Rightarrow (c = 1,\ h = 3,\ e = -1,\ d = 2(28\sqrt[3]{4} - 16\sqrt[3]{2} - 23),$ $f = 2(4\sqrt[3]{4} - 8\sqrt[3]{2} + 3))$ (if $\eta = 0$).

1.1.2.2.2: $\mathcal{F}_1 = 0$. By Theorem 6.2 the weak saddle is of order at least two and its order depends on $\mathcal{F}_2$.

1. $\mathcal{F}_2 \neq 0$. We have a weak saddle of order two, and we arrive at the next three global configurations, for which to construct examples we use the canonical systems

$$\dot{x} = x + ax^2 + bxy + cy^2, \quad \dot{y} = -y - kx^2 - lxy - my^2 : \quad (10.20)$$

   - $s^{(2)}, n, f; \overline{\binom{1}{1}}SN, ©, ©$ $\Rightarrow (a = 0,\ b = 1,\ c = 0,\ m = 0,\ k = 5,\ l = 4)$ (if $\eta < 0$);

- $s^{(2)}, n, f; \overline{\binom{1}{1}}SN, S, N^{\infty} \Rightarrow (a=0,\ b=1,\ c=0,\ m=0,\ k=5,\ l=6)$ (if $\eta > 0$);
- $s^{(2)}, n, f; \overline{\binom{1}{1}}SN,\ \overline{\binom{0}{2}}SN \Rightarrow (a=0,\ b=1,\ c=0,\ m=0,\ k=5,\ l=2\sqrt{5})$ (if $\eta = 0$).

2. $\mathcal{F}_2 = 0$. According to Theorem 7.3 we could not have an integrable saddle apart from a node and a focus. So we have a weak saddle of order three, and we get the next three global configurations:
   - $s^{(3)}, n, f; \overline{\binom{1}{1}}SN, \copyright, \copyright \Rightarrow (a=2/5,\ b=1,\ c=(27\sqrt{201}-227)/250,$ $m=-2,\ k=1,\ l=-1/5)$ (if $\eta < 0$);
   - $s^{(3)}, n, f; \overline{\binom{1}{1}}SN, S, N^{\infty} \Rightarrow (a=4/5,\ b=1,\ c=(33\sqrt{51}-233)/125,$ $m=-2,\ k=1,\ l=-2/5)$ (if $\eta > 0$);
   - $s^{(3)}, n, f; \overline{\binom{1}{1}}SN,\ \overline{\binom{0}{2}}SN \Rightarrow (a=\sqrt{2},\ b=1,\ c=(5\sqrt{5}-11)/\sqrt{2},\ m=-2,$ $k=(5\sqrt{5}+11)/4,\ l=-1/\sqrt{2})$ (if $\eta = 0$).

1.1.3: $\mathcal{T}_4 = 0$, $\mathcal{T}_3 = 0$. Then considering (10.18) we conclude that two traces are zero, but the third one is different from zero (as one point is a node). So according to Theorem 6.2 (see statement (c)) we have either a weak focus and a weak saddle both of order one (if $\mathcal{F}_1 \neq 0$), or a center and an integrable saddle (if $\mathcal{F}_1 = 0$). However by Theorem 7.3 we could not have, apart from a node, an integrable saddle and a center. So we get the following three global configurations:

- $s^{(1)}, n, f^{(1)}; \overline{\binom{1}{1}}SN, \copyright, \copyright \Rightarrow (c=-1/10,\ h=1/10,\ e=1,\ d=2,\ f=1/10)$ (if $\eta < 0$);
- $s^{(1)}, n, f^{(1)}; \overline{\binom{1}{1}}SN, S, N^{\infty} \Rightarrow (c=-7/5,\ h=7/5,\ e=1,\ d=2,\ f=7/5)$ (if $\eta > 0$);
- $s^{(1)}, n, f^{(1)}; \overline{\binom{1}{1}}SN,\ \overline{\binom{0}{2}}SN \Rightarrow (c=-4/3,\ h=4/3,\ e=1,\ d=2,\ f=4/3)$ (if $\eta = 0$).

1.2: $W_4 > 0$. By Theorem 6.1 (see Table 6.2, lines 115,121 and 122), apart from the saddle, we have either two foci or two nodes, and this situation if governed by the invariant polynomials W_3 and W_2.

1.2.1: $W_3 < 0$ *or* $(W_3 > 0, W_2 < 0)$. In this case we have two foci.

1.2.1.1: $\mathcal{T}_4 \neq 0$. Then by Theorem 6.2 all three singularities are strong, and according to Lemma 10.1 we get the following three global configurations:

- $s, f, f; \overline{\binom{1}{1}}SN, \copyright, \copyright \Rightarrow (c=1,\ h=3,\ e=-1,\ d=7,\ f=-3)$ (if $\eta < 0$);
- $s, f, f; \overline{\binom{1}{1}}SN, S, N^{\infty} \Rightarrow (c=1,\ h=3,\ e=-1,\ d=40,\ f=12)$ (if $\eta > 0$);
- $s, f, f; \overline{\binom{1}{1}}SN,\ \overline{\binom{0}{2}}SN \Rightarrow (c=1,\ h=3,\ e=-1,\ d=40,\ f=2(4\sqrt{5}-3))$ (if $\eta = 0$).

1.2.1.2: $\mathcal{T}_4 = 0$, $\mathcal{T}_3 \neq 0$. Then at least one singularity is weak.

1.2.1.2.1: $\mathcal{T}_3\mathcal{F} < 0$. In this case by Theorem 6.2 we have one weak focus, the order of which is governed by the polynomials $\mathcal{F}_i$, $i = 1, 2, 3, 4$.

1. $\mathcal{F}_1 \neq 0$. The focus is of order one, and we arrive at the next three global configurations:

   - $s, f, f^{(1)}; \overline{\binom{1}{1}}SN, ©, ©$ $\Rightarrow (c = 1,\ h = 1,\ e = -1,\ d = 5/4,\ f = -1)$ (if $\eta < 0$);
   - $s, f, f^{(1)}; \overline{\binom{1}{1}}SN, S, N^{\infty}$ $\Rightarrow (c = 1,\ h = -2,\ e = -5/11,\ d = 11/4,\ f = -1)$ (if $\eta > 0$);
   - $s, f, f^{(1)}; \overline{\binom{1}{1}}SN,\ \overline{\binom{0}{2}}SN$ $\Rightarrow (c = 1, h = -5/4, e = -40/49, d = 49/32, f = -1)$ (if $\eta = 0$).

2. $\mathcal{F}_1 = 0$. In this case the weak focus is of order two or three because by Theorem 7.4 a quadratic system could not have a center apart from a focus and a saddle. Additionally according to Theorem 7.3 systems with three real finite singularities could not have a weak focus of order three and a strong focus. Moreover according to Theorem 7.7 we could have the singularities $s, f, f^{(2)}$ only in the case $\eta < 0$. So we get the global configuration (see systems (10.19))

   - $s, f, f^{(2)}; \overline{\binom{1}{1}}SN, ©, ©$ $\Rightarrow (b = -3\sqrt[3]{4},\ v = -3,\ a = 1)$.

1.2.1.2.2: $\mathcal{T}_3\mathcal{F} > 0$. By Theorem 6.2 in this case we have a weak saddle.

1. $\mathcal{F}_1 \neq 0$. The weak saddle is of order one, and we arrive at the next three global configurations:

   - $s^{(1)}, f, f; \overline{\binom{1}{1}}\ SN, ©, ©$ $\Rightarrow (c = -11/10, h = 1/10, e = -42/47, d = 47/200,$
     $f = 11/10)$ (if $\eta < 0$);
   - $s^{(1)}, f, f; \overline{\binom{1}{1}}SN, S, N^{\infty}$ $\Rightarrow (c = -21/20,\ h = 7/5,\ e = -82/145,$ $d = 29/160,\ f = 21/20)$ (if $\eta > 0$);
   - $s^{(1)}, f, f; \overline{\binom{1}{1}}SN,\ \overline{\binom{0}{2}}SN$ $\Rightarrow (c = -(1 + \sqrt{226})/15,\ h = 1/10,$ $e = -8(1 + \sqrt{226})/(121 + 4\sqrt{226}),\ d = (121 + 4\sqrt{226})/900,$ $f = (1 + \sqrt{226})/15)$ (if $\eta = 0$).

2. $\mathcal{F}_1 = 0$. In this case the weak saddle is of order two or three. However according to Theorem 7.9 we could not have two foci apart from a weak saddle of order three. Moreover, examining the graphics of the functions $W_i = 0$ $(i = 4, 3, 2)$, $\mathcal{T}_4 = 0$, $\mathcal{F}_1 = 0$ and $\eta = 0$, it can be proved that in the case under consideration, we could not have $\eta \geq 0$. So we get the global configuration

   - $s^{(2)}, f, f; \overline{\binom{1}{1}}SN, ©, ©$ $\Rightarrow (c = -2,\ h = 1,\ e = -6/5,\ d = 5/2,\ f = 2)$.

1.2.1.2.3: $\mathcal{T}_3 = 0$. We note that in this case the systems could not be Hamiltonian, otherwise we get two centers and an integrable saddle, which is impossible (see Theorem 7.4). Additionally according to Theorem 6.2 in the case of non-Hamiltonian systems, we could have only two weak singularities, and this implies $\mathcal{T}_2 \neq 0$. So we could apply the statement (c) of Theorem 6.2, which says that we have two weak foci if $\mathcal{T}_2 < 0$ and a weak focus and a weak saddle if $\mathcal{T}_2 > 0$.

1. $\mathcal{T}_2 < 0$. By Theorem 6.2 we have two weak foci (each one of order one) if $\mathcal{F}_1 \neq 0$ and two centers if $\mathcal{F}_1 = 0$. However according to Theorem 7.4 the systems (10.16) could not have as finite singularities c, c, s. Thus in this case we arrive at the following three global configurations:

   - $s, f^{(1)}, f^{(1)}; \overline{\binom{1}{1}}SN, ©, ©$ $\Rightarrow (c = -1,\ h = 1,\ e = -4/7,\ d = 7/2,\ f = 1)$ (if $\eta < 0$);
   - $s, f^{(1)}, f^{(1)}; \overline{\binom{1}{1}}SN, S, N^{\infty}$ $\Rightarrow (c{=}-1,\ h = 1,\ e{=}-306/313,\ d = 313/288,\ f = 1)$ (if $\eta > 0$);
   - $s, f^{(1)}, f^{(1)}; \overline{\binom{1}{1}}SN,\ \overline{\binom{0}{2}}SN$ $\Rightarrow (c{=}-1,\ h = 1,\ e{=}-17/18, d = 9/8,\ f = 1)$ (if $\eta = 0$).

2. $\mathcal{T}_2 > 0$. Then by the statement (c) of Theorem 6.2, we have either a weak focus and a weak saddle (each of order one) if $\mathcal{F}_1 \neq 0$, or we have a center and an integrable saddle if $\mathcal{F}_1 = 0$. However according to Theorem 7.4 in the second case we could not have a focus. So we get the following three global configurations:

   - $s^{(1)}, f, f^{(1)}; \overline{\binom{1}{1}}SN, ©, ©$ $\Rightarrow (c = -1,\ h = 1,\ e = -3/4,\ d = 5/4,\ f = 1)$ (if $\eta < 0$);
   - $s^{(1)}, f, f^{(1)}; \overline{\binom{1}{1}}SN, S, N^{\infty}$ $\Rightarrow (c{=}-1, h{=}1, e{=}-192/221, d{=}221/200, f{=}1)$ (if $\eta > 0$);
   - $s^{(1)}, f, f^{(1)}; \overline{\binom{1}{1}}SN,\ \overline{\binom{0}{2}}SN$ $\Rightarrow (c = -1,\ h = 1,\ e = -5/6,\ d = 9/8,\ f = 1)$ (if $\eta = 0$).

1.2.2: $W_3 > 0, W_2 > 0$. By Theorem 6.1 (see Table 6.2, line 115), apart from a saddle, we have two nodes, which are generic due to $W_4 \neq 0$.

1.2.2.1: $\mathcal{T}_4 \neq 0$. Then the saddle is strong, and we get the following three global configurations:

- $s, n, n; \overline{\binom{1}{1}}SN, ©, ©$ $\Rightarrow (c = 1,\ h = 1,\ e = 1/100,\ d = 3,\ f = 2)$ (if $\eta < 0$);
- $s, n, n; \overline{\binom{1}{1}}SN, S, N^{\infty}$ $\Rightarrow (c = 1,\ h = 1,\ e = 1/3,\ d = 6,\ f = 5)$ (if $\eta > 0$);
- $s, n, n; \overline{\binom{1}{1}}SN,\ \overline{\binom{0}{2}}SN$ $\Rightarrow (c = 1,\ h = 1,\ e = 1/2,\ d = 3 + 2\sqrt{2},\ f = 2 + 2\sqrt{2})$ (if $\eta = 0$).

1.2.2.2: $\mathcal{T}_4 = 0$. In this case we have $\mathcal{T}_3 \neq 0$ because systems (10.16) possess two nodes apart from a saddle. Then we have a weak saddle, the order of which is governed by the invariant polynomials $\mathcal{F}_i$, $i = 1, 2, 3, 4$.

1.2.2.2.1: $\mathcal{F}_1 \neq 0$. By Theorem 6.2 the weak saddle is of order one, and we arrive at the next three global configurations:

- $s^{(1)}, n, n; \overline{\binom{1}{1}}SN, ©, ©\ \Rightarrow (c = 1,\ h = 2,\ e = 1,\ d = 49/40,\ f = -1)$ (if $\eta < 0$);
- $s^{(1)}, n, n; \overline{\binom{1}{1}}SN, S, N^{\infty}\ \Rightarrow (c = 1,\ h = 2,\ e = 1,\ d = 1/8,\ f = -1)$ (if $\eta > 0$);
- $s^{(1)}, n, n; \overline{\binom{1}{1}}SN,\ \overline{\binom{0}{2}}SN\ \Rightarrow (c = 1,\ h = 2,\ e = 1,\ d = 9/8,\ f = -1)$ (if $\eta = 0$).

1.2.2.2.2: $\mathcal{F}_1 = 0$. Then the order of the weak saddle is greater than or equal to two.

1. $\mathcal{F}_2 \neq 0$. In this case we have a weak saddle of order two, and this leads to the next three global configurations (see systems (10.20)):
   - $s^{(2)}, n, n; \overline{\binom{1}{1}}SN, ©, ©\ \Rightarrow (a = 1,\ b = -9/10,\ c = -27/100,\ m = 1,$ $l = -9/10,\ k = (127\sqrt{21} - 581)/54)$ (if $\eta < 0$);
   - $s^{(2)}, n, n; \overline{\binom{1}{1}}SN, S, N^{\infty}\ \Rightarrow a = 1,\ b = -9/10,\ c = -1/4,\ m = 1,$ $l = -9/10,\ k = (9\sqrt{181} - 121)/10)$ (if $\eta > 0$);
   - $s^{(2)}, n, n; \overline{\binom{1}{1}}SN,\ \overline{\binom{0}{2}}SN\ \Rightarrow a = 1,\ b = -9/10,\ c = -(87\sqrt{29}$ $+ 419)/3380,\ m = 1,\ l = -9/10,\ k = 13(-5089747 - 317811\sqrt{29}$ $+ 261\sqrt{58(7312891 + 821733\sqrt{29})}\ )/(10(419 + 87\sqrt{29})^2))$ (if $\eta = 0$).
2. $\mathcal{F}_2 = 0$. Since $\mathcal{T}_3 \neq 0$, according to Theorem 7.3 we could not have an integrable saddle with two nodes. Additionally by Theorem 7.9 in this case we could only have one global configuration with a weak saddle of order three (here we use systems (10.20))
   - $s^{(3)}, n, n; \overline{\binom{1}{1}}SN, S, N^{\infty}\ \Rightarrow (a = -2,\ b = 1,\ c = (5 - \sqrt{17})/2,\ k = 2,\ l = 1,$ $m = -2)$.

1.3: $W_4 = 0$. Considering (10.18) we obtain $\tau_1\tau_2\tau_3 = 0$, i.e. at least one discriminant vanishes, and this leads to a node with coinciding eigenvalues.

1.3.1: $W_3 < 0$. In this case by Theorem 6.1 (see Table 6.2, line 119), apart from the node, systems (10.16) possess a focus and a saddle. Since $W_3 < 0$, according to Lemma 10.3 the systems (10.16) could not have a star node, and therefore we have a node n^d.

1.3.1.1: $\mathcal{T}_4 \neq 0$. Then by Theorem 6.2 we do not have weak singularities and this leads to the following three global configurations:

- $s, n^d, f; \overline{\binom{1}{1}}SN, ©, ©\ \Rightarrow (c = 2,\ h = -1,\ e = -1/4,\ d = 1,\ f = 1)$ (if $\eta < 0$);
- $s, n^d, f; \overline{\binom{1}{1}}SN, S, N^{\infty}\ \Rightarrow (c = 2,\ h = -2,\ e = -1/4,\ d = 1,\ f = 1)$ (if $\eta > 0$);

- $s, n^d, f; \overline{\binom{1}{1}}SN, \overline{\binom{0}{2}}SN \Rightarrow (c = 2,\ h = -(1+2\sqrt{2})/2,\ e = -1/4,\ d = 1,\ f = 1)$ (if $\eta = 0$).

1.3.1.2: $\mathcal{T}_4 = 0$, $\mathcal{T}_3 \neq 0$. We have one weak singularity.

1.3.1.2.1: $\mathcal{T}_3\mathcal{F} < 0$. By Theorem 6.2 in this case we have one weak focus, the order of which is governed by the polynomials $\mathcal{F}_i$, $i = 1, 2, 3, 4$.

1. $\mathcal{F}_1 \neq 0$. The weak focus is of order one, and we arrive at the next three global configurations:

   - $s, n^d, f^{(1)}; \overline{\binom{1}{1}}SN, ©, © \Rightarrow (c = -3/2,\ h = 5/4,\ e = -25/32,\ d{=}2,\ f{=}1)$ (if $\eta < 0$);
   - $s, n^d, f^{(1)}; \overline{\binom{1}{1}}SN, S, N^{\infty} \Rightarrow (c = -9/4,\ h = 13/8,\ e = -169/128,\ d = 2,$ $f = 1)$ (if $\eta > 0$);
   - $s, n^d, f^{(1)}; \overline{\binom{1}{1}}SN, \overline{\binom{0}{2}}SN \Rightarrow (c = -2,\ h = 3/2,\ e = -9/8,\ d = 2,\ f = 1)$ (if $\eta = 0$).

2. $\mathcal{F}_1 = 0$. Then by Theorem 6.2 the weak focus is of order at least two. However according to Theorems 7.8 and 7.4, systems with three real finite singularities could have neither the finite configuration $s, n^d, f^{(3)}$ nor s, n^d, c. Additionally by Theorem 7.7 in the case of the weak focus of order two, only the case $\eta < 0$ is realizable. So we get the following global configuration (the example below is based on systems (10.19))

   - $s, n^d, f^{(2)}; \overline{\binom{1}{1}}SN, ©, © \Rightarrow (a = 1,\ \{\mathcal{F}_1^{-1}(0), W_4^{-1}(0)\} \ni (b, v) \approx$ $(-4.1394, -1.64395))$.

1.3.1.2.2: $\mathcal{T}_3\mathcal{F} > 0$. In this case we have one weak saddle.

1. $\mathcal{F}_1 \neq 0$. By Theorem 6.2 the weak saddle is of order one, and we obtain three global configurations:

   - $s^{(1)}, n^d, f; \overline{\binom{1}{1}}SN, ©, © \Rightarrow (c = -27/20,\ h = 1,\ e = -58/45,\ d = 45/32,$ $f = 27/20)$ (if $\eta < 0$);
   - $s^{(1)}, n^d, f; \overline{\binom{1}{1}}SN, S, N^{\infty} \Rightarrow (c = -13/10,\ h = 1,\ e = -84/67,\ d = 67/50,$ $f = 13/10)$ (if $\eta > 0$);
   - $s^{(1)}, n^d, f; \overline{\binom{1}{1}}SN, \overline{\binom{0}{2}}SN \Rightarrow (c = -(10+\sqrt{103})/15,\ h = 1,$ $e = -2(17+2\sqrt{103})(23+2\sqrt{103})/(1703+80\sqrt{103}),$ $d = (1703+80\sqrt{103})/1800,\ f = (10+\sqrt{103})/15)$ (if $\eta = 0$).

2. $\mathcal{F}_1 = 0$. The order of the weak saddle is ≥ 2. However by Theorem 7.3 in the case under examination, we could not have the finite configuration $\$, n^d, f$. By Theorem 7.9 we could not have the configuration $s^{(3)}, n^d, f$. Moreover examining the graphics of the functions $W_i = 0$ $(i = 4, 3, 2)$, $\mathcal{T}_4 = 0$, $\mathcal{F}_1 = 0$

and $\eta = 0$, it can be proved from these graphics that we could not have $\eta \geq 0$. So considering Lemma 10.1 we get the global configuration

- $s^{(2)}, n^d, f; \overline{\binom{1}{1}}SN, ©, ©$ $\Rightarrow (c = -2,\ h = 1,\ e = -14/9,\ d = 9/4,\ f = 2)$.

1.3.1.2.3: $\mathcal{T}_3 = 0$. In this case considering (10.18) two traces vanish, and we have two weak singularities: a focus and a saddle. Additionally by Theorem 7.3 we could not have the finite configuration $\$, n^d, c$. So according to Theorem 6.2 both weak singularities are of order one, and we get three global configurations:

- $s^{(1)}, n^d, f^{(1)}; \overline{\binom{1}{1}}SN, ©, ©$ $\Rightarrow (c = -1,\ h = 9/5,\ e = -7/9,\ d = 729/275,\ f = 1)$ (if $\eta < 0$);
- $s^{(1)}, n^d, f^{(1)}; \overline{\binom{1}{1}}SN, S, N^\infty$ $\Rightarrow (c = -1,\ h = 89/50,\ e = -139/178,\ f = 1,$ $d = 704969/271250)$ (if $\eta > 0$);
- $s^{(1)}, n^d, f^{(1)}; \overline{\binom{1}{1}}SN,\ \overline{\binom{0}{2}}SN$ $\Rightarrow (c = -1,\ h = (3 + \sqrt{17})/4,\ f = 1,$ $e = -(7 + \sqrt{17})/(6 + 2\sqrt{17}),\ d = (45 + 11\sqrt{17})/(10 + 6\sqrt{17}))$ (if $\eta = 0$).

1.3.2: $W_3 > 0$. In this case by Theorem 6.1 (see Table 6.2, line 116), systems (10.16) possess one saddle and two nodes (one generic and another one with coinciding eigenvalues).

1.3.2.1: $U_3 \neq 0$. Then by Lemma 10.3 we have a node n^d.

1.3.2.1.1: $\mathcal{T}_4 \neq 0$. By Theorem 6.2 the saddle is strong, and we obtain three global configurations:

- $s, n, n^d; \overline{\binom{1}{1}}SN, ©, ©$ $\Rightarrow (c = 33/8,\ h = 1,\ e = -2,\ d = 625/512,\ f = 1)$ (if $\eta < 0$);
- $s, n, n^d; \overline{\binom{1}{1}}SN, S, N^\infty$ $\Rightarrow (c = 31/8,\ h = 1,\ e = -2,\ d = 529/512,\ f = 1)$ (if $\eta > 0$);
- $s, n, n^d; \overline{\binom{1}{1}}SN,\ \overline{\binom{0}{2}}SN$ $\Rightarrow (c = 4,\ h = 1,\ e = -2,\ d = 9/8,\ f = 1)$ (if $\eta = 0$).

1.3.2.1.2: $\mathcal{T}_4 = 0$. Then we have a weak saddle.

1. $\mathcal{F}_1 \neq 0$. By Theorem 6.2 the weak saddle is of order one, and we arrive at the following three global configurations:

   - $s^{(1)}, n, n^d; \overline{\binom{1}{1}}SN, ©, ©$ $\Rightarrow (c = -2/5,\ h = 7/10,\ e = -2441/3600,\ f = 1,$ $d = 1764/2441)$ (if $\eta < 0$);

   - $s^{(1)}, n, n^d; \overline{\binom{1}{1}}SN, S, N^\infty$ $\Rightarrow (c = -2/5,\ h = 7/10,\ e = -1243/1800,\ f = 1,$ $d = 882/1243)$ (if $\eta > 0$);

   - $s^{(1)}, n, n^d; \overline{\binom{1}{1}}SN,\ \overline{\binom{0}{2}}SN$ $\Rightarrow (c = -2/5,\ h = 7/10,\ e = -49/72,\ f = 1,$ $d = 18/25)$ (if $\eta = 0$).

2. $\mathcal{F}_1 = 0$, $\mathcal{F}_2 \neq 0$. According to Theorem 6.2 we have a weak saddle of order two, and this leads to the following three global configurations:

- $s^{(2)}, n, n^d; \overline{\binom{1}{1}}SN, ©, ©$ $\Rightarrow$ $(c = 1 - 2\xi,\ h = \xi,\ e = -11,\ f = 1,$ $d = \xi^2/11)$ (if $\eta < 0$) (where $\{\mathcal{F}_1^{-1}(0)\} \ni \xi \approx -0.8549$);
- $s^{(2)}, n, n^d; \overline{\binom{1}{1}}SN, S, N^\infty$ $\Rightarrow$ $(c = 1 - 2\xi,\ h = \xi,\ e = -12,\ f = 1,$ $d = \xi^2/12)$ (if $\eta > 0$) (where $\{\mathcal{F}_1^{-1}(0)\} \ni \xi \approx -0.8557$);
- $s^{(2)}, n, n^d; \overline{\binom{1}{1}}SN, \overline{\binom{0}{2}}SN$ $\Rightarrow$ $(c = 1 - 2\xi,\ h = \xi,\ d = -\xi^2/e,\ f = 1)$ (if $\eta = 0$) (where $\{\mathcal{F}_1^{-1}(0), \eta^{-1}(0)\} \ni (e, \xi) \approx (-11.5890,\ -0.8553)$).

3. $\mathcal{F}_1 = 0$, $\mathcal{F}_2 = 0$. By Theorem 7.3 in this case we could not have an integrable saddle. Additionally by Theorem 7.9 the systems (10.16) have finite singularities $s^{(3)}, n, n^d$ only in the case $\eta > 0$. So we get the global configuration

   - $s^{(3)}, n, n^d; \overline{\binom{1}{1}}SN, S, N^\infty$ $\Rightarrow$ $(c = 1 - 2\xi,\ h = \xi,\ d = -\xi^2/e,\ f = 1)$ (where $\{\mathcal{F}_1^{-1}(0), \mathcal{F}_2^{-1}(0)\} \ni (e, \xi) \approx (-4.7882,\ -2.9346)$).

1.3.2.2: $\mathcal{U}_3 = 0$. Then by Lemma 10.3 we have a star node. Moreover from the proof of Lemma 10.3, it follows that in the case $\mathcal{W}_3 \neq 0$ the condition $\mathcal{W}_4 = \mathcal{U}_3 = 0$ yields $d = e = 0$ and $f = c$. In this case we calculate $\eta = 4(c + 2h)^2$ and $\mathbf{D} = -192c^8(c + 2h)^2$, and therefore the condition $\mathbf{D} < 0$ implies $\eta > 0$. Since by Lemma 10.3 the condition $\mathcal{T}_4 \neq 0$ holds (i.e. the saddle is strong), we could only have one global configuration

- $s, n, n^*; \overline{\binom{1}{1}}SN, S, N^\infty$ $\Rightarrow$ $(c = 1,\ h = -2,\ e = 0,\ d = 0,\ f = 1)$.

1.3.3: $\mathcal{W}_3 = 0$. In this case by Theorem 6.1 (see Table 6.2, line 116) systems (10.16) possess one saddle and two nodes, each one with coinciding eigenvalues.

1.3.3.1: $\mathcal{U}_3 \neq 0$. Then by Lemma 10.3 we have two nodes n^d.

1.3.3.1.1: $\mathcal{T}_4 \neq 0$. According to Theorem 6.2 the saddle is strong, and we obtain the following three global configurations:

- $s, n^d, n^d; \overline{\binom{1}{1}}SN, ©, ©$ $\Rightarrow$ $(c = 1,\ h = -2,\ e = 3/(\sqrt{6} - 4),\ d = (4 - \sqrt{6})/2,$ $f = 1 + \sqrt{6})$ (if $\eta < 0$);
- $s, n^d, n^d; \overline{\binom{1}{1}}SN, S, N^\infty$ $\Rightarrow$ $(c = 1,\ h = -2,\ e = (2 - \sqrt{6})^2/(3\sqrt{6} - 8),$ $d = (8 - 3\sqrt{6})/2,\ f = \sqrt{6} - 1)$ (if $\eta > 0$);
- $s, n^d, n^d; \overline{\binom{1}{1}}SN, \overline{\binom{0}{2}}SN$ $\Rightarrow$ $(c = 1,\ h = -2,\ e = (7 - 2\sqrt{6})/(4\sqrt{6} - 11),$ $d = (11 - 4\sqrt{6})/4,\ f = \sqrt{6})$ (if $\eta = 0$).

1.3.3.1.2: $\mathcal{T}_4 = 0$. In this case we have a weak saddle.

1. $\mathcal{F}_1 \neq 0$. By Theorem 6.2 the weak saddle is of order one, and we arrive at the following three global configurations:

   - $s^{(1)}, n^d, n^d; \overline{\binom{1}{1}}SN, ©, ©$ $\Rightarrow$ $(c = 1,\ h = -(8 + 5f^2)/20,$ $e = -200(f - 1)^2/(104 - 160f + 180f^2 - 100f^3 + 25f^4),\ d = (104 - 160f + 180f^2 - 100f^3 + 25f^4)/800)$ (if $\eta < 0$) (where $\{\mathcal{T}_4^{-1}(0)\} \ni f \approx -3.2840$);

- $s^{(1)}, n^d, n^d; \overline{\binom{1}{1}}SN, S, N^\infty \Rightarrow (c = 1,\ h = -2,\ e = 0,\ d = 1/2,\ f = 1)$ (if $\eta > 0$);

- $s^{(1)}, n^d, n^d; \overline{\binom{1}{1}}SN,\ \overline{\binom{0}{2}}SN \Rightarrow (c = 1,\ h = -(2 + f^2)/4,$ $e = -8(f-1)^2/(2-2f+f^2)^2,\ d = (2-2f+f^2)^2/32,$ $f = (\sqrt{2(5+\sqrt{41})} - 1 - \sqrt{41})/4)$ (if $\eta = 0$).

2. $\mathcal{F}_1 = 0$. In this case the weak saddle is of order ≥ 2. According to Theorem 7.9 we could not have the finite configuration $s^{(3)}, n^d, n^d$. By Theorem 7.6 we could only obtain the finite configuration $s^{(2)}, n^d, n^d$ with $\eta < 0$. Therefore we arrive at the global configuration

   - $s^{(2)}, n^d, n^d; \overline{\binom{1}{1}}SN, ©, © \Rightarrow (h = 1,\ e = -(c-f)^2/[(c+1)^2 + (f+1)^2],$ $d = [(c+1)^2 + (f+1)^2]/4),$ where $\{\mathcal{F}_1^{-1}(0), \mathcal{T}_4^{-1}(0)\} \ni (c, f) \approx (-1.9245, 0.7926)$.

1.3.3.2: $U_3 = 0$. According to Lemma 10.3 the conditions $W_4 = W_3 = U_3 = 0$ imply $\mathcal{T}_4 = 16c^6 \neq 0$ and $\eta > 0$. So the saddle is strong, and at infinity we have three real singularities. This leads to the global configuration

- $s, n^d, n^*; \overline{\binom{1}{1}}SN, S, N^\infty \Rightarrow (c = 1,\ h = -1,\ e = 0,\ d = 0,\ f = 1)$.

2: $\kappa > 0$. According to Remark 10.5 the condition $\kappa > 0$ implies $\widetilde{K} < 0$, and by Theorem 6.1 (see Table 6.2, lines 110–113) systems (10.16) possess two saddles and one anti-saddle. Additionally considering (10.18) the condition $\kappa > 0$ implies $\gamma < 0$ and then $\eta > 0$. By Lemma 10.1 this means that at infinity we only have the configuration $\overline{\binom{1}{1}}SN, N^f, N^f$.

We observe that due to the condition $G_9 \neq 0$ according to Table 6.2 (see Theorem 6.1) the type of the anti-saddle is governed by the invariant polynomials W_i, $i = 4, 3, 2$.

2.1: $W_4 < 0$. The anti-saddle is a focus. If $\mathcal{T}_4 \neq 0$ then we do not have weak singularities, and hence we get only one global configuration

- $s, s, f; \overline{\binom{1}{1}}SN, N^f, N^f \Rightarrow (c = 1,\ h = 2,\ e = 1,\ d = -1/3,\ f = 2)$.

Assume now $\mathcal{T}_4 = 0$, i.e. we have at least one weak singularity. We consider two cases: $\mathcal{T}_3 \neq 0$ and $\mathcal{T}_3 = 0$.

2.1.1: $\mathcal{T}_3 \neq 0$. Then by Theorem 6.2 the systems (10.16) have exactly one weak singularity. If $\mathcal{F}_1 \neq 0$ or $\mathcal{F}_1 = 0$ and $\mathcal{F}_2 \neq 0$, considering Theorem 6.2 we conclude that there are four possible global configurations, and they indeed are all realizable:

- $s, s, f^{(1)}; \overline{\binom{1}{1}}SN, N^f, N^f \Rightarrow (c = 0,\ h = 1,\ e = 1,\ d = -1,\ f = 0)$ (if $\mathcal{T}_3\mathcal{F} < 0$, $\mathcal{F}_1 \neq 0$);
- $s, s^{(1)}, f; \overline{\binom{1}{1}}SN, N^f, N^f \Rightarrow (c = -1,\ h = 2,\ e = 2,\ d = 0,\ f = 1)$ (if $\mathcal{T}_3\mathcal{F} > 0$, $\mathcal{F}_1 \neq 0$);

- $s, s, f^{(2)}; \overline{\binom{1}{1}}SN, N^f, N^f \Rightarrow (c = -1/2,\ h = 1,\ e = 2,\ d = -1/3,\ f = 1/2)$ (if $\mathcal{T}_3\mathcal{F} < 0,\ \mathcal{F}_1 = 0,\ \mathcal{F}_2 \neq 0$);
- $s, s^{(2)}, f; \overline{\binom{1}{1}}SN, N^f, N^f \Rightarrow (c = -1,\ h = 6/5,\ e = 1,\ d = 3/5,\ f = 1)$ (if $\mathcal{T}_3\mathcal{F} > 0,\ \mathcal{F}_1 = 0,\ \mathcal{F}_2 \neq 0$).

Suppose now $\mathcal{F}_1 = \mathcal{F}_2 = 0$, i.e. we have a single weak singularity of order ≥ 2. By Theorems 7.4 and 7.3, in the case under consideration, neither could we have a center, nor could we have an integrable saddle. Moreover as $\kappa > 0$, by Theorem 7.9 a weak saddle of order three is also impossible. Therefore considering the only possible configuration here for infinite singularities, we arrive at the global configuration

- $s, s, f^{(3)}; \overline{\binom{1}{1}}SN, N^f, N^f \Rightarrow (c = -14/11,\ h = 1,\ e = -7/4,\ d = 6496/5203,\ f = 14/11)$.

2.1.2: $\mathcal{T}_3 = 0$. Then at least two singularities are weak, and since $\sigma = c + f + 2x - 2(f - h)y \neq 0$ (i.e. systems (10.16) could not be Hamiltonian), we conclude that we could only have two weak singularities. This implies the condition $\mathcal{T}_2 \neq 0$. Moreover as it was mentioned above, neither could we have a center, nor could we have an integrable saddle, and in this case by Theorem 6.2 (see statement (c)) the condition $\mathcal{F}_1 \neq 0$ is satisfied. So we obtain the following two global configurations:

- $s^{(1)}, s^{(1)}, f; \overline{\binom{1}{1}}SN, N^f, N^f \Rightarrow (c = 2,\ h = -2,\ e = 1,\ d = 1,\ f = -2)$ (if $\mathcal{T}_2 < 0$);
- $s, s^{(1)}, f^{(1)}; \overline{\binom{1}{1}}SN, N^f, N^f \Rightarrow (c = -2,\ h = 2,\ e = -5,\ d = 1,\ f = 2)$ (if $\mathcal{T}_2 > 0$).

2.2: $W_4 > 0$. In this case apart from two saddles, systems (10.16) possess a node, which is generic. By Theorem 7.3, in the case under consideration, we could not have an integrable saddle. Moreover by Theorem 7.9 it is impossible to have a weak saddle of order three together with a saddle and a node. This leads to the next four global configurations:

- $s, s, n; \overline{\binom{1}{1}}SN, N^f, N^f \Rightarrow (c = 1,\ h = 2,\ e = 1,\ d = 0,\ f = 2)$ (if $\mathcal{T}_4 \neq 0$);
- $s, s^{(1)}, n; \overline{\binom{1}{1}}SN, N^f, N^f \Rightarrow (c = -1,\ h = 2,\ e = -3/4,\ d = 0,\ f = 1)$ (if $\mathcal{T}_4 = 0,\ \mathcal{T}_3 \neq 0,\ \mathcal{F}_1 \neq 0$);
- $s, s^{(2)}, n; \overline{\binom{1}{1}}SN, N^f, N^f \Rightarrow (c = -1,\ h = 251/200,\ e = 1,\ d = 49/100,\ f = 1)$ (if $\mathcal{T}_4 = 0,\ \mathcal{T}_3 \neq 0,\ \mathcal{F}_1 = 0$);
- $s^{(1)}, s^{(1)}, n; \overline{\binom{1}{1}}SN, N^f, N^f \Rightarrow (c = 2,\ h = -2,\ e = -4/5,\ d = 1,\ f = -2)$ (if $\mathcal{T}_4 = \mathcal{T}_3 = 0$).

2.3: $W_4 = 0$. Then the anti-saddle is a node with coinciding eigenvalues. Moreover by Lemma 10.3 this node is a star node if and only if $U_3 = 0$, and in this case the condition $\mathcal{T}_4 \neq 0$ holds. By Theorem 7.9 we could not have a weak saddle of order three apart from a saddle and apart from a node of any type. So we get the following five global configurations:

- $s, s, n^d; \overline{\binom{1}{1}}SN, N^f, N^f \Rightarrow (c = 2,\ h = 2,\ e = -1/4,\ d = 1,\ f = 1)$ (if $\mathcal{T}_4 \neq 0,\ U_3 \neq 0$);

- $s, s, n^*; \overline{\binom{1}{1}}SN, N^f, N^f \Rightarrow (c=1,\ h=1,\ e=0,\ d=0,\ f=1)$ (if $\mathcal{T}_4 \neq 0,\ U_3=0$);
- $s, s^{(1)}, n^d; \overline{\binom{1}{1}}SN, N^f, N^f \Rightarrow (c=-7/2,\ h=9/4,\ e=-81/32,\ d=2,\ f=1)$ (if $\mathcal{T}_4=0,\ \mathcal{T}_3\neq 0,\ \mathcal{F}_1\neq 0$);
- $s, s^{(2)}, n^d; \overline{\binom{1}{1}}SN, N^f, N^f \Rightarrow (c=-3/4,\ h=1,\ e=-2(3+\sqrt{39})/(23+\sqrt{39}),$ $d=(23+\sqrt{39})/40,\ f=3/4)$ (if $\mathcal{T}_4=0,\ \mathcal{T}_3\neq 0,\ \mathcal{F}_1=0$);
- $s^{(1)}, s^{(1)}, n^d; \overline{\binom{1}{1}}SN, N^f, N^f \Rightarrow (c=-1,\ h=4/5,\ e=-9/8,\ d=128/175,$ $f=1)$ (if $\mathcal{T}_4=\mathcal{T}_3=0$).

3: $\kappa=0$. In this case for systems (10.15) we get $m=0$, and we have $\mu_1 = 4efh^2y\neq 0$. Therefore we may assume $e=h=1$ due to the rescaling $(x,y,t)\mapsto(hx/e, y, t/h)$, and this leads to the 3-parameter family of systems:

$$\dot{x}=cx+dy+2xy-dy^2,\quad \dot{y}=x+fy-fy^2. \tag{10.21}$$

These systems possess the singular points

$$M_1(0,0),\quad M_2(0,1),\quad M_3\Big(\frac{(d-cf)(d-cf-2f)}{4f},\ \frac{d-cf}{2f}\Big),$$

for which we have

$$\begin{gathered}\rho_1=c+f,\ \ \Delta_1=cf-d;\quad \rho_2=c-f+2,\\ \Delta_2=d-2f-cf;\ \ \rho_3=f(c+1)+d(1-f)/f,\\ \Delta_3=(cf-d)(d-2f-cf)/(2f);\ \ \tau_i=\rho_i^2-4\Delta_i,\ \ i=1,2,3.\end{gathered}$$

Then for the systems above we calculate

$$\begin{gathered}\kappa=\eta=\mu_0=0,\ G_9=0,\ \mathcal{T}_i=0=W_i,\ i=1,2,3,4,\ \mu_1=4fy\neq 0,\\ \widetilde{K}=-4fy^2,\ \widetilde{N}=-4f(1+f)y^2,\ \widetilde{M}=-8(2+f)^2y^2,\ \mathcal{B}_1=2f\rho_1\rho_2\rho_3,\\ \mathcal{B}_2=2f(f-1)^2(\rho_1\rho_2+\rho_1\rho_3+\rho_2\rho_3),\ W_5=1444f^2(\tau_1+\tau_2+\tau_3),\\ \tilde{L}=8f(2+f)y^2,\ W_6=4f^2(\tau_1\tau_2+\tau_1\tau_3+\tau_2\tau_3),\ W_7=12f^2\tau_1\tau_2\tau_3.\end{gathered} \tag{10.22}$$

We shall examine step by step the global configurations of singularities of systems (10.21). We observe that the condition $\mu_1\neq 0$ implies $\widetilde{K}\neq 0$ and $\operatorname{sign}(\widetilde{K})=-\operatorname{sign}(f)$.

Remark 10.6. For systems (10.21) the condition $\tilde{L}=0$ implies $\widetilde{M}=0$. Moreover if $\tilde{L}\widetilde{N}\neq 0$ then we have $\operatorname{sign}(\tilde{L})=\operatorname{sign}(f(2+f))$ and $\operatorname{sign}(\widetilde{N})=-\operatorname{sign}(f(1+f))$.

3.1: $\widetilde{K}<0$. According to Theorem 6.1, systems (10.21) have two saddles and one anti-saddle.

3.1.1: $W_7<0$. Then the anti-saddle is a focus or a center.

3.1.1.1: $\mathcal{B}_1 \neq 0$. We have $\rho_1\rho_2\rho_3 \neq 0$, i.e. all three singularities are strong. Additionally we observe that the condition $\widetilde{K} < 0$ implies $f > 0$ and then $\widetilde{M} \neq 0$. So considering Lemma 10.1 we obtain the global configuration

- $s, s, f;\ \widehat{\binom{1}{2}}\ \widehat{P_\lambda} E \widehat{P_\lambda} - H,\ N^f \ \Rightarrow (c = -1,\ d = -2,\ f = 3)$.

3.1.1.2: $\mathcal{B}_1 = 0$. Then $\rho_1\rho_2\rho_3 = 0$ and at least one of the traces is zero. Therefore without loss of generality we may assume $\rho_1 = c + f = 0$, i.e. $c = -f$, and calculations yield

$$\begin{aligned} \mathcal{B}_1 = \mathcal{H} = 0,\ \mathcal{B}_2 = 4(f-1)^4(d+f^2),\ \sigma = 2(1-f)y,\\ \mathbf{D} = -192(d+f^2)^2(d-2f+f^2)^2,\ \mathcal{F}_1 = 2(1-f)(d+2f+f^2),\\ W_7 = 192(1-d)(d+f^2)^2(d-3f^2+df^2+f^4),\ C_2 = (2+f)xy^2 - dy^3. \end{aligned} \tag{10.23}$$

1. $\mathcal{B}_2 \neq 0$. Since by Theorem 7.3 we could not have an integrable saddle, we arrive at the following three global configurations:

   - $s, s, f^{(1)};\ \widehat{\binom{1}{2}}\ \widehat{P_\lambda} E \widehat{P_\lambda} - H,\ N^f \ \Rightarrow (c = -5/4,\ d = 2,\ f = 3/4)$ (if $\mathcal{B}_2 < 0,\ \mathcal{F}_1 \neq 0$);
   - $s, s, c;\ \widehat{\binom{1}{2}}\ \widehat{P_\lambda} E \widehat{P_\lambda} - H,\ N^f \ \Rightarrow (c = -2,\ d = -8,\ f = 2)$ (if $\mathcal{B}_2 < 0,\ \mathcal{F}_1 = 0$);
   - $s, s^{(1)}, f;\ \widehat{\binom{1}{2}}\ \widehat{P_\lambda} E \widehat{P_\lambda} - H,\ N^f \ \Rightarrow (c = -1/2,\ d = 2,\ f = 3/2)$ (if $\mathcal{B}_2 > 0$).

2. $\mathcal{B}_2 = 0$. In this case $(f-1)(d+f^2) = 0$, and due to $W_7 \neq 0$ we get $f = 1$, i.e. $\sigma = 0$. So the systems become Hamiltonian, and we arrive at the global configuration

   - $\$, \$, c;\ \widehat{\binom{1}{2}}\ \widehat{P_\lambda} E \widehat{P_\lambda} - H,\ N^f \ \Rightarrow (c = -1,\ d = 2,\ f = 1)$.

3.1.2: $W_7 > 0$. Then the anti-saddle is a node, which is generic.

1. $\mathcal{B}_1 \neq 0$. In this case both saddles are strong, and as $\widetilde{K} < 0$, considering Lemma 10.1 we get the global configuration

   - $s, s, n;\ \widehat{\binom{1}{2}}\ \widehat{P_\lambda} E \widehat{P_\lambda} - H,\ N^f \ \Rightarrow (c = -1,\ d = -2,\ f = 9/4)$.

2. $\mathcal{B}_1 = 0$. We set again $\rho_1 = 0$, which implies $c = -f$. We claim that the conditions $W_7 > 0$ and $\widetilde{K} < 0$ imply $\mathcal{B}_2 > 0$ and $\mathcal{F}_1 \neq 0$. We first show that we could not have $\mathcal{F}_1 = 0$ nor $\mathcal{B}_2 = 0$. Supposing the contrary, by (10.23) we obtain either

   $$(f-1)(d+f^2) = 0 \quad \text{or} \quad (1-f)(d+2f+f^2) = 0.$$

   Since the condition $W_7 > 0$ implies $d + f^2 \neq 0$, we must have either $f = 1$ or $d = -f(f+2)$, and we obtain either $W_7 = -384(d^2-1)^2$ or $W_7 = -1536f^3(1+f)^4$. The condition $\widetilde{K} = -4fy^2 < 0$ gives $f > 0$, and therefore in both cases we obtain the relation $W_7 < 0$, which contradicts the assumption.

Assume now $\mathcal{B}_2 < 0$. Then $d + f^2 < 0$ and we can set $d = -f^2 - u^2$ (with $u \neq 0$), and this leads to the relation $W_7 = -192u^4(1 + f^2 + u^2)(4f^2 + u^2 + f^2u^2) < 0$. So our claim is proved.

Thus $\mathcal{B}_2 > 0$ and $\mathcal{F}_1 \neq 0$, and considering Theorem 6.2 and Lemma 10.1, we get the global configuration

- $s, s^{(1)}, n;\ \widehat{\binom{1}{2}}\ \widehat{P}_{\curlywedge} E \widehat{P}_{\curlywedge} - H,\ N^f \ \Rightarrow (c = -3/2,\ d = 7/13,\ f = 3/2).$

3.1.3: $W_7 = 0$. Then the anti-saddle is a node with coinciding eigenvalues. For systems (10.21) we have $U_1 = -4(1 + f)y^2 \neq 0$ and $Y_1 = 1 + f \neq 0$ due to $\widetilde{K} < 0$ (i.e. $f > 0$), and by Theorem 7.1 we could not have a star node.

1. $\mathcal{B}_1 \neq 0$. In this case both saddles are strong, and considering Lemma 10.1 we obtain the global configuration
   - $s, s, n^d;\ \widehat{\binom{1}{2}}\ \widehat{P}_{\curlywedge} E \widehat{P}_{\curlywedge} - H,\ N^f \ \Rightarrow (c = 1,\ d = -1/4,\ f = 2).$
2. $\mathcal{B}_1 = 0$. Then one saddle is weak, and without loss of generality we may assume that it is located at the origin and the node n^d is at the point $M_2(0, 1)$. Then $\rho_1 = 0$ gives $c = -f$, whereas $\tau_2 = 4(1 - d) = 0$ implies $d = 1$. In this case calculations yield
   $$\mathcal{B}_1 = \mathcal{H} = 0,\quad \mathcal{B}_2 = 4(f-1)^4(1+f^2),\quad \sigma = 2(1-f)y,$$
   $$W_7 = 0,\quad \mathcal{F}_1 = 2(1-f)(f+1)^2,\quad \mathbf{D} = -192(f-1)^4(1+f^2)^2.$$
   Since for the family of systems (10.21) we have $\mathbf{D} < 0$, this implies $\mathcal{F}_1\mathcal{B}_2\sigma \neq 0$. Then by Theorem 6.2 (see statement (e_1)) the weak saddle is of order one. Therefore we get the global configuration
   - $s, s^{(1)}, n^d;\ \widehat{\binom{1}{2}}\ \widehat{P}_{\curlywedge} E \widehat{P}_{\curlywedge} - H,\ N^f \ \Rightarrow (c = -2,\ d = 1,\ f = 2).$

3.2: $\widetilde{K} > 0$. According to Theorem 6.1 systems (10.21) possess one saddle and two anti-saddles, the types of which are governed by the invariant polynomial W_7.

3.2.1: $W_7 < 0$. Then apart from the saddle, we have a node (which is generic) and a focus. We observe that the condition $W_7 < 0$ implies $C_2 \neq 0$. Indeed, otherwise setting $C_2 = 0$ (i.e. $f = -2$ and $d = 0$), we get $W_7 = 192c^2(c + 2)^2 \geq 0$.

1. $\mathcal{B}_1 \neq 0$. Then we could not have weak singularities, and considering the condition $C_2 \neq 0$, Remark 10.6 and Lemma 10.1, we arrive at the next five global configurations:
   - $s, n, f;\ \widehat{\binom{1}{2}}\ \widehat{P}_{\curlywedge} E \widehat{P}_{\curlywedge} - H,\ S \Rightarrow (c = 5/4,\ d = 1,\ f = -1/2)$ (if $\tilde{L} < 0,\ \widetilde{N} < 0$);
   - $s, n, f;\ \widehat{\binom{1}{2}}\ \widehat{P}_{\curlywedge} H \widehat{P}_{\curlywedge} - E,\ S \ \Rightarrow (c = 9/4,\ d = 1,\ f = -3/2)$ (if $\tilde{L} < 0,\ \widetilde{N} > 0$);

- $s, n, f;\ \widehat{\binom{1}{2}}\, H - E,\ S\ \Rightarrow (c = 2,\ d = 3,\ f = -1)$ (if $\tilde{L} < 0,\ \widetilde{N} = 0$);
- $s, n, f;\ \widehat{\binom{1}{2}}\, H_{\curlywedge} H H_{\curlywedge} - H,\ N^{\infty}\ \Rightarrow (c = 4,\ d = -7,\ f = -3)$ (if $\tilde{L} > 0$);
- $s, n, f;\ \widehat{\binom{1}{3}}\, H_{\curlywedge} H \widehat{P}_{\curlywedge} - \widehat{P}\ \Rightarrow (c = -5,\ d = 1,\ f = -2)$ (if $\tilde{L} = 0$).

2. $\mathcal{B}_1 = 0$. Then $\rho_1\rho_2\rho_3 = 0$ and at least one of the traces (which corresponds to the saddle or to the focus) is zero. Without loss of generality we may assume $\rho_1 = c + f = 0$, i.e. $c = -f$, and we arrive at the relations given in (10.23). Therefore we deduce that the conditions $\widetilde{K} > 0$ (i.e. $f < 0$) and $W_7 < 0$ imply $\mathcal{B}_2\sigma \neq 0$. Moreover we have $\mathcal{F}_1 \neq 0$, otherwise setting $d = -f(f+2)$ we obtain $W_7 = -1536f^3(1+f)^4 \geq 0$ due to $f < 0$.

   Thus by Lemma 10.1, considering the condition $C_2 \neq 0$, we get in the case $\mathcal{B}_2 < 0$ the following five global configurations:

   - $s, n, f^{(1)};\ \widehat{\binom{1}{2}}\, \widehat{P}_{\curlywedge} E \widehat{P}_{\curlywedge} - H,\ S\ \Rightarrow (c = 1/2,\ d = -1,\ f = -1/2)$ (if $\tilde{L} < 0,\ \widetilde{N} < 0$);
   - $s, n, f^{(1)};\ \widehat{\binom{1}{2}}\, \widehat{P}_{\curlywedge} H \widehat{P}_{\curlywedge} - E,\ S\ \Rightarrow (c = 3/2,\ d = -3,\ f = -3/2)$ (if $\tilde{L} < 0,\ \widetilde{N} > 0$);
   - $s, n, f^{(1)};\ \widehat{\binom{1}{2}}\, H - E,\ S\ \Rightarrow (c = 1,\ d = -2,\ f = -1)$ (if $\tilde{L} < 0,\ \widetilde{N} = 0$);
   - $s, n, f^{(1)};\ \widehat{\binom{1}{2}}\, H_{\curlywedge} H H_{\curlywedge} - H,\ N^{\infty}\ \Rightarrow (c = 3,\ d = -10,\ f = -3)$ (if $\tilde{L} > 0$);
   - $s, n, f^{(1)};\ \widehat{\binom{1}{3}}\, H_{\curlywedge} H \widehat{P}_{\curlywedge} - \widehat{P}\ \Rightarrow (c = 2,\ d = -5,\ f = -2)$ (if $\tilde{L} = 0$).

   In the case $\mathcal{B}_2 > 0$ we obtain the next five global configurations:

   - $s^{(1)}, n, f;\ \widehat{\binom{1}{2}}\, \widehat{P}_{\curlywedge} E \widehat{P}_{\curlywedge} - H,\ S\ \Rightarrow (c = 1/2,\ d = 1/4,\ f = -1/2)$ (if $\tilde{L} < 0,\ \widetilde{N} < 0$);
   - $s^{(1)}, n, f;\ \widehat{\binom{1}{2}}\, \widehat{P}_{\curlywedge} H \widehat{P}_{\curlywedge} - E,\ S\ \Rightarrow (c = 3/2,\ d = 1/4,\ f = -3/2)$ (if $\tilde{L} < 0,\ \widetilde{N} > 0$);
   - $s^{(1)}, n, f;\ \widehat{\binom{1}{2}}\, H - E,\ S\ \Rightarrow (c = 1,\ d = 0,\ f = -1)$ (if $\tilde{L} < 0,\ \widetilde{N} = 0$);
   - $s^{(1)}, n, f;\ \widehat{\binom{1}{2}}\, H_{\curlywedge} H H_{\curlywedge} - H,\ N^{\infty}\ \Rightarrow (c = 3,\ d = -8,\ f = -3)$ (if $\tilde{L} > 0$);
   - $s^{(1)}, n, f;\ \widehat{\binom{1}{3}}\, H_{\curlywedge} H \widehat{P}_{\curlywedge} - \widehat{P}\ \Rightarrow (c = 2,\ d = -3,\ f = -2)$ (if $\tilde{L} = 0$).

3.2.2: $W_7 > 0$. Then by Theorem 6.1 (see Table 6.2, line 117), apart from the saddle, we have two nodes, which are generic due to $W_7 \neq 0$.

3.2.2.1: $\mathcal{B}_1 \neq 0$. In this case the saddle is strong, and considering Remark 10.6 and Lemma 10.1, we arrive at the next six global configurations:

- $s, n, n;\ \widehat{\binom{1}{2}}\ \widehat{P}_{\curlywedge} E \widehat{P}_{\curlywedge} - H,\ S\ \Rightarrow (c = 1,\ d = 1,\ f = -1/2)$ (if $\tilde{L} < 0,\ \widetilde{N} < 0$);
- $s, n, n;\ \widehat{\binom{1}{2}}\ \widehat{P}_{\curlywedge} H \widehat{P}_{\curlywedge} - E,\ S\ \Rightarrow (c = 2,\ d = 1,\ f = -3/2)$ (if $\tilde{L} < 0,\ \widetilde{N} > 0$);
- $s, n, n;\ \widehat{\binom{1}{2}}\ H - E,\ S\ \Rightarrow (c = 2,\ d = 2,\ f = -1)$ (if $\tilde{L} < 0,\ \widetilde{N} = 0$);
- $s, n, n;\ \widehat{\binom{1}{2}}\ H_{\curlywedge} H H_{\curlywedge} - H,\ N^{\infty}\ \Rightarrow (c = 1,\ d = -7/2,\ f = -3)$ (if $\tilde{L} > 0$);
- $s, n, n;\ \widehat{\binom{1}{3}}\ H_{\curlywedge} H \widehat{P}_{\curlywedge} - \widehat{P}\ \Rightarrow (c = 1,\ d = -7/2,\ f = -2)$ (if $\tilde{L} = 0,\ C_2 \neq 0$);
- $s, n, n;\ [\infty; \emptyset]\ \Rightarrow (c = 3,\ d = 0,\ f = -2)$ (if $\tilde{L} = 0,\ C_2 = 0$).

3.2.2.2: $\mathcal{B}_1 = 0$. Then the saddle is weak, we may assume that it is located at the origin, i.e. $c = -f$, and we get the relations (10.23). We observe that due to $f < 0$ and $\mathbf{D} < 0$, we obtain $\mathcal{B}_2 \neq 0$. Moreover the condition $\mathcal{B}_2 > 0$ has to be satisfied, otherwise we have $d + f^2 < 0$, and setting $d = -f^2 - u^2$ (where $u \neq 0$), we obtain $W_7 = -192u^4(1 + f^2 + u^2)(4f^2 + u^2 + f^2u^2) < 0$.

Additionally we observe that in the case under examination we have $\widetilde{N} \neq 0$, otherwise the condition $\widetilde{N} = 0$ yields $f = -1$ (due to $\mu_1 \neq 0$), and then we arrive at the contradiction $W_7 = -384(d^2 - 1)^2 \leq 0$.

1. $\mathcal{F}_1 \neq 0$. Then $C_2 \neq 0$ because the condition $C_2 = 0$ implies $d = f + 2 = 0$ and then $\mathcal{F}_1 = 0$. So considering Remark 10.6 and Lemma 10.1, we arrive at the next four global configurations:
   - $s^{(1)}, n, n;\ \widehat{\binom{1}{2}}\ \widehat{P}_{\curlywedge} E \widehat{P}_{\curlywedge} - H,\ S\ \Rightarrow (c = 1/2,\ d = 4/5,\ f = -1/2)$ (if $\tilde{L} < 0,\ \widetilde{N} < 0$);
   - $s^{(1)}, n, n;\ \widehat{\binom{1}{2}}\ \widehat{P}_{\curlywedge} H \widehat{P}_{\curlywedge} - E,\ S\ \Rightarrow (c = 3/2,\ d = 4/5,\ f = -3/2)$ (if $\tilde{L} < 0,\ \widetilde{N} > 0$);
   - $s^{(1)}, n, n;\ \widehat{\binom{1}{2}}\ H_{\curlywedge} H H_{\curlywedge} - H,\ N^{\infty} \Rightarrow (c = 3,\ d = 1/2,\ f = -3)$ (if $\tilde{L} > 0$);
   - $s^{(1)}, n, n;\ \widehat{\binom{1}{3}}\ H_{\curlywedge} H \widehat{P}_{\curlywedge} - \widehat{P}\ \Rightarrow (c = 2,\ d = 1/2,\ f = -2)$ (if $\tilde{L} = 0$).
2. $\mathcal{F}_1 = 0$. Then by Theorem 6.2 (see statement $(e_3)[\beta]$ we have an integrable saddle. Considering (10.23) we have $d = -f(f + 2)$, and calculations yield
$$C_2 = (f + 2)y^2(x - fy),\quad \tilde{L} = 8f(f + 2)y^2.$$
   Since $f \neq 0$ the condition $\tilde{L} = 0$ is equivalent to $C_2 = 0$, and considering Remark 10.6 and Lemma 10.1, we arrive at the next four global configurations:
   - $\$, n, n;\ \widehat{\binom{1}{2}}\ \widehat{P}_{\curlywedge} E \widehat{P}_{\curlywedge} - H,\ S\ \Rightarrow (c = 1/2,\ d = 3/4,\ f = -1/2)$ (if $\tilde{L} < 0,\ \widetilde{N} < 0$);
   - $\$, n, n;\ \widehat{\binom{1}{2}}\ \widehat{P}_{\curlywedge} H \widehat{P}_{\curlywedge} - E,\ S\ \Rightarrow (c = 3/2,\ d = 3/4,\ f = -3/2)$ (if $\tilde{L} < 0,\ \widetilde{N} > 0$);
   - $\$, n, n;\ \widehat{\binom{1}{2}}\ H_{\curlywedge} H H_{\curlywedge} - H,\ N^{\infty}\ \Rightarrow (c = 3,\ d = -3,\ f = -3)$ (if $\tilde{L} > 0$);
   - $\$, n, n;\ [\infty; \emptyset]\ \Rightarrow (c = 2,\ d = 0,\ f = -2)$ (if $\tilde{L} = 0$).

3.2.3: $W_7 = 0$. Then apart from the saddle, we have two nodes, and at least one of them has coinciding eigenvalues. Without loss of generality we may assume that this point is located at the origin, and therefore the condition $\tau_1 = (c-f)^2+4d = 0$ gives $d = -(c-f)^2/4$. Then calculations yield

$$U_1 = -4(1+f)y^2, \quad Y_1 = 1+f, \quad U_6 = -4y^2, \quad \tilde{L} = 8f(f+2)y^2,$$
$$\mathbf{D} = -3/4(c+f)^4(c^2+8f+2cf+f^2)^2, \quad C_2 = (2+f)xy^2 + 1/4(c-f)^2y^3,$$

and considering Theorem 7.1 we conclude that systems (10.21) could not possess a star node. Moreover $C_2 \neq 0$, otherwise the conditions $c = f = -2$ imply $\mathbf{D} = 0$.

3.2.3.1: $W_6 \neq 0$. Considering (10.22) we conclude that only one discriminant vanishes, i.e. in this case we could not have two nodes n^d.

1. $\mathcal{B}_1 \neq 0$. Then the saddle is strong, and considering Remark 10.6 and Lemma 10.1, we arrive at the next five global configurations:
   - $s, n, n^d;\ \widehat{\binom{1}{2}}\ \widehat{P}_\curlywedge E\widehat{P}_\curlywedge - H,\ S\ \Rightarrow (c=1,\ d=-9/16,\ f=-1/2)$ (if $\tilde{L}<0,\ \widetilde{N}<0$);
   - $s, n, n^d;\ \widehat{\binom{1}{2}}\ \widehat{P}_\curlywedge H\widehat{P}_\curlywedge - E,\ S\ \Rightarrow (c=1,\ d=-25/16,\ f=-3/2)$ (if $\tilde{L}<0,\ \widetilde{N}>0$);
   - $s, n, n^d;\ \widehat{\binom{1}{2}}\ H - E,\ S\ \Rightarrow (c=2,\ d=-9/4,\ f=-1)$ (if $\tilde{L}<0,\ \widetilde{N}=0$);
   - $s, n, n^d;\ \widehat{\binom{1}{2}}\ H_\curlywedge HH_\curlywedge - H,\ N^\infty\ \Rightarrow (c=1,\ d=-4,\ f=-3)$ (if $\tilde{L}>0$);
   - $s, n, n^d;\ \widehat{\binom{1}{3}}\ H_\curlywedge H\widehat{P}_\curlywedge - \widehat{P}\ \Rightarrow (c=1,\ d=-9/4,\ f=-2)$ (if $\tilde{L}=0$).
2. $\mathcal{B}_1 = 0$. In this case we have $\rho_1\rho_2\rho_3 = 0$ and $\rho_1 \neq 0$ (due to $\tau_1 = 0$). So without loss of generality we may assume $\rho_2 = 2+c-f = 0$, and this gives $c = f-2$. Then we have

$$\mathcal{B}_1 = \mathcal{H} = 0,\ \mathcal{B}_2 = 4(f-1)^4(1+f^2),\ \sigma = 2(1-f)(y-1),\ \widetilde{N} = -4(f+1)y^2,$$
$$W_7 = 0,\quad \mathcal{F}_1 = 2(1-f)(f+1)^2,\quad \mathbf{D} = -192(f-1)^4(1+f^2)^2,$$

   and hence the condition $\mathbf{D} < 0$ implies $\mathcal{B}_2 > 0$. We note that in this case $\mathcal{F}_1 \neq 0$ and $\widetilde{N} \neq 0$, otherwise we get $f=-1$, $c=-3$ and this implies $W_6 = 0$. Therefore by Theorem 6.2 (see statement (e_1)) we have a weak saddle of first order. Since $C_2\widetilde{N} \neq 0$, considering Remark 10.6 and Lemma 10.1, we arrive at the next four global configurations:
   - $s^{(1)}, n, n^d;\ \widehat{\binom{1}{2}}\ \widehat{P}_\curlywedge E\widehat{P}_\curlywedge - H,\ S\ \Rightarrow (c=-5/2,\ d=-1,\ f=-1/2)$ (if $\tilde{L}<0,\ \widetilde{N}<0$);
   - $s^{(1)}, n, n^d;\ \widehat{\binom{1}{2}}\ \widehat{P}_\curlywedge H\widehat{P}_\curlywedge - E,\ S\ \Rightarrow (c=-7/2,\ d=-1,\ f=-3/2)$ (if $\tilde{L}<0,\ \widetilde{N}>0$);

- $s^{(1)}, n, n^d;\ \widehat{\binom{1}{2}}\ H_\curlywedge HH_\curlywedge - H,\ N^\infty \Rightarrow (c=-5,\ d=-1,\ f=-3)$ (if $\tilde{L}>0$);
- $s^{(1)}, n, n^d;\ \widehat{\binom{1}{3}}\ H_\curlywedge H\widehat{P}_\curlywedge - \widehat{P} \Rightarrow (c=-4,\ d=-1,\ f=-2)$ (if $\tilde{L}=0$).

3.2.3.2: $W_6 = 0$. Then apart from the node n^d located at the origin, we have another node n^d, and we can assume that this is the singularity $M_2(0,1)$. Then we have $\tau_2 = 2\big[(c+1)^2 + (f+1)^2\big] = 0$, which implies $c = f = -1$. In this case we calculate

$$\mathcal{B}_1 = \mathcal{H} = \mathcal{F}_1 = 0,\ \ \mathcal{B}_2 = 32 > 0,\ \ \sigma = 2(2y-1) \neq 0,$$
$$W_7 = W_6 = 0,\ \ \widetilde{K} = 4y^2 > 0,\ \ \tilde{L} = -8y^2 < 0,\ \ \widetilde{N} = 0.$$

So by Theorem 6.2 (see statement $(e_3)[\beta]$) the saddle is integrable, and considering Lemma 10.1 we get the global configuration

- $\$, n^d, n^d;\ \widehat{\binom{1}{2}}\ H - E,\ S \Rightarrow (c=-1,\ d=0,\ f=-1)$.

As all the cases are examined, we have constructed all 147 possible global configurations for the family of quadratic systems with exactly three real distinct finite singularities. □

10.3.2 Systems with zero-cycle $\mathcal{D}_{\mathbb{C}^2}(S) = p + q^c + r^c$

Theorem 10.4. (a) *We consider here all the geometrical global configurations of singularities of quadratic vector fields with exactly three distinct finite singularities: one real and two complex. These configurations are classified in Diagram* 10.5 *according to the geometrical equivalence relation of configurations. The total number of geometrically distinct global configurations of singularities is* 48.

(b) *Necessary and sufficient conditions for each one of the* 48 *different geometrical equivalence classes can be assembled from this diagram in terms of* 23 *invariant polynomials with respect to the action of the affine group and time rescaling given in Section* 5.4.

(c) *Diagram* 10.5 *actually contains the global bifurcation diagram in the* 12*-dimensional space of parameters of the global geometrical global configurations of singularities, of this family of quadratic differential systems and provides an algorithm for deciding for any given system which is its respective geometrical global configuration.*

Proof. Assume that the quadratic systems (8.1) possess one real and two complex elemental finite singularities. Then according to Proposition 6.2 (see Table 6.1) via an affine transformation, these systems could be brought to the systems 11. So we consider here the family of systems:

$$\begin{aligned}\dot{x} &= 2(h-gu)x + g(u^2+1)y + gx^2 - 2hxy,\\ \dot{y} &= 2(m-lu)x + l(u^2+1)y + lx^2 - 2mxy,\end{aligned} \tag{10.24}$$

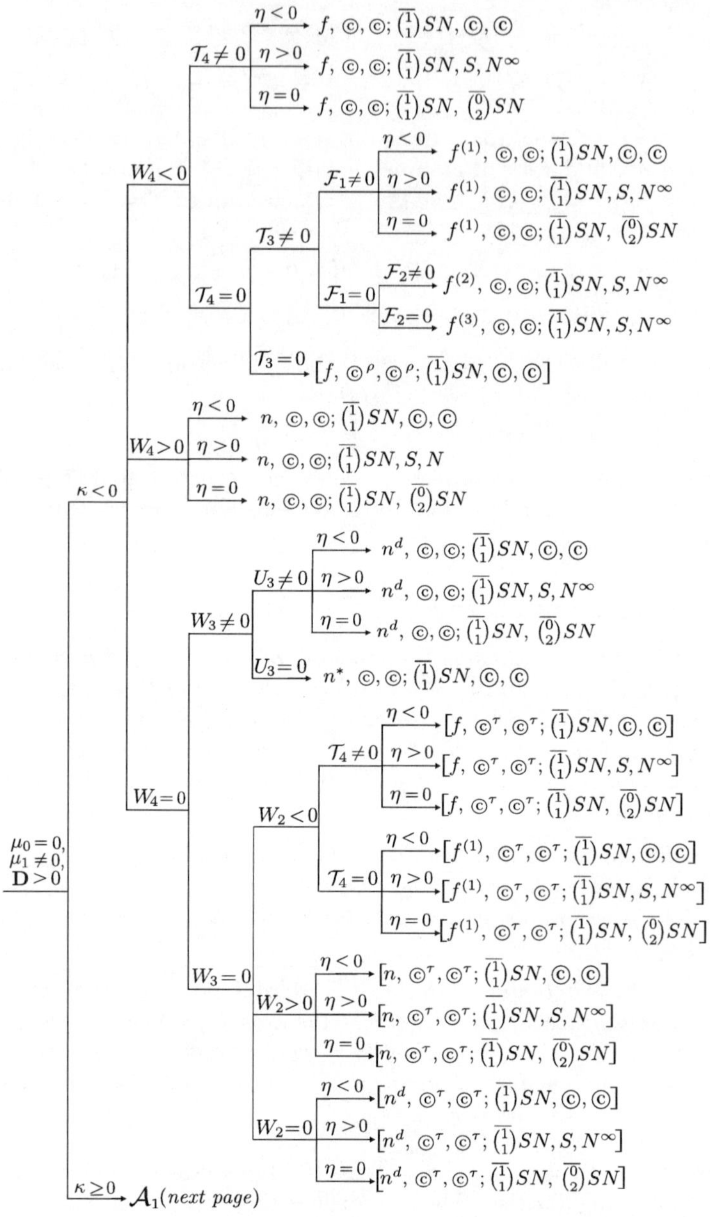

Diagram 10.5: Global configurations: the case $\mu_0=0$, $\mu_1\neq 0$, $\mathbf{D}>0$.

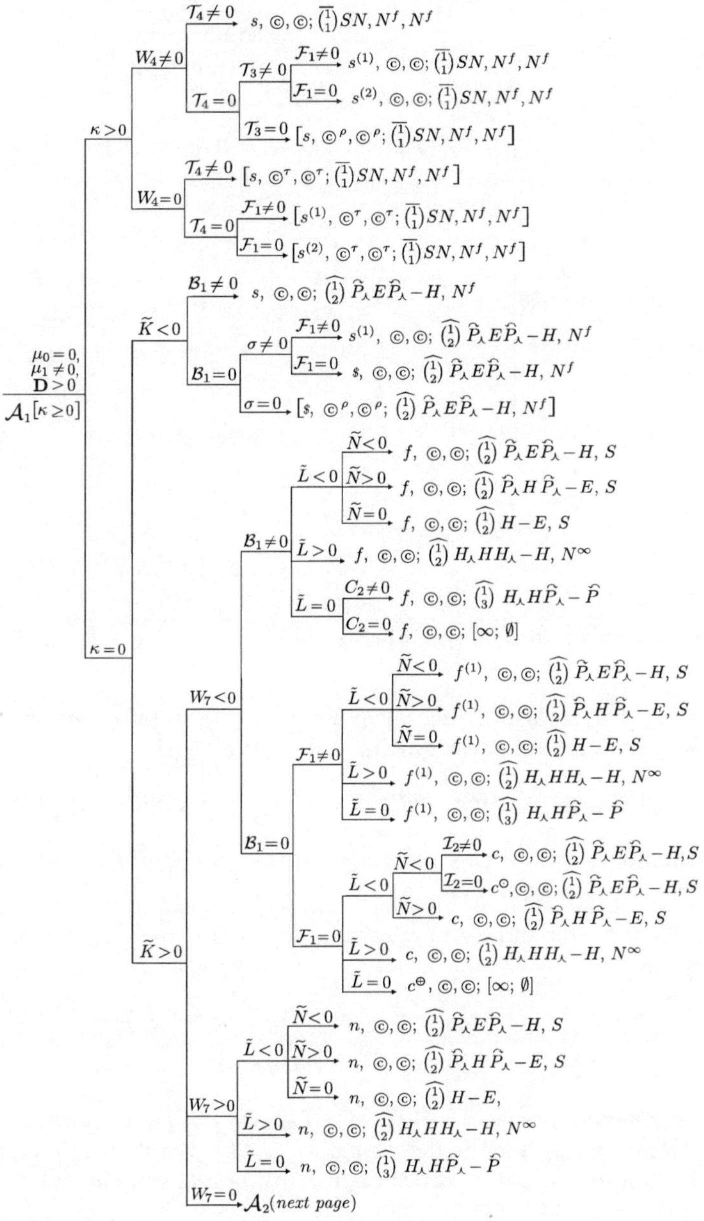

Diagram 10.5 (*cont.*): Global configurations: the case $\mu_0 = 0$, $\mu_1 \neq 0$, $\mathbf{D} > 0$.

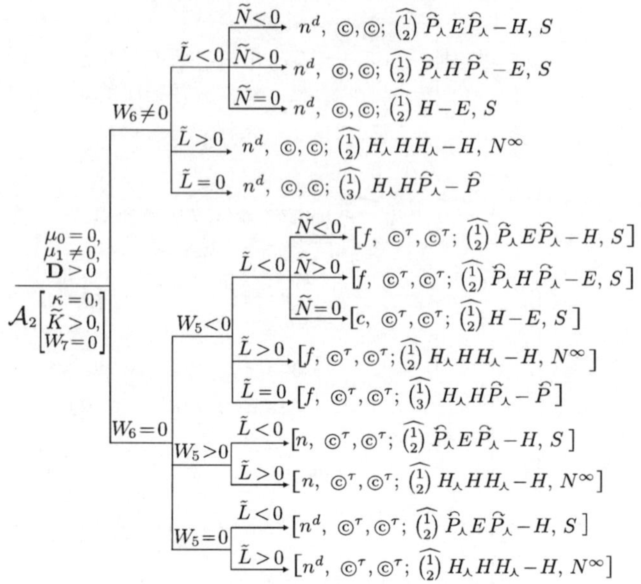

Diagram 10.5 (*cont.*): Global configurations: the case $\mu_0 = 0$, $\mu_1 \neq 0$, $\mathbf{D} > 0$.

possessing the real elemental singular point $M_1(0,0)$ and the two complex singularities $M_{2,3}(u \pm i, 1)$. For these systems calculations yield

$$\mu_0 = 0, \quad \mu_1 = 4(gm - hl)^2(1+u^2)x, \quad \kappa = 128h^2(gm - hl).$$

Since for the above systems we have $\mu_1 \neq 0$ (i.e. $gm - hl \neq 0$), we observe that the condition $\kappa = 0$ is equivalent to $h = 0$.

1: $\kappa \neq 0$. Then $h \neq 0$ and we may assume $h = 1$ due to a time rescaling. So we get the systems:

$$\begin{aligned} \dot{x} &= 2(1 - gu)x + g(u^2+1)y + gx^2 - 2xy, \\ \dot{y} &= 2(m - lu)x + l(u^2+1)y + lx^2 - 2mxy, \end{aligned} \tag{10.25}$$

and for the singular points $M_1(0,0)$ and $M_{2,3}(u \pm i, 1)$, we have the following values for the traces ρ_i, for the determinants Δ_i, for the discriminants τ_i and for the linearization matrix $\mathcal{M}_1$ (corresponding to the real singularity):

$$\mathcal{M}_1 = \begin{pmatrix} 2(1-gu) & g(1+u^2) \\ 2(m-lu) & l(1+u^2) \end{pmatrix}, \quad \rho_1 = 2 + l - 2gu + lu^2,$$

$$\Delta_1 = 2(l - gm)(1+u^2), \quad \tau_1 = \rho_1^2 - 4\Delta_1; \quad \tau_{2,3} = \rho_{2,3}^2 - 4\Delta_{2,3},$$

$$\rho_{2,3} = l - 2mu + lu^2 \pm 2i(g - m), \quad \Delta_{2,3} = 4(gm - l)(1 \mp iu).$$

Then for systems (10.24) we calculate

$$\kappa = -64\Delta_1/(1+u^2), \quad \widetilde{K} = 2\Delta_1 x^2/(1+u^2), \quad G_9 = \Delta_1/(2(1+u^2)), \tag{10.26}$$

$$\begin{gathered}
\mathcal{T}_4 = 2\rho_1\rho_2\rho_3\Delta_1^2/(1+u^2), \quad \mathcal{T}_3 = 2(\rho_1\rho_2+\rho_1\rho_3+\rho_2\rho_3)\Delta_1^2/(1+u^2),\\
\mathcal{T}_2 = 2(\rho_1+\rho_2+\rho_3)\Delta_1^2/(1+u^2), \quad \eta = 4\big[(g+2m)^2-8l\big],\\
\widetilde{M} = -8\big[(g+2m)^2-6l\big]x^2 + 16(g+2m)xy - 32y^2,\\
W_4 = 4\tau_1\tau_2\tau_3\Delta_1^4/(1+u^2)^2, \quad W_2 = 4(\tau_1+\tau_2+\tau_3)\Delta_1^4/(1+u^2)^2,\\
W_3 = 4(\tau_1\tau_2+\tau_1\tau_3+\tau_2\tau_3)\Delta_1^4/(1+u^2)^2.
\end{gathered} \tag{10.27}$$

Remark 10.7. We observe that $\widetilde{M} \neq 0$ and the condition $\kappa \neq 0$ implies $\widetilde{K}G_9 \neq 0$. Moreover we have sign $(\widetilde{K}) = -$sign (κ).

1.1: $\kappa < 0$. This implies $\widetilde{K} > 0$, and according to Theorem 6.1 (see Table 6.2, lines 124–133) the elemental real singular point is an anti-saddle and its type is governed by the invariant polynomials W_i, $i = 4, 3, 2$ (as in the case we consider we have $G_9 \neq 0$).

1.1.1: $W_4 < 0$. Clearly for the complex singularities $M_{2,3}$ we have $\tau_3 = \bar{\tau}_2$ (also $\rho_3 = \bar{\rho}_2$) and then $\tau_2\tau_3 > 0$. So considering (10.26) we have $\tau_1 < 0$, i.e. the real singular point is a focus or a center.

1.1.1.1: $\mathcal{T}_4 \neq 0$. Then we have a strong focus, and considering Lemma 10.1 and the condition $\widetilde{M} \neq 0$ (see Remark 10.7), we arrive at the following three global configurations:

- f, ©, ©; $\overline{\binom{1}{1}}SN$, ©, © $\Rightarrow (g = -2,\ l = 1,\ m = 1,\ u = 0)$ (if $\eta < 0$);
- f, ©, ©; $\overline{\binom{1}{1}}SN, S, N^\infty$ $\Rightarrow (g = -2,\ l = -1,\ m = 1,\ u = 0)$ (if $\eta > 0$);
- f, ©, ©; $\overline{\binom{0}{2}}SN$, $\overline{\binom{1}{1}}SN$ $\Rightarrow (g = -2,\ l = 0,\ m = 1,\ u = 0)$ (if $\eta = 0$).

1.1.1.2: $\mathcal{T}_4 = 0$. We examine two possibilities: $\mathcal{T}_3 \neq 0$ and $\mathcal{T}_3 = 0$.

1.1.1.2.1: $\mathcal{T}_3 \neq 0$. Considering (10.26) we have $\rho_1 = 0$ (i.e. $l = 2(gu-1)/(1+u^2)$), and we calculate

$$W_4 = -\frac{16\Delta_1^5\tau_2\tau_3}{(1+u^2)^2}, \quad \mathcal{T}_3 = \frac{2\rho_2\rho_3\Delta_1^2}{(1+u^2)}, \quad \mathcal{F} = -\frac{\rho_2\rho_3\Delta_1}{4(1+u^2)},$$
$$\mathcal{F}_1 = 2\Delta_1\big[2g^2u - 5g - 2m(gu-1) + 4(g+u)/(1+u^2)\big].$$

1. $\mathcal{F}_1 \neq 0$. As $\mathcal{T}_3\mathcal{F} < 0$, considering Theorem 6.2 (see statement (b_1)) we have a first order weak focus. So by Lemma 10.1 and Remark 10.7, we get the next three global configurations:

   - $f^{(1)}$, ©, ©; $\overline{\binom{1}{1}}SN$, ©, © $\Rightarrow (g = 3,\ l = 2,\ m = -3,\ u = 1)$ (if $\eta < 0$);

- $f^{(1)}, ©, ©; \overline{\binom{1}{1}}SN, S, N^{\infty} \Rightarrow (g=3,\ l=2,\ m=-4,\ u=1)$ (if $\eta>0$);
- $f^{(1)}, ©, ©; \binom{0}{2}SN,\ \overline{\binom{1}{1}}SN \Rightarrow (g=3,\ l=2,\ m=-7/2,\ u=1)$ (if $\eta=0$).

2. $\mathcal{F}_1=0$. Then the order of the weak focus is at least two. According to Theorem 7.4 we could not have a center in this case. Moreover by Theorems 7.7 and 7.8, we could have only the next two global configurations, which by Theorem 6.2 are distinguished by the invariant polynomial $\mathcal{F}_2$:

   - $f^{(2)}, ©, ©; \overline{\binom{1}{1}}SN, S, N^{\infty} \Rightarrow (g=2/3,\ l=-1/3,\ m=-4/3,\ u=1)$ (if $\mathcal{F}_2\neq 0$);
   - $f^{(3)}, ©, ©; \overline{\binom{1}{1}}SN, S, N^{\infty} \Rightarrow (g=4/5,\ l=-1/5,\ m=-11/5,\ u=1)$ (if $\mathcal{F}_2=0$).

1.1.1.2.2: $\mathcal{T}_3=0$. Considering (10.26) it is clear that the condition $\mathcal{T}_4=\mathcal{T}_3=0$ gives $\rho_2=\rho_3=0$, and this implies $g=m$ and $l=2mu/(1+u^2)$. Then calculations yield

$$W_4=256\Delta_1^6\tau_1/(1+u^2)^3,\quad \kappa=-64\Delta_1/(1+u^2),$$
$$\eta=4m(9m-16u+9mu^2)/(1+u^2),\quad \rho_2=\rho_3=0,\quad \rho_1=2.$$

So considering (10.26) we have $\mathcal{T}_2\neq 0$ due to $\rho_1\neq 0$. We claim that in this case the condition $W_4<0$ implies $\eta<0$. Indeed, we observe that the curves $\eta=0$ and $\tau_1=0$ have only complex points of intersection. So it is sufficient to determine the sign of the polynomial η at a point where $\tau_1<0$. At the point $(m,u)=(-1/2,-2)$ we have $\tau_1=-2<0$ and $\eta=-19/5<0$, and this proves our claim.

Thus due to $\rho_1\neq 0$ the focus is strong, and we get the global configuration

- $f, ©^{\rho}, ©^{\rho}; \overline{\binom{1}{1}}SN, ©, © \Rightarrow (g=1/2,\ l=6/13,\ m=1/2,\ u=3/2)$.

1.1.2: $W_4>0$. In this case by Theorem 6.1 (see Table 6.2, line 124), the elemental singularity is a generic node, and we arrive at the next three global configurations:

- $n, ©, ©; \overline{\binom{1}{1}}SN, ©, © \Rightarrow (g=1,\ l=2,\ m=1,\ u=0)$ (if $\eta<0$);
- $n, ©, ©; \overline{\binom{1}{1}}SN, S, N^{\infty} \Rightarrow (g=1,\ l=17/16,\ m=1,\ u=0)$ (if $\eta>0$);
- $n, ©, ©; \binom{0}{2}SN,\ \overline{\binom{1}{1}}SN \Rightarrow (g=1,\ l=9/8,\ m=1,\ u=0)$ (if $\eta=0$).

1.1.3: $W_4=0$. We consider two cases: $W_3\neq 0$ and $W_3=0$.

1. $W_3\neq 0$. Considering (10.26) we obtain $\tau_1=0$, and we have a node with coinciding eigenvalues. According to Lemma 10.3 this node will be a star node if and only if $U_3=0$. We observe that in the case $U_3=0$ we obtain $\eta=-64/(1+u^2)^2<0$. So according to Lemma 10.1 we get the following four global configurations:

   - $n^d, ©, ©; \overline{\binom{1}{1}}SN, ©, © \Rightarrow (g=-1/4,\ l=4,\ m=2,\ u=0)$ (if $U_3\neq 0$, $\eta<0$);

- n^d, ©, ©; $\overline{\binom{1}{1}}SN, S, N^{\infty}$ $\Rightarrow (g = -1/16,\ l = 1,\ m = 2,\ u = 0)$ (if $U_3 \neq 0$, $\eta > 0$);
- n^d, ©, ©; $\binom{0}{2}SN$, $\overline{\binom{1}{1}}SN \Rightarrow (g = 0,\ l = 2,\ m = 2,\ u = 0)$ (if $U_3 \neq 0, \eta{=}0$);
- n^*, ©, ©; $\overline{\binom{1}{1}}SN$, ©, © $\Rightarrow (g = 0,\ l = 2,\ m = 0,\ u = 0)$ (if $U_3 = 0$).

2. $W_3 = 0$. Considering (10.26) we get $\tau_2 = \tau_3 = 0$. However for systems (10.25) these conditions lead to a quite complicated system of polynomial equations. So we decide to construct a more convenient canonical form, and this will be done in Lemma 10.4 (see page 386). According to this lemma, in the case $\kappa < 0$ we obtain 12 geometrically distinct global configurations and the corresponding invariant conditions that characterize them.

1.2: $\kappa > 0$. Then $\Delta_1 < 0$ and the real singular point is a saddle. Additionally considering (10.26) we obtain $\eta > 0$, and according to Lemma 10.1, at infinity we get the unique configuration $\overline{\binom{1}{1}}SN, N^f, N^f$.

1.2.1: $W_4 \neq 0$. In this case the discriminants of the complex points are different from zero.

1.2.1.1: $\mathcal{T}_4 \neq 0$. In this case the saddle is strong, and we arrive at the global configuration

- s, ©, ©; $\overline{\binom{1}{1}}SN, N^f, N^f$ $\Rightarrow (g = 1,\ l = -1,\ m = 1,\ u = 0)$.

1.2.1.2: $\mathcal{T}_4 = 0$. We consider two possibilities: $\mathcal{T}_3 \neq 0$ and $\mathcal{T}_3 = 0$.

1.2.1.2.1: $\mathcal{T}_3 \neq 0$. Then $\rho_1 = 0$, i.e. we have a weak saddle whose order depends on the polynomial $\mathcal{F}_1$.

1. $\mathcal{F}_1 \neq 0$. The weak saddle has order one, and we get the global configuration
   - $s^{(1)}$, ©, ©; $\overline{\binom{1}{1}}SN, N^f, N^f$ $\Rightarrow (g = 2,\ l = 1,\ m = 1,\ u = 1)$.
2. $\mathcal{F}_1 = 0$. Then we have a weak saddle whose order is at least two. By Theorem 7.9 in this case we could not have a weak saddle of order three. Thus we obtain the unique global configuration
   - $s^{(2)}$, ©, ©; $\overline{\binom{1}{1}}SN, N^f, N^f$ $\Rightarrow (g = 0,\ l = -2,\ m = 0,\ u = 0)$.

1.2.1.2.2: $\mathcal{T}_3 = 0$. Considering (10.26) it is clear that the condition $\mathcal{T}_4 = \mathcal{T}_3 = 0$ gives $\rho_2 = \rho_3 = 0$, and this implies $g = m$ and $l = 2mu/(1 + u^2)$. Then we have $\rho_1 = 2 \neq 0$, i.e. the saddle is strong and we get the global configuration

- s, $©^{\rho}$, $©^{\rho}$; $\overline{\binom{1}{1}}SN, N^f, N^f$ $\Rightarrow (g = 1,\ l = 0,\ m = 1,\ u = 0)$.

1.2.1: $W_4 = 0$. Since the real singularity is a saddle, we conclude that in this case both complex points have zero discriminants (i.e. $W_4 = W_3 = 0$). In this case by Lemma 10.4 (see page 386) we get the three global configurations (Cf_{13}), (Cf_{14}) and (Cf_{15}) and the corresponding invariant conditions that characterize them.

2: $\kappa = 0$. Then $h = 0$ and we have $\mu_1 = 4g^2m^2(1+u^2)x \neq 0$. Therefore we may assume $g = 1$ due to a time rescaling, and we get the 3-parameter family of systems:

$$\begin{aligned} \dot{x} &= -2ux + (u^2+1)y + x^2, \\ \dot{y} &= 2(m-lu)x + l(u^2+1)y + lx^2 - 2mxy. \end{aligned} \tag{10.28}$$

For the singular points $M_1(0,0)$ and $M_{2,3}(u \pm i, 1)$ we have the following values for the traces ρ_i, for the determinants Δ_i, for the discriminants τ_i and for the linearization matrix $\mathcal{M}_1$ (corresponding to the real singularity):

$$\begin{aligned} \mathcal{M}_1 = \begin{pmatrix} -2u & 1+u^2 \\ 2(m-lu) & l(1+u^2) \end{pmatrix}, \quad \rho_1 = l - 2u + lu^2, \ \Delta_1 = -2m(1+u^2), \\ \tau_1 = \rho_1^2 - 4\Delta_1; \quad \rho_{2,3} = l - 2mu + lu^2 \pm 2i(1-m), \\ \Delta_{2,3} = 4m(1 \mp iu), \ \tau_{2,3} = \rho_{2,3}^2 - 4\Delta_{2,3}. \end{aligned} \tag{10.29}$$

Then for the above systems we calculate

$$\begin{aligned} \mu_1 = 4m^2(1+u^2)x, \quad \widetilde{K} = -4mx^2, \quad G_9 = 0, \\ W_7 = 3\tau_1\tau_2\tau_3\Delta_1^6/(1+u^2)^2/16, \quad W_5 = 9(\tau_1+\tau_2+\tau_3)\Delta_1^6/(1+u^2)^2/4, \\ W_6 = \big[\tau_1(\tau_2+\tau_3) + \tau_2\tau_3\big]\Delta_1^6/(1+u^2)^2/16, \quad \mathcal{T}_4 = \mathcal{T}_3 = \mathcal{T}_2 = \mathcal{T}_1 = 0, \\ \sigma = \rho_1 - 2(m-1)x, \quad \mathcal{B}_1 = \rho_1\rho_2\rho_3\Delta_1, \\ \mathcal{B}_2 = -(m-1)^2(1+u^2)(\rho_1\rho_2 + \rho_1\rho_3 + \rho_2\rho_3)\Delta_1. \end{aligned} \tag{10.30}$$

We observe that the condition $\mu_1 \neq 0$ implies $\widetilde{K} \neq 0$, and by (10.30) we have $\operatorname{sign}(\widetilde{K}) = \operatorname{sign}(\Delta_1)$. According to Theorems 6.1 and 6.2, the above polynomials are responsible for the types of the finite singularities of systems (10.28). In order to describe the infinite singularities, considering Lemma 10.1 we calculate the additional invariant polynomials:

$$\begin{aligned} \kappa = \eta = 0, \quad \widetilde{M} = -8(1+2m)^2x^2, \quad \tilde{L} = 8(1+2m)x^2, \\ C_2 = -lx^3 + (1+2m)x^2y, \quad \widetilde{N} = -4m(1+m)x^2. \end{aligned} \tag{10.31}$$

So by Lemma 10.1 we have the next remark.

Remark 10.8. Systems (10.28) could possess at infinity only one of the following seven global configurations:

$$\begin{array}{lll}
\widetilde{M}\neq 0,\ \widetilde{K}<0 & \Rightarrow & \widehat{\binom{1}{2}}\ \widehat{P}_{\curlywedge}E\widehat{P}_{\curlywedge}-H,\ N^{f};\\
\widetilde{M}\neq 0,\ \widetilde{K}>0,\ \tilde{L}<0,\ \widetilde{N}<0 & \Rightarrow & \widehat{\binom{1}{2}}\ \widehat{P}_{\curlywedge}E\widehat{P}_{\curlywedge}-H,\ S;\\
\widetilde{M}\neq 0,\ \widetilde{K}>0,\ \tilde{L}<0,\ \widetilde{N}>0 & \Rightarrow & \widehat{\binom{1}{2}}\ \widehat{P}_{\curlywedge}H\widehat{P}_{\curlywedge}-E,\ S;\\
\widetilde{M}\neq 0,\ \widetilde{K}>0,\ \tilde{L}<0,\ \widetilde{N}=0 & \Rightarrow & \widehat{\binom{1}{2}}\ H-E,\ S;\\
\widetilde{M}\neq 0,\ \widetilde{K}>0,\ \tilde{L}>0 & \Rightarrow & \widehat{\binom{1}{2}}\ H_{\curlywedge}HH_{\curlywedge}-H,\ N^{\infty};\\
\widetilde{M}=0,\ C_2\neq 0 & \Rightarrow & \widehat{\binom{1}{3}}\ H_{\curlywedge}H\widehat{P}_{\curlywedge}-\widehat{P};\\
\widetilde{M}=C_2=0 & \Rightarrow & [\infty;\ \emptyset].
\end{array}$$

2.1: $\widetilde{K}<0$. Then $\Delta_1<0$ and the real singularity is a saddle. We observe that the condition $\widetilde{K}<0$ implies $m>0$ and then $\widetilde{M}\neq 0$.

2.1.1: $\mathcal{B}_1\neq 0$. In this case $\rho_1\neq 0$ and the saddle is strong. So considering Remark 10.8 we obtain the global configuration

- s, ©, ©; $\widehat{\binom{1}{2}}\ \widehat{P}_{\curlywedge}E\widehat{P}_{\curlywedge}-H,\ N^{f}\ \Rightarrow (l=1,\ m=1,\ u=0)$.

2.1.2: $\mathcal{B}_1=0$. Then $\rho_1\rho_2\rho_3=0$ and we consider two cases: $\mathcal{B}_2\neq 0$ and $\mathcal{B}_2=0$.

1. $\mathcal{B}_2\neq 0$. As $\rho_3=\bar{\rho}_2$, considering (10.30) we have $\rho_2\rho_3\neq 0$ and then $\rho_1=0$, i.e. $l=2u/(1+u^2)$, and the saddle is weak. We calculate

$$\mathcal{B}_1=\mathcal{H}=0,\quad \mathcal{B}_2=8m(m-1)^4(1+u^2)^3,$$
$$\sigma=2(1-m)x,\quad \mathcal{F}_1=8m(m-1)u(1+u^2),$$

and clearly the conditions $\mathcal{B}_2\neq 0$ and $\widetilde{K}<0$ imply $\sigma\neq 0$ and $\mathcal{B}_2>0$. So according to Theorem 6.2 (see statements (e_1) and $(e_3)[\beta]$) the weak saddle is of order one if $\mathcal{F}_1\neq 0$ and is integrable if $\mathcal{F}_1=0$. Considering Remark 10.8 we get the next two global configurations:

- $s^{(1)}$, ©, ©; $\widehat{\binom{1}{2}}\ \widehat{P}_{\curlywedge}E\widehat{P}_{\curlywedge}-H,\ N^{f}\Rightarrow (l=1, m=2, u=1)$ (if $\mathcal{F}_1\neq 0$);
- $\$$, ©, ©; $\widehat{\binom{1}{2}}\ \widehat{P}_{\curlywedge}E\widehat{P}_{\curlywedge}-H,\ N^{f}\ \Rightarrow (l=0,\ m=2,\ u=0)$ (if $\mathcal{F}_1=0$).

2. $\mathcal{B}_2=0$. We claim that the condition $\mathcal{B}_1=\mathcal{B}_2=0$ implies $\sigma=0$. Indeed, we observe that the condition $m=1$ has to be satisfied, otherwise we must have $\rho_2=\rho_3=0$ and this implies $m=1$. Therefore as $m=1$ we get $\rho_1=\rho_2=\rho_3$, and then the condition $\mathcal{B}_1=\rho_1^3\Delta_1=0$ implies $\sigma=\rho_1=0$, i.e. our claim is proved.

Thus $\sigma=0$ and we get Hamiltonian systems. So we arrive at the global configuration

- $\$$, $©^{\rho}$, $©^{\rho}$; $\widehat{\binom{1}{2}}\ \widehat{P}_{\curlywedge}E\widehat{P}_{\curlywedge}-H,\ N^{f}\ \Rightarrow (l=0,\ m=1,\ u=0)$.

Remark 10.9. Considering (10.31) we observe that the condition $\tilde{L}=0$ is equivalent to $\widetilde{M}=0$.

2.2: $\widetilde{K} > 0$. In this case $\Delta_1 > 0$, i.e. the real singular point is an anti-saddle, and as $G_9 = 0$, according to Theorem 6.1 (see Table 6.2, lines 127–129, 132 and 133) its type is governed by the invariant polynomials W_7, W_6 and W_5.

2.2.1: $W_7 < 0$. By (10.30) due to $\tau_2\tau_3 > 0$ we have $\tau_1 < 0$, and we conclude that the real singularity is a focus or a center.

2.2.1.1: $\mathcal{B}_1 \neq 0$. Then $\rho_1 \neq 0$ and we have a strong focus. Considering Remarks 10.8 and 10.9, we arrive at the next six global configurations:

- f, ©, ©; $\widehat{\binom{1}{2}}\ \widehat{P}_{\curlywedge} E \widehat{P}_{\curlywedge} - H,\ S \Rightarrow (l = 1,\ m = -3/2,\ u = 0)$ (if $\tilde{L} < 0,\ \widetilde{N} < 0$);
- f, ©, ©; $\widehat{\binom{1}{2}}\ \widehat{P}_{\curlywedge} H \widehat{P}_{\curlywedge} - E,\ S \Rightarrow (l = 1,\ m = -2/3,\ u = 0)$ (if $\tilde{L} < 0,\ \widetilde{N} > 0$);
- f, ©, ©; $\widehat{\binom{1}{2}}\ H - E,\ S \Rightarrow (l = 1,\ m = -1,\ u = 0)$ (if $\tilde{L} < 0,\ \widetilde{N} = 0$);
- f, ©, ©; $\widehat{\binom{1}{2}}\ H_{\curlywedge} H H_{\curlywedge} - H,\ N^{\infty} \Rightarrow (l = 1,\ m = -1/3,\ u = 0)$ (if $\tilde{L} > 0$);
- f, ©, ©; $\widehat{\binom{1}{3}}\ H_{\curlywedge} H \widehat{P}_{\curlywedge} - \widehat{P} \Rightarrow (l = 1,\ m = -1/2,\ u = 0)$ (if $\tilde{L} = 0,\ C_2 \neq 0$).
- f, ©, ©; $[\infty;\ \emptyset] \Rightarrow (l = 0,\ m = -1/2,\ u = 1)$ (if $\tilde{L} = 0,\ C_2 = 0$).

2.2.1.2: $\mathcal{B}_1 = 0$. Then $\rho_1\rho_2\rho_3 = 0$, and as $\Im(\rho_2) = -\Im(\rho_3) = 2(1 - m)$, we obtain $\rho_2\rho_3 \neq 0$ due to $\widetilde{K} > 0$ (which implies $m < 0$). So we get $\rho_1 = 0$, and this gives $l = 2u/(u^2 + 1)$. Then following Theorem 6.2 we calculate

$$\begin{aligned}&\mathcal{B}_1 = \mathcal{H} = 0,\quad \sigma = 2(1 - m)x,\quad \mathcal{F}_1 = 8mu(m - 1)(1 + u^2),\quad \widetilde{K} = -4mx^2,\\ &\mathcal{B}_2 = 8(m - 1)^4 m(1 + u^2)^3,\quad C_2 = x\big[-2ux + (1 + u)^2(1 + 2m)y\big]/(1 + u)^2,\end{aligned}$$

and clearly the condition $\widetilde{K} > 0$ implies $\mathcal{B}_2 < 0$ and $\sigma \neq 0$. So according to Theorem 6.2 we have a weak focus of the first order if $\mathcal{F}_1 \neq 0$ (statement (e_2)) and a center if $\mathcal{F}_1 = 0$ (statement $(e_4)[\beta]$).

1. $\mathcal{F}_1 \neq 0$. Then $u \neq 0$ (this implies $C_2 \neq 0$), and we arrive at the next five global configurations:

   - $f^{(1)}$, ©, ©; $\widehat{\binom{1}{2}}\ \widehat{P}_{\curlywedge} E \widehat{P}_{\curlywedge} - H,\ S \Rightarrow (l = 1,\ m = -3/2,\ u = 1)$ (if $\tilde{L} < 0,\ \widetilde{N} < 0$);
   - $f^{(1)}$, ©, ©; $\widehat{\binom{1}{2}}\ \widehat{P}_{\curlywedge} H \widehat{P}_{\curlywedge} - E,\ S \Rightarrow (l = 1,\ m = -2/3,\ u = 1)$ (if $\tilde{L} < 0,\ \widetilde{N} > 0$);
   - $f^{(1)}$, ©, ©; $\widehat{\binom{1}{2}}\ H - E,\ S \Rightarrow (l = 1,\ m = -1,\ u = 1)$ (if $\tilde{L} < 0,\ \widetilde{N} = 0$);
   - $f^{(1)}$, ©, ©; $\widehat{\binom{1}{2}}\ H_{\curlywedge} H H_{\curlywedge} - H,\ N^{\infty} \Rightarrow (l = 1,\ m = -1/3,\ u = 1)$ (if $\tilde{L} > 0$);
   - $f^{(1)}$, ©, ©; $\widehat{\binom{1}{3}}\ H_{\curlywedge} H \widehat{P}_{\curlywedge} - \widehat{P} \Rightarrow (l = 1,\ m = -1/2,\ u = 1)$ (if $\tilde{L} = 0$).

2. $\mathcal{F}_1 = 0$. Then $u = 0$ and hence $l = 0$, and this leads to the 1-parameter family of systems:

$$\dot{x} = y + x^2,\quad \dot{y} = 2mx - 2mxy. \tag{10.32}$$

This family of systems possesses for $m < 0$ a center and two complex singularities. We claim that the system (S_2) as well as the system (S_3) (see Lemma 7.4) belong to this family for some specific values of the parameter m.

Indeed, it is easy to observe that for $m = -1/2$, after the rescaling $(x, y, t) \mapsto (-x, -y, -t)$ we get exactly the system (S_2). Considering Lemmas 7.4 and 7.5, we conclude that the system (10.32) corresponding to $m = -1/2$ has a uniformly isochronous center. Considering (10.31) we deduce that the condition $m = -1/2$ is equivalent to $\tilde{L} = 0$.

Secondly we examine the case of the system (S_3). We observe that this system possesses the same two distinct real infinite singularities as systems (10.32), and these singularities are located at the points $N_1[0:1:0]$ and $N_2[1:0:0]$. Therefore in order to bring a system (10.32) to the system (S_3), we can use only a rescaling $(x, y, t) \mapsto (\alpha\, x, \beta\, y, \tau\, t)$. Since $m < 0$ we put $m = -2v^2 < 0$, and then for $\alpha = -8v/3$, $\beta = 16v^2/3$ and $\tau = 1/(2v)$, we get the systems:

$$\dot{x} = -y - 4x^2/3, \quad \dot{y} = x - 16v^2 xy/3.$$

Evidently the system (S_3) belongs to this family if and only if $v^2 = 1$, i.e. $m = -2$. Additionally for systems (10.32) we calculate

$$\mathcal{I}_2 = 3m(1-m)(2+m), \quad \tilde{L} = 8(1+2m)x^2, \quad C_2 = (1+2m)x^2y,$$
$$\widetilde{N} = -4m(1+m)x^2, \quad W_7 = 1536m^7(1+m)^4.$$

So the condition $W_7 < 0$ implies $m < 0$ and $m + 1 \neq 0$. Therefore we have $\widetilde{N} \neq 0$, and the condition $m = -2$ is equivalent to $\mathcal{I}_2 = 0$. Moreover in this case the condition $\tilde{L} = 0$ implies also $C_2 = 0$, and if $\mathcal{I}_2 = 0$ (i.e. $m = -2$) we have $\widetilde{N} < 0$. So according to Lemma 10.1 we arrive at the next five global configurations:

- c, ©, ©; $\widehat{\binom{1}{2}}\ \widehat{P}_{\curlywedge} E \widehat{P}_{\curlywedge} - H,\ S \Rightarrow (l = 0,\ m = -3/2,\ u = 0)$
  (if $\tilde{L} < 0$, $\widetilde{N} < 0$, $\mathcal{I}_2 \neq 0$);

- $c^{\odot}$, ©, ©; $\widehat{\binom{1}{2}}\ \widehat{P}_{\curlywedge} E \widehat{P}_{\curlywedge} - H,\ S \Rightarrow (l = 0,\ m = -2,\ u = 0)$
  (if $\tilde{L} < 0$, $\widetilde{N} < 0$, $\mathcal{I}_2 = 0$);

- c, ©, ©; $\widehat{\binom{1}{2}}\ \widehat{P}_{\curlywedge} H \widehat{P}_{\curlywedge} - E,\ S \Rightarrow (l = 0, m = -2/3, u = 0)$
  (if $\tilde{L} < 0, \widetilde{N} > 0$);

- c, ©, ©; $\widehat{\binom{1}{2}}\ H_{\curlywedge} H H_{\curlywedge} - H,\ N^{\infty} \Rightarrow (l = 0,\ m = -1/3,\ u = 0)$ (if $\tilde{L} > 0$);

- $c^{\oplus}$, ©, ©; $[\infty; \emptyset] \Rightarrow (l = 0,\ m = -1/2,\ u = 0)$ (if $\tilde{L} = 0$).

2.2.2: $W_7 > 0$. In this case we have $\tau_1 > 0$ and the anti-saddle is a generic node.

Remark 10.10. We observe that the condition $C_2 = 0$ (i.e. $l = 0$ and $m = -1/2$) implies $W_7 = -3(1+u^2)^6/4 < 0$.

By this remark we conclude that the condition $W_7 > 0$ implies $C_2 \neq 0$. So considering Remarks 10.8 and 10.9, we arrive at the next five global configurations:

- n, ©, ©; $\widehat{\binom{1}{2}}\, \widehat{P}_\curlywedge E \widehat{P}_\curlywedge - H,\ S \Rightarrow (l=4,\ m=-3/2,\ u=0)$ (if $\tilde{L} < 0,\ \widetilde{N} < 0$);
- n, ©, ©; $\widehat{\binom{1}{2}}\, \widehat{P}_\curlywedge H \widehat{P}_\curlywedge - E,\ S \Rightarrow (l=4,\ m=-2/3,\ u=0)$ (if $\tilde{L} < 0,\ \widetilde{N} > 0$);
- n, ©, ©; $\widehat{\binom{1}{2}}\, H - E,\ S \Rightarrow (l=4,\ m=-1,\ u=0)$ (if $\tilde{L} < 0,\ \widetilde{N} = 0$);
- n, ©, ©; $\widehat{\binom{1}{2}}\, H_\curlywedge H H_\curlywedge - H,\ N^\infty \Rightarrow (l=4,\ m=-1/3,\ u=0)$ (if $\tilde{L} > 0$);
- n, ©, ©; $\widehat{\binom{1}{3}}\, H_\curlywedge H \widehat{P}_\curlywedge - \widehat{P} \Rightarrow (l=4,\ m=-1/2,\ u=0)$ (if $\tilde{L} = 0$).

2.2.2: $W_7 = 0$. Considering (10.30) we obtain $\tau_1\tau_2\tau_3 = 0$.

2.2.2.1: $W_6 \neq 0$. Then $\tau_2\tau_3 \neq 0$ and we get $\tau_1 = 0$, i.e. the real singularity is a node with coinciding eigenvalues. Considering the corresponding linearization matrix $\mathcal{M}_1$ given by (10.29), we conclude that this node could not be a star node. Thus considering Remarks 10.8, 10.10 and 10.9, we arrive at the next five global configurations:

- n^d, ©, ©; $\widehat{\binom{1}{2}}\, \widehat{P}_\curlywedge E \widehat{P}_\curlywedge - H,\ S \Rightarrow (l=4,\ m=-2,\ u=0)$ (if $\tilde{L} < 0, \widetilde{N} < 0$);
- n^d, ©, ©; $\widehat{\binom{1}{2}}\, \widehat{P}_\curlywedge H \widehat{P}_\curlywedge - E,\ S \Rightarrow (l=5/2,\ m=-25/32,\ u=0)$
  (if $\tilde{L} < 0, \widetilde{N} > 0$);
- n^d, ©, ©; $\widehat{\binom{1}{2}}\, H - E,\ S \Rightarrow (l=2\sqrt{2},\ m=-1,\ u=0)$ (if $\tilde{L} < 0, \widetilde{N} = 0$);
- n^d, ©, ©; $\widehat{\binom{1}{2}}\, H_\curlywedge H H_\curlywedge - H,\ N^\infty \Rightarrow (l=1,\ m=-1/8,\ u=0)$ (if $\tilde{L} > 0$);
- n^d, ©, ©; $\widehat{\binom{1}{3}}\, H_\curlywedge H \widehat{P}_\curlywedge - \widehat{P} \Rightarrow (l=2,\ m=-1/2,\ u=0)$ (if $\tilde{L} = 0$).

2.2.2.2: $W_6 = 0$. By (10.30) the conditions $W_7 = W_6 = 0$ give $\tau_2 = \tau_3 = 0$, and this implies for systems (10.28)

$$\Re(\tau_2) = (l - 2mu + lu^2)^2 - 4(1+m)^2 = 0,$$
$$\Im(\tau_2) = 4\big[l(m-1)(1+u^2) - 2m(1+m)u\big] = 0.$$

Since $\widetilde{K} \neq 0$ implies $m \neq 0$, the above equations yield

$$l = \varepsilon\frac{4m^2(1+m)^2}{(m^2+1)^2}, \quad u = \varepsilon\frac{m^2-1}{2m}, \quad \varepsilon = \pm 1.$$

Then we calculate

$$W_7 = W_6 = 0, \quad W_5 = \frac{9(1+m^2)^{10}(1+4m+m^2)}{16m^4} = 9(m^2+1)^4\Delta_1^2\tau_1/4,$$
$$\mathcal{B}_1 = -4\varepsilon(m+1)(1+m^2)^4/m^2 = 8(1+m^2)\Delta_1\rho_1,$$
$$\widetilde{N} = -4m(1+m)x^2, \quad \widetilde{K} = -4mx^2, \quad \tilde{L} = 8(1+2m)x^2.$$

1. $W_5 < 0$. Then $\tau_1 < 0$ and the real singular point is a focus or a center. Clearly we have a strong focus if $\rho_1 \neq 0$ (i.e. $\mathcal{B}_1 \neq 0$) and a weak one if $\rho_1 = 0$. In the second case we have $m = -1$, and then calculations yield

$$\mathcal{B}_1 = \mathcal{H} = \mathcal{F}_1 = 0, \quad \sigma = 4x, \quad \mathcal{B}_2 = -128 < 0.$$

According to Theorem 6.2 (see statement $(e_4)[\beta]$) we have a center (in the case $m = -1$). Additionally, due to $\widetilde{K} \neq 0$, the condition $\mathcal{B}_1 = 0$ (i.e. $m = -1$) is equivalent to $\widetilde{N} = 0$. So considering Remarks 10.8, 10.10 and 10.9, we arrive at the next five global configurations:

   - f, $©^{\tau}, ©^{\tau}$; $\widehat{\binom{1}{2}}\ \widehat{P}_{\curlywedge} E \widehat{P}_{\curlywedge} - H,\ S$ $\Rightarrow (l = 16/25,\ m = -2,\ u = -3/4)$ (if $\tilde{L} < 0$, $\widetilde{N} < 0$);
   - f, $©^{\tau}, ©^{\tau}$; $\widehat{\binom{1}{2}}\ \widehat{P}_{\curlywedge} H \widehat{P}_{\curlywedge} - E,\ S$ $\Rightarrow (l = 16/169,\ m = -2/3,\ u = 5/12)$ (if $\tilde{L} < 0$, $\widetilde{N} > 0$);
   - c, $©^{\tau}, ©^{\tau}$; $\widehat{\binom{1}{2}}\ H - E,\ S$ $\Rightarrow (l = 0,\ m = -1,\ u = 0)$ (if $\tilde{L} < 0, \widetilde{N} = 0$);
   - f, $©^{\tau}, ©^{\tau}$; $\widehat{\binom{1}{2}}\ H_{\curlywedge} H H_{\curlywedge} - H,\ N^{\infty}$ $\Rightarrow (l = 4/25, m = -1/3, u = 4/3)$ (if $\tilde{L} > 0$);
   - f, $©^{\tau}, ©^{\tau}$; $\widehat{\binom{1}{3}}\ H_{\curlywedge} H \widehat{P}_{\curlywedge} - \widehat{P}$ $\Rightarrow (l = 4/25,\ m = -1/2,\ u = 3/4)$ (if $\tilde{L} = 0$).

2. $W_5 > 0$. Then $\tau_1 > 0$ and the real singular point is a generic node. Additionally the conditions $\widetilde{K} > 0$ and $\tau_1 = 1 + 4m + m^2 > 0$ imply either $m < -2 - \sqrt{3}$ or $\sqrt{3} - 2 < m < 0$. So the condition $\tilde{L} < 0$ gives $m < -1/2$, and then we have $m < -2 - \sqrt{3}$, which implies $\widetilde{N} < 0$. We also observe that in this case $\tilde{L} \neq 0$. Thus by Remark 10.8 we get the next two global configurations:

   - n, $©^{\tau}, ©^{\tau}$; $\widehat{\binom{1}{2}}\ \widehat{P}_{\curlywedge} E \widehat{P}_{\curlywedge} - H,\ S$ $\Rightarrow (l = 576/289,\ m = -4,\ u = -15/8)$ (if $\tilde{L} < 0$);
   - n, $©^{\tau}, ©^{\tau}$; $\widehat{\binom{1}{2}}\ H_{\curlywedge} H H_{\curlywedge} - H,\ N^{\infty}$ $\Rightarrow (l = 36/289,\ m = -1/4,\ u = 15/8)$ (if $\tilde{L} > 0$).

3. $W_5 = 0$. Then $\tau_1 = 0$, and considering the corresponding matrix we get a node with one direction. We obtain $m_{1,2} = -2 \pm \sqrt{3}$ $(m_2 < m_1 < 0)$, and we get $\tilde{L} < 0$ for $m = m_2$ and $\tilde{L} > 0$ for $m = m_1$. So denoting

$$l_i = 4m_i^2(1+m_i)^2/(1+m_i^2)^2, \quad u_i = (m_i^2 - 1)/(2m_i), \quad i = 1, 2,$$

we arrive at the following two global configurations:

   - n^d, $©^{\tau}, ©^{\tau}$; $\widehat{\binom{1}{2}}\ \widehat{P}_{\curlywedge} E \widehat{P}_{\curlywedge} - H,\ S$ $\Rightarrow (l = l_2,\ m = m_2,\ u = u_2)$ (if $\tilde{L} < 0$);
   - n^d, $©^{\tau}, ©^{\tau}$; $\widehat{\binom{1}{2}}\ H_{\curlywedge} H H_{\curlywedge} - H,\ N^{\infty}$ $\Rightarrow (l = l_1,\ m = m_1,\ u = u_1)$ (if $\tilde{L} > 0$).

Thus the case $\kappa = 0$ is completely examined.

In what follows we consider the family of systems (10.25) with the following special property: the discriminants corresponding to the complex points vanish. However for these systems the discriminants are polynomials that are quite large, which imply complicated computations. In order to avoid such computations, we will use a new normal form of this family of systems.

Assume that a non-degenerate system (8.1) possesses singular points $M_{1,2}(\alpha \pm i\beta, \gamma \pm i\delta)$ $(\beta^2+\delta^2 \neq 0)$. Via the substitution $x \leftrightarrow y$ we may assume $\beta \neq 0$, and then we can consider $\beta = 1$ due to the change $x \to x/\beta$. Therefore the affine transformation $\tilde{x} = -\delta x + y + \alpha\delta - \gamma$, $\tilde{y} = x - \alpha$, moves the points $M_{1,2}(\alpha \pm i, \gamma \pm i\delta)$ to the points $\widetilde{M}_{1,2}(0, \pm i)$. Then from the identities $p(0,i) = q(0,i) = 0$ we obtain: $d = f = 0$, $k = a$ and $n = b$, and introducing some new parameters we obtain the systems:

$$\dot{x} = a + cx + gx^2 + 2hxy + ay^2, \quad \dot{y} = b + ex + lx^2 + 2mxy + by^2, \tag{10.33}$$

with $a^2 + b^2 \neq 0$, having the singular points $M_{1,2}(0, \pm i)$. For these systems the conditions $\mu_0 = 0$ and $\mu_1 \neq 0$ must be fulfilled (i.e. only one real finite singular point must be at infinity).

Additionally the invariant polynomial μ_0 is the discriminant of the binary form $\widetilde{K}(\tilde{a}, x, y)$. So in this case the homogeneous quadratic polynomial $\widetilde{K}$ has the form $\widetilde{K} = (ux + vy)^2$. Moreover the linear form $ux + vy$ is a common factor of the quadratic parts of systems (8.1), and hence the fourth finite singularity has coalesced with the infinite singular point $N[-v : u : 0]$. We observe that the condition $v \neq 0$ has to be satisfied, otherwise the common factor of the quadratic parts of systems (10.33) will be x, which implies $a = b = 0$, and we get degenerate systems. So $v \neq 0$ and via the transformation $x_1 = x$ and $y_1 = ux/v + y$, which preserves the singular points $M_{1,2}(0, \pm i)$, we get $\widetilde{K}(\tilde{a}, x, y) = \delta y^2$. This means that the common factor of the homogeneous quadratic parts of systems (10.33) will be y, and therefore the applied transformation implies the conditions $g = l = 0$.

Thus we arrive at the family of systems:

$$\dot{x} = a + cx + 2hxy + ay^2, \quad \dot{y} = b + ex + 2mxy + by^2, \tag{10.34}$$

possessing three distinct finite singularities: $M_{1,2}(0, \pm i)$ and one real singularity M_3.

Lemma 10.4. *Assume that for a system (10.34) the conditions $\kappa \neq 0$ and $W_4 = W_3 = 0$ hold, i.e. $\tau_1 = \tau_2 = 0$. Then this system could possess only one of the following 15 global configurations:*

- $f,\ ©^{\tau},\ ©^{\tau};\ \overline{\binom{1}{1}}SN,\ ©,\ ©$ *(if $\kappa < 0$, $W_2 < 0$, $\mathcal{T}_4 \neq 0$, $\eta < 0$);* (Cf_1)
- $f,\ ©^{\tau},\ ©^{\tau};\ \overline{\binom{1}{1}}SN,\ S,\ N^{\infty}$ *(if $\kappa < 0$, $W_2 < 0$, $\mathcal{T}_4 \neq 0$, $\eta > 0$);* (Cf_2)
- $f,\ ©^{\tau},\ ©^{\tau};\ \overline{\binom{1}{1}}SN,\ \overline{\binom{0}{2}}SN$ *(if $\kappa < 0$, $W_2 < 0$, $\mathcal{T}_4 \neq 0$, $\eta = 0$);* (Cf_3)

- $f^{(1)}$, $©^{\tau}, ©^{\tau}; \overline{\binom{1}{1}} SN, ©, ©$ (*if* $\kappa < 0,\ W_2 < 0,\ \mathcal{T}_4 = 0,\ \eta < 0$); ($Cf_4$)
- $f^{(1)}$, $©^{\tau}, ©^{\tau}; \overline{\binom{1}{1}} SN, S, N^{\infty}$ (*if* $\kappa < 0,\ W_2 < 0,\ \mathcal{T}_4 = 0,\ \eta > 0$); ($Cf_5$)
- $f^{(1)}$, $©^{\tau}, ©^{\tau}; \overline{\binom{1}{1}} SN,\ \overline{\binom{0}{2}} SN$ (*if* $\kappa < 0,\ W_2 < 0,\ \mathcal{T}_4 = 0,\ \eta = 0$); ($Cf_6$)
- n, $©^{\tau}, ©^{\tau}; \overline{\binom{1}{1}} SN, ©, ©$ (*if* $\kappa < 0,\ W_2 > 0,\ \eta < 0$); ($Cf_7$)
- n, $©^{\tau}, ©^{\tau}; \overline{\binom{1}{1}} SN, S, N^{\infty}$ (*if* $\kappa < 0,\ W_2 > 0,\ \eta > 0$); ($Cf_8$)
- n, $©^{\tau}, ©^{\tau}; \overline{\binom{1}{1}} SN,\ \overline{\binom{0}{2}} SN$ (*if* $\kappa < 0,\ W_2 > 0,\ \eta = 0$); ($Cf_9$)
- n^d, $©^{\tau}, ©^{\tau}; \overline{\binom{1}{1}} SN, ©, ©$ (*if* $\kappa < 0,\ W_2 = 0,\ \eta < 0$); ($Cf_{10}$)
- n^d, $©^{\tau}, ©^{\tau}; \overline{\binom{1}{1}} SN, S, N^{\infty}$ (*if* $\kappa < 0,\ W_2 = 0,\ \eta > 0$); ($Cf_{11}$)
- n^d, $©^{\tau}, ©^{\tau}; \overline{\binom{1}{1}} SN,\ \overline{\binom{0}{2}} SN$ (*if* $\kappa < 0,\ W_2 = 0,\ \eta = 0$); ($Cf_{12}$)
- s, $©^{\tau}, ©^{\tau}; \overline{\binom{1}{1}} SN, N^f, N^f$ (*if* $\kappa > 0,\ \mathcal{T}_4 \neq 0$); ($Cf_{13}$)
- $s^{(1)}$, $©^{\tau}, ©^{\tau}; \overline{\binom{1}{1}} SN, N^f, N^f$ (*if* $\kappa > 0,\ \mathcal{T}_4 = 0,\ \mathcal{F}_1 \neq 0$); ($Cf_{14}$)
- $s^{(2)}$, $©^{\tau}, ©^{\tau}; \overline{\binom{1}{1}} SN, N^f, N^f$ (*if* $\kappa > 0,\ \mathcal{T}_4 = 0,\ \mathcal{F}_1 = 0$). ($Cf_{15}$)

Proof. For systems (10.34) we calculate

$$\tau_{1,2} = c^2 - 16am - 4(b-h)^2 \pm 4i(2ae - bc + ch) = 0.$$

Hence we get the conditions $c^2 - 16am - 4(b-h)^2 = 0 = 2ae - bc + ch$, and we shall consider two possibilities: $a \neq 0$ and $a = 0$.

1: $a \neq 0$. Then we may assume $a = 1$ due to a time rescaling, and we calculate $e = c(b-h)/2$, $m = \left[c^2 - 4(b-h)^2\right]/16$. So we get the family of systems:

$$\begin{aligned} \dot{x} &= 1 + cx + 2hxy + y^2, \\ \dot{y} &= b + c(b-h)x/2 + \left[c^2 - 4(b-h)^2\right]xy/8 + by^2, \end{aligned} \tag{10.35}$$

possessing the complex singular points $M_{1,2}(0, \pm i)$ and one real elemental singularity M_3. For these systems we calculate

$$\begin{aligned} \mu_0 &= 0,\ \ \mu_1 = -\frac{c\,\omega_1\omega_2}{64}\, y, \quad \widetilde{K} = \frac{\omega_1}{4}\, y^2, \quad \kappa = -\frac{\omega_1\omega_3^2}{32}, \\ G_9 &= \frac{\omega_1\omega_3^2}{4096}, \quad W_4 = W_3 = 0, \quad W_2 = \frac{c^2\omega_1^2\omega_2^2\omega_3^2}{2^{18}}\,\tau_3, \ \ \eta = -\omega_3^2(\omega_2 - 2b^2)/128 \\ \mathcal{T}_4 &= \frac{c\omega_1\omega_2\omega_3}{512}\,\rho_1\rho_2\rho_3, \quad \mathcal{T}_3 = \frac{c\omega_1\omega_2\omega_3}{512}\,(\rho_1\rho_2 + \rho_1\rho_3 + \rho_2\rho_3), \end{aligned}$$

where

$$\omega_1 = 4(b+h)^2 - c^2, \quad \omega_2 = 4(b^2 - h^2) - c^2, \quad \omega_3 = 4(b-h)^2 - c^2.$$

1.1: $\kappa < 0$. Then $\widetilde{K} > 0$ and according to Theorem 6.1 (see the Table 6.2, lines 124–133), the elemental real singular point is an anti-saddle and its type is governed by the invariant polynomial W_2, as in the case under consideration we have $G_9 \neq 0$ and $W_4 = W_3 = 0$.

1.1.1: $W_2 < 0$. In this case we obtain $\tau_3 < 0$, i.e. the real singularity of systems (10.35) is a focus, which could be either strong or weak.

1. $\mathcal{T}_4 \neq 0$. Then the focus is strong, and we arrive at the following three global configurations:

   •(Cf_1) if $\eta < 0 \Rightarrow (b = 1,\ c = 1/2,\ h = 0)$;

   •(Cf_2) if $\eta > 0 \Rightarrow (b = 1,\ c = 1/5,\ h = 4/5)$;

   •(Cf_3) if $\eta = 0 \Rightarrow (b = 1,\ c = 1/5,\ h = 7/10)$.

2. $\mathcal{T}_4 = 0$. As $\tau_1 = \tau_2 = 0$ it is clear that $\rho_1\rho_2 \neq 0$, otherwise we get degenerate systems. Therefore $\rho_3 = 0$ (i.e. focus is weak), and in this case $\mathcal{T}_3 \neq 0$ due to $\rho_2\rho_3 \neq 0$.

   Additionally according to [19] we have the next remark.

   Remark 10.11. In the class of quadratic systems with a weak focus of order two, the condition $\mu_0 = 0 = W_4$ implies the existence of exactly three distinct real finite singularities.

   So by this remark, systems (10.35) could not have a weak focus of the second order. Therefore in the case considered, we arrive at the following three global configurations:

   •(Cf_4) if $\eta < 0 \Rightarrow (b = 1,\ c = 2/\sqrt{7},\ h = 0)$;

   •(Cf_5) if $\eta > 0 \Rightarrow (b = 1,\ c = \frac{2}{5}\sqrt{(76 - \sqrt{5209})/7},\ h = 4/5)$;

   •(Cf_6) if $\eta = 0 \Rightarrow (b = 1,\ c = \sqrt{(\sqrt{41} - 5)/8},\ h = (\sqrt{41} - 1)/8)$.

1.1.2: $W_2 > 0$. Then $\tau_3 > 0$, i.e. the real singular point is a generic node. In this case we get the next three global configurations:

•(Cf_7) if $\eta < 0 \Rightarrow (b = 1,\ c = 1/5,\ h = 0)$;

•(Cf_8) if $\eta > 0 \Rightarrow (b = 1,\ c = 1/5,\ h = -4/5)$;

•(Cf_9) if $\eta = 0 \Rightarrow (b = 1,\ c = 1/5,\ h = -7/10)$.

1.1.3: $W_2 = 0$. Then $\tau_3 = 0$ and the real singularity is a node with coinciding eigenvalues. According to Theorem 7.2 a finite star node could not coexist with two complex singularities with zero discriminants.

Thus we get a node n^d, and we obtain the following three global configurations:

• (Cf_{10}) if $\eta < 0 \Rightarrow (b = 1,\ c = 2(4 - \sqrt{7})/9,\ h = 0)$;

- (Cf_{11}) if $\eta > 0 \Rightarrow (b = 1,\ c = 2(4 + \sqrt{7})/9,\ h = 0)$;
- (Cf_{12}) if $\eta = 0 \Rightarrow (b = 1, c = -(\alpha^{1/3} - 1)^2/(2\alpha^{1/3}), h = (\alpha^{4/3} + 1)/(4\alpha^{2/3})$ where $\alpha = 2 - \sqrt{3}$.

1.2: $\kappa > 0$. Then $\widetilde{K} < 0$ and according to Theorem 6.1 (see the Table 6.2, line 123), the elemental real singular point is a saddle. In this case the total index of the infinite singularities must be $+2$, and hence at infinity, apart from a saddle-node, we have two nodes (i.e. $\eta > 0$).

1.2.1: $\mathcal{T}_4 \neq 0$. Then the saddle is strong, and we get the global configuration
- $(Cf_{13}) \Rightarrow (b = 1,\ c = 2,\ h = -1)$.

1.2.2: $\mathcal{T}_4 = 0$. Then $\rho_3 = 0$ and the saddle is weak and its order is governed by the invariant polynomial $\mathcal{F}_1$ (see Theorem 6.2).

1. $\mathcal{F}_1 \neq 0$. In this case the we have a weak saddle of the first order
   - $(Cf_{14}) \Rightarrow (b = 1,\ c = \sqrt{(256 + 4\sqrt{1009})/175},\ h = -2/5)$.
2. $\mathcal{F}_1 = 0$. In this case we have a weak saddle of order two, and by Theorem 7.9 it could not be of order three. So we arrive at the global configuration
   - $(Cf_{15}) \Rightarrow \big(h = 1, (b, c) \approx (-1.9835, 2.0633) \in \{\mathcal{T}_4^{-1}(0), \mathcal{F}_1^{-1}(0)\}\big)$.

2: $a = 0$. In this case for systems (10.34) we get $\tau_{1,2} = -(2b - 2h \pm ic)^2 = 0$, which implies $c = 0$ and $h = b$. Hence we get the family of systems:

$$\dot{x} = 2bxy, \quad \dot{y} = b + ex + 2mxy + by^2,$$

for which we have $\mu_1 = -4eb^3y \neq 0$. Therefore we can consider $e = 1 = b$ due to the rescaling $(x, y, t) \mapsto (bx/e, y, t/b)$, and then we calculate

$$\kappa = -128m^2, \quad \mathcal{T}_4 = 256m^2, \quad W_2 = 256m^2(m^2 - 2), \quad \eta = 4m^2.$$

Since $\kappa \neq 0$ we get $\kappa < 0$, $\eta > 0$, $\mathcal{T}_4 \neq 0$ and $\operatorname{sign}(W_2) = \operatorname{sign}(m^2 - 2)$. So we could only have the global configurations (Cf_2), (Cf_8) and (Cf_{11}), which are already detected in the case $a \neq 0$. This completes the proof of Lemma 10.4. □

As all the cases are examined, we have constructed all 48 possible global configurations for the family of quadratic systems with exactly one real and two complex finite singularities all being elemental, and hence Theorem 10.4 is proved. □

Chapter 11

Quadratic systems with finite singularities of total multiplicity four

Consider real the quadratic systems (8.1). According to Proposition 5.1 for a quadratic system (8.1) to have finite singularities of total multiplicity four (i.e. $m_f = 4$), the condition $\mu_0 \neq 0$ must be satisfied. Therefore according to Theorem 6.4 the following lemma is valid.

Lemma 11.1. *The geometrical configurations of singularities at infinity of the family of quadratic systems possessing finite singularities of total multiplicity four (i.e. $\mu_0 \neq 0$) are classified in Diagram* 11.1 *according to the geometrical equivalence relation. Necessary and sufficient conditions for each one of the* 19 *different equivalence classes can be assembled from these diagrams in terms of nine invariant polynomials with respect to the action of the affine group and time rescaling.*

Considering Theorem 7.1 we have the following lemma.

Lemma 11.2. *Assume that a quadratic system possesses finite singularities of total multiplicity four (i.e. $\mu_0 \neq 0$). Then this system possesses exactly one star node if and only if the condition $(\mathcal{S}^1_{n^*})$ of Theorem* 7.1 *is satisfied.*

Proof. In order to prove this lemma, it is sufficient to show that the condition

$$(\mathcal{S}^2_{n^*})\ \ U_1 = U_4 = U_5 = 0, \ \ Y_2 > 0, \ \ U_6 = 0$$

of Theorem 7.1 could not be satisfied for a quadratic system with $\mu_0 \neq 0$. Indeed, by a straightforward calculation it could be detected that for a quadratic system (8.1) we have $\text{Discrim}[U_6, x] = 16\mu_0 y^2$, and clearly the condition $\mu_0 \neq 0$ implies $U_6 \neq 0$. This proves that the condition $(\mathcal{S}^2_{n^*})$ could not be satisfied. □

J. C. Artés et al., *Geometric Configurations of Singularities of Planar Polynomial Differential Systems*, https://doi.org/10.1007/978-3-030-50570-7_11

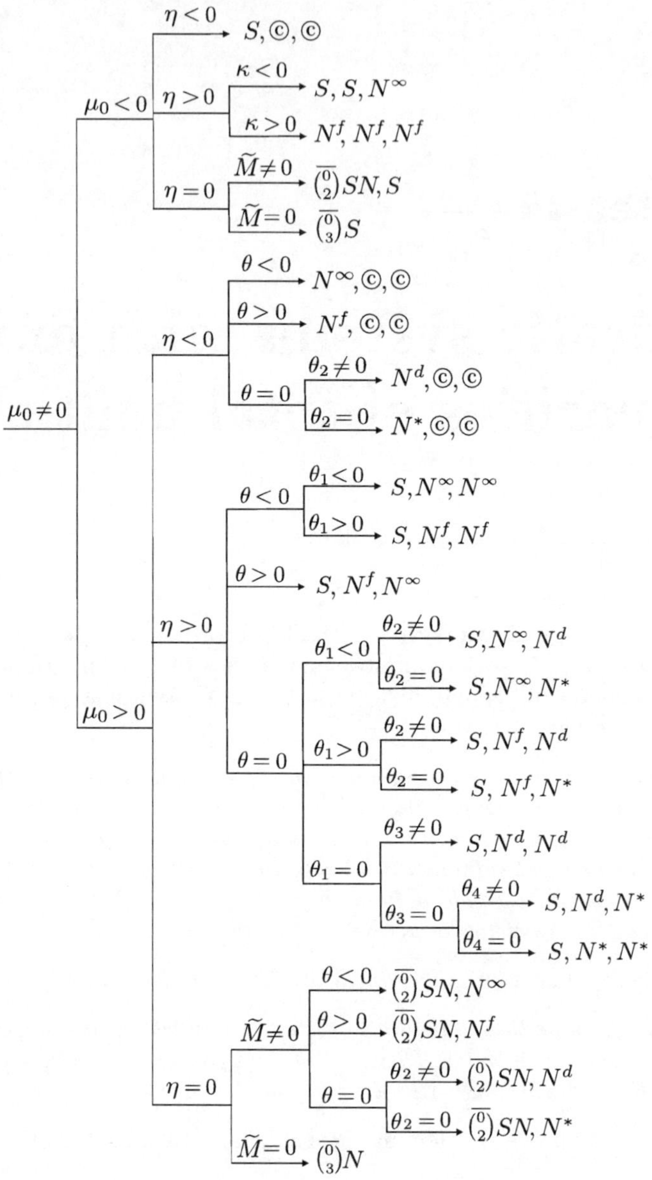

Diagram 11.1: Configurations of infinite singularities in the case $\mu_0 \neq 0$.

11.1 Exactly one singularity

Theorem 11.1. (a) *We consider here all the configurations of singularities, finite and infinite, of quadratic vector fields possessing exactly one finite singularity* *that is of multiplicity four. These configurations are classified in Diagram* 11.2 *according to the geometrical equivalence relation of configurations. We have a total of* 35 *geometrically distinct configurations of singularities, finite and infinite.*

(b) *Necessary and sufficient conditions for each one of the* 35 *different geometrical equivalence classes can be assembled from this diagram in terms of* 12 *invariant polynomials with respect to the action of the affine group and time rescaling. These invariant polynomials, which appear in Diagram* 11.2, *are given in Section* 5.4.

(c) *Diagram* 11.2 *actually contains the global bifurcation diagram in the* 12-*dimensional space of parameters of the global configurations of singularities, finite and infinite, of this family of quadratic differential systems and provides an algorithm for finding for any given system in the family considered, its respective geometrical configuration of singularities.*

First of all we observe that by Proposition 5.1 (see Table 5.1 on page 123) the quadratic systems (8.1) possess one finite singularity that is real and of multiplicity four if and only if the conditions $\mu_0 \neq 0$ and $\mathbf{D} = \mathbf{T} = \mathbf{P} = \mathbf{R} = 0$ are satisfied. Then according to Proposition 6.2 via an affine transformation and time rescaling these systems could be brought to systems of the following kinds: 9a or 9b or 9c (see Table 6.1). However we need the affine invariant conditions to distinguish these three families of systems, and therefore we begin with proving the next lemma (see also [338]).

Lemma 11.3. *Assume that the quadratic systems* (8.1) *possess one finite singularity that is real and of multiplicity four, i.e. the conditions* $\mu_0 \neq 0$ *and* $\mathbf{D} = \mathbf{T} = \mathbf{P} = \mathbf{R} = 0$ *hold. Then via an affine transformation and time rescaling, this system could be brought to the canonical form indicated in Table* 6.1 *if and only if the following corresponding conditions are satisfied, respectively:*

(9a) $\dot{x} = y^2,\ \ \dot{y} = y + x^2 + 2mxy + ny^2 \quad \Leftrightarrow \quad \widetilde{D} \neq 0,\ \mathcal{T}_4 \neq 0;$

(9b) $\dot{x} = y + x^2 + ky^2,\ \ \dot{y} = y^2 \quad \Leftrightarrow \quad \widetilde{D} \neq 0,\ \mathcal{T}_4 = 0;$

(9c) $\dot{x} = gx^2 + 2hxy,\ \ \dot{y} = lx^2 + 2mxy + ny^2 \quad \Leftrightarrow \quad \widetilde{D} = 0.$

Proof. Consider a quadratic system (8.1) possessing one real finite singularity of multiplicity four, i.e. the corresponding conditions provided by the lemma are fulfilled. Then via a translation we may locate this point at the origin of coordinates. Moreover as $M_1(0,0)$ has multiplicity greater than two, via a linear transformation this system can be brought to the canonical form

$$\dot{x} = cx + gx^2 + 2hxy + ky^2, \quad \dot{y} = ex + lx^2 + 2mxy + ny^2.$$

For these systems we calculate

$$\widetilde{D} = (ex - cy)^2(nx - ky), \quad \mu_0 = (lk - gn)^2 - 4(km - hn)(lh - gm),$$

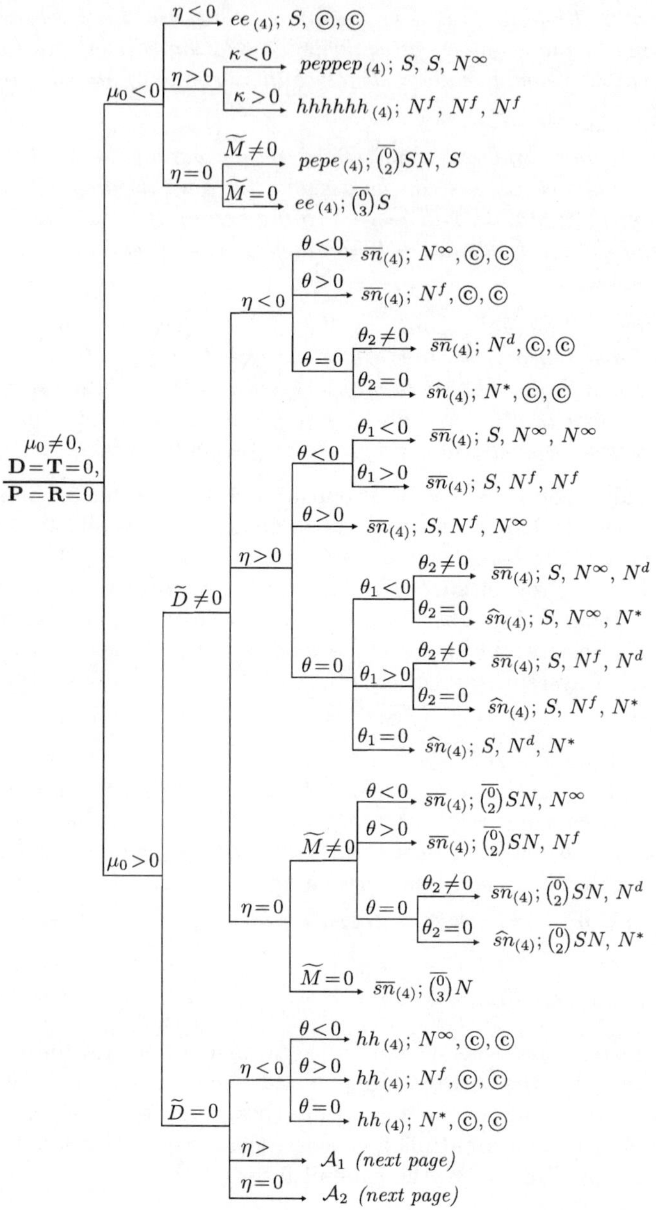

Diagram 11.2: Global configurations in the case $\mu_4=\mu_3=\mu_2=\mu_1=0,\ \mu_0\neq 0$.

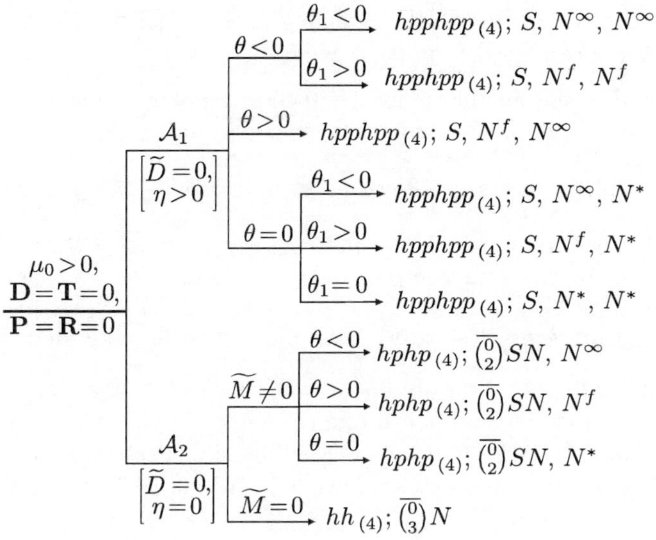

Diagram 11.2(*cont.*): Global configurations in the case $\mu_4=\mu_3=\mu_2=\mu_1=0$, $\mu_0\neq 0$.

and since $\mu_0 \neq 0$ we deduce that the condition $\widetilde{D} = 0$ is equivalent to $e = c = 0$. In other words in the case under examination, the above systems become homogeneous if and only if $\widetilde{D} = 0$. Additionally at infinity we have at least one real singularity, and hence in the case $\widetilde{D} = 0$ applying a rotation we may place this singularity at the point $N_1[0:1:0]$ (i.e. we force the condition $k = 0$), and we arrive at the canonical systems 9c.

Assume now $\widetilde{D} \neq 0$. Clearly in this case we could have either a semi-elemental (with one non-zero eigenvalue) or a nilpotent (with two zero eigenvalues) singular point.

1. In the case of a semi-elemental singular point, it is known (see, for example, [2]) that via a linear transformation and time rescaling a quadratic system can be transformed to the canonical form

$$\dot{x} = gx^2 + 2hxy + ky^2, \quad \dot{y} = y + lx^2 + 2mxy + ny^2.$$

By Lemma 5.2 the singular point $M_1(0,0)$ has multiplicity four if and only if $\mu_4 = \mu_3 = \mu_2 = \mu_1 = 0$ and $\mu_0 \neq 0$. For the above systems we calculate $\mu_4 = \mu_3 = 0$, $\mu_2 = g(gx^2 + 2hxy + ky^2)$, and hence the condition $\mu_2 = 0$ yields $g = 0$. Then we have:

$$\mu_1 = 2hl(2hx + ky) = 0, \quad \mu_0 = l(k^2l - 4hkm + 4h^2n) \neq 0,$$

which implies $h = 0$, and therefore we obtain $\mu_0 = k^2l^2 \neq 0$. Then we may assume $k = l = 1$ due to the rescaling $(x, y) \mapsto (k^{-\frac{1}{3}}l^{-\frac{2}{3}}x,\ k^{-\frac{2}{3}}l^{-\frac{1}{3}}y)$. This leads to the canonical form 9a.

2. Assume that the singular point $M_1(0,0)$ is nilpotent. Then via a linear transformation and time rescaling, these systems can be brought to the form

$$\dot{x} = y + gx^2 + 2hxy + ky^2, \quad \dot{y} = lx^2 + 2mxy + ny^2.$$

Considering Lemma 5.2 we calculate $\mu_4 = \mu_3 = 0$, $\mu_2 = l(lx^2+2mxy+ny^2)$, and therefore the condition $\mu_2 = 0$ yields $l = 0$. Then we have

$$\mu_1 = 2gm(2mx + ny) = 0, \quad \mu_0 = g(4km^2 - 4hmn + gn^2),$$

and due to $\mu_0 \neq 0$ we obtain $m = 0$. So $\mu_0 = g^2n^2 \neq 0$, and we may consider $g = n = 1$ and $h = 0$ due to the transformation $(x, y, t) \mapsto \big((nx - hy)/n^2, gy/n^2, nt/g\big)$. Thus we obtain the canonical form 9b. It remains to remark that for systems 9a we have $\mathcal{T}_4 = 1 \neq 0$, whereas for systems 9b we have $\mathcal{T}_4 = 0$.

This completes the proof of Lemma 11.3. □

Proof of Theorem 11.1. We consider the family of systems with one finite singularity of multiplicity four, and as $\mu_0 \neq 0$ we examine two cases: $\mu_0 < 0$ and $\mu_0 > 0$.

1: $\mu_0 < 0$. Since for systems 9a as well as for systems 9b we have $\mu_0 = 1 > 0$, we have to examine the canonical form 9c, i.e. generic homogeneous systems, for which the condition $\widetilde{D} = 0 = F_1$ holds. Note that $\widetilde{D} \equiv F_1$ from Table 6.2.

We observe that in the case $\mu_0 < 0$ the conditions for the classification of finite singularity of Theorem 6.1 (see Table 6.2, lines 105–109) are identical to the ones for classification of the singularities at infinity provided by Lemma 11.1 (see Diagram 11.1, case $\mu_0 < 0$). We have a total of five global configurations, and each one of them is realizable:

- $ee_{(4)}$; S, ©, © $\Rightarrow (0,0,0,0,1,0), (0,0,0,-5,2,1)$ (if $\eta < 0$);
- $ppeppe_{(4)}$; S, S, $N^\infty \Rightarrow (0,0,0,0,1,0), (0,0,0,-3,2,1)$ (if $\eta > 0$, $\kappa < 0$);
- $hhhhhh_{(4)}$; $N^f, N^f, N^f \Rightarrow (0,0,0,0,1,0), (0,0,0,1,1,-1)$ (if $\eta > 0$, $\kappa > 0$);
- $pepe_{(4)}$; $\overline{\binom{0}{2}}SN$, $S \Rightarrow (0,0,0,0,1,0), (0,0,0,-1,1,2)$ (if $\eta = 0$, $\widetilde{M} \neq 0$);
- $ee_{(4)}$; $\overline{\binom{0}{3}}S \Rightarrow (0,0,0,0,1,0), (0,0,0,-1,0,2)$ (if $\eta = 0$, $\widetilde{M} = 0$).

2: $\mu_0 > 0$. We consider two subcases: $\widetilde{D} \neq 0$ and $\widetilde{D} = 0$.

2.1: $\widetilde{D} \neq 0$. According to Lemma 11.3 the systems (8.1) possessing one finite singularity that is real and of multiplicity four could be brought via an affine transformation and time rescaling to the form either 9a (if $\mathcal{T}_4 \neq 0$) or 9b (if $\mathcal{T}_4 = 0$). We examine both these possibilities.

2.1.1: $\mathcal{T}_4 \neq 0$. In this case we consider systems 9a possessing a semi-elemental finite singular point of multiplicity four that is a saddle-node. For these systems we calculate

$$\theta = 64(1 - m^3 + mn), \quad \theta_2 = -1,$$
$$\theta_1 = 64(10m^3 + m^2n^2 - n^3 - 9 - 11mn),$$

and we claim that the condition $\theta = 0$ implies $\theta_1 \neq 0$. Indeed assuming $\theta = 0$ we get $m \neq 0$ (otherwise $\theta = 64 \neq 0$), and then $n = (m^3 - 1)/m$. Therefore we obtain $\theta_1 = 64/m^3 \neq 0$, and our claim is proved.

Additionally by Lemma 11.1 (see Diagram 11.1, case $\mu_0 > 0$) at infinity we can have 19 configurations. However considering the above claim and the condition $\theta_2 = -1 \neq 0$, we conclude that seven possibilities could not be realizable. It remains to present the following examples that prove the realizability of the remaining 12 global configurations:

- $\overline{sn}_{(4)}$; $N^\infty, ©, © \Rightarrow (0,0,0,0,0,1), (0,0,1,1,-1,3)$ (if $\eta < 0$, $\theta < 0$);
- $\overline{sn}_{(4)}$; $N^f, ©, © \Rightarrow (0,0,0,0,0,1), (0,0,1,1,-1,1)$ (if $\eta < 0$, $\theta > 0$);
- $\overline{sn}_{(4)}$; $N^d, ©, © \Rightarrow (0,0,0,0,0,1), (0,0,1,1,-1,2)$ (if $\eta < 0$, $\theta = 0$);
- $\overline{sn}_{(4)}$; S, N^∞, $N^\infty \Rightarrow (0,0,0,0,0,1), (0,0,1,1,-157/100,63/20)$ (if $\eta > 0$, $\theta < 0$, $\theta_1 < 0$);
- $\overline{sn}_{(4)}$; S, N^f, $N^f \Rightarrow (0,0,0,0,0,1), (0,0,1,1,1/2,-2)$ (if $\eta > 0$, $\theta < 0$, $\theta_1 > 0$);
- $\overline{sn}_{(4)}$; S, N^f, $N^\infty \Rightarrow (0,0,0,0,0,1), (0,0,1,1,-3,5)$ (if $\eta > 0$, $\theta < 0$, $\theta_1 > 0$);
- $\overline{sn}_{(4)}$; S, N^∞, $N^d \Rightarrow (0,0,0,0,0,1), (0,0,1,1,-2,9/2)$ (if $\eta > 0$, $\theta = 0$, $\theta_1 < 0$);
- $\overline{sn}_{(4)}$; S, N^f, $N^d \Rightarrow (0,0,0,0,0,1), (0,0,1,1,1,0)$ (if $\eta > 0$, $\theta = 0$, $\theta_1 > 0$);
- $\overline{sn}_{(4)}$; $\overline{\binom{0}{2}}SN$, $N^\infty \Rightarrow (0,0,0,0,0,1)$, $(0,0,1,1,-38/25,\xi)$ (if $\eta = 0$, $\widetilde{M} \neq 0$, $\theta < 0$) (where $\{\eta^{-1}(0)\} \ni \xi \approx 3.0433$);
- $\overline{sn}_{(4)}$; $\overline{\binom{0}{2}}SN$, $N^f \Rightarrow (0,0,0,0,0,1)$, $(0,0,1,1,0,-3/\sqrt[3]{4})$ (if $\eta = 0$, $\widetilde{M} \neq 0$, $\theta > 0$);
- $\overline{sn}_{(4)}$; $\overline{\binom{0}{2}}SN$, $N^d \Rightarrow (0,0,0,0,0,1)$, $(0,0,1,1,-\sqrt[3]{4},5/\sqrt[3]{4})$ (if $\eta = 0$, $\widetilde{M} \neq 0$, $\theta = 0$);
- $\overline{sn}_{(4)}$; $\overline{\binom{0}{3}}N \Rightarrow (0,0,0,0,0,1)$, $(0,0,1,1,-3/2,3)$ (if $\eta = 0$, $\widetilde{M} = 0$).

2.1.2: $\mathcal{T}_4 = 0$. According to Lemma 11.3 in this case we consider systems 9b possessing a nilpotent finite singular point of multiplicity four, which is a saddle-node. For these systems we calculate

$$\mu_0 = 1, \quad \eta = 1 - 4k, \quad \widetilde{M} = -8x^2 + 8xy + 8(3k - 1)y^2,$$
$$\theta = \theta_2 = \theta_3 = 0, \quad \theta_1 = -64k, \quad \theta_4 = -1.$$

So we have $\mu_0 > 0$, and by Lemma 11.1 in the case $\eta < 0$ we obtain a single configuration since $\theta = \theta_2 = 0$, which is

• $\widehat{sn}_{(4)}$; $N^*, ©, ©$: $\Rightarrow (0,0,1,1,0,1),\ (0,0,0,0,0,1)$.

Assume now $\eta > 0$, i.e. $k < 1/4$, and considering the value of the invariant polynomial θ_1 by Lemma 11.1, we arrive at the following three global configurations:

• $\widehat{sn}_{(4)}$; $S,\ N^\infty,\ N^* \Rightarrow (0,0,1,1,0,1/5),\ (0,0,0,0,0,1)$ (if $\theta_1 < 0$);

• $\widehat{sn}_{(4)}$; $S,\ N^f,\ N^* \Rightarrow (0,0,1,1,0,-1),\ (0,0,0,0,0,1)$ (if $\theta_1 > 0$);

• $\widehat{sn}_{(4)}$; $S,\ N^d,\ N^* \Rightarrow (0,0,1,1,0,0),\ (0,0,0,0,0,1)$ (if $\theta_1 = 0$).

Since $\widetilde{M} \neq 0$ in the case $\eta = 0$ (i.e. $k = 1/4$) by Lemma 11.1 due to $\theta = \theta_2 = 0$, we arrive at the global configuration

• $\widehat{sn}_{(4)}$; $\binom{0}{2}SN,\ N^* \Rightarrow (0,0,1,1,0,1/4),\ (0,0,0,0,0,1)$.

We observe that in the case $\tilde{D} \neq 0$ (i.e. when the systems are non-homogeneous) for systems 9a we have $\mathcal{T}_4 = 1 \neq 0$ and $\theta_2 = -1 \neq 0$, whereas for systems 9b we have $\mathcal{T}_4 = \theta_2 = 0$. So we conclude that the condition $\mathcal{T}_4 \neq 0$ could be substituted by the condition $\theta_2 \neq 0$. Hence the global configurations are governed only by the invariant polynomials μ_0, η, $\widetilde{M}$, θ and θ_i as they are presented in Diagram 11.2 for the case $\mu_0 > 0$ and $\tilde{D} \neq 0$.

2.2: $\widetilde{D} = 0$. According to Lemma 11.3 the systems (8.1) with a finite singularity of multiplicity four become homogeneous, and we consider canonical form 9c. For these systems we calculate

$$
\begin{aligned}
\mu_0 &= g^2n^2 - 4hn(gm - hl),\\
\kappa &= 16\big[8lh^3 - g^2n^2 + 4hn(3gm - 2m^2 + 2ln) - 4h^2(2gm + 5ln)\big],\\
\eta &= (2h-n)^2(g^2 + 8lh - 4gm + 4m^2 - 4ln),\\
\widetilde{M} &= -8(g^2 + 6lh - 4gm + 4m^2 - 3ln)x^2 - 8(g-2m)(2h-n)xy\\
&\quad - 8(2h-n)^2y^2,\\
\theta &= -64h\big[l(h-n)^2 - mg(h-n) - m^2n\big], \quad \theta_2 = \theta_3 = \theta_4 = 0,\\
\theta_1 &= 64\big[(h-n)^2(g^2 + 10lh - 6gm + 4m^2 + 3ln)\\
&\quad + (h-n)(g^2n - 5gmn + 2m^2n + 4ln^2) - m^2n^2\big].
\end{aligned}
\tag{11.1}
$$

1. $\eta < 0$. By Theorem 6.1 (see Table 6.2, line 104) the finite singularity will be of the type $hh_{(4)}$. Since $\theta_2 = 0$ considering Lemma 11.1, in this case we obtain the following three global configurations:

    • $hh_{(4)}$; $N^\infty, ©, © \Rightarrow (0,0,0,1,1,0),\ (0,0,0,118/55,-1,31/10)$ (if $\theta < 0$);

    • $hh_{(4)}$; $N^f, ©, © \Rightarrow (0,0,0,1/10,-1/10,0),\ (0,0,0,67/2200,-1/10,9/10)$ (if $\theta > 0$);

    • $hh_{(4)}$; $N^*, ©, © \Rightarrow (0,0,0,1,0,0),\ (0,0,0,2,0,1)$ (if $\theta = 0$).

2. $\eta > 0$. According to Theorem 6.1 (see Table 6.2, line 101), the finite singularity is of the type $phpphp_{(4)}$. As we can see from (11.1), the conditions $\theta_2 = \theta_3 = \theta_4 = 0$ are satisfied. Therefore by Lemma 11.1 in the case $\eta > 0$, at infinity we could have six configurations at infinity, and each one of them is realizable:

   - $phpphp_{(4)}$; S, N^∞, $N^\infty \Rightarrow (0,0,0,1/10,-2,0)$, $(0,0,0,-43/440,1/10,-29/10)$ (if $\theta < 0$, $\theta_1 < 0$);

   - $phpphp_{(4)}$; S, N^f, $N^f \Rightarrow (0,0,0,1/10,-1/5,0)$, $(0,0,0,317/440,1,7/10)$ (if $\theta < 0$, $\theta_1 > 0$);

   - $phpphp_{(4)}$; S, N^f, $N^\infty \Rightarrow (0,0,0,1/10,1/5,0)$, $(0,0,0,317/440,1,3/2)$ (if $\theta > 0$);

   - $phpphp_{(4)}$; S, N^∞, $N^* \Rightarrow (0,0,0,1,1,0)$, $(0,0,0,0,0,3)$ (if $\theta = 0$, $\theta_1 < 0$);

   - $phpphp_{(4)}$; S, N^f, $N^* \Rightarrow (0,0,0,1,2,0)$, $(0,0,0,0,0,1)$ (if $\theta = 0$, $\theta_1 > 0$);

   - $phpphp_{(4)}$; S, N^*, $N^* \Rightarrow (0,0,0,1,0,0)$, $(0,0,0,0,0,1)$ (if $\theta = \theta_1 = 0$).

3. $\eta = 0$. According to Theorem 6.1 (see Table 6.2, lines 102 and 103), the finite singularity is of the type $hphp_{(4)}$ if $\widetilde{M} \neq 0$ and of the type $hh_{(4)}$ if $\widetilde{M} = 0$.

   Additionally by Lemma 11.1 due to $\theta_2 = 0$, at infinity we could have three configurations in the case $\widetilde{M} \neq 0$ and one in the case $\widetilde{M} = 0$. So we arrive at the following four global configurations:

   - $hphp_{(4)}$; $\overline{\binom{0}{2}}SN$, $N^\infty \Rightarrow (0,0,0,1,1/10,0)$, $(0,0,0,5/16,0,1)$ (if $\widetilde{M} \neq 0$, $\theta < 0$);

   - $hphp_{(4)}$; $\overline{\binom{0}{2}}SN$, $N^f \Rightarrow (0,0,0,1,-1/10,0)$, $(0,0,0,5/24,0,1)$ (if $\widetilde{M} \neq 0, \theta > 0$);

   - $hphp_{(4)}$; $\overline{\binom{0}{2}}SN$, $N^* \Rightarrow (0,0,0,-1,0,0)$, $(0,0,0,9/8,1,2)$ (if $\widetilde{M} \neq 0$, $\theta = 0$);

   - $hh_{(4)}$; $\overline{\binom{0}{3}}N \Rightarrow (0,0,0,0,1,0)$, $(0,0,0,1,0,2)$ (if $\widetilde{M} = 0$).

As all the cases are examined, we have constructed all 35 possible global configurations for the family of quadratic systems with one real finite singularity that is of multiplicity four. □

11.2 Exactly two distinct singularities

We consider here the three subclasses of quadratic differential systems with $m_f = 4$ possessing exactly two finite singularities, namely

- systems with one triple and one simple real singularities ($\mu_0 \neq 0$, $\mathbf{D} = \mathbf{T} = \mathbf{P} = 0$, $\mathbf{R} \neq 0$),
- systems with two double real singularities ($\mu_0 \neq 0$, $\mathbf{D} = \mathbf{T} = 0$, $\mathbf{PR} > 0$),
- systems with two double complex singularities ($\mu_0 \neq 0$, $\mathbf{D} = \mathbf{T} = 0$, $\mathbf{PR} < 0$).

11.2.1 One triple and one simple real singularities

Theorem 11.2. (a) *We consider here all the configurations of singularities, finite and infinite, of quadratic vector fields with finite singularities of total multiplicity $m_f = 4$ possessing exactly two distinct real finite singularities: one elemental and one of multiplicity three. These configurations are classified in Diagram* 11.3 *according to the geometrical equivalence relation of configurations. We have a total of* 130 *geometrically distinct configurations of singularities, finite and infinite.*

(b) *Necessary and sufficient conditions for each one of the* 130 *different geometrical equivalence classes can be assembled from this diagram in terms of* 20 *invariant polynomials with respect to the action of the affine group and time rescaling. These invariant polynomials, which appear in Diagram* 11.3, *are given in Section* 5.4.

(c) *The Diagram* 11.3 *actually contains the global bifurcation diagram in the* 12*-dimensional space of parameters of the global geometrical configurations of singularities, finite and infinite, of this subclass of quadratic differential systems and provides an algorithm for finding for any given system in this family, its respective geometrical configuration of singularities.*

Proof. Assume that systems (8.1) possess one triple and one simple real finite singularities. According to Proposition 6.2 by applying an affine transformation, these systems could be brought to the canonical form 8 (see Table 6.1). This canonical form is

$$\begin{aligned} \dot{x} &= cx + cuy - cx^2 + 2hxy + (2hu + cv)y^2, \\ \dot{y} &= ex + euy - ex^2 + 2mxy + (2mu + ev)y^2, \end{aligned} \tag{11.2}$$

and it has a triple singularity at the origin. This form has the invariants

$$\begin{aligned} \mu_0 = 4e^2(h+mu)^2(u^2-v) \neq 0, \ \mathcal{T}_4 = W_4 = 0, \\ Y_1 = e(h+mu)(h+mu+eu^2-ev), \ E_3 = 2e^3(h+mu)^3(u^2-v). \end{aligned} \tag{11.3}$$

First we prove the next lemma.

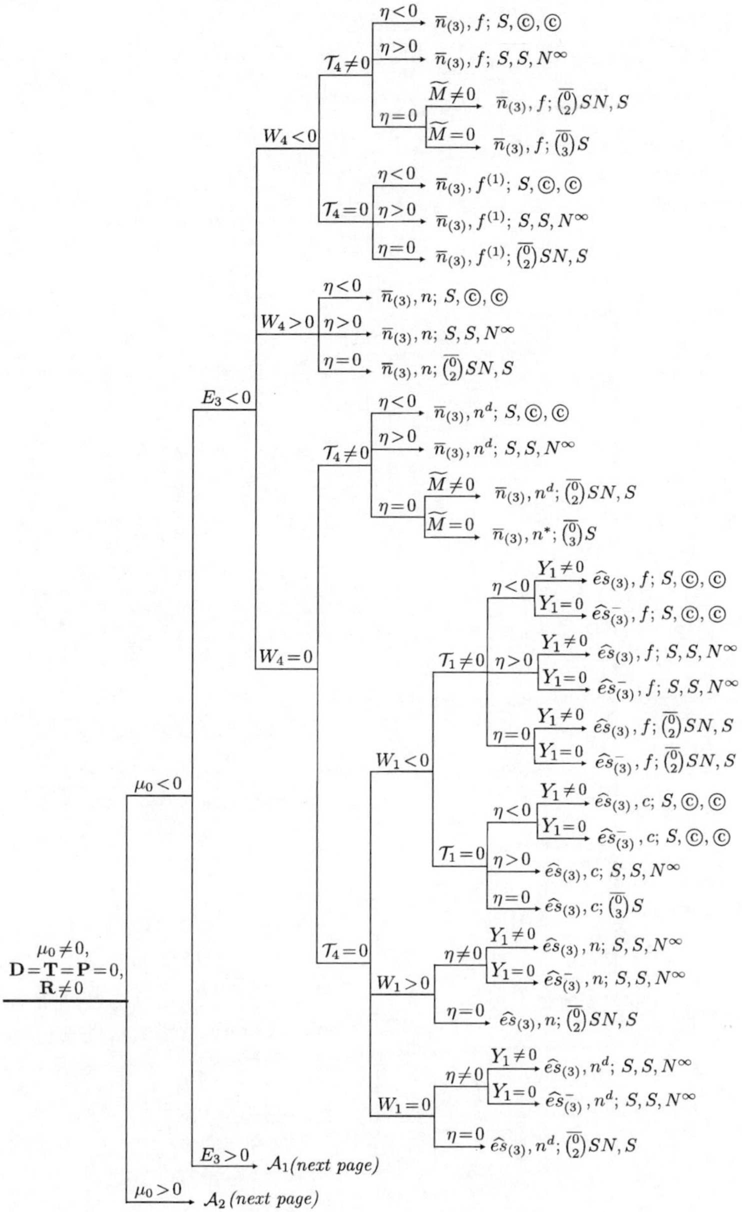

Diagram 11.3: Global configurations in the case $\mu_0 \neq 0$, $\mathbf{D} = \mathbf{T} = \mathbf{P} = 0$, $\mathbf{R} \neq 0$.

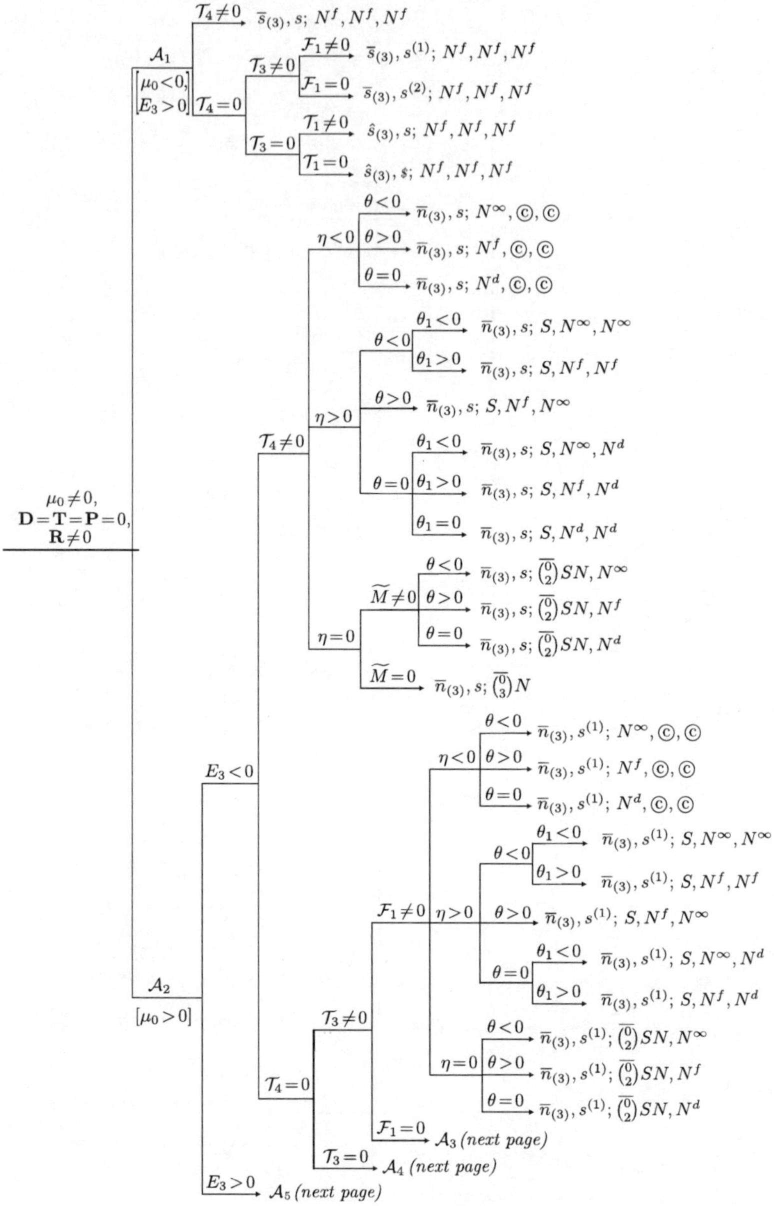

Diagram 11.3 (*cont.*): Global configurations in the case $\mu_0\neq0$, $\mathbf{D}=\mathbf{T}=\mathbf{P}=0$, $\mathbf{R}\neq0$.

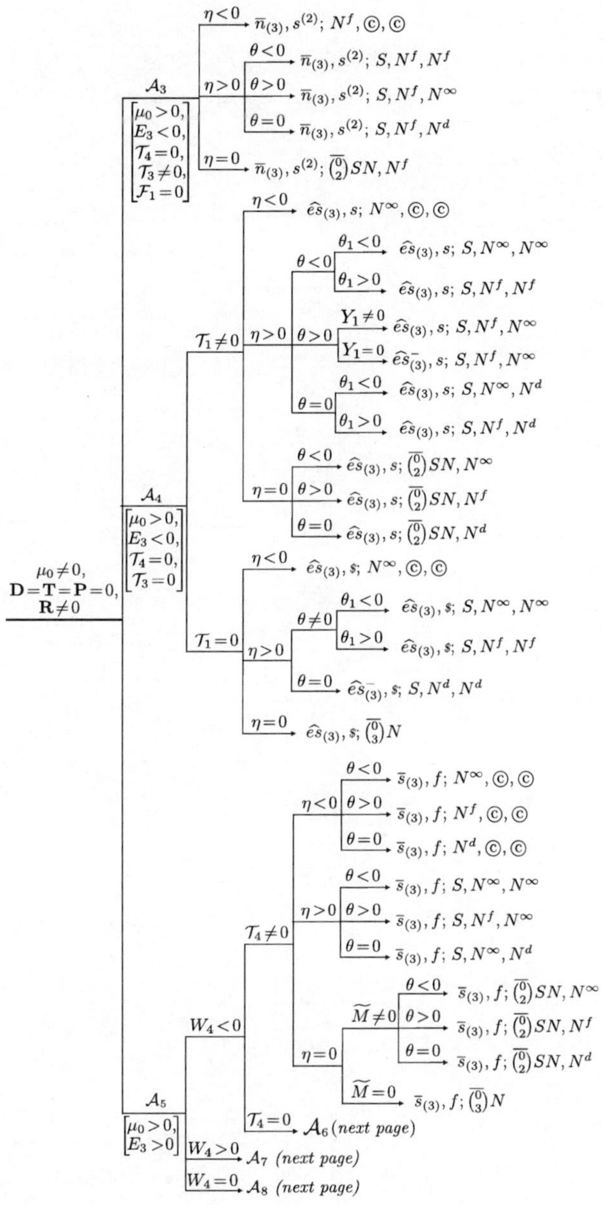

Diagram 11.3 (*cont.*): Global configurations in the case $\mu_0\neq 0$, $\mathbf{D}=\mathbf{T}=\mathbf{P}=0$, $\mathbf{R}\neq 0$.

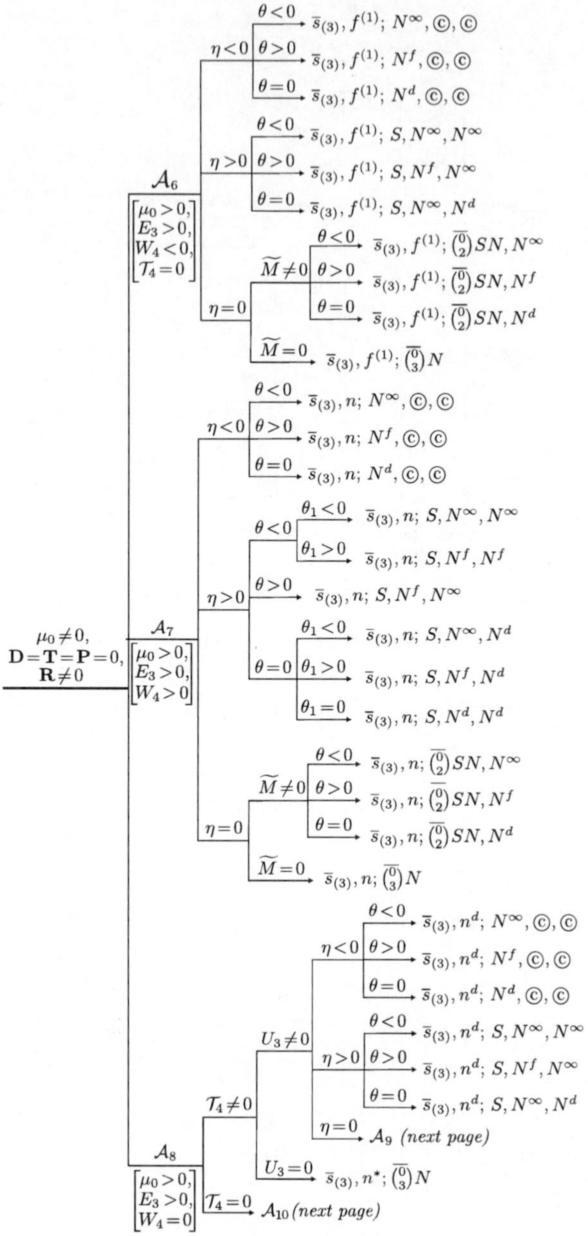

Diagram 11.3 (*cont.*): Global configurations in the case $\mu_0 \neq 0$, $\mathbf{D}=\mathbf{T}=\mathbf{P}=0$, $\mathbf{R}\neq 0$.

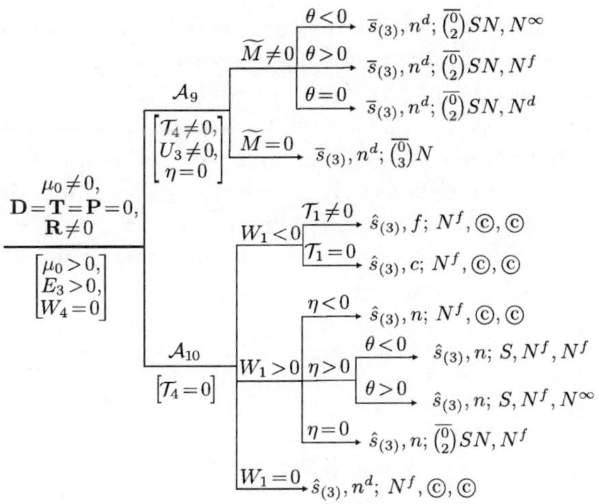

Diagram 11.3 (*cont.*): Global configurations in the case $\mu_0 \neq 0$, $\mathbf{D}=\mathbf{T}=\mathbf{P}=0$, $\mathbf{R}\neq 0$.

Lemma 11.4. *If a system of the family* (11.2) *has an elliptic-saddle, then it is generic if* $Y_1 \neq 0$. *Otherwise it is non-generic.*

Proof. We first force that a system of the family (11.2) has a nilpotent singularity at the origin by making its trace zero, that is, $c = -eu$. Since this system also has a singularity at $(1,0)$, the line $y = 0$ cannot be a characteristic direction at $(0,0)$. Indeed, following the blow-up technique described in Section 3.4, the polynomial of characteristic directions is $PCD(x,y) = e(x+uy)^2$ and $e \neq 0$ for the conditions of canonical form 8. We make a change $(x,y) \to (y,x)$ to be sure that we can make a vertical directional blow-up, and we do it. After this we have an intricate singularity at $(0,-1/u)$, we move it to the origin and since the vertical axis is invariant, we do a 1-twist followed by a second vertical blow-up. Then we get three singularities on the axis $x = 0$, which are $(0,0)$, $(0,1)$ and $(0,(h+mu+eu^2-ev)/(h+mu+eu^2+eu^3-ev))$ (in the case $h+mu+eu^2+eu^3-ev = 0$ we just make a 2-twist instead of a 1-twist, and we arrive at the same result). If the three singularities are different, the original elliptic-saddle is generic and has four borsecs. Then, the condition to obtain a non-generic elliptic-saddle (with just two borsecs) is that the third singularity of the blow-up coincides with $(0,0)$ (clearly it cannot coincide with $(0,1)$). And the expression $h+mu+eu^2+eu^3-ev$ is governed by the invariant Y_1 from (11.3) since the other factors of Y_1 cannot be zero due to the condition $E_3 < 0$, which is needed to have an elliptic-saddle. □

We observe that systems (11.2) depend on six parameters, and it is more convenient to split this canonical form into four different systems each one depending on fewer parameters.

Consider systems (8.1) possessing one triple and one simple real finite singularities. Then via an affine transformation we may assume that these systems possess as singularities the points $M_1(0,0)$ and $M_2(1,0)$ and the singular point M_0 is triple. Moreover, as in this case we have $C_2 \neq 0$ (otherwise we get $\mu_0 = 0$), we may consider that there exists at least one isolated real infinite singularity. We shall consider two possibilities: (a) there exists a real infinite singular point that does not coincide with $N[1:0:0]$ (i.e. the end of the axis $y = 0$), and (b) there exists a unique real infinite singularity that is located at the point $N[1:0:0]$.

In the case (b) we get the systems:

$$\dot{x} = cx + dy - cx^2 + 2hxy + ky^2, \quad \dot{y} = fy + 2mxy + ny^2, \tag{11.4}$$

for which we calculate

$$\mu_0 = c(cn^2 - 4km^2 + 4hmn), \quad \Delta_1 = cf.$$

Since we have to force the singular point $M_1(0,0)$ to be triple, we get $\Delta_1 = 0$ (i.e. the corresponding determinant vanishes). Since $\mu_0 \neq 0$ (i.e. $c \neq 0$) we have $f = 0$, and then we calculate

$$\mu_4 = \mu_3 = 0, \quad \mu_2 = c(cn - 2dm)y(2mx + ny).$$

So according to Lemma 5.2 the singular point $M_1(0,0)$ of systems (11.4) is of multiplicity three if and only if $\mu_2 = 0$, which is equivalent to $c(cn - 2dm) = 0$. As $c \neq 0$ we may assume $c = 1$ due to a time rescaling, and then we obtain $n = 2dm$. Thus we arrive at the family of systems:

$$\dot{x} = x + dy - x^2 + 2hxy + ky^2, \quad \dot{y} = 2my(x + dy), \quad d \in \{0,1\}, \tag{11.5}$$

since in the case $d \neq 0$ we apply the rescaling $y \to y/d$.

Remark 11.1. We remark that since we have assumed that the singularity $N[1:0:0]$ is the unique real infinite singularity of this family of systems, then the condition either $\eta < 0$ or $\widetilde{M} = 0$ must hold for these systems.

We shall construct the normal form of systems in the case (a), i.e. when there exists a real infinite singular point that does not coincide with $N[1:0:0]$. Then via a linear transformation (which keeps the singularities $M_1(0,0)$ and $M_2(1,0)$), we may assume that this real point is located at the point $N_1[0:1:0]$ (i.e. on the end of the axis $x = 0$), and we obtain the systems:

$$\dot{x} = cx + dy - cx^2 + 2hxy, \quad \dot{y} = ex + fy - ex^2 + 2mxy + ny^2.$$

For the determinant corresponding to the point $M_1(0,0)$, which we want to be triple, we have $\Delta_1 = cf - de = 0$, and we shall consider two cases: $e \neq 0$ and $e = 0$.

1. $e \neq 0$. Then we assume $e = 1$ (due to a time rescaling), and therefore the condition $\Delta_1 = 0$ gives $d = cf$. Herein we calculate

$$\mu_4 = \mu_3 = 0, \quad \mu_2 = (2fh - 2cfm + cn)y(2cmx - 2hx + cny),$$

and we split the examination into two subcases: $f \neq 0$ and $f = 0$.

   (a) $f \neq 0$. In this case via the rescaling $(x, y, t) \mapsto (x, y/f, t/f)$ we may assume $f = 1$, and then the condition $\mu_2 = 0$ is equivalent to $h = c(2m - n)/2$. So we arrive at the family of systems:

$$\begin{aligned} \dot{x} &= cx + cy - cx^2 + c(2m - n)xy, \\ \dot{y} &= x + y - x^2 + 2mxy + ny^2. \end{aligned} \tag{11.6}$$

   (b) $f = 0$. Then the condition $\mu_2 = 0$ implies $cn = 0$, and as $\mu_0 = n(4chm + c^2n - 4h^2) \neq 0$ (i.e. $n \neq 0$), we get $c = 0$. This leads to the family of systems:

$$\dot{x} = 2hxy, \quad \dot{y} = x - x^2 + 2mxy + ny^2, \quad m \in \{0, 1\} \tag{11.7}$$

   since in the case $m \neq 0$ we apply the rescaling $(x, y, t) \mapsto (x, y/m, t/m)$.

2. $e = 0$. Then the condition $\Delta_1 = 0$ implies $cf = 0$, and as $\mu_0 = cn(4hm+cn) \neq 0$ (i.e. $c \neq 0$), we get $f = 0$. In this case we calculate

$$\mu_4 = \mu_3 = 0, \quad \mu_2 = c(cn - 2dm)y(2mx + ny), \quad \mu_0 = cn(4hm + cn) \neq 0.$$

As $c \neq 0$ we may assume $c = 1$ due to a time rescaling, and then the condition $\mu_2 = 0$ gives $n = 2dm \neq 0$. Thus we may assume $d = 1$ due to the rescaling $y \to y/d$. So we arrive at the 2-parameter family of systems:

$$\dot{x} = x + y - x^2 + 2hxy, \quad \dot{y} = 2my(x + y). \tag{11.8}$$

As a result we constructed four canonical forms of systems possessing one triple and one elemental real finite singularities: (11.6), (11.7), (11.8) and (11.5). In what follows we examine each one of these families of systems.

1: The family of systems (11.6). This family of systems possesses the triple singular point $M_{1,2,3}(0, 0)$ and the elemental one $M_4(1, 0)$, for which we have the following values for the traces ρ_i, for the determinants Δ_i, for the discriminants τ_i and for the linearization matrices $\mathcal{M}_{1,2,3}$ and $\mathcal{M}_4$:

$$\begin{aligned} &\mathcal{M}_1 = \begin{pmatrix} c & c \\ 1 & 1 \end{pmatrix}, \ \mathcal{M}_4 = \begin{pmatrix} -c & c(1 + 2m - n) \\ -1 & 1 + 2m \end{pmatrix}, \ \rho_1 = 1 + c, \ \Delta_1 = 0; \\ &\rho_4 = 1 - c + 2m, \ \Delta_4 = -cn; \ \tau_i = \rho_i^2 - 4\Delta_i, \ i = 1, 4. \end{aligned} \tag{11.9}$$

For this family we calculate

$$\mu_0 = \alpha\Delta_4^2,\ \eta = (2cm - n - cn)^2\big[(c - 2m)^2 + 4n(c+1)\big],$$
$$E_3 = \alpha\Delta_4^3/4,\ G_{10} = -\alpha\rho_1^3\Delta_4^3/8,\ \mathcal{F}_1 = 3\alpha\rho_1\Delta_4,$$
$$\mathcal{T}_1 = \alpha\Delta_4^2(3\rho_1 + \rho_4),\ \mathcal{T}_2 = 3\alpha\Delta_4^2\rho_1(\rho_1 + \rho_4),\ \mathcal{T}_3 = \alpha\Delta_4^2\rho_1^2(\rho_1 + 3\rho_4), \tag{11.10}$$
$$\widetilde{M} = -8(c^2 - 2cm + 4m^2 + 3n\rho_1)x^2 + 8(c+2m)(2cm - n\rho_1)xy$$
$$-8(2cm - n\rho_1)^2y^2,\ \mathcal{T}_4 = \alpha\Delta_4^2\rho_1^3\rho_4,\ W_4 = \alpha^2\rho_1^6\Delta_4^4\tau_4,$$

where $\alpha = 1 + 2m - n$.

Lemma 11.5. *For systems (11.6) the following are true:*

(i) *If the condition $\widetilde{M} = 0$ holds, then $\rho_1 \neq 0$ and*

(i_1) *the condition $\mu_0 < 0$ implies $E_3 < 0$, $W_4 < 0$ and $\mathcal{T}_4 \neq 0$;*

(i_2) *the condition $\mathcal{T}_4 = 0$ implies $E_3 > 0$.*

(ii) *If the conditions $\eta < 0$ and $\rho_1 \neq 0$ are satisfied, then* sign (μ_0) = sign $(c+1)$ *and* sign (E_3) = sign (c).

(iii) *If the conditions $\theta = \theta_2 = 0$ hold, then for these systems we have $\mu_0 = \eta$. Moreover in this case the condition $\mu_0 > 0$ implies $W_4 \geq 0$.*

(iv) *If the condition $\theta = 0$ is fulfilled, then the condition $\theta_1 = 0$ is equivalent to $\theta_2 = 0$. Moreover if $\theta_1 = 0$ then $\theta_3 \neq 0$.*

Proof. (i) First we observe that in the case $\widetilde{M} = 0$ the condition $\rho_1 = c + 1 \neq 0$ must hold. Indeed, admitting $c = -1$ we have Coefficient$[\widetilde{M}, y^2] = -32m^2 = 0$ and then $\widetilde{M} = -8x^2 \neq 0$. So $c + 1 \neq 0$ and considering (11.10) the condition $\widetilde{M} = 0$ yields $m = -c/2$ and $n = -c^2/(c+1)$. Then we calculate

$$\mu_0 = \frac{c^6}{(c+1)^3},\quad E_3 = \frac{c^9}{4(c+1)^4},\quad W_4 = \frac{c^{12}(1-3c)}{c+1},\quad \mathcal{T}_4 = c^6(1-2c),$$

and hence the condition $\mu_0 < 0$ gives $c < -1$, which obviously implies the validity of the statement (i_1).

Additionally we observe that the condition $\mathcal{T}_4 = 0$ implies $c = 1/2$ and then $E_3 > 0$. This completes the proof of the statement (i).

(ii) Assume $\eta < 0$. Then considering (11.10) we conclude that the second factor of η is negative, and due to the condition $\rho_1 = c + 1 \neq 0$, we may assume

$$(c - 2m)^2 + 4n(c+1) = -v^2 \ \Rightarrow\ n = -\big[(c-2m)^2 + v^2\big]/[4(1+c)].$$

Therefore we calculate

$$\mu_0 = \frac{c^2\big[(c-2m)^2 + v^2\big]^2\big[(c+2m+2)^2 + v^2\big]}{64(1+c)^3},$$
$$E_3 = \frac{c^3\big[(c-2m)^2 + v^2\big]^3\big[(c+2m+2)^2 + v^2\big]}{1024(1+c)^4},$$

and obviously the statement (ii) of the lemma is proved.

(iii) Assume now that for systems (11.6) the conditions $\theta = \theta_2 = 0$ are satisfied. For these systems we have

$$\theta = -8c(2m-n)n(4cm+2c^2m-4m^2-4n-4cn-c^2n),$$
$$\theta_2 = -cn(4+2c+6m+2cm-2n-cn)/4,$$

and as $\mu_0 \neq 0$ (i.e. $cn \neq 0$), we get the relations

$$(2m-n)\big[2m(2c+c^2-2m)-n(c+2)^2\big] = 0 = (2+c)(2+2m-n)+2m.$$

We consider two subcases: $c+2 \neq 0$ and $c+2=0$.

(a) Assume first $c+2 \neq 0$. Then from the second equation we get $n = 2(2+c+3m+cm)/(2+c)$, and we have

$$\theta_2 = 0, \quad \theta = -\frac{64c(2+c+m)^2(2+c+2m)(2+c+3m+cm)}{(2+c)^2},$$
$$\mu_0 = -\frac{4c^2(2+c+2m)(2+c+3m+cm)^2}{(2+c)^3}.$$

As $\mu_0 \neq 0$ the condition $\theta = 0$ gives $m = -2-c$. Therefore we obtain

$$\theta_2 = \theta = 0, \quad \mu_0 = 4c^2(2+c)^2 = \eta, \quad W_4 = 16c^4(1+c)^6(2+c)^4(9+2c+c^2),$$

and we observe that $W_4 \geq 0$.

(b) Suppose now $c = -2$. Then we calculate

$$\theta = -64m^2(2m-n)n, \quad \theta_2 = mn, \quad \mu_0 = 4(1+2m-n)n^2,$$

and as $\mu_0 \neq 0$ (i.e. $n \neq 0$) the condition $\theta_2 = 0$ yields $m = 0$. Then we obtain:

$$\theta_2 = \theta = 0, \quad \mu_0 = 4(1-n)n^2 = \eta, \quad W_4 = 16(1-n)^2n^4(9-8n),$$

and it is clear that the condition $\mu_0 > 0$ implies $W_4 > 0$. This completes the proof of the statement (iii) of the lemma.

(iv) Assume $\theta = 0$. As $\mu_0 \neq 0$ the condition $\theta = 0$ implies one of the following three relations: (a) $n = 2m$, (b) $c+2 \neq 0$ and $n = 2\big[cm(c+2)-2m^2\big]/(2+c)^2$ or (c) $c = -2$ and $m = 0$.

(a) We calculate

$$\mu_0 = 4c^2m^2, \quad \theta_1 = 256m^3(2+c+m), \quad \theta_2 = -cm(2+c+m),$$

and clearly due to $\mu_0 \neq 0$ the condition $\theta_1 = 0$ is equivalent to $\theta_2 = 0$, which yields $m = -(2+c)$. However in this case we get $\theta_3 = -2c^3(2+c)^3 \neq 0$ since $\mu_0 = 4c^2(2+c)^2 \neq 0$.

(b) We have

$$\mu_0 = \frac{4c^2(2c+c^2-2m)^2m^2(2+c+2m)^2}{(2+c)^6},$$
$$\theta_1 = \frac{256c^3m^3(2+c+m)(2+c+2m)^2}{(2+c)^5},$$
$$\theta_2 = -\frac{c(2c+c^2-2m)m(2+c+m)(2+c+2m)}{(2+c)^3},$$

and since $\mu_0 \neq 0$ we obtain that the condition $\theta_1 = 0$ is equivalent to $\theta_2 = 0$, which again implies $m = -(2+c)$. Then we calculate $\theta_3 = -2c^3(2+c)^3 \neq 0$ because $\mu_0 = 4c^2(2+c)^2 \neq 0$.

(c) In the last case (i.e. when $c = -2$ and $m = 0$), we obtain

$$\theta_1 = \theta_2 = 0, \quad \theta_3 = 2(n-1)n^3, \quad \mu_0 = -4(n-1)n^2 \neq 0.$$

This completes the proof of the lemma. □

Now we consider the two cases $\mu_0 < 0$ and $\mu_0 > 0$.

1.1: $\mu_0 < 0$. Then $\alpha < 0$ and by (11.10) this implies $\text{sign}(E_3) = -\text{sign}(\Delta_4)$.

1.1.1: $E_3 < 0$. In this case the elemental singularity M_4 is an anti-saddle (i.e. a node, a focus or a center). By Theorem 6.1 the type of this point is governed by the invariant polynomials W_4 or W_1.

1.1.1.1: $W_4 < 0$. Then from (11.10) it follows that $G_{10} \neq 0$, and according to Theorem 6.1 (see Table 6.2, line 96) we have a focus and a semi-elemental triple node. And by (11.10) the focus could be weak only if $\mathcal{T}_4 = 0$.

1. $\mathcal{T}_4 \neq 0$. In this case we have a strong focus, and considering Lemma 11.1 and the condition $C_2 \neq 0$, we arrive at the following four global configurations:
   - $\overline{n}_{(3)}, f;\ S, ©, © \Rightarrow (c=-2,\ m=-1,\ n=1)$ (if $\eta < 0$);
   - $\overline{n}_{(3)}, f;\ S, S, N^\infty \Rightarrow (c=-6/5,\ m=-1/10,\ n=1)$ (if $\eta > 0$);
   - $\overline{n}_{(3)}, f;\ \overline{\binom{0}{2}}SN, S \Rightarrow (c=-5/4, m=-1/8, n=1)$ (if $\eta = 0,\ \widetilde{M} \neq 0$);
   - $\overline{n}_{(3)}, f;\ \overline{\binom{0}{3}}S \Rightarrow (c=-2,\ m=1,\ n=4)$ (if $\eta = 0,\ \widetilde{M} = 0$).
2. $\mathcal{T}_4 = 0$. Since by (11.10) the condition $W_4 \neq 0$ implies $\mathcal{F}_1 \neq 0$ and $\rho_1 \neq 0$, we obtain $\rho_4 = 0$, i.e. we have a first order weak focus. According to Lemma 11.5 in this case the condition $\widetilde{M} \neq 0$ must hold. So considering Lemma 11.1 we get the following three global configurations:
   - $\overline{n}_{(3)}, f^{(1)};\ S, ©, © \Rightarrow (c=-3,\ m=-2,\ n=2)$ (if $\eta < 0$);
   - $\overline{n}_{(3)}, f^{(1)};\ S, S, N^\infty \Rightarrow (c=-2,\ m=-3/2,\ n=1/10)$ (if $\eta > 0$);
   - $\overline{n}_{(3)}, f^{(1)};\ \overline{\binom{0}{2}}SN, S \Rightarrow (c=-3,\ m=-2,\ n=1/8)$ (if $\eta = 0$).

1.1.1.2: $W_4 > 0$. Since $G_{10} \neq 0$, according to Theorem 6.1 (see Table 6.2, line 95) we have a node (which is generic due to $W_4 \neq 0$) and a semi-elemental triple node.

Since by Lemma 11.5 the condition $\widetilde{M} \neq 0$ has to be satisfied, by Lemma 11.1, we arrive at the following three global configurations:
- $\overline{n}_{(3)}, n;\ S, ©, © \Rightarrow (c = -2,\ m = -2/3,\ n = 1/4)$ (if $\eta < 0$);
- $\overline{n}_{(3)}, n;\ S, S, N^{\infty} \Rightarrow (c = 1,\ m = -4,\ n = -2)$ (if $\eta > 0$);
- $\overline{n}_{(3)}, n;\ \overline{\binom{0}{2}}SN, S \Rightarrow (c = 2,\ m = -2,\ n = -8/3)$ (if $\eta = 0$).

1.1.1.3: $W_4 = 0$. In this case due to $\mu_0 \neq 0$, we get $\rho_1 \tau_4 = 0$, and we consider two cases: $\mathcal{T}_4 \neq 0$ and $\mathcal{T}_4 = 0$.

1.1.1.3.1: $\mathcal{T}_4 \neq 0$. Then $\rho_1 \neq 0$ and hence we have $\tau_4 = 0$, i.e. we have a node with coinciding eigenvalues. Considering the corresponding linear matrix $\mathcal{M}_4$, from (11.9), we conclude that this node is not a star node. Therefore by Lemmas 11.1 and 11.5, we arrive at the following three global configurations:
- $\overline{n}_{(3)}, n^d;\ S, ©, © \Rightarrow (c = -4,\ m = 0,\ n = 25/16)$ (if $\eta < 0$);
- $\overline{n}_{(3)}, n^d;\ S, S, N^{\infty} \Rightarrow (c = -1/4,\ m = 0,\ n = 25/16)$ (if $\eta > 0$);
- $\overline{n}_{(3)}, n^d;\ \overline{\binom{0}{2}}SN, S \Rightarrow (c = -2,\ m = 9/2,\ n = 18)$ (if $\eta = 0$).

1.1.1.3.2: $\mathcal{T}_4 = 0$. Then $\rho_1 \rho_4 = 0$, and considering the condition $\rho_1 \tau_4 = 0$, we obtain $\rho_1 = 0$ (otherwise we get $\rho_4 = \tau_4 = 0$, which is impossible for an elemental singular point). Thus $\rho_1 = 0$ (i.e. $c = -1$), and we calculate

$$\begin{gathered}\mu_0 = (1 + 2m - n)n^2, \quad \eta = 4m^2(1 + 2m)^2, \quad G_{10} = \mathcal{F}_1 = 0, \\ E_3 = (1 + 2m - n)n^3/4, \quad \mathcal{T}_4 = \mathcal{T}_3 = \mathcal{T}_2 = 0, \quad Y_1 = (2 + 4m - n)n/4, \\ \mathcal{T}_1 = (1 + 2m - n)n^2 \rho_4, \quad W_4 = 0, \quad W_1 = 4(1 + 2m - n)^2 n^4 \tau_4.\end{gathered} \tag{11.11}$$

Since $G_{10} = 0$ by Theorem 6.1 (see Table 6.2, lines 97 and 98) the triple point is an elliptic-saddle and the elemental one is an anti-saddle, the type of which is governed by W_1.

1. $W_1 < 0$. Then M_4 is a focus (or a center). More exactly, it is a strong focus if $\mathcal{T}_1 \neq 0$ and a weak focus (or a center) if $\mathcal{T}_1 = 0$. Moreover in the second case, i.e. when $\rho_1 = \rho_4 = 0$ (which implies $c = -1 = m$), we calculate

$$\begin{gathered}\mu_0 = -n^2(1 + n), \quad E_3 = -n^3(1 + n)/4, \quad \eta = 4, \\ \theta = 8n(2 + n)^2, \quad \theta_1 = 16(2 + n)(2 - 3n), \quad Y_1 = -(2 + n)n/4, \\ \mathcal{T}_4 = \mathcal{T}_3 = \mathcal{T}_2 = \mathcal{T}_1 = \mathcal{F} = 0, \quad \sigma = (2 + 3n)y, \\ \mathcal{H} = -n(2 + 3n)^2/2, \quad \mathcal{B} = -(2 + 3n)^4/8.\end{gathered} \tag{11.12}$$

So the conditions $\mu_0 < 0$ and $E_3 < 0$ imply $n > 0$, and this gives $\sigma \neq 0$, $\mathcal{H} < 0$ and $\mathcal{B} < 0$. By Theorem 6.2 (see statement $(e_4)[\alpha]$) we obtain a center.

If $\mathcal{T}_1 = 0$ then $E_3 < 0$ implies $Y_1 \neq 0$, and by Lemma 11.4 the elliptic-saddle is generic. Otherwise, both possibilities are realizable.

Thus considering Lemma 11.1 we get the following five global configurations:

- $\widehat{es}_{(3)}, f;\ S, S, N^\infty \Rightarrow (c = -1,\ m = -1/3,\ n = 1)$ (if $\eta > 0$, $\mathcal{T}_1 \neq 0$, $Y_1 \neq 0$);
- $\widehat{es}_{(3)}^-, f;\ S, S, N^\infty \Rightarrow (c = -1, m = -1/3, n = 2/3)$ (if $\eta > 0, \mathcal{T}_1 \neq 0, Y_1 = 0$);
- $\widehat{es}_{(3)}, c;\ S, S, N^\infty \Rightarrow (c = -1,\ m = -1,\ n = 1)$ (if $\eta > 0$, $\mathcal{T}_1 = 0$);
- $\widehat{es}_{(3)}, f;\ \overline{\binom{0}{2}} SN, S \Rightarrow (c = -1,\ m = -1/2,\ n = 1)$ (if $\eta = 0$, $Y_1 \neq 0$),
- $\widehat{es}_{(3)}^-, f;\ \overline{\binom{0}{2}} SN, S \Rightarrow (c = -1,\ m = 0,\ n = 2)$ (if $\eta = 0$, $Y_1 = 0$).

2. $W_1 > 0$. Then M_4 is a generic node. If $\eta = 0$ then $Y_1 \neq 0$, and by Lemma 11.4 the elliptic-saddle is generic. Otherwise $\eta > 0$ and both possibilities are realizable. We get the next three global configurations:
   - $\widehat{es}_{(3)}, n;\ S, S, N^\infty \Rightarrow (c = -1,\ m = -3,\ n = 2)$ (if $\eta > 0$, $Y_1 \neq 0$);
   - $\widehat{es}_{(3)}^-, n;\ S, S, N^\infty \Rightarrow (c = -1,\ m = -7/16,\ n = 1/4)$ (if $\eta > 0$, $Y_1 = 0$);
   - $\widehat{es}_{(3)}, n;\ \overline{\binom{0}{2}} SN, S \Rightarrow (c = -1,\ m = -1/2,\ n = 1/5)$ (if $\eta = 0$).
3. $W_1 = 0$. Then M_4 is a one-direction node. Again if $\eta = 0$ then $Y_1 \neq 0$, and by Lemma 11.4 the elliptic-saddle is generic. Otherwise $\eta > 0$ and both possibilities are available. We get the next three global configurations:
   - $\widehat{es}_{(3)}, n^d;\ S, S, N^\infty \Rightarrow (c = -1,\ m = -1/3,\ n = 4/9)$ (if $\eta > 0$, $Y_1 \neq 0$);
   - $\widehat{es}_{(3)}^-, n^d;\ S, S, N^\infty \Rightarrow (c = -1,\ m = 1-\sqrt{2},\ n = 6-4\sqrt{2})$ (if $\eta > 0, Y_1 = 0$);
   - $\widehat{es}_{(3)}, n^d;\ \overline{\binom{0}{2}} SN, S \Rightarrow (c = -1,\ m = -1/2,\ n = 1/4)$ (if $\eta = 0$).

1.1.2: $E_3 > 0$. In this case the elemental singularity M_4 is a saddle, whereas M_1 is a triple saddle, which could be semi-elemental or nilpotent. So at infinity we have three nodes.

1.1.2.1: $\mathcal{T}_4 \neq 0$. Then $\rho_1\rho_4 \neq 0$ and both saddles have non-zero traces. In this case we arrive at the global configuration

- $\overline{s}_{(3)}, s;\ N^f, N^f, N^f \Rightarrow (c = -2,\ m = -2,\ n = -2)$.

1.1.2.2: $\mathcal{T}_4 = 0$. We consider two cases: $\mathcal{T}_3 \neq 0$ and $\mathcal{T}_3 = 0$.

1. $\mathcal{T}_3 \neq 0$. According to (11.10) this implies $\rho_1 \neq 0$, and then $\mathcal{F}_1 \neq 0$ and $\rho_4 = 0$, i.e. $c = 2m + 1$. Then we have a first order weak saddle, and we get the configuration
   - $\overline{s}_{(3)}, s^{(1)};\ N^f, N^f, N^f \Rightarrow (c = -3,\ m = -2,\ n = -2)$.

2. $\mathcal{T}_3 = 0$. This implies $\rho_1 = 0$ (i.e. $c = -1$) and the triple saddle is nilpotent. So we get the expressions (11.11), and in this case the conditions $\mu_0 < 0$ and $E_3 > 0$ imply $1 + 2m < n < 0$. Therefore depending on the value of the invariant polynomial $\mathcal{T}_1$, we arrive at the following two global configurations:
   - $\hat{s}_{(3)}, s;\ N^f, N^f, N^f \Rightarrow (c = -1,\ m = -2,\ n = -2)$ (if $\mathcal{T}_1 \neq 0$);
   - $\hat{s}_{(3)}, \$;\ N^f, N^f, N^f \Rightarrow (c = -1,\ m = -1,\ n = -1/2)$ (if $\mathcal{T}_1 = 0$).

1.2: $\mu_0 > 0$. Then considering (11.10) we have $\alpha > 0$, and this implies sign $(E_3) =$ sign (Δ_4).

1.2.1: $E_3 < 0$. In this case the elemental singularity M_4 is a saddle. Additionally by Theorem 6.1 the triple point could be either a semi-elemental triple node or a nilpotent elliptic-saddle.

1.2.1.1: $\mathcal{T}_4 \neq 0$. Then by (11.10) we get $\rho_1\rho_4 \neq 0$ and $G_{10} \neq 0$. In this case by Theorem 6.1 (see Table 6.2, line 87) the triple point is a semi-elemental triple node and the saddle is strong. Therefore considering Lemmas 11.1 and 11.5, we examine three cases: $\eta < 0$, $\eta > 0$ and $\eta = 0$.

1. $\eta < 0$. By Lemma 11.1 the configurations of infinite singularities are governed by the invariant polynomials θ and θ_2. According to Lemma 11.5 in this case the condition $\theta = 0$ implies $\theta_2 \neq 0$, otherwise we get $\mu_0 = \eta$, which contradicts $\eta < 0$ and $\mu_0 > 0$. So considering Lemma 11.1, we arrive at the following three global configurations:
   - $\overline{n}_{(3)}, s;\ N^\infty, ©, © \Rightarrow (c = -1/2,\ m = 1,\ n = -29/8)$ (if $\theta < 0$);
   - $\overline{n}_{(3)}, s;\ N^f, ©, © \Rightarrow (c = -1/2,\ m = -1/2,\ n = -1/4)$ (if $\theta > 0$);
   - $\overline{n}_{(3)}, s;\ N^d, ©, © \Rightarrow (c = -1/2,\ m = -5/4,\ n = -5/2)$ (if $\theta = 0$).
2. $\eta > 0$. From Lemmas 11.5 and 11.1 we get the following six global configurations:
   - $\overline{n}_{(3)}, s;\ S, N^\infty, N^\infty \Rightarrow (c = -3/2,\ m = 0,\ n = -3)$ (if $\theta < 0,\ \theta_1 < 0$);
   - $\overline{n}_{(3)}, s;\ S, N^f, N^f \Rightarrow (c = 2,\ m = 7/4,\ n = 1/2)$ (if $\theta < 0,\ \theta_1 > 0$);
   - $\overline{n}_{(3)}, s;\ S, N^f, N^\infty \Rightarrow (c = 2,\ m = 5/2,\ n = 2)$ (if $\theta > 0$);
   - $\overline{n}_{(3)}, s;\ S, N^\infty, N^d \Rightarrow (c=-3/2, m=-1/12, n=-1/6)$ (if $\theta=0,\ \theta_1<0$);
   - $\overline{n}_{(3)}, s;\ S, N^f, N^d \Rightarrow (c = -4,\ m = -1/2,\ n = -1)$ (if $\theta = 0,\ \theta_1 > 0$);
   - $\overline{n}_{(3)}, s;\ S, N^d, N^d \Rightarrow (c = -3/2,\ m = -1/2,\ n = -1)$ (if $\theta = \theta_1 = 0$).
3. $\eta = 0$. According to Lemma 11.5 in this case the conditions $\theta = 0$ and $\theta_2 = 0$ are incompatible (otherwise we get $\eta = \mu_0 \neq 0$). So considering Lemma 11.1 we get the following four global configurations:

- $\overline{n}_{(3)}, s; \overline{\binom{0}{2}}SN, N^\infty \Rightarrow (c{=}-3/2, m{=}-1/6, n{=}-1)$ (if $\widetilde{M}\neq 0,\ \theta<0$);
- $\overline{n}_{(3)}, s; \overline{\binom{0}{2}}SN, N^f$ $(c = 1,\ m = 1,\ n = 1)$ (if $\widetilde{M} \neq 0,\ \theta > 0$);
- $\overline{n}_{(3)}, s; \overline{\binom{0}{2}}SN, N^d \Rightarrow (c{=}-1/2, m{=}3/4, n{=}-3/2)$ (if $\widetilde{M}\neq 0,\ \theta=0$);
- $\overline{n}_{(3)}, s; \overline{\binom{0}{3}}N \Rightarrow (c = -1/2,\ m = 1/4,\ n = -1/2)$ (if $\widetilde{M} = 0$).

1.2.1.2: $\mathcal{T}_4 = 0$. Then by (11.10) we get $\rho_1\rho_4 = 0$, and we consider two cases: $\mathcal{T}_3 \neq 0$ and $\mathcal{T}_3 = 0$.

1.2.1.2.1: $\mathcal{T}_3 \neq 0$. Then $\rho_1 \neq 0$ and by (11.10) this implies $\mathcal{F}_1 G_{10} \neq 0$. Therefore we get $\rho_4 = 1 - c + 2m = 0$, i.e. the finite saddle is weak of order one (due to $\mathcal{F}_1 \neq 0$).

1. $\eta < 0$. So considering Lemmas 11.1 and 11.5, we arrive at the following three global configurations:
   - $\overline{n}_{(3)}, s^{(1)}; N^\infty, ©, © \Rightarrow (c = -1/2,\ m = -3/4,\ n = -2)$ (if $\theta < 0$);
   - $\overline{n}_{(3)}, s^{(1)}; N^f, ©, © \Rightarrow (c = -1/2,\ m = -3/4,\ n = -1)$ (if $\theta > 0$);
   - $\overline{n}_{(3)}, s^{(1)}; N^d, ©, © \Rightarrow (c = -1/2,\ m = -3/4,\ n = -3/2)$ (if $\theta = 0$).
2. $\eta > 0$. From Lemmas 11.5 and 11.1 we can get up to six global configurations. Examples for five of them are as follows:
   - $\overline{n}_{(3)}, s^{(1)}; S, N^\infty, N^\infty \Rightarrow (c = -21/20,\ m = -41/40,\ n = -3)$ (if $\theta < 0,\ \theta_1 < 0$);
   - $\overline{n}_{(3)}, s^{(1)}; S, N^f, N^f \Rightarrow (c = 1/2,\ m = -1/4,\ n = 1/3)$ (if $\theta < 0,\ \theta_1 > 0$);
   - $\overline{n}_{(3)}, s^{(1)}; S, N^f, N^\infty \Rightarrow (c = -2,\ m = -3/2,\ n = -4)$ (if $\theta > 0$);
   - $\overline{n}_{(3)}, s^{(1)}; S, N^\infty, N^d \Rightarrow (c = -9/10,\ m = -19/20,\ n = -19/10)$ (if $\theta = 0,\ \theta_1 < 0$);
   - $\overline{n}_{(3)}, s^{(1)}; S, N^f, N^d \Rightarrow (c = -2,\ m = -3/2,\ n = -3)$ (if $\theta = 0,\ \theta_1 > 0$).

   The sixth configuration corresponding to the case $\theta = \theta_1 = 0$ cannot be realizable. Indeed setting $\rho_4 = 1 - c + 2m = 0$ (i.e. $m = (c-1)/2$), we found that the common solutions (c, n) of the equations $\theta = 0$ and $\theta_1 = 0$ are the following ones: $\{(-1, 2),\ (-1/2, -1/2),\ (0, -1/4),\ (0, 0),\ (1, 0)\}$. However all these solutions contradict the condition $\mu_0 = c^2n^2(c - n) > 0$.
3. $\eta = 0$. According to Lemma 11.5 in this case the conditions $\theta = 0$ and $\theta_2 = 0$ are incompatible (otherwise we get $\eta = \mu_0 \neq 0$). Moreover since $\mathcal{T}_4 = 0$ and $E_3 < 0$, by Lemma 11.5 in this case we have $\widetilde{M} \neq 0$. So considering Lemma 11.1 we get the following three global configurations:
   - $\overline{n}_{(3)}, s^{(1)}; \overline{\binom{0}{2}}SN, N^\infty \Rightarrow (c{=}-22/25, m{=}-47/50, n{=}-25/12)$ (if $\theta{<}0$);

- $\overline{n}_{(3)}, s^{(1)}; \overline{\binom{0}{2}}SN, N^f \Rightarrow (c = -4/5, m = -9/10, n = -5/4)$ (if $\theta > 0$);
- $\overline{n}_{(3)}, s^{(1)}; \overline{\binom{0}{2}}SN, N^d \Rightarrow (c = -\sqrt{3}/2, m = -(2+\sqrt{3})/4, n = -1-\sqrt{3}/2)$ (if $\theta = 0$).

1.2.1.2.2: $\mathcal{T}_3 = 0$. This implies $\rho_1 = 0$ (i.e. $c = -1$) and then $G_{10} = 0$. Therefore by Theorem 6.1 (see Table 6.2, line 88) the triple singularity is an elliptic-saddle (nilpotent). Additionally the elemental saddle could be a weak one, and we shall consider two subcases: $\mathcal{T}_1 \neq 0$ and $\mathcal{T}_1 = 0$.

1.2.1.2.2.1: $\mathcal{T}_1 \neq 0$. Then we have a strong saddle, and by (11.11) the condition $\eta \geq 0$ holds.

1. $\eta > 0$. We detect that the condition $Y_1 = (2 + 4m - n)n/4 = 0$ implies that $n = 2 + 4m \neq 0$ (by $\mu_0 \neq 0$), and then we have $\theta = 64(1+m)^2(1+2m)^2 > 0$ since $\mathcal{T}_1 = -8(1+m)(1+2m)^3 \neq 0$. So by Lemma 11.4 the only possibility for non-generic elliptic-saddle is when $\theta > 0$. Additionally, considering Lemmas 11.1 and 11.5, we can get up to six configurations at infinity. Five of them are realizable by normal form (11.6). So we get the following six global configurations:

   - $\widehat{es}_{(3)}, s;\ S, N^\infty, N^\infty \Rightarrow (c = -1,\ m = -1/3,\ n = -1)$ (if $\theta < 0,\ \theta_1 < 0$);
   - $\widehat{es}_{(3)}, s;\ S, N^f, N^f \Rightarrow (c = -1,\ m = -21/25,\ n = -1)$ (if $\theta < 0,\ \theta_1 > 0$);
   - $\widehat{es}_{(3)}, s;\ S, N^f, N^\infty \Rightarrow (c = -1,\ m = -2,\ n = -5)$ (if $\theta > 0,\ Y_1 \neq 0$);
   - $\widehat{es}_{\overline{(3)}}, s;\ S, N^f, N^\infty \Rightarrow (c = -1,\ m = -5/4,\ n = -3)$ (if $\theta > 0,\ Y_1 = 0$);
   - $\widehat{es}_{(3)}, s;\ S, N^\infty, N^d \Rightarrow (c = -1,\ m = -1/4,\ n = -1/2)$ (if $\theta = 0,\ \theta_1 < 0$);
   - $\widehat{es}_{(3)}, s;\ S, N^f, N^d \Rightarrow (c = -1,\ m = -3/2,\ n = -3)$ (if $\theta = 0,\ \theta_1 > 0$).

   The sixth configuration at infinity corresponding to the case $\theta = \theta_1 = 0$ cannot be realized. Indeed, considering (11.11) the conditions $\mu_0 > 0$ and $E_3 < 0$ imply $n < 0$. Additionally supposing $\theta = 0 = \theta_1$, we get the unique solution with $n < 0$ and namely $(m, n) = (-1, -2)$. However in this case we get $\mathcal{T}_1 = 0$, which enters in contradiction with the case under examination.

2. $\eta = 0$. In this case according to Lemma 11.5 the condition $\theta = 0$ implies $\theta_2 \neq 0$, and since $\rho_1 = 0$, by Lemma 11.5 we have $\widetilde{M} \neq 0$. Moreover since $\eta = 4m^2(1+2m)^2 = 0$, either $m = -1/2$, which implies $\mu_0 = -n^3$ and $Y_1 = -n^2/4 \neq 0$, or $m = 0$, which implies $Y_1 = n(n-2)$ and $\mu_0 = (1-n)n^2$. In both cases we have that $Y_1 = 0$ is incompatible with $\mu_0 > 0$. Then by Lemma 11.4 the elliptic-saddle is always generic. So considering Lemma 11.1 we get the following three global configurations:

   - $\widehat{es}_{(3)}, s;\ \overline{\binom{0}{2}}SN, N^\infty \Rightarrow (c = -1,\ m = 0,\ n = -1)$ (if $\theta < 0$);

- $\widehat{es}_{(3)}, s; \overline{\binom{0}{2}}SN, N^f \Rightarrow (c = -1,\ m = -1/2,\ n = -1/2)$ (if $\theta > 0$);
- $\widehat{es}_{(3)}, s; \overline{\binom{0}{2}}SN, N^d \Rightarrow (c = -1,\ m = -1/2,\ n = -1)$ (if $\theta = 0$).

1.2.1.2.2.2: $\mathcal{T}_1 = 0$. In this case the saddle is weak, and the conditions $\rho_1 = \rho_4 = 0$ yield $c = -1 = m$. Then we arrive at the formulas (11.12). Clearly the condition $\mu_0 > 0$ implies $n < -1$ (then $E_3 < 0$), and therefore we obtain $\sigma \neq 0$, $\mathcal{H} > 0$ and $\mathcal{B} < 0$. By Theorem 6.2 (see statement $(e_3)[\alpha]$) we obtain an integrable saddle. Moreover, in this case $Y_1 = -n(2+n)/4$ and $\mu_0 = -n^2(1+n)$, so get $Y_1 = 0$ only if $n = -2$ (which implies $\mu_0 > 0$). Thus by Lemma 11.4 the elliptic-saddle is generic if $n \neq -2$ and non-generic if $n = -2$. Moreover, since $\theta = 8n(2+n)^2$ and $\theta_1 = -16(2+n)(3n-2)$, the case of the non-generic elliptic-saddle implies $\theta = \theta_1 = 0$. Therefore by Lemma 11.1 the infinite if formed by a saddle and two N^d.

Additionally from formulas (11.12) we observe that the condition $\theta \leq 0$ holds. Since $\eta > 0$ considering Lemmas 11.1 and 11.5, we arrive at the following three global configurations:

- $\widehat{es}_{(3)}, \$; S, N^\infty, N^\infty \Rightarrow (c = -1,\ m = -1,\ n = -3)$ (if $\theta \neq 0$, $\theta_1 < 0$);
- $\widehat{es}_{(3)}, \$; S, N^f, N^f \Rightarrow (c = -1,\ m = -1,\ n = -3/2)$ (if $\theta \neq 0$, $\theta_1 > 0$);
- $\widehat{es}_{\overline{(3)}}, \$; S, N^d, N^d \Rightarrow (c = -1,\ m = -1,\ n = -2)$ (if $\theta = 0$).

1.2.2: $E_3 > 0$. In this case the elemental singularity M_4 is an anti-saddle. Additionally by Theorem 6.1 the singular point $M_1(0,0)$ is a triple saddle (semi-elemental or nilpotent).

1.2.2.1: $W_4 < 0$. Then from (11.10) it follows that $G_{10} \neq 0$, and according to Theorem 6.1 (see Table 6.2, line 91), we have a focus (or a center) and a semi-elemental triple saddle. When $\mathcal{T}_4 = 0$ the elemental singularity can be a weak focus or a center.

1.2.2.1.1: $\mathcal{T}_4 \neq 0$. In this case we have a strong focus.

1. The subcase $\eta < 0$. By Lemma 11.1 the configurations of infinite singularities are governed by the invariant polynomials θ and θ_2. According to Lemma 11.5, in this case the condition $\theta = 0$ implies $\theta_2 \neq 0$, and considering Lemma 11.1, we arrive at the following three global configurations:
   - $\overline{s}_{(3)}, f;\ N^\infty, ©, © \Rightarrow (c = 2,\ m = -11/8,\ n = -2)$ (if $\theta < 0$);
   - $\overline{s}_{(3)}, f;\ N^f, ©, © \Rightarrow (c = 2,\ m = -1/2,\ n = -2)$ (if $\theta > 0$);
   - $\overline{s}_{(3)}, f;\ N^d, ©, © \Rightarrow (c = 2,\ m = -1,\ n = -2)$ (if $\theta = 0$).
2. $\eta > 0$. We claim that in this case instead of the 10 possible configurations indicated in Diagram 11.1, there could be realized only three. More precisely we claim that if $\mu_0 > 0$, $W_4 < 0$ and $\eta > 0$, then the condition $\theta \leq 0$ implies $\theta_1 < 0$, and in the case $\theta = 0$ we have $\theta_2 \neq 0$.

Indeed, it can be proved directly that if the conditions $\mu_0 > 0$, $W_4 < 0$ (i.e. $\tau_4 < 0$) and $\eta > 0$ hold, then the surfaces $\theta = 0$ and $\theta_1 = 0$ do not intersect in real points. So it is sufficient to take any slice in the three-dimensional space of the parameters (c, m, n) that intersects the region where the conditions hold, and to check that at a point where $\theta_1 = 0$ the polynomial θ is positive.

It remains to observe that in the case $\theta = 0$ by Lemma 11.5(iii) the condition $\theta_2 \neq 0$ must hold. This completes the proof of our claim.

Remark 11.2. We point out that the above arguments hold regardless of the value of $\mathcal{T}_4$. So the same reasons could be used when $\mathcal{T}_4 = 0$.

Thus by Lemma 11.1 the remaining three global configurations are:

- $\overline{s}_{(3)}, f;\ S, N^{\infty}, N^{\infty} \Rightarrow (c = 20,\ m = -17/2,\ n = -65/4)$ (if $\theta < 0$);
- $\overline{s}_{(3)}, f;\ S, N^{f}, N^{\infty} \Rightarrow (c = 545/64,\ m = 127/128,\ n = -65/64)$ (if $\theta > 0$);
- $\overline{s}_{(3)}, f;\ S, N^{\infty}, N^{d} \Rightarrow (c = 137/16,\ m = -65/32,\ n = -65/16)$ (if $\theta = 0$).

3. $\eta = 0$. According to Lemma 11.5 in this case the condition $\theta = 0$ implies $\theta_2 \neq 0$, and considering Lemma 11.1, we arrive at the following four global configurations:

- $\overline{s}_{(3)}, f;\ \overline{\binom{0}{2}}SN, N^{\infty} \Rightarrow (c = 5/4,\ m = -7/8,\ n = -1)$ (if $\widetilde{M} \neq 0$, $\theta < 0$);
- $\overline{s}_{(3)}, f;\ \overline{\binom{0}{2}}SN, N^{f} \Rightarrow (c = 65/16, m = 47/32, n = -1/16)$ (if $\widetilde{M} \neq 0, \theta > 0$);
- $\overline{s}_{(3)}, f;\ \overline{\binom{0}{2}}SN, N^{d} \Rightarrow (c = 5/4,\ m = -1/8,\ n = -1/4)$ (if $\widetilde{M} \neq 0$, $\theta = 0$);
- $\overline{s}_{(3)}, f;\ \overline{\binom{0}{3}}N \Rightarrow (c = 5/11,\ m = -5/22,\ n = -25/176)$ (if $\widetilde{M} = 0$).

1.2.2.1.2: $\mathcal{T}_4 = 0$. Then by (11.10) we get $\rho_1\rho_4 = 0$, and as $W_4 \neq 0$ implies $\rho_1\mathcal{F}_1 \neq 0$, we obtain $\rho_4 = 0$ (i.e. $1 - c + 2m = 0$). So we have a first order weak focus. Taking into consideration Remark 11.2, we split the study according to η.

1. $\eta < 0$. Then we have the next three global configurations:
   - $\overline{s}_{(3)}, f^{(1)};\ N^{\infty}, ©, © \Rightarrow (c = 1/2,\ m = -1/4,\ n = -1/3)$ (if $\theta < 0$);
   - $\overline{s}_{(3)}, f^{(1)};\ N^{f}, ©, © \Rightarrow (c = 1/2,\ m = -1/4,\ n = -1)$ (if $\theta > 0$);
   - $\overline{s}_{(3)}, f^{(1)};\ N^{d}, ©, © \Rightarrow (c = 1/2,\ m = -1/4,\ n = -1/2)$ (if $\theta = 0$).
2. $\eta > 0$. Considering Lemmas 11.1 and 11.5, we arrive at the following three global configurations:
   - $\overline{s}_{(3)}, f^{(1)};\ S, N^{\infty}, N^{\infty} \Rightarrow (c = 1/2,\ m = -1/4,\ n = -8/50)$ (if $\theta < 0$);
   - $\overline{s}_{(3)}, f^{(1)};\ S, N^{f}, N^{\infty} \Rightarrow (c = 1/2, m = -1/4, n = -6/50)$ (if $\theta > 0$);

- $\overline{s}_{(3)}, f^{(1)}; S, N^{\infty}, N^{d} \Rightarrow (c = 1/2,\ m = -1/4,\ n = -7/50)$ (if $\theta = 0$).

Considering the graphic $\theta_1 = 0$, we observe that we cannot have any branch inside the region defined by the conditions $\mu_0 > 0$, $E_3 > 0$ and $\theta < 0$.

3. $\eta = 0$. Then we have the next four global configurations:

- $\overline{s}_{(3)}, f^{(1)}; \overline{\binom{0}{2}}SN, N^{\infty} \Rightarrow (c = 1/3, m = -1/3, n = -3/16)$ (if $\widetilde{M} \neq 0, \theta < 0$);
- $\overline{s}_{(3)}, f^{(1)}; \overline{\binom{0}{2}}SN, N^{f} \Rightarrow (c = 1,\ m = 0,\ n = -1/8)$ (if $\widetilde{M} \neq 0,\ \theta > 0$);
- $\overline{s}_{(3)}, f^{(1)}; \overline{\binom{0}{2}}SN, N^{d} \Rightarrow (c = \sqrt{3}/2,\ m = (\sqrt{3} - 2)/4,\ n = (\sqrt{3} - 2)/2)$ (if $\widetilde{M} \neq 0,\ \theta = 0$);
- $\overline{s}_{(3)}, f^{(1)}; \overline{\binom{0}{3}}N \Rightarrow (c = 1/2,\ m = -1/4,\ n = -1/6)$ (if $\widetilde{M} = 0$).

1.2.2.2: $W_4 > 0$. In this case $G_{10} \neq 0$, and according to Theorem 6.1 (see Table 6.2, line 89), we have a generic node and a semi-elemental triple saddle.

1. $\eta < 0$. Therefore considering Lemmas 11.1 and 11.5(iii), we arrive at the following three global configurations:

- $\overline{s}_{(3)}, n; N^{\infty}, ©, © \Rightarrow (c = 5/2,\ m = -23/8,\ n = -5)$ (if $\theta < 0$);
- $\overline{s}_{(3)}, n; N^{f}, ©, © \Rightarrow (c = 1,\ m = 5/4,\ n = -1/2)$ (if $\theta > 0$);
- $\overline{s}_{(3)}, n; N^{d}, ©, © \Rightarrow (c = 1,\ m = -5/2,\ n = -5)$ (if $\theta = 0$).

2. $\eta > 0$. So by Lemmas 11.1 and 11.5(iv), we obtain the following six global configurations:

- $\overline{s}_{(3)}, n; S, N^{\infty}, N^{\infty} \Rightarrow (c = 1,\ m = -14849/5000,\ n = -59/10)$ (if $\theta < 0,\ \theta_1 < 0$);
- $\overline{s}_{(3)}, n; S, N^{f}, N^{f} \Rightarrow (c = 1,\ m = -65/18,\ n = -8)$ (if $\theta < 0,\ \theta_1 > 0$);
- $\overline{s}_{(3)}, n; S, N^{f}, N^{\infty} \Rightarrow (c = 1,\ m = -29/50,\ n = -1/5)$ (if $\theta > 0$);
- $\overline{s}_{(3)}, n; S, N^{\infty}, N^{d} \Rightarrow (c = 1,\ m = -59/20,\ n = -59/10)$ (if $\theta = 0, \theta_1 < 0$);
- $\overline{s}_{(3)}, n; S, N^{f}, N^{d} \Rightarrow (c = 1,\ m = -61/20,\ n = -61/10)$ (if $\theta = 0, \theta_1 > 0$);
- $\overline{s}_{(3)}, n; S, N^{d}, N^{d} \Rightarrow (c = 1,\ m = -3,\ n = -6)$ (if $\theta = 0,\ \theta_1 = 0$).

3. $\eta = 0$. In this case considering Lemmas 11.1 and 11.5, we get the following four global configurations:

- $\overline{s}_{(3)}, n; \overline{\binom{0}{2}}SN, N^{\infty} \Rightarrow (c = 1/5,\ m = -1/30,\ n = -1/90)$ (if $\widetilde{M} \neq 0, \theta < 0$);
- $\overline{s}_{(3)}, n; \overline{\binom{0}{2}}SN, N^{f} \Rightarrow (c = 1/5,\ m = -31/120,\ n = -31/360)$ (if $\widetilde{M} \neq 0,\ \theta > 0$);

- $\overline{s}_{(3)}, n; \binom{0}{2}SN, N^d \Rightarrow (c = 1/5,\ m = -11/60,\ n = -11/180)$ (if $\widetilde{M} \neq 0$, $\theta = 0$);
- $\overline{s}_{(3)}, n; \binom{0}{3}N \Rightarrow (c = 1/4,\ m = -1/8,\ n = -1/20)$ (if $\widetilde{M} = 0$).

1.2.2.3: $W_4 = 0$. In this case due to $\mu_0 \neq 0$ we get $\rho_1\tau_4 = 0$, and we consider two cases: $\mathcal{T}_4 \neq 0$ and $\mathcal{T}_4 = 0$.

1.2.2.3.1: $\mathcal{T}_4 \neq 0$. Then $\rho_1 \neq 0$ and hence we obtain $\tau_4 = 0$, i.e. we have a node with coinciding eigenvalues, which by (11.9) is not a star node.

1. $\eta < 0$. In this case considering Lemmas 11.1 and 11.5(iii), we arrive at the following three global configurations:
   - $\overline{s}_{(3)}, n^d;\ N^\infty, ©, © \Rightarrow (c = 1,\ m = -3/2,\ n = -9/4)$ (if $\theta < 0$);
   - $\overline{s}_{(3)}, n^d;\ N^f, ©, © \Rightarrow (c = 1,\ m = -3,\ n = -9)$ (if $\theta > 0$);
   - $\overline{s}_{(3)}, n^d;\ N^d, ©, © \Rightarrow (c = 1,\ m = -2,\ n = -4)$ (if $\theta = 0$).
2. $\eta > 0$. Then by Lemmas 11.1 and 11.5 (statements (iii) and (iv)), we could have five possible global configurations. For three of them, we give corresponding three global configurations:
   - $\overline{s}_{(3)}, n^d;\ S, N^\infty, N^\infty \Rightarrow (c = 5/2, m = -1/2, n = -5/8)$ (if $\theta < 0$, $\theta_1 < 0$);
   - $\overline{s}_{(3)}, n^d;\ S, N^f, N^\infty \Rightarrow (c = 1,\ m = -1/2,\ n = -1/4)$ (if $\theta > 0$);
   - $\overline{s}_{(3)}, n^d;\ S, N^\infty, N^d \Rightarrow (c = 9/4,\ m = -1/8,\ n = -1/4)$ (if $\theta = 0$, $\theta_1 < 0$).

   We claim that the remaining two configurations (defined respectively by the conditions $\theta < 0$, $\theta_1 > 0$ and $\theta = 0$, $\theta_1 > 0$) could not be realizable. More exactly we prove below that the condition $\theta \leq 0$ implies in this case $\theta_1 < 0$.

   Indeed, as $\Delta_4 \neq 0$ (i.e. $cn \neq 0$) the condition $\tau_4 = 0$ yields $n = -(1 - c + 2m)^2/(4c)$, and then the invariant polynomials $\theta = \theta(c, m)$ and $\theta_1 = \theta_1(c, m)$ depend on the two parameters c and m. Considering the system of the equations $\theta(c, m) = 0$ and $\theta_1(c, m) = 0$, we detect that the only real solutions of this system are $(-2, 0)$ and $(1, 0)$. However in these points we have $\mu_0 = -81/128 < 0$ and $\mu_0 = 0$, respectively.

   Additionally in this case we calculate

$$\mu_0 = (1 - c + 2m)^4(1 + c + 2m)^2/(64c),$$
$$\eta = (1 + c + 2m)^2\big[(c-1)^2 + 2(1 + c)m\big]^2\big[c^2 + c - (1 + 2m)^2\big]/(16c^3),$$

   and as $\mu_0 > 0$ we get $c > 0$. So it is enough to check in the region defined by $c > 0$ and $c^2 + c - (1 + 2m)^2 > 0$ the signs of θ and θ_1, and to verify that in any point of this region where $\theta \leq 0$ and $\theta_1 \neq 0$ we have $\theta_1 < 0$.

In order to check the conditions, we need to find the intersections of the graphic $\theta_1 = 0$ with the component $c^2 + c - 1 - 4m - 4m^2 = 0$. We obtain only one point (c, m) with $c > 0$, namely $(1/3, -1/6)$, and this is a contact point of $\theta_1 = 0$ and $\eta = 0$, which does not produce any new region inside $\eta > 0$.

3. $\eta = 0$. In this case considering Lemmas 11.1 and 11.5, we get the following four global configurations:

- $\overline{s}_{(3)}, n^d; \overline{\binom{0}{2}}SN, N^\infty \Rightarrow (c = 9/16, m = -31/32, n = -1)$ (if $\widetilde{M} \neq 0, \theta < 0$);
- $\overline{s}_{(3)}, n^d; \overline{\binom{0}{2}}SN, N^f \Rightarrow (c = 49/72, m = 5/144, n = -1/18)$ (if $\widetilde{M} \neq 0, \theta > 0$);
- $\overline{s}_{(3)}, n^d; \overline{\binom{0}{2}}SN, N^d \Rightarrow (c = 9/16, m = -1/32, n = -1/16)$ (if $\widetilde{M} \neq 0, \theta = 0$);
- $\overline{s}_{(3)}, n^d; \overline{\binom{0}{3}}N \Rightarrow (c = 1/3,\ m = -1/6,\ n = -1/12)$ (if $\widetilde{M} = 0$).

1.2.2.3.2: $\mathcal{T}_4 = 0$. Then $\rho_1\rho_4 = 0$, and considering the condition $\rho_1\tau_4 = 0$ (as $W_4 = 0$), we obtain $\rho_1 = 0$ (otherwise we get $\rho_4 = \tau_4 = 0$, which is impossible for an elemental singular point). Thus $\rho_1 = 0$ (i.e. $c = -1$) and we arrive at the relations (11.11). As $G_{10} = 0$, by Theorem 6.1 the triple point is a nilpotent saddle and the elemental one is an anti-saddle. We claim that the condition $\mu_0 > 0$ implies $\tau_4 > 0$. Indeed considering (11.11) the condition $\mu_0 > 0$ gives $1 + 2m - n > 0$, and therefore $\tau_4 = (1 + 2m - n) + m^2 > 0$. So our claim is proved, and hence the elemental singular point is a generic node.

We observe also that the conditions $\mu_0 > 0$ and $E_3 > 0$ imply $0 < n < 2m+1$. Then considering the graphics $\theta = 0$ and $\theta_1 = 0$, it is easy to detect that they do not intersect for $n > 0$ and that the condition $\theta < 0$ implies $\theta_1 > 0$.

Additionally by (11.11) we have $\eta \geq 0$. Moreover due to the condition $E_3 > 0$, we could have $\eta = 0$ only if $m = 0$ (otherwise in the case $m = -1/2$ we get $E_3 = -n^4/4 < 0$). And in the case $m = 0$ we obtain $\widetilde{M} = -8x^2 \neq 0$ and $\theta = 8n^3 > 0$ as $n > 0$. Thus considering Lemma 11.1, we arrive at the following three global configurations:

- $\hat{s}_{(3)}, n;\ S, N^f, N^f \Rightarrow (c = -1,\ m = 2,\ n = 1)$ (if $\eta > 0,\ \theta < 0$);
- $\hat{s}_{(3)}, n;\ S, N^f, N^\infty \Rightarrow (c = -1,\ m = 1,\ n = 9/4)$ (if $\eta > 0,\ \theta > 0$);
- $\hat{s}_{(3)}, n;\ \overline{\binom{0}{2}}SN, N^f \Rightarrow (c = -1,\ m = 0,\ n = 1/2)$ (if $\eta = 0$).

2: The family of systems (11.7). So we consider the family of systems:

$$\dot{x} = 2hxy, \quad \dot{y} = x - x^2 + 2mxy + ny^2, \quad m \in \{0, 1\}, \tag{11.7}$$

which possess the triple nilpotent singular point $M_1(0, 0)$ and the elemental one $M_4(1, 0)$. For the second singularity we have

$$\mathcal{M}_4 = \begin{pmatrix} 0 & 2h \\ -1 & 2m \end{pmatrix}, \quad \rho_4 = 2m, \ \ \Delta_4 = 2h; \ \ \tau_4 = 4(m^2 - 2h). \tag{11.13}$$

For this family we calculate

$$\begin{aligned} \mu_0 = -n\Delta_4^2, \quad \eta = 4(2h-n)^2(m^2+n-2h), \quad E_3 = -n\Delta_4^3/4, \\ G_{10} = 0, \quad \mathcal{T}_4 = \mathcal{T}_3 = \mathcal{T}_2 = 0, \quad \mathcal{T}_1 = -n\Delta_4^2\rho_4, \quad \mathcal{F}_1 = 0, \\ \mathcal{F} = m(h+n)^2\Delta_4 \quad W_4 = 0, \quad W_1 = n^2\Delta_4^4\tau_4, \\ \widetilde{M} = 8(6h-4m^2-3n)x^2 + 16m(2h-n)xy - 8(2h-n)^2y^2. \end{aligned} \tag{11.14}$$

Lemma 11.6. *For a system* (11.7) *the following are true:*

(i) *If the condition* $\widetilde{M} = 0$ *holds, then* $\mathcal{T}_1 = \theta_1 = 0$*,* $E_3 < 0$ *and* $\operatorname{sign}(\theta) = -\operatorname{sign}(\mu_0)$.

(ii) *If the conditions* $\theta = 0$ *and* $\mathcal{T}_1 \neq 0$ *hold, then for this system we have* $\eta \geq 0$*,* $\theta_1\theta_2 \neq 0$.

Proof. (i) By (11.14) we obtain that the condition $\widetilde{M} = 0$ is equivalent to $n-2h = m = 0$, and we obtain

$$\mathcal{T}_1 = \theta_1 = 0, \quad E_3 = -4h^4, \quad \mu_0 = -8h^3, \quad \theta = 64h^3.$$

Evidently statement (i) follows immediately.

(ii) Setting a new parameter $u = h - n$, we have $\theta = 64(n+u)(m^2n+u^2)$ and $\mu_0 = -4n(n+u)^2 \neq 0$. So due to $m \neq 0$ (as $\mathcal{T}_1 \neq 0$), the condition $\theta = 0$ gives $n = -u^2/m^2$. Then we calculate

$$\begin{aligned} \eta = 4(m^2-u)^2(2m^2-u)^2u^2/m^6, \quad \theta_1 = 256(m^2-u)^3u^2/m^4, \\ \theta_2 = (m^2-u)u^2/m^2, \quad \mu_0 = 4(m^2-u)^2u^4/m^6, \end{aligned}$$

and this completes the proof of the lemma as $\mu_0 \neq 0$. □

We observe that for systems (11.7) the parameter $m \in \{0,1\}$, and as the condition $m = 0$ is equivalent to $\mathcal{T}_1 = 0$, we consider two cases.

2.1: $\mathcal{T}_1 \neq 0$. Then $m = 1$ and the condition $\mu_0 \neq 0$ implies $E_3 \neq 0$.

2.1.1: $\mu_0 < 0$. Therefore $n > 0$ and we obtain $\operatorname{sign}(E_3) = -\operatorname{sign}(\Delta_4)$.

2.1.1.1: $E_3 < 0$. In this case by Theorem 6.1 (see Table 6.2, lines 97 and 98), the nilpotent point is an elliptic-saddle, whereas the elemental singularity M_4 is an anti-saddle. Therefore since the sum of the indexes of the finite singularities equals +2, we conclude that the sum of the indices of the infinite singularities must be −1.

Additionally by Theorem 6.1 the type of the elemental singularity is governed by the invariant polynomial W_1.

1. $W_1 < 0$. Then the elemental singular point is a strong focus, and considering Lemmas 11.1 and 11.6, we should now study the cases $\eta < 0$, $\eta > 0$ and $\eta = 0$. But the last two cases have already produced all possible configurations of singularities (with an elliptic-saddle and a strong focus) in the family of systems (11.6). So we only need to study the case $\eta < 0$, for which we obtain the next two global configurations:

   - $\widehat{es}_{(3)}, f;\ S, ©, © \Rightarrow (h = 2,\ m = 1,\ n = 1)$ (if $\eta < 0$, $Y_1 \neq 0$);
   - $\widehat{es}^{-}_{(3)}, f;\ S, ©, © \Rightarrow (h = 2,\ m = 1,\ n = 2)$ (if $\eta < 0$, $Y_1 = 0$).

2. $W_1 > 0$. Under this condition we have a generic node. Moreover the above conditions imply $n > 0$, $h > 0$ and $h < 1/2$, and this gives $\eta > 0$. So by Lemma 11.1 we get the global configurations $\widehat{es}_{(3)}, n;\ S, S, N^\infty$ and $\widehat{es}^{-}_{(3)}, n;\ S, S, N^\infty$, which already have been obtained for the previous family.

3. $W_1 = 0$. Then $\tau_4 = 0$, and considering the matrix $\mathcal{M}_4$ from (11.13), we have a node n^d. We get $h = 1/2$, which implies $\eta > 0$, and we get the global configurations $\widehat{es}_{(3)}, n^d;\ S, S, N^\infty$ and $\widehat{es}^{-}_{(3)}, n^d;\ S, S, N^\infty$, both obtained earlier.

2.1.1.2: $E_3 > 0$. In this case by Theorem 6.1 we have a nilpotent triple saddle and a saddle. So this implies at infinity the existence of three nodes, and according to Lemma 11.1 we get the configuration $\hat{s}_{(3)}, s;\ N^f, N^f, N^f$, which already has been obtained.

2.1.2: $\mu_0 > 0$. Therefore $n < 0$ and we obtain $\operatorname{sign}(E_3) = \operatorname{sign}(\Delta_4)$.

2.1.2.1: $E_3 < 0$. Then $h < 0$ and in this case the elemental singularity M_4 is a saddle (which is strong due to $m \neq 0$). By Theorem 6.1 (see Table 6.2, line 88), the triple point is a nilpotent elliptic-saddle.

1. $\eta < 0$. According to Lemma 11.6 the condition $\theta \neq 0$ holds, otherwise we get $\eta \geq 0$. Moreover we observe that the condition $\eta < 0$ implies $\theta < 0$. Indeed, as $\eta < 0$ (i.e. $n - 2h + 1 < 0$), we set $n - 2h + 1 = -u^2$ (i.e. $n = 2h - 1 - u^2$), and then we calculate $\theta = 64h\big[(u^2 - h)^2 + u^2\big] < 0$ due to $h < 0$. Moreover $Y_1 = h(h - n)$, h cannot be zero by $\mu_0 = -4h^2 n > 0$ and $h - n \neq 0$, otherwise $\mu_0 = -4n^3 > 0$, implying $n < 0$, and then $\eta = 4n^2(m^2 - n)$ cannot be negative. So by Lemma 11.4 the elliptic-saddle is generic, and by Lemma 11.1 we get one unique global configuration, which is a new one

   - $\widehat{es}_{(3)}, s;\ N^\infty, ©, © \Rightarrow (h = -1/4,\ m = 1,\ n = -2)$.

2. $\eta > 0$. It was proved for the previous family (see the case defined by the conditions $\mu_0 > 0$, $E_3 > 0$, $\mathcal{T}_4 = \mathcal{T}_3 = 0$ and $\mathcal{T}_1 \neq 0$) that there could exist only six configurations. More precisely the possibilities with $\theta = 0$ and either $\theta_1 \neq 0$ and $\theta_2 = 0$ or $\theta_1 = \theta_3 = 0$ cannot be realizable. However according to Lemma 11.6 none of these cases can be realizable for the family (11.7).

3. $\eta = 0$. Since $\widetilde{M} \neq 0$, in this case according to Lemma 11.1, at infinity we can have in this case four configurations. We have proved (see the same case above mentioned) that for the previous family (11.6) all cases except the possibility $\theta = \theta_2 = 0$ are realizable, as the condition $\theta = 0$ implies $\theta_2 \neq 0$. We observe that for systems (11.7) by Lemma 11.6 the condition $\theta = 0$ also implies $\theta_2 \neq 0$, and hence we cannot obtain new global configurations.

2.1.2.2: $E_3 > 0$. As $\mu_0 > 0$, considering (11.14) we get $n < 0$ and $h > 0$. In this case by Theorem 6.1 (see Table 6.2, lines 90 and 92), the triple singular point is a saddle, whereas the elemental singularity M_4 is an anti-saddle, the type of which is governed by the invariant polynomial W_1.

1. $W_1 < 0$. Then $\tau_4 = 4(1 - 2h) < 0$, and the elemental singular point is a strong focus. Additionally considering (11.14) the conditions $n < 0$ and $h > 1/2$ imply $\eta < 0$ and $\theta > 0$. So by Lemma 11.1 we arrive at the new global configuration

   - $\hat{s}_{(3)}, f;\ N^f, \copyright, \copyright \Rightarrow (h = 1,\ m = 1,\ n = -2)$.

2. $W_1 > 0$. We have a generic node, and considering the conditions $n < 0$, $0 < h < 1/2$ and the graphics $2h - 1 - n = 0$ (for η) and $(h - n)^2 + n = 0$ (for θ), we obviously obtain that the condition $\eta < 0$ implies $\theta > 0$. Moreover, the conditions $\eta > 0$ and $\theta < 0$ give $\theta_1 > 0$, and the condition $\eta = 0$ (i.e. $n = 2h - 1$) yields $\theta = 64h^3 > 0$. So considering Lemma 11.1 we get the global configurations $\hat{s}_{(3)}, n;\ S, N^f, N^f$ (if $\eta > 0$, $\theta < 0$), $\hat{s}_{(3)}, n;\ S, N^f, N^\infty$ (if $\eta > 0$, $\theta > 0$) and $\hat{s}_{(3)}, n;\ \overline{\binom{0}{2}}SN, N^f$ (if $\eta = 0$), which already have been obtained for the previous family. In addition we get the following new global configuration

   - $\hat{s}_{(3)}, n;\ N^f, \copyright, \copyright \Rightarrow (h = 1/5,\ m = 1,\ n = -1)$ (if $\eta < 0$),

   that is realizable if the conditions $\mu_0 > 0$, $E_3 > 0$, $W_4 = 0$, $\mathcal{T}_4 = 0$, $W_1 > 0$, $\mathcal{T}_1 \neq 0$ and $\eta < 0$ hold. This allows us to include it in the diagram in the corresponding place.

3. $W_1 = 0$. Then $h = 1/2$ (in this case we have a node with one direction), and we calculate

$$\mu_0 = -n = 4E_3, \quad \eta = 4n(n-1)^2, \quad \theta = 8(1 + 4n^2) > 0.$$

   Therefore as the condition $\mu_0 > 0$ implies $\eta < 0$, we only get one global configuration that is new

   - $\hat{s}_{(3)}, n^d;\ N^f, \copyright, \copyright \Rightarrow (h = 1/2,\ m = 1,\ n = -1)$.

2.2: $\mathcal{T}_1 = 0$. Then $m = 0$, and since $n \neq 0$ due to the rescaling $(x, y, t) \mapsto (x, |n|^{-1/2}y, |n|^{-1/2}t)$, we may assume $n \in \{-1, 1\}$.

2.2.1: $\mu_0 < 0$. Then $n > 0$, and assuming $n = 1$, we get the family of systems:

$$\dot{x} = 2hxy, \quad \dot{y} = x - x^2 + y^2,$$

for which we have

$$\mu_0 = -4h^2, \ E_3 = -2h^3, \ \eta = 4(1-2h)^3, \ W_1 = -128h^5,$$
$$\mathcal{T}_4 = \mathcal{T}_3 = \mathcal{T}_2 = \mathcal{T}_1 = \mathcal{F} = \mathcal{F}_1 = 0, \ \sigma = 2(h+1)y,$$
$$\mathcal{H} = -4h(h+1)^2, \ \mathcal{B} = -2(h+1)^4, \ \widetilde{M} = 8(2h-1)(3x^2 - 2hy^2 + y^2).$$

1. $E_3 < 0$. Then $h > 0$ and this implies $\sigma \neq 0$, $\mathcal{H} < 0$, $\mathcal{B} < 0$. We observe that the condition $\eta = 0$ implies $\widetilde{M} = 0$. So by Theorems 6.2 and 6.1, we have a center and an elliptic-saddle. In this case $Y_1 = h(h-n) = 0$ implies $\eta < 0$, so by Lemma 11.4 we can have only a non-generic elliptic-saddle if $\eta < 0$. Considering Lemma 11.1, if $\eta > 0$ we get the unique global configuration, $\widehat{es}_{(3)}, c;\ S, S, N^\infty$, which already has been obtained for the previous family.

   If $\eta \leq 0$ we have the following three new global configurations:

   - $\widehat{es}_{(3)}, c;\ S,$ ©, © $\Rightarrow (h = 2,\ m = 0,\ n = 1)$ (if $\eta < 0,\ Y_1 \neq 0$);
   - $\widehat{es}^{-}_{(3)}, c;\ S,$ ©, © $\Rightarrow (h = 2,\ m = 0,\ n = 2)$ (if $\eta < 0,\ Y_1 = 0$);
   - $\widehat{es}_{(3)}, c;\ \overline{\binom{0}{3}}S \Rightarrow (h = 1/2,\ m = 0,\ n = 1)$ (if $\eta = 0$).

   These configurations are defined by the conditions $\mu_0 < 0$, $E_3 < 0$, $W_4 = 0$, $\mathcal{T}_4 = 0$, $W_1 < 0$, $\mathcal{T}_1 = 0$ and either $\eta < 0$ or $\eta = 0$. In the case $\eta < 0$ there are two subcases for whether Y_1 is zero or not.

2. $E_3 > 0$. In this case by Theorem 6.1 we have a triple nilpotent saddle and a weak saddle, which is integrable. Since the condition $h < 0$ implies $\eta > 0$, we only get one configuration, $\hat{s}_{(3)}, \$;\ N^f, N^f, N^f$, which is already obtained for the previous family.

2.2.2: $\mu_0 > 0$. Then $n < 0$, and assuming $n = -1$, we get the family of systems:

$$\dot{x} = 2hxy, \quad \dot{y} = x - x^2 - y^2,$$

for which we have

$$\mu_0 = 4h^2, \ E_3 = 2h^3, \ \eta = -4(1+2h)^3, \ W_1 = -128h^5,$$
$$\mathcal{T}_4 = \mathcal{T}_3 = \mathcal{T}_2 = \mathcal{T}_1 = \mathcal{F} = \mathcal{F}_1 = 0, \ \sigma = 2(h-1)y,$$
$$\theta = 64h(1+h)^2, \ \theta_1 = -64(1+h)(1+2h)(1+5h), \ \theta_3 = -h(5h^2 + 2h - 1),$$
$$\mathcal{H} = -4h(h-1)^2, \ \mathcal{B} = -2(h-1)^4, \ \widetilde{M} = 8(2h+1)(3x^2 - 2hy^2 - y^2).$$

1. $E_3 < 0$. Then $h < 0$ and the elemental singularity is an integrable saddle, whereas the triple point is an elliptic-saddle. We observe that in this case we

have $\theta \le 0$, and the condition $\eta = 0$ implies $\widetilde{M} = 0$. Moreover $Y_1 = h(n-h) = 0$ implies $h = n < 0$ since $\mu_0 = -4h^2 n > 0$, and then $\theta = 64h(h-n)^2 = 0$. Thus the only possibility with non-generic elliptic-saddle (see Lemma 11.4) has already appeared in the family of systems (11.6). So if $\eta > 0$ we arrive at the following three global configurations: $\widehat{es}_{(3)}, \$; S, N^\infty, N^\infty$ (if $\eta > 0$, $\theta \neq 0$, $\theta_1 < 0$), $\widehat{es}_{(3)}, \$; S, N^f, N^f$ (if $\eta > 0$, $\theta \neq 0$, $\theta_1 > 0$) and $\widehat{es}_{\overline{(3)}}, \$; S, N^d, N^d$ (if $\eta > 0$, $\theta = 0$), which already have been obtained for the previous family.

If $\eta \le 0$ we get the following two new global configurations:

- $\widehat{es}_{(3)}, \$; N^\infty, ©, © \Rightarrow (h = -1/3,\ m = 0,\ n = -1)$ (if $\eta < 0$);
- $\widehat{es}_{(3)}, \$; \overline{\binom{0}{3}} N \Rightarrow (h = -1/2,\ m = 0,\ n = -1)$ (if $\eta = 0$).

2. $E_3 > 0$. In this case $h > 0$ and the elemental singular point is a center, which exists apart from a nilpotent triple saddle. We observe that the condition $h > 0$ implies $\eta < 0$, $\theta > 0$, and we get the new unique global configuration
   - $\widehat{s}_{(3)}, c; N^f, ©, © \Rightarrow (h = 1,\ m = 0,\ n = -1)$.

3: The family of systems (11.8). So we consider the family of systems

$$\dot{x} = x + y - x^2 + 2hxy, \quad \dot{y} = 2my(x+y), \tag{11.8}$$

which possess the triple semi-elemental singular point $M_{1,2,3}(0,0)$ and the elemental one $M_4(1,0)$. For the second singularity we have

$$\mathcal{M}_4 = \begin{pmatrix} -1 & 1+2h \\ 0 & 2m \end{pmatrix}, \quad \rho_4 = 2m-1, \ \ \Delta_4 = -2m; \ \ \tau_4 = (2m+1)^2. \tag{11.15}$$

For this family we calculate

$$\begin{gathered} \mu_0 = (1+2h)\Delta_4^2, \quad \eta = 4(h-m)^2(2m+1)^2, \\ E_3 = (1+2h)\Delta_4^3/4, \quad G_{10} = -(1+2h)\Delta_4^3/8, \quad \mathcal{F}_1 = 0; \\ \mathcal{T}_4 = (1+2h)\Delta_4^2\rho_4, \quad \mathcal{T}_3 = 2(1+2h)\Delta_4^2(3m-1), \\ \mathcal{T}_2 = -3(1+2h)\Delta_4^3, \quad W_4 = (1+2h)^2\Delta_4^4\tau_4, \\ \widetilde{M} = -8(1+2m)^2x^2 + 16(h-m)(1+2m)xy - 32(h-m)^2y^2, \\ \theta = 32h(h-2m-2m^2)\Delta_4, \quad \theta_2 = (1+h+m)\Delta_4/2, \\ \theta_1 = 64(h^2 - 2hm + 6h^2m - 14hm^2 + 4h^2m^2 + 4m^3 - 12hm^3 + 4m^4), \\ \theta_3 = (1+2h)(m^2+2m-1)\Delta_4/2. \end{gathered} \tag{11.16}$$

Lemma 11.7. *For a system* (11.8) *with* $\mu_0 \neq 0$ *the following statements hold:*

(i) *The conditions* $E_3 G_{10}\widetilde{M} \neq 0$, $\eta \ge 0$ *and* $W_4 \ge 0$ *are fulfilled.*

(ii) *If* $\theta = 0$ *then the condition* $\theta_1 = 0$ *is equivalent to* $\theta_2 = 0$ *and in the last case* $\theta_3 \neq 0$.

Proof. (i) The conditions $E_3G_{10} \neq 0$, $\eta \geq 0$ and $W_4 \geq 0$ follow directly from (11.16). Assume $\widetilde{M} = 0$. By (11.16) this implies $m = -1/2$, and then we calculate $\widetilde{M} = -8(1+2h)^2y^2 \neq 0$ due to $\mu_0 \neq 0$.

(ii) Assume $\theta = 0$. Then we obtain either (a) $h = 0$ or (b) $h = 2m(m+1)$. Then the calculations yield

$$\theta_1 = 256m^3(1+m), \quad \theta_2 = -m(1+m), \quad \theta_3 = -m(m^2+2m-1), \quad \mu_0 = 4m^2,$$

in the case (a) and

$$\theta_1 = 256m^3(1+m)(1+2m)^2, \quad \theta_2 = -m(1+m)(1+2m),$$
$$\theta_3 = -m(m^2+2m-1)(1+2m)^2, \quad \mu_0 = 4m^2(1+2m)^2,$$

in the case (b). Clearly in both cases due to $\mu_0 \neq 0$, the condition $\theta_1 = 0$ is equivalent to $\theta_2 = 0$, and in the case $\theta_1 = 0$ we get $\theta_3 \neq 0$. □

3.1: $\mu_0 < 0$. According to Table 6.2 (see Theorem 6.1) we consider two subcases: $E_3 < 0$ and $E_3 > 0$.

3.1.1: $E_3 < 0$. Since by Lemma 11.7 we have $G_{10} \neq 0$ and $W_4 \geq 0$, according to Theorem 6.1 the triple singularity is a semi-elemental node and the elemental one is a node.

1. $W_4 \neq 0$. Then we have a generic node. As by Lemma 11.7 we have $\eta \geq 0$, according to Lemma 11.1 this leads to the two configurations that were previously detected for the family (11.6).
2. $W_4 = 0$. In this case $\tau_4 = 0$, which implies $m = -1/2$. So we have a node with two coinciding eigenvalues, which could not be a star node because $\mu_0 \neq 0$ (i.e. $2h+1 \neq 0$ and this appears in the matrix $\mathcal{M}_4$ from (11.15)). In this case we have $\eta = 0$, and since by Lemma 11.7 the condition $\widetilde{M} \neq 0$ holds, we get the unique global configuration $\overline{n}_{(3)}, n^d; \overline{\binom{0}{3}}S$, which was previously detected by the same invariant conditions.

3.1.2: $E_3 > 0$. By Theorem 6.1 we have a semi-elemental triple saddle and a saddle that could be a weak one. At infinity we have three nodes (i.e. $\eta > 0$), and then in the case $\mathcal{T}_4 \neq 0$ we obtain the same configuration previously found. Assuming $\mathcal{T}_4 = 0$ we get $m = 1/2$, and then we calculate

$$\mathcal{T}_4 = \mathcal{F}_1 = 0, \quad \mathcal{T}_3\mathcal{F} = (1+2h)^2/8, \quad \mathcal{F}_2 = -3(1+2h)^2/2. \qquad (11.17)$$

Therefore due to $\mu_0 \neq 0$ we have $\mathcal{T}_3\mathcal{F} > 0$ and $\mathcal{F}_2 \neq 0$, and according to [338] we have a weak saddle of order two. Considering the condition $\eta > 0$ and Lemma 11.1, we get only one global configuration and it is new

• $\overline{s}_{(3)}, s^{(2)}; N^f, N^f, N^f \Rightarrow (h=-1,\ m=1/2)$.

3.2: $\mu_0 > 0$. We consider two possibilities: $E_3 < 0$ and $E_3 > 0$.

3.2.1: $E_3 < 0$. By Theorem 6.1 we have a triple semi-elemental node and an elemental saddle.

1. $\mathcal{T}_4 \neq 0$. In this case the saddle is strong. It was shown earlier (see page 413) that in the case $\mu_0 > 0$, $E_3 < 0$, $\mathcal{T}_4 \neq 0$ and $\eta \geq 0$, for the family (11.6) all the possible configurations for the infinite singularities are realizable except the cases $\theta = \theta_2 = 0$, $\theta_1 \neq 0$ and $\theta = \theta_1 = \theta_3 = 0$. However by Lemma 11.7 none of these cases can be realizable for the family (11.8).
2. $\mathcal{T}_4 = 0$. We obtain $m = 1/2$ and we arrive at the relations (11.17). According to Theorem 6.2 (see statement (b_2)), we have a weak saddle of order two. In this case we have

$$\theta = 16h(3-2h),\quad \theta_1 = 16(3-24h+20h^2),\quad \theta_2 = -(3+2h)/4,$$
$$E_3 = -(1+2h)/4,\quad \eta = 4(2h-1)^2.$$

We observe that the condition $\theta \leq 0$ implies $\theta_1 > 0$, and in the case $\theta = 0$ we have $\theta_2 \neq 0$. So considering Lemmas 11.7 and 11.1, we get the following three new global configurations in the case $\eta \neq 0$ (i.e. $\eta > 0$):

• $\overline{n}_{(3)}, s^{(2)}; S, N^f, N^f \Rightarrow (h=2,\ m=1/2)$ (if $\theta < 0$);

• $\overline{n}_{(3)}, s^{(2)}; S, N^f, N^\infty \Rightarrow (h=1,\ m=1/2)$ (if $\theta > 0$);

• $\overline{n}_{(3)}, s^{(2)}; S, N^f, N^d \Rightarrow (h=0,\ m=1/2)$ (if $\theta = 0$).

In the case $\eta = 0$ we have $h = 1/2$, and this implies $\theta > 0$. So we get the new global configuration

• $\overline{n}_{(3)}, s^{(2)}; \overline{\binom{0}{2}}SN, N^f \Rightarrow (h=1/2,\ m=1/2)$.

3.2.2: $E_3 > 0$. By Theorem 6.1 the triple point is a semi-elemental saddle and the elemental singular point is an anti-saddle, which must be a node due to $\tau_4 \geq 0$.

It was shown earlier (see page 418) that in the case $\mu_0 > 0$, $E_3 > 0$, $\eta \geq 0$ and $W_4 > 0$ for the family (11.6), all the possible configurations for the infinite singularities are realizable except the cases $\theta = \theta_2 = 0$ and $\theta = \theta_1 = \theta_3 = 0$. However by Lemma 11.7 none of these cases can be realizable for the family (11.8).

In the case $W_4 = 0$ we get $m = -1/2$ and then $\eta = 0$ and $\theta^2 + \theta_2^2 \neq 0$, and again no new global configurations appear.

4: The family of systems (11.5). So we consider the family of systems

$$\dot{x} = x + dy - x^2 + 2hxy + ky^2,\quad \dot{y} = 2my(x+dy),\quad d \in \{0,1\}, \tag{11.5}$$

which possess the triple semi-elemental singular point $M_{1,2,3}(0,0)$ and the elemental one $M_4(1,0)$. For the second singularity we have

$$\mathcal{M}_4 = \begin{pmatrix} -1 & d+2h \\ 0 & 2m \end{pmatrix}, \quad \rho_4 = 2m-1, \ \ \Delta_4 = -2m; \ \ \tau_4 = (2m+1)^2. \quad (11.18)$$

For this family we calculate

$$\begin{aligned} &\mu_0 = \alpha\Delta_4^2, \ \ \eta = 4(1+2m)^2\big[(h-dm)^2 + k(2m+1)\big], \ \ E_3 = \alpha\Delta_4^3/4, \\ &\quad G_{10} = -\alpha\Delta_4^3/8, \ \mathcal{F}_1 = 0, \ \ \mathcal{T}_4 = \alpha\Delta_4^2\rho_4, \ \mathcal{T}_3 = 2\alpha\Delta_4^2(3m-1), \\ &\quad W_4 = \alpha^2\Delta_4^4\tau_4, \ \ \widetilde{M} = -8(1+2m)^2x^2 + 16(1+2m)(h-dm)xy \\ &\qquad\qquad -8(4h^2+3k-8dhm+6km+4d^2m^2)y^2, \end{aligned} \quad (11.19)$$

where $\alpha = (d^2+2dh-k)$. We observe that due to $\mu_0 \neq 0$ for the above systems the conditions $E_3G_{10} \neq 0$ and $W_4 \geq 0$ are satisfied.

According to Remark 11.1 we have to consider only two cases: $\eta < 0$ or $\widetilde{M} = 0$.

4.1: $\eta < 0$. Then $1+2m \neq 0$ (i.e. $\tau_4 > 0$) and the elemental singular point is either a generic node or a saddle.

4.1.1: $\mu_0 < 0$. Then $E_3 < 0$, otherwise by Theorem 6.1 the systems possess two finite saddles (one triple and one elemental), which implies the existence of three nodes at infinity, i.e. $\eta > 0$. So $E_3 < 0$, and according to Theorem 6.1 the systems (11.5) possess a semi-elemental triple node and a generic node. Considering the condition $\eta < 0$ by Lemma 11.1, we arrive at the unique global configuration $\overline{n}_{(3)}, n;\ S, ©, ©$, previously detected.

4.1.2: $\mu_0 > 0$. Since we are interested in the case $\eta < 0$, according to Lemma 11.1 in this case the invariant polynomial θ is necessary to be taken into consideration. We calculate

$$\theta = -64m\big[k(m+1)^2 - 2hdm(m+1) + h^2\big],$$

and we claim that the conditions $\mu_0 \neq 0$ and $\eta < 0$ imply $\theta \neq 0$. Indeed, if $m+1 \neq 0$ then setting $\theta = 0$ we obtain $k = \big[2hdm(m+1) - h^2\big]/(m+1)^2$. However in this case we get

$$\eta = 4m^2(1+2m)^2(d+h+dm)^2/(1+m)^2 \geq 0,$$

which contradicts $\eta < 0$.

Assume now $m = -1$. Then the condition $\theta = 64h^2 = 0$ gives $h = 0$, and we obtain $\eta = 4(d^2-k) = \mu_0 > 0$. The contradiction obtained completes the proof of our claim, and in what follows we assume $\theta \neq 0$.

1. $E_3 < 0$. According to Theorem 6.1 the systems (11.5) have a semi-elemental triple node and an elemental saddle.

(a) $\mathcal{T}_4 \neq 0$. Then the saddle is strong, and by Theorem 6.1 we obtain two global configurations: $\overline{n}_{(3)}, s;\ N^{\infty}, ⓒ, ⓒ$ if $\theta < 0$ and $\overline{n}_{(3)}, s;\ N^f, ⓒ, ⓒ$ if $\theta > 0$. However both configurations were previously constructed.

(b) $\mathcal{T}_4 = 0$. Then $\rho_4 = 0$ (i.e. $m = 1/2$), and we obtain:

$$\mathcal{T}_4 = \mathcal{F}_1 = 0, \quad \mathcal{T}_3\mathcal{F} = (d^2 + 2dh - k)^2/8, \quad \mathcal{F}_2 = -3(d^2 + 2dh - k)^2/2,$$
$$\mu_0 = d^2 + 2dh - k, \quad \eta = 4\big[(d - 2h)^2 + 8k\big], \quad \theta = 8(6dh - 4h^2 - 9k).$$

So the condition $\mu_0 \neq 0$ implies $\mathcal{T}_3\mathcal{F} > 0$ and $\mathcal{F}_2 \neq 0$, and by Theorem 6.2 (see statement (b_2)) we have a weak saddle of the second order. Since for (11.5) we have $d \in \{0, 1\}$, we observe that in both cases the functions $\eta = 0$ and $\theta = 0$ represent two parabolas having a contact point of multiplicity two. Moreover the region $\{\eta = 0\} \subset \{\theta \geq 0\}$ and the region $\{\theta = 0\} \subset \{\eta \geq 0\}$, and hence the condition $\eta < 0$ implies $\theta > 0$. So considering Theorem 6.1, we arrive at the following new global configuration

• $\overline{n}_{(3)}, s^{(2)};\ N^f, ⓒ, ⓒ \Rightarrow (d = 0,\ h = 0,\ k = -1, m = 1/2)$.

2. $E_3 > 0$. By Theorem 6.1 the systems (11.5) have a semi-elemental triple saddle and an elemental node. We observe that due to $\eta < 0$ the condition $\tau_4 \neq 0$ (i.e. $W_4 \neq 0$) holds. Then the node is generic, and we can have only the following two global configurations: $\overline{s}_{(3)}, n;\ N^{\infty}, ⓒ, ⓒ$ if $\theta < 0$ and $\overline{s}_{(3)}, s;\ N^f, ⓒ, ⓒ$ if $\theta > 0$. However these global configurations were previously detected.

4.2: $\widetilde{M} = 0$. Considering (11.19) the condition $\widetilde{M} = 0$ is equivalent to $m = -1/2$ and $d = -2h$, and by (11.18) the elemental singular point is a star node. Since in this case we have $\mu_0 = -k$ and $E_3 = -k/4$, according to Theorem 6.1 and Lemma 11.1, we get the following two new global configurations:

• $\overline{n}_{(3)}, n^*;\ \overline{\binom{0}{3}}S \Rightarrow (d = 0,\ h = 0,\ k = 1, m = -1/2)$ (if $\mu_0 < 0$);

• $\overline{s}_{(3)}, n^*;\ \overline{\binom{0}{3}}N \Rightarrow (d = 0,\ h = 0,\ k = -1, m = -1/2)$ (if $\mu_0 > 0$).

In both cases we have $U_3 = 0$, whereas for the family (11.6) with $\widetilde{M} = 0$ (which is equivalent to $m = -c/2$ and $n = -c^2/(1 + c)$), we have that Coefficient$[U_3, y^5] = 3c^7/(1 + c)^3 \neq 0$. Therefore the invariant polynomial U_3 will distinguish in both cases the existence of a star node from the existence of a node n^d.

We remark that the condition $U_3 = 0$ is equivalent to $\widetilde{M} = 0$ for systems (11.5). Moreover it is not difficult to check that for the families of systems (11.6), (11.7) and (11.8), the condition $U_3 = 0$ cannot be satisfied.

As all the cases are examined, we have constructed all 130 possible configurations for the family of quadratic systems possessing exactly two finite singularities, both real: one elemental and one of multiplicity three. Therefore Theorem 11.2 is completely proved. □

11.2.2 Two double real singularities

Theorem 11.3. (a) *We consider here all the configurations of singularities, finite and infinite, of quadratic vector fields possessing exactly two distinct finite singularities both real and of multiplicity two. These configurations are classified in Diagram* 11.4 *according to the geometrical equivalence relation of configurations. We have a total of* 33 *geometrically distinct configurations of singularities, finite and infinite.*

(b) *Necessary and sufficient conditions for each one of the* 33 *different geometrical equivalence classes can be assembled from these diagrams in terms of* 20 *invariant polynomials with respect to the action of the affine group and time rescaling. These invariant polynomials, which appear in Diagram* 11.4, *are given in Section* 5.4.

(c) *Diagram* 11.4 *actually contains the global bifurcation diagram in the* 12*-dimensional space of parameters of the global configurations of singularities, finite and infinite, of this family of quadratic differential systems and provides an algorithm for finding for any given system in the family considered, its respective geometrical configuration of singularities.*

Proof. Assume that systems (8.1) possess two double real finite singularities. Then according to Proposition 6.2 (see Table 6.1), via an affine transformation these systems could be brought to the systems 6, i.e. to the systems of the form:

$$\begin{aligned} \dot{x} &= cx + cuy - cx^2 + 2cvxy + ky^2, \\ \dot{y} &= ex + euy - ex^2 + 2evxy + ny^2, \end{aligned} \tag{11.20}$$

with $cn - ek \neq 0$. These systems possess two finite double distinct singularities: $M_{1,2}(0,0)$ and $M_{3,4}(1,0)$.

Following Theorem 6.1 for this family of systems we calculate

$$\begin{aligned} \mu_0 &= (cn-ek)^2, \quad E_2 = -e(ek-cn)^2(u+v), \\ G_1 &= (cn-ek)^2(c+eu)^2(c-eu-2ev)^2, \end{aligned} \tag{11.21}$$

and hence $\mu_0 > 0$. Moreover according to Theorem 6.1 (see Table 6.2, lines 84–86), systems (11.20) possess two saddle-nodes if $G_1 \neq 0$, one saddle-node and one cusp if $G_1 = 0$ and $E_2 \neq 0$ and two cusps if $G_1 = E_2 = 0$.

Lemma 11.8. *For systems* (11.20) *the following statements are valid:*

(i) *The conditions $\theta = \theta_1 = 0$ imply $\theta_3 = 0$.*

(ii) *The condition $E_2 \neq 0$ implies $\theta_2 \neq 0$; moreover if in addition the condition $\theta = 0$ is satisfied, then the condition $\theta_1 \neq 0$ holds.*

(iii) *The condition $G_1 = E_2 = 0$ implies $\theta > 0$ and $\widetilde{M} \neq 0$.*

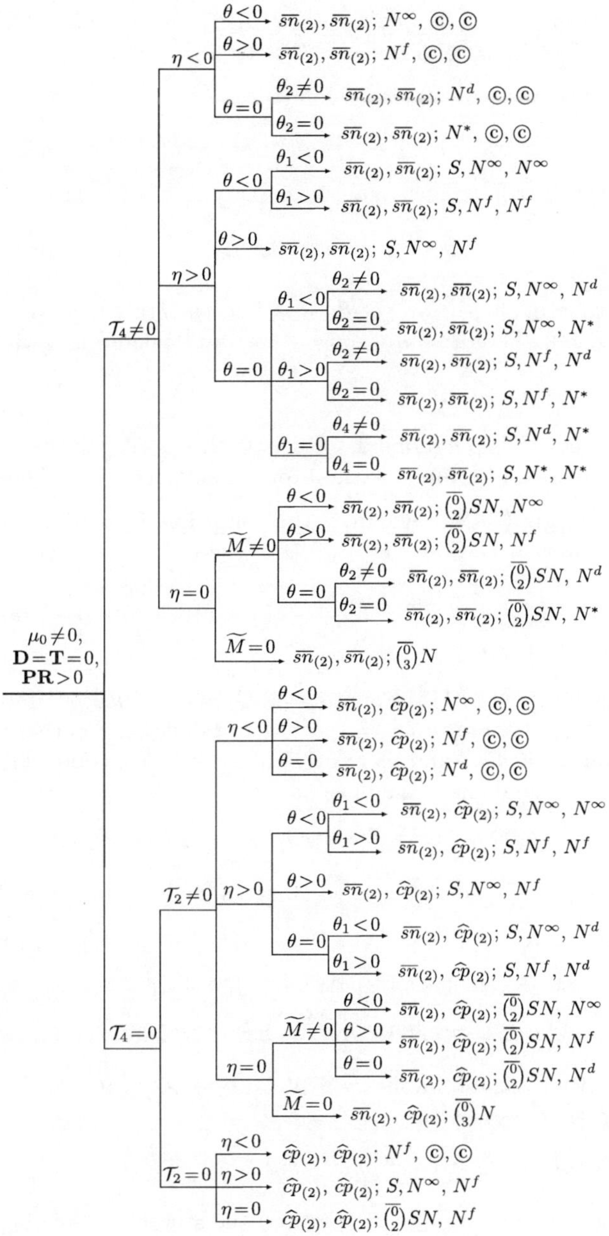

Diagram 11.4: Global configurations in the case $\mu_0 \neq 0$, $\mathbf{D} = \mathbf{T} = 0$, $\mathbf{PR} > 0$.

(iv) *The condition* $G_1 = 0$ *is equivalent to* $\mathcal{T}_4 = 0$*, and if* $G_1 = 0$ *then the condition* $E_2 = 0$ *is equivalent to* $\mathcal{T}_2 = 0$.

Proof. (i) For systems (11.20) we calculate

$$\begin{aligned} \theta &= 64e(ek-cn)(k-nv+cv^2-ev^3), \quad \theta_2 = e(ek-cn)(u+v), \\ \theta_3 &= e(ek-cn)\widehat{U}(c,e,k,n,u,v), \quad E_2 = -e(ek-cn)^2(u+v), \end{aligned} \tag{11.22}$$

where $\widehat{U}(c,e,k,n,u,v)$ is a polynomial. As $\mu_0 \neq 0$ the condition $\theta = 0$ gives $e(k-nv+cv^2-ev^3) = 0$.

If $e = 0$ then evidently we get $\theta_3 = 0$, and the statement (i) holds.

Assume $e \neq 0$. Then $k = nv - cv^2 + ev^3$, and calculations yield

$$\theta = 0, \quad \theta_1 = 64e(n+ev^2)^3, \quad \theta_3 = e(c-ev)^2(n+ev^2)\widehat{V}(c,e,n,u,v),$$

where $\widehat{V}(c,e,n,u,v)$ is a polynomial. Clearly in this case again the condition $\theta_1 = 0$ implies $\theta_3 = 0$, and this completes the proof of statement (i) of the lemma.

(ii) Considering (11.22) evidently the condition $E_2 \neq 0$ implies $\theta_2 \neq 0$.

Assume now $\theta = 0$. Since $E_2 \neq 0$ this yields $k = nv - cv^2 + ev^3$. Then we have

$$\theta_1 = 64e(n+ev^2)^3, \quad E_2 = -e(u+v)(c-ev)^2(n+ev^2)^2,$$

and clearly the condition $E_2 \neq 0$ gives $\theta_1 \neq 0$.

(iii) Considering (11.21) the conditions $E_2 = 0$ and $\mu_0 \neq 0$ imply $e(u+v) = 0$. We observe that the condition $u = -v$ has to be satisfied, otherwise in the case $e = 0$ we obtain $G_1 = c^6n^2$ and $\mu_0 = c^2n^2$, and hence the condition $\mu_0 \neq 0$ implies $G_1 \neq 0$. So $u = -v$ and then calculations yield

$$G_1 = (ek-cn)^2(c-ev)^4, \quad \mu_0 = (ek-cn)^2.$$

Therefore as $\mu_0 \neq 0$ the condition $G_1 = 0$ gives $c = ev$, and we obtain

$$\begin{aligned} G_1 = E_2 = 0, \quad \theta &= 64e^2(k-nv)^2, \quad \mu_0 = e^2(k-nv)^2, \\ \widetilde{M}/8 &= -3e(n+ev^2)x^2 + 3e(3k-nv+2ev^3)xy \\ &\quad + (4env^2 - n^2 - 9ekv - 4e^2v^4)y^2. \end{aligned}$$

Hence the condition $\mu_0 \neq 0$ implies $\theta > 0$. Additionally the condition $\widetilde{M} = 0$ gives $n = -ev^2$, and then we obtain

$$\widetilde{M} = 72e(k+ev^3)y(x-vy), \quad \mu_0 = e^2(k+ev^3)^2.$$

So the condition $\mu_0 \neq 0$ implies $\widetilde{M} \neq 0$, and the statement (iii) is proved.

(iv) For systems (11.20) calculations confirm that $\mathcal{T}_4 = G_1$. Assuming $G_1 = 0$ (i.e. $\mathcal{T}_4 = 0$) due to $\mu_0 \neq 0$, we obtain either

$$E_2 = -e^3(k+nu)^2(u+v), \quad \mathcal{T}_2 = 4e^4(k+nu)^2(u+v)^2$$

if $c = -eu$ or

$$E_2 = -e^3(u+v)(k - nu - 2nv)^2, \quad \mathcal{T}_2 = 4e^4(u+v)^2(k - nu - 2nv)^2$$

if $c = e(u + 2v)$. In both cases we obtain that the condition $E_2 = 0$ is equivalent to $\mathcal{T}_2 = 0$, and this completes the proof of Lemma 11.8. □

Considering Theorem 6.1 and the above lemma, we get the next remark.

Remark 11.3. Systems (11.20) possess two saddle-nodes if $\mathcal{T}_4 \neq 0$, one saddle-node and one cusp if $\mathcal{T}_4 = 0$ and $\mathcal{T}_2 \neq 0$ and two cusps if $\mathcal{T}_4 = \mathcal{T}_2 = 0$.

1: $\mathcal{T}_4 \neq 0$. By the above remark both double finite singular points are saddle-nodes.

1.1: $\eta < 0$. In this case systems (11.20) possess one real and two complex infinite singular points, and according to Lemma 11.1 there could be only four distinct configurations at infinity. It remains to construct corresponding global configurations:

- $\overline{sn}_{(2)}, \overline{sn}_{(2)};\ N^{\infty}, ©, © \Rightarrow (c = -1, e = 1, k = 0, n = -3, u = 0, v = 1)$ (if $\theta < 0$);
- $\overline{sn}_{(2)}, \overline{sn}_{(2)};\ N^{f}, ©, © \Rightarrow (c = 1,\ e = 1,\ k = 0,\ n = -3,\ u = 0,\ v = 1)$ (if $\theta > 0$);
- $\overline{sn}_{(2)}, \overline{sn}_{(2)};\ N^{d}, ©, © \Rightarrow (c = 1,\ e = -1,\ k = 0,\ n = 1,\ u = 2,\ v = 0)$ (if $\theta = 0, \theta_2 \neq 0$);
- $\overline{sn}_{(2)}, \overline{sn}_{(2)};\ N^{*}, ©, © \Rightarrow (c = 1,\ e = 0,\ k = -1,\ n = 1,\ u = 0,\ v = 0)$ (if $\theta = 0, \theta_2 = 0$).

1.2: $\eta > 0$. In this case systems (11.20) possess three real infinite singular points, and taking into consideration Lemma 11.8(i) and the condition $\mu_0 > 0$, by Lemma 11.1 we could have at infinity only nine distinct configurations. The corresponding global configurations are as follows:

- $\overline{sn}_{(2)}, \overline{sn}_{(2)};\ S, N^{\infty},\ N^{\infty} \Rightarrow (c = -4,\ e = 1,\ k = -1,\ n = -8,\ u = 0,\ v = 1)$ (if $\theta < 0, \theta_1 < 0$);
- $\overline{sn}_{(2)}, \overline{sn}_{(2)};\ S, N^{f},\ N^{f} \Rightarrow (c = -1,\ e = 1,\ k = 0,\ n = 1,\ u = 0,\ v = 1)$ (if $\theta < 0, \theta_1 > 0$);
- $\overline{sn}_{(2)}, \overline{sn}_{(2)};\ S, N^{\infty},\ N^{f} \Rightarrow (c = 1,\ e = 1,\ k = 0,\ n = 1,\ u = 0,\ v = 1)$ (if $\theta > 0$);
- $\overline{sn}_{(2)}, \overline{sn}_{(2)};\ S, N^{\infty},\ N^{d} \Rightarrow (c = 3,\ e = 1,\ k = 0,\ n = -1,\ u = 1,\ v = 0)$ (if $\theta = 0, \theta_1 < 0, \theta_2 \neq 0$);
- $\overline{sn}_{(2)}, \overline{sn}_{(2)};\ S, N^{\infty},\ N^{*} \Rightarrow (c = 1,\ e = 0,\ k = 1,\ n = 3,\ u = 0,\ v = 1)$ (if $\theta = 0, \theta_1 < 0, \theta_2 = 0$);
- $\overline{sn}_{(2)}, \overline{sn}_{(2)};\ S, N^{f},\ N^{d} \Rightarrow (c = 2,\ e = 1,\ k = 0,\ n = 1,\ u = 1,\ v = 1)$ (if $\theta = 0, \theta_1 > 0, \theta_2 \neq 0$);
- $\overline{sn}_{(2)}, \overline{sn}_{(2)};\ S, N^{f},\ N^{*} \Rightarrow (c = 1,\ e = 0,\ k = 1,\ n = 1,\ u = 0,\ v = 1)$ (if $\theta = 0, \theta_1 > 0, \theta_2 = 0$);
- $\overline{sn}_{(2)}, \overline{sn}_{(2)};\ S, N^{d},\ N^{*} \Rightarrow (c = 1,\ e = 0,\ k = 0,\ n = 1,\ u = 1,\ v = 0)$ (if $\theta = 0, \theta_1 = 0, \theta_4 \neq 0$);

• $\overline{sn}_{(2)},\overline{sn}_{(2)}$; S, N^*, $N^* \Rightarrow (c=1,\ e=0,\ k=-2,\ n=1,\ u=1,\ v=-1)$ (if $\theta=\theta_1=\theta_4=0$).

1.3: $\eta=0$. In this case systems (11.20) possess at infinity either one double and one simple real singular points (if $\widetilde{M}\neq 0$) or one triple real singularity (if $\widetilde{M}=0$). So by Lemma 11.1 we could have at infinity exactly five configurations which lead to the global configurations:

• $\overline{sn}_{(2)},\overline{sn}_{(2)}$; $\overline{\binom{0}{2}}SN$, $N^\infty \Rightarrow (c=-3, e=1, k=0, n=-6, u=0, v=1)$ (if $\widetilde{M}\neq 0$, $\theta<0$);

• $\overline{sn}_{(2)},\overline{sn}_{(2)}$; $\overline{\binom{0}{2}}SN$, $N^f \Rightarrow (c=1,\ e=1,\ k=0,\ n=2,\ u=0,\ v=1)$ (if $\widetilde{M}\neq 0$, $\theta>0$);

• $\overline{sn}_{(2)},\overline{sn}_{(2)}$; $\overline{\binom{0}{2}}SN$, $N^d \Rightarrow (c=-1, e=1, k=0, n=-2, u=0, v=1)$ (if $\widetilde{M}\neq 0$, $\theta=0, \theta_2\neq 0$);

• $\overline{sn}_{(2)},\overline{sn}_{(2)}$; $\overline{\binom{0}{2}}SN$, $N^* \Rightarrow (c=-1, e=1, k=0, n=-2, u=-1, v=1)$ (if $\widetilde{M}\neq 0$, $\theta=\theta_2=0$),

• $\overline{sn}_{(2)},\overline{sn}_{(2)}$; $\overline{\binom{0}{3}}N \Rightarrow (c=3, e=3, k=-1/9, n=-1, u=0, v=0)$ (if $\widetilde{M}=0$).

2: $\mathcal{T}_4=0$. In this case we have at least one cusp, and considering Remark 11.3 we consider two subcases: $\mathcal{T}_2\neq 0$ and $\mathcal{T}_2=0$.

2.1: $\mathcal{T}_2\neq 0$. Then by Remark 11.3 the systems (11.20) possess one saddle-node and one cusp.

1. $\eta<0$. In this case systems (11.20) possess one real and two complex infinite singular points. Considering Lemma 11.8 (see statements (ii) and (iv)) and the condition $\mu_0>0$, by Lemma 11.1 there could be only three distinct configurations at infinity which provide the corresponding global configurations:

   • $\overline{sn}_{(2)},\widehat{cp}_{(2)}$; N^∞, ©, © $\Rightarrow (c=1,\ e=3,\ k=0,\ n=-3,\ u=1,\ v=-1/3)$ (if $\theta<0$);

   • $\overline{sn}_{(2)},\widehat{cp}_{(2)}$; N^f, ©, © $\Rightarrow (c=-2,\ e=-1,\ k=0,\ n=1,\ u=0,\ v=1)$ (if $\theta>0$);

   • $\overline{sn}_{(2)},\widehat{cp}_{(2)}$; N^d, ©, © $\Rightarrow (c=-1,\ e=-1,\ k=0,\ n=1,\ u=1,\ v=0)$ (if $\theta=0$).

2. $\eta>0$. In this case systems (11.20) possess three real infinite singular points, and taking into consideration Lemma 11.8 (see statements (ii) and (iv)) and the condition $\mu_0>0$, by Lemma 11.1 there could be only five distinct configurations at infinity which provide the corresponding global configurations:

   • $\overline{sn}_{(2)},\widehat{cp}_{(2)}$; S, N^∞, $N^\infty \Rightarrow (c=-5,\ e=1,\ k=0,\ n=-11,\ u=5,\ v=1)$ (if $\theta<0, \theta_1<0$);

   • $\overline{sn}_{(2)},\widehat{cp}_{(2)}$; S, N^f, $N^f \Rightarrow (c=3,\ e=1,\ k=0,\ n=1,\ u=-3,\ v=1)$ (if $\theta<0, \theta_1>0$);

- $\overline{sn}_{(2)}, \widehat{cp}_{(2)}$; $S, N^{\infty}, N^f \Rightarrow (c=1,\ e=1,\ k=0,\ n=1,\ u=-1,\ v=2)$ (if $\theta>0$);
- $\overline{sn}_{(2)}, \widehat{cp}_{(2)}$; $S, N^{\infty}, N^d \Rightarrow (c=-3,,e=1,k=0,n=-1,u=3,v=0)$ (if $\theta=0, \theta_1<0$);
- $\overline{sn}_{(2)}, \widehat{cp}_{(2)}$; $S, N^f, N^d \Rightarrow (c=1,\ e=1,\ k=0,\ n=1,\ u=-1,\ v=0)$ (if $\theta=0, \theta_1>0$).

3. $\eta=0$. In this case systems (11.20) possess at infinity either one double and one simple real singular points (if $\widetilde{M}\neq 0$) or one triple real singularity (if $\widetilde{M}=0$). So by Lemma 11.8 (see statements (ii) and (iv)) and Lemma 11.1, we could have at infinity exactly four distinct configurations and they lead to the corresponding global configurations:
   - $\overline{sn}_{(2)}, \widehat{cp}_{(2)}$; $\binom{\overline{0}}{2}SN, N^{\infty} \Rightarrow (c=-3,e=1,k=0,n=-6,u=3,v=1)$ (if $\widetilde{M}\neq 0,\ \theta<0$);
   - $\overline{sn}_{(2)}, \widehat{cp}_{(2)}$; $\binom{\overline{0}}{2}SN, N^f \Rightarrow (c=2,\ e=1,\ k=0,\ n=4,\ u=-2,\ v=1)$ (if $\widetilde{M}\neq 0,\ \theta>0$);
   - $\overline{sn}_{(2)}, \widehat{cp}_{(2)}$; $\binom{\overline{0}}{2}SN, N^d \Rightarrow (c=-1,e=1,k=0,n=-2,u=1,v=1)$ (if $\widetilde{M}\neq 0,\ \theta=0$);
   - $\overline{sn}_{(2)}, \widehat{cp}_{(2)}$; $\binom{\overline{0}}{3}N \Rightarrow (c=3,\ e=3,\ k=-1/9,\ n=-1,\ u=-1,\ v=0)$ (if $\widetilde{M}=0$).

2.2: $\mathcal{T}_2=0$. Then by Remark 11.3 the systems (11.20) possess two nilpotent cusps. Additionally by Lemma 11.8(iii) the conditions $\theta>0$ and $\widetilde{M}\neq 0$ are satisfied. Therefore according to Lemma 11.1 in this case we could have only three global configurations:

- $\widehat{cp}_{(2)}, \widehat{cp}_{(2)}$; N^f, ©, © $\Rightarrow (c=0,\ e=1,\ k=1,\ n=0,\ u=0,\ v=0)$ (if $\eta<0$);
- $\widehat{cp}_{(2)}, \widehat{cp}_{(2)}$; $S, N^{\infty}, N^f \Rightarrow (c=-1,e=1,k=0,n=1,u=1,v=-1)$ (if $\eta>0$);
- $\widehat{cp}_{(2)}, \widehat{cp}_{(2)}$; $\binom{\overline{0}}{2}SN, N^f \Rightarrow (c=1,\ e=1,\ k=0,\ n=2,\ u=-1,\ v=1)$ (if $\eta=0$).

As all the cases are examined, we have detected all 33 possible global configurations for the family of quadratic systems possessing exactly two distinct finite singularities both real and of multiplicity two. So Theorem 11.3 is completely proved. □

11.2.3 Two double complex singularities

Theorem 11.4. (a) *We consider here all the configurations of singularities, finite and infinite, of quadratic vector fields possessing two distinct finite singularities both complex and each of multiplicity two. These configurations are* *classified in Diagram* 11.5 *according to the geometrical equivalence relation of configurations. We have* 18 *geometrically distinct configurations of singularities, finite and infinite.*

(b) *Necessary and sufficient conditions for each one of the* 18 *different geometrical equivalence classes can be assembled from these diagrams in terms of* 11 *invariant polynomials with respect to the action of the affine group and time rescaling. These invariant polynomials, which appear in Diagram* 11.5, *are given in Section* 5.4.

(c) *Diagram* 11.5 *actually contains the global bifurcation diagram in the* 12-*dimensional space of parameters of the global configurations of singularities, finite and infinite, of this family of quadratic differential systems and provides an algorithm for finding for any given system in the family considered, its respective geometrical configuration of singularities.*

Proof. Assume that systems (8.1) have two double complex finite singularities. In this case according to Proposition 6.2 (see Table 6.1) via an affine transformation, these systems could be brought to the systems 7, i.e. to the systems of the form:

$$\begin{aligned} \dot{x} &= a + aux + gx^2 + 2avxy + ay^2, \\ \dot{y} &= b + bux + lx^2 + 2bvxy + by^2, \end{aligned} \tag{11.23}$$

with $al-bg\neq 0$, possessing the following two double distinct singularities: $M_{1,2}(0,i)$ and $M_{3,4}(0,-i)$.

Lemma 11.9. *The condition* $\theta=\theta_1=0$ *implies for systems* (11.23) $\theta_3=0$.

Proof. For the above systems we have

$$\theta = 64a(al-bg)(l+gv-bv^2-av^3),\ \ \mu_0=(al-bg)^2,\ \ \theta_3=a(al-bg)\widehat{U}(a,b,g,l,u,v),$$

where $\widehat{U}(a,b,g,l,u,v)$ is a polynomial. As $\mu_0\neq 0$ the condition $\theta=0$ gives $a(l+gv-bv^2-av^3)=0$.

If $a=0$ then evidently we get $\theta_3=0$, and the statement of the lemma is valid.

Assume $a\neq 0$. Then $l=-v(g-bv-av^2)$, and calculations yield

$$\theta=0,\quad \theta_1=64a(-g+av^2)^3,\quad \theta_3=a(b+av)^2(g-av^2)\widehat{V}(a,b,g,u,v),$$

where $\widehat{V}(a,b,g,u,v)$ is a polynomial. Clearly in this case the condition $\theta_1=0$ implies again $\theta_3=0$, and this completes the proof of the lemma. □

We continue the proof of Theorem 11.4.

1: $\eta<0$. Then systems (11.23) possess one real and two complex infinite singular points, and according to Lemma 11.1 there could be only four distinct configurations at infinity which provide the corresponding global configurations:

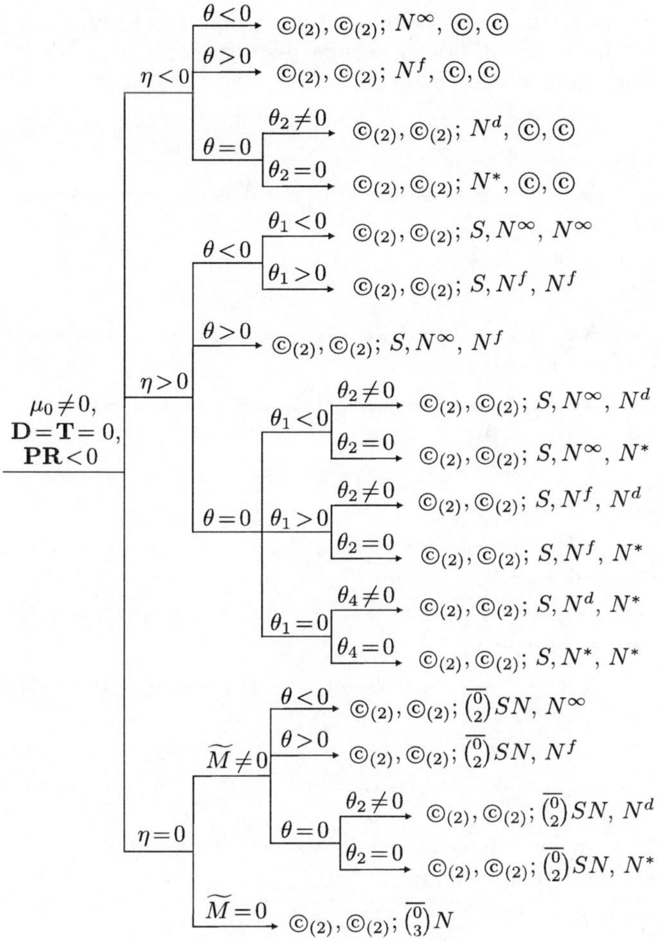

Diagram 11.5: Global configurations in the case $\mu_0 \neq 0$, $\mathbf{D} = \mathbf{T} = 0$, $\mathbf{PR} < 0$.

- $©_{(2)}, ©_{(2)};\ N^\infty,\ ©, © \Rightarrow (a = 1,\ b = 1,\ g = 3,\ l = 0,\ u = 0,\ v = 1)$ (if $\theta < 0$);
- $©_{(2)}, ©_{(2)};\ N^f,\ ©, © \Rightarrow (a = 1,\ b = -1,\ g = 1,\ l = 0,\ u = 0,\ v = 1)$ (if $\theta > 0$);
- $©_{(2)}, ©_{(2)};\ N^d,\ ©, © \Rightarrow (a = 1,\ b = 1,\ g = 1,\ l = 0,\ u = 1,\ v = 0)$ (if $\theta = 0, \theta_2 \neq 0$);
- $©_{(2)}, ©_{(2)};\ N^*,\ ©, © \Rightarrow (a = 0,\ b = 1,\ g = 1,\ l = 1,\ u = 0,\ v = 0)$ (if $\theta = 0, \theta_2 = 0$).

2: $\eta > 0$. In this case systems (11.23) possess three real infinite singular points, and taking into consideration Lemma 11.9 and the condition $\mu_0 > 0$, by Lemma 11.1 we could have at infinity only nine distinct configurations. Corresponding Global configurations are as follows:

- $©_{(2)}, ©_{(2)}; \, S, N^\infty, \, N^\infty \Rightarrow (a = 1, b = 1, g = 35/16, l = 0, u = 0, v = 1)$ (if $\theta < 0, \theta_1 < 0$);
- $©_{(2)}, ©_{(2)}; \, S, N^f, \, N^f \Rightarrow (a = 1, \, b = 1, \, g = 5/4, \, l = 0, \, u = 0, \, v = -2)$ (if $\theta < 0, \theta_1 > 0$);
- $©_{(2)}, ©_{(2)}; \, S, N^\infty, \, N^f \Rightarrow (a = 1, \, b = 1, \, g = 5/4, \, l = 0, \, u = 0, \, v = 1)$ (if $\theta > 0$);
- $©_{(2)}, ©_{(2)}; \, S, N^\infty, \, N^d \Rightarrow (a = 1, \, b = 1, \, g = 6, \, l = 0, \, u = 1, \, v = 2)$ (if $\theta = 0, \theta_1 < 0, \theta_2 \neq 0$);
- $©_{(2)}, ©_{(2)}; \, S, N^\infty, \, N^* \Rightarrow (a = 0, \, b = 1, \, g = 3, \, l = 0, \, u = 0, \, v = 1)$ (if $\theta = 0, \theta_1 < 0, \theta_2 = 0$);
- $©_{(2)}, ©_{(2)}; \, S, N^f, \, N^d \Rightarrow (a = 1, \, b = 1, \, g = 2, \, l = 0, \, u = 1, \, v = -2)$ (if $\theta = 0, \theta_1 > 0, \theta_2 \neq 0$);
- $©_{(2)}, ©_{(2)}; \, S, N^f, \, N^* \Rightarrow (a = 0, \, b = 1, \, g = -1, \, l = 0, \, u = 0, \, v = 1)$ (if $\theta = 0, \theta_1 > 0, \theta_2 = 0$);
- $©_{(2)}, ©_{(2)}; \, S, N^d, \, N^* \Rightarrow (a = 0, \, b = 1, \, g = 1, \, l = 0, \, u = 1, \, v = 1)$ (if $\theta = 0, \theta_1 = 0, \theta_4 \neq 0$);
- $©_{(2)}, ©_{(2)}; \, S, N^*, \, N^* \Rightarrow (a = 0, \, b = 1, \, g = 1, \, l = 0, \, u = 0, \, v = 1)$ (if $\theta = \theta_1 = \theta_4 = 0$).

2: $\eta = 0$. In this case systems (11.23) possess at infinity either one double and one simple real singular points (if $\widetilde{M} \neq 0$) or one triple real singularity (if $\widetilde{M} = 0$). So by Lemma 11.1 we could have at infinity exactly five configurations which provide the corresponding global configurations:

- $©_{(2)}, ©_{(2)}; \, \overline{\binom{0}{2}} SN, \, N^\infty \Rightarrow (a = 1, \, b = 3, \, g = 6, \, l = 0, \, u = 0, \, v = 1)$ (if $\widetilde{M} \neq 0, \, \theta < 0$);
- $©_{(2)}, ©_{(2)}; \, \overline{\binom{0}{2}} SN, \, N^f \Rightarrow (a = -1, \, b = 1, \, g = 2, \, l = 0, \, u = 0, \, v = 1)$ (if $\widetilde{M} \neq 0, \, \theta > 0$);
- $©_{(2)}, ©_{(2)}; \, \overline{\binom{0}{2}} SN, \, N^d \Rightarrow (a = 1, \, b = 1, \, g = 2, \, l = 0, \, u = 1, \, v = 1)$ (if $\widetilde{M} \neq 0, \, \theta = 0, \theta_2 \neq 0$);
- $©_{(2)}, ©_{(2)}; \, \overline{\binom{0}{2}} SN, \, N^* \Rightarrow (a = 1, \, b = 1, \, g = 2, \, l = 0, \, u = 0, \, v = 1)$ (if $\widetilde{M} \neq 0, \, \theta = \theta_2 = 0$);
- $©_{(2)}, ©_{(2)}; \, \overline{\binom{0}{3}} N \Rightarrow (a = 1, \, b = 2, \, g = 4, \, l = 0, \, u = 0, \, v = 1)$ (if $\widetilde{M} = 0$).

This completes the proof of Theorem 11.4. □

11.3 Exactly three distinct singularities

According to Proposition 5.1 for a quadratic system (8.1) to have finite singularities of total multiplicity four (i.e. $m_f = 4$), the condition $\mu_0 \neq 0$ must be satisfied. We consider here the two subclasses of quadratic differential systems with $m_f = 4$ possessing exactly three finite singularities, namely

- systems with one double and two simple real finite singularities ($\mu_0 \neq 0$, $\mathbf{D} = 0$, $\mathbf{T} < 0$),
- systems with one double real and two simple complex singularities ($\mu_0 \neq 0$, $\mathbf{D} = 0$, $\mathbf{T} > 0$).

We shall need the next result proved in [272].

Lemma 11.10 ([272]). *Consider the equation*

$$az^4 + 4bz^3 + 6cz^2 + 4dz + e = 0$$

and the associated polynomials:

$$P_\diamond = ae - 4bd + 3c^2, \quad Q_\diamond = (b^2 - ac)e + ad^2 + (c^2 - 2bd)c, \quad D_\diamond = 27Q_\diamond^2 - P_\diamond^3,$$
$$R_\diamond = b^2 - ac, \quad S_\diamond = 12R_\diamond^2 - a^2P_\diamond, \quad T_\diamond = 3aQ_\diamond = 2P_\diamond R_\diamond, \quad U_\diamond = 2d^2 - 3ce.$$

These polynomials completely determine the number of distinct roots, real and complex, and their multiplicities. More precisely, in the case $a \neq 0$ we have

- *four real simple roots* $\Leftrightarrow$ $D_\diamond < 0, R_\diamond > 0, S_\diamond > 0$;
- *two real and two complex simple roots* $\Leftrightarrow$ $D_\diamond > 0$;
- *four complex simple roots* $\Leftrightarrow$ $D_\diamond < 0$ *and either* $R_\diamond \leq 0$ *or* $S_\diamond < 0$;
- *three real roots, one double and two simple* $\Leftrightarrow$ $D_\diamond = 0, T_\diamond < 0$;
- *one real double and two complex simple roots* $\Leftrightarrow$ $D_\diamond = 0, T_\diamond > 0$;
- *two real roots, one triple and one simple* $\Leftrightarrow$ $D_\diamond = P_\diamond = 0, R_\diamond \neq 0$;
- *two real roots, both double* $\Leftrightarrow$ $D_\diamond = T_\diamond = 0, P_\diamond R_\diamond > 0$;
- *two complex roots, both double* $\Leftrightarrow$ $D_\diamond = T_\diamond = 0, P_\diamond R_\diamond < 0$;
- *one real root of multiplicity four* $\Leftrightarrow$ $D_\diamond = P_\diamond = R_\diamond = 0$.

11.3.1 One double and two elemental real singularities

Theorem 11.5. (a) *We consider here all the configurations of singularities, finite and infinite, of quadratic vector fields possessing exactly three distinct real finite singularities: one double and two elemental. These configurations are classified in Diagram* 11.6 *according to the geometrical equivalence relation of configurations. We have a total of* 265 *geometrically distinct configurations of singularities, finite and infinite.*

(b) *Necessary and sufficient conditions for each one of the* 265 *different geometrical equivalence classes can be assembled from this diagram in terms of* 26 *invariant polynomials with respect to the action of the affine group and time rescaling. These invariant polynomials, which appear in Diagram* 11.6*, are given in Section* 5.4.

(c) *Diagram* 11.6 *actually contains the global bifurcation diagram in the* 12*-dimensional space of parameters of the global configurations of singularities, finite and infinite, of this family of quadratic differential systems and provides an algorithm for finding for any given system in the family considered, its respective geometrical configuration of singularities.*

Proof. Assume that systems (8.1) possess one double and two simple real finite singularities. Then from Proposition 6.2 (see Table 6.1) via an affine transformation these systems could be brought to the systems 4, i.e. to the systems of the form

$$\begin{aligned} \dot{x} &= cx + cuy - cx^2 + 2hxy - cuy^2, \\ \dot{y} &= ex + euy - ex^2 + 2mxy - euy^2, \end{aligned} \tag{11.24}$$

with $u(cm - eh) \neq 0$, possessing the next three distinct singularities: $M_{1,2}(0,0)$, $M_3(1,0)$ and $M_4(0,1)$. For them we get the next values for the traces ρ_i, for the determinants Δ_i and for the linearization matrices $\mathcal{M}_3$ and $\mathcal{M}_4$:

$$\begin{gathered} \mathcal{M}_3 = \begin{pmatrix} -c & 2h+cu \\ -e & 2m+eu \end{pmatrix}, \quad \mathcal{M}_4 = \begin{pmatrix} c+2h & -cu \\ e+2m & -eu \end{pmatrix}, \\ \rho_1 = c+eu, \;\; \rho_3 = -c+2m+eu, \;\; \rho_4 = c+2h-eu; \\ \Delta_1 = 0, \;\; \Delta_3 = 2(eh-cm), \;\; \Delta_4 = -2(eh-cm)u, \end{gathered} \tag{11.25}$$

having the discriminants $\tau_i = \rho_i^2 - 4\Delta_i$ for $i = 1,3,4$. Then for the above systems we calculate

$$\begin{gathered} \mu_0 = 4(eh-cm)^2u = -\Delta_3\Delta_4, \;\; \widetilde{K} = 2\Delta_3(x^2-uy^2), \;\; E_1 = -\Delta_3^2\Delta_4^2\rho_1/2, \\ \eta = -4(\mathcal{N}_2^2 - \mathcal{N}_1\mathcal{N}_3)/3, \;\; \widetilde{M} = -8(\mathcal{N}_1x^2 - 2\mathcal{N}_2xy + \mathcal{N}_3y^2), \\ \mathcal{T}_4 = -\Delta_3\Delta_4\rho_1^2\rho_3\rho_4, \;\; \mathcal{T}_3 = -\Delta_3\Delta_4\rho_1\big[\rho_1(\rho_3+\rho_4) + 2\rho_3\rho_4\big], \\ \mathcal{T}_2 = -\Delta_3\Delta_4\big[\rho_1^2 + 2\rho_1(\rho_3+\rho_4) + \rho_3\rho_4\big], \;\; \mathcal{T}_1 = -\Delta_3\Delta_4(2\rho_1+\rho_3+\rho_4), \\ W_4 = \Delta_3^2\Delta_4^2\rho_1^4\tau_3\tau_4, \;\; W_3 = \Delta_3^2\Delta_4^2\rho_1^2\big[\rho_1^2(\tau_3+\tau_4) + 2\tau_3\tau_4\big], \\ W_2 = \Delta_3^2\Delta_4^2\big[\rho_1^4 + 2\rho_1^2(\tau_3+\tau_4) + \tau_3\tau_4\big], \;\; W_1 = \Delta_3^2\Delta_4^2\big[2\rho_1^2 + \tau_3 + \tau_4\big], \end{gathered} \tag{11.26}$$

where

$$\begin{gathered} \mathcal{N}_1 = c^2 - 6eh + 4cm + 4m^2 - 3e^2u, \;\; \mathcal{N}_2 = ch + 2hm - 4ceu + emu, \\ \mathcal{N}_3 = 4h^2 - 3c^2u + 4ehu - 6cmu + e^2u^2. \end{gathered}$$

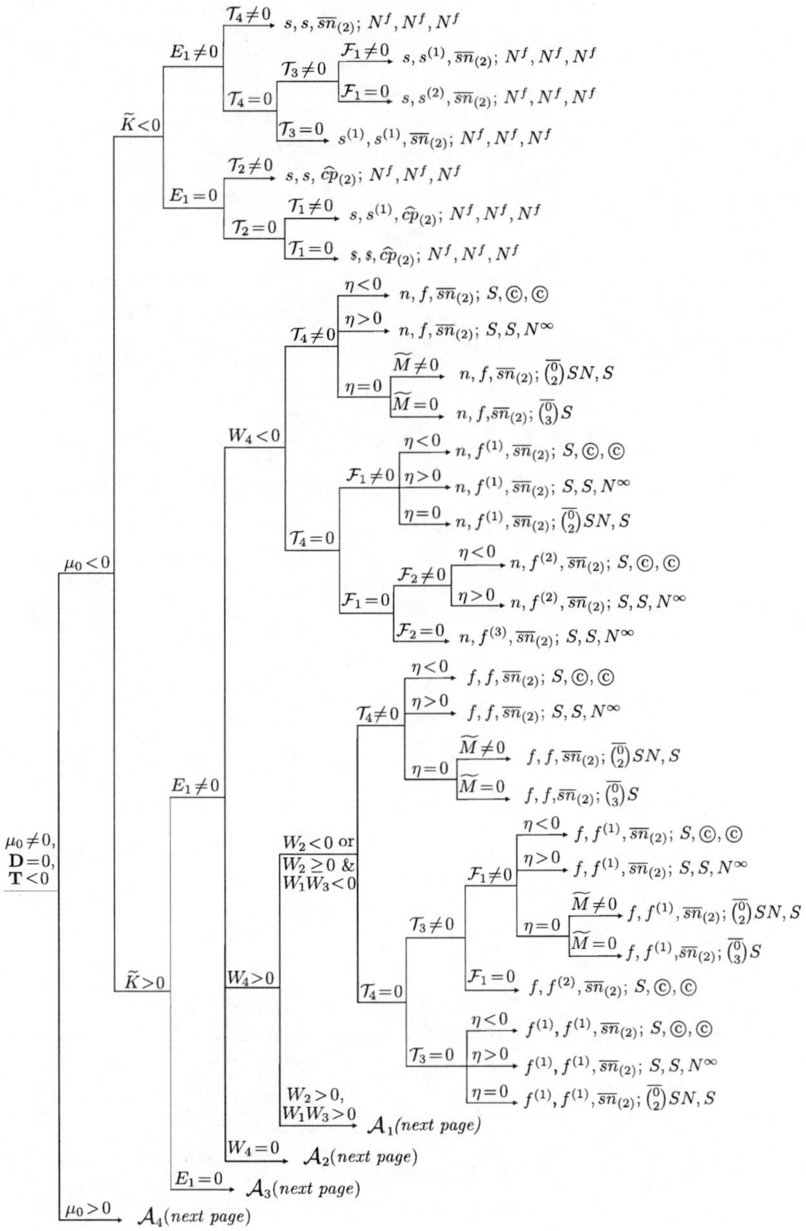

Diagram 11.6: Global configurations in the case $\mu_0 \neq 0, \mathbf{D} = 0, \mathbf{T} < 0$.

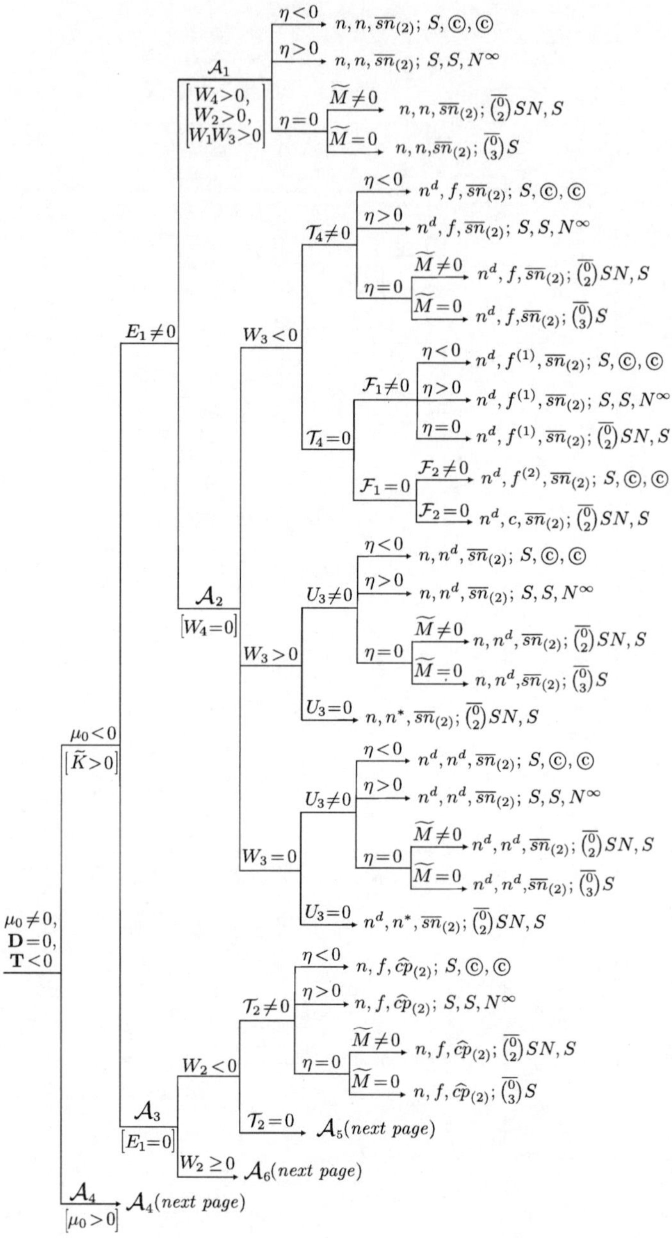

Diagram 11.6 (*cont.*): Global configurations in the case $\mu_0 \neq 0, \mathbf{D} = 0, \mathbf{T} < 0$.

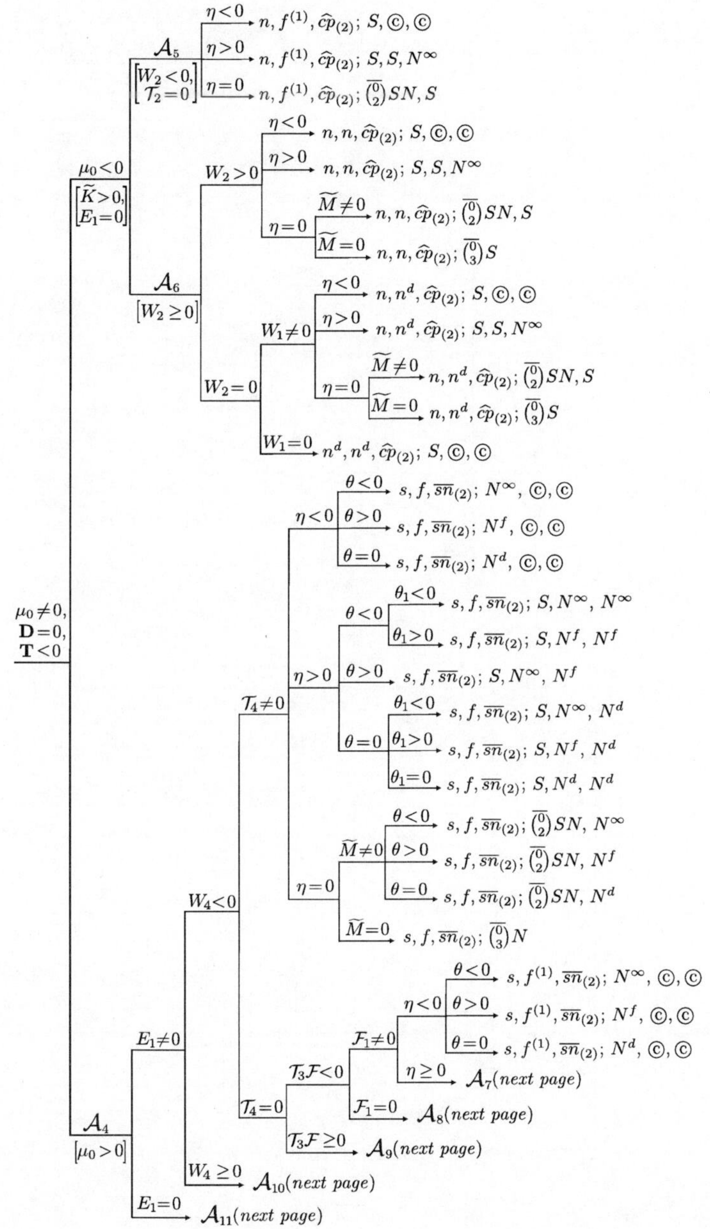

Diagram 11.6 (*cont.*): Global configurations in the case $\mu_0\neq0, \mathbf{D}=0, \mathbf{T}<0$.

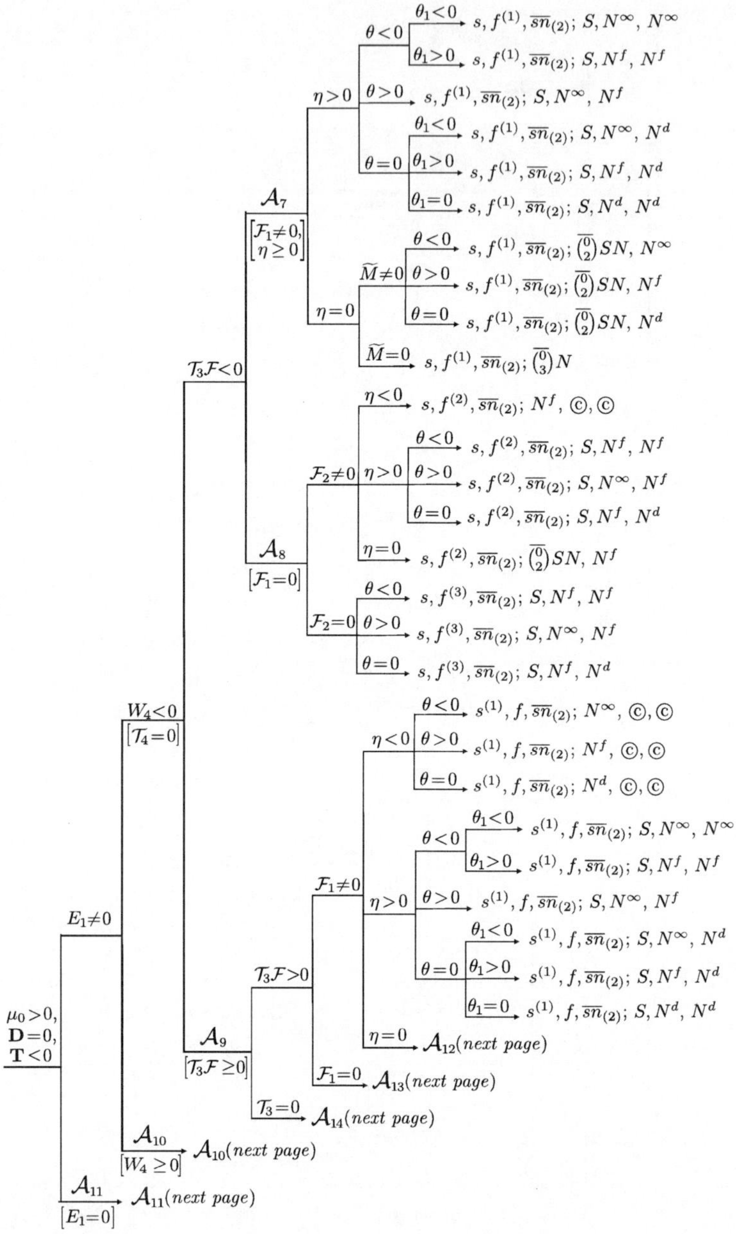

Diagram 11.6 (*cont.*): Global configurations in the case $\mu_0 \neq 0, \mathbf{D} = 0, \mathbf{T} < 0$.

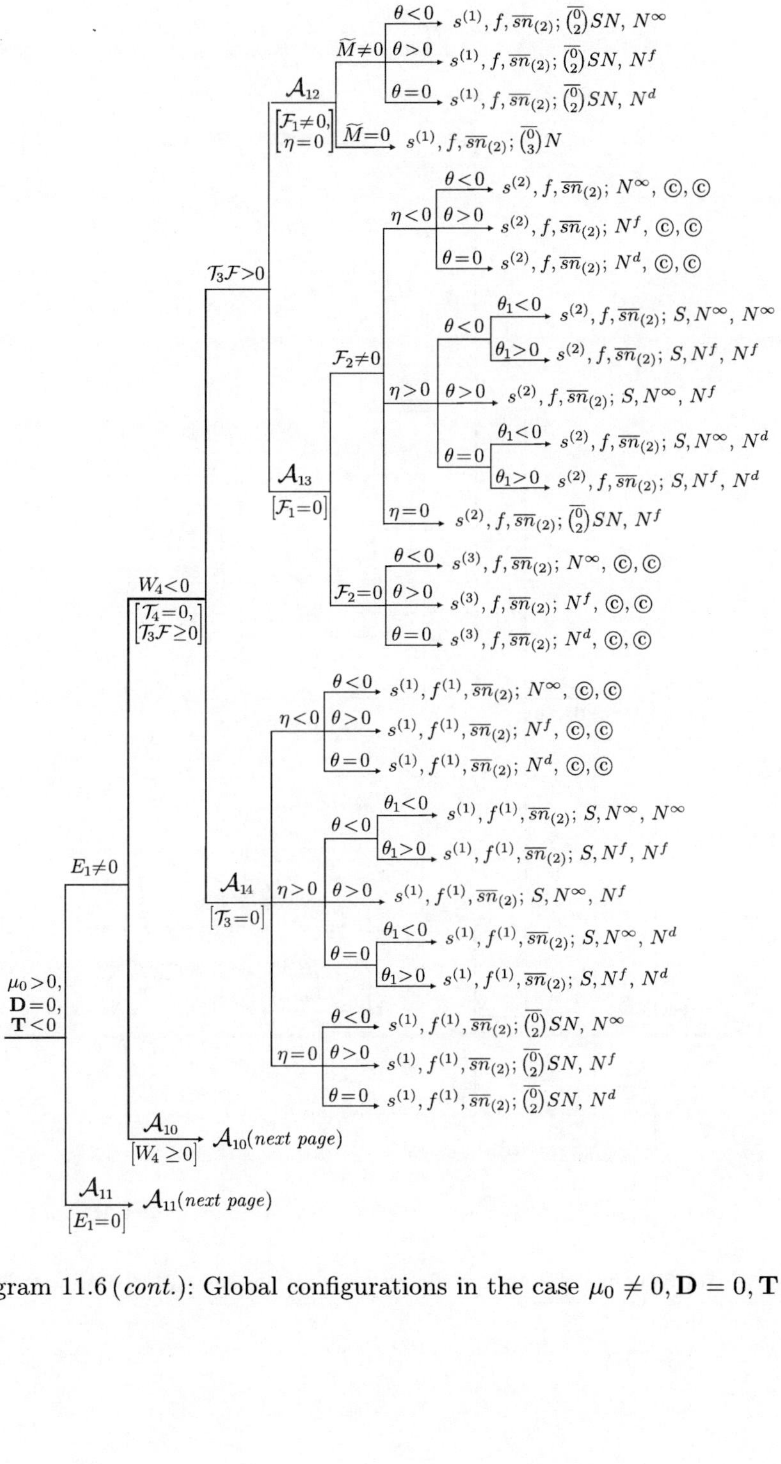

Diagram 11.6 (*cont.*): Global configurations in the case $\mu_0 \neq 0, \mathbf{D} = 0, \mathbf{T} < 0$.

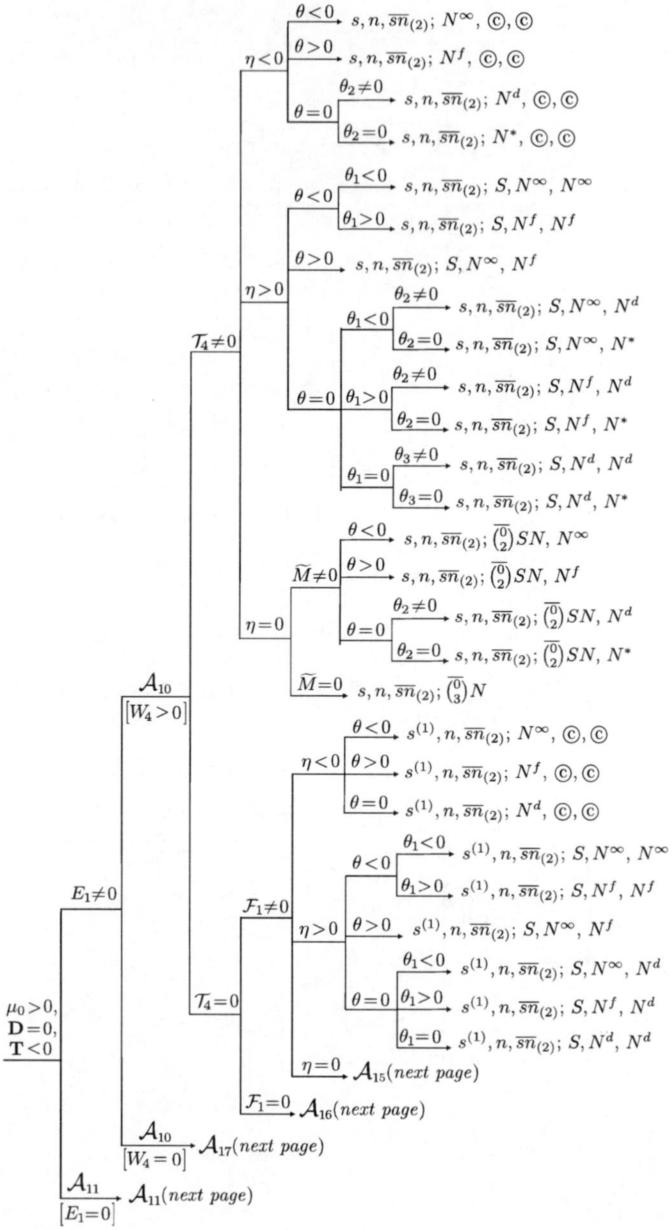

Diagram 11.6 (*cont.*): Global configurations in the case $\mu_0 \neq 0, \mathbf{D} = 0, \mathbf{T} < 0$.

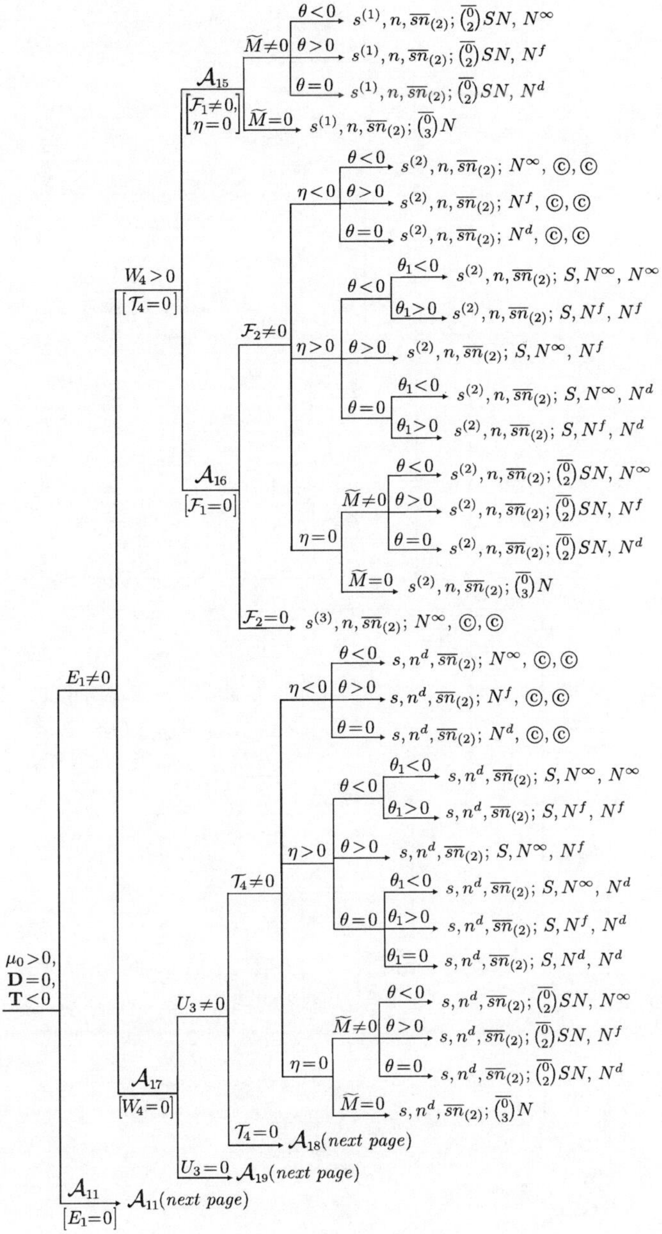

Diagram 11.6 (*cont.*): Global configurations in the case $\mu_0 \neq 0, \mathbf{D} = 0, \mathbf{T} < 0$.

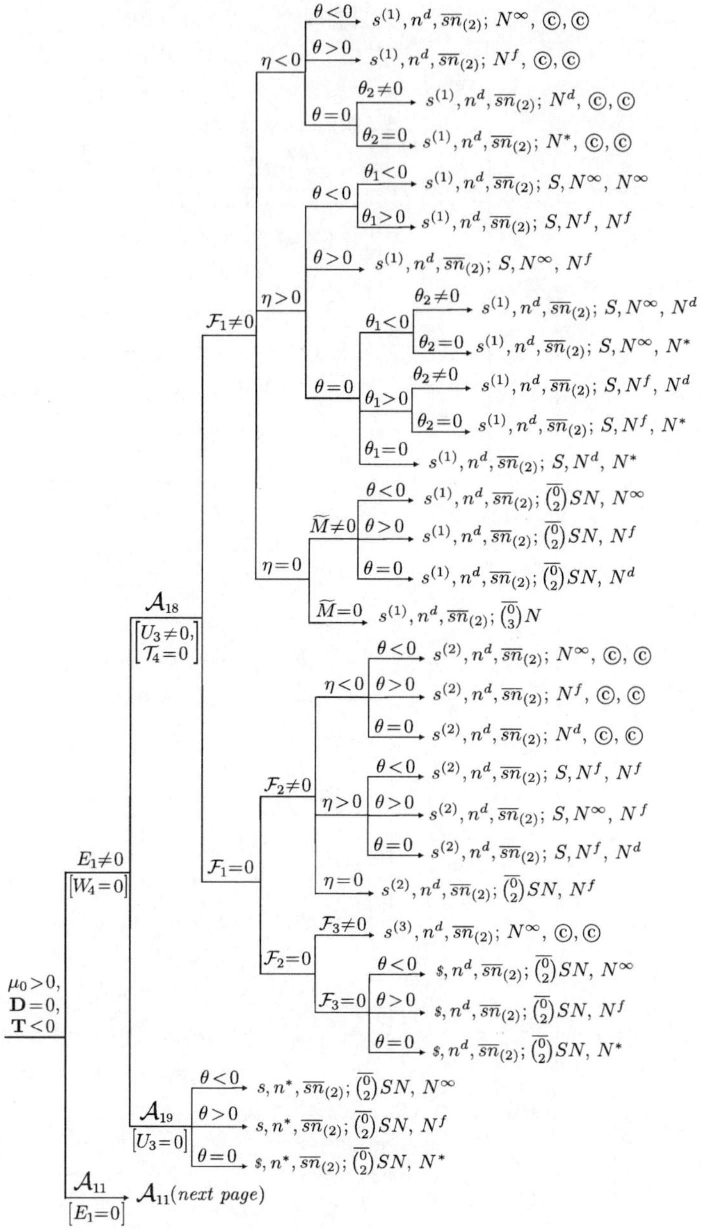

Diagram 11.6 (*cont.*): Global configurations in the case $\mu_0 \neq 0, \mathbf{D} = 0, \mathbf{T} < 0$.

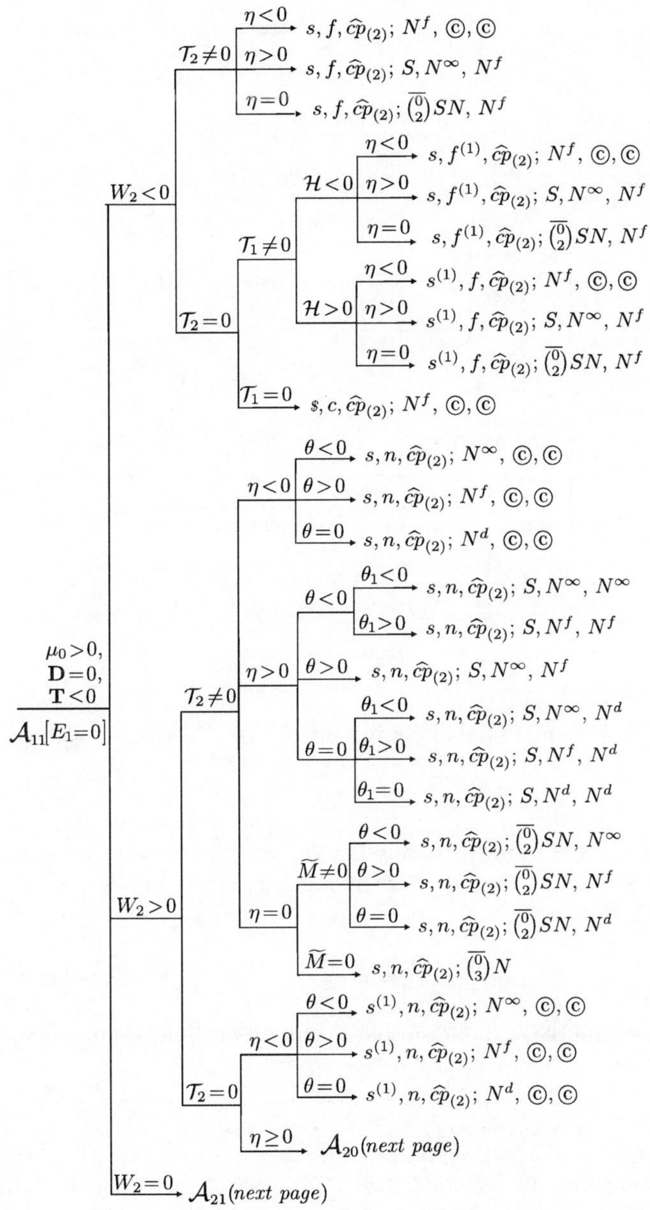

Diagram 11.6 (*cont.*): Global configurations in the case $\mu_0 \neq 0, \mathbf{D} = 0, \mathbf{T} < 0$.

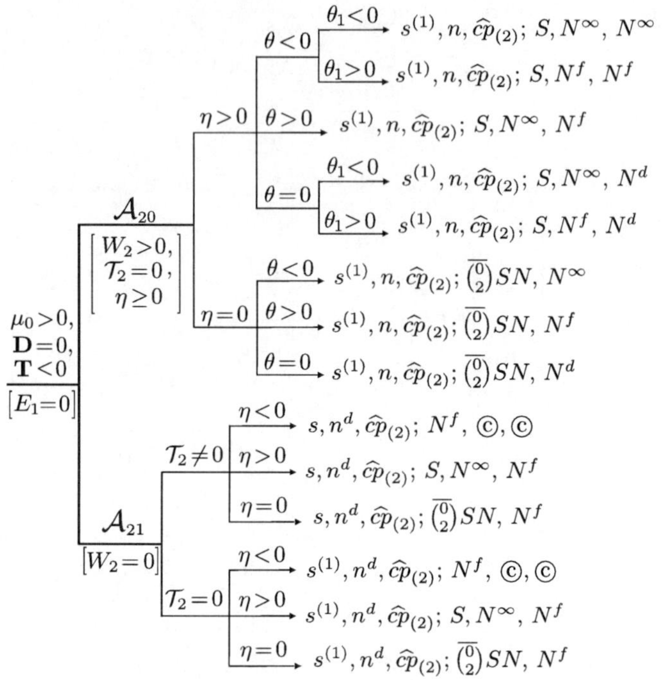

Diagram 11.6 (*cont.*): Global configurations in the case $\mu_0 \neq 0, \mathbf{D} = 0, \mathbf{T} < 0$.

Remark 11.4. In order to construct the examples or to prove nonexistence of some configurations, apart from the family (11.24) we shall use here another normal form of quadratic systems, associated with singularities at infinity. So we will use the family of systems:

$$(\tilde{\mathbf{S}}_{III})\quad \dot{x} = cx + dy + gx^2 + hxy, \quad \dot{y} = ex + fy + (g-1)xy + hy^2,$$

which have one double and one simple real distinct infinite singularities (i.e. $\eta = 0$, $\widetilde{M} \neq 0$).

Lemma 11.11. *For a system* (11.24) *the following are true:*

(i) *If the condition* $\widetilde{M} = 0$ *holds, then the conditions* $\mu_0 < 0$ *and* $\rho_3\rho_4 = 0$ *imply* $\tau_3\tau_4 > 0$.

(ii) *If the conditions* $\rho_3\rho_4 = \mathcal{F}_1 = 0$, $\mu_0 < 0$ *and* $W_4 \neq 0$ *are satisfied, then* $\eta \neq 0$. *Moreover if* $\mathcal{F}_2 = 0$ *then* $\eta > 0$ *and* $\mathcal{F}_3\mathcal{F}_4 \neq 0$.

(iii) *If the condition* $\theta = 0$ *holds, then the following hold:*

(iii$_1$) *The conditions* $\theta_1 \neq 0$ *and* $\theta_2 = 0$ *imply* $\tau_3\tau_4 \geq 0$*, and we have* $\tau_3\tau_4 = 0$ *if and only if* $\rho_3\rho_4 = 0$*. In the case* $\tau_3\tau_4 = 0$ *the condition* $\eta = 0$ *is equivalent to* $\mathcal{F}_1 = 0$*.*

(iii$_2$) *The condition* $\theta_1 = 0$ *implies* $\mu_0 = \eta$ *and* $\theta_2 = 0$*, and furthermore if* $E_1 = 0$ *then the condition* $\theta_2 = 0$ *is equivalent to* $\theta_1 = 0$*. In addition* $\theta_1 = \theta_2 = 0$ *implies* $\theta_3 \neq 0$*; if* $\theta_3 = 0$ *then* $\tau_3\tau_4 \geq 0$*,* $\theta_4 \neq 0$*, and we have* $\tau_3\tau_4 = 0$ *if and only if* $\rho_3\rho_4 = 0$*.*

(iv) *If the conditions* $\rho_3\rho_4 = \tau_3\tau_4 = \mathcal{F}_1 = 0$ *and* $\mu_0 < 0$ *hold, then either* $\mathcal{F}_2 \neq 0$ *and* $\eta < 0$ *or* $\mathcal{F}_2 = \mathcal{F}_3 = \eta = 0$*.*

(v) *If the conditions* $\rho_3\rho_4 = \theta = \theta_1 = 0$ *hold, then* $\mathcal{F}_1 \neq 0$*. Moreover the condition* $\tau_3\tau_4 = 0$ *is equivalent to* $\theta_3 = 0$*.*

Proof. (i) Assume that for a system (11.24) the conditions $\widetilde{M} = 0$, $\mu_0 < 0$ and $\rho_3\rho_4 = 0$ are fulfilled. We claim that in this case the condition $ce \neq 0$ holds. Indeed, considering (11.26) we detect that in the case $ce = 0$ the conditions $\mathcal{N}_1 = \mathcal{N}_2 = \mathcal{N}_3 = 0$ yield either $c = m = 2h + eu = 0$ or $c + 2m = e = h = 0$. In the first case we obtain $\mu_0 = e^4u^3$, $\rho_3\rho_4 = -2e^2u^2$, and in the second one we get $\mu_0 = 16m^4u$, $\rho_3\rho_4 = -8m^2$. Therefore in both cases the condition $\mu_0 \neq 0$ implies $\rho_3\rho_4 \neq 0$. The contradiction obtained proves our claim.

Thus $ce \neq 0$ and the relations $\mathcal{N}_1 = \mathcal{N}_2 = \mathcal{N}_3 = 0$ are equivalent to

$$h = \frac{(4c - m)(c + 2m)^2}{27ce}, \quad u = \frac{(c + 2m)^3}{27ce^2}.$$

We calculate

$$\rho_3 = \frac{(c + 2m)^3 - 27ce(c - 2m)}{27ce}, \quad \rho_4 = \frac{(7c - 4m)(c + 2m)^2 + 27c^2e}{27ce}.$$

Since the condition $\rho_3\rho_4 = 0$ holds, without loss of generality we may assume $\rho_4 = 0$, i.e. $e = (4m - 7c)(c + 2m)^2/(27c^2) \neq 0$, and then we calculate

$$\tau_3\tau_4 = \frac{2^{10}c(5c - 8m)(c - m)^6}{27(7c - 4m)^4}, \quad \mu_0 = \frac{64c(c - m)^6}{27(7c - 4m)^2(c + 2m)}.$$

So the condition $\mu_0 < 0$ implies $c(c + 2m) < 0$ and then $cm < 0$. Therefore $c(5c - 8m) > 0$, which implies $\tau_3\tau_4 > 0$, and this completes the proof of the statement (i) of the lemma.

(ii) Assume that for a system (11.24) the conditions $\rho_3\rho_4 = 0$ and $\mathcal{F}_1 = 0$ hold. Without loss of generality we may assume $\rho_3 = 0$ (i.e. $m = (c - eu)/2$), and then we calculate

$$\mathcal{F}_1 = u(ceu - c^2 + 2eh)(2c^2 + 3ce + 2eh + 2ceu + e^2u),$$
$$\mu_0 = u(ceu - c^2 + 2eh)^2.$$

So due to $\mu_0 \neq 0$ the condition $\mathcal{F}_1 = 0$ yields $2c^2 + e(3c+2h+2cu+eu) = 0$, and we observe that $e \neq 0$; otherwise we get $c = 0$, which implies $\mu_0 = 0$. Therefore we obtain $h = -(2c^2 + 3ce + 2ceu + e^2u)/(2e)$, and we calculate

$$\eta = 4c(c+e)\Psi_1(c,e,u)/e^2, \quad W_4 = c(c+eu)^4\Psi_2(c,e,u)/e^2,$$
$$\mu_0 = u(c+e)^2(3c+eu)^2, \ \mathcal{F}_2 = -c(c+e)^4u^2(c+eu)^4(3c+eu)^2(6c+eu)/e^2,$$

where $\Psi_2(c,e,u)$ is a polynomial and

$$\Psi_1(c,e,u) = 12c^4 + 4c^3e(9+7u) + 3c^2e^2(9+6u+7u^2) + 3ce^3u^2(3+2u) + u^4e^4.$$

Suppose that the conditions $\mu_0 < 0$ and $W_4 \neq 0$ are satisfied. Then the condition $\eta = 0$ is equivalent to $\Psi_1(c,e,u) = 0$. According to Lemma 11.10, for this quartic equation with respect to c, we calculate

$$D_\diamond = 19683e^{12}u^5(u-2)^3(4u-9)(1+2u+2u^2)/16,$$
$$R_\diamond = e^2(27+90u+7u^2), \quad S_\diamond = -48e^4u(8u^3 - 477u^2 - 1998u - 972).$$

We observe that $\mu_0 < 0$ and $\Psi_1(c,e,u)$=0 imply $u < 0$ and $e \neq 0$. Then $D_\diamond < 0$, and checking the roots of the polynomials $R_\diamond|_{e=1}$ and $S_\diamond|_{e=1}$, it is easy to determine that the possibility $R_\diamond > 0$ and $S_\diamond > 0$ (simultaneously) cannot be realized.

Thus by Lemma 11.10 the polynomial $\Psi_1(c,e,u)$ does not have real roots, i.e. $\eta \neq 0$.

Next, if we impose the condition $\mathcal{F}_2 = 0$ (i.e. $c = -eu/6$) to be fulfilled, then we obtain:

$$\eta = e^4(u-6)u^3(81 - 126u + 50u^2)/972, \quad \mu_0 = e^4(u-6)^2u^3/144,$$
$$\mathcal{F}_3\mathcal{F}_4 = 5^7e^{20}(u-6)^8u^{19}/(2^{17}3^{19}),$$

and clearly the condition $\mu_0 < 0$ implies $\eta > 0$ and $\mathcal{F}_3\mathcal{F}_4 \neq 0$. This completes the proof of the statement (ii).

(iii) Assume now that for systems (11.24) the condition $\theta = 0$ is satisfied. For these systems we have

$$\theta = 64(eh - cm)\big[(h+eu)^2 - (c+m)^2u\big], \quad \mu_0 = 4u(eh-cm)^2,$$

and as $\mu_0 \neq 0$ the condition $\theta = 0$ implies $(h+eu)^2 - (c+m)^2u = 0$.

(iii$_1$) Assume first $\theta_1 \neq 0$. We observe that in this case the condition $c+m \neq 0$ must hold, otherwise we get $c+m = h+eu = 0$, which implies $\theta_1 = 0$. So $c+m \neq 0$, and setting a new parameter $v = h+eu$ (then $h = v - eu$), the condition $\theta = 0$ gives $u = v^2/(c+m)^2$. Then we calculate

$$\theta = 0, \quad \theta_1 = 256v^2(cm+m^2-ev)^3/(c+m)^4,$$
$$\mu_0 = 4v^2(c^2+cm-ev)^2(cm+m^2-ev)^2/(c+m)^6, \qquad (11.27)$$
$$\theta_2 = -v(c+m+v)(c^2+cm-ev)(cm+m^2-ev)/(c+m)^3,$$

and due to $\mu_0 \neq 0$ the condition $\theta_2 = 0$ yields $c+m+v=0$ (i.e. $v=-c-m$), which gives

$$\tau_3\tau_4 = \rho_3^2\rho_4^2 \geq 0, \quad \mu_0 = 4(c+e)^2(e+m)^2.$$

So obviously the condition $\tau_3\tau_4 = 0$ is equivalent to $\rho_3\rho_4 = 0$. Moreover if we suppose that $\rho_3\rho_4 = 0$, then we may assume $\rho_3 = 0$ (i.e. $c = e+2m$). We calculate

$$\mathcal{F}_1 = -16(e+m)^3(e+2m), \quad \eta = 32(e+m)^3(e+2m),$$

and evidently the condition $\mathcal{F}_1 = 0$ is equivalent to $\eta = 0$.

(iii$_2$) Suppose now $\theta_1 = 0$. Then $c+m = 0$, otherwise we obtain (11.27) and therefore the condition $\mu_0 \neq 0$ implies $\theta_1 \neq 0$. So assuming $m = -c$ and $h = -eu$, we obtain

$$\theta = \theta_1 = \theta_2 = 0, \quad E_1 = -8u^2(c+eu)(c^2-e^2u)^4,$$
$$\mu_0 = 4u(c^2-e^2u)^2 = \eta, \quad \theta_3 = 2u(1-u)(c^2-e^2u)^3.$$

(a) Assume $E_1 = 0$. Due to $\mu_0 \neq 0$ this condition is equivalent to $c+eu = 0$, i.e. $c = -eu$, and then we get $\theta_3 = 2e^6(u-1)^4u^4 \neq 0$ due to $\mu_0 = 4e^4(u-1)^2u^3 \neq 0$.

(b) Suppose now $\theta_3 = 0$. Since $\mu_0 \neq 0$ we get $u = 1$, and this gives

$$\tau_3\tau_4 = \rho_3^2\rho_4^2, \quad \theta_4 = -4(c-e)^4(c+e), \quad \mu_0 = 4(c-e)^2(c+e)^2.$$

Clearly $\theta_4 \neq 0$ due to $\mu_0 \neq 0$, and the condition $\tau_3\tau_4 = 0$ is equivalent to $\rho_3\rho_4 = 0$. This completes the proof of the statement (iii) of the lemma.

(iv) Assume that for a system (11.24) the conditions $\rho_3\rho_4 = 0$ and $\mathcal{F}_1 = 0$ hold. As it was shown in the proof of the statement (ii), supposing $\rho_3 = \mathcal{F}_1 = 0$ we arrive at the following relations

$$m = (c-eu)/2, \quad h = -(2c^2+3ce+2ceu+e^2u)/(2e),$$

where $e \neq 0$ due to $\mu_0 \neq 0$. Since we choose $\rho_3 = 0$, the condition $\tau_3\tau_4 = 0$ gives $\tau_4 = 0$ (as the singular point $M_3(1,0)$ is elemental). We calculate

$$\tau_4 = \frac{4c(c+e)}{e^2}\left[(c+eu)^2 + e(c-eu)\right] \equiv \frac{4c(c+e)}{e^2}\,\xi(c,e,u),$$
$$\mu_0 = (c+e)^2u(3c+eu)^2,$$

and as $\mu_0 \neq 0$, the condition $\tau_4 = 0$ yields $c\xi = 0$.

If $c = 0$ then calculations yield $\mathcal{F}_2 = \mathcal{F}_3 = \eta = 0$, i.e. in this case the statement (iv) is valid.

Assume now $c \neq 0$. Then $\xi(c, e, u) = 0$, and as Discrim $[\xi, c] = e^2(1 + 8u)$, to factorize this polynomial, we set a new variable v as follows: $1 + 8u = v^2 \geq 0$, i.e. $u = (v^2 - 1)/8$. The we obtain

$$\xi = (8c + 3e - 4ev + ev^2)(8c + 3e + 4ev + ev^2)/64 = 0,$$

and we may assume $(8c + 3e - 4ev + ev^2) = 0$, as the second possibility can be obtained from the first one by substituting v with $-v$.

Thus we obtain $c = e(3 - v)(v - 1)/8$, and we calculate

$$\begin{aligned} \mu_0 &= 2^{-13}e^4(v-5)^4(v^2-1)^3, \\ \mathcal{F}_2 &= 2^{-32}e^{10}(v-5)^6(3-v)(v-1)^{10}(1+v)^6(5v-19), \\ \eta &= 2^{-13}e^4(v-5)^2(3-v)(v-1)^2(v^2-1)(205+67v-65v^2+9v^3). \end{aligned}$$

We observe that the equation $205 + 67v - 65v^2 + 9v^3 = 0$ possesses a single real root $v_0 < 1.25$. Therefore obviously the condition $\mu_0 < 0$ (i.e. $|v| < 1$) implies $\eta < 0$ and $\mathcal{F}_2 \neq 0$. This completes the proof of the statement (iv) of Lemma 11.11.

(v) Assume that for a system (11.24) the condition $\theta = \theta_1 = 0$ holds. As it was shown in the proof of the statement (iii), in this case the conditions $m = -c$ and $h = -eu$ are fulfilled, and we obtain $\rho_3\rho_4 = (eu - 3c)(c - 3eu) = 0$. As it was mentioned earlier, we may assume $\rho_3 = 0$ (i.e. $c = eu/3$), and we calculate

$$\begin{aligned} \mathcal{F}_1 &= 16e^4u^4(u-9)/81, \quad & \theta_3 &= 2e^6(u-9)^3(u-1)u^4/729, \\ \mu_0 &= 4e^4u^3(u-9)^2/81, \quad & \tau_3\tau_4 &= -64e^4(u-9)(u-1)u^3/81. \end{aligned}$$

Clearly the condition $\mu_0 \neq 0$ implies $\mathcal{F}_1 \neq 0$, and the condition $\tau_3\tau_4 = 0$ is equivalent to $\theta_3 = 0$. This completes the proof of the statement (v) and also the proof of Lemma 11.11. □

Lemma 11.12. *A system* (11.24) *possesses a finite star node if and only if the condition* $(\mathcal{S}^1_{n^*})$ *(see Theorem* 7.1*) holds, and in this case the star node is unique. Moreover, for a system* (11.24) *the condition* $(\mathcal{S}^1_{n^*})$ *implies* $\eta = \theta_2 = 0$ *and* $E_1\widetilde{M} \neq 0$.

Proof. We claim that neither the condition $(\mathcal{S}^2_{n^*})$ nor $(\mathcal{S}^3_{n^*})$ of Theorem 7.1 could be satisfied for systems (11.24). To prove this it is sufficient to show that the condition $U_1 = U_4 = 0$ does not hold for these systems.

Indeed, suppose that for systems (11.24) we have $U_1 = U_4 = 0$. We calculate

$$U_1 = 4(eh - cm - m^2 + e^2u)x^2 + 8(hm - ceu)xy - 4(h^2 - c^2u + ehu - cmu)y^2,$$

and we consider two possibilities: $e \neq 0$ and $e = 0$.

1. If $e \neq 0$ then we may assume $e = 1$ (due to a time rescaling), and therefore the condition $U_1 = 0$ gives either

$$\text{(i) } h = cm + m^2 - u,\ m = -c, \text{ or } \text{(ii) } h = cm + m^2 - u,\ u = m^2.$$

However in the case (ii) we get $\mu_0 = 0$, and therefore we examine the first case. We calculate Coefficient$[U_4, x^3] = 6(u-c) = 0$, which implies $u = c$, and then we have

$$U_4 = 12c(1-c)xy(x+cy), \quad \mu_0 = 4c^3(1-c)^2.$$

Evidently the condition $\mu_0 \neq 0$ implies $U_4 \neq 0$.

2. Assuming $e = 0$ we obtain

$$U_1 = -4m(c+m)x^2 + 8hmxy - 4(h^2 - c^2u - cmu)y^2, \; \mu_0 = 4c^2m^2u,$$

and due to $\mu_0 \neq 0$ the condition $U_1 = 0$ gives $h = 0$, $m = -c$. Then we calculate $U_4 = -6c^2y(x^2 + 2uxy + uy^2) \neq 0$ due to $\mu_0 \neq 0$, and this completes the proof of our claim.

Thus considering Theorem 7.1 we conclude that a system (11.24) has a finite star node if and only if the condition $(\mathcal{S}^1_{n^*})$ is satisfied.

Assume now that the condition $(\mathcal{S}^1_{n^*})$ holds, i.e. a system (11.24) possess a finite star node. Then without loss of generality we may consider that such a point is $M_2(1,0)$, and considering (11.25) we obtain $c = 0$, $m = -c/2$ and $h = -cu/2$. Therefore calculations yield

$$\mu_0 = c^4u, \quad \eta = \theta_2 = 0, \quad E_1 = -c^9u^2/2, \quad \widetilde{M} = -8c^2u^2y^2,$$

and clearly the condition $\mu_0 \neq 0$ gives $E_1\widetilde{M} \neq 0$. This completes the proof of the lemma. □

Lemma 11.13. *Assume that for systems* (11.24) *the conditions* $\mu_0 > 0$, $E_1 \neq 0$, $\mathcal{T}_4 = \mathcal{F}_1 = 0$ *and* $W_4 < 0$ *hold. Then we have:*

(i) $\widetilde{M} \neq 0$.

(ii) *The condition* $\eta = 0$ *implies* $\theta > 0$.

(iii) *If in addition* $\mathcal{T}_3\mathcal{F} < 0$, *then*

  (iii_1) *the condition* $\eta < 0$ *implies* $\theta > 0$,

  (iii_2) *the conditions* $\eta > 0$ *and* $\theta \leq 0$ *imply* $\theta_1 > 0$.

Proof. Since $\mu_0 E_1 \neq$ the condition $\mathcal{T}_4 = 0$ gives $\rho_3\rho_4 = 0$, and we may consider $\rho_3 = 0$. Assuming $\mathcal{F}_1 = 0$ we claim that in this case the condition $e \neq 0$ holds. Indeed, since the condition $\rho_3 = 0$ gives $m = (c - eu)/2$, supposing $e = 0$ we obtain $\mu_0 = c^4u > 0$ and then $\mathcal{F}_1 = -2c^4u \neq 0$.

So $e \neq 0$ and due to a time rescaling we may assume $e = 1$. Therefore we calculate

$$\mathcal{F}_1 = u(cu - c^2 + 2h)(3c + 2c^2 + 2h + u + 2cu), \quad \mu_0 = u(cu - c^2 + 2h)^2,$$

and as $\mu_0 \neq 0$ the condition $\mathcal{F}_1 = 0$ implies $h = -(3c + 2c^2 + u + 2cu)/2$. Then we have

$$\begin{aligned} \mu_0 &= u(c+1)^2(3c+u)^2, \ \eta = 4c(c+1)\Psi_2(c,u), \\ \mathcal{F}_2 &= -c(1+c)^4u^2(c+u)^4(3c+u)^2(6c+u), \\ \theta &= -8(1+c)(3c+u)\Psi_3(c,u), \ \tau_4 = 4c(1+c)[(c+u)^2+c-u], \\ \tau_3 &= 4(c+1)(3c+u), \ \ E_1 = -(1+c)^4u^2(c+u)(3c+u)^4/2, \end{aligned} \tag{11.28}$$

where

$$\begin{aligned} \Psi_2(c,u) &= 12c^4 + u^4 + 3cu^2(3+2u) + 4c^3(9+7u) + 3c^2(9+6u+7u^2), \\ \Psi_3(c,u) &= 4c^4 + 2c(u-3)u - (u-1)u^2 + 4c^3(3+2u) + c^2(9-u+4u^2). \end{aligned}$$

(i) Suppose that the condition $\widetilde{M} = 0$ holds. We calculate Coefficient$[\widetilde{M}, xy] = -16c(3+2c-u)(c+u)xy$ and as $E_1\tau_4 \neq 0$ the condition $\widetilde{M} = 0$ implies $3+2c-u = 0$, i.e. $u = 3+2c$. Then we obtain $\widetilde{M} = -72(1+c)(1+2c)(x^2+3cy^2+2c^2y^2)$. Hence the condition $\widetilde{M} = 0$ yields $c = -1/2$ and this implies $\tau_3\tau_4 = 1/4 > 0$, i.e. $W_4 > 0$ and we get a contradiction.

(ii) Assume that the condition $\eta = 0$ is fulfilled. The only intersection of the curves $\eta = 0$ and $\theta = 0$ outside the union $\{\mu_0 = 0, E_1 = 0, W_4 = 0\}$ is the point $(c_0, u_0) \approx (-0.5745, 2.1564)$ for which $W_4 > 0$. In any other open subset of the region $\mathfrak{R}$ defined by $\{\mu_0 > 0, E_1 \neq 0, W_4 < 0\}$ these curves do not intersect. So when $\eta = 0$ the polynomial θ has a fixed sign which could be different if $\mathfrak{R}$ is disconnected. Checking the sign of θ on the points of the curve $\eta = 0$ in any subset of $\mathfrak{R}$ we detect that θ is always positive.

(iii) Assume now that the condition $\mathcal{T}_3\mathcal{F} < 0$ holds. Due to $\rho_3 = 0$ this is equivalent to $\tau_3 < 0$ (as by Theorem 6.2 we have a weak focus).

(iii$_1$) As we mentioned above the only intersection of the curves $\eta = 0$ and $\theta = 0$ outside the union $\{\mu_0 = 0, E_1 = 0, W_4 = 0\}$ is in the domain $W_4 > 0$ where we also have $\tau_3 > 0$. Additionally inside the intersection of the region $\mathfrak{R}$ with the region defined by $\tau_3 < 0$ we can have either $\eta > 0$ or $\eta \leq 0$ (respectively $\theta > 0$ or $\theta \leq 0$). But since there is no intersection of these curves it means that some combinations of signs is not possible. It remains to observe that in the domain $\eta < 0$ we have $\theta > 0$.

(iii$_2$) Considering the intersection points of the curve $\theta_1 = 0$ with $\theta = 0$ (respectively with $\eta = 0$) we detect that they are also in the complement of the region of $\tau_3 < 0$ and $\mu_0 > 0$. Moreover in this region the curve $\theta_1 = 0$ is located on the domain where $\eta > 0$ and $\theta > 0$. It remains to observe that in the region where $\theta < 0$ we have $\theta_1 > 0$. This completes the proof of Lemma 11.13. □

In what follows we determine the geometrical configurations of systems (11.24).

1: $\mu_0 < 0$. Then $u < 0$ and hence $\text{sign}(\widetilde{K}) = \text{sign}(\Delta_3) = \text{sign}(\Delta_4)$.

1.1: $\widetilde{K} < 0$. Then the elemental singular points are both saddles, and by Theorem 6.1 the type of the double singular point is governed by the invariant polynomial E_1. Additionally at infinity we must have three nodes as the sum of the indices of the finite singularities equals -2.

1.1.1: $E_1 \neq 0$. In this case $\rho_1 \neq 0$, and apart from the two saddles, we have a semi-elemental saddle-node.

1.1.1.1: $\mathcal{T}_4 \neq 0$. Then $\rho_3\rho_4 \neq 0$ and both saddles are strong. So we arrive at the global configuration

- $s, s, \overline{sn}_{(2)}; N^f, N^f, N^f \Rightarrow ((11.24): c = 1, e = 0, h = 0, m{=}1, u{=}-1)$.

1.1.1.2: $\mathcal{T}_4 = 0$. By (11.26) we get $\rho_3\rho_4 = 0$, and we consider two subcases: $\mathcal{T}_3 \neq 0$ and $\mathcal{T}_3 = 0$.

1. $\mathcal{T}_3 \neq 0$. Then by Theorem 6.2 only one saddle is weak.
   (a) $\mathcal{F}_1 \neq 0$. In this case according to Theorem 6.2 the weak saddle is of order one, and we get the global configuration
   - $s, s^{(1)}, \overline{sn}_{(2)}; N^f, N^f, N^f \Rightarrow ((11.24): c = 1, e = 3, h = 0, m = 2, u = -1)$.

   (b) $\mathcal{F}_1 = 0$. Then by Theorem 6.2 the weak saddle has order ≥ 2. However by Theorems 7.9 and 7.3, we could not have a weak saddle of order three or an integrable saddle apart from a saddle and a saddle-node. Thus the weak saddle has order two, and this leads to the global configuration
   - $s, s^{(2)}, \overline{sn}_{(2)}; N^f, N^f, N^f \Rightarrow ((11.24): c = 1, e = 1/3, h = -10/3, m = 2/3, u = -1)$.
2. $\mathcal{T}_3 = 0$. Considering (11.26) and the condition $E_1 \neq 0$ (i.e. $\rho_1 \neq 0$), we obtain $\rho_3 = \rho_4 = 0$ and $\mathcal{T}_2 \neq 0$. Then by Theorem 6.2 we have two weak saddles, which could be both of order one (if $\mathcal{F}_1 \neq 0$), and they are both integrable if ($\mathcal{F}_1 = 0$). However by Theorem 7.3 we could not have two integrable saddles apart from a saddle-node, and we obtain the global configuration
   - $s^{(1)}, s^{(1)}, \overline{sn}_{(2)}; N^f, N^f, N^f \Rightarrow ((11.24): c = -3,\ e = 1,\ h = 1,\ m = -1,\ u = -1)$.

1.1.2: $E_1 = 0$. In this case $\rho_1 = 0$, and apart from the two saddles, we have a cusp. Then $c = -eu$ and by (11.26) we obtain

$$\mathcal{T}_4 = \mathcal{T}_3 = 0, \quad \mathcal{T}_2 = -\Delta_3\Delta_4\rho_3\rho_4, \quad \mathcal{T}_1 = -\Delta_3\Delta_4(\rho_3 + \rho_4).$$

1. $\mathcal{T}_2 \neq 0$. Then $\rho_3\rho_4 \neq 0$ and both saddles are strong. So we arrive at the global configuration
   - $s, s, \widehat{cp}_{(2)}; N^f, N^f, N^f \Rightarrow ((11.24): c{=}1, e{=}1, h{=}1, m = 2, u{=}-1)$.

2. $\mathcal{T}_2 = 0$. This implies $\rho_3\rho_4 = 0$, and we consider two subcases: $\mathcal{T}_1 \neq 0$ and $\mathcal{T}_1 = 0$.

   (a) $\mathcal{T}_1 \neq 0$. Then only one saddle is weak. According to Theorem 6.2 (see statement (d)), in the case $\mathcal{T}_4 = \mathcal{T}_3 = \mathcal{T}_2 = 0$ and $\mathcal{T}_1 \neq 0$, we could have a weak singularity only of order one. So the weak saddle is of order one, and we get the global configuration
   - $s, s^{(1)}, \widehat{cp}_{(2)};\ N^f, N^f, N^f \Rightarrow ((11.24): c = 1,\ e = 1,\ h = -2,\ m = 1,\ u = -1)$.

   (b) $\mathcal{T}_1 = 0$. We observe that in this case all the traces vanish (this implies $\sigma = 0$), and we arrive at the Hamiltonian systems. So we obtain the global configuration
   - $\$, \$, \widehat{cp}_{(2)};\ N^f, N^f, N^f \Rightarrow ((11.24): c = -1,\ e = -1,\ h = 1,\ m = -1,\ u = -1)$.

1.2: $\widetilde{K} > 0$. Then according to Theorem 6.1 the elemental singular points are both anti-saddles, and the type of the double singular point is governed by the invariant polynomial E_1.

1.2.1: $E_1 \neq 0$. In this case $\rho_1 \neq 0$, and apart from the two anti-saddles, we have a semi-elemental saddle-node.

1.2.1.1: $W_4 < 0$. According to Theorem 6.1 we have a node and a focus (or a center). Moreover the node is generic, whereas the type of the focus depends on the invariant polynomial $\mathcal{T}_4$.

1.2.1.1.1: $\mathcal{T}_4 \neq 0$. Then we have a strong focus. Since the total index of the finite singularities equals $+2$, we deduce that at infinity we must have singular points of a total index -1. So considering Lemma 11.1, we arrive at the following four global configurations:

- $n, f, \overline{sn}_{(2)};\ S, ©, © \Rightarrow ((11.24): c = 1,\ e = 47/20,\ h = 1,\ m = 1/10,\ u = -1)$ (if $\eta < 0$);
- $n, f, \overline{sn}_{(2)};\ S, S, N^\infty \Rightarrow ((11.24): c = 1, e = 5, h = 1, m = 1, u = -1)$ (if $\eta > 0$);
- $n, f, \overline{sn}_{(2)};\ \overline{\binom{0}{2}}SN, S \Rightarrow ((11.24): c{=}-2, e{=}-1, h = 1, m = 1, u{=}-1)$ (if $\eta = 0$, $\widetilde{M} \neq 0$);
- $n, f, \overline{sn}_{(2)};\ \overline{\binom{0}{3}}S \Rightarrow ((11.24): c = 1,\ e = 1,\ h = 5/27,\ m = -1,\ u = -1/27)$ (if $\eta = 0$, $\widetilde{M} = 0$).

1.2.1.1.2: $\mathcal{T}_4 = 0$. Considering (11.26) and the condition $E_1 \neq 0$, we get $\rho_3\rho_4 = 0$, i.e. the focus is weak. Then without loss of generality we may assume that the singularity $M_3(1,0)$ is a weak focus, i.e. $m = (c - eu)/2$. In this case we obtain

$$\mu_0 = u(-c^2 + 2eh + ceu)^2, \quad \mathcal{T}_4 = 0,$$
$$\mathcal{F}_1 = u(ceu - c^2 + 2eh)(2c^2 + 3ce + 2eh + 2ceu + e^2u).$$

By Lemma 11.11(i) in this case the condition $\widetilde{M} \neq 0$ holds, and we consider two possibilities: $\mathcal{F}_1 \neq 0$ and $\mathcal{F}_1 = 0$.

1. $\mathcal{F}_1 \neq 0$. By Theorem 6.2 the weak focus has order one, and considering Lemma 11.1 we arrive at the following three global configurations:
   - $n, f^{(1)}, \overline{sn}_{(2)};\ S, ©, © \Rightarrow ((11.24) : c = 1, e = 2, h = 2, m = 3/2, u = -1)$ (if $\eta < 0$);
   - $n, f^{(1)}, \overline{sn}_{(2)};\ S, S, N^{\infty} \Rightarrow ((11.24) : c = 1, e = 2, h = 1, m = 3/2, u = -1)$ (if $\eta > 0$);
   - $n, f^{(1)}, \overline{sn}_{(2)};\ \overline{\binom{0}{2}}SN, S \Rightarrow ((11.24) : c=0, e = 8/5, h=1, m = 4/5, u=-1)$ (if $\eta = 0$).
2. $\mathcal{F}_1 = 0$. In this case we have a weak focus of order at least two. According to Lemma 11.11(ii) in this case the condition $\eta \neq 0$ is verified.
   (a) $\mathcal{F}_2 \neq 0$. By Theorem 6.2 the weak focus has order two, and considering Lemma 11.1 and the condition $\eta \neq 0$, we arrive at the following two global configurations:
   - $n, f^{(2)}, \overline{sn}_{(2)};\ S, ©, © \Rightarrow ((11.24) : c = 1, e = -2, h = -1, m = 7/16,$ $u = -1/16)$ (if $\eta < 0$);
   - $n, f^{(2)}, \overline{sn}_{(2)};\ S, S, N^{\infty} \Rightarrow ((11.24) : c = 1, e = 4, h = 5/4, m = 5/2,$ $u = -1)$ (if $\eta > 0$).

   (b) $\mathcal{F}_2 = 0$. Since by Lemma 11.11(ii) we have $\mathcal{F}_3\mathcal{F}_4 \neq 0$ and $\eta > 0$, according to Theorem 6.2 the weak focus has order three, and we get one global configuration
   - $n, f^{(3)}, \overline{sn}_{(2)};\ S, S, N^{\infty} \Rightarrow ((11.24) : c = 1/6, e = 1, h=7/18, m=7/12,$ $u = -1)$.

1.2.1.2: $W_4 > 0$. According to Theorem 6.1 in this case we have two foci if either $W_2 < 0$ or ($W_2 \geq 0$ and $W_1W_3 < 0$); and we have two nodes if $W_2 > 0$ and $W_1W_3 > 0$.

1.2.1.2.1: $W_2 < 0$ or ($W_2 \geq 0, W_1W_3 < 0$). We have two foci, and for the existence of at least one weak focus, the condition $\mathcal{T}_4 = 0$ is necessary.

1.2.1.2.1.1: $\mathcal{T}_4 \neq 0$. Then both foci are strong. So considering Lemma 11.1, we arrive at the following four global configurations:
- $f, f, \overline{sn}_{(2)};\ S, ©, © \Rightarrow ((11.24) : c = -2,\ e = 2/5,\ h = 1,\ m = 0,\ u = -2)$ (if $\eta < 0$);
- $f, f, \overline{sn}_{(2)};\ S, S, N^{\infty} \Rightarrow ((11.24) : c = -2,\ e = 1/5,\ h = 1,\ m = 0,\ u = -2)$ (if $\eta > 0$);
- $f, f, \overline{sn}_{(2)};\ \overline{\binom{0}{2}}SN, S \Rightarrow ((11.24) : c = -2,\ e = \xi,\ h = 1,\ m = 0,\ u = -2)$ (where $\{\eta^{-1}(0)\} \ni \xi \approx 0.3825$); (if $\eta = 0, \widetilde{M} \neq 0$);

- $f, f, \overline{sn}_{(2)}; \overline{\binom{0}{3}}S \Rightarrow ((11.24): c = -1,\ e = 1/2,\ h = 10/27,\ m = 1,\ u = -4/27)$ (if $\eta = 0,\ \widetilde{M} = 0$).

1.2.1.2.1.2: $\mathcal{T}_4 = 0$. Considering the condition $E_1 \neq 0$ and (11.26), we get $\rho_3\rho_4 = 0$. Then at least one focus is weak, and without loss of generality we may assume that the singularity $M_3(1,0)$ is a weak focus, i.e. $m = (c - eu)/2$. In this case we obtain

$$\mathcal{T}_4 = 0,\quad \mathcal{T}_3 = -\Delta_3\Delta_4\rho_1^2\rho_4,\quad \mu_0 = u(ceu - c^2 + 2eh)^2,$$
$$\mathcal{F}_1 = u(ceu - c^2 + 2eh)(2c^2 + 3ce + 2eh + 2ceu + e^2u).$$

1. $\mathcal{T}_3 \neq 0$. Then only the focus $M_3(1,0)$ is weak.
   (a) $\mathcal{F}_1 \neq 0$. By Theorem 6.2 the weak focus has order one, and considering Lemma 11.1 we arrive at the following four global configurations:
   - $f, f^{(1)}, \overline{sn}_{(2)}; S, ©, © \Rightarrow ((11.24): c{=}-1, e = 3, h{=}1, m = 1, u{=}-1)$ (if $\eta < 0$);
   - $f, f^{(1)}, \overline{sn}_{(2)}; S, S, N^{\infty} \Rightarrow ((11.24): c = -1,\ e = 8,\ h = 1, m = 7/2,$ $u = -1)$ (if $\eta > 0$);
   - $f, f^{(1)}, \overline{sn}_{(2)}; \overline{\binom{0}{2}}SN, S \Rightarrow ((11.24): c = -1,\ e = \xi,\ h = 1,\ u = -1,$ $m = (\xi - 1)/2)$ (where $\{\eta^{-1}(0)\} \ni \xi \approx 7.7549$) (if $\eta = 0,\ \widetilde{M} \neq 0$);
   - $f, f^{(1)}, \overline{sn}_{(2)}; \overline{\binom{0}{3}}S \Rightarrow ((11.24): c = 1, e = -5, h = -2/5, m = -2,$ $u = -1/25)$ (if $\eta = 0,\ \widetilde{M} = 0$).

   (b) $\mathcal{F}_1 = 0$. According to Theorem 7.8 (respectively Theorem 7.4) systems (11.24) could not have a weak focus of order three (respectively a center) with a focus and a saddle-node, i.e. in this case the condition $\mathcal{F}_2 \neq 0$ holds. Moreover by Theorem 7.7 in the case under examination, we could have a weak focus of order two only if at infinity we have one real and two complex singularities (i.e. $\eta < 0$). By Theorem 6.2 the weak focus has order two. Thus considering Lemma 11.1 we get the global configuration
   - $f, f^{(2)}, \overline{sn}_{(2)}; S, ©, © \Rightarrow ((11.24): c = -1,\ e = 4,\ h = 9/4,\ m = 3/2,$ $u = -1)$.

2. $\mathcal{T}_3 = 0$. Then $\rho_3 = \rho_4 = 0$ and both elemental singularities are weak singularities (foci or centers). The condition $\rho_3 = \rho_4 = 0$ implies $m = (c - eu)/2$ and $h = (eu - c)/2$, and we calculate

$$\mu_0 = u(c+e)^2(c - eu)^2,\quad \mathcal{F}_1 = -2u(c+e)^2(c - eu)(c + eu),$$
$$E_1 = -u^2(c+e)^4(c - eu)^4(c + eu)/2.$$

So we arrive at the next remark.

Remark 11.5. If for a system (11.24) the conditions $\rho_3 = \rho_4 = 0$ and $E_1 \neq 0$ hold, then $\mathcal{F}_1\widetilde{M} \neq 0$.

Indeed, from the above expressions it immediately follows that $\mathcal{F}_1 \neq 0$. Additionally considering the relations $m = (c - eu)/2$ and $h = (eu - c)/2$, we calculate Coefficient$[\widetilde{M}, xy] = -16(c + eu)^2 \neq 0$ due to $E_1 \neq 0$.

Therefore considering Lemma 11.1, we arrive at the following three global configurations of singularities:

- $f^{(1)}, f^{(1)}, \overline{sn}_{(2)};\ S, ©, © \Rightarrow ((11.24): c = 1,\ e = -2,\ h = -3/8,\ m = 3/8,\ u = -1/8)$ (if $\eta < 0$);
- $f^{(1)}, f^{(1)}, \overline{sn}_{(2)};\ S, S, N^{\infty} \Rightarrow ((11.24): c = 4/15,\ e = -2,\ h = -1/120,\ m = 1/120,\ u = -1/8)$ (if $\eta > 0$);
- $f^{(1)}, f^{(1)}, \overline{sn}_{(2)};\ \overline{\binom{0}{2}}SN, S \Rightarrow ((11.24): c = 5/16,\ e = -2,\ h = -1/32,\ m = 1/32,\ u = -1/8)$ (if $\eta = 0$).

1.2.1.2.2: $W_2 > 0$ and $W_1W_3 > 0$. We have two nodes that are generic due to $W_4 \neq 0$. So considering Lemma 11.1, we arrive at the following four global configurations:

- $n, n, \overline{sn}_{(2)};\ S, ©, © \Rightarrow ((11.24): c = 1,\ e = 1,\ h = 1,\ m = 0,\ u = -2)$ (if $\eta < 0$);
- $n, n, \overline{sn}_{(2)};\ S, S, N^{\infty} \Rightarrow ((11.24): c = 1,\ e = 3,\ h = 1,\ m = 0,\ u = -2)$ (if $\eta > 0$);
- $n, n, \overline{sn}_{(2)};\ \overline{\binom{0}{2}}SN, S \Rightarrow ((11.24): c = 1,\ e = \xi,\ h = 1,\ m = 0,\ u = -2)$ (where $\{\eta^{-1}(0)\} \ni \xi \approx 2.7192$) (if $\eta = 0$, $\widetilde{M} \neq 0$);
- $n, n, \overline{sn}_{(2)};\ \overline{\binom{0}{3}}S \Rightarrow ((11.24): c = -1/2,\ e = -3/5,\ h = -5/6,\ u = -25/35,\ m = 1)$ (if $\eta = 0$, $\widetilde{M} = 0$).

1.2.1.3: $W_4 = 0$. Since $E_1 \neq 0$ (i.e. $\rho_1 \neq 0$), by (11.26) we obtain $\tau_3\tau_4 = 0$. Therefore at least one elemental singular point is a node with coinciding eigenvalues, and we may assume that such a singular point is $M_3(1, 0)$ (i.e. $\tau_3 = 0$). Considering Theorem 6.1 we examine three subcases: $W_3 < 0$, $W_3 > 0$ and $W_3 = 0$.

1.2.1.3.1: $W_3 < 0$. According to Theorem 6.1 the second elemental singularity is a focus or a center, and then by Theorem 7.2 in this case we could not have a star node.

1.2.1.3.1.1: $\mathcal{T}_4 \neq 0$. In this case we have a strong focus, and considering Lemma 11.1 we get the following four global configurations:

- $n^d, f, \overline{sn}_{(2)};\ S, ©, © \Rightarrow ((11.24): c = 1,\ e = -1,\ h = 1/2,\ m = -1,\ u = -1)$ (if $\eta < 0$);
- $n^d, f, \overline{sn}_{(2)};\ S, S, N^{\infty} \Rightarrow ((11.24): c = 2,\ e = -1,\ h = -1/8,\ m = 0,\ u = -1)$ (if $\eta > 0$);
- $n^d, f, \overline{sn}_{(2)};\ \overline{\binom{0}{2}}SN, S \Rightarrow ((11.24): c = 3,\ e = 0,\ h = 1,\ m = -3/2,\ u = -1)$ (if $\eta = 0$, $\widetilde{M} \neq 0$);

- $n^d, f, \overline{sn}_{(2)}; \overline{\binom{0}{3}}S \Rightarrow ((11.24): c = -2,\ e = 0,\ h = 0,\ m = 1,\ u = -1)$ (if $\eta = 0,\ \widetilde{M} = 0$).

1.2.1.3.1.2: $\mathcal{T}_4 = 0$. Since $\tau_3 = 0$ we must have $\rho_3 \neq 0$, and then the condition $\mathcal{T}_4 = 0$ implies $\rho_4 = 0$. In this case we have a weak focus or a center.

1. $\mathcal{F}_1 \neq 0$. By Theorem 6.2 we have a weak focus of order one. Since $W_4 = 0$, by Lemma 11.11(i) we have $\widetilde{M} \neq 0$. So considering Lemma 11.1 we get the following three global configurations:
   - $n^d, f^{(1)}, \overline{sn}_{(2)};\ S, ©, © \Rightarrow ((11.24): c = -1,\ e = 1,\ h = 0,\ m = 2,\ u = -1)$ (if $\eta < 0$);
   - $n^d, f^{(1)}, \overline{sn}_{(2)};\ S, S, N^\infty \Rightarrow ((11.24): c = -5, e = 1, h = 2, m = 0, u = -1)$ (if $\eta > 0$);
   - $n^d, f^{(1)}, \overline{sn}_{(2)}; \overline{\binom{0}{2}}SN, S \Rightarrow ((11.24): c = -(16\xi^2 + 1)/4,\ h = \xi,$ $e = (4\xi - 1)^2/4,\ m = (1-2\xi)/2,\ u = -1)$ (where $\{\eta^{-1}(0)\} \ni \xi \approx -0.9159$) (if $\eta = 0$).
2. $\mathcal{F}_1 = 0,\ \mathcal{F}_2 \neq 0$. Then we have a weak focus of order two. According to Lemma 11.11(iv) in this case the condition $\eta < 0$ holds, and according to Lemma 11.1 we get the global configuration
   - $n^d, f^{(2)}, \overline{sn}_{(2)};\ S, ©, © \Rightarrow ((11.24): c = -2/5,\ e = 3/20,\ h = -2/5,$ $m = 7/5,\ u = -8)$.
3. $\mathcal{F}_1 = \mathcal{F}_2 = 0$. By Lemma 11.11(iv) in this case the conditions $\mathcal{F}_3 = 0$ and $\eta = 0$ are satisfied. Therefore by Theorem 6.2 (see statement (b_4)) we have a center, and considering Theorem 7.4 (see Table 7.1 on page 201, line γ_5) we obtain the global configuration
   - $n^d, c, \overline{sn}_{(2)}; \overline{\binom{0}{2}}SN, S \Rightarrow ((11.24): c = 0, e = 2, h = 1, m = 1, u = -1)$.

1.2.1.3.2: $W_3 > 0$. According to Theorem 6.1 the second elemental singularity is a generic node. Applying Lemma 11.12 we consider two possibilities: the condition $(\mathcal{S}^1_{n^*})$ holds, and the condition $(\mathcal{S}^1_{n^*})$ is not satisfied (we denote this by $\neg(\mathcal{S}^1_{n^*})$).

1. $\neg(\mathcal{S}^1_{n^*})$. In this case by Lemma 11.12 we cannot have a finite star node, and considering Lemma 11.1 we get the following four global configurations:
   - $n, n^d, \overline{sn}_{(2)};\ S, ©, © \Rightarrow ((11.24): c = 2,\ e = 1,\ h = 9/8,\ m = 0,\ u = -1)$ (if $\eta < 0$);
   - $n, n^d, \overline{sn}_{(2)};\ S, S, N^\infty \Rightarrow ((11.24): c = 0,\ e = 1,\ h = 1/8,\ m = 0,\ u = -1)$ (if $\eta > 0$);
   - $n, n^d, \overline{sn}_{(2)}; \overline{\binom{0}{2}}SN, S \Rightarrow ((11.24): c = -2,\ e = 0,\ h = 4,\ m = 1,\ u = -1)$ (if $\eta = 0,\ \widetilde{M} \neq 0$);

- $n, n^d, \overline{sn}_{(2)}; \overline{\binom{0}{3}}S \Rightarrow ((11.24): c = -2,\ e = 0,\ h = 0,\ m = 1,\ u = -1/5)$ (if $\eta = 0$, $\widetilde{M} = 0$).

2. $(\mathcal{S}^1_{n^*})$. By Lemma 11.12 we have one finite star node (which is unique), and in this case the condition $\widetilde{M} \neq 0$ holds. Hence by Lemma 11.1 we get the global configuration

- $n, n^*, \overline{sn}_{(2)}; \overline{\binom{0}{2}}SN, S \Rightarrow ((11.24):\ c = 2,\ e = 0,\ h = 2,\ m = -1,\ u = -2)$.

1.2.1.3.3: $W_3 = 0$. In this case we have $\tau_3 = \tau_4 = 0$, i.e. each one of the nodes has coinciding eigenvalues.

1. $\neg(\mathcal{S}^1_{n^*})$. In this case by Lemma 11.12 we cannot have a star node, and considering Lemma 11.1 we get the following four global configurations:

- $n^d, n^d, \overline{sn}_{(2)};\ S, ©, © \Rightarrow ((11.24):\ c = 2,\ e = -2,\ h = -5,\ m = 1,\ u = -4)$ (if $\eta < 0$);
- $n^d, n^d, \overline{sn}_{(2)};\ S, S, N^\infty \Rightarrow ((11.24):\ c = 2,\ e = -1/2, h = 3/4, m = -3/4, u = -1)$ (if $\eta > 0$);
- $n^d, n^d, \overline{sn}_{(2)}; \overline{\binom{0}{2}}SN, S \Rightarrow ((11.24):\ c = -1,\ e = 0,\ h = -1,\ m = 1/2, u = -9/4)$ (if $\eta = 0$, $\widetilde{M} \neq 0$);
- $n^d, n^d, \overline{sn}_{(2)}; \overline{\binom{0}{3}}S \Rightarrow ((11.24): c = 1,\ e = 0,\ h = 0,\ m = -1/2,\ u = -1/4)$ (if $\eta = 0$, $\widetilde{M} = 0$).

2. $(\mathcal{S}^1_{n^*})$. By Lemma 11.12 only one of the nodes is a star node, and considering Theorem 7.2 we get the unique global configuration

- $n^d, n^*, \overline{sn}_{(2)}; \overline{\binom{0}{2}}SN, S \Rightarrow ((11.24):\ c = -1,\ e = 0,\ h = -1/2,\ m = 1/2, u = -1)$.

1.2.2: $E_1 = 0$. In this case $\rho_1 = 0$, and apart from two anti-saddles, we have a cusp. Then $c = -eu$ and by (11.26) we obtain

$$W_4 = W_3 = 0,\ \ W_2 = \Delta_3^2\Delta_4^2\tau_3\tau_4,\ \ W_1 = \Delta_3^2\Delta_4^2(\tau_3 + \tau_4),$$
$$\mathcal{T}_4 = \mathcal{T}_3 = 0,\ \ \mathcal{T}_2 = -\Delta_3\Delta_4\rho_3\rho_4,\ \ \mathcal{T}_1 = -\Delta_3\Delta_4(\rho_3 + \rho_4).$$

1.2.2.1: $W_2 < 0$. According to Theorem 6.1 (see Table 6.2, line 74) we have a node and a focus. Moreover the node is generic, whereas the type of the focus depends on the invariant polynomial $\mathcal{T}_2$.

1. $\mathcal{T}_2 \neq 0$. The focus is strong, and considering Lemma 11.1, we arrive at the following four global configurations:

- $n, f, \widehat{cp}_{(2)};\ S, ©, © \Rightarrow ((11.24):\ c = 1,\ e = 1,\ h = 1,\ m = 0,\ u = -1)$ (if $\eta < 0$);

- $n, f, \widehat{cp}_{(2)}$; $S, S, N^\infty \Rightarrow$ ((11.24) : $c = -3/2$, $e = -1$, $h = 2$, $m = 2$, $u = -3/2$) (if $\eta > 0$);

- $n, f, \widehat{cp}_{(2)}$; $\overline{\binom{0}{2}}SN, S \Rightarrow$ ((11.24) : $c = \xi$, $e = -1$, $h = 2$, $m = 2$, $u = \xi$) (where $\{\eta^{-1}(0)\} \ni \xi \approx -1.5278$) (if $\eta = 0$, $\widetilde{M} \neq 0$);

- $n, f, \widehat{cp}_{(2)}$; $\overline{\binom{0}{3}}S \Rightarrow$ ((11.24) : $c = 8$, $e = 1$, $h = 28$, $m = -10$, $u = -8$) (if $\eta = 0$, $\widetilde{M} = 0$).

2. $\mathcal{T}_2 = 0$. Then the focus is weak, and since we have a node, the condition $\mathcal{T}_1 \neq 0$ holds. Therefore according to Theorem 6.2 (see statement (d)), in the case $\mathcal{T}_4 = \mathcal{T}_3 = \mathcal{T}_2 = 0$ and $\mathcal{T}_1 \neq 0$, we could have a weak singularity only of order one. So the weak focus is of order one. Additionally since by Lemma 11.11(i) the condition $\widetilde{M} \neq 0$ holds, according to Lemma 11.1 we arrive at the following three global configurations:

- $n, f^{(1)}, \widehat{cp}_{(2)}$; $S, ©, © \Rightarrow$ ((11.24) : $c = 1$, $e = 1$, $h = 2$, $m = 1$, $u = -1$) (if $\eta < 0$);

- $n, f^{(1)}, \widehat{cp}_{(2)}$; $S, S, N^\infty \Rightarrow$ ((11.24) : $c = 1$, $e = 1$, $h = 6/5$, $m = 1$, $u = -1$) (if $\eta > 0$);

- $n, f^{(1)}, \widehat{cp}_{(2)}$; $\overline{\binom{0}{2}}SN, S \Rightarrow$ ((11.24) : $c = 1$, $e = 1$, $h = \xi$, $m = 1$, $u = -1$) (where $\{\eta^{-1}(0)\} \ni \xi \approx 1.3112$) (if $\eta = 0$).

1.2.2.2: $W_2 > 0$. According to Theorem 6.1 (see Table 6.2, line 73) we have two nodes that are both generic (due to $W_2 \neq 0$), and considering Lemma 11.1 we arrive at the following four global configurations:

- $n, n, \widehat{cp}_{(2)}$; $S, ©, © \Rightarrow$ ((11.24) : $c = 3$, $e = 3$, $h = 3/2$, $m = 1$, $u = -1$) if $\eta < 0$);
- $n, n, \widehat{cp}_{(2)}$; $S, S, N^\infty \Rightarrow$ ((11.24) : $c = 3$, $e = 3$, $h = 21/20$, $m = 1$, $u = -1$) (if $\eta > 0$);
- $n, n, \widehat{cp}_{(2)}$; $\overline{\binom{0}{2}}SN, S \Rightarrow$ ((11.24) : $c = 3$, $e = 3$, $h = \xi$, $m = 1$, $u = -1$) (where $\{\eta^{-1}(0)\} \ni \xi \approx 1.0904$) (if $\eta = 0$, $\widetilde{M} \neq 0$);
- $n, n, \widehat{cp}_{(2)}$; $\overline{\binom{0}{3}}S \Rightarrow$ ((11.24) : $c = 1$, $e = 1$, $h = 2$, $m = -2$, $u = -1$) (if $\eta = 0$, $\widetilde{M} = 0$).

1.2.2.3: $W_2 = 0$. By Theorem 6.1 (see Table 6.2, line 73) we have two nodes, and at least one is with coinciding eigenvalues. Since $E_1 = 0$, by Lemma 11.12 we could not have a finite star node.

We assume that the singular point $M_3(1, 0)$ is a point with coinciding eigenvalues. So we impose $\tau_3 = 0$, and since $\rho_1 = 0$ (due to $E_1 = 0$), by (11.26) we obtain $W_1 = \Delta_3^2\Delta_4^2\tau_4$.

1. $W_1 \neq 0$. Then $\tau_4 \neq 0$ and hence the second node is generic. So considering Lemma 11.1, we arrive at the following four global configurations:

- $n, n^d, \widehat{cp}_{(2)}$; $S, ©, © \Rightarrow ((11.24): c = 1,\ e = 1,\ h = 1/2,\ m = 0,\ u = -1)$ (if $\eta < 0$);
- $n, n^d, \widehat{cp}_{(2)}$; $S, S, N^\infty \Rightarrow ((11.24): c = 1,\ e = 1,\ h = 5/2,\ m = 2,\ u = -1)$ (if $\eta > 0$);
- $n, n^d, \widehat{cp}_{(2)}$; $\overline{\binom{0}{2}} SN, S \Rightarrow ((11.24): c = 1,\ e = 1,\ h = (1+\xi^2)/2,\ m = \xi,$ $u = -1)$ (where $\{\eta^{-1}(0)\} \ni \xi \approx 0.5694$) (if $\eta = 0,\ \widetilde{M} \neq 0$);
- $n, n^d, \widehat{cp}_{(2)}$; $\overline{\binom{0}{3}} S \Rightarrow ((11.24): c = 64/125,\ e = 1,\ h = 544/625,\ u = -4/5,$ $m = -152/125)$ (if $\eta = 0,\ \widetilde{M} = 0$).

2. $W_1 = 0$. Then $\tau_4 = 0$ and hence the singular point $M_4(0,1)$ is a node with coinciding eigenvalues. It was shown above that none of the nodes could be a star node.

   We claim that in this case the condition $\eta < 0$ holds. Indeed considering the relations $c = -eu$ and $h = (m^2 + e^2u^2)/(2e)$, we calculate

$$\tau_4 = (m^4 + 4e^4u^2 + 8e^3mu^2 + 2e^2m^2u^2 + e^4u^4)/e^2 \equiv \phi(e, m, u)/e^2.$$

   Consider the equation $\phi(e, m, u) = 0$. We observe that the polynomial ϕ is homogeneous of degree 4 with respect to c and m, and it is bi-quadratic in u. So denoting $m/e = z$, we calculate Discrim $[\phi(1, z, u), u^2] = 16(1+z)^2(1+2z)$, and clearly the condition $(1+z)^2(1+2z) \geq 0$ must hold.

   Assume $1 + z \neq 0$. Then $1 + 2z \geq 0$, and setting a new variable w as $1 + 2z = w^2 \geq 0$ (i.e. $z = (w^2 - 1)/2$), we calculate

$$\phi(1, z, u) = \left[4u^2 + (w-1)^4\right]\left[4u^2 + (w+1)^4\right].$$

   It is clear that the condition $u \neq 0$ implies $\phi(1, z, u) \neq 0$.

   Suppose now $z = -1$. This yields $m = -e$, and then we have

$$\phi(e, -e, u) = e^2(u^2 - 1)^2 = 0.$$

   Since $\mu_0 < 0$ (i.e. $u < 0$), we get $u = -1$, and then we calculate

$$\tau_3 = \tau_4 = 0, \quad \eta = -16e^4 = \mu_0.$$

   As $\mu_0 < 0$ this completes the proof of our claim.

   Thus in the case $W_2 = W_1 = 0$ we get the unique global configuration

- $n^d, n^d, \widehat{cp}_{(2)}$; $S, ©, © \Rightarrow ((11.24): c = 1,\ e = 1,\ h = 1,\ m = -1,\ u = -1)$.

2: $\mu_0 > 0$. Following Theorem 6.1 we shall consider two subcases: $E_1 \neq 0$ and $E_1 = 0$.

2.1: $E_1 \neq 0$. Then the double singular point is a semi-elemental saddle-node.

2.1.1: $W_4 < 0$. According to Theorem 6.1, apart from the saddle-node, we have a saddle and a focus. Moreover by Theorem 6.2 their types depend on the invariant polynomials $\mathcal{T}_i$ and $\mathcal{F}_i$ $(i = 1, \ldots, 4)$.

2.1.1.1: $\mathcal{T}_4 \neq 0$. Then $\rho_3\rho_4 \neq 0$ and both elemental singularities are strong.

1. $\eta < 0$. Then systems (11.24) possess one real and two complex infinite singular points, and according to Lemmas 11.1 and 11.11(iii), there can only be three distinct configurations at infinity which provide the global configurations:

   - $s, f, \overline{sn}_{(2)};\ N^{\infty}, ©, © \Rightarrow ((11.24): c = 2,\ e = 1/2,\ h = 1,\ m = 0,\ u = 1)$ (if $\theta < 0$);
   - $s, f, \overline{sn}_{(2)};\ N^{f}, ©, © \Rightarrow ((11.24): c = 0,\ e = 1,\ h = 1,\ m = 0,\ u = 1)$ (if $\theta > 0$);
   - $s, f, \overline{sn}_{(2)};\ N^{d}, ©, © \Rightarrow ((11.24): c = 2,\ e = 1,\ h = 1,\ m = 0,\ u = 1)$ (if $\theta = 0$).

2. $\eta > 0$. In this case systems (11.24) possess three real infinite singular points. Since for these systems the condition $\mu_0 > 0$ holds, taking into consideration Lemmas 11.1 and 11.11(iii), we could have at infinity only six distinct configurations which provide the following global configurations:

   - $s, f, \overline{sn}_{(2)};\ S, N^{\infty}, N^{\infty} \Rightarrow ((11.24): c = 2,\ e = 2/5,\ h = -2/5,\ m = -8/5,\ u = 1)$ (if $\theta < 0, \theta_1 < 0$);
   - $s, f, \overline{sn}_{(2)};\ S, N^{f}, N^{f} \Rightarrow ((11.24): c = 1/2,\ e = 1,\ h = 0,\ m = -2,\ u = 2)$ (if $\theta < 0, \theta_1 > 0$);
   - $s, f, \overline{sn}_{(2)};\ S, N^{\infty}, N^{f} \Rightarrow ((11.24): c = -3/2,\ e = 1/2,\ h = 0,\ m = -2,\ u = 2)$ (if $\theta > 0$);
   - $s, f, \overline{sn}_{(2)};\ S, N^{\infty}, N^{d} \Rightarrow ((11.24): c = 6/5,\ e = 1/5,\ h = 1,\ m = 0,\ u = 1)$ (if $\theta = 0, \theta_1 < 0$);
   - $s, f, \overline{sn}_{(2)};\ S, N^{f}, N^{d} \Rightarrow ((11.24): c = 1/4,\ e = 1/2,\ h = -1/4,\ m = 0,\ u = 1/4)$ (if $\theta = 0, \theta_1 > 0$);
   - $s, f, \overline{sn}_{(2)};\ S, N^{d}, N^{d} \Rightarrow ((11.24): c = -3,\ e = -1,\ h = 2,\ m = 3,\ u = 2)$ (if $\theta = 0, \theta_1 = 0$).

3. $\eta = 0$. In this case systems (11.24) possess at infinity either one double and one simple real singular points (if $\widetilde{M} \neq 0$) or one triple real singularity (if $\widetilde{M} = 0$). So by Lemmas 11.1 and 11.11(iii), we have the following three global configurations if $\widetilde{M} \neq 0$ (here we use the canonical systems $(\mathbf{S}_{III})$ and $(\mathbf{S}_{IV})$ both with $a = b = 0$, see Lemma 6.1):

- $s, f, \overline{sn}_{(2)};\ \overline{\binom{0}{2}}SN,\ N^{\infty} \Rightarrow ((\mathbf{S}_{III}) : c = 0, d = -1, e = 1, f = 1, g = 2, h = 1)$ (if $\theta < 0$);
- $s, f, \overline{sn}_{(2)};\ \overline{\binom{0}{2}}SN,\ N^{f} \Rightarrow ((\mathbf{S}_{III}) : c = 1, d = 0, e = 1, f = 0, g = 1/2, h = 1)$ (if $\theta > 0$);
- $s, f, \overline{sn}_{(2)};\ \overline{\binom{0}{2}}SN,\ N^{d} \Rightarrow ((\mathbf{S}_{III}) : (c = 1, d = 0,\ e = 1, f = 0, g = 1, h = 1)$ (if $\theta = 0$).

And we have one global configuration if $\widetilde{M} = 0$:

- $s, f, \overline{sn}_{(2)};\ \overline{\binom{0}{3}}N \Rightarrow \quad ((\mathbf{S}_{IV}) : c = 0,\ d = -1,\ e = 0,\ f = 2,\ g = 1,\ h = -1)$.

2.1.1.2: $\mathcal{T}_4 = 0$. Considering the condition $E_1 \neq 0$ and (11.26), we get $\rho_3\rho_4 = 0$. Then at least one singularity is weak, and without loss of generality we may assume that such a singularity is $M_3(1, 0)$.

2.1.1.2.1: $\mathcal{T}_3\mathcal{F} < 0$. According to Theorem 6.2 (see statement (b)), the weak singularity is a focus.

2.1.1.2.1.1: $\mathcal{F}_1 \neq 0$. Then the order of the weak focus is one.

1. $\eta < 0$. According to Lemma 11.11(iii), the condition $\theta = \theta_2 = 0$ could not be satisfied. So by Lemma 11.1 there can only be three distinct configurations at infinity which lead to the following global configurations:
   - $s, f^{(1)}, \overline{sn}_{(2)};\ N^{\infty},\ ©, © \Rightarrow ((11.24) : c = -3, e = -2, h = -1, m = -1/2, u = 1)$ (if $\theta < 0$);
   - $s, f^{(1)}, \overline{sn}_{(2)};\ N^{f},\ ©, © \Rightarrow ((11.24) : c = -1, e = -3, h = 0, m = 1, u = 1)$ (if $\theta > 0$);
   - $s, f^{(1)}, \overline{sn}_{(2)};\ N^{d},\ ©, © \Rightarrow ((11.24) : c = 1, e = 2, h = 1/2, m = 3/2, u = 1)$ (if $\theta = 0$).
2. $\eta > 0$. In this case systems (11.24) possess three real infinite singular points. Since for these systems the condition $\mu_0 > 0$ holds, taking into consideration Lemmas 11.1 and 11.11(iii), we could have at infinity only six configurations which provide the following global configurations:
   - $s, f^{(1)}, \overline{sn}_{(2)};\ S, N^{\infty},\ N^{\infty} \Rightarrow ((11.24) : c = 7/2, e=1, h = 4, m=1, u = 3/2)$ (if $\theta < 0, \theta_1 < 0$);
   - $s, f^{(1)}, \overline{sn}_{(2)};\ S, N^{f},\ N^{f} \Rightarrow ((11.24) : c = -1/2, e = 1, h = 3/2, m = -5/2, u = 9/2)$ (if $\theta < 0, \theta_1 > 0$);
   - $s, f^{(1)}, \overline{sn}_{(2)};\ S, N^{\infty},\ N^{f} \Rightarrow ((11.24) : c = -3/2,\ e = 1,\ h = 4,\ u = 7/2$ $m = -5/2)$ (if $\theta > 0$);
   - $s, f^{(1)}, \overline{sn}_{(2)};\ S, N^{\infty},\ N^{d} \Rightarrow ((11.24) : c = 1/4, e = 7/4, h = 3/32, u = 1/4, m = 13/16)$ (if $\theta = 0, \theta_1 < 0$);

- $s, f^{(1)}, \overline{sn}_{(2)}; \ S, N^f, \ N^d \Rightarrow ((11.24): c = 1/4, e = -1/4, h = -5/32, m = 3/16, u = 1/4)$ (if $\theta = 0, \theta_1 > 0$);
- $s, f^{(1)}, \overline{sn}_{(2)}; \ S, N^d, \ N^d \Rightarrow ((11.24): c = 3/10, \ e = 1, \ h = -1/10, m = -3/10, \ u = 1/10)$ (if $\theta = 0, \theta_1 = 0$).

3. $\eta = 0$. In this case systems (11.24) possess at infinity either one double and one simple real singular points (if $\widetilde{M} \neq 0$) or one triple real singularity (if $\widetilde{M} = 0$). So by Lemmas 11.1 and 11.11(iii), we have the following three global configurations if $\widetilde{M} \neq 0$:

- $s, f^{(1)}, \overline{sn}_{(2)}; \ \overline{\binom{0}{2}}SN, \ N^\infty \Rightarrow ((11.24): c = 3/2, e = 1/42, h = -7, m = -1, u = 147)$ (if $\theta < 0$);
- $s, f^{(1)}, \overline{sn}_{(2)}; \ \overline{\binom{0}{2}}SN, \ N^f \Rightarrow ((11.24): c = 50, \ e = \xi, \ h = -58, m = (50 - 8\xi)/2, \ u = 8)$ (where $\{\eta^{-1}(0)\} \ni \xi \approx 44.6351$) (if $\theta > 0$);
- $s, f^{(1)}, \overline{sn}_{(2)}; \ \overline{\binom{0}{2}}SN, \ N^d \Rightarrow ((11.24): c = \xi, \ e = \chi, \ u = 8, h = (1 + 4\xi^2 - 32\xi\chi/(8\chi), \ m = (\xi - 8\chi)/2)$ (where $\{\eta^{-1}(0), \theta^{-1}(0)\} \ni (\xi, \chi) \approx (1.1777, 0.1072)$) (if $\theta = 0$).

And we have one global configuration if $\widetilde{M} = 0$:

- $s, f^{(1)}, \overline{sn}_{(2)}; \ \overline{\binom{0}{3}}N \Rightarrow ((11.24): c = -3, \ e = -1/405, \ h = -65, \ m = 1, u = 2025)$.

2.1.1.2.1.2: $\mathcal{F}_1 = 0$. We consider two subcases: $\mathcal{F}_2 \neq 0$ and $\mathcal{F}_2 = 0$.

1. $\mathcal{F}_2 \neq 0$. In this case the weak focus is of order two, and by Theorem 7.7 we could have only the following five global configurations:

- $s, f^{(2)}, \overline{sn}_{(2)}; \ N^f, \ ©, © \Rightarrow ((11.24): c = -1/2, e = 1, h = 1/2, m = -3/8, u = 1/4)$ (if $\eta < 0$);
- $s, f^{(2)}, \overline{sn}_{(2)}; \ S, N^f, \ N^f \Rightarrow ((11.24): c = -3/2, \ e = 1, \ h = 5, \ m = -13/4, u = 5)$ (if $\eta > 0, \ \theta < 0$);
- $s, f^{(2)}, \overline{sn}_{(2)}; \ S, N^\infty, \ N^f \Rightarrow ((11.24): c = -3, \ e = 1, \ h = 41/2, \ u = 10, m = -13/2)$ (if $\eta > 0, \ \theta > 0$);
- $s, f^{(2)}, \overline{sn}_{(2)}; \ S, N^f, \ N^d \Rightarrow ((11.24): c = \xi, e = 1, h = -(2\xi^2 + 23\xi + 10)/2, m = (\xi - 10)/2, \ u = 10)$ (where $\{\theta^{-1}(0)\} \ni \xi \approx -2.6027$) (if $\eta > 0, \theta = 0$);
- $s, f^{(2)}, \overline{sn}_{(2)}; \ \overline{\binom{0}{2}}SN, \ N^f \Rightarrow ((11.24): c = -4/5, \ e = 1, \ m = -(4 + 5\xi)/10, h = (28 + 15\xi)/50, \ u = \xi)$ (where $\{\eta^{-1}(0)\} \ni \xi \approx 2.2364$) (if $\eta = 0$).

2. $\mathcal{F}_2 = 0$. By Theorem 6.2 we could have either a weak focus of order three or a center. However by Theorem 7.4 we could not have a center in the case under examination. Therefore considering Theorem 7.8, we arrive at the following three global configurations:

- $s, f^{(3)}, \overline{sn}_{(2)}$; $S, N^f, N^f \Rightarrow ((11.24): c=-2,\ e=1,\ h=17,\ m=-7,\ u=12)$ (if $\theta<0$);
- $s, f^{(3)}, \overline{sn}_{(2)}$; $S, N^\infty, N^f \Rightarrow ((11.24): c=-3,\ e=1,\ h=81/2,\ u=18,\ m=-21/2)$ (if $\theta>0$);
- $s, f^{(3)}, \overline{sn}_{(2)}$; $S, N^f, N^d \Rightarrow ((11.24): c=-\xi/6,\ e=1,\ h=\xi(5\xi-9)/36,\ m=-7\xi/12,\ u=\xi)$ (where $\xi=27(17+3\sqrt{21})/50$) (if $\theta=0$).

2.1.1.2.2: $\mathcal{T}_3\mathcal{F}>0$. According to Theorem 6.2 the weak singularity is a saddle.

2.1.1.2.2.1: $\mathcal{F}_1\neq 0$. Then the order of the weak saddle is one.

1. $\eta<0$. According to Lemmas 11.11(iii) and 11.1, there can only be three distinct configurations at infinity which lead to the corresponding global configurations:
   - $s^{(1)}, f, \overline{sn}_{(2)}$; N^∞, ©, © $\Rightarrow ((11.24): c=2,\ e=4,\ h=-2,\ m=-1,\ u=1)$ (if $\theta<0$);
   - $s^{(1)}, f, \overline{sn}_{(2)}$; N^f, ©, © $\Rightarrow ((11.24): c=2,\ e=2,\ h=-2,\ m=0,\ u=1)$ (if $\theta>0$);
   - $s^{(1)}, f, \overline{sn}_{(2)}$; N^d, ©, © $\Rightarrow ((11.24): c=2,\ e=10/3,\ h=-2,\ m=-2/3,\ u=1)$ (if $\theta=0$).
2. $\eta>0$. In this case systems (11.24) possess three real infinite singular points. Since for these systems the condition $\mu_0>0$ holds, taking into consideration Lemmas 11.1 and 11.11(iii), we could have at infinity six configurations which lead to the corresponding global configurations:
   - $s^{(1)}, f, \overline{sn}_{(2)}$; $S, N^\infty, N^\infty \Rightarrow ((11.24): c=3, e=17/2, h=-8, m=-11/4,\ u=1)$ (if $\theta<0, \theta_1<0$);
   - $s^{(1)}, f, \overline{sn}_{(2)}$; $S, N^f, N^f \Rightarrow ((11.24): c=-5/2,\ e=-1,\ h=3,\ m=-1,\ u=1/2)$ (if $\theta<0, \theta_1>0$);
   - $s^{(1)}, f, \overline{sn}_{(2)}$; $S, N^\infty, N^f \Rightarrow ((11.24): c=3,\ e=8,\ h=-8,\ m=-5/2,\ u=1)$ (if $\theta>0$);
   - $s^{(1)}, f, \overline{sn}_{(2)}$; $S, N^\infty, N^d \Rightarrow ((11.24): c=2,\ e=40,\ h=-9,\ m=-4,\ u=1/4)$ (if $\theta=0, \theta_1<0$);
   - $s^{(1)}, f, \overline{sn}_{(2)}$; $S, N^f, N^d \Rightarrow ((11.24): c=4,\ e=40,\ h=-21/2,\ m=-3,\ u=1/4)$ (if $\theta=0, \theta_1>0$);
   - $s^{(1)}, f, \overline{sn}_{(2)}$; $S, N^d, N^d \Rightarrow ((11.24): c=-1,\ e=-12,\ h=3,\ m=1,\ u=1/4)$ (if $\theta=0, \theta_1=0$).
3. $\eta=0$. In this case systems (11.24) possess at infinity either one double and one simple real singular points (if $\widetilde{M}\neq 0$) or one triple real singularity (if

$\widetilde{M} = 0$). So by Lemmas 11.1 and 11.11(iii), we have the following three global configurations if $\widetilde{M} \neq 0$:

- $s^{(1)}, f, \overline{sn}_{(2)}; \overline{\binom{0}{2}}SN,\ N^{\infty} \Rightarrow ((11.24): c = 1/2,\ e = 27/10,\ h = -5/3,$ $m = -1,\ u = 25/27)$ (if $\theta < 0$);
- $s^{(1)}, f, \overline{sn}_{(2)}; \overline{\binom{0}{2}}SN,\ N^{f} \Rightarrow ((11.24): c = \xi,\ e = 1,\ h = -2, m = (\xi - 1)/2,$ $u = 1)$ (where $\{\eta^{-1}(0)\} \ni \xi \approx 1.1373$) (if $\theta > 0$);
- $s^{(1)}, f, \overline{sn}_{(2)}; \overline{\binom{0}{2}}SN,\ N^{d} \Rightarrow ((11.24): c = \xi,\ e = \chi,\ h = -2, m = (\xi - \chi)/2,$ $u = 1)$ (where $\{\eta^{-1}(0), \theta^{-1}(0)\} \ni (\xi, \chi) \approx (0.8240,\ 2.1574)$) (if $\theta = 0$).

And we have one global configuration if $\widetilde{M} = 0$:

- $s^{(1)}, f, \overline{sn}_{(2)}; \overline{\binom{0}{3}}N \Rightarrow ((11.24): c = 3/4, e = -1331/1620, h = -10/11, m = 1,$ $u = 2025/1331)$.

2.1.1.2.2.2: $\mathcal{F}_1 = 0$. Then the order of the weak saddle is at least two. In this case by Lemma 11.13(i), we have $\widetilde{M} \neq 0$ (i.e. at infinity we could not have a triple singularity), and the condition $\eta = 0$ implies $\theta > 0$ (see statement (ii) of this lemma). We consider two subcases: $\mathcal{F}_2 \neq 0$ and $\mathcal{F}_2 = 0$.

1. $\mathcal{F}_2 \neq 0$. According to Theorem 6.2 the weak saddle is of order two. As $\widetilde{M} \neq 0$, by Lemmas 11.1 and 11.11(v), we arrive at the following nine global configurations:
   - $s^{(2)}, f, \overline{sn}_{(2)}; N^{\infty}, ©, © \Rightarrow ((11.24): c = -4/25, e = 1, h = -172/625,$ $m = -4/5,\ u = 36/25)$ (if $\eta < 0,\ \theta < 0$);
   - $s^{(2)}, f, \overline{sn}_{(2)}; N^{f}, ©, © \Rightarrow ((11.24): c = -2,\ e = 1,\ h = 1/2,\ m = -3/2,$ $u = 1)$ (if $\eta < 0,\ \theta > 0$);
   - $s^{(2)}, f, \overline{sn}_{(2)}; N^{d}, ©, © \Rightarrow ((11.24): c = -1/5,\ e = 1,\ h = (13 - 15\xi)/50,$ $m = -(1 + 5\xi)/10,\ u = \xi)$ (where $\{\theta^{-1}(0)\} \ni \xi \approx 1.5686$) (if $\eta < 0,\ \theta = 0$);
   - $s^{(2)}, f, \overline{sn}_{(2)}; S, N^{\infty},\ N^{\infty} \Rightarrow ((11.24): c = 1/20,\ e = 1,$ $h = -3/16, m = -3/40,\ u = 1/5)$ (if $\eta > 0,\ \theta < 0, \theta_1 < 0$);
   - $s^{(2)}, f, \overline{sn}_{(2)}; S, N^{f},\ N^{f} \Rightarrow ((11.24): c = -9/8,\ e = 1,\ h = 1/2,$ $m = -5/8,\ u = 1/8)$ (if $\eta > 0,\ \theta < 0, \theta_1 > 0$);
   - $s^{(2)}, f, \overline{sn}_{(2)}; S, N^{\infty},\ N^{f} \Rightarrow ((11.24): c = 1/10,\ e = 1,\ h = -2/5,$ $m = -3/20,\ u = 2/5)$ (if $\eta > 0,\ \theta > 0$);
   - $s^{(2)}, f, \overline{sn}_{(2)}; S, N^{\infty},\ N^{d} \Rightarrow ((11.24): c = 1/20, e = 1,\ m = (1 - 20\xi)/40,$ $h = -(31 + 220\xi)/400,\ u = \xi)$ (where $\{\theta^{-1}(0)\} \ni \xi \approx 0.1935$) (if $\eta > 0,\ \theta = 0, \theta_1 < 0$);

- $s^{(2)}, f, \overline{sn}_{(2)}$; $S, N^f, N^d \Rightarrow ((11.24): c = (1+8\xi)/4,\ e = 1,\ m = \xi,$ $h = -3(1+4\xi)/8,\ u = 1/4)$ (where $\xi = -(3+\sqrt{5})/8$) (if $\eta > 0,\ \theta = 0, \theta_1 > 0$);
- $s^{(2)}, f, \overline{sn}_{(2)}$; $\overline{\binom{0}{2}}SN, N^f \Rightarrow ((11.24): c = -6/5,\ e = 1,\ h = (35\xi + 18)/50,$ $m = -(5\xi + 6)/10, u = \xi)$ (where $\{\eta^{-1}(0)\} \ni \xi \approx 0.07382$) (if $\eta = 0$).

2. $\mathcal{F}_2 = 0$. By Theorem 7.3 we cannot have an integrable saddle plus a focus and a saddle–node. According to Theorem 7.9 we could have only the following three global configurations:
   - $s^{(3)}, f, \overline{sn}_{(2)}$; N^∞, ©, © $\Rightarrow ((11.24): c = -17/60, e = 1, h = -17/720,$ $m = -119/120,\ u = 17/10)$ (if $\theta < 0$);
   - $s^{(3)}, f, \overline{sn}_{(2)}$; N^f, ©, © $\Rightarrow ((11.24): c = -1/3,\ e = 1,\ h = 1/18,\ u = 2$ $m = -7/6)$ (if $\theta > 0$);
   - $s^{(3)}, f, \overline{sn}_{(2)}$; N^d, ©, © $\Rightarrow ((11.24): c = -\xi/6,\ e = 1,\ h = \xi(5\xi - 9)/36,$ $m = -7\xi/12,\ u = \xi)$ (where $\xi = 27(17 - 3\sqrt{21}\,)/50$) (if $\theta = 0$).

2.1.1.2.3: $\mathcal{T}_3 = 0$. In this case we have $\rho_3 = \rho_4 = 0$, and hence both elemental singularities are weak. Considering Remark 11.5 we have $\mathcal{F}_1 \neq 0$ and $\widetilde{M} \neq 0$, i.e. both singularities have order one, and at infinity there could not be a triple singularity. Moreover, we claim that in this case the condition $\theta = \theta_1 = 0$ could not be satisfied. Indeed, suppose the contrary, that $\theta = \theta_1 = 0$. As it was shown in the proof of the statement (iii) of Lemma 11.11 (see page 453), in this case for systems (11.24) the conditions $m = -c$ and $h = -eu$ must hold. However this implies

$$\rho_3 = eu - 3c, \quad \rho_4 = c - eu, \quad \mu_0 = 4u(c^2 - e^2u)^2,$$

and evidently the condition $\rho_3 = \rho_4 = 0$ yields $\mu_0 = 0$. This contradiction proves our claim.

1. $\eta < 0$. Then systems (11.24) possess one real and two complex infinite singular points, and according to Lemmas 11.1 and 11.11(iii), there can only be three configurations at infinity which lead to the corresponding global configurations:
   - $s^{(1)}, f^{(1)}, \overline{sn}_{(2)}$; N^∞, ©, © $\Rightarrow ((11.24): c = 2,\ e = 1,\ h = 1/4,\ m = -1/4,$ $u = 5/2)$ (if $\theta < 0$);
   - $s^{(1)}, f^{(1)}, \overline{sn}_{(2)}$; N^f, ©, © $\Rightarrow ((11.24): c = 1,\ e = 0,\ h = -1/2,\ m = 1/2,$ $u = 1)$ (if $\theta > 0$);
   - $s^{(1)}, f^{(1)}, \overline{sn}_{(2)}$; N^d, ©, © $\Rightarrow ((11.24): c = 2,\ e = 1,\ h = \xi/2 - 1,$ $m = 1 - \xi/2,\ u = \xi)$ (where $\{\theta^{-1}(0)\} \ni \xi \approx 2.5098$) (if $\theta = 0$).
2. $\eta > 0$. In this case systems (11.24) possess three real infinite singular points. Taking into consideration the conditions $\mu > 0$, $\theta^2 + \theta_1^2 \neq 0$ and Lemmas

11.1 and 11.11(iii), we could have at infinity only five configurations which lead to the corresponding global configurations:

- $s^{(1)}, f^{(1)}, \overline{sn}_{(2)}$; $S, N^\infty, N^\infty \Rightarrow$ ((11.24) : $c = 619/5$, $e = 1$, $h = 1001/10$, $m = -1001/10$, $u = 324$) (if $\theta < 0, \theta_1 < 0$);
- $s^{(1)}, f^{(1)}, \overline{sn}_{(2)}$; $S, N^f, N^f \Rightarrow$ ((11.24) : $c = 1$, $e = 0$, $h = -1/2$, $m = 1/2$, $u = 1/10$) (if $\theta < 0, \theta_1 > 0$);
- $s^{(1)}, f^{(1)}, \overline{sn}_{(2)}$; $S, N^\infty, N^f \Rightarrow$ ((11.24) : $c = 1$, $e = 0$, $h = -1/2$, $m = 1/2$, $u = 3/25$) (if $\theta > 0$);
- $s^{(1)}, f^{(1)}, \overline{sn}_{(2)}$; $S, N^\infty, N^d \Rightarrow$ ((11.24) : $c = 6804/55$, $e = 1$, $h = 5508/55$, $m = -5508/55$, $u = 324$) (if $\theta = 0, \theta_1 < 0$);
- $s^{(1)}, f^{(1)}, \overline{sn}_{(2)}$; $S, N^f, N^d \Rightarrow$ ((11.24) : $c = 1$, $e = 0$, $h = -1/2$, $m = 1/2$, $u = 1/9$) (if $\theta = 0, \theta_1 > 0$).

3. $\eta = 0$. Since by Lemma 11.13 we have $\widetilde{M} \neq 0$, in this case systems (11.24) possess at infinity one double and one simple real singular points. So by Lemmas 11.1 and 11.11(iii), we have the following three global configurations:

- $s^{(1)}, f^{(1)}, \overline{sn}_{(2)}$; $\binom{\overline{0}}{2}SN, N^\infty \Rightarrow$ ((11.24) : $c = \xi$, $e = 1$, $h = (324 - \xi)/2$, $m = (\xi - 324)/2$, $u = 324$) (where $\{\eta^{-1}(0)\} \ni \xi \approx 123.842$) (if $\theta < 0$);
- $s^{(1)}, f^{(1)}, \overline{sn}_{(2)}$; $\binom{\overline{0}}{2}SN, N^f \Rightarrow$ ((11.24) : $c = 1$, $e = 0$, $h = -1/2$, $m = 1/2$, $u = 1/8$) (if $\theta > 0$);
- $s^{(1)}, f^{(1)}, \overline{sn}_{(2)}$; $\binom{\overline{0}}{2}SN, N^d \Rightarrow$ ((11.24) : $c = \xi$, $e = 1$, $h = (\chi - \xi)/2$, $m = (\xi - \chi)/2$, $u = \chi$) (where $\{\eta^{-1}(0), \theta^{-1}(0)\} \ni (\xi, \chi) = (123 + 55\sqrt{5})/2, 161 + 72\sqrt{5})$ (if $\theta = 0$).

2.1.2: $W_4 > 0$. According to Theorem 6.1 (see Table 6.2, line 75), apart from the semi-elemental saddle-node, we have a saddle and a node, and this node is generic (due to $W_4 \neq 0$, i.e. $\tau_3\tau_4 \neq 0$).

2.1.2.1: $\mathcal{T}_4 \neq 0$. Then $\rho_3\rho_4 \neq 0$ and the saddle is strong.

1. $\eta < 0$. According to Lemma 11.1 we obtain four global configurations:

- $s, n, \overline{sn}_{(2)}$; N^∞, ©, © $\Rightarrow$ ((11.24) : $c = -3$, $e = 0$, $h = 1$, $m = 1$, $u = 1$) (if $\theta < 0$);
- $s, n, \overline{sn}_{(2)}$; N^f, ©, © $\Rightarrow$ ((11.24) : $c = 4$, $e = 0$, $h = 2$, $m = 1$, $u = 1$) (if $\theta > 0$);
- $s, n, \overline{sn}_{(2)}$; N^d, ©, © $\Rightarrow$ ((11.24) : $c = 3$, $e = 1$, $h = 4$, $m = 1$, $u = 4$) (if $\theta = 0$, $\theta_2 \neq 0$);
- $s, n, \overline{sn}_{(2)}$; N^*, ©, © $\Rightarrow$ ((11.24) : $c = -2, e = 1, h = 0, m = 1, u = 1$) (if $\theta = 0$, $\theta_2 = 0$).

2. $\eta > 0$. Taking into consideration Lemmas 11.1 and 11.11(iii), we could have at infinity only nine configurations which lead to the corresponding global configurations:

   - $s, n, \overline{sn}_{(2)};\ S, N^\infty,\ N^\infty \Rightarrow ((11.24): c = -3/2,\ e = 0,\ h = 1/3,\ m = 1,$ $u = 1)\quad (\text{if } \theta < 0, \theta_1 < 0)$;
   - $s, n, \overline{sn}_{(2)};\ S, N^f,\ N^f \Rightarrow ((11.24): c = -1/2, e = 0, h = 1/3, m = 1, u = 1)$ $(\text{if } \theta < 0, \theta_1 > 0)$;
   - $s, n, \overline{sn}_{(2)};\ S, N^\infty,\ N^f \Rightarrow ((11.24): c = -1/2,\ e = 0,\ h = 2,\ m = 1,\ u = 1)$ $(\text{if } \theta > 0)$;
   - $s, n, \overline{sn}_{(2)};\ S, N^\infty,\ N^d \Rightarrow ((11.24): c = -3/2, e = 0, h = 1/2, m = 1, u = 1)$ $(\text{if } \theta = 0, \theta_1 < 0, \theta_2 \neq 0)$;
   - $s, n, \overline{sn}_{(2)};\ S, N^\infty,\ N^* \Rightarrow ((11.24): c = 4,\ e = 1,\ h = -3,\ m = -2,\ u = 1)$ $(\text{if } \theta = 0, \theta_1 < 0, \theta_2 = 0)$;
   - $s, n, \overline{sn}_{(2)};\ S, N^f,\ N^d \Rightarrow ((11.24): c = -1/2, e = 0, h = 1/2, m = 1, u = 1)$ $(\text{if } \theta = 0, \theta_1 > 0, \theta_2 \neq 0)$;
   - $s, n, \overline{sn}_{(2)};\ S, N^f,\ N^* \Rightarrow ((11.24): c = -2,\ e = 1,\ h = 3,\ m = -2,\ u = 1)$ $(\text{if } \theta = 0, \theta_1 > 0, \theta_2 = 0)$;
   - $s, n, \overline{sn}_{(2)};\ S, N^d,\ N^d \Rightarrow ((11.24): c = 0,\ e = -1/2,\ h = 1,\ m = 0,\ u = 2)$ $(\text{if } \theta = \theta_1 = 0, \theta_3 \neq 0)$;
   - $s, n, \overline{sn}_{(2)};\ S, N^d,\ N^* \Rightarrow ((11.24): c = 2,\ e = 1,\ h = -1,\ m = -2,\ u = 1)$ $(\text{if } \theta = \theta_1 = 0, \theta_3 = 0)$.

3. $\eta = 0$. In this case systems (11.24) possess at infinity either one double and one simple real singular points (if $\widetilde{M} \neq 0$) or one triple real singularity (if $\widetilde{M} = 0$). By Lemma 11.1 we could have at infinity exactly five configurations. So we have the following four global configurations if $\widetilde{M} \neq 0$:

   - $s, n, \overline{sn}_{(2)};\ \overline{\binom{0}{2}}SN,\ N^\infty \Rightarrow ((11.24): c = 1,\ e = 0,\ h = 1/2,\ m = -3/8,$ $u = 1)\ \ (\text{if } \theta < 0)$;
   - $s, n, \overline{sn}_{(2)};\ \overline{\binom{0}{2}}SN,\ N^f \Rightarrow ((11.24): c = 1,\ e = 0,\ h = 2,\ m = 3/2,\ u = 1)$ $(\text{if } \theta > 0)$;
   - $s, n, \overline{sn}_{(2)};\ \overline{\binom{0}{2}}SN,\ N^d \Rightarrow ((11.24): c = 1,\ e = 1,\ h = -2(7 + 5\sqrt{2}),$ $m = 0,\ u = 17 + 12\sqrt{2})\quad (\text{if } \theta = 0, \theta_2 \neq 0)$;
   - $s, n, \overline{sn}_{(2)};\ \overline{\binom{0}{2}}SN,\ N^* \Rightarrow ((11.24): c = -4,\ e = 1,\ h = 7/2,\ m = -1/2,$ $u = 1)\ (\text{if } \theta = 0, \theta_2 = 0)$.

   And we have one global configuration if $\widetilde{M} = 0$:

   - $s, n, \overline{sn}_{(2)};\ \overline{\binom{0}{3}}N \Rightarrow ((11.24): c = 1,\ e = 1,\ h = 40/27,\ u = 64/27,$ $m = 3/2)$.

2.1.2.2: $\mathcal{T}_4 = 0$. By (11.26) we get $\rho_3\rho_4 = 0$, and hence the saddle is weak. We consider two subcases: $\mathcal{F}_1 \neq 0$ and $\mathcal{F}_1 = 0$.

2.1.2.2.1: $\mathcal{F}_1 \neq 0$. Then by Theorem 6.2 the weak saddle has order one.

1. $\eta < 0$. Considering Lemmas 11.1 and 11.11(iii), there can only be three configurations at infinity which lead to the corresponding global configurations:
   - $s^{(1)}, n, \overline{sn}_{(2)};\ N^{\infty}$, ©, © $\Rightarrow$ ((11.24) : $c = 1, e = 1, h = -2, m = -1, u = 3$) (if $\theta < 0$);
   - $s^{(1)}, n, \overline{sn}_{(2)};\ N^{f}$, ©, © $\Rightarrow$ ((11.24) : $c = 2,\ e = 0,\ h = 2,\ m = 1,\ u = 1$) (if $\theta > 0$);
   - $s^{(1)}, n, \overline{sn}_{(2)};\ N^{d}$, ©, © $\Rightarrow$ ((11.24) : $c = -3,\ e = 1,\ h = 7/2,\ m = -3/4,$ $u = 4$) (if $\theta = 0$).
2. $\eta > 0$. Since for these systems the condition $\mu_0 > 0$ holds, taking into consideration Lemma 11.1 and Lemma 11.11(iii), we could have at infinity only six configurations which lead to the corresponding global configurations:
   - $s^{(1)}, n, \overline{sn}_{(2)};\ S, N^{\infty},\ N^{\infty} \Rightarrow$ ((11.24) : $c = 0,\ e = 1,\ h = -1/2,\ m = -3/4,$ $u = 3/2$) (if $\theta < 0, \theta_1 < 0$);
   - $s^{(1)}, n, \overline{sn}_{(2)};\ S, N^{f},\ N^{f} \Rightarrow$ ((11.24) : $c = 1,\ e = 0,\ h = 2,\ m = 1/2,\ u = 1$) (if $\theta < 0, \theta_1 > 0$);
   - $s^{(1)}, n, \overline{sn}_{(2)};\ S, N^{\infty},\ N^{f} \Rightarrow$ ((11.24) : $c = 1,\ e = 0,\ h = 2,\ m = 1/2,$ $u = 9/5$) (if $\theta > 0$);
   - $s^{(1)}, n, \overline{sn}_{(2)};\ S, N^{\infty},\ N^{d} \Rightarrow$ ((11.24) : $c = 2/3,\ e = 1,\ h = -2,\ m = -5/3,$ $u = 4$) (if $\theta = 0, \theta_1 < 0$);
   - $s^{(1)}, n, \overline{sn}_{(2)};\ S, N^{f},\ N^{d} \Rightarrow$ ((11.24) : $c = 0,\ e = -1,\ h = -2,\ m = -3,$ $u = 4$) (if $\theta = 0, \theta_1 > 0$);
   - $s^{(1)}, n, \overline{sn}_{(2)};\ S, N^{d},\ N^{d} \Rightarrow$ ((11.24) : $c = 2/3,\ e = 1,\ h = -2,\ u = 2,$ $m = -2/3,$) (if $\theta = 0, \theta_1 = 0$).
3. $\eta = 0$. In this case systems (11.24) possess at infinity either one double and one simple real singular points (if $\widetilde{M} \neq 0$) or one triple real singularity (if $\widetilde{M} = 0$). So by Lemmas 11.1 and 11.11(iii) we have the following three global configurations if $\widetilde{M} \neq 0$:
   - $s^{(1)}, n, \overline{sn}_{(2)};\ \overline{\binom{0}{2}}SN,\ N^{\infty} \Rightarrow$ ((11.24) : $c = 1,\ e = 2,\ h = -3,\ m = -3/2,$ $u = 2$) (if $\theta < 0$);
   - $s^{(1)}, n, \overline{sn}_{(2)};\ \overline{\binom{0}{2}}SN,\ N^{f} \Rightarrow$ ((11.24) : $c = 1,\ e = 0,\ h = 2,\ m = 1/2,\ u = 2$) (if $\theta > 0$);
   - $s^{(1)}, n, \overline{sn}_{(2)};\ \overline{\binom{0}{2}}SN,\ N^{d} \Rightarrow$ ((11.24) : $c = \xi^2 - 2,\ e = 1, h = \xi(3 - \xi - \xi^2),$ $m = -1,\ u = \xi^2$) (where $\{\eta^{-1}(0)\} \ni \xi \approx 1.3816$) (if $\theta = 0$).

And we have one global configuration if $\widetilde{M} = 0$:

- $s^{(1)}, n, \overline{sn}_{(2)};\ \binom{0}{3}N \Rightarrow ((11.24): c = 17-3\sqrt{33},\ e = 1, h = (155-27\sqrt{33})/2,$ $m = -1,\ u = 19 - 3\sqrt{33})$.

2.1.2.2.2: $\mathcal{F}_1 = 0$, $\mathcal{F}_2 \neq 0$. In this case according to Theorem 6.2 the weak saddle is of order two. We claim that for $\mathcal{F}_1 = 0$ we could not have $\theta = \theta_1 = 0$. Indeed, suppose the contrary. As it was shown in the proof of the statement (iii) of Lemma 11.11 (see page 453), in this case for systems (11.24) the conditions $m = -c$ and $h = -eu$ must hold. Then we have $\rho_3 = eu - 3c = 0$ (i.e. $c = eu/3$), which implies

$$\mathcal{F}_1 = 16e^4u^4(u-9)/81, \quad \mu_0 = 4e^4u^3(u-9u)^2/81,$$

and evidently the condition $\mu_0 \neq 0$ gives $\mathcal{F}_1 \neq 0$. The contradiction we have obtained proves our claim.

1. $\eta < 0$. According to Lemmas 11.1 and 11.11(iii), there can only be three configurations at infinity which lead to the corresponding global configurations:
   - $s^{(2)}, n, \overline{sn}_{(2)};\ N^{\infty},$ ©, © $\Rightarrow ((11.24): c = -1/3,\ e = 1,\ h = 1/6,\ u = 4/3,$ $m = -5/6)$ (if $\theta < 0$);
   - $s^{(2)}, n, \overline{sn}_{(2)};\ N^{f},$ ©, © $\Rightarrow ((11.24): c = -3,\ e = 1,\ h = -13/4,\ u = 1/2,$ $m = -7/4)$ (if $\theta > 0$);
   - $s^{(2)}, n, \overline{sn}_{(2)};\ N^{d},$ ©, © $\Rightarrow ((11.24): c = (841 + 800\xi)/400,\ e = 1,\ m = \xi,$ $h = -29(1421 + 1200\xi)/8000,\ u = 841/400)$ (where $\xi = (-2253 + \sqrt{23281})/1600$) (if $\theta = 0$).
2. The case $\eta > 0$. Thus as for these systems the condition $\mu_0 > 0$ holds, taking into consideration Lemmas 11.1 and 11.11(iii), we could have at infinity only five distinct configurations. The corresponding global configurations are as follows:
   - $s^{(2)}, n, \overline{sn}_{(2)};\ S, N^{\infty}, N^{\infty} \Rightarrow ((11.24): c = -2/3,\ e = 1,\ h = 25/27,$ $m = -13/9,\ u = 20/9)$ (if $\theta < 0, \theta_1 < 0$);
   - $s^{(2)}, n, \overline{sn}_{(2)};\ S, N^{f}, N^{f} \Rightarrow ((11.24): c = 1,\ e = 1,\ h = -4, m = 0,\ u = 1)$ (if $\theta < 0, \theta_1 > 0$);
   - $s^{(2)}, n, \overline{sn}_{(2)};\ S, N^{\infty}, N^{f} \Rightarrow ((11.24): c = 1/2,\ e = 1,\ h = -4,\ m = -5/4,$ $u = 3)$ (if $\theta > 0$);
   - $s^{(2)}, n, \overline{sn}_{(2)};\ S, N^{\infty}, N^{d} \Rightarrow ((11.24): c = -3/5,\ e = 1,\ h = (27 + 5\xi)/50,$ $m = -(3 + 5\xi)/10,\ u = \xi)$ (where $\{\theta^{-1}(0)\} \ni \xi \approx 2.1794$) (if $\theta = 0, \theta_1 < 0$);
   - $s^{(2)}, n, \overline{sn}_{(2)};\ S, N^{f}, N^{d} \Rightarrow ((11.24): c = -2,\ e = 1, h = (3\xi - 2)/2,$ $m = -(2+\xi)/2\, u = \xi)$ (where $\{\theta^{-1}(0)\} \ni \xi \approx 0.07264)$ (if $\theta = 0, \theta_1 > 0$).

3. $\eta = 0$. In this case systems (11.24) possess at infinity either one double and one simple real singular points (if $\widetilde{M} \neq 0$) or one triple real singularity (if $\widetilde{M} = 0$). So by Lemmas 11.1 and 11.11(iii), we have the following three global configurations if $\widetilde{M} \neq 0$:
   - $s^{(2)}, n, \overline{sn}_{(2)};\ \overline{\binom{0}{2}}SN,\ N^{\infty} \Rightarrow ((11.24): c = \xi,\ e = 1,\ h = -(43 + 146\xi + 40\xi^2)/40,\ m = (20\xi - 43)/40,\ u = 43/20)$ (where $\{\eta^{-1}(0)\} \ni \xi \approx -0.5715$) (if $\theta < 0$);
   - $s^{(2)}, n, \overline{sn}_{(2)};\ \overline{\binom{0}{2}}SN,\ N^{f} \Rightarrow ((11.24): c = -2(5 + \sqrt{13})/3,\ e = 1,\ h = -(56 + 19\sqrt{13})/9,\ m = -(8 + \sqrt{13})/3,\ u = 2)$ (if $\theta > 0$);
   - $s^{(2)}, n, \overline{sn}_{(2)};\ \overline{\binom{0}{2}}SN,\ N^{d} \Rightarrow ((11.24): c = \xi,\ e = 1,\ h = -(2\xi^2 + 3\xi + 2\xi\chi + \chi)/2,\ m = (\chi - \xi)/2,\ u = \chi)$ (where $\{\eta^{-1}(0), \theta^{-1}(0)\} \ni (\xi, \chi) \approx (-0.5746,\ 2.1565)$) (if $\theta = 0$).

   And we have one global configuration if $\widetilde{M} = 0$:
   - $s^{(2)}, n, \overline{sn}_{(2)};\ \overline{\binom{0}{3}}N \Rightarrow ((11.24): c = -1/2, e = 1, h = 1/2, m = -5/4, u = 2)$.

2.1.2.2.3: $\mathcal{F}_1 = \mathcal{F}_2 = 0$. Then by Theorem 6.2 we have either a weak saddle of order three or an integrable saddle. However by Theorem 7.3 we could not have an integrable saddle. Additionally according to Theorems 7.3 and 7.9 we have a single global configuration
- $s^{(3)}, n, \overline{sn}_{(2)};\ N^{\infty},\ ©, © \Rightarrow ((11.24): c = -1/6,\ e = 1,\ h = -1/9,\ m = -7/12,\ u = 1)$.

2.1.3: $W_4 = 0$. Since $E_1 \neq 0$ (i.e. $\rho_1 \neq 0$) by (11.26) we obtain $\tau_3\tau_4 = 0$, and therefore we have a node with coinciding eigenvalues. Clearly we may assume that such a singular point is $M_3(1, 0)$ (i.e. $\tau_3 = 0$). Additionally considering Theorem 6.1 (see Table 6.2, line 75), we conclude that, apart from this node, we have a semi-elemental saddle-node $M_{1,2}(0, 0)$ and a saddle $M_4(0, 1)$.

Considering Lemma 11.12 we have to examine two cases: the condition $(\mathcal{S}^1_{n^*})$ does not hold and the condition $(\mathcal{S}^1_{n^*})$ is satisfied.

2.1.3.1: $\neg(\mathcal{S}^1_{n^*})$. Then the node $M_3(1, 0)$ is a one-direction node.

2.1.3.1.1: $\mathcal{T}_4 \neq 0$. We obtain $\rho_3\rho_4 \neq 0$ and the saddle is strong.

1. $\eta < 0$. According to Lemmas 11.1 and 11.11(iii), there can only be three configurations at infinity which lead to the corresponding global configurations:
   - $s, n^d, \overline{sn}_{(2)};\ N^{\infty},\ ©, © \Rightarrow ((11.24): c = 2,\ e = 1,\ h = 1/8,\ m = 0,\ u = 1)$ (if $\theta < 0$);
   - $s, n^d, \overline{sn}_{(2)};\ N^{f},\ ©, © \Rightarrow ((11.24): c = 0,\ e = 1,\ h = 1/8,\ m = 0,\ u = 1)$ (if $\theta > 0$);
   - $s, n^d, \overline{sn}_{(2)};\ N^{d},\ ©, © \Rightarrow ((11.24): c = -11/4,\ e = 1,\ h = 9/8,\ m = 0,\ u = 1/4)$ (if $\theta = 0$).

2. $\eta > 0$. Since for these systems the condition $\mu_0 > 0$ holds, taking into consideration Lemmas 11.1 and 11.11(iii), as we could have at infinity only six configurations. The corresponding global configurations are as follows:
   - $s, n^d, \overline{sn}_{(2)};\ S, N^\infty,\ N^\infty \Rightarrow ((11.24): c = -4,\ e = -1,\ h = 15/8,\ m = 3,\ u = 1)$ (if $\theta < 0, \theta_1 < 0$);
   - $s, n^d, \overline{sn}_{(2)};\ S, N^f,\ N^f \Rightarrow ((11.24): c = 0, e = -1, h = -1/2, m = 2, u = 2)$ (if $\theta < 0, \theta_1 > 0$);
   - $s, n^d, \overline{sn}_{(2)};\ S, N^\infty,\ N^f \Rightarrow ((11.24): c = 0, e{=}-1, h{=}-9/8, m = 2, u = 1)$ (if $\theta > 0$);
   - $s, n^d, \overline{sn}_{(2)};\ S, N^\infty,\ N^d \Rightarrow ((11.24): c = -6,\ e = -1,\ h = 10,\ m = 3,\ u = 4)$ (if $\theta = 0, \theta_1 < 0$);
   - $s, n^d, \overline{sn}_{(2)};\ S, N^f,\ N^d \Rightarrow ((11.24): c = 0,\ e = 1,\ h = 8,\ m = -6, u = 4)$ (if $\theta = 0, \theta_1 > 0$);
   - $s, n^d, \overline{sn}_{(2)};\ S, N^d,\ N^d \Rightarrow ((11.24): c = 10,\ e = 1,\ h = -2, m = -10,\ u = 2)$ (if $\theta = 0, \theta_1 = 0$).
3. $\eta = 0$. In this case systems (11.24) possess at infinity either one double and one simple real singular points (if $\widetilde{M} \neq 0$) or one triple real singularity (if $\widetilde{M} = 0$). So by Lemmas 11.1 and 11.11(iii), we have the following three global configurations if $\widetilde{M} \neq 0$:
   - $s, n^d, \overline{sn}_{(2)};\ \overline{\binom{0}{2}}SN,\ N^\infty \Rightarrow ((11.24): c = 1,\ e = 0,\ h = 1/2,\ m = -1/2,\ u = 2)$ (if $\theta < 0$);
   - $s, n^d, \overline{sn}_{(2)};\ \overline{\binom{0}{2}}SN,\ N^f \Rightarrow ((11.24): c = 1,\ e = 0,\ h = 1,\ m = -1/2,\ u = 1)$ (if $\theta > 0$);
   - $s, n^d, \overline{sn}_{(2)};\ \overline{\binom{0}{2}}SN,\ N^d \Rightarrow ((11.24): c = 1, e = 0, h = 1/2, m = -1/2, u = 1)$ (if $\theta = 0$).

   And we have one global configuration if $\widetilde{M} = 0$:
   - $s, n^d, \overline{sn}_{(2)};\ \overline{\binom{0}{3}}N \Rightarrow ((11.24): c = 1,\ e = 0,\ h = 0,\ m = -1/2,\ u = 1)$.

2.1.3.1.2: $\mathcal{T}_4 = 0$. We obtain $\rho_3\rho_4 = 0$ (then $\rho_4 = 0$ due to $\tau_3 = 0$), and the saddle is weak.

2.1.3.1.2.1: $\mathcal{F}_1 \neq 0$. In this case we have a weak saddle of order one.

1. $\eta < 0$. According to Lemmas 11.1 and 11.11(iii), there can be four configurations at infinity which lead to the corresponding global configurations:
   - $s^{(1)}, n^d, \overline{sn}_{(2)};\ N^\infty,$ ©, © $\Rightarrow ((11.24): c = 4,\ e = 1,\ h = 2,\ m = 0,\ u = 8)$ (if $\theta < 0$);

- $s^{(1)}, n^d, \overline{sn}_{(2)}$; N^f, ©, © $\Rightarrow$ ((11.24) : $c = -1$, $e = 1$, $h = 2$, $m = 0$, $u = 3$) (if $\theta > 0$);
- $s^{(1)}, n^d, \overline{sn}_{(2)}$; N^d, ©, © $\Rightarrow$ ((11.24) : $c = (\xi^2 - 48)/32$, $e = 1$, $h = 2 + \xi/4$, $m = -1$, $u = (\xi^2 + 16\xi + 80)/32$) (where $\{\theta^{-1}(0)\} \ni \xi \approx -9.4251$) (if $\theta = 0$, $\theta_2 \neq 0$);
- $s^{(1)}, n^d, \overline{sn}_{(2)}$; N^*, ©, © $\Rightarrow$ ((11.24) : $c = -3$, $e = 1$, $h = 2$, $m = 0$, $u = 1$) (if $\theta = 0$, $\theta_2 = 0$).

2. $\eta > 0$. Since $\mu_0 > 0$, by Lemma 11.1 there are 10 possibilities. However according to Lemma 11.11 (statements (iii$_2$) and (v)), at infinity we cannot have in this case the configurations S, N^d, N^d or S, N^*, N^*.

   Thus at infinity we could only have eight configurations. The corresponding global configurations are as follows:

   - $s^{(1)}, n^d, \overline{sn}_{(2)}$; S, N^∞, $N^\infty \Rightarrow$ ((11.24) : $c = 17/4$, $e = 1/4$, $h = -2$, $m = -2$, $u = 1$) (if $\theta < 0, \theta_1 < 0$);
   - $s^{(1)}, n^d, \overline{sn}_{(2)}$; S, N^f, $N^f \Rightarrow$ ((11.24) : $c = -7/6$, $e = 1$, $h = 4$, $m = -3$, $u = 41/6$) (if $\theta < 0, \theta_1 > 0$);
   - $s^{(1)}, n^d, \overline{sn}_{(2)}$; S, N^∞, $N^f \Rightarrow$ ((11.24) : $c = -23/32$, $e = 1$, $h = 3/4$, $m = -1$, $u = 25/32$) (if $\theta > 0$);
   - $s^{(1)}, n^d, \overline{sn}_{(2)}$; S, N^∞, $N^d \Rightarrow$ ((11.24) : $c = (12 - \sqrt{287}\,)/156$, $e = -12(14 + \sqrt{287}\,)/13$, $h = (11 + \sqrt{287}\,)/144$, $m = 1/12$, $u = 1/144$) (if $\theta = 0, \theta_1 < 0, \theta_2 \neq 0$);
   - $s^{(1)}, n^d, \overline{sn}_{(2)}$; S, N^∞, $N^* \Rightarrow$ ((11.24) : $c = 1$, $e = 5$, $h = -4$, $m = -2$, $u = 1$) (if $\theta = 0, \theta_1 < 0, \theta_2 = 0$);
   - $s^{(1)}, n^d, \overline{sn}_{(2)}$; S, N^f, $N^d \Rightarrow$ ((11.24) : $c = (2 - \sqrt{7}\,)/6$, $e = 2(4 - \sqrt{7}\,)/3$, $h = (-1 + \sqrt{7}\,)/4$, $m = -1/2$, $u = 1/4$) (if $\theta = 0, \theta_1 > 0$, $\theta_2 \neq 0$);
   - $s^{(1)}, n^d, \overline{sn}_{(2)}$; S, N^f, $N^* \Rightarrow$ ((11.24) : $c = 1$, $e = 1$, $h = -2$, $m = 0$, $u = 1$) (if $\theta = 0, \theta_1 > 0$, $\theta_2 = 0$);
   - $s^{(1)}, n^d, \overline{sn}_{(2)}$; S, N^d, $N^* \Rightarrow$ ((11.24) : $c = 1/3$, $e = 1$, $h = -1$, $m = -1/3$, $u = 1$) (if $\theta = \theta_1 = 0$).

3. $\eta = 0$. In this case systems (11.24) possess at infinity either one double and one simple real singular points (if $\widetilde{M} \neq 0$) or one triple real singularity (if $\widetilde{M} = 0$).

   So by Lemmas 11.1 and 11.11(iii), we have the following three global configurations if $\widetilde{M} \neq 0$:

   - $s^{(1)}, n^d, \overline{sn}_{(2)}$; $\overline{\binom{0}{2}}SN$, $N^\infty \Rightarrow$ ((11.24) : $c = 1 - 2\xi$, $e = 1$, $h = \xi$, $m = \xi$, $u = 1$) (where $\{\eta^{-1}(0)\} \ni \xi \approx -7.4138$) (if $\theta < 0$);

- $s^{(1)}, n^d, \overline{sn}_{(2)}; \overline{\binom{0}{2}}SN,\ N^f \Rightarrow$ ((11.24) : $c = -2\xi + 2\sqrt{1+2\xi}$, $e = -1$, $h = -1 + \xi - \sqrt{1+2\xi}$, $m = \xi$, $u = 2$) (where $\{\eta^{-1}(0)\} \ni \xi \approx 1.4293$) (if $\theta > 0$);
- $s^{(1)}, n^d, \overline{sn}_{(2)}; \overline{\binom{0}{2}}SN,\ N^d \Rightarrow$ ((11.24) : $c = \dfrac{3\xi - \chi}{\chi - 1}$, $e = \dfrac{\xi + 2\xi\chi - \chi}{\chi^2(\chi - 1)}$, $h = \xi, m = 1, u = \chi^2$) (where $\{\eta^{-1}(0), W_4^{-1}(0)\} \ni (\xi, \chi) \approx$ $(-2.5850,\ 36.9003)$) (if $\theta = 0$).

And we have one global configuration if $\widetilde{M} = 0$:

- $s^{(1)}, n^d, \overline{sn}_{(2)}; \overline{\binom{0}{3}}N \Rightarrow$ ((11.24) : $c = 32/27$, $e = -1$, $h = -8/9, m = 20/27$, $u = 16/27$).

2.1.3.1.2.2: $\mathcal{F}_1 = 0$. In this case we have a weak saddle of order at least two. We consider two cases: $\mathcal{F}_2 \neq 0$ and $\mathcal{F}_2 = 0$.

1. $\mathcal{F}_2 \neq 0$. Then the weak saddle has order two. We prove now that in this case we could have only seven configurations.

   Assume that for systems (11.24) the conditions $\mu_0 E_1 \neq 0$ and $\mathcal{T}_4 = \mathcal{F}_1 = W_4 = 0$ are satisfied. As it was shown earlier (see page 456), the conditions $\mathcal{T}_4 = 0$ (we assume $\rho_3 = 0$) and $\mathcal{F}_1 = 0$ imply the relations (11.28). Therefore the condition $W_4 = 0$ (which in this case is equivalent to $\tau_4 = 0$) gives

$$\begin{aligned}
E_1 &= -(1+c)^4 u^2 (c+u)(3c+u)^4/2,\\
\mathcal{F}_2 &= -c(1+c)^4 u^2 (c+u)^4 (3c+u)^2 (6c+u),\\
\mathcal{F}_3 &= -c(1+c)^4 u^2 (c+u)^4 (2c+u)(3c+u)^2,\\
\tau_4 &= 4c(1+c)[(c+u)^2 + c - u] \equiv 4c(1+c)\psi(c,u),
\end{aligned}$$

   and as $\mathcal{F}_2 \neq 0$ the condition $\tau_4 = 0$ implies $\psi(c,u) = 0$. Since Discriminant$[\psi, c]$ $= 1+8u$, in order to have a real solution we set $1+8u = v^2$ (i.e. $u = (v^2-1)/8$), and then we obtain $\psi(c, v^2-1) = (3+8c-4v+v^2)(3+8c+4v+v^2)/64 = 0$. We may consider only the first factor because the second factor is obtained from the first one by replacing v by $-v$, and we arrive at the same result.

   So $c = (4v - v^2 - 3)/8$ and we calculate

$$\begin{aligned}
\mathcal{T}_4 = \mathcal{F}_1 = W_4 = 0, \quad &\mu_0 = 2^{-13}(v-5)^4(v^2-1)^3,\\
\mathcal{F}_2 &= -2^{-32}(v-5)^6(v-3)(v-1)^{10}(1+v)^6(5v-19),\\
\eta &= -2^{-13}(v-5)^2(v-3)(v-1)^3(1+v)(205+67v-65v^2+9v^3),
\end{aligned}$$

$$\begin{aligned}
\theta &= (v-5)^2(v-3)(v-1)^3(1+v)(11+v-5v^2+v^3)/256,\\
\theta_1 &= -(v-5)^2(v-1)^3(1+v)(38v-361+166v^2-66v^3+7v^4)/256,\\
\theta_2 &= 2^{-11}(v-5)^2(v-3)(v-1)^2(1+v)(v^2-v-4).
\end{aligned}$$

Since we have only one parameter v (which satisfies $|v| > 1$, due to $\mu_0 > 0$), we could only obtain the following seven global configurations:

- $s^{(2)}, n^d, \overline{sn}_{(2)}; N^\infty$, ©, © $\Rightarrow$ ((11.24) : $c = -3/8$, $e = 1$, $h = 3/16$, $m = -9/8$, $u = 15/8$) (if $\eta < 0, \theta < 0$);
- $s^{(2)}, n^d, \overline{sn}_{(2)}; N^f$, ©, © $\Rightarrow$ ((11.24) : $c = -15/8$, $e = 1$, $h = -3/16$, $m = -9/8$, $u = 3/8$) (if $\eta < 0, \theta > 0$);
- $s^{(2)}, n^d, \overline{sn}_{(2)}; N^d$, ©, © $\Rightarrow$ ((11.24) : $c = (3-\xi)(\xi-1)/8$, $e = 1$, $h = (\xi-1)(\xi^2-2\xi-7)/16$, $m = -(\xi-1)^2/8$, $u = (\xi^2-1)/8$) (where $\{\theta^{-1}(0)\} \ni \xi \approx 4.1028$) (if $\eta < 0, \theta = 0$);
- $s^{(2)}, n^d, \overline{sn}_{(2)}; S, N^f, N^f \Rightarrow$ ((11.24) : $c = 1/8$, $e = 1$, $h = -7/16$, $m = -1/8$, $u = 3/8$) (if $\eta > 0, \theta < 0$);
- $s^{(2)}, n^d, \overline{sn}_{(2)}; S, N^\infty, N^f \Rightarrow$ ((11.24) : $c = 3/32$, $e = 1$, $h = -69/128$, $m = -9/32$, $u = 21/32$) (if $\eta > 0, \theta > 0$);
- $s^{(2)}, n^d, \overline{sn}_{(2)}; S, N^\infty, N^d \Rightarrow$ ((11.24) : $c = (3-\xi)(\xi-1)/8$, $e = 1$, $h = (\xi-1)(\xi^2-2\xi-7)/16$, $m = -(\xi-1)^2/8$, $u = (\xi^2-1)/8$) (where $\{\theta^{-1}(0)\} \ni \xi \approx -1.2491$) (if $\eta > 0, \theta = 0$);
- $s^{(2)}, n^d, \overline{sn}_{(2)}; \overline{\binom{0}{2}}SN, N^f \Rightarrow$ ((11.24) : $c = (3-\xi)(\xi-1)/8$, $e = 1$, $h = (\xi-1)(\xi^2-2\xi-7)/16$, $m = -(\xi-1)^2/8$, $u = (\xi^2-1)/8$) (where $\{\eta^{-1}(0)\} \ni \xi \approx -1.2577$) (if $\eta = 0$).

2. $\mathcal{F}_2 = 0$. Then by Theorem 6.2 we have either a weak saddle of order three (if $\mathcal{F}_3\mathcal{F}_4 \neq 0$) or an integrable saddle. Considering Theorems 7.3 and 7.9 we get a single global configuration

   - $s^{(3)}, n^d, \overline{sn}_{(2)}; N^\infty$, ©, © $\Rightarrow$ ((11.24) : $c = -7/25$, $e = 1$, $h = -7/250$, $m = -49/50$, $u = 42/25$).

   Additionally by Theorem 7.3 we arrive at the following three global configurations:

   - $\$, n^d, \overline{sn}_{(2)}; \overline{\binom{0}{2}}SN, N^\infty \Rightarrow$ ((11.24) : $c = 0$, $e = 1$, $h = -1/4$, $m = -1/4$, $u = 1/2$) (if $\theta < 0$);
   - $\$, n^d, \overline{sn}_{(2)}; \overline{\binom{0}{2}}SN, N^f \Rightarrow$ ((11.24) : $c = 0$, $e = 1$, $h = -1$, $m = -1$, $u = 2$) (if $\theta > 0$);
   - $\$, n^d, \overline{sn}_{(2)}; \overline{\binom{0}{2}}SN, N^* \Rightarrow$ ((11.24) : $c = 0$, $e = 1$, $h = -1/2$, $m = -1/2$, $u = 1$) (if $\theta = 0$).

2.1.3.2: $(\mathcal{S}^1_{n^*})$. Then by Lemma 11.12 the node $M_3(1,0)$ is a star node. Considering the corresponding matrix from (11.25), we obtain $e = 0$, $h = -cu/2$,

$m = -c/2$ and $c \neq 0$, otherwise we get degenerate systems. So we may assume $c = 1$ (due to time rescaling), and we arrive at the family of systems:

$$\dot{x} = x + uy - x^2 - uxy - uy^2, \quad \dot{y} = -xy, \tag{11.29}$$

for which we calculate

$$\eta = \mathcal{F}_1 = \mathcal{F}_4 = \theta_2 = 0, \quad \mu_0 = u, \quad \widetilde{M} = -8u^2y^2,$$
$$\mathcal{T}_4 = 2u(u-1), \quad \theta = 8u(u-1), \quad \mathcal{F}_2 = u^2(u-1)(9+2u)/2.$$

Therefore the conditions $\mathcal{T}_4 = 0$ and $\theta = 0$ (respectively $\mu = 0$ and $\widetilde{M} = 0$) are equivalent. Moreover, the condition $\theta = 0$ implies $\mathcal{F}_2 = 0$.

Thus considering these implications, by Lemma 11.1 we could only have the following three global configurations:

- $s, n^*, \overline{sn}_{(2)};\ \overline{\binom{0}{2}}SN,\ N^\infty \Rightarrow \quad ((11.29): u = 1/2) \quad (\text{if } \theta < 0);$
- $s, n^*, \overline{sn}_{(2)};\ \overline{\binom{0}{2}}SN,\ N^f \Rightarrow \quad ((11.29): u = 2) \quad (\text{if } \theta > 0);$
- $\$, n^*, \overline{sn}_{(2)};\ \overline{\binom{0}{2}}SN,\ N^* \Rightarrow \quad ((11.29): u = 1) \quad (\text{if } \theta = 0).$

2.2: $E_1 = 0$. In this case $\rho_1 = 0$, and apart from the two elemental singularities, we have a cusp. Moreover, since $\mu_0 > 0$ from (11.26) it follows that $\Delta_3\Delta_4 < 0$, i.e. the two elemental singularities are a saddle and an anti-saddle. The condition $E_1 = 0$ gives $c = -eu$ and then $e \neq 0$, otherwise we get degenerate systems. So due to a rescaling systems (11.24) become

$$\begin{aligned} \dot{x} &= -ux - u^2y + ux^2 + 2hxy + u^2y^2, \\ \dot{y} &= x + uy - x^2 + 2mxy - uy^2, \end{aligned} \tag{11.30}$$

for which calculations yield

$$\begin{aligned} \mu_0 = 4u(h+mu)^2, \quad \mathcal{T}_4 = \mathcal{T}_3 = 0, \quad \mathcal{T}_2 = \mu_0\rho_3\rho_4, \quad \mathcal{T}_1 = \mu_0(\rho_3+\rho_4), \\ W_4 = W_3 = 0, \quad W_2 = \mu_0^2\tau_3\tau_4, \quad W_1 = \mu_0^2(\tau_3+\tau_4), \\ \theta = 64(h+mu)\big[(h+u)^2 - u(m-u)^2\big], \\ \eta = -4(\mathcal{N}_2^2 - \mathcal{N}_1\mathcal{N}_3)/3, \quad \widetilde{M} = 8(\mathcal{N}_1x^2 - 2\mathcal{N}_2xy - \mathcal{N}_3y^2), \end{aligned} \tag{11.31}$$

where

$$\rho_3 = 2(m+u), \quad \rho_4 = 2(h-u), \quad \tau_3 = 4(m^2+u^2) - 8h,$$
$$\tau_4 = 4(h^2+u^2) + 8mu^2, \quad \mathcal{N}_1 = (2m-u)^2 - 6h - 3u,$$
$$\mathcal{N}_2 = 2hm - hu + mu + 4u^2, \quad \mathcal{N}_3 = (2h+u)^2 + 6mu^2 - 3u^3.$$

Lemma 11.14. *If for a system* (11.30) *the condition* $\mu_0 > 0$ *holds, then*

(i) *the condition* $W_2 \leq 0$ *implies* $\theta > 0$*;*

(ii) *the condition* $\widetilde{M} = 0$ *implies* $W_2 > 0$ *and* $\mathcal{T}_2 \neq 0$.

Proof. (i) Assume that the condition $W_2 \leq 0$ holds. This implies $\tau_3\tau_4 \leq 0$, and we may assume $\tau_3 \leq 0$, i.e. the singular point $M_3(1,0)$ is either a focus or a node (with coinciding eigenvalues), and then for the saddle we have $\tau_4 > 0$. We set a new parameter v as $\tau_3 = -v^2 \leq 0$, and we get $h = (4m^2+4u^2+v^2)/8$. Calculations yield

$$\theta = 2\big[(4(m+u)^2+v^2)\big](\tau_4+uv^2), \quad \mu_0 = u[(4(m+u)^2+v^2)]^2/16.$$

Hence due to the condition $\mu_0 > 0$ and $\tau_4 > 0$, we obtain $\theta > 0$ for any value of the parameter v. So the statement (i) of the lemma is proved.

(ii) According to (11.31) the condition $\widetilde{M} = 0$ implies $\mathcal{N}_1 = \mathcal{N}_2 = \mathcal{N}_3 = 0$. The equality $\mathcal{N}_1 = 0$ gives $h = \big[(2m-u)^2 - 3u\big]/6$, and then we obtain $\mathcal{N}_2 = (2m-u)^3 + 27u^2 = 0$. Since $u \neq 0$ (due to $\mu_0 \neq 0$), we set two new parameters v and w as follows: $v = 2m - u \neq 0$ (i.e. $m = (u+v)/2$) and $u = vw$. Then we get $\mathcal{N}_2 = v^2(v+27w^2)/6 = 0$, which yields $v = -27w^2$. So we obtain $m = -27w^2(1+w)/2$, $u = -27w^3$ and $h = 27w^3(1+9w)/2$, and for these values of the parameters of systems (11.30) we get $\widetilde{M} = 0$ and

$$W_2 = -3^{27}w^{27}(1+3w)^{16}(12w-5)(15w-4),$$
$$\mathcal{T}_2 = 3^{16}w^{14}(1+3w)^8, \quad \mu_0 = -3^9w^9(1+3w)^6.$$

Clearly due to $\mu_0 > 0$ we have $W_2 > 0$ and $\mathcal{T}_2 \neq 0$ and this completes the proof of Lemma 11.14. □

2.2.1: $W_2 < 0$. In this case the anti-saddle is a focus or a center, and the existence of weak singularities depends on the invariant polynomial $\mathcal{T}_2$.

2.2.1.1: $\mathcal{T}_2 \neq 0$. Then $\rho_3\rho_4 \neq 0$ and the saddle as well as the focus are strong ones. Therefore considering Lemmas 11.14 and 11.1, we could only obtain the following three global configurations:

- $s, f; \widehat{cp}_{(2)};\ N^f, ©, ©$ $\Rightarrow$ ((11.30): $h=2,\ m=0,\ u=1$) (if $\eta < 0$);
- $s, f; \widehat{cp}_{(2)};\ S, N^\infty, N^f$ $\Rightarrow$ ((11.30): $h=4,\ m=-5/2,\ u=1$) (if $\eta > 0$);
- $s, f; \widehat{cp}_{(2)};\ \binom{0}{2}SN, N^f$ $\Rightarrow$ ((11.30): $h=(5+4\xi^2)/8,\ m=\xi,\ u=1$) (where $\{\eta^{-1}(0)\} \ni \xi \approx -1.4744$) (if $\eta = 0$).

2.2.1.2: $\mathcal{T}_2 = 0$. In this case we have at least one weak singularity.

2.2.1.2.1: $\mathcal{T}_1 \neq 0$. So by (11.31) only one singularity is weak, and by Theorem 6.2 its type is governed by the invariant polynomial $\mathcal{H}$.

1. $\mathcal{H} < 0$. According to Theorem 6.2 (see statement (d)) we have a weak focus of order one, and we obtain the following three global configurations:

   - $s, f^{(1)}, \widehat{cp}_{(2)};\ N^f, ©, ©$ $\Rightarrow$ ((11.30): $h=1,\ m=-2,\ u=1$) (if , $\eta < 0$);

- $s, f^{(1)}, \widehat{cp}_{(2)}$; S, N^∞, N^f $\Rightarrow$ ((11.30): $h = 1/12$, $m = -13/12$, $u = 1/12$) (if $\eta > 0$);
- $s, f^{(1)}, \widehat{cp}_{(2)}$; $\overline{\binom{0}{2}}SN$, N^f $\Rightarrow$ ((11.30): $h = (1 + 8\xi^2)/8$, $m = -\xi$, $u = \xi$) (where $\{\eta^{-1}(0)\} \ni \xi \approx 1.7627$) (if $\eta = 0$).

2. $\mathcal{H} > 0$. Then by Theorem 6.2 the weak singularity is a saddle of order one, and we arrive at the following three global configurations:
   - $s^{(1)}, f, \widehat{cp}_{(2)}$; N^f, ©, © $\Rightarrow$ ((11.30): $h = 1$, $m = 0$, $u = 1$) (if $\eta < 0$);
   - $s^{(1)}, f, \widehat{cp}_{(2)}$; S, N^∞, N^f $\Rightarrow$ ((11.30): $h = 567/400$, $m = -6/5, u = 6/5$) (if $\eta > 0$);
   - $s^{(1)}, f, \widehat{cp}_{(2)}$; $\overline{\binom{0}{2}}SN$, N^f $\Rightarrow$ $((\tilde{\mathbf{S}}_{III}):$ $b = 27/4$, $c = -3/2$, $d = -3/4, e = 3$, $f = 3/2$, $g = 5/4$, $h = 1$) (if $\eta = 0$).

2.2.1.2.2: $\mathcal{T}_1 = 0$. Then $\rho_3 = \rho_4 = 0$, and this implies $h = 0$ and $m = -u$. In this case for systems (11.30) we calculate

$$\sigma = \frac{\partial P}{\partial x} + \frac{\partial Q}{\partial y} = 0, \quad \eta = -108(u-1)^2u^3, \quad \mu_0 = 4(u-1)^2u^3.$$

So we get Hamiltonian systems that apart from a cusp, have a center and an integrable saddle. Since the condition $\mu_0 > 0$ implies $\eta < 0$, we obtain the unique global configuration

- $\$, c, \widehat{cp}_{(2)}$; N^f, ©, © $\Rightarrow$ ((11.30): $h = 2$, $m = -2$, $u = 2$).

2.2.2: $W_2 > 0$. In this case the anti-saddle is a generic node, and we consider two cases: $\mathcal{T}_2 \neq 0$ and $\mathcal{T}_2 = 0$.

2.2.2.1: $\mathcal{T}_2 \neq 0$. The saddle is strong.

1. $\eta < 0$. According to Lemmas 11.1 and 11.11(iii), there can only be three global configurations:
   - $s, n, \widehat{cp}_{(2)}$; N^∞, ©, © $\Rightarrow$ ((11.30): $h = -1$, $m = 1$, $u = 2$) (if $\theta < 0$);
   - $s, n, \widehat{cp}_{(2)}$; N^f, ©, © $\Rightarrow$ ((11.30): $h = -1$, $m = 0$, $u = 1$) (if $\theta > 0$);
   - $s, n, \widehat{cp}_{(2)}$; N^d, ©, © $\Rightarrow$ ((11.30): $h = 2(\sqrt{2} - 1)$, $m = 0$, $u = 2$) (if $\theta = 0$).
2. $\eta > 0$. According to Lemmas 11.1 and 11.11(iii), there can only be six global configurations:
   - $s, n, \widehat{cp}_{(2)}$; S, N^∞, N^∞ $\Rightarrow$ ((11.30): $h = -11/6$, $m = 5/3$, $u = 7/4$) (if $\theta < 0, \theta_1 < 0$);
   - $s, n, \widehat{cp}_{(2)}$; S, N^f, N^f $\Rightarrow$ ((11.30): $h = 0$, $m = 3$, $u = 1$) (if $\theta < 0, \theta_1 > 0$);

- $s, n, \widehat{cp}_{(2)}$; $S, N^{\infty}, N^f \Rightarrow$ ((11.30): $h=2,\ m=3,\ u=1$) (if $\theta>0$);
- $s, n, \widehat{cp}_{(2)}$; $S, N^{\infty}, N^d \Rightarrow$ ((11.30): $h=(10\sqrt{13}-13)/4,\ m=-7/4,$ $u=13/4$) (if $\theta=0, \theta_1<0$);
- $s, n, \widehat{cp}_{(2)}$; $S, N^f, N^d \Rightarrow$ ((11.30): $h=0,\ m=2,\ u=1$) (if $\theta=0, \theta_1>0$);
- $s, n, \widehat{cp}_{(2)}$; $S, N^d, N^d \Rightarrow$ ((11.30): $h=-2,\ m=2,\ u=2$) (if $\theta=\theta_1=0$).

3. $\eta=0$. In this case systems (11.24) possess at infinity either one double and one simple real singular points (if $\widetilde{M}\neq 0$) or one triple real singularity (if $\widetilde{M}=0$). So by Lemmas 11.1 and 11.11(iii), we have the following three global configurations if $\widetilde{M}\neq 0$:

- $s, n, \widehat{cp}_{(2)}$; $\overline{\binom{0}{2}}SN, N^{\infty} \Rightarrow ((\tilde{\mathbf{S}}_{III}):\ b=-1,\ c=-2,\ d=4,\ e=-1, f=2,$ $g=-3,\ h=1)$ (if $\theta<0$);
- $s, n, \widehat{cp}_{(2)}$; $\overline{\binom{0}{2}}SN, N^f \Rightarrow ((\tilde{\mathbf{S}}_{III}):\ b=-4,\ c=1,\ d=1,\ e=-1, f=-1,$ $g=-3,\ h=1)$ (if $\theta>0$);
- $s, n, \widehat{cp}_{(2)}$; $\overline{\binom{0}{2}}SN, N^d \Rightarrow ((\tilde{\mathbf{S}}_{III}):\ b=-1/3, c=-2, d=4/3, e=-3, f=$ 2,
  $g=-1/3,\ h=1)$ (if $\theta=0$).

And we have one global configuration if $\widetilde{M}=0$

- $s, n, \widehat{cp}_{(2)}$; $\overline{\binom{0}{3}}N \Rightarrow ((11.30):\ h=108,\ m=0,\ u=27)$.

2.2.2.2: $\mathcal{T}_2=0$. Then the saddle is weak, and we claim that in this case:

(i) the weak saddle could only be of order one and the condition $\widetilde{M}\neq 0$ holds;

(ii) the conditions $\theta=\theta_1=0$ and $\mu_0\neq 0$ are incompatible.

Indeed, the condition $\mathcal{T}_2=0$ implies $\rho_3\rho_4=0$, and we may consider that $\rho_3=0$, i.e. for systems (11.30) we have $m=-u$. Then we calculate

$$\mu_0=4u(h-u^2)^2,\quad \mathcal{F}_1=4(h-u)u(h-u^2),$$
$$\theta=64(h-u^2)\big[(h+u)^2-4u^3\big],\quad \text{Coefficient}[\widetilde{M},xy]=-48(h-u)u,$$

and since $\mu_0\neq 0$ the condition $\mathcal{F}_1=0$, as well as the condition $\widetilde{M}=0$, implies $h=u$ and then $W_2=-1024(-1+u)^6u^9\leq 0$ due to $\mu_0>0$, and this contradicts with $W_2>0$, i.e. the claim (i) is proved.

Assume now that $\theta=0$. Since $u\neq 0$ we set a new parameter v as $h+u=2uv$, and then the condition $\theta=0$ gives $u=v^2$. In this case we calculate

$$\theta=0,\quad \mu_0=4(v-1)^4v^6,\quad \theta_1=512(v-1)^3v^7,$$

and since the condition $\mu_0\neq 0$ implies $\theta_1\neq 0$, the claim (ii) is proved.

1. $\eta < 0$. According to Lemmas 11.1 and 11.11(iii), there can only be three global configurations:

   - $s^{(1)}, n, \widehat{cp}_{(2)};\ N^{\infty}$, ©, © $\Rightarrow$ ((11.30): $h = 2,\ m = -5/6,\ u = 2$) (if $\theta < 0$);
   - $s^{(1)}, n, \widehat{cp}_{(2)};\ N^{f}$, ©, © $\Rightarrow$ ((11.30): $h = 2,\ m = 1,\ u = 2$) (if $\theta > 0$);
   - $s^{(1)}, n, \widehat{cp}_{(2)};\ N^{d}$, ©, © $\Rightarrow$ ((11.30): $h = 4,\ m = 0,\ u = 4$) (if $\theta = 0$).

2. $\eta > 0$. As it was mentioned above, in the case $\theta = 0$ we have $\theta_1 \neq 0$, and according to Lemmas 11.1 and 11.11(iii), there can be only five global configurations:

   - $s^{(1)}, n, \widehat{cp}_{(2)};\ S, N^{\infty}, N^{\infty} \Rightarrow$ ((11.30): $h = 100,\ m = 79,\ u = 100$) (if $\theta < 0, \theta_1 < 0$);
   - $s^{(1)}, n, \widehat{cp}_{(2)};\ S, N^{f}, N^{f} \Rightarrow$ ((11.30): $h = 1,\ m = 4,\ u = 1$) (if $\theta < 0, \theta_1 > 0$);
   - $s^{(1)}, n, \widehat{cp}_{(2)};\ S, N^{\infty}, N^{f} \Rightarrow$ ((11.30): $h = 2,\ m = 4,\ u = 2$) (if $\theta > 0$);
   - $s^{(1)}, n, \widehat{cp}_{(2)};\ S, N^{\infty}, N^{d} \Rightarrow$ ((11.30): $h = 100,\ m = 80,\ u = 100$) (if $\theta = 0, \theta_1 < 0$);
   - $s^{(1)}, n, \widehat{cp}_{(2)};\ S, N^{f}, N^{d} \Rightarrow$ ((11.30): $h = 1/4,\ m = -3/4,\ u = 1/4$) (if $\theta = 0, \theta_1 > 0$).

3. $\eta = 0$. Since $\widetilde{M} \neq 0$ in this case systems (11.24) possess at infinity one double and one simple real singular points. So by Lemmas 11.1 and 11.11(iii), we have the following three global configurations:

   - $s^{(1)}, n, \widehat{cp}_{(2)};\ \overline{\binom{0}{2}}SN,\ N^{\infty} \Rightarrow$ (($\tilde{\mathbf{S}}_{III}$) : $b = -9/2,\ c = 1,\ d = 1/2,\ e = -2,$ $f = -1,\ g = -11/2,\ h = 4$) (if $\theta < 0$);
   - $s^{(1)}, n, \widehat{cp}_{(2)};\ \overline{\binom{0}{2}}SN,\ N^{f} \Rightarrow$ (($\tilde{\mathbf{S}}_{III}$) : $b = -9/2,\ c = 1,\ d = 1/2,\ e = -2,$ $f = -1,\ g = -5/2,\ h = 1$) (if $\theta > 0$);
   - $s^{(1)}, n, \widehat{cp}_{(2)};\ \overline{\binom{0}{2}}\,SN,\ N^{d} \Rightarrow$ (($\tilde{\mathbf{S}}_{III}$) : $b = -3/2,\ c = 1/3,\ d = 1/6,\ e = -2/3,$ $f = -1/3,\ g = -3/2,\ h = 1$) (if $\theta = 0$).

2.2.3: $W_2 = 0$. We have a node with coinciding eigenvalues, and we observe that it could not be a star node because the corresponding linear matrices for the elemental singularities are

$$\mathcal{M}_3 = \begin{pmatrix} u & 2h - u^2 \\ -1 & 2m + u \end{pmatrix}, \quad \mathcal{M}_4 = \begin{pmatrix} 2h - u & u^2 \\ 1 + 2m & -u \end{pmatrix},$$

with $u \neq 0$.

1. $\mathcal{T}_2 \neq 0$. So according to Lemmas 11.14 and 11.1, there can only be three global configurations:
   - $s, n^d, \widehat{cp}_{(2)}; N^f$, ©, © $\Rightarrow$ ((11.30): $h = 1/2$, $m = 0$, $u = 1$) (if $\eta < 0$);
   - $s, n^d, \widehat{cp}_{(2)}; S, N^\infty, N^f \Rightarrow$ ((11.30): $h = 5/2$, $m = -2$, $u = 1$) (if $\eta > 0$);
   - $s, n^d, \widehat{cp}_{(2)}; \overline{\binom{0}{2}}SN, N^f \Rightarrow ((\tilde{\mathbf{S}}_{III}) : b = 4,\ c = -1,\ d = -1,\ e = 1,\ f = 1,$ $g = 3/4,\ h = 5/4)$ (if $\eta = 0$).
2. $\mathcal{T}_2 = 0$. Then by Lemmas 11.14 and 11.1 we obtain only three global configurations:
   - $s^{(1)}, n^d, \widehat{cp}_{(2)}; N^f$, ©, © $\Rightarrow$ ((11.30): $h = -1$, $m = -1$, $u = 1$) (if $\eta < 0$);
   - $s^{(1)}, n^d, \widehat{cp}_{(2)}; S, N^\infty, N^f \Rightarrow$ ((11.30): $h = 4/5$, $m = -2\sqrt{6}/5$, $u = 4/5$) (if $\eta > 0$);
   - $s^{(1)}, n^d, \widehat{cp}_{(2)}; \overline{\binom{0}{2}}SN, N^f \Rightarrow$ ((11.30): $h = 81/125$, $m = -117/125$, $u = 81/125$) (if $\eta = 0$).

Thus we have examined all the possibilities for the family of systems possessing one double and two elemental real finite singularities, and we proved the existence of 265 geometrically distinct configurations. We have proved Theorem 11.5. □

11.3.2 One double real and two elemental complex singularities

Theorem 11.6. (a) *We consider here all the configurations of singularities, finite and infinite, of quadratic vector fields possessing exactly three distinct finite singularities: one double real and two elemental complex. These configurations are classified in Diagram* 11.7 *according to the geometrical equivalence relation of configurations. We have a total of* 31 *geometrically distinct configurations of singularities, finite and infinite.*

(b) *Necessary and sufficient conditions for each one of the* 31 *different geometrical equivalence classes can be assembled from this diagram in terms of* 10 *invariant polynomials with respect to the action of the affine group and time rescaling. These invariant polynomials, which appear in Diagram* 11.7, *are given in Section* 5.4.

(c) *Diagram* 11.7 *actually contains the global bifurcation diagram in the* 12-*dimensional space of parameters of the global configurations of singularities, finite and infinite, of this family of quadratic differential systems and provides an algorithm for finding for any given system in the family considered, its respective geometrical configuration of singularities.*

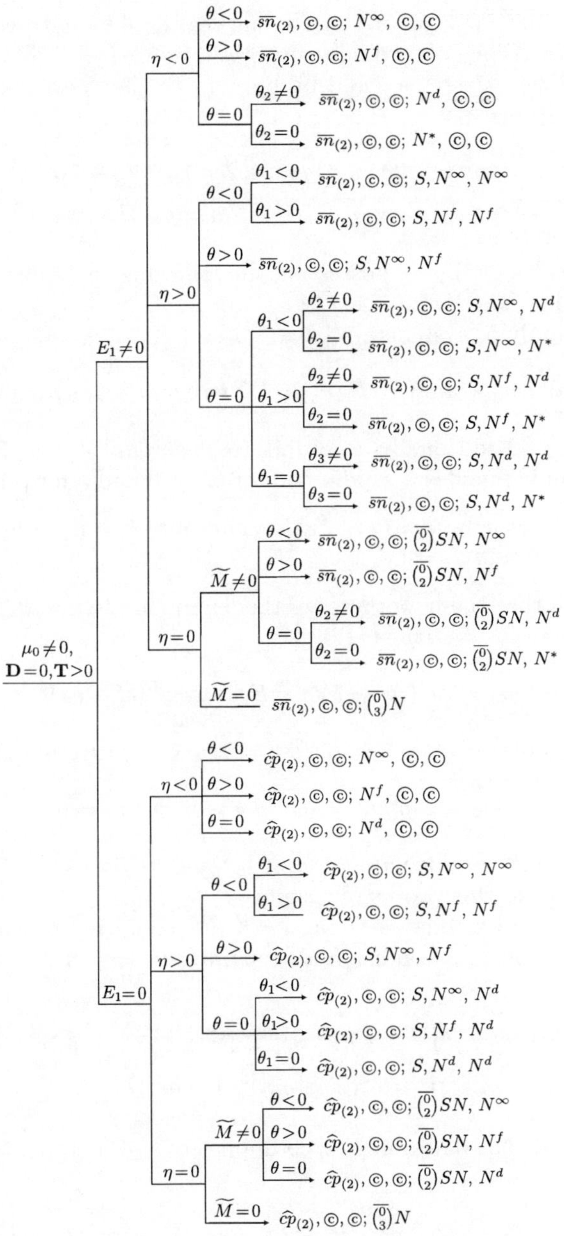

Diagram 11.7: Global configurations in the case $\mu_0 \neq 0, \mathbf{D} = 0, \mathbf{T} > 0$.

Proof. Assume that systems (8.1) have one real double and two simple complex finite singularities. Then according to Proposition 6.2 (see Table 6.1) via an affine transformation these systems could be brought to the systems 5. So we consider the family of systems:

$$\begin{aligned}\dot{x} &= cm\,x + 2cn\,y + g\,x^2 - 2cn\,xy + (g+cm)\,y^2,\\ \dot{y} &= em\,x + 2en\,y + l\,x^2 - 2en\,xy + (l+em)\,y^2,\end{aligned} \tag{11.32}$$

with $(cl-eg)(m^2+4n^2) \neq 0$, possessing the following three distinct singularities: $M_{1,2}(0,0)$ (double) and $M_{3,4}(1,\pm i)$.

For this family of systems we have

$$\mu_0 = (cl-eg)^2(m^2+4n^2), \quad E_1 = 2(cl-eg)^4(cm+2en)(m^2+4n^2)^2 \tag{11.33}$$

and hence $\mu_0 > 0$. Additionally according to Theorem 6.1 (see Table 6.2, lines 82 and 83) the double point is a saddle-node if $E_1 \neq 0$ and a cusp if $E_1 = 0$.

Lemma 11.15. *If for a system* (11.32) *the conditions* $\theta = E_1 = \theta_1\theta_2 = 0$ *hold, then* $\theta_1 = \theta_2 = 0$, $\eta > 0$ *and* $\theta_3 \neq 0$.

Proof. We claim that the hypotheses of the lemma imply $n \neq 0$. Indeed, assuming $n = 0$ we calculate for systems (11.32)

$$\mu_0 = m^2(cl-eg)^2, \quad E_1 = 2cm^5(cl-eg)^4.$$

Therefore the condition $E_1 = 0$ implies $c = 0$, and then we have

$$\mu_0 = e^2g^2m^2 \neq 0, \quad \theta = -64eg^2lm = 0.$$

So we get $l = 0$, and we calculate $\theta_1 = -64g^4$ and $\theta_2 = -eg^2m^2$. Clearly $\theta_1\theta_2 \neq 0$ (due to $\mu_0 \neq 0$), and this proves our claim.

Thus $n \neq 0$ and we may assume $n = 1$ due to a time rescaling. Considering (11.33) we deduce that the condition $E_1 = 0$ implies $e = -cm/2$. Then calculations yield

$$\begin{aligned}\theta &= 8c(2l+gm)(4c^2 - 4g^2 + 8cl + 4l^2 + 4glm + c^2m^2 + 2clm^2),\\ \mu_0 &= c^2(2l+gm)^2(4+m^2)/4,\end{aligned}$$

and due to $\mu_0 \neq 0$ the condition $E_1 = 0$ implies $c = -l \pm (lm-2g)/\sqrt{4+m^2}$. We calculate

$$\begin{aligned}\theta_1 &= -32(2l+gm)^3(lm-2g)(m \pm \sqrt{4+m^2})/(4+m^2),\\ \theta_2 &= -\frac{(2l+gm)(lm-2g)}{(4+m^2)}\big[2g + l(-m \pm \sqrt{4+m^2}\,)\big](4+m^2 \mp m\sqrt{4+m^2}),\\ \mu_0 &= (2l+gm)^2\big[2g + l(-m \pm \sqrt{4+m^2})\big]^2/4.\end{aligned}$$

We observe that $(m \pm \sqrt{4+m^2})(4+m^2 \mp m\sqrt{4+m^2}) \neq 0$ (we have only complex roots), and therefore due to $\mu_0 \neq 0$ we get that $\theta_1 = 0$ is equivalent to $\theta_2 = 0$, which implies $g = lm/2$. Then we calculate

$$\eta = c^4(4+m^2)^3/16 = \mu_0, \quad \theta_3 = c^6(4+m^2)^4/32,$$

and since $\mu_0 \neq 0$ we get $\theta_3 \neq 0$ and $\eta > 0$. This completes the proof of the Lemma 11.15. □

1: $E_1 \neq 0$. Then the double finite singular point is a semi-elemental saddle-node.

1.1: $\eta < 0$. Then systems (11.32) possess one real and two complex infinite singular points, and according to Lemma 11.1 there can be only four configurations at infinity. It remains to construct the corresponding global configurations:

- $\overline{sn}_{(2)}, ©, ©;\ N^{\infty},\ ©, © \Rightarrow ((11.32): c=1,\ e=-2,\ g=5,\ l=0,\ m=1,\ n=1)$ (if $\theta<0$);
- $\overline{sn}_{(2)}, ©, ©;\ N^{f},\ ©, © \Rightarrow ((11.32): c=-1,\ e=2,\ g=5,\ l=0, m=1,\ n=1)$ (if $\theta>0$);
- $\overline{sn}_{(2)}, ©, ©;\ N^{d},\ ©, © \Rightarrow ((11.32): c=1,\ e=0,\ g=-5/2, l=1,\ m=3/2,$ $n=1)$ (if $\theta=0, \theta_2\neq 0$);
- $\overline{sn}_{(2)}, ©, ©;\ N^{*},\ ©, © \Rightarrow ((11.32):\ c=1,\ e=0,\ g=-1,\ l=1, m=1,\ n=0)$ (if $\theta=0, \theta_2=0$).

1.2: $\eta > 0$. In this case systems (11.32) possess three real infinite singular points. Since for these systems the condition $\mu_0 > 0$ holds, taking into consideration Theorem 7.5 and Lemma 11.1, we can have at infinity only nine configurations. The corresponding global configurations are as follows:

- $\overline{sn}_{(2)}, ©, ©;\ S, N^{\infty},\ N^{\infty} \Rightarrow ((11.32): c=1,\ e=1,\ g=-21/20, l=-3,\ m=0,$ $n=2)$ (if $\theta<0, \theta_1<0$);
- $\overline{sn}_{(2)}, ©, ©;\ S, N^{f},\ N^{f} \Rightarrow ((11.32): c=1,\ e=2,\ g=-1,\ l=0, m=1,\ n=1)$ (if $\theta<0, \theta_1>0$);
- $\overline{sn}_{(2)}, ©, ©;\ S, N^{\infty},\ N^{f} \Rightarrow ((11.32): c=1,\ e=1,\ g=-1,\ l=0, m=1,\ n=1)$ (if $\theta>0$);
- $\overline{sn}_{(2)}, ©, ©;\ S, N^{\infty},\ N^{d} \Rightarrow ((11.32): c=1, e=1, g=-1, l=-3m=0, n=2)$ (if $\theta=0, \theta_1<0, \theta_2\neq 0$);
- $\overline{sn}_{(2)}, ©, ©;\ S, N^{\infty},\ N^{*} \Rightarrow ((11.32): c=3, e=1, g=-3, l=-2, m=1, n=0)$ (if $\theta=0, \theta_1<0, \theta_2=0$);
- $\overline{sn}_{(2)}, ©, ©;\ S, N^{f},\ N^{d} \Rightarrow ((11.32): c=2, e=1, g=6, l=0,$ $m=-3, n=2)$ (if $\theta=0, \theta_1>0, \theta_2\neq 0$);
- $\overline{sn}_{(2)}, ©, ©;\ S, N^{f},\ N^{*} \Rightarrow ((11.32): c=1, e=1, g=2, l=1m=-2, n=0)$ (if $\theta=0, \theta_1>0, \theta_2=0$);
- $\overline{sn}_{(2)}, ©, ©;\ S, N^{d},\ N^{d} \Rightarrow ((11.32): c=1, e=1, g=-2, l=-1, m=1, n=1)$ (if $\theta=0, \theta_1=0, \theta_3\neq 0$);

- $\overline{sn}_{(2)}, ©, ©; S, N^d, N^* \Rightarrow ((11.32): c=1,\ e=1,\ g=-1,\ l=0, m=1,\ n=0)$ (if $\theta=\theta_1=\theta_3=0$).

1.3: $\eta=0$. In this case systems (11.32) possess at infinity either one double and one simple real singular points (if $\widetilde{M}\neq 0$) or one triple real singularity (if $\widetilde{M}=0$). So by Lemma 11.1 we could have at infinity exactly five configurations. We have the following four global configurations if $\widetilde{M}\neq 0$:

- $\overline{sn}_{(2)}, ©, ©; \overline{\binom{0}{2}}SN,\ N^\infty \Rightarrow ((11.32): c=1, e=1, g=-1, l=-3, m=1, n=1)$ (if $\theta<0$);
- $\overline{sn}_{(2)}, ©, ©; \overline{\binom{0}{2}}SN,\ N^f \Rightarrow ((11.32): c=1,\ e=1,\ g=1,\ l=-1, m{=}-1,\ n{=}1)$ (if $\theta>0$);
- $\overline{sn}_{(2)}, ©, ©; \overline{\binom{0}{2}}SN,\ N^d \Rightarrow ((11.32): c=1,\ e=1,\ g=0,\ l=-2, m=0,\ n=1)$ (if $\theta=0, \theta_2\neq 0$);
- $\overline{sn}_{(2)}, ©, ©; \overline{\binom{0}{2}}SN,\ N^* \Rightarrow ((11.32): c=1,\ e=0,\ g=-2,\ l=1, m=2,\ n=0)$ (if $\theta=\theta_2=0$).

And we have one global configuration if $\widetilde{M}=0$

- $\overline{sn}_{(2)}, ©, ©; \overline{\binom{0}{3}}N \Rightarrow ((11.32): c=0,\ e=1,\ g=1/4,\ l=3\sqrt{3}/4, m=0,\ n=1)$.

2: $E_1=0$. Then by Theorem 6.1 the double finite singular point is a cusp. As $\mu_0\neq 0$, considering (11.33) we get the relation $cm+2en=0$.

2.1: $\eta<0$. Then systems (11.32) possess one real and two complex infinite singular points, and considering Lemmas 11.15 and 11.1, there could be only three configurations at infinity which lead to the corresponding global configurations:

- $\widehat{cp}_{(2)}, ©, ©; N^\infty,\ ©, © \Rightarrow ((11.32): c=2,\ e=1,\ g=5,\ l=0, m=1,\ n=-1)$ (if $\theta<0$);
- $\widehat{cp}_{(2)}, ©, ©; N^f,\ ©, © \Rightarrow ((11.32): c=2,\ e=1,\ g=3,\ l=0, m=-1,\ n=1)$ (if $\theta>0$);
- $\widehat{cp}_{(2)}, ©, ©; N^d,\ ©, © \Rightarrow ((11.32): c=-2,\ e=1, g=(3+\sqrt{5})/2,\ l=3,$ $m=1,\ n=1)$ (if $\theta=0$).

2.2: $\eta>0$. In this case systems (11.32) possess three real infinite singular points. Since for these systems the condition $\mu_0>0$ holds, taking into consideration Lemmas 11.15 and 11.1, we could have at infinity only six configurations. The corresponding global configurations are as follows:

- $\widehat{cp}_{(2)}, ©, ©; S, N^\infty,\ N^\infty \Rightarrow ((11.32): c=2,\ e=1,\ g=12,\ l=0,\ m=5,$ $n=-5)$ (if $\theta<0, \theta_1<0$);
- $\widehat{cp}_{(2)}, ©, ©; S, N^f,\ N^f \Rightarrow ((11.32): c=-2,\ e=1,\ g=2,\ l=0, m=1,\ n=1)$ (if $\theta<0, \theta_1>0$);
- $\widehat{cp}_{(2)}, ©, ©; S, N^\infty,\ N^f \Rightarrow ((11.32): c=-1,\ e=1,\ g=2,\ l=1,$ $m=2,\ n=1)$ (if $\theta>0$);
- $\widehat{cp}_{(2)}, ©, ©; S, N^\infty,\ N^d \Rightarrow ((11.32): c=-2,\ e=1,\ l=-17/10,$ $g=(-17+37\sqrt{5})/20, m=1, n=1)$ (if $\theta{=}0, \theta_1{<}0$);

- $\widehat{cp}_{(2)}, \copyright, \copyright;\ S, N^f,\ N^d \Rightarrow ((11.32):\ c = -\sqrt{3},\ e = 1,\ g = 2,\ l = 0,$ $m = 2/\sqrt{3},\ n = 1)$ (if $\theta = 0, \theta_1 > 0$);
- $\widehat{cp}_{(2)}, \copyright, \copyright;\ S, N^d,\ N^d \Rightarrow ((11.32):\ c = 1,\ e = 0,\ g = 0,\ l = -1,$ $m = 0,\ n = 1)$ (if $\theta = \theta_1 = 0$).

2.1: $\eta = 0$. In this case systems (11.32) possess at infinity either one double and one simple real singular points (if $\widetilde{M} \neq 0$) or one triple real singularity (if $\widetilde{M} = 0$). So by Lemmas 11.15 and 11.1, we could have at infinity exactly four configurations. We have the following three global configurations if $\widetilde{M} \neq 0$:

- $\widehat{cp}_{(2)}, \copyright, \copyright;\ \overline{\binom{0}{2}}SN,\ N^\infty \Rightarrow ((11.32):\ c = 1,\ e = -2,\ g = 4,\ l = 0, m = 4,\ n = 1)$ (if $\theta < 0$);
- $\widehat{cp}_{(2)}, \copyright, \copyright;\ \overline{\binom{0}{2}}SN,\ N^f \Rightarrow ((11.32):\ c = 2,\ e = -1,\ g = 2,\ l = 0, m = 1,\ n = 1)$ (if $\theta > 0$);
- $\widehat{cp}_{(2)}, \copyright, \copyright;\ \overline{\binom{0}{2}}SN,\ N^d \Rightarrow ((11.32):\ c = 1,\ e = -1/\sqrt{3},\ g = 1, l = 0,\ m = 1,$ $n = \sqrt{3}/2)$ (if $\theta = 0$).

And we have one global configuration if $\widetilde{M} = 0$

- $\widehat{cp}_{(2)}, \copyright, \copyright;\ \overline{\binom{0}{3}}N \Rightarrow ((11.32):\ c = 1,\ e = 0,\ g = 0,\ l = -2,\ m = 0, n = 1)$.

This completes the proof of Theorem 11.6. □

11.4 Exactly four distinct finite singularities

We consider here the three subclasses of quadratic differential systems with finite singularities of total multiplicity four (i.e. $m_f = 4$) possessing exactly four distinct finite singularities, namely

- systems with four real elemental finite singularities ($\mu_0 \neq 0$, $\mathbf{D} < 0$, $\mathbf{R} > 0$, $\mathbf{S} > 0$);
- systems with two real and two complex elemental singularities ($\mu_0 \neq 0$, $\mathbf{D} > 0$);
- systems with four complex elemental finite singularities ($\mu_0 \neq 0$, $\mathbf{D} < 0$, ($\mathbf{R} \leq 0$ or $\mathbf{S} \leq 0$)).

11.4.1 Four real elemental singularities

Theorem 11.7. (a) *We consider here all the configurations of singularities, finite and infinite, of quadratic vector fields possessing exactly four real elemental finite singularities. These configurations are classified in Diagram* 11.8 *according to the geometrical equivalence relation of configurations. We obtain a total of* 396 *geometrically distinct configurations of singularities, finite and infinite. There could be up to eight more configurations sharing the same finite part* $s, s^{(2)}, f, f$ *and three singularities at infinity, which we conjecture not to be realizable (see Conjecture* 11.1*).*

(b) *Necessary and sufficient conditions for each one of the* 396 *different geometrical equivalence classes can be assembled from this diagram in terms of* 29 *invariant polynomials with respect to the action of the affine group and time rescaling. These invariant polynomials, which appear in Diagram* 11.8, *are given in Section* 5.4.

(c) *Diagram* 11.8 *actually contains the global bifurcation diagram in the* 12-*dimensional space of parameters of the global configurations of singularities, finite and infinite, of this family of quadratic differential systems and provides an algorithm for finding for any given system in the family considered, its respective geometrical configuration of singularities.*

Proof. Assume that systems (8.1) have four real elemental finite singularities. Then according to Proposition 6.2 (see Table 6.1) via an affine transformation these systems could be brought to the systems 1. So we consider the family of systems:

$$\begin{aligned} \dot{x} &= cx + dy - cx^2 + 2hxy - dy^2, \\ \dot{y} &= ex + fy - ex^2 + 2mxy - fy^2, \end{aligned} \tag{11.34}$$

which possess the following four singularities:

$$M_1(0,0),\quad M_2(1,0),\quad M_3(0,1),\quad M_4\Big(\frac{\beta(\beta-2\gamma)}{\beta^2-4\alpha\gamma},\ \frac{\beta(\beta-2\alpha)}{\beta^2-4\alpha\gamma}\Big),$$

where $\alpha = eh - cm$, $\beta = cf - de$, $\gamma = dm - fh$, and $\beta^2 - 4\alpha\gamma \neq 0$.

First of all we prove several lemmas that we need for proving Theorem 11.7.

Lemma 11.16. *If for a quadratic system with four real finite distinct singularities the conditions $\mathcal{T}_4 = \mathcal{T}_3 = \mathcal{T}_2 = 0$ are satisfied, then this system is Hamiltonian.*

Proof. For the singularities M_i $(i = 1, 2, 3, 4)$ of systems 11.34 we have, respectively,

$$\begin{aligned} &\rho_1 = c + f, \quad \rho_2 = -c + f + 2m, \quad \rho_3 = c - f + 2h, \\ &\rho_4 = c + f + 2(m - c)\frac{\beta(\beta-2\gamma)}{\beta^2-4\alpha\gamma} + 2(h - f)\frac{\beta(\beta-2\alpha)}{\beta^2-4\alpha\gamma}. \end{aligned}$$

Considering Lemma 5.3(i) it is clear that the conditions $\mathcal{T}_4 = \mathcal{T}_3 = \mathcal{T}_2 = 0$ imply the vanishing of three traces. Then without lost of generality we may assume $\rho_1 = \rho_2 = \rho_3 = 0$. These equalities give us the conditions $f = h = -c$ and $m = c$, and therefore we obtain $\mathcal{T}_1 = \sigma = 0$, i.e. the systems are Hamiltonian. □

Lemma 11.17. *Assume that for a quadratic system with four real finite distinct singularities the condition $\widetilde{M} = 0$ holds, i.e. this system possesses a triple infinite singularity. Then for this system the following are true:*

(i) *If the conditions $\mathcal{T}_4 = 0$ and $\mathcal{T}_3 \neq 0$ hold, then $\mathcal{F}_2 \neq 0$. Moreover, the following hold:*

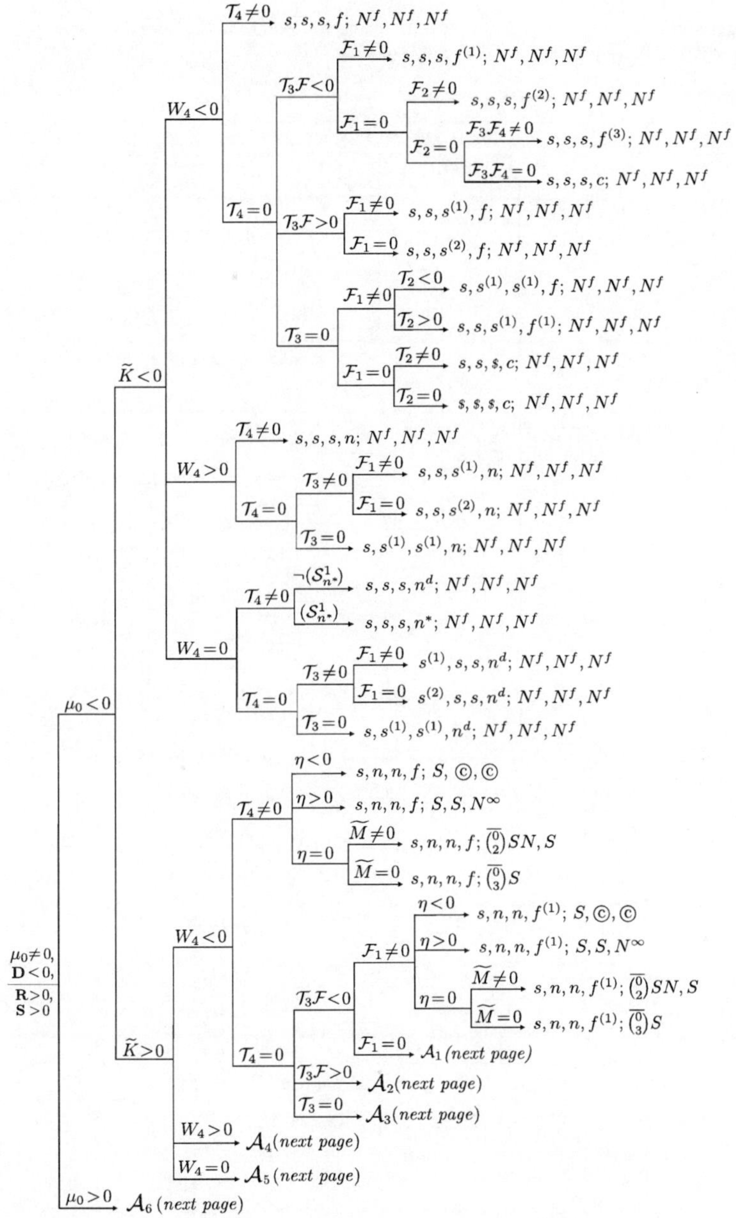

Diagram 11.8: Global configurations in the case $\mu_0 \neq 0, \mathbf{D} < 0, \mathbf{R} > 0, \mathbf{S} > 0$.

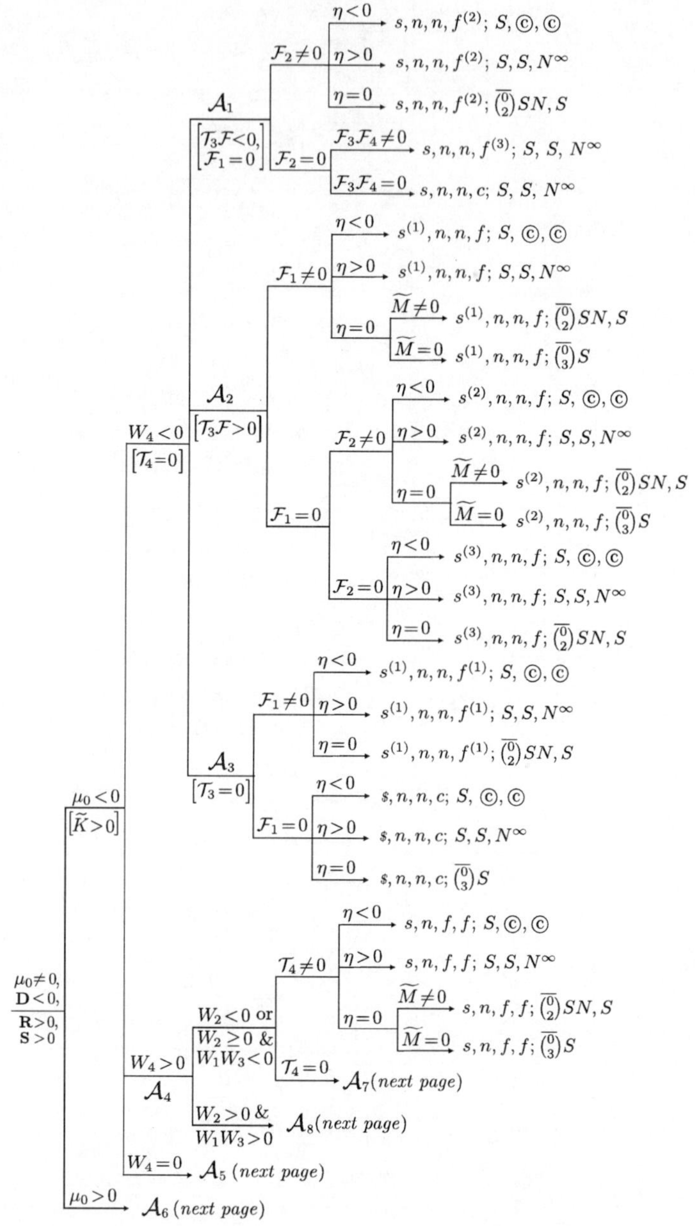

Diagram 11.8 (*cont.*): Global configurations in the case $\mu_0 \neq 0, \mathbf{D} < 0, \mathbf{R} > 0, \mathbf{S} > 0$.

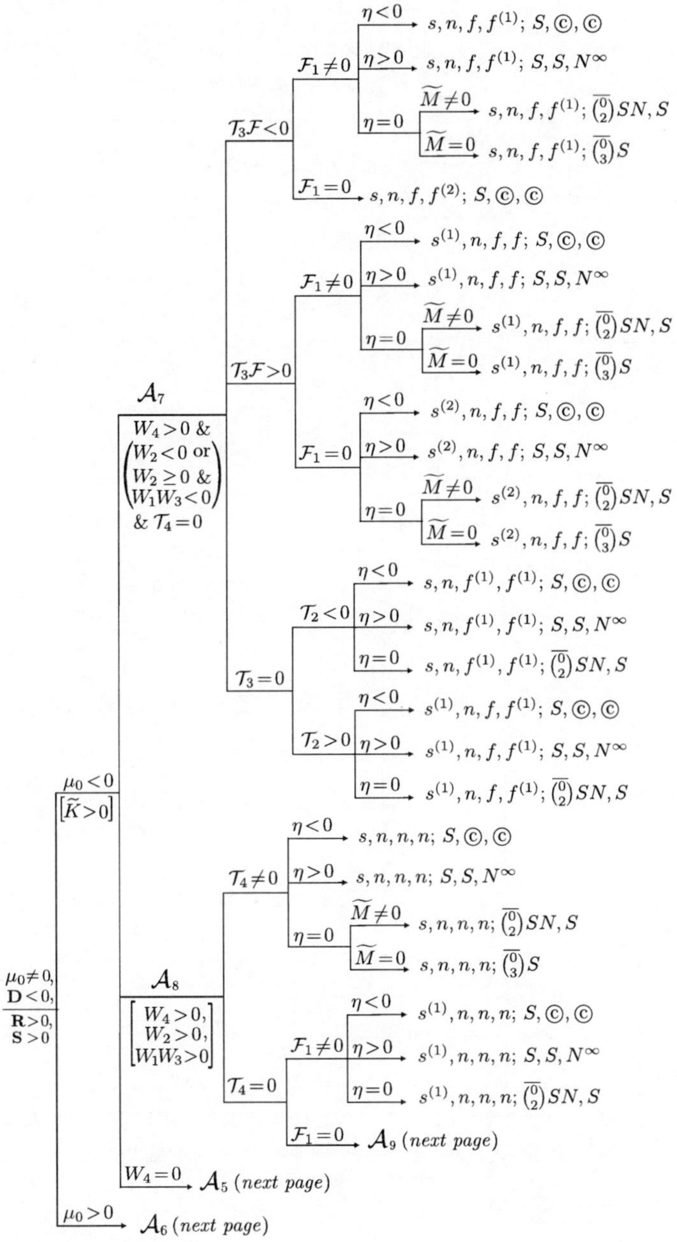

Diagram 11.8 (*cont.*): Global configurations in the case $\mu_0 \neq 0, \mathbf{D} < 0, \mathbf{R} > 0, \mathbf{S} > 0$.

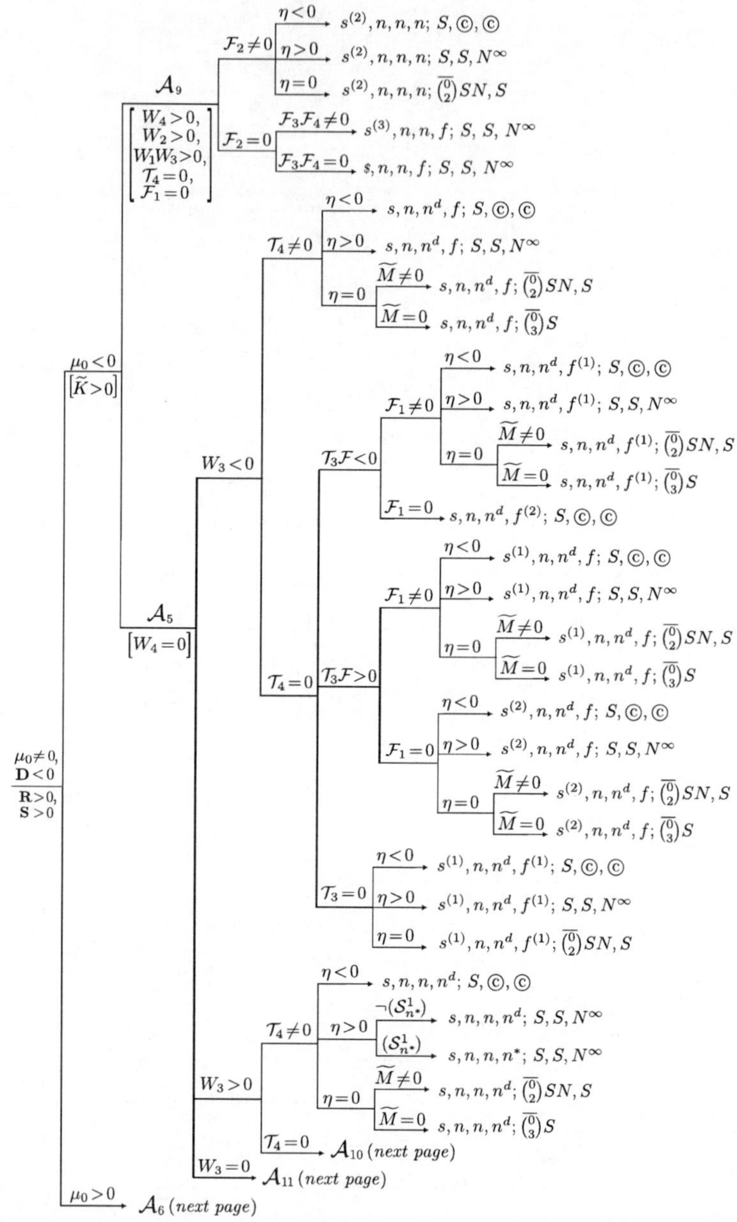

Diagram 11.8 (*cont.*): Global configurations in the case $\mu_0 \neq 0, \mathbf{D} < 0, \mathbf{R} > 0, \mathbf{S} > 0$.

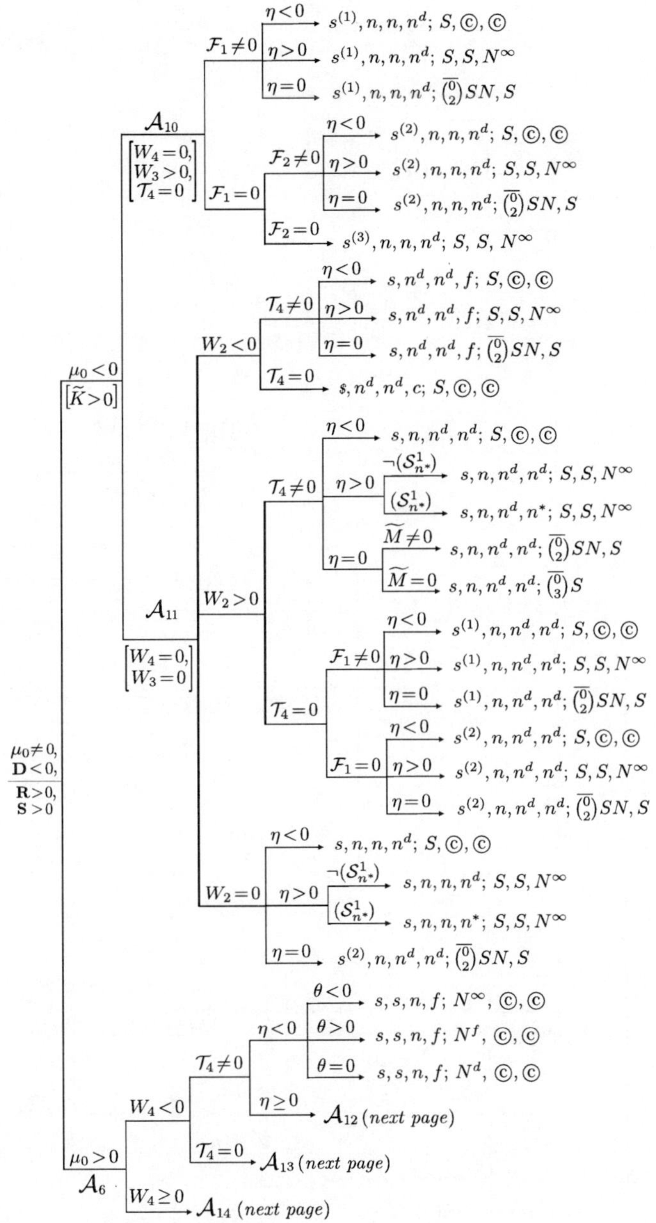

Diagram 11.8 (*cont.*): Global configurations in the case $\mu_0 \neq 0, \mathbf{D} < 0, \mathbf{R} > 0, \mathbf{S} > 0$.

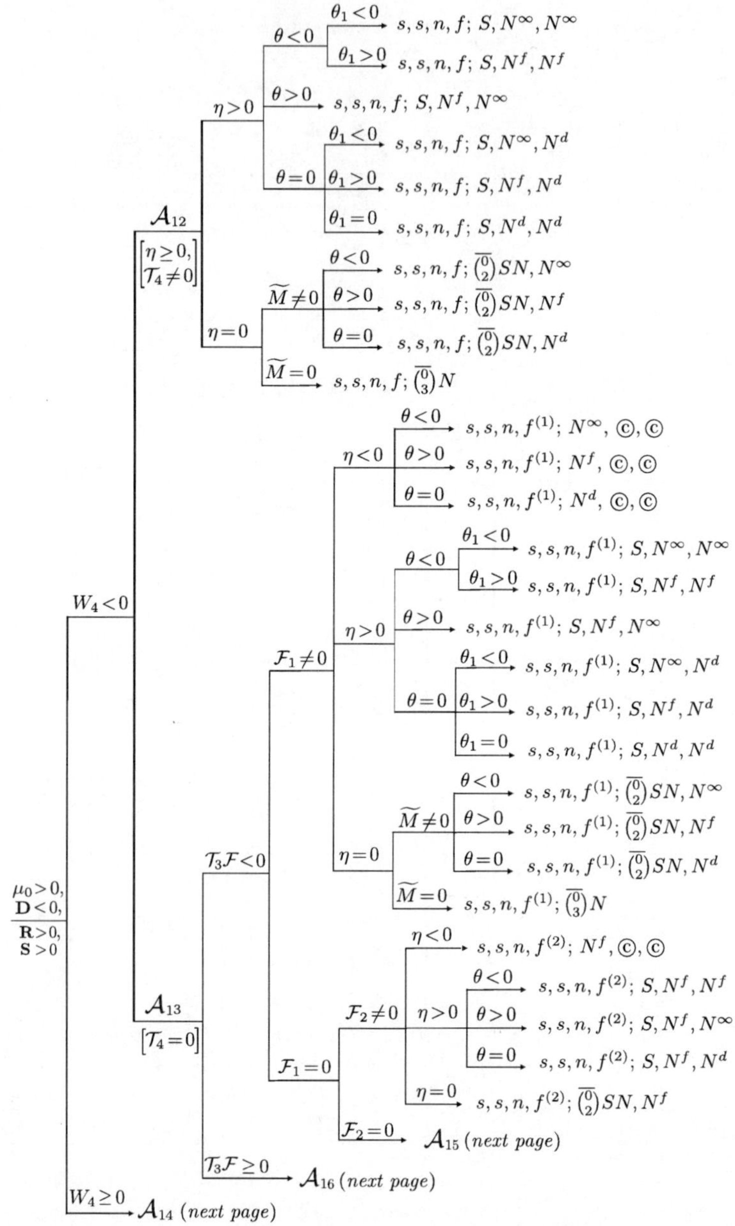

Diagram 11.8 (*cont.*): Global configurations in the case $\mu_0\neq 0, \mathbf{D}<0, \mathbf{R}>0, \mathbf{S}>0$.

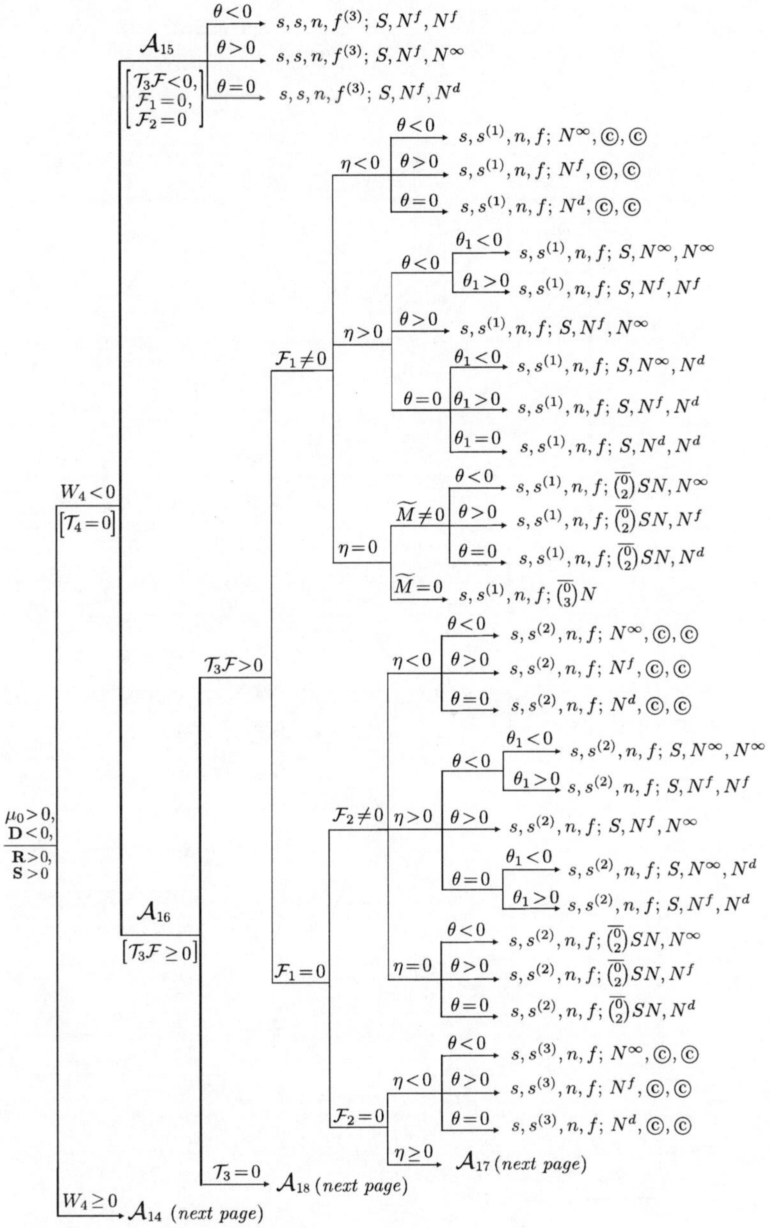

Diagram 11.8 (*cont.*): Global configurations in the case $\mu_0 \neq 0, \mathbf{D} < 0, \mathbf{R} > 0, \mathbf{S} > 0$.

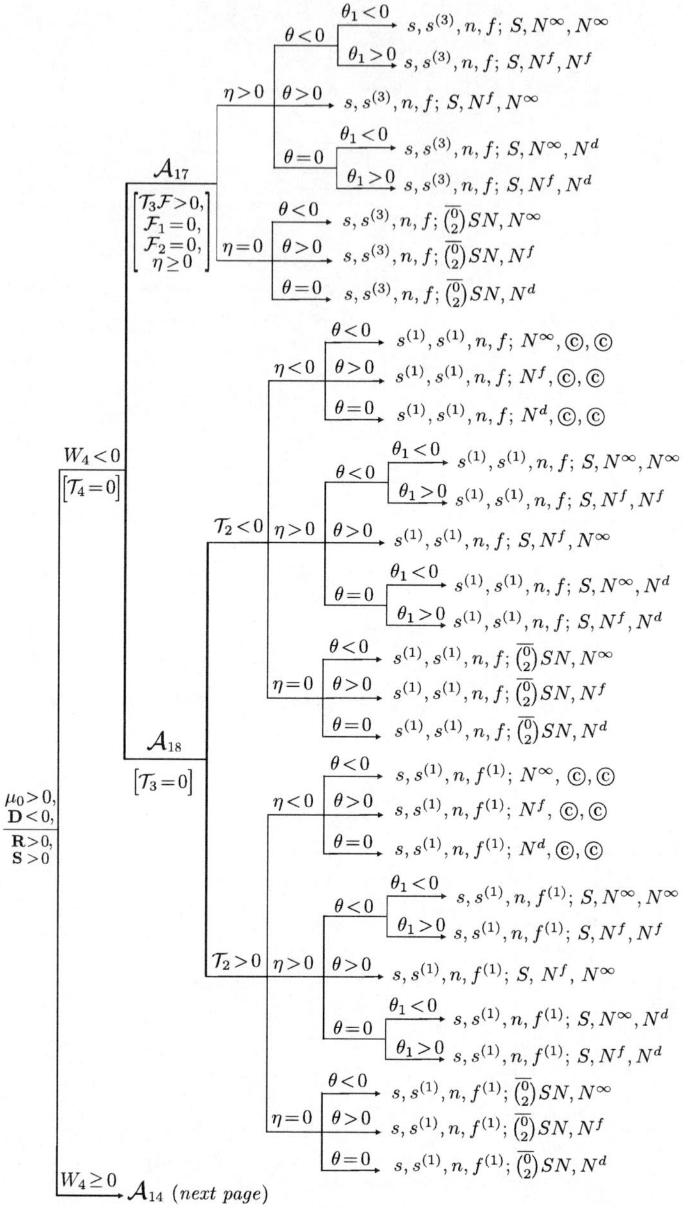

Diagram 11.8 (*cont.*): Global configurations in the case $\mu_0 \neq 0, \mathbf{D} < 0, \mathbf{R} > 0, \mathbf{S} > 0$.

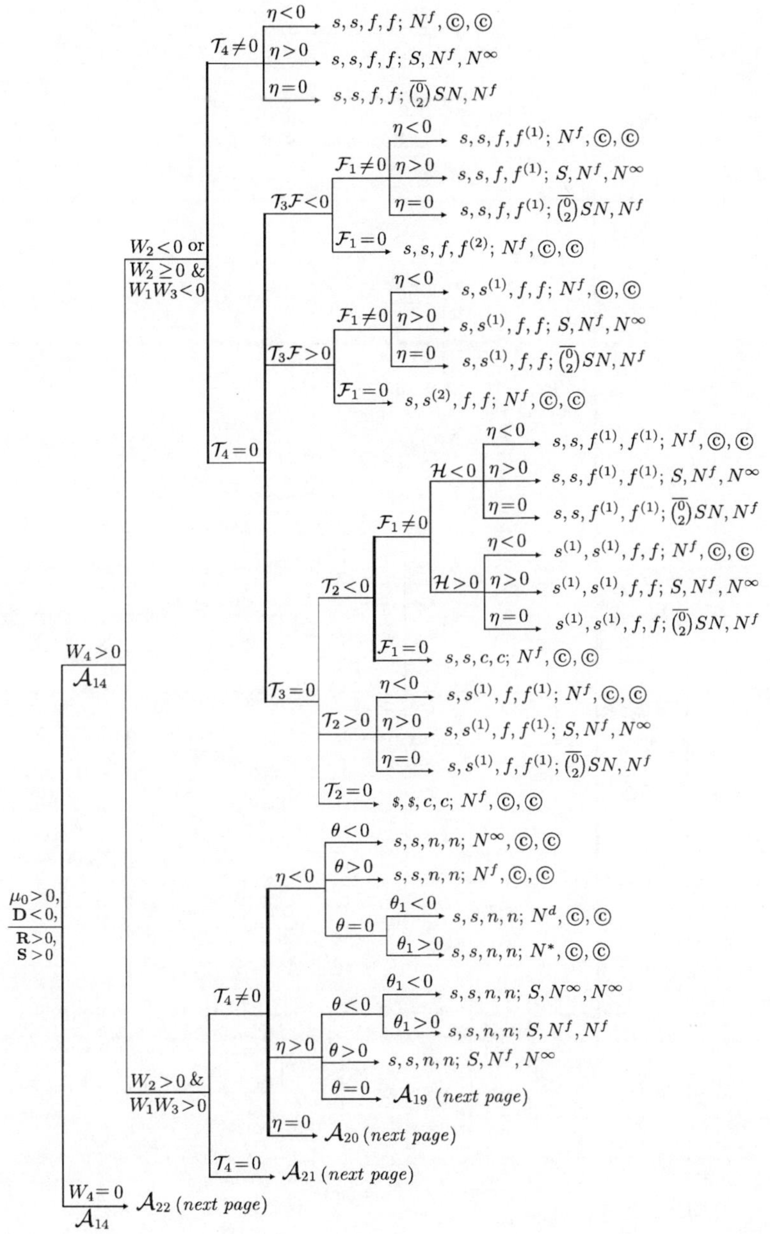

Diagram 11.8 (*cont.*): Global configurations in the case $\mu_0 \neq 0, \mathbf{D} < 0, \mathbf{R} > 0, \mathbf{S} > 0$.

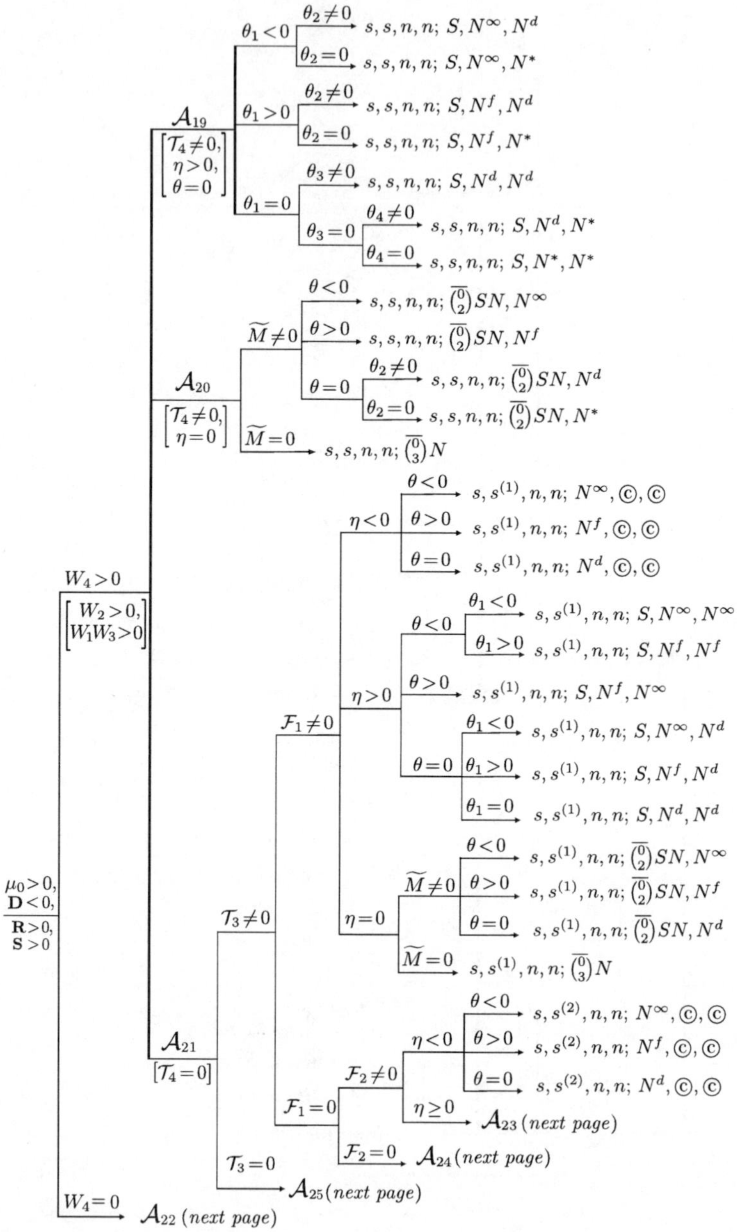

Diagram 11.8 (*cont.*): Global configurations in the case $\mu_0 \neq 0, \mathbf{D} < 0, \mathbf{R} > 0, \mathbf{S} > 0$.

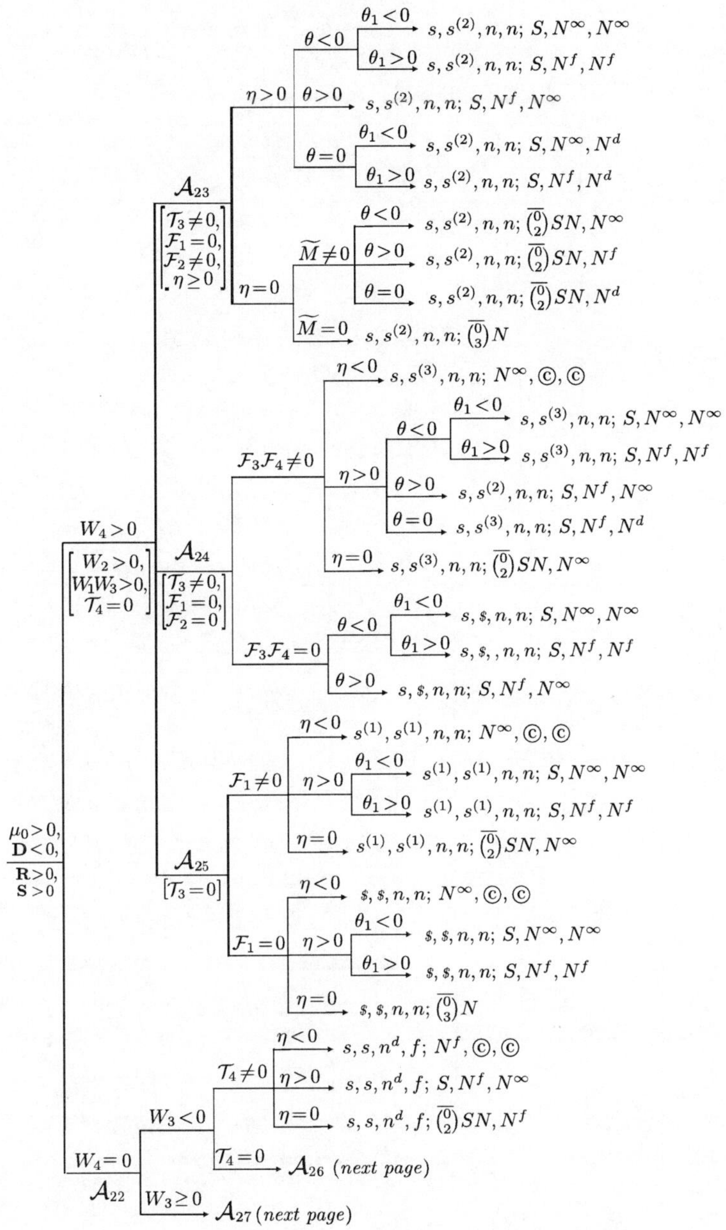

Diagram 11.8 (*cont.*): Global configurations in the case $\mu_0 \neq 0, \mathbf{D} < 0, \mathbf{R} > 0, \mathbf{S} > 0$.

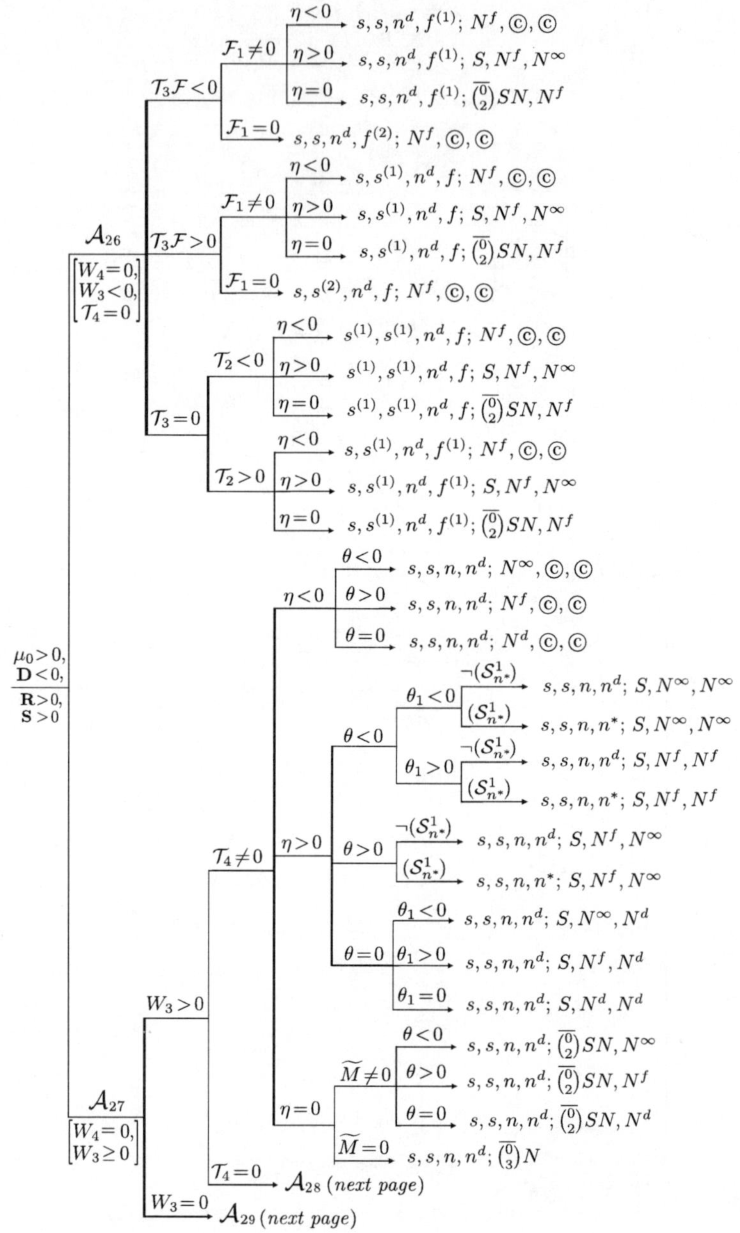

Diagram 11.8 (*cont.*): Global configurations in the case $\mu_0 \neq 0, \mathbf{D} < 0, \mathbf{R} > 0, \mathbf{S} > 0$.

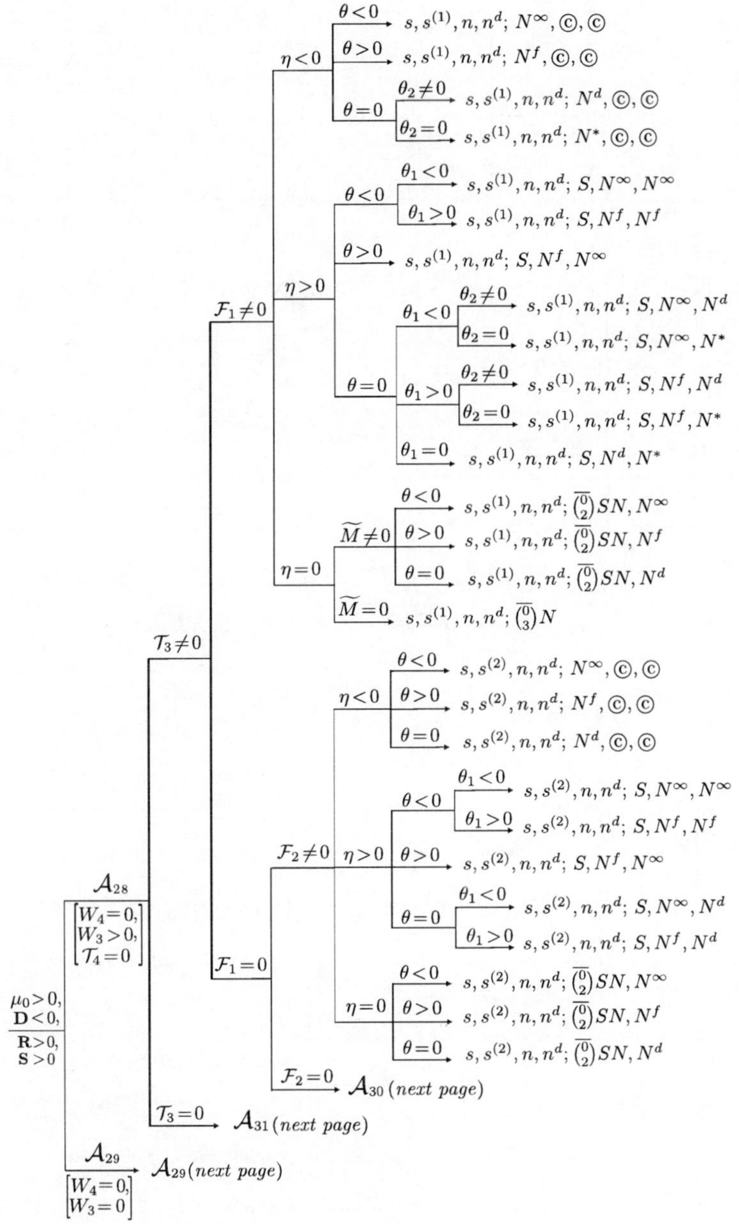

Diagram 11.8 (*cont.*): Global configurations in the case $\mu_0\neq0, \mathbf{D}<0, \mathbf{R}>0, \mathbf{S}>0$.

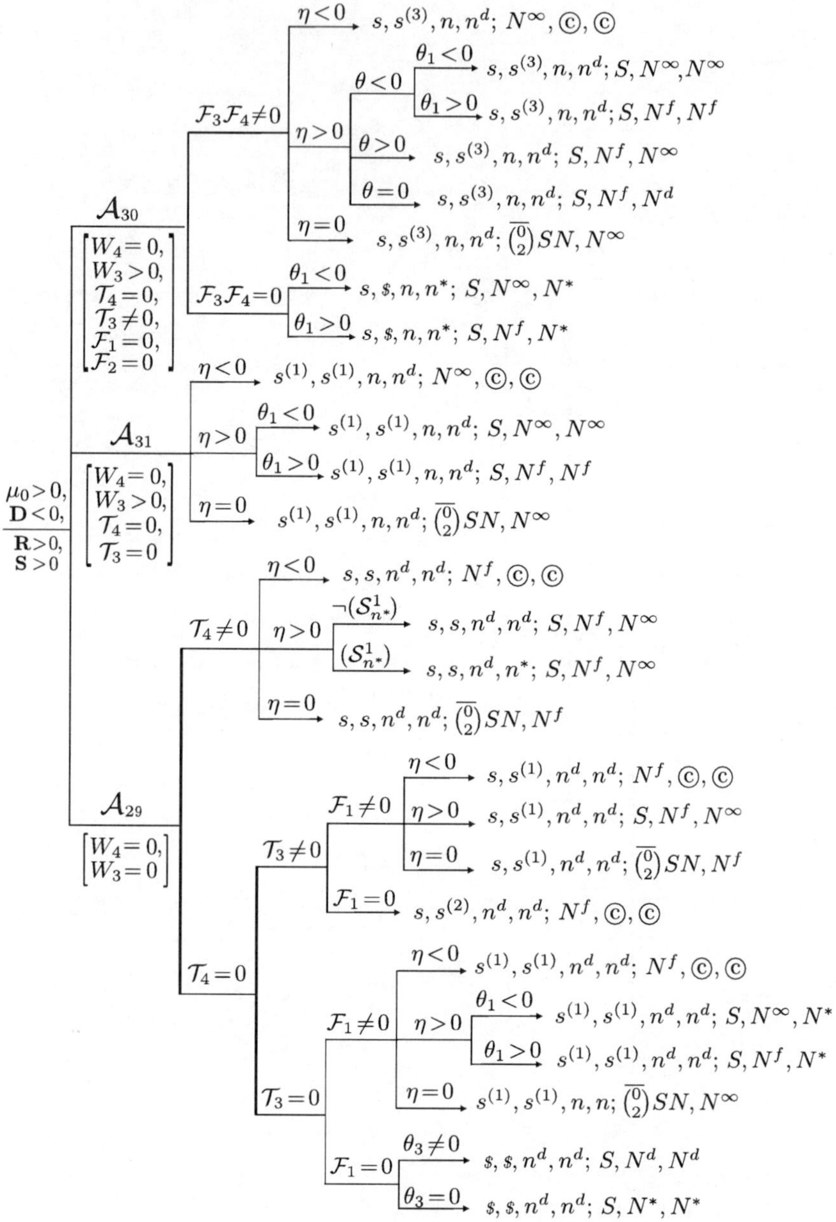

Diagram 11.8 (*cont.*): Global configurations in the case $\mu_0 \neq 0, \mathbf{D} < 0, \mathbf{R} > 0, \mathbf{S} > 0$.

(i$_1$) *If the condition $\mathcal{F}_1 = 0$ holds, then the condition $\mu_0 > 0$ implies $W_4 > 0$.*

(i$_2$) *If the condition $W_4 = 0$ is satisfied, then $\mu_0 W_3 \geq 0$.*

(ii) *The conditions $\mathcal{T}_4 = \mathcal{T}_3 = 0$ and $\mathcal{F}_1 \neq 0$ are incompatible.*

(iii) *If the conditions $W_4 = W_3 = 0$ hold (i.e. this system possesses two nodes with coinciding eigenvalues), then it could not have a focus.*

(iv) *The conditions $\mu_0 > 0$, $W_4 = 0$ and $W_3 < 0$ are incompatible.*

(v) *The conditions $\mu_0 > 0$, $W_4 > 0$ and $(W_2 < 0) \bigwedge (W_2 > 0, W_1 W_3 < 0)$ are incompatible.*

(vi) *The conditions $\mu_0 > 0$ and $W_4 = W_3 = 0$ are incompatible.*

Proof. First of all we construct a suitable canonical form of the systems possessing simultaneously four real finite and a triple infinite singularities. Consider systems (11.34), and assume that these systems possess a triple infinite singular point. Then clearly we must have $C_2 = (kx + ny)^3$ with $k^2 + n^2 \neq 0$.

Additionally for systems (11.34) we have

$$C_2 = ex^3 - (c + 2m)x^2y + (f + 2h)xy^2 - dy^3 = (kx + ny)^3$$

and we get the following relations:

$$e = k^3, \;\; d = -n^3, \;\; h = (3kn^2 - f)/2, \;\; m = -(c + 3k^2n)/2.$$

Therefore we arrive at the following canonical form:

$$\begin{aligned} \dot{x} &= cx - n^3y - cx^2 + (3kn^2 - f)xy + n^3y^2, \\ \dot{y} &= k^3x + fy - k^3x^2 - (c + 3k^2n)xy - fy^2. \end{aligned} \tag{11.35}$$

(i) Assume that for systems (11.35) the conditions $\mathcal{T}_4 = 0$ and $\mathcal{T}_3 \neq 0$ are satisfied. In this case we have exactly one trace zero, and we may consider $\rho_1 = c + f = 0$, i.e. $f = -c$. Then we calculate

$$\begin{aligned} \mathcal{T}_4 = 0, \quad \mu_0 &= -(ck + cn + 2k^2n^2)^3, \\ \mathcal{T}_3 &= 27(k - n)(c + k^2n)(c + kn^2)(ck + cn + 2k^2n^2)^2(k^3n^3 - c^2), \\ \mathcal{F}_1 &= -3k(k - n)n(c + k^2n)(c + kn^2)(2c + k^2n + kn^2), \\ \mathcal{F}_2 &= -81(k - n)^2(c + k^2n)^2(c + kn^2)^2(ck + cn + 2k^2n^2)^4/2, \end{aligned} \tag{11.36}$$

and due $\mathcal{T}_3 \neq 0$ evidently we have $\mathcal{F}_2 \neq 0$.

(i$_1$) Suppose now that in addition $\mathcal{F}_1 = 0$. Since $\mathcal{T}_3 \neq 0$ the condition $\mathcal{F}_1 = 0$ yields $kn(2c + k^2n + kn^2) = 0$. Examining the corresponding three possibilities, we get

$\mu_0 = -c^3n^3$ and $W_4 = -4c^{13}n^3(c-4n^3)(16c-n^3)$ if $k=0$;
$\mu_0 = -c^3k^3$ and $W_4 = -4c^{13}k^3(c-4k^3)(16c-k^3)$ if $n=0$;
$\mu_0 = k^3n^3(k-n)^6/8$ and $W_4 = k^{13}n^{13}(k-n)^{20}(8k+n)(k+8n)/4096$ if $c=-kn(k+n)/2$.

We observe that the condition $\mu_0 > 0$ implies $W_4 > 0$.

(i_2) Assume now that in addition the condition $W_4 = 0$ holds, i.e. one of the singular points is a node with coinciding eigenvalues. Without loss of generality we may assume $M_2(1,0)$ to be such a node. Then considering the relation $f=-c$, we have

$$\tau_2 = c^2 - 2ck^2(2k-3n) - k^3n^2(3k-4n) = 0,$$

and we observe that $k \neq 0$, otherwise we get $c=0$, which implies $\mu_0 = 0$. Since Discriminant$[\tau_2, c] = 16k^3(k-n)^3$, we set a new parameter u as $k(k-n) = u^2 \geq 0$, and since $k \neq 0$ we find that $n = (k^2-u^2)/k$. Then we obtain

$$\tau_2 = (c+k^3-3ku^2-2u^3)(c+k^3-3ku^2+2u^3) = 0,$$

and due to the change u by $-u$, we may assume that the first factor vanishes, i.e. $c = (2u-k)(k+u)^2$. Then calculations yield $W_4 = 0$ and

$$\mu_0 W_3 = 36(3k-5u)^2(3k-2u)^6u^{28}(k+u)^{26}(9k^2+3ku+2u^2) \times (9k^2-15ku+8u^2)/k^{12},$$

and it easily could be observed that the condition $\mu_0 W_3 \geq 0$ holds. This completes the proof of the statement (i).

(ii) Suppose now that for systems (11.35), apart from the condition $\mathcal{T}_4 = 0$, we have in addition $\mathcal{T}_3 = 0$ and $\mathcal{F}_1 \neq 0$. Then considering (11.36), due to $\mu_0 \neq 0$ we get the unique factor $k^3n^3 - c^2 = 0$. However for systems (11.35) with $f=-c$, we calculate

$$\mathbf{D} = -48(c^2-k^3n^3)^2(2c^2+ck^3+3ck^2n+3k^4n^2-k^3n^3)^2 \times (2c^2+3ckn^2+cn^3-k^3n^3+3k^2n^4)^2 = 0,$$

and as $\mathbf{D} < 0$ the contradiction we obtained completes the proof of the statement (ii) of the lemma.

(iii) Suppose the contrary, that a quadratic system with a triple point at infinity and four real finite distinct singularities possesses two nodes with coinciding eigenvalues and a focus. Considering systems (11.35), without loss of generality we may assume that $M_1(0,0)$ and $M_2(1,0)$ are two such nodes (i.e. $\tau_1 = \tau_2 = 0$), whereas $M_3(0,1)$ is a focus (i.e. $\tau_3 < 0$). For systems (11.35) we have $\tau_1 = (c-f)^2 - 4k^3n^3 = 0$, and we observe that the condition $kn \neq 0$

must hold. Indeed supposing $kn = 0$, we may consider $k = 0$ due to the change $(x, y, c, f, k, n) \mapsto (y, x, f, c, -n, -k)$, and then we get $f = c$. However in this case we obtain a contradiction: $\tau_2 = c^2 = 0$ and $\tau_3 = c(c - 4n^3) < 0$.

Thus $kn \neq 0$ and we set a new parameter u as follows: $c - f = 2ku \neq 0$. Then $f = c - 2ku$, and the condition $\tau_1 = -4k^2(kn^3 - u^2) = 0$ gives $k = u^2/n^3$. Herein we calculate

$$\begin{aligned}
\tau_2 &= \left[c^2 n^{12} - 2cn^3u^3(2n^6 + 3n^4u - 2u^3) + u^6(8n^6 + 12n^4u - 3n^2u^2 - 8u^3)\right]/n^{12}\\
&\equiv \psi_2(c, n, u)/n^{12},\\
\tau_3 &= \left[c^2n^6 - 2cn^5(2n^4 - 3u^2) - u^4(3n^4 - 4u^2)\right]/n^6 \equiv \psi_3(c, n, u)/n^6,\\
\tilde{D}_2 &\equiv \text{Discriminant}[\psi_2(c, n, u), c] = -16n^6u^6(n^4 - u^2)^3,\\
\tilde{D}_3 &\equiv \text{Discriminant}[\psi_3(c, n, u), c] = 16n^6(n^4 - u^2)^3.
\end{aligned}$$

Clearly in order to force the condition $\tau_3 < 0$, it is necessary that $\tilde{D}_3 > 0$. However in this case we get $\tilde{D}_2 < 0$, hence the equation $\tau_2 = 0$ could not have real solutions. This completes the proof of the statement (iii) of the lemma.

(iv)–(v) In order to prove these two statements, it is convenient to construct a new canonical form of systems, possessing a triple point at infinity and at least two real finite singularities.

We consider the systems $(\mathbf{S}_{IV})$ (see Lemma 6.1) with $\widetilde{M} = 0$ and $C_2 = x^3$, i.e. the triple infinite singularity is located at the point $N[0:1:0]$. Then applying a translation and a linear transformation (which keeps the point $N[0:1:0]$), as well as a time rescaling, we could locate the two real finite singularities at the points $M_1(0, 0)$ and $M_2(1, 1)$. As a result we get the family of systems:

$$\begin{aligned}
\dot{x} &= cx - (c + g + h)y + gx^2 + hxy,\\
\dot{y} &= (1 - f - g - h)x + fy - x^2 + gxy + hy^2.
\end{aligned} \tag{11.37}$$

Assume that the conditions (iv) (respectively, (v)) are satisfied for these systems. According to Theorem 6.1 in this case systems (11.37) possess a focus and a node with coinciding eigenvalues (respectively, two foci). Then without loss of generality we may assume that the singular point $M_2(1, 1)$ is a focus (i.e. the condition $\tau_2 < 0$ holds), whereas $M_1(0, 0)$ is either a focus or a node with coinciding eigenvalues (i.e. the condition $\tau_1 \leq 0$ holds). Since

$$\tau_1 = (c + f + 2g + 2h)^2 - 4(c + g + h),$$

we set a new parameter $u = c + f$ (i.e. $f = u - c$), and we impose the condition $\tau_1 = -v^2 \leq 0$, where v is a new parameter, too. The last equality gives $c = (g + h)^2 + (g + h)(u - 1) + (u^2 + v^2)/4$, and then we calculate

$$\mu_0 = -h^3, \ \tau_1 = -v^2, \ \tau_2 = 5g^2 + 2g(5h + 3u) - 4h + 5h^2 + 6hu + 2u^2 + v^2.$$

We observe that Discriminant$[\tau_2, g] = 4(20h - u^2 - 5v^2) < 0$ due to the condition $\mu_0 > 0$ (i.e. $h < 0$). Therefore $\tau_2 > 0$, i.e. M_2 could not be a focus, and this contradiction completes the proof of the statements (iv) and (v).

(vi) We consider systems (11.37). Since the conditions $W_4 = W_3 = 0$ imply the existence of two nodes with coinciding eigenvalues, we may assume that $M_1(0,0)$ and $M_2(1,1)$ are such points, i.e. the conditions

$$\tau_1 = (c + f + 2g + 2h)^2 - 4(c + g + h) = 0,$$
$$\tau_2 = (c + f + g + h)^2 + 4(c + g) = 0$$

must hold. Clearly the condition $c + g \leq 0$ has to be satisfied, and we set $c + g = -u^2$, i.e. $g = -(c+u^2)$, and this yields $\tau_2 = (f+h-2u-u^2)(f+h+2u-u^2) = 0$. Due to the change $u \to -u$, we may assume $f = u(2+u) - h$, and we calculate

$$\tau_1 = (c - h - 2u)^2 + u^2(u-2)^2 - 2h(2+u^2) + 2cu^2 = 0,$$
$$\text{Discrim}[\tau_1, c] = 16(h - u^2) \geq 0, \quad \mu_0 = -h^3 > 0.$$

Therefore the above conditions are incompatible, and this proves the statement (vi) as well as Lemma 11.17. □

Lemma 11.18. *If a quadratic system* (8.1) *possesses a node* N^d *at infinity, then it could not have one finite focus and a node with coinciding eigenvalues.*

Proof. Assuming that a system (8.1) possesses a node N^d at infinity, we can locate it via a rotation at the point $N_1[0:1:0]$ (i.e. we force the condition $k = 0$). Additionally suppose that we have a focus and a node with coinciding eigenvalues. Due to a translation we may consider the focus located at the origin of coordinates (i.e. at the point $M_1(0,0)$). Then the second singularity could not be located on the y-axis, otherwise this line becomes invariant, which contradicts the existence of the focus $M_1(0,0)$. Therefore the second singular point could be located at $M_2(1,0)$ due to a linear change of coordinates. Thus we arrive at the family of systems:

$$\dot{x} = cx + dy - cx^2 + 2hxy, \quad \dot{y} = ex + fy - ex^2 + 2mxy + ny^2, \tag{11.38}$$

for which the linear matrix corresponding to the infinite singularity $N_1[0:1:0]$ has the form:

$$\begin{pmatrix} h-n & d \\ 0 & -n \end{pmatrix}.$$

Clearly $N_1[0:1:0]$ is a node N^d if and only if $h = 0$ and $dn \neq 0$, and we may assume $d = 1$ due to a rescaling. Then for systems (11.38) with $h = 0$ and $d = 1$, we have $\tau_1 = c^2 + 4e - 2cf + f^2$. Since $M_1(0,0)$ is a focus, we set $\tau_1 = -u^2 < 0$, and we obtain $e = -\left[u^2 + (c-f)^2\right]/4$. Therefore we obtain

$$\tau_1 = -u^2 < 0, \quad \tau_2 = 2(c+m)^2 + 2(f+m)^2 + u^2 > 0,$$

due to $u \neq 0$. So $M_2(1,0)$ could not be a node with coinciding eigenvalues, and this contradiction proves the lemma. □

Lemma 11.19. *Assume that for a quadratic system possessing four real finite distinct singularities, the conditions* $\theta = \theta_1 = \mathcal{T}_4 = 0$ *and* $\eta > 0$ *hold. Then for this system*

(i) *the condition* $W_4 \neq 0$ *implies* $\mathcal{T}_3\mathcal{F}_1 \neq 0$,

(ii) *the conditions* $W_4 = 0$ *and* $\theta_3 \neq 0$ *imply* $\mathcal{T}_3 = \mathcal{F}_1 = W_3 = 0$.

Proof. Since $\eta > 0$ we consider the systems $(\mathbf{S}_I)$ (see Lemma 6.1), and applying a translation we locate one finite singularity at the origin of coordinates. Since $\mathcal{T}_4 = 0$ we may assume that $M_1(0,0)$ is a weak singularity, i.e. $\rho_1 = c + f = 0$, and this gives $f = -c$. So we arrive at the family of systems:

$$\dot{x} = cx + gx^2 + dy + (h-1)xy, \quad \dot{y} = ex - cy + (g-1)xy + hy^2,$$

for which we calculate

$$\theta = -8(g-1)(h-1)(g+h), \quad \theta_1 = 16\big[(g-1)(h-1) - (g-1)(g+h) - (h-1)(g+h)\big].$$

Clearly the condition $\theta = \theta_1 = 0$ implies the vanishing of two of the factors of θ, and we get $(g,h) \in \big\{(1,1),(1,-1),(-1,1)\big\}$. Due to the change $(x,y,c,d,e,g,h) \mapsto (y,x,-c,e,d,h,g)$, it is sufficient to consider $(g,h) \in \big\{(1,1),(1,-1)\big\}$.

In the case $g = h = 1$ we have

$$\begin{gathered}\mathcal{T}_3 = 8(d+e)\Delta_1, \quad \mathcal{F}_1 = 2c(d+e), \quad \theta_3 = -2de,\\ W_4 = 256de(d+e)^2\Delta_1^2, \quad W_3 = 64(d+e)^2\Delta_1^2, \quad \mathbf{D}|_{c=0} = 1296d^4e^4 > 0,\end{gathered}$$

whereas in the case $g = -h = 1$ we obtain

$$\begin{gathered}\mathcal{T}_3 = -8(2c+d)\Delta_1, \quad \mathcal{F}_1 = 2(2c+d)(c-e),\\ \theta_3 = -2e(2c+d-e), \quad W_4 = 256e(2c+d-e)(2c+d)^2\Delta_1^2,\\ W_3 = 64(2c+d)^2\Delta_1^2, \quad \mathbf{D}|_{c=e} = 1296e^4(d+e)^4 > 0.\end{gathered}$$

It is not difficult to observe that in both cases the implication (i) as well as the implication (ii) of the lemma are valid. □

Lemma 11.20. *Assume that a quadratic system possesses four real finite distinct singularities. Then the condition* $\mathcal{T}_4 = \mathcal{T}_3 = \theta = 0$ *implies* $W_4 \leq 0$.

Proof. Assume that for a quadratic system the conditions $\mathcal{T}_4 = \mathcal{T}_3 = 0$ and $\theta = 0$ are satisfied. We shall construct some new convenient canonical systems possessing the required number of infinite singularities as well as two fixed finite singular points. We will discuss in detail only one case, when at infinity there exists three distinct real singularities (i.e. when $\eta > 0$). The same arguments could be applied in the other three cases, defined by the condition $\eta \leq 0$.

So consider the case $\eta > 0$. Then due to a linear transformation we may assume $C_2 = xy(x+wy)$ with $w \neq 0$, and considering the relation $C_2 = yp_2(x,y) - xq_2(x,y)$, we detect that the homogenous parts $p_2(x,y)$ and $q_2(x,y)$ have the forms $p_2(x,y) = x(gx+ny+wy)$ and $q_2(x,y) = y(gx-x+ny)$. Additionally since we have four real finite singularities due to a translation and additional rescaling, we can locate two of the singularities at $M_1(0,0)$ and $M_2(1,1)$. So we get the 5-parameter family of systems:

$$\begin{aligned} \dot{x} &= cx - (c+g+n+w)y + gx^2 + (n+w)xy, \\ \dot{y} &= (1-f-g-n)x + fy + (g-1)xy + ny^2, \end{aligned} \tag{11.39}$$

for which we calculate $\rho_1 = c+f$ and $\rho_2 = -1+c+f+3g+3n+w$.

Following the same algorithm as above, we construct the canonical systems:

$$\begin{aligned} \dot{x} &= cx - (1+c+g+h)y + gx^2 + (1+h)xy, \\ \dot{y} &= (1-f-g-h)x + fy - x^2 + gxy + hy^2, \end{aligned} \tag{11.40}$$

(with $\rho_1 = c+f$ and $\rho_2 = 1+c+f+3g+3h$) in the case $\eta < 0$ and the systems:

$$\begin{aligned} \dot{x} &= cx - (c+g+h)y + gx^2 + hxy, \\ \dot{y} &= (1-f-g-h)x + fy + (g-1)xy + hy^2, \end{aligned} \tag{11.41}$$

(with $\rho_1 = c+f$ and $\rho_2 = -1+c+f+3g+3h$) in the case $\eta = 0$ and $\widetilde{M} \neq 0$.

We observe that for the case $\eta = \widetilde{M} = 0$, the canonical form was constructed earlier (see systems (11.37)), and for these systems the condition $\mathcal{T}_4 = \mathcal{T}_3 = 0$ implies the existence of two weak singularities. Then without loss of generality we may assume $\rho_1 = c+f = 0$ and $\rho_2 = c+f+3g+3h = 0$, which yields $f = -c$, $g = -h$. However in this case for systems (11.37) we calculate $\theta = 8h^3 = -8\mu_0 \neq 0$, i.e. we could not have $\theta = 0$.

Next we compute the invariant W_4 for each one of the above three systems.

1. *Systems* (11.39). The condition $\rho_1 = \rho_2 = 0$ implies $f = -c$, $w = 1 - 3g - 3n$, and then we calculate $\theta = 8(g-1)(3g+2n-1)(3g^2+3gn+n-g) = 0$. Therefore we obtain

$$\begin{aligned} W_4 &= -256(c-n)^2(1+n)^2(1+2n)^2(c-n+3cn-4n^2)^2(c-n+3cn-2n^2)^2 \\ &\quad \text{if } g = 1; \\ W_4 &= -(g^2-1)^2(c+g)^2(3g-1)^2(c+g+3cg+g^2)^2(c+g+3cg+5g^2)^2/4 \\ &\quad \text{if } n = (1-3g)/2; \\ W_4 &= -\frac{16(g^2-1)^2}{(1+3g)^{14}}(3g-1)^2(1+2c+6cg-g^2)^2(1+2c+6cg+3g^2)^2 \\ &\quad \times (1+2c+6cg+7g^2)^2 \quad \text{if } n = (g-3g^2)/(3g+1), \end{aligned}$$

as in the case $3g^2 - g + n(3g+1) = 0$ the condition $3g+1 \neq 0$ holds.

2. *Systems* (11.40). The condition $\rho_1 = \rho_2 = 0$ implies $f = -c$, $g = -(1 + 3h)/3$, and then we calculate $\theta = 16(1+h)(5-6h+9h^2)/9 = 0$. This yields $h = -1$, and we obtain

$$W_4 = -409600(2+3c)^2(4+9c)^2(8+9c)^2/4782969.$$

3. *Systems* (11.41). The condition $\rho_1 = \rho_2 = 0$ gives $f = -c$, $g = (1-3h)/3$, and then we calculate $\theta = 8h^2(2+3h)/3 = 0$ and $\mu_0 = h^2(1-3h)/3 \neq 0$. This implies $h = -2/3$, and we obtain

$$W_4 = -256(2+3c)^2(2+9c)^2(10+9c)^2/4782969.$$

Since for all three systems we have $W_4 \leq 0$, we conclude that Lemma 11.20 is proved. □

Lemma 11.21. *Assume that a quadratic system possesses four real finite distinct singularities. Then this system possesses exactly one star node if and only if the condition ($\mathcal{S}^1_{n^*}$) of Theorem 7.1 is satisfied. Moreover if this system possesses two star nodes, then the conditions $\mu_0 > 0$, $\eta > 0$, $\mathcal{T}_4 = \mathcal{T}_3 = \mathcal{F}_1 = \mathcal{F}_2 = \mathcal{F}_3 = \mathcal{F} = 0$ and $\theta = \theta_1 = \theta_3 = \theta_4 = 0$ are fulfilled.*

Proof. The first part of the lemma follows directly from Lemma 11.2.

Assume now that we have two star nodes. According to Theorem 7.2 we could have only one global configuration in this case, namely (α_{11}) $\$, \$, n^*, n^*; S, N^*, N^*$. As it was shown in the proof of Theorem 7.2, in this case the conditions (7.13) must be satisfied. From these conditions immediately follow the conditions stated in the lemma. This completes the proof of Lemma 11.21. □

Lemma 11.22. *Assume that a quadratic system possesses four real finite distinct singularities. Then the conditions from each one of the following sets are incompatible:*

(i) $\mu_0 > 0$, $W_4 > 0$, $(W_2 < 0) \vee (W_2 > 0, W_1W_3 < 0)$ *and* $\theta \leq 0$;

(ii) $\mu_0 > 0$, $W_4 > 0$, $(W_2 < 0) \vee (W_2 > 0, W_1W_3 < 0)$, $\mathcal{T}_4 = \mathcal{F}_1 = 0$, $\mathcal{T}_3\mathcal{F} > 0$ *and* $\eta = 0$;

(iii) $\mu_0 > 0$, $W_4 > 0$, $W_2 > 0$, $W_1W_3 > 0$, $\mathcal{T}_4 = \mathcal{T}_3 = 0$, $\eta \leq 0$ *and* $\theta \geq 0$;

(iv) $\mu_0 > 0$, $W_4 = 0$, $W_3 < 0$ *and* $\theta \leq 0$;

(v) $\mu_0 > 0$, $W_4 = 0$, $W_3 < 0$, $\mathcal{T}_4 = \mathcal{F}_1 = \eta = 0$ *and* $\theta > 0$;

(vi) $\mu_0 > 0$, $W_4 = W_3 = 0$ *and either* $\theta < 0$ *or* $\theta = 0$ *and* $\mathcal{T}_4^2 + \mathcal{T}_3^2 \neq 0$;

(vii) $\mu_0 > 0$, $W_4 > 0$, $W_2 > 0$, $W_1W_3 > 0$, $\mathcal{T}_4 = \mathcal{T}_3 = 0$, $\eta > 0$ *and* $\theta > 0$;

(viii) $\mu_0 < 0$, $W_4 > 0$, $W_2 > 0$ *and* $\mathcal{T}_4 = \widetilde{M} = 0$;

(ix) $\mu_0 > 0$, $W_4 = 0$, $W_3 < 0$, $\mathcal{T}_4 = \mathcal{F}_1 = 0$, $\mathcal{T}_3\mathcal{F} > 0$ *and* $\eta > 0$.

Proof. We have not been able to find a direct proof of these statements. However the Mathematica tools "Reduce", "Assuming" and/or "Find" reply for each one of the above sets of conditions that are incompatible. □

Conjecture 11.1. *Assume that a quadratic system possesses four finite distinct real singularities. Then the conditions* $\mu_0 > 0$, $W_4 > 0$, $(W_2 < 0) \vee (W_2 > 0$, $W_1 W_3 < 0)$, $\mathcal{T}_4 = \mathcal{F}_1 = 0$, $\mathcal{T}_3\mathcal{F} > 0$ *and* $\eta > 0$ *are incompatible.*

The reason why we leave this statement as a conjecture is that the computer algebra system Mathematica (and other systems) collapses by lack of sufficient memory when trying to prove this incompatibility. We mention that a massive numerical search for an example with these conditions has given no result. We note that if these conditions could be realizable, then this would lead to a configuration of finite singularities $s, s^{(2)}, f, f$ with three real infinite singularities. But since by Theorem 7.5 a system with a star node at infinite and some special singularity, cannot occur with a weak saddle of order two, the total number of possibilities that appear if the conjecture is false is at most eight.

Next we examine the global configurations of singularities of the family of systems under consideration.

1: $\mu_0 < 0$. We consider two possibilities: $\widetilde{K} < 0$ and $\widetilde{K} > 0$.

1.1: $\widetilde{K} < 0$. In this case by Theorem 6.1 (see Table 6.2, lines 1 and 2) systems (8.1) possess three saddles and one anti-saddle. This implies the existence of three nodes at infinity; more precisely, according to Lemma 11.1, at infinity we have N^f, N^f, N^f, i.e. the condition $\eta > 0$ holds.

1.1.1: $W_4 < 0$. According to Theorem 6.1, apart from the saddles, we have a focus or a center.

1. $\mathcal{T}_4 \neq 0$. By Theorem 6.2 the systems could not have weak singularities. So we get the global configuration
   - $s, s, s, f;\ N^f, N^f, N^f \Rightarrow (0, 0, 1, 0, -2, -1), (0, 2, -2, -2, 0, 2)$.
2. $\mathcal{T}_4 = 0$, $\mathcal{T}_3 \neq 0$. According to Theorem 6.2 we have one weak singularity that is a focus (or a center) if $\mathcal{T}_3\mathcal{F} < 0$ and a saddle if $\mathcal{T}_3\mathcal{F} > 0$.
   (a) $\mathcal{T}_3\mathcal{F} < 0$. So we have a weak focus, and its order is defined by the invariant polynomials $\mathcal{F}_i$, $i \in \{1, 2, 3, 4\}$. Thus we arrive at the following four global configurations:
   - $s, s, s, f^{(1)};\ N^f, N^f, N^f \Rightarrow (0, -1, 1, 1, 2, -1), (0, -2, 1, 2, 0, -1)$ (if $\mathcal{F}_1 \neq 0$);
   - $s, s, s, f^{(2)};\ N^f, N^f, N^f \Rightarrow (0, 0, -1, -1, 1/2, 1), (0, 1, 0, 1, 1, 0)$ (if $\mathcal{F}_1 = 0$, $\mathcal{F}_2 \neq 0$);
   - $s, s, s, f^{(3)};\ N^f, N^f, N^f \Rightarrow (0, 0, -1, -1, 5/2, 1), (0, 1, 0, 1, 1, 0)$ (if $\mathcal{F}_1 = \mathcal{F}_2 = 0$, $\mathcal{F}_3\mathcal{F}_4 \neq 0$);
   - $s, s, s, c;\ N^f, N^f, N^f \Rightarrow (0, 0, -1, 0, 1, 0), (0, 1, 0, 1, 1/2, -1)$ (if $\mathcal{F}_1 = \mathcal{F}_2 = \mathcal{F}_3\mathcal{F}_4 = 0$).

(b) $\mathcal{T}_3\mathcal{F} > 0$. In accordance with Theorem 6.2 we have a weak saddle. According to Theorems 7.9 and 7.3, in this case we could have neither an integrable saddle nor a weak saddle of order three. So we get the following two global configurations:
- $s, s, s^{(1)}, f;\ N^f, N^f, N^f \Rightarrow (0,-1,0,1,2,0), (0,-2,1,2,0,-1)$ (if $\mathcal{F}_1 \neq 0$);
- $s, s, s^{(2)}, f;\ N^f, N^f, N^f \Rightarrow (0,1,0,2,-3,3), (0,0,-1,0,-3/2,4)$ (if $\mathcal{F}_1 = 0$).

3. $\mathcal{T}_4 = \mathcal{T}_3 = 0$, $\mathcal{T}_2 \neq 0$. According to Theorem 6.2 we could have two, and only two, weak singularities. Additionally by Theorem 7.3, in the case under consideration, quadratic systems could not have two integrable saddles. Therefore we arrive at the following three global configurations:

   - $s, s^{(1)}, s^{(1)}, f;\ N^f, N^f, N^f \Rightarrow (0,1,0,-4,3/2,0), (0,0,-1,-5,7/2,-1)$ (if $\mathcal{F}_1 \neq 0$, $\mathcal{T}_2 < 0$);
   - $s, s, s^{(1)}, f^{(1)};\ N^f, N^f, N^f \Rightarrow (0,1,0,1/4,-7/8,1/2), (0,0,-1,-1/4,5/8,0)$ (if $\mathcal{F}_1 \neq 0$, $\mathcal{T}_2 > 0$);
   - $s, s, \$, c;\ N^f, N^f, N^f \Rightarrow (0,1,0,0,-5/8,1/4), (0,0,-1,-1/4,5/8,0)$ (if $\mathcal{F}_1 = 0$).

4. $\mathcal{T}_4 = \mathcal{T}_3 = \mathcal{T}_2 = 0$. By Lemma 11.16 we obtain Hamiltonian systems, and by Theorem 7.3 we arrive at the global configuration

   - $\$, \$, \$, c;\ N^f, N^f, N^f \Rightarrow (0,1,0,-3,1,0), (0,0,-1,-4,3,-1)$.

1.1.2: $W_4 > 0$. According to Theorem 6.1 (see Table 6.2, line 1), apart from the three saddles, we have a node.

1. $\mathcal{T}_4 \neq 0$. By Theorem 6.2 the systems could not have weak singularities, and we get the global configuration

   - $s, s, s, n;\ N^f, N^f, N^f \Rightarrow (0,0,1,0,-2,-1), (0,3,-1,-3,0,1)$.

2. $\mathcal{T}_4 = 0$, $\mathcal{T}_3 \neq 0$. According to Theorem 6.2 we have one weak singularity that evidently is a saddle, and its order is defined by the invariant polynomials $\mathcal{F}_i$, $i \in \{1,2,3,4\}$. We remark that by Theorems 7.9 and 7.3 we neither could have an integrable saddle nor an $s^{(3)}$.

   Thus we arrive at the following two global configurations:

   - $s, s, s^{(1)}, n;\ N^f, N^f, N^f \Rightarrow (0,-1,-2,1,2,2), (0,-2,1,2,1,-1)$ (if $\mathcal{F}_1 \neq 0$);
   - $s, s, s^{(2)}, n;\ N^f, N^f, N^f \Rightarrow (0,1,0,0,1/2,0), (0,0,-1,1,0,-2)$ (if $\mathcal{F}_1 = 0$).

3. $\mathcal{T}_4 = \mathcal{T}_3 = 0$. We observe that in this case $\mathcal{T}_2 \neq 0$, otherwise by Lemma 11.16 the systems become Hamiltonian, which is impossible due to the existence of a node. Therefore according to Theorem 6.2 we could have two, and only two, weak singularities that clearly must be saddles. Since by Theorem 7.3

we cannot have in this case two integrable saddles, we arrive at the global configuration

- $s, s^{(1)}, s^{(1)}, n$; $N^f, N^f, N^f \Rightarrow (0,-1,0,1,2/5,0),(0,0,1,0,-1,-1)$.

1.1.3: $W_4 = 0$. According to Theorem 6.1, apart from the three saddles, we have a node with coinciding eigenvalues.

1. $\mathcal{T}_4 \neq 0$. By Theorem 6.2 the systems could not have weak singularities. So considering Lemma 11.21 we get the following two global configurations:

   - s, s, s, n^d; $N^f, N^f, N^f \Rightarrow (0,1,-1,-1,1,1),(0,0,1,0,1,-1)$ (if $\neg(\mathcal{S}^1_{n^*})$);
   - s, s, s, n^*; $N^f, N^f, N^f \Rightarrow (0,1,0,-1,1,0),(0,0,1,0,1,-1)$ (if $(\mathcal{S}^1_{n^*})$).

2. $\mathcal{T}_4 = 0$, $\mathcal{T}_3 \neq 0$. In this case we could not have a star node with a weak saddle (see Theorem 7.2). Additionally by Theorems 7.9 and 7.3 we could have neither a weak saddle of order three nor an integrable saddle. So we arrive at the following two global configurations:

   - $s^{(1)}, s, s, n^d$; $N^f, N^f, N^f \Rightarrow (0,1,-1,-1,0,1),(0,0,1,0,1,-1)$ (if $\mathcal{F}_1 \neq 0$);
   - $s^{(2)}, s, s, n^d$; $N^f, N^f, N^f \Rightarrow (0,2,-9,-2,-3,9),\big(0,1,-4,-1,(59-\sqrt{7913}\,)/24,4\big)$ (if $\mathcal{F}_1 = 0$).

3. $\mathcal{T}_4 = \mathcal{T}_3 = 0$. As it was mentioned earlier, in this case $\mathcal{T}_2 \neq 0$, otherwise by Lemma 11.16 systems become Hamiltonian, which is impossible due to the existence of a node. Therefore according to Theorem 6.2 we could have two, and only two, weak singularities that clearly must be saddles. Since by Theorem 7.3 we cannot have in this case two integrable saddles, we arrive at the global configuration

   - $s, s^{(1)}, s^{(1)}, n^d$; $N^f, N^f, N^f \Rightarrow (0,-2,1,2,1,-1),(0,-1,0,1,-1,0)$.

1.2: $\widetilde{K} > 0$. In this case systems (8.1) possess one saddle and three anti-saddles.

1.2.1: $W_4 < 0$. According to Theorem 6.1 (see Table 6.2, line 15), apart from the saddle, we have a focus (or a center) and two nodes (which are generic).

1.2.1.1: The case $\mathcal{T}_4 \neq 0$. By Theorem 6.2 the systems could not have weak singularities. We observe that the total index of infinite singularities must be -1, and therefore considering Lemma 11.1 we obtain the following four global configurations:

- s, n, n, f; S, ©, © $\Rightarrow (0,0,1,0,2,-1),(0,2,-1,-2,0,1)$ (if $\eta < 0$);
- s, n, n, f; S, S, $N^\infty \Rightarrow (0,0,1,0,4,-1),(0,2,-1,-2,4,1)$ (if $\eta > 0$);
- s, n, n, f; $\overline{\binom{0}{2}}SN$, $S \Rightarrow (0,2,2,-2,-1,0),(0,6,-1,0,-3/2,-2)$ (if $\eta = 0$, $\widetilde{M} \neq 0$);
- s, n, n, f; $\overline{\binom{0}{3}}S \Rightarrow (0,-1,1,-1,1/2,0),(0,2,-1,-1,-1/2,1)$ (if $\eta = 0$, $\widetilde{M} = 0$).

1.2.1.2: $\mathcal{T}_4 = 0$, $\mathcal{T}_3 \neq 0$. By Theorem 6.2 the systems have one weak singularity that is a saddle (if $\mathcal{T}_3\mathcal{F} > 0$) or a focus (or a center) (if $\mathcal{T}_3\mathcal{F} < 0$). Moreover its order is defined by the invariant polynomials $\mathcal{F}_i$, $i \in \{1,2,3,4\}$.

1. $\mathcal{T}_3\mathcal{F} < 0$. In accordance with Theorem 6.2 we have a weak focus or a center.

   (a) $\mathcal{F}_1 \neq 0$. Then we have a weak focus of order one, and considering Lemma 11.1 we obtain the following four global configurations:
   - $s, n, n, f^{(1)};\ S,\ ©, © \Rightarrow (0,-1,3,1,-1,-3), (0,-2,1,2,1,-1)$ (if $\eta < 0$);
   - $s, n, n, f^{(1)};\ S,\ S,\ N^{\infty} \Rightarrow (0,-1,1,1,-3,-1), (0,-2,1,2,4,-1)$ (if $\eta > 0$);
   - $s, n, n, f^{(1)};\ \overline{\binom{0}{2}}SN,\ S \Rightarrow (0,2,-2,-1,1/2,0), (0,3,-2,0,-1,1)$ (if $\eta = 0$, $\widetilde{M} \neq 0$);
   - $s, n, n, f^{(1)};\ \overline{\binom{0}{3}}S \Rightarrow (0,1,-5/2,-3/2,3/2,0), (0,1/2,-1,-1,-3/4,3)$ (if $\eta = 0$, $\widetilde{M} = 0$).

   (b) $\mathcal{F}_1 = 0, \mathcal{F}_2 \neq 0$. In this case we have a weak focus of order two. Since by Theorem 7.7 we could not have in this case $\widetilde{M} = 0$, we arrive at the next three global configurations:
   - $s, n, n, f^{(2)};\ S,\ ©, © \Rightarrow (0,0,-1,2/25,4/5,1), (0,1,0,-6/5,-4/5,0)$ (if $\eta < 0$);
   - $s, n, n, f^{(2)};\ S,\ S,\ N^{\infty} \Rightarrow (0,0,-1,-19/5,3/5,1), (0,1,0,2/5,-2/5,0)$ (if $\eta > 0$);
   - $s, n, n, f^{(2)};\ \overline{\binom{0}{2}}SN,\ S \Rightarrow (0,0,-1,-5/6,1/2,1), (0,1,0,1/2,-5/12,-1/2)$ (if $\eta = 0$).

   (c) $\mathcal{F}_1 = \mathcal{F}_2 = 0, \mathcal{F}_3\mathcal{F}_4 \neq 0$. According to Theorem 7.8 in this case we get the unique global configuration
   - $s, n, n, f^{(3)};\ S,\ S,\ N^{\infty} \Rightarrow (0,0,-1,-7/4,1/4,1), (0,1,0,1/4,-1/8,-1/4)$.

   (d) $\mathcal{F}_1 = \mathcal{F}_2 = \mathcal{F}_3\mathcal{F}_4 = 0$. In this case we have a center, and according to Theorem 7.4 we arrive at the global configuration
   - $s, n, n, c;\ S,\ S,\ N^{\infty} \Rightarrow (0,0,-1,-1,1,1), (0,1,0,1,-5/4,-1)$.

2. $\mathcal{T}_3\mathcal{F} > 0$. By Theorem 6.2 we have a weak saddle.

   (a) $\mathcal{F}_1 \neq 0$. Then the weak saddle has order one, and considering Lemma 11.1 we obtain the following four global configurations:
   - $s^{(1)}, n, n, f;\ S,\ ©, © \Rightarrow (0,1,0,1,-1/2,-1), (0,0,-1,-1,1,0)$ (if $\eta < 0$);
   - $s^{(1)}, n, n, f;\ S,\ S,\ N^{\infty} \Rightarrow (0,1,0,2,-1,-1), (0,0,-1,0,1/2,0)$ (if $\eta > 0$);
   - $s^{(1)}, n, n, f;\ \overline{\binom{0}{2}}SN,\ S \Rightarrow (0,-1,1,-1,1/2,0), (0,0,1,0,-1,1)$ (if $\eta = 0$, $\widetilde{M} \neq 0$);

- $s^{(1)}, n, n, f;\ \overline{\binom{0}{3}}S \Rightarrow (0,-2,-1,1,1,0), (0,-4,2,-1,1/2,2)$ (if $\eta = 0,\ \widetilde{M} = 0$).

(b) $\mathcal{F}_1 = 0, \mathcal{F}_2 \neq 0$. In this case we have a weak saddle of order two. So we arrive at the next four global configurations:

- $s^{(2)}, n, n, f;\ S,$ ©, © $\Rightarrow (0,1,0,0,-1,1), (0,0,-1,2,0,-1)$ (if $\eta < 0$);
- $s^{(2)}, n, n, f;\ S,\ S,\ N^{\infty} \Rightarrow (0,1,0,-2,1/2,0), (0,0,-1,9/2,-2,1/2)$ (if $\eta > 0$);
- $s^{(2)}, n, n, f;\ \overline{\binom{0}{2}}SN,\ S \Rightarrow (0, 1+\sqrt{2}/2, -\sqrt{2}/2, -2, 1/2, 0), (0, 3+\sqrt{2}/2, -1-\sqrt{2}/2, 0, -3/2, 1)$ (if $\eta = 0,\ \widetilde{M} \neq 0$);
- $s^{(2)}, n, n, f;\ \overline{\binom{0}{3}}S \Rightarrow (0,-3,0,2,1/2,0), (0,-5,3,-1,1,1)$ (if $\eta = 0,\ \widetilde{M} = 0$).

(c) $\mathcal{F}_1 = \mathcal{F}_2 = 0$. We observe that in this case the condition $\mathcal{F}_3\mathcal{F}_4 \neq 0$ is satisfied, since by Theorem 7.3 we could not have an integrable saddle with two nodes and a focus. Moreover, by Theorem 7.9 we could have the following three global configurations:

- $s^{(3)}, n, n, f;\ S,$ ©, © $\Rightarrow (0,1,0,-11/10,0,1/10), (0,0,-1,31/20, -11/40,0)$ (if $\eta < 0$);
- $s^{(3)}, n, n, f;\ S,\ S,\ N^{\infty} \Rightarrow (0,1,0,9/5,0,-14/5), (0,0,-1,1/10,9/20,0)$ (if $\eta > 0$);
- $s^{(3)}, n, n, f;\ \overline{\binom{0}{2}}SN,\ S \Rightarrow (0,1,0,0,1/2,1), (0,0,-1,-4/27,0,2)$ (if $\eta = 0$).

3. $\mathcal{T}_4 = \mathcal{T}_3 = 0$. Since we have two nodes, we could not have $\mathcal{T}_2 = 0$. So according to Theorem 6.2 we could have two, and only two, weak singularities, which clearly must be a focus (or a center) and a saddle.

(a) $\mathcal{F}_1 \neq 0$. Then both weak singularities have order one. Since by Lemma 11.17(ii) we could not have a triple infinite singularity, considering Lemma 11.1 we obtain the following three global configurations:

- $s^{(1)}, n, n, f^{(1)};\ S,$ ©, © $\Rightarrow (0,1,0,3,-1,-2), (0,0,-1,3,0,-2)$ (if $\eta < 0$);
- $s^{(1)}, n, n, f^{(1)};\ S,\ S,\ N^{\infty} \Rightarrow (0,1,0,3,-3,2), (0,0,-1,0,1,-1)$ (if $\eta > 0$);
- $s^{(1)}, n, n, f^{(1)};\ \overline{\binom{0}{2}}SN,\ S \Rightarrow (0,-1,2/3,-1,2/3,0), (0,-1/3,1,0,-1,4/3)$ (if $\eta = 0$).

(b) $\mathcal{F}_1 = 0$. In this case according to Theorem 6.2 we have an integrable saddle and a center. So by Theorem 7.3 we arrive at the next three global configurations:

- $\$, n, n, c;\ S,$ ©, © $\Rightarrow (0,1,0,1,-1/2,-1), (0,0,-1,1,1/2,-1)$ (if $\eta < 0$);
- $\$, n, n, c;\ S,\ S,\ N^{\infty} \Rightarrow (0,1,0,1,-1,0), (0,0,-1,0,1,-1)$ (if $\eta > 0$);

- $\$, n, n, c;\ \overline{\binom{0}{3}}S \Rightarrow (0,-1,1,-1,1/2,0),(0,0,1,-1,-1/2,1)$ (if $\eta = 0 \ \Rightarrow \widetilde{M} = 0$).

1.2.2: $W_4 > 0$, $(W_2 < 0) \cup (W_2 \geq 0, W_1W_3 < 0)$. According to Theorem 6.1 (see Table 6.2, lines 20 and 21), apart from the saddle, we have a node and two foci.

1.2.2.1: $\mathcal{T}_4 \neq 0$. By Theorem 6.2 the systems could not have weak singularities. So considering Lemma 11.1 we obtain four global configurations:

- $s, n, f, f;\ S,\ ©, © \Rightarrow (0,0,1,0,1,-1),(0,2,-1,-2,0,1)$ (if $\eta < 0$);
- $s, n, f, f;\ S,\ S,\ N^\infty \Rightarrow (0,0,1,0,3,-1),(0,2,-1,-2,3,1)$ (if $\eta > 0$);
- $s, n, f, f;\ \overline{\binom{0}{2}}SN,\ S \Rightarrow (0,0,1,-1,1/2,0),(0,1,-1,0,-1,1)$ (if $\eta = 0$, $\widetilde{M} \neq 0$);
- $s, n, f, f;\ \overline{\binom{0}{3}}S \Rightarrow (0,1,-2,0,1/2,0),(0,-1,1,-1,0,1)$ (if $\eta = 0$, $\widetilde{M} = 0$).

1.2.2.2: $\mathcal{T}_4 = 0$, $\mathcal{T}_3 \neq 0$. By Theorem 6.2 the systems have one weak singularity that is a saddle (if $\mathcal{T}_3\mathcal{F} > 0$) or a focus (if $\mathcal{T}_3\mathcal{F} < 0$). Moreover its order is defined by the invariant polynomials $\mathcal{F}_i$, $i \in \{1,2,3,4\}$.

1.2.2.2.1: $\mathcal{T}_3\mathcal{F} < 0$. In accordance with Theorem 6.2 we have a weak focus.

1. $\mathcal{F}_1 \neq 0$. In this case the weak focus has order one, and considering Lemma 11.1 we obtain the following four global configurations:
   - $s, n, f, f^{(1)};\ S,\ ©, © \Rightarrow (0,-1,1,1,-1,-1),(0,-2,1,2,0,-1)$ (if $\eta < 0$);
   - $s, n, f, f^{(1)};\ S,\ S,\ N^\infty \Rightarrow (0,-1,1,1,-5,-1),(0,-2,1,2,-4,-1)$ (if $\eta > 0$);
   - $s, n, f, f^{(1)};\ \overline{\binom{0}{2}}SN,\ S \Rightarrow (0,3,-2,-2,1/2,0),(0,5,-3,0,-3/2,1)$ (if $\eta = 0$, $\widetilde{M} \neq 0$);
   - $s, n, f, f^{(1)};\ \overline{\binom{0}{3}}S \Rightarrow (0,1,-2,0,1/2,0),(0,1,-1,-1,0,1)$ (if $\eta = 0$, $\widetilde{M} = 0$).
2. $\mathcal{F}_1 = 0$. According to Theorem 7.8 (respectively, Theorem 7.4) in this case we could not have a weak focus of order three (respectively, a center). So we have a weak focus of order two, and by Theorem 7.7 we get the unique global configuration
   - $s, n, f, f^{(2)};\ S,\ ©, © \Rightarrow (0,0,-1,0,1/2,1),(0,1,0,-1/2,-1,0)$.

1.2.2.2.2: $\mathcal{T}_3\mathcal{F} > 0$. In this case the weak singularity is the saddle.

1. $\mathcal{F}_1 \neq 0$. Then the weak saddle has order one, and considering Lemma 11.1 we obtain the following four global configurations:
   - $s^{(1)}, n, f, f;\ S,\ ©, © \Rightarrow (0,-1,1,1,-1,-1),(0,0,1,0,1,-1)$ (if $\eta < 0$);
   - $s^{(1)}, n, f, f;\ S,\ S,\ N^\infty \Rightarrow (0,0,5/4,-2,3/8,0),(0,2,0,0,-3/2,1)$ (if $\eta > 0$);

- $s^{(1)}, n, f, f; \binom{\overline{0}}{2}SN, S \Rightarrow (0,0,1,-2,1/2,0), (0,2,0,0,-3/2,1)$ (if $\eta = 0$, $\widetilde{M} \neq 0$);
- $s^{(1)}, n, f, f; \binom{\overline{0}}{3}S \Rightarrow (0,-2,1,0,1/2,0), (0,-2,2,-1,0,1)$ (if $\eta = 0$, $\widetilde{M} = 0$).

2. $\mathcal{F}_1 = 0$. According to Theorems 7.9 and 7.3, in this case we could have neither a weak saddle of order three nor an integrable saddle. So we have a weak saddle of order two, and considering Lemma 11.1 we get the following four global configurations:
   - $s^{(2)}, n, f, f; S, ©, © \Rightarrow (0,1,0,2,-5/2,2), (0,0,-1,10,-5,1)$ (if $\eta < 0$);
   - $s^{(2)}, n, f, f; S, S, N^{\infty} \Rightarrow (0,1,0,0,-1/2,0), (0,0,-1,5/2,0,-3/2)$ (if $\eta > 0$);
   - $s^{(2)}, n, f, f; \binom{\overline{0}}{2}SN, S \Rightarrow (0,1-\sqrt{2}/2,\sqrt{2}/2,-2,1/2,0)$, $(0,3-\sqrt{2}/2,-1+\sqrt{2}/2,0,-3/2,1)$ (if $\eta = 0$, $\widetilde{M} \neq 0$);
   - $s^{(2)}, n, f, f; \binom{\overline{0}}{3}S \Rightarrow (0,-1,0,0,1/2,0), (0,-1,1,-1,0,1)$ (if $\eta = 0$, $\widetilde{M} = 0$).

1.2.2.2.3: $\mathcal{T}_3 = 0$. By Theorem 6.2 we have two weak singularities. More exactly (see statement (c)), we have two weak foci if $\mathcal{T}_2 < 0$, and we have a weak saddle and a weak focus if $\mathcal{T}_2 > 0$. We observe that by Theorem 7.4 in this case we could have neither one nor two centers. Therefore the condition $\mathcal{F}_1 \neq 0$, holds and according to Lemma 11.17(ii) this implies $\widetilde{M} \neq 0$. Thus we get the following six global configurations:

- $s, n, f^{(1)}, f^{(1)}; S, ©, © \Rightarrow (0,0,-1,-1,0,1), (0,1,0,5,-3,0)$ (if $\mathcal{T}_2 < 0$, $\eta < 0$);
- $s, n, f^{(1)}, f^{(1)}; S, S, N^{\infty} \Rightarrow (0,0,-1,-32/5,0,1), (0,1,0,4,-4/5,0)$ (if $\mathcal{T}_2 < 0$, $\eta > 0$);
- $s, n, f^{(1)}, f^{(1)}; \binom{\overline{0}}{2}SN, S \Rightarrow (0,-2/25,-19/75,-1/25,14/75,0)$, $(0,44/75,2/25,0,-13/25,28/75)$ (if $\mathcal{T}_2 < 0$, $\eta = 0$);
- $s^{(1)}, n, f, f^{(1)}; S, ©, © \Rightarrow (0,0,-1,-2/5,0,1), (0,1,0,2/5,-2/5,0)$ (if $\mathcal{T}_2 > 0$, $\eta < 0$);
- $s^{(1)}, n, f, f^{(1)}; S, S, N^{\infty} \Rightarrow (0,0,-1,-16/5,0,1), (0,1,0,6/5,-2/5,0)$ (if $\mathcal{T}_2 > 0$, $\eta > 0$);
- $s^{(1)}, n, f, f^{(1)}; \binom{\overline{0}}{2}SN, S \Rightarrow (0,-1/4,-1/12,-1/4,7/24,0)$, $(0,5/12,1/4,0,-5/8,7/12)$ (if $\mathcal{T}_2 > 0$, $\eta = 0$).

1.2.3: $W_4 > 0$, $W_2 > 0$ and $W_1W_3 > 0$. In this case apart from the saddle, we have three nodes that all are generic due to $W_4 \neq 0$.

1.2.3.1: $\mathcal{T}_4 \neq 0$. Then by Theorem 6.2 the saddle is strong, and considering Lemma 11.1 we obtain the following four global configurations:

- $s, n, n, n;\ S,$ ©, © $\Rightarrow (0,0,1,0,8/5,-1), (0,2,-2,-2,-2,2)$ (if $\eta < 0$);
- $s, n, n, n;\ S,\ S,\ N^\infty \Rightarrow (0,0,1,0,3/2,-1), (0,2,-2,-2,-2,2)$ (if $\eta > 0$);
- $s, n, n, n;\ \overline{\binom{0}{2}}SN,\ S \Rightarrow (0,-1,1,-1,1/2,0), (0,2,-1,0,-1,1)$ (if $\eta = 0,\ \widetilde{M} \neq 0$);
- $s, n, n, n;\ \overline{\binom{0}{3}}S \Rightarrow (0,2,-2,-1,1/2,0), (0,0,1,-1,-1/2,1)$ (if $\eta = 0,\ \widetilde{M} = 0$).

1.2.3.2: $\mathcal{T}_4 = 0$. Then the saddle is weak, and its order is governed by the invariant polynomials $\mathcal{F}_i$ $(i = 1,2,3,4)$.

1. $\mathcal{F}_1 \neq 0$. By Theorem 6.2 the weak saddle has order one, and considering Lemma 11.22(viii) we deduce that in this case the condition $\widetilde{M} \neq 0$ holds. So by Lemma 11.1 we arrive at the following three global configurations:
   - $s^{(1)}, n, n, n;\ S,$ ©, © $\Rightarrow (0,-1,-2/3,1,-1,2/3), (0,-1/3,1,1/3,1,-1)$ (if $\eta < 0$);
   - $s^{(1)}, n, n, n;\ S,\ S,\ N^\infty \Rightarrow (0,-1,-2/3,1,-1,2/3), (0,0,1,0,1,-1)$ (if $\eta > 0$);
   - $s^{(1)}, n, n, n;\ \overline{\binom{0}{2}}SN,\ S \Rightarrow (0,1/5,1,-1/5,-1/2,0), (0,12/5,-1/5,0,-3/5,-1)$ (if $\eta = 0$).
2. $\mathcal{F}_1 = 0,\ \mathcal{F}_2 \neq 0$. Then the weak saddle has order two, and by Lemma 11.22(viii) in this case the condition $\widetilde{M} \neq 0$ holds. So we get the next three global configurations:
   - $s^{(2)}, n, n, n;\ S,$ ©, © $\Rightarrow (0,1,0,2,-13/10,-2/5), (0,0,-1,-32/15,13/6,-6/5)$ (if $\eta < 0$);
   - $s^{(2)}, n, n, n;\ S,\ S,\ N^\infty \Rightarrow (0,1,0,-1/2,-1/4,0), (0,0,-1,8/3,-1/12,-3/2)$ (if $\eta > 0$);
   - $s^{(2)}, n, n, n;\ \overline{\binom{0}{2}}SN,\ S \Rightarrow (0,1,0,-2/3,-3/2,5/9), (0,0,-1,-1/9,-1/2,-2)$ (if $\eta = 0$).
3. $\mathcal{F}_1 = \mathcal{F}_2 = 0$. In this case we have either a weak saddle of order three or an integrable saddle. According to Theorems 7.9 and 7.3, we arrive at the next two global configurations:
   - $s^{(3)}, n, n, n;\ S,\ S,\ N^\infty \Rightarrow (0,0,1,0,-2,-7/5), (0,1,0,2,7/2,2)$ (if $\mathcal{F}_3\mathcal{F}_4 \neq 0$);
   - $\$, n, n, n;\ S,\ S,\ N^\infty \Rightarrow (0,1,0,0,-1/2,0), (0,0,-1,3,0,-2)$ (if $\mathcal{F}_3\mathcal{F}_4 = 0$).

1.2.4: $W_4 = 0$. In this case we have at least one node with coinciding eigenvalues.

1.2.4.1: $W_3 < 0$. Then apart from the saddle, we have one focus (or a center) and two nodes, one of them being n^d or n^*. However by Theorem 7.2 a star node could not coexist with a focus (or a center). So we could have only a node n^d.

1.2.4.1.1: $\mathcal{T}_4 \neq 0$. Therefore by Theorem 6.2 we do not have weak singularities. Considering Lemma 11.1 we obtain the following four global configurations:

- $s, n, n^d, f;\ S,\ ©, © \Rightarrow (0,1,-2,-1,1,2), (0,0,1,0,-2,-1)$ (if $\eta < 0$);
- $s, n, n^d, f;\ S,\ S,\ N^\infty \Rightarrow (0,1,-2,-1,-3,2), (0,0,1,0,-2,-1)$ (if $\eta > 0$);
- $s, n, n^d, f;\ \overline{\binom{0}{2}}SN,\ S \Rightarrow (0,3,-1,-3,1/2,0), (0,0,3,0,-2,1)$ (if $\eta = 0,\ \widetilde{M} \neq 0$);
- $s, n, n^d, f;\ \overline{\binom{0}{3}}S \Rightarrow (0,2,-1,-2,1/2,0), (0,0,2,-1,-1,1)$ (if $\eta = 0,\ \widetilde{M} = 0$).

1.2.4.1.2: $\mathcal{T}_4 = 0$, $\mathcal{T}_3 \neq 0$. By Theorem 6.2 the systems have one weak singularity that is a saddle (if $\mathcal{T}_3\mathcal{F} > 0$) or a focus (if $\mathcal{T}_3\mathcal{F} < 0$). Moreover its order is defined by the invariant polynomials $\mathcal{F}_i$, $i \in \{1,2,3,4\}$.

1. $\mathcal{T}_3\mathcal{F} < 0$. So we have a weak focus.

   (α) If $\mathcal{F}_1 \neq 0$ the weak focus has order one, and taking into account Lemma 11.1, we arrive at the following four global configurations:

   - $s, n, n^d, f^{(1)};\ S,\ ©, © \Rightarrow (0,1,-1,-1,0,1), (0,0,1,0,-2,-1)$ (if $\eta < 0$);
   - $s, n, n^d, f^{(1)};\ S,\ S,\ N^\infty \Rightarrow (0,1,-8/5,-1,0,8/5), (0,0,1,0,-2/5,-1)$ (if $\eta > 0$);
   - $s, n, n^d, f^{(1)};\ \overline{\binom{0}{2}}SN,\ S \Rightarrow (0,2,-1,-4,3/2,0), (0,0,2,0,-5/2,3)$ (if $\eta = 0,\ \widetilde{M} \neq 0$);
   - $s, n, n^d, f^{(1)};\ \overline{\binom{0}{3}}S \Rightarrow (0,-3/4,-1/4,-1,1,0), (0,9/4,-9/4,-1,-1/2,2)$ (if $\eta = 0,\ \widetilde{M} = 0$).

   (β) If $\mathcal{F}_1 = 0$ then according to Theorems 7.8 and 7.4, we could not have a weak focus of order three or a center (i.e. the condition $\mathcal{F}_2 \neq 0$ must hold). So we have a weak focus of order two, and by Theorem 7.7 we get the unique global configuration

   - $s, n, n^d, f^{(2)};\ S,\ ©, © \Rightarrow (0,0,-1,-(9+2\xi)/6,1,1), (0,1,0,1,(4\xi-9)/12,-1)$, where $\{W_4^{-1}(0)\} \ni \xi \approx -1.6284$.

2. $\mathcal{T}_3\mathcal{F} > 0$. In this case we have a weak saddle.

   (α) If $\mathcal{F}_1 \neq 0$ the weak saddle has order one, and taking into account Lemma 11.1 we get the next four global configurations:

   - $s^{(1)}, n, n^d, f;\ S,\ ©, © \Rightarrow (0,1,1,-1,2,-1), (0,4,-1,-4,0,1)$ (if $\eta < 0$);
   - $s^{(1)}, n, n^d, f;\ S,\ S,\ N^\infty \Rightarrow (0,-1,2,1,-2,-2), (0,0,1,0,1,-1)$ (if $\eta > 0$);
   - $s^{(1)}, n, n^d, f;\ \overline{\binom{0}{2}}SN,\ S \Rightarrow (0,1,0,-2,1/2,0), (0,3,-1,0,-3/2,1)$ (if $\eta = 0,\ \widetilde{M} \neq 0$);
   - $s^{(1)}, n, n^d, f;\ \overline{\binom{0}{3}}S \Rightarrow (0,-1/4,-3/4,0,1/2,0), (0,-1/4,1/4,-1,0,1)$ (if $\eta = 0,\ \widetilde{M} = 0$).

(β) Assume now that $\mathcal{F}_1 = 0$. By Theorems 7.9 and 7.3 we could not have in this case a weak saddle of order three or an integrable saddle. So the condition $\mathcal{F}_2 \neq 0$ must hold. Hence the weak saddle has order two, and considering Lemma 11.1 we could have four configurations. For the first three we construct examples using the same system, depending on one parameter, ξ:

$$(0,1,0,1,-1-\chi/2,\chi),\,(0,0,-1,-\nu,(1+\nu+\xi)/2,-\xi), \quad \text{where}$$

$$\chi = \frac{-2}{(\xi-1)^2}\big(1+\xi^3-2\xi\sqrt{2-2\xi+\xi^3}\,\big),$$
$$\nu = -\frac{1}{(\xi-1)^2}\big(5-3\xi+\xi^2+\xi^3-4\sqrt{2-2\xi+\xi^3}\,\big).$$

Then depending on ξ we have the following three global configurations:

- $s^{(2)}, n, n^d, f;\ S,\ \copyright, \copyright \Rightarrow \xi = 3 \quad (\text{if } \eta < 0)$;
- $s^{(2)}, n, n^d, f;\ S,\ S,\ N^\infty \Rightarrow \xi = 16/5 \quad (\text{if } \eta > 0)$;
- $s^{(2)}, n, n^d, f;\ \overline{\binom{0}{2}}SN,\ S \Rightarrow \{\eta^{-1}(0)\} \ni \xi \approx 3.1655 \quad (\text{if } \eta = 0,\ \widetilde{M} \neq 0)$.

The last global configuration is

- $s^{(2)}, n, n^d, f;\ \overline{\binom{0}{3}}S \Rightarrow (0,-2,0,1,1/2,0),\,(0,-3,2,-1,1/2,1)$ $(\text{if } \eta = 0,\ \widetilde{M} = 0)$.

1.2.4.1.3: $\mathcal{T}_3 = 0$. In this case we have two weak singularities, which must be a saddle and a focus. According to Theorem 7.3 we could not have an integrable saddle, and hence by Theorem 6.2 (see statement (c)) the condition $\mathcal{F}_1 \neq 0$ holds. Additionally by Lemma 11.17(ii) we could not have a triple infinite singularity. So we arrive at the following three global configurations:

- $s^{(1)}, n, n^d, f^{(1)};\ S,\ \copyright, \copyright \Rightarrow (0,0,-5,-1,0,5),\ \left(0,5,0,\sqrt{9\sqrt{5}/5-4},-2,0\right)$ $(\text{if } \eta < 0)$;
- $s^{(1)}, n, n^d, f^{(1)};\ S,\ S,\ N^\infty \Rightarrow (0,0,-1,-2,0,1)$, $\left(0,1,0,\tfrac{2}{5}\sqrt{(9\sqrt{89}-77)/5}-2/5,0\right) \quad (\text{if } \eta > 0)$;
- $s^{(1)}, n, n^d, f^{(1)};\ \overline{\binom{0}{2}}SN, S \Rightarrow (0,0,-1,-11/10,0,1)$, $(0,1,0,$ $\sqrt{(11+20\xi)^3}/(15\sqrt{30}),\xi,0)$, where $\{W_4^{-1}(0)\} \ni \xi \approx -0.4991$ $(\text{if } \eta = 0)$.

1.2.4.2: $W_3 > 0$. Then apart from the saddle, we have three nodes, one of them being n^d or n^*. We remark that by Theorem 7.2 we could have a star node (defined by the conditions $(\mathcal{S}^1_{n^*})$, see Lemma 11.21) only with a strong saddle and with three real infinite singularities.

1. $\mathcal{T}_4 \neq 0$. Therefore by Theorem 6.2 the saddle is strong, and considering the above observation and Lemma 11.1, we arrive at the following five global configurations:
   - $s, n, n, n^d;\ S,\ ©, © \Rightarrow (0, 1, -3/2, -1, 3/2, 3/2), (0, 0, 1, 0, -2, -1)$ (if $\eta < 0$);
   - $s, n, n, n^d;\ S,\ S,\ N^\infty \Rightarrow (0, 1, -2, -1, 2, 2), (0, 0, 1, 0, -2, -1)$ (if $\eta > 0,\ \neg(\mathcal{S}^1_{n^*})$);
   - $s, n, n, n^*;\ S,\ S,\ N^\infty \Rightarrow (0, 1, 0, -1, -2, 0), (0, 0, 1, 0, -2, -1)$ (if $\eta > 0,\ (\mathcal{S}^1_{n^*})$);
   - $s, n, n, n^d;\ \overline{\binom{0}{2}}SN,\ S \Rightarrow (0, 1, -1, -1/2, 1/4, 0), (0, 0, 1, 0, -3/4, 1/2)$ (if $\eta = 0,\ \widetilde{M} \neq 0$);
   - $s, n, n, n^d;\ \overline{\binom{0}{3}}S \Rightarrow (0, 1, -1, -1/2, 1/4, 0), (0, 0, 1, -1, -1/4, 1/2)$ (if $\eta = 0,\ \widetilde{M} = 0$).
2. $\mathcal{T}_4 = 0$. Then the saddle is weak, and its order is governed by the invariant polynomials $\mathcal{F}_i$ $(i = 1, 2, 3, 4)$.
   (a) $\mathcal{F}_1 \neq 0$. Then the weak saddle has order one, and since $\mu_0 W_3 < 0$, by Lemma 11.17(i$_2$) we conclude that in this case the condition $\widetilde{M} \neq 0$ holds. So by Lemma 11.1 we get the following three global configurations:
   - $s^{(1)}, n, n, n^d;\ S,\ ©, © \Rightarrow (0, -2, -4, 2, 3, 4), (0, 1, -6, -1, 2, 6)$ (if $\eta < 0$);
   - $s^{(1)}, n, n, n^d;\ S,\ S,\ N^\infty \Rightarrow (0, -1, -16, 1, 8, 16), (0, 1, 7, -1, -4, -7)$ (if $\eta > 0$);
   - $s^{(1)}, n, n, n^d;\ \overline{\binom{0}{2}}SN,\ S \Rightarrow (0, 5, -5, -4, 2, 0), (0, 5, 15, 0, -12, 4)$ (if $\eta = 0$).

   (b) $\mathcal{F}_1 = 0,\ \mathcal{F}_2 \neq 0$. Then the weak saddle has order two. Since by Lemma 11.17(i$_2$) we could not have $\widetilde{M} = 0$, considering Lemma 11.1 we arrive at the following three global configurations:
   - $s^{(2)}, n, n, n^d;\ S,\ ©, © \Rightarrow (0, 0, 1, 14, 3, -1), (0, 1, 0, -5/2, 4, 0)$ (if $\eta < 0$);
   - $s^{(2)}, n, n, n^d;\ S,\ S,\ N^\infty \Rightarrow (0, 0, 1, 1/2, 3, -1), (0, 1, 0, -1, 4, 0)$ (if $\eta > 0$);
   - $s^{(2)}, n, n, n^d;\ \overline{\binom{0}{2}}SN,\ S \Rightarrow (0, 0, 1, \xi, -\sqrt{3}, -1), (0, 1, 0, \sqrt{3}, 1, 0)$, where $\{\eta^{-1}(0)\} \ni \xi \approx 7.3246$ (if $\eta = 0$).

   (c) $\mathcal{F}_1 = \mathcal{F}_2 = 0$. According to Theorem 7.3 we could not have an integrable saddle apart from three nodes, one of them being n^d. So $\mathcal{F}_3\mathcal{F}_4 \neq 0$, and in accordance with Theorem 7.9 we get the unique global configuration
   - $s^{(3)}, n, n, n^d;\ S,\ S,\ N^\infty \Rightarrow (0, 0, 1, 1, 3, -1), (0, 1, 0, -6/5, 4, 0)$.

1.2.4.3: $W_3 = 0$. Then we have at least two nodes with coinciding eigenvalues.

1.2.4.3.1: $W_2 < 0$. In this case apart from a saddle and two nodes, we have a focus (or a center), and by Theorem 7.2 we could not have a star node, i.e. both nodes are n^d.

1. $\mathcal{T}_4 \neq 0$. We do not have weak singularities. Since by Lemma 11.17(iii) the condition $\widetilde{M} \neq 0$ holds, considering Lemma 11.1 we get the next three global configurations:
   - $s, n^d, n^d, f;\ S,\ ©, © \Rightarrow (0, 1, -1, -1, -1, 1), (0, 0, 1, 0, -1, -1)$ (if $\eta < 0$);
   - $s, n^d, n^d, f;\ S,\ S,\ N^\infty \Rightarrow (0, 1, -1, -1, -2, 1), (0, 0, 1, 0, -1, -1)$ (if $\eta > 0$);
   - $s, n^d, n^d, f;\ \overline{\binom{0}{2}}SN,\ S \Rightarrow (0, 1, -3, -3, 5/2, 0), (0, 0, 1, 0, -3, 5)$ (if $\eta = 0$).
2. $\mathcal{T}_4 = 0$. We have at least one weak singularity. As there also exist two n^d's, according to Theorem 7.6 we could not have two n^d's with either $s^{(i)}, f$ or $s, f^{(i)}$ $(i = 1, 2, 3)$ or $s^{(1)}, f^{(1)}$. Additionally by Theorem 7.3 (or Theorem 7.4) the only possibility is to have an integrable saddle and a center, and in this case at infinity we must have one real and two complex singularities (i.e. $\eta < 0$). So we get the unique global configuration
   - $\$, n^d, n^d, c;\ S,\ ©, © \Rightarrow (0, 1, -1, -1, 0, 1), (0, 0, 1, 0, -1, -1)$.

1.2.4.3.2: $W_2 > 0$. In this case apart from a saddle, we have three nodes, two of them having coinciding eigenvalues.

Remark 11.6. (i) A weak saddle of order three or an integrable saddle could not coexist with three nodes, among which two have coinciding eigenvalues (see Theorems 7.9 and 7.3).

(ii) In the case we consider, (a) if we have a star node, then it is unique, the saddle is strong and at infinity we must have three real singularities (see Theorem 7.2); (b) if we have two nodes n^d and a weak singularity, then we could not have a triple singular point at infinity (see Theorem 7.6).

1. $\mathcal{T}_4 \neq 0$. The saddle is strong, and considering the above remark and Lemma 11.1, we arrive at the next five global configurations:
   - $s, n, n^d, n^d;\ S,\ ©, © \Rightarrow (0, 1, -1/10, -1, -1/2, 1/10), (0, 0, 1, 0, -1, -1)$ (if $\eta < 0$);
   - $s, n, n^d, n^d;\ S,\ S,\ N^\infty \Rightarrow (0, 1, 1, -1, -2, -1), (0, 0, 1, 0, -1, -1)$ (if $\eta > 0$, $\neg(\mathcal{S}^1_{n^*})$);
   - $s, n, n^d, n^*;\ S,\ S,\ N^\infty \Rightarrow (0, 1, 0, -1, -2, 0), (0, 0, 1, 0, -1, -1)$ (if $\eta > 0$, $(\mathcal{S}^1_{n^*})$);
   - $s, n, n^d, n^d;\ \overline{\binom{0}{2}}SN,\ S \Rightarrow (0, 1, -81/100, -1, 2/5, 81/100), (0, 0, 1, 0, -1, -1)$ (if $\eta = 0$, $\widetilde{M} \neq 0$);

- $s, n, n^d, n^d; \overline{\binom{0}{3}}S \Rightarrow (0,1,-1,-2,1,0), (0,0,1,-1,-1,2)$ (if $\eta = 0, \widetilde{M} = 0$).

2. $\mathcal{T}_4 = 0$. The saddle is weak, and its order is governed by the invariant polynomials $\mathcal{F}_i$, $i \in \{1,2,3,4\}$.

(α) $\mathcal{F}_1 \neq 0$. Then we have a weak saddle of order one. So considering Remark 11.6 and Lemma 11.1, we obtain the following three global configurations:

- $s^{(1)}, n, n^d, n^d; S$, ©, © $\Rightarrow (0,4,1/2,-4,-2,-1/2), \big(0,-8,0,8,2(\sqrt{7}-1),0\big)$ (if $\eta < 0$);
- $s^{(1)}, n, n^d, n^d; S, S, N^\infty \Rightarrow (0,3,1,-3,-2,-1), (0,-4,-1,4,2\sqrt{3}-1,1)$ (if $\eta > 0$);
- $s^{(1)}, n, n^d, n^d; \overline{\binom{0}{2}}SN, S \Rightarrow \big(0,(41-104\xi+16\xi^2)/64,-(4\xi-5)^2/64,-1/4, \xi/2,0\big), \big(0,(4\xi+3)^2/64,(71-88\xi-16\xi^2)/64,0,-5/8,\xi\big)$, where $\{W_3^{-1}(0)\} \ni \xi \approx 0.2301$ (if $\eta = 0$).

(β) $\mathcal{F}_1 = 0$. By Remark 11.6 the condition $\mathcal{F}_2 \neq 0$ must hold, and hence the weak saddle has order two. Moreover by Remark 11.6 we could not have a triple singularity at infinity. So we arrive at three configurations listed further below. For the first two we construct the corresponding examples using the same canonical systems, depending on one parameter ξ:

$$\big(0,\ \xi,\ -169/2500,\ -\xi,\ 13/100,\ 169/2500\big),$$
$$\big(0,\ 1/4,\ 13/50+\xi,\ -1/4,\ -(13+\sqrt{481}\,)/100-\xi, -13/50-\xi\big).$$

Then we have the following two global configurations:

- $s^{(2)}, n, n^d, n^d; S$, ©, © $\Rightarrow$ with $\{\mathcal{F}_1^{-1}(0)\} \ni \xi \approx -0.1699$ (if $\eta < 0$),
- $s^{(2)}, n, n^d, n^d; S, S, N^\infty \Rightarrow$ with $\{\mathcal{F}_1^{-1}(0)\} \ni \xi \approx -0.1888$ (if $\eta > 0$).

The third configuration is

- $s^{(2)}, n, n^d, n^d; \overline{\binom{0}{2}}SN, S \Rightarrow \big(0,\xi,-\chi^2,-\xi,\chi/2,\chi^2\big), \big(0,1/4,\xi+\chi,-1/4, -(2\xi+\chi+\sqrt{(1-\xi)\xi}\,)/2,-\xi-\chi\big)$, where $\{\mathcal{F}_1^{-1}(0),\eta^{-1}(0)\} \ni (\xi,\chi) \approx \approx (-0.2460, 0.4499)$ (if $\eta = 0$).

1.2.4.3.3: $W_2 = 0$. The condition $W_4 = W_3 = W_2 = 0$ implies the existence of three nodes with coinciding eigenvalues, and hence the fourth singularity is a saddle. So by Theorem 7.6 and Lemmas 11.21 and 11.1, we arrive at the next four global configurations:

- $s, n^d, n^d, n^d; S$, ©, © $\Rightarrow (0,1,-9/8,-1,1/2,9/8), (0,0,1,0,-1,-1)$ (if $\eta < 0$);
- $s, n^d, n^d, n^d; S, S, N^\infty \Rightarrow (0,1,-9/2,-1,2,9/2), (0,0,1,0,-1,-1)$ (if $\eta > 0, \neg(\mathcal{S}_{n^*}^1)$);
- $s, n^d, n^d, n^*; S, S, N^\infty \Rightarrow (0,1,0,-1,-1,0), (0,0,1,0,-1,-1)$ (if $\eta > 0, (\mathcal{S}_{n^*}^1)$);

- $s, n^d, n^d, n^d; \overline{\binom{0}{2}}SN, S \Rightarrow (0, 1, -(3+2\sqrt{2})/4, -1, \sqrt{2}/2, (3+2\sqrt{2})/4)$, $(0, 0, 1, 0, -1, -1)$ (if $\eta = 0$).

2: $\mu_0 > 0$. In this case by Theorem 6.1 the systems (8.1) possess two saddles and two anti-saddles.

Remark 11.7. According to Theorem 7.5 if a quadratic system possesses an infinite star node, then it could not possess a focus.

2.1: $W_4 < 0$. Then the anti-saddles are one node (which is generic) and one focus, and we consider two possibilities: $\mathcal{T}_4 \neq 0$ and $\mathcal{T}_4 = 0$.

2.1.1: $\mathcal{T}_4 \neq 0$. In this case by Theorem 6.2 we do not have weak singularities, and hence the configuration of finite singularities is well defined. Regarding the infinite singularities, according to Lemma 11.1 we need to distinguish not only their number and types (which are governed by the invariant polynomials η and $\widetilde{M}$), but also the type of elemental nodes. And in these cases we will apply the invariant polynomials θ and θ_i $(i = 1, 2, 3, 4)$.

1. $\eta < 0$. According to Remark 11.7 and Lemma 11.1, in this case we get only three global configurations:
   - $s, s, n, f;\ N^\infty,\ ©, © \Rightarrow (0, 0, 1, 0, -1, -1), (0, 2, 5/2, -2, 0, -5/2)$ (if $\theta < 0$);
   - $s, s, n, f;\ N^f,\ ©, © \Rightarrow (0, 0, 1, 0, -1, -1), (0, 2, 1, -2, 0, -1)$ (if $\theta > 0$);
   - $s, s, n, f;\ N^d,\ ©, © \Rightarrow (0, 0, -2, 2, 0, 0), (0, -1, 1, -1, 1, -1)$ (if $\theta = 0$).
2. $\eta > 0$. By Lemma 11.1 in this case there are 10 possible configurations. However considering Remark 11.7, we arrive at the next six global configurations:
   - $s, s, n, f;\ S,\ N^\infty,\ N^\infty \Rightarrow (0, 0, 1, 0, -7/2, -1), (0, 2, 9/2, -2, 0, -9/2)$ (if $\theta < 0$, $\theta_1 < 0$);
   - $s, s, n, f;\ S,\ N^f,\ N^f \Rightarrow (0, 0, 1, 0, 1, -1), (0, 2, 4, -2, 4, -4)$ (if $\theta < 0, \theta_1 > 0$);
   - $s, s, n, f;\ S,\ N^f,\ N^\infty \Rightarrow (0, 0, 1, 0, -3/2, -1), (0, 2, 1/2, -2, 0, -1/2)$ (if $\theta > 0$);
   - $s, s, n, f;\ S,\ N^\infty,\ N^d \Rightarrow (0, -2, -3, 1, 2, 0), (0, -1, -2, 0, 0, 3)$ (if $\theta = 0$, $\theta_1 < 0$);
   - $s, s, n, f;\ S,\ N^f,\ N^d \Rightarrow (0, 0, -4, 1, 3/2, 0), (0, 1, -2, 0, 0, 1)$ (if $\theta = 0, \theta_1 > 0$);
   - $s, s, n, f;\ S,\ N^d,\ N^d \Rightarrow (0, -2, 1, 1, 0, 0), (0, -1, -1, 0, 0, 2)$ (if $\theta = \theta_1 = 0$).

3. $\eta = 0$. By Lemma 11.1, at infinity we could have either one double and one simple (if $\widetilde{M} \neq 0$) or one triple (if $\widetilde{M} = 0$) real singularities. Considering Remark 11.7 we obtain the following four global configurations:

   - $s, s, n, f; \overline{\binom{0}{2}}SN,\ N^\infty \Rightarrow (0,-4,1,2,1/2,0), (0,-2,0,0,1/2,1)$ (if $\widetilde{M} \neq 0,\ \theta < 0$);
   - $s, s, n, f; \overline{\binom{0}{2}}SN,\ N^f \Rightarrow (-4,-2,2,2,0), (0,-2,0,0,-1,4)$ (if $\widetilde{M} \neq 0,\ \theta > 0$);
   - $s, s, n, f; \overline{\binom{0}{2}}SN,\ N^d \Rightarrow (0,-2,-1,1,1,0), (0,-1,-1,0,0,2)$ (if $\widetilde{M} \neq 0,\ \theta = 0$);
   - $s, s, n, f; \overline{\binom{0}{3}}N \Rightarrow (0,1,-2,2,-1/2,0), (0,-2,2,-1,1,-1)$ (if $\widetilde{M} = 0$).

2.1.2: $\mathcal{T}_4 = 0,\ \mathcal{T}_3 \neq 0$. In this case we have at least one weak singularity.

2.1.2.1: $\mathcal{T}_3\mathcal{F} < 0$. According to Theorem 6.2 there exists one weak focus, and its order is governed by the invariant polynomials $\mathcal{F}_i$ $(i = 1,2,3,4)$.

1. $\mathcal{F}_1 \neq 0$. In this case we have a weak focus of order one, and considering Remark 11.7, according to Lemma 11.1 we arrive at the next 13 global configurations:

   - $s, s, n, f^{(1)}; N^\infty,$ ©, © $\Rightarrow (0,-3,-10,3,1,10), (0,1,3,-1,0,-3)$ (if $\eta < 0,\ \theta < 0$);
   - $s, s, n, f^{(1)}; N^f,$ ©, © $\Rightarrow (0,0,-1,0,2,1), (0,1,0,-1,2,0)$ (if $\eta < 0,\ \theta > 0$);
   - $s, s, n, f^{(1)}; N^d,$ ©, © $\Rightarrow (0,0,-15,2,0,0), (0,1,0,1,1,1)$ (if $\eta < 0,\ \theta = 0$);
   - $s, s, n, f^{(1)}; S,\ N^\infty,\ N^\infty \Rightarrow (0,-3,-10,3,-4,10), (0,1,3,-1,5/2,-3)$ (if $\eta > 0,\ \theta < 0,\ \theta_1 < 0$);
   - $s, s, n, f^{(1)}; S,\ N^f,\ N^f \Rightarrow (0,-1,-2,1,-2,2), (0,1,1,-1,1,-1)$ (if $\eta > 0,\ \theta < 0,\ \theta_1 > 0$);
   - $s, s, n, f^{(1)}; S,\ N^f,\ N^\infty \Rightarrow (0,-1,-2,1,0,2), (0,1,1,-1,2,-1)$ (if $\eta > 0,\ \theta > 0$);
   - $s, s, n, f^{(1)}; S,\ N^\infty,\ N^d \Rightarrow (0,2,-9/2,1,3/4,0), (0,1,-2,0,0,1)$ (if $\eta > 0,\ \theta = 0,\ \theta_1 < 0$);
   - $s, s, n, f^{(1)}; S,\ N^f,\ N^d \Rightarrow (0,2,-6,1,3/2,0), (0,1,-2,0,0,1)$ (if $\eta > 0,\ \theta = 0,\ \theta_1 > 0$);
   - $s, s, n, f^{(1)}; S,\ N^d,\ N^d \Rightarrow (0,-4,3,1,0,0), (0,-6,4,0,0,2)$ (if $\eta > 0,\ \theta = 0,\ \theta_1 = 0$);

- $s, s, n, f^{(1)}$; $\overline{\binom{0}{2}}SN$, $N^\infty \Rightarrow (0,10,-18,6,1,0),(0,6,-10,0,1,2)$ (if $\eta = 0$, $\widetilde{M} \neq 0$, $\theta < 0$);
- $s, s, n, f^{(1)}$; $\overline{\binom{0}{2}}SN$, $N^f \Rightarrow (0,1,-2,1/2,1/4,0),(0,1,-1,0,-1/4,1/2)$ (if $\eta = 0$, $\widetilde{M} \neq 0$, $\theta > 0$);
- $s, s, n, f^{(1)}$; $\overline{\binom{0}{2}}SN$, $N^d \Rightarrow (0,3,-5,1,1/2,0),(0,2,-3,0,0,1)$ (if $\eta = 0$, $\widetilde{M} \neq 0$, $\theta = 0$);
- $s, s, n, f^{(1)}$; $\overline{\binom{0}{3}}N \Rightarrow (0,1,-5/2,2,-1/4,0),(0,1/2,-1,-1,1,-1/2)$ (if $\eta = 0$, $\widetilde{M} = 0$).

2. $\mathcal{F}_1 = 0$, $\mathcal{F}_2 \neq 0$. We have a weak focus of order two, and therefore by Theorem 7.7 we have the next five global configurations:
   - $s, s, n, f^{(2)}$; N^f, ©, © $\Rightarrow (0,0,-1,4/3,1,1),(0,1,0,1,-1/3,-1)$ (if $\eta < 0$);
   - $s, s, n, f^{(2)}$; S, N^f, $N^f \Rightarrow (0,0,-1,1,1,1),(0,1,0,1,-1/4,-1)$ (if $\eta > 0$, $\theta < 0$);
   - $s, s, n, f^{(2)}$; S, N^f, $N^\infty \Rightarrow (0,0,-1,-1/3,1,1),(0,1,0,1,-11/12,-1)$ (if $\eta > 0$, $\theta > 0$);
   - $s, s, n, f^{(2)}$; S, N^f, $N^d \Rightarrow \big(0,0,-1,(3-2\xi)/6,1,1\big),\big(0,1,0,1,(3+4\xi)/12,-1\big)$, where $\{\theta^{-1}(0)\} \ni \xi \approx -1.5873$ (if $\eta > 0$, $\theta = 0$);
   - $s, s, n, f^{(2)}$; $\overline{\binom{0}{2}}SN$, $N^f \Rightarrow (0,0,-1,-4/9,\xi,1),(0,1,0,\xi,-1/18,-\xi)$, where $\xi = \sqrt{3}/9$ (if $\eta = 0$).
3. $\mathcal{F}_1 = \mathcal{F}_2 = 0$. By Theorem 7.4 we could not have a center and a node together with two saddles. So the condition $\mathcal{F}_3\mathcal{F}_4 \neq 0$ holds (i.e. we have a weak focus of order three), and then by Theorem 7.8 we get the next three global configurations:
   - $s, s, n, f^{(3)}$; S, N^f, $N^f \Rightarrow (0,0,-1,2,5,1),(0,1,0,2,11/2,0)$ (if $\theta < 0$);
   - $s, s, n, f^{(3)}$; S, N^f, $N^\infty \Rightarrow (0,0,-1,-9/4,75/32,1),(0,1,0,15/16,-7/8,0)$ (if $\theta > 0$);
   - $s, s, n, f^{(3)}$; S, N^f, $N^d \Rightarrow (0,0,-1,-11/4,-15/4,1),(0,1,0,-3/2,-13/8,0)$ (if $\theta = 0$).

2.1.2.2: $\mathcal{T}_3\mathcal{F} > 0$. According to Theorem 6.2 there exists one weak saddle.

1. $\mathcal{F}_1 \neq 0$. In this case we have a weak saddle of order one, and considering Remark 11.7, by Lemma 11.1 we arrive at the next 13 global configurations:
   - $s, s^{(1)}, n, f$; N^∞, ©, © $\Rightarrow (0,-6,-5,6,3,5),(0,4,6,-4,-3,-6)$ (if $\eta < 0$, $\theta < 0$);

- $s, s^{(1)}, n, f;\ N^f,\ ©, © \Rightarrow (0,0,1,0,-1,-1), (0,1,0,-1,1,0)$ (if $\eta < 0,\ \theta > 0$);
- $s, s^{(1)}, n, f;\ N^d,\ ©, © \Rightarrow (0,0,-2,3,0,0), (0,-1,0,1,3/2,1)$ (if $\eta < 0,\ \theta = 0$);
- $s, s^{(1)}, n, f;\ S,\ N^\infty,\ N^\infty \Rightarrow (0,-3,-8,3,-5/2,8), (0,1,3,-1,3,-3)$ (if $\eta > 0,\ \theta < 0,\ \theta_1 < 0$);
- $s, s^{(1)}, n, f;\ S,\ N^f,\ N^f \Rightarrow (0,-1,0,1,-2,0), (0,1,1,-1,0,-1)$ (if $\eta > 0,\ \theta < 0,\ \theta_1 > 0$);
- $s, s^{(1)}, n, f;\ S,\ N^f,\ N^\infty \Rightarrow (0,-1,0,1,-1,0), (0,1,1,-1,-2,-1)$ (if $\eta > 0,\ \theta > 0$);
- $s, s^{(1)}, n, f;\ S,\ N^\infty,\ N^d \Rightarrow (0,-1,0,-1,3/2,0), (0,1,1,0,0,2)$ (if $\eta > 0,\ \theta = 0,\ \theta_1 < 0$);
- $s, s^{(1)}, n, f;\ S,\ N^f,\ N^d \Rightarrow (0,-1,0,-2,3/2,0), (0,1,1,0,0,1)$ (if $\eta > 0,\ \theta = 0,\ \theta_1 > 0$);
- $s, s^{(1)}, n, f;\ S,\ N^d,\ N^d \Rightarrow (0,-2,1,1,0,0), (0,-3,2,0,0,1)$ (if $\eta > 0,\ \theta = 0,\ \theta_1 = 0$);
- $s, s^{(1)}, n, f;\ \overline{\binom{0}{2}}SN,\ N^\infty \Rightarrow (0,2,-3,2,1/2,0), (0,1,-2,0,1/2,1)$ (if $\eta = 0,\ \widetilde{M} \neq 0,\ \theta < 0$);
- $s, s^{(1)}, n, f;\ \overline{\binom{0}{2}}SN,\ N^f \Rightarrow (0,2,6,2,4,0), (0,2,-2,0,-1,8)$ (if $\eta = 0,\ \widetilde{M} \neq 0,\ \theta > 0$);
- $s, s^{(1)}, n, f;\ \overline{\binom{0}{2}}SN,\ N^d \Rightarrow (0,1,0,1,1/2,0), (0,1,-1,0,0,1)$ (if $\eta = 0,\ \widetilde{M} \neq 0,\ \theta = 0$);
- $s, s^{(1)}, n, f;\ \overline{\binom{0}{3}}N \Rightarrow (0,2,3/2,1,-1/2,0), (0,-2,-2,-1,1/2,-1)$ (if $\eta = 0,\ \widetilde{M} = 0$).

2. $\mathcal{F}_1 = 0,\ \mathcal{F}_2 \neq 0$. We have a weak saddle of order two.

   (a) $\eta < 0$. Since by Remark 11.7 we could not have an infinite star node, considering Lemma 11.1 we obtain the next three global configurations:
   - $s, s^{(2)}, n, f;\ N^\infty,\ ©, © \Rightarrow (0,1,0,1,4/25,1/5), (0,0,-1,-1/5,1/5,4/5)$ (if $\theta < 0$);
   - $s, s^{(2)}, n, f;\ N^f,\ ©, © \Rightarrow (0,1,0,1,1/10,1/5), (0,0,-1,-1/5,1/10,1)$ (if $\theta > 0$);
   - $s, s^{(2)}, n, f;\ N^d,\ ©, ©: \Rightarrow (0,1,0,1,1/10,1/5), \big(0,0,-1,(247-27\sqrt{89})/80,1/10,1\big)$ (if $\theta = 0$).

   (b) $\eta > 0$. According to Remark 11.7, Lemma 11.19(i) and Lemma 11.1, we could have at infinity neither a star node nor two one-direction nodes. All the remaining possibilities are realizable and the corresponding global configurations are:

- $s, s^{(2)}, n, f;\ S,\ N^{\infty},\ N^{\infty} \Rightarrow (0,1,0,1,1/25,0), (0,0,-1,-1/5,1/10, 2/5)$ (if $\theta < 0,\ \theta_1 < 0$);
- $s, s^{(2)}, n, f;\ S,\ N^{f},\ N^{f} \Rightarrow (0,1,0,1,6/25,-1/5), (0,0,-1,-2/5,3/10, 4/5)$ (if $\theta < 0,\ \theta_1 > 0$);
- $s, s^{(2)}, n, f;\ S,\ N^{f},\ N^{\infty} \Rightarrow (0,1,0,1,1/16,0), (0,0,-1,-1/4,1/8,1/2)$ (if $\theta > 0$);
- $s, s^{(2)}, n, f;\ S,\ N^{\infty},\ N^{d} \Rightarrow (0,1,0,1,5/16,0), \big(0,0,-1,-30/121, 5/16,1\big)$ (if $\theta = 0,\ \theta_1 < 0$);
- $s, s^{(2)}, n, f;\ S,\ N^{f},\ N^{d} \Rightarrow (0,1,0,1,6,0), (0,0,-1,66/25,6,1)$ (if $\theta = 0,\ \theta_1 > 0$).

(c) $\eta = 0$. By Lemma 11.17(i_1) we could not have at infinity a triple point, i.e. $\widetilde{M} \neq 0$. So considering Lemma 11.1 we obtain the next three global configurations:

- $s, s^{(2)}, n, f;\ \overline{\binom{0}{2}}SN,\ N^{\infty} \Rightarrow (0,-3/4,-7/16,65/63,1/2,0), (0,-1,3/4, 0,1/63,1)$ (if $\theta < 0$);
- $s, s^{(2)}, n, f;\ \overline{\binom{0}{2}}SN,\ N^{f} \Rightarrow (0,1,-3/5,1/9,1/10,0), (0,0,-1,0,-4/9, 1/5)$ (if $\theta > 0$);
- $s, s^{(2)}, n, f;\ \overline{\binom{0}{2}}SN,\ N^{d} \Rightarrow \big(0,\sqrt{3}/2,1/2,5/(4\sqrt{3}),1/4,0\big),\ \big(0,1/2, -\sqrt{3}/2,0,0,1/2\big)$ (if $\theta = 0$).

3. $\mathcal{F}_1 = \mathcal{F}_2 = 0$. By Theorem 7.3 we could not have an integrable saddle and a focus in the case $\mu_0 \neq 0$. So the condition $\mathcal{F}_3\mathcal{F}_4 \neq 0$ holds, and we get a weak saddle of order three. Therefore by Theorem 7.9 we get the next 11 global configurations:

- $s, s^{(3)}, n, f;\ N^{\infty},\ ©, © \Rightarrow (0,1,0,-8/5,-1/2,2/5), (0,0,-1,-2/5,-2/5, -2)$ (if $\eta < 0,\ \theta < 0$);
- $s, s^{(3)}, n, f;\ N^{f},\ ©, © \Rightarrow (0,1,0,-2,-1/4,1), (0,0,-1,-1,-1/2,-1)$ (if $\eta < 0,\ \theta > 0$);
- $s, s^{(3)}, n, f;\ N^{d},\ ©, © \Rightarrow (0,1,0,6,-1/4,-3/88), (0,0,-1,-\sqrt{77},3/2,-1)$ (if $\eta < 0,\ \theta = 0$);
- $s, s^{(3)}, n, f;\ S,\ N^{\infty},\ N^{\infty} \Rightarrow (0,1,0,-2,3/8,0), (0,0,-1,-1/2,-1/2,3/2)$ (if $\eta > 0,\ \theta < 0,\ \theta_1 < 0$);
- $s, s^{(3)}, n, f;\ S,\ N^{f},\ N^{f} \Rightarrow (0,1,0,-3/5,-1/4,3/5), (0,0,-1,1/10,-3/20, -1)$ (if $\eta > 0,\ \theta < 0,\ \theta_1 > 0$);
- $s, s^{(3)}, n, f;\ S,\ N^{f},\ N^{\infty} \Rightarrow (0,1,0,-2,1/4,0), (0,0,-1,-1,-1/2,1)$ (if $\eta > 0,\ \theta > 0$);
- $s, s^{(3)}, n, f;\ S,\ N^{\infty},\ N^{d} \Rightarrow (0,1,0,-1,-3/8,1/2), (0,0,-1,0,-1/4,-3/2)$ (if $\eta > 0,\ \theta = 0,\ \theta_1 < 0$);

- $s, s^{(3)}, n, f;\ S,\ N^f,\ N^d \Rightarrow (0,1,0,-1,-1/2,3/2), \big(0,0,-1,(5-\sqrt{3})/24, -1/4,-2\big)$ (if $\eta>0,\ \theta=0,\ \theta_1>0$);
- $s, s^{(3)}, n, f;\ \overline{\binom{0}{2}}SN,\ N^\infty \Rightarrow (0,4/5,9/25,11/36,1/10,0),(0,1,-4/5,0, 1/72,1/5)$ (if $\eta=0,\ \theta<0$);
- $s, s^{(3)}, n, f;\ \overline{\binom{0}{2}}SN,\ N^f \Rightarrow (0,15,45/2,2,15/2,0),(0,30,-15,0,-7,15)$ (if $\eta=0,\ \theta>0$);
- $s, s^{(3)}, n, f;\ \overline{\binom{0}{2}}SN,\ N^d \Rightarrow (0,\sqrt{3/5},2/5,\sqrt{5/3},1/2,0),(0,1,-\sqrt{3/5},0,0,1)$ (if $\eta=0,\ \theta=0$).

2.1.2: $\mathcal{T}_4=\mathcal{T}_3=0$. Then the condition $\mathcal{T}_2\neq 0$ must hold, otherwise by Lemma 11.16 we get Hamiltonian systems, which is impossible due to the existence of a node. So according to Theorem 6.2 we have two weak singularities, which could be either a focus and a saddle or two saddles. Moreover by Theorem 7.3, in the case under consideration, we could not have an integrable saddle, and hence both weak singularities have order one.

1. $\mathcal{T}_2<0$. Then we have two weak saddles. By Remark 11.7, Lemma 11.19(i) and Lemma 11.1, we could have at infinity neither a star node nor two one-direction nodes. So all the remaining possibilities are realizable, and we get the next 11 global configurations:
   - $s^{(1)}, s^{(1)}, n, f;\ N^\infty,$ ©, © $\Rightarrow (0,-5,-1,5,0,1),(0,18,5,-18,-5,-5)$ (if $\eta<0,\ \theta<0$);
   - $s^{(1)}, s^{(1)}, n, f;\ N^f,$ ©, © $\Rightarrow (0,-3,-1,3,0,1),(0,8,3,-8,-3,-3)$ (if $\eta<0,\ \theta>0$);
   - $s^{(1)}, s^{(1)}, n, f;\ N^d,$ ©, © $\Rightarrow \big(0,-1,(\sqrt{2}-4)/16,1,1/4,(4-\sqrt{2})/16\big)$ $(0,2,1,-2,-1,-1)$ (if $\eta<0,\ \theta=0$);
   - $s^{(1)}, s^{(1)}, n, f;\ S,\ N^\infty,\ N^\infty \Rightarrow (0,-11,-1,11,0,1),(0,25,11,-25,-11,-11)$ (if $\eta>0,\ \theta<0,\ \theta_1<0$);
   - $s^{(1)}, s^{(1)}, n, f;\ S,\ N^f,\ N^f \Rightarrow (0,1,0,-1,-1,0),(0,1,-1,-1,-2,1)$ (if $\eta>0,\ \theta<0,\ \theta_1>0$);
   - $s^{(1)}, s^{(1)}, n, f;\ S,\ N^f,\ N^\infty \Rightarrow (0,-1,0,1,1,0),(0,1,1,-1,3,-1)$ (if $\eta>0,\ \theta>0$);
   - $s^{(1)}, s^{(1)}, n, f;\ S,\ N^\infty,\ N^d \Rightarrow \big(0,1,2(5-\sqrt{13}),-1,-1,2(\sqrt{13}-5)\big)$, $(0,1/16,-1,-1/16,-1/4,1)$ (if $\eta>0,\ \theta=0,\ \theta_1<0$);
   - $s^{(1)}, s^{(1)}, n, f;\ S,\ N^f,\ N^d \Rightarrow \big(0,2,4(\sqrt{3}-1),-2,-2,4(1-\sqrt{3})\big)$, $(0,1,-2,-1,0,2)$ (if $\eta>0,\ \theta=0,\ \theta_1>0$);
   - $s^{(1)}, s^{(1)}, n, f;\ \overline{\binom{0}{2}}SN,\ N^\infty \Rightarrow (0,-6/5,13/15,4/3,-1/2,0)$, $(0,-8/15,6/5,0,1/6,-1)$ (if $\eta=0,\ \theta<0$);

- $s^{(1)}, s^{(1)}, n, f; \overline{\binom{0}{2}}SN,\ N^f \Rightarrow (0,-9,5,10,-3,0), (0,-1,9,0,-1,-6)$ (if $\eta = 0,\ \theta > 0$);
- $s^{(1)}, s^{(1)}, n, f; \overline{\binom{0}{2}}SN,\ N^d \Rightarrow (0,-1,2/3,1,-1/3,0), (0,-1/3,1,0,0,-2/3)$ (if $\eta = 0,\ \theta = 0$).

2. $\mathcal{T}_2 > 0$. In this case we have one weak saddle and one weak focus, both of order one. By the same arguments as those mentioned above (in the case of two weak saddles), we arrive at the next possible 11 global configurations:

- $s, s^{(1)}, n, f^{(1)}; N^\infty, ©, © \Rightarrow (0,-6,-1,6,0,1), (0,40,6,-40,-6,-6)$ (if $\eta < 0,\ \theta < 0$)
- $s, s^{(1)}, n, f^{(1)}; N^f, ©, © \Rightarrow (0,-1,-1,1,0,1), (0,4,1,-4,-1,-1)$ (if $\eta < 0,\ \theta > 0$);
- $s, s^{(1)}, n, f^{(1)}; N^d, ©, © \Rightarrow (0,-16,-3,16,0,3), (0,128,16,-128,-16,-16)$ (if $\eta < 0,\ \theta = 0$);
- $s, s^{(1)}, n, f^{(1)}; S,\ N^\infty,\ N^\infty \Rightarrow (0,-1,-1/512,1,-8/9,1/512)$, $(0,528,1,-528,-1,-1)$ (if $\eta > 0,\ \theta < 0,\ \theta_1 < 0$);
- $s, s^{(1)}, n, f^{(1)}; S,\ N^f,\ N^f \Rightarrow (0,-1,3/2,1,0,-3/2), (0,1,1,-1,-1,-1)$ (if $\eta > 0,\ \theta < 0,\ \theta_1 > 0$);
- $s, s^{(1)}, n, f^{(1)}; S,\ N^f,\ N^\infty \Rightarrow (0,1,0,-1,1,0), (0,1,-1,-1,1,1)$ (if $\eta > 0,\ \theta > 0$);
- $s, s^{(1)}, n, f^{(1)}; S,\ N^\infty,\ N^d \Rightarrow (0,-1,-1/512,1,-7/8,1/512)$, $(0,32(44+29\sqrt{2}),1,-32(44+29\sqrt{2}),-1,-1)$ (if $\eta > 0,\ \theta = 0,\ \theta_1 < 0$);
- $s, s^{(1)}, n, f^{(1)}; S,\ N^f,\ N^d \Rightarrow (0,-9,4,9,-9,-4), (0,-27/4,9,27/4,$ $-9,-9)$ (if $\eta > 0,\ \theta = 0,\ \theta_1 > 0$);
- $s, s^{(1)}, n, f^{(1)}; \overline{\binom{0}{2}}SN,\ N^\infty \Rightarrow (0,-185/128,427/384,4/3,-1/2,0)$, $(0,-299/384,185/128,0,1/6,-1)$ (if $\eta = 0,\ \theta < 0$);
- $s, s^{(1)}, n, f^{(1)}; \overline{\binom{0}{2}}SN,\ N^f \Rightarrow (0,-2,5/3,19/48,-1/32,0)$, $(0,-4/3,2,0,-29/96,-1/16)$ (if $\eta = 0,\ \theta > 0$);
- $s, s^{(1)}, n, f^{(1)}; \overline{\binom{0}{2}}SN,\ N^d \Rightarrow (0,-11/50,-17/150,1,-1/3,0)$, $(0,67/150,11/50,0,0,-2/3)$ (if $\eta = 0,\ \theta = 0$).

2.2: $W_4 > 0$. Then Theorem 6.1 (see Table 6.2, lines 4, 8 and 9), apart from the two saddles, we have either two foci, two nodes or two centers, because by Theorem 7.4 we could not have in this case a focus and a center.

2.2.1: $W_2 < 0$ or $W_2 > 0$ and $W_1W_3 < 0$. In this case by Theorem 6.1 we have two foci.

Remark 11.8. Since we have two foci, by Lemmas 11.22(i) and 11.17(v), we conclude that the conditions $\theta > 0$ and $\widetilde{M} \neq 0$ must hold.

We consider two cases: $\mathcal{T}_4 \neq 0$ and $\mathcal{T}_4 = 0$.

2.2.1.1: $\mathcal{T}_4 \neq 0$. Then by Theorem 6.2 we do not have weak singularities, and hence the configuration of finite singularities is well defined. So considering the above remark and Lemma 11.1, we get the next three global configurations:
- $s, s, f, f;\ N^f,$ ©, © $\Rightarrow (0,0,1,0,0,-1), (0,2,1,-2,0,-1)$ (if $\eta < 0$);
- $s, s, f, f;\ S,\ N^f,\ N^\infty \Rightarrow (0,0,1,0,-1/5,-1), (0,2,-2,-2,0,2)$ (if $\eta > 0$);
- $s, s, f, f;\ \overline{\binom{0}{2}}SN,\ N^f \Rightarrow (0,8,-6,2,-2,0), (0,6,4,0,-3,-4)$ (if $\eta = 0$).

2.2.1.2: $\mathcal{T}_4 = 0$, $\mathcal{T}_3 \neq 0$. According to Theorem 6.2 we have one weak singularity and its type depends on the sign of the product $\mathcal{T}_3\mathcal{F}$.

2.2.1.2.1: $\mathcal{T}_3\mathcal{F} < 0$. Then we have a weak focus.

1. $\mathcal{F}_1 \neq 0$. In this case the weak focus has order one, and since by Remark 11.8 the conditions $\theta > 0$ and $\widetilde{M} \neq 0$ hold, we arrive at the next three global configurations:
   - $s, s, f, f^{(1)};\ N^f,$ ©, © $\Rightarrow (0,0,1,0,-2,-1), (0,-1,0,1,-1,0)$ (if $\eta < 0$);
   - $s, s, f, f^{(1)};\ S,\ N^f,\ N^\infty \Rightarrow (0,0,-1,-7/2,-1/4,1), (0,1,0,2,0,0)$ (if $\eta > 0$);
   - $s, s, f, f^{(1)};\ \overline{\binom{0}{2}}SN,\ N^f \Rightarrow (0,8,-7,1,-1,0), (0,17,-8,0,-7/2,-2)$ (if $\eta = 0$).
2. $\mathcal{F}_1 = 0$. According to Theorems 7.8 and 7.4, we could have neither a weak focus of order three nor a center, i.e. the condition $\mathcal{F}_2 \neq 0$ must hold. According to Theorem 6.2 the weak focus has order two, and considering Theorem 7.7 we get the global configuration
   - $s, s, f, f^{(2)};\ N^f,$ ©, © $\Rightarrow (0,0,-1,4/5,1,1), (0,1,0,1,-3/5,-1)$.

2.2.1.2.2: $\mathcal{T}_3\mathcal{F} > 0$. By Theorem 6.2 we have a weak saddle the order of which is governed by the invariant polynomials $\mathcal{F}_i$ $(i = 1,2,3,4)$.

1. $\mathcal{F}_1 \neq 0$. In this case the weak saddle has order one, and considering Lemma 11.1 and Remark 11.8, we obtain the following three global configurations:
   - $s, s^{(1)}, f, f;\ N^f,$ ©, © $\Rightarrow (0,0,2,0,-2,-2), (0,1,0,-1,1,0)$ (if $\eta < 0$);
   - $s, s^{(1)}, f, f;\ S,\ N^f,\ N^\infty \Rightarrow (0,1,0,-1,1/2,-1), (0,0,-1,0,5/2,-4)$ (if $\eta > 0$);
   - $s, s^{(1)}, f, f;\ \overline{\binom{0}{2}}SN,\ N^f \Rightarrow (0,-4,-2,2,2,0), (0,2,4,0,-5,4)$ (if $\eta = 0$).

2. $\mathcal{F}_1 = 0$. We remark that according to Theorems 7.9 and 7.3, apart from two foci, we could have neither a weak saddle of order three nor an integrable saddle, i.e. the condition $\mathcal{F}_2 \neq 0$ holds. Then by Theorem 6.2 the weak saddle has order two, and by Lemma 11.22(ii) and Conjecture 11.1, the condition $\eta < 0$ must hold. So considering Remark 11.8 and Lemma 11.1, we arrive at the global configuration
   - $s, s^{(2)}, f, f;\ N^f,$ ©, © $\Rightarrow (0,4,0,0,2,1), (0,0,-4,-1,0,0)$.

2.2.1.2.3: $\mathcal{T}_3 = 0$, $\mathcal{T}_2 \neq 0$. According to Theorem 6.2 we have two weak singularities, which could be either two foci or two saddles or a focus and a saddle.

1. $\mathcal{F}_1 \neq 0$. Then by Theorem 6.2 we have two weak singularities both of order one.
   (a) $\mathcal{T}_2 < 0$. In this case we could have either two weak foci (if $\mathcal{H} < 0$) or two weak saddles (if $\mathcal{H} > 0$). So considering Lemma 11.1 and Remark 11.8, we obtain the following six global configurations:
   - $s, s, f^{(1)}, f^{(1)};\ N^f,$ ©, © $\Rightarrow (0,0,-1,0,1,1), (0,1,0,-1,0,0)$ (if $\mathcal{H} < 0$, $\eta < 0$);
   - $s, s, f^{(1)}, f^{(1)};\ S,\ N^f,\ N^\infty \Rightarrow (0,1,-7/10,-1,5/8,7/10)$, $(0,2,-1,-2,1,1)$ (if $\mathcal{H} < 0$, $\eta > 0$);
   - $s, s, f^{(1)}, f^{(1)};\ \overline{\binom{0}{2}}SN,\ N^f \Rightarrow (0,-6,-4,1,9/2,0), (0,14,6,0,-29/2,9)$ (if $\mathcal{H} < 0$, $\eta = 0$);
   - $s^{(1)}, s^{(1)}, f, f;\ N^f,$ ©, © $\Rightarrow (0,0,1,0,-1,-1), (0,2,0,-2,0,0)$ (if $\mathcal{H} > 0$, $\eta < 0$);
   - $s^{(1)}, s^{(1)}, f, f;\ S,\ N^f,\ N^\infty \Rightarrow (0,19,7/16,-19,139/32,-7/16)$, $(0,8,-19,-8,19,19)$ (if $\mathcal{H} > 0$, $\eta > 0$);
   - $s^{(1)}, s^{(1)}, f, f;\ \overline{\binom{0}{2}}SN,\ N^f \Rightarrow (0,-6,-2,5,3/2,0), (0,10,6,0,-19/2,3)$ (if $\mathcal{H} > 0$, $\eta = 0$).

   (b) $\mathcal{T}_2 > 0$. Then by Theorem 6.2 we have one weak focus and one weak saddle both of order one. Therefore considering the same remark and lemma as above, we arrive at the next three global configurations:
   - $s, s^{(1)}, f, f^{(1)};\ N^f,$ ©, © $\Rightarrow (0,0,2,0,1,-2), (0,1,0,-1,0,0)$ (if $\eta < 0$);
   - $s, s^{(1)}, f, f^{(1)};\ S,\ N^f,\ N^\infty \Rightarrow (0,-1,-3/2,1/17,-1/2,12)$, $(0,-2,1,2/17,-1/17,-8)$ (if $\eta > 0$);
   - $s, s^{(1)}, f, f^{(1)};\ \overline{\binom{0}{2}}SN,\ N^f \Rightarrow (0,-6,-2,2,3,0), (0,10,6,0,-11,6)$ (if $\eta = 0$).

2. $\mathcal{F}_1 = 0$. In this case according to Theorem 6.2 we could have either two centers or two integrable saddles or one center and one integrable saddle. However by Theorem 7.3, in the case under consideration, we could not have a focus with an integrable saddle. So we must have two centers (i.e. the

conditions $\mathcal{T}_2 < 0$ and $\mathcal{H} < 0$ must hold), and by Theorem 7.4 we obtain the global configuration

- $s, s, c, c;\ N^f,\ ©, © \Rightarrow (0,-1,-2,1,2,2), (0,2,1,-2,-1,-1)$.

2.2.1.2.4: $\mathcal{T}_3 = \mathcal{T}_2 = 0$. By Lemma 11.16 systems become Hamiltonian, i.e. we have two centers and two integrable saddles. According to Theorem 7.3 we get the unique global configuration

- $\$, \$, c, c;\ N^f,\ ©, © \Rightarrow (0,1,1,-1,-1,-1), (0,0,-1,0,1,1)$.

2.2.2: $W_2 > 0$ and $W_1W_3 > 0$. In this case by Theorem 6.1 (see Table 6.2, line 4), apart from two saddles, we have two nodes that are generic due to $W_4 \neq 0$. We consider two cases: $\mathcal{T}_4 \neq 0$ and $\mathcal{T}_4 = 0$.

2.2.2.1: $\mathcal{T}_4 \neq 0$. Then by Theorem 6.2 we do not have weak singularities, and hence the configuration of finite singularities is well defined. According to Lemma 11.1, at infinity we could have 19 distinct configurations of singularities, and all these possibilities are realizable as shown by the next global configurations:

- $s, s, n, n;\ N^\infty,\ ©, © \Rightarrow (0,0,3,0,-2,-3), (0,6,10,-6,-2,-10)$ (if $\eta < 0,\ \theta < 0$);
- $s, s, n, n;\ N^f,\ ©, © \Rightarrow (0,0,1,0,-1,-1), (0,2,3/2,-2,-1,-3/2)$ (if $\eta < 0,\ \theta > 0$);
- $s, s, n, n;\ N^d,\ ©, © \Rightarrow (0,-1,-3/2,5/2,0,0), (0,-1/2,0,-1,5/4,-1)$ (if $\eta < 0,\ \theta = 0,\ \theta_2 \neq 0$);
- $s, s, n, n;\ N^*,\ ©, © \Rightarrow (0,2,0,-2,0,0), (0,3,1,-1,-1,-1)$ (if $\eta < 0,\ \theta = 0,\ \theta_2 = 0$);
- $s, s, n, n;\ S,\ N^\infty,\ N^\infty \Rightarrow (0,0,3,0,-8,-3), (0,6,10,-6,-2,-10)$ (if $\eta > 0,\ \theta < 0,\ \theta_1 < 0$);
- $s, s, n, n;\ S,\ N^f,\ N^f \Rightarrow (0,0,1,0,-2,-1), (0,2,1,-2,-1,-1)$ (if $\eta > 0,\ \theta < 0,\ \theta_1 > 0$);
- $s, s, n, n;\ S,\ N^f,\ N^\infty \Rightarrow (0,0,1,0,1,-1), (0,2,-2,-2,-2,2)$ (if $\eta > 0,\ \theta > 0$);
- $s, s, n, n;\ S,\ N^\infty,\ N^d \Rightarrow (0,-1,3,-2,0,0), (0,4,0,0,-3/2,-1)$ (if $\eta > 0,\ \theta = 0,\ \theta_1 < 0,\ \theta_2 \neq 0$);
- $s, s, n, n;\ S,\ N^\infty,\ N^* \Rightarrow (0,2,0,-2,0,0), (0,3,1,0,-3/2,-1)$ (if $\eta > 0,\ \theta = 0,\ \theta_1 < 0,\ \theta_2 = 0$);
- $s, s, n, n;\ S,\ N^f,\ N^d \Rightarrow (0,-4,2,2,0,0), (0,6,0,0,-1,-4)$ (if $\eta > 0,\ \theta = 0,\ \theta_1 > 0,\ \theta_2 \neq 0$);
- $s, s, n, n;\ S,\ N^f,\ N^* \Rightarrow (0,2,0,-2,0,0), (0,6,4,0,-3,-4)$ (if $\eta > 0,\ \theta = 0,\ \theta_1 > 0,\ \theta_2 = 0$);
- $s, s, n, n;\ S,\ N^d,\ N^d \Rightarrow (0,-2,1,1,0,0), (0,1,0,0,0,-1)$ (if $\eta > 0,\ \theta = 0,\ \theta_1 = 0,\ \theta_3 \neq 0$);
- $s, s, n, n;\ S,\ N^d,\ N^* \Rightarrow (0,-1,0,1,0,0), (0,3,-2,0,0,-1)$ (if $\eta > 0,\ \theta = 0,\ \theta_1 = \theta_3 = 0,\ \theta_4 \neq 0$);
- $s, s, n, n;\ S,\ N^*,\ N^* \Rightarrow (0,-1,0,1,0,0), (0,0,2,0,0,-2)$ (if $\eta > 0,\ \theta = 0,\ \theta_1 = \theta_3 = \theta_4 = 0$);

- $s, s, n, n; \overline{\binom{0}{2}}SN,\ N^{\infty} \Rightarrow (0,-1,1,2,-1,0), (0,1,0,0,1/2,-2)$ (if $\eta = 0$, $\widetilde{M} \neq 0$, $\theta < 0$);
- $s, s, n, n; \overline{\binom{0}{2}}SN,\ N^{f} \Rightarrow (0,1,1/2,1/2,-1,0), (0,5/2,0,0,-1/4,-2)$ (if $\eta = 0$, $\widetilde{M} \neq 0$, $\theta > 0$);
- $s, s, n, n; \overline{\binom{0}{2}}SN,\ N^{d} \Rightarrow (0,-1,-2,1,1,0), (0,-2,0,0,0,2)$ (if $\eta = 0$, $\widetilde{M} \neq 0$, $\theta = 0$, $\theta_2 \neq 0$);
- $s, s, n, n; \overline{\binom{0}{2}}SN,\ N^{*} \Rightarrow (0,1,-4,1,1,0), (0,0,-2,0,0,2)$ (if $\eta = 0$, $\widetilde{M} \neq 0$, $\theta = 0$, $\theta_2 = 0$);
- $s, s, n, n; \binom{0}{3}N \Rightarrow (0,1,1,0,-1,0), (0,3,0,-1,0,-2)$ (if $\eta = 0$, $\widetilde{M} = 0$).

2.2.2.2: $\mathcal{T}_4 = 0$, $\mathcal{T}_3 \neq 0$. According to Theorem 6.2 we have only one weak singularity, which evidently could be only a saddle.

1. $\mathcal{F}_1 \neq 0$. Then the weak saddle has order one. By Theorem 7.5 we could not have a star node at infinity, and considering Lemma 11.1 we arrive at the next 13 global configurations:

   - $s, s^{(1)}, n, n;\ N^{\infty},\ ©, © \Rightarrow (0,-28,5,28,-14,-5), (0,10,28,-10,7,-28)$ (if $\eta < 0$, $\theta < 0$);
   - $s, s^{(1)}, n, n;\ N^{f},\ ©, © \Rightarrow (0,-2,1,2,-8/3,-1), (0,2,2,-2,2,-2)$ (if $\eta < 0$, $\theta > 0$);
   - $s, s^{(1)}, n, n;\ N^{d},\ ©, © \Rightarrow (0,-1,1/8,7/8,0,0), (0,1/8,1,-1,7/16,-1)$ (if $\eta < 0$, $\theta = 0$);
   - $s, s^{(1)}, n, n;\ S,\ N^{\infty},\ N^{\infty} \Rightarrow (0,-5,1,5,-5/2,-1), (0,2,5,-2,0,-5)$ (if $\eta > 0$, $\theta < 0$, $\theta_1 < 0$);
   - $s, s^{(1)}, n, n;\ S,\ N^{f},\ N^{f} \Rightarrow (0,0,1,0,-2,-1), (0,2,0,-2,1/2,0)$ (if $\eta > 0$, $\theta < 0$, $\theta_1 > 0$);
   - $s, s^{(1)}, n, n;\ S,\ N^{f},\ N^{\infty} \Rightarrow (0,0,1,0,-3,-1), (0,2,0,-2,5/2,0)$ (if $\eta > 0$, $\theta > 0$);
   - $s, s^{(1)}, n, n;\ S,\ N^{\infty},\ N^{d} \Rightarrow (0,0,2,-2,0,0), (0,2,0,0,-3/2,1)$ (if $\eta > 0$, $\theta = 0$, $\theta_1 < 0$);
   - $s, s^{(1)}, n, n;\ S,\ N^{f},\ N^{d} \Rightarrow (0,-1,2,-1,0,0), (0,3,1,0,-3/2,-1)$ (if $\eta > 0$, $\theta = 0$, $\theta_1 > 0$);
   - $s, s^{(1)}, n, n;\ S,\ N^{d},\ N^{d} \Rightarrow (0,-3/2,1/2,1,0,0), (0,1/2,3/2,0,0,-2)$ (if $\eta > 0$, $\theta = 0$, $\theta_1 = 0$);
   - $s, s^{(1)}, n, n;\ \overline{\binom{0}{2}}SN,\ N^{\infty} \Rightarrow (0,-1,1,2,-1,0), (0,0,1,0,1/2,-2)$ (if $\eta = 0$, $\widetilde{M} \neq 0$, $\theta < 0$);
   - $s, s^{(1)}, n, n;\ \overline{\binom{0}{2}}SN,\ N^{f} \Rightarrow (0,0,1,1,-1,0), (0,3,0,0,-1/2,-2)$ (if $\eta = 0$, $\widetilde{M} \neq 0$, $\theta > 0$);

- $s, s^{(1)}, n, n; \overline{\binom{0}{2}}SN,\ N^d \Rightarrow (0,0,1,1,-1,0), (0,2,0,0,0,-2)$ (if $\eta = 0,\ \widetilde{M} \neq 0,\ \theta = 0$);

- $s, s^{(1)}, n, n; \overline{\binom{0}{3}}N \Rightarrow (0,1,1,0,-1,0), (0,4,-1,-1,0,-2)$ (if $\eta = 0,\ \widetilde{M} = 0$).

2. $\mathcal{F}_1 = 0,\ \mathcal{F}_2 \neq 0$. By Theorem 6.2 the weak saddle has order two, and by Theorem 7.5 we again could not have a star node at infinity.

Additionally by Lemma 11.19(i) in this case the conditions $\eta > 0,\ \theta = \theta_1 = 0$ are incompatible. So considering Lemma 11.1 we conclude that there could exist 12 global configurations, and their realization is confirmed by the next examples:

- $s, s^{(2)}, n, n; N^\infty$, ©, © $\Rightarrow (0,-1,6,0,-5/2,0), (0,6,1,-1,0,-6)$ (if $\eta < 0,\ \theta < 0$);

- $s, s^{(2)}, n, n; N^f$, ©, © $\Rightarrow (0,25,0,22,0,2), (0,0,-25,0,-12,0)$ (if $\eta < 0,\ \theta > 0$);

- $s, s^{(2)}, n, n; N^d$, ©, © $\Rightarrow (0,-1,3,3,0,0), (0,0,1,2/15,1,3)$ (if $\eta < 0,\ \theta = 0$);

- $s, s^{(2)}, n, n; S,\ N^\infty,\ N^\infty \Rightarrow (0,-9,21,8,-10,0), (0,3,9,0,2,-4)$ (if $\eta > 0,\ \theta < 0,\ \theta_1 < 0$);

- $s, s^{(2)}, n, n; S,\ N^f,\ N^f \Rightarrow (0,-15,-40,16,39/2,0), (0,-6,15,0,-8,7)$ (if $\eta > 0,\ \theta < 0,\ \theta_1 > 0$);

- $s, s^{(2)}, n, n; S,\ N^f,\ N^\infty \Rightarrow (0,4,4,-4,4,-4), (0,17,-4,-17,-18,4)$ (if $\eta > 0,\ \theta > 0$);

- $s, s^{(2)}, n, n; S,\ N^\infty,\ N^d \Rightarrow (0,15,-7,-8,0,0), (0,24,-15,0,-5,1)$ (if $\eta > 0,\ \theta = 0,\ \theta_1 < 0$);

- $s, s^{(2)}, n, n; S,\ N^f,\ N^d \Rightarrow (0,7,-4,-3,0,0), (0,6,-7,0,-7/2,8)$ (if $\eta > 0,\ \theta = 0,\ \theta_1 > 0$);

- $s, s^{(2)}, n, n; \overline{\binom{0}{2}}SN,\ N^\infty \Rightarrow \big(0,\xi,-(\xi+3),2,1/2,0\big), (0,\xi-2,-\xi,0,1/2,1)$, where $\xi = -(2+\sqrt{7})/2$ (if $\eta = 0,\ \theta < 0$);

- $s, s^{(2)}, n, n; \overline{\binom{0}{2}}SN,\ N^f \Rightarrow (0,\xi,1/2-\xi,1/2,-1/2,0), (0,\xi+3/2,-\xi,0,-1/4,-1)$, where $\xi = (\sqrt{145}-5)/20$ (if $\eta = 0,\ \theta > 0$);

- $s, s^{(2)}, n, n; \overline{\binom{0}{2}}SN,\ N^d \Rightarrow \big(0,\xi,-(\xi+3),1,1,0\big), (0,\xi-2,-\xi,0,0,2)$ where $\xi = -(\sqrt{7}+1)/2$ (if $\eta = 0,\ \theta = 0$);

- $s, s^{(2)}, n, n; \overline{\binom{0}{3}}N \Rightarrow (0,1,1,-1,1/2,-1), (0,0,-1,0,-1/2,1)$ (if $\eta = 0,\ \widetilde{M} = 0$).

3. $\mathcal{F}_1 = \mathcal{F}_2 = 0$. Then by Theorem 6.2 the weak saddle could either be of order three or be an integrable saddle. So considering Theorems 7.9 and 7.3, in this case we could have only the following six global configurations if $\mathcal{F}_3\mathcal{F}_4 \neq 0$:
   - $s, s^{(3)}, n, n;\ N^\infty$, ©, © $\Rightarrow (0, 1, 0, -2, 1/2, -2/3), (0, 0, -1, 1/2, -1/2, 2)$ (if $\eta < 0$);
   - $s, s^{(3)}, n, n;\ S,\ N^\infty,\ N^\infty \Rightarrow (0, 10, 0, -20, 5, 1), (0, 0, -10, -2, -5, 20)$ (if $\eta > 0,\ \theta < 0,\ \theta_1 < 0$);
   - $s, s^{(3)}, n, n;\ S,\ N^f,\ N^f \Rightarrow (0, 10, 0, -20, 5, 7), (0, 0, -10, -8, -5, 20)$, (if $\eta > 0,\ \theta < 0,\ \theta_1 > 0$);
   - $s, s^{(3)}, n, n;\ S,\ N^f,\ N^\infty \Rightarrow (0, 1, 0, -2, 1/2, 3/5), (0, 0, -1, -7/5, -1/2, 2)$ (if $\eta > 0,\ \theta > 0$);
   - $s, s^{(3)}, n, n;\ S,\ N^f,\ N^d \Rightarrow (0, 1, 0, -2, 1/2, 5/8), (0, 0, -1, -1, -1/2, 2)$ (if $\eta > 0,\ \theta = 0$);
   - $s, s^{(3)}, n, n;\ \overline{\binom{0}{2}}SN,\ N^\infty \Rightarrow (0, 1, 0, -2, 1/2, -1/4), (0, 0, -1, 0, -1/2, 2)$ (if $\eta = 0$).

   In the case $\mathcal{F}_3\mathcal{F}_4 = 0$ we have the next three global configurations:
   - $s, \$, n, n;\ S,\ N^\infty,\ N^\infty \Rightarrow (0, 1, 0, 9/2, 0, 1), (0, 0, -1, -1, 0, -4)$ (if $\eta > 0,\ \theta < 0,\ \theta_1 < 0$);
   - $s, \$, n, n;\ S,\ N^f,\ N^f \Rightarrow (0, 1, 0, -2, 0, 1), (0, 0, -1, -1, 0, 1)$ (if $\eta > 0,\ \theta < 0,\ \theta_1 > 0$);
   - $s, \$, n, n;\ S,\ N^f,\ N^\infty \Rightarrow (0, 1, 0, -2, 0, 1), (0, 0, -1, -1, 0, 0)$ (if $\eta > 0,\ \theta > 0$).

2.2.2.3: $\mathcal{T}_4 = \mathcal{T}_3 = 0$. In this case the condition $\mathcal{T}_2 \neq 0$ must hold due to the existence of nodes. Therefore by Theorem 6.2 we have two weak singularities, which are weak saddles of order one if $\mathcal{F}_1 \neq 0$ and integrable saddles if $\mathcal{F}_1 = 0$.

1. $\mathcal{F}_1 \neq 0$. So we have two weak saddles of order one, and considering Lemma 11.22 (see statements (iii) and (vii)) we conclude that the condition $\theta < 0$ must hold.

   Additionally by Lemma 11.17(ii) we could not have a triple infinite singularity. So according to Lemma 11.1 there remain four possible global configurations, and all these possibilities are realizable:
   - $s^{(1)}, s^{(1)}, n, n;\ N^\infty$, ©, © $\Rightarrow (0, -9, 0, 9, -19/48, 0), (0, 24, 9, -24, -9, -9)$ (if $\eta < 0$);
   - $s^{(1)}, s^{(1)}, n, n;\ S,\ N^\infty,\ N^\infty \Rightarrow (0, -4, 0, 4, -9/20, 0), (0, 10, 4, -10, -4, -4)$, (if $\eta > 0,\ \theta_1 < 0$);

- $s^{(1)}, s^{(1)}, n, n;\ S,\ N^f,\ N^f \Rightarrow (0,0,1,0,-5,2), (0,1,0,-5,0,0)$ (if $\eta > 0,\ \theta_1 > 0$);
- $s^{(1)}, s^{(1)}, n, n;\ \overline{\binom{0}{2}}SN,\ N^\infty \Rightarrow (0,-1,2/3,4/3,-1/2,0), (0,-1/3,1,0,1/6,-1)$ (if $\eta = 0$).

2. $\mathcal{F}_1 = 0$. So we have two integrable saddles, and by Theorem 7.3 we arrive at the following four global configurations:

   - $\$, \$, n, n;\ N^\infty,\ ©, © \Rightarrow (0,-2,-1,2,13/12,1), (0,3,2,-3,-2,-2)$ (if $\eta < 0$);
   - $\$, \$, n, n;\ S,\ N^\infty,\ N^\infty \Rightarrow (0,-1,0,1,-1/6,0), (0,3,1,-3,-1,-1)$ (if $\eta > 0,\ \theta_1 < 0$);
   - $\$, \$, n, n;\ S,\ N^f,\ N^f \Rightarrow (0,-1,1,1,-1,-1), (0,2,1,-2,-1,-1)$ (if $\eta > 0,\ \theta_1 > 0$);
   - $\$, \$, n, n;\ \overline{\binom{0}{3}}N \Rightarrow (0,-1,1,2,-1,0), (0,0,1,-1,1,-2)$ (if $\eta = 0$).

2.3: $W_4 = 0$. In this case apart from two saddles and a node with coinciding eigenvalues, we have an anti-saddle that could be either a focus or a node.

2.3.1: $W_3 < 0$. Then by Theorem 6.1 (see Table 6.2, line 7) the anti-saddle is a focus, and since by Theorem 7.2 we could not have a star node apart from a focus, we conclude that the node with coinciding eigenvalues must be n^d.

Remark 11.9. We observe that by Lemma 11.22(iv), in the case under consideration, the condition $\theta > 0$ must hold. Moreover by Lemma 11.17(iv) at infinity we could not have a triple singularity, i.e. the condition $\widetilde{M} \neq 0$ is satisfied.

We consider two cases: $\mathcal{T}_4 \neq 0$ and $\mathcal{T}_4 = 0$.

2.3.1.1: $\mathcal{T}_4 \neq 0$. In this case by Theorem 6.2 we do not have weak singularities, and hence the configuration of finite singularities is well defined. So considering the above remark and Lemma 11.1, there remain three possible global configurations, and all these possibilities are realizable:

- $s, s, n^d, f;\ N^f,\ ©, © \Rightarrow (0,1,1,-1,0,-1), (0,-1/4,0,1/4,1,0)$ (if $\eta < 0$);
- $s, s, n^d, f;\ S,\ N^f,\ N^\infty \Rightarrow (0,-2,1,2,-1,-1), (0,-1/4,-1,1/4,0,1)$ (if $\eta > 0$);
- $s, s, n^d, f;\ \overline{\binom{0}{2}}SN,\ N^f \Rightarrow (0,-2,-1,1,1,0), (0,9,-8,0,-3/2,2)$ (if $\eta = 0$).

2.3.1.2: $\mathcal{T}_4 = 0,\ \mathcal{T}_3 \neq 0$. According to Theorem 6.2 we have only one weak singularity the type of which is governed by the invariant polynomial $\mathcal{T}_3\mathcal{F}$.

2.3.1.2.1: $\mathcal{T}_3\mathcal{F} < 0$. Then we have a weak focus.

1. $\mathcal{F}_1 \neq 0$. In this case the weak focus has order one. Considering Remark 11.9 and Lemma 11.1, we arrive at three global configurations:

   - $s, s, n^d, f^{(1)};\ N^f,\ ©, © \Rightarrow (0,0,-1,0,2,1), (0,1,2,-1,-1,-2)$ (if $\eta < 0$);

- $s, s, n^d, f^{(1)}$; S, N^f, $N^\infty \Rightarrow (0,3,-9,-3,6,9), (0,4,-9,-4,6,9)$ (if $\eta > 0$);
- $s, s, n^d, f^{(1)}$; $\overline{\binom{0}{2}}SN$, $N^f \Rightarrow (0,-151/16,-9/16,2,4,0)$, $(0,169/16,-73/16,0,-7,8)$ (if $\eta = 0$).

2. $\mathcal{F}_1 = 0$. We observe that by Theorems 7.8 and 7.4, apart from two saddles and a node n^d, we could have neither a weak focus of order three nor a center. So the condition $\mathcal{F}_2 \neq 0$ holds, i.e. by Theorem 6.2 the weak focus has order two, and considering Theorem 7.7 we get the unique global configuration
   - $s, s, n^d, f^{(2)}$; N^f, ©, © $\Rightarrow \big(0,0,-1,(2+\xi)/3,1,1\big), \big(0,1,0,1,(\xi-4)/6,-1\big)$ where $\{W_4^{-1}(0)\} \ni \xi \approx 1.7251$.

2.3.1.2.2: $\mathcal{T}_3\mathcal{F} > 0$. Then by Theorem 6.2 we have one weak saddle.

1. $\mathcal{F}_1 \neq 0$. In this case the weak saddle has order one, and considering Remark 11.9 and Lemma 11.1, we arrive at the next three global configurations:
   - $s, s^{(1)}, n^d, f$; N^f, ©, © $\Rightarrow (0,4,1,-4,0,-1), (0,-1,2,1,1,-2)$ (if $\eta < 0$);
   - $s, s^{(1)}, n^d, f$; S, N^f, $N^\infty \Rightarrow (0,10,-1,-10,1,1), (0,9,4,-9,3,-4)$ (if $\eta > 0$);
   - $s, s^{(1)}, n^d, f$; $\overline{\binom{0}{2}}SN$, $N^f \Rightarrow (0,49,-25,8,-16,0), (0,1,39,0,-4,-32)$ (if $\eta = 0$).
2. $\mathcal{F}_1 = 0$. First of all we remark that by Theorems 7.9 and 7.3, in the case under consideration, we could have neither a weak saddle of order three nor an integrable saddle. So the condition $\mathcal{F}_2 \neq 0$ holds, i.e. by Theorem 6.2 the weak saddle has order two.

   Additionally, by Remark 11.9 the conditions $\theta > 0$ and $\widetilde{M} \neq 0$ hold. Moreover according to Lemma 11.22 (statements (v) and (ix)), the condition $\eta < 0$ must hold. As a result we get the unique global configuration
   - $s, s^{(2)}, n^d, f$; N^f, ©, © $\Rightarrow (0,1,0,0,0,-1), (0,0,-1,-1,1,0)$.

2.3.1.3: $\mathcal{T}_4 = \mathcal{T}_3 = 0$. In this case the condition $\mathcal{T}_2 \neq 0$ must hold, otherwise by Lemma 11.16 the systems become Hamiltonian, which is impossible due to the existence of a node.

Therefore by Theorem 6.2 we have two weak singularities that could be both saddles or one saddle and one focus, depending on the sign of the invariant polynomial $\mathcal{T}_2$. Moreover by Theorem 7.3 we could not have an integrable saddle apart from a node n^d and a focus (or a center). Therefore according to Theorem 6.2 in this case the condition $\mathcal{F}_1 \neq 0$ is satisfied, and hence we could have only two weak singularities of order one.

1. $\mathcal{T}_2 < 0$. So we have two weak saddles of order one, and considering Remark 11.9 (i.e. the conditions $\theta > 0$ and $\widetilde{M} \neq 0$ are fulfilled), according to Lemma 11.1 we get the next three global configurations:

- $s^{(1)}, s^{(1)}, n^d, f$; N^f, ©, © $\Rightarrow (0, 40, -144, -40, 12, 144)$, $(0, 1, 64, -1, -12, -64)$ (if $\eta < 0$);
- $s^{(1)}, s^{(1)}, n^d, f$; S, N^f, $N^\infty \Rightarrow (0, 109/2, -225, -109/2, 15, 225)$, $(0, 1, 169/2, -1, -15, -169/2)$ (if $\eta > 0$);
- $s^{(1)}, s^{(1)}, n^d, f$; $\overline{\binom{0}{2}}SN$, $N^f \Rightarrow (0, 6\sqrt{3} - 1, -1, 1 - 6\sqrt{3}, 1, 1)$, $(0, 1, 1 + 6\sqrt{3}, -1, -1, -1 - 6\sqrt{3})$ (if $\eta = 0$).

2. $\mathcal{T}_2 > 0$. In this case we have a weak saddle and a weak focus both of order one, and again considering Remark 11.9, we arrive at the next three global configurations:
   - $s, s^{(1)}, n^d, f^{(1)}$; N^f, ©, © $\Rightarrow (0, 109/2, -256, -109/2, 16, 256)$, $(0, 1, 173/2, -1, -16, -173/2)$ (if $\eta < 0$);
   - $s, s^{(1)}, n^d, f^{(1)}$; S, N^f, $N^\infty \Rightarrow (0, 60, -289, -60, 17, 289)$, $(0, 1, 94, -1, -17, -94)$ (if $\eta > 0$);
   - $s, s^{(1)}, n^d, f^{(1)}$; $\overline{\binom{0}{2}}SN$, $N^f \Rightarrow (0, \xi, -256, -\xi, 16, 256)$, $(0, 1, 32 + \xi, -1, -16, -32 - \xi)$ where $\{\eta^{-1}(0)\} \ni \xi \approx 55.7133$ (if $\eta = 0$).

2.3.2: $W_3 > 0$. In this case apart from two saddles and a node with coinciding eigenvalues, we have a generic node.

Remark 11.10. We remark that by Theorem 7.2 a quadratic system (8.1) with four real distinct finite singularities could possess a finite star node only if this system has three real infinite singularities, i.e. the condition $\eta > 0$ must hold in this case. Moreover, if we have a finite star node and $\mathcal{T}_4 \neq 0$, then at infinity we could not have a node with coinciding eigenvalues, i.e. the condition $\theta \neq 0$ must hold.

We consider two cases: $\mathcal{T}_4 \neq 0$ and $\mathcal{T}_4 = 0$.

2.3.2.1: $\mathcal{T}_4 \neq 0$. Then according to Theorem 6.2 we have no weak singularities, and by Theorem 7.5 we could not have an infinite star node together with a node with coinciding eigenvalues and two strong saddles. So considering the above remark and Lemma 11.1, we could have 16 global configurations, and all these possibilities are realizable as shown by the next examples:

- s, s, n, n^d; N^∞, ©, © $\Rightarrow (0, 2, 1, -2, 0, -1), (0, -1/4, 1, 1/4, -1, -1)$ (if $\eta < 0$, $\theta < 0$);
- s, s, n, n^d; N^f, ©, © $\Rightarrow (0, 2, 1, -2, 0, -1), (0, -9/4, -1, 9/4, 1, 1)$ (if $\eta < 0$, $\theta > 0$);
- s, s, n, n^d; N^d, ©, © $\Rightarrow \big(0, 2(1 - \sqrt{2}), -3 + 2\sqrt{2}, 1, 0, 0\big), (0, 1, 0, -1, 1/2, -1)$ (if $\eta < 0$, $\theta = 0$);
- s, s, n, n^d; S, N^∞, $N^\infty \Rightarrow (0, 6, 4, -6, 0, -4), (0, -1, 2, 1, -4, -2)$ (if $\eta > 0$, $\theta < 0$, $\theta_1 < 0$, $\neg(\mathcal{S}^1_{n^*})$);

- $s, s, n, n^*;\ S,\ N^\infty,\ N^\infty \Rightarrow (0,-3,0,3,1,0), (0,0,-3,0,1,3)$ (if $\eta>0,\ \theta<0,\ \theta_1<0,\ (\mathcal{S}^1_{n^*})$);
- $s, s, n, n^d;\ S,\ N^f,\ N^f \Rightarrow (0,1,1,-1,0,-1), (0,-1/4,0,1/4,-2,0)$ (if $\eta>0,\ \theta<0,\ \theta_1>0,\ \neg(\mathcal{S}^1_{n^*})$);
- $s, s, n, n^*;\ S,\ N^f,\ N^f \Rightarrow (0,3,0,-3,1,0), (0,0,3,0,1,-3)$ (if $\eta>0,\ \theta<0,\ \theta_1>0,\ (\mathcal{S}^1_{n^*})$);
- $s, s, n, n^d;\ S,\ N^f,\ N^\infty \Rightarrow (0,1,1,-1,0,-1), (0,-1/4,0,1/4,-1,0)$ (if $\eta>0,\ \theta>0,\ \neg(\mathcal{S}^1_{n^*})$);
- $s, s, n, n^*;\ S,\ N^f,\ N^\infty \Rightarrow (0,1,0,-1,2,0), (0,0,1,0,-2,-1)$ (if $\eta>0,\ \theta>0,\ (\mathcal{S}^1_{n^*})$);
- $s, s, n, n^d;\ S,\ N^\infty,\ N^d \Rightarrow (0,1,-9/4,1,1/8,0), (0,1,-2,0,0,1)$ (if $\eta>0,\ \theta=0,\ \theta_1<0$);
- $s, s, n, n^d;\ S,\ N^f,\ N^d \Rightarrow (0,1,1/4,1,-9/8,0), (0,-1,0,0,0,1)$ (if $\eta>0,\ \theta=0,\ \theta_1>0$);
- $s, s, n, n^d;\ S,\ N^d,\ N^d \Rightarrow (0,0,-1,1,0,0), (0,1/4,1,0,0,-5/4)$ (if $\eta>0,\ \theta=0,\ \theta_1=0$);
- $s, s, n, n^d;\ \overline{\binom{0}{2}}SN,\ N^\infty \Rightarrow (0,1,-1,2,-1,0), (0,0,1,0,1/2,-2)$ (if $\eta=0,\ \widetilde{M}\neq 0,\ \theta<0$);
- $s, s, n, n^d;\ \overline{\binom{0}{2}}SN,\ N^f \Rightarrow (0,-4,-4,2,3,0), (0,0,-4,0,-1,6)$ (if $\eta=0,\ \widetilde{M}\neq 0,\ \theta>0$);
- $s, s, n, n^d;\ \overline{\binom{0}{2}}SN,\ N^d \Rightarrow \big(0,2(2-\sqrt{2}),2\sqrt{2}-3,1,-1,0\big), (0,-2,0,0,0,2)$ (if $\eta=0,\ \widetilde{M}\neq 0,\ \theta=0$);
- $s, s, n, n^d;\ \overline{\binom{0}{3}}N \Rightarrow \big(0,2(2-\sqrt{2}),2\sqrt{2}-3,0,-1/2,0\big), (0,2,0,-1,0,-1)$ (if $\eta=0,\ \widetilde{M}=0$).

2.3.2.2: $\mathcal{T}_4=0,\ \mathcal{T}_3\neq 0$. Then we have one weak saddle, the order of which is governed by the invariant polynomials $\mathcal{F}_j=0,\ j=1,2,3,4$.

1. $\mathcal{F}_1\neq 0$. Then the weak saddle has order one. We remark that by Theorem 7.2, in the case under consideration, we could not have a finite star node, whereas by Theorem 7.5 we could have only one infinite star node and only in the case $\eta\neq 0$. We also observe that by Lemma 11.19(ii) we could not have two N^d's. So there remain 16 possibilities of global configurations, and all are realizable:

   - $s, s^{(1)}, n, n^d;\ N^\infty,\ ©, © \Rightarrow (0,4,4,-4,0,-4),\ (0,-1/4,6,1/4,-1,-6)$ (if $\eta<0,\ \theta<0$);
   - $s, s^{(1)}, n, n^d;\ N^f,\ ©, © \Rightarrow (0,8,4,-8,0,-4), (0,-9/4,2,9/4,3,-2)$ (if $\eta<0,\ \theta>0$);
   - $s, s^{(1)}, n, n^d;\ N^d,\ ©, © \Rightarrow (0,-7,-1,8,0,0), (0,9,-1,-8,4,-8)$ (if $\eta<0,\ \theta=0,\ \theta_2\neq 0$);
   - $s, s^{(1)}, n, n^d;\ N^*,\ ©, © \Rightarrow (0,1,0,-1,0,0), (0,2,1,-1,-1/2,-1)$ (if $\eta<0,\ \theta=0,\ \theta_2=0$);

- $s, s^{(1)}, n, n^d$; S, N^∞, $N^\infty \Rightarrow (0, 6, 4, -6, -8, -4), (0, -1/4, 8, 1/4, -1, -8)$ (if $\eta > 0$, $\theta < 0$, $\theta_1 < 0$);
- $s, s^{(1)}, n, n^d$; S, N^f, $N^f \Rightarrow (0, 0, 1, 0, 0, -1), (0, -1/4, 1, 1/4, -1/2, -1)$ (if $\eta > 0$, $\theta < 0$, $\theta_1 > 0$);
- $s, s^{(1)}, n, n^d$; S, N^f, $N^\infty \Rightarrow (0, 1, 1, -1, 0, -1), (0, -1/4, 2, 1/4, -1/2, -2)$ (if $\eta > 0$, $\theta > 0$);
- $s, s^{(1)}, n, n^d$; S, N^∞, $N^d \Rightarrow (0, -31, -1, 8, 12, 0), (0, 25, -41, 0, 0, 16)$ (if $\eta > 0$, $\theta = 0$, $\theta_1 < 0$, $\theta_2 \neq 0$);
- $s, s^{(1)}, n, n^d$; S, N^∞, $N^* \Rightarrow (0, 2, -2, 1, -1/2, 0), (0, 0, 2, 0, 0, -2)$ (if $\eta > 0$, $\theta = 0$, $\theta_1 < 0$, $\theta_2 = 0$);
- $s, s^{(1)}, n, n^d$; S, N^f, $N^d \Rightarrow (0, -47, -1, 12, 18, 0), (0, 25, -37, 0, 0, 12)$ (if $\eta > 0$, $\theta = 0$, $\theta_1 > 0$, $\theta_2 \neq 0$);
- $s, s^{(1)}, n, n^d$; S, N^f, $N^* \Rightarrow (0, 1, -3, 1, 1/2, 0), (0, 0, 1, 0, 0, -1)$ (if $\eta > 0$, $\theta = 0$, $\theta_1 > 0$, $\theta_2 = 0$);
- $s, s^{(1)}, n, n^d$; S, N^d, $N^* \Rightarrow (0, 1, -2, 1, 0, 0), (0, 0, 3, 0, 0, -3)$ (if $\eta > 0$, $\theta = 0$, $\theta_1 = 0$);
- $s, s^{(1)}, n, n^d$; $\overline{\binom{0}{2}} SN$, $N^\infty \Rightarrow (0, -7, -9, 8, 4, 0), (0, 25, -37, 0, 2, 8)$ (if $\eta = 0$, $\widetilde{M} \neq 0$, $\theta < 0$);
- $s, s^{(1)}, n, n^d$; $\overline{\binom{0}{2}} SN$, $N^f \Rightarrow (0, -31, -9, 8, 16, 0), (0, 49, -73, 0, -4, 32)$ (if $\eta = 0$, $\widetilde{M} \neq 0$, $\theta > 0$);
- $s, s^{(1)}, n, n^d$; $\overline{\binom{0}{2}} SN$, $N^d \Rightarrow (0, -7, -1, 4, 2, 0), (0, 9, -13, 0, 0, 4)$ (if $\eta = 0$, $\widetilde{M} \neq 0$, $\theta = 0$);
- $s, s^{(1)}, n, n^d$; $\overline{\binom{0}{3}} N \Rightarrow (0, 65, -25, -32, -4, 0), (0, 1, 55, -16, -16, -8)$ (if $\eta = 0$, $\widetilde{M} = 0$).

2. $\mathcal{F}_1 = 0$, $\mathcal{F}_2 \neq 0$. By Theorem 6.2 the weak saddle has order two. We observe that by Theorem 7.2 (respectively Theorem 7.5) we could not have a finite (respectively an infinite) star node. We also observe that by Lemma 11.19(ii) we could not have two N^d's.

   Additionally by Lemma 11.17(i_1) we could not have a triple infinite singularity. So according to Lemma 11.1 there remain 11 possibilities, and all these global configurations are realizable as shown in the examples:

- $s, s^{(2)}, n, n^d$; N^∞, ©, © $\Rightarrow \big(0, 10, 10, -10, -(61 + 13\sqrt{737})/192, -10\big)$, $(0, -9/10, 16, 9/10, -3, -16)$ (if $\eta < 0$, $\theta < 0$);
- $s, s^{(2)}, n, n^d$; N^f, ©, © $\Rightarrow \big(0, 8, 16, -8, -(15 + \sqrt{865})/4, -16\big)$, $(0, -1, 16, 1, -4, -16)$ (if $\eta < 0$, $\theta > 0$);

- $s, s^{(2)}, n, n^d;\ N^d,$ ©, © $\Rightarrow (0, 71, -25, 184/21, -25, 25), (0, 25, 21, 0, 0, -21)$ (if $\eta < 0,\ \theta = 0$);
- $s, s^{(2)}, n, n^d;\ S,\ N^\infty,\ N^\infty \Rightarrow \big(0, 10, 1, -100, (13\sqrt{737} - 61)/192, -1\big)$, $(0, -9, 16, 90, -30, -16)$ (if $\eta > 0,\ \theta < 0,\ \theta_1 < 0$);
- $s, s^{(2)}, n, n^d;\ S,\ N^f,\ N^f \Rightarrow \big(0, 8, 4, -8, (\sqrt{2361} - 27)/12, -4\big)$, $(0, -1, 12, 1, -2, -12)$ (if $\eta > 0,\ \theta < 0,\ \theta_1 > 0$);
- $s, s^{(2)}, n, n^d;\ S,\ N^f,\ N^\infty \Rightarrow \big(0, 6, 6, -6, (5\sqrt{154} - 84)/14, -6\big)$, $(0, -8/3, 14, 8/3, -4, -14)$ (if $\eta > 0,\ \theta > 0$);
- $s, s^{(2)}, n, n^d;\ S,\ N^\infty,\ N^d \Rightarrow \big(0, -(8 + 20\xi + 3\xi^2)/(4(1+\xi)), -(2-\xi)^2/(4(1+\xi)), 1, (2+\xi)/2, 0\big)$, $\big(0, (4+\xi)^2/(4(1+\xi)), -(24 + 16\xi + \xi^2)/(4(1+\xi)), 0, 0, 2\big)$ where $\{\mathcal{F}_1^{-1}(0)\} \ni \xi \approx 0.6577$ (if $\eta > 0,\ \theta = 0,\ \theta_1 < 0$);
- $s, s^{(2)}, n, n^d;\ S,\ N^f,\ N^d \Rightarrow (0, 7, 8, 3, -9, 0), (0, -2, -1, 0, 0, 3)$ (if $\eta > 0,\ \theta = 0,\ \theta_1 > 0$);
- $s, s^{(2)}, n, n^d;\ \overline{\binom{0}{2}}SN,\ N^\infty \Rightarrow \big(0, -(4 + 4\xi - \xi^2), -\xi^2, 4\xi, 2, 0\big), \big(0, (2+\xi)^2, -(4 + 8\xi + \xi^2), 0, 2(\xi - 1), 4\big)$ where $\{\mathcal{F}_1^{-1}(0)\} \ni \xi \approx 1.1129$ (if $\eta = 0,\ \theta < 0$);
- $s, s^{(2)}, n, n^d;\ \overline{\binom{0}{2}}SN,\ N^f \Rightarrow \big(0, 8 - 280\xi + 49\xi^2, -(6 - 7\xi)^2, 196\xi, 14, 0\big)$, $\big(0, (8 + 7\xi)^2, 104 - 308\xi - 49\xi^2, 0, 98(\xi - 1), 28\big)$ where $\{\mathcal{F}_1^{-1}(0)\} \ni \xi \approx 0.009473$ (if $\eta = 0,\ \theta > 0$);
- $s, s^{(2)}, n, n^d;\ \overline{\binom{0}{2}}SN,\ N^d \Rightarrow (0, \xi^2 - 4\xi - 4, -\xi^2, 4, 2\xi, 0), \big(0, (2+\xi)^2, -4 - 8\xi - \xi^2, 0, 0, 4\xi\big)$ where $\{\mathcal{F}_1^{-1}(0)\} \ni \xi \approx 1.4713$ (if $\eta = 0,\ \theta = 0$).

3. $\mathcal{F}_1 = \mathcal{F}_2 = 0$. By Theorem 6.2 in this case we have either a weak saddle of order three or an integrable saddle. Considering Theorems 7.9 and 7.3 as well as the condition $W_3 \neq 0$ (i.e. we have only one n^d), we conclude that there exist the following six global configurations if $\mathcal{F}_3\mathcal{F}_4 \neq 0$:

   - $s, s^{(3)}, n, n^d;\ N^\infty,$ ©, © $\Rightarrow (0, 1, 0, -2, 1/2, -1/2), (0, 0, -1, \xi, -1/2, 2)$ where $\{W_4^{-1}(0)\} \ni \xi \approx 0.2812$ (if $\eta < 0$);
   - $s, s^{(3)}, n, n^d;\ S,\ N^\infty,\ N^\infty \Rightarrow (0, 1, 0, -2, 1/2, 0), \big(0, 0, -1, 6(11\sqrt{6} - 27), -1/2, 2\big)$ (if $\eta > 0,\ \theta < 0,\ \theta_1 < 0$);
   - $s, s^{(3)}, n, n^d;\ S,\ N^f,\ N^f \Rightarrow (0, 1, 0, -2, 1/2, \xi), (0, 0, -1, -3/5, -1/2, 2)$ where $\{W_4^{-1}(0)\} \ni \xi \approx 0.5553$ (if $\eta > 0,\ \theta < 0,\ \theta_1 > 0$);
   - $s, s^{(3)}, n, n^d;\ S,\ N^f,\ N^\infty \Rightarrow (0, 1, 0, -2, 1/2, \xi), (0, 0, -1, -6/5, -1/2, 2)$ where $\{W_4^{-1}(0)\} \ni \xi \approx 0.4482$ (if $\eta > 0,\ \theta > 0$);
   - $s, s^{(3)}, n, n^d;\ S,\ N^f,\ N^d \Rightarrow (0, 1, 0, -2, 1/2, \xi), (0, 0, -1, -\nu, -1/2, 2)$ where $\{\theta^{-1}(0), W_4^{-1}(0)\} \ni (\xi, \nu) \approx (0.5503, 0.7749)$ (if $\eta > 0,\ \theta = 0$);

- $s, s^{(3)}, n, n^d$; $\overline{\binom{0}{2}}SN$, $N^\infty \Rightarrow (0,1,0,-2,1/2,\xi), (0,0,-1,-\nu,-1/2,2)$
  where $\{\eta^{-1}(0), W_4^{-1}(0)\} \ni (\xi,\nu) \approx (0.3062, -0.2209)$ (if $\eta = 0$).

  Additionally in the case $\mathcal{F}_3\mathcal{F}_4 = 0$ we get two global configurations:

- $s, \$, n, n^*$; S, N^∞, $N^* \Rightarrow (0,1,0,-1,0,0), (0,0,1,0,-1/3,-1)$ $(\theta_1 < 0)$,
- $s, \$, n, n^*$; S, N^f, $N^* \Rightarrow (0,1,0,-1,0,0), (0,0,1,0,1,-1)$ $(\theta_1 > 0)$.

2.3.2.3: $\mathcal{T}_4 = \mathcal{T}_3 = 0$. In this case the condition $\mathcal{T}_2 \neq 0$ must hold, otherwise by Lemma 11.16 systems become Hamiltonian, which is impossible due to the existence of nodes.

Therefore according to Theorem 6.2 we have two weak saddles of order one if $\mathcal{F}_1 \neq 0$ and two integrable saddles if $\mathcal{F}_1 = 0$. However by Theorem 7.3 we could not have only one n^d apart from two integrable saddles. So the condition $\mathcal{F}_1 \neq 0$ must be satisfied, i.e. we have two weak saddles of order one. Since apart from these saddles we have one node n^d, by Theorem 7.6 we arrive at the following four global configurations:

- $s^{(1)}, s^{(1)}, n, n^d$; N^∞, ©, © $\Rightarrow (0,-13,1/4,13,2,-1/4)$, $(0,-16,-9,16,-2,9)$ (if $\eta < 0$);
- $s^{(1)}, s^{(1)}, n, n^d$; S, N^∞, N^∞: $\Rightarrow (0,-9,1/2,9,1,-1/2), (0,-2,-7,2,-1,7)$ (if $\eta > 0$, $\theta_1 < 0$);
- $s^{(1)}, s^{(1)}, n, n^d$; S, N^f, $N^f \Rightarrow (0,-9,1,9,1,-1), (0,-1,-7,1,-1,7)$ (if $\eta > 0$, $\theta_1 > 0$);
- $s^{(1)}, s^{(1)}, n, n^d$; $\overline{\binom{0}{2}}SN$, $N^\infty \Rightarrow (0,-11,1/4,11,2,-1/4), (0,-16,-7,16,-2,7)$ (if $\eta = 0$).

2.3.2: $W_3 = 0$. It is clear that in this case, apart from two saddles, we have two nodes with coinciding eigenvalues.

Remark 11.11. According to Lemma 11.22(vi) in this case the condition $\theta \geq 0$ must hold, and in the case $\theta = 0$ we have $\mathcal{T}_4 = \mathcal{T}_3 = 0$. Moreover by Lemma 11.17(vi) the condition $\widetilde{M} \neq 0$ is satisfied, i.e. we could not have a triple singularity at infinity.

1. $\mathcal{T}_4 \neq 0$. Then according to Theorem 6.2 both saddles are strong, and by the above remark the conditions $\theta > 0$ and $\widetilde{M} \neq 0$ are fulfilled. Additionally by Lemma 11.21 we could have exactly one star node if and only if the condition $(\mathcal{S}^1_{n^*})$ holds. Moreover by the same lemma we could not have two star nodes in the case $\mathcal{T}_4 \neq 0$. So considering Remark 11.10 and Lemmas 11.21 and 11.1, we conclude that, in the case under consideration, we could have only four global configurations, and each one of these possibilities is realizable:
   - s, s, n^d, n^d; N^f, ©, © $\Rightarrow (0,4,4,-4,0,-4), (0,-1,0,1,1,0)$ (if $\eta < 0$);
   - s, s, n^d, n^d; S, N^f, $N^\infty \Rightarrow (0,-4,4,4,4,-4), (0,-1,0,1,1,0)$ (if $\eta > 0$, $\neg(\mathcal{S}^1_{n^*})$);

- s, s, n^d, n^*; S, N^f, $N^\infty \Rightarrow (0,1,0,-1,-1,0), (0,0,1,0,1,-1)$ (if $\eta > 0$, $(\mathcal{S}^1_{n^*})$);
- s, s, n^d, n^d; $\overline{\binom{0}{2}}SN$, $N^f \Rightarrow (0, 123-\sqrt{10}\chi-\xi, \xi+2\sqrt{10}\chi-283/2, 5/2, 8, 0)$, $(0, \chi/2, \xi, 0, -155/4, 16)$, where $\chi = 4\sqrt{123-2\xi}$, $\xi = (58240-1329\sqrt{239})/2560$ (if $\eta = 0$).

2. $\mathcal{T}_4 = 0$. Then we have at least one weak saddle, and since we have also two finite nodes with coinciding eigenvalues, considering Theorems 7.6, 7.9 and 7.3, we conclude that there could be only 10 global configurations. More exactly we obtain the global configurations
   - $s, s^{(1)}, n^d, n^d$; N^f, ©, © $\Rightarrow (0,-2,-1,2,0,1), (0,1,0,-1,-1,0)$ (if $\mathcal{F}_1 \neq 0$, $\eta < 0$);
   - $s, s^{(1)}, n^d, n^d$; S, N^f, $N^\infty \Rightarrow (0, 3-\sqrt{2}, 1, \sqrt{2}-3, -2, -1)$, $(0, \sqrt{2}-3/2, 1, 3/2-\sqrt{2}, (2-\sqrt{2})/2, -1)$ (if $\mathcal{F}_1 \neq 0$, $\eta > 0$);
   - $s, s^{(1)}, n^d, n^d$; $\overline{\binom{0}{2}}SN$, $N^f \Rightarrow \big(0, -(13+2\xi+\xi^2), 21+6\xi+\xi^2, 4(1-\xi), -6, 0\big)$, $\big(0, -(\xi-1)^2, 13+2\xi+\xi^2, 0, -2\xi, -12\big)$, where $\{W_3^{-1}(0)\} \ni \xi \approx 0.6432$ (if $\mathcal{F}_1 \neq 0$, $\eta = 0$);
   - $s, s^{(2)}, n^d, n^d$; N^f, ©, © $\Rightarrow (0, 5\xi, -16/5, -5\xi, 2, 16/5)$, $\big(0, 5/4, (4+5\xi), -5/4, -(3+5\xi), -(4+5\xi)\big)$, where $\{\mathcal{F}_1^{-1}(0)\} \ni \xi \approx -0.2150$ (if $\mathcal{F}_1 = 0$)

   in the case $\mathcal{T}_3 \neq 0$ and the global configurations
   - $s^{(1)}, s^{(1)}, n^d, n^d$; N^*, ©, © $\Rightarrow (0,1,1,-1,0,-1), (0,0,1,0,0,-1)$ (if $\mathcal{F}_1 \neq 0$, $\eta < 0$);
   - $s^{(1)}, s^{(1)}, n^d, n^d$; S, N^∞, $N^* \Rightarrow (0,1,1/8,-1,0,-1/8), (0,0,1,0,0,-1)$ (if $\mathcal{F}_1 \neq 0$, $\eta > 0$, $\theta_1 < 0$);
   - $s^{(1)}, s^{(1)}, n^d, n^d$; S, N^f, $N^* \Rightarrow (0,-1,1,1,0,-1), (0,0,-1,0,0,1)$ (if $\mathcal{F}_1 \neq 0$, $\eta > 0$ $\theta_1 > 0$);
   - $s^{(1)}, s^{(1)}, n^d, n^d$; $\overline{\binom{0}{2}}SN$, $N^* \Rightarrow (0,4,1,-4,0,-1), (0,0,4,0,0,-4)$ (if $\mathcal{F}_1 \neq 0$, $\eta = 0$);
   - $\$, \$, n^d, n^d$; S, N^d, $N^d \Rightarrow (0,-7,1/2,7,1,-1/2), (0,-2,-5,2,-1,5)$ (if $\mathcal{F}_1 = 0$, $\theta_3 \neq 0$);
   - $\$, \$, n^*, n^*$; S, N^*, $N^* \Rightarrow (0,1,0,-1,0,0), (0,0,1,0,0,-1)$ (if $\mathcal{F}_1 = 0$, $\theta_3 = 0$)

   in the case $\mathcal{T}_3 = 0$.

Thus all the possibilities was examined, and hence we proved that the family of quadratic systems with four distinct real finite singularities possesses 396 global configurations. This completes the proof of Theorem 11.7. □

11.4.2 Two real and two complex elemental singularities

Theorem 11.8. (a) *We consider here all the configurations of singularities, finite and infinite, of quadratic vector fields possessing exactly four elemental finite singularities: two real and two complex. These configurations are classified in Diagram* 11.9 *according to the geometrical equivalence relation of configurations. We have a total of* 239 *geometrically distinct configurations of singularities, finite and infinite.*

(b) *Necessary and sufficient conditions for each one of the* 239 *different geometrical equivalence classes can be assembled from this diagram in terms of* 42 *invariant polynomials with respect to the action of the affine group and time rescaling. These invariant polynomials, which appear in Diagram* 11.9, *are given in Section* 5.4.

(c) *Diagram* 11.9 *actually contains the global bifurcation diagram in the* 12*-dimensional space of parameters of the global configurations of singularities, finite and infinite, of this family of quadratic differential systems and provides an algorithm for finding for any given system in the family considered, its respective geometrical configuration of singularities.*

Remark 11.12. From Table 6.2 of Theorem 6.1 and conditions (6.2), it follows without difficulty that a quadratic system with two complex and two real elemental singularities possesses

(i) two foci if and only if the condition

$$\big(W_4 > 0 \wedge (\mathcal{C}_{ff})\big) \vee \big(W_4 = W_3 = 0 \wedge (\mathcal{C}^{\mathcal{T}}_{ff})\big)$$

are satisfied,

(ii) two nodes if and only if the condition

$$\big(W_4 > 0 \wedge (\mathcal{C}_{nn})\big) \vee \big[(W_4 = W_3 = 0) \wedge ((\mathcal{C}^{\mathcal{T}}_{nn}) \vee (\mathcal{C}^{\mathcal{T}}_{n^d n^d}))\big]$$

hold, where

$$\begin{aligned}
(\mathcal{C}_{ff}) : {} & (V_1 > 0 \wedge W_1 < 0) \vee (V_1 < 0 \wedge V_2 < 0) \vee (V_1 = 0 \wedge V_2 V_3 \neq 0 \wedge W_1 < 0) \\
& \vee (V_1 = V_3 = 0 \wedge V_2 < 0) \vee \big[(V_1 \leq 0 \wedge V_2 = 0) \wedge \big((H_1 \neq 0 \wedge W_{10} \leq 0) \\
& \vee (H_1 = 0 \wedge H_4 H_5 < 0) \vee (H_1 = H_4 = 0 \wedge H_2 < 0)\big)\big]; \\
(\mathcal{C}^{\mathcal{T}}_{ff}) : {} & (V_4 > 0 \wedge W_1 < 0) \vee (V_4 = 0 \wedge H_1 \neq 0 \wedge W_1 < 0 \wedge W_{10} < 0) \\
& \vee (V_4 = 0 \wedge H_1 = 0 \wedge W_1 < 0 \wedge H_4 H_5 < 0); \\
(\mathcal{C}_{nn}) : {} & (V_1 > 0 \wedge W_1 > 0) \vee (V_1 < 0 \wedge V_2 > 0) \vee (V_1 = 0 \wedge V_2 V_3 \neq 0 \wedge W_1 > 0) \\
& \vee (V_1 = V_3 = 0 \wedge V_2 > 0) \vee \big[(V_1 \leq 0 \wedge V_2 = 0) \wedge \big((H_1 \neq 0 \wedge W_{10} > 0) \\
& \vee (H_1 = 0 \wedge H_4 H_5 > 0) \vee (H_1 = H_4 = 0 \wedge H_2 > 0)\big)\big];
\end{aligned}$$

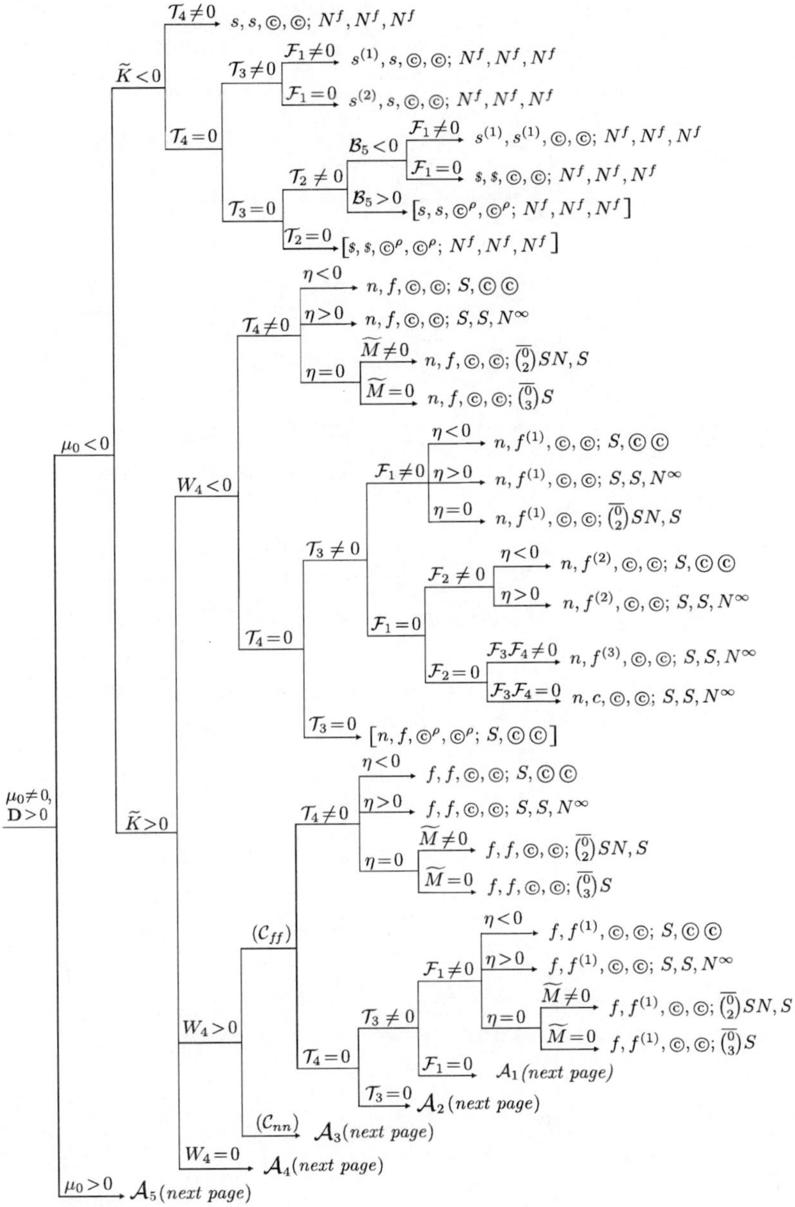

Diagram 11.9: Global configurations in the case $\mu_0\neq 0, \mathbf{D}>0$.

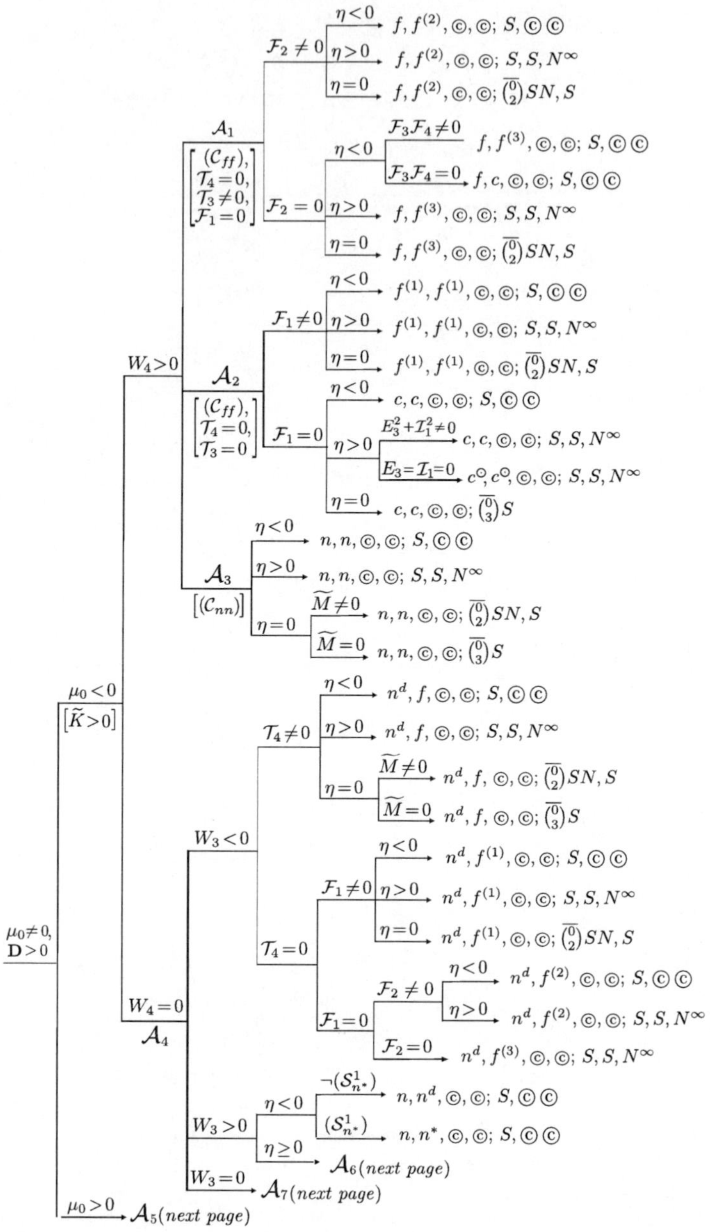

Diagram 11.9 (*cont.*): Global configurations in the case $\mu_0 \neq 0, \mathbf{D} > 0$.

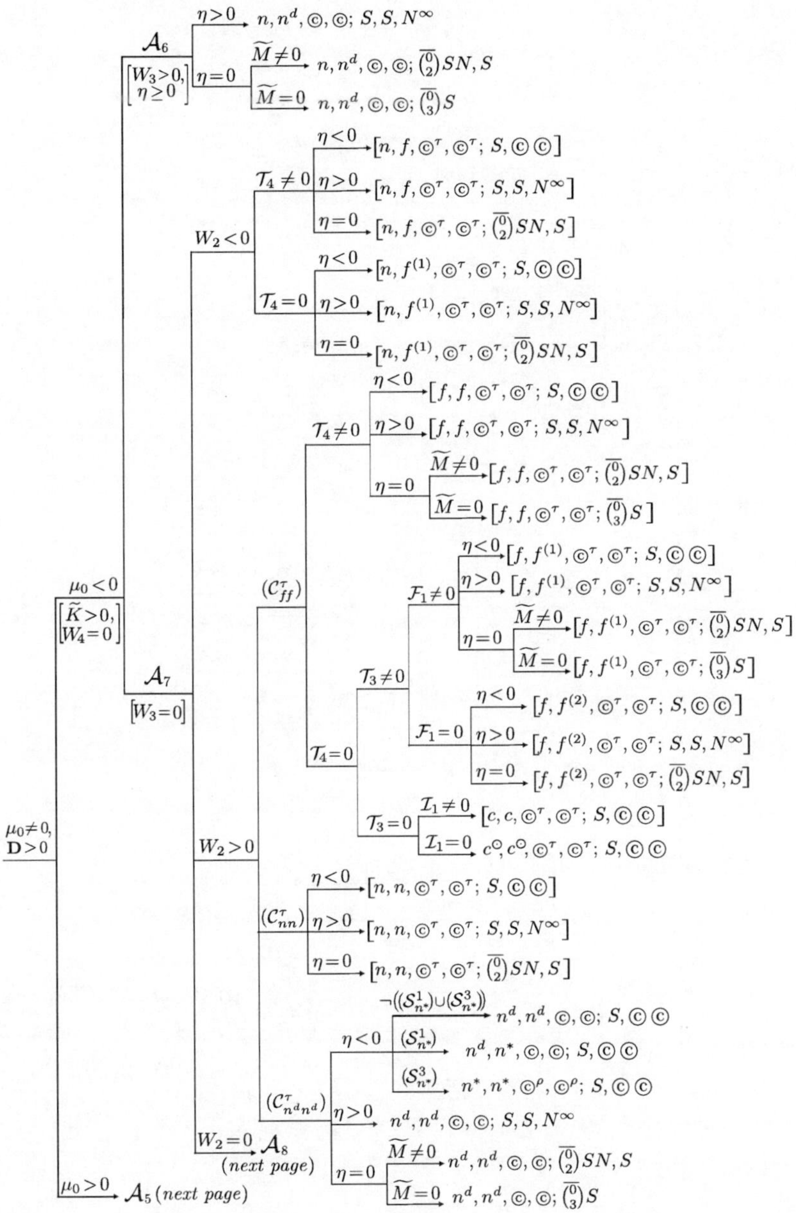

Diagram 11.9 (*cont.*): Global configurations in the case $\mu_0 \neq 0, \mathbf{D} > 0$.

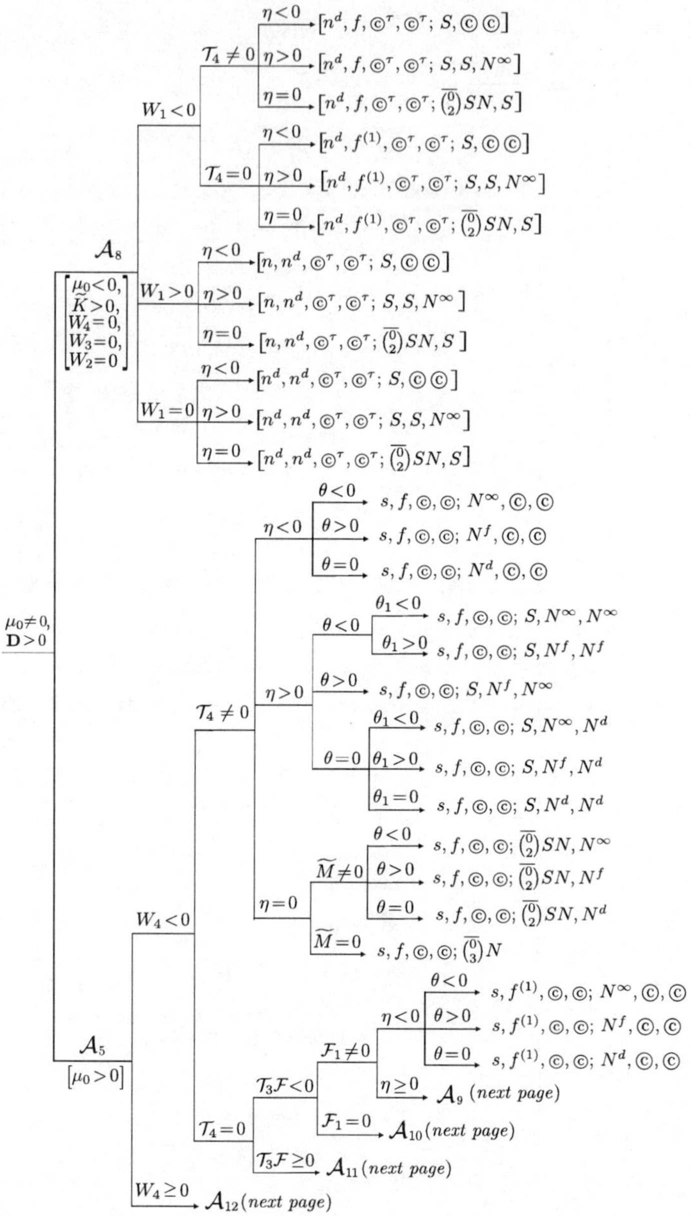

Diagram 11.9 (*cont.*): Global configurations in the case $\mu_0 \neq 0, \mathbf{D} > 0$.

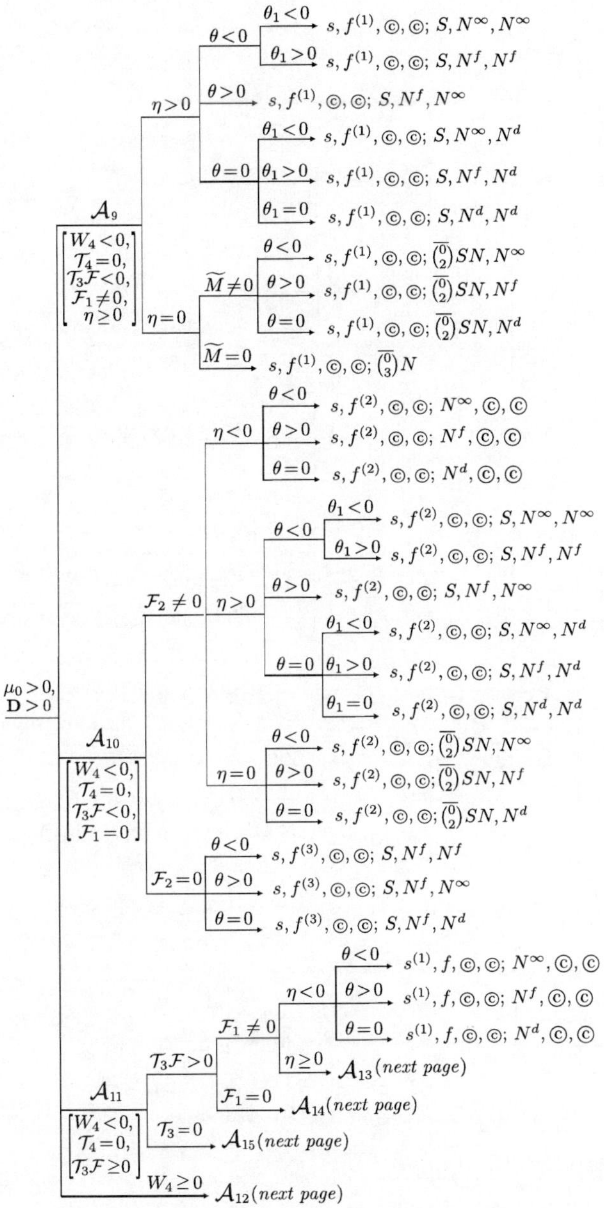

Diagram 11.9 (*cont.*): Global configurations in the case $\mu_0\neq0, \mathbf{D}>0$.

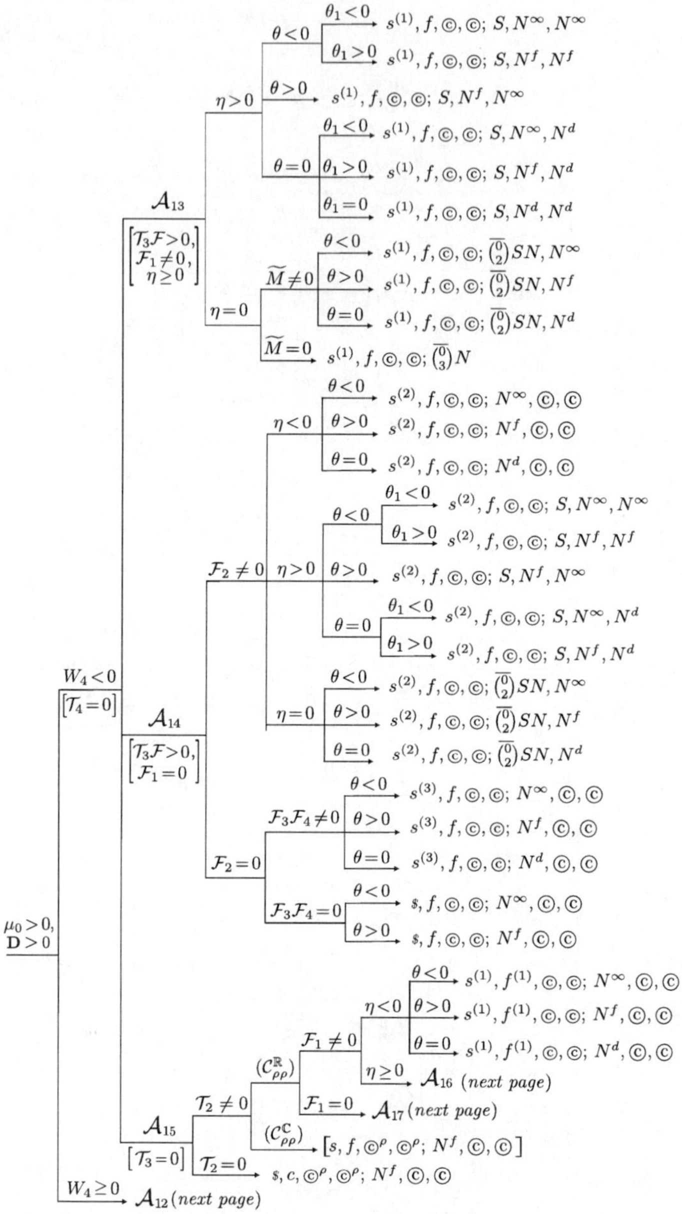

Diagram 11.9 (*cont.*): Global configurations in the case $\mu_0 \neq 0, \mathbf{D} > 0$.

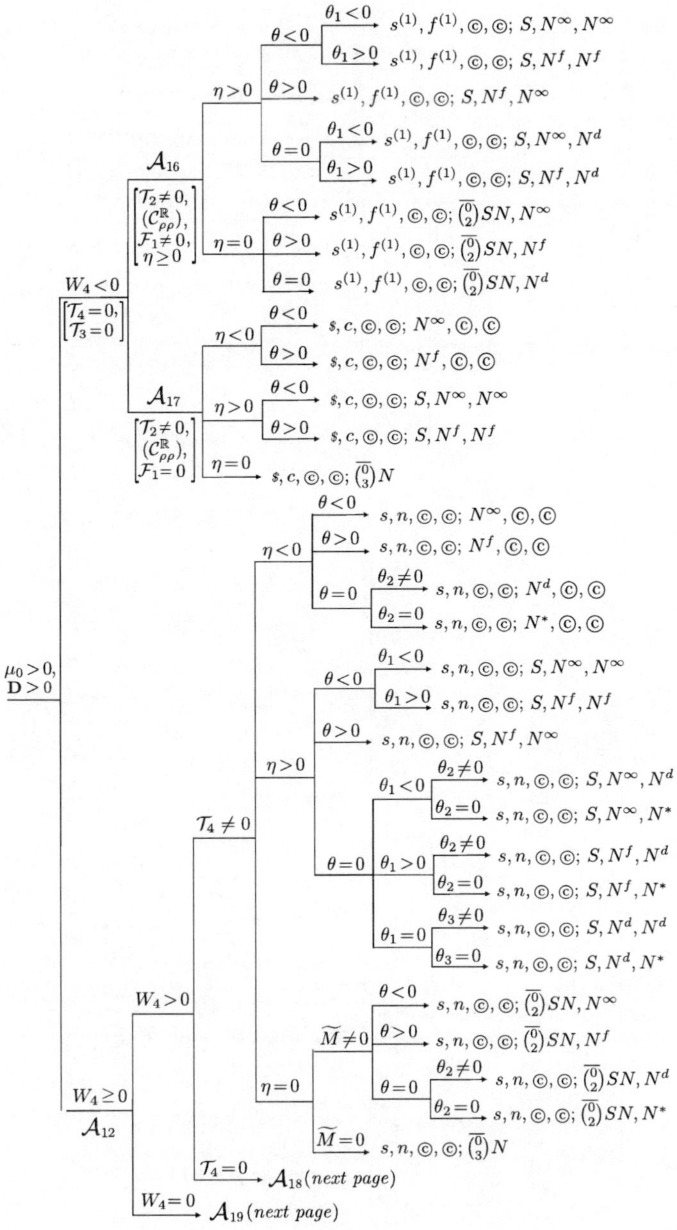

Diagram 11.9 (*cont.*): Global configurations in the case $\mu_0 \neq 0, \mathbf{D} > 0$.

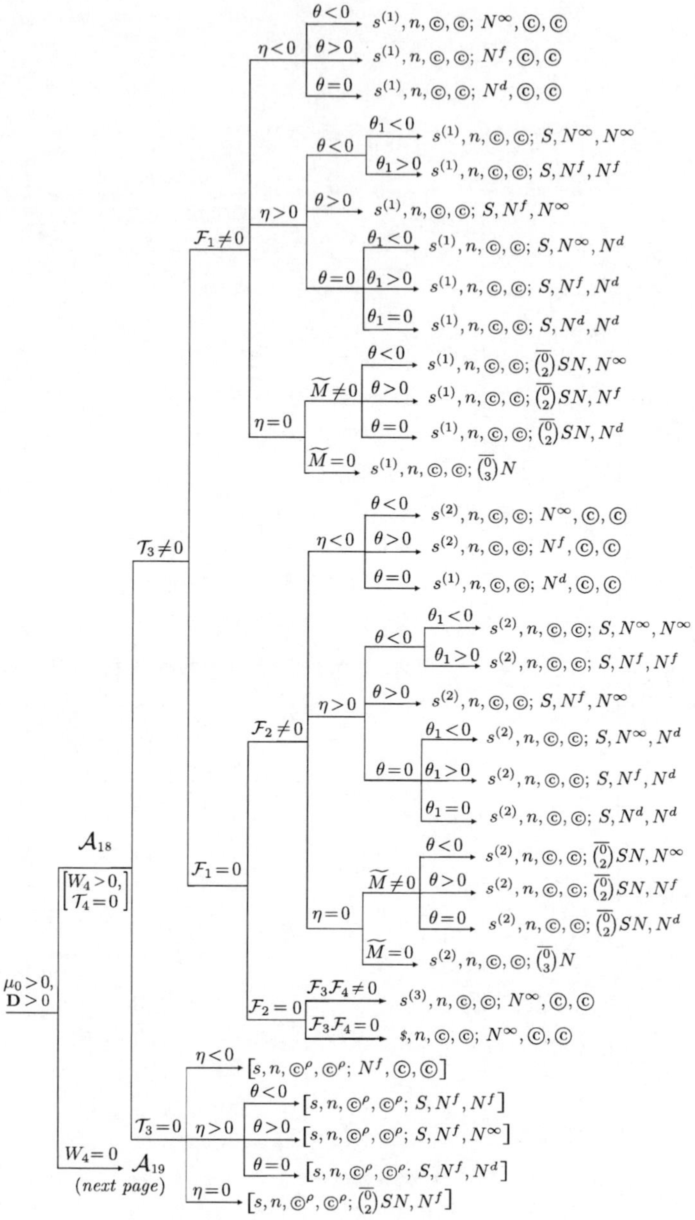

Diagram 11.9 (*cont.*): Global configurations in the case $\mu_0 \neq 0, \mathbf{D} > 0$.

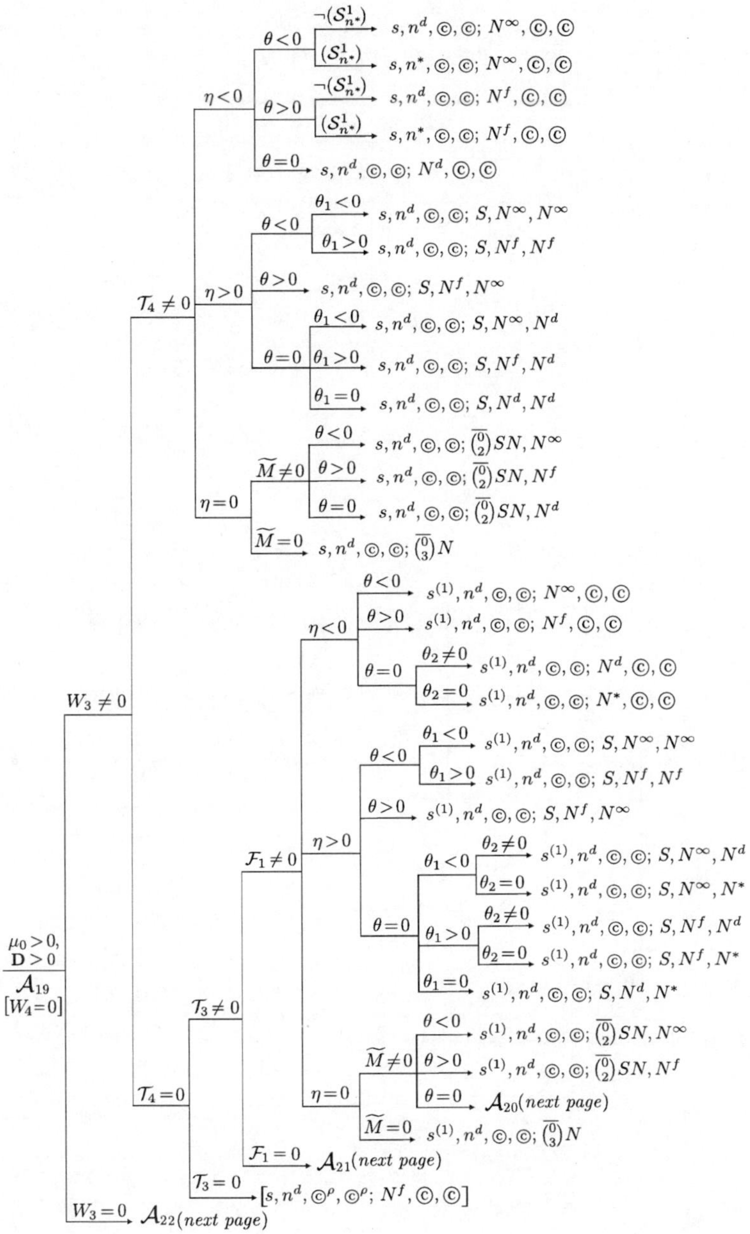

Diagram 11.9 (*cont.*): Global configurations in the case $\mu_0\neq 0, \mathbf{D}>0$.

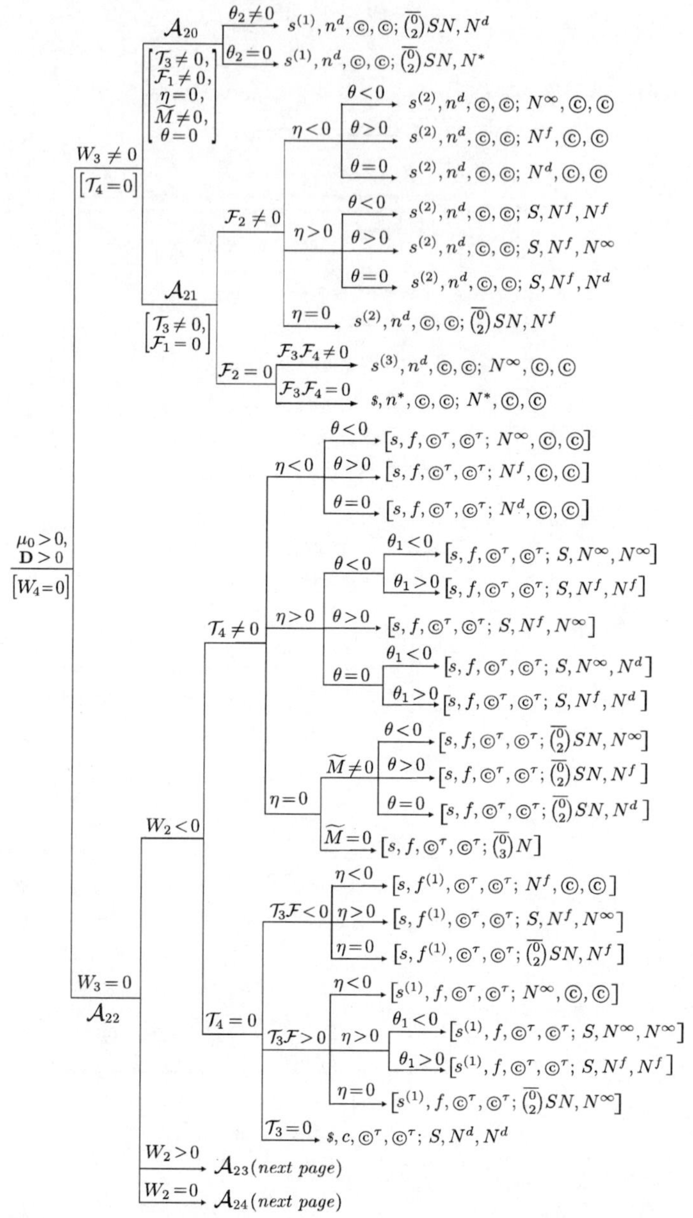

Diagram 11.9 (*cont.*): Global configurations in the case $\mu_0\neq 0, \mathbf{D}>0$.

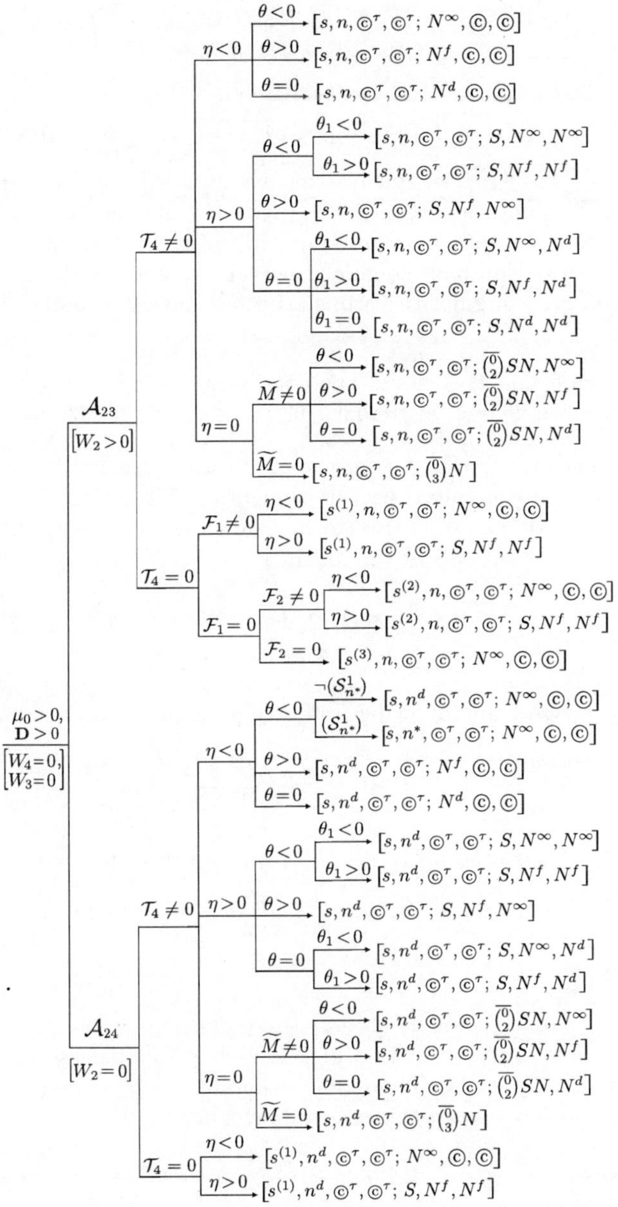

Diagram 11.9 (*cont.*): Global configurations in the case $\mu_0 \neq 0, \mathbf{D} > 0$.

$$(\mathcal{C}^{\mathcal{T}}_{nn}) : (V_4 > 0 \wedge W_1 > 0) \vee (V_4 = 0 \wedge H_1 \neq 0 \wedge W_1 > 0 \wedge W_{10} > 0)$$
$$\vee (V_4 = 0 \wedge H_1 = 0 \wedge W_1 > 0 \wedge H_4H_5 > 0);$$
$$(\mathcal{C}^{\mathcal{T}}_{n^dn^d}) : (V_4 < 0) \vee (V_4 = 0 \wedge H_1 \neq 0 \wedge W_1 > 0 \wedge W_{10} = 0)$$
$$\vee (V_4 = 0 \wedge H_1 \neq 0 \wedge W_1W_{10} < 0) \vee (V_4 = 0 \wedge H_1 = 0 \wedge W_1H_4H_5 < 0).$$

The notation $(\mathcal{C}_{jj})$, $j \in \{f, n\}$, means that we do not have singularities with zero discriminants. The notation $(\mathcal{C}^{\mathcal{T}}_{jj})$, $j \in \{f, n, n^d\}$, means that we have at least two singularities with zero discriminants. More exactly in the case $j \in \{f, n\}$ the discriminants of complex singularities vanish, and in the case $j = n^d$ the discriminants of real singularities vanish, i.e. we have two nodes with coinciding eigenvalues.

We point out that the sets of the conditions $(\mathcal{R}^{\mathbb{R}}_{\rho\rho})$ and $(\mathcal{C}^{\mathbb{R}}_{\rho\rho})$ that are used in Diagram 11.9 are defined in Remark 11.13 on the page 562.

Proof of Theorem 11.8. Assume that the quadratic systems (8.1) possess two real and two complex finite singularities (all elemental). Then according to Proposition 6.2 (see Table 6.1) via an affine transformation these systems could be brought to systems 2, i.e. to the systems of the form:

$$\begin{aligned} \dot{x} &= a - (a+g)x + gx^2 + 2hxy + ay^2, \\ \dot{y} &= b - (b+l)x + lx^2 + 2mxy + by^2. \end{aligned} \tag{11.42}$$

These systems possess the following finite singularities:

$$M_{1,2}(0, \pm i), \quad M_3(1, 0), \quad M_4\Big(\frac{\beta^2 + 4\gamma^2}{\beta^2 - 4\alpha\gamma}, \frac{2\beta(\alpha+\gamma)}{4\alpha\gamma - \beta^2}\Big),$$

where $\alpha = gm - hl$, $\beta = bg - al$, $\gamma = bh - am$ and $\beta^2 - 4\alpha\gamma \neq 0$.

Lemma 11.23. *If for a quadratic system with two real and two complex finite singularities (all distinct) the conditions $\mathcal{T}_4 = \mathcal{T}_3 = \mathcal{T}_2 = 0$ are satisfied, then this system is Hamiltonian.*

Proof. Assume that for systems (11.42) the conditions $\mathcal{T}_4 = \mathcal{T}_3 = \mathcal{T}_2 = 0$ holds. Then at least three traces vanish, and this means that necessarily the traces of complex singularities are zero. Therefore the conditions $\rho_{1,2} = -(a+g) - 2i(b+h) = 0$ give $a = -g$ and $b = -h$. In this case we calculate

$$\mu_0 = g^2(l-h)^2 + 4(gm - hl)(h^2 - gm), \quad \sigma = 2(g+m)x,$$
$$\mathcal{T}_4 = \mathcal{T}_3 = 0, \quad \mathcal{T}_2 = 4(g+m)^2\big[g^2(l-h)^2 + 4(h^2 - gm)^2\big],$$
$$\mathcal{T}_1 = 4(l-h)(g+m)\big[g^2(l-h) - 2h(h^2 - gm)\big].$$

We observe that due to $\mu_0 \neq 0$ the condition $\mathcal{T}_2 = 0$ implies $g + m = 0$ and then $\mathcal{T}_1 = \sigma = 0$, i.e. systems become Hamiltonian. □

Lemma 11.24. *Assume that for a quadratic system with two real and two complex finite distinct singularities the conditions* $\mathcal{T}_4 = \mathcal{T}_3 = 0$ *and* $\mathcal{T}_2 \neq 0$ *hold, i.e. this system has two finite singularities (real or complex) with the corresponding traces zero. In addition we assume that the condition* $\mathcal{B}_5 = 0$ *is fulfilled. Then the condition* $\mu_0 > 0$ *is satisfied, and moreover the vanishing traces correspond to real singularities if* $\mathcal{B} < 0$ *and to complex singularities if* $\mathcal{B} \geq 0$.

Proof. First of all we recall that according to Lemma 5.3 we have the following relations

$$\mathcal{T}_4 = \mu_0\, \rho_1\rho_2\rho_3\rho_4, \quad \mathcal{T}_3 = \mu_0(\rho_1\rho_2\rho_3 + \rho_1\rho_2\rho_4 + \rho_1\rho_3\rho_4 + \rho_2\rho_3\rho_4),$$
$$\mathcal{T}_2 = \mu_0(\rho_1\rho_2 + \rho_1\rho_3 + \rho_1\rho_4 + \rho_2\rho_3 + \rho_2\rho_4 + \rho_3\rho_4), \quad \mathcal{T}_1 = \mu_0(\rho_1 + \rho_2 + \rho_3 + \rho_4),$$

and therefore by Viète's theorem the traces ρ_i are the solutions of the equation

$$z^4 - \frac{\mathcal{T}_1}{\mu_0} z^3 + \frac{\mathcal{T}_2}{\mu_0} z^2 - \frac{\mathcal{T}_3}{\mu_0} z + \frac{\mathcal{T}_4}{\mu_0} = 0.$$

Hence in the case $\mathcal{T}_4 = \mathcal{T}_3 = 0$ the other two non-zero traces will be the roots of the equation of the second degree

$$z^2 - \frac{\mathcal{T}_1}{\mu_0} z + \frac{\mathcal{T}_2}{\mu_0} = 0, \tag{11.43}$$

and the type of these roots depends of the discriminant of this equation, i.e.

$$\frac{\mathcal{T}_1^2}{\mu_0^2} - 4\frac{\mathcal{T}_2}{\mu_0} = \frac{\mathcal{T}_1^2 - 4\mu_0\mathcal{T}_2}{\mu_0^2}.$$

Since $\mathcal{B}_5 = \mathcal{T}_1^2 - 4\mu_0\mathcal{T}_2$ (see Section 5.4), we conclude that the condition $\mathcal{B}_5 = 0$ implies the existence of a double root of the above equation.

Now we consider systems (11.42) and assume that the condition $\mathcal{T}_4 = \mathcal{T}_3 = 0$ holds. We examine both possibilities, when the traces corresponding either (1) to complex singularities or (2) to real singularities, vanish.

1. As it was shown above, in this case the conditions $\rho_{1,2} = -(a+g) - 2i(b+h) = 0$ give $g = -a$ and $h = -b$. Then calculations yield

$$\mathcal{B}_5 = 64(a-m)^2(b^2+am)^2(b^2-bl+2am)^2,$$
$$\mathcal{T}_2 = 4(a-m)^2\big[a^2(l+b)^2 + 4(b^2+am)^2\big],$$

   and due to $\mathcal{T}_2 \neq 0$ the condition $\mathcal{B}_5 = 0$ implies $(b^2+am)(b^2-bl+2am) = 0$. We consider two cases: $a \neq 0$ and $a = 0$.

   (a) If $a \neq 0$ then we may assume $a = 1$ due to a time rescaling, and therefore we get either $m = -b^2$ or $m = b(l-b)/2$. In the first case we have

$$\mathcal{B}_5 = 0, \quad \mathcal{B} = 8(1+b^2)^4, \quad \mu_0 = (b+l)^2,$$

whereas in the second one we obtain

$$\mathcal{B}_5 = 0, \ \ \mathcal{B} = (2 + b^2 - bl)^4/2, \ \ \mu_0 = (1 + b^2)(b + l)^2,$$
$$\mathcal{T}_2 = (1 + b^2)(b + l)^2(2 + b^2 - bl)^2 \neq 0.$$

We observe that in both cases we get $\mathcal{B} > 0$ and $\mu_0 > 0$.

(b) Assume now $a = 0$. Then we calculate

$$\mathcal{B}_5 = 64b^6(b - l)^2m^2, \ \ \mathcal{B} = 0, \ \ \mu_0 = 4b^3l, \ \ \mathcal{T}_2 = 16b^4m^2 \neq 0.$$

So clearly due to $\mathcal{T}_2 \neq 0$ the condition $\mathcal{B}_5 = 0$ gives $l = b$, and then we obtain $\mu_0 > 0$.

2. Assume now that the traces corresponding to real singularities vanish, i.e. $\rho_3 = \rho_4 = 0$. In this case the condition $\mathcal{B}_5 = 0$ implies $\rho_1 = \rho_2$ (to have a double root of the indicated above second-degree equation in z), and then for systems (11.42) we have

$$\rho_3 = g + 2m - a = 0, \quad \rho_1 - \rho_2 = -4i(b + h) = 0.$$

Therefore we get $g = a - 2m$ and $b = -h$, and then

$$\mathcal{T}_3 = 32(a - m)^3(h^2 + am)(lh + h^2 + 2m^2), \quad \mathcal{T}_2 = (a - m)^2\Psi_1(a, h, l, m),$$
$$\mu_0 = a^2\left[4m^2 + (h + l)^2\right] - 4(hl + m^2)(h^2 + 2am),$$
$$\mathbf{D} = (lh + h^2 + 2m^2)^2\Psi_2(a, h, l, m),$$

where $\Psi_i(a, h, l, m)$ are polynomials. So due to $\mathcal{T}_2\mathbf{D} \neq 0$ the condition $\mathcal{T}_3 = 0$ implies $h^2 + am = 0$. In this case $a \neq 0$, otherwise we get $a = h = 0$, which implies $\mu_0 = 0$. Then due to a time rescaling we may consider $a = 1$, and we get $m = -h^2$, which gives

$$\mathcal{B}_5 = 0, \ \ \mathcal{B} = -8h^2(1+h^2)^4, \ \ \mu_0 = (cl+h+2h^3)^2, \ \ \mathbf{D} = 768h^2(cl+h+2h^3)^6.$$

So since $\mathbf{D} \neq 0$ we get $\mathcal{B} < 0$ and $\mu_0 > 0$, and we conclude that Lemma 11.24 is proved. □

Remark 11.13. Considering the above lemma, we conclude that in the case $\mu_0 > 0$ if the conditions $\mathcal{T}_4 = \mathcal{T}_3 = 0$ are satisfied, then systems (11.42) possess two finite singularities (real or complex) with the corresponding traces zero and the vanishing traces correspond to real singularities if the conditions

$$(\mathcal{C}_{\rho\rho}^{\mathbb{R}}): \quad (\mathcal{B}_5 < 0) \cup (\mathcal{B}_5 = 0 \cap \mathcal{B} < 0)$$

hold and to the complex singularities if the conditions

$$(\mathcal{C}_{\rho\rho}^{\mathbb{C}}): \quad (\mathcal{B}_5 > 0) \cup (\mathcal{B}_5 = 0 \cap \mathcal{B} \geq 0)$$

are satisfied.

Lemma 11.25. *Assume that for a quadratic system with two real and two complex finite distinct singularities, the condition $\widetilde{M} = 0$ holds, i.e. this system possesses a triple infinite singularity. Then for this system*

(i) *the conditions $\mathcal{T}_3 = \mathcal{T}_4 = 0$ and $(\mathcal{C}_{\rho\rho}^{\mathbb{C}})$ imply $\mu_0 < 0$ and $W_4 > 0$;*

(ii) *the conditions $\mu_0 < 0$ and $\mathcal{T}_4 = 0$ imply $W_4 \geq 0$;*

(iii) *the conditions $\mathcal{T}_3 = \mathcal{T}_4 = 0$ and $(\mathcal{C}_{\rho\rho}^{\mathbb{R}})$ imply $\mathcal{F}_1 = 0$;*

(iv) *the conditions $W_4 = 0$ and $W_3 < 0$ imply $\mathcal{T}_4 \neq 0$;*

(v) *the conditions $\mu_0 < 0$, $W_4 = W_3 = 0$ and $W_2 \leq 0$ are incompatible;*

(vi) *the conditions $\mu_0 < 0$, $\widetilde{K} > 0$, $W_4 = W_3 = 0$, $W_2 > 0$ and $(\mathcal{C}_{nn}^{\mathcal{T}})$ are incompatible;*

(vii) *the conditions $\mu_0 > 0$, $W_4 < 0$, $\mathcal{T}_4 = 0$, $\mathcal{T}_3 \neq 0$ and $\mathcal{F}_1 = 0$ imply $\mathcal{F}_2 = 0$;*

(viii) *the conditions $W_4 = \mathcal{T}_4 = \mathcal{F}_1 = 0$ are incompatible;*

(ix) *the conditions $\mu_0 > 0$ and $W_4 = W_3 = \mathcal{T}_4 = 0$ are incompatible.*

Proof. First of all we construct a suitable canonical form of the systems possessing simultaneously two real and two complex finite singularities as well as a triple infinite singularity. Consider systems (11.42) and assume that these systems possess a triple infinite singular point. Then clearly we must have $C_2 = (vx + wy)^3$ with $v^2 + w^2 \neq 0$. Additionally for systems (11.42) we have

$$C_2 = -lx^3 + (g - 2m)x^2y + (-b + 2h)xy^2 + ay^3 = (vx + wy)^3,$$

and we get the following relations:

$$l = -v^3, \quad a = w^3, \quad b = 2h - 3vw^2, \quad g = 2m + 3v^2w.$$

Therefore we arrive at the canonical systems:

$$\begin{aligned} \dot{x} &= w^3 - (2m + 3v^2w + w^3)x + (2m + 3v^2w)x^2 + 2hxy + w^3y^2, \\ \dot{y} &= 2h - 3vw^2 - (2h - v^3 - 3vw^2)x - v^3x^2 + 2mxy + (2h - 3vw^2)y^2. \end{aligned} \tag{11.44}$$

(i) According to the assumption, both traces corresponding to complex singularities vanish. So we have $\rho_{1,2} = -2m - 3v^2w - w^3 \pm 6i(h - vw^2) = 0$, i.e. $m = -w(3v^2 + w^2)/2$, $h = vw^2$, and calculations yield

$$\mu_0 = -w^6(v^2 + w^2)^3, \quad W_4 = 64w^{22}(v^2 + w^2)^{13}.$$

Clearly due to the condition $\mu_0 \neq 0$ we get $\mu_0 < 0$ and $W_4 > 0$, and this proves the statement (i) of the lemma.

(ii) For the above systems the condition $\mathcal{T}_4 = 0$ implies the existence of at least one singularity with zero trace. If the traces corresponding to complex singularities vanish, then by statement (i) we deduce that statement (ii) of the lemma is valid.

Assume now that a real singularity has zero trace. Then without loss of generality we may consider $\rho_3 = 4m + 3v^2w - w^3 = 0$ (i.e. $m = w(w^2 - 3v^2)/4$); then for systems (11.44) we have $\mu_0 = (-4hv+5v^2w^2+w^4)^3/8$, and due to $\mu_0 < 0$ we get $v \neq 0$. Then setting $\mu_0 = -z^2$ we obtain $h = (5v^2w^2 + w^4 + z^2)/(4v)$, and calculations yield

$$\begin{aligned}\tau_4 = {} & \frac{\tau_3}{4v^4z^6}\big[2w^6(v^2+w^2)^5 + w^4(v^2+w^2)^3(3v^2+8w^2)z^2 \\ & + 12w^4(v^2+w^2)^2z^4 + w^2(7v^2+8w^2)z^6 + 2z^8\big], \\ \tau_3 = {} & -\big[w^2(v^2+w^2)^3 + 2(v^2+w^2)^2z^2 + z^4\big]/v^2.\end{aligned} \tag{11.45}$$

Since we have two complex singularities, it is clear that sign $(W_4) =$ sign $(\tau_3\tau_4)$ in the case $W_4 \neq 0$. Additionally considering (11.45) we see that sign $(\tau_3) =$ sign (τ_4). So we conclude that $W_4 \geq 0$, and this completes the proof of the statement (ii) of Lemma 11.25.

(iii) According to the conditions provided by this statement, both real singularities have the corresponding traces zero.

As it was shown in the proof of the statement (ii), setting $\rho_3 = 0$ we obtain $m = w(w^2 - 3v^2)/4$, and then we calculate

$$\Delta_3 = -\big[w^2(3v^2-w^2)^2 - 16h^2 - 8hv(v^2-3w^2)\big]/4, \quad \rho_4 = \frac{-6w\Delta_3}{5v^2w^2 + w^4 - 4hv},$$
$$\mu_0 = (5v^2w^2 + w^4 - 4hv)^3/8, \quad \mathcal{F}_1 = w\varphi(h,v,w),$$

where $\varphi(h,v,w)$ is a polynomial. Clearly $\Delta_3 \neq 0$ (due to $\rho_3 = 0$), and hence the condition $\rho_4 = 0$ implies $w = 0$. Then we obtain $\mathcal{F}_1 = 0$, and thus the statement (iii) of the lemma is proved.

(iv) Assume the contrary, that the condition $\mathcal{T}_4 = 0$ holds, i.e. at least one singularity has zero trace. Since $W_4 = 0$, by statement (i) of this lemma this singularity must be real, and without loss of generality we may consider $\rho_3 = 0$. Due to the conditions $W_4 = 0$ and $W_3 < 0$, the real singularity M_3 is a weak focus, and M_4 is a node with coinciding eigenvalues.

Therefore as we have two anti-saddles, by Theorem 6.1 (see Table 6.2) in this case the condition $\mu_0 < 0$ holds. Then setting $\mu_0 = -z^2$, after the corresponding calculations we arrive at the expressions (11.45) for τ_3 and τ_4. Since $W_3 \neq 0$ the condition $W_4 = 0$ implies $\tau_3\tau_4 = 0$, and due to $\rho_3 = 0$ (then $\tau_3 \neq 0$), we obtain $\tau_4 = 0$. However from (11.45) it is clear that the condition $\tau_4 = 0$ gives $w = z = 0$, which leads to $\mu_0 = 0$, and this contradicts our assumption, thus completing the proof of the statement (iv) of the lemma.

(v)–(vi) To prove these statements we first construct more convenient canonical systems with $\widetilde{M} = 0$ and $\mu_0 < 0$. Assume that a quadratic system (8.1) possesses two complex singularities $M_{1,2}(p \pm iq, r \pm is)$ $(q^2 + s^2 \neq 0)$ and in addition the condition $\mu_0 < 0$ is satisfied. Via the substitution $x \leftrightarrow y$ we may assume $q \neq 0$, and then we can consider $q = 1$ via the change $x \to x/q$. Therefore the affine transformation $\tilde{x} = -sx + y + ps - r$, $\tilde{y} = x - p$, moves the points $M_{1,2}(p \pm i, r \pm is)$ to the points $M'_{1,2}(0, \pm i)$. Then from the identities $p(0, i) = q(0, i) = 0$ we obtain $a + id - k = 0$, $b + if - n = 0$. Thus, $d = f = 0$, $k = a$ and $n = b$, and we obtain the canonical system

$$\dot{x} = a + cx + gx^2 + 2hxy + ay^2, \qquad \dot{y} = b + ex + lx^2 + 2mxy + by^2, \tag{11.46}$$

for which we calculate

$$\widetilde{K} = 4(d_{45}x^2 + d_{46}xy + d_{56}y^2), \ \mu_0 = d_{46}^2 - 4d_{45}d_{56} \equiv \frac{1}{16}\text{Discrim}(\widetilde{K})$$

where $d_{45} = gm - lh$, $d_{46} = bg - al$ and $d_{56} = bh - am$. Since $\mu_0 < 0$ then $d_{45}d_{56} \neq 0$ and applying the transformation

$$q: \begin{cases} x_1 = \sqrt{-\mu_0}x/(2d_{56}), \\ y_1 = d_{46}x_1/(2d_{56}) + y, \end{cases} \quad \Rightarrow \quad q^{-1}: \begin{cases} x = 2d_{56}\, x_1/\sqrt{-\mu_0}, \\ y = -d_{46}\, x_1/\sqrt{-\mu_0} + y_1, \end{cases}$$

directly to the quadratic binary form $\widetilde{K}(x, y)$ produces $\widetilde{K}'(x_1, y_1) = 2d_{56}(x_1^2 + y_1^2)$. We note that this transformation conserves the singular points $M_{1,2}(0, \pm i)$, and applying it to systems (11.46) we remain in the same class of systems.

Additionally, if we apply the same transformation q to the initial system and calculate for the transformed system its polynomial $\widetilde{K}(\tilde{a}(q), x_1, y_1)$ (we denote here $\tilde{a} = (a, b, c, e, g, h, l, m)$), according to Definition 5.1 the following relation holds:

$$\widetilde{K}(\tilde{a}(q), x_1, y_1) = \det(q)\, \widetilde{K}(\tilde{a}, x, y) = \det(q)\, \widetilde{K}(\tilde{a},\, q^{-1}(x_1, y_1)).$$

Thus the coefficients of the transformed comitant $\widetilde{K}'(x_1, y_1)$ are proportional to the coefficients of the comitant calculated for the transformed systems (see also [316]). In other words we must have the following relations: $d_{46} = 0$ and $d_{45} = d_{56}$, i.e. for systems (11.46) we obtain $bg - al = 0$ and $gm - hl = bh - am$. The determinant of these linear equations with respect to (g, l) equals $am - bh = -d_{56} \neq 0$ since $\mu_0 < 0$. Therefore we obtain the unique solution $g = -a$ and $l = -b$, and this leads to the family of systems:

$$\dot{x} = a + cx - ax^2 + 2hxy + ay^2, \quad \dot{y} = b + ex - bx^2 + 2mxy + by^2. \tag{11.47}$$

These systems possess the complex singularities $M_{1,2}(0, \pm i)$ and two other arbitrary ones; moreover for these systems the condition $\mu_0 < 0$ is satisfied. As done earlier we force the invariant polynomial $C_2(x, y)$ to have for these systems the form $(vx + wy)^3$ with $v^2 + w^2 \neq 0$. We obtain

$$C_2 = bx^3 - (a + 2m)x^2y + (2h - b)xy^2 + ay^3 = (vx + wy)^3,$$

and we get the following relations:

$$a = w^3, \quad b = v^3, \quad m = -w(3v^2 + w^2)/2, \quad h = v(v^2 + 3w^2)/2.$$

Therefore we arrive at the required canonical systems:

$$\dot{x} = w^3 + cx - w^3x^2 + v(v^2 + 3w^2)xy + w^3y^2,$$
$$\dot{y} = v^3 + ex - v^3x^2 - w(3v^2 + w^2)xy + v^3y^2.$$

According to the assumption of the statement (v), for these systems the conditions $\mu_0 < 0$, $W_4 = W_3 = 0$ and $W_2 \leq 0$ hold, i.e. at least two discriminants corresponding to finite singularities of the systems vanish. Moreover since the sign of W_2 (if $W_2 \neq 0$) coincides with the sign of the product of the remaining two non-zero discriminants, we conclude that in the case $W_2 \leq 0$ the discriminants corresponding to complex singularities necessarily vanish.

Additionally the conditions $(\mathcal{C}^{\tau}_{nn})$ provided by statement (vi) also imply the vanishing of the discriminants corresponding to complex singularities, i.e. $\tau_{1,2} = 0$. So calculations yield

$$\tau_{1,2} = c^2 - (v^2 - 8w^2)(v^2 + w^2)^2 \pm i\big[8ew^3 - 2c(v^3 - 3vw^2)\big], \quad \mu_0 = -(v^2 + w^2)^6,$$

and we impose the conditions $\tau_{1,2} = 0$ to be satisfied. We observe that $w \neq 0$, otherwise we get $c = v = 0$, which implies $\mu_0 = 0$.

So $w \neq 0$ and in this case the conditions $\tau_{1,2} = 0$ gives

$$c = \pm(v^2 + w^2)\sqrt{v^2 - 8w^2}, \quad e = \pm v(v^2 + w^2)(v^2 - 3w^2)\sqrt{v^2 - 8w^2})/(4w^3).$$

For both these solutions calculations yield

$$W_2 = \frac{1}{256w^6}(5v^2 - 4w^2)^2(v^2 + w^2)^{16}(12v^6 - 135v^4w^2 + 408v^2w^4 + 16w^6),$$
$$W_1 = -\frac{1}{16w^6}(v^2 + w^2)^{14}(4v^8 - 53v^6w^2 + 222v^4w^4 - 192v^2w^6 + 96w^8),$$
$$\mathbf{D} = \frac{3}{64w^6}(v - 2w)^3(v + 2w)^3(5v^2 - 4w^2)^4(v^2 + w^2)^{14},$$

and since $\mathbf{D} > 0$ the condition $W_2 \leq 0$ is equivalent to $(12v^6 - 135v^4w^2 + 408v^2w^4 + 16w^6) \leq 0$. We get a contradiction, because the equation $12z^6 - 135z^4 + 408z^2 + 16 = 0$ (where $z = v/w$) has only complex solutions. This completes the proof of the statement (v) of the lemma.

Thus $W_2 > 0$ and we observe that the equation $4z^8 - 53z^6 + 222z^4 - 192z^2 + 96 = 0$ also has only complex solutions, i.e. $W_1 < 0$. This means that both discriminants corresponding to real singularities are negative, i.e. we could not have two nodes, and this contradicts the conditions $(\mathcal{C}^{\tau}_{nn})$. So we conclude that the statement (vi) of the lemma is also proved.

(vii) We consider systems (11.44) for which the conditions $\mathcal{T}_4 = 0$ and $\mathcal{T}_3 \neq 0$ imply that only one (real) singularity possesses the corresponding trace zero. We assume $\rho_3 = 0$, and by what was shown earlier, this leads to the condition $m = w(w^2 - 3v^2)/4$. Then for systems (11.44) we have $\mu_0 = (-4hv + 5v^2w^2 + w^4)^3/8$. We claim that in the case we consider the condition $v \neq 0$ must hold. Indeed, supposing $v = 0$ we get

$$\mathcal{F}_1 = -3hw^3(16h^2 + w^6)/2, \quad \mu_0 = w^{12}/8,$$

and due to $\mu_0 \neq 0$ the condition $\mathcal{F}_1 = 0$ gives $h = 0$. In this case we arrive at the contradiction $W_4 = 49w^{48}/4096 > 0$.

So $v \neq 0$ and since $\mu_0 > 0$ we set $\mu_0 = z^2$, which gives $h = (5v^2w^2 + w^4 - z^2)/(4v)$. Then calculations yield

$$\mathcal{F}_1 = -\frac{3w}{8v^3}(v^2 + w^2)(v^2w^2 + w^4 - z^2)\psi(v, w, z), \quad \mathcal{F}_2 = -\frac{81w^2z^8}{512v^4}\psi(v, w, z)^2,$$

where $\psi(v, w, z) = w^2(v^2 + w^2)^3 - 2w^2(v^2 + w^2)z^2 + z^4$. Supposing $\mathcal{F}_2 \neq 0$, the condition $\mathcal{F}_1 = 0$ gives $z^2 = w^2(v^2 + w^2)$, and then calculations yield

$$W_4 = w^{20}(v^2 + w^2)^{13}(81v^2 + 49w^2)/4096 > 0.$$

The contradiction we have obtained completes the proof of statement (vii) of the lemma.

(viii) Assume $\mathcal{T}_4 = 0$, i.e. at least one singularity has the corresponding trace zero. Since $W_4 = 0$, by the statement (i) of this lemma, this singularity must be real, and without loss of generality we may consider $\rho_3 = 0$. Then $m = w(w^2 - 3v^2)/4$, and we calculate

$$\begin{aligned}
\mathcal{F}_1 &= -\frac{3w}{2}(v^2 + w^2)(h - vw^2)(16h^2 - 32hvw^2 + v^4w^2 + 18v^2w^4 + w^6) \\
&\equiv -\frac{3w}{2}(v^2 + w^2)(h - vw^2)\phi(h, v, w), \\
\mu_0 &= \frac{(-4hv + 5v^2w^2 + w^4)^3}{8} \neq 0.
\end{aligned}$$

We observe that $\text{Discrim}[\phi, h] = -64w^2(v^2 + w^2)^2$, and clearly in order to have a real solution of the equation $\phi(h, v, w) = 0$, the condition $w = 0$ is necessary. So the condition $\mathcal{F}_1 = 0$ is equivalent to $w(h - vw^2) = 0$. If $w = 0$ (then $v \neq 0$ by $\mu_0 \neq 0$) we get

$$W_4 = 2^{17}h^{13}v^3(2h + v^3)^2, \quad \mathbf{D} = 196608h^{10}(2h + v^3)^2,$$

whereas in the case $h = vw^2$ we obtain

$$W_4 = 2^{-12}w^{20}(v^2 + w^2)^{13}(81v^2 + 49w^2), \quad \mathbf{D} = 3w^{16}(v^2 + w^2)^{10}/4.$$

We observe that in both cases due to $\mathbf{D} > 0$ we have $W_4 \neq 0$, and this completes the proof of the statement (viii).

(ix) We consider systems (11.44) for which the condition $\mathcal{T}_4 = 0$ implies that at least one singularity has the corresponding trace zero. Since $\mu_0 > 0$ we have a saddle and an anti-saddle, and hence the conditions $W_4 = W_3 = 0$ imply the vanishing of two discriminants corresponding to the complex singularities.

So assuming $\rho_3 = 0$ (i.e. $m = w(w^2 - 3v^2)/4$), we calculate

$$\begin{aligned}\tau_{1,2} &= (96hvw^2 - 16h^2 + 9v^4w^2 - 78v^2w^4 - 7w^6)/4\\ &\quad \pm 2iw\big[h(3v^2 - 5w^2) - vw^2(5v^2 - 3w^2)\big],\end{aligned}$$

and we force $\tau_{1,2} = 0$. We observe that in this case the condition $w \neq 0$ must hold, otherwise we get $\tau_{1,2} = -4h^2 \neq 0$ due to $\mu_0 = -8h^3v^3 \neq 0$. Moreover the condition $3v^2 - 5w^2 \neq 0$ also has to be satisfied, because in the case $w^2 = 3v^2/5 \neq 0$, the conditions $\tau_{1,2} = 0$ imply $v = 0$, and we again get a contradiction.

So $w(3v^2 - 5w^2) \neq 0$, and the imaginary part of $\tau_{1,2}$ being zero yields $h = vw^2(5v^2 - 3w^2)/(3v^2 - 5w^2)$. However in this case we calculate

$$\mu_0 = -\frac{125w^6(v^2 + w^2)^6}{8(3v^2 - 5w^2)^3} < 0,$$

and this contradicts $\mu_0 > 0$. So we conclude that the statement (ix) is valid, and this completes the proof of Lemma 11.25. □

Lemma 11.26. *Assume that a quadratic system possesses two real and two complex finite distinct singularities. Then the conditions of each one of the following sets are incompatible:*

(i) $\mu_0 < 0,\ \widetilde{K} > 0,\ W_4 < 0,\ \mathcal{T}_4 = \mathcal{T}_3 = 0,\ \eta \geq 0$;

(ii) $W_4 = 0,\ W_3 < 0,\ \mathcal{T}_4 = \mathcal{T}_3 = 0$;

(iii) $W_4 = W_3 = 0,\ (\mathcal{C}^{\mathbb{R}}_{\rho\rho}),\ \mathcal{T}_4 = \mathcal{T}_3 = 0,\ \mathcal{F}_1 \neq 0$;

(iv) $\mu_0 > 0,\ W_4 < 0,\ \mathcal{T}_4 = 0,\ \mathcal{T}_3\mathcal{F} > 0,\ \mathcal{F}_1 = 0,\ \mathcal{F}_2 \neq 0,\ \theta = \theta_1 = 0$;

(v) $\mu_0 > 0,\ \mathcal{T}_4 = \mathcal{T}_3 = 0,\ (\mathcal{C}^{\mathbb{C}}_{\rho\rho}),\ \eta \leq 0,\ \theta \leq 0$;

(vi) $W_4 < 0,\ \mathcal{T}_4 = \mathcal{T}_3 = 0,\ \mathcal{B}_5 < 0,\ \eta > 0,\ \theta = \theta_1 = 0$;

(vii) $W_4 = \mathcal{T}_4 = 0,\ W_3\mathcal{T}_3 \neq 0,\ \eta > 0,\ \theta = 0,\ \theta_1 = 0,\ (\theta_3 \neq 0) \vee (\theta_3 = \theta_4 = 0)$;

(viii) $W_4 = \mathcal{T}_4 = \mathcal{F}_1 = 0,\ W_3\mathcal{T}_3\mathcal{F}_2 \neq 0,\ \eta > 0,\ \theta = 0,\ \theta_1 \leq 0$;

(ix) $W_4 = \mathcal{T}_4 = \mathcal{F}_1 = 0,\ W_3\mathcal{T}_3\mathcal{F}_2 \neq 0,\ \theta = \theta_2 = 0$;

(x) $W_4 = \mathcal{T}_4 = \mathcal{T}_3 = 0,\ W_3 \neq 0,\ \eta \geq 0$;

(xi) $W_4 = W_3 = 0,\ W_2 < 0,\ \mathcal{T}_4 \neq 0,\ \eta > 0,\ \theta = \theta_1 = 0$;

(xii) $\mu_0 > 0,\ W_4 = W_3 = W_2 = \mathcal{T}_4 = 0,\ \theta \geq 0$;

(xiii) $\mu_0 > 0,\ W_4 = W_3 = W_2 = \mathcal{T}_4 = 0,\ \eta > 0,\ \theta < 0,\ \theta_1 < 0;$

(xiv) $\mu_0 > 0,\ W_4 = W_3 = W_2 = \theta = \theta_1 = 0.$

Proof. (i) Since $\mu_0 < 0$ we consider systems (11.47) for which the conditions $W_4 < 0$ and $\mathcal{T}_4 = \mathcal{T}_3 = 0$ imply the vanishing of the traces corresponding to complex singularities. So for these systems the conditions $\rho_{1,2} = c \pm 2i(b+h) = 0$ imply $c = 0$, $b = -h$, and then we calculate $\widetilde{K} = -4(h^2 + am)(x^2 + y^2) > 0$. So $am \neq 0$, and we may assume $a = 1$ (due to a time rescaling) and $h^2 + m = -z^2$ (i.e. $m = -(h^2 + z^2)$). Then calculations yield

$$\begin{aligned} W_4 &= 2^{10}(e^2 + 4z^4)^2\Big[4z^4\big[(1+h^2+z^2)^2 - 4z^2\big]^2 + e^2\Phi(e,h,z)\Big],\\ \Phi &= \big[(1+h^2)^3(1+h^2-4z^2) - 2(5h^4+2h^2-3)z^4 - 4(1+h^2)z^6 + z^8\big],\\ \eta &= 4\big[(1+h^2)^3 - 6(1+h^2)^2z^2 + 3(4-5h^2)z^4 - 8z^6\big]. \end{aligned}$$

We claim that for $\eta = 0$ the polynomial $\Phi(e,h,z)$ is positive. Indeed we observe that this polynomial can be written in the form

$$\Phi = \eta(1+h^2+2z^2)/4 + z^4\big[17z^4 + 2(17h^2-10)z^2 + (1+h^2)(6+17h^2)\big],$$

and to complete the proof of our claim, it is sufficient to observe that Discrim $[17z^4 + 2(17h^2-10)z^2 + (1+h^2)(6+17h^2), z^2] = -4(2+731h^2) < 0$. So in the case $\eta = 0$, we get $\Phi(e,h,z) > 0$, and this implies $W_4 > 0$, which contradicts the assumption (i) of the lemma, i.e. the statement (i) is proved.

(ii) Since $W_4 = 0$ and $W_3 < 0$, we deduce that the conditions $\mathcal{T}_4 = \mathcal{T}_3 = 0$ imply the vanishing of the traces corresponding to complex singularities. So we consider systems (11.42) for which the conditions $\rho_{1,2} = -(a+g) - 2i(b+h) = 0$ give $a = -g$ and $b = -h$. Since one discriminant corresponding to a real singularity must vanish, without loss of generality we may assume $\tau_3 = 4[(g-m)^2 + 2h^2 + 2hl] = 0$. We observe that $h \neq 0$, otherwise we get $m = g$, which implies $W_3 = 2048l^2g^8(l^2+4g^2)^2 \geq 0$.

Thus $h \neq 0$ and due to a time rescaling we may assume $h = 1$, and the condition $\tau_3 = 0$ gives $l = -1 - (g-m)^2/2$. However in this case calculations yield

$$\begin{aligned} W_3 = 8\big[&16 + 16g^2 + 8g^4 + g^6 - 4g(8+4g^2+g^4)m + 6g^2(4+g^2)m^2 - 4g^3m^3\\ &+ g^2m^4\big]^2(g+m)^2(2+g^2-gm)^2\big[4+(g-m)^2\big] \geq 0, \end{aligned}$$

and the contradiction we obtained proves the statement (ii) of the lemma.

(iii) According to the conditions provided by this statement, we must have two traces corresponding to real singularities zero as well as two zero discriminants corresponding to complex singularities. So we consider systems (11.42), and we examine two cases: $a = 0$ and $a \neq 0$.

If $a=0$ then we calculate $\tau_{1,2}=g^2-4(b-h)^2\pm 4ig(b-h)$, and hence the conditions $\tau_{1,2}=0$ give $g=0$ and $h=b$, which imply $\mathcal{F}_1=0$, i.e. the statement (iii) is valid.

Assume now $a\neq 0$. We claim that in this case the conditions $\tau_1=\tau_2=\rho_3=\rho_4=0$ lead to degenerate systems. Indeed, as $a\neq 0$, due to a time rescaling we may consider $a=1$, and then we obtain

$$\tau_{1,2}=(1+g)^2-4(b-h)^2-16m\pm 4i\big[b(g-1)-2l-h(1+g)\big]=0 \;\Rightarrow$$
$$l=(bg-b-h-gh)/2,\quad m=\big[(1+g)^2-4(b-h)^2\big]/16.$$

So we calculate

$$\rho_3=\big[g(10+g)-4(b-h)^2-7\big]/8\equiv\varphi_3(b,g,h)/8,$$
$$\rho_4=\frac{1}{512\mu_0}\big[(1+g)^2+4(b+h)^2\big]\varphi_4(b,g,h),\quad \mu_0\neq 0,$$

where

$$\varphi_4(b,g,h)=32b^2(h^2-2+g+g^2)-16b^4+1+12g+14g^2-4g^3-7g^4$$
$$+16b(-11-2g+g^2)h-128h^2-96gh^2-32g^2h^2-16h^4.$$

We have to force $\rho_3=0$ and $\rho_4=0$, which is equivalent to $\varphi_3=\varphi_4=0$ (otherwise we get $g=-1$, $b=-h$ and then $\varphi_3=-16(1+h^2)\neq 0$). We remark that the common solution of these two equations must depend on b because we have the term $4b^2$ in ρ_3. So we calculate

$$\text{Res}_b(\varphi_3,\varphi_4)=2^{18}\big[2(1-g)^2+8h^2-h(1+g)^2\big]\big[2(1-g)^2+8h^2+h(1+g)^2\big](1+g)^2,$$

and we observe that the second factor could be obtained from the first due to the change $h\to -h$. So we will examine only the case $h>0$ as the case $h<0$ could be examined similarly.

Thus setting $h=2z^2$ means that only the first factor could be zero, and this relation implies

$$\big[g(z+1)-1-3z-4z^2\big]\big[g(z-1)+1-3z+4z^2\big]=0.$$

We observe that the first factor could be reduced (up to the sign) to the second one via the change $z\to -z$ and hence we may assume $z+1\neq 0$. Then forcing the first factor to be zero, we get $g=(1+3z+4z^2)/(1+z)$, and as $h=2z^2$ we obtain that the common factor (which depends on b) of the polynomials φ_3 and φ_4 is $b(1+z)-1-4z-4z^2-2z^3$. However the same factor is contained in the determinant Δ_3, and hence the condition $\tau_3=\Delta_3=0$ leads to degenerate systems. So our claim is proved, and this completes the proof of the statement (iii).

(iv) According to the assumption and Theorem 6.2 (see statement (b)), we have a weak saddle of order two, and by Lemma 11.1 at infinity there exists at least one node $N[x_0 : y_0 : 0]$ with coinciding eigenvalues. So we consider the family of systems:

$$\dot{x} = x + gx^2 + 2hxy + ky^2, \quad \dot{y} = -y + lx^2 + 2mxy + ny^2, \tag{11.48}$$

possessing a weak saddle $M_1(0,0)$. If for the node $N[x_0 : y_0 : 0]$ with coinciding eigenvalues the condition $x_0y_0 \neq 0$ holds, then due to the rescaling $(x,y) \mapsto (x_0x, y_0y)$ we may assume that this singularity is located at the point $N[1:1:0]$. In the case $x_0y_0 = 0$ we may assume that $N[x_0 : y_0 : 0]$ is located at $N[1:0:0]$ due to the change $(x,y,t,g,h,k,l,m,n) \mapsto (-y,-x,-t,n,m,l,k,h,g)$. So we consider two positions of the node at infinity: $N[1:0:0]$ and $N[1:1:0]$.

1. If we have the node $N[1 : 0 : 0]$, then clearly the condition $l = 0$ must hold. In this case we detect the following Jacobian matrix corresponding to this singularity: $\begin{pmatrix} g-2m & 0 \\ 0 & g \end{pmatrix}$. Evidently this singularity has coinciding eigenvalues if and only if $m = 0$ (and in this case we get a star node).

   Additionally for systems (11.48) with $l = m = 0$, we calculate

   $$\mathcal{T}_4 = 0, \quad \mathcal{F}_1 = -4gh, \quad \mathcal{F}_2 = 24g^4hn(gk - h^2 - hn),$$

   and therefore the condition $\mathcal{F}_1 = 0$ implies $\mathcal{F}_2 = 0$. So the contradiction obtained confirms the validity of the statement (iv) in this case.

2. Assume now that systems (11.48) possess the infinite node $N[1:1:0]$. In this case the invariant polynomial $C_2 = xp_2(x,y) - yq_2(x,y)$ must possess as a factor $x-y$, and this implies the condition $g = l-2h-k+2m+n$. Therefore we obtain the following Jacobian matrix corresponding to this singularity: $\begin{pmatrix} l+2h+2k-n & 2 \\ 0 & l+2m+n \end{pmatrix}$. Hence we have a node N^d if and only if $m = h+k-n$. Then we calculate

   $$\begin{aligned} \mathcal{F}_1 &= -4(lh + hk - 2hn - kn + n^2), \\ \theta_1 &= -64(l + 2h + 2k - n)^2(lk - h^2 - hk + hn), \end{aligned}$$

   and we have to force the conditions $\mathcal{F}_1 = 0$ and $\theta_1 = 0$. We consider two cases: $h = 0$ and $h \neq 0$. If $h = 0$ then we calculate

   $$\mathcal{F}_1 = 4n(k-n), \quad \mu_0 = (k-n)^2(l+2k-n)^2, \quad \theta_1 = -64lk(l+2k-n)^2,$$

   and due to $\mu_0 \neq 0$ the condition $\mathcal{F}_1 = \theta_1 = 0$ yields $n = l = 0$. However in this case we get $W_4 = 16384k^8 > 0$, which contradicts $W_4 < 0$, i.e. in the case $h = 0$ the statement is valid.

   Assume now $h \neq 0$. Then the condition $\mathcal{F}_1 = 0$ gives $l = -(hk - 2hn - kn + n^2)/h$, and calculations yield

$$\theta_1 = \frac{64}{h^3}(h-n)(h+n)^2(2h+k-n)^2(h^2+hk+k^2-kn),$$
$$\mathcal{F}_2 = \frac{16}{h^5}(h^2+hk+k^2-kn)^2\varphi(h,k,n),$$
$$\mu_0 = (k-n)^2(h+n)^2(2h+k-n)^2/h^2,$$

where $\varphi(h,k,n)$ is a polynomial. So due to $\mu_0\mathcal{F}_2 \neq 0$ the condition $\theta_1 = 0$ gives $h = n$, and then we obtain $W_4 = 16384(k-n)^2(k+n)^2(k^2+n^2)^2 \geq 0$. The obtained contradiction completes the proof of the statement (iv) of Lemma 11.26.

(v) Considering systems (11.42) and setting $\rho_{1,2} = -(a+g) - 2i(b+h) = 0$, we get $g = -a$ and $h = -b$. Then calculations yield

$$\theta = 64\big[a^2l^2 + 2bl(a^2+2b^2+2am) - m(a^3+2ab^2+2a^2m+b^2m+am^2)\big],$$
$$\eta = -27\theta/64 + (a-m)^2\big[4(a^2+b^2)+5(b^2+am)\big],$$
$$\text{Discrim}[\theta,l] = 2^{14}(b^2+am)(a^2+2b^2+am)^2$$

and supposing $\theta \leq 0$ we must have $\text{Discrim}[\theta,l] \geq 0$, i.e. $b^2+am \geq 0$. In this case we get $\eta \geq 0$, and to have the equality, the condition $a-m=0=\theta$ must hold. However for $m=a$ we get $\theta = 64\mu_0 \neq 0$, and we conclude that the statement (v) is valid.

(vi) According to Remark 11.13 the conditions provided by this statement imply the vanishing of the traces corresponding to real singularities. This needs canonical systems with simpler traces of real singularities. So via an affine transformation we fix two real singularities at the points $M_1(0,0)$ and $M_2(1,0)$, and we arrive at the following systems:

$$\dot{x} = cx - cx^2 + dy + 2hxy + ky^2, \quad \dot{y} = ex - ex^2 + fy + 2mxy + ny^2. \qquad (11.49)$$

Since the condition $\eta > 0$ must hold, we force now the existence of three simple real singularities at infinity. In other words we force the invariant polynomial $C_2(x,y)$ to be of the form $(x+uy)(wx+ry)(sx+y)$ with $(u^2+r^2)(s^2+w^2) \neq 0$. We obtain

$$C_2 = ex^3 - (c+2m)x^2y + (2h-n)xy^2 + ky^3 = (x+uy)(wx+ry)(sx+y),$$

and we get the relations

$$e = sw, \quad c = -(2m+rs+w+suw), \quad k = ru, \quad h = (n+r+rsu+uw)/2,$$

which lead to the canonical systems:

$$\begin{aligned}\dot{x} &= -(2m+rs+w+suw)x + dy + (2m+rs+w+suw)x^2\\ &\quad + (n+r+rsu+uw)xy + ruy^2,\\ \dot{y} &= rv^2x + fy - rv^2x^2 + 2mxy + ny^2.\end{aligned} \qquad (11.50)$$

Then we calculate

$$\begin{aligned}\theta &= 8\gamma_1\gamma_2\gamma_3, \quad \eta = (1-su)^2(rs-w)^2(uw-r)^2,\\ \theta_1 &= 16\big[\gamma_1\gamma_2(uw-r)(w-rs) + \gamma_1\gamma_3(1-su)(uw-r)\\ &\quad +\gamma_2\gamma_3(1-su)(w-rs)\big],\end{aligned} \tag{11.51}$$

where

$$\gamma_1 = 2m - ns + rs - rs^2u + suw, \quad \gamma_2 = n + r - 2mu - rsu - uw,$$
$$\gamma_3 = 2mr - nw + rw + rsuw - uw^2,$$

and due to the condition $\eta \neq 0$, we obtain that $\theta = \theta_1 = 0$ if and only if two of the factors γ_1, γ_2 and γ_3 vanish.

Additionally the real singularities must have zero traces. We calculate $\rho_1 = f - 2m - rs - w - suw$, $\rho_2 = f + 4m + rs + w + suw$, and therefore the conditions $\rho_1 = \rho_2 = 0$ yield

$$f = (rs + w + suw)/3, \quad m = -(rs + w + suw)/3.$$

Assume $\gamma_1 = \gamma_2 = 0$. Considering the above relations, we obtain

$$\gamma_1 = -(3ns - rs + 3rs^2u + 2w - suw)/3, \; \gamma_2 = (3n + 3r - rsu - uw + 2su^2w)/3,$$

and we calculate $\mathrm{Res}_r(\gamma_1, \gamma_2) = 2(1-su)(2ns + w + s^2u^2w)/3 = 0$. Due to $\eta \neq 0$ we have $w = -2ns/(1+s^2u^2)$, and then γ_1 and γ_2 possess as a common factor $\psi = (n+r+nsu+rs^2u^2)$. Therefore the condition $\psi = 0$ gives $r = n(1+su)/(1+s^2u^2)$. However in this case we get $W_4 = 0$, which contradicts the assumption of the statement.

In an analogous way, $\gamma_1 = \gamma_3 = 0$ and $\gamma_2 = \gamma_3 = 0$ could be the examined cases, and in both these cases we get the same contradiction: $W_4 = 0$. This completes the proof of the statement (vi) of the lemma.

(vii) Since $W_3\mathcal{T}_3 \neq 0$ we deduce that the conditions $W_4 = 0$ and $\mathcal{T}_4 = 0$ imply the vanishing of one trace and one discriminant corresponding to different real singularities.

In order to prove this statement, we consider systems (11.50) for which we have the values of the invariant polynomials θ and θ_1 given in (11.51). As it was mentioned earlier, $\theta = \theta_1 = 0$ if and only if two among the three polynomials γ_1, γ_2 and γ_3 vanish. We calculate $\rho_1 = f - 2m - rs - w - suw$, and therefore the condition $\rho_1 = 0$ yields $f = 2m + rs + w + suw$.

We consider the case when the condition $\gamma_1 = \gamma_2 = 0$ (which implies $\theta = \theta_1 = 0$) is satisfied. We calculate $\mathrm{Res}_m(\gamma_1, \gamma_2) = 2(1-su)(n + r + rsu - uw) = 0$. Due to $\eta \neq 0$ we have $n = -r - rsu + uw$, and then γ_1 and γ_2 possess as a common factor $\psi_1 = m + rs$. Therefore the condition $\psi_1 = 0$ gives $m = -rs$, and calculations yield

$$\tau_2 = 4s(r^2s - dw - 2uw^2), \quad \mathcal{T}_3 = (r^2s - dw - 2uw^2)\Phi_1, \quad \theta_3 = s\Phi_2,$$

where $\Phi_i(d,r,s,u,w)$ are polynomials. So due to $\mathcal{T}_3 \neq 0$ the condition $\tau_2 = 0$ gives $s = 0$, and this implies $\theta_3 = 0$. Moreover, forcing the condition $s = 0$ (i.e. $\theta_3 = 0$), we obtain

$$\theta_4 = w^3(r - uw)(d + 2uw), \quad \mathcal{T}_3 = 8w^5(r - uw)(d + 2uw),$$

and due to $\mathcal{T}_3 \neq 0$ we get $\theta_4 \neq 0$.

The remaining two cases ($\gamma_1 = \gamma_3 = 0$ and $\gamma_2 = \gamma_3 = 0$) could be examined in an analogous way, and we arrive at the same conclusions as above. This completes the proof of the statement (vii) of the lemma.

(viii)–(ix) We consider systems (11.49) that possess the real singularities $M_1(0,0)$ and $M_2(1,0)$. Since these statements provide the condition $\theta = 0$, by Lemma 11.1 at infinity we must have a node $N[x_0 : y_0 : 0]$ with coinciding eigenvalues. If $y_0 \neq 0$ then we may assume $x_0 = 0$ due to the change $x_1 = x - x_0 y/y_0$, $y_1 = y$, which keeps the singularities $M_1(0,0)$ and $M_2(1,0)$ (due to an additional rescaling $x_1 \to \lambda x_1$). Hence we consider the two possibilities of when this node is located either at $N_1[1:0:0]$ or at $N_2[0:1:0]$.

1. In the case of the point $N_1[1:0:0]$, for systems (11.49) the polynomial $C_2 = ex^3 - (c+2m)x^2y + (2h-n)xy^2 + ky^3$ must have the factor y, and this implies the condition $e = 0$. In this case we detect the following Jacobian matrix corresponding to $N_1[1:0:0]$: $\begin{pmatrix} -c-2m & 0 \\ 0 & -c \end{pmatrix}$. Evidently this singularity is a node with coinciding eigenvalues if and only if $m = 0$ (moreover, in this case we have a star node).

   Additionally the conditions of both statements guarantee the existence of a weak saddle and of a finite n^d (or n^*), i.e. we could assume that the conditions $\rho_1 = \tau_2 = 0$ are satisfied. For systems (11.49) with $e = m = 0$, we calculate $\rho_1 = c + f = 0$, $\tau_2 = (c+f)^2 = 0$, which imply $f = -c$. However in this case we get

$$\mathcal{F}_1 = 2c^3(d+2h), \quad \mathcal{F}_2 = -12c^6(d+2h)n(h^2 + ck + dn + hn),$$

   and clearly the condition $\mathcal{F}_1 = 0$ implies $\mathcal{F}_2 = 0$, which contradicts the condition $\mathcal{F}_2 \neq 0$ provided by both statements.

2. Assume now that the infinite star node is located at the point $N_2[0:1:0]$. In this case the polynomial $C_2(x,y)$ has as a factor x, i.e. the condition $k = 0$ holds and the corresponding Jacobian matrix has the form $\begin{pmatrix} n-2h & -d \\ 0 & n \end{pmatrix}$. So $N_1[0:1:0]$ is a node with coinciding eigenvalues if and only if $h = 0$. Again assuming that the conditions $\rho_1 = \tau_2 = 0$ are satisfied, we calculate $\rho_1 = c + f = 0$, i.e. $f = -c$, and this gives $\tau_2 = 4(m^2 - de) = 0$. If $d = 0$ we get $m = 0$, and this implies $\mathcal{F}_2 = 0$, i.e. we conclude that both statements of the lemma are valid.

Assuming $d \neq 0$ we may consider $d = 1$ due to a time rescaling, and then the condition $\tau_2 = 0$ gives $e = m^2 \neq 0$; otherwise we get $e = m = 0$, and then the condition $\mathcal{F}_1 = 2c^3 = 0$ implies again $\mathcal{F}_2 = -12c^6n^2 = 0$. So calculations yield

$$\mu_0 = c^2n^2, \quad \mathcal{F}_1 = 2(c+m+mn)(c^2 - m^2 + cmn),$$
$$\mathcal{T}_3 = 8n(c-m)^2(c^2+m^2)(c+m+mn),$$

and due to $\mu_0\mathcal{T}_3 \neq 0$ and $m \neq 0$, the condition $\mathcal{F}_1 = 0$ gives $n = (m^2 - c^2)/(cm)$ (here $c \neq 0$ due to $m \neq 0$). Then we obtain

$$\theta_1 = 64(c-m)^2(c+m)^3/c^3, \quad \theta_2 = -m^2(c+m)/c,$$
$$\mu_0 = (c-m)^2(c+m)^2/m^2,$$

and due to $\mu_0 \neq 0$ we get $\theta_1\theta_2 \neq 0$, which proves the validity of the statement (viii) (as well as the statement (ix)) in the case $\theta_1 = 0$.

Assume now $\theta_1 < 0$. Then $(c+m)/c < 0$, and we may set $(c+m)/c = -z^2 \neq 0$, i.e. $m = -c(1+z^2)$, which implies

$$\eta = c^2z^4(2+z^2)^2(-1+2z+2z^3)(1+2z+2z^3)/(1+z^2)^2,$$
$$\mathbf{D} = 48c^{10}(2+z^2)^4(2+2z^2+z^4)^3(-2+6z^2+11z^4+4z^6)/(1+z^2)^2.$$

We observe that the sign of η is defined by the sign of the polynomial $\omega_1 = (-1+2z+2z^3)(1+2z+2z^3)$, whereas the sign of $\mathbf{D}$ is defined by the sign of $\omega_2 = (-2+6z^2+11z^4+4z^6)$. These polynomials depend on a single variable, and one can find the roots of each one of them. Checking that the intervals formed by this set of roots are properly ordered, it is easy to verify that the case $\eta > 0$ and $\mathbf{D} > 0$ is not realizable. This completes the proof of the statement (ix).

(x) Since $W_3 \neq 0$ we have a node with coinciding eigenvalues and complex singularities with zero traces. So we consider systems (11.42) for which the conditions $\rho_{1,2} = -(a+g) - 2i(b+h) = 0$ give $g = -a$ and $h = -b$. Then we calculate $\tau_3 = 4\big[(a+m)^2 + 2b(b-l)\big]$, and we consider two cases: $b = 0$ and $b \neq 0$.

If $b = 0$ then the condition $\tau_3 = 0$ gives $m = -a$, and then $\eta = -a^2(4a^2 + 27l^2) < 0$ due to $\mu_0 = a^2(l^2 - 4a^2) \neq 0$.

Assume now $b \neq 0$. Then we get $l = \big[(a+m)^2 + 2b^2\big]/(2b)$, and calculations yield

$$\begin{aligned}\eta = &-\frac{1}{4b^2}\big[27a^2m^4 + 4a(27a^2+22b^2)m^3 + 6(27a^4+76a^2b^2+12b^4)m^2\\ &+ 12a(9a^4+46a^2b^2+60b^4)m + 27a^6 + 200a^4b^2 + 504a^2b^4 + 432b^6\big]\\ \equiv &-\frac{1}{4b^2}\Psi.\end{aligned}$$

We observe that in the case $a = 0$ we have $\Psi = 72b^4(6b^2 + m^2) > 0$, and this implies $\eta < 0$.

Assume $a \neq 0$. Then Ψ is a polynomial of degree 4 with respect to the parameter m. Considering Lemma 11.10 we calculate

$$D_\diamond = -2^{12}3^9a^2b^6(a^2+b^2)^8, \quad R_\diamond = -32a^2b^2(27a^2 - 5b^2),$$
$$S_\diamond = -48a^4b^2(6561a^6 - 166941a^4b^2 + 88803a^2b^4 + 161b^6).$$

Since $D_\diamond < 0$ in order to have real roots, by Lemma 11.10 the conditions $R_\diamond > 0$ and $S_\diamond > 0$ must hold. So assuming $R_\diamond > 0$ we set $b^2 = (27a^2 + z^2)/5$, and then we get

$$S_\diamond = -\frac{48a^2}{625}(27a^2+z^2)\big[2^{10}3^8a^4(32a^2+3z^2) + 457056a^2z^4 + 161z^6\big] < 0,$$

i.e. the polynomial Ψ has no real roots. Therefore we conclude that we could not have $\eta \geq 0$, and this proves the statement (x) of the lemma.

(xi) According to the conditions provided by this statement, both discriminants corresponding to complex singularities must vanish since W_2 is negative, and the other two discriminants must have opposite signs. So in order to prove this statement, we construct a more convenient family of systems.

First of all we force the generic family of systems to possess the singular points $M_{1,2}(0, \pm i)$, and in such a way we get the conditions $d = f = 0$, $k = a$ and $n = b$. Moreover since $\eta > 0$ we have three real infinite singularities, and via a linear transformation of the form $x_1 = x$, $y_1 = \alpha x + y$, we may assume that one of these singularities is located at the point $[1:0:0]$, i.e. at the "end" of the x-axis. This implies $l = 0$, and we arrive at the family of systems:

$$\dot{x} = a + cx + gx^2 + 2hxy + ay^2, \quad \dot{y} = b + ex + 2mxy + by^2,$$

for which we calculate

$$\tau_{1,2} = c^2 - 4(b-h)^2 - 16am \pm 4i(2ae - bc + ch).$$

We observe that in the case $a = 0$ the conditions $\tau_{1,2} = -(2b \pm ic - 2h)^2 = 0$ yield $c = 0$ and $b = h$, and then we get $\mathcal{T}_4 = 4h^4\theta = 256h^6m^2 \neq 0$, which contradicts the assumption.

So $a \neq 0$, and we may assume $a = 1$ due to the rescaling $x \to ax$. Then the conditions $\tau_{1,2} = 0$ give $e = c(b-h)/2$ and $m = \big[c^2 - 4(b-h)^2\big]/16$. By Lemma 11.1 the condition $\theta = 0$ implies the existence at infinity of a node with coinciding eigenvalues. Without loss of generality we may assume that the singular point $N_1[1:0:0]$ is such a node. So we detect the following Jacobian matrix corresponding to this singularity:

$$\begin{pmatrix} \big[8g - c^2 + 4(b-h)^2\big]/8 & c(h-b)/2 \\ 0 & g \end{pmatrix}.$$

Clearly the coincidence of the eigenvalues gives

$$\left[8g - c^2 + 4(b-h)^2\right]/8 - g = (2b - c - 2h)(2b + c - 2h)/8 = 0.$$

Since the rescaling $(x, y, t) \mapsto (-x, -y, -t)$ changes the sign of the parameter c, we will examine only the condition given by the first factor, i.e. $c = 2(b-h)$. Then we have

$$\theta_1 = -64g^2(g + bh - h^2), \quad \mathcal{T}_4 = -32g(b-h)^2(b^2+h^2)^2(b^2 - g - 2bh + h^2),$$

and due to $g \neq 0$ (as $\mathcal{T}_4 \neq 0$), the condition $\theta_1 = 0$ gives $g = h(h-b)$. However in this case we obtain $W_2 = 16b^2h^2(-b+h)^4(b^2+h^2)^4 \geq 0$, which contradicts the assumption, and hence the statement (xi) of the lemma is valid.

(xii)–(xiv) According to the conditions $W_4 = W_3 = W_2 = 0$ provided by each one of these statements, we must have three discriminants zero. So we consider systems (11.42), and we examine two cases: $a = 0$ and $a \neq 0$.

If $a = 0$ then we calculate $\tau_{1,2} = -(2b + 2h \pm gi)^2$. Hence the conditions $\tau_{1,2} = 0$ give $g = 0$ and $h = b$, and we calculate $\mathcal{T}_4 = 256b^6m^2$, $\theta = 64b^2m^2$ and $\mu_0 = 4b^3l$. Since $\mu_0 \neq 0$ the condition either $\mathcal{T}_4 = 0$ or $\theta = 0$ gives $m = 0$, and this implies $\mathbf{D} = 12288b^{10}(b-l)^2$, $W_2 = -1024b^9l(b-l)^2$. Therefore the condition $\mu_0\mathbf{D} \neq 0$ implies $W_2 \neq 0$, i.e. the statements (xii)–(xiv) are valid in this case.

Assume now $a \neq 0$. Then due to a time rescaling we may consider $a = 1$, and then we obtain

$$\tau_{1,2} = (1+g)^2 - 4(b-h)^2 - 16m \pm 4i\left[b(g-1) - 2l - h(1+g)\right] = 0 \;\Rightarrow$$
$$l = (bg - b - h - gh)/2, \quad m = \left[(1+g)^2 - 4(b-h)^2\right]/16.$$

So we calculate $\tau_3 = \varphi_1(b, g, h)\varphi_2(b, g, h)$, where

$$\varphi_1(b, g, h) = (2b + g - 3)^2 - 4(2b + g + 5)h + 4h^2,$$
$$\varphi_2(b, g, h) = (2b - g + 3)^2 - 4(2b - g - 5)h + 4h^2.$$

We have $\text{Discrim}[\varphi_1, b] = 512h$ and $\text{Discrim}[\varphi_2, b] = -512h$, and we conclude that $\tau_3 = 0$ is equivalent to $\varphi_1 = 0$ if $h > 0$ and to $\varphi_2 = 0$ if $h < 0$. So we present here only the case $\varphi_1 = 0$ (i.e. $h \geq 0$), as the case $\varphi_2 = 0$ could be examined analogously and leads to the same results.

Thus $h \geq 0$ and setting $h = 2z^2$ we obtain

$$\varphi_1 = (2b - 3 + g - 8z - 4z^2)(2b - 3 + g + 8z - 4z^2) = 0.$$

Since the second factor could be brought to the first one via the change $z \to -z$, we shall consider the condition given by the first factor. So we get $b = (3 - g + 8z + 4z^2)/2$, and calculations yield

$$\mathcal{T}_4 = -\nu_1^2\nu_2^2\nu_3/2, \quad \theta = 8\nu_1\nu_2\nu_4, \quad \mathbf{D} = 3\nu_1^4\nu_2^4/4,$$
$$\nu_1(g, z) = -1 - 3z - 4z^2 + g(1+z), \quad \nu_4(g, z) = -2 - 3z + gz - 4z^2,$$
$$\nu_2(g, z) = 5 + g^2 + 24z + 56z^2 + 64z^3 + 32z^4 - 2g(1+2z)^2,$$
$$\nu_3(g, z) = g^2(1+2z) + 2g(1+2z) - 7 - 30z - 64z^2 - 64z^3 - 32z^4.$$

We remark that we must have either $\mathcal{T}_4 = 0$ (statements (xii) and (xiii)) or $\theta = 0$ (statement (xiv)).

1. Assume first $\mathcal{T}_4 = 0$. Since $\mathbf{D} \neq 0$ the condition $\mathcal{T}_4 = 0$ implies $\nu_3 = 0$, and we calculate

$$\text{Discrim}[\nu_3, g] = 32(1+2z)(1+2z+2z^2)^2 \geq 0,$$

which is equivalent to $2z+1 \geq 0$. So we set $2z+1 = 2u^2$ (i.e. $z = (2u^2-1)/2$), and then we obtain

$$\nu_3 = -2(1-u-gu+4u^4)(1+u+gu+4u^4) = 0.$$

We remark that if we change z by $-z$ and follow the same algorithm, then we get the equality $\nu_3(g,-z)\big|_{z=(1-2u^2)/2} = \nu_3(g,z)\big|_{z=(2u^2-1)/2}$.

So we consider the condition $\nu_3 = 0$. Since by the change $z \to -z$ we arrived at the same value of ν_3 and the second factor could be obtained from the first one via the change $u \to -u$, we shall only consider the condition given by the first factor. We observe that $u \neq 0$, and hence we have $g = (1-u+4u^4)/u$. Then calculations yield

$$\begin{aligned}
\mu_0 &= -(2u^2-2u-1)(1-2u+2u^2)^6(1+2u+2u^2)^4/(16u^4),\\
\eta &= (2u^2-2u-1)(1-2u+2u^2)^6(1+2u+2u^2)^3\\
&\quad \times (2-3u+6u^2+2u^3)/(16u^5),\\
\theta &= 2(2u^2-2u-1)(1-2u+2u^2)^6(1+2u+2u^2)^4/u^4,\\
\theta_1 &= 2(1+2u-2u^2)(1-2u+2u^2)^6(1+2u+2u^2)^3(3u+2u^3-1))/u^5.
\end{aligned}$$

These functions depend on a single variable, and one can find the roots of each one of them. Checking that the intervals formed by this set of roots are properly ordered, it is easy to verify that the case $\mu_0 > 0$, $\eta > 0$, $\theta < 0$ and $\theta_1 < 0$ is not realizable. Then we have $\mu_0\theta < 0$. Apart from this, the condition $\theta = 0$ implies $\mu_0 = 0$. This completes the proof of the statements (xii) and (xiii).

2. Suppose now $\theta = 0$, i.e. due to $\mathbf{D} \neq 0$ we have $\nu_4 = 0$. Then $z \neq 0$ and we obtain $g = (2+3z+4z^2)/z$. Calculations yield

$$\begin{aligned}
\theta_1 &= -64(-1-2z+2z^2)^2(1-2z+2z^2)(1+2z+2z^2)^3/z^5,\\
\mu_0 &= (-1-2z+2z^2)^2(1+2z+2z^2)^4/z^4.
\end{aligned}$$

Since the condition $\mu_0 \neq 0$ implies $\theta_1 \neq 0$ we conclude that the statement (xiv) is valid, and this completes the proof of Lemma 11.26. □

Lemma 11.27. *Assume that for a quadratic system with two real and two complex finite distinct singularities, the conditions $\mu_0 < 0$, $\widetilde{K} > 0$, $W_4 > 0$ and $(\mathcal{C}_{ff})$ hold,* *i.e. this system possesses two foci. If in addition we have $\mathcal{T}_4 = \mathcal{T}_3 = 0$, then the traces corresponding to complex singularities could not vanish, and hence we must have two weak foci (or centers).*

Proof. Consider the family of systems (11.42) and suppose the contrary, that $\rho_{1,2} = -(a+g) \pm 2i(b+h) = 0$, i.e. $a = -g$ and $b = -h$. In this case calculations yield $\mathcal{T}_4 = \mathcal{T}_3 = 0$ and $\tau_3 = 4(g-m)^2 + 8h(l+h)$, and since the real singularities must be foci, it is necessary that $\tau_3 < 0$. This implies $h \neq 0$, and we may assume $h = 1$ due to a time rescaling. So setting a new parameter v as $\tau_3 = -4v^2$, we obtain $l = -\big[(g-m)^2 + v^2 + 2)\big]/2$, and we calculate

$$\begin{aligned} W_4 &= -16v^2\big[16(gm-1)^2 + g^2(4+g^2-2gm+m^2+v^2)^2\big]^2\Psi(g,m,v),\\ \Psi &= 2(g+m)^2(2+g^2-gm)^2\big[4+(g-m)^2\big] + \psi(g,m)v^2 + 4\big[(gm-1)^2\\ &\quad + (1+g^2)^2\big]v^4 + g^2v^6, \end{aligned}$$

where

$$\psi(g,m) = 5g^2m^4 - 4g(4+g^2)m^3 + 2(8+4g^2-g^4)m^2 - 4g^5m + 16 + 32g^2 + 24g^4 + 5g^6.$$

We claim that in this case we have $W_4 < 0$. To prove this claim, it is sufficient to show that $\psi(g,m) > 0$. As this polynomial is of degree 4 with respect to the parameter m, considering Lemma 11.10 we calculate the corresponding polynomials on coefficients of ψ, which depend on the parameter g. So we calculate

$$\begin{aligned} D_\diamond &= -4096g^2(1+g^2)^6\big[4(1+g^2)^2 - g^2\big]^2,\\ S_\diamond &= -16g^4(4(1+g^2)^2 - g^2)(7+26g^2+7g^4), \end{aligned}$$

and since $g \neq 0$ (otherwise we get $\mu_0 = 2(2+m^2+v^2) > 0$), we deduce that $D_\diamond < 0$ and $S_\diamond < 0$. Therefore by Lemma 11.10 the polynomial $\psi(g,m)$ could not have real roots, and hence it is positive for any values of the parameters g and m. This completes the proof of our claim as well as the proof of Lemma 11.27. □

Lemma 11.28. *If a quadratic system* (8.1) *possesses an infinite star node N^*, then the condition $W_4 = 0$ is equivalent to $\mathcal{T}_4 = 0$.*

Proof. Consider a quadratic system (8.1) and assume that it has an infinite star node. Without lost of generality we may suppose that this point is located at $N[0:1:0]$, i.e. the parameter $k = 0$. So we detect the following Jacobian matrix corresponding to this singularity: $\begin{pmatrix} 2h-n & d\\ 0 & -n \end{pmatrix}$. Evidently we have a star node if and only if $h = d = 0$. Then we obtain $W_4 = \mathcal{T}_4^2$, and this proves the lemma. □

Lemma 11.29. *Consider a quadratic system possessing two real and two complex finite distinct singularities. Assume that the complex singularities have vanishing discriminants (i.e. $W_4 = W_3 = 0$) and at least one real singularity is weak (i.e. $\mathcal{T}_4 = 0$). Then* (i) *in the case $\mathcal{T}_3 \neq 0$ the condition $\theta\mathcal{T}_3\mathcal{F} < 0$ must hold, and* (ii) *the condition $\theta = 0$ implies $\mathcal{T}_3 = 0$.*

Proof. We consider systems (11.42) for which we force the conditions $\tau_{1,2} = 0$ and $\rho_3 = 0$ to be satisfied, and we examine two cases: $g = 0$ and $g \neq 0$.

If $g = 0$ then the condition $\rho_3 = 2m - a = 0$ gives $a = 2m$, and we calculate $\tau_{1,2} = 4[(b-h)^2 + m^2 \pm 2im(b + 2cl + h)]$. Therefore the conditions $\tau_{1,2} = 0$ give $m = 0$ and $h = b$, and we obtain $\mathcal{T}_4 = 0$ and $\theta = \mathcal{T}_3 = 0$, i.e. the statement (ii) of the lemma is valid in this case.

Assume now $g \neq 0$. Then due to a time rescaling we may consider $g = 1$, and the condition $\rho_3 = 1 - a + 2m = 0$ gives $a = 2m + 1$. Herein we calculate

$$\tau_{1,2} = 4\big[(1 - 2m - 7m^2) - (b-h)^2\big] \pm 8i\big[l(2m+1) + h + m(b+h)\big] = 0,$$

and we observe that $2m + 1 \neq 0$; otherwise we get $m = -1/2$, and then we have $\Im(\tau_{1,2}) = 4(h - b) = 0$ (i.e. $b = h$), which implies $\Re(\tau_{1,2}) = 1 \neq 0$.

So in what follows we assume that the condition $2m + 1 \neq 0$ holds. Then $\tau_{1,2} = 0$ gives

$$b = h \pm \sqrt{1 - 2m - 7m^2}, \quad l = -\frac{h(1+2m) \pm m\sqrt{1 - 2m - 7m^2}}{1 + 2m},$$

and we calculate

$$\begin{aligned} \theta = \frac{64(1+m)^2}{1+2m}\Phi_\pm, \quad \mathcal{T}_3 &= -\frac{32(1+m)}{1+2m}\Phi_\pm \Psi_\pm^2, \\ \theta \mathcal{T}_3 \mathcal{F} &= -\frac{4096}{(1+2m)^2}(1+m)^4 \Phi_\pm^2 \Psi_\pm^4, \end{aligned} \tag{11.52}$$

where

$$\begin{aligned} \Phi_\pm &= 2(h^2 - m^2)(2m+1) \pm h(1+3m)\sqrt{1 - 2m - 7m^2}, \\ \Psi_\pm &= 1 - 3m^2 + 2h(h \pm \sqrt{1 - 2m - 7m^2}\,). \end{aligned}$$

Evidently if $\mathcal{T}_3 \neq 0$ the condition $\theta \mathcal{T}_3 \mathcal{F} < 0$ is satisfied, i.e. the statement (i) of the lemma is valid. It remains to show that the condition $\theta = 0$ implies $\mathcal{T}_3 = 0$ which follows directly from (11.52). This completes the proof of Lemma 11.29. □

Lemma 11.30. *Assume that a quadratic system possesses two real and two complex finite distinct singularities. Then the conditions in each one of the following sets are incompatible:*

(i) $W_4 > 0,\ \mathcal{T}_4 = \mathcal{T}_3 = 0,\ \eta > 0,\ \theta \le 0,\ \theta_1 \le 0;$

(ii) $W_4 = \mathcal{T}_4 = \mathcal{F}_1 = 0,\ (\eta = 0 \wedge \theta \le 0) \vee (\eta > 0 \wedge \theta < 0 \wedge \theta_1 < 0);$

(iii) $\mu_0 > 0,\ W_4 = W_3 = \mathcal{T}_4 = 0,\ \mathcal{T}_3\mathcal{F} > 0,\ \mathcal{F}_1 = 0;$

(iv) $\mu_0 > 0,\ W_4 = W_3 = \mathcal{T}_4 = 0,\ W_2 > 0,\ (\theta \ge 0) \vee (\eta = 0)$
$\vee\ (\eta > 0 \wedge \theta < 0 \wedge \theta_1 < 0);$

(v) $W_4 < 0,\ \mathcal{T}_4 = \mathcal{T}_3 = 0,\ (\mathcal{C}^{\mathbb{C}}_{pp}),\ \eta \ge 0;$

(vi) $\mu_0 > 0$, $W_4 = W_3 = W_2 = \mathcal{T}_4 = \mathcal{F}_1\eta = 0$.

Proof. We have not been able to prove these statements by direct computation. However the Mathematics tools "Reduce", "Assuming" and/or "Find" give incompatibility of each one of the above sets of the conditions. □

Now we examine the possible global configurations of singularities for the family of systems under consideration.

1: $\mu_0 < 0$. Then by Theorem 6.1 (see Table 6.2) we have either two saddles or two anti-saddles.

1.1: $\widetilde{K} < 0$. In this case the real finite singularities are both saddles, and this implies the existence at infinity of three nodes. By Lemma 11.1, at infinity we have N^f, N^f, N^f.

1.1.1: $\mathcal{T}_4 \neq 0$. Then both saddles are strong, and we get the global configuration

- $s, s, ©, ©; N^f, N^f, N^f \Rightarrow (0, -1, 0, 0, -1, 0), (1, 0, 0, -1, -1, 1)$.

1.1.2: $\mathcal{T}_4 = 0$. In this case at least one saddle is weak.

1.1.2.1: $\mathcal{T}_3 \neq 0$. Then we have only one weak saddle the order of which is governed by the invariant polynomials $\mathcal{F}_i$, $i = 1, 2, 3, 4$.

1. $\mathcal{F}_1 \neq 0$. The weak saddle is of the first order. We get a unique global configuration
   - $s^{(1)}, s, ©, ©; N^f, N^f, N^f \Rightarrow (0, -1, 0, 0, -1, 0), (1, 1, 0, -1, -1, 1)$.
2. $\mathcal{F}_1 = 0$. The weak saddle is of order ≥ 2. However according to Theorems 7.9 and 7.3, we could have neither a weak saddle of the third order nor an integrable saddle. So we arrive at the global configuration
   - $s^{(2)}, s, ©, ©; N^f, N^f, N^f \Rightarrow (0, 1, 0, 1, 2, 1), (0, 0, -1, 2, -1/2, -4)$.

1.1.2.2: $\mathcal{T}_3 = 0$. In this case at least two traces vanish.

1. $\mathcal{T}_2 \neq 0$. Then only two singularities have zero traces, and we need to detect whether these singular points are real or complex. Since $\mu_0 < 0$, by Lemma 11.24 the condition $\mathcal{B}_5 \neq 0$ holds, and hence the sign of this polynomial distinguishes these two possibilities.
   (a) $\mathcal{B}_5 < 0$. This means that the non-zero roots of the equation (11.43) are complex, and hence the traces of the real singularities vanish. By Theorem 6.2 (see statement (c)) we have two weak saddles, which either both have order one or both are integrable depending on the value of the invariant polynomial $\mathcal{F}_1$. So we arrive at the following two global configurations:
   - $s^{(1)}, s^{(1)}, ©, ©; N^f, N^f, N^f \Rightarrow (0, 1, 0, -1, 0, 1), (0, -1, -1, 1, 1, -1)$ (if $\mathcal{F}_1 \neq 0$);
   - $\$, \$, ©, ©; N^f, N^f, N^f \Rightarrow (0, -1, 0, 1, 1, -4), (0, -1, 1, 1, -1, 0)$ (if $\mathcal{F}_1 = 0$).

(b) $\mathcal{B}_5 > 0$. In this case we deduce that the traces of the complex singularities vanish, and we get the global configuration
- $s, s, ©^\rho, ©^\rho;\ N^f, N^f, N^f \Rightarrow (1,0,0,-1,1,1), (-1,0,0,1,-1/2,-1)$.

2. $\mathcal{T}_2 = 0$. According to Lemma 11.23, the systems become Hamiltonian (and hence all the traces are zero), and we arrive at the global configuration
   - $\$, \$, ©^\rho, ©^\rho;\ N^f, N^f, N^f \Rightarrow (1,0,0,-1,-1,1), (1,-1,0,0,1,1)$.

1.2: $\widetilde{K} > 0$. In this case by [29] we have two anti-saddles. Additionally the sum of the indices of infinite singularities must be -1, and this means that we could only have four configurations of singularities at infinity (see Diagram 11.1).

1.2.1: $W_4 < 0$. Then systems (8.1) possess one node (which is generic due to $W_4 \neq 0$) and one focus (or a center).

1.2.1.1: $\mathcal{T}_4 \neq 0$. In this case we have a strong focus, and considering Lemma 11.1 we get the following four global configurations:
- $n, f, ©, ©;\ S, ⓒⓒ \Rightarrow (2,0,0,-2,1,2), (1,1,0,-1,0,1)$ (if $\eta < 0$);
- $n, f, ©, ©;\ S, S, N^\infty \Rightarrow (3,0,0,-3,5,3), (5,5,0,-5,5,5)$ (if $\eta > 0$);
- $n, f, ©, ©;\ \overline{\binom{0}{2}}SN, S \Rightarrow (1,6,0,-1,-9/8,1), (-5,0,0,5,0,-5)$ (if $\eta = 0$, $\widetilde{M} \neq 0$);
- $n, f, ©, ©;\ \overline{\binom{0}{3}}S \Rightarrow (25,-24,0,-25,38/5,25), (1/5,-18,0,-1/5,-14,1/5)$ (if $\eta = 0$, $\widetilde{M} = 0$).

1.2.1.2: $\mathcal{T}_4 = 0$, $\mathcal{T}_3 \neq 0$. In this case exactly one (real) singularity is weak, and clearly this singularity is a focus.

1. $\mathcal{F}_1 \neq 0$. By Theorem 6.2 the weak focus is of order one. Since by Lemma 11.25(ii) in this case the condition $\widetilde{M} = 0$ is impossible, considering Lemma 11.1 we arrive at the next three global configurations:
   - $n, f^{(1)}, ©, ©;\ S, ⓒⓒ \Rightarrow (1/8,-9/8,0,1,1/2,1/8), (0,2,0,-2,-7/16,0)$ (if $\eta < 0$);
   - $n, f^{(1)}, ©, ©;\ S, S, N^\infty \Rightarrow (-1/17,18/17,0,-1,-3/10,-1/17), (0,-2,0, 2,8/17,0)$ (if $\eta > 0$);
   - $n, f^{(1)}, ©, ©;\ \overline{\binom{0}{2}}SN, S \Rightarrow (2,-207/4,0,199/4,-64,2), (-157,403/4,0, 225/4,-191/8,-157)$ (if $\eta = 0$).
2. $\mathcal{F}_1 = 0$. Then the weak focus is of order greater than one.

   (a) $\mathcal{F}_2 \neq 0$. So by Theorem 6.2 the weak focus has order two, and by Theorem 7.7 the condition $\eta \neq 0$ must be satisfied. Thus we have the next two global configurations:
   - $n, f^{(2)}, ©, ©;\ S, ⓒⓒ \Rightarrow (0,0,-1,-4/35,1/100,1), (0,1,0,1/100, -17/35,-1/100)$ (if $\eta < 0$);
   - $n, f^{(2)}, ©, ©;\ S, S, N^\infty \Rightarrow (0,0,-1,-9,5/4,1), (0,1,0,1,-1,0)$ (if $\eta > 0$).

(b) $\mathcal{F}_2 = 0$. In this case we have a weak focus of order three if $\mathcal{F}_3\mathcal{F}_4 \neq 0$ and a center if $\mathcal{F}_3\mathcal{F}_4 = 0$. Additionally by Theorems 7.8 and 7.4, in both cases the condition $\eta > 0$ must hold, and we get the following two global configurations:
- $n, f^{(3)}, ©, ©;\ S, S, N^\infty \Rightarrow (0,0,-1,-11/4,1,1), (0,1,0,1,-13/8,-1)$ (if $\mathcal{F}_3\mathcal{F}_4 \neq 0$);
- $n, c, ©, ©;\ S, S, N^\infty \Rightarrow (0,0,-1,-3,1,1), (0,1,0,1,-2,-1)$ (if $\mathcal{F}_3\mathcal{F}_4 = 0$).

1.2.1.3: $\mathcal{T}_4 = 0$, $\mathcal{T}_3 = 0$. In this case at least two traces vanish. However by Lemma 11.23 the condition $\mathcal{T}_2 \neq 0$ must hold (as we have a node). So we have exactly two traces zero, and since we have a node, they must be the traces corresponding to complex singularities. Additionally by Lemma 11.26(i) the condition $\eta < 0$ must hold, and we get the global configuration
- $n, f, ©^\rho, ©^\rho;\ S, ⓒⓒ \Rightarrow (1,0,0,-1,-1,1), (1,-1365/32,0,1333/32,-10,1)$.

1.2.2: $W_4 > 0$. Then systems (8.1) possess either two foci or two nodes, and according to Theorem 6.1 (see Table 6.2) and Remark 11.12, we consider two cases: the conditions $(\mathcal{C}_{ff})$ are satisfied and the conditions $(\mathcal{C}_{nn})$ hold.

1.2.2.1: $(\mathcal{C}_{ff})$ hold. So we have two foci or two centers or one focus and one center.

1.2.2.1.1: $\mathcal{T}_4 \neq 0$. Then we have two foci which are strong, and considering Lemma 11.1 we get the following four global configurations:
- $f, f, ©, ©;\ S, ⓒⓒ \Rightarrow (1,-2,0,1,3,1), (1,0,0,-1,1,1)$ (if $\eta < 0$);
- $f, f, ©, ©;\ S, S, N^\infty \Rightarrow (1,-2,0,1,9/4,1), (1,0,0,-1,-5/4,1)$ (if $\eta > 0$);
- $f, f, ©, ©;\ \overline{\binom{0}{2}}SN, S \Rightarrow (1,4,0,-5,-1,1), (-3,2,0,1,-2,-3)$ (if $\eta = 0$, $\widetilde{M} \neq 0$);
- $f, f, ©, ©;\ \overline{\binom{0}{3}}S \Rightarrow (-1,2,0,-1,-2,-1), (-1,0,0,1,1,-1)$ (if $\eta = 0$, $\widetilde{M} = 0$).

1.2.2.1.2: $\mathcal{T}_4 = 0$, $\mathcal{T}_3 \neq 0$. This means that we have exactly one weak singularity, and this singularity is a focus.

1. $\mathcal{F}_1 \neq 0$. So the weak focus is of order one. Considering Lemma 11.1, we arrive at the following four global configurations:
   - $f, f^{(1)}, ©, ©;\ S, ⓒⓒ \Rightarrow (1,-2,0,1,2,1), (1,0,0,-1,0,1)$ (if $\eta < 0$);
   - $f, f^{(1)}, ©, ©;\ S, S, N^\infty \Rightarrow (0,2,0,-2,3,0), (8,0,0,-8,1,8)$ (if $\eta > 0$);
   - $f, f^{(1)}, ©, ©;\ \overline{\binom{0}{2}}SN, S \Rightarrow (0,1,0,-1,1,0), (1,0,0,-1,1/2,1)$ (if $\eta = 0$, $\widetilde{M} \neq 0$);
   - $f, f^{(1)}, ©, ©;\ \overline{\binom{0}{3}}S \Rightarrow (1,-3,0,2,2,1), (1,0,0,-1,-1/2,1)$ (if $\eta = 0$, $\widetilde{M} = 0$).
2. $\mathcal{F}_1 = 0$. Then we have a weak focus of order at least two.

(a) $\mathcal{F}_2 \neq 0$. By Theorem 6.2 (see statement (b_2)) the weak focus has order two. Then according to Theorem 7.7 in this case we arrive at the next three global configurations:
- $f, f^{(2)}, ©, ©;\ S, \copyright\copyright \Rightarrow (0,0,-1,-5/6,1/2,1), (0,1,0,1/2,-2/3,-1/2)$ (if $\eta < 0$);
- $f, f^{(2)}, ©, ©;\ S, S, N^\infty \Rightarrow (0,0,-1,-4/3,1/3,1), (0,1,0,1/3,-1/2,-1/3)$ (if $\eta > 0$);
- $f, f^{(2)}, ©, ©;\ \overline{\binom{0}{2}}SN, S \Rightarrow (0,0,-1,-(5+\xi)/3,1,1), (0,1,0,1,(2\xi-5)/6,-1)$, where $\{\eta^{-1}(0)\} \ni \xi \approx -1.6224$ (if $\eta = 0$).

(b) $\mathcal{F}_2 = 0$. According to Theorems 7.4 and 7.8, we arrive at the following four global configurations:
- $f, f^{(3)}, ©, ©;\ S, \copyright\copyright \Rightarrow (0,0,-1,-7/2,1,1), (0,1,0,1,-11/4,-1)$ (if $\eta < 0$, $\mathcal{F}_3\mathcal{F}_4 \neq 0$);
- $f, c, ©, ©;\ S, \copyright\copyright \Rightarrow (0,0,-1,-1,1,1), (0,1,0,1,-3/2,-1)$ (if $\eta < 0$, $\mathcal{F}_3\mathcal{F}_4 = 0$);
- $f, f^{(3)}, ©, ©;\ S, S, N^\infty \Rightarrow (0,0,-1,-13/4,1,1), (0,1,0,1,-19/8,-1)$ (if $\eta > 0$);
- $f, f^{(3)}, ©, ©;\ \overline{\binom{0}{2}}SN, S \Rightarrow \big(0,0,-1,-4,-\sqrt{(1+\sqrt{5})/2},1\big), \big(0,1,0,-\sqrt{(1+\sqrt{5})/2},-7/2,\sqrt{(1+\sqrt{5})/2}\big)$ (if $\eta = 0$).

1.2.2.1.2: $\mathcal{T}_4 = 0$, $\mathcal{T}_3 = 0$. According to Lemma 11.23, we could not have $\mathcal{T}_2 = 0$, and moreover by Lemma 11.27 exactly the two traces corresponding to the real singularities vanish. So there exist two weak foci, which by Theorem 6.2 (see statement (c)) are of order one if $\mathcal{F}_1 \neq 0$, or two centers if $\mathcal{F}_1 = 0$.

1. $\mathcal{F}_1 \neq 0$. So the weak foci are both of order one. According to Lemma 11.25(iii) in this case the condition $\widetilde{M} \neq 0$ holds, and considering Lemma 11.1 we arrive at the following three global configurations:
   - $f^{(1)}, f^{(1)}, ©, ©;\ S, \copyright\copyright \Rightarrow (0,-1,2,1,-2,0), (0,-1,1,1,-1,-1)$ (if $\eta < 0$);
   - $f^{(1)}, f^{(1)}, ©, ©;\ S, S, N^\infty \Rightarrow (0,-1,3/2,1,-3/2,-1/4), (0,-1,1,1,-1,-1)$ (if $\eta > 0$);
   - $f^{(1)}, f^{(1)}, ©, ©;\ \overline{\binom{0}{2}}SN, S \Rightarrow (0,1/3,1/8,-1/3,-1/4,0), (0,-1,-1/3,1,1/3,-1/2)$ (if $\eta = 0$).
2. $\mathcal{F}_1 = 0$. Then the real singularities are centers, and according to Theorem 7.4 we have the following three global configurations:
   - $c, c, ©, ©;\ S, \copyright\copyright \Rightarrow (0,1,-1,-1,1,-5/8), (0,2,-1,-2,1,-1/4)$ (if $\eta < 0$);
   - $c, c, ©, ©;\ S, S, N^\infty \Rightarrow (0,1,-1,-1,1,0), (0,2,-1,-2,1,1)$ (if $\eta > 0$);
   - $c, c, ©, ©;\ \overline{\binom{0}{3}}S \Rightarrow (0,1,3/2,-1,-3/2,-1), (0,-1,-1,1,1,0)$ (if $\eta = 0$).

1.2.2.2: $(\mathcal{C}_{nn})$. So we have two nodes. Since $W_4 \neq 0$ both nodes are generic. As a result taking into account Lemma 11.1, we arrive at the following four global configurations:

- $n, n, ©, ©;\ S, ©\,©\ \Rightarrow (-2,1,0,2,1,-21/10), (-1,-2,0,1,1,-1)$ (if $\eta < 0$);
- $n, n, ©, ©;\ S, S, N^{\infty} \Rightarrow (-2,1,0,2,1,-19/10), (-1,-2,0,1,1,-1)$ (if $\eta > 0$);
- $n, n, ©, ©;\ \overline{\binom{0}{2}}SN, S \Rightarrow (-2,1,0,2,1,-2), (-1,-2,0,1,1,-1)$ (if $\eta = 0$, $\widetilde{M} \neq 0$);
- $n, n, ©, ©;\ \overline{\binom{0}{3}}S \Rightarrow (-64, 3971/64, 0, 125/64, 1, -64), (-1, 4097/4096, 0, -1/4096, 1, -1)$ (if $\eta = 0$, $\widetilde{M} = 0$).

1.2.3: $W_4 = 0$. In this case at least one singularity (real or complex) of systems (8.1) has discriminant zero.

1.2.3.1: $W_3 < 0$. Then by Theorem 6.1 the systems (8.1) possess a node and a focus (or a center). Moreover since only one discriminant vanishes, the node must have coinciding eigenvalues, and by Theorem 7.2 it could not be a star node.

1. $\mathcal{T}_4 \neq 0$. According to Theorem 6.2 we have a strong focus, and considering Lemma 11.1 we arrive at the following four global configurations:

   - $n^d, f, ©, ©;\ S, ©\,©\ \Rightarrow (81\xi-80, -2\xi, 0, 80-79\xi, 3\xi, 81\xi-80), (246\xi-245, 1-\xi, 0, 245-246\xi, 94-93\xi, 246\xi-245)$, where $\{W_4^{-1}(0)\} \ni \xi \approx 0.9984$ (if $\eta < 0$);

   - $n^d, f, ©, ©;\ S, S, N^{\infty} \Rightarrow \big((1+3\xi)/4, -2\xi, 0, (5\xi-1)/4, 9\xi/4, (1+3\xi)/4\big), \big(4\xi-3, 1-\xi, 0, 3-4\xi, -(4+\xi)/4, 4\xi-3\big)$, where $\{W_4^{-1}(0)\} \ni \xi \approx 0.9751$ (if $\eta > 0$);

   - $n^d, f, ©, ©;\ \overline{\binom{0}{2}}SN, S \Rightarrow (0,1,1,-29,1,0), (0,0,1,-231/2,-11,2)$ (if $\eta = 0$, $\widetilde{M} \neq 0$);

   - $n^d, f, ©, ©;\ \overline{\binom{0}{3}}S \Rightarrow \big(-3\sqrt{3}, 6(\sqrt{3}-5), 0, 3(10-\sqrt{3}), 0, -3\sqrt{3}\big), (9,-10,0,1,15,9)$ (if $\eta = 0$, $\widetilde{M} = 0$).

2. $\mathcal{T}_4 = 0$. We have at least one singularity with zero trace. However by Lemma 11.26(ii), in this case the condition $\mathcal{T}_3 \neq 0$ holds. Therefore we have exactly one weak singularity, and this singularity is a focus.

   (a) $\mathcal{F}_1 \neq 0$. By Theorem 6.2 the weak focus is of order one, and according to Lemma 11.25(iv), which says that in this case the condition $\widetilde{M} \neq 0$ holds, and Lemma 11.1, we get the next three global configurations:
   - $n^d, f^{(1)}, ©, ©;\ S, ©\,©\ \Rightarrow (0,-1,-2,1,1,-1), (0,1,1,-1,0,2)$ (if $\eta < 0$);
   - $n^d, f^{(1)}, ©, ©;\ S, S, N^{\infty} \Rightarrow (0,-1,-2,1,1,-1), (0,1,1,-1,0,4)$ (if $\eta > 0$);
   - $n^d, f^{(1)}, ©, ©;\ \overline{\binom{0}{2}}SN, S \Rightarrow (0,-1,-2,1,1,-1), (0,1,1,-1,0,3)$ (if $\eta = 0$).

(b) $\mathcal{F}_1 = 0$, $\mathcal{F}_2 \neq 0$. By Theorem 6.2 we have a weak focus of second order, and according to Theorem 7.7 we get the global configurations:
- $n^d, f^{(2)}, ©, ©;\ S, ©©$ $\Rightarrow$ $(0, 0, -1, -(6+5\xi)/15, 1/100, 1)$, $(0, 1, 0, 1/100, (5\xi - 3)/15, -1/100)$, where $\{W_4^{-1}(0)\} \ni \xi \approx -0.8725$ (if $\eta < 0$);
- $n^d, f^{(2)}, ©, ©;\ S, S, N^\infty \Rightarrow (0, 0, -1, (17+100\xi)/300, 1, 1)$, $(0, 1, 0, 1, (50\xi - 17)/300, -1)$, where $\{W_4^{-1}(0)\} \ni \xi \approx -14.9953$ (if $\eta > 0$).

(c) $\mathcal{F}_1 = 0$, $\mathcal{F}_2 = 0$. According to Theorem 7.4 we could not have a center (i.e. the condition $\mathcal{F}_3\mathcal{F}_4 \neq 0$ holds). Hence we have a weak focus of order three, and considering Theorem 7.8 we arrive at the global configuration
- $n^d, f^{(3)}, ©, ©;\ S, S, N^\infty \Rightarrow (0, 0, -1, (\xi - 5)/2, 1, 1)$, $(0, 1, 0, 1, (3\xi - 5)/4, -1)$, where $\{W_4^{-1}(0)\} \ni \xi \approx -1.0400$.

1.2.3.2: $W_3 > 0$. By Theorem 6.1 the systems (8.1) possess two nodes, and since only one discriminant vanishes, one node must have coinciding eigenvalues. According to Theorem 7.1 and Lemma 11.2, this node is a star node if and only if the conditions $(\mathcal{S}^1_{n^*})$ are satisfied. Moreover by Theorem 7.2, in the case under consideration, the existence of a star node implies the condition $\eta < 0$. So considering Lemma 11.1 we arrive at the following five global configurations:
- $n, n^d, ©, ©;\ S, ©© \Rightarrow (-2(1+39\xi), 1-\xi, 0, 2(1+39\xi), 1-\xi, -(21+779\xi)/10)$, $(-1-244\xi, -2+3\xi, 0, 1+244\xi, 1+93\xi, -1-244\xi)$, where $\{W_4^{-1}(0)\} \ni \xi \approx -0.004322$ (if $\eta < 0$, $\neg(\mathcal{S}^1_{n^*})$);
- $n, n^*, ©, ©;\ S, ©© \Rightarrow (0, 1, 0, 0, 1, 2), (0, 0, 1, -1, -1, 0)$ (if $\eta < 0$, $(\mathcal{S}^1_{n^*})$);
- $n, n^d, ©, ©;\ S, S, N^\infty \Rightarrow ((1-9\xi)/4, \xi, 0, (9\xi - 1)/4, \xi, (5-43\xi)/20)$, $(-3+2\xi, 1-3\xi, 0, 3-2\xi, -1+2\xi, -3+2\xi)$, where $\{W_4^{-1}(0)\} \ni \xi \approx 0.01709$ (if $\eta > 0$);
- $n, n^d, ©, ©;\ \overline{\binom{0}{2}} SN, S \Rightarrow (0, 1, 1, -59/2, 1, 0), (0, 0, 1, -117, -11, 2)$ (if $\eta = 0$, $\widetilde{M} \neq 0$);
- $n, n^d, ©, ©;\ \overline{\binom{0}{3}} S \Rightarrow (-3\sqrt{3}, -2(32-3\sqrt{3}), 0, 64-3\sqrt{3}, 0, -3\sqrt{3}), (9, -10, 0, 1, 32, 9)$ (if $\eta = 0$, $\widetilde{M} = 0$).

1.2.3.3: $W_3 = 0$. In this case at least two singularities have zero discriminants.

1.2.3.3.1: $W_2 < 0$. By Theorem 6.1 the systems (8.1) possess a node and a focus (or a center). Therefore the vanishing discriminants correspond to complex singularities, and hence the node is generic. Additionally by Lemma 11.25(v) we could not have in this case $\widetilde{M} = 0$.

1. $\mathcal{T}_4 \neq 0$. Then we have a strong focus, and considering Lemma 11.1 we arrive at the following three global configurations:
   - $n, f, ©^{\tau}, ©^{\tau};\ S, ©© \Rightarrow (1, -1, 0, 0, 1, 1), (0, 1/2, 0, -1/2, -3/16, 0)$ (if $\eta < 0$);

- $n, f, ©^{\tau}, ©^{\tau}$; $S, S, N^{\infty} \Rightarrow (1,1,0,-2,4,1), (1,-3/2,0,1/2,-35/16,1)$ (if $\eta > 0$);

- $n, f, ©^{\tau}, ©^{\tau}$; $\overline{\binom{0}{2}}SN, S \Rightarrow (1,\xi,0,-1,-1,1), \big(-8,-7\xi/2,0,8,(\xi^2-196)/16, -8\big)$, where $\{\eta^{-1}(0)\} \ni \xi \approx 11.6655$ (if $\eta = 0$).

2. $\mathcal{T}_4 = 0$. By Theorems 7.7, 7.8 and 7.4, we could not have a weak focus of order greater than or equal to two nor a center together with a node and two complex finite singularities. Therefore the condition $\mathcal{F}_1 \neq 0$ holds, i.e. we have a weak focus of order one. Additionally by Lemma 11.25(v) the condition $\widetilde{M} \neq 0$ holds. So considering Lemma 11.1 we get the following three global configurations:

   - $n, f^{(1)}, ©^{\tau}, ©^{\tau}$; $S, ©©$ $\Rightarrow (-1,4,0,-3,-3,-1), (-3+2\sqrt{2}, -4\sqrt{2}, 0, 3+2\sqrt{2}, 1, -3+2\sqrt{2})$ (if $\eta < 0$);

   - $n, f^{(1)}, ©^{\tau}, ©^{\tau}$; $S, S, N^{\infty} \Rightarrow (1471/16384, 14913/8192, 0, -31297/16384, -8, 1471/16384), \big((-131072+\sqrt{125994113})/16384, 14913\sqrt{125994113}/24100864, 0, (11768-\sqrt{125994113})/1471, 1, (-131072+\sqrt{125994113})/16384\big)$ (if $\eta > 0$);

   - $n, f^{(1)}, ©^{\tau}, ©^{\tau}$; $\overline{\binom{0}{2}}SN, S \Rightarrow \big(2+\xi, -2(1+\xi), 0, \xi, 325/64, 2+\xi\big)$, $\big(\chi+325/64, -\chi(1+\xi)/(2+\xi), 0, -(650+325\xi+64\chi)/(64(2+\xi)), 1, \chi+325/64\big)$, where$\{\eta^{-1}(0)\} \ni \xi \approx -2.5319$ and $\chi = \sqrt{\xi^2-2\xi-7}$ (if $\eta = 0$).

1.2.3.3.2: $W_2 > 0$. Then systems (11.42) possess two anti-saddles the types of which depend on the sets of conditions $(\mathcal{C}^{\tau}_{ff})$, $(\mathcal{C}^{\tau}_{nn})$ and $(\mathcal{C}^{\tau}_{n^d n^d})$ (see Remark 11.12).

1.2.3.3.2.1: $(\mathcal{C}^{\tau}_{ff})$. Then we have either two foci, or two centers, or one focus and one center, and the discriminants of complex singularities vanish (due to $W_3 = W_4 = 0$).

1. $\mathcal{T}_4 \neq 0$. By Theorem 6.2 we have two strong foci, and considering Lemma 11.1 we arrive at the following four global configurations:

   - $f, f, ©^{\tau}, ©^{\tau}$; $S, ©©$ $\Rightarrow (0,0,0,0,1,0), (1,1,0,-2,1,1)$ (if $\eta < 0$);

   - $f, f, ©^{\tau}, ©^{\tau}$; $S, S, N^{\infty} \Rightarrow (0,0,0,0,1,0), (1,1,0,-2,2,1)$ (if $\eta > 0$);

   - $f, f, ©^{\tau}, ©^{\tau}$; $\overline{\binom{0}{2}}SN, S \Rightarrow (0,0,0,0,1,0), (1,1,0,-1,1,1)$ (if $\eta=0, \widetilde{M}\neq 0$);

   - $f, f, ©^{\tau}, ©^{\tau}$; $\overline{\binom{0}{3}}S \Rightarrow (1,10,0,-1,18,1), (27,45,0,-27,-14,27)$ (if $\eta = 0, \widetilde{M} = 0$).

2. $\mathcal{T}_4 = 0$, $\mathcal{T}_3 \neq 0$. In this case by Theorem 6.2 only one singularity has zero trace, and we have one weak focus.

(a) $\mathcal{F}_1 \neq 0$. Then the weak focus has order one, and taking into account Lemma 11.1 we get the following four global configurations:

- $f, f^{(1)}, ©^{\tau}, ©^{\tau}; S, ©\,© \Rightarrow \big(1, 2(2+\sqrt{849}), 0, -5-2\sqrt{849}, -32, 1\big),$ $\big(-61, -29(2+\sqrt{849}), 0, 119+29\sqrt{849}, 3+\sqrt{849}, -61\big)$ (if $\eta < 0$);

- $f, f^{(1)}, ©^{\tau}, ©^{\tau}; S, S, N^{\infty} \Rightarrow \big(1, 4(1+\sqrt{902}), 0, -5-4\sqrt{902}, -32, 1\big),$ $\big(-92, -120(1+\sqrt{902}), 0, 4(53+30\sqrt{902}), 3+2\sqrt{902}, -92\big)$ (if $\eta > 0$);

- $f, f^{(1)}, ©^{\tau}, ©^{\tau}; \overline{\binom{0}{2}}SN, S \Rightarrow \big(1, 2(2+\chi), 0, -5-2\chi, -32, 1\big),$ $\big(\xi, (32+\xi)(2+\chi), 0, -64-3\xi-(32+\xi)\chi, 3+\chi, \xi\big)$, where $\{\eta^{-1}(0)\} \ni \xi \approx -74.5876$ and $\chi = \sqrt{\xi^2+64\xi+1032}$ (if $\eta = 0$, $\widetilde{M} \neq 0$);

- $f, f^{(1)}, ©^{\tau}, ©^{\tau}; \overline{\binom{0}{3}}S \Rightarrow (1, -128, 0, 101, -79\sqrt{7}, 1), (-113\sqrt{7}, 2176\sqrt{7}/27, 0,$ $875\sqrt{7}/27, -37, -113\sqrt{7})$ (if $\eta = 0$, $\widetilde{M} = 0$).

(b) $\mathcal{F}_1 = 0$. According to Theorems 7.8 and 7.4, we could have neither a weak focus of order three nor a center, i.e. the condition $\mathcal{F}_2 \neq 0$ is satisfied, and we have a weak focus of order two. Therefore by Theorem 7.7 we could only have three global configurations corresponding to $\eta < 0$, $\eta > 0$ and $\eta = 0$, respectively. In order to construct the corresponding examples, we shall use the next lemma.

Lemma 11.31. *If $\mu_0 \neq 0$ and $W_4 = W_3 = 0$, then each derivative of the polynomial W_4 with respect to each parameter vanishes.*

Proof. According to Lemma 5.4 for any quadratic system with $\mu_0 \neq 0$, we have the identities

$$W_4 = \mu_0^2 \tau_1\tau_2\tau_3\tau_4, \quad W_3 = \mu_0^2\big(\tau_1\tau_2\tau_3 + \tau_1\tau_3\tau_4 + \tau_2\tau_3\tau_4\big),$$

where τ_i $(i = 1, \ldots, 4)$ are the four discriminants of the finite singularities (which could be real or complex, distinct or coinciding). The condition $W_4 = W_3 = 0$ implies the vanishing of at least two discriminants. Additionally for any arbitrary parameter, say a, we have

$$\begin{aligned} \frac{\partial W_4}{\partial a} &= 2\mu_0\tau_1\tau_2\tau_3\tau_4\frac{\partial \mu_0}{\partial a} \\ &\quad + \mu_0^2\left[\tau_2\tau_3\tau_4\frac{\partial \tau_1}{\partial a} + \tau_1\tau_3\tau_4\frac{\partial \tau_2}{\partial a} + \tau_1\tau_2\tau_4\frac{\partial \tau_3}{\partial a} + \tau_1\tau_2\tau_3\frac{\partial \tau_4}{\partial a}\right], \end{aligned}$$

and since we have at least two discriminants zero, we deduce that $\dfrac{\partial W_4}{\partial a} = 0$. This completes the proof of the lemma. □

So to construct the required three examples, we shall use the canonical systems:

$$\dot{x} = -y + (v-b)x^2/3 + 2axy - ny^2,$$
$$\dot{y} = x + ax^2 - (b+2v)xy/3 - ay^2,$$

possessing a weak focus of order two at the origin of coordinates. For these systems we calculate

$$\begin{aligned}W_4 = \frac{-4}{2187}\Big\{&6^{10}a^8 - 6^7a^6(7b^2+24bn+36n^2-8bv-168nv-44v^2)\\&+108a^4\big[203b^4+16b^3(99n-26v)+24b^2(225n^2-558nv-91v^2)\\&+64b(3n-v)(45n^2-210nv-43v^2)-16(729n^4+540n^3v-4914n^2v^2\\&-4788nv^3-479v^4)\big]+9a^2\big[7b^6+4b^5(54n-25v)+8b^4(285n^2-75nv\\&+61v^2)+48b^3(207n^3-46n^2v+59nv^2-16v^3)+32b^2(648n^4-297n^3v\\&-342n^2v^2-585nv^3-16v^4)+256b(3n-v)(27n^4-18n^3v-69n^2v^2\\&-72nv^3-4v^4)-256v(81n^5-324n^3v^2-516n^2v^3-201nv^4-8v^5)\big]\\&+4n(b-4v)^4(b+3n+2v)^3\Big\},\end{aligned}$$

and considering Lemma 11.31 in order to find out the needed examples, we shall consider the common solutions of the polynomial equations

$$W_4 = 0,\ \partial W_4/\partial a = 0,\ \partial W_4/\partial b = 0,\ \partial W_4/\partial n = 0,\ \partial W_4/\partial v = 0. \quad (11.53)$$

Then we take the coefficients $(0,0,-1,(v_0-b_0)/3,a_0,-n_0),(0,1,0,a_0,-(b_0+2v_0)/6,-a_0)$, where (a_0,b_0,n_0,v_0) is a common solution of equations (11.53), and we construct the next three global configurations:

- $f, f^{(2)}, ©^{\tau}, ©^{\tau};\ S, ©©\Rightarrow$ for $(a_0,b_0,n_0,v_0)\approx(0.97515934, 4v_0, -3.82479968, 6.64073493)$ (if $\eta<0$);
- $f, f^{(2)}, ©^{\tau}, ©^{\tau};\ S, S, N^{\infty}\Rightarrow$ for $(a_0,b_0,n_0,v_0)\approx(0.26881531, -4.97826473, 1, 7/16)$ (if $\eta>0$);
- $f, f^{(2)}, ©^{\tau}, ©^{\tau};\ \overline{\binom{0}{2}}SN, S\Rightarrow$ for $(a_0,b_0,n_0,v_0)\approx(0.32252959, -5.18361596, 1, 0.26660841)$ (if $\eta=0$).

3. $\mathcal{T}_4=0$, $\mathcal{T}_3=0$. Then clearly by Theorem 6.2 we have two weak foci of order one (if $\mathcal{F}_1\neq0$) or two centers (if $\mathcal{F}_1=0$). However by Lemma 11.26(iii) the condition $\mathcal{F}_1=0$ must hold, i.e. we have two centers. In this case by Theorem 7.4 we get the unique global configuration

- $c, c, ©^{\tau}, ©^{\tau};\ S, ©©\Rightarrow(0,0,0,0,1,0),(1,0,0,-1,0,1)$.

1.2.3.3.2.2: $(\mathcal{C}^{\tau}_{nn})$. So by Theorem 6.1 and Remark 11.12, apart from the complex singularities with zero discriminants, we have two generic nodes. Since by Lemma 11.25(vi) the condition $\widetilde{M} \neq 0$ must hold, considering Lemma 11.1 we obtain the following three global configurations:

- $n, n, ©^{\tau}, ©^{\tau}$; S, ⓒⓒ $\Rightarrow (1,0,0,-1,0,1), (1,0,0,-1,-1/4,1)$ (if $\eta < 0$);
- $n, n, ©^{\tau}, ©^{\tau}$; $S, S, N^{\infty} \Rightarrow (1,-1,0,0,-2,1), (1,-3/2,0,1/2,-35/16,1)$ (if $\eta > 0$);
- $n, n, ©^{\tau}, ©^{\tau}$; $\overline{\binom{0}{2}}SN, S \Rightarrow (1,-1,0,-1,3/2,1), (0,3/4,0,0,-1/2,0)$ (if $\eta = 0$).

1.2.3.3.2.3: $(\mathcal{C}^{\tau}_{n^d n^d})$. In this case the discriminants corresponding to the real singularities vanish, i.e. we have two nodes with coinciding eigenvalues. Moreover by Theorem 7.1 and Lemma 11.2, we have one star node if and only if the conditions $(\mathcal{S}^1_{n^*})$ are satisfied, and we have two star nodes if and only if the set of conditions $(\mathcal{S}^3_{n^*})$ holds. Considering Theorem 7.2 we deduce that the existence of at least one finite star node implies the condition $\eta < 0$. So considering Lemma 11.1, we arrive at the following six global configurations:

- n^d, n^d, ©, ©; S, ⓒⓒ $\Rightarrow (0,-1,0,1,841/2048,-1/9), (0,1,-1,-1,3/32,7/8)$ (if $\eta < 0$, $\neg\big((\mathcal{S}^1_{n^*}) \vee (\mathcal{S}^3_{n^*})\big)$);
- n^d, n^*, ©, ©; S, ⓒⓒ $\Rightarrow (0,1,0,-1,-1,2), (0,0,1,0,-1,0)$ (if $\eta < 0$, $(\mathcal{S}^1_{n^*})$);
- n^*, n^*, ©, ©; S, ⓒⓒ $\Rightarrow (0,-1,0,1,0,-1), (0,0,-1,0,1,1)$ (if $\eta < 0$, $(\mathcal{S}^3_{n^*})$);
- n^d, n^d, ©, ©; $S, S, N^{\infty} \Rightarrow (0,-1,0,1,25/2,-12), (0,1,-1,-1,-4,75/8)$ (if $\eta > 0$);
- n^d, n^d, ©, ©; $\overline{\binom{0}{2}}SN, S \Rightarrow (0,1,0,-1,225/2,108), (0,1,1,-1,-16,113)$ (if $\eta = 0$, $\widetilde{M} \neq 0$);
- n^d, n^d, ©, ©; $\overline{\binom{0}{3}}S \Rightarrow (0,4,-4,-4,25/8,-1), (0,1,0,-1,-1/2,13/4)$ (if $\eta = 0$, $\widetilde{M} = 0$).

1.2.3.3.3: $W_2 = 0$. So at least three discriminants vanish, and this means that necessarily two of them correspond to the complex singularities and at least one to a real one. So we have one node with coinciding eigenvalues, and by Theorem 7.2 it could not be a star node. Moreover by Lemma 11.25(v) the condition $\widetilde{M} = 0$ is impossible in this case. Additionally the type of the second real singularity is governed by the invariant polynomial W_1.

1. $W_1 < 0$. Then apart from n^d, we have a focus or a center.
   (a) $\mathcal{T}_4 \neq 0$. In this case we have a strong focus, and considering Lemma 11.1 we obtain the following three global configurations:
   - $n^d, f, ©^{\tau}, ©^{\tau}$; S, ⓒⓒ $\Rightarrow (1,1,0,-2,-1/2,1), (-1,-1/4,0,5/4,0,-1)$ (if $\eta < 0$);
   - $n^d, f, ©^{\tau}, ©^{\tau}$; $S, S, N^{\infty} \Rightarrow (1,-1,0,0,-9/2,1), (0,-9/4,0,9/4,-5,0)$ (if $\eta > 0$);
   - $n^d, f, ©^{\tau}, ©^{\tau}$; $\overline{\binom{0}{2}}SN, S \Rightarrow \big(1,-2(128+\xi),0,255+2\xi,-98,1\big)$, $\big(\xi,-(98+\xi)(128+\xi),0,12544+225\xi+\xi^2,15(113+\xi),\xi\big)$, where $\{\eta^{-1}(0)\} \ni \xi \approx -376.768$ (if $\eta = 0$).

(b) $\mathcal{T}_4 = 0$. By Theorem 6.2 we have a weak focus. Considering Theorems 7.7, 7.8 and 7.4 we deduce that we could not have a center and that this weak focus could only be of order one. So considering Lemma 11.1, we arrive at the following three global configurations:

- $n^d, f^{(1)}, ©^\tau, ©^\tau; S, ©\,© \Rightarrow (1,-2,0,1,-1-\sqrt{2}/2,1), (-\sqrt{2}/2,-1,0, 1+\sqrt{2}/2, 0, -\sqrt{2}/2)$ (if $\eta < 0$);
- $n^d, f^{(1)}, ©^\tau, ©^\tau; S, S, N^\infty \Rightarrow \big(1, 15, 0, -16, (165-19\sqrt{89}+\xi)/16, 1\big), \big((165-11\sqrt{89}+\xi)/16, 15\sqrt{89}/4, 0, -(165+49\sqrt{89}+\xi)/16, 17/2, (165-11\sqrt{89}+\xi)/16\big)$, where $\xi = 15\sqrt{2(173-15\sqrt{89})}$ (if $\eta > 0$);
- $n^d, f^{(1)}, ©^\tau, ©^\tau; \overline{\binom{0}{2}}SN, S \Rightarrow \big(1, -1-\chi, 0, \chi, (\xi+\omega/2), 1\big), \big(\xi, (1+\chi)\omega/4, 0, -\xi-(1+\chi)\omega/4, (1-\chi)/2, \xi\big)$, where $\{W_2^{-1}(0), \eta^{-1}(0)\} \ni (\xi, \chi) \approx (-13.2759, 4.64856)$ and $\omega = \sqrt{\chi^2+10\chi-7}$ (if $\eta = 0$).

2. $W_1 > 0$. Then apart from n^d, we have a node, which is generic due to $W_1 \neq 0$. As it was mentioned above, in the case under consideration, we could not have $\widetilde{M} = 0$, and therefore we get the next three global configurations:

- $n, n^d, ©^\tau, ©^\tau; S, ©\,© \Rightarrow (1,-3,0,2,-9/2,1), (1,-33/4,0,29/4,-7,1)$ (if $\eta < 0$);
- $n, n^d, ©^\tau, ©^\tau; S, S, N^\infty \Rightarrow (1,1,0,-2,-9/2,1), (-1,7/4,0,-3/4,-3,-1)$ (if $\eta > 0$);
- $n, n^d, ©^\tau, ©^\tau; \overline{\binom{0}{2}}SN, S \Rightarrow \big(1, -(9025+128\xi)/64, 0, (8961+128\xi)/64, -6241/128, 1\big), \big(\xi, -(6241+128\xi)(9025+128\xi)/16384, 0, (56325025 + 1937664\xi + 16384\xi^2)/16384, 87(7633+128\xi)/1024, \xi\big)$, where $\{\eta^{-1}(0)\} \ni \xi \approx -69.0074$ (if $\eta = 0$).

3. $W_1 = 0$. In this case all the discriminants vanish, and since we could not have a star node and the condition $\widetilde{M} \neq 0$ holds, we arrive at the following three global configurations:

- $n^d, n^d, ©^\tau, ©^\tau; S, ©\,© \Rightarrow (5,0,0,-5,2\sqrt{10}-11,1), (-1,0,0,1,2\sqrt{10}-3,-1)$ (if $\eta < 0$);
- $n^d, n^d, ©^\tau, ©^\tau; S, S, N^\infty \Rightarrow (0,0,0,0,1,0), (1,0,0,-1,2,1)$ (if $\eta > 0$);
- $n^d, n^d, ©^\tau, ©^\tau; \overline{\binom{0}{2}}SN, S \Rightarrow \big(1,0,0,-1,(91\sqrt{7}-256)/9,1\big), \big(-\xi/18, 0, 0, \xi/18, -(7\sqrt{7}-16)^2/9, -\xi/18\big)$, where $\xi = (7\sqrt{7}-19)^2$ (if $\eta = 0$).

2: $\mu_0 > 0$. Then according to Theorem 6.1 (see Table 6.2, lines 23–27) we have one saddle and one anti-saddle.

Remark 11.14. According to Theorem 7.5 if a quadratic system possesses an infinite star node, then it could not possess a focus.

2.1: $W_4 < 0$. Then the anti-saddle is a focus (or a center), and we consider two possibilities: $\mathcal{T}_4 \neq 0$ and $\mathcal{T}_4 = 0$.

2.1.1: $\mathcal{T}_4 \neq 0$. In this case by Theorem 6.2 we do not have weak singularities, and hence the configuration of finite singularities is well defined. Regarding the infinite singularities, according to Lemma 11.1 we need to distinguish not only their number and types (which are governed by the invariant polynomials η and $\widetilde{M}$), but also the type of elemental nodes if we have them. And in these cases we will apply the invariant polynomials θ and θ_i $(i = 1, 2, 3)$.

1. $\eta < 0$. According to Remark 11.14 and Lemma 11.1, in this case we get only three global configurations:
   - $s, f, \copyright, \copyright;\ N^{\infty},\ \copyright, \copyright \Rightarrow (0,0,0,0,1/4,0), (1,-3,0,2,1/4,1)$ (if $\theta < 0$);
   - $s, f, \copyright, \copyright;\ N^{f},\ \copyright, \copyright \Rightarrow (0,0,0,0,-2,0), (1,-3,0,2,1,1)$ (if $\theta > 0$);
   - $s, f, \copyright, \copyright;\ N^{d},\ \copyright, \copyright \Rightarrow (2,-6,0,4,2,2), (3,-2,0,-1,1,3)$ (if $\theta = 0$).
2. $\eta > 0$. By Lemma 11.1 in this case there are 10 possible configurations. However considering Remark 11.14, we arrive at the next six global configurations:
   - $s, f, \copyright, \copyright;\ S, N^{\infty},\ N^{\infty} \Rightarrow (0,-1/4,0,1/4,3/4,0), (1,-5/4,0,1/4,0,1)$ (if $\theta < 0$, $\theta_1 < 0$);
   - $s, f, \copyright, \copyright;\ S, N^{f},\ N^{f} \Rightarrow (0,-1,0,1,2,0), (1,-3,0,2,0,1)$ (if $\theta < 0$, $\theta_1 > 0$);
   - $s, f, \copyright, \copyright;\ S, N^{f},\ N^{\infty} \Rightarrow (0,-1/2,0,1/2,1/4,0), (1,-3/4,0,-1/4,0,1)$ (if $\theta > 0$);
   - $s, f, \copyright, \copyright;\ S, N^{\infty},\ N^{d} \Rightarrow (1/2,-23/8,0,19/8,1,1/2), (0,3/4,0,-3/4,0,0)$ (if $\theta = 0$, $\theta_1 < 0$);
   - $s, f, \copyright, \copyright;\ S, N^{f},\ N^{d} \Rightarrow (-2,0,0,2,1,-2), (3,-2,0,-1,0,3)$ (if $\theta = 0$, $\theta_1 > 0$);
   - $s, f, \copyright, \copyright;\ S, N^{d},\ N^{d} \Rightarrow (0,-1,0,1,1,0), (1,-1,0,0,0,1)$ (if $\theta = 0$, $\theta_1 = 0$).
3. $\eta = 0$. By Lemma 11.1, at infinity we could have either one double and one simple (if $\widetilde{M} \neq 0$) or one triple (if $\widetilde{M} = 0$) real singularities. Considering Remark 11.14 we obtain the following four global configurations:
   - $s, f, \copyright, \copyright;\ \overline{\binom{0}{2}}SN,\ N^{\infty} \Rightarrow (-9/2,21/2,0,-6,0,-9/2), (15,-19,0,4,4,15)$ (if $\widetilde{M} \neq 0$, $\theta < 0$);
   - $s, f, \copyright, \copyright;\ \overline{\binom{0}{2}}SN,\ N^{f} \Rightarrow (9,-11,0,2,0,9), (-15/2,11/2,0,2,2,-15/2)$ (if $\widetilde{M} \neq 0$, $\theta > 0$);

- $s, f, ©, ©; \binom{\overline{0}}{2}SN, N^d \Rightarrow (-2, 6, 0, -4, 0, -2), (-9, 13, 0, -4, 4, -9)$ (if $\widetilde{M} \neq 0$, $\theta = 0$);
- $s, f, ©, ©; \binom{\overline{0}}{3}N \Rightarrow (1, -9, 0, 8, 0, 1), (-6, 14, 0, -8, -2, -6)$ (if $\widetilde{M} = 0$).

2.1.2: $\mathcal{T}_4 = 0$, $\mathcal{T}_3 \neq 0$. In this case we have exactly one singularity with zero trace.

2.1.2.1: $\mathcal{T}_3\mathcal{F} < 0$. According to Theorem 6.2 (see statement (b)) there exists one weak focus (or a center), and the order of the focus is governed by the invariant polynomials $\mathcal{F}_i$ ($i = 1, 2, 3, 4$).

1. $\mathcal{F}_1 \neq 0$. In this case we have a weak focus of order one, and considering Remark 11.14, according to Lemma 11.1 we arrive at the next 13 global configurations:
   - $s, f^{(1)}, ©, ©; N^\infty, ©, © \Rightarrow (1, -2, 0, 1, 1, 1), (3, -4, 0, 1, 0, 3)$ (if $\eta < 0$, $\theta < 0$);
   - $s, f^{(1)}, ©, ©; N^f, ©, © \Rightarrow (5, -35/2, 0, 25/2, 8, 5), (18, -26, 0, 8, -15/4, 18)$ (if $\eta < 0$, $\theta > 0$);
   - $s, f^{(1)}, ©, ©; N^d, ©, © \Rightarrow (1, -2, 0, 1, 1, 1), (2, -2, 0, 0, 0, 2)$ (if $\eta < 0$, $\theta = 0$);
   - $s, f^{(1)}, ©, ©; S, N^\infty, N^\infty \Rightarrow (0, 3/2, 0, -3/2, 1, 0), (4, -5, 0, 1, 3/4, 4)$ (if $\eta > 0$, $\theta < 0$, $\theta_1 < 0$);
   - $s, f^{(1)}, ©, ©; S, N^f, N^f \Rightarrow (0, 1, 0, -1, 2, 0), (1, -1, 0, 0, 1/2, 1)$ (if $\eta > 0$, $\theta < 0$, $\theta_1 > 0$);
   - $s, f^{(1)}, ©, ©; S, N^f, N^\infty \Rightarrow (0, 1, 0, -1, 2, 0), (6, -6, 0, 0, 1/2, 6)$ (if $\eta > 0$, $\theta > 0$);
   - $s, f^{(1)}, ©, ©; S, N^\infty, N^d \Rightarrow (0, 4, 0, -4, 7, 0), (21, -22, 0, 1, 2, 21)$ (if $\eta > 0$, $\theta = 0$, $\theta_1 < 0$);
   - $s, f^{(1)}, ©, ©; S, N^f, N^d \Rightarrow (0, 2, 0, -2, 1, 0), (1/2, 3/2, 0, -2, 1, 1/2)$ (if $\eta > 0$, $\theta = 0$, $\theta_1 > 0$);
   - $s, f^{(1)}, ©, ©; S, N^d, N^d \Rightarrow (-2, 2, 0, 0, 2, -2), (3, -4, 0, 1, -1, 3)$ (if $\eta > 0$, $\theta = 0$, $\theta_1 = 0$);
   - $s, f^{(1)}, ©, ©; \binom{\overline{0}}{2}SN, N^\infty \Rightarrow (0, -1, 0, 1, 1, 0), (3, -4, 0, 1, -1/2, 3)$ (if $\eta = 0$, $\widetilde{M} \neq 0$, $\theta < 0$);
   - $s, f^{(1)}, ©, ©; \binom{\overline{0}}{2}SN, N^f \Rightarrow (2, -3, 0, 1, 1, 2), (5, -4, 0, -1, 1/2, 5)$ (if $\eta = 0$, $\widetilde{M} \neq 0$, $\theta > 0$);
   - $s, f^{(1)}, ©, ©; \binom{\overline{0}}{2}SN, N^d \Rightarrow (1, -2, 0, 1, 1, 1), (4, -4, 0, 0, 0, 4)$ (if $\eta = 0$, $\widetilde{M} \neq 0$, $\theta = 0$);

- $s, f^{(1)}, ©, ©; \overline{\binom{0}{3}}N \Rightarrow (1, -3/2, 0, 1/2, 1, 1), (2, -2, 0, 0, 1/4, 2)$ (if $\eta = 0$, $\widetilde{M} = 0$).

2. $\mathcal{F}_1 = 0$, $\mathcal{F}_2 \neq 0$. We have a weak focus of order two, and therefore by Theorem 7.7 and Lemma 11.1, we have the next 12 global configurations:

- $s, f^{(2)}, ©, ©; N^{\infty}, ©, © \Rightarrow (0, 0, -4, 11/3, 1, 4), (0, 4, 0, 1, 11/6, -1)$ (if $\eta < 0$, $\theta < 0$);
- $s, f^{(2)}, ©, ©; N^{f}, ©, © \Rightarrow (0, 0, -1, 5/3, 1, 1), (0, 1, 0, 1, -1/6, -1)$ (if $\eta < 0$, $\theta > 0$);
- $s, f^{(2)}, ©, ©; N^{d}, ©, © \Rightarrow (0, 0, -1, 15/8, \xi, 1), (0, 1, 0, \xi, 3/8, -\xi)$, where $\xi = \frac{1}{16}\sqrt{(289 - 5\sqrt{1129})/2}$ (if $\eta < 0$, $\theta = 0$);
- $s, f^{(2)}, ©, ©; S, N^{\infty}, N^{\infty} \Rightarrow (0, 0, -5, 13/3, 1, 5), (0, 5, 0, 1, 19/6, -1)$ (if $\eta > 0$, $\theta < 0$, $\theta_1 < 0$);
- $s, f^{(2)}, ©, ©; S, N^{f}, N^{f} \Rightarrow (0, 0, -1, 1, 1, 1), (0, 1, 0, 1, 1/2, -1)$ (if $\eta > 0$, $\theta < 0$, $\theta_1 > 0$);
- $s, f^{(2)}, ©, ©; S, N^{f}, N^{\infty} \Rightarrow (0, 0, -3, 4/3, 1, 3), (0, 3, 0, 1, 7/6, -1)$ (if $\eta > 0$, $\theta > 0$);
- $s, f^{(2)}, ©, ©; S, N^{\infty}, N^{d} \Rightarrow (0, 0, -1, 7/3, \xi, 1), (0, 1, 0, \xi, 5/3, -\xi)$, where $\xi = \frac{1}{3}\sqrt{(9 - \sqrt{21})/2}$ (if $\eta > 0$, $\theta = 0$, $\theta_1 < 0$);
- $s, f^{(2)}, ©, ©; S, N^{f}, N^{d} \Rightarrow (0, 0, -1, -11, \xi, 1), (0, 1, 0, \xi, -5, -\xi)$, where $\xi = \sqrt{3(7 + \sqrt{69})/2}$ (if $\eta > 0$, $\theta = 0$, $\theta_1 > 0$);
- $s, f^{(2)}, ©, ©; S, N^{d}, N^{d} \Rightarrow (0, 0, -1, 4, \sqrt{3/2}, 1), (0, 1, 0, \sqrt{3/2}, 1, -\sqrt{3/2})$ (if $\eta > 0$, $\theta = 0$, $\theta_1 = 0$);
- $s, f^{(2)}, ©, ©; \overline{\binom{0}{2}}SN, N^{\infty} \Rightarrow (0, 0, -1, 5/3, 1/2, 1), (0, 1, 0, 1/2, 5/6, -1/2)$ (if $\eta = 0$, $\theta < 0$);
- $s, f^{(2)}, ©, ©; \overline{\binom{0}{2}}SN, N^{f} \Rightarrow (0, 0, -6, 8, 3, 6), (0, 6, 0, 3, 4, -3)$ (if $\eta = 0$, $\theta > 0$);
- $s, f^{(2)}, ©, ©; \overline{\binom{0}{2}}SN, N^{d} \Rightarrow (0, 0, -2, 3, 1, 2), (0, 2, 0, 1, 3/2, -1)$ (if $\eta = 0$, $\theta = 0$).

3. $\mathcal{F}_1 = \mathcal{F}_2 = 0$. By Theorem 7.4 we could not have a center in this case. So the condition $\mathcal{F}_3\mathcal{F}_4 \neq 0$ holds (i.e. we have a weak focus of order three), and then by Theorem 7.8 the condition $\eta > 0$ is fulfilled. So we get the next three global configurations:

- $s, f^{(3)}, ©, ©; S, N^{f}, N^{f} \Rightarrow (0, 0, -1, 2, 5/2, 1), (0, 1, 0, 1, 11/2, 0)$ (if $\theta < 0$);

- $s, f^{(3)}, \copyright, \copyright; S, N^f, N^\infty \Rightarrow (0,0,-1,4,5,1), (0,1,0,2,17/2,0)$ (if $\theta > 0$);
- $s, f^{(3)}, \copyright, \copyright; S, N^f, N^d \Rightarrow (0,0,-1,3,5/\sqrt{2},1), (0,1,0,\sqrt{2},7,0)$ (if $\theta = 0$).

2.1.2.2: $\mathcal{T}_3\mathcal{F} > 0$. According to Theorem 6.2 we have a weak saddle.

1. $\mathcal{F}_1 \neq 0$. In this case the weak saddle has order one, and considering Remark 11.14, by Lemma 11.1, we arrive at the next 13 global configurations:
   - $s^{(1)}, f, \copyright, \copyright; N^\infty, \copyright, \copyright \Rightarrow (1,-7/2,0,5/2,1,1), (3,-7,0,4,-3/4,3)$ (if $\eta < 0$, $\theta < 0$);
   - $s^{(1)}, f, \copyright, \copyright; N^f, \copyright, \copyright \Rightarrow (2,-7,0,5,1,2), (3,-7,0,4,-3/2,3)$ (if $\eta < 0$, $\theta > 0$);
   - $s^{(1)}, f, \copyright, \copyright; N^d, \copyright, \copyright \Rightarrow (4,-13/2,0,5/2,1,4), (-1,7/4,0,-3/4,3/4,-1)$ (if $\eta < 0$, $\theta = 0$);
   - $s^{(1)}, f, \copyright, \copyright; S, N^\infty, N^\infty \Rightarrow (0,3/2,0,-3/2,1,0), (3,-5,0,2,3/4,3)$ (if $\eta > 0$, $\theta < 0$, $\theta_1 < 0$);
   - $s^{(1)}, f, \copyright, \copyright; S, N^f, N^f \Rightarrow (0,1,0,-1,8,0), (4,-8,0,4,1/2,4)$ (if $\eta > 0$, $\theta < 0$, $\theta_1 > 0$);
   - $s^{(1)}, f, \copyright, \copyright; S, N^f, N^\infty \Rightarrow (0,-1,0,1,2,0), (2,-6,0,4,-1/2,2)$ (if $\eta > 0$, $\theta > 0$);
   - $s^{(1)}, f, \copyright, \copyright; S, N^\infty, N^d \Rightarrow (-6,-8,0,14,16,-6), (27,-67,0,40,-10,27)$ (if $\eta > 0$, $\theta = 0$, $\theta_1 < 0$);
   - $s^{(1)}, f, \copyright, \copyright; S, N^f, N^d \Rightarrow (0,-8,0,8,4,0), (3,-19,0,16,-4,3)$ (if $\eta > 0$, $\theta = 0$, $\theta_1 > 0$);
   - $s^{(1)}, f, \copyright, \copyright; S, N^d, N^d \Rightarrow (-1,-4,0,5,4,-1), (6,-18,0,12,-3,6)$ (if $\eta > 0$, $\theta = 0$, $\theta_1 = 0$);
   - $s^{(1)}, f, \copyright, \copyright; \overline{\binom{0}{2}} SN, N^\infty \Rightarrow (-2,5,0,-3,1,-2), (-3,4,0,-1,1/2,-3)$ (if $\eta = 0$, $\widetilde{M} \neq 0$, $\theta < 0$);
   - $s^{(1)}, f, \copyright, \copyright; \overline{\binom{0}{2}} SN, N^f \Rightarrow (-4,7,0,-3,1,-4), (-5,4,0,1,-1/2,-5)$ (if $\eta = 0$, $\widetilde{M} \neq 0$, $\theta > 0$);
   - $s^{(1)}, f, \copyright, \copyright; \overline{\binom{0}{2}} SN, N^d \Rightarrow (-3,6,0,-3,1,-3), (-4,4,0,0,0,-4)$ (if $\eta = 0$, $\widetilde{M} \neq 0$, $\theta = 0$);
   - $s^{(1)}, f, \copyright, \copyright; \overline{\binom{0}{3}} N \Rightarrow (1,-15/2,0,13/2,1,1), (-4,12,0,-8,-11/4,-4)$ (if $\eta = 0$, $\widetilde{M} = 0$).
2. $\mathcal{F}_1 = 0$, $\mathcal{F}_2 \neq 0$. We have a weak saddle of order two.

(a) $\eta < 0$. Since by Remark 11.14 we could not have an infinite star node, considering Lemma 11.1 we obtain the next three global configurations:
- $s^{(2)}, f, ©, ©;\ N^{\infty},\ ©, © \Rightarrow (0,0,4,4,-1,6), (0,4,0,1,0,1)$ (if $\theta < 0$);
- $s^{(2)}, f, ©, ©;\ N^{f},\ ©, © \Rightarrow (0,0,1,1,-1,3/2), (0,1,0,1,0,1)$ (if $\theta > 0$);
- $s^{(2)}, f, ©, ©;\ N^{d},\ ©, © \Rightarrow (0,0,1,1,-\sqrt{3}/4,3/2), (0,1,0,\sqrt{3}/4,0,\sqrt{3}/4)$ (if $\theta = 0$).

(b) $\eta > 0$. According to Remark 11.14 and Lemmas 11.26(iv) and 11.1, at infinity we could have neither a star node nor two one-direction nodes. So all the remaining possibilities are realizable, and the corresponding global configurations are:
- $s^{(2)}, f, ©, ©;\ S, N^{\infty},\ N^{\infty} \Rightarrow (0,0,16,16,-1,24), (0,16,0,1,12,1)$ (if $\theta < 0,\ \theta_1 < 0$);
- $s^{(2)}, f, ©, ©;\ S, N^{f},\ N^{f} \Rightarrow (0,0,1,1,-1/40,2), (0,1,0,1/40,17/16,1/40)$ (if $\theta < 0,\ \theta_1 > 0$);
- $s^{(2)}, f, ©, ©;\ S, N^{f},\ N^{\infty} \Rightarrow (0,0,8,8,-1,12), (0,8,0,1,7,1)$ (if $\theta > 0$);
- $s^{(2)}, f, ©, ©;\ S, N^{\infty},\ N^{d} \Rightarrow (0,0,1,1,-\xi,2), (0,1,0,\xi,7/8,\xi)$, where $\xi = \sqrt{(23\sqrt{497}-511)/2}/16$ (if $\theta > 0$) (if $\theta = 0,\ \theta_1 < 0$);
- $s^{(2)}, f, ©, ©;\ S, N^{f},\ N^{d} \Rightarrow (0,0,1,1,-\xi,2), (0,1,0,\xi,9/8,\xi)$, where $\xi = \sqrt{3(25\sqrt{73}-213)/2}/16$ (if $\theta = 0,\ \theta_1 > 0$).

(c) $\eta = 0$. By Lemma 11.25(vii) we could not have at infinity a triple point, i.e. $\widetilde{M} \neq 0$. So, considering Lemma 11.1 we obtain the next three global configurations:
- $s^{(2)}, f, ©, ©;\ \overline{\binom{0}{2}}SN,\ N^{\infty} \Rightarrow \big(0,0,1,12,-1,2(5+3\sqrt{3})\big), (0,1,0,1,9,1)$ (if $\theta < 0$);
- $s^{(2)}, f, ©, ©;\ \overline{\binom{0}{2}}SN,\ N^{f} \Rightarrow (0,0,1,1,-1,45+32\sqrt{2}), (0,1,0,1,5,1)$ (if $\theta > 0$);
- $s^{(2)}, f, ©, ©;\ \overline{\binom{0}{2}}SN,\ N^{d} \Rightarrow (0,0,1,10,-1,10-\xi/2), (0,1,0,1,\chi-10,1)$, where $\{\eta^{-1}(0), \theta^{-1}(0)\} \ni (\xi,\chi) \approx (-15.3275, 17.8981)$ (if $\theta = 0$).

3. $\mathcal{F}_1 = \mathcal{F}_2 = 0$. Then we have a weak saddle whose order is at least three.

(a) $\mathcal{F}_3\mathcal{F}_4 \neq 0$. In this case the weak saddle is of order three and by Theorem 7.9 the condition $\eta < 0$ must hold. So considering this theorem and Lemma 11.1, we get the next three global configurations:
- $s^{(3)}, f, ©, ©;\ N^{\infty},\ ©, © \Rightarrow (0,1,0,-2,1/2,-1), (0,0,-1,4,-1/2,2)$ (if $\theta < 0$);
- $s^{(3)}, f, ©, ©;\ N^{f},\ ©, © \Rightarrow (0,1,0,-2,1/2,-5), (0,0,-1,4,-1/2,2)$ (if $\theta > 0$);
- $s^{(3)}, f, ©, ©;\ N^{d},\ ©, © \Rightarrow (0,1,0,-2,1/2,-2), (0,0,-1,-11/32,$

$-1/2, 2)$ (if $\theta = 0$).

(b) $\mathcal{F}_3\mathcal{F}_4 = 0$. By Theorem 6.2 we have an integrable saddle. So considering Theorem 7.3 and Lemma 11.1, we arrive at the next two global configurations:
- $\$, f, ©, ©; N^{\infty}, ©, © \Rightarrow (0,1,0,2,0,1), (0,0,-1,-1,0,-1)$ (if $\theta < 0$);
- $\$, f, ©, ©; N^{f}, ©, © \Rightarrow (0,1,0,1,0,1), (0,0,-1,-1,0,0)$ (if $\theta > 0$).

2.1.3: $\mathcal{T}_4 = \mathcal{T}_3 = 0$. Then at least two of the singularities have zero traces.

2.1.3.1: $\mathcal{T}_2 \neq 0$. Then the traces corresponding to exactly two singularities vanish, and we need to distinguish whether these singularities are real or complex. Considering Remark 11.13 we examine two subcases, when the conditions $(\mathcal{C}_{\rho\rho}^{\mathbb{R}})$ or $(\mathcal{C}_{\rho\rho}^{\mathbb{C}})$ are satisfied.

1. $(\mathcal{C}_{\rho\rho}^{\mathbb{R}})$. Then the vanishing traces correspond to real singularities, and hence we have a weak saddle and a weak focus (or a center).

(a) $\mathcal{F}_1 \neq 0$. By Theorem 6.2 (see statement (c)) both weak singularities have order one. By Remark 11.14, at infinity we could not have a star node. Moreover by Lemma 11.26(vi) we could not have two one-direction nodes at infinity, and by Lemma 11.25(iii) in this case the condition $\widetilde{M} \neq 0$ must hold. So considering Lemma 11.1, we arrive at the next 11 global configurations:
- $s^{(1)}, f^{(1)}, ©, ©; N^{\infty}, ©, © \Rightarrow (0,-2,6,2,3,6), (0,-6,2,6,-2,12)$ (if $\eta < 0$, $\theta < 0$);
- $s^{(1)}, f^{(1)}, ©, ©; N^{f}, ©, © \Rightarrow (0,-1,3,1,0,3), (0,-3,1,3,-1,3)$ (if $\eta < 0$, $\theta > 0$);
- $s^{(1)}, f^{(1)}, ©, ©; N^{d}, ©, © \Rightarrow (0,-1,3,1,1,3)$, $(0,-3,1,3,-1,5)$ (if $\eta < 0$, $\theta = 0$);
- $s^{(1)}, f^{(1)}, ©, ©; S, N^{\infty}, N^{\infty} \Rightarrow (0,-1,-1,1,1/6,3/20)$, $(0,2,1,-2,-1,-1)$ (if $\eta > 0$, $\theta < 0$, $\theta_1 < 0$);
- $s^{(1)}, f^{(1)}, ©, ©; S, N^{f}, N^{f} \Rightarrow (0,-1,-1,1,-5,-6), (0,2,1,-2,-1,-1)$ (if $\eta > 0$, $\theta < 0$, $\theta_1 > 0$);
- $s^{(1)}, f^{(1)}, ©, ©; S, N^{f}, N^{\infty} \Rightarrow (0,-1,-9,1,-5,-27/2), (0,2,1,-2,-1,-1)$ (if $\eta > 0$, $\theta > 0$);
- $s^{(1)}, f^{(1)}, ©, ©; S, N^{\infty}, N^{d} \Rightarrow (0,1,-2,-1,0,-1/3)$, $(0,-1,-1,1,1,13/6)$ (if $\eta > 0$, $\theta = 0$, $\theta_1 < 0$);
- $s^{(1)}, f^{(1)}, ©, ©; S, N^{f}, N^{d} \Rightarrow (0,1,-2,-1,0,0), (0,-1,-1,1,1,3/2)$ (if $\eta > 0$, $\theta = 0$, $\theta_1 > 0$);
- $s^{(1)}, f^{(1)}, ©, ©; \overline{\binom{0}{2}}SN, N^{\infty} \Rightarrow (0,-1/3,-7/8,1/3,3/8,0), (0,-1,1/3,1,-1/3,3/4)$ (if $\eta = 0$, $\theta < 0$);
- $s^{(1)}, f^{(1)}, ©, ©; \overline{\binom{0}{2}}SN, N^{f} \Rightarrow (0,1,-3,-1,3/4,0), (0,-3,-1,3,1,3/2)$ (if $\eta = 0$, $\theta > 0$);

- $s^{(1)}, f^{(1)}, ©, ©; \overline{\binom{0}{2}}SN,\ N^d \Rightarrow (0,-1,-3,1,-1,0), (0,3,1,-3,-1,-2)$ (if $\eta = 0,\ \theta = 0$).

(b) $\mathcal{F}_1 = 0$. In this case by Theorem 6.2 we have an integrable saddle and a center. So considering Theorem 7.4 and Lemma 11.1, we get the following five global configurations:

- $\$, c, ©, ©;\ N^\infty,\ ©, © \Rightarrow (0,1,-1,-1,1/4,-5/18), (0,3,-1,-3,1,-1)$ (if $\eta < 0,\ \theta < 0$);
- $\$, c, ©, ©;\ N^f,\ ©, © \Rightarrow (0,1,-1,-1,1/2,-4/9), (0,3,-1,-3,1,-1)$ (if $\eta < 0,\ \theta > 0$);
- $\$, c, ©, ©;\ S, N^\infty,\ N^\infty \Rightarrow (0,1,-1,-1,-7/8,59/1156), (0,17,-1,-17,1,-1)$ (if $\eta > 0,\ \theta_1 < 0$);
- $\$, c, ©, ©;\ S, N^f,\ N^f \Rightarrow (0,1,-1,-1,-2,2), (0,2,-1,-2,1,-1)$ (if $\eta > 0,\ \theta_1 > 0$);
- $\$, c, ©, ©;\ \overline{\binom{0}{3}}N \Rightarrow (0,-1,0,1,-3/4,1), (0,-1,1,1,-1,3/2)$ (if $\eta = 0,\ \widetilde{M} = 0$).

2. ($\mathcal{C}^{\mathbb{C}}_{\rho\rho}$). In this case we could not have weak singularities, and this implies the existence of complex singularities of the type $©^\rho, ©^\rho$. Additionally by Lemma 11.30(v) the condition $\eta < 0$ must be satisfied. Moreover by Lemma 11.26(v) in this case the condition $\theta > 0$ holds, and hence we arrive at the unique global configuration

   - $s, f, ©^\rho, ©^\rho;\ N^f,\ ©, © \Rightarrow (3,0,0,-3,2,3), (-2,26,0,-24,-15/2,-2)$.

2.1.3.2: $\mathcal{T}_2 = 0$. By Lemma 11.23 the systems become Hamiltonian (then all the traces vanish), and hence we have an integrable saddle and a center. So by Theorem 7.3 we arrive at the global configuration

- $\$, c, ©^\rho, ©^\rho;\ N^f,\ ©, © \Rightarrow (-1,0,0,1,-1,-1), (1,10,0,-11,-1,1)$.

2.2: $W_4 > 0$. Then the anti-saddle is a (generic) node, and we consider two possibilities: $\mathcal{T}_4 \neq 0$ and $\mathcal{T}_4 = 0$.

2.2.1: $\mathcal{T}_4 \neq 0$. In this case by Theorem 6.2 we do not have weak singularities, and hence we have a strong saddle and a generic node.

1. $\eta < 0$. According to Lemma 11.1 in this case we obtain all four global configurations:

   - $s, n, ©, ©;\ N^\infty,\ ©, © \Rightarrow (0,-1,0,1,1/4,0), (1,-3,0,2,1,1)$ (if $\theta < 0$);
   - $s, n, ©, ©;\ N^f,\ ©, © \Rightarrow (0,5,0,-5,-1,0), (1,-3,0,2,-1,1)$ (if $\theta > 0$);
   - $s, n, ©, ©;\ N^d,\ ©, © \Rightarrow (1,-3,0,2,0,1), (0,-7/8,0,7/8,1/4,0)$ (if $\theta = 0,\ \theta_2 \neq 0$);
   - $s, n, ©, ©;\ N^*,\ ©, © \Rightarrow (0,5,0,-5,0,0), (1,-3,0,2,-2,1)$ (if $\theta = 0,\ \theta_2 = 0$).

2. $\eta > 0$. By Lemma 11.1 in this case there are 10 possible configurations. Additionally according to Theorem 7.5, in the case $\mathbf{D} > 0$ we could not have two infinite star nodes. So we arrive at the following nine global configurations:

- $s, n, ©, ©;\ S, N^{\infty},\ N^{\infty} \Rightarrow (0,-4,0,4,1/2,0),(2,-3,0,1,0,2)$ (if $\theta < 0,\ \theta_1 < 0$);
- $s, n, ©, ©;\ S, N^{f},\ N^{f} \Rightarrow (0,-4,0,4,-2,0),(2,-1,0,-1,0,2)$ (if $\theta < 0,\ \theta_1 > 0$);
- $s, n, ©, ©;\ S, N^{f},\ N^{\infty} \Rightarrow (0,-6,0,6,-2,0),(2,-3,0,1,0,2)$ (if $\theta > 0$);
- $s, n, ©, ©;\ S, N^{\infty},\ N^{d} \Rightarrow (0,-2,0,2,1/4,0),(1,-1,0,0,0,1)$ (if $\theta = 0,\ \theta_1 < 0,\ \theta_2 \neq 0$);
- $s, n, ©, ©;\ S, N^{\infty},\ N^{*} \Rightarrow (0,-1,0,1,0,0),(1,-9/8,0,1/8,1,1)$ (if $\theta = 0,\ \theta_1 < 0,\ \theta_2 = 0$);
- $s, n, ©, ©;\ S, N^{f},\ N^{d} \Rightarrow (0,-1,0,1,-6,0),(1,-1,0,0,0,1)$ (if $\theta = 0,\ \theta_1 > 0,\ \theta_2 \neq 0$);
- $s, n, ©, ©;\ S, N^{f},\ N^{*} \Rightarrow (0,-1,0,1,0,0),(1,0,0,-1,1,1)$ (if $\theta = 0,\ \theta_1 > 0,\ \theta_2 = 0$);
- $s, n, ©, ©;\ S, N^{d},\ N^{d} \Rightarrow (0,-7,0,7,1,0),(1,-1,0,0,0,1)$ (if $\theta = 0,\ \theta_1 = 0,\ \theta_3 \neq 0$);
- $s, n, ©, ©;\ S, N^{d},\ N^{*} \Rightarrow (0,1,0,-1,0,0),(1,-3,0,2,1,1)$ (if $\theta = 0,\ \theta_1 = 0,\ \theta_3 = 0$).

3. $\eta = 0$. By Lemma 11.1, at infinity we could have either one double and one simple (if $\widetilde{M} \neq 0$) or one triple (if $\widetilde{M} = 0$) real singularities. Considering Lemma 11.1 we obtain the following five global configurations:

- $s, n, ©, ©;\ \overline{\binom{0}{2}}SN,\ N^{\infty} \Rightarrow (0,-8,0,8,2,0),(4,-5,0,1,0,4)$ (if $\widetilde{M} \neq 0,\ \theta < 0$);
- $s, n, ©, ©;\ \overline{\binom{0}{2}}SN,\ N^{f} \Rightarrow (0,-3,0,3,1/2,0),(1,0,0,-1,0,1)$ (if $\widetilde{M} \neq 0,\ \theta > 0$);
- $s, n, ©, ©;\ \overline{\binom{0}{2}}SN,\ N^{d} \Rightarrow (0,-\sqrt{5},0,\sqrt{5},1/2,0),(1,-1,0,0,0,1)$ (if $\widetilde{M} \neq 0,\ \theta = 0,\ \theta_2 \neq 0$);
- $s, n, ©, ©;\ \overline{\binom{0}{2}}SN,\ N^{*} \Rightarrow (0,4,0,-4,0,0),(1,-2,0,1,-1,1)$ (if $\widetilde{M} \neq 0,\ \theta = 0,\ \theta_2 = 0$);
- $s, n, ©, ©;\ \overline{\binom{0}{3}}N \Rightarrow (-1,14,0,-13,0,-1),(-3,4,0,-1,-5,-3)$ (if $\widetilde{M} = 0$).

2.2.2: $\mathcal{T}_4 = 0$. In this case at least one finite singularity has trace zero, and by Lemma 11.28 we could not have a star node at infinity.

2.2.2.1: $\mathcal{T}_3 \neq 0$. Then we have exactly one weak singularity, and it must be the saddle.

1. $\mathcal{F}_1 \neq 0$. Then the weak saddle has order one.

   (a) $\eta < 0$. Since we could not have a star node at infinity, considering Lemma 11.1 we arrive at the following three global configurations:
   - $s^{(1)}, n, ©, ©; N^\infty, ©, © \Rightarrow (0,1,0,1,-1/2,2), (0,0,-1,-1/8,0,-1)$ (if $\theta < 0$);
   - $s^{(1)}, n, ©, ©; N^f, ©, © \Rightarrow (0,1,0,1,-1/2,2), (0,0,-1,-1/4,0,1)$ (if $\theta > 0$);
   - $s^{(1)}, n, ©, ©; N^d, ©, © \Rightarrow (0,1,0,1,-1/2,2), (0,0,-1,-17/128,0,$ $-3/4)$ (if $\theta = 0$).

   (b) $\eta > 0$. Taking into account the impossibility of a star node at infinity (see Lemma 11.28), we obtain the following six global configurations:
   - $s^{(1)}, n, ©, ©; S, N^\infty, N^\infty \Rightarrow (0,1,0,1,-1/2,-1/8), (0,0,-1,-1,0,-1)$ (if $\theta < 0$, $\theta_1 < 0$);
   - $s^{(1)}, n, ©, ©; S, N^f, N^f \Rightarrow (0,1,0,1,-1/2,-257/128), (0,0,-1,-1,$ $0,1)$ (if $\theta < 0$, $\theta_1 > 0$);
   - $s^{(1)}, n, ©, ©; S, N^f, N^\infty \Rightarrow (0,1,0,1,-1/2,-33/16), (0,0,-1,-1,0,1)$ (if $\theta > 0$);
   - $s^{(1)}, n, ©, ©; S, N^\infty, N^d \Rightarrow (0,1,0,1,-1/2,-7/32), (0,0,-1,-16/49,$ $0,-1)$ (if $\theta = 0$, $\theta_1 < 0$);
   - $s^{(1)}, n, ©, ©; S, N^f, N^d \Rightarrow (0,1,0,1,-1/2,-3), (0,0,-1,-5/8,0,1)$ (if $\theta = 0$, $\theta_1 > 0$);
   - $s^{(1)}, n, ©, ©; S, N^d, N^d \Rightarrow (0,1,0,1,-1/2,0), (0,0,-1,-1,0,-1/2)$ (if $\theta = 0$, $\theta_1 = 0$).

   (c) $\eta = 0$. By Lemma 11.1, at infinity we could have either one double and one simple (if $\widetilde{M} \neq 0$) or one triple (if $\widetilde{M} = 0$) real singularities. Considering Lemma 11.28 (i.e. we could not have a node N^*), we get the following four global configurations:
   - $s^{(1)}, n, ©, ©; \binom{0}{2}SN, N^\infty \Rightarrow (0,1,0,1,-1/2,0), (0,0,-1,-1,0,-1)$ (if $\widetilde{M} \neq 0$, $\theta < 0$);
   - $s^{(1)}, n, ©, ©; \binom{0}{2}SN, N^f \Rightarrow (0,2,0,2,-1,-(25+27\sqrt{2})/27), (0,0,-2,$ $-2,0,1/3)$ (if $\widetilde{M} \neq 0$, $\theta > 0$);
   - $s^{(1)}, n, ©, ©; \binom{0}{2}SN, N^d \Rightarrow (0,2,0,2,-1,-2), (0,0,-2,-17/64,0,$ $-21/4)$ (if $\widetilde{M} \neq 0$, $\theta = 0$);
   - $s^{(1)}, n, ©, ©; \binom{0}{3}N \Rightarrow (0,2,1,-2,-1/4,0), (0,1,-2,-1,-1,-1/2)$ (if $\widetilde{M} = 0$).

(d) $\mathcal{F}_1 = 0$, $\mathcal{F}_2 \neq 0$. Then By Theorem 6.2 we have a weak saddle of order two. So considering Lemma 11.1 we arrive at the following 13 global configurations:

- $s^{(2)}, n, ©, ©;\ N^{\infty}, ©, © \Rightarrow (0,0,1,-5/3,-1,-7/6), (0,1,0,1,2/3,1)$ (if $\eta < 0$, $\theta < 0$);
- $s^{(2)}, n, ©, ©;\ N^{f}, ©, ©, \Rightarrow (0,0,32,-79,-32,-15), (0,32,0,32,47,32)$ (if $\eta < 0$, $\theta > 0$);
- $s^{(2)}, n, ©, ©;\ N^{d}, ©, © \Rightarrow (0,0,4,-7,-4,-4), (0,4,0,4,3,4)$ (if $\eta < 0$, $\theta = 0$);
- $s^{(2)}, n, ©, ©;\ S, N^{\infty}, N^{\infty} \Rightarrow (0,0,9,-16,-9,-443/62), (0,9,0,9,7,9)$ (if $\eta > 0$, $\theta < 0$, $\theta_1 < 0$);
- $s^{(2)}, n, ©, ©;\ S, N^{f}, N^{f} \Rightarrow (0,0,16,-39,-16,-7), (0,16,0,16,23,16)$ (if $\eta > 0$, $\theta < 0$, $\theta_1 > 0$);
- $s^{(2)}, n, ©, ©;\ S, N^{f}, N^{\infty} \Rightarrow (0,0,128,-313,-128,-57), (0,128,0,128,185,128)$ (if $\eta > 0$, $\theta > 0$);
- $s^{(2)}, n, ©, ©;\ S, N^{\infty}, N^{d} \Rightarrow (0,0,4,-7,-4,-13/4), (0,4,0,4,3,4)$ (if $\eta > 0$, $\theta = 0$, $\theta_1 < 0$);
- $s^{(2)}, n, ©, ©;\ S, N^{f}, N^{d} \Rightarrow \big(0,0,1,-2,-1,(1+\sqrt{5})/2\big), (0,1,0,1,1,1)$ (if $\eta > 0$, $\theta = 0$, $\theta_1 > 0$);
- $s^{(2)}, n, ©, ©;\ S, N^{d}, N^{d} \Rightarrow (0,0,2,7,-2,1), (0,2,0,2,-1,2)$ (if $\eta > 0$, $\theta = 0$, $\theta_1 = 0$);
- $s^{(2)}, n, ©, ©;\ \overline{\binom{0}{2}}SN, N^{\infty} \Rightarrow (0,0,1,-2,-1,-23/27), (0,1,0,1,5/8,1)$ (if $\eta = 0$, $\widetilde{M} \neq 0$, $\theta < 0$);
- $s^{(2)}, n, ©, ©;\ \overline{\binom{0}{2}}SN, N^{f} \Rightarrow \big(0,0,1,-2,-1,-2(10+7\sqrt{7})/27\big), (0,1,0,1,-3,1)$ (if $\eta = 0$, $\widetilde{M} \neq 0$, $\theta > 0$);
- $s^{(2)}, n, ©, ©;\ \overline{\binom{0}{2}}SN, N^{d} \Rightarrow \big(0,0,1,\xi,-1,-\xi(46+42\xi+9\xi^2)/9\big), (0,1,0,1,-2-\xi,1)$, where $\{\theta^{-1}(0)\} \ni \xi \approx -2.6140$ (if $\eta = 0$, $\widetilde{M} \neq 0$, $\theta = 0$);
- $s^{(2)}, n, ©, ©;\ \overline{\binom{0}{3}}N \Rightarrow (0,0,3,-3,1,-1), (0,3,0,-1,0,-1)$ (if $\eta = 0$, $\widetilde{M} = 0$).

(e) $\mathcal{F}_1 = \mathcal{F}_2 = 0$. So the weak saddle has order three (if $\mathcal{F}_3\mathcal{F}_4 \neq 0$), or it is an integrable saddle (if $\mathcal{F}_3\mathcal{F}_4 = 0$). Considering Theorems 7.9 and 7.3, we arrive at the following two global configurations:

- $s^{(3)}, n, ©, ©;\ N^{\infty}, ©, © \Rightarrow (0,2,0,-4,1,-2), (0,0,-2,0,-1,4)$ (if $\mathcal{F}_3\mathcal{F}_4 \neq 0$);
- $\$, n, ©, ©;\ N^{\infty}, ©, © \Rightarrow (0,2,0,-4,1,-4), (0,0,-2,1,-1,4)$ (if $\mathcal{F}_3\mathcal{F}_4 = 0$).

2.2.2.2: $\mathcal{T}_3 = 0$. Then at least two singularities have vanishing traces, and due to the existence of a finite node, it is clear that the complex singularities have zero traces. Moreover by Lemma 11.23 the condition $\mathcal{T}_2 \neq 0$ must hold, i.e. in this case we have a strong saddle. Additionally we have the following implications

- By Lemma 11.26(v) the condition $\eta \leq 0$ implies $\theta > 0$.
- By Lemma 11.30(i) the conditions $\eta > 0$ and $\theta \leq 0$ imply $\theta_1 > 0$.
- By Lemma 11.25(i) the condition $\mu_0 > 0$ implies $\widetilde{M} \neq 0$.

As a result considering Lemma 11.1 and the fact that in this case by Lemma 11.28 we could not have an infinite star node, we get the following five global configurations:

- $s, n, ©^{\rho}, ©^{\rho}; N^f, ©, © \Rightarrow (6,0,0,-6,4,6), (-4,7,0,-3,-15/4,-4)$ (if $\eta < 0$);
- $s, n, ©^{\rho}, ©^{\rho}; S, N^f, N^f \Rightarrow (0,0,0,0,2,0), (-2,8,0,-6,-15/2,-2)$ (if $\eta > 0$, $\theta < 0$);
- $s, n, ©^{\rho}, ©^{\rho}; S, N^f, N^{\infty} \Rightarrow (0,0,0,0,1,0), (-1,7,0,-6,-9/2,-1)$ (if $\eta > 0$, $\theta > 0$);
- $s, n, ©^{\rho}, ©^{\rho}; \underline{S}, N^f, N^d \Rightarrow (0,0,0,0,-1,0), (1,-10,0,9,6,1)$ (if $\eta > 0$, $\theta = 0$);
- $s, n, ©^{\rho}, ©^{\rho}; \overline{\binom{0}{2}}SN, N^f \Rightarrow (0,0,0,0,-1,0), (1,-4,0,3,3,1)$ (if $\eta = 0$).

2.3: $W_4 = 0$. Then at least one finite singularity has the corresponding discriminant zero.

2.3.1: $W_3 \neq 0$. Then the anti-saddle is a node with coinciding eigenvalues, and it can be either a node n^d or a star node. We consider two cases: $\mathcal{T}_4 \neq 0$ and $\mathcal{T}_4 = 0$.

2.3.1.1: $\mathcal{T}_4 \neq 0$. Then by Theorem 6.2 the saddle is strong.

1. $\eta < 0$. By Lemma 11.28 in this case we could not have an infinite star node. Considering Theorem 7.2 and Lemmas 11.2 and 11.1, we arrive at the following five global configurations:

   - $s, n^d, ©, ©; N^{\infty}, ©, ©: \Rightarrow (0,0,-1,0,-2,0), (0,4,4,-4,0,-8)$ (if $\theta < 0$, $\neg(\mathcal{S}^1_{n^*})$);
   - $s, n^*, ©, ©; N^{\infty}, ©, © \Rightarrow (0,-6,0,6,1,2), (0,0,-6,0,2,2)$ (if $\theta < 0$, $(\mathcal{S}^1_{n^*})$);
   - $s, n^d, ©, ©; N^f, ©, © \Rightarrow (0,0,-9/4,0,1/2,0), (0,64/9,8,-64/9,0,-55/9)$ (if $\theta > 0$, $\neg(\mathcal{S}^1_{n^*})$);
   - $s, n^*, ©, ©; N^f, ©, © \Rightarrow (0,-2,0,2,0,1), (0,0,-2,0,-1,0)$ (if $\theta > 0$, $(\mathcal{S}^1_{n^*})$);
   - $s, n^d, ©, ©; N^d, ©, © \Rightarrow (0,2,-25/8,-2,0,0), (0,8,-8,-8,-1,-8)$ (if $\theta = 0$).

2. $\eta > 0$. By Theorem 7.2 and Lemma 11.28, we could have neither a finite star node nor an infinite one. So considering Lemma 11.1 we get the following six global configurations:

   - $s, n^d, ©, ©; S, N^{\infty}, N^{\infty} \Rightarrow (0,-4,4,4,1/2,0), (0,-1,0,1,-1,6)$ (if $\theta < 0$, $\theta_1 < 0$);

- $s, n^d, ©, ©; S, N^f, N^f \Rightarrow (0,-4,4,4,-3/2,0), (0,-1,0,1,-1,2)$ (if $\theta < 0$, $\theta_1 > 0$);
- $s, n^d, ©, ©; S, N^f, N^\infty \Rightarrow (0,3,9/4,-3,2,0), (0,-1,0,1,-1,2)$ (if $\theta > 0$);
- $s, n^d, ©, ©; S, N^\infty, N^d \Rightarrow (0,-30,225/4,30,0,0), (0,-4,0,4,-3,32)$ (if $\theta = 0$, $\theta_1 < 0$);
- $s, n^d, ©, ©; S, N^f, N^d \Rightarrow (0,-13,-169/40,13,0,0), (0,10,0,-10,-6,-10)$ (if $\theta = 0$, $\theta_1 > 0$);
- $s, n^d, ©, ©; S, N^d, N^d \Rightarrow (0,-9,-81/40,9,0,0), (0,10,0,-10,-1,-1)$ (if $\theta = 0$, $\theta_1 = 0$).

3. $\eta = 0$. Since we could not have a finite star node as well as an infinite one, considering Lemma 11.1 we obtain the following four global configurations:
   - $s, n^d, ©, ©; \overline{\binom{0}{2}} SN, N^\infty \Rightarrow (0,-12,-25/8,12,-1,0), (0,8,-22,-8,-2,-10)$ (if $\widetilde{M} \neq 0$, $\theta < 0$);
   - $s, n^d, ©, ©; \overline{\binom{0}{2}} SN, N^f \Rightarrow (0,-13,-324,13,1/2,0), (0,1/4,-31,-1/4,-1,-224)$ (if $\widetilde{M} \neq 0$, $\theta > 0$);
   - $s, n^d, ©, ©; \overline{\binom{0}{2}} SN, N^d \Rightarrow (0,-2,-841/4,2,0,0), (0,1,-31,-1,-1,-4)$ (if $\widetilde{M} \neq 0$, $\theta = 0$);
   - $s, n^d, ©, ©; \overline{\binom{0}{3}} N \Rightarrow (0,0,-1,0,-2,0), (0,4,4,-4,0,-4)$ (if $\widetilde{M} = 0$).

2.3.1.2: $\mathcal{T}_4 = 0$, $\mathcal{T}_3 \neq 0$. Then exactly one finite singularity has the corresponding trace zero, and hence the saddle is weak.

1. $\mathcal{F}_1 \neq 0$. By Theorem 6.2 the weak saddle has order one. Since by Lemma 11.26(vii) in this case if $\eta > 0$ then the condition $\theta = \theta_1 = 0$ implies $\theta_3 = 0$ and $\theta_4 \neq 0$, considering Lemma 11.1 we arrive at the following 17 global configurations:
   - $s^{(1)}, n^d, ©, ©; N^\infty, ©, © \Rightarrow (0,-2,-1,2,-1/2,0), (0,1,-4,-1,1,-2)$ (if $\eta < 0$, $\theta < 0$);
   - $s^{(1)}, n^d, ©, ©; N^f, ©, © \Rightarrow (0,-4,-4,4,1/2,0), (0,1,-8,-1,2,-3)$ (if $\eta < 0$, $\theta > 0$);
   - $s^{(1)}, n^d, ©, ©; N^d, ©, © \Rightarrow (0,-2,-1,2,0,0), (0,1,-4,-1,1,-1)$ (if $\eta < 0$, $\theta = 0$, $\theta_2 \neq 0$);
   - $s^{(1)}, n^d, ©, ©; N^*, ©, © \Rightarrow (0,-1,-1,1,-1/2,1), (0,0,-1,0,0,-1)$ (if $\eta < 0$, $\theta = 0$, $\theta_2 = 0$);
   - $s^{(1)}, n^d, ©, ©; S, N^\infty, N^\infty \Rightarrow (0,-11,-4,11,-5/2,0), (0,4,-19,-4,4,-4)$ (if $\eta > 0$, $\theta < 0$, $\theta_1 < 0$);

- $s^{(1)}, n^d, ©, ©; S, N^f, N^f \Rightarrow (0,4,-4/3,-4,-2,0), (0,3,0,-3,2,-1)$ (if $\eta > 0$, $\theta < 0$, $\theta_1 > 0$);
- $s^{(1)}, n^d, ©, ©; S, N^f, N^\infty \Rightarrow (0,-10,4,10,-3/2,0), (0,-1,-14,1,2,2)$ (if $\eta > 0$, $\theta > 0$);
- $s^{(1)}, n^d, ©, ©; S, N^\infty, N^d \Rightarrow (0,-20,4,20,0,0), (0,-1,-24,1,2,15)$ (if $\eta > 0$, $\theta = 0$, $\theta_1 < 0$, $\theta_2 \neq 0$);
- $s^{(1)}, n^d, ©, ©; S, N^\infty, N^* \Rightarrow (0,10,1,-10,-4,-5), (0,0,10,0,0,7)$ (if $\eta > 0$, $\theta = 0$, $\theta_1 < 0$, $\theta_2 = 0$);
- $s^{(1)}, n^d, ©, ©; S, N^f, N^d \Rightarrow (0,-3,4,3,0,0), (0,-1,1,1,-2,6)$ (if $\eta > 0$, $\theta = 0$, $\theta_1 > 0$, $\theta_2 \neq 0$);
- $s^{(1)}, n^d, ©, ©; S, N^f, N^* \Rightarrow (0,6,1,-6,-167/32,-5), (0,0,6,0,0,9/16)$ (if $\eta > 0$, $\theta = 0$, $\theta_1 > 0$, $\theta_2 = 0$);
- $s^{(1)}, n^d, ©, ©; S, N^d, N^* \Rightarrow (0,1,1,-1,0,0), (0,0,1,0,0,1)$ (if $\eta > 0$, $\theta = 0$, $\theta_1 = 0$);
- $s^{(1)}, n^d, ©, ©; \overline{\binom{0}{2}}SN, N^\infty \Rightarrow (0,10,-16,-10,-2,0), (0,1,18,-1,-4,-5)$ (if $\eta = 0$, $\widetilde{M} \neq 0$, $\theta < 0$);
- $s^{(1)}, n^d, ©, ©; \overline{\binom{0}{2}}SN, N^f \Rightarrow (0,12,-8,-12,1/2,0), (0,2,20,-2,-4,-1)$ (if $\eta = 0$, $\widetilde{M} \neq 0$, $\theta > 0$);
- $s^{(1)}, n^d, ©, ©; \overline{\binom{0}{2}}SN, N^d \Rightarrow (0,4,-1,-4,0,0), (0,1,6,-1,-1,-1)$ (if $\eta = 0$, $\widetilde{M} \neq 0$, $\theta = 0$, $\theta_2 \neq 0$);
- $s^{(1)}, n^d, ©, ©; \overline{\binom{0}{2}}SN, N^* \Rightarrow (0,-1,-1,1,1/2,1), (0,0,-1,0,0,-1)$ (if $\eta = 0$, $\widetilde{M} \neq 0$, $\theta = 0$, $\theta_2 = 0$);
- $s^{(1)}, n^d, ©, ©; \overline{\binom{0}{3}}N \Rightarrow (0,-2,-1,2,-1/2,0), (0,1,-4,-1,1,-1)$ (if $\eta = 0$, $\widetilde{M} = 0$).

2. $\mathcal{F}_1 = 0$. In this case the order of the weak saddle is ≥ 2.
   (a) $\mathcal{F}_2 \neq 0$. Then apart from a node n^d (by Theorem 7.2 we could not have a star node), systems (11.42) possess a weak saddle of order two, i.e. the configuration of the finite singularities is defined. Regarding infinite singularities we remark that, in the case under consideration, the following implications hold
      - By Lemma 11.26(ix) the condition $\theta = 0$ implies $\theta_2 \neq 0$.
      - By Lemmas 11.26(viii) and 11.30(ii), the conditions $\eta > 0$ and $\theta \leq 0$ imply $\theta_1 > 0$.
      - By Lemma 11.30(ii) the condition $\eta = 0$ implies $\theta > 0$.
      - By Lemma 11.25(viii) the condition $\eta = 0$ implies $\widetilde{M} \neq 0$.

Therefore considering Lemma 11.1 we get the following seven global configurations:

- $s^{(2)}, n^d, ©, ©; N^\infty, ©, © \Rightarrow (0,-4,-1,4,17/11,2), (0,1,-6,-1,1, 23/11)$ (if $\eta < 0$, $\theta < 0$);
- $s^{(2)}, n^d, ©, ©; N^f, ©, © \Rightarrow (0,-3,-1,3,5/3,1), (0,1,-5,-1,1,7/3)$ (if $\eta < 0$, $\theta > 0$);
- $s^{(2)}, n^d, ©, ©; N^d, ©, © \Rightarrow (0,-4,-1,4,2,7), (0,1,-6,-1,1,1/2)$ (if $\eta < 0$, $\theta = 0$);
- $s^{(2)}, n^d, ©, ©; S, N^f, N^f \Rightarrow (0,0,16,-17/4,4,-16), (0,16,0,4, -15/2,0)$(if $\eta > 0$, $\theta < 0$);
- $s^{(2)}, n^d, ©, ©; S, N^f, N^\infty \Rightarrow (0,0,32,-89/4,24,-32), (0,32,0,8,-7,0)$ (if $\eta > 0$, $\theta > 0$);
- $s^{(2)}, n^d, ©, ©; S, N^f, N^d \Rightarrow \big(0,0,8,(20-11\sqrt{7})/3,4,-8\big), \big(0,8,0, 2(\sqrt{7}-1),-3,0\big)$ (if $\eta > 0$, $\theta = 0$);
- $s^{(2)}, n^d, ©, ©; \overline{\binom{0}{2}}SN, N^f \Rightarrow \big(0,0,2,(1-4\xi-\xi^2)/(1+2\xi),2\xi,-2\big), (0,2,0,1,\xi^2-1,0)$, where $\{\eta^{-1}(0)\} \ni \xi \approx 0.4130$ (if $\eta = 0$).

(b) $\mathcal{F}_2 = 0$. Then we have either a weak saddle of order three (if $\mathcal{F}_3\mathcal{F}_4 \neq 0$) or an integrable saddle (if $\mathcal{F}_3\mathcal{F}_4 = 0$). Considering Theorems 7.9 and 7.3, we arrive at the following two global configurations:

- $s^{(3)}, n^d, ©, ©; N^\infty, ©, © \Rightarrow (0,2,0,-4,1,-4), (0,0,-2,19/16,-1,4)$ (if $\mathcal{F}_3\mathcal{F}_4 \neq 0$);
- $\$, n^*, ©, ©; N^*, ©, © \Rightarrow (0,1,0,1,0,0), (0,0,-1,-1,0,-1)$ (if $\mathcal{F}_3\mathcal{F}_4 = 0$).

2.3.1.3: $\mathcal{T}_4 = \mathcal{T}_3 = 0$. Since we have a finite node, the systems could not be Hamiltonian, i.e. by Lemma 11.23 the condition $\mathcal{T}_2 \neq 0$ is satisfied. Therefore only the traces corresponding to the complex finite singularities vanish, and we have a strong finite saddle.

Additionally by Lemma 11.26 (see statements (v) and (x)) in this case the conditions $\eta < 0$ and $\theta > 0$ must hold. So considering Lemma 11.1 we get the unique global configuration

- $s, n^d, ©^\rho, ©^\rho; N^f, ©, © \Rightarrow (12,0,0,-12,8,12), \big(-8,32(5-2\sqrt{5}),0, 8(8\sqrt{5}-19),4(4\sqrt{5}-11),-8\big)$.

2.3.2: $W_3 = 0$. Since we have one saddle, it results that none of the real singularities could have coinciding eigenvalues, and hence the two discriminants of complex singularities vanish. We consider three cases: $W_2 < 0$, $W_2 > 0$ and $W_2 = 0$.

2.3.2.1: $W_2 < 0$. Then by Theorem 6.1 (see Table 6.2, line 27) the anti-saddle is a focus (or a center).

2.3.2.1.1: $\mathcal{T}_4 \neq 0$. By Theorem 6.2 the systems do not have weak singularities. According to Lemma 11.28 in this case we could not have an infinite star node.

Additionally by Lemma 11.26(xi) the conditions $\eta > 0$ and $\theta = 0$ imply $\theta_1 \neq 0$. So considering Lemma 11.1 we arrive at the following 12 global configurations:

- $s, f, \copyright^{\tau}, \copyright^{\tau};\ N^{\infty},\ \copyright, \copyright \Rightarrow (1,-3,0,2,0,1), (-2,3,0,-1,-7/16,-2)$ (if $\eta < 0,\ \theta < 0$);
- $s, f, \copyright^{\tau}, \copyright^{\tau};\ N^{f},\ \copyright, \copyright \Rightarrow (8,-8,0,0,-2,8), (-4,1,0,3,3/8,-4)$ (if $\eta < 0,\ \theta > 0$);
- $s, f, \copyright^{\tau}, \copyright^{\tau};\ N^{d},\ \copyright, \copyright \Rightarrow (1,-4,0,3,-1,1), (-1,0,0,1,1,-1)$ (if $\eta < 0,\ \theta = 0$);
- $s, f, \copyright^{\tau}, \copyright^{\tau};\ S, N^{\infty},\ N^{\infty} \Rightarrow (-5/2,8,0,4,-11,-5/2), (-16,8,0,0,9/10,-16)$ (if $\eta > 0,\ \theta < 0,\ \theta_1 < 0$);
- $s, f, \copyright^{\tau}, \copyright^{\tau};\ S, N^{f},\ N^{f} \Rightarrow (4,8,0,-12,-4,4), (-5,-1,0,6,15/16,-5)$ (if $\eta > 0,\ \theta < 0,\ \theta_1 > 0$);
- $s, f, \copyright^{\tau}, \copyright^{\tau};\ S, N^{f},\ N^{\infty} \Rightarrow (-11/2,64,0,-128,-53,-11/2)$, $(-64,64,0,0,-903/22,-64)$ (if $\eta > 0,\ \theta > 0$);
- $s, f, \copyright^{\tau}, \copyright^{\tau};\ S, N^{\infty},\ N^{d} \Rightarrow (1,4,0,7/8,0,1), (2,4,0,0,0,2)$ (if $\eta > 0,\ \theta = 0,\ \theta_1 < 0$);
- $s, f, \copyright^{\tau}, \copyright^{\tau};\ S, N^{f},\ N^{d} \Rightarrow (-1,4,0,-3,4,-1), (2,4,0,0,0,2)$ (if $\eta > 0,\ \theta = 0,\ \theta_1 > 0$);
- $s, f, \copyright^{\tau}, \copyright^{\tau};\ \overline{\binom{0}{2}}SN,\ N^{\infty} \Rightarrow \big((\sqrt{6}-3)/4, 1, 0, -1, (1-\sqrt{6})/3, (\sqrt{6}-3)/4\big)$, $\big(-1,1,0,0,(3-2\sqrt{6})/6,-1\big)$ (if $\eta = 0,\ \widetilde{M} \neq 0,\ \theta < 0$);
- $s, f, \copyright^{\tau}, \copyright^{\tau};\ \overline{\binom{0}{2}}SN,\ N^{f} \Rightarrow \big(1-\sqrt{2}, 4, 0, -4, (8\sqrt{2}-17)/4, 1-\sqrt{2}\big)$, $(-9/4,4,0,0,-2,-9/4)$ (if $\eta = 0,\ \widetilde{M} \neq 0,\ \theta > 0$);
- $s, f, \copyright^{\tau}, \copyright^{\tau};\ \overline{\binom{0}{2}}SN,\ N^{d} \Rightarrow (1,4,0,1,0,1), (2,4,0,0,0,2)$ (if $\eta = 0,\ \widetilde{M} \neq 0,\ \theta = 0$);
- $s, f, \copyright^{\tau}, \copyright^{\tau};\ \overline{\binom{0}{3}}N \Rightarrow (1,10,0,21/2,2,1), (4,10,0,0,21/4,4)$ (if $\eta = 0,\ \widetilde{M} = 0$).

2.3.2.1.2: $\mathcal{T}_4 = 0,\ \mathcal{T}_3 \neq 0$. So exactly one singularity has zero trace, and this means that there exists only one weak (real) singular point.

1. $\mathcal{T}_3\mathcal{F} < 0$. By Theorem 6.2 we have a weak focus of order ≥ 1. However according to Theorems 7.7, 7.8 and 7.4, in the case we consider we could have neither a weak focus of order two, a weak focus of order three, nor a center, i.e. the condition $\mathcal{F}_1 \neq 0$ is satisfied. Additionally by Lemma 11.29 the condition $\theta\mathcal{T}_3\mathcal{F} < 0$ must hold. Therefore we get $\theta > 0$, and since by Lemma 11.25(ix) we get $\widetilde{M} \neq 0$, considering Lemma 11.1 we obtain the following three global configurations:

   - $s, f^{(1)}, \copyright^{\tau}, \copyright^{\tau};\ N^{f},\ \copyright, \copyright \Rightarrow (36,-60,0,24,23,36), (29,-5,0,-24,6,29)$ (if $\eta < 0$);

   - $s, f^{(1)}, \copyright^{\tau}, \copyright^{\tau};\ S, N^{f},\ N^{\infty} \Rightarrow (4,-36,0,32,7\sqrt{137}-12,4)$, $\big(3(3\sqrt{137}-4), -9\sqrt{137}, 0, 12, -14, 3(3\sqrt{137}-4)\big)$ (if $\eta > 0$);

   - $s, f^{(1)}, \copyright^{\tau}, \copyright^{\tau};\ \overline{\binom{0}{2}}SN,\ N^{f} \Rightarrow \big(4\xi, -4(1+\xi), 0, 4, \chi/\xi - 2 - \chi, 4\xi\big)$,

$(\chi/\xi - 2 + \chi, -\chi/\xi - 1, 0, 2, -2 + 2\xi, \chi/\xi - 2 + \chi)$, where
$\nu \in \{\eta^{-1}(0)\}$, $\xi = 1 + 2\nu \approx 0.1401$, $\chi = \sqrt{1 + 10\xi - 7\xi^2}$ (if $\eta = 0$).

2. $\mathcal{T}_3\mathcal{F} > 0$. In this case by Theorem 6.2 we have a weak saddle. We remark that by Lemma 11.30(iii) the condition $\mathcal{F}_1 \neq 0$ is satisfied, i.e. we have a weak saddle of order one. Additionally by Lemma 11.29 the condition $\theta\mathcal{T}_3\mathcal{F} < 0$ must hold, which implies $\theta < 0$. Moreover by Lemma 11.25(ix) we get $\widetilde{M} \neq 0$, and therefore considering Lemma 11.1 we obtain the following four global configurations:

   - $s^{(1)}, f, ©^{\tau}, ©^{\tau};\ N^{\infty},\ ©, © \Rightarrow (2, -6, 0, 4, 0, 2), (\sqrt{17}, -3\sqrt{17}/2, 0, \sqrt{17}/2, -1, \sqrt{17})$ (if $\eta < 0$);
   - $s^{(1)}, f, ©^{\tau}, ©^{\tau};\ S, N^{\infty},\ N^{\infty} \Rightarrow (2, -34, 0, 32, 58, 2), \big(58 + \sqrt{409}, -17\sqrt{409}/2, 0, (15\sqrt{409} - 116)/2, -15, 58 + \sqrt{409}\big)$ (if $\eta > 0,\ \theta_1 < 0$);
   - $s^{(1)}, f, ©^{\tau}, ©^{\tau};\ S, N^{f},\ N^{f} \Rightarrow (-6, -58, 0, 64, -32, -6), (-31, 29/6, 0, 157/6, -35, -31)$ (if $\eta > 0,\ \theta_1 > 0$);
   - $s^{(1)}, f, ©^{\tau}, ©^{\tau};\ \overline{\binom{0}{2}}SN,\ N^{\infty} \Rightarrow \big(4\xi, -4(1 + \xi), 0, 4, \chi/\xi - 8 - \chi, 4\xi\big), \big(\chi/\xi - 8 + \chi, -\chi/\xi - 1, 0, 8, 2(\xi - 1), \chi/\xi - 8 + \chi\big)$, where $\nu \in \{\eta^{-1}(0)\}$, $\xi = 1 + 2\nu \approx 0.086307$, $\chi = \sqrt{1 + 10\xi - 7\xi^2}$ (if $\eta = 0$).

2.3.2.1.3: $\mathcal{T}_4 = \mathcal{T}_3 = 0$. Since the systems are non-degenerate, it is clear that the traces corresponding to real singularities vanish. So the saddle and the focus are weak. By Lemma 11.26(iii) in this case the condition $\mathcal{F}_1 = 0$ must hold, and hence by Theorem 6.2 we have an integrable saddle and a center. Therefore considering Theorem 7.3 we get the global configuration

- $\$, c, ©^{\tau}, ©^{\tau};\ S, N^{d},\ N^{d} \Rightarrow (0, 0, 0, 0, 1, 0), (1, -3, 0, 2, 0, 1)$.

2.3.2.2: $W_2 > 0$. According to Theorem 6.1 (see Table 6.2, line 25) the anti-saddle is a node, which is generic due to $W_2 \neq 0$.

2.3.2.2.1: $\mathcal{T}_4 \neq 0$. Then the saddle is strong, and since by Lemma 11.28 in this case we could not have an infinite star node, considering Lemma 11.1 we arrive at the following 13 global configurations:

- $s, n, ©^{\tau}, ©^{\tau};\ N^{\infty},\ ©, © \Rightarrow (2, -3, 0, 1, 0, 2), (0, 0, 0, 0, 9/32, 0)$ (if $\eta < 0,\ \theta < 0$);
- $s, n, ©^{\tau}, ©^{\tau};\ N^{f},\ ©, © \Rightarrow (8, -4, 0, -4, -8, 8), (12, -5, 0, -7, -99/8, 12)$ (if $\eta < 0,\ \theta > 0$);
- $s, n, ©^{\tau}, ©^{\tau};\ N^{d},\ ©, © \Rightarrow (20, -22, 0, 2, 20\xi, 20), \big(0, 11\xi, 0, -11\xi, (121 - 400\xi^2)/80, 0\big)$, where $\{\theta^{-1}(0)\} \ni \xi \approx -0.01089$ (if $\eta < 0,\ \theta = 0$);
- $s, n, ©^{\tau}, ©^{\tau};\ S, N^{\infty},\ N^{\infty} \Rightarrow (10, -11, 0, 1, 0, 10), (0, 0, 0, 0, 121/160, 0)$ (if $\eta > 0,\ \theta < 0,\ \theta_1 < 0$);
- $s, n, ©^{\tau}, ©^{\tau};\ S, N^{f},\ N^{f} \Rightarrow (40, -44, 0, 4, -40, 40), (35, -165/4, 0, 25/4, -5141/160, 35)$ (if $\eta > 0,\ \theta < 0,\ \theta_1 > 0$);

- $s, n, ©^{\tau}, ©^{\tau}; S, N^f, N^{\infty} \Rightarrow (8,-16,0,8,16,8), (-23,39,0,-16,-1457/32,-23)$ (if $\eta > 0$, $\theta > 0$);
- $s, n, ©^{\tau}, ©^{\tau}; S, N^{\infty}, N^d \Rightarrow (10,-11,0,1,-1,10), \big(10\xi, -11(1+10\xi)/20, 0, (11-90\xi)/20, (9-20\xi)(13+20\xi)/160, 10\xi\big)$, where $\{\theta^{-1}(0)\} \ni \xi \approx 0.1409$ (if $\eta > 0$, $\theta = 0$, $\theta_1 < 0$);
- $s, n, ©^{\tau}, ©^{\tau}; S, N^f, N^d \Rightarrow (1,-2,0,1,\xi,1), \big(-2, 2+\xi, 0, -\xi, -(1+\xi)(3+\xi)/4, -2\big)$, where $\{\theta^{-1}(0)\} \ni \xi \approx 2.5098$ (if $\eta > 0$, $\theta = 0$, $\theta_1 > 0$);
- $s, n, ©^{\tau}, ©^{\tau}; S, N^d, N^d \Rightarrow (1,4,0,2,-1,1), (1,4,0,0,0,1)$ (if $\eta > 0$, $\theta = 0$, $\theta_1 = 0$);
- $s, n, ©^{\tau}, ©^{\tau}; \overline{\binom{0}{2}} SN, N^{\infty} \Rightarrow (16,-32,0,16,16\xi,16), \big(-21, 21+16\xi, 0, -16\xi, -(5+16\xi)(37+16\xi)/64, -21\big)$, where $\{\eta^{-1}(0)\} \ni \xi \approx 0.8620$ (if $\eta = 0$, $\widetilde{M} \neq 0$, $\theta < 0$);
- $s, n, ©^{\tau}, ©^{\tau}; \overline{\binom{0}{2}} SN, N^f: \Rightarrow (1,-2,0,1,\xi,1), \big(-2, 2+\xi, 0, -\xi, -(1+\xi)(3+\xi)/4, -2\big)$
  where $\{\eta^{-1}(0)\} \ni \xi \approx 1.3484$ (if $\eta = 0$, $\widetilde{M} \neq 0$, $\theta > 0$);
- $s, n, ©^{\tau}, ©^{\tau}; \overline{\binom{0}{2}} SN, N^d \Rightarrow (1,-2,0,1,\xi,1), \big(\chi, -\chi+\xi, 0, -\xi, (1-(\chi-\xi)^2)/4, \chi\big)$, where $\{\eta^{-1}(0), \theta^{-1}(0)\} \ni (\xi, \chi) \approx (-0.8823, 1.3216)$ (if $\eta = 0$, $\widetilde{M} \neq 0$, $\theta = 0$);
- $s, n, ©^{\tau}, ©^{\tau}; \overline{\binom{0}{3}} N \Rightarrow (1/2, 32, 0, 255, 1, 1/2), (2, 32, 0, 0, 255/2, 2)$ (if $\eta = 0$, $\widetilde{M} = 0$).

2.3.2.2.2: $\mathcal{T}_4 = 0$. In this case only the trace corresponding to the saddle could vanish.

1. $\mathcal{F}_1 \neq 0$. By Theorem 6.2 we have a weak saddle of order one. According to Lemma 11.30(iv) in this case the conditions $\theta < 0$ and $\eta \neq 0$ hold. Moreover in the case $\eta > 0$ and $\theta < 0$, we must have $\theta_1 > 0$. So considering Lemma 11.1 we obtain the following two global configurations:
   - $s^{(1)}, n, ©^{\tau}, ©^{\tau}; N^{\infty}, ©, © \Rightarrow (2,-34,0,32,0,2), (\sqrt{409}, -17\sqrt{409}/2, 0, 15\sqrt{409}/2, -15, \sqrt{409})$ (if $\eta < 0$);
   - $s^{(1)}, n, ©^{\tau}, ©^{\tau}; S, N^f, N^f \Rightarrow (-2,-30,0,32,-16,-2), \big(\sqrt{89}-16, 15\sqrt{89}/2, 0, (32-17\sqrt{89})/2, -17, \sqrt{89}-16\big)$ (if $\eta > 0$).
2. $\mathcal{F}_1 = 0$. In this case we have a weak saddle of order at least two.
   (a) $\mathcal{F}_2 \neq 0$. Then the weak saddle has order two, and considering the same lemmas as above, we get the same two global configurations with the only difference being that instead of a saddle of order one we have a saddle of order two:
   - $s^{(2)}, n, ©^{\tau}, ©^{\tau}; N^{\infty}, ©, © \Rightarrow (0,0,343,2173/2,18\sqrt{2},572), (0,343,0,-18\sqrt{2},571/2,-18\sqrt{2})$ (if $\eta < 0$);
   - $s^{(2)}, n, ©^{\tau}, ©^{\tau}; S, N^f, N^f \Rightarrow \big(0,0,65863,21547/2,288\sqrt{2},109568\big), \big(0,65863,0,-288\sqrt{2},110179/2,-288\sqrt{2}\big)$ (if $\eta > 0$).

(b) $\mathcal{F}_2 = 0$. So by Theorem 7.3 we could not have an integrable saddle in this configuration, and according to Theorem 7.9 we have the global configuration
   - $s^{(3)}, n, ©^{\tau}, ©^{\tau}; N^{\infty}, ©, ©: \Rightarrow \big(0, 6, 0, -12, 3, -9-4\sqrt{3}\big), \big(0, 0, -6, 9-4\sqrt{3}, -3, 12\big)$.

2.3.2.3: $W_2 = 0$. By Theorem 6.1 the anti-saddle is a node, which has coinciding eigenvalues due to $W_2 = 0$. According to Theorem 7.2 and Lemma 11.1, we deduce that in the case of the existence of a star node (i.e. by Lemma 11.2 the conditions $(\mathcal{S}^1_{n^*})$ hold), we must have $\mathcal{T}_4 \neq 0$, $\eta < 0$ and $\theta < 0$.

1. $\mathcal{T}_4 \neq 0$. Then the saddle is strong, and we note that by Lemma 11.28 in this case we could not have an infinite star node. Moreover, by Lemma 11.26(xiv) in the case $\theta = 0$, the condition $\theta_1 \neq 0$ holds, i.e. at infinity we could not have two nodes N^d. Therefore considering Lemma 11.1 as well as the above remark about the existence of a finite star node, we arrive at the following 13 global configurations:
   - $s, n^d, ©^{\tau}, ©^{\tau}; N^{\infty}, ©, © \Rightarrow (1, -2, 0, 1, 3-2\sqrt{2}, 1), (0, 3-2\sqrt{2}, 0, -3+2\sqrt{2},$
     $-4+3\sqrt{2}, 0)$ (if $\eta < 0$, $\theta < 0$, $\neg(\mathcal{S}^1_{n^*})$);
   - $s, n^*, ©^{\tau}, ©^{\tau}; N^{\infty}, ©, © \Rightarrow (1, -4, 0, 3, 0, 1), (0, 0, 0, 0, 1, 0)$ (if $\eta < 0$, $\theta < 0$, $(\mathcal{S}^1_{n^*})$);
   - $s, n^d, ©^{\tau}, ©^{\tau}; N^f, ©, © \Rightarrow (4, -4, 0, 0, 2, 4), (0, 1, 0, -1, 0, 0)$ (if $\eta < 0$, $\theta > 0$);
   - $s, n^d, ©^{\tau}, ©^{\tau}; N^d, ©, © \Rightarrow (2, -20, 0, 18, 1, 2), (-1, 10, 0, -9, 12, -1)$ (if $\eta < 0$, $\theta = 0$);
   - $s, n^d, ©^{\tau}, ©^{\tau}; S, N^{\infty}, N^{\infty} \Rightarrow (8, -78, 0, 70, 16, 8), (25, -351/8, 0, 151/8, 45, 25)$ (if $\eta > 0$, $\theta < 0$, $\theta_1 < 0$);
   - $s, n^d, ©^{\tau}, ©^{\tau}; S, N^f, N^f \Rightarrow (4, 12, 0, -16, 2, 4), (8, 9, 0, -17, 0, 8)$ (if $\eta > 0$, $\theta < 0$, $\theta_1 > 0$);
   - $s, n^d, ©^{\tau}, ©^{\tau}; S, N^f, N^{\infty} \Rightarrow (0, 0, 0, 0, 5, 0), (5, -32/5, 0, 7/5, 6, 5)$ (if $\eta > 0$, $\theta > 0$);
   - $s, n^d, ©^{\tau}, ©^{\tau}; S, N^{\infty}, N^d \Rightarrow (1, -10, 0, 9, 2, 1), (3, -5, 0, 2, 6, 3)$ (if $\eta > 0$, $\theta = 0$, $\theta_1 < 0$);
   - $s, n^d, ©^{\tau}, ©^{\tau}; S, N^f, N^d \Rightarrow (1, 2, 0, -3, 2, 1), (1, -1, 0, 0, 0, 1)$ (if $\eta > 0$, $\theta = 0$, $\theta_1 > 0$);
   - $s, n^d, ©^{\tau}, ©^{\tau}; \binom{0}{2}SN, N^{\infty} \Rightarrow (4, -30, 0, 26, 2, 4), (3, -15/4, 0, 3/4, 14, 3)$ (if $\eta = 0$, $\widetilde{M} \neq 0$, $\theta < 0$);

- $s, n^d, ©^\tau, ©^\tau; \overline{\binom{0}{2}}SN,\ N^f \Rightarrow \big(4, -4(1+\xi), 0, 4\xi, 8, 4\big), \big(4\chi, 2(2-\chi)(1+\xi), 0, 2(\chi\xi - \chi - 2\xi - 2), (5 - 2\chi + \xi)(2\chi + \xi - 3)/4, 4\chi\big)$, where $\{\eta^{-1}(0), W_2^{-1}(0)\} \ni (\xi, \chi) \approx (12.5203, 1.2399)$ (if $\eta = 0,\ \widetilde{M} \neq 0,\ \theta > 0$);

- $s, n^d, ©^\tau, ©^\tau; \overline{\binom{0}{2}}SN,\ N^d \Rightarrow \big(1, -2(2 + \chi + \xi + 2\sqrt{2\xi}), 0, 3 + 2\chi + 2\xi + 4\sqrt{2\xi}, -\xi, 1\big), \big(\chi, -(2 + \chi + \xi + 2\sqrt{2\xi})(\xi + \chi), 0, \chi + \chi^2 + 2\chi\sqrt{2\xi} + 2\xi + 2\chi\xi + 2\xi\sqrt{2\xi} + \xi^2, (1 + \sqrt{2\xi})(1 + \chi + \xi + \sqrt{2\xi}), \chi\big)$, where $\{\eta^{-1}(0), \theta^{-1}(0)\} \ni (\xi, \chi) \approx (0.9442, -0.8630)$ (if $\eta = 0,\ \widetilde{M} \neq 0,\ \theta = 0$);

- $s, n^d, ©^\tau, ©^\tau; \overline{\binom{0}{3}}N \Rightarrow \big(1, 2(\chi - \xi - 2 - 2\sqrt{2\xi}), 0, 3 - 2\chi + 2\xi + 4\sqrt{2\xi}, \xi, 1\big), \big(\chi, (\chi - \xi)(\chi - \xi - 2 - 2\sqrt{2\xi}), 0, \chi - \chi^2 + 2\chi\sqrt{2\xi} - 2\xi + 2\chi\xi - 2\xi\sqrt{2\xi} - \xi^2, (1 + \sqrt{2\xi})(1 - \chi + \sqrt{2\xi} + \xi), \chi\big)$, where $\{\widetilde{M}^{-1}(0)\} \ni (\xi, \chi) \approx (0.29428650, 0.41619378)$ (if $\eta = 0,\ \widetilde{M} = 0$).

2. $\mathcal{T}_4 = 0$. Then the saddle is weak and the node is n^d. Moreover by Lemma 11.30(vi) the condition $\mathcal{F}_1 \neq 0$ holds (i.e. the weak saddle is of order one), and the condition $\eta \neq 0$ is satisfied.

   Regarding the types of infinite singularities, we remark that in the case under consideration, according to Lemma 11.26(xii), the condition $\theta < 0$ must hold. Moreover by Lemma 11.26(xiii) the conditions $\eta > 0$ and $\theta < 0$ imply $\theta_1 > 0$. Therefore considering Lemma 11.1 we get the following two global configurations:

   - $s^{(1)}, n^d, ©^\tau, ©^\tau;\ N^\infty,\ ©, © \Rightarrow \big(1, -2, 0, 1, (1-\sqrt{3})/2, 1\big), \big((3-\sqrt{3})/2, -1, 0, (\sqrt{3} - 1)/2, 0, (3 - \sqrt{3})/2\big)$ (if $\eta < 0$);

   - $s^{(1)}, n^d, ©^\tau, ©^\tau;\ S, N^f,\ N^f \Rightarrow (1, 4, 0, -5, \xi, 1), \big(4+\xi-2\sqrt{2\xi}, 4(2-\sqrt{2\xi}), 0, 6\sqrt{2\xi} - \xi - 12, (3 - \sqrt{2\xi})(\sqrt{2\xi} - 1), 4 + \xi - 2\sqrt{2\xi}\big)$, where $\{\mathcal{T}_4^{-1}(0)\} \ni \xi \approx 0.3791389$ (if $\eta > 0$).

Thus all the possibilities have been examined, and hence we proved that the family of quadratic systems with two real and two complex all distinct finite singularities possesses 239 global configurations. □

11.4.3 Four complex elemental finite singularities

Theorem 11.9. (a) *We consider here all the configurations of singularities, finite and infinite, of quadratic vector fields possessing exactly four elemental finite singularities that are complex. These configurations are classified in Diagram* 11.10 *according to the geometrical equivalence relation of configurations. We have a total of* 19 *geometrically distinct configurations of singularities, finite and infinite.*

(b) *Necessary and sufficient conditions for each one of the* 19 *different geometrical equivalence classes can be assembled from this diagram in terms of* 11 *invariant polynomials with respect to the action of the affine group and time rescaling. These invariant polynomials, which appear in Diagram* 11.10*, are given in Section* 5.4.

(c) *Diagram* 11.10 *actually contains the global bifurcation diagram in the* 12*-dimensional space of parameters of the global configurations of singularities, finite and infinite, of this family of quadratic differential systems and provides an algorithm for finding for any given system in the family considered, its respective geometrical configuration of singularities.*

Proof. As it was mentioned at the beginning of this section a quadratic system possesses four complex elemental finite singularities if and only if $\mu_0 \neq 0$, $\mathbf{D} < 0$ and either $\mathbf{R} \leq 0$ or $\mathbf{S} \leq 0$. Since we do not have real finite singularities, it is clear that the total index of the finite singularities equals zero, and hence the total index of infinite singularities must be $+1$. Considering Lemma 11.1 (see Diagram 11.1) this could happen only in the case $\mu_0 > 0$, and we have exactly 19 possibilities of geometrically distinct configurations at infinity with total index one. Moreover all these configurations are realizable, and to complete the proof it only remains to present the corresponding examples. We give here in the examples, only the configurations of infinite singularities, taking into consideration that in all the cases we must add the same configuration of finite singularities, namely ©, ©, ©, ©.

1. In the case $\eta < 0$, we have four global configurations:
   - N^{∞}, ©, © $\Rightarrow (2,0,0,3,1,1),\ (1,0,0,0,1,1)$ (if $\theta < 0$);
   - N^{f}, ©, © $\Rightarrow (2,0,0,1,1,1),\ (-1,0,0,0,-1,-1)$ (if $\theta > 0$);
   - N^{d}, ©, © $\Rightarrow (2,1,0,1,0,1),\ (1,1,0,0,0,1)$ (if $\theta = 0,\ \theta_2 \neq 0$);
   - N^{*}, ©, © $\Rightarrow (1,0,0,1,0,0),\ (0,0,0,1,0,1)$ (if $\theta = \theta_2 = 0$).
2. In the case $\eta > 0$, we have 10 global configurations:
   - S, N^{∞}, N^{∞} $\Rightarrow (2,0,0,35/16,1,1),\ (1,0,0,0,1,1)$ (if $\theta < 0,\ \theta_1 < 0$);
   - S, N^{f}, N^{f} $\Rightarrow (2,0,0,5/4,-2,1),\ (1,0,0,0,-2,1)$ (if $\theta < 0,\ \theta_1 > 0$);
   - S, N^{f}, N^{∞} $\Rightarrow (2,0,0,5/4,1,1),\ (1,0,0,0,1,1)$ (if $\theta > 0$);
   - S, N^{∞}, N^{d} $\Rightarrow (2,1,0,6,2,1),\ (1,1,0,0,2,1)$ (if $\theta = 0,\ \theta_1 < 0,\ \theta_2 \neq 0$);
   - S, N^{∞}, N^{*} $\Rightarrow (1,0,0,3,0,0),\ (1,0,0,0,1,1)$ (if $\theta = 0,\ \theta_1 < 0,\ \theta_2 = 0$);
   - S, N^{f}, N^{d} $\Rightarrow (2,1,0,2,-2,1),\ (1,1,0,0,-2,1)$ (if $\theta = 0,\ \theta_1 > 0,\ \theta_2 \neq 0$);
   - S, N^{f}, N^{*} $\Rightarrow (1,0,0,-1,0,0),\ (2,0,0,0,1,1)$ (if $\theta = 0,\ \theta_1 > 0,\ \theta_2 = 0$);
   - S, N^{d}, N^{d} $\Rightarrow (0,0,1/2,1,0,0),\ (1,1,0,0,1,1)$ (if $\theta = \theta_1 = 0,\ \theta_3 \neq 0$);

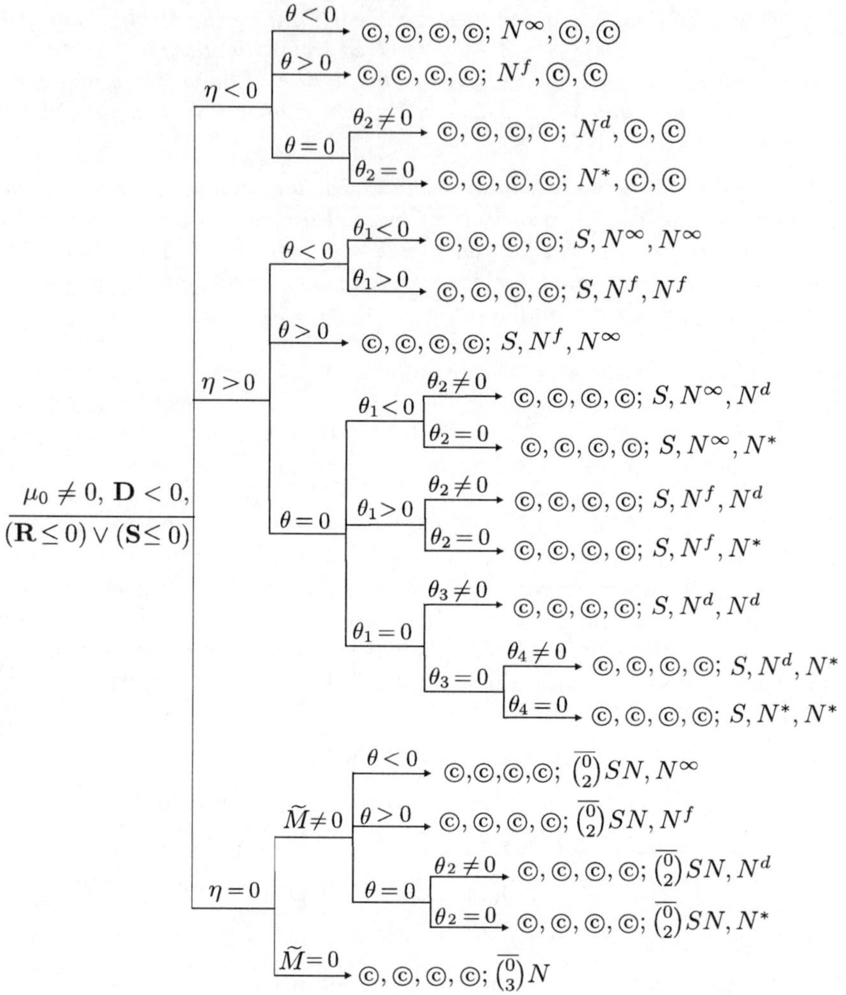

Diagram 11.10: Global configurations: the case $\mu_0 \neq 0, \mathbf{D} < 0, (\mathbf{R} \leq 0) \vee (\mathbf{S} \leq 0)$.

- S, N^d, N^* $\Rightarrow (1,0,0,1,0,0)$, $(1,1,0,0,1,1)$ (if $\theta = \theta_1 = \theta_3 = 0$, $\theta_4 \neq 0$);
- S, N^*, N^* $\Rightarrow (1,0,0,1,0,0)$, $(1,0,0,0,1,1)$ (if $\theta = \theta_1 = \theta_3 = \theta_4 = 0$).

3. In the case $\eta = 0$, $\widetilde{M} \neq 0$, we have four global configurations:

- $\overline{\binom{0}{2}}SN$, N^∞ $\Rightarrow (2,0,0,6,1,1)$, $(3,0,0,0,3,3)$ (if $\theta < 0$);
- $\overline{\binom{0}{2}}SN$, N^f $\Rightarrow (0,0,0,2,-1,-1)$, $(1,0,0,0,1,1)$ (if $\theta > 0$);

- $\overline{\binom{0}{2}}SN,\ N^d \ \Rightarrow (2,1,0,2,1,1),\ (1,1,0,0,1,1)\ \ (\text{if } \theta=0,\ \theta_2\neq 0);$
- $\overline{\binom{0}{2}}SN,\ N^* \ \Rightarrow (2,0,0,2,1,1),\ (1,0,0,0,1,1)\ \ (\text{if } \theta=\theta_2=0).$

4. In the case $\eta=\widetilde{M}=0$, we have a single configuration

- $\overline{\binom{0}{3}}N \ \Rightarrow (2,0,0,4,1,1),\ (2,0,0,0,2,2).$

Thus we have shown that each one of the 19 geometrically distinct configurations is realizable, and this completes the proof of Theorem 11.9. □

Chapter 12

Degenerate quadratic systems ($m_f = \infty$)

A study of configurations of singularities for the whole family of quadratic systems cannot be considered complete unless we also add the degenerate cases. These are the systems that have an infinite number of finite singular points (real or complex) located either on a line or on a conic, and for which there is a common factor between the polynomials $p(x, y)$ and $q(x, y)$ of (8.1).

According to Lemma 5.2 for the quadratic systems to be degenerate, the conditions $\mu_i(\tilde{a}, x, y) = 0$ in $\mathbb{R}[x, y]$ for every $i = 0, 1, 2, 3, 4$ must be satisfied.

So by Theorem 6.4 the following lemma is valid.

Lemma 12.1. *The configurations of singularities at infinity of the family of degenerate quadratic systems (i.e.* $\mu_i(\tilde{a}, x, y) = 0$ *for every* $i = 0, 1, 2, 3, 4$*) are classified in Diagram* 12.1 *according to the geometrical equivalence relation. Necessary and sufficient conditions for each one of the* 35 *different equivalence classes can be assembled from these diagrams in terms of* 19 *invariant polynomials with respect to the action of the affine group and time rescaling, given in Section* 5.4.

We remark that the notations for the configurations of singularities in the case of degenerate systems are given in Section 3.7.4.2 (see page 86).

Theorem 12.1. (a) *We consider here all the configurations of singularities, finite and infinite, of degenerate quadratic vector fields. These configurations are classified in Diagram* 12.2 *according to the geometrical equivalence relation of configurations. We have a total of* 41 *geometrically distinct configurations of singularities, finite and infinite.*

(b) *Necessary and sufficient conditions for each one of the* 41 *different equivalence classes can be assembled from this diagram in terms of* 21 *invariant*

J. C. Artés et al., *Geometric Configurations of Singularities of Planar Polynomial Differential Systems*, https://doi.org/10.1007/978-3-030-50570-7_12

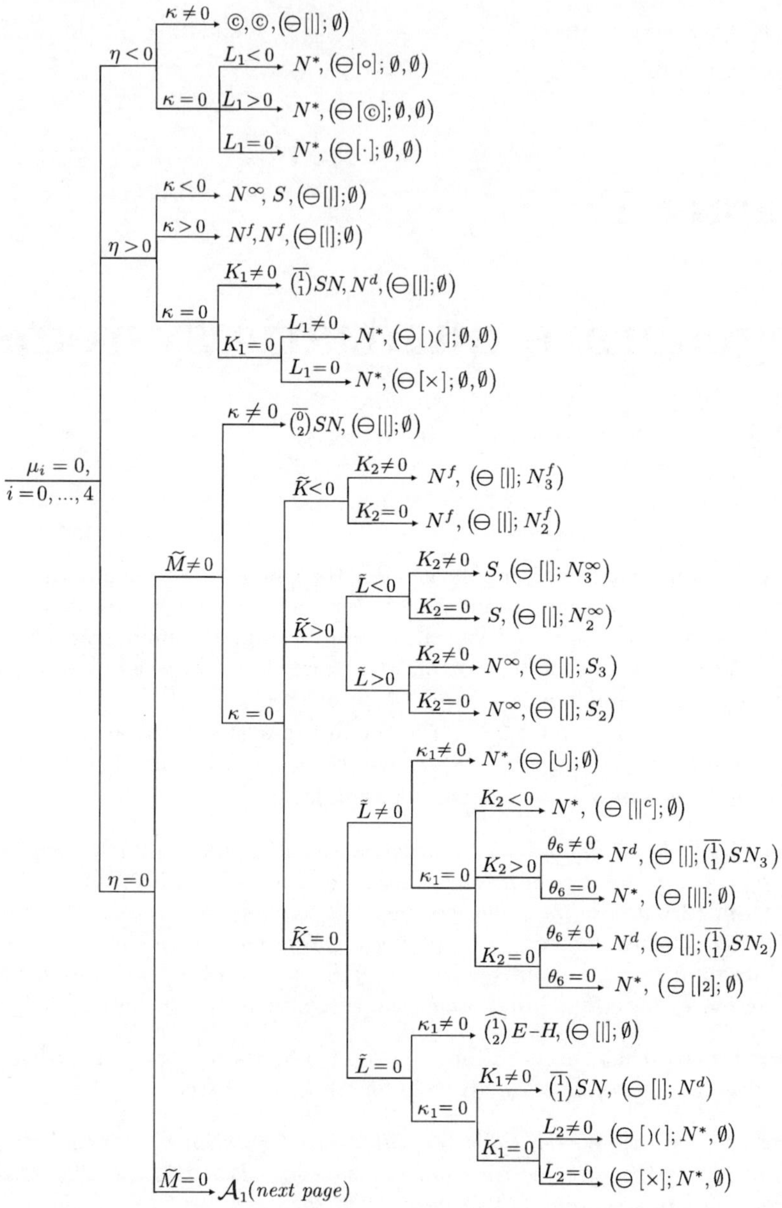

Diagram 12.1: Configurations of infinite singularities in the case $\mu_i = 0$ for $i = 0, \ldots, 4$.

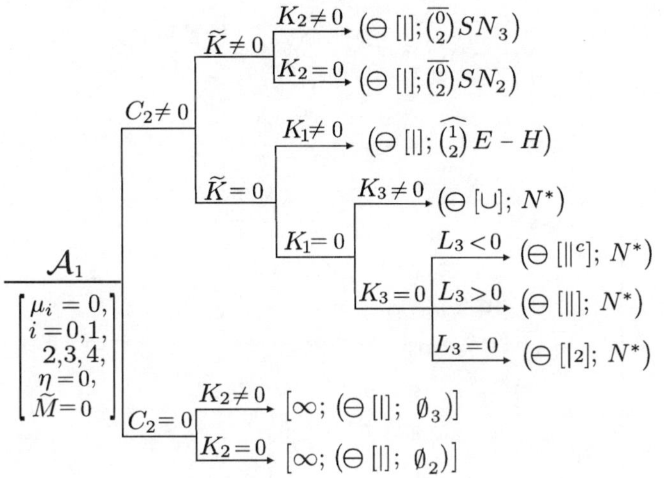

Diagram 12.1 (*cont.*): Configurations of infinite singularities in the case $\mu_i = 0$ for $i = 0, \ldots, 4$.

polynomials with respect to the action of the affine group and time rescaling. These invariant polynomials, which appear in Diagram 12.2*, are given in Section* 5.4.

(c) *Diagram* 12.2 *actually contains the global bifurcation diagram in the* 12*-dimensional space of parameters of the global configurations of singularities, finite and infinite, of this family of degenerate quadratic differential systems and provides an algorithm for finding for any given system in the family considered, its respective geometrical configuration of singularities.*

Proof. We assume that the conditions $\mu_i(\tilde{a}, x, y) = 0$ in $\mathbb{R}[x, y]$ for every $i = 0, 1, 2, 3, 4$ are fulfilled, and we examine each one of the canonical systems $(\mathbf{S}_I)$ – $(\mathbf{S}_V)$ provided by Lemma 6.1.

1: Three real distinct infinite singularities. According to Lemma 6.1 we consider the family of systems:

$$\begin{aligned} \dot{x} &= a + cx + dy + gx^2 + (h-1)xy, \\ \dot{y} &= b + ex + fy + (g-1)xy + hy^2, \end{aligned} \tag{12.1}$$

for which we calculate $\mu_0 = gh(g+h-1)$, and hence the condition $\mu_0 = 0$ implies $gh(g + h - 1) = 0$. Then without loss of generality we may assume just that $g = 0$. Indeed, if $h = 0$ (respectively, $g + h - 1 = 0$), we can apply the linear transformation that will replace the straight line $y = 0$ with $x = 0$ (respectively, $y = 0$ with $y = x$). Let $g = 0$. By using the translation $x = x_1 + f + 2eh$, $y = y_1 + e$,

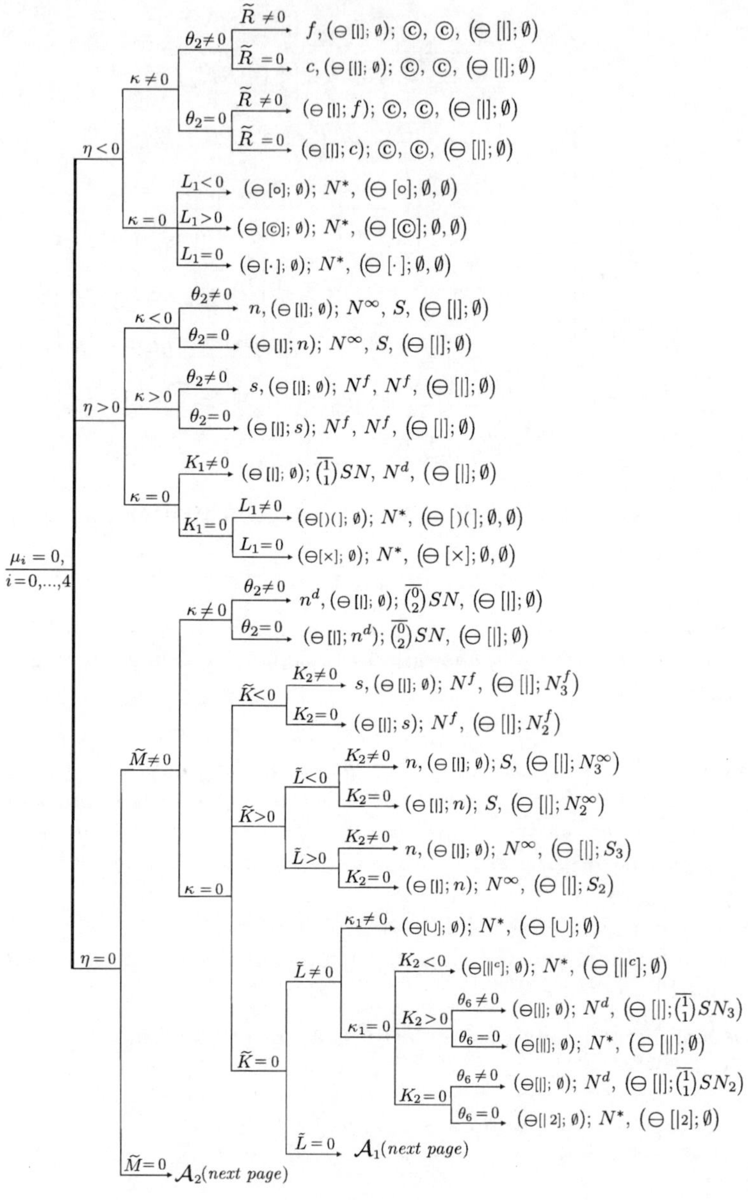

Diagram 12.2: Global configurations in the case $\mu_i = 0$, $i = 0, \dots, 4$.

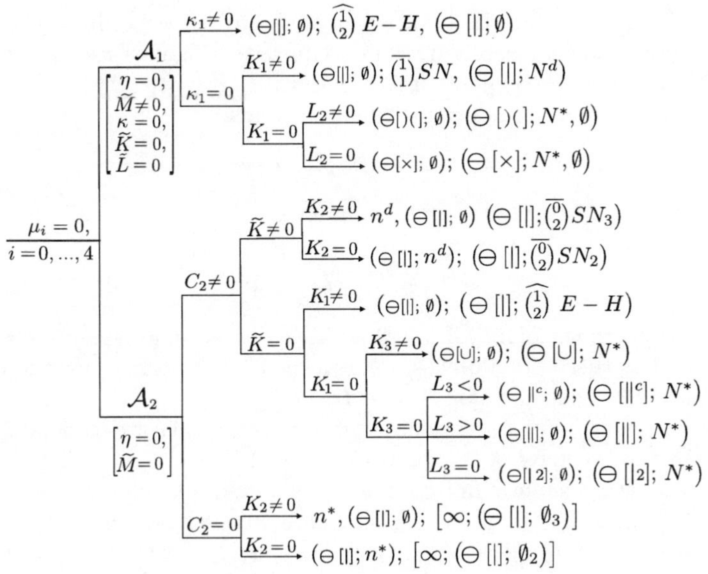

Diagram 12.2 (*cont.*): Global configurations in the case $\mu_i = 0$, $i = 0, \ldots, 4$.

we may assume $e = f = 0$. In this way the systems (12.1) with $g = 0$ have the following canonical form:

$$\dot{x} = a + cx + dy + (h-1)xy, \quad \dot{y} = b - xy + hy^2,$$

for which we have

$$\mu_1 = ch(1-h)y, \quad \kappa = 16h(1-h).$$

We consider two cases: $\kappa \neq 0$ and $\kappa = 0$.

1.1: $\kappa \neq 0$. Then $h(1-h) \neq 0$ and the condition $\mu_1 = 0$ gives $c = 0$. So we obtain

$$\mu_2 = h(h-1)(a - b + bh)y^2 = 0,$$

and clearly we get $a = b(1-h)$. In this case calculations yield

$$\mu_0 = \mu_1 = \mu_2 = 0, \quad \mu_3 = bdh(1-h)y^3, \quad \mu_4 = -by^3\left[d^2(x - hy) - bh^2(1-h)^2y\right],$$

and evidently due to $\kappa \neq 0$ the condition $\mu_3 = \mu_4 = 0$ is equivalent to $b = 0$. As a result we arrive at the family of degenerate systems:

$$\dot{x} = y(d - x + hx), \quad \dot{y} = -y(x - hy), \tag{12.2}$$

for which $\kappa = 16h(1-h) \neq 0$. These systems possess the invariant singular line $y = 0$, and the corresponding linear systems have the matrix $\begin{pmatrix} h-1 & 0 \\ -1 & h \end{pmatrix}$.

As $\kappa \neq 0$ considering the notation of singularities, according to Lemma 12.1 we obtain the following configurations of infinite singularities of the quadratic systems (12.2): N^∞, S, $(\ominus\,[|];\emptyset)$ if $\kappa < 0$ and N^f, N^f, $(\ominus\,[|];\emptyset)$ if $\kappa > 0$.

Let us examine the underlying linear systems:

$$\dot{x} = d - x + hx, \quad \dot{y} = -x + hy,$$

which due to $\kappa \neq 0$ possess the finite singularity

$$M_1\Big(\frac{d}{1-h}, \frac{d}{h(1-h)}\Big).$$

This singular point is isolated if $d \neq 0$ and lies on the singular line $y = 0$ if $d = 0$. We observe that these possibilities are governed by the affine invariant θ_2, which for systems (12.2) has the value $\theta_2 = -d/4$.

If the singular point M_0 is isolated (i.e. $\theta_2 \neq 0$), then we calculate $\lambda_1 = h - 1$ and $\lambda_2 = h$. So we have $\lambda_1\lambda_2 = h(h-1)$ and hence $\text{sign}\,(\lambda_1\lambda_2) = -\text{sign}\,(\kappa)$. Therefore we have a generic finite node if $\kappa < 0$ and a saddle if $\kappa > 0$.

Thus we arrive at the following four global configurations:

- $n, (\ominus\,[|];\,\emptyset);\ N^\infty,\ S,\ (\ominus\,[|];\emptyset) \Rightarrow \big((12.2):\ d = 1,\ h = -1\big)$ (if $\theta_2 \neq 0,\ \kappa < 0$);
- $s, (\ominus\,[|];\,\emptyset);\ N^f,\ N^f,\ (\ominus\,[|];\emptyset) \Rightarrow \big((12.2):\ d = 1,\ h = 1/2\big)$ (if $\theta_2 \neq 0,\ \kappa > 0$);
- $(\ominus\,[|];n);\ N^\infty,\ S,\ (\ominus\,[|];\emptyset) \Rightarrow \big((12.2):\ d = 0,\ h = -1\big)$ (if $\theta_2 = 0,\ \kappa < 0$);
- $(\ominus\,[|];s);\ N^f,\ N^f,\ (\ominus\,[|];\emptyset) \Rightarrow \big((12.2):\ d = 0,\ h = 1/2\big)$ (if $\theta_2 = 0,\ \kappa > 0$).

1.2: $\kappa = 0$. Then $h(h-1) = 0$, and without loss of generality we may consider $h = 0$ in the systems (12.1) with $g = 0$ (due to a linear transformation that keeps the line $y = 0$ fixed and replaces the line $y = x$ with $x = 0$). Moreover since $g = h = 0$ (making a translation), we may assume $d = e = 0$, and systems (12.1) become

$$\dot{x} = a + cx - xy, \quad \dot{y} = b + fy - xy, \tag{12.3}$$

for which we calculate

$$\mu_0 = \mu_1 = 0, \quad \mu_2 = cfxy = 0.$$

Therefore we get $cf = 0$, and we can assume $f = 0$ due to the change $(x, y, a, b, c, f) \mapsto (y, x, b, a, f, c)$, which conserves systems (12.3). We calculate $\mu_0 = \mu_1 = \mu_2 = 0$ and

$$\mu_3 = (b-a)cx^2y, \quad \mu_4 = x^2y\big[-bc^2x + (a-b)^2y\big], \quad \theta_2 = -c/4.$$

1. $\theta_2 \neq 0$. Then $c \neq 0$ and the condition $\mu_3 = \mu_4 = 0$ implies $b = a = 0$. Via the rescaling $(x, y, t) \mapsto (cx, cy, t/c)$, we arrive at the degenerate system

$$\dot{x} = x(1-y), \quad \dot{y} = -xy.$$

Clearly the underlying linear system has no finite singular point, and according to Lemma 12.1 we get the global configuration

- $(\ominus\,[|];\,\emptyset);\ \overline{\binom{1}{1}}SN,\ N^d,\ (\ominus\,[|];\emptyset).$

2. $\theta_2 = 0$. Then $c = 0$ (which implies $\mu_3 = 0$, and the condition $\mu_4 = 0$ gives $b = a$. So we arrive at the degenerate systems:

$$\dot{x} = a - xy, \quad \dot{y} = a - xy. \tag{12.4}$$

These systems possess an invariant hyperbola $xy - a = 0$ filled with singularities, which splits into two lines if $a = 0$. As for these systems we have $L_1 = 12a(x - y)^2$, according to Lemma 12.1 we arrive at the following two global configurations:

- $(\ominus[)(]; \emptyset)$; N^*, $\big(\ominus\,[)(]; \emptyset, \emptyset\big) \;\Rightarrow \big((12.4):\;\; a = 1\big)$ (if $L_1 \neq 0$);
- $(\ominus[\times]; \emptyset)$; N^*, $\big(\ominus\,[\times]; \emptyset, \emptyset\big) \;\Rightarrow \big((12.4):\;\; a = 0\big)$ (if $L_1 = 0$).

2: One real and two complex infinite singularities. So according to Lemma 6.1 we consider the family of systems:

$$\begin{aligned} \dot{x} &= a + cx + dy + gx^2 + (h+1)xy, \\ \dot{y} &= b + ex + fy - x^2 + gxy + hy^2, \end{aligned} \tag{12.5}$$

for which we calculate

$$\mu_0 = -h\big[(h+1)^2 + g^2\big], \quad \kappa = -16\big[g^2 + (h+1)(1-3h)\big]. \tag{12.6}$$

We again examine two cases: $\kappa \neq 0$ and $\kappa = 0$.

2.1: $\kappa \neq 0$. Then for systems (12.5) the conditions $\mu_0 = 0$ and $\kappa \neq 0$ imply $h = 0$. In this case we may assume $c = d = 0$ (making a translation), and then the condition $\mu_1 = -f(1 + g^2)x = 0$ gives $f = 0$. Therefore we get the systems:

$$\dot{x} = a + gx^2 + xy, \quad \dot{y} = b + ex - x^2 + gxy,$$

for which the condition $\mu_2 = (ag - b)(1 + g^2)x^2 = 0$ yields $b = ag$. Then we calculate

$$\mu_0 = \mu_1 = \mu_2 = 0, \quad \mu_3 = ae(1+g^2)x^3, \quad \mu_4 = ax^3\big[\big(e^2g + a(1+g^2)^2\big)x + e^2y\big],$$

and evidently the condition $\mu_3 = \mu_4 = 0$ is equivalent to $a = 0$. So we get the 2-parameter family of degenerate systems:

$$\dot{x} = x(gx + y), \quad \dot{y} = x(e - x + gy). \tag{12.7}$$

These systems possess the invariant line $x = 0$ filled with singularities. The corresponding linear systems have complex infinite points, and hence according to Lemma 12.1, at infinity we get the configuration ©, ©, $\big(\ominus\,[|]; \emptyset\big)$.

We examine now the corresponding underlying linear systems:

$$\dot{x} = gx + y, \quad \dot{y} = e - x + gy,$$

possessing the finite singularity $M_1\left(\dfrac{e}{1+g^2}, \dfrac{-eg}{1+g^2}\right)$. This singular point is isolated if $e \neq 0$ and lies on the singular line $x = 0$ if $e = 0$. We observe that this condition is again governed by the affine invariant polynomial θ_2, which for systems (12.7) has the value $\theta_2 = e/4$. Calculating for M_1 the corresponding eigenvalues, we obtain $\lambda_{1,2} = g \pm i$. Therefore the singular point M_1 of the above systems is a focus if $g \neq 0$ and a center if $g = 0$. Since for systems (12.7) we have $\widetilde{R} = 16gx(gx+y)$, we deduce that the above two possibilities are governed by this invariant polynomial.

So in the case $\kappa \neq 0$ we obtain the following four geometrically distinct global configurations:

- $f, (\ominus[||]; \emptyset)$; ©, ©, $(\ominus[||]; \emptyset)$ $\Rightarrow$ $\big((12.7):\; e=1,\; g=1\big)$ (if $\theta_2 \neq 0,\ \widetilde{R} \neq 0$);
- $c, (\ominus[||]; \emptyset)$; ©, ©, $(\ominus[||]; \emptyset)$ $\Rightarrow$ $\big((12.7):\; e=1,\; g=0\big)$ (if $\theta_2 \neq 0,\ \widetilde{R} = 0$);
- $(\ominus[||]; f)$; ©, ©, $(\ominus[||]; \emptyset)$ $\Rightarrow$ $\big((12.7):\; e=0,\; g=1\big)$ (if $\theta_2 = 0,\ \widetilde{R} \neq 0$);
- $(\ominus[||]; c)$; ©, ©, $(\ominus[||]; \emptyset)$ $\Rightarrow$ $\big((12.7):\; e=0,\; g=0\big)$ (if $\theta_2 = 0,\ \widetilde{R} = 0$).

2.2: $\kappa = 0$. In this case considering (12.6) we conclude that the conditions $\mu_0 = 0$ and $\kappa = 0$ imply $g = 0$ and $h = -1$. Then for systems (12.5) we may assume $e = f = 0$ by a translation, and we get the family of systems:

$$\dot{x} = a + cx + dy, \quad \dot{y} = b - x^2 - y^2.$$

For these systems we calculate

$$\mu_0 = \mu_1 = 0, \quad \mu_2 = (c^2 + d^2)(x^2 + y^2),$$

and clearly the condition $\mu_2 = 0$ gives $c = d = 0$. This implies $\mu_3 = 0$ and $\mu_4 = a^2(x^2 + y^2)^2 = 0$. So $a = 0$ and we arrive at the degenerate systems:

$$\dot{x} = 0, \quad \dot{y} = b - x^2 - y^2 \tag{12.8}$$

possessing the invariant conics $x^2 + y^2 = b$ filled with singularities. For these systems we have $L_1 = -48bx^2$, i.e. sign $(b) = -$sign (L_1) if $b \neq 0$. Therefore by Lemma 12.1 we obtain the following three global configurations:

- $(\ominus[\circ]; \emptyset)$; N^*, $(\ominus[\circ]; \emptyset, \emptyset)$ $\Rightarrow$ $\big((12.8):\; b=1\big)$ (if $\tilde{L} < 0$);
- $(\ominus[©]; \emptyset)$; N^*, $(\ominus[©]; \emptyset, \emptyset)$ $\Rightarrow$ $\big((12.8):\; b=-1\big)$ (if $\tilde{L} > 0$);
- $(\ominus[\cdot]; \emptyset)$; N^*, $(\ominus[\cdot]; \emptyset, \emptyset)$ $\Rightarrow$ $\big((12.8):\; b=0\big)$ (if $\tilde{L} = 0$).

3: Two real distinct infinite singularities. In this case according to Lemma 6.1 we consider the family of systems $(\mathbf{S}_{III})$, i.e.

$$\begin{aligned} \dot{x} &= a + cx + dy + gx^2 + hxy, \\ \dot{y} &= b + ex + fy + (g-1)xy + hy^2. \end{aligned} \tag{12.9}$$

For these systems we calculate

$$\mu_0 = gh^2, \quad \kappa = -16h^2,$$

and evidently the condition $\mu_0 = 0$ implies $gh = 0$.

3.1: $\kappa \neq 0$. Then we have $g = 0$, and since $\kappa \neq 0$ (i.e. $h \neq 0$), we may assume $h = 1$ and $c = d = 0$ due to the affine transformation $x_1 = x + d/h$, $y_1 = hy + c$. So we arrive at the systems:

$$\dot{x} = a + xy, \quad \dot{y} = b + ex + fy - xy + y^2,$$

for which $\mu_0 = 0$, and the condition $\mu_1 = -ey = 0$ yields $e = 0$. Then we have $\mu_2 = (a+b)y^2 = 0$, which gives $b = -a$. We calculate

$$\mu_3 = afy^3, \quad \mu_4 = ay^3(f^2x + ay).$$

Clearly the condition $\mu_3 = \mu_4 = 0$ is equivalent to $a = 0$, and hence we obtain the degenerate systems:

$$\dot{x} = xy, \quad \dot{y} = y(f - x + y). \tag{12.10}$$

These systems possess the invariant line $y = 0$ filled with singularities, and since $\kappa \neq 0$, by Lemma 12.1 we arrive at the configuration of infinite singularities $\overline{\binom{0}{2}}SN$, $(\ominus[|];\emptyset)$.

Additionally the underlying linear systems:

$$\dot{x} = x, \quad \dot{y} = (f - x + y),$$

possess the finite singular point $M_1(0, -f)$, which is isolated if $f \neq 0$ and lies on the singular line $y = 0$ if $f = 0$. Since for systems (12.10) we have $\theta_2 = -f/4$, it is clear that θ_2 governs this situation.

It remains to observe that M_1 is a one-direction node. So we arrive at the following two geometrically distinct global configurations:

- $n^d, (\ominus[|];\emptyset)$; $\overline{\binom{0}{2}}SN$, $(\ominus[|];\emptyset) \Rightarrow \big((12.10):\ f = 1\big)$ (if $\theta_2 \neq 0$);
- $(\ominus[|]; n^d)$; $\overline{\binom{0}{2}}SN$, $(\ominus[|];\emptyset) \Rightarrow \big((12.10):\ f = 0\big)$ (if $\theta_2 = 0$).

3.2: $\kappa = 0$. Then $h = 0$ (which implies $\mu_0 = 0$), and for systems (12.9) we calculate

$$\mu_1 = dg(g-1)^2x, \quad \widetilde{K} = 2g(g-1)x^2, \quad \tilde{L} = 8gx^2. \tag{12.11}$$

We consider two subcases: $\widetilde{K} \neq 0$ and $\widetilde{K} = 0$.

3.2.1: $\widetilde{K} \neq 0$. In this case $g - 1 \neq 0$, and by a translation we may assume $e = f = 0$. Since the condition $\mu_1 = 0$ implies $d = 0$, the condition $\mu_2 = a(g-1)^2gx^2 = 0$ yields $a = 0$. As a result we get the systems:

$$\dot{x} = cx + gx^2, \quad \dot{y} = b - xy + gxy,$$

for which we have

$$\mu_0 = \mu_1 = \mu_2 = 0, \quad \mu_3 = bc(1-g)gx^3, \quad \mu_4 = bx^3\big[bg^2x + c^2(g-1)y\big].$$

Clearly the condition $\mu_3 = \mu_4 = 0$ is equivalent to $b = 0$, and we arrive at the 2-parameter family of degenerate systems:

$$\dot{x} = x(c + gx), \quad \dot{y} = (g-1)xy. \tag{12.12}$$

These systems possess the invariant line $x = 0$ filled with singularities. We have

$$\widetilde{K} = 2g(g-1)x^2 \neq 0, \quad \tilde{L} = 8gx \quad K_2 = 48c^2(2 - g + g^2)x^2,$$

and according to Lemma 12.1 we obtain the following configurations of infinite singularities:

$$\begin{array}{lcll}
\widetilde{K} < 0,\, K_2 \neq 0 & \Rightarrow & N^f,\, (\ominus\,[|];N_3^f) & QD_7^\infty; \\
\widetilde{K} < 0,\, K_2 = 0 & \Rightarrow & N^f,\, (\ominus\,[|];N_2^f) & QD_8^\infty; \\
\widetilde{K} > 0,\, \tilde{L} < 0 \neq K_2 & \Rightarrow & S,\, (\ominus\,[|];N_3^\infty) & QD_9^\infty; \\
\widetilde{K} > 0,\, \tilde{L} < 0 = K_2 & \Rightarrow & S,\, (\ominus\,[|];N_2^\infty) & QD_{10}^\infty; \\
\widetilde{K} > 0,\, \tilde{L} > 0 \neq K_2 & \Rightarrow & N^\infty,\, (\ominus\,[|];S_3) & QD_{11}^\infty; \\
\widetilde{K} > 0,\, \tilde{L} > 0 = K_2 & \Rightarrow & N^\infty,\, (\ominus\,[|];S_2) & QD_{12}^\infty.
\end{array}$$

Additionally we observe that the underlying linear systems:

$$\dot{x} = c + gx, \quad \dot{y} = (g-1)y,$$

due to $\widetilde{K} \neq 0$ (i.e. $g(g-1) \neq 0$) possess the finite singular point $M_1(-c/g,\, 0)$. Moreover this singularity is isolated if $c \neq 0$ and lies on the singular line $x = 0$ if $c = 0$. We observe that these possibilities are governed by the invariant polynomial K_2 (see its value above).

For the finite singularity $M_1(-c/g, 0)$ we obtain $\lambda_1\lambda_2 = g(g-1) \neq 0$ and hence sign $(\lambda_1\lambda_2) =$ sign $(\widetilde{K})$. So we get a saddle if $\widetilde{K} < 0$ and a generic node if $\widetilde{K} > 0$. We recall that this singularity is isolated if and only if $K_2 \neq 0$. Considering the above detected configurations at infinity, we arrive at the following six global configurations:

- $s, (\ominus\,[|];\,\emptyset);\; N^f,\, (\ominus\,[|];N_3^f) \Rightarrow \big((12.12):\; c = 1,\; g = 1/2\big)$ (if $\widetilde{K} < 0,\, K_2 \neq 0$);
- $(\ominus\,[|];s);\; N^f,\, (\ominus\,[|];N_2^f) \Rightarrow \big((12.12):\; c = 0,\; g = 1/2\big)$ (if $\widetilde{K} < 0,\, K_2 = 0$);
- $n, (\ominus\,[|];\,\emptyset);\, S,\, (\ominus\,[|];N_3^\infty) \Rightarrow \big((12.12):\, c = 1,\, g = -1\big)$ (if $\widetilde{K} > 0,\, \tilde{L} < 0,\, K_2 \neq 0$);
- $(\ominus\,[|];n);\; S,\, (\ominus\,[|];N_2^\infty) \Rightarrow \big((12.12):\; c = 0,\; g = -1\big)$ (if $\widetilde{K} > 0,\, \tilde{L} < 0,\, K_2 = 0$);
- $n, (\ominus\,[|];\,\emptyset);\; N^\infty,\, (\ominus\,[|];S_3) \Rightarrow \big((12.12):\; c = 1,\; g = 2\big)$ (if $\widetilde{K} > 0,\, \tilde{L} > 0,\, K_2 \neq 0$);
- $(\ominus\,[|];n);\; N^\infty,\, (\ominus\,[|];S_2) \Rightarrow \big((12.12):\; c = 0,\; g = 2\big)$ (if $\widetilde{K} > 0,\, \tilde{L} > 0,\, K_2 = 0$).

3.2.2: $\widetilde{K} = 0$. Then we have $g(g-1) = 0$, and we examine two possibilities: $\tilde{L} \neq 0$ and $\tilde{L} = 0$.

1. $\tilde{L} \neq 0$. In this case we get $g = 1$, and we may assume $c = 0$ due to a translation. So we obtain the systems:

$$\dot{x} = a + dy + x^2, \quad \dot{y} = b + ex + fy,$$

for which the condition $\mu_2 = f^2x^2 = 0$ implies $f = 0$. Then calculations yield

$$\mu_0 = \mu_1 = \mu_2 = 0, \quad \mu_3 = de^2x^3, \quad \mu_4 = x^3\big[(b^2 + ae^2)x - bdey\big], \quad \kappa_1 = -32d,$$

and we examine two cases: $\kappa_1 \neq 0$ and $\kappa_1 = 0$.

(a) $\kappa_1 \neq 0$. Then $d \neq 0$ and the condition $\mu_3 = \mu_4 = 0$ is equivalent to $e = b = 0$. So doing the transformation $(x, y, t) \mapsto \big(x, (y - a)/d, t\big)$, we get the degenerate systems:

$$\dot{x} = y + x^2, \quad \dot{y} = 0$$

possessing the invariant parabola $y = -x^2$ filled with singularities. Considering Lemma 12.1, we arrive at the global configuration

- $(\ominus[\cup];\, \emptyset)$; N^*, $\big(\ominus\, [\cup]; \emptyset\big)$.

(b) $\kappa_1 = 0$. Then $d = 0$ (which implies $\mu_3 = 0$), and we obtain the systems:

$$\dot{x} = a + x^2, \quad \dot{y} = b + ex, \tag{12.13}$$

for which calculations yield

$$\mu_0 = \mu_1 = \mu_2 = \mu_3 = 0, \quad \mu_4 = (b^2 + ae^2)x^4, \quad \theta_6 = 8ex^4.$$

i. $\theta_6 \neq 0$. Then $e \neq 0$ and we could assume $e = 1$ due to the rescaling $y \to ey$. Therefore the condition $\mu_4 = 0$ gives $a = -b^2$. So we get the degenerate systems:

$$\dot{x} = (b + x)(x - b), \quad \dot{y} = b + x, \tag{12.14}$$

possessing the invariant line $x = -b$ filled with singularities. We observe that for these systems $K_2 = 384b^2x^2 \geq 0$ and $\theta_6 = 8x^4 \neq 0$. Since the underlying linear systems do not have finite singularities, according to Lemma 12.1 we arrive at the following two global configurations:

- $(\ominus[|];\, \emptyset)$; N^d, $\big(\ominus\, [|]; \overline{\binom{1}{1}} SN_3\big) \Rightarrow \big((12.14):\ b = 1\big)$ (if $K_2 > 0$);
- $(\ominus[|];\, \emptyset)$; N^d, $\big(\ominus\, [|]; \overline{\binom{1}{1}} SN_2\big) \Rightarrow \big((12.14):\ b = 0\big)$ (if $K_2 = 0$).

ii. $\theta_6 = 0$. Then for systems (12.13) we have $e = 0$ and the condition $\mu_4 = 0$ gives $b = 0$. So we get the degenerate systems:

$$\dot{x} = a + x^2, \quad \dot{y} = 0 \tag{12.15}$$

with $K_2 = -384ax^2$. Since for these systems the condition $\theta_6 = 0$ holds, by Lemma 12.1 we get the following three global configurations:

- $(\ominus[||^c]; \emptyset)$; N^*, $(\ominus\,[||^c]; \emptyset) \Rightarrow \big((12.15):\ a = 1\big)$ (if $K_2 < 0$);
- $(\ominus[|||]; \emptyset)$; N^*, $(\ominus\,[|||]; \emptyset) \Rightarrow \big((12.15):\ a = -1\big)$ (if $K_2 > 0$);
- $(\ominus[|\,2]; \emptyset)$; N^*, $(\ominus\,[|2]; \emptyset) \Rightarrow \big((12.15):\ a = 0\big)$ (if $K_2 = 0$).

2. $\tilde{L} = 0$. Then for systems (12.11) we have $g = 0$ (which implies $\mu_1 = 0 = \widetilde{K}$), and we may assume $e = f = 0$ by a translation. So we get the family of systems:

$$\dot{x} = a + cx + dy, \quad \dot{y} = b - xy, \tag{12.16}$$

for which we have

$$\mu_0 = \mu_1 = 0, \quad \mu_2 = -cdxy, \quad \mu_3 = axy(-cx + dy), \quad \kappa_1 = -32d.$$

Therefore the condition $\mu_2 = 0$ implies $cd = 0$, and we consider two cases: $\kappa_1 \neq 0$ and $\kappa_1 = 0$.

(a) $\kappa_1 \neq 0$. Then $d \neq 0$ and the condition $\mu_2 = \mu_3 = 0$ is equivalent to $c = a = 0$. Moreover via a rescaling we may assume $d = 1$, and we obtain $\mu_4 = -bxy^3 = 0$, which implies $b = 0$. Thus we arrive at the degenerate system

$$\dot{x} = y, \quad \dot{y} = -xy,$$

possessing the invariant line $y = 0$ filled with singularities. The underlying linear system does not have finite singularities, and considering Lemma 12.1 we get the global configuration

- $(\ominus[|]; \emptyset)$; $\widehat{\binom{1}{2}}\, E\!-\!H$, $(\ominus\,[|]; \emptyset)$.

(b) $\kappa_1 = 0$. Then $d = 0$ and hence for systems (12.16) with this condition, we obtain

$$\begin{aligned} \mu_0 &= \mu_1 = \mu_2 = 0, \quad \mu_3 = -acx^2y, \\ \mu_4 &= x^2y(-bc^2x + a^2y), \quad K_1 = -cx^2y. \end{aligned} \tag{12.17}$$

i. $K_1 \neq 0$. In this case $c \neq 0$ (then due to a rescaling we may assume $c = 1$), and hence the condition $\mu_3 = \mu_4 = 0$ is equivalent to $a = b = 0$. As a result we arrive at the degenerate system

$$\dot{x} = x, \quad \dot{y} = -xy,$$

possessing the invariant line $x = 0$ filled with singularities. Clearly the underlying linear system does not have finite singular points. Therefore according to Lemma 12.1 we obtain the global configuration

- $(\ominus[|]; \emptyset)$; $\overline{\binom{1}{1}} SN$, $(\ominus\,[|]; N^d)$.

ii. $K_1 = 0$. Considering (12.17) we obtain $c = 0$ (which implies $\mu_3 = 0$), and then the condition $\mu_4 = a^2x^2y^2 = 0$ gives $a = 0$. Therefore we get the degenerate systems:

$$\dot{x} = 0, \quad \dot{y} = b - xy, \tag{12.18}$$

for which $L_2 = -3bx^4$. These systems possess the invariant conic $xy = b$ filled with singularities. Clearly this conic splits into two lines if $b = 0$, and this situation is governed by the invariant polynomial L_2. Then according to Lemma 12.1 we obtain the following two global configurations:

- $(\ominus[)(];\, \emptyset)$; $\big(\ominus\, [)(]; N^*, \emptyset\big) \Rightarrow \big((12.18):\ \ b = 1\big)$ (if $L_2 \neq 0$);
- $(\ominus[\times];\, \emptyset)$; $\big(\ominus\, [\times]; N^*, \emptyset\big) \Rightarrow \big((12.18):\ \ b = 0\big)$ (if $L_2 = 0$).

4: One unique real infinite singularity. According to Lemma 6.1 in this case we consider the family of systems $(\mathbf{S}_{IV})$, i.e.

$$\begin{aligned} \dot{x} &= a + cx + dy + gx^2 + hxy, \\ \dot{y} &= b + ex + fy - x^2 + gxy + hy^2, \end{aligned} \tag{12.19}$$

for which we have $\mu_0 = -h^3$, and hence the condition $\mu_0 = 0$ gives $h = 0$. Then for these systems we calculate

$$\mu_0 = 0, \quad \mu_1 = dg^3x, \quad \widetilde{K} = 2g^2x^2,$$

and we examine two cases: $\widetilde{K} \neq 0$ and $\widetilde{K} = 0$.

4.1: $\widetilde{K} \neq 0$. Then $g \neq 0$ and hence the condition $\mu_1 = 0$ implies $d = 0$. Moreover due to $g \neq 0$, we may assume $e = f = 0$ (by a translation) and $g = 1$ (by a rescaling). So we get the systems:

$$\dot{x} = a + cx + x^2, \quad \dot{y} = b - x^2 + xy,$$

for which the condition $\mu_2 = ax^2 = 0$ yields $a = 0$. Then for these systems we calculate

$$\mu_0 = \mu_1 = \mu_2 = 0, \quad \mu_3 = -bcx^3, \quad \mu_4 = bx^3\big[(b - c^2)x + c^2y\big].$$

Evidently the condition $\mu_3 = \mu_4 = 0$ is equivalent to $b = 0$, and we arrive at the degenerate systems:

$$\dot{x} = x(c + x), \quad \dot{y} = x(y - x), \tag{12.20}$$

which possess the invariant line $x = 0$ filled with singularities. Since for these systems we have $K_2 = 48c^2x^2$, according to Lemma 12.1, at infinity we have the configuration $\big(\ominus\, [|];\overline{\binom{0}{2}}SN_3\big)$ if $K_2 \neq 0$ and $\big(\ominus\, [|];\overline{\binom{0}{2}}SN_2\big)$ if $K_2 = 0$.

Additionally we observe that the corresponding underlying linear systems possess the finite singular point $M_1(-c,\, -c)$. Moreover this singularity is isolated if $c \neq 0$ and lies on the singular line $x = 0$ if $c = 0$, and clearly these possibilities are governed by the invariant polynomial K_2.

It remains to observe that M_1 is a one-direction node. So considering Lemma 12.1, we arrive at the following two global configurations:

- $n^d, (\ominus\,[|]; \emptyset)\ (\ominus\,[|]; \overline{\binom{0}{2}}SN_3) \Rightarrow ((12.20):\ \ c = 1)\quad (\text{if } K_2 \neq 0);$
- $(\ominus\,[|]; n^d);\ (\ominus\,[|]; \overline{\binom{0}{2}}SN_2) \Rightarrow ((12.20):\ \ c = 0)\quad (\text{if } K_2 = 0).$

4.2: $\widetilde{K} = 0$. Then for systems (12.19) with $h = 0$ (as $\mu_0 = 0$), we get $g = 0$, which implies $\mu_1 = 0$. Then by a translation we may assume $e = 0$, and therefore for systems (12.19) the condition $\mu_2 = d^2x^2 = 0$ gives $d = 0$. So we arrive at the systems

$$\dot{x} = a + cx, \quad \dot{y} = b + fy - x^2, \tag{12.21}$$

for which we calculate

$$\mu_0 = \mu_1 = \mu_2 = 0, \quad \mu_3 = -c^2fx^3, \quad \mu_4 = x^3\big[(a^2 - bc^2)x + acfy\big], \quad K_1 = -cx^3.$$

1. $K_1 \neq 0$. Then $c \neq 0$ and we may assume $c = 1$ due to a rescaling. Hence the condition $\mu_3 = \mu_4 = 0$ is equivalent to $f = 0 = a^2 - b$. So $b = a^2$, and we get the degenerate systems

   $$\dot{x} = a + x, \quad \dot{y} = (a + x)(a - x),$$

   possessing the invariant line $x = -a$ filled with singularities. It is clear that the underlying linear systems do not have finite singularity. If $a \neq 0$ then via the transformation $(x, y, t) \mapsto \big(a(2x - 1), -4a^2y, t\big)$, the above systems can be brought to the system $\dot{x} = x$, $\dot{y} = x(x - 1)$. So considering Lemma 12.1 in both cases (i.e. $a = 0$ or $a = 1$) we obtain the same global configuration

   - $(\ominus[|]; \emptyset);\ (\ominus\,[|]; \widehat{\binom{1}{2}}\ E - H).$

   The unique difference is that in the case $a = 0$ the singular line $x = 0$ is a line of symmetry for the corresponding system.

2. $K_1 = 0$. In this case for systems (12.21) we get $c = 0$ (which implies $\mu_3 = 0$), and then the condition $\mu_4 = 0$ implies $a = 0$. So we get the degenerate systems:

   $$\dot{x} = 0, \quad \dot{y} = b + fy - x^2, \tag{12.22}$$

   possessing the invariant conic $b + fy - x^2 = 0$ filled with singularities. Clearly the type of this conic depends on the parameters b and f. More exactly, if $f \neq 0$ we have a parabola. In the case $f = 0$ this conic splits into two parallel lines, which are real if $b > 0$, complex if $b < 0$ and a double real line if $b = 0$. It is clear that the underlying constant system does not have finite singularities.

Additionally for systems (12.22) we have $K_3 = -6f^2x^6$, and in the case $f = 0$ (i.e. $K_3 = 0$) we obtain $L_3 = 4bx^4$. So it is clear that the above conditions are governed by these two invariant polynomials.

So considering Lemma 12.1 we arrive at the following four global configurations:

- $(\ominus[\cup];\, \emptyset)$; $\big(\ominus\,[\cup];\, N^*\big) \Rightarrow \big((12.22):\ \ b=1,\ f=1\big)$ (if $K_3 \neq 0$);
- $(\ominus\,\|^c;\, \emptyset)$; $\big(\ominus\,[\|^c];\, N^*\big) \Rightarrow \big((12.22):\ \ b=-1,\ f=0\big)$ (if $K_3 = 0,\ L_3 < 0$);
- $(\ominus[|||];\, \emptyset)$; $\big(\ominus\,[|||];\, N^*\big) \Rightarrow \big((12.22):\ \ b=1,\ f=0\big)$ (if $K_3 = 0,\ L_3 > 0$);
- $(\ominus[|\,2];\, \emptyset)$; $\big(\ominus\,[|2];\, N^*\big) \Rightarrow \big((12.22):\ \ b=0,\ f=0\big)$ (if $K_3 = 0,\ L_3 = 0$).

5: A line of singularities at infinity. According to Lemma 6.1 in this case we consider the family of systems ($\mathbf{S}_V$), i.e.

$$\dot{x} = a + cx + dy + x^2, \quad \dot{y} = b + ex + fy + xy,$$

for which $\mu_0 = 0$ and $\mu_1 = dx$. So the condition $\mu_1 = 0$ implies $d = 0$. Moreover due to a translation we may assume $e = f = 0$, and we get the family of systems:

$$\dot{x} = a + cx + x^2, \quad \dot{y} = b + xy,$$

for which we have $\mu_2 = ax^2 = 0$ (which gives $a = 0$). Then calculations yield

$$\mu_0 = \mu_1 = \mu_2 = 0, \quad \mu_3 = -bcx^3, \quad \mu_4 = bx^3(bx + c^2y).$$

Evidently the condition $\mu_3 = \mu_4 = 0$ is equivalent to $b = 0$, and we arrive at the degenerate systems:

$$\dot{x} = x(c + x), \quad \dot{y} = xy. \tag{12.23}$$

These systems possess the invariant line $x = 0$ filled with singularities.

Additionally the underlying linear systems:

$$\dot{x} = (c + x), \quad \dot{y} = y,$$

possess the finite singular point $M_1(-c, 0)$, which is isolated if $c \neq 0$ and lies on the singular line $x = 0$ if $c = 0$. Since for systems (12.23) we have $K_2 = 48c^2x^2$, it is clear that the invariant polynomial K_2 governs this situation.

Since the singular point M_1 is a star node, considering Lemma 12.1 we arrive at the following two global configurations:

- $n^*, (\ominus\,[|];\, \emptyset)$; $\big[\infty;\, (\ominus\,[|];\, \emptyset_3)\big] \Rightarrow \big((12.23):\ \ c = 1\big)$ (if $K_2 \neq 0$);
- $(\ominus\,[|]; n^*)$; $\big[\infty;\, (\ominus\,[|];\, \emptyset_2)\big] \Rightarrow \big((12.23):\ \ c = 0\big)$ (if $K_2 = 0$).

Thus all the possibilities have been considered, and this completes the proof of Theorem 12.1. □

Chapter 13

Conclusions

The goals of this chapter are firstly to sum up and highlight the core contributions of this book and secondly to cast a view towards the future indicating ways opened up by these contributions.

13.1 New concepts

This book is on singularities of quadratic differential systems. However in this book we introduced a number of concepts that are general, i.e. can be applied for any polynomial differential system. For example we introduced a new concept of *sector* of a singularity, the concept of *borsec* (which is completely new), the notion of *geometrical configuration of singularities* and the notion of *equivalence relation of geometrical configurations of singularities* as well as several other new concepts.

The classical notion of *hyperbolic singular point* was introduced more than a century ago because of its role in determining the stability of the singular point. But this concept is simply not adequate for classifying the global schemes of singularities in a geometric way. Indeed, while weak foci and weak saddles are sort of *dual geometric concepts*, each playing an important role in the bifurcation diagram of the class of quadratic systems, the weak foci are topologically unstable singularities (non-hyperbolic) while weak saddles are always stable (hyperbolic). Thus the old concept of hyperbolic singular point turns out to be unfit for a classification with geometric meaning. So we have defined the concept of *elemental* point, and we no longer use the classical *elementary* point. All the elementary points are elemental, but *elemental* gathers now all simple singularities, that is, all those of multiplicity one. Also the notion of *semi-elementary* singularity that has been used in many papers is replaced by the notion of *semi-elemental* singularity.

J. C. Artés et al., *Geometric Configurations of Singularities of Planar Polynomial Differential Systems*, https://doi.org/10.1007/978-3-030-50570-7_13

13.2 The classical versus the new way

The new concepts were used to obtain the global geometrical configurations of singularities and their bifurcation diagrams for the particular case of quadratic systems. Usually in studying a family of quadratic systems, we start by computing the singularities and the local phase portraits around them. We provide a way of obtaining directly these local phase portraits but the information we give goes way beyond this, by adding information such as for example order of foci and saddles, multiplicity, types of nodes and tangency information, etc. This ampler information contains essential items for studying how the systems bifurcate. Moreover, a substantial part of the global tools we use (such as for example the affine comitants μ_i, etc.) have geometrical meaning, unlike the coefficients of the normal forms.

The classical way to study a family of quadratic differential systems is to start by calculating the coordinates of the singularities of the systems (finite and infinite) with respect to an adequate normal form for the family.

Next, we obtain the types of singularities in the respective normal forms, and their variation depending on the parameters involved. Finally the separatrix connection, limit cycles and types of graphics are studied. The expressions of the coordinates of the singularities for quadratic systems involve iterated square roots or roots of quartic polynomials, and for singularities at infinity cubic roots and iterated cubic roots, and the calculations involved in producing the phase portraits could be quite tricky. For certain cases such as the family **QW3** of quadratic systems with a third order weak focus, the calculations were simple. But for the family **QW2** of quadratic systems with a second order weak focus and its set of limiting points within **QS**, which includes the sets **QW3** and **QC** (quadratic systems with a center), another normal form needed to be used because of the complexity of the algebraic hypersurfaces on which the systems with centers were located in the first normal form. In the second normal form these hypersurfaces turn out to be just planes. This new normal form did not provide an easy way to calculate the singularities. So the use of invariants to find the bifurcation diagram was a step in the good direction (used in [19] for the first time).

The class **QS** of quadratic differential systems modulo the action of the group of affine transformations and time rescaling is five-dimensional. Just as for the case of **QW2**, in general while studying a subfamily of **QS** and **QS** itself, several normal forms are usually needed to cover the whole family. Clearly we need to be able to pass from one normal form to another normal form since two distinct normal forms may have a non-empty intersection in **QS**. In the literature such a passage is seldom considered, bifurcation diagrams being provided only for the fragments of the family described by the respective normal forms without specifying any procedure as to how the results in the two normal forms are to be glued.

Classical works have dealt with high codimension cases leading to normal forms with few parameters. In the first part of these works, algebraic calculations were involved for determining the kind of singularities, while the second part contained an analytic study. But these classical tools fail when trying to study more generic cases. In particular no four-dimensional subfamily of **QS** has so far been studied.

The new way to begin the study of a subfamily of **QS** is by applying the algorithm devised in this work which renders redundant the calculation of singularities and their kinds. This algorithm is expressed in terms of affine invariant polynomials and this makes it independent of the particular presentation of the systems.

13.3 An algorithm to study the singularities of quadratic differential systems

There are two parts involved in the study of families of quadratic differential systems. The first part is the study of singularities, and the second part deals with the separatrix connections and the limit cycles. The first part is a completely algebraic study, while the second part involves analytic and numerical methods.

In this book we studied the global schemes of singularities for the whole class **QS**. This work provides us with an *algorithm* for calculating the geometrical configuration of singularities for any subfamily of **QS** given with respect to any normal form, which allows us to skip the effective calculations of singularities in the specific normal forms. This algorithm is given in terms of invariant polynomials, which must be calculated for the respective family of systems. By applying the algorithm given in this book, we obtain not only the geometrical configurations of singularities but also their bifurcation diagram for the specific family of quadratic systems.

Due to the numerous of calculations involved in studying families of quadratic systems, it often happens that certain cases are missed. Since the calculation of singularities were done in this work for all quadratic systems, skipping cases can no longer occur.

In order to be able to use the algorithm, we provide all the necessary invariant polynomials as well as the Mathematica tools for searching the specific configurations of singularities we look for (see Appendix B).

If for a family several normal forms are needed in order to study it, then because the algorithm is done in terms of invariant polynomials, the transfer of information from one normal form to another can easily be done. This is an affinely invariant classification because it is done using conditions expressed in terms of invariant polynomials, conditions that are invariant with respect to the affine group and time rescaling.

13.4 The topological configurations of singularities

We have here the global classification of geometrical configurations of singularities for the whole class **QS**. Each such configuration contains much more information than the topological configuration of singularities. It contains, for example, the distinction between foci and nodes, which the topological configuration ignores. It also contains information about multiplicities, order of weak foci or saddles, distinction among nodes etc., all this being completely ignored in the topological configurations of singularities. However, for the bifurcation diagram of the class **QS**, this information is essential. The topological configurations of singularities could be obtained from the geometrical configurations, and this has already been done in [27]. From the 1765 geometrically distinct geometrical configurations we have obtained exactly 208 topological configurations.

13.5 The study of the quadratic differential systems modulo limit cycles

Since the study of limit cycles is in general a very difficult problem, it is advisable to separate the study of limit cycles and consider the problem of giving the complete set of phase portraits of quadratic differential systems modulo limit cycles. We first recall that a quadratic differential system has at most two eyes of limit cycles and that inside a limit cycle (or an eye of limit cycles) there is only one singularity, which is a focus. Consider a quadratic system (S), and let $TP(S)$ be its topological phase portrait on the Poincaré disc. We define below the notion of *phase portrait modulo limit cycles* of a quadratic system (S).

We first consider a quadratic system (S) without limit cycles. In this case the phase portrait modulo limit cycles of (S) is just $TP(S)$. If (S) has limit cycles, we introduce an equivalence relation on the phase space of $TP(S)$ as follows: Any point outside the two eyes of limit cycles is equivalent only to itself, and any two points inside an eye of limit cycles or on the largest limit cycle of the corresponding eye are equivalent. This equivalence relation yields a quotient topological space from the phase space, call it $TP(S)/\sim$ and a quotient map $TP(S) \to TP(S)/\sim$ transferring the oriented orbits of $TP(S)$ onto oriented orbits on $TP(S)/\sim$. The topological object obtained in this way is called the *phase portrait modulo limit cycles* of the quadratic system (S). The notion of topological equivalence between phase portraits modulo limit cycles is clearly easy to define. Two phase portraits modulo limit cycles of systems (S) and (S') are *topologically equivalent modulo limit cycles* if there is a homeomorphism $f : TP(S)/\sim \to TP(S')/\sim$ that carries oriented orbits to oriented orbits, conserving the orientation.

By a *structurally stable quadratic system modulo limit cycles* we mean a system $(S) \in$ **QS** that possesses a neighborhood in the space $\mathbb{R}^{12}$ of coefficients of quadratic differential systems such that for any system (S') in this neighborhood, (S) and (S') have topologically equivalent phase portraits modulo limit cycles.

The study of structurally stable quadratic systems modulo limit cycles has a long history. A first attempt by Tavares dos Santos [290] to obtain all such phase portraits led to a total of 25 such phase portraits. But after more work done by other mathematicians, more such phase portraits modulo limit cycles were discovered. The final and correct result was obtain by Artés, Kooij and Llibre [12] who obtained 44 phase portraits modulo limit cycles for the systems in **QS** that are structurally stable systems. For this result they mainly used topological and combinatorial methods. First they used such tools to make a list of possibilities for such phase portraits. Then for each one of these possibilities, they needed to produce examples showing that the corresponding phase portrait modulo limit cycles can be realized in the quadratic class. In the case that such examples turned out to be impossible to find, they proved (mainly using the bound of the number of contact points of orbits with straight lines) that the corresponding possibility cannot actually occur.

Artés and Llibre continued this study and devised a program for obtaining all phase portraits modulo limit cycles of systems in **QS**. After the work on structurally stable systems modulo limit cycles, they continued this work with the study of the class of codimension one structurally unstable quadratic differential systems modulo limit cycles [17]. The compilation of all possible phase portraits was relatively easy, but the location of examples for each one of them, or proofs of their impossibility, were not so easy. The study of many new families of quadratic systems (obtained mainly thanks to polynomial invariants) has been needed to obtain a wide collection of realizable phase portraits, and so reduce the number of those that needed to be proven impossible. A. C. Rezende also contributed for part of this work. A system (S) is a *codimension one structurally unstable system modulo limit cycles* if it possesses a neighborhood with the property that any system (S') in this neighborhood is either structurally stable or topologically equivalent modulo limit cycles to (S).

This work was also mainly topological and combinatorial. The proofs to show the impossibility of some phase portraits modulo limit cycles of codimension one proceed by showing that at least one of the structurally stable systems modulo limit cycles that appear in perturbations of the unstable element, is not realizable. The unstable element is either a saddle-node or a separatrix connection. By perturbing this unstable element we may obtain two (sometimes just one by symmetry) of the topologically possible 72 structurally stable phase portraits of quadratic systems (modulo limit cycles). But 28 of them are not realizable. And the non-realizability of the 28 structurally stable phase portraits implies the non-realizability of many more topologically possible phase portraits of codimension one. Also the bound of contact points continued to be an important tool to use. We may have graphics in this class, actually only three types of graphics: a homoclinic loop through a saddle or a graphic with two singularities at infinity on the Poincaré disc, formed by a part of the line at infinity connecting the two singularities and an orbit that may happen to be a straight line or not. This work has recently been published [17].

The next step in this program is in the process of being done by Artés, Llibre, Oliveira and Rezende, and it concerns *codimension two structurally unstable modulo limit cycles* quadratic systems (S), which are systems possessing a neighborhood containing only structurally stable, codimension one unstable systems or systems that are topologically equivalent modulo limit cycles to (S). For the study of this class, the topological/combinatorial and the geometric methods need to be combined since some subfamilies need to be studied with algebro-geometric and analytic tools while others need a topological/combinatorial study. The possible codimension two bifurcations that could be produced are either the existence of a codimension two singularity like a cusp (nilpotent), a triple node or a triple saddle (both semi-elemental) or combinations of two phenomena from the following list: saddle-nodes finite or infinite, finite saddle to saddle connections, finite-infinite saddle connections, connections of two opposite saddles at infinity and connections of two non-opposite saddle-nodes at infinity. One of the two possible cases of a combination of a finite saddle-node and an infinite saddle-node was completely studied by Artés, Oliveira and Rezende [35].

Up to codimension two, the singularities involved are sufficiently easy so that there is no need of a deep geometric study. But starting at codimension three and higher, our book on geometrical configurations of singularities provides a "backbone" from which to assemble the separatrix connections and obtain all possible combinations of singularities and connections up to the codimension being considered.

It is clear that the realization of the program to obtain all phase portraits of systems in **QS** modulo limit cycles is a long-term project. But we think that the tools presented in this book along with methods used in [17] are sufficient for completing this program.

Appendix A

Table of notation

In Table A we list the categories of singularities, starting with the elemental ones, in the first column, the notation for the various singularities in the second column and then the description of the sectors for each one of the singularities in the third column. We next give in the fourth column the Jacobian matrix of the respective singularity, and lastly in the fifth column we give the drawing of the local phase portrait around the singularity. In some cases, if the singularity is at infinity, we present the local phase portrait on both sides of the line at infinity, if we consider it to be helpful. The list is complete for some cases, but it is only a sketch of some examples for an intricate singularity at infinity or for a degenerate system since there are too many possibilities to consider. The given examples however may be enough for understanding those cases that are not included in this table. The reader is advised to print or copy the pages of this table and have them handy to be consulted whenever necessary.

J. C. Artés et al., *Geometric Configurations of Singularities of Planar Polynomial Differential Systems*, https://doi.org/10.1007/978-3-030-50570-7

Table A

Notation for Singular Points				
	lowercase are finite singularities			
	UPPERCASE ARE INFINITE SINGULARITIES			
CLASS	NOTATION	DESCRIPTION	JACOBIAN	PHASE PORTRAIT
ELEMENTAL	s	STRONG SADDLE	$\begin{pmatrix} -1 & 0 \\ 0 & 2 \end{pmatrix}$	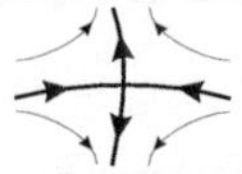
	n	GENERIC NODE	$\begin{pmatrix} 1 & 0 \\ 0 & 2 \end{pmatrix}$	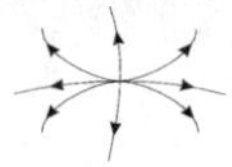
	f	STRONG FOCUS	$\begin{pmatrix} 1 & 2 \\ -2 & 1 \end{pmatrix}$	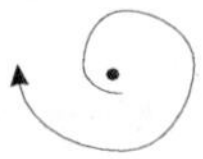
	n^d	ONE–DIRECTION NODE	$\begin{pmatrix} 1 & 1 \\ 0 & 1 \end{pmatrix}$	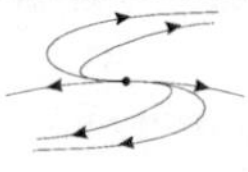
	n^*	STAR NODE	$\begin{pmatrix} 1 & 0 \\ 0 & 1 \end{pmatrix}$	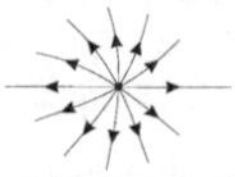
	N^f	INFINITE NODE WITH $\lambda_1 > \lambda_2$	$\begin{pmatrix} 2 & 0 \\ 0 & 1 \end{pmatrix}$	
	N^∞	INFINITE NODE WITH $\lambda_1 < \lambda_2$	$\begin{pmatrix} 1 & 0 \\ 0 & 2 \end{pmatrix}$	
	$s^{(i)}$	WEAK SADDLE OF ORDER $i = 1, 2, 3$	$\begin{pmatrix} -1 & 0 \\ 0 & 1 \end{pmatrix}$	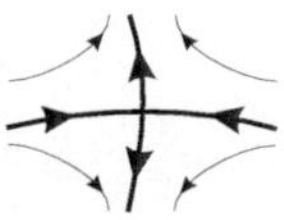
	$	INTEGRABLE SADDLE		
	$f^{(i)}$	WEAK FOCUS OF ORDER $i = 1, 2, 3$	$\begin{pmatrix} 0 & 1 \\ -1 & 0 \end{pmatrix}$	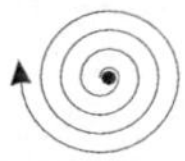
	c	CENTER		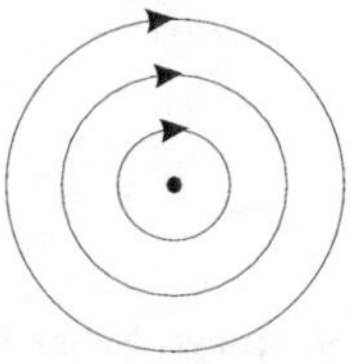
	$c^{\odot}$	ISOCHRONUS CENTER		
	$c^{\oplus}$	UNIFORM ISOCHRONUS CENTER		

Table A *(continued)*

CLASS	NOTATION	DESCRIPTION	JACOBIAN	PHASE PORTRAIT
	$\overline{sn}_{(2)}$ or $\overline{sn}_{(4)}$	SADDLE-NODE		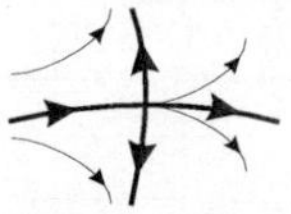
	$\overline{n}_{(3)}$	NODE	$\begin{pmatrix} 0 & 0 \\ 0 & 1 \end{pmatrix}$	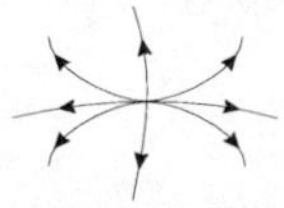
	$\overline{s}_{(3)}$	SADDLE		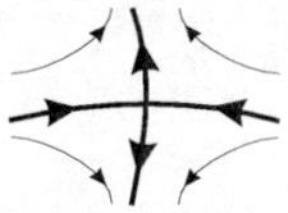
	$\overline{\binom{0}{2}}\,SN$	INFINITE SADDLE-NODE		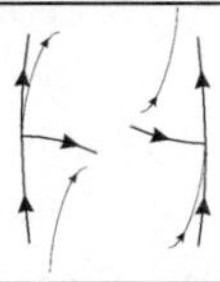
SEMI-ELEMENTAL	$\overline{\binom{0}{3}}\,S$	INFINITE SADDLE	$\begin{pmatrix} 0 & 0 \\ 0 & 1 \end{pmatrix}$	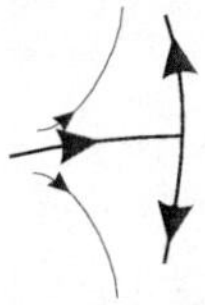
	$\overline{\binom{0}{3}}\,N$	INFINITE NODE		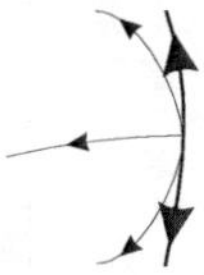
	$\overline{\binom{1}{1}}\,SN$	INFINITE SADDLE-NODE		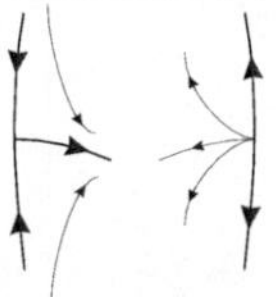
	$\overline{\binom{2}{1}}\,S$	INFINITE SADDLE	$\begin{pmatrix} 1 & 0 \\ 0 & 0 \end{pmatrix}$	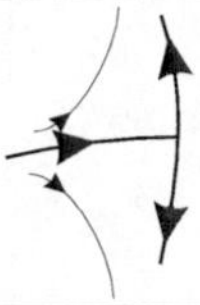
	$\overline{\binom{2}{1}}\,N$	INFINITE NODE		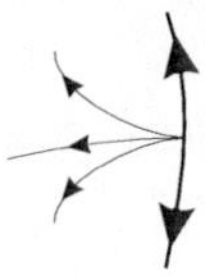

Table A *(continued)*

CLASS	NOTATION	DESCRIPTION	JACOBIAN	PHASE PORTRAIT
	$\widehat{cp}_{(2)}$	CUSP		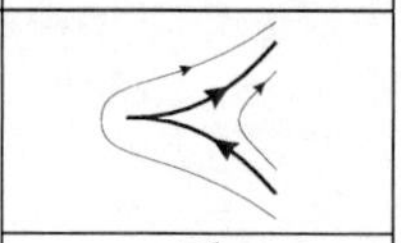
	$\widehat{es}_{(3)}$ or $\widehat{es}^{-1}_{(3)}$	ELLIPTIC-SADDLE		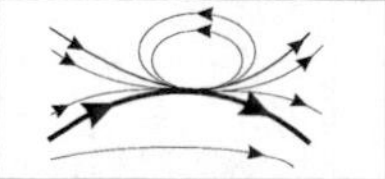
	$\hat{s}_{(3)}$	SADDLE		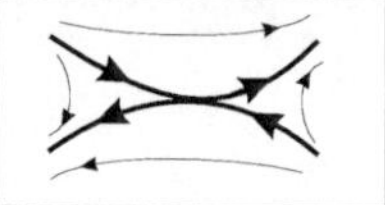
	$\hat{n}_{(3)}$	NODE		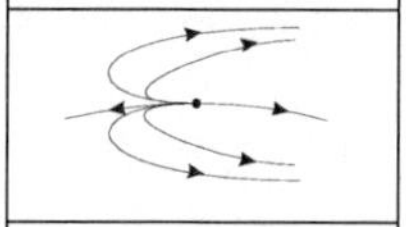
	$\widehat{sn}_{(4)}$	SADDLE-NODE		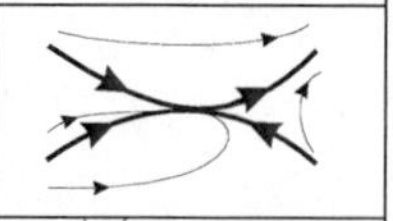
NILPOTENT	$\widehat{\binom{1}{2}}\ H_\lambda H H_\lambda - H$	INFINITE SADDLE	$\begin{pmatrix} 0 & 1 \\ 0 & 0 \end{pmatrix}$	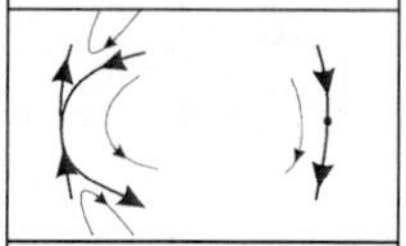
	$\widehat{\binom{1}{2}}\ \widehat{P}_\lambda H \widehat{P}_\lambda - E$	INFINITE ELLIPTIC-SADDLE		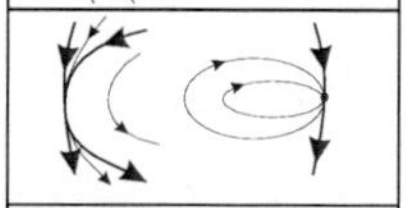
	$\widehat{\binom{1}{2}}\ \widehat{P}_\lambda E \widehat{P}_\lambda - H$	INFINITE ELLIPTIC-SADDLE		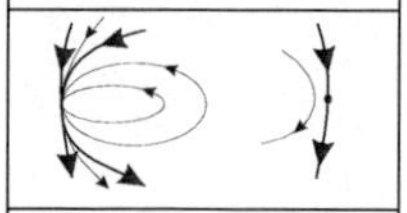
	$\widehat{\binom{1}{2}}\ H - E$	INFINITE ELLIPTIC-SADDLE		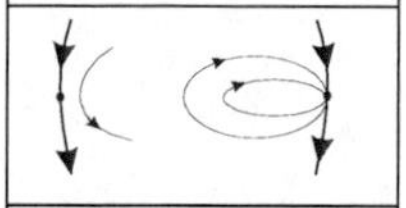
	$\widehat{\binom{2}{2}}\ \widehat{P}_\lambda \widehat{P} H_\lambda - H$	INFINITE SADDLE-NODE		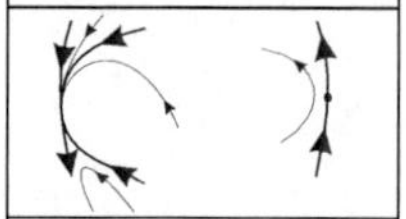
	$\widehat{\binom{1}{3}}\ H_\lambda H \widehat{P}_\lambda - \widehat{P}$	INFINITE SADDLE-NODE		

Table A *(continued)*

CLASS	NOTATION	DESCRIPTION	JACOBIAN	PHASE PORTRAIT
	H	HYPERBOLIC		
	E	ELLIPTIC		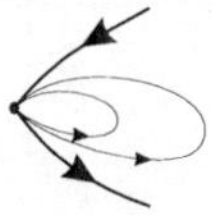
	$\widetilde{P}$	PARABOLIC		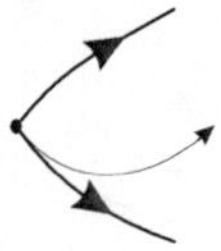
	$\widehat{P}$	PARABOLIC		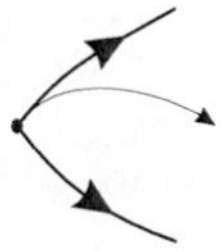
SECTORS	$H_{\curlywedge}$ or $H_{\curlywedge}^{*}$	CUSP-LIKE HYPERBOLIC or STAR-LIKE HYPERBOLIC		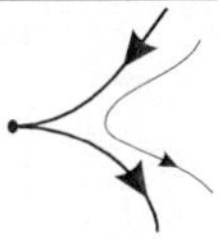
	$E_{\curlywedge}$	CUSP-LIKE ELLIPTIC		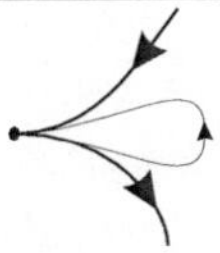
	$\widetilde{P}_{\curlywedge}$	CUSP-LIKE PARABOLIC		
	$\widehat{P}_{\curlywedge}$	CUSP-LIKE PARABOLIC		
	E^{*}	STAR-LIKE ELLIPTIC		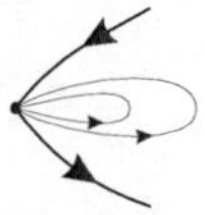
	P^{*}	STAR-LIKE PARABOLIC		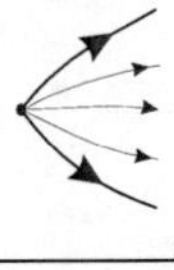

Table A *(continued)*

CLASS	NOTATION	DESCRIPTION	JACOBIAN	PHASE PORTRAIT
	$pphpph_{(4)}$	TWO HYPERBOLIC FOUR PARABOLIC		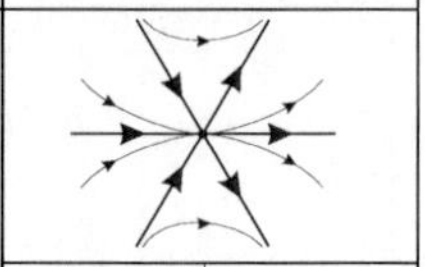
	$phph_{(4)}$	TWO HYPERBOLIC TWO PARABOLIC		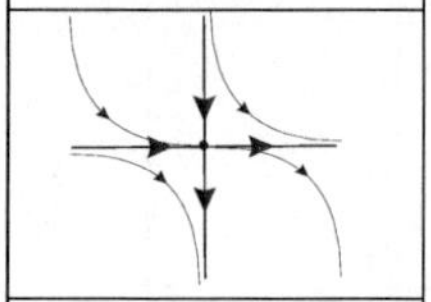
	$hh_{(4)}$	TWO HYPERBOLIC "FLAT" SINGULARITY		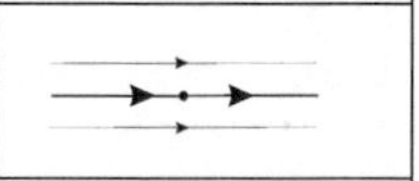
INTRICATE	$hhhhhh_{(4)}$	SIX HYPERBOLIC	$\begin{pmatrix} 0 & 0 \\ 0 & 0 \end{pmatrix}$	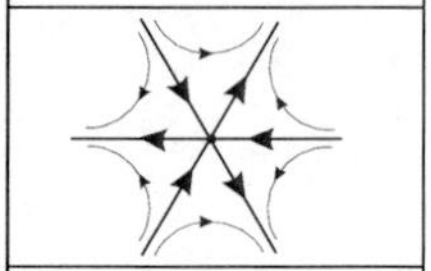
	$ppeppe_{(4)}$	TWO ELLIPTIC FOUR PARABOLIC		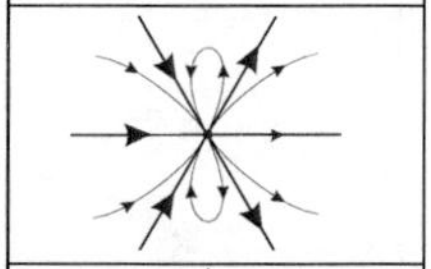
	$pepe_{(4)}$	TWO ELLIPTIC TWO PARABOLIC		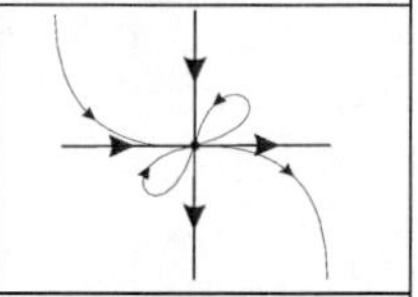
	$ee_{(4)}$	TWO ELLIPTIC		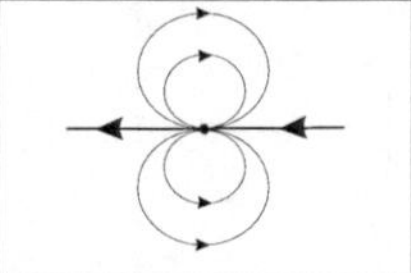

Table A *(continued)*

Class	Notation	Description	Jacobian	Phase Portrait
	$\binom{3}{3}\ \widehat{P}_{\curlywedge} E E \widehat{P}_{\curlywedge} - \widehat{P}\,\widehat{P}$			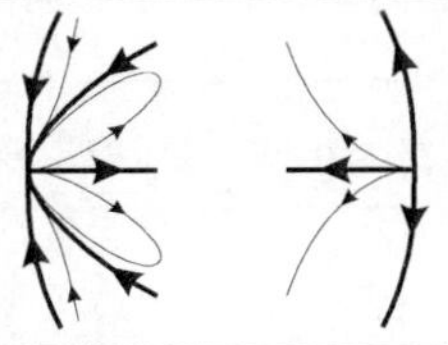
	$\binom{3}{3}\ H_{\curlywedge} \widehat{P}\,\widehat{P} H_{\curlywedge} - \widehat{P}\,\widehat{P}$			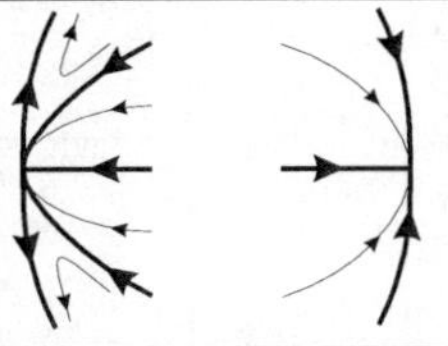
	$\binom{3}{3}\ \widehat{P}_{\curlywedge} \widehat{P}\,\widehat{P}\,\widehat{P}_{\curlywedge} - H H$			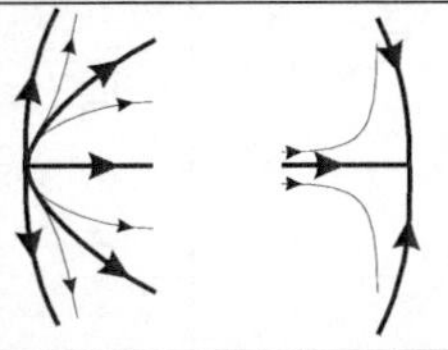
INTRICATE	$\binom{4}{3}\ \widehat{P}_{\curlywedge} \widehat{P} - \widehat{P}\,\widehat{P}_{\curlywedge}$		$\begin{pmatrix} 0 & 0 \\ 0 & 0 \end{pmatrix}$	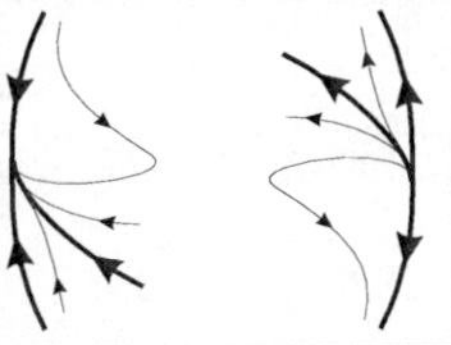
	$\binom{4}{3}\ E \widehat{P}_{\curlywedge} H - H \widehat{P}_{\curlywedge} E$			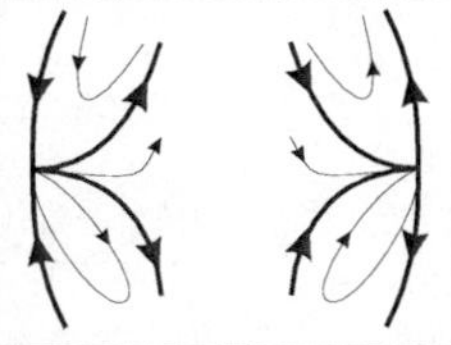
	$\binom{4}{2}\ H - H$			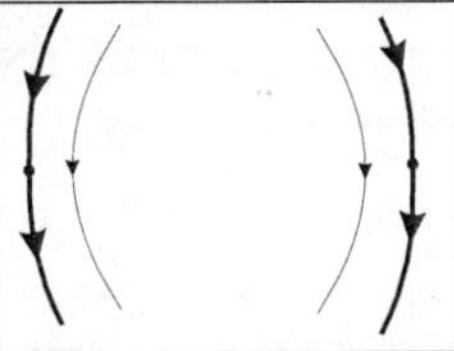
	$\binom{3}{2}\ \widehat{P} H \widehat{P} - \widehat{P}\,\widehat{P} E$			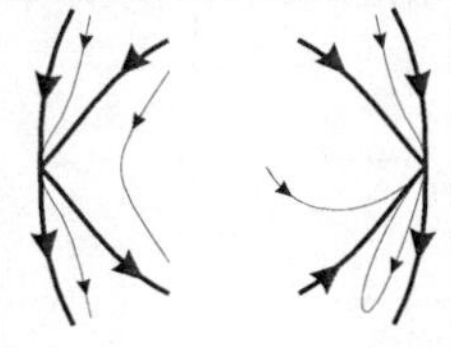

Table A *(continued)*

Class	Notation	Description	Jacobian	Phase Portrait
	$[\infty;\ \emptyset]$	NO SINGULARITY "UNDERNEATH" INFINITY		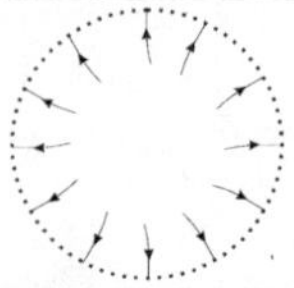
	$[\infty;\ N]$	NODE SINGULARITY "UNDERNEATH" INFINITY		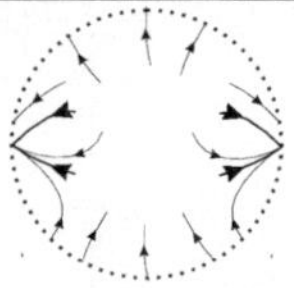
	$[\infty;\ N^d]$	N^d SINGULARITY "UNDERNEATH" INFINITY		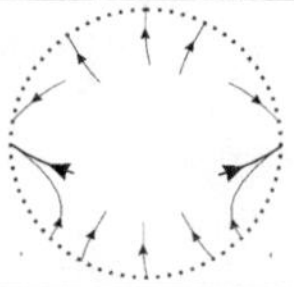
$C_2 \equiv 0$	$[\infty;\ S]$	SADDLE SINGULARITY "UNDERNEATH" INFINITY		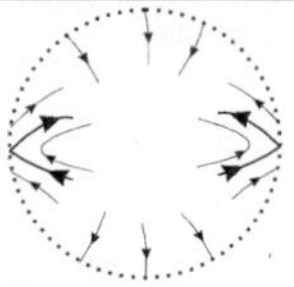
	$[\infty;\ C]$	CENTER (or FOCUS) "UNDERNEATH" INFINITY		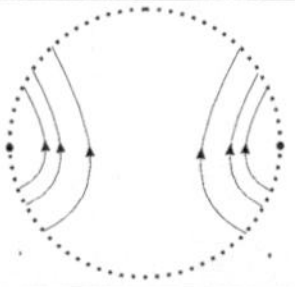
	$[\infty;\ \overline{\binom{2}{0}}SN]$	SADDLE–NODE "UNDERNEATH" INFINITY		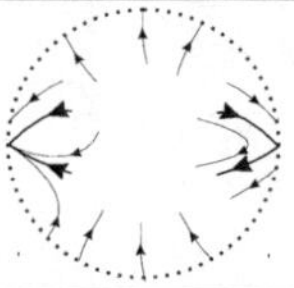
	$[\infty;\ \widehat{\binom{3}{0}}ES]$	ELLIPTIC-SADDLE "UNDERNEATH" INFINITY		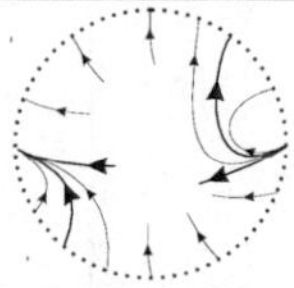

Table A *(continued)*

CLASS	NOTATION	DESCRIPTION	JACOBIAN	PHASE PORTRAIT
	$f, (\ominus\,[\,|\,];\emptyset)$	LINE FILLED WITH SINGULARITIES AND FOCUS OUTSIDE		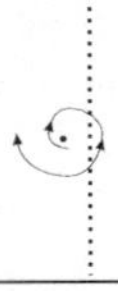
	$(\ominus\,[\,|\,]; f)$	LINE FILLED WITH SINGULARITIES AND FOCUS ON IT		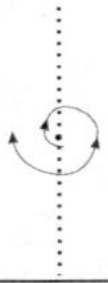
	$(\ominus\,[\,||\,];\emptyset)$	TWO PARALLEL LINES FILLED WITH SINGULARITIES		
DEGENERATE	$(\ominus\,[\times];\emptyset)$	TWO INTERSECTING LINES FILLED WITH SINGULARITIES		
	$(\ominus\,[\circ];\emptyset)$	REAL ELLIPSE FILLED WITH SINGULARITIES		
	$(\ominus\,[\cup];\emptyset)$	PARABOLA FILLED WITH SINGULARITIES		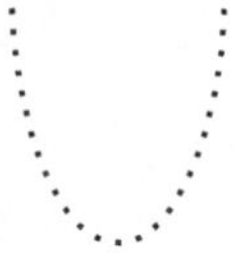
	$(\ominus\,[)(];\emptyset)$	HYPERBOLA FILLED WITH SINGULARITIES		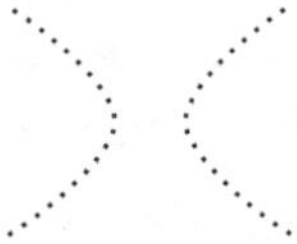
	$(\ominus\,[\cdot];\emptyset)$	TWO COMPLEX INTERSECTING LINES FILLED WITH SINGULARITIES		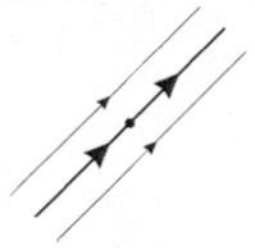

Appendix B

Manual of Mathematica tools

The set of all invariants and the diagrams given in this book provide a complete algorithm that, if followed correctly, gives the global geometrical configurations of singularities for any family of quadratic systems, in whatever normal form it may be presented. If all the coefficients of a quadratic system are known, when applied to just this system, the algorithm provides the entire configuration of singularities without any further intervention of the user.

In some cases of a family of quadratic systems depending on a few parameters, we may end up with just one configuration of singularities for the whole family. This may occur if some restrictions among the parameters are known. When the restrictions upon parameters are not enough to uniquely determine a single configuration, the values of some of the affinely invariant polynomials (polynomial invariants/comitants) depending on the parameters will give an algebraic bifurcation diagram that will classify the possible configurations of singularities according to the different values of the corresponding parameters.

Since this task may be quite hard to do by hand, both the computation of the 83 invariant polynomials needed and the path through all the diagrams to arrive at the different possibilities that a family of systems may have, we have created an interactive program to help the user research the configuration (in the case that it is unique) or the possible configurations and their bifurcation diagram (in the case that there are several such configurations).

The program is written in Mathematica, and it will be offered freely through the link http://mat.uab.es/∼artes/articles/config.nb (for unlimited time in principle). Some previous knowledge of Mathematica is recommended.

The program contains a complete set of instructions to compute all the invariant polynomials needed, as well as some examples (from 0– to 2–parameter families of quadratic systems).

We give here a short description of how the program works, as well as some important concepts and indications about how to use the program along with some examples. We warn the reader that the given examples do not cover all the

J. C. Artés et al., *Geometric Configurations of Singularities of Planar Polynomial Differential Systems*, https://doi.org/10.1007/978-3-030-50570-7

possibilities that may occur in some of the branches of the diagrams. The user is advised to experiment with the program and to create new functions that may improve the ones that we propose. If the reader detects some gaps, we will be thankful to be informed so as to improve the program.

We want to point out that it was not our intention to produce a program, commercial or not. This is simply the code produced after many years of computations and diagrams, and we think it is worthwhile to offer it in this form to the mathematical community so that each user will not have to recode it all. We offer it freely as it is. We know that it is far from perfect, not on the level of the standard performance of a commercial program.

B.1 How to initiate the program

The user must start by loading the file and executing the first block of instructions (Block to be activated). This block contains thousands of lines of code preparing all invariant polynomials, the functions to study families of systems and many other tools to make changes of variables of systems. It also contains hundreds of different normal forms. We do not describe here all these tools. We leave to the reader the opportunity to explore them.

After the block has been activated, there will appear on the screen an introductory section called "Preliminary steps", which helps to understand the notation used. Next come four examples that illustrate the way to use the tools. It is advisable that the user opens a new window in Mathematica so as to develop his/her own examples. There is no need to add in this new window the whole block to every example to be done. The block can be executed in one window and the example in a different window, but always under the same kernel of Mathematica.

B.2 Notation

We now highlight some important details that the user must take into account when using the Mathematica files for the algorithm.

The variables used to describe the systems will always be of the type "x[i][t]", where "x" is the name of the variable and the typical letters to use here are "x", "y", "z" or "u", the integer "i" is normally set to be 1 or 0 (if we need to do some changes of variables, it may be modified) and "t" always represents the time. The time "t" is the variable with respect to which we differentiate, and this symbol cannot be changed. Normally the two working variables can easily be accessed with the variables "var1" and "var2" in the program.

The parameters of a quadratic system are initially ("a00", "a10", "a01", "a20", "a11", "a02", "b00", "b10", "b01", "b20", "b11", "b02"), but they can be changed by numbers, or by other parameters, or even by algebraic formulas involving parameters. It is advisable to use letters for the parameters starting always with “c”, for example “ca”, “cc” or “ch”. This advise cannot be ignored since if you start using letters or words that have already been used in the main block, the calculations will become mixed up and the program will not work at all, or it will give wrong results. We have avoided the use of words starting with “c” or “d”, so that they may be used as parameters. However the reader is advised not to use words like “change” or “determinant” as parameters since they clearly have other meanings. It is also important to recall the tensor form in which the family of quadratic systems is written. So the coefficients of the monomial “x[1][t]*y[1][t]” in the two equations are “2a11” and “2b11”.

The equations of the family of systems are normally stored in the variables “eq1” and “eq2”: these variables do not just contain the polynomials defining the differential equations, but also include the left side of these equations involving the differentiation of the variables. Then, if one needs to access only the right parts of the equation, one must type “eq1[[2]]” and “eq2[[2]]”.

So, a general family of systems may look like the following:

```
In:=    eq1
Out:=   x[1]'[t] == a00 + a10 x[1][t] + a20 x[1][t]^2
        + a01 y[1][t] + 2 a11 x[1][t] y[1][t] + a02 y[1][t]^2
In:=    eq2
Out:=   y[1]'[t] == b00 + b10 x[1][t] + b20 x[1][t]^2 +
        b01 y[1][t] + 2 b11 x[1][t] y[1][t] + b02 y[1][t]^2
```

In this book we mostly use the following normal form, which is stored in the Mathematica variable “changea00toca”:

```
In:=  change=changea00toca
Out:= {a00 → ca, a10 → cc, a01 → cd, a20 → cg, a11 → ch, a02 → ck,
       b00 → cb, b10 → ce, b01 → cf, b20 → cl, b11 → cm, b02 → cn}
In:=  eq1[[2]]/. change
Out:= ca + cc x[1][t] + cg x[1][t]^2 + cd y[1][t] + 2 ch x[1][t] y[1][t] + ck y[1][t]^2
In:=  eq2[[2]]/. change
Out:= cb + ce x[1][t] + cl x[1][t]^2 + cf y[1][t] + 2 cm x[1][t] y[1][t] + cn y[1][t]^2
```

We recall that the variables are stored in “var1” and “var2”. If it is too long to read the equations in this way, one can always do the following at printing time:

```
In:=    var1
Out:=   x[1][t]
In:=    var2
Out:=   y[1][t]
```

In:= **renamevar = {var1 → x, var2 → y}**
Out:= {x[1][t] → x, y[1][t] → y}
In:= **eq1[[2]]/. change/. renamevar**
Out:= ca + cc x + cg x^2 + cd y + 2 ch x y + ck y^2
In:= **eq2[[2]]/. change/. renamevar**
Out:= cb + ce x + cl x^2 + cf y + 2 cm x y + cn y^2

It is important to respect all these rules, otherwise the program will not work since the code will not be able to extract the important items from the equations that we propose.

B.3 Examples

Example B.1. We start with a simple, concrete quadratic system to get acquainted with the tools. Consider the system $\dot{x} = -3x^2$, $\dot{y} = 2y$. We need to introduce this system in our Mathematica variable "change", where the system to be studied is always stored.

In:= **change={a00 → 0, a10 → 0, a01 → 0, a20 → -3, a11 → 0, a02 →0, b00 → 0, b10 → 0, b01 → 2, b20 → 0, b11 → 0, b02 → 0}**
Out:= {a00 → 0, a10 → 0, a01 → 0, a20 → -3, a11 → 0, a02 → 0, b00 → 0, b10 → 0, b01 → 2, b20 → 0, b11 → 0, b02 → 0}

We now ask our main function "comp" to evaluate all the invariant polynomials on this system.

In:= **comp**
Out:= {Null}

The function makes no output, but now all 83 needed invariant polynomials (and even some additional ones) are loaded with their corresponding values. For example, for the invariants W_3, $\widetilde{M}$, $\mathbf{T}$ and μ_2, we have the following:

In:= **{WW3, WM, T, μ2}**
Out:= {0, -72 x[1][t]^2, 0, 36 x[1][t]^2}

Although we made an effort to use exactly the same set of characters to describe the polynomial invariants in TEX and in Mathematica, some features could not be exactly reproduced. Simply there is not always exact coincidence between the font used in TEX and the one offered by Mathematica. Some of our invariant polynomials have remained as we programmed them some years ago. For example, for faster typing, we never use subindex in Mathematica, and we prefer to type the subindex as a number at the end, for example μ2, which stands for μ_2. Mathematica variables do not use the tilde that we have in some TEX variables, and many of our Mathematica variables start with "W". We include, at this point of the Mathematica file, a complete list with the names of the 83 variables that we need (see (5.13)) in the same order as they appear in this book.

However, since this way to show several invariant polynomials may not be adequate when they are complicated (or too many) expressions, we suggest the following structure that can be adapted to as many invariants as needed:

In:= **Print[Column**[{{"μ0=", **Factor**[μ0]}, {"μ2=", **Factor**[μ2]}, {"η=",**Factor**[η]}, {"**WK1**=", **Factor[WK1]**}}]]
{μ0= 0}
{μ2= 36 x[1][t]^2}
{η= 0}
{WK1= 6 x[1][t]^2 y[1][t]}

We are now ready to obtain our first configuration of singularities. We are going to use the function CSP[{},{},{}]; (CSP stands for Configuration of Singular Points). For the moment we leave empty the three possible parameters that we may add, but we must set the brackets "{}" with nothing inside. It is important to add the ";" at the end, otherwise you obtain a lot of unnecessary stuff.

In:= **CSP**[{},{},{}];
{μ_0= 0}
{μ_1= 0}
{μ_2= 36 x[1][t]^2}
{**U**= 0}
{κ= 0}
{$\widetilde{K}$= 0}
{$\tilde{L}$= 72 x[1][t]^2}
{$\mathcal{T}_4$= 0}
{η= 0}
{κ_1= 0}

Finite singularities : $\overline{sn}_{(2)}$

{η= 0}
{$\widetilde{M}$= -72 x[1][t]^2}
{μ_0= 0}
{μ_1= 0}
{μ_2= 36 x[1][t]^2}
{κ= 0}
{$\widetilde{K}$= 0}
{$\tilde{L}$= 72 x[1][t]^2}
{κ_1= 0}
{K_2= 0}
{θ_6= 0}

Infinite singularities : $\binom{2}{2}\ \widehat{P}H-\widehat{P}H,\ N^*$ (Config. 19)

In fact " CSP[{},{},{}];" is equivalent to two independent functions finiteGeo[{},{},{}]; and infiniteGeo[{},{},{}]; each one of which makes the study of singularities in the respective part. This is why some information about invariants may appear repeated.

The information given by each one of the two "Geo" functions is formed by all the invariants that are relevant for the result. This information is followed by the concrete configuration of singularities, finite and infinite. If one takes the values of the invariants obtained and looks for the precise diagram where these invariants are used, one arrives at the same configuration of singularities, finite and infinite. The programs have moved along all the branches of the diagrams, checking if the invariants were positive, negative, zero or non-zero, until they obtained a branch that is described completely by the deployed invariants. The invariants that are not needed for this branch are then not shown. The expression "(Config. 19)" that appears accompanying the configuration of infinite singularities indicates the topological phase portrait at infinity according to Figure 6.1. The reader is reminded that several different geometrical configurations may lead to the same topological configuration.

Example B.2. Let us make a small change in the studied system by adding one parameter. Let us study the family of systems $\dot{x} = -3x^2$, $\dot{y} = 2y + ay^2$. So we introduce the following:

In:= **change={a00 → 0, a10 → 0, a01 → 0, a20 → -3, a11 → 0, a02 →0, b00 → 0, b10 → 0, b01 → 2, b20 → 0, b11 → 0, b02 → ca}; comp**
You are working with at most 1 different parameters, or quite complicated parameters
Depending on the position of the parameters, in the worst case, this should last less than 1 minute
Out:= {Null}

The program detects that there is one parameter present and warns you that computations may take a little longer than in the previous case. The number of parameters that the program detects is not always right, since it may be confused by complicated algebraic formulas with the parameters, or even by radical numbers. It will try to make the calculations anyway. Only in the case that it thinks that there are more than four parameters, the program will tell you that it is not advisable to work with the version "comp", and it will suggest other ways that may imply numerical calculations that reduce the precision of the results. Even though the program "comp" as well as others can deal with the 12 parameters of a quadratic system, it is not advisable to use so many parameters for several reasons. First, the results for some invariants may be huge with polynomial expressions of up to degree 14 in 14 variables (12 parameters plus the variables x and y), and so they are difficult to manipulate. Secondly, the computation time needed may be very great. Thirdly, if you have so many parameters, the most probable outcome is that the family of systems may also have many different configurations, and so it will be very hard, even with the help of the computer, to produce the bifurcation diagram that gives you all possibilities. The programs we offer work well when operating on systems with two free parameters or less, but it can also work, with some effort, with three or four parameters.

Let us now return to our system. If we try to find the configuration of singularities, we proceed as before:

In:= **CSP**[{},{},{}];
μ0= 9 ca^2 =0??
μ1= 36 ca x[1][t]=0??
μ0= 9 ca^2 =0??
...

If one gets no definitive conclusion, and instead of that just gets a set of questions marks, whether some invariant is or is not zero, or is or is not positive, this means that the final decision may depend on that invariant. The list of questions posed may be very long, as in this example, or can be quite short. There can also appear repetitive questions as shown above. The reader should not be bothered by that. Normally one just needs to deal with the first question, offer a restriction on the parameters that solves it, and run the input again to see if something else is needed. If in one of the following examples, a long list of questions appear as here, we will just type the first one.

In this case, the first question is whether μ_0 is zero or not. It is clear that if $a \neq 0$ $(= 0)$, then $\mu_0 \neq 0$ (respectively $= 0$). The user will need to help the program at this point. Now is the moment to explain the reason of the three elements of the function "CSP[{},{},{}]" that we have left empty up to now. These are three lists that can contain an unlimited number of polynomial expressions in the parameters and variables, or directly the name of some invariant polynomials, which we fix either to be zero, to be positive, or to be non-zero, respectively. It does not matter at all the order in which you put expressions in the same list. This may be easily understood with an example. We set $a \neq 0$.

In:= **CSP**[{},{},{**ca**}];
−**PR**= -1679616 ca^2 x[1][t]^6 >0??
{η= 9 ca^2}
{μ_0= 9 ca^2}
{θ= 0}
{θ_1= 0}
{θ_3= 0}
{θ_4= 0}
Infinite singularities : S, N^*, N^* (Config. 5)

Here we may notice that the program is still asking one question related to the finite singular points for which no configuration is given. But with the current restrictions, the program is already able to know the configuration at infinity. Now, it is important to notice here the difference between an invariant and a comitant.

An invariant only depends on the parameters, but a comitant is a polynomial expression that depends on the parameters but also the polynomial variables x and y. The computer is not able to distinguish between the parameters and the variables x and y. For the program all of them are free parameters. And when the program asks if a certain comitant is zero, or if it is positive, it asks if it is so generically, i.e. not just for some concrete values of x and y. As in the example above, it is not valid to say that **PR** is going to be zero if $x = 0$. One can play with the parameters but not with the variables. So, with the current conditions, it is already known that **PR** cannot be zero, but we must inform the computer that x cannot be zero. In general it is advisable to always add the variables "var1" and "var2" to the list of expressions that cannot be zero. In others cases, one may need to add, for example, that "var1+var2" is not zero, or any other homogeneous polynomial expression in x and y.

In:= **CSP[{},{},{ca,var1,var2}];**
{μ_0= 9 ca^2}
{**D**= 0}
{**T**= 0}
{**PR**= -1679616 ca^2 x[1][t]^6}
{$\mathcal{T}_4$= 144 ca^2}

Finite singularities : $\overline{sn}_{(2)}$, $\overline{sn}_{(2)}$

{η= 9 ca^2}
{μ_0= 9 ca^2}
{θ= 0}
{θ_1= 0}
{θ_3= 0}
{θ_4= 0}

Infinite singularities : $\mathrm{S}, \mathrm{N}^*, \mathrm{N}^*$ (Config. 5)

As we see, we have completed the study obtaining the configuration of singularities with the restriction $a \neq 0$.

If we now set $a = 0$ by "CSP[{ca},{},{var1,var2}];" this leads to the system of Example B.1 and so we will obtain the same configuration as there.

In:= **CSP[{ca},{},{var1,var2}];**
{μ_0= 9 ca^2}
{μ_1= 36 ca x[1][t]}
{μ_2= 36 x[1][t]^2}
{**U**= 0}
{κ= -144 ca^2}
{$\widetilde{K}$= -12 ca x[1][t]y[1][t]}
{$\tilde{L}$= 8 (9 x[1][t]^2+9 ca x[1][t] y[1][t]+ca^2 y[1][t]^2)}

$\{\mathcal{T}_4$= 144 ca$^2\}$
$\{\eta$= 9 ca$^2\}$
$\{\kappa_1$= -96 ca$\}$

Finite singularities : $\overline{sn}_{(2)}$

$\{\eta$= 9 ca$^2\}$
$\{\widetilde{M}$= -8 (9 x[1][t]^2+3 ca x[1][t] y[1][t]+ca^2 y[1][t]^2))$\}$
$\{\mu_0$= 9 ca$^2\}$
$\{\mu_1$= 36 ca x[1][t]$\}$
$\{\mu_2$= 36 x[1][t]$^2\}$
$\{\kappa$= -144 ca$^2\}$
$\{\widetilde{K}$= -12 ca x[1][t]y[1][t]$\}$
$\{\tilde{L}$= 8 (9 x[1][t]^2+9 ca x[1][t] y[1][t]+ca^2 y[1][t]^2))$\}$
$\{\kappa_1$= -96 ca$\}$
$\{K_2$= 384 ca^2 x[1][t]$^2\}$
$\{\theta_6$= -48 x[1][t]^{2}y[1][t]$^2\}$

Infinite singularities : $\binom{2}{2}\ \widehat{P}H - \widehat{P}H,\ N^*$ (Config. 19)

Note in this case that even though we have set $a = 0$, the program has not substituted the value but has continued to operate with it, and whenever an expression that contains the parameter appeared, the program already applied the restriction and decided its value. Another option would have been to set directly in the variable "change" where the system is stored, the value $a = 0$. If there are several parameters, it is advisable to follow this technique because then you reduce a lot the time and effort of computation, but if there is only one parameter that you may wish to try with different values, it is better the way we have it done here.

The program may prompt two types of questions: whether some invariant is zero or not, or whether it is positive or not. In the first case, the user simply confirms one of the possibilities (if there is only one) or proposes both in independent tests as to obtain all possible answers. In the second case, up to three possibilities (> 0, < 0 and $= 0$) may need to be tested. There are some invariants whose only relevance is whether they are zero or not and their sign is always irrelevant (for example $\mathcal{T}_4$). But there are other invariants whose sign may be relevant in some cases. The user need not memorize which invariants have a relevance in their sign. Just provide the help required by the program to decide exactly the answer to what it asks. With time and experience the user may gather knowledge about some of the most common invariants.

Example B.3. Let us consider another simple 1–parameter family. We study now the systems $\dot{x} = a - 3x^2$, $\dot{y} = 2y$. So we introduce the following:

In:= **change={a00 → ca, a10 → 0, a01 → 0, a20 → -3, a11 → 0, a02 →0, b00 → 0, b10 → 0, b01 → 2, b20 → 0, b11 → 0, b02 → 0}; comp**
You are working with at most 1 different parameters, or quite complicated parameters
Depending on the position of the parameters, in the worst case, this should last less than 1 minute
Out:= {Null}
In:= **CSP**[{},{},{**var1,var2**}];
−**U**= -1728 ca x[1][t]^4 y[1][t]^2 >0??
$-K2$= -10368 ca x[1][t]^2 >0??
$-K2$= -10368 ca x[1][t]^2 >0??

Now the first question is whether **U** is positive or not. We have already set by default that x and y cannot be zero. It is clear that if $a > 0$ (respectively < 0, $= 0$), then $\mathbf{U} < 0$ (respectively > 0, $= 0$).

In:= **CSP**[{},{**ca**},{**var1,var2**}];
$\mathcal{B}_3$= -2592 (-1+3 ca) x[1][t]^2 = 0??
{η= 0}
{$\widetilde{M}$= -72 x[1][t]^2}
{μ_0= 0}
{μ_1= 0}
{μ_2= 36 x[1][t]^2}
{κ= 0}
{$\widetilde{K}$= 0}
{$\tilde{L}$= 72 x[1][t]^2}
{κ_1= 0}
{K_2= 10368 ca x[1][t]^2}
{θ_6= 0}

Infinite singularities : $\binom{2}{2}\ \widehat{P}\widehat{P}H - \widehat{P}\widehat{P}H,\ N^*$ (Config. 19)

Again the conditions are enough to know about infinite singularities but not yet about finite singularities. We must decide if $\mathcal{B}_3$ could or could not be zero, and this depends on $-1 + 3a$. Let us first fix it to be non-zero.

In:= **CSP**[{},{**ca**},{**var1,var2,3ca-1**}];
{μ_0= 0}
{μ_1= 0}
{μ_2= 36 x[1][t]^2}
{**U**= 1728 ca x[1][t]^4 y[1][t]^2}
{κ= 0}
{$\widetilde{K}$= 0}
{W_4= 0}
{$\mathcal{T}_4$= 0}

$\{\eta = 0\}$
$\{\tilde{L} = 72\ x[1][t]^2\}$
$\{W_8 = 0\}$
$\{\kappa_1 = 0\}$
$\{\mathcal{B}_3 = -2592\ (-1+3\ ca)\ x[1][t]^2\}$

Finite singularities : s, n

$\{\eta = 0\}$
$\{\widetilde{M} = -72\ x[1][t]^2\}$
$\{\mu_0 = 0\}$
$\{\mu_1 = 0\}$
$\{\mu_2 = 36\ x[1][t]^2\}$
$\{\kappa = 0\}$
$\{\widetilde{K} = 0\}$
$\{\tilde{L} = 72\ x[1][t]^2\}$
$\{\kappa_1 = 0\}$
$\{K_2 = 10368\ ca\ x[1][t]^2\}$
$\{\theta_6 = 0\}$

Infinite singularities : $\binom{2}{2}\ \widehat{P}\widehat{P}H-\widehat{P}\widehat{P}H,\ N^*$ (Config. 19)

Now the configuration of singularities is fully exposed.

Next we must consider the other options. First let us check what happens if $3a - 1 = 0$. Notice that now we do not even need to say that $a > 0$ since this is already forced: $a = 1/3 > 0$.

In:= **CSP[{3ca-1},{},{var1,var2}];**
$\{\mu_0 = 0\}$
$\{\mu_1 = 0\}$
$\{\mu_2 = 36\ x[1][t]^2\}$
$\{\mathbf{U} = 1728\ ca\ x[1][t]^4\ y[1][t]^2\}$
$\{\kappa = 0\}$
$\{\widetilde{K} = 0\}$
$\{W_4 = 0\}$
$\{\mathcal{T}_4 = 0\}$
$\{\eta = 0\}$
$\{\tilde{L} = 72\ x[1][t]^2\}$
$\{W_8 = 0\}$
$\{\kappa_1 = 0\}$
$\{\mathcal{B}_3 = -2592\ (-1+3\ ca)\ x[1][t]^2\}$
$\{\theta_6 = 0\}$

Finite singularities : $\$$, n^*

$\{\eta = 0\}$
$\{\widetilde{M} = -72\ x[1][t]^2\}$
$\{\mu_0 = 0\}$
$\{\mu_1 = 0\}$
$\{\mu_2 = 36\ x[1][t]^2\}$
$\{\kappa = 0\}$
$\{\widetilde{K} = 0\}$
$\{\tilde{L} = 72\ x[1][t]^2\}$
$\{\kappa_1 = 0\}$
$\{K_2 = 10368\ ca\ x[1][t]^2\}$
$\{\theta_6 = 0\}$

Infinite singularities : $\binom{2}{2}\ \widehat{P}\widehat{P}H - \widehat{P}\widehat{P}H,\ N^*$ (Config. 19)

So when $a > 0$ and tends to $1/3$, the generic node and saddle turn into the star node and the integrable saddle, respectively.

What happens now if $a < 0$? There is no list assigned to expressions that are negative, however a negative sign can be added to variables in the list for positive.

In:= **CSP**[{},**{-ca}**,**{var1,var2}**];
$\{\mu_0 = 0\}$
$\{\mu_1 = 0\}$
$\{\mu_2 = 36\ x[1][t]^2\}$
$\{\mathbf{U} = 1728\ ca\ x[1][t]^4\ y[1][t]^2\}$

Finite singularities : ©, ©

$\{\eta = 0\}$
$\{\widetilde{M} = -72\ x[1][t]^2\}$
$\{\mu_0 = 0\}$
$\{\mu_1 = 0\}$
$\{\mu_2 = 36\ x[1][t]^2\}$
$\{\kappa = 0\}$
$\{\widetilde{K} = 0\}$
$\{\tilde{L} = 72\ x[1][t]^2\}$
$\{\kappa_1 = 0\}$
$\{K_2 = 10368\ ca\ x[1][t]^2\}$
$\{\theta_6 = 0\}$

Infinite singularities : $\binom{2}{2}\ H - H,\ N^*$ (Config. 8)

Notice now that the program has not asked about $\mathcal{B}_3$ as before. For $a < 0$ the program already knows that $\mathcal{B}_3$ cannot be zero, or simply $\mathcal{B}_3$ is not neeed.

If we set $a = 0$, " CSP[{ca},{},{var1,var2}];", we will be in the system of

Example B.3, and so we will obtain the same configuration as then.

Please notice that if you try to cheat the program and enter some set of contradictory statements, the program will detect it. This function is heavily based on the Mathematica function "Assuming", which is able to detect algebraic contradictions. Just try what happens if you enter, for example, "CSP[{3ca-1},{-ca},{var1,var2}];".

With this we have completed another study of a 1–parameter family. Let us now consider a 2–parameter family where the power of this tool will be even more appreciated.

Example B.4. Now we are going to study the family of systems:

$$\begin{aligned} \frac{dx}{dt} &= -1 + 2bx + 3x^2 + 2axy - y^2, \\ \frac{dy}{dt} &= -1 - 2by + ax^2 - 2xy - 3ay^2. \end{aligned} \tag{B.1}$$

In:= **change={a00 → -1, a10 → 2cb, a01 → 0, a20 → 3, a11 → ca, a02 →-1, b00 → -ca, b10 → 0, b01 → -2cb, b20 → ca, b11 → -1, b02 → -3ca}; comp**
You are working with at most 3 different parameters, or quite complicated parameters
Depending on the position of the parameters, in the worst case, this should last less than 1 minute
Out:= {Null}

The program mentions that we are working with at most three parameters when we are just working with two. In any case, it does not say anything wrong. As mentioned, some expressions in the parameters may look different to the program, and it may consider that there are more parameters, but this has not handicapped its power to compute invariants anyway. We start the study:

In:= **CSP[{},{},{var1,var2}];**
μ_0= -12 (-1+ca)^2 (1+ca)^2=0??
...??

The program is asking us if μ_0 is zero or not. So, let us assume first that it is not zero. One can simply put invariant μ_0 on the third list, but if there is a simple equivalent algebraic expression, we will help the program since this same expression may appear in other invariants and thus will shorten the time it needs to run. So, we set $a^2 - 1 \neq 0$.

In:= **CSP[{},{},{var1,var2,ca^2-1}];**
D= 49152 (-1+ca)^2 (1+ca)^2 (-3+3 ca^2-cb^2) (-1+ca^2+cb^2)^3=0??
...??

Now it needs to know if **D** is zero or not, and this may depend both on the hyperbola $-3 + 3a^2 - b^2 = 0$ and/or on the circle $-1 + a^2 + b^2 = 0$. So let us set both defining expressions non-zero. Normally one first considers the generic possibilities, and later goes to the non-generic ones. But it can be done otherwise with no problem.

In:= **CSP[{},{},{var1,var2,ca^2-1,(-3+3 ca^2-cb^2)(-1+ca^2+cb^2)}];**
D= 49152 (-1+ca)^2 (1+ca)^2 (-3+3 ca^2-cb^2) (-1+ca^2+cb^2)^3 >0??
...??

And now it asks if **D** is positive or not. Why did it not say it before? The program follows step by step the diagrams of the configurations of singularities from this book and tests one by one all the options. If at some moment it needs to know if one invariant is zero or not, then it will prompt this question. If once this is answered, the program enters in a branch of the diagram where the sign is also needed, then it will prompt for the sign. Maybe one could think that it would be better to check the sign of every invariant before entering into the diagram, but in this case the program would ask for conditions one after the other until it had perfectly controlled all 83 invariants though it may not really need them all.

So, we check first the case $(-3+3a^2-b^2)(-1+a^2+b^2)>0$.

In:= **CSP[{},{(-3+3 ca^2-cb^2)(-1+ca^2+cb^2)},{var1,var2,ca^2-1}];**
$-\widetilde{K}$= 4 (3 x[1][t]^2+ca^2 x[1][t]^2+8 ca x[1][t] y[1][t]+y[1][t]^2+
3 ca^2 y[1][t]^2) > 0??
$\mathcal{T}_4$= -1024 (1+ca^2)^2 (-4 ca^2-cb^2+ca^2 cb^2) >0??
...

Now, $\widetilde{K}$ is a comitant that is only relevant when $\mu_0<0$, as we previously fixed by saying that $a^2-1\neq 0$. And in this case it is always sure that $\widetilde{K}$ cannot be decomposed into factors of smaller degree, that is, the discriminant of this homogeneous polynomial in x and y must be negative. We have to help the program by telling it if the comitant is always positive or negative, and the simplest way to do this is by checking the sign of the coefficient of x or y at the maximum power. In this case we have that the coefficient of x^2 is $4(3+a^2)$, and since the expression we get is for $-\widetilde{K}$, we can affirm that $\widetilde{K}$ is negative. It is not worth trying to describe all the terms of $\widetilde{K}$ in "CSP[{},{},{}];". It is better to simply say that $\widetilde{K}$ is negative. The variable that contains this comitant is **WK**. So we input the following:

In:= **CSP[{},{(-3+3 ca^2-cb^2)(-1+ca^2+cb^2),-WK},{var1,var2,ca^2-1}];**
$\mathcal{T}_4$= -1024 (1+ca^2)^2 (-4 ca^2-cb^2+ca^2 cb^2) =0??
...

Here we need to decide also if $\mathcal{T}_4$ may be zero or not. Let us start by assuming that it is not zero:

In:= **CSP[{},{(-3+3 ca^2-cb^2)(-1+ca^2+cb^2),-WK},{var1,var2,ca^2-1, (-4 ca^2-cb^2+ca^2 cb^2)}];**
{μ_0= -12 (-1+ca)^2 (1+ca)^2}
{**D**= 49152 (-1+ca)^2 (1+ca)^2 (-3+3 ca^2-cb^2) (-1+ca^2+cb^2)^3}
{$\widetilde{K}$= 4 (3 x[1][t]^2+ca^2 x[1][t]^2+8 ca x[1][t] y[1][t]+y[1][t]^2+3 ca^2 y[1][t]^2}
{$\mathcal{T}_4$= -1024 (1+ca^2)^2 (-4 ca^2-cb^2+ca^2 cb^2)}

Finite singularities : s, s, ©, ©

{η= 4 (125+262 ca^2+125 ca^4)}
{μ_0= -12 (-1+ca)^2 (1+ca)^2}
{κ= 64 (35+58 ca^2+35 ca^4)}

Infinite singularities : N^f, N^f, N^f (Config. 1)

And now assume that $\mathcal{T}_4$ is zero:

In:= **CSP[{(-4 ca^2-cb^2+ca^2 cb^2)},{(-3+3 ca^2-cb^2)(-1+ca^2+cb^2),-WK}, {var1,var2,ca^2-1}];**
$\mathcal{T}_3$= -512 (-1+ca) (1+ca) (1+ca^2) cb (-1+ca^2+cb^2) =0??
...

Now we must decide whether $\mathcal{T}_3$ is zero or not. The only way for $\mathcal{T}_3 = 0$ in the current conditions is if $b = 0$. So, let us try both possibilities:

In:= **CSP[{(-4 ca^2-cb^2+ca^2 cb^2)},{(-3+3 ca^2-cb^2)(-1+ca^2+cb^2),-WK}, {var1,var2,ca^2-1,cb}];**
{μ_0= -12 (-1+ca)^2 (1+ca)^2}
{**D**= 49152 (-1+ca)^2 (1+ca)^2 (-3+3 ca^2-cb^2) (-1+ca^2+cb^2)^3}
{$\widetilde{K}$= 4 (3 x[1][t]^2+ca^2 x[1][t]^2+8 ca x[1][t] y[1][t]+y[1][t]^2+3 ca^2 y[1][t]^2}
{$\mathcal{T}_4$= -1024 (1+ca^2)^2 (-4 ca^2-cb^2+ca^2 cb^2)}
{$\mathcal{T}_3$= -512 (-1+ca) (1+ca) (1+ca^2) cb (-1+ca^2+cb^2)}
{$\mathcal{F}_1$= 192 ca (1+ca^2)}

Finite singularities : $s^{(1)}$, s, ©, ©

{η= 4 (125+262 ca^2+125 ca^4)}
{μ_0= -12 (-1+ca)^2 (1+ca)^2}
{κ= 64 (35+58 ca^2+35 ca^4)}

Infinite singularities : N^f, N^f, N^f (Config. 1)

In:= **CSP[{(-4 ca^2-cb^2+ca^2 cb^2),cb},{(-3+3 ca^2-cb^2)(-1+ca^2+cb^2),-WK}, {var1,var2,ca^2-1}];**
{μ_0= -12 (-1+ca)^2 (1+ca)^2}

{**D**= 49152 (-1+ca)^2 (1+ca)^2 (-3+3 ca^2-cb^2) (-1+ca^2+cb^2)^3}
{$\widetilde{K}$= 4 (3 x[1][t]^2+ca^2 x[1][t]^2+8 ca x[1][t] y[1][t]+y[1][t]^2+3 ca^2 y[1][t]^2}
{$\mathcal{T}_4$= -1024 (1+ca^2)^2 (-4 ca^2-cb^2+ca^2 cb^2)}
{$\mathcal{T}_3$= -512 (-1+ca) (1+ca) (1+ca^2) cb (-1+ca^2+cb^2)}
{$\mathcal{T}_2$= -64 (-1+11 ca^2-11 ca^4+ca^6+7 cb^2+2 ca^2 cb^2+7 ca^4 cb^2)}
{$\mathcal{F}$= -16 (-1+ca) (1+ca) (1+ca^2) cb}
{$\mathcal{B}$= -128 (-1+11 ca^2-11 ca^4+ca^6+4 ca^2 cb^2)}

Finite singularities : s, s, $©^\rho$, $©^\rho$

{η= 4 (125+262 ca^2+125 ca^4)}
{μ_0= -12 (-1+ca)^2 (1+ca)^2}
{κ= 64 (35+58 ca^2+35 ca^4)}

Infinite singularities : N^f, N^f, N^f (Config. 1)

Even though it does not seem that we have continuity between the geometrical configurations of the cases $b = 0$ and $b \neq 0$, note that $b = \mathcal{T}_4 = 0$ also implies $a = 0$. So this is an isolated component of the quartic, formed by just one point. Another part of the quartic is when $b \neq 0$, which implies $a = \pm b/\sqrt{b^2-4}$, and so this is only defined if $b^2 > 4$. So, since the quartic has more than one connected components, the problem of the inconsistency in continuity is solved.

In the previous case, it may have been easier for the computer if we had set $b = 0$ in the original system and if we recalculated again all invariants (with the command "comp") with just the parameter a. But then, in order to continue our study, we would have had to start again with both free parameters. If the computations become too complicated, it is advisable to use a reduction of conditions. Otherwise, the way in which we have proceeded is faster.

Now we consider the other case **D** < 0, that is, $(-3+3a^2-b^2)(-1+a^2+b^2) < 0$.

In:= **CSP[{},{-(-3+3 ca^2-cb^2)(-1+ca^2+cb^2)},{var1,var2,ca^2-1}];**
−**R**= -384 (-1+ca)^2 (1+ca)^2 (3 x[1][t]^2...-3 ca^2 cb^2 y[1][t]^2) >0??
R= -384 (-1+ca)^2 (1+ca)^2 (3 x[1][t]^2...-3 ca^2 cb^2 y[1][t]^2) =0??
...??

We draw attention to the fact that in the expressions of **R** and **S** we introduce "..." so as not to give their long full expressions. The program is asking if **R** is positive, but we see below that it is also interested in knowing if it may be zero. And it is also asking the same questions for **S**. It is important now to recall here the geometric meaning of the comitants **R** and **S**. When the comitant **D** > 0 they have no meaning. But if it is negative (and $\mu_0 \neq 0$), then if both **R** and **S** are positive, there are four real finite singular points and if at least one of them is negative, then the system has four finite complex singular points. The comitants **R** and **S** are usually quite large. In fact in this example, if you follow the Mathematica file, the comitants **R** and **S** that you see here are relatively small compared to other cases. Due to this reason, it is not advisable to introduce them directly in the function "CSP[{},{},{}];" without a previous checking, as it is further explained. Maybe some combinations can be proved to be impossible in an easier way, and we avoid the computation time of testing all possible configurations. So, we will simply ask Mathematica using its own function " Assuming" if there are some incompatibilities or not. And then, for those that are compatible, we continue the work.

In:= **Assuming[{ca^2-1 $\neq$ 0,-(-3+3 ca^2-cb^2)(-1+ca^2+cb^2) $>$ 0, R $>$ 0, S $>$ 0, var1 $\neq$ 0, var2 $\neq$ 0 }, Print[1]]**
1

Since there is no problem, it means that the conditions are compatible. However, if one tries to do the remaining combinations of signs for **R** and **S**, a contradiction appears. Remember that we must ask not only if **R** and **S** can be negative, but also if they can be zero, and we can do this as follows:

In:= **Assuming[{ca^2-1 $\neq$ 0,-(-3+3 ca^2-cb^2)(-1+ca^2+cb^2) $>$ 0, R $\leq$ 0, var1 $\neq$ 0, var2 $\neq$ 0 }, Print[1]]**
1

that yields a contradiction. We do the same with $\mathbf{S} \leq 0$ and get another contradiction. So we do not need to try them in "CSP[{},{},{}];" and we apply only the combination that is possible:

In:= **CSP[{},{-(-3+3 ca^2-cb^2)(-1+ca^2+cb^2),R, S},{var1,var2,ca^2-1}];**
$\mathcal{T}_4$= -1024 (1+ca^2)^2 (-4 ca^2-cb^2+ca^2 cb^2)=0??
. . . ??

We are questioned again on $\mathcal{T}_4$. Assume first that it is non-zero:

In:= **CSP[{},{-(-3+3 ca^2-cb^2)(-1+ca^2+cb^2),R, S}, {var1,var2,ca^2-1,(-4 ca^2-cb^2+ca^2 cb^2)}];**
{μ_0= -12 (-1+ca)^2 (1+ca)^2}
{**D**= 49152 (-1+ca)^2 (1+ca)^2 (-3+3 ca^2-cb^2) (-1+ca^2+cb^2)^3}
{$\widetilde{K}$= 4 (3 x[1][t]^2+ca^2 x[1][t]^2+8 ca x[1][t] y[1][t]+y[1][t]^2+3 ca^2 y[1][t]^2}
. . .

Finite singularities : s, s, s, n

{η= 4 (125+262 ca^2+125 ca^4)}
{μ_0= -12 (-1+ca)^2 (1+ca)^2}
{κ= 64 (35+58 ca^2+35 ca^4)}

Infinite singularities : N^f, N^f, N^f (Config. 1)

Next assume that $\mathcal{T}_4 = 0$:

In:= **CSP[{(-4 ca^2-cb^2+ca^2 cb^2)},{-(-3+3 ca^2-cb^2)(-1+ca^2+cb^2),R, S},**
{var1,var2,ca^2-1}];
{μ_0= -12 (-1+ca)^2 (1+ca)^2}
{**D**= 49152 (-1+ca)^2 (1+ca)^2 (-3+3 ca^2-cb^2) (-1+ca^2+cb^2)^3}
{$\widetilde{K}$= 4 (3 x[1][t]^2+ca^2 x[1][t]^2+8 ca x[1][t] y[1][t]+y[1][t]^2+3 ca^2 y[1][t]^2}
...

Finite singularities : $s^{(1)}$, s, s, n

{η= 4 (125+262 ca^2+125 ca^4)}
{μ_0= -12 (-1+ca)^2 (1+ca)^2}
{κ= 64 (35+58 ca^2+35 ca^4)}

Infinite singularities : N^f, N^f, N^f (Config. 1)

With what we have done up to now, we have covered already all the generic possibilities. So, one can already perform the bifurcation diagram in two dimensions, gathering the important polynomial expressions that were needed to decide about their sign, in our case the couple of parallel lines $a^2 - 1 = 0$, the circle $a^2+b^2-1 = 0$, the hyperbola $-3+3a^2-b^2 = 0$ and the quartic $-4a^2-b^2+a^2cb^2 = 0$. Now it is time to study what happens on the bifurcation curves.

So, let us see what happens if $a^2 + b^2 - 1 = 0$. Since in this case the double symmetry ($a \to -a$ and $b \to -b$) can easily be observed, we can restrict ourselves to the first quadrant of the parameter space. So we set $b = \sqrt{1-a^2}$ in the original system. When one sets a radical condition of even degree, it is worthwhile to fix also the sign of whatever is inside the radical. In this case $1 - a^2 > 0$.

In:= **change={a00 → -1, a10 → 2cb, a01 → 0, a20 → 3, a11 → ca, a02 →-1,**
b00 → -ca, b10 → 0, b01 → -2cb, b20 → ca, b11 → -1, b02 → -3ca}
/.cb– >Sqrt[1-ca^2]; comp

You are working with at most 2 different parameters, or quite complicated parameters
Depending on the position of the parameters, in the worst case, this should last less than 1 minute

Out:= {Null}
In:= **CSP[{},{1-ca^2},{var1,var2}];**
R= -768 (-1+ca)^3 (1+ca)^3 (ca x[1][t]+y[1][t])^2 =0??
...

We must confirm either that $ax + y \neq 0$ or simply that $\mathbf{R} \neq 0$:

In:= **CSP[{},{1-ca^2},{var1,var2,R}];**
{μ_0= -12 (-1+ca)^2 (1+ca)^2}
{**D**= 0}

{**T**= 0}
{**P**= 0}
{**R**= -768 (-1+ca)^3 (1+ca)^3 (ca x[1][t]+y[1][t])^2
{E_3= -32 (-1+ca)^2 (1+ca)^2}
{$\mathcal{T}_4$= 1024 (1+ca^2)^4}

Finite singularities : $\overline{s}_{(3)}$, s

{η= 4 (125+262 ca^2+125 ca^4)}
{μ_0= -12 (-1+ca)^2 (1+ca)^2}
{κ= 64 (35+58 ca^2+35 ca^4)}

Infinite singularities : N^f, N^f, N^f (Config. 1)

If we are exactly at the point $(a, b) = (1, 0)$, we simply input the following:

```
In:=    change={a00 → -1, a10 → 2cb, a01 → 0, a20 → 3, a11 → ca,
         a02 →-1, b00 → -ca, b10 → 0, b01 → -2cb, b20 → ca, b11 → -1,
        b02 → -3ca}/.cb− >0/.ca− >1; comp
Out:=   {Null}
In:=    CSP[{},{},{var1,var2}];
        μ4= -4 (x[1][t]+y[1][t])^4 =0??
        ...
```

Now we need to explain to the computer that the expression $x + y$ is not zero.

In:= **CSP**[{},{},{**var1,var2,var1+var2**}];
{μ_0= 0}
{μ_1= 0}
{μ_2= 0}
{μ_3= 0}
{μ_4= -4 (x[1][t]+y[1][t])^4 }

No finite singularity

{η= 2048)}
{μ_0= 0}
{μ_1= 0}
{μ_2= 0}
{μ_3= 0}
{μ_4= -4 (x[1][t]+y[1][t])^4 }
{κ= 8192}

Infinite singularities : $\overline{\binom{4}{1}}S$, N^f, N^f (Config. 5)

Next we must consider the case $a = \pm 1$ and $b \neq 0$. As we have said, by means of a symmetry we can consider just the positive case. It is usually useful to keep the expressions that appear in the preceding comitants (for example $x + y$) since they may easily come up again.

In:= **change={a00 → -1, a10 → 2cb, a01 → 0, a20 → 3, a11 → ca, a02 →-1, b00 → -ca, b10 → 0, b01 → -2cb, b20 → ca, b11 → -1, b02 → -3ca}/.ca− >1; comp**
Out:= {Null}
In:= **CSP[{},{},{var1,var2,var1+var2,cb}];**
U= 64 cb^2 (x[1][t]+y[1][t])^4 (-2 x[1][t]+cb^2 x[1][t]-2 y[1][t] -cb^2 y[1][t])^2 =0??
...

Here the comitant **U** must be set as different from zero in order to continue. Check that it cannot be zero since $(-2x + b^2x - 2y - b^2y) = (b^2 - 2)x - (2 + b^2)y$ cannot vanish.

In:= **CSP[{},{},{var1,var2,var1+var2,cb,U}];**
{μ_0= 0}
{μ_1= 0}
{μ_2= -16 cb^2 (x[1][t]+y[1][t])^2}
{**U**= 64 cb^2 (x[1][t]+y[1][t])^4 (-2 x[1][t]+cb^2 x[1][t]-2 y[1][t]-cb^2 y[1][t])^2}
{κ= 8192}
{$\mathcal{T}_4$= 16384}

Finite singularities : s, s

{η= 2048)}

{μ_0= 0}
{μ_1= 0}
{μ_2= -16 cb^2 (x[1][t]+y[1][t])^2}
{κ= 8192}

Infinite singularities : $\overline{\binom{2}{1}}N$, N^f, N^f (Config. 5)

We finally must study the configurations on the hyperbola $-3+3a^2-b^2 = 0$, and just as above, we only need to study the behavior in one branch. We set $a^2 - 1 > 0$ to fix the sign of the radical.

In:= **change={a00 → -1, a10 → 2cb, a01 → 0, a20 → 3, a11 → ca, a02 →-1, b00 → -ca, b10 → 0, b01 → -2cb, b20 → ca, b11 → -1, b02 → -3ca} /.cb− >Sqrt[-3+3ca^2]; comp**
Out:= {Null}

In:= **CSP[{},{ca^2-1},{var1,var2,var1+var2}];**
T= -1769472 (-1+ca)^5 (1+ca)^5 (x[1][t]+ca y[1][t])^2 (-3 x[1][t]^2+
+ca^2 x[1][t]^2-4 ca x[1][t] y[1][t]+y[1][t]^2-3 ca^2 y[1][t]^2)^2=0??
...

We must set $\mathbf{T} \neq 0$ since we already have $a^2 - 1 > 0$. Notice that the factor $(-3x^2 + a^2x^2 - 4axy + y^2 - 3a^2y^2)$ cannot vanish.

In:= **CSP[{},{ca^2-1},{var1,var2,var1+var2,T}];**
$\mathcal{T}_4$= -1024 (-3+ca^2) (1+ca^2)^2 (-1+3 ca^2)=0??
...

We are close to the end. Let us try both possibilities for $\mathcal{T}_4$. First, $\mathcal{T}_4 \neq 0$.

In:= **CSP[{},{ca^2-1},{var1,var2,var1+var2,-T,$\mathcal{T}_4$}];**
{μ_0= -12(-1+ca)^2(1+ca)^2
{**D**= 0
{**T**= -1769472 (-1+ca)^5 (1+ca)^5 (x[1][t]+ca y[1][t])^2 (-3 x[1][t]^2+
+ca^2 x[1][t]^2-4 ca x[1][t] y[1][t]+y[1][t]^2-3 ca^2 y[1][t]^2)^2
{$\widetilde{K}$= -4 (3 x[1][t]^2+ca^2 x[1][t]^2+8 ca x[1][t] y[1][t]+y[1][t]^2+3 ca^2 y[1][t]^2)
{E_1= -2048 $\sqrt{3}$ (-1+ca)^3 (1+ca)^3 $\sqrt{-1+ca^2}$(1+ca^2)}
{$\mathcal{T}_4$= -1024 (-3+ca^2) (1+ca^2)^2 (-1+3 ca^2)}

Finite singularities : $\overline{sn}_{(2)}$, s, s

{η= 4 (125+262 ca^2+125 ca^4)}
{μ_0= -12 (-1+ca)^2 (1+ca)^2}

{κ= 64 (35+58 ca^2+35 ca^4)}

Infinite singularities : N^f, N^f, N^f (Config. 1)

And then $\mathcal{T}_4 = 0$.

In:= **CSP[{$\mathcal{T}_4$},{ca^2-1},{var1,var2,var1+var2,-T}];**
{μ_0= -12(-1+ca)^2(1+ca)^2
{**D**= 0
{**T**= -1769472 (-1+ca)^5 (1+ca)^5 (x[1][t]+ca y[1][t])^2 (-3 x[1][t]^2+
ca^2 x[1][t]^2-4 ca x[1][t] y[1][t]+y[1][t]^2-3 ca^2 y[1][t]^2)^2
{$\widetilde{K}$= -4 (3 x[1][t]^2+ca^2 x[1][t]^2+8 ca x[1][t] y[1][t]+y[1][t]^2+3 ca^2 y[1][t]^2)}
{E_1= -2048 $\sqrt{3}$ (-1+ca)^3 (1+ca)^3 $\sqrt{-1+ca^2}$(1+ca^2)}
$\mathcal{T}_4$= -1024 (-3+ca^2) (1+ca^2)^2 (-1+3 ca^2)

Finite singularities : $\overline{sn}_{(2)}$, s, $s^{(1)}$

{η= 4 (125+262 ca^2+125 ca^4)}
{μ_0= -12 (-1+ca)^2 (1+ca)^2}
{κ= 64 (35+58 ca^2+35 ca^4)}

Infinite singularities : N^f, N^f, N^f (Config. 1)

We include an example of another function made by us that we call "implicit-plot[{fun},var1,var2,xa,xb,xc,xd,grid]" and that is quite similar to "CountourPlot" but works in a slightly different way. It can plot up to seven different functions "{fun}' in two variables "var1" and "var2", in a square "(xa,xc)→(xb,cd)" defined by the user and using a grid (vertical and horizontal) of "grid" points. We provide an example with seven lines followed by our bifurcation diagram with the circle, the hyperbola, the vertical straight line and the quartic restricted to the first quadrant. We include here this algebraic bifurcation diagram (see Figure B.1), indicating the configurations of singularities at every region.

We hope that this brief manual may be useful to the reader. As you may see, the reader will need to acquire quite a bit of practice experimenting with it. If you find any gap in the formula "CSP[{},{},{}];" we ask that you be so kind as to contact the authors so we may improve it. Also, if you are an expert in Mathematica and find a way to improve the function or the routines involved, we will be very happy to incorporate your improvements and give you credit for that.

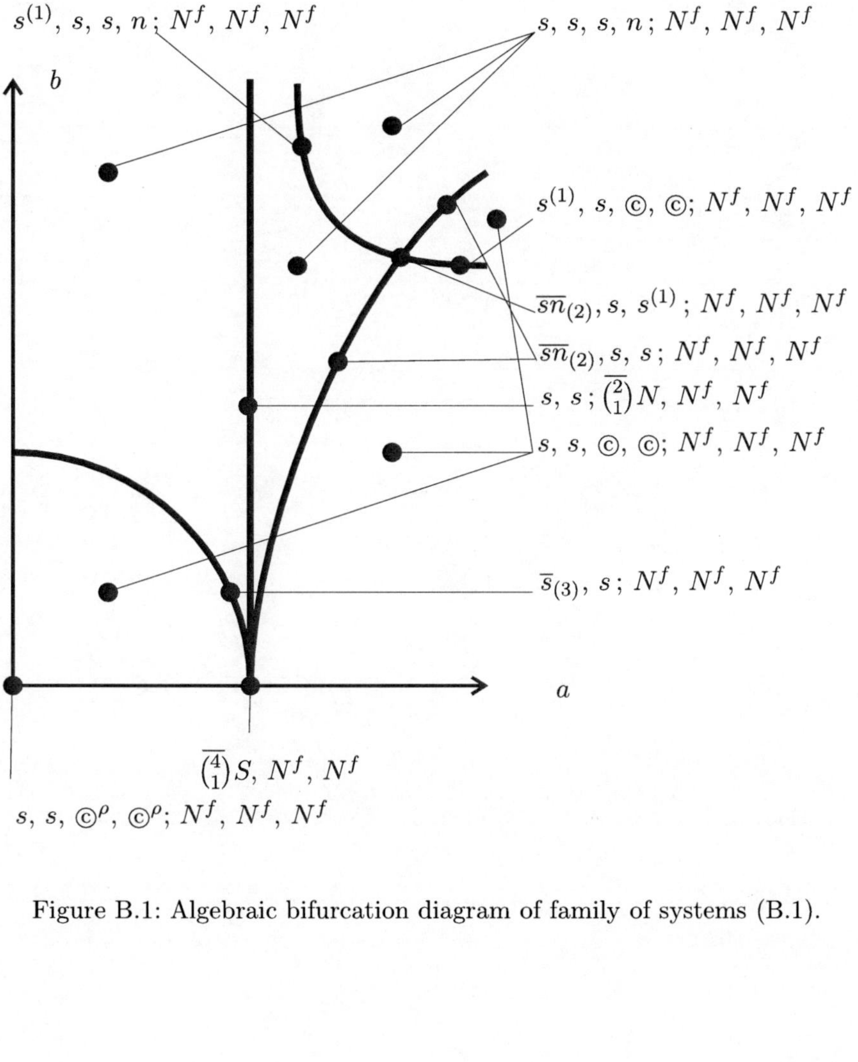

Figure B.1: Algebraic bifurcation diagram of family of systems (B.1).

Bibliography

[1] A. A. Andronov and E. A. Leontovich, On the generation of limit cycles from a loop of a separatrix and from the separatrix of equilibrium of saddle-node type (in Russian), *Mat. Sb. (N.S).*, **48, no. 90** (1959), 335–376.

[2] A. A. Andronov, E. A. Leontovich, I. I. Gordon and A. G. Maier, *Qualitative theory of second-order dynamic systems*, translated from Russian by D. Louvish, Halsted Press, New York; Israel Program for Scientific Translations, Jerusalem, 1973, xxiii+524 pp.

[3] A. A. Andronov, E. A. Leontovich, I. I. Gordon and A. G. Maier, *Theory of bifurcations of dynamic systems on a plane*, translated from Russian. Halsted Press, New York; Israel Program for Scientific Translations, Jerusalem, 1973, xiv+482 pp.

[4] A. A. Andronov, A. A. Vitt and S. E. Khaikin, *Teoriya kolebanii*, [The theory of oscillations], Corrected reprint of the 1937 original, with a preface by L. I. Mandel'shtam, "Nauka", Moscow, 1981, 568 pp.

[5] A. A. Andronov, A. A. Vitt and S. E. Khaikin, *Theory of oscillators*, translated from Russian by F. Immirzi, reprint of the 1966 translation, Dover Publications, Inc., New York, 1987, xxxiv+815 pp.

[6] P. Appell, Sur les invariants de quelques équations différentielles, *Journ. Math. Pures et Appl.*, **5** (1889), 361–423.

[7] V. I. Arnold, *Ordinary differential equations* (in Russian), Izdatelstvo Nauka, Moscow, 1971.

[8] V. I. Arnold, *Les Méthodes mathématiques de la mécanique classique* (in Russian) Editions Mir, Moscow, 1974. (French translation by Mir 1976.)

[9] V. I. Arnold, *Chapitres supplémentaires de la théorie des équations différentielles ordinaires* (in Russian), Editions Mir, Moscow, 1978. (French translation by Mir 1980.)

[10] V. I. Arnold, *Geometrical methods in the theory of ordinary differential equations.* Fundamental Principles of Mathematical Science, vol. 250, Springer-Verlag, Berlin, 1983, xi+334 pp.

J. C. Artés et al., *Geometric Configurations of Singularities of Planar Polynomial Differential Systems*, https://doi.org/10.1007/978-3-030-50570-7

[11] V. I. Arnold, V. V. Kozlov and A. I. Neishtadt, *Mathematical aspects of classical and celestial Mechanics* Springer-Verlag, Berlin, 2006, 518 pp.

[12] J. C. Artés, R. Kooij and J. Llibre, *Structurally stable quadratic vector fields*, Memoires of the American Mathematical Society, vol 134, no. 639, American Mathematical Society, 1998, 108 pp.

[13] J. C. Artés and J. Llibre, Quadratic Hamiltonian vector fields, *J. Differential Equations*, **107** (1994), 80–95.

[14] J. C. Artés and J. Llibre, Phase portraits for quadratic systems having a focus and one anti-saddle, *Rocky Mountain J. Math.*, **24** (1994), 875–889.

[15] J. C. Artés and J. Llibre, Corrigendum: Quadratic Hamiltonian vector fields, *J. Differential Equations*, **129** (1996), 559–560.

[16] J. C. Artés and J. Llibre, Quadratic vector fields with a weak focus of third order, *Publ. Mat.*, **41** (1997), 7–39.

[17] J. C. Artés, J. Llibre and A. C. Rezende, *Structurally unstable quadratic vector fields of codimension one*, Birkhäuser/Springer, Cham, 2018, vi+267 pp.

[18] J. C. Artés, J. Llibre, A. C. Rezende, D. Schlomiuk and N. Vulpe, Global configurations of singularities for quadratic differential systems with exactly two finite singularities of total multiplicity four, *Electron. J. Qual. Theory Differ. Equ.*, **2014** (2014), No. 60, 43 pp.

[19] J. C. Artés, J. Llibre and D. Schlomiuk, The geometry of quadratic differential systems with a weak focus of second order, *Internat. J. Bifur. Chaos Appl. Sci. Engrg.*, **16** (2006), 3127–3194.

[20] J. C. Artés, J. Llibre and D. Schlomiuk, The geometry of quadratic polynomial differential systems with a weak focus and an invariant straight line, *Internat. J. Bifur. Chaos Appl. Sci. Engrg.*, **20**, no. 11 (2010), 3627–3662.

[21] J. C. Artés, J. Llibre, D. Schlomiuk and N. Vulpe, Configurations of singularities for quadratic differential systems with total finite multiplicity lower than 2, *Bul. Acad. Ştiinţe Repub. Mold. Mat.*, **1 (71)** 2013, 72–124.

[22] J. C. Artés, J. Llibre, D. Schlomiuk and N. Vulpe, Geometric configurations of singularities for quadratic differential systems with three distinct real simple finite singularities, *J. Fixed Point Theory Appl.*, **14**, no. 2 (2013), 555–618.

[23] J. C. Artés, J. Llibre, D. Schlomiuk and N. Vulpe, Geometric configurations of singularities for quadratic differential systems with total finite multiplicity $m_f = 2$, *Electron. J. Differ. Equ.*, **2014** (2014), No. 159, 79 pp.

[24] J. C. Artés, J. Llibre, D. Schlomiuk and N. Vulpe, Global configurations of singularities for quadratic differential systems with total finite multiplicity

three and at most two real singularities, *Qual. Theory Dyn. Syst.*, **13**, no. 2 (2014), 305–351.

[25] J. C. Artés, J. Llibre, D. Schlomiuk and N. Vulpe, Global configurations of singularities for quadratic differential systems with exactly three finite singularities of total multiplicity four, *Electron. J. Qual. Theory Differ. Equ.*, **2015** (2015), No. 49, 60 pp.

[26] J. C. Artés, J. Llibre, D. Schlomiuk and N. Vulpe, From topological to geometric equivalence in the classification of singularities at infinity for quadratic vector fields, *Rocky Mountain J. of Math.*, **45**, no. 1 (2015), 29–113.

[27] J. C. Artés, J. Llibre, D. Schlomiuk and N. Vulpe, Global topological configurations of singularities for the whole family of quadratic differential systems, *Qual. Theory Dyn. Syst.*, **19**(51), (2020), pp. 32.

[28] J. C. Artés, J. Llibre and N. Vulpe, Quadratic systems with a rational first integral of degree 2: A complete classification in the coefficient space $\mathbb{R}^{12}$, *Rend. Circ. Mat. Palermo*, **56**, no. 3 (2007), 417–444.

[29] J. C. Artés, J. Llibre and N. Vulpe, Singular points of quadratic systems: A complete classification in the coefficient space $\mathbb{R}^{12}$, *International J. of Bifurcation and Chaos*, **18** (2008), 313–362.

[30] J. C. Artés, J. Llibre and N. Vulpe, Quadratic systems with a polynomial first integral: A complete classification in the coefficient space $\mathbb{R}^{12}$, *J. Differential Equations*, **246** (2009), 3535–3558.

[31] J. C. Artés, J. Llibre and N. Vulpe, Complete geometric invariant study of two classes of quadratic systems, *Electron. J. Differ. Equ.*, **2012** (2012) No. 9, 35pp.

[32] J. C. Artés, J. Llibre and N. Vulpe, Quadratic systems with an integrable saddle: A complete classification in the coefficient space $\mathbb{R}^{12}$, *Nonlinear Anal.*, **75**, no. 14 (2012), 5416–5447.

[33] J. C. Artés, A. C. Rezende and R. Oliveira, Global phase portraits of quadratic polynomial differential systems with a semi-elemental triple node, *Internat. J. Bifur. Chaos Appl. Sci. Engrg.*, **23**, no. 8 (2013), 1350140, 21 pp.

[34] J. C. Artés, A. C. Rezende and R. Oliveira, The geometry of quadratic polynomial differential systems with a finite and an infinite saddle-node (A,B), *Internat. J. Bifur. Chaos Appl. Sci. Engrg.*, **24**, no. 4 (2014), 1450044, 30 pp.

[35] J. C. Artés, A. C. Rezende and R. D. S. Oliveira, The geometry of quadratic polynomial differential systems with a finite and an infinite saddle-node (C), *Internat. J. Bifur. Chaos Appl. Sci. Engrg.*, **25**, no. 3 (2015), 1530009, 111 pp.

[36] M. Aschenbrenner, L. van den Dries and J. van der Hoeven, Asymptotic differential algebra: Annalyzable functions and applications, *Contemp. Math.*, **373** (2005), 49–85.

[37] M. Aschenbrenner, L. van den Dries and J. van der Hoeven, Toward a model theory for transseries, *Notre Dame J. Form. Log.*, **54**, no. 3–4 (2013), 279–310.

[38] M. Aschenbrenner, L. van den Dries and J. van der Hoeven, *Asymptotic differential algebra and model theory of transseries*, Annals of Mathematics Studies, 195, Princeton University Press, Princeton, NJ, 2017, xxi+849 pp.

[39] V. Baltag and I. Calin, The transvectants and the integrals for Darboux systems of differential equations (English summary), *Bul. Acad. Ştiinţe Repub. Mold. Mat.*, **1 (56)** (2008), 4–18.

[40] V. Baltag and N. Vulpe, The number and multiplicity of singular points of a quadratic differential system (in Russian), *Dokl. Akad. Nauk*, **323**, no. 1 (1992), 9–12. (Translation in *Russian Acad. Sci. Dokl. Math.*, **45**, no. 2 (1993), 235–238.

[41] V. A. Baltag and N. I. Vulpe, Affine-invariant conditions for determining the number and multiplicity of singular points of quadratic differential systems, *Izv. Akad. Nauk Respub. Moldova Mat.*, **57**, no. 1 (1993), 39–48.

[42] V. Baltag and N. Vulpe, Total multiplicity of all finite critical points of the polynomial differential system: Planar nonlinear dynamical systems (Delft, 1995), *Differential Equations Dynam. Systems*, **5**, no. 3–4 (1997), 455–471.

[43] V. Baltag and N. Vulpe, The differential operators and multiplicity of singular points for polynomial differential system, *Bul. Acad. Ştiinţe Repub. Mold. Mat.*, **1 (26)** (1998), 81–94.

[44] R. Bamon, Quadratic vector fields in the plane have a finite number of limit cycles, *Inst. Hautes Études Sci. Publ. Math.*, **64** (1986), 111–142.

[45] N. N. Bautin, Du nombre de cycles limites naissant en cas de variation des coefficients d'un état d'équilibre du type foyer ou centre, *C. R. (Doklady) Acad. Sci. URSS (N. S.)*, **24** (1939), 669–672.

[46] N. N. Bautin, On the number of limit cycles which appear with the variation of coefficients from an equilibrium position of focus or center type (in Russian), *Mat. Sb. (N.S)*, **30**, no. (7)2 (1952), 181–196. (Translation in *Transl. Amer. Math. Soc. Ser. (1)*, **100** (1954), 397–413.)

[47] L. N. Belyustin, On the conditions of the existence of the center (in Russian), *Prik. Mat. i Mek.*, **18** (1954), 511.

[48] A. Bendixon, Sur les courbes définies par des équations différentielles, *Acta Math.*, **24** (1901), 1–88.

[49] A. Berlinskii, On the coexistence of singular points of different types (in Russian), *Izv. Vyss. Ucebn. Zaved. Matematika*, **18**, no. 5 (1960), 27–32.

[50] G. Binyamini and G. Dor, An explicit linear estimate for the number of zeros of Abelian integrals, *Nonlinearity*, **25** (2012), 1931–1946.

[51] G. Binyamini, D. Novikov and S. Yakovenko, On the number of zeros of Abelian integrals (English summary), *Invent. Math.*, **181**, no. 2 (2010), 227–289.

[52] Y. Bolaños, J. Llibre and C. Valls, Liouvillian first integrals for quadratic systems with an integrable saddle, *Rocky Mountain J. Math.*, **45** (2015), 1765–1779.

[53] G. Boole, Exposition of a general theory of linear transformations, *Cambridge Math. J.*, **3** (1841–1842), 1–20, 106–119.

[54] D. Boularas, Affine invariants of differential systems (in Russian), *Algebraic invariants of dynamical systems. Mat. Issled.*, **55** (1980), 30–36.

[55] D. Boularas, Computation of affine covariants of quadratic bivariate differential systems (English summary), *J. Complexity*, **16**, no. 4 (2000), 691–715.

[56] D. Boularas, A new classification of planar homogeneous quadratic systems (English summary), *Qual. Theory Dyn. Syst.*, **2**, no. 1 (2001), 93–110.

[57] D. Boularas, A. Braicov and M. Popa, Hilbert series for comitants algebra of a full quadratic differential system and its applications, *Bul. Acad. Ştiinţe Repub. Mold. Mat.*, **3 (31)** (1999), 81–94.

[58] D. Boularas, Iu. Calin, L. Timochouk and N. I. Vulpe, *T-comitants of quadratic systems: A study via the translation invariants*, Delft University of Technology, Faculty of Technical Mathematics and Informatics, Report no. 96-90, 1996.

[59] D. Boularas and D. Dali, Sur les bases de concomitants centro-affines des systèmes différentiels plans (with English summary) [On the bases of centro-affine concomitants of plane differential systems], *Cahiers Math.*, **2** (1987), 25–30.

[60] D. Boularas, N. I. Vulpe and K. S. Sibirschi, The problem of the center "in the large" for a general quadratic system (in Russian), *Dokl. Akad. Nauk SSSR*, **311**, no. 4 (1990), 777–780. (Translation in *Soviet Math. Dokl.*, **41**, no. 2 (1991), 287–290.)

[61] W. Büchel, Zur topologie der durch eine gewöhnliche differentialgleichung erster ordnung und ersten grades definierten kurvenschar, *Mitteil. der Math. Gesellsch. in Hamburg*, **4** (1904), 33–68.

[62] C. Bujac, One new class of cubic systems with maximum number of invariant lines omitted in the classification of J. Llibre and N. Vulpe, *Bul. Acad. Ştiinţe Repub. Mold. Mat.*, **2 (75)** (2014), 102–105.

[63] C. Bujac, J. Llibre and N. Vulpe, First integrals and phase portraits of planar polynomial differential cubic systems with the maximum number of invariant straight lines, *Qual. Theory Dyn. Syst.*, **15**, no. 2 (2016), 327–348

[64] C. Bujac and N. Vulpe, Cubic systems with invariant lines of total multiplicity eight and with four distinct infinite singularities, *J. Math. Anal. Appl.*, **423**, no. 2 (2015), 1025–1080.

[65] C. Bujac and N. Vulpe, Cubic systems with invariant lines of total multiplicity eight and with three distinct infinite singularities, *Qual. Theory Dyn. Syst.*, **14**, no. 1 (2015), 109–137.

[66] C. Bujac and N. Vulpe, Classification of cubic differential systems with invariant straight lines of total multiplicity eight and two distinct infinite singularities, *Electron. J. Qual. Theory Differ. Equ.*, **2015** (2015), No. 74, 38 pp.

[67] C. Bujac and N. Vulpe, Cubic differential systems with invariant straight lines of total multiplicity eight and possessing one infinite singularity, *Qual. Theory Dyn. Syst.*, **16**, no. 1 (2017), 1–30.

[68] W. Burnside, *Theory of groups of finite order*, First Edition, Cambridge University Press, Cambridge, 1897, 388 pp. (Second Edition, 1911.)

[69] N. V. Butenin, Maintained vibrating systems of gyroscopic forces (R), *Prikl. Mat. Meh.*, **6** (1942), 327–346.

[70] Cai, Sui Lin, The weak saddle and separatrix cycle of a quadratic system (in Chinese), *Acta Math. Sinica*, **30**, no. 4 (1987), 553–559.

[71] L. Cairo, H. Giacomini and J. Llibre, Liouvillian first integrals for the planar Lotka–Volterra systems, *Rend. Circ. Mat. Palermo (2)*, **52** (2003), 389–418.

[72] L. Cairo and J. Llibre, Phase portraits of planar semi-homogeneous vector fields, *Nonlinear Anal.*, **29** (1997), 783–811.

[73] L. Cairo and J. Llibre, Integrability of the 2D Lotka–Volterra system via polynomial first integrals and polynomial inverse integrating factors, *J. Phys. A*, **33** (2000), 2407–2417.

[74] L. Cairo and J. Llibre, Phase portraits of quadratic polynomial vector fields having a rational first integral of degree 2, *Nonlinear Anal.*, **67** (2007), 327–348.

[75] Iu. Calin, On rational bases of $GL(2,\mathbb{R})$-comitants of planar polynomial systems of differential equations, *Bul. Acad. Ştiinţe Repub. Mold. Mat.*, **2 (42)** (2003), 69–86.

[76] C. Camacho and L. H. de Figueiredo, The dynamics of the Jouanolou foliation on the complex projective 2-space. *Ergodic Theory Dynam. Systems*, **21**, no. 3 (2001), 757–766.

[77] F. Cao and J. Jiang, The classification on the global phase portraits of two-dimensional Lotka–Volterra system, *J. Dynam. Differential Equations*, **20**, no. 4 (2008), 797–830.

[78] M. Carnicer, The Poincaré problem in the nondicritical case, *Ann. of Math.* **140**, no. 2 (1994), 289–294.

[79] A. Cayley, On the theory of linear transformations, *Camb. Math. J.*, **4** (1845), 193–209. (Also in *The Collected Mathematical Papers*, vol. 1, Cambridge University Press, Cambridge, England, 1889, pp. 95–112.)

[80] S. Chandrasekhar, *An introduction to the study of stellar structure*, Chapter 4. Chicago Univ. Press, 1939. (Reprinted by Dover, New York, 1957, 512 pp.)

[81] J. Chavarriga, H. Giacomini and J. Llibre, Uniqueness of algebraic limit cycles for quadratic systems, *J. Math. Anal. Appl.*, **261** (2001), 85–99.

[82] J. Chavarriga, J. Llibre and J. Moulin Ollagnier, On a result of Darboux, *LMS J. Comput. Math.*, **4** (2001), 197–210.

[83] J. Chavarriga, J. Llibre and J. Sorolla, Algebraic limit cycles of degree 4 for quadratic systems, *J. Differential Equations*, **200** (2004), 206–244.

[84] J. Chavarriga and M. Sabatini, A survey of isochronous centers, *Qual. Theory Dyn. Syst.*, **1**, no. 1 (1999), 1–70.

[85] F. Chen, C. Li, J. Llibre and Z. Zhang, A unified proof on the weak Hilbert's 16th problem for $n = 2$, *J. Differential Equations*, **221** (2006), 309–342.

[86] L. Chen, X. Ma, J. Zhang and C. Li, Cyclicity of several quadratic reversible systems with center of genus one, *J. Appl. Anal. Comput.*, **1**, no. 4 (2011), 439–437.

[87] L. S. Chen and M. S. Wang, The relative position and number of limit cycles of a quadratic differential system, *Acta Math. Sinica*, **22** (1979), 751–768.

[88] C. Chicone and D. Shafer, Separatrix and limit cycles of quadratic systems and Dulac's theorem, *Trans. Amer. Math. Soc.*, **278**, no. 2 (1983), 585–612.

[89] C. Chicone and Tian Jinghuang, On general properties of quadratic systems, *The American Mathematical Monthly*, **89**, no. 3 (Mar. 1982), 167–178.

[90] C. Christopher and C. Li, *Limit cycles of differential equations*, Advanced Courses in Mathematics, CRM Barcelona, Birkhäuser Verlag, Basel, 2007, 171 pp.

[91] C. Christopher, J. Llibre and J. Pereira, Multiplicity of invariant algebraic curves in polynomial vector fields, *Pacific J. Math.*, **29**, no. 1 (2007), 63–117.

[92] C. Christopher, J. Llibre and G. Świrszcs, Invariant algebraic curves of large degree for quadratic system, *J. Math. Anal. Appl.*, **303** (2005), 450–461.

[93] A. Cima, A. Gasull and F. Mañosas, Some applications of the Euler-Jacobi formula to differential equations, *Proc. Amer. Math. Soc.*, **118** (1993), 151–163.

[94] B. Coll, *Qualitative study of some classes of vector fields in the plane*, Ph. D. thesis, Universitat Autònoma de Barcelona, 1987, 34 pp.

[95] B. Coll, A. Ferragut and J. Llibre, Phase portraits of the quadratic systems with a polynomial inverse integrating factor, *Internat. J. Bifur. Chaos Appl. Sci. Engrg.*, **19** (2009), 765–783.

[96] B. Coll, A. Ferragut and J. Llibre, Polynomial inverse integrating factors for quadratic differential systems, *Nonlinear Anal.*, **73** (2010), 881–914.

[97] B. Coll, A. Gasull and J. Llibre, Some theorems on the existence, uniqueness and non existence of limit cycles for quadratic systems, *J. Differential Equations*, **67** (1987), 372–399.

[98] B. Coll, A. Gasull and J. Llibre, Quadratic systems with a unique finite rest point, *Publ. Mat.*, **32** (1988), 199–259.

[99] B. Coll, C. Li and R. Prohens, Quadratic perturbations of a class of quadratic reversible systems with two centers, *Discrete Contin. Dyn. Syst.*, **24** (2009), 699–729.

[100] B. Coll and J. Llibre, Limit cycles for a quadratic system with an invariant straight line and some evolution of phase portraits, *Colloquia Mathematica Societatis Janos Bolyai*, **53** (1988), pp. 111–123.

[101] R. Conti, Uniformly isochronous centers of polynomial systems in $\mathbb{R}^2$, *Lecture Notes in Pure and Appl. Math.*, **152** (1994), 21–31.

[102] W. A. Coppel, On a differential equation of boundary-layer theory, *Phil. Trans. Roy, Soc. Ser. A*, **253** (1961), 101–136.

[103] W. A. Coppel, A survey of quadratic systems, *J. Differential Equations*, **2** (1966), 293–304.

[104] W. A. Coppel, Some quadratic systems with at most one limit cycle, in Y. Kirchgraber and H. O. Walther (eds.), Dynamics reported: A series in dynamical systems and their applications, vol. 2, Wiley, Chichester, 1989, pp. 61–88.

[105] D. Cozma, The problem of the centre for cubic systems with two parallel invariant straight lines and one invariant conic, *NoDEA Nonlinear Differential Equations Appl.*, **16**, no. 2 (2009), 213–234.

[106] D. Cozma, Darboux integrability in the cubic differential systems with three invariant straight lines, *ROMAI J.*, **5**, no. 1 (2009), 45–61.

[107] D. Cozma, The problem of the centre for cubic differential systems with two homogeneous invariant straight lines and one invariant conic, *Ann. Differential Equations*, **26**, no. 4 (2010), 385–399.

[108] D. Cozma, Center problem for a class of cubic systems with a bundle of two invariant straight lines and one invariant conic, *Bul. Acad. Ştiinţe Repub. Mold. Mat.*, **3 (64)** (2010), 51–66.

[109] D. Cozma, Center problem for cubic systems with a bundle of two invariant straight lines and one invariant conic, *Bul. Acad. Ştiinţe Repub. Mold. Mat.*, **1 (68)** (2012), 32–49.

[110] D. V. Kozma and A. S. Shuba, Center conditions of a cubic system with four integral lines (in Russian), *Izv. Akad. Nauk Respub. Moldova Mat.*, **56**, no. 3 (1992), 62–67.

[111] D. V. Kozma and A. S. Shuba, Conditions for the presence of four integral straight lines in a cubic differential system in the case of a center or focus (in Russian), *Izv. Akad. Nauk Repub. Moldova Mat.*, **57**, no. 3 (1993), 54–62.

[112] D. Cozma and A. S. Şubă, Conditions of center for some cubic systems with three linear particular integrals, *Scr. Sci. Math.*, **1**, no. 1 (1997), 82–94.

[113] D. Cozma and A. Şubă, The solution of the problem of center for cubic differential systems with four invariant straight lines, *An. Ştiinţ. Univ. Al. I. Cuza Iaşi. Mat. (N.S.)*, **44**, suppl. (1998), 517–530.

[114] D. Cozma and A. Şubă, Solution of the problem of the centre for a cubic differential system with three invariant straight lines, *Qual. Theory Dyn. Syst.* **2**, no. 1 (2001), 129–143.

[115] C. W. Curtis, *Pioneers of representation theory: Frobenius, Burnside, Schur and Brauer*, American Mathematical Society, Providence, RI, 1999, 287 pp.

[116] Dang Din' Bik, Affine invariants of an autonomous second order differential system with cubic homogeneous nonlinearities (in Russian), *Mat. Issled.*, **9**, no. 3(33) (1974), 64–81.

[117] Dang Din' Bik, Comitants of a cubic system (in Russian), *Mat. Issled.*, **10**, no. 3(37) (1975), 44–55.

[118] Dang Din' Bik and K. S. Sibirschi, Affine classification of a cubic differential system (in Russian), in *Studies in algebra, mathematical analysis, and their applications* (in Russian), *Izdat. "Ştiinţa", Kishinev*, 1977, pp. 43–52, 121.

[119] G. Darboux, Mémoire sur les équations différentielles du premier ordre et du premier degré, *Bulletin de Sciences Mathématiques*, 2me série, **2**, no. 1 (1878), 60–96, 123–144, 151–200.

[120] T. Date, Classification and analysis of two-dimensional real homogeneous quadratic differential equation systems, *J. Differential Equations*, **32** (1979), 311–334.

[121] M. Demers, *Action de groupe, formes normales et systèmes quadratiques à foyer faible dórdere trois*, Master's thesis, Université de Montréal, 2015, 91 pp.

[122] J. Dieudonné, *Invariant theory old and new*, Academic Press, New York, 1970, 84 pp.

[123] H. Dulac, Détermination et integration d'une certaine classe d'equations différentielle ayant par point singulier un centre, *Bull. Sci. Math. Sèr (2)*, **32** (1908), 230–252.

[124] H. Dulac, Sur les cycles limites, *Bulletin de la Société Mathématique de France*, **51** (1923), 45–188.

[125] F. Dumortier, Singularities of vector fields on the plane, *J. Differential Equations*, **23** (1977), 53–106.

[126] F. Dumortier, C. Herssens and L. Perko, Local bifurcations and a survey of bounded quadratic systems, *J. Differential Equations*, **165** (2000), 430–467.

[127] F. Dumortier, C. Li and Z. Zhang, Unfolding of a quadratic integrable system with two centers and two unbounded heteroclinic loops, *J. Differential Equations*, **139**, no. 1 (1997), 146–193.

[128] F. Dumortier, J. Llibre and J. C. Artés, *Qualitative theory of planar differential systems*, Universitext, Springer-Verlag, New York, 2008, 298 pp.

[129] F. Dumortier, R. Roussarie and C. Rousseau, Hilbert's 16th problem for quadratic vector fields, *J. Differential Equations*, **110**, no. 1 (1994), 86–133.

[130] J. Ecalle, Finitude des cycles limites et accéléro-sommation de l'application de retour, in *Bifurcation of Planar Vector Fields*, Lecture Notes in Mathematics, vol. 1455, Springer, New York, 1990, pp. 74–159.

[131] J. Ecalle, *Introduction aux fonctions analysables et preuve constructive de la conjecture du Dulac*, Actualités Mathématiques, Hermann, Paris, 1992, 337 pp.

[132] J. Ecalle, J. Martinet, R. Moussu, and J. P. Ramis, Non-accumulation de cycles-limites, *C. R. Acad. Sci. Paris Sér. I Math.*, **304**, no. 13 (1987), 375–378, (Part I); no. 14, 431–434 (Part II).

[133] R. M. Evdokimenko, Construction of algebraic trajectories and a qualitative investigation in the large of the behavior of the integral curves of a certain system of differential equations (in Russian), *Differentsial'nye Uravneniya*, **6** (1970), 1780–1791.

[134] R. M. Evdokimenko, Behavior of the integral curves of a certain dynamical system (in Russian), *Differentsial'nye Uravneniya*, **9** (1974), 1557–1567.

[135] R. M. Evdokimenko, Investigation in the large of a dynamical system in the presence of a given integral curve (in Russian), *Differentsial'nye Uravneniya*, **15**, no. 2 (1979), 215–221.

[136] V. F. Filiptsov, Algebraic limit cycles, *Differentsial'nye Uravneniya*, **9** (1973), 1281–1288.

[137] J. P. Françoise and C. C. Pugh, Keeping track of limit cycles, *J. Differential Equations*, **65**, no. 2 (1986), 139–157.

[138] F. G. Frobenius, Über die darstellung der endlichen gruppen durch lineare substitutionen, *Sber. Akad. Wiss. Berlin* (1897), 944–1015; Gesammelte Abhandlungen III, 82–103.

[139] M. Frommer, Über das Auftreten von Wirbeln und Strudeln (geschlossener und spiraliger Integralcurven) in der Umgebung rationaler Unbestimmtheitsstellen, *Math. Ann.*, **109** (1934), 395–424.

[140] B. Garcia, J. Llibre and J. Pérez del Rio, Phase portraits of the quadratic vector fields with a polynomial first integral, *Rend. Circ. Mat. Palermo (2)*, **55**, no. 3 (2006), 420–440.

[141] E. F. Gasinskaja and K. S. Sibirschi, Affine invariants of a system with quadratic nonlinearities (in Russian), *Differencial'nye Uravnenija*, **9** (1973), 1371–1382.

[142] A. Gasull, S. Li Ren and J. Llibre, Chordal quadratic systems, *Rocky Mountain J. Math.*, **16**, no. 4 (1986), 751–782.

[143] A. Gasull and J. Llibre, On the nonsingular quadratic differential equations in the plane, *Proc. Amer. Math. Soc.*, **104**, no. 3 (1988), 793–794.

[144] A. Gasull and R. Prohens, Quadratic and cubic systems with degenerate infinity, *J. Math. Anal. Appl.*, **198** (1996), 25–34.

[145] G. F. Gause, *The struggle for existence*, Williams and Wilkins, Baltimore, 1934, 176 pp.

[146] C. F. Gauss, *Disquisitiones arithmeticae*, translated and with a preface by Arthur A. Clarke, revised by William C. Waterhouse, Cornelius Greither and A. W. Grootendorst and with a preface by Waterhouse, Springer-Verlag, New York, 1986, xx+472 pp.

[147] L. Gavrilov, The infinitesimal 16th Hilbert problem in the quadratic case, *Invent. Math.*, **143**, no. 3 (2001), 449–497.

[148] L. Gavrilov and E. Horozov, Limit cycles of perturbations of quadratic Hamiltonian vector fields, *J. Math. Pures Appl.*, **72**, no. 2 (1993), 2013–238.

[149] R. M. Georgescu, *Group theory applications to the bifurcation study for some dynamical models* (in Romanian), Ph.D. thesis, University of Pitesti, Romania, 2007.

[150] E. A. González Velasco, Generic properties of polynomial vector fields at infinity, *Trans. Amer. Math. Soc.*, **143** (1969), 201–222.

[151] P. Gordan, Beweis, dass jede covariante und invariante einer binaren form eine ganze function mit numerische coefficienten einer endlichen anzahl solchen formen ist, *Jour. r. ang. Math.*, **69** (1868), 323–354.

[152] J. H. Grace and A. Young, *The algebra of invariants*, Stechert, New York, 1941, 392 pp.

[153] J. Guckenheimer and P. Holmes, *Nonlinear oscillations, dynamical systems, and bifurcations of vector fields*, Springer-Verlag, New York, 1983, 453 pp.

[154] G. Guderley and H. Yoshihara, An axial-symmetric transonic flow pattern, *Quart. Appl. Math.*, **8** (1950), 333–339.

[155] G. B. Gurevich, *Foundations of the theory of algebraic invariants*, translated from Russian by J. R. M. Radok and A. J. M. Spencer, P. Noordhoff Ltd., Groningen, The Netherlands, 1964, 429 pp.

[156] G. H. Halphen, *Sur la réduction des équations différentielles linéaires aux formes intégrables*, Mém. prés. par divers savants à l'Acad. des Sci., Imprimerie nationale, Paris, 1884, 397 pp. (Also in *Oeuvres de G. H. Halphen*, vol 3. Gauthier-Villars, Paris, 1921, pp. 1–260.)

[157] R. S. Hamilton, Three-manifolds with positive Ricci curvature, *J. Differential Geom.*, **17**, no. 2 (1982), 255–306.

[158] T. Hawking, *Emergence of the theory of Lie groups*, Springer, New York, 2000, 564 pp.

[159] D. Hilbert, The theory of algebraic forms, *Math. Ann.*, **36** (1890), 473–534.

[160] D. Hilbert, Über die vollen invariantensysteme, *Math. Ann.*, **42** (1893), 313–373. (Ges. Abh. Vol. 2, pp. 287–344.)

[161] D. Hilbert, Mathematical problems, *Bull. Amer. Mat. Soc.*, **8** (1902), 437–479. (Reprinted in *Bull. Amer. Math.* **37** (2000), 407–436.)

[162] C. Hong and L. Zou, Abelian integrals and critical periods for a class of reversible quadratic systems, *Sichuan Daxue Xuebao*, **49**, no. 5 (2012), 960–964.

[163] X. Hong, R. Ma and B. Wang, Estimation of the number of zeros of Abelian integrals of a quadratic reversible system, *Math. Pract. Theory*, **45**, no. 5 (2015), 270–276.

[164] X. Hong, S. Xie and L. Chen, Estimating the number of zeros for Abelian integrals of quadratic reversible centers with orbits formed by higher order curves, *Internat. J. Bifur. Chaos Appl. Sci. Engrg.*, **26**, no. 2 (2016), 1650020, 16 pp.

[165] X. Hong, S. Xie and R. Ma, On the Abelian integrals of quadratic reversible centers with orbits formed by genus 1 curves of higher degree, *J. Math. Anal. Appl.*, **429**, no. 2 (2015), 924–941.

[166] X. Huang and J. W. Reyn, On the limit cyle distribution over two nests in quadratic systems, *Bull. Austral. Math. Soc.*, **52**, no. 3 (1995), 461–474.

[167] I. D. Iliev, The cyclicity of the period annulus of the quadratic Hamiltonian triangle, *J. Differential Equations*, **128**, no. 1 (1996), 309–326.

[168] I. D. Iliev, Perturbations of quadratic centers, *Bull. Sci. Math.*, **122**, no. 2 (1998), 107–161.

[169] I. D. Iliev, C. Li and J. Yu, Bifurcations of limit cycles in a reversible quadratic system with a center, a saddle and two nodes, *Commun. Pure Appl. Anal.*, **9**, no. 3 (2010), 583–610.

[170] Y. Ilyashenko, Singular points and limit cycles of differential equations in the real and complex plane, Preprint NIVTS AN SSSR, Pushchino (1982), 38 pp.

[171] Y. Ilyashenko, Limit cycles of polynomial vector fields with non-degenerate singular points on the real plane (in Russian), *Funktsional. Anal. i Prilozhen.*, **18** (1984), 32–42. (Translation in *Funct. Anal. and its Appl.*, **18** (1985), 199–209.)

[172] Y. Ilyashenko, Dulac's memoir "On limit cycles" and related questions of the local theory of differential equations (in Russian), *Uspekhi Mat. Nauk*, **40**, no. 6 (1985), 41–78.

[173] Y. Ilyashenko, Finiteness theorems for limit cycles (in Russian), *Uspekhi Mat. Nauk*, **42**, no. 3 (1987), 223.

[174] Y. Ilyashenko, *Finiteness theorems for limit cycles*, Translation of Mathematical Monographs, vol. 94, American Mathematical Society, Providence, RI, 1991, 288 pp.

[175] Y. Ilyashenko, Editor, *Nonlinear Stokes phenomena*, Advances in Soviet Mathematics, vol. 14, American Mathematical Society, Providence, RI, 1993, 287 pp.

[176] Y. Ilyashenko, Centennial history of Hilbert's 16th problem, *Bull. Amer. Math. Soc. (N.S.)*, **39**, no. 3 (2002), 301–354.

[177] Y. Ilyashenko and C. Rousseau, Editors, *Normal forms, bifurcations and finiteness problems in differential equations*, NATO Science Series II: Mathematics, Physics and Chemistry, vol. 137, Kluwer Academic Publishers, Dordrecht, 2004, xxiv+513 pp.

[178] Y. Ilyashenko and S. Yakovenko, Editors, *Concerning the Hilbert 16th problem*, AMS Translations, Series 2, vol. 165, American Mathematical Society, Providence, RI, 1995, 219 pp.

[179] Y. Ilyashenko and S. Yakovenko, *Lectures on analytic differential equations*, Graduate Studies in Mathematics, vol. 86, American Mathematical Society, Providence, RI, 2008, xiv+625 pp.

[180] P. de Jager, Phase portraits for quadratic systems with a higher order singularity with two zero eigenvalues, *J. Differential Equations*, **87** (1990), 169–204.

[181] X. Jarque, J. Llibre and D. S. Shafer, Structurally stable quadratic foliations, *Rocky Mountain J. Math.*, **38**, no. 2 (2008), 489–530.

[182] Q. Jiang and J. Llibre, Qualitative classification of singular points, *Qual. Theory Dyn. Syst.*, **6**, no. 1 (2005), 87–167.

[183] C. W. Jones, On reducible non-linear differential equations occurring in mechanics, *Proc. Roy. Soc. Ser. A*, **217** (1953), 327–343.

[184] C. Jordan, *Traité des substitutions et des équations algébriques*, Gauthier-Villars, Paris, 1870, xvi+670 pp.

[185] P. Joyal and C. Rousseau, Saddle quantities and applications, *J. Differential Equations*, **78**, no. 2 (1989), 374–399.

[186] T. Kaiser, J. P. Rolin and P. Speissegger, Transition maps at non-resonant hyperbolic singularities are o-minimal, *J. Reine Angew. Math.*, **636** (2009), 1–45.

[187] Y. F. Kalin and N. I. Vulpe, Affine invariant conditions for the topological discrimination of quadratic Hamiltonian differential systems (in Russian), *Differ. Uravn.*, **34**, no. 3 (1998), 298–302. (Translation in *Differential Equations*, **34**, no. 3 (1998), 297–301.)

[188] V. Kaloshin, The existential Hilbert 16-th problem and an estimate for cyclicity of elementary polycycles, *Invent. Mat.*, **151**, no. 3 (2003), 451–512.

[189] W. Kapteyn, On the midpoints of integral curves of differential equations of the first order and the first degree (in Dutch), *Nederl. Acad. Wetensch. Verslangen, Afd. Natuurkunde Koninkl. Nederland*, **XIX**, no. 2 (1911) 1446–1457.

[190] W. Kapteyn, New investigations on the midpoints of integrals of differential equations of the first order and the first degree (in Dutch), *Nederl. Acad. Wetensch. Verslangen, Afd. Natuurkunde*, **XX**, no. 2 (1912) 1354–1365.

[191] K. Kestin and S. K. Zaremba, Adiabatic one-dimensional flow of a perfect gas through a rotating tube of uniform cross section, *Aeronaut. Quart.*, **4** (1954), 373–399.

[192] A. G. Khovanskii, Real analytic manifolds with the property of finiteness, and complex Abelian integrals, *Funktsional. Anal. i Prilozhen.*, **18**, no. 2 (1984), 40–50.

[193] A. G. Khovanskii, *Fewnomials*, Translations of Mathematical Monographs, vol. 88, American Mathematical Society, Providence, RI, 1991, viii+139 pp.

[194] E. F. Kirnickaja and K. S. Sibirschi, Conditions for two centers for a quadratic differential system (in Russian), *Differentsial'nye Uravneniya*, **14**, no. 9 (1978), 1589–1593. (Translation in *Differential Equations*, **14**, no. 9 (1978), 1129–1133.)

[195] F. Klein, *Le programme d'Erlangen, Considérations comparatives sur les recherches géométriques modernes*, translated from German by H. Padé, with preface by J. Dieudonné and postface by François Russo, Collection "Discours de la Méthode", Gauthier-Villars Éditeur, Paris 1974, xiv+72 pp.

[196] D. Kotschick, Gauge theory is dead!–Long live Gauge theory!, *Notices Amer. Math. Soc.*, **42**, no. 3 (1995), 335–338.

[197] I. S. Kukles and M. Hasanova, On the distribution of critical points of the first and second group (in Russian), *Izv. Vysš. Učebn. Zaved. Matematika*, **6** (1964), 88–97.

[198] E. Laguerre, Sur les équations différentielles du troisième ordre, *C. r. Acad. Sci.*, **88** (1879), 116–119.

[199] E. Laguerre, Sur quelques invariants des équations différentielles linéaires, *C. r. Acad. Sci.*, **88** (1879), 224–227.

[200] S. Lefschetz, *Differential equations: Geometric theory*, Second Edition, Dover Publications Inc., New York, 1977, x+390 pp. (This is an unabridged republication of the second (1963) edition of the work originally published by Interscience Publishers in 1957 as Volume VI of the Interscience Pure and Applied Mathematics series.)

[201] E. A. Leontovich, On the generation of limit cycles from separatrices, *Doklady Akad. Nauk SSSR (N.S)*, **78** (1951), 641–644.

[202] C. Li, Non-existence of limit cycles around a weak focus of order three for any quadratic system, *Chinese Ann. Math. Series B*, **7**, no. 2 (1986), 174–190.

[203] C. Li, Abelian integrals and limit cycles, *Qual. Theory Dyn. Syst.*, **11**, no. 1 (2012), 111–128.

[204] C. Li and J. Llibre, The cyclicity of period annulus of a quadratic reversible Lotka–Volterra system, *Nonlinearity* **22**, no. 12 (2009), 2971–2979.

[205] C. Li and J. Llibre, Quadratic perturbations of a quadratic reversible Lotka–Volterra system, *Qual. Theory Dyn. Syst.*, **9**, no. 1–2 (2010), 235–239.

[206] C. Li, J. Llibre and Z. Zhang, Abelian integrals of quadratic Hamiltonian vector fields with an invariant straight line, *Publ. Math.*, **39**, no. 2 (1995), 355–366.

[207] C. Li, J. Llibre and Z. Zhang, Weak focus, limit cycles and bifurcations for bounded quadratic systems, *J. Differential Equations*, **115**, no. 1 (1995), 193–223.

[208] C. Li and R. Roussarie, The cyclicity of the elliptic segment loops of the reversible quadratic Hamiltonian systems under quadratic perturbations, *J. Differential Equations*, **205**, no. 2 (2004), 488–520.

[209] Jibin Li, Hilbert's 16th problem and bifurcations of planar polynomial vector fields, *Internat. J. Bifur. Chaos Appl. Sci. Engrg.*, **13**, no. 1 (2003), 47–106.

[210] A. Lins Neto, Polynomial differential equations with many real ovals in the same algebraic complex solution, *Publ. Mat.*, **55**, no. 2 (2011), 379–399.

[211] R. Liouville, Sur certaines équations différentielles du premier ordre, *C. r. Acad. Sci.*, **103** (1886), 476–479.

[212] C. Liu, Limit cycles bifurcated from some reversible quadratic centers with a non-algebraic first integral, *Nonlinearity*, **25**, no. 6 (2012), 1653–1660.

[213] C. Liu, The cyclicity of period annuli of a class of quadratic reversible systems with two centers, *J. Differential Equations*, **252**, no. 10 (2012), 5260–5273.

[214] Liu Yirong, Li Jibin and Huang Wentao, *Singular point values, center problem and bifurcations of limit cycles of two dimensional autonomous systems*, Science Press, Beijing, 2008, 228 pp.

[215] Liu Yirong, Li Jibin and Huang Wentao, *Planar dynamical systems*, Selected classical problems, De Gruyter, Berlin; Science Press, New York, 2014, 371 pp.

[216] J. Llibre, *Integrability of polynomial differential systems*, in A. Cañada, P. Drabek and A. Fonda, Elsevier (eds.), Handbook of differential equations: Ordinary differential equations, vol. 1, 2004, pp. 437–533.

[217] J. Llibre and J. C. Medrado, Darboux integrability and reversible quadratic vector fields, *Rocky Mountain J. Math.*, **35**, no. 6 (2005), 1999–2057.

[218] J. Llibre and R. Oliveira, Phase portraits of quadratic polynomial vector fields having a rational first integral of degree 3, *Nonlinear Anal.*, **70**, no. 10 (2009), 3549–3560.

[219] J. Llibre and R. Oliveira, Erratum to "Phase portraits of quadratic polynomial vector fields having a rational first integral of degree 3", *Nonlinear Anal.* **70** (2009), 3549–3560, *Nonlinear Anal.*, **71** (2009), 6378–6379.

[220] J. Llibre, R. Ramirez and A. Sadovskaia, On the 16th Hilbert problem for algebraic limit cycles, *J. Differential Equations*, **248**, no. 6 (2010), 1401–1409.

[221] J. Llibre, R. Ramirez and A. Sadovskaia, On the 16th Hilbert problem for limit cycles on nonsingular algebraic curves, *J. Differential Equations*, **250**, no. 2 (2011), 983–999.

[222] J. Llibre and D. Schlomiuk, Geometry of quadratic differential systems with a weak focus of third order, *Canad. J. of Math.*, **56**, no. 2 (2004), 310–343.

[223] J. Llibre and G. Swirszcz, Relationships between limit cycles and algebraic invariant curves for quadratic systems, *J. Differential Equations*, **229**, no. 2 (2006), 529–537.

[224] J. Llibre and G. Swirszcz, Classification of quadratic systems admitting the existence of an algebraic limit cycle, *Bull. Sci. Math.*, **131**, no. 5 (2007), 405–421.

[225] J. Llibre and N. Vulpe, Planar cubic polynomial differential systems with the maximum number of invariant straight lines, *Rocky Mountain J. Math.*, **36**, no. 4 (2006), 1301–1373.

[226] J. Llibre and X. Zhang, Topological phase portraits of planar semi-linear quadratic vector fields, *Houston J. Math.*, **27**, no. 2 (2001), 247–296.

[227] J. Llibre and X. Zhang, Polynomial first integrals of quadratic systems, *Rocky Mountain J. Math.*, **31**, (2001), 1317–1371.

[228] A. J. Lotka, *Elements of physical biology*, Waverly Press by the Williams and Wilkins Company, Baltimore, 1925.

[229] W. S. Loud, Behavior of the period of solutions of certain plane autonomous systems near centers, *Contributions to Differential Equations*, **3** (1964), 21–36.

[230] V. A. Lunkevich and K. S. Sibirschi, Integrals of general quadratic systems in cases of a center (in Russian), *Differentsial'nye Uravneniya*, **8**, no. 5 (1982), 786–792. (Translation in *Differential Equations*, **8**, no. 5 (1982), 563–568.)

[231] M. Lupan and N. Vulpe, Classification of quadratic systems with a symmetry center and simple infinite singular points, *Bul. Acad. Ştiinţe Repub. Mold. Mat.*, **1 (41)** (2003), 102–119. (In memory of C. S. Sibirschi.)

[232] L. S. Lyagina, Integral'nye kriv'ye uravnenya $y' = (ax^2 + bxy + cy^2)/(dx^2 + exy + fy^2)$, *Uspehi Matem. Nauk (N.S.)*, **6**, no. 2(42) (1951), 171–183.

[233] A. M. Lyapunov, *Problème général de la stabilité du mouvement*, Annals of Mathematics Studies, vol. 17, Princeton University Press, Princeton, NJ, 1947, 272 pp. (Reproduction of the French translation in 1907 of the Russian mémoire dated 1892).

[234] L. Markus, Global structure of ordinary differential equations in the plane, *Trans. Amer. Math. Soc.*, **76** (1954), 127–148.

[235] L. Markus, Quadratic differential equations and non-associative algebras in *Contributions to the theory of nonlinear oscillations*, vol. V, Annals of Mathematics Studies, vol. 45, Princeton University Press, Princeton, NJ, 1960, pp. 185–213.

[236] A. Mayer, On the theory of coupled vibrations of two self-excited generators, *Tech. Phys. URSS*, **2** (1935), 465–481.

[237] A. Mayer, A contribution to the theory of forced oscillations in a generator with two degrees of freedom, *Tech. Phys. URSS*, **3** (1936), 1056–1071.

[238] R. Moussu, Le problème de la finitude du nombre de cycles limites, (Séminaire Bourbaki, 4me année, 1985-86, no. 655), *Astérisque*, **145–146** (1987), 89–101.

[239] R. Moussu and C. Roche, Théorie de Hovansky et problème de Dulac, *Invent. Math.*, **105**, no. 2 (1991), 431–441.

[240] D. Mumford, J. Fogarty and F. Kirwan, *Geometric invariant theory*, Third Edition, Ergebnisse der Mathematik und ihrer Grenzgebiete (2) [Results in Mathematics and Related Areas (2)], 34. Springer-Verlag, Berlin, 1994, xiv+292 pp. (The first edition was authored only by Mumford, D. (1965).)

[241] V. V. Nemytskii and V. V. Stepanov, *Qualitative theory of differential equations* (in Russian), OGIZ, Moscow-Leningrad, 1947, 448p. (Translation to English by Princeton Mathematical Series, vol. 22, Princeton University Press, Princeton, NJ, 1960, viii+523 pp.)

[242] D. Neumann, Classification of continuous flows on 2–manifolds, *Proc. Amer. Math. Soc.*, **48** (1975), 73–81.

[243] I. V. Nikolaev and N. I. Vulpe, Topological classification of quadratic systems with a unique finite second order singularity with two zero eigenvalues, *Izv. Akad. Nauk Respub. Moldova Mat.*, **57**, no. 1 (1993), 3–8.

[244] I. Nikolaev and N. Vulpe, Topological classification of quadratic systems at infinity, *J. London Math. Soc. (2)*, **55** (1997), 473–488.

[245] E. Noether, Über die bildung des formensystems der ternären biquadratischen form, *Journal fur Reine und Andgewandte Mathematik*, **134** (1908), 23–90. (Also in *Collected papers of Emmy Noether*, edited by Nathan Jacobson, Springer-Verlag, Berlin, 1983, pp 31–99.)

[246] E. Noether, Der endlichkeitssatz der invarianten endlicher gruppen, *Math. Ann.*, **77**, no. 1 (1915), 89–92.

[247] P. J. Olver, *Classical invariant theory*, London Mathematical Society Student Texts, vol. 44, Cambridge University Press, Cambridge, England, 1999, xxii+280 pp.

[248] J. Pal and D. Schlomiuk, Summing up the dynamics of quadratic Hamiltonian systems with a center, *Canad. J. Math.*, **49**, no. 3 (1997), 583–599.

[249] M. M. Peixoto, On the classification on flows on 2–manifolds, in *Dynamical systems: Proceedings of a symposium held at the University of Bahia, Salvador, Brasil, July 26–August 14, 1971*, Academic Press, New York, 1973, pp. 389–419.

[250] L. Peng, Quadratic perturbations of a quadratic reversible center of genus one, *Front. Math. China*, **6**, no. 5 (2011), 911–930.

[251] L. Peng, Z. Feng and C. Liu, Quadratic perturbations of a quadratic reversible Lotka-Volerra system with two centers, *Discrete Contin. Dyn. Syst.*, **34**, no. 11 (2014), 4807–4826.

[252] L. Peng and Y. Lei, Bifurcations of limit cycles from a quadratic reversible center with the unbounded elliptic separatrix, *Bull. Iranian Math. Soc.*, **39**, no. 6 (2013), 1225–1248.

[253] L. Peng and Y. Li, On the limit cycles bifurcating from a quadratic reversible center of genus one, *Mediterr. J. Math.*, **11**, no. 2 (2014), 373–392.

[254] G. Perelman, The entropy formula for the Ricci flow and its geometric applications, Preprint, arXiv:math/0211159 [math.DG] 11 Nov. 2002, or arXiv:math/0211159v1 [math.DG], 39 pp.

[255] G. Perelman, Ricci flow with surgery on three-manifolds, Preprint, arXiv:-math/0303109 [math.DG], 10 Mar 2003, 22 pp.

[256] G. S. Petrov, On the non-oscillation of elliptic integrals, *Funktsional. Anal. i Prilozhen.*, **31**, no. 4 (1997), 47–51. (Translation in *Funkt. Ana. Appl.*, **31**, no. 4 (1997), 262–265.)

[257] I. Petrovski and E. Landis, On the number of limit cycles of the equation $dy/dx = P(x,y)/Q(x,y)$, where P and Q are polynomials of 2nd degree (in Russian), *Mat. Sb. N.S.*, **37 (79)** (1955), 209–250.

[258] I. Petrovski and E. Landis, On the number of limit cycles of the equation $dy/dx = M(x,y)/N(x,y)$, where the M and N are polynomials of second degree (in Russian), *Dokl. Akad. Nauk SSSR (N.S.)*, **102**, (1955), 29–32.

[259] I. Petrovski and E. Landis, On the number of limit cycles of the equation $dy/dx = P(x,y)/Q(x,y)$, where P and Q are polynomials of the second degree, *American Mathematical Society Translations, Ser. 2*, **10** (1958), 177–221.

[260] I. Petrovski and E. Landis, Corrections to the articles "On the number of limit cycles of the equations $dy/dx = P(x,y)/Q(x,y)$, where P and Q are polynomials of 2nd degree" and "On the number of limit cycles of the equation $dy/dx = P(x,y)/Q(x,y)$, where P and Q are polynomials" (in Russian), *Mat. Sb. (N.S.)*, **48 (90)** (1959), 253–255.

[261] I. Petrovski and E. Landis, A letter to the editors (in Russian), *Mat. Sb. (N.S.)*, **73 (115)** (1967), 160.

[262] H. Poincaré, Mémoire sur les courbes définies par une équation differentielle, *Journal de Mathématiques*, 3e *série*, **t. 7** (1881), 375–422. (Oeuvres de Henri Poincaré, Edited by P. Appel and J. Drach, Volume I, Paris: Gauthier-Villar 1916. 3–84.)

[263] H. Poincaré, Sur les courbes définies par les équations differentielles, *Journal de Mathématiques Pures et Appliquées*, 4e *série*, **t. 1** (1885), 167–244. (Oeuvres de Henri Poincaré, Edited by P. Appel and J. Drach, Volume I, Paris: Gauthier-Villar 1916. 90–158.)

[264] H. Poincaré, Sur l'intégration algébrique des équations différentielles du premier ordre et du premier degré I, *Rendiconti del Circolo Matematico di Palermo*, **5** (1891), 161–191.

[265] H. Poincaré, Sur l'intégration algébrique des équations différentielles du premier ordre et du premier degré II, *Rendiconti del Circolo Matematico di Palermo*, **11** (1897), 193–239.

[266] H. Poincaré, *Sur les courbes définies par une équation différentielle*, Oeuvres, Vol. 1. (1880–1890), Gauthier–Villar, Paris.

[267] M. Popa, *Applications of algebras to differential systems* (in Russian), The Academy of Sciences of Moldova, Chişinău, 2001, 224 pp.

[268] M. Popa, *Methods with algebras in differential systems* (in Romanian), Seria Matematica Aplicata şi Industriala, vol. 15, The Flower Power, University of Piteşti, Romania, 2004, 340 pp.

[269] M. Popa and V. Pricop, Applications of algebraic methods in solving the center-focus problem, *Bul. Acad. Ştiinţe Repub. Mold. Mat.*, **1 (71)** (2013), 45–71.

[270] M. N. Popa and V. V. Pricop, *Problema tsentra i fokusa: algebraicheskie resheniya i gipotezy* [The center-focus problem: algebraic solutions and hypotheses], Akademiya Nauk Respubliki Moldova, Institut Matematiki i Informatiki, Chişinău, 2018, 255 pp.

[271] M. N. Popa and K. S. Sibirschi, Affine classification of a system with quadratic nonlinearities and not single valued canonical form (in Russian), *Differencial'nye Uravnenija*, **14**, no. 6 (1978), 1028–1033. (Translation in *Differential equations*, **14**, no. 6 (1978), 732–735.)

[272] M. N. Popa and K. S. Sibirschi, Conditions for the existence of a homogeneous linear partial integral of a differential system (in Russian), *Differencial'nye Uravnenija*, **23**, no. 8 (1987), 1324–1331. (Translation in *Differential Equations*, **23**, no. 8 (1987), 894–900.)

[273] M. Popov, Modern developments in invariant theory, *Proc. Int. Congr. Math., Berkeley, Calif.*, 1986, Vol. 1,2, 394–406.

[274] M. Popov and E. B. Vinberg, Invariant theory, *Itogi Nauki i Tekhniki, Sovrem. Probl. Mat. Fund. Naprav.*, **55** (1989) 137–314. (Translation in A. N. Parshin and I. R. Shafarevich (eds.), *Algebraic geometry IV*, Encyclopaedia of Mathematical Sciences, vol. 55, Springer-Verlag, Berlin, 1994, 125–284.)

[275] Qin Yuan–Xun, On the algebraic limit cycles of degree 2 of the differential equation $dy/dx = \sum_{0\le i+j\le 2} a_{ij}x^iy^j / \sum_{0\le i+j\le 2} b_{ij}x^iy^j$, *Sci. Sinica*, **7** (1958), 934–945.

[276] J. W. Reyn, Phase portraits of a quadratic system of differential equations occurring frequently in applications, *Nieuw Arch. Wisk. (4)*, **5**, no. 2 (1987), 107–151.

[277] J. W. Reyn, Classes of quadratic systems of differential equations in the plane, in *Dynamical systems*, Nankai Series in Pure, Applied Mathematics and Theoretical Physics, vol. 4, World Scientific Publishers, Rivers Edge, NJ, 1993, pp. 146–180.

[278] J. W. Reyn, *A bibliography of the qualitative theory of quadratic systems of differential equations in the plane*, Third edition, Faculty of Technical Mathematics and Informatics, **2**, Delft University of Technology, Delft, The Netherlands, 1994.

[279] J. W. Reyn, *Phase portraits of planar quadratic systems*, Mathematics and its Applications, vol. 583, Springer, New York, 2007, xvi+334 pp.

[280] L. F. Richardson, *Generalized foreign politics*, British Journal of Psychology Monograph Supplements, vol. 23, Cambridge University Press, Cambridge, England, 1936, viii+91 pp.

[281] V. G. Romanovski and D. Shafer, *The center and cyclicity problems: A computational algebra approach*, Birkhäuser Boston, Boston, 2009, xvi+330 pp.

[282] R. Roussarie, On the number of limit cycles which appear by perturbation of separatrix loop of planar vector fields, *Bol. Soc. Brasil. Mat.*, **17**, no. 2 (1986), 67–101.

[283] R. Roussarie, A note on finite cyclicity and Hilbert's 16th problem, in *Dynamical systems*, Lecture Notes in Mathematics, vol. 1331, Springer, New York, 1988, pp. 161–168.

[284] R. Roussarie, *Bifurcation of planar vector fields and Hilbert's sixteenth problem*, Progress in Mathematics, vol. 164, Birkhäuser Verlag, Basel, 1998, 204 pp.

[285] R. Roussarie and D. Schlomiuk, On the Geometric Structure of the Class of Planar Quadratic Differential Systems, *Qualitative Theory of Dynamical Systems*, **3**, (2002), 93-121.

[286] C. Rousseau, Bifurcation methods in polynomial systems, in *Bifurcation and periodic orbits of vector fields (Montreal, PQ, 1992)*, NATO Advanced Science Institutes Series, Series C: Mathematical and Physical Sciences, **408**, Kluwer Academic Publishers, Dordrecht, 1993, pp. 383–428.

[287] C. Rousseau, Normal forms, bifurcations and finiteness properties of vector fields, *Normal forms, bifurcations and finiteness problems in differential equations*, NATO Science Series II: Mathematics, Physics and Chemistry, **137**, Kluwer Academic Publishers, Dordrecht, 2004, pp. 431–470.

[288] C. Rousseau, C. Shan and H. Zhu, Finite cyclicity of some graphics through a nilpotent point of saddle type inside quadratic systems, *Qual. Theory Dyn. Syst.*, **15**, no. 1 (2016), 237–256.

[289] N. A. Sakharnikov, On Frommer's conditions for the existence of a center (in Russian), *Prikl. Mat. Mek.*, **12** (1948), 669–670.

[290] G. T. dos Santos, Classification of generic quadratic vector fields with no limit cycles, *Lecture Notes in Mathematics*, vol. 597, Springer, New York, 1977, pp. 605–640.

[291] D. Schlomiuk, Algebraic particular integrals, integrability and the problem of the center, *Trans. Amer. Math. Soc.*, **338** (1993), 799–841.

[292] D. Schlomiuk, Elementary first integrals and algebraic invariant curves of differential equations, *Exposition. Math.*, **11**, no. 5 (1993), 433–454.

[293] D. Schlomiuk, Algebraic and geometric aspects of the theory of polynomial vector fields, in *Bifurcations and periodic Orbits of Vector Fields*, D. Schlomiuk (ed.), *NATO Advanced Science Institutes Series, Series C: Mathematical and Physical Sciences*, vol. 408, Kluwer Academic Publishers, Dordrecht, 1993, pp.429–467.

[294] D. Schlomiuk Editor, *Bifurcations and periodic orbits of vector fields*, NATO Advanced Science Institutes Series, Series C: Mathematical and Physical Sciences, vol. 408, Kluwer Academic Publishers, Dordrecht, 1993, 472 pp.

[295] D. Schlomiuk, Aspects of planar polynomial vector fields: global versus local, real versus complex, analytic versus algebraic and geometric, in *Normal forms, bifurcations and finiteness problems in differential equations*, NATO Science Series II: Mathematics, Physics and Chemistry, vol. 137, Kluwer Academic Publishers, Dordrecht, 2004, pp. 471–509.

[296] D. Schlomiuk, Topological and polynomial invariants, moduli spaces, in classification problems of polynomial vector fields, *Publ. Mat.*, **58**, suppl. (2014), 461–496.

[297] D. Schlomiuk, A. A. Bolibrukh, S. Yakovenko, V. Kaloshin and A. Buium, *On finiteness in differential equations and Diophantine geometry*, edited by Schlomiuk, CRM Monograph Series, vol. 24, American Mathematical Society, Providence, RI, 2005, x+182 pp.

[298] D. Schlomiuk, J. Guckenheimer and R. Rand, Integrability of plane quadratic vector fields, *Exposition. Math.*, **8**, no. 1 (1990), 3–25.

[299] D. Schlomiuk and J. Pal, On the geometry in the neighborhood of infinity of quadratic differential systems with a weak focus, *Qual. Theory Dyn. Syst.*, **2** (2001), 1–43.

[300] D. Schlomiuk and N. Vulpe, Planar quadratic differential systems with invariant straight lines of at least five total multiplicity, *Qual. Theory Dyn. Syst.*, **5**, no. 1 (2004), 135–194.

[301] D. Schlomiuk and N. Vulpe, Geometry of quadratic differential systems in the neighborhood of infinity, *J. Differential Equations*, **215**, no. 2 (2005), 357–400.

[302] D. Schlomiuk and N. Vulpe, Integrals and phase portraits of planar quadratic differential systems with invariant lines of at least five total multiplicity, *Rocky Mountain J. Math.*, **38**, no. 6 (2008), 2015–2075.

[303] D. Schlomiuk and N. Vulpe, Integrals and phase portraits of planar quadratic differential systems with invariant lines of total multiplicity four, *Bul. Acad. Ştiinţe Repub. Mold. Mat.*, **1 (56)** (2008), 27–83.

[304] D. Schlomiuk and N. Vulpe, Planar quadratic differential systems with invariant straight lines of total multiplicity four, *Nonlinear Anal.*, **68**, no. 4 (2008), 681–715.

[305] D. Schlomiuk and N. Vulpe, The full study of planar quadratic differential systems possessing a line of singularities at infinity, *J. Dynam. Differential Equations*, **20**, no. 4 (2008), 737–775.

[306] D. Schlomiuk and N. Vulpe, Global classification of the planar Lotka–Volterra differential systems according to their configurations of invariant straight lines, *J. Fixed Point Theory Appl.*, **8**, no. 1 (2010), 177–245.

[307] D. Schlomiuk and N. Vulpe, Global topological classification of Lotka–Volterra quadratic differential systems, *Electron. J. Differential Equations*, **2012** (2012), No. 64, 69 pp.

[308] D. Schlomiuk and X. Zhang, Quadratic differential systems with complex conjugate invariant lines meeting at a finite point, *J. Differential Equations*, **265**, no. 8 (2018), 3650–3684.

[309] H. Seifert and W. Threlfall, *A textbook of topology*, Academic Press, New York, 1980, 437 pp.

[310] Y. Shao and A. Chunxiang, Quadratic perturbations of a class of quadratic reversible Lotka–Volterra systems, *Internat. J. Bifur. Chaos Appl. Sci. Engrg.*, **23**, no. 8 (2013), 1350137, 8 pp.

[311] S. Shi, A concrete example of the existence of four limit cycles for plane quadratic systems, *Sci. Sinica*, **23**, no. 2 (1980), 153–158.

[312] K. S. Sibirschi, On conditions for the presence of a center and a focus (in Russian), *Kisinev. Gos. Unic. Uc. Zap.*, **11** (1954), 115–117.

[313] K. S. Sibirschi, The principle of symmetry and the problem of the center (in Russian), *Kishinev. Gos. Unic. Uc. Zap.*, **17** (1955), 27–34.

[314] K. S. Sibirschi, *The method of invariants in the qualitative theory of differential equations* (in Russian), Kishinev. Stiinta, The Academy of Sciences of Moldova, Chişinău, 1968, 184 pp.

[315] K. S. Sibirschi, *Algebraic invariants of differential equations and matrices* (in Russian), Kishinev, Stiinta, The Academy of Sciences of Moldova, Chişinău, 1976, 268 pp.

[316] K. S. Sibirschi, *Introduction to the algebraic theory of invariants of differential equations*, Nonlinear Science: Theory and Applications, Manchester University Press, Manchester, 1988. (This is a translation of the Russian original ["Shtiintsa", Kishinev, 1982; MR0716501].)

[317] S. Smale, Mathematical problems for the next century, *Mathematics: Frontiers and perspectives*, American Mathematical Society, Providence, RI, 2000, pp. 271–294.

[318] J. Sotomayor, *Curvas definidas por equaçoes diferenciais no plano*, Instituto de Matemática Pura e Aplicada, Rio de Janeiro, 1981, iii+166 pp.

[319] J. Sotomayor and R. Paterlini, Quadratic vector fields with finitely many periodic orbits, in *Geometric dynamics*, Lecture Notes in Mathematics, vol. 1007, Springer, Berlin, 1983, pp. 753–766.

[320] T. A. Springer, *Invariant theory*, Lecture Notes in Mathematics, vol. 585, Springer-Verlag, Berlin, 1977, iv+112 pp.

[321] A. Şubă and D. Cozma, Solution of the problem of the center for cubic system with two homogeneous and one nonhomogeneous invariant straight lines, *Bul. Acad. Ştiinţe Repub. Mold. Mat.*, **1 (29)** (1999), 37–44.

[322] A. Şubă and D. Cozma, Solution of the problem of the centre for cubic differential system with three invariant straight lines two of which are parallel, *Bul. Acad. Ştiinţe Repub. Mold. Mat.*, **2 (36)** (2001), 75–86.

[323] A. Şubă and D. Cozma, Solution of the problem of the centre for cubic differential system with three invariant straight lines in generic position, *Qual. Theory Dyn. Syst.*, **6**, no. 1 (2005), 45–58.

[324] A. Şubă, V. Repeşco and V. Puţuntică, Cubic systems with seven invariant straight lines of configuration (3,3,1), *Bul. Acad. Ştiinţe Repub. Mold. Mat.*, **2 (69)** (2012), 81–98.

[325] A. Şubă, V. Repeşco and V. Puţuntică. Cubic systems with invariant affine straight lines of total parallel multiplicity seven, *Electron. J. Differential Equations*, **2013** (2013), No. 274, 22 pp.

[326] G. Suo and Y. Chen, The real quadratic systems with two conjugate imaginary straight line solutions, *Ann. Differential Equations*, **2**, no. 2 (1986), 197–207.

[327] C. H. Taubes, Differential topology, in *The Princeton Companion to Mathematics*, T. Gowers (ed.), Princeton University Press, Princeton, NJ, 2008, pp. 396–408.

[328] G. I. Taylor, Recent work on the flow of compressible fluids, *J. London Math. Soc.*, **5** (1930), 224–240.

[329] R. Taylor and A. Wiles, Ring-theoretic properties of certain Hecke algebras, *Ann. of Math. (2)*, **141**, no. 3 (1995), 553–572.

[330] A. N. Varchenko, Estimation of the number of zeros of an abelian integral depending on a parameter and limit cycles (in Russian), *Funktsional. Anal. i Prilozhen.*, **18**, no. 2 (1984), 14–25.

[331] M. Voldman, Iu. T. Calin and N. I. Vulpe, Affine invariant conditions for the topological distinction of quadratic systems with a critical point of the 4th multiplicity, *Publ. Mat.*, **40**, no. 2 (1996), 431–441.

[332] A. Voldman and N. Vulpe, Affine invariant conditions for topologically distinguishing quadratic systems without finite critical points, *Izv. Akad. Nauk Respub. Moldova Mat.*, **59**, no. 2–3 (1995), 100–112.

[333] A. Voldman and N. Vulpe, Affine invariant conditions for topologically distinguishing quadratic systems with $m_f = 1$, *Nonlinear Anal.*, **31**, no. 1–2 (1998), 171–179.

[334] V. Volterra, Variazione flutuazioni del numero d'individus in specie animali conventi, *Mem. R. Acad. Naz. dei Lincei*, **2** (1926), 31–113.

[335] N. I. Vulpe, A polynomial basis of center-affine comitants of a differential system (in Russian), *Dokl. Akad. Nauk SSSR*, **250**, no. 5 (1980), 1033–1037. (Translation in *Soviet Math. Dokl.*, **21**, no. 1 (1980), 234–239.)

[336] N. I. Vulpe, Affine–invariant conditions for the topological distinction of quadratic systems with a center (in Russian), *Differentsial'nye Uravneniya*, **19**, no. 3 (1983), 371–379. (Translation in *Differential Equations*, **19** (1983), 273–280.)

[337] N. I. Vulpe, *Polynomial bases of comitants of differential systems and their applications in qualitative theory* (in Russian), Kishinev, Stiinta, The Academy of Sciences of Moldova, Chişinău, 1986, 172 pp.

[338] N. Vulpe, Characterization of the finite weak singularities of quadratic systems via invariant theory, *Nonlinear Anal.*, **74**, no. 17 (2011), 6553–6582.

[339] N. I. Vulpe and A. Yu. Likhovetski, Coefficient conditions for the topological discrimination of quadratic systems of Darboux type (in Russian), *Mat. Issled., Differ. Uravneniya i Mat. Fizika*, **106** (1989), 34–49.

[340] N. Vulpe and M. Lupan, Quadratic systems with dicritical points, *Izv. Akad. Nauk Respub. Moldova Mat.*, **58**, no. 3 (1994), 52–60.

[341] N. I. Vulpe and I. V. Nikolaev, Topological classification of QS with a unique third order singular point, *Izv. Akad. Nauk Respub. Moldova Mat.*, **92**, no. 2 (1992), 37–44.

[342] N. I. Vulpe and I. V. Nikolaev, Topological classification of a quadratic system with a four-fold singular point (in Russian), *Differentsial'nye Uravneniya*, **29**, no. 10 (1993), 1669–1674. (Translation in *Differential Equations*, **29** (1993), 1449–1453.)

[343] N. I. Vulpe and K. S. Sibirschi, Affine classification of a quadratic system (in Russian), *Differencial'nye Uravnenija*, **10** (1974), 2111–2124.

[344] N. I. Vulpe and K. S. Sibirschi, Affinely invariant coefficient conditions for the topological distinctness of quadratic systems, *Mat. Issled.*, **10**, no. 3(37) (1975), 15–28.

[345] N. I. Vulpe and K. S. Sibirschi, Geometric classification of a quadratic differential system (in Russian), *Differentsial'nye Uravneniya*, **13**, no. 5 (1977), 803–814. (Translation in *Differential Equations*, **13**, no. 5 (1977), 548–556.)

[346] H. Weyl, *The classical groups*, Princeton University Press, Princeton, NJ, 1939, 320 pp. (Second edition published in 1946.)

[347] H. Weyl, On the differential equations of the simplest boundary-layer problems, *Ann. of Math.*, **43** (1942), 381–407.

[348] H. Weyl, David Hilbert and his mathematical work, *Bull. Amer. Math. Soc.*, **50** (1944), 612–654.

[349] A. Wiles, Modular elliptic curves and Fermat's last theorem, *Ann. of Math. (2)*, **141**, no. 3 (1995), 443–551.

[350] A. Wörz-Buserkros, A complete classification of all two-dimensional Lotka–Volterra systems, *Differential Equations Dynam. Systems*, **1**, no. 2 (1993), 101–118.

[351] J. Wu, L. Peng and C. Li, On the number of limit cycles in perturbations of quadratic reversible center, *J. Aust. Math. Soc.*, **92**, no. 3 (2012), 409–423.

[352] A. I. Yablonskii, Limit cycles of a certain differential equations (in Russian), *Differentsial'nye Uravneniya*, **2** (1966), 335–344.

[353] S. Yakovenko, Review on [147], MathSciNet review MR1817642 (2002g:-34069). (2002).

[354] S. Yakovenko, Quantitative theory of ordinary differential equations and the tangential Hilbert 16th problem, in *On finiteness in differential equations and Diophantine geometry*, CRM Monograph Series, vol. 24, American Mathematical Society, Providence, RI, 2005, pp. 41–109. MR2180125 (2006g:34062).

[355] Y. Q. Ye et al., *Theory of limit cycles*, translated by Y. L. Chi, Translations of Mathematical Monographs, vol. 66, American Mathematical Society, Providence, RI, 1986, xi+435 pp.

[356] J. Ch. Yoccoz, Non-accumulation des cycles limites, *Astérisque (Séminaire Bourbaki, no. 655)*, **161–162**, no. 3 (1988), 87–103.

[357] A. Zegeling and R. E. Kooij, The distribution of limit cycles in quadratic systems with four finite singularities, *J. Differential Equations*, **151**, no. 2 (1999), 373–385.

[358] Pingguang Zhang, On the distribution and number of limit cycles for quadratic system with two foci (in Chinese), *Acta Math. Sinica (Chin. Ser.)*, **44**, no. 1 (2001), 37–44.

[359] Pingguang Zhang, On the distribution and number of limit cycles for quadratic systems with two foci, *Qual. Theory Dyn. Syst.*, **3**, no. 2 (2002), 437–463.

[360] X. Zhang, The 16th Hilbert problem on algebraic limit cycles, *J. Differential Equations*, **251**, no. 7 (2011), 1778–1789.

[361] X. Zhang, *Integrability of Dynamical Systems: Algebra and Analysis*, Developments in Mathematics, Springer, 2017, 380 pp.

[362] Z. F. Zhang, T. R. Ding, W. Z. Huang and Z. X. Dong, *Qualitative theory of differential equations*, translated from the Chinese by Anthony Wing Kwok Leung, Translations of Mathematical Monographs, 101, American Mathematical Society, Providence, RI, 1992, xiv+461 pp.

[363] Y. Zhao and H. Zhu, Bifurcations of limit cycles from a non-Hamiltonian quadratic integrable system with homoclinic loop, in *Infinite dimensional dynamical systems,* Fields Institute Communications, vol. 64, Springer, New York, 2013, pp. 445–479.

[364] H. Żołądek, Quadratic systems with center and their perturbations, *J. Differential Equations*, **109** (1994), 223–273.

[365] H. Żołądek, *The monodromy group*, Instytut Matematyczny Polskiej Akademii Nauk. Monografie Matematyczne [Mathematics Institute of the Polish Academy of Sciences Mathematical Monographs], vol. 67, Birkhäuser, Basel, 2006, xi+583 pp.

Index

k-twist, 51

Abelian integral, 6, 32
affine
  blow-up, 48
  transformation, 86
angle
  well defined, 44

blow-down, 48
blow-up
  affine, 48
  directional, 48
  line, 51
  one-directional, 50
  polar, 45
  technique, 45
  vertical, 52
blowing down, 46
borsec, 66

canonical region, 97
center
  center-focus problem, 6
  isochronus, 42
  problem of, 6
  uniformly isochronous, 42
central projection, 38
characteristic
  angle, 44
    special, 44
  direction, 44, 51
    special, 44, 63
  orbit, 44
  polynomial, 44, 51

comitant, 110
  CT-comitant, 111
  GL-comitant, 111
  T-comitant, 111
  affine, 105, 111
compactification
  on the projective plane, 8, 41
  Poincaré, 37
configuration
  invariant lines, 24
  singularities, 78
covariant, 100
cusp, 43
cyclicity, 28
  finite, 28

Darboux, 3
degree, 37
  of a system, 3
  of the representation, 100
Desingularization Theorem, 50
differential systems
  polynomial, 3
  quadratic, 3
Dulac, 8

elliptic-saddle, 43
equivalence
  blow-up, 76, 78
    one-step, 77
  geometrical, 35, 76, 78
  group action, 92, 95
  modulo limit cycles, 11, 95, 97
  order, 76
  qualitative, 35, 75, 95

J. C. Artés et al., *Geometric Configurations of Singularities of Planar Polynomial Differential Systems*, https://doi.org/10.1007/978-3-030-50570-7

tangent, 75
topological, 35, 75
Erlangen program, 99

finiteness
problem
general for limit cycles, 9, 27
individual, 97
theorem general, 8

geometric, 99
classification, 36
invariant theory, 104
local sector, 65
property, 99
geometrical equivalence
of configurations, 13, 36
of singularities, 10, 35, 79
global finiteness conjecture, 9, 10, 28
graphic, 27
degenerate, 27
generic, 28
irreducible, 28
group action, 92
left, 92

Hilbert number, 9
Hilbert's 16th problem, 7
infinitesimal, 10
Hilbert–Arnold problem, 10, 33
hyperbolicity ratio, 28

invariant, 91, 100
absolute, 100
complete set, 91
group, 92
relative, 100
topological, 97
isochronous
center, 42
level, 76

limit
cycle, 4
configuration of, 22
multiple, 97
periodic set, 27
degenerate, 27

multiplicity, 76

node
one-direction, 65
star, 65

orbit
characteristic, 44
half, 67
tangent to a half line, 44
order
of weak focus, 12, 42
of weak saddle, 12, 42

phase portrait, 94
Poincaré, 4
compactification, 8, 37
first return map, 4, 28
problem, 7, 94
sphere, 37
polycycle, 28
problem
16th Hilbert
existential, 9
infinitesimal, 10
second part, 7
tangential, 10
weak, 10
on algebraic integrability, 6, 7
center-focus, 45
classification, 91
Hilbert's 16th, 7
Hilbert–Arnold, 10
projective plane, 4, 37

QS, 3
quadratic
differential system, 37
bounded, 25
qualitative study
differential systems, 4

regular point, 41
representation linear, 100

saddle-node, 42
sector, 42
    cusp-like, 44, 74, 85
    elliptic, 42, 50, 53
    hyperbolic, 42, 50, 53
    parabolic, 42, 50, 53
    star-like, 85
sectorial decomposition, 50
separatrix
    connection, 632
    configurations, 97
    connection, 11
    local, 43
singular point, 3, 41, 43
    center, 42
        isochronous, 42
        isochronous uniform, 42
    degenerate, 41
    elemental, 43
    elementary, 42
    finite, 12
    focus, 42
        stable, 42
        strong, 42
        unstable, 42
        weak, 22, 42
    hyperbolic, 41
    infinite, 37
    intricate, 44
    linearly zero, 42
    multiple, 76
    nilpotent, 42, 44
    node, 42
        stable, 42
        unstable, 42
    saddle, 42
        integrable, 25, 82, 133, 145
        strong, 42
        weak, 42
    semi-elemental, 43
    semi-hyperbolic, 42
    star, 76
    star-like, 44, 63
syzygy, 101

topological local sector, 65
trace polynomial, 123

well defined angle, 44